# Contents

a LANGE medical book



# Harper's Illustrated Biochemistry

twenty-sixth edition

**Robert K. Murray, MD, PhD**
*Professor (Emeritus) of Biochemistry*
*University of Toronto*
*Toronto, Ontario*

**Daryl K. Granner, MD**
*Joe C. Davis Professor of Biomedical Science*
*Director, Vanderbilt Diabetes Center*
*Professor of Molecular Physiology and Biophysics*
*and of Medicine*
*Vanderbilt University*
*Nashville, Tennessee*

**Peter A. Mayes, PhD, DSc**
*Emeritus Professor of Veterinary Biochemistry*
*Royal Veterinary College*
*University of London*
*London*

**Victor W. Rodwell, PhD**
*Professor of Biochemistry*
*Purdue University*
*West Lafayette, Indiana*

**Lange Medical Books/McGraw-Hill**
Medical Publishing Division

New York Chicago San Francisco Lisbon London Madrid Mexico City
Milan New Delhi San Juan Seoul Singapore Sydney Toronto

The McGraw-Hill Companies

***Harper's Illustrated Biochemistry, Twenty-Sixth Edition***



1 2 3 4 5 6 7 8 9 0 DOC/DOC 0 9 8 7 6 5 4 3

ISBN 0-07-138901-6
ISSN 1043-9811

## Notice

Medicine is an ever-changing science. As new research and clinical experience broaden our knowledge, changes in treatment and drug therapy are required. The authors and the publisher of this work have checked with sources believed to be reliable in their efforts to provide information that is complete and generally in accord with the standards accepted at the time of publication. However, in view of the possibility of human error or changes in medical sciences, neither the authors nor the publisher nor any other party who has been involved in the preparation or publication of this work warrants that the information contained herein is in every respect accurate or complete, and they disclaim all responsibility for any errors or omissions or for the results obtained from use of the information contained in this work. Readers are encouraged to confirm the information contained herein with other sources. For example and in particular, readers are advised to check the product information sheet included in the package of each drug they plan to administer to be certain that the information contained in this work is accurate and that changes have not been made in the recommended dose or in the contraindications for administration. This recommendation is of particular importance in connection with new or infrequently used drugs.

This book was set in Garamond by Pine Tree Composition
The editors were Janet Foltin, Jim Ransom, and Janene Matragrano Oransky.
The production supervisor was Phil Galea.
The illustration manager was Charissa Baker.
The text designer was Eve Siegel.
The cover designer was Mary McKeon.
The index was prepared by Kathy Pitcoff.
RR Donnelley was printer and binder.

This book is printed on acid-free paper.

ISBN-0-07-121766-5 (International Edition)

# Authors


### David A. Bender, PhD

Sub-Dean Royal Free and University College Medical School, Assistant Faculty Tutor and Tutor to Medical Students, Senior Lecturer in Biochemistry, Department of Biochemistry and Molecular Biology, University College London

### Kathleen M. Botham, PhD, DSc

Senior Lecturer in Biochemistry, Royal Veterinary College, University of London

### Daryl K. Granner, MD

Joe C. Davis Professor of Biomedical Science, Director, Vanderbilt Diabetes Center, Professor of Molecular Physiology and Biophysics and of Medicine, Vanderbilt University, Nashville, Tennessee

### Frederick W. Keeley, PhD

Associate Director and Senior Scientist, Research Institute, Hospital for Sick Children, Toronto, and Professor, Department of Biochemistry, University of Toronto

### Peter J. Kennelly, PhD

Professor of Biochemistry, Virginia Polytechnic Institute and State University, Blacksburg, Virginia

### Peter A. Mayes, PhD, DSc

Emeritus Professor of Veterinary Biochemistry, Royal Veterinary College, University of London

### Robert K. Murray, MD, PhD

Professor (Emeritus) of Biochemistry, University of Toronto

### Margaret L. Rand, PhD

Scientist, Research Institute, Hospital for Sick Children, Toronto, and Associate Professor, Departments of Laboratory Medicine and Pathobiology and Department of Biochemistry, University of Toronto

### Victor W. Rodwell, PhD

Professor of Biochemistry, Purdue University, West Lafayette, Indiana

### P. Anthony Weil, PhD

Professor of Molecular Physiology and Biophysics, Vanderbilt University School of Medicine, Nashville, Tennessee

# Preface

The authors and publisher are pleased to present the twenty-sixth edition of *Harper's Illustrated Biochemistry. Review of Physiological Chemistry* was first published in 1939 and revised in 1944, and it quickly gained a wide readership. In 1951, the third edition appeared with Harold A. Harper, University of California School of Medicine at San Francisco, as author. Dr. Harper remained the sole author until the ninth edition and co-authored eight subsequent editions. Peter Mayes and Victor Rodwell have been authors since the tenth edition, Daryl Granner since the twentieth edition, and Rob Murray since the twenty-first edition. Because of the increasing complexity of biochemical knowledge, they have added co-authors in recent editions.

Fred Keeley and Margaret Rand have each co-authored one chapter with Rob Murray for this and previous editions. Peter Kennelly joined as a co-author in the twenty-fifth edition, and in the present edition has co-authored with Victor Rodwell all of the chapters dealing with the structure and function of proteins and enzymes. The following additional co-authors are very warmly welcomed in this edition: Kathleen Botham has co-authored, with Peter Mayes, the chapters on bioenergetics, biologic oxidation, oxidative phosphorylation, and lipid metabolism. David Bender has co-authored, also with Peter Mayes, the chapters dealing with carbohydrate metabolism, nutrition, digestion, and vitamins and minerals. P. Anthony Weil has co-authored chapters dealing with various aspects of DNA, of RNA, and of gene expression with Daryl Granner. We are all very grateful to our co-authors for bringing their expertise and fresh perspectives to the text.

## CHANGES IN THE TWENTY-SIXTH EDITION

A major goal of the authors continues to be to provide both medical and other students of the health sciences with a book that both describes the basics of biochemistry and is user-friendly and interesting. A second major ongoing goal is to reflect the most significant advances in biochemistry that are important to medicine. However, a third major goal of this edition was to achieve a substantial reduction in size, as feedback indicated that many readers prefer shorter texts.

To achieve this goal, all of the chapters were rigorously edited, involving their amalgamation, division, or deletion, and many were reduced to approximately one-half to two-thirds of their previous size. This has been effected without loss of crucial information but with gain in conciseness and clarity.

Despite the reduction in size, there are many new features in the twenty-sixth edition. These include:

- A new chapter on amino acids and peptides, which emphasizes the manner in which the properties of biologic peptides derive from the individual amino acids of which they are comprised.
- A new chapter on the primary structure of proteins, which provides coverage of both classic and newly emerging "proteomic" and "genomic" methods for identifying proteins. A new section on the application of mass spectrometry to the analysis of protein structure has been added, including comments on the identification of covalent modifications.
- The chapter on the mechanisms of action of enzymes has been revised to provide a comprehensive description of the various physical mechanisms by which enzymes carry out their catalytic functions.
- The chapters on integration of metabolism, nutrition, digestion and absorption, and vitamins and minerals have been completely re-written.
- Among important additions to the various chapters on metabolism are the following: update of the information on oxidative phosphorylation, including a description of the rotary ATP synthase; new insights into the role of GTP in gluconeogenesis; additional information on the regulation of acetyl-CoA carboxylase; new information on receptors involved in lipoprotein metabolism and reverse cholesterol transport; discussion of the role of leptin in fat storage; and new information on bile acid regulation, including the role of the farnesoid X receptor (FXR).
- The chapter on membrane biochemistry in the previous edition has been split into two, yielding two new chapters on the structure and function of membranes and intracellular traffic and sorting of proteins.
- Considerable new material has been added on RNA synthesis, protein synthesis, gene regulation, and various aspects of molecular genetics.
- Much of the material on individual endocrine glands present in the twenty-fifth edition has been replaced with new chapters dealing with the diversity of the endocrine system, with molecular mechanisms of hormone action, and with signal transduction.

- The chapter on plasma proteins, immunoglobulins, and blood coagulation in the previous edition has been split into two new chapters on plasma proteins and immunoglobulins and on hemostasis and thrombosis.
- New information has been added in appropriate chapters on lipid rafts and caveolae, aquaporins, connexins, disorders due to mutations in genes encoding proteins involved in intracellular membrane transport, absorption of iron, and conformational diseases and pharmacogenomics.
- A new and final chapter on "The Human Genome Project" (HGP) has been added, which builds on the material covered in Chapters 35 through 40. Because of the impact of the results of the HGP on the future of biology and medicine, it appeared appropriate to conclude the text with a summary of its major findings and their implications for future work.
- As initiated in the previous edition, references to useful Web sites have been included in a brief Appendix at the end of the text.

## ORGANIZATION OF THE BOOK

The text is divided into two introductory chapters ("Biochemistry & Medicine" and "Water & pH") followed by six main sections.

**Section I** deals with the structures and functions of proteins and enzymes, the workhorses of the body. Because almost all of the reactions in cells are catalyzed by enzymes, it is vital to understand the properties of enzymes before considering other topics.

**Section II** explains how various cellular reactions either utilize or release energy, and it traces the pathways by which carbohydrates and lipids are synthesized and degraded. It also describes the many functions of these two classes of molecules.

**Section III** deals with the amino acids and their many fates and also describes certain key features of protein catabolism.

**Section IV** describes the structures and functions of the nucleotides and nucleic acids, and covers many major topics such as DNA replication and repair, RNA synthesis and modification, and protein synthesis. It also discusses new findings on how genes are regulated and presents the principles of recombinant DNA technology.

**Section V** deals with aspects of extracellular and intracellular communication. Topics covered include membrane structure and function, the molecular bases of the actions of hormones, and the key field of signal transduction.

**Section VI** consists of discussions of eleven special topics: nutrition, digestion, and absorption; vitamins and minerals; intracellular traffic and sorting of proteins; glycoproteins; the extracellular matrix; muscle and the cytoskeleton; plasma proteins and immunoglobulins; hemostasis and thrombosis; red and white blood cells; the metabolism of xenobiotics; and the Human Genome Project.

## ACKNOWLEDGMENTS

The authors thank Janet Foltin for her thoroughly professional approach. Her constant interest and input have had a significant impact on the final structure of this text. We are again immensely grateful to Jim Ransom for his excellent editorial work; it has been a pleasure to work with an individual who constantly offered wise and informed alternatives to the sometimes primitive text transmitted by the authors. The superb editorial skills of Janene Matragrano Oransky and Harriet Lebowitz are warmly acknowledged, as is the excellent artwork of Charissa Baker and her colleagues. The authors are very grateful to Kathy Pitcoff for her thoughtful and meticulous work in preparing the Index. Suggestions from students and colleagues around the world have been most helpful in the formulation of this edition. We look forward to receiving similar input in the future.

Robert K. Murray, MD, PhD
Daryl K. Granner, MD
Peter A. Mayes, PhD, DSc
Victor W. Rodwell, PhD

Toronto, Ontario
Nashville, Tennessee
London
West Lafayette, Indiana
March 2003

# Biochemistry & Medicine

1

*Robert K. Murray, MD, PhD*

## INTRODUCTION

Biochemistry can be defined as *the science concerned with the chemical basis of life* (Gk *bios* "life"). The **cell** is the structural unit of living systems. Thus, biochemistry can also be described as *the science concerned with the chemical constituents of living cells and with the reactions and processes they undergo.* By this definition, biochemistry encompasses large areas of **cell biology,** of **molecular biology,** and of **molecular genetics.**

### The Aim of Biochemistry Is to Describe & Explain, in Molecular Terms, All Chemical Processes of Living Cells

The major objective of biochemistry is the complete understanding, at the molecular level, of all of the chemical processes associated with living cells. To achieve this objective, biochemists have sought to isolate the numerous molecules found in cells, determine their structures, and analyze how they function. Many techniques have been used for these purposes; some of them are summarized in Table 1–1.

### A Knowledge of Biochemistry Is Essential to All Life Sciences

The biochemistry of the nucleic acids lies at the heart of genetics; in turn, the use of genetic approaches has been critical for elucidating many areas of biochemistry. Physiology, the study of body function, overlaps with biochemistry almost completely. Immunology employs numerous biochemical techniques, and many immunologic approaches have found wide use by biochemists. Pharmacology and pharmacy rest on a sound knowledge of biochemistry and physiology; in particular, most drugs are metabolized by enzyme-catalyzed reactions. Poisons act on biochemical reactions or processes; this is the subject matter of toxicology. Biochemical approaches are being used increasingly to study basic aspects of pathology (the study of disease), such as inflammation, cell injury, and cancer. Many workers in microbiology, zoology, and botany employ biochemical approaches almost exclusively. These relationships are not surprising, because life as we know it depends on biochemical reactions and processes. In fact, the old barriers among the life sciences are breaking down, and biochemistry is increasingly becoming their common language.

### A Reciprocal Relationship Between Biochemistry & Medicine Has Stimulated Mutual Advances

The two major concerns for workers in the health sciences—and particularly physicians—are the understanding and maintenance of health and the understanding and effective treatment of diseases. Biochemistry impacts enormously on both of these fundamental concerns of medicine. In fact, the interrelationship of biochemistry and medicine is a wide, two-way street. Biochemical studies have illuminated many aspects of health and disease, and conversely, the study of various aspects of health and disease has opened up new areas of biochemistry. Some examples of this two-way street are shown in Figure 1–1. For instance, a knowledge of protein structure and function was necessary to elucidate the single biochemical difference between normal hemoglobin and sickle cell hemoglobin. On the other hand, analysis of sickle cell hemoglobin has contributed significantly to our understanding of the structure and function of both normal hemoglobin and other proteins. Analogous examples of reciprocal benefit between biochemistry and medicine could be cited for the other paired items shown in Figure 1–1. Another example is the pioneering work of Archibald Garrod, a physician in England during the early 1900s. He studied patients with a number of relatively rare disorders (alkaptonuria, albinism, cystinuria, and pentosuria; these are described in later chapters) and established that these conditions were genetically determined. Garrod designated these conditions as **inborn errors of metabolism.** His insights provided a major foundation for the development of the field of human biochemical genetics. More recent efforts to understand the basis of the genetic disease known as **familial hypercholesterolemia,** which results in severe atherosclerosis at an early age, have led to dramatic progress in understanding of cell receptors and of mechanisms of uptake of cholesterol into cells. Studies of **oncogenes** in cancer cells have directed attention to the molecular mechanisms involved in the control of normal cell growth. These and many other examples emphasize how the study of

***Table 1–1.*** The principal methods and preparations used in biochemical laboratories.

**Methods for Separating and Purifying Biomolecules**[1]
- Salt fractionation (eg, precipitation of proteins with ammonium sulfate)
- Chromatography: Paper; ion exchange; affinity; thin-layer; gas-liquid; high-pressure liquid; gel filtration
- Electrophoresis: Paper; high-voltage; agarose; cellulose acetate; starch gel; polyacrylamide gel; SDS-polyacrylamide gel
- Ultracentrifugation

**Methods for Determining Biomolecular Structures**
- Elemental analysis
- UV, visible, infrared, and NMR spectroscopy
- Use of acid or alkaline hydrolysis to degrade the biomolecule under study into its basic constituents
- Use of a battery of enzymes of known specificity to degrade the biomolecule under study (eg, proteases, nucleases, glycosidases)
- Mass spectrometry
- Specific sequencing methods (eg, for proteins and nucleic acids)
- X-ray crystallography

**Preparations for Studying Biochemical Processes**
- Whole animal (includes transgenic animals and animals with gene knockouts)
- Isolated perfused organ
- Tissue slice
- Whole cells
- Homogenate
- Isolated cell organelles
- Subfractionation of organelles
- Purified metabolites and enzymes
- Isolated genes (including polymerase chain reaction and site-directed mutagenesis)

[1]Most of these methods are suitable for analyzing the components present in cell homogenates and other biochemical preparations. The sequential use of several techniques will generally permit purification of most biomolecules. The reader is referred to texts on methods of biochemical research for details.

disease can open up areas of cell function for basic biochemical research.

The relationship between medicine and biochemistry has important implications for the former. As long as medical treatment is firmly grounded in a knowledge of biochemistry and other basic sciences, the practice of medicine will have a rational basis that can be adapted to accommodate new knowledge. This contrasts with unorthodox health cults and at least some "alternative medicine" practices, which are often founded on little more than myth and wishful thinking and generally lack any intellectual basis.

## NORMAL BIOCHEMICAL PROCESSES ARE THE BASIS OF HEALTH

The World Health Organization (WHO) defines health as a state of "complete physical, mental and social well-being and not merely the absence of disease and infirmity." From a strictly biochemical viewpoint, health may be considered that situation in which all of the many thousands of intra- and extracellular reactions that occur in the body are proceeding at rates commensurate with the organism's maximal survival in the physiologic state. However, this is an extremely reductionist view, and it should be apparent that caring for the health of patients requires not only a wide knowledge of biologic principles but also of psychologic and social principles.

### Biochemical Research Has Impact on Nutrition & Preventive Medicine

One major prerequisite for the maintenance of health is that there be optimal dietary intake of a number of chemicals; the chief of these are **vitamins,** certain **amino acids,** certain **fatty acids,** various **minerals,** and **water.** Because much of the subject matter of both biochemistry and nutrition is concerned with the study of various aspects of these chemicals, there is a close relationship between these two sciences. Moreover, more emphasis is being placed on systematic attempts to maintain health and forestall disease, ie, on **preventive medicine.** Thus, nutritional approaches to—for example—the prevention of atherosclerosis and cancer are receiving increased emphasis. Understanding nutrition depends to a great extent on a knowledge of biochemistry.

### Most & Perhaps All Disease Has a Biochemical Basis

We believe that most if not all diseases are manifestations of abnormalities of molecules, chemical reactions, or biochemical processes. The major factors responsible for causing diseases in animals and humans are listed in Table 1–2. All of them affect one or more critical chemical reactions or molecules in the body. Numerous examples of the biochemical bases of diseases will be encountered in this text; the majority of them are due to causes 5, 7, and 8. In most of these conditions, biochemical studies contribute to both the diagnosis and treatment. Some major uses of biochemical investigations and of laboratory tests in relation to diseases are summarized in Table 1–3.

Additional examples of many of these uses are presented in various sections of this text.

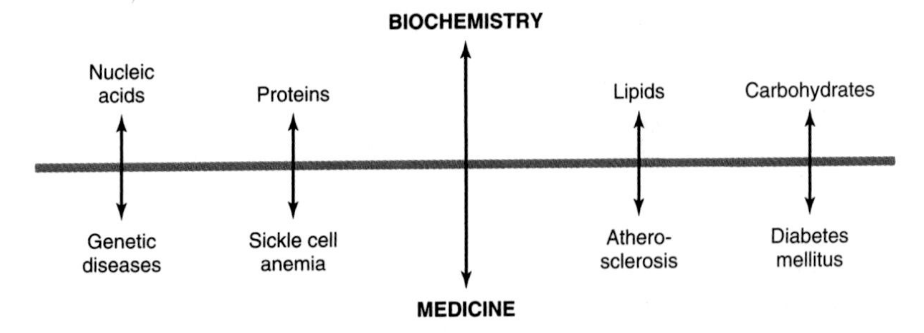


***Figure 1–1.*** Examples of the two-way street connecting biochemistry and medicine. Knowledge of the biochemical molecules shown in the top part of the diagram has clarified our understanding of the diseases shown in the bottom half—and conversely, analyses of the diseases shown below have cast light on many areas of biochemistry. Note that sickle cell anemia is a genetic disease and that both atherosclerosis and diabetes mellitus have genetic components.

## Impact of the Human Genome Project (HGP) on Biochemistry & Medicine

Remarkable progress was made in the late 1990s in sequencing the human genome. This culminated in July 2000, when leaders of the two groups involved in this effort (the International Human Genome Sequencing Consortium and Celera Genomics, a private company) announced that over 90% of the genome had been sequenced. Draft versions of the sequence were published

***Table 1–2.*** The major causes of diseases. All of the causes listed act by influencing the various biochemical mechanisms in the cell or in the body.[1]

1. Physical agents: Mechanical trauma, extremes of temperature, sudden changes in atmospheric pressure, radiation, electric shock.
2. Chemical agents, including drugs: Certain toxic compounds, therapeutic drugs, etc.
3. Biologic agents: Viruses, bacteria, fungi, higher forms of parasites.
4. Oxygen lack: Loss of blood supply, depletion of the oxygen-carrying capacity of the blood, poisoning of the oxidative enzymes.
5. Genetic disorders: Congenital, molecular.
6. Immunologic reactions: Anaphylaxis, autoimmune disease.
7. Nutritional imbalances: Deficiencies, excesses.
8. Endocrine imbalances: Hormonal deficiencies, excesses.

[1]Adapted, with permission, from Robbins SL, Cotram RS, Kumar V: *The Pathologic Basis of Disease*, 3rd ed. Saunders, 1984.

***Table 1–3.*** Some uses of biochemical investigations and laboratory tests in relation to diseases.

| Use | Example |
|---|---|
| 1. To reveal the fundamental causes and mechanisms of diseases | Demonstration of the nature of the genetic defects in cystic fibrosis. |
| 2. To suggest rational treatments of diseases based on (1) above | A diet low in phenylalanine for treatment of phenylketonuria. |
| 3. To assist in the diagnosis of specific diseases | Use of the plasma enzyme creatine kinase MB (CK-MB) in the diagnosis of myocardial infarction. |
| 4. To act as screening tests for the early diagnosis of certain diseases | Use of measurement of blood thyroxine or thyroid-stimulating hormone (TSH) in the neonatal diagnosis of congenital hypothyroidism. |
| 5. To assist in monitoring the progress (eg, recovery, worsening, remission, or relapse) of certain diseases | Use of the plasma enzyme alanine aminotransferase (ALT) in monitoring the progress of infectious hepatitis. |
| 6. To assist in assessing the response of diseases to therapy | Use of measurement of blood carcinoembryonic antigen (CEA) in certain patients who have been treated for cancer of the colon. |

in early 2001. It is anticipated that the entire sequence will be completed by 2003. The implications of this work for biochemistry, all of biology, and for medicine are tremendous, and only a few points are mentioned here. Many previously unknown genes have been revealed; their protein products await characterization. New light has been thrown on human evolution, and procedures for tracking disease genes have been greatly refined. The results are having major effects on areas such as proteomics, bioinformatics, biotechnology, and pharmacogenomics. Reference to the human genome will be made in various sections of this text. The Human Genome Project is discussed in more detail in Chapter 54.

## SUMMARY

- Biochemistry is the science concerned with studying the various molecules that occur in living cells and organisms and with their chemical reactions. Because life depends on biochemical reactions, biochemistry has become the basic language of all biologic sciences.
- Biochemistry is concerned with the entire spectrum of life forms, from relatively simple viruses and bacteria to complex human beings.
- Biochemistry and medicine are intimately related. Health depends on a harmonious balance of biochemical reactions occurring in the body, and disease reflects abnormalities in biomolecules, biochemical reactions, or biochemical processes.
- Advances in biochemical knowledge have illuminated many areas of medicine. Conversely, the study of diseases has often revealed previously unsuspected aspects of biochemistry. The determination of the sequence of the human genome, nearly complete, will have a great impact on all areas of biology, including biochemistry, bioinformatics, and biotechnology.
- Biochemical approaches are often fundamental in illuminating the causes of diseases and in designing appropriate therapies.
- The judicious use of various biochemical laboratory tests is an integral component of diagnosis and monitoring of treatment.
- A sound knowledge of biochemistry and of other related basic disciplines is essential for the rational practice of medical and related health sciences.

## REFERENCES

Fruton JS: *Proteins, Enzymes, Genes: The Interplay of Chemistry and Biology.* Yale Univ Press, 1999. (Provides the historical background for much of today's biochemical research.)

Garrod AE: Inborn errors of metabolism. (Croonian Lectures.) Lancet 1908;2:1, 73, 142, 214.

International Human Genome Sequencing Consortium. Initial sequencing and analysis of the human genome. Nature 2001:409;860. (The issue [15 February] consists of articles dedicated to analyses of the human genome.)

Kornberg A: Basic research: The lifeline of medicine. FASEB J 1992;6:3143.

Kornberg A: Centenary of the birth of modern biochemistry. FASEB J 1997;11:1209.

McKusick VA: *Mendelian Inheritance in Man. Catalogs of Human Genes and Genetic Disorders,* 12th ed. Johns Hopkins Univ Press, 1998. [Abbreviated MIM]

Online Mendelian Inheritance in Man (OMIM): Center for Medical Genetics, Johns Hopkins University and National Center for Biotechnology Information, National Library of Medicine, 1997. http://www.ncbi.nlm.nih.gov/omim/

(The numbers assigned to the entries in MIM and OMIM will be cited in selected chapters of this work. Consulting this extensive collection of diseases and other relevant entries—specific proteins, enzymes, etc—will greatly expand the reader's knowledge and understanding of various topics referred to and discussed in this text. The online version is updated almost daily.)

Scriver CR et al (editors): *The Metabolic and Molecular Bases of Inherited Disease,* 8th ed. McGraw-Hill, 2001.

Venter JC et al: The Sequence of the Human Genome. Science 2001;291:1304. (The issue [16 February] contains the Celera draft version and other articles dedicated to analyses of the human genome.)

Williams DL, Marks V: *Scientific Foundations of Biochemistry in Clinical Practice,* 2nd ed. Butterworth-Heinemann, 1994.

# Water & pH

# 2

*Victor W. Rodwell, PhD, & Peter J. Kennelly, PhD*

## BIOMEDICAL IMPORTANCE

Water is the predominant chemical component of living organisms. Its unique physical properties, which include the ability to solvate a wide range of organic and inorganic molecules, derive from water's dipolar structure and exceptional capacity for forming hydrogen bonds. The manner in which water interacts with a solvated biomolecule influences the structure of each. An excellent nucleophile, water is a reactant or product in many metabolic reactions. Water has a slight propensity to dissociate into hydroxide ions and protons. The acidity of aqueous solutions is generally reported using the logarithmic pH scale. Bicarbonate and other buffers normally maintain the pH of extracellular fluid between 7.35 and 7.45. Suspected disturbances of acid-base balance are verified by measuring the pH of arterial blood and the $CO_2$ content of venous blood. Causes of acidosis (blood pH < 7.35) include diabetic ketosis and lactic acidosis. Alkalosis (pH > 7.45) may, for example, follow vomiting of acidic gastric contents. Regulation of water balance depends upon hypothalamic mechanisms that control thirst, on antidiuretic hormone (ADH), on retention or excretion of water by the kidneys, and on evaporative loss. Nephrogenic diabetes insipidus, which involves the inability to concentrate urine or adjust to subtle changes in extracellular fluid osmolarity, results from the unresponsiveness of renal tubular osmoreceptors to ADH.

## WATER IS AN IDEAL BIOLOGIC SOLVENT

### Water Molecules Form Dipoles

A water molecule is an irregular, slightly skewed tetrahedron with oxygen at its center (Figure 2–1). The two hydrogens and the unshared electrons of the remaining two $sp^3$-hybridized orbitals occupy the corners of the tetrahedron. The 105-degree angle between the hydrogens differs slightly from the ideal tetrahedral angle, 109.5 degrees. Ammonia is also tetrahedral, with a 107-degree angle between its hydrogens. Water is a **dipole,** a molecule with electrical charge distributed asymmetrically about its structure. The strongly electronegative oxygen atom pulls electrons away from the hydrogen nuclei, leaving them with a partial positive charge, while its two unshared electron pairs constitute a region of local negative charge.

Water, a strong dipole, has a high **dielectric constant.** As described quantitatively by Coulomb's law, the strength of interaction F between oppositely charged particles is inversely proportionate to the dielectric constant ε of the surrounding medium. The dielectric constant for a vacuum is unity; for hexane it is 1.9; for ethanol it is 24.3; and for water it is 78.5. Water therefore greatly decreases the force of attraction between charged and polar species relative to water-free environments with lower dielectric constants. Its strong dipole and high dielectric constant enable water to dissolve large quantities of charged compounds such as salts.

### Water Molecules Form Hydrogen Bonds

An unshielded hydrogen nucleus covalently bound to an electron-withdrawing oxygen or nitrogen atom can interact with an unshared electron pair on another oxygen or nitrogen atom to form a **hydrogen bond.** Since water molecules contain both of these features, hydrogen bonding favors the self-association of water molecules into ordered arrays (Figure 2–2). Hydrogen bonding profoundly influences the physical properties of water and accounts for its exceptionally high viscosity, surface tension, and boiling point. On average, each molecule in liquid water associates through hydrogen bonds with 3.5 others. These bonds are both relatively weak and transient, with a half-life of about one microsecond. Rupture of a hydrogen bond in liquid water requires only about 4.5 kcal/mol, less than 5% of the energy required to rupture a covalent O—H bond.

Hydrogen bonding enables water to dissolve many organic biomolecules that contain functional groups which can participate in hydrogen bonding. The oxygen atoms of aldehydes, ketones, and amides provide pairs of electrons that can serve as hydrogen acceptors. Alcohols and amines can serve both as hydrogen acceptors and as donors of unshielded hydrogen atoms for formation of hydrogen bonds (Figure 2–3).

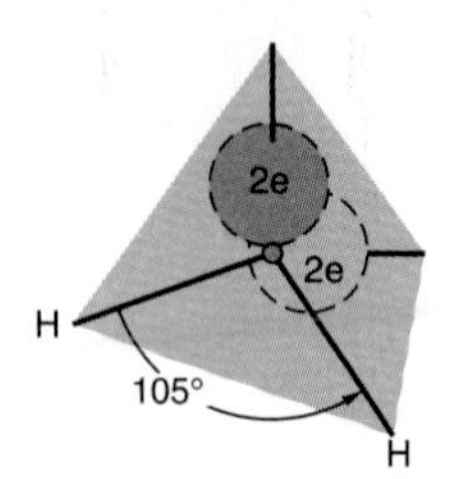


***Figure 2–1.*** The water molecule has tetrahedral geometry.

# INTERACTION WITH WATER INFLUENCES THE STRUCTURE OF BIOMOLECULES

## Covalent & Noncovalent Bonds Stabilize Biologic Molecules

The covalent bond is the strongest force that holds molecules together (Table 2–1). Noncovalent forces, while of lesser magnitude, make significant contributions to the structure, stability, and functional competence of macromolecules in living cells. These forces, which can be either attractive or repulsive, involve interactions both within the biomolecule and between it and the water that forms the principal component of the surrounding environment.

## Biomolecules Fold to Position Polar & Charged Groups on Their Surfaces

Most biomolecules are **amphipathic;** that is, they possess regions rich in charged or polar functional groups as well as regions with hydrophobic character. Proteins tend to fold with the R-groups of amino acids with hydrophobic side chains in the interior. Amino acids with charged or polar amino acid side chains (eg, arginine, glutamate, serine) generally are present on the surface in contact with water. A similar pattern prevails in a phospholipid bilayer, where the charged head groups of phosphatidyl serine or phosphatidyl ethanolamine contact water while their hydrophobic fatty acyl side chains cluster together, excluding water. This pattern maximizes the opportunities for the formation of energetically favorable charge-dipole, dipole-dipole, and hydrogen bonding interactions between polar groups on the biomolecule and water. It also minimizes energetically unfavorable contact between water and hydrophobic groups.

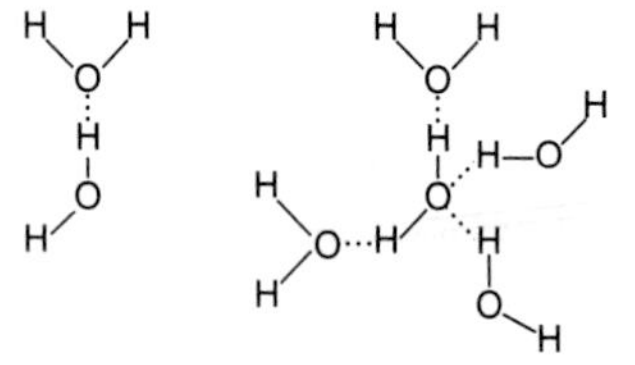

***Figure 2–2.*** **Left:** Association of two dipolar water molecules by a hydrogen bond (dotted line). **Right:** Hydrogen-bonded cluster of four water molecules. Note that water can serve simultaneously both as a hydrogen donor and as a hydrogen acceptor.

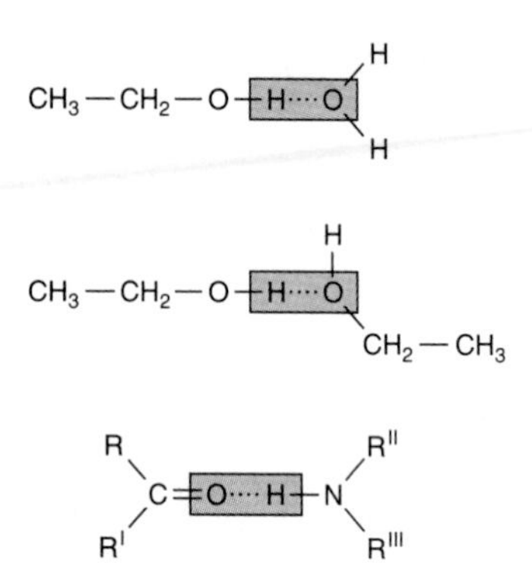


***Figure 2–3.*** Additional polar groups participate in hydrogen bonding. Shown are hydrogen bonds formed between an alcohol and water, between two molecules of ethanol, and between the peptide carbonyl oxygen and the peptide nitrogen hydrogen of an adjacent amino acid.

## Hydrophobic Interactions

Hydrophobic interaction refers to the tendency of nonpolar compounds to self-associate in an aqueous environment. This self-association is driven neither by mutual attraction nor by what are sometimes incorrectly referred to as "hydrophobic bonds." Self-association arises from the need to minimize energetically unfavorable interactions between nonpolar groups and water.

***Table 2–1.*** Bond energies for atoms of biologic significance.

| Bond Type | Energy (kcal/mol) | Bond Type | Energy (kcal/mol) |
|---|---|---|---|
| O—O | 34 | O═O | 96 |
| S—S | 51 | C—H | 99 |
| C—N | 70 | C═S | 108 |
| S—H | 81 | O—H | 110 |
| C—C | 82 | C═C | 147 |
| C—O | 84 | C═N | 147 |
| N—H | 94 | C═O | 164 |

While the hydrogens of nonpolar groups such as the methylene groups of hydrocarbons do not form hydrogen bonds, they do affect the structure of the water that surrounds them. Water molecules adjacent to a hydrophobic group are restricted in the number of orientations (degrees of freedom) that permit them to participate in the maximum number of energetically favorable hydrogen bonds. Maximal formation of multiple hydrogen bonds can be maintained only by increasing the order of the adjacent water molecules, with a corresponding decrease in entropy.

It follows from the second law of thermodynamics that the optimal free energy of a hydrocarbon-water mixture is a function of both maximal enthalpy (from hydrogen bonding) and minimum entropy (maximum degrees of freedom). Thus, nonpolar molecules tend to form droplets with minimal exposed surface area, reducing the number of water molecules affected. For the same reason, in the aqueous environment of the living cell the hydrophobic portions of biopolymers tend to be buried inside the structure of the molecule, or within a lipid bilayer, minimizing contact with water.

### Electrostatic Interactions

Interactions between charged groups shape biomolecular structure. Electrostatic interactions between oppositely charged groups within or between biomolecules are termed **salt bridges.** Salt bridges are comparable in strength to hydrogen bonds but act over larger distances. They thus often facilitate the binding of charged molecules and ions to proteins and nucleic acids.

### Van der Waals Forces

Van der Waals forces arise from attractions between transient dipoles generated by the rapid movement of electrons on all neutral atoms. Significantly weaker than hydrogen bonds but potentially extremely numerous, van der Waals forces decrease as the sixth power of the distance separating atoms. Thus, they act over very short distances, typically 2–4 Å.

### Multiple Forces Stabilize Biomolecules

The DNA double helix illustrates the contribution of multiple forces to the structure of biomolecules. While each individual DNA strand is held together by covalent bonds, the two strands of the helix are held together exclusively by noncovalent interactions. These noncovalent interactions include hydrogen bonds between nucleotide bases (Watson-Crick base pairing) and van der Waals interactions between the stacked purine and pyrimidine bases. The helix presents the charged phosphate groups and polar ribose sugars of the backbone to water while burying the relatively hydrophobic nucleotide bases inside. The extended backbone maximizes the distance between negatively charged backbone phosphates, minimizing unfavorable electrostatic interactions.

## WATER IS AN EXCELLENT NUCLEOPHILE

Metabolic reactions often involve the attack by lone pairs of electrons on electron-rich molecules termed **nucleophiles** on electron-poor atoms called **electrophiles.** Nucleophiles and electrophiles do not necessarily possess a formal negative or positive charge. Water, whose two lone pairs of $sp^3$ electrons bear a partial negative charge, is an excellent nucleophile. Other nucleophiles of biologic importance include the oxygen atoms of phosphates, alcohols, and carboxylic acids; the sulfur of thiols; the nitrogen of amines; and the imidazole ring of histidine. Common electrophiles include the carbonyl carbons in amides, esters, aldehydes, and ketones and the phosphorus atoms of phosphoesters.

Nucleophilic attack by water generally results in the cleavage of the amide, glycoside, or ester bonds that hold biopolymers together. This process is termed **hydrolysis.** Conversely, when monomer units are joined together to form biopolymers such as proteins or glycogen, water is a product, as shown below for the formation of a peptide bond between two amino acids.

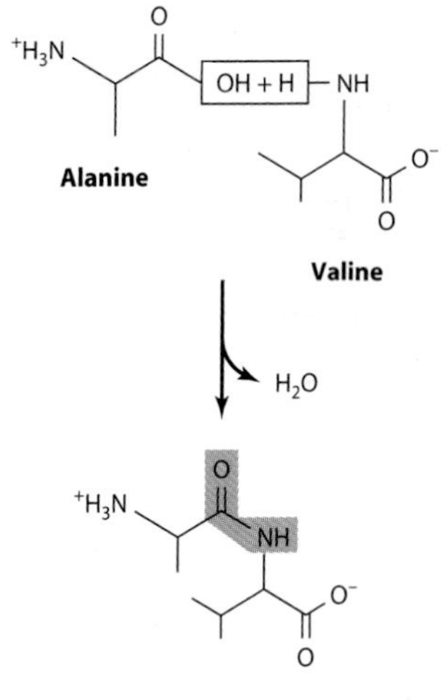

Alanine
Valine
$H_2O$

While hydrolysis is a thermodynamically favored reaction, the amide and phosphoester bonds of polypeptides and oligonucleotides are stable in the aqueous environment of the cell. This seemingly paradoxic behavior reflects the fact that the thermodynamics governing the equilibrium of a reaction do not determine the rate at which it will take place. In the cell, protein catalysts called **enzymes** are used to accelerate the rate of hydrolytic reactions when needed. **Proteases** catalyze the hydrolysis of proteins into their component amino acids, while **nucleases** catalyze the hydrolysis of the phosphoester bonds in DNA and RNA. Careful control of the activities of these enzymes is required to ensure that they act only on appropriate target molecules.

## Many Metabolic Reactions Involve Group Transfer

In group transfer reactions, a group G is transferred from a donor D to an acceptor A, forming an acceptor group complex A–G:

$$D-G + A = A-G + D$$

The hydrolysis and phosphorolysis of glycogen represent group transfer reactions in which glucosyl groups are transferred to water or to orthophosphate. The equilibrium constant for the hydrolysis of covalent bonds strongly favors the formation of split products. The biosynthesis of macromolecules also involves group transfer reactions in which the thermodynamically unfavored synthesis of covalent bonds is coupled to favored reactions so that the overall change in free energy favors biopolymer synthesis. Given the nucleophilic character of water and its high concentration in cells, why are biopolymers such as proteins and DNA relatively stable? And how can synthesis of biopolymers occur in an apparently aqueous environment? Central to both questions are the properties of enzymes. In the absence of enzymic catalysis, even thermodynamically highly favored reactions do not necessarily take place rapidly. Precise and differential control of enzyme activity and the sequestration of enzymes in specific organelles determine under what physiologic conditions a given biopolymer will be synthesized or degraded. Newly synthesized polymers are not immediately hydrolyzed, in part because the active sites of biosynthetic enzymes sequester substrates in an environment from which water can be excluded.

## Water Molecules Exhibit a Slight but Important Tendency to Dissociate

The ability of water to ionize, while slight, is of central importance for life. Since water can act both as an acid and as a base, its ionization may be represented as an intermolecular proton transfer that forms a hydronium ion ($H_3O^+$) and a hydroxide ion ($OH^-$):

$$H_2O + H_2O \quad H_3O^+ + OH^-$$

The transferred proton is actually associated with a cluster of water molecules. Protons exist in solution not only as $H_3O^+$, but also as multimers such as $H_5O_2^+$ and $H_7O_3^+$. The proton is nevertheless routinely represented as $H^+$, even though it is in fact highly hydrated.

Since hydronium and hydroxide ions continuously recombine to form water molecules, an *individual* hydrogen or oxygen cannot be stated to be present as an ion or as part of a water molecule. At one instant it is an ion. An instant later it is part of a molecule. Individual ions or molecules are therefore not considered. We refer instead to the *probability* that at any instant in time a hydrogen will be present as an ion or as part of a water molecule. Since 1 g of water contains $3.46 \times 10^{22}$ molecules, the ionization of water can be described statistically. To state that the probability that a hydrogen exists as an ion is 0.01 means that a hydrogen atom has one chance in 100 of being an ion and 99 chances out of 100 of being part of a water molecule. The actual probability of a hydrogen atom in pure water existing as a hydrogen ion is approximately $1.8 \times 10^{-9}$. The probability of its being part of a molecule thus is almost unity. Stated another way, for every hydrogen ion and hydroxyl ion in pure water there are 1.8 billion or $1.8 \times 10^9$ water molecules. Hydrogen ions and hydroxyl ions nevertheless contribute significantly to the properties of water.

For dissociation of water,

$$K = \frac{[H^+][OH^-]}{[H_2O]}$$

where brackets represent molar concentrations (strictly speaking, molar activities) and $K$ is the **dissociation constant.** Since one mole (mol) of water weighs 18 g, one liter (L) (1000 g) of water contains $1000 \times 18 = 55.56$ mol. Pure water thus is 55.56 molar. Since the probability that a hydrogen in pure water will exist as a hydrogen ion is $1.8 \times 10^{-9}$, the molar concentration of $H^+$ ions (or of $OH^-$ ions) in pure water is the product of the probability, $1.8 \times 10^{-9}$, times the molar concentration of water, 55.56 mol/L. The result is $1.0 \times 10^{-7}$ mol/L.

We can now calculate $K$ for water:

$$K = \frac{[H^+][OH^-]}{[H_2O]} = \frac{[10^{-7}][10^{-7}]}{[55.56]}$$
$$= 0.018 \times 10^{-14} = 1.8 \times 10^{-16} \text{mol/L}$$

The molar concentration of water, 55.56 mol/L, is too great to be significantly affected by dissociation. It therefore is considered to be essentially constant. This constant may then be incorporated into the dissociation constant $K$ to provide a useful new constant $K_w$ termed the **ion product** for water. The relationship between $K_w$ and $K$ is shown below:

$$K = \frac{[H^+][OH^-]}{[H_2O]} = 1.8 \times 10^{-16} \text{ mol/L}$$

$$K_w = (K)[H_2O] = [H^+][OH^-]$$
$$= (1.8 \times 10^{-16} \text{ mol/L})(55.56 \text{ mol/L})$$
$$= 1.00 \times 10^{-14} (\text{mol/L})^2$$

Note that the dimensions of $K$ are moles per liter and those of $K_w$ are moles$^2$ per liter$^2$. As its name suggests, the ion product $K_w$ is numerically equal to the product of the molar concentrations of $H^+$ and $OH^-$:

$$K_w = [H^+][OH^-]$$

At 25 °C, $K_w = (10^{-7})^2$, or $10^{-14}$ (mol/L)$^2$. At temperatures below 25 °C, $K_w$ is somewhat less than $10^{-14}$; and at temperatures above 25 °C it is somewhat greater than $10^{-14}$. Within the stated limitations of the effect of temperature, **$K_w$ equals $10^{-14}$ (mol/L)$^2$ for all aqueous solutions, even solutions of acids or bases.** We shall use $K_w$ to calculate the pH of acidic and basic solutions.

## pH IS THE NEGATIVE LOG OF THE HYDROGEN ION CONCENTRATION

The term **pH** was introduced in 1909 by Sörensen, who defined pH as the negative log of the hydrogen ion concentration:

$$pH = -\log [H^+]$$

This definition, while not rigorous, suffices for many biochemical purposes. To calculate the pH of a solution:

1. Calculate hydrogen ion concentration $[H^+]$.
2. Calculate the base 10 logarithm of $[H^+]$.
3. pH is the negative of the value found in step 2.

For example, for pure water at 25°C,

$$pH = -\log [H^+] = -\log 10^{-7} = -(-7) = 7.0$$

Low pH values correspond to high concentrations of $H^+$ and high pH values correspond to low concentrations of $H^+$.

Acids are **proton donors** and bases are **proton acceptors. Strong acids** (eg, HCl or $H_2SO_4$) completely dissociate into anions and cations even in strongly acidic solutions (low pH). **Weak acids** dissociate only partially in acidic solutions. Similarly, **strong bases** (eg, KOH or NaOH)—but not **weak bases** (eg, $Ca[OH]_2$)—are completely dissociated at high pH. Many biochemicals are weak acids. Exceptions include phosphorylated intermediates, whose phosphoryl group contains two dissociable protons, the first of which is strongly acidic.

The following examples illustrate how to calculate the pH of acidic and basic solutions.

***Example 1:*** What is the pH of a solution whose hydrogen ion concentration is $3.2 \times 10^{-4}$ mol/L?

$$pH = -\log [H^+]$$
$$= -\log (3.2 \times 10^{-4})$$
$$= -\log (3.2) - \log (10^{-4})$$
$$= -0.5 + 4.0$$
$$= 3.5$$

***Example 2:*** What is the pH of a solution whose hydroxide ion concentration is $4.0 \times 10^{-4}$ mol/L? We first define a quantity **pOH** that is equal to $-\log [OH^-]$ and that may be derived from the definition of $K_w$:

$$K_w = [H^+][OH^-] = 10^{-14}$$

Therefore:

$$\log [H^+] + \log [OH^-] = \log 10^{-14}$$

or

$$pH + pOH = 14$$

To solve the problem by this approach:

$$[OH^-] = 4.0 \times 10^{-4}$$
$$pOH = -\log [OH^-]$$
$$= -\log (4.0 \times 10^{-4})$$
$$= -\log (4.0) - \log (10^{-4})$$
$$= -0.60 + 4.0$$
$$= 3.4$$

Now:

$$pH = 14 - pOH = 14 - 3.4$$
$$= 10.6$$

***Example 3:*** What are the pH values of (a) $2.0 \times 10^{-2}$ mol/L KOH and of (b) $2.0 \times 10^{-6}$ mol/L KOH? The $OH^-$ arises from two sources, KOH and water. Since pH is determined by the total $[H^+]$ (and pOH by the total $[OH^-]$), both sources must be considered. In the first case (a), the contribution of water to the total $[OH^-]$ is negligible. The same cannot be said for the second case (b):

| | Concentration (mol/L) | |
|---|---|---|
| | (a) | (b) |
| Molarity of KOH | $2.0 \times 10^{-2}$ | $2.0 \times 10^{-6}$ |
| $[OH^-]$ from KOH | $2.0 \times 10^{-2}$ | $2.0 \times 10^{-6}$ |
| $[OH^-]$ from water | $1.0 \times 10^{-7}$ | $1.0 \times 10^{-7}$ |
| Total $[OH^-]$ | $2.00001 \times 10^{-2}$ | $2.1 \times 10^{-6}$ |

Once a decision has been reached about the significance of the contribution by water, pH may be calculated as above.

The above examples assume that the strong base KOH is completely dissociated in solution and that the concentration of $OH^-$ ions was thus equal to that of the KOH. This assumption is valid for dilute solutions of strong bases or acids but not for weak bases or acids. Since weak electrolytes dissociate only slightly in solution, we must use the **dissociation constant** to calculate the concentration of $[H^+]$ (or $[OH^-]$) produced by a given molarity of a weak acid (or base) before calculating total $[H^+]$ (or total $[OH^-]$) and subsequently pH.

## Functional Groups That Are Weak Acids Have Great Physiologic Significance

Many biochemicals possess functional groups that are weak acids or bases. Carboxyl groups, amino groups, and the second phosphate dissociation of phosphate esters are present in proteins and nucleic acids, most coenzymes, and most intermediary metabolites. Knowledge of the dissociation of weak acids and bases thus is basic to understanding the influence of intracellular pH on structure and biologic activity. Charge-based separations such as electrophoresis and ion exchange chromatography also are best understood in terms of the dissociation behavior of functional groups.

We term the protonated species (eg, HA or $R—NH_3^+$) the **acid** and the unprotonated species (eg, $A^-$ or $R—NH_2$) its **conjugate base.** Similarly, we may refer to a **base** (eg, $A^-$ or $R—NH_2$) and its **conjugate acid** (eg, HA or $R—NH_3^+$). Representative weak acids (left), their conjugate bases (center), and the $pK_a$ values (right) include the following:

$$
\begin{array}{lll}
R—CH_2—COOH & R—CH_2—COO^- & pK_a = 4-5 \\
R—CH_2—NH_3^+ & R—CH_2—NH_2 & pK_a = 9-10 \\
H_2CO_3 & HCO_3^- & pK_a = 6.4 \\
H_2PO_4^- & HPO_4^{-2} & pK_a = 7.2
\end{array}
$$

We express the relative strengths of weak acids and bases in terms of their dissociation constants. Shown below are the expressions for the dissociation constant ($K_a$) for two representative weak acids, R—COOH and $R—NH_3^+$.

$$R—COOH \quad R—COO^- + H^+$$

$$K_a = \frac{[R—COO^-][H^+]}{[R—COOH]}$$

$$R—NH_3^+ \quad R—NH_2 + H^+$$

$$K_a = \frac{[R—NH_2][H^+]}{[R—NH_3^+]}$$

Since the numeric values of $K_a$ for weak acids are negative exponential numbers, we express $K_a$ as $pK_a$, where

$$pK_a = -\log K$$

Note that $pK_a$ is related to $K_a$ as pH is to $[H^+]$. The stronger the acid, the lower its $pK_a$ value.

$pK_a$ is used to express the relative strengths of both acids and bases. For any weak acid, its conjugate is a strong base. Similarly, the conjugate of a strong base is a weak acid. The relative strengths of bases are expressed in terms of the $pK_a$ of their conjugate acids. For polyproteic compounds containing more than one dissociable proton, a numerical subscript is assigned to each in order of relative acidity. For a dissociation of the type

$$R—NH_3^+ \rightarrow R—NH_2$$

the $pK_a$ is the pH at which the concentration of the acid $R—NH_3^+$ equals that of the base $R—NH_2$.

From the above equations that relate $K_a$ to $[H^+]$ and to the concentrations of undissociated acid and its conjugate base, when

$$[R—COO^-] = [R—COOH]$$

or when

$$[R—NH_2] = [R—NH_3^+]$$

then

$$K_a = [H^+]$$

Thus, when the associated (protonated) and dissociated (conjugate base) species are present at equal concentrations, the prevailing hydrogen ion concentration $[H^+]$ is numerically equal to the dissociation constant, $K_a$. If the logarithms of both sides of the above equation are taken and both sides are multiplied by −1, the expressions would be as follows:

$$K_a = [H^+]$$
$$-\log K_a = -\log [H^+]$$

Since $-\log K_a$ is defined as $pK_a$, and $-\log [H^+]$ defines pH, the equation may be rewritten as

$$pK_a = pH$$

ie, the $pK_a$ of an acid group is the pH at which the protonated and unprotonated species are present at equal concentrations. The $pK_a$ for an acid may be determined by adding 0.5 equivalent of alkali per equivalent of acid. The resulting pH will be the $pK_a$ of the acid.

## The Henderson-Hasselbalch Equation Describes the Behavior of Weak Acids & Buffers

The Henderson-Hasselbalch equation is derived below.

A weak acid, HA, ionizes as follows:

$$HA \quad H^+ + A^-$$

The equilibrium constant for this dissociation is

$$K_a = \frac{[H^+][A^-]}{[HA]}$$

Cross-multiplication gives

$$[H^+][A^-] = K_a[HA]$$

Divide both sides by $[A^-]$:

$$[H^+] = K_a \frac{[HA]}{[A^-]}$$

Take the log of both sides:

$$\log [H^+] = \log \left( K_a \frac{[HA]}{[A^-]} \right)$$
$$= \log K_a + \log \frac{[HA]}{[A^-]}$$

Multiply through by −1:

$$-\log [H^+] = -\log K_a - \log \frac{[HA]}{[A^-]}$$

Substitute pH and $pK_a$ for $-\log [H^+]$ and $-\log K_a$, respectively; then:

$$pH = pK_a - \log \frac{[HA]}{[A^-]}$$

Inversion of the last term removes the minus sign and gives the Henderson-Hasselbalch equation:

$$pH = pK_a + \log \frac{[A^-]}{[HA]}$$

The Henderson-Hasselbalch equation has great predictive value in protonic equilibria. For example,

(1) When an acid is exactly half-neutralized, $[A^-] = [HA]$. Under these conditions,

$$pH = pK_a + \log \frac{[A^-]}{[HA]} = pK_a + \log \frac{1}{1} = pK_a + 0$$

Therefore, at half-neutralization, $pH = pK_a$.

(2) When the ratio $[A^-]/[HA] = 100:1$,

$$pH = pK_a + \log \frac{[A^-]}{[HA]}$$
$$pH = pK_a + \log 100/1 = pK_a + 2$$

(3) When the ratio $[A^-]/[HA] = 1:10$,

$$pH = pK_a + \log 1/10 = pK_a + (-1)$$

If the equation is evaluated at ratios of $[A^-]/[HA]$ ranging from $10^3$ to $10^{-3}$ and the calculated pH values are plotted, the resulting graph describes the titration curve for a weak acid (Figure 2–4).

## Solutions of Weak Acids & Their Salts Buffer Changes in pH

Solutions of weak acids or bases and their conjugates exhibit buffering, the ability to resist a change in pH following addition of strong acid or base. Since many metabolic reactions are accompanied by the release or uptake of protons, most intracellular reactions are buffered. Oxidative metabolism produces $CO_2$, the anhydride of carbonic acid, which if not buffered would produce severe acidosis. Maintenance of a constant pH involves buffering by phosphate, bicarbonate, and proteins, which accept or release protons to resist a change

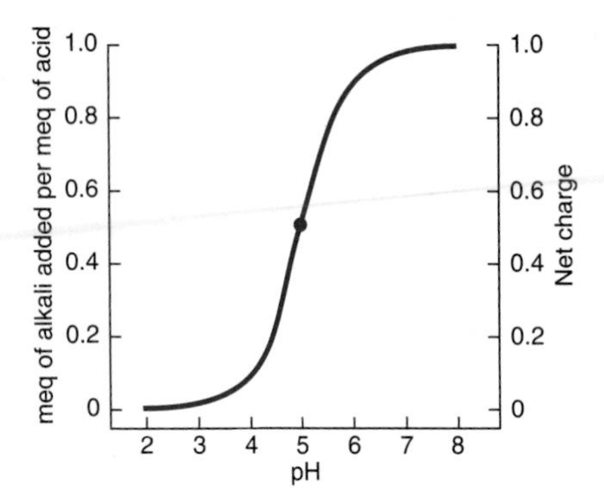


***Figure 2–4.*** Titration curve for an acid of the type HA. The heavy dot in the center of the curve indicates the p$K_a$ 5.0.

in pH. For experiments using tissue extracts or enzymes, constant pH is maintained by the addition of buffers such as MES ([2-*N*-morpholino]ethanesulfonic acid, p$K_a$ 6.1), inorganic orthophosphate (p$K_{a2}$ 7.2), HEPES (*N*-hydroxyethylpiperazine-*N'*-2-ethanesulfonic acid, p$K_a$ 6.8), or Tris (tris[hydroxymethyl] aminomethane, p$K_a$ 8.3). The value of p$K_a$ relative to the desired pH is the major determinant of which buffer is selected.

Buffering can be observed by using a pH meter while titrating a weak acid or base (Figure 2–4). We can also calculate the pH shift that accompanies addition of acid or base to a buffered solution. In the example, the buffered solution (a weak acid, p$K_a$ = 5.0, and its conjugate base) is initially at one of four pH values. We will calculate the pH shift that results when 0.1 meq of KOH is added to 1 meq of each solution:

| | | | | |
|---|---|---|---|---|
| Initial pH | 5.00 | 5.37 | 5.60 | 5.86 |
| $[A^-]_{initial}$ | 0.50 | 0.70 | 0.80 | 0.88 |
| $[HA]_{initial}$ | 0.50 | 0.30 | 0.20 | 0.12 |
| $([A^-]/[HA])_{initial}$ | 1.00 | 2.33 | 4.00 | 7.33 |
| Addition of 0.1 meq of KOH produces | | | | |
| $[A^-]_{final}$ | 0.60 | 0.80 | 0.90 | 0.98 |
| $[HA]_{final}$ | 0.40 | 0.20 | 0.10 | 0.02 |
| $([A^-]/[HA])_{final}$ | 1.50 | 4.00 | 9.00 | 49.0 |
| $\log([A^-]/[HA])_{final}$ | 0.176 | 0.602 | 0.95 | 1.69 |
| Final pH | 5.18 | 5.60 | 5.95 | 6.69 |
| ΔpH | 0.18 | 0.60 | 0.95 | 1.69 |

Notice that the change in pH per milliequivalent of $OH^-$ added depends on the initial pH. The solution resists changes in pH most effectively at pH values close to the p$K_a$. A solution of a weak acid and its conjugate base buffers most effectively in the pH range p$K_a$ ± 1.0 pH unit.

Figure 2–4 also illustrates the net charge on one molecule of the acid as a function of pH. A fractional charge of −0.5 does not mean that an individual molecule bears a fractional charge, but the *probability* that a given molecule has a unit negative charge is 0.5. Consideration of the net charge on macromolecules as a function of pH provides the basis for separatory techniques such as ion exchange chromatography and electrophoresis.

## Acid Strength Depends on Molecular Structure

Many acids of biologic interest possess more than one dissociating group. The presence of adjacent negative charge hinders the release of a proton from a nearby group, raising its p$K_a$. This is apparent from the p$K_a$ values for the three dissociating groups of phosphoric acid and citric acid (Table 2–2). The effect of adjacent charge decreases with distance. The second p$K_a$ for succinic acid, which has two methylene groups between its carboxyl groups, is 5.6, whereas the second p$K_a$ for glu-

***Table 2–2.*** Relative strengths of selected acids of biologic significance. Tabulated values are the p$K_a$ values (−log of the dissociation constant) of selected monoprotic, diprotic, and triprotic acids.

| | | |
|---|---|---|
| **Monoprotic Acids** | | |
| Formic | p$K$ | 3.75 |
| Lactic | p$K$ | 3.86 |
| Acetic | p$K$ | 4.76 |
| Ammonium ion | p$K$ | 9.25 |
| **Diprotic Acids** | | |
| Carbonic | p$K_1$ | 6.37 |
| | p$K_2$ | 10.25 |
| Succinic | p$K_1$ | 4.21 |
| | p$K_2$ | 5.64 |
| Glutaric | p$K_1$ | 4.34 |
| | p$K_2$ | 5.41 |
| **Triprotic Acids** | | |
| Phosphoric | p$K_1$ | 2.15 |
| | p$K_2$ | 6.82 |
| | p$K_3$ | 12.38 |
| Citric | p$K_1$ | 3.08 |
| | p$K_2$ | 4.74 |
| | p$K_3$ | 5.40 |

taric acid, which has one additional methylene group, is 5.4.

## p$K_a$ Values Depend on the Properties of the Medium

The p$K_a$ of a functional group is also profoundly influenced by the surrounding medium. The medium may either raise or lower the p$K_a$ depending on whether the undissociated acid or its conjugate base is the charged species. The effect of dielectric constant on p$K_a$ may be observed by adding ethanol to water. The p$K_a$ of a carboxylic acid *increases,* whereas that of an amine *decreases* because ethanol decreases the ability of water to solvate a charged species. The p$K_a$ values of dissociating groups in the interiors of proteins thus are profoundly affected by their local environment, including the presence or absence of water.

## SUMMARY


- Water forms hydrogen-bonded clusters with itself and with other proton donors or acceptors. Hydrogen bonds account for the surface tension, viscosity, liquid state at room temperature, and solvent power of water.
- Compounds that contain O, N, or S can serve as hydrogen bond donors or acceptors.
- Macromolecules exchange internal surface hydrogen bonds for hydrogen bonds to water. Entropic forces dictate that macromolecules expose polar regions to an aqueous interface and bury nonpolar regions.
- Salt bonds, hydrophobic interactions, and van der Waals forces participate in maintaining molecular structure.
- pH is the negative log of [$H^+$]. A low pH characterizes an acidic solution, and a high pH denotes a basic solution.
- The strength of weak acids is expressed by p$K_a$, the negative log of the acid dissociation constant. Strong acids have low p$K_a$ values and weak acids have high p$K_a$ values.
- Buffers resist a change in pH when protons are produced or consumed. Maximum buffering capacity occurs ± 1 pH unit on either side of p$K_a$. Physiologic buffers include bicarbonate, orthophosphate, and proteins.


## REFERENCES


Segel IM: *Biochemical Calculations.* Wiley, 1968.

Wiggins PM: Role of water in some biological processes. Microbiol Rev 1990;54:432.

# SECTION I

## Structures & Functions of Proteins & Enzymes

# Amino Acids & Peptides 3

*Victor W. Rodwell, PhD, & Peter J. Kennelly, PhD*

## BIOMEDICAL IMPORTANCE

In addition to providing the monomer units from which the long polypeptide chains of proteins are synthesized, the L-α-amino acids and their derivatives participate in cellular functions as diverse as nerve transmission and the biosynthesis of porphyrins, purines, pyrimidines, and urea. Short polymers of amino acids called peptides perform prominent roles in the neuroendocrine system as hormones, hormone-releasing factors, neuromodulators, or neurotransmitters. While proteins contain only L-α-amino acids, microorganisms elaborate peptides that contain both D- and L-α-amino acids. Several of these peptides are of therapeutic value, including the antibiotics bacitracin and gramicidin A and the antitumor agent bleomycin. Certain other microbial peptides are toxic. The cyanobacterial peptides microcystin and nodularin are lethal in large doses, while small quantities promote the formation of hepatic tumors. Neither humans nor any other higher animals can synthesize 10 of the 20 common L-α-amino acids in amounts adequate to support infant growth or to maintain health in adults. Consequently, the human diet must contain adequate quantities of these nutritionally essential amino acids.

## PROPERTIES OF AMINO ACIDS

### The Genetic Code Specifies 20 L-α-Amino Acids

Of the over 300 naturally occurring amino acids, 20 constitute the monomer units of proteins. While a nonredundant three-letter genetic code could accommodate more than 20 amino acids, its redundancy limits the available codons to the 20 L-α-amino acids listed in Table 3–1, classified according to the polarity of their R groups. Both one- and three-letter abbreviations for each amino acid can be used to represent the amino acids in peptides (Table 3–1). Some proteins contain additional amino acids that arise by modification of an amino acid already present in a peptide. Examples include conversion of peptidyl proline and lysine to 4-hydroxyproline and 5-hydroxylysine; the conversion of peptidyl glutamate to γ-carboxyglutamate; and the methylation, formylation, acetylation, prenylation, and phosphorylation of certain aminoacyl residues. These modifications extend the biologic diversity of proteins by altering their solubility, stability, and interaction with other proteins.

### Only L-α-Amino Acids Occur in Proteins

With the sole exception of glycine, the α-carbon of amino acids is chiral. Although some protein amino acids are dextrorotatory and some levorotatory, all share the absolute configuration of L-glyceraldehyde and thus are L-α-amino acids. Several free L-α-amino acids fulfill important roles in metabolic processes. Examples include ornithine, citrulline, and argininosuccinate that participate in urea synthesis; tyrosine in formation of thyroid hormones; and glutamate in neurotransmitter biosynthesis. D-Amino acids that occur naturally include free D-serine and D-aspartate in brain tissue, D-alanine and D-glutamate in the cell walls of gram-positive bacteria, and D-amino acids in some nonmammalian peptides and certain antibiotics.

**Table 3–1.** L-α-Amino acids present in proteins.

| Name | Symbol | Structural Formula | $pK_1$ | $pK_2$ | $pK_3$ |
|---|---|---|---|---|---|
| | | | α-COOH | $\alpha\text{-}NH_3^+$ | R Group |
| **With Aliphatic Side Chains** | | | | | |
| Glycine | Gly [G] | $H-CH(NH_3^+)-COO^-$ | 2.4 | 9.8 | |
| Alanine | Ala [A] | $CH_3-CH(NH_3^+)-COO^-$ | 2.4 | 9.9 | |
| Valine | Val [V] | $(H_3C)_2CH-CH(NH_3^+)-COO^-$ | 2.2 | 9.7 | |
| Leucine | Leu [L] | $(H_3C)_2CH-CH_2-CH(NH_3^+)-COO^-$ | 2.3 | 9.7 | |
| Isoleucine | Ile [I] | $CH_3-CH_2-CH(CH_3)-CH(NH_3^+)-COO^-$ | 2.3 | 9.8 | |
| **With Side Chains Containing Hydroxylic (OH) Groups** | | | | | |
| Serine | Ser [S] | $CH_2(OH)-CH(NH_3^+)-COO^-$ | 2.2 | 9.2 | about 13 |
| Threonine | Thr [T] | $CH_3-CH(OH)-CH(NH_3^+)-COO^-$ | 2.1 | 9.1 | about 13 |
| Tyrosine | Tyr [Y] | See below. | | | |
| **With Side Chains Containing Sulfur Atoms** | | | | | |
| Cysteine | Cys [C] | $CH_2(SH)-CH(NH_3^+)-COO^-$ | 1.9 | 10.8 | 8.3 |
| Methionine | Met [M] | $CH_3-S-CH_2-CH_2-CH(NH_3^+)-COO^-$ | 2.1 | 9.3 | |
| **With Side Chains Containing Acidic Groups or Their Amides** | | | | | |
| Aspartic acid | Asp [D] | $^-OOC-CH_2-CH(NH_3^+)-COO^-$ | 2.0 | 9.9 | 3.9 |
| Asparagine | Asn [N] | $H_2N-C(=O)-CH_2-CH(NH_3^+)-COO^-$ | 2.1 | 8.8 | |
| Glutamic acid | Glu [E] | $^-OOC-CH_2-CH_2-CH(NH_3^+)-COO^-$ | 2.1 | 9.5 | 4.1 |
| Glutamine | Gln [Q] | $H_2N-C(=O)-CH_2-CH_2-CH(NH_3^+)-COO^-$ | 2.2 | 9.1 | |

(continued)

***Table 3–1.*** L-α-Amino acids present in proteins. (continued)

| Name | Symbol | Structural Formula | $pK_1$ | $pK_2$ | $pK_3$ |
|---|---|---|---|---|---|
| **With Side Chains Containing Basic Groups** | | | α-COOH | α-$NH_3^+$ | R Group |
| Arginine | Arg [R] | 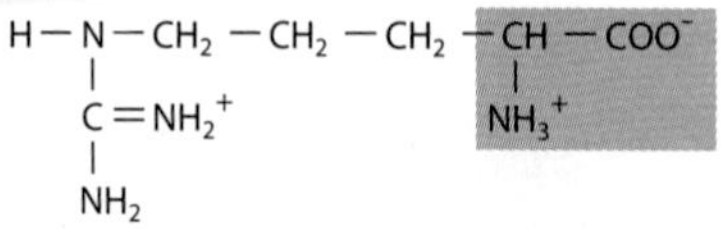 $H-N(-C(=NH_2^+)-NH_2)-CH_2-CH_2-CH_2-CH(NH_3^+)-COO^-$ | 1.8 | 9.0 | 12.5 |
| Lysine | Lys [K] | 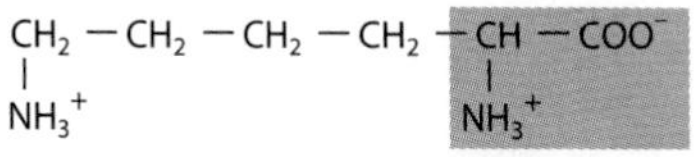 $CH_2(NH_3^+)-CH_2-CH_2-CH_2-CH(NH_3^+)-COO^-$ | 2.2 | 9.2 | 10.8 |
| Histidine | His [H] | 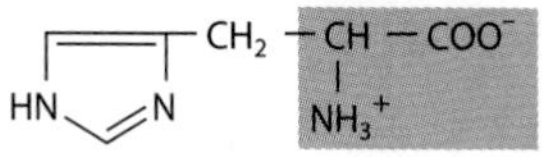 (imidazole ring, HN, N)$-CH_2-CH(NH_3^+)-COO^-$ | 1.8 | 9.3 | 6.0 |
| **Containing Aromatic Rings** | | | | | |
| Histidine | His [H] | See above. | | | |
| Phenylalanine | Phe [F] | 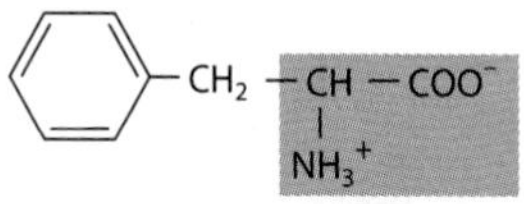 (benzene ring)$-CH_2-CH(NH_3^+)-COO^-$ | 2.2 | 9.2 | |
| Tyrosine | Tyr [Y] | 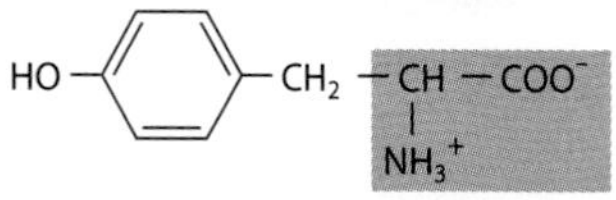 $HO-$(benzene ring)$-CH_2-CH(NH_3^+)-COO^-$ | 2.2 | 9.1 | 10.1 |
| Tryptophan | Trp [W] | 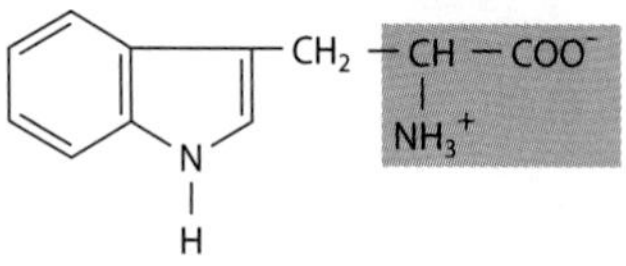 (indole ring, N—H)$-CH_2-CH(NH_3^+)-COO^-$ | 2.4 | 9.4 | |
| **Imino Acid** | | | | | |
| Proline | Pro [P] | 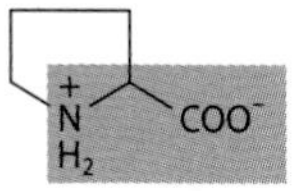 (pyrrolidine ring, $\overset{+}{N}H_2$)$-COO^-$ | 2.0 | 10.6 | |

## Amino Acids May Have Positive, Negative, or Zero Net Charge

Charged and uncharged forms of the ionizable —COOH and —$NH_3^+$ weak acid groups exist in solution in protonic equilibrium:

$$R-COOH = R-COO^- + H^+$$

$$R-NH_3^+ = R-NH_2 + H^+$$

While both R—COOH and R—$NH_3^+$ are weak acids, R—COOH is a far stronger acid than R—$NH_3^+$. At physiologic pH (pH 7.4), carboxyl groups exist almost entirely as R—$COO^-$ and amino groups predominantly as R—$NH_3^+$. Figure 3–1 illustrates the effect of pH on the charged state of aspartic acid.

Molecules that contain an equal number of ionizable groups of opposite charge and that therefore bear no net charge are termed **zwitterions.** Amino acids in blood and most tissues thus should be represented as in A, below.

A: $R-CH(NH_3^+)-C(=O)-O^-$

B: $R-CH(NH_2)-C(=O)-OH$

Structure B cannot exist in aqueous solution because at any pH low enough to protonate the carboxyl group the amino group would also be protonated. Similarly, at any pH sufficiently high for an uncharged amino

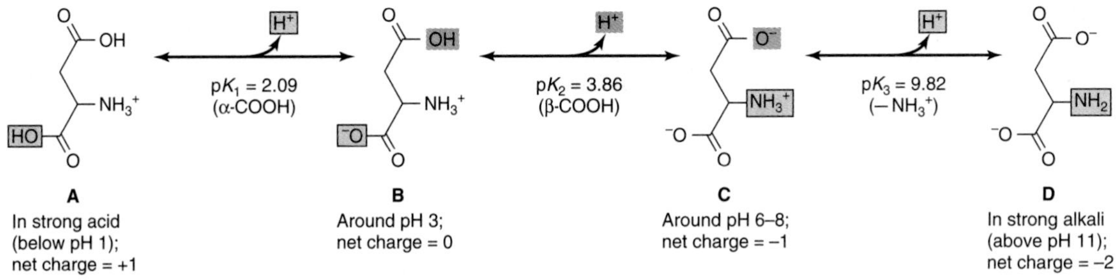

$H^+$ $pK_1 = 2.09$ ($\alpha$-COOH) $H^+$ $pK_2 = 3.86$ ($\beta$-COOH) $H^+$ $pK_3 = 9.82$ ($-NH_3^+$)

**A** In strong acid (below pH 1); net charge = +1

**B** Around pH 3; net charge = 0

**C** Around pH 6–8; net charge = −1

**D** In strong alkali (above pH 11); net charge = −2

***Figure 3–1.*** Protonic equilibria of aspartic acid.

group to predominate, a carboxyl group will be present as $R—COO^-$. The uncharged representation B (above) is, however, often used for reactions that do not involve protonic equilibria.

## p$K_a$ Values Express the Strengths of Weak Acids

The acid strengths of weak acids are expressed as their **p$K_a$** (Table 3–1). The imidazole group of histidine and the guanidino group of arginine exist as resonance hybrids with positive charge distributed between both nitrogens (histidine) or all three nitrogens (arginine) (Figure 3–2). The net charge on an amino acid—the algebraic sum of all the positively and negatively charged groups present—depends upon the p$K_a$ values of its functional groups and on the pH of the surrounding medium. Altering the charge on amino acids and their derivatives by varying the pH facilitates the physical separation of amino acids, peptides, and proteins (see Chapter 4).

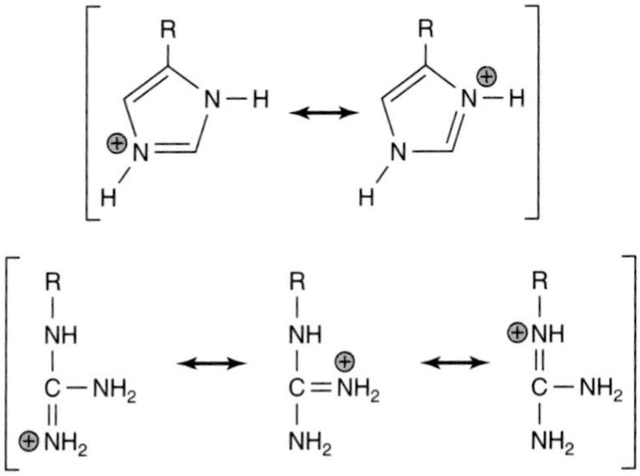

***Figure 3–2.*** Resonance hybrids of the protonated forms of the R groups of histidine and arginine.

## At Its Isoelectric pH (pI), an Amino Acid Bears No Net Charge

The **isoelectric** species is the form of a molecule that has an equal number of positive and negative charges and thus is electrically neutral. The isoelectric pH, also called the pI, is the pH midway between p$K_a$ values on either side of the isoelectric species. For an amino acid such as alanine that has only two dissociating groups, there is no ambiguity. The first p$K_a$ (R—COOH) is 2.35 and the second p$K_a$ ($R—NH_3^+$) is 9.69. The isoelectric pH (pI) of alanine thus is

$$pI = \frac{pK_1 + pK_2}{2} = \frac{2.35 + 9.69}{2} = 6.02$$

For polyfunctional acids, pI is also the pH midway between the p$K_a$ values on either side of the isoionic species. For example, the pI for aspartic acid is

$$pI = \frac{pK_1 + pK_2}{2} = \frac{2.09 + 3.96}{2} = 3.02$$

For lysine, pI is calculated from:

$$pI = \frac{pK_2 + pK_3}{2}$$

Similar considerations apply to all polyprotic acids (eg, proteins), regardless of the number of dissociating groups present. In the clinical laboratory, knowledge of the pI guides selection of conditions for electrophoretic separations. For example, electrophoresis at pH 7.0 will separate two molecules with pI values of 6.0 and 8.0 because at pH 8.0 the molecule with a pI of 6.0 will have a net positive charge, and that with pI of 8.0 a net negative charge. Similar considerations apply to understanding chromatographic separations on ionic supports such as DEAE cellulose (see Chapter 4).

### p$K_a$ Values Vary With the Environment

The environment of a dissociable group affects its p$K_a$. The p$K_a$ values of the R groups of free amino acids in aqueous solution (Table 3–1) thus provide only an approximate guide to the p$K_a$ values of the same amino acids when present in proteins. A polar environment favors the charged form (R—$COO^-$ or R—$NH_3^+$), and a nonpolar environment favors the uncharged form (R—COOH or R—$NH_2$). A nonpolar environment thus *raises* the p$K_a$ of a carboxyl group (making it a weaker acid) but *lowers* that of an amino group (making it a stronger acid). The presence of adjacent charged groups can reinforce or counteract solvent effects. The p$K_a$ of a functional group thus will depend upon its location within a given protein. Variations in p$K_a$ can encompass whole pH units (Table 3–2). p$K_a$ values that diverge from those listed by as much as three pH units are common at the active sites of enzymes. An extreme example, a buried aspartic acid of thioredoxin, has a p$K_a$ above 9—a shift of over six pH units!

### The Solubility and Melting Points of Amino Acids Reflect Their Ionic Character

The charged functional groups of amino acids ensure that they are readily solvated by—and thus soluble in—polar solvents such as water and ethanol but insoluble in nonpolar solvents such as benzene, hexane, or ether. Similarly, the high amount of energy required to disrupt the ionic forces that stabilize the crystal lattice account for the high melting points of amino acids (> 200 °C).

Amino acids do not absorb visible light and thus are colorless. However, tyrosine, phenylalanine, and especially tryptophan absorb high-wavelength (250–290 nm) ultraviolet light. Tryptophan therefore makes the major contribution to the ability of most proteins to absorb light in the region of 280 nm.

***Table 3–2.*** Typical range of p$K_a$ values for ionizable groups in proteins.

| Dissociating Group | p$K_a$ Range |
|---|---|
| α-Carboxyl | 3.5–4.0 |
| Non-α COOH of Asp or Glu | 4.0–4.8 |
| Imidazole of His | 6.5–7.4 |
| SH of Cys | 8.5–9.0 |
| OH of Tyr | 9.5–10.5 |
| α-Amino | 8.0–9.0 |
| ε-Amino of Lys | 9.8–10.4 |
| Guanidinium of Arg | ~12.0 |

## THE α-R GROUPS DETERMINE THE PROPERTIES OF AMINO ACIDS

Since glycine, the smallest amino acid, can be accommodated in places inaccessible to other amino acids, it often occurs where peptides bend sharply. The hydrophobic R groups of alanine, valine, leucine, and isoleucine and the aromatic R groups of phenylalanine, tyrosine, and tryptophan typically occur primarily in the interior of cytosolic proteins. The charged R groups of basic and acidic amino acids stabilize specific protein conformations via ionic interactions, or salt bonds. These bonds also function in "charge relay" systems during enzymatic catalysis and electron transport in respiring mitochondria. Histidine plays unique roles in enzymatic catalysis. The p$K_a$ of its imidazole proton permits it to function at neutral pH as either a base or an acid catalyst. The primary alcohol group of serine and the primary thioalcohol (—SH) group of cysteine are excellent nucleophiles and can function as such during enzymatic catalysis. However, the secondary alcohol group of threonine, while a good nucleophile, does not fulfill an analogous role in catalysis. The —OH groups of serine, tyrosine, and threonine also participate in regulation of the activity of enzymes whose catalytic activity depends on the phosphorylation state of these residues.

## FUNCTIONAL GROUPS DICTATE THE CHEMICAL REACTIONS OF AMINO ACIDS

Each functional group of an amino acid exhibits all of its characteristic chemical reactions. For carboxylic acid groups, these reactions include the formation of esters, amides, and acid anhydrides; for amino groups, acylation, amidation, and esterification; and for —OH and —SH groups, oxidation and esterification. The most important reaction of amino acids is the formation of a peptide bond (shaded blue).

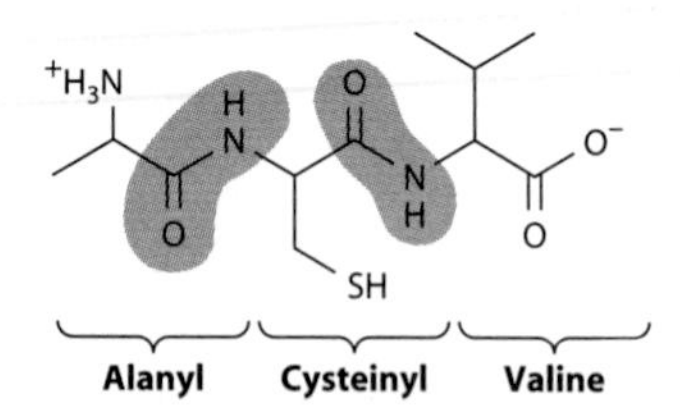


### Amino Acid Sequence Determines Primary Structure

The number and order of all of the amino acid residues in a polypeptide constitute its primary structure. Amino acids present in peptides are called aminoacyl residues and are named by replacing the *-ate* or *-ine* suffixes of free amino acids with *-yl* (eg, alan*yl*, aspart*yl*, tyros*yl*). Peptides are then named as derivatives of the carboxyl terminal aminoacyl residue. For example, Lys-Leu-Tyr-Gln is called lys*yl*-leuc*yl*-tyros*yl*-glutam*ine*. The *-ine* ending on glutamine indicates that its α-carboxyl group is *not* involved in peptide bond formation.

## Peptide Structures Are Easy to Draw

Prefixes like *tri-* or *octa-* denote peptides with three or eight **residues,** respectively, not those with three or eight peptide **bonds.** By convention, peptides are written with the residue that bears the free α-amino group at the left. To draw a peptide, use a zigzag to represent the main chain or backbone. Add the main chain atoms, which occur in the repeating order: α-nitrogen, α-carbon, carbonyl carbon. Now add a hydrogen atom to each α-carbon and to each peptide nitrogen, and an oxygen to the carbonyl carbon. Finally, add the appropriate R groups (shaded) to each α-carbon atom.

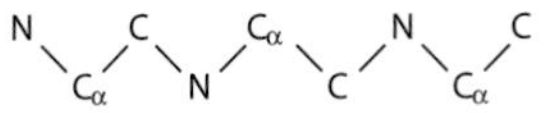

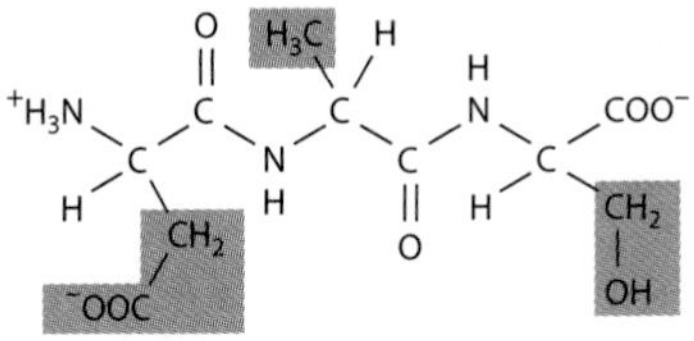

Three-letter abbreviations linked by straight lines represent an unambiguous primary structure. Lines are omitted for single-letter abbreviations.

Glu - Ala - Lys - Gly - Tyr - Ala

E A K G Y A

Where there is uncertainty about the order of a portion of a polypeptide, the questionable residues are enclosed in brackets and separated by commas.

Glu - Lys - (Ala, Gly, Tyr) - His - Ala

## Some Peptides Contain Unusual Amino Acids

In mammals, peptide hormones typically contain only the α-amino acids of proteins linked by standard peptide bonds. Other peptides may, however, contain nonprotein amino acids, derivatives of the protein amino acids, or amino acids linked by an atypical peptide bond. For example, the amino terminal glutamate of glutathione, which participates in protein folding and in the metabolism of xenobiotics (Chapter 53), is linked to cysteine by a non-α peptide bond (Figure 3–3). The amino terminal glutamate of thyrotropin-releasing hormone (TRH) is cyclized to pyroglutamic acid, and the carboxyl group of the carboxyl terminal prolyl residue is amidated. Peptides elaborated by fungi, bacteria, and lower animals can contain nonprotein amino acids. The antibiotics tyrocidin and gramicidin S are cyclic polypeptides that contain D-phenylalanine and ornithine. The heptapeptide opioids dermorphin and deltophorin in the skin of South American tree frogs contain D-tyrosine and D-alanine.

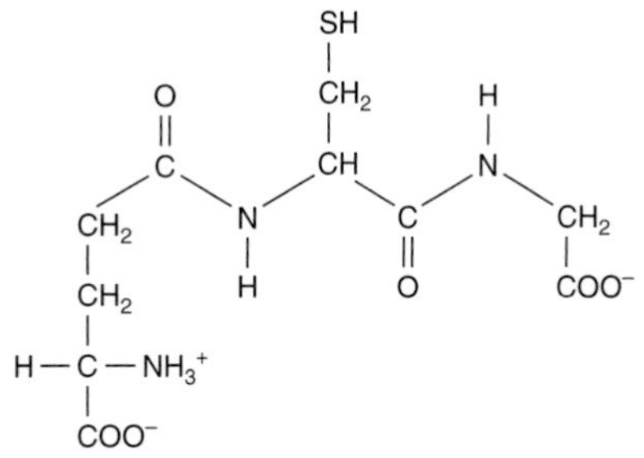

***Figure 3–3.*** Glutathione (γ-glutamyl-cysteinyl-glycine). Note the non-α peptide bond that links Glu to Cys.

## Peptides Are Polyelectrolytes

The peptide bond is uncharged at any pH of physiologic interest. Formation of peptides from amino acids is therefore accompanied by a net loss of one positive and one negative charge per peptide bond formed. Peptides nevertheless are charged at physiologic pH owing to their carboxyl and amino terminal groups and, where present, their acidic or basic R groups. As for amino acids, the net charge on a peptide depends on the pH of its environment and on the $pK_a$ values of its dissociating groups.

## The Peptide Bond Has Partial Double-Bond Character

Although peptides are written as if a single bond linked the α-carboxyl and α-nitrogen atoms, this bond in fact exhibits partial double-bond character:

There thus is no freedom of rotation about the bond that connects the carbonyl carbon and the nitrogen of a peptide bond. Consequently, all four of the colored atoms of Figure 3–4 are coplanar. The imposed semirigidity of the peptide bond has important conse-

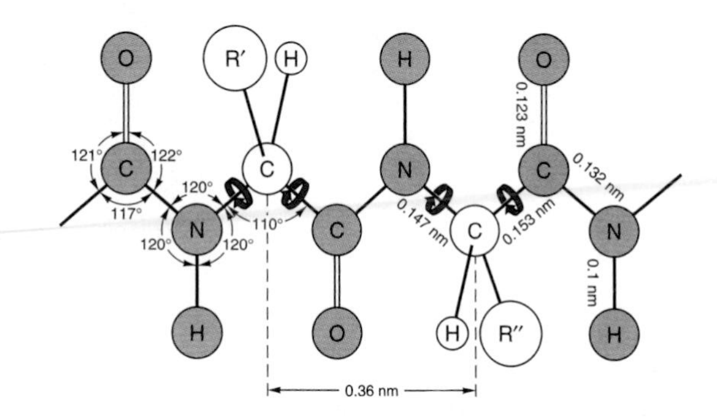


***Figure 3–4.*** **Dimensions of a fully extended polypeptide chain. The four atoms of the peptide bond (colored blue) are coplanar. The unshaded atoms are the α-carbon atom, the α-hydrogen atom, and the α-R group of the particular amino acid. Free rotation can occur about the bonds that connect the α-carbon with the α-nitrogen and with the α-carbonyl carbon (blue arrows). The extended polypeptide chain is thus a semirigid structure with two-thirds of the atoms of the backbone held in a fixed planar relationship one to another. The distance between adjacent α-carbon atoms is 0.36 nm (3.6 Å). The interatomic distances and bond angles, which are not equivalent, are also shown.** (Redrawn and reproduced, with permission, from Pauling L, Corey LP, Branson HR: The structure of proteins: Two hydrogen-bonded helical configurations of the polypeptide chain. Proc Natl Acad Sci U S A 1951;37:205.)

quences for higher orders of protein structure. Encircling arrows (Figure 3–4) indicate free rotation about the remaining bonds of the polypeptide backbone.

### Noncovalent Forces Constrain Peptide Conformations

Folding of a peptide probably occurs coincident with its biosynthesis (see Chapter 38). The physiologically active conformation reflects the amino acid sequence, steric hindrance, and noncovalent interactions (eg, hydrogen bonding, hydrophobic interactions) between residues. Common conformations include α-helices and β-pleated sheets (see Chapter 5).

## ANALYSIS OF THE AMINO ACID CONTENT OF BIOLOGIC MATERIALS

In order to determine the identity and quantity of each amino acid in a sample of biologic material, it is first necessary to hydrolyze the peptide bonds that link the amino acids together by treatment with hot HCl. The resulting mixture of free amino acids is then treated with 6-amino-*N*-hydroxysuccinimidyl carbamate, which reacts with their α-amino groups, forming fluorescent derivatives that are then separated and identified using high-pressure liquid chromatography (see Chapter 5). Ninhydrin, also widely used for detecting amino acids, forms a purple product with α-amino acids and a yellow adduct with the imine groups of proline and hydroxyproline.

## SUMMARY

- Both D-amino acids and non-α-amino acids occur in nature, but only L-α-amino acids are present in proteins.
- All amino acids possess at least two weakly acidic functional groups, R—$NH_3^+$ and R—COOH. Many also possess additional weakly acidic functional groups such as —OH, —SH, guanidino, or imidazole groups.
- The $pK_a$ values of all functional groups of an amino acid dictate its net charge at a given pH. pI is the pH at which an amino acid bears no net charge and thus does not move in a direct current electrical field.
- Of the biochemical reactions of amino acids, the most important is the formation of peptide bonds.
- The R groups of amino acids determine their unique biochemical functions. Amino acids are classified as basic, acidic, aromatic, aliphatic, or sulfur-containing based on the properties of their R groups.
- Peptides are named for the number of amino acid residues present, and as derivatives of the carboxyl terminal residue. The primary structure of a peptide is its amino acid sequence, starting from the amino-terminal residue.
- The partial double-bond character of the bond that links the carbonyl carbon and the nitrogen of a peptide renders four atoms of the peptide bond coplanar and restricts the number of possible peptide conformations.

## REFERENCES

Doolittle RF: Reconstructing history with amino acid sequences. Protein Sci 1992;1:191.

Kreil G: D-Amino acids in animal peptides. Annu Rev Biochem 1997;66:337.

Nokihara K, Gerhardt J: Development of an improved automated gas-chromatographic chiral analysis system: application to non-natural amino acids and natural protein hydrolysates. Chirality 2001;13:431.

Sanger F: Sequences, sequences, and sequences. Annu Rev Biochem 1988;57:1.

Wilson NA et al: Aspartic acid 26 in reduced *Escherichia coli* thioredoxin has a $pK_a$ greater than 9. Biochemistry 1995;34:8931.

# Proteins: Determination of Primary Structure

4

*Victor W. Rodwell, PhD, & Peter J. Kennelly, PhD*

## BIOMEDICAL IMPORTANCE

Proteins perform multiple critically important roles. An internal protein network, the cytoskeleton (Chapter 49), maintains cellular shape and physical integrity. Actin and myosin filaments form the contractile machinery of muscle (Chapter 49). Hemoglobin transports oxygen (Chapter 6), while circulating antibodies search out foreign invaders (Chapter 50). Enzymes catalyze reactions that generate energy, synthesize and degrade biomolecules, replicate and transcribe genes, process mRNAs, etc (Chapter 7). Receptors enable cells to sense and respond to hormones and other environmental cues (Chapters 42 and 43). An important goal of molecular medicine is the identification of proteins whose presence, absence, or deficiency is associated with specific physiologic states or diseases. The primary sequence of a protein provides both a molecular fingerprint for its identification and information that can be used to identify and clone the gene or genes that encode it.

## PROTEINS & PEPTIDES MUST BE PURIFIED PRIOR TO ANALYSIS

Highly purified protein is essential for determination of its amino acid sequence. Cells contain thousands of different proteins, each in widely varying amounts. The isolation of a specific protein in quantities sufficient for analysis thus presents a formidable challenge that may require multiple successive purification techniques. Classic approaches exploit differences in relative solubility of individual proteins as a function of pH (isoelectric precipitation), polarity (precipitation with ethanol or acetone), or salt concentration (salting out with ammonium sulfate). Chromatographic separations partition molecules between two phases, one mobile and the other stationary. For separation of amino acids or sugars, the stationary phase, or matrix, may be a sheet of filter paper (paper chromatography) or a thin layer of cellulose, silica, or alumina (thin-layer chromatography; TLC).

### Column Chromatography

Column chromatography of proteins employs as the stationary phase a column containing small spherical beads of modified cellulose, acrylamide, or silica whose surface typically has been coated with chemical functional groups. These stationary phase matrices interact with proteins based on their charge, hydrophobicity, and ligand-binding properties. A protein mixture is applied to the column and the liquid mobile phase is percolated through it. Small portions of the mobile phase or eluant are collected as they emerge (Figure 4–1).

### Partition Chromatography

Column chromatographic separations depend on the relative affinity of different proteins for a given stationary phase and for the mobile phase. Association between each protein and the matrix is weak and transient. Proteins that interact more strongly with the stationary phase are retained longer. The length of time that a protein is associated with the stationary phase is a function of the composition of both the stationary and mobile phases. Optimal separation of the protein of interest from other proteins thus can be achieved by careful manipulation of the composition of the two phases.

### Size Exclusion Chromatography

Size exclusion—or gel filtration—chromatography separates proteins based on their **Stokes radius,** the diameter of the sphere they occupy as they tumble in solution. The Stokes radius is a function of molecular mass and shape. A tumbling elongated protein occupies a larger volume than a spherical protein of the same mass. Size exclusion chromatography employs porous beads (Figure 4–2). The pores are analogous to indentations in a river bank. As objects move downstream, those that enter an indentation are retarded until they drift back into the main current. Similarly, proteins with Stokes radii too large to enter the pores (excluded proteins) remain in the flowing mobile phase and emerge before proteins that can enter the pores (included proteins).

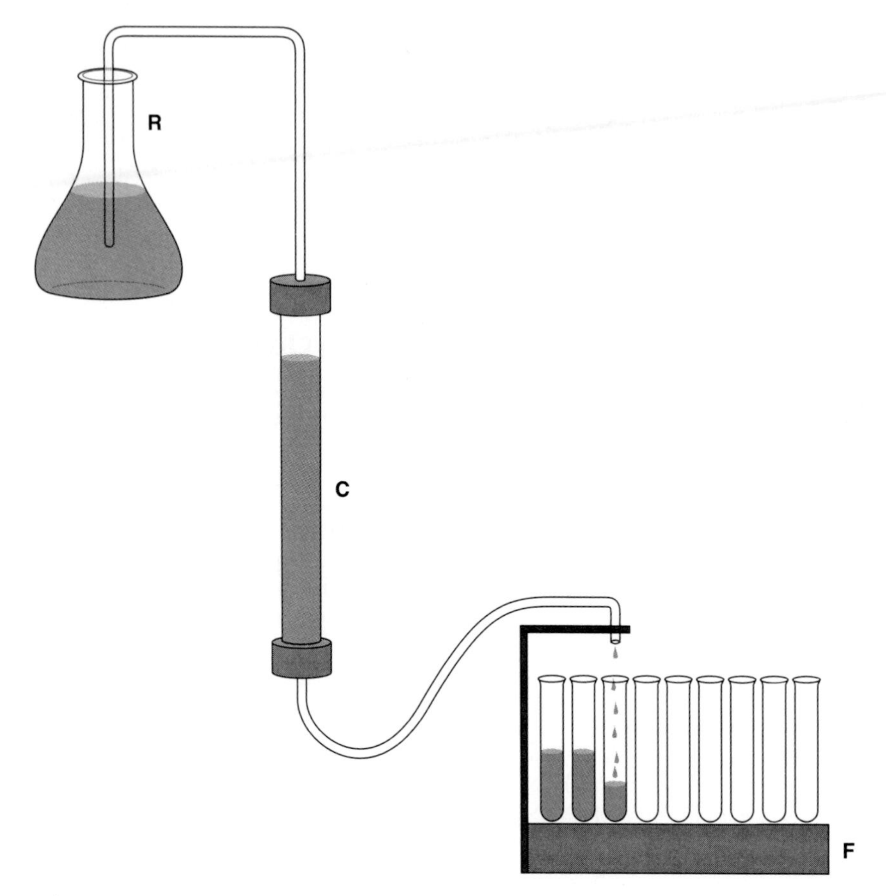


***Figure 4–1.*** Components of a simple liquid chromatography apparatus. **R:** Reservoir of mobile phase liquid, delivered either by gravity or using a pump. **C:** Glass or plastic column containing stationary phase. **F:** Fraction collector for collecting portions, called fractions, of the eluant liquid in separate test tubes.

Proteins thus emerge from a gel filtration column in descending order of their Stokes radii.

## Absorption Chromatography

For absorption chromatography, the protein mixture is applied to a column under conditions where the protein of interest associates with the stationary phase so tightly that its partition coefficient is essentially unity. Nonadhering molecules are first eluted and discarded. Proteins are then sequentially released by disrupting the forces that stabilize the protein-stationary phase complex, most often by using a gradient of increasing salt concentration. The composition of the mobile phase is altered gradually so that molecules are selectively released in descending order of their affinity for the stationary phase.

## Ion Exchange Chromatography

In ion exchange chromatography, proteins interact with the stationary phase by charge-charge interactions. Proteins with a net positive charge at a given pH adhere to beads with negatively charged functional groups such as carboxylates or sulfates (cation exchangers). Similarly, proteins with a net negative charge adhere to beads with positively charged functional groups, typically tertiary or quaternary amines (anion exchangers). Proteins, which are polyanions, compete against monovalent ions for binding to the support—thus the term "ion exchange." For example, proteins bind to diethylaminoethyl (DEAE) cellulose by replacing the counter-ions (generally $Cl^-$ or $CH_3COO^-$) that neutralize the protonated amine. Bound proteins are selectively displaced by gradually raising the concentration of monovalent ions in

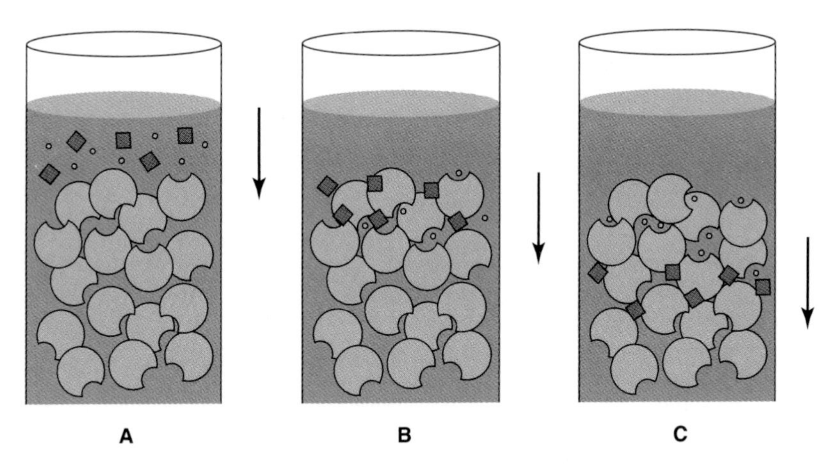


***Figure 4–2.*** Size-exclusion chromatography. **A:** A mixture of large molecules (diamonds) and small molecules (circles) are applied to the top of a gel filtration column. **B:** Upon entering the column, the small molecules enter pores in the stationary phase matrix from which the large molecules are excluded. **C:** As the mobile phase flows down the column, the large, excluded molecules flow with it while the small molecules, which are temporarily sheltered from the flow when inside the pores, lag farther and farther behind.

the mobile phase. Proteins elute in inverse order of the strength of their interactions with the stationary phase.

Since the net charge on a protein is determined by the pH (see Chapter 3), sequential elution of proteins may be achieved by changing the pH of the mobile phase. Alternatively, a protein can be subjected to consecutive rounds of ion exchange chromatography, each at a different pH, such that proteins that co-elute at one pH elute at different salt concentrations at another pH.

## Hydrophobic Interaction Chromatography

Hydrophobic interaction chromatography separates proteins based on their tendency to associate with a stationary phase matrix coated with hydrophobic groups (eg, phenyl Sepharose, octyl Sepharose). Proteins with exposed hydrophobic surfaces adhere to the matrix via hydrophobic interactions that are enhanced by a mobile phase of high ionic strength. Nonadherent proteins are first washed away. The polarity of the mobile phase is then decreased by gradually lowering the salt concentration. If the interaction between protein and stationary phase is particularly strong, ethanol or glycerol may be added to the mobile phase to decrease its polarity and further weaken hydrophobic interactions.

## Affinity Chromatography

Affinity chromatography exploits the high selectivity of most proteins for their ligands. Enzymes may be purified by affinity chromatography using immobilized substrates, products, coenzymes, or inhibitors. In theory, only proteins that interact with the immobilized ligand adhere. Bound proteins are then eluted either by competition with soluble ligand or, less selectively, by disrupting protein-ligand interactions using urea, guanidine hydrochloride, mildly acidic pH, or high salt concentrations. Stationary phase matrices available commercially contain ligands such as $NAD^+$ or ATP analogs. Among the most powerful and widely applicable affinity matrices are those used for the purification of suitably modified recombinant proteins. These include a $Ni^{2+}$ matrix that binds proteins with an attached polyhistidine "tag" and a glutathione matrix that binds a recombinant protein linked to glutathione *S*-transferase.

## Peptides Are Purified by Reversed-Phase High-Pressure Chromatography

The stationary phase matrices used in classic column chromatography are spongy materials whose compressibility limits flow of the mobile phase. High-pressure liquid chromatography (HPLC) employs incompressible silica or alumina microbeads as the stationary phase and pressures of up to a few thousand psi. Incompressible matrices permit both high flow rates and enhanced resolution. HPLC can resolve complex mixtures of lipids or peptides whose properties differ only slightly. Reversed-phase HPLC exploits a hydrophobic stationary phase of aliphatic polymers 3–18 carbon atoms in length. Peptide mixtures are eluted using a gradient of a water-miscible organic solvent such as acetonitrile or methanol.

### Protein Purity Is Assessed by Polyacrylamide Gel Electrophoresis (PAGE)

The most widely used method for determining the purity of a protein is SDS-PAGE—polyacrylamide gel electrophoresis (PAGE) in the presence of the anionic detergent sodium dodecyl sulfate (SDS). Electrophoresis separates charged biomolecules based on the rates at which they migrate in an applied electrical field. For SDS-PAGE, acrylamide is polymerized and cross-linked to form a porous matrix. SDS denatures and binds to proteins at a ratio of one molecule of SDS per two peptide bonds. When used in conjunction with 2-mercaptoethanol or dithiothreitol to reduce and break disulfide bonds (Figure 4–3), SDS separates the component polypeptides of multimeric proteins. The large number of anionic SDS molecules, each bearing a charge of −1, on each polypeptide overwhelms the charge contributions of the amino acid functional groups. Since the charge-to-mass ratio of each SDS-polypeptide complex is approximately equal, the physical resistance each peptide encounters as it moves through the acrylamide matrix determines the rate of migration. Since large complexes encounter greater resistance, polypeptides separate based on their relative molecular mass ($M_r$). Individual polypeptides trapped in the acrylamide gel are visualized by staining with dyes such as Coomassie blue (Figure 4–4).

### Isoelectric Focusing (IEF)

Ionic buffers called ampholytes and an applied electric field are used to generate a pH gradient within a polyacrylamide matrix. Applied proteins migrate until they reach the region of the matrix where the pH matches their isoelectric point (pI), the pH at which a peptide's net charge is zero. IEF is used in conjunction with SDS-PAGE for two-dimensional electrophoresis, which separates polypeptides based on pI in one dimension and based on $M_r$ in the second (Figure 4–5). Two-dimensional electrophoresis is particularly well suited for separating the components of complex mixtures of proteins.

## SANGER WAS THE FIRST TO DETERMINE THE SEQUENCE OF A POLYPEPTIDE

Mature insulin consists of the 21-residue A chain and the 30-residue B chain linked by disulfide bonds. Frederick Sanger reduced the disulfide bonds (Figure 4–3),

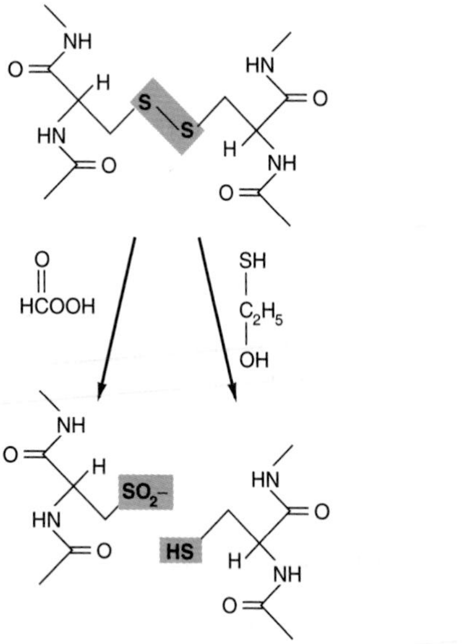

***Figure 4–3.*** Oxidative cleavage of adjacent polypeptide chains linked by disulfide bonds (shaded) by performic acid **(left)** or reductive cleavage by β-mercaptoethanol **(right)** forms two peptides that contain cysteic acid residues or cysteinyl residues, respectively.

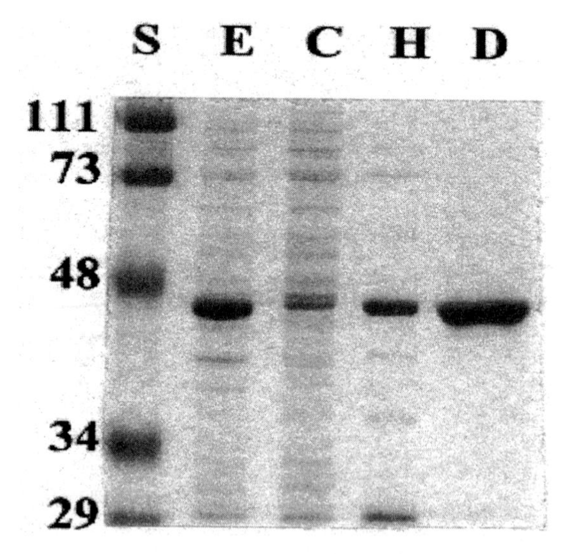


***Figure 4–4.*** Use of SDS-PAGE to observe successive purification of a recombinant protein. The gel was stained with Coomassie blue. Shown are protein standards (lane **S**) of the indicated mass, crude cell extract **(E)**, high-speed supernatant liquid **(H)**, and the DEAE-Sepharose fraction **(D)**. The recombinant protein has a mass of about 45 kDa.

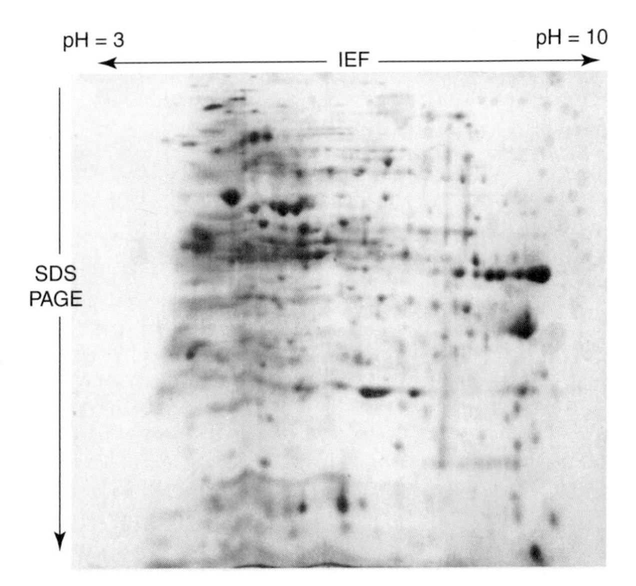


***Figure 4–5.*** Two-dimensional IEF-SDS-PAGE. The gel was stained with Coomassie blue. A crude bacterial extract was first subjected to isoelectric focusing (IEF) in a pH 3–10 gradient. The IEF gel was then placed horizontally on the top of an SDS gel, and the proteins then further resolved by SDS-PAGE. Notice the greatly improved resolution of distinct polypeptides relative to ordinary SDS-PAGE gel (Figure 4–4).

separated the A and B chains, and cleaved each chain into smaller peptides using trypsin, chymotrypsin, and pepsin. The resulting peptides were then isolated and treated with acid to hydrolyze peptide bonds and generate peptides with as few as two or three amino acids. Each peptide was reacted with 1-fluoro-2,4-dinitrobenzene (Sanger's reagent), which derivatizes the exposed α-amino group of amino terminal residues. The amino acid content of each peptide was then determined. While the ε-amino group of lysine also reacts with Sanger's reagent, amino-terminal lysines can be distinguished from those at other positions because they react with 2 mol of Sanger's reagent. Working backwards to larger fragments enabled Sanger to determine the complete sequence of insulin, an accomplishment for which he received a Nobel Prize in 1958.

## THE EDMAN REACTION ENABLES PEPTIDES & PROTEINS TO BE SEQUENCED

Pehr Edman introduced phenylisothiocyanate (Edman's reagent) to selectively label the amino-terminal residue of a peptide. In contrast to Sanger's reagent, the phenylthiohydantoin (PTH) derivative can be removed under mild conditions to generate a new amino terminal residue (Figure 4–6). Successive rounds of derivatization with Edman's reagent can therefore be used to sequence many residues of a single sample of peptide. Edman sequencing has been automated, using a thin film or solid matrix to immobilize the peptide and HPLC to identify PTH amino acids. Modern gas-phase sequencers can analyze as little as a few picomoles of peptide.

### Large Polypeptides Are First Cleaved Into Smaller Segments

While the first 20–30 residues of a peptide can readily be determined by the Edman method, most polypeptides contain several hundred amino acids. Consequently, most polypeptides must first be cleaved into smaller peptides prior to Edman sequencing. Cleavage also may be necessary to circumvent posttranslational modifications that render a protein's α-amino group "blocked", or unreactive with the Edman reagent.

It usually is necessary to generate several peptides using more than one method of cleavage. This reflects both inconsistency in the spacing of chemically or enzymatically susceptible cleavage sites and the need for sets of peptides whose sequences overlap so one can infer the sequence of the polypeptide from which they derive (Figure 4–7). Reagents for the chemical or enzymatic cleavage of proteins include cyanogen bromide (CNBr), trypsin, and *Staphylococcus aureus* V8 protease (Table 4–1). Following cleavage, the resulting peptides are purified by reversed-phase HPLC—or occasionally by SDS-PAGE—and sequenced.

## MOLECULAR BIOLOGY HAS REVOLUTIONIZED THE DETERMINATION OF PRIMARY STRUCTURE

Knowledge of DNA sequences permits deduction of the primary structures of polypeptides. DNA sequencing requires only minute amounts of DNA and can readily yield the sequence of hundreds of nucleotides. To clone and sequence the DNA that encodes a partic-

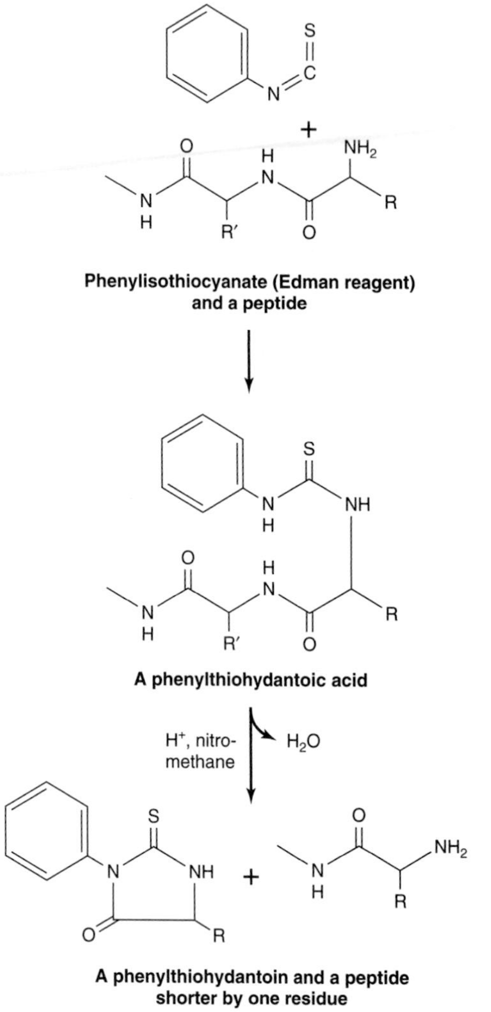

**Phenylisothiocyanate (Edman reagent) and a peptide**

**A phenylthiohydantoic acid**

$H^+$, nitromethane

$H_2O$

**A phenylthiohydantoin and a peptide shorter by one residue**

***Figure 4–6.*** The Edman reaction. Phenylisothiocyanate derivatizes the amino-terminal residue of a peptide as a phenylthiohydantoic acid. Treatment with acid in a nonhydroxylic solvent releases a phenylthiohydantoin, which is subsequently identified by its chromatographic mobility, and a peptide one residue shorter. The process is then repeated.

ular protein, some means of identifying the correct clone—eg, knowledge of a portion of its nucleotide sequence—is essential. A hybrid approach thus has emerged. Edman sequencing is used to provide a partial amino acid sequence. Oligonucleotide primers modeled on this partial sequence can then be used to identify clones or to amplify the appropriate gene by the polymerase chain reaction (PCR) (see Chapter 40). Once an authentic DNA clone is obtained, its oligonucleotide sequence can be determined and the genetic code used to infer the primary structure of the encoded polypeptide.

Peptide X Peptide Y

Peptide Z

Carboxyl terminal portion of peptide X

Amino terminal portion of peptide Y

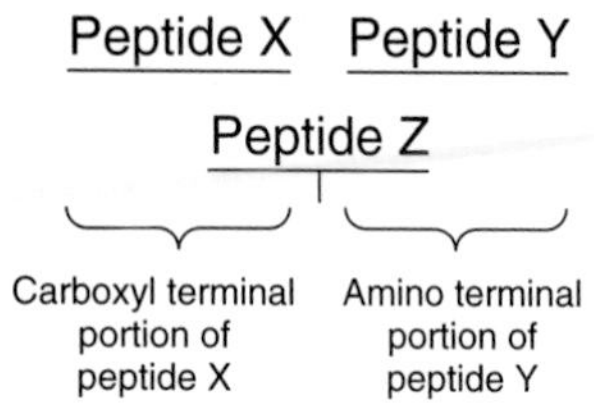

***Figure 4–7.*** The overlapping peptide Z is used to deduce that peptides X and Y are present in the original protein in the order X → Y, not Y ← X.

The hybrid approach enhances the speed and efficiency of primary structure analysis and the range of proteins that can be sequenced. It also circumvents obstacles such as the presence of an amino-terminal blocking group or the lack of a key overlap peptide. Only a few segments of primary structure must be determined by Edman analysis.

DNA sequencing reveals the order in which amino acids are added to the nascent polypeptide chain as it is synthesized on the ribosomes. However, it provides no information about posttranslational modifications such as proteolytic processing, methylation, glycosylation, phosphorylation, hydroxylation of proline and lysine, and disulfide bond formation that accompany maturation. While Edman sequencing can detect the presence of most posttranslational events, technical limitations often prevent identification of a specific modification.

***Table 4–1.*** Methods for cleaving polypeptides.

| Method | Bond Cleaved |
|---|---|
| CNBr | Met-X |
| Trypsin | Lys-X and Arg-X |
| Chymotrypsin | Hydrophobic amino acid-X |
| Endoproteinase Lys-C | Lys-X |
| Endoproteinase Arg-C | Arg-X |
| Endoproteinase Asp-N | X-Asp |
| V8 protease | Glu-X, particularly where X is hydrophobic |
| Hydroxylamine | Asn-Gly |
| *o*-Iodosobenzene | Trp-X |
| Mild acid | Asp-Pro |

## MASS SPECTROMETRY DETECTS COVALENT MODIFICATIONS

Mass spectrometry, which discriminates molecules based solely on their mass, is ideal for detecting the phosphate, hydroxyl, and other groups on posttranslationally modified amino acids. Each adds a specific and readily identified increment of mass to the modified amino acid (Table 4–2). For analysis by mass spectrometry, a sample in a vacuum is vaporized under conditions where protonation can occur, imparting positive charge. An electrical field then propels the cations through a magnetic field which deflects them at a right angle to their original direction of flight and focuses them onto a detector (Figure 4–8). The magnetic force required to deflect the path of each ionic species onto the detector, measured as the current applied to the electromagnet, is recorded. For ions of identical net charge, this force is proportionate to their mass. In a time-of-flight mass spectrometer, a briefly applied electric field accelerates the ions towards a detector that records the time at which each ion arrives. For molecules of identical charge, the velocity to which they are accelerated—and hence the time required to reach the detector—will be inversely proportionate to their mass.

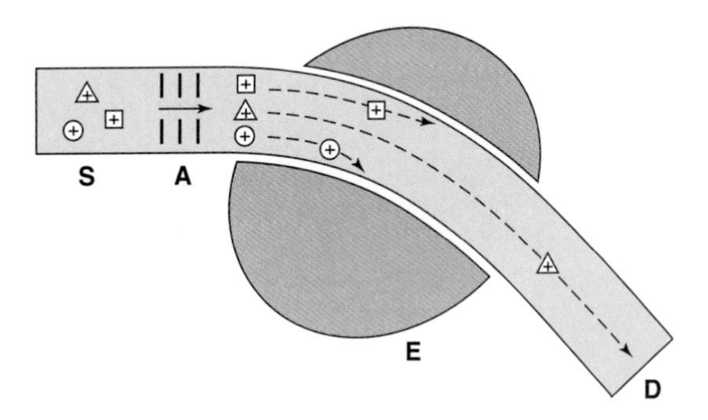


***Figure 4–8.*** Basic components of a simple mass spectrometer. A mixture of molecules is vaporized in an ionized state in the sample chamber **S.** These molecules are then accelerated down the flight tube by an electrical potential applied to accelerator grid **A.** An adjustable electromagnet, **E,** applies a magnetic field that deflects the flight of the individual ions until they strike the detector, **D.** The greater the mass of the ion, the higher the magnetic field required to focus it onto the detector.

Conventional mass spectrometers generally are used to determine the masses of molecules of 1000 Da or less, whereas time-of-flight mass spectrometers are suited for determining the large masses of proteins. The analysis of peptides and proteins by mass spectometry initially was hindered by difficulties in volatilizing large organic molecules. However, matrix-assisted laser-desorption (MALDI) and electrospray dispersion (eg, nanospray) permit the masses of even large polypeptides (> 100,000 Da) to be determined with extraordinary accuracy (± 1 Da). Using electrospray dispersion, peptides eluting from a reversed-phase HPLC column are introduced directly into the mass spectrometer for immediate determination of their masses.

Peptides inside the mass spectrometer are broken down into smaller units by collisions with neutral helium atoms (collision-induced dissociation), and the masses of the individual fragments are determined. Since peptide bonds are much more labile than carbon-carbon bonds, the most abundant fragments will differ from one another by units equivalent to one or two amino acids. Since—with the exception of leucine and isoleucine—the molecular mass of each amino acid is unique, the sequence of the peptide can be reconstructed from the masses of its fragments.

***Table 4–2.*** Mass increases resulting from common posttranslational modifications.

| Modification | Mass Increase (Da) |
|---|---|
| Phosphorylation | 80 |
| Hydroxylation | 16 |
| Methylation | 14 |
| Acetylation | 42 |
| Myristylation | 210 |
| Palmitoylation | 238 |
| Glycosylation | 162 |

### Tandem Mass Spectrometry

Complex peptide mixtures can now be analyzed without prior purification by tandem mass spectrometry, which employs the equivalent of two mass spectrometers linked in series. The first spectrometer separates individual peptides based upon their differences in mass. By adjusting the field strength of the first magnet, a single peptide can be directed into the second mass spectrometer, where fragments are generated and their masses determined. As the sensitivity and versatility of mass spectrometry continue to increase, it is displacing Edman sequencers for the direct analysis of protein primary structure.

## GENOMICS ENABLES PROTEINS TO BE IDENTIFIED FROM SMALL AMOUNTS OF SEQUENCE DATA

Primary structure analysis has been revolutionized by genomics, the application of automated oligonucleotide sequencing and computerized data retrieval and analysis to sequence an organism's entire genetic complement. The first genome sequenced was that of *Haemophilus influenzae,* in 1995. By mid 2001, the complete genome sequences for over 50 organisms had been determined. These include the human genome and those of several bacterial pathogens; the results and significance of the Human Genome Project are discussed in Chapter 54. Where genome sequence is known, the task of determining a protein's DNA-derived primary sequence is materially simplified. In essence, the second half of the hybrid approach has already been completed. All that remains is to acquire sufficient information to permit the open reading frame (ORF) that encodes the protein to be retrieved from an Internet-accessible genome database and identified. In some cases, a segment of amino acid sequence only four or five residues in length may be sufficient to identify the correct ORF.

Computerized search algorithms assist the identification of the gene encoding a given protein and clarify uncertainties that arise from Edman sequencing and mass spectrometry. By exploiting computers to solve complex puzzles, the spectrum of information suitable for identification of the ORF that encodes a particular polypeptide is greatly expanded. In peptide mass profiling, for example, a peptide digest is introduced into the mass spectrometer and the sizes of the peptides are determined. A computer is then used to find an ORF whose predicted protein product would, if broken down into peptides by the cleavage method selected, produce a set of peptides whose masses match those observed by mass spectrometry.

## PROTEOMICS & THE PROTEOME

### The Goal of Proteomics Is to Identify the Entire Complement of Proteins Elaborated by a Cell Under Diverse Conditions

While the sequence of the human genome is known, the picture provided by genomics alone is both static and incomplete. Proteomics aims to identify the entire complement of proteins elaborated by a cell under diverse conditions. As genes are switched on and off, proteins are synthesized in particular cell types at specific times of growth or differentiation and in response to external stimuli. Muscle cells express proteins not expressed by neural cells, and the type of subunits present in the hemoglobin tetramer undergo change pre- and postpartum. Many proteins undergo posttranslational modifications during maturation into functionally competent forms or as a means of regulating their properties. Knowledge of the human genome therefore represents only the beginning of the task of describing living organisms in molecular detail and understanding the dynamics of processes such as growth, aging, and disease. As the human body contains thousands of cell types, each containing thousands of proteins, the proteome—the set of all the proteins expressed by an individual cell at a particular time—represents a moving target of formidable dimensions.

### Two-Dimensional Electrophoresis & Gene Array Chips Are Used to Survey Protein Expression

One goal of proteomics is the identification of proteins whose levels of expression correlate with medically significant events. The presumption is that proteins whose appearance or disappearance is associated with a specific physiologic condition or disease will provide insights into root causes and mechanisms. Determination of the proteomes characteristic of each cell type requires the utmost efficiency in the isolation and identification of individual proteins. The contemporary approach utilizes robotic automation to speed sample preparation and large two-dimensional gels to resolve cellular proteins. Individual polypeptides are then extracted and analyzed by Edman sequencing or mass spectroscopy. While only about 1000 proteins can be resolved on a single gel, two-dimensional electrophoresis has a major advantage in that it examines the proteins themselves. An alternative and complementary approach employs gene arrays, sometimes called DNA chips, to detect the expression of the mRNAs which encode proteins. While changes in the expression of the mRNA encoding a protein do not necessarily reflect comparable changes in the level of the corresponding protein, gene arrays are more sensitive probes than two-dimensional gels and thus can examine more gene products.

### Bioinformatics Assists Identification of Protein Functions

The functions of a large proportion of the proteins encoded by the human genome are presently unknown. Recent advances in bioinformatics permit researchers to compare amino acid sequences to discover clues to potential properties, physiologic roles, and mechanisms of action of proteins. Algorithms exploit the tendency of nature to employ variations of a structural theme to perform similar functions in several proteins (eg, the Rossmann nucleotide binding fold to bind NAD(P)H, nuclear targeting sequences, and EF hands to bind $Ca^{2+}$). These domains generally are detected in the primary structure by conservation of particular amino acids at key positions. Insights into the properties and physiologic role of a newly discovered protein thus may be inferred by comparing its primary structure with that of known proteins.

## SUMMARY

- Long amino acid polymers or polypeptides constitute the basic structural unit of proteins, and the structure of a protein provides insight into how it fulfills its functions.
- The Edman reaction enabled amino acid sequence analysis to be automated. Mass spectrometry provides a sensitive and versatile tool for determining primary structure and for the identification of post-translational modifications.
- DNA cloning and molecular biology coupled with protein chemistry provide a hybrid approach that greatly increases the speed and efficiency for determination of primary structures of proteins.
- Genomics—the analysis of the entire oligonucleotide sequence of an organism's complete genetic material—has provided further enhancements.
- Computer algorithms facilitate identification of the open reading frames that encode a given protein by using partial sequences and peptide mass profiling to search sequence databases.
- Scientists are now trying to determine the primary sequence and functional role of every protein expressed in a living cell, known as its proteome.
- A major goal is the identification of proteins whose appearance or disappearance correlates with physiologic phenomena, aging, or specific diseases.

## REFERENCES


Deutscher MP (editor): *Guide to Protein Purification.* Methods Enzymol 1990;182. (Entire volume.)

Geveart K, Vandekerckhove J: Protein identification methods in proteomics. Electrophoresis 2000;21:1145.

Helmuth L: Genome research: map of the human genome 3.0. Science 2001;293:583.

Khan J et al: DNA microarray technology: the anticipated impact on the study of human disease. Biochim Biophys Acta 1999;1423:M17.

McLafferty FW et al: Biomolecule mass spectrometry. Science 1999;284:1289.

Patnaik SK, Blumenfeld OO: Use of on-line tools and databases for routine sequence analyses. Anal Biochem 2001;289:1.

Schena M et al: Quantitative monitoring of gene expression patterns with a complementary DNA microarray. Science 1995;270:467.

Semsarian C, Seidman CE: Molecular medicine in the 21st century. Intern Med J 2001;31:53.

Temple LK et al: Essays on science and society: defining disease in the genomics era. Science 2001;293:807.

Wilkins MR et al: High-throughput mass spectrometric discovery of protein post-translational modifications. J Mol Biol 1999;289:645.

# Proteins: Higher Orders of Structure 5

Victor W. Rodwell, PhD, & Peter J. Kennelly, PhD

## BIOMEDICAL IMPORTANCE

Proteins catalyze metabolic reactions, power cellular motion, and form macromolecular rods and cables that provide structural integrity to hair, bones, tendons, and teeth. In nature, form follows function. The structural variety of human proteins therefore reflects the sophistication and diversity of their biologic roles. Maturation of a newly synthesized polypeptide into a biologically functional protein requires that it be folded into a specific three-dimensional arrangement, or **conformation.** During maturation, **posttranslational modifications** may add new chemical groups or remove transiently needed peptide segments. Genetic or nutritional deficiencies that impede protein maturation are deleterious to health. Examples of the former include Creutzfeldt-Jakob disease, scrapie, Alzheimer's disease, and bovine spongiform encephalopathy (mad cow disease). Scurvy represents a nutritional deficiency that impairs protein maturation.

## CONFORMATION VERSUS CONFIGURATION

The terms configuration and conformation are often confused. **Configuration** refers to the geometric relationship between a given set of atoms, for example, those that distinguish L- from D-amino acids. Interconversion of *configurational* alternatives requires breaking covalent bonds. **Conformation** refers to the spatial relationship of every atom in a molecule. Interconversion between conformers occurs without covalent bond rupture, with retention of configuration, and typically via rotation about single bonds.

## PROTEINS WERE INITIALLY CLASSIFIED BY THEIR GROSS CHARACTERISTICS

Scientists initially approached structure-function relationships in proteins by separating them into classes based upon properties such as solubility, shape, or the presence of nonprotein groups. For example, the proteins that can be extracted from cells using solutions at physiologic pH and ionic strength are classified as **soluble.** Extraction of **integral membrane proteins** requires dissolution of the membrane with detergents. **Globular proteins** are compact, are roughly spherical or ovoid in shape, and have **axial ratios** (the ratio of their shortest to longest dimensions) of not over 3. Most enzymes are globular proteins, whose large internal volume provides ample space in which to construct cavities of the specific shape, charge, and hydrophobicity or hydrophilicity required to bind substrates and promote catalysis. By contrast, many structural proteins adopt highly extended conformations. These **fibrous proteins** possess axial ratios of 10 or more.

**Lipoproteins** and **glycoproteins** contain covalently bound lipid and carbohydrate, respectively. Myoglobin, hemoglobin, cytochromes, and many other proteins contain tightly associated metal ions and are termed **metalloproteins.** With the development and application of techniques for determining the amino acid sequences of proteins (Chapter 4), more precise classification schemes have emerged based upon similarity, or **homology,** in amino acid sequence and structure. However, many early classification terms remain in common use.

## PROTEINS ARE CONSTRUCTED USING MODULAR PRINCIPLES

Proteins perform complex physical and catalytic functions by positioning specific chemical groups in a precise three-dimensional arrangement. The polypeptide scaffold containing these groups must adopt a conformation that is both functionally efficient and physically strong. At first glance, the biosynthesis of polypeptides comprised of tens of thousands of individual atoms would appear to be extremely challenging. When one considers that a typical polypeptide can adopt $\geq 10^{50}$ distinct conformations, folding into the conformation appropriate to their biologic function would appear to be even more difficult. As described in Chapters 3 and 4, synthesis of the polypeptide backbones of proteins employs a small set of common building blocks or modules, the amino acids, joined by a common linkage, the peptide bond. A stepwise modular pathway simplifies the folding and processing of newly synthesized polypeptides into mature proteins.

## THE FOUR ORDERS OF PROTEIN STRUCTURE

The modular nature of protein synthesis and folding are embodied in the concept of orders of protein structure: **primary structure,** the sequence of the amino acids in a polypeptide chain; **secondary structure,** the folding of short (3- to 30-residue), contiguous segments of polypeptide into geometrically ordered units; **tertiary structure,** the three-dimensional assembly of secondary structural units to form larger functional units such as the mature polypeptide and its component domains; and **quaternary structure,** the number and types of polypeptide units of oligomeric proteins and their spatial arrangement.

## SECONDARY STRUCTURE

### Peptide Bonds Restrict Possible Secondary Conformations

Free rotation is possible about only two of the three covalent bonds of the polypeptide backbone: the α-carbon (Cα) to the carbonyl carbon (Co) bond and the Cα to nitrogen bond (Figure 3–4). The partial double-bond character of the peptide bond that links Co to the α-nitrogen requires that the carbonyl carbon, carbonyl oxygen, and α-nitrogen remain coplanar, thus preventing rotation. The angle about the Cα—N bond is termed the phi (Φ) angle, and that about the Co—Cα bond the psi (Ψ) angle. For amino acids other than glycine, most combinations of phi and psi angles are disallowed because of steric hindrance (Figure 5–1). The conformations of proline are even more restricted due to the absence of free rotation of the N—Cα bond.

Regions of ordered secondary structure arise when a series of aminoacyl residues adopt similar phi and psi angles. Extended segments of polypeptide (eg, loops) can possess a variety of such angles. The angles that define the two most common types of secondary structure, the **α helix** and the **β sheet,** fall within the lower and upper left-hand quadrants of a Ramachandran plot, respectively (Figure 5–1).

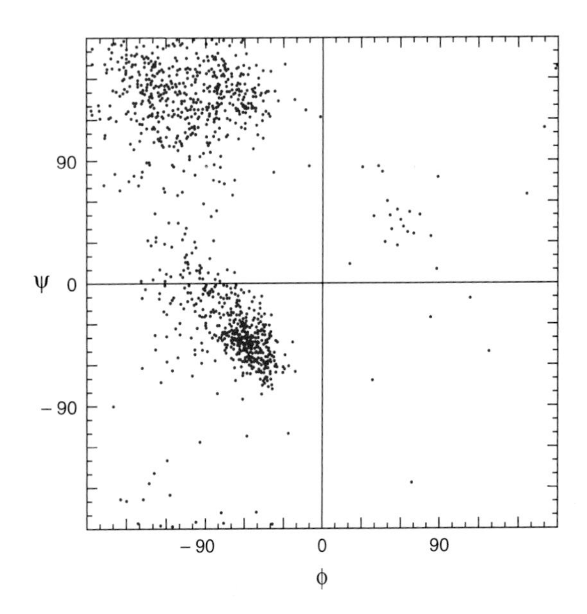


***Figure 5–1.*** Ramachandran plot of the main chain phi (Φ) and psi (Ψ) angles for approximately 1000 nonglycine residues in eight proteins whose structures were solved at high resolution. The dots represent allowable combinations and the spaces prohibited combinations of phi and psi angles. (Reproduced, with permission, from Richardson JS: The anatomy and taxonomy of protein structures. Adv Protein Chem 1981;34:167.)

### The Alpha Helix

The polypeptide backbone of an α helix is twisted by an equal amount about each α-carbon with a phi angle of approximately −57 degrees and a psi angle of approximately −47 degrees. A complete turn of the helix contains an average of 3.6 aminoacyl residues, and the distance it rises per turn (its pitch) is 0.54 nm (Figure 5–2). The R groups of each aminoacyl residue in an α helix face outward (Figure 5–3). Proteins contain only L-amino acids, for which a right-handed α helix is by far the more stable, and only right-handed α helices occur in nature. Schematic diagrams of proteins represent α helices as cylinders.

The stability of an α helix arises primarily from hydrogen bonds formed between the oxygen of the peptide bond carbonyl and the hydrogen atom of the peptide bond nitrogen of the fourth residue down the polypeptide chain (Figure 5–4). The ability to form the maximum number of hydrogen bonds, supplemented by van der Waals interactions in the core of this tightly packed structure, provides the thermodynamic driving force for the formation of an α helix. Since the peptide bond nitrogen of proline lacks a hydrogen atom to contribute to a hydrogen bond, proline can only be stably accommodated within the first turn of an α helix. When present elsewhere, proline disrupts the conformation of the helix, producing a bend. Because of its small size, glycine also often induces bends in α helices.

Many α helices have predominantly hydrophobic R groups on one side of the axis of the helix and predominantly hydrophilic ones on the other. These **amphipathic helices** are well adapted to the formation of interfaces between polar and nonpolar regions such as the hydrophobic interior of a protein and its aqueous envi-

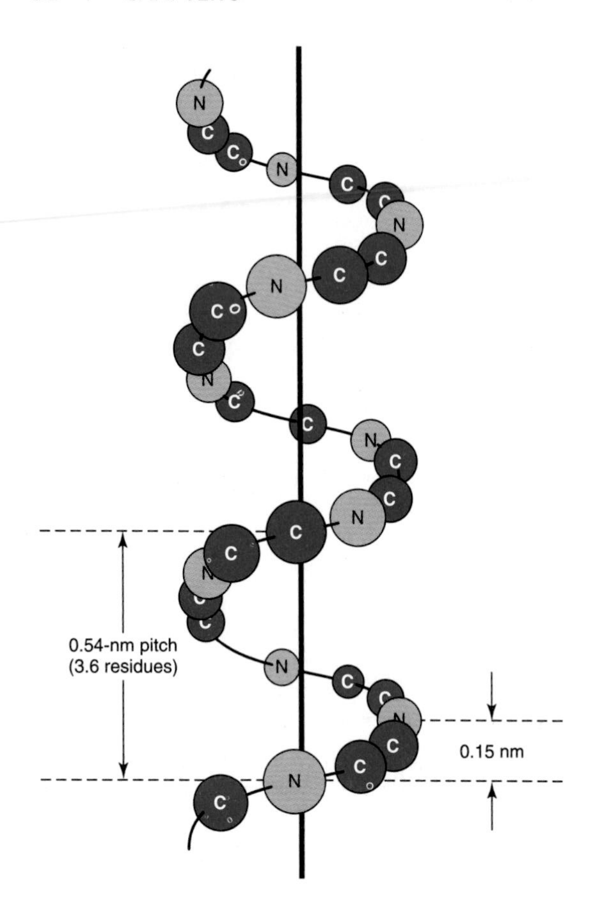


***Figure 5–2.*** Orientation of the main chain atoms of a peptide about the axis of an α helix.

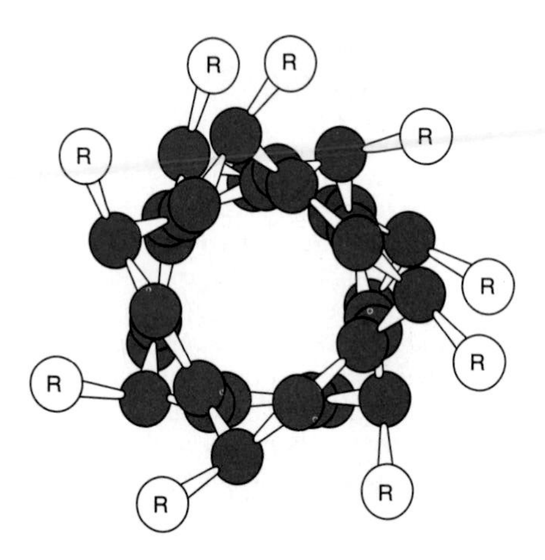

***Figure 5–3.*** View down the axis of an α helix. The side chains (R) are on the outside of the helix. The van der Waals radii of the atoms are larger than shown here; hence, there is almost no free space inside the helix. (Slightly modified and reproduced, with permission, from Stryer L: *Biochemistry,* 3rd ed. Freeman, 1995. Copyright © 1995 by W.H. Freeman and Co.)

ronment. Clusters of amphipathic helices can create a channel, or pore, that permits specific polar molecules to pass through hydrophobic cell membranes.

## The Beta Sheet

The second (hence "beta") recognizable regular secondary structure in proteins is the β sheet. The amino acid residues of a β sheet, when viewed edge-on, form a zigzag or pleated pattern in which the R groups of adjacent residues point in opposite directions. Unlike the compact backbone of the α helix, the peptide backbone of the β sheet is highly extended. But like the α helix, β sheets derive much of their stability from hydrogen bonds between the carbonyl oxygens and amide hydrogens of peptide bonds. However, in contrast to the α helix, these bonds are formed with adjacent segments of β sheet (Figure 5–5).

Interacting β sheets can be arranged either to form a **parallel** β sheet, in which the adjacent segments of the polypeptide chain proceed in the same direction amino to carboxyl, or an **antiparallel** sheet, in which they proceed in opposite directions (Figure 5–5). Either configuration permits the maximum number of hydrogen bonds between segments, or strands, of the sheet. Most β sheets are not perfectly flat but tend to have a right-handed twist. Clusters of twisted strands of β sheet form the core of many globular proteins (Figure 5–6). Schematic diagrams represent β sheets as arrows that point in the amino to carboxyl terminal direction.

## Loops & Bends

Roughly half of the residues in a "typical" globular protein reside in α helices and β sheets and half in loops, turns, bends, and other extended conformational features. Turns and bends refer to short segments of amino acids that join two units of secondary structure, such as two adjacent strands of an antiparallel β sheet. A β turn involves four aminoacyl residues, in which the first residue is hydrogen-bonded to the fourth, resulting in a tight 180-degree turn (Figure 5–7). Proline and glycine often are present in β turns.

Loops are regions that contain residues beyond the minimum number necessary to connect adjacent regions of secondary structure. Irregular in conformation, loops nevertheless serve key biologic roles. For many enzymes, the loops that bridge domains responsible for binding substrates often contain aminoacyl residues

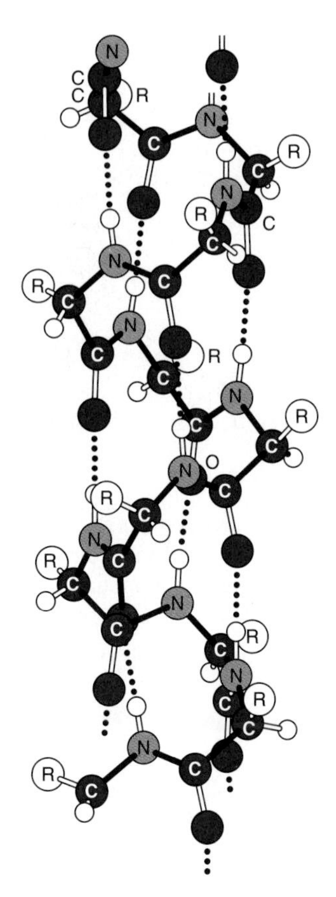

***Figure 5–4.*** Hydrogen bonds (dotted lines) formed between H and O atoms stabilize a polypeptide in an α-helical conformation. (Reprinted, with permission, from Haggis GH et al: *Introduction to Molecular Biology.* Wiley, 1964.)

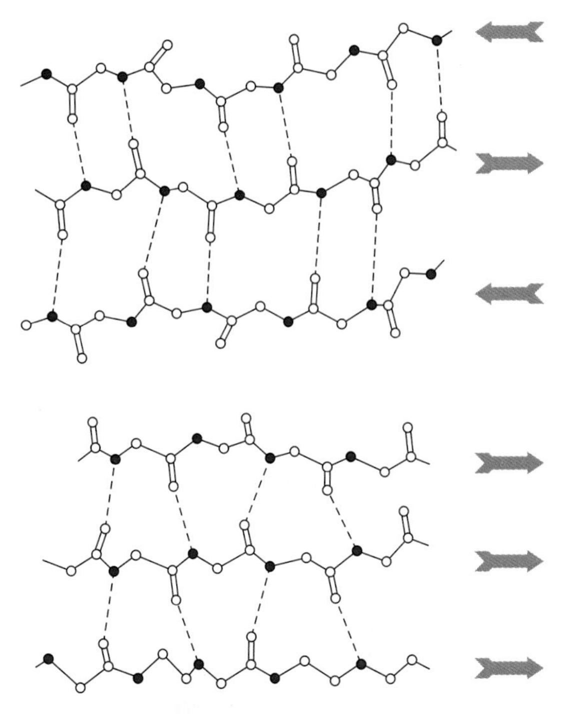

***Figure 5–5.*** Spacing and bond angles of the hydrogen bonds of antiparallel and parallel pleated β sheets. Arrows indicate the direction of each strand. The hydrogen-donating α-nitrogen atoms are shown as blue circles. Hydrogen bonds are indicated by dotted lines. For clarity in presentation, R groups and hydrogens are omitted. **Top:** Antiparallel β sheet. Pairs of hydrogen bonds alternate between being close together and wide apart and are oriented approximately perpendicular to the polypeptide backbone. **Bottom:** Parallel β sheet. The hydrogen bonds are evenly spaced but slant in alternate directions.

that participate in catalysis. **Helix-loop-helix motifs** provide the oligonucleotide-binding portion of DNA-binding proteins such as repressors and transcription factors. Structural motifs such as the helix-loop-helix motif that are intermediate between secondary and tertiary structures are often termed **supersecondary structures.** Since many loops and bends reside on the surface of proteins and are thus exposed to solvent, they constitute readily accessible sites, or **epitopes,** for recognition and binding of antibodies.

While loops lack apparent structural regularity, they exist in a specific conformation stabilized through hydrogen bonding, salt bridges, and hydrophobic interactions with other portions of the protein. However, not all portions of proteins are necessarily ordered. Proteins may contain "disordered" regions, often at the extreme amino or carboxyl terminal, characterized by high conformational flexibility. In many instances, these disordered regions assume an ordered conformation upon binding of a ligand. This structural flexibility enables such regions to act as ligand-controlled switches that affect protein structure and function.

## Tertiary & Quaternary Structure

The term "tertiary structure" refers to the entire three-dimensional conformation of a polypeptide. It indicates, in three-dimensional space, how secondary structural features—helices, sheets, bends, turns, and loops—assemble to form **domains** and how these domains relate spatially to one another. A domain is a section of protein structure sufficient to perform a particular chemical or physical task such as binding of a substrate

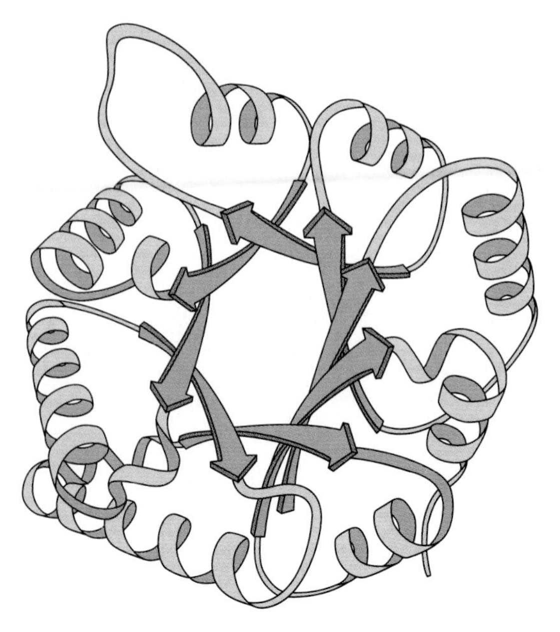

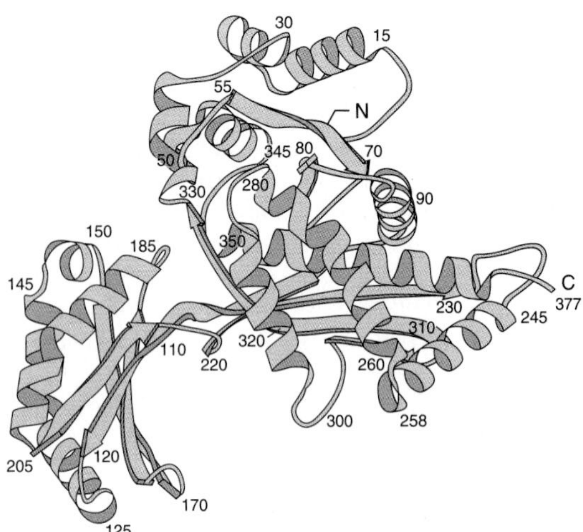


***Figure 5–6.*** Examples of tertiary structure of proteins. **Top:** The enzyme triose phosphate isomerase. Note the elegant and symmetrical arrangement of alternating β sheets and α helices. (Courtesy of J Richardson.) **Bottom:** Two-domain structure of the subunit of a homodimeric enzyme, a bacterial class II HMG-CoA reductase. As indicated by the numbered residues, the single polypeptide begins in the large domain, enters the small domain, and ends in the large domain. (Courtesy of C Lawrence, V Rodwell, and C Stauffacher, Purdue University.)

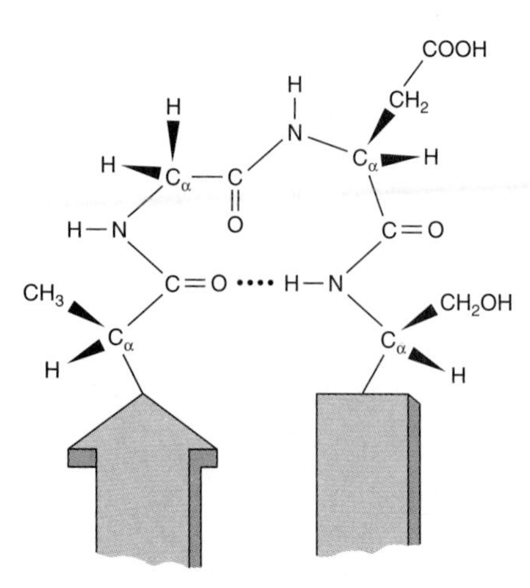


***Figure 5–7.*** A β-turn that links two segments of antiparallel β sheet. The dotted line indicates the hydrogen bond between the first and fourth amino acids of the four-residue segment Ala-Gly-Asp-Ser.

or other ligand. Other domains may anchor a protein to a membrane or interact with a regulatory molecule that modulates its function. A small polypeptide such as triose phosphate isomerase (Figure 5–6) or myoglobin (Chapter 6) may consist of a single domain. By contrast, protein kinases contain two domains. Protein kinases catalyze the transfer of a phosphoryl group from ATP to a peptide or protein. The amino terminal portion of the polypeptide, which is rich in β sheet, binds ATP, while the carboxyl terminal domain, which is rich in α helix, binds the peptide or protein substrate (Figure 5–8). The groups that catalyze phosphoryl transfer reside in a loop positioned at the interface of the two domains.

In some cases, proteins are assembled from more than one polypeptide, or protomer. Quaternary structure defines the polypeptide composition of a protein and, for an oligomeric protein, the spatial relationships between its subunits or protomers. **Monomeric** proteins consist of a single polypeptide chain. **Dimeric** proteins contain two polypeptide chains. Homodimers contain two copies of the same polypeptide chain, while in a heterodimer the polypeptides differ. Greek letters (α, β, γ etc) are used to distinguish different subunits of a heterooligomeric protein, and subscripts indicate the number of each subunit type. For example, $\alpha_4$ designates a homotetrameric protein, and $\alpha_2\beta_2\gamma$ a protein with five subunits of three different types.

Since even small proteins contain many thousands of atoms, depictions of protein structure that indicate the position of every atom are generally too complex to be readily interpreted. Simplified schematic diagrams thus are used to depict key features of a protein's ter-

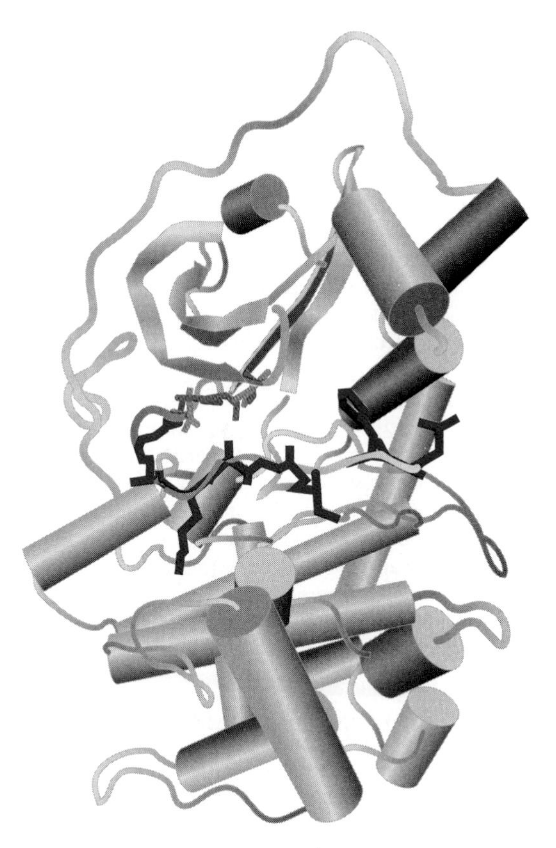

***Figure 5–8.*** Domain structure. Protein kinases contain two domains. The upper, amino terminal domain binds the phosphoryl donor ATP (light blue). The lower, carboxyl terminal domain is shown binding a synthetic peptide substrate (dark blue).

tiary and quaternary structure. Ribbon diagrams (Figures 5–6 and 5–8) trace the conformation of the polypeptide backbone, with cylinders and arrows indicating regions of α helix and β sheet, respectively. In an even simpler representation, line segments that link the α carbons indicate the path of the polypeptide backbone. These schematic diagrams often include the side chains of selected amino acids that emphasize specific structure-function relationships.

## MULTIPLE FACTORS STABILIZE TERTIARY & QUATERNARY STRUCTURE

Higher orders of protein structure are stabilized primarily—and often exclusively—by noncovalent interactions. Principal among these are hydrophobic interactions that drive most hydrophobic amino acid side chains into the interior of the protein, shielding them from water. Other significant contributors include hydrogen bonds and salt bridges between the carboxylates of aspartic and glutamic acid and the oppositely charged side chains of protonated lysyl, argininyl, and histidyl residues. While individually weak relative to a typical covalent bond of 80–120 kcal/mol, collectively these numerous interactions confer a high degree of stability to the biologically functional conformation of a protein, just as a Velcro fastener harnesses the cumulative strength of multiple plastic loops and hooks.

Some proteins contain covalent disulfide (S—S) bonds that link the sulfhydryl groups of cysteinyl residues. Formation of disulfide bonds involves oxidation of the cysteinyl sulfhydryl groups and requires oxygen. Intrapolypeptide disulfide bonds further enhance the stability of the folded conformation of a peptide, while interpolypeptide disulfide bonds stabilize the quaternary structure of certain oligomeric proteins.

## THREE-DIMENSIONAL STRUCTURE IS DETERMINED BY X-RAY CRYSTALLOGRAPHY OR BY NMR SPECTROSCOPY

### X-Ray Crystallography

Since the determination of the three-dimensional structure of myoglobin over 40 years ago, the three-dimensional structures of thousands of proteins have been determined by x-ray crystallography. The key to x-ray crystallography is the precipitation of a protein under conditions in which it forms ordered crystals that diffract x-rays. This is generally accomplished by exposing small drops of the protein solution to various combinations of pH and precipitating agents such as salts and organic solutes such as polyethylene glycol. A detailed three-dimensional structure of a protein can be constructed from its primary structure using the pattern by which it diffracts a beam of monochromatic x-rays. While the development of increasingly capable computer-based tools has rendered the analysis of complex x-ray diffraction patterns increasingly facile, a major stumbling block remains the requirement of inducing highly purified samples of the protein of interest to crystallize. Several lines of evidence, including the ability of some crystallized enzymes to catalyze chemical reactions, indicate that the vast majority of the structures determined by crystallography faithfully represent the structures of proteins in free solution.

### Nuclear Magnetic Resonance Spectroscopy

Nuclear magnetic resonance (NMR) spectroscopy, a powerful complement to x-ray crystallography, measures the absorbance of radio frequency electromagnetic energy by certain atomic nuclei. "NMR-active" isotopes of biologically relevant atoms include $^{1}H$, $^{13}C$, $^{15}N$, and $^{31}P$. The frequency, or chemical shift, at which a particular nucleus absorbs energy is a function of both the functional group within which it resides and the proximity of other NMR-active nuclei. Two-dimensional NMR spectroscopy permits a three-dimensional representation of a protein to be constructed by determining the proximity of these nuclei to one another. NMR spectroscopy analyzes proteins in aqueous solution, obviating the need to form crystals. It thus is possible to observe changes in conformation that accompany ligand binding or catalysis using NMR spectroscopy. However, only the spectra of relatively small proteins, $\leq$ 20 kDa in size, can be analyzed with current technology.

### Molecular Modeling

An increasingly useful adjunct to the empirical determination of the three-dimensional structure of proteins is the use of computer technology for molecular modeling. The types of models created take two forms. In the first, the known three-dimensional structure of a protein is used as a template to build a model of the probable structure of a homologous protein. In the second, computer software is used to manipulate the static model provided by crystallography to explore how a protein's conformation might change when ligands are bound or when temperature, pH, or ionic strength is altered. Scientists also are examining the library of available protein structures in an attempt to devise computer programs that can predict the three-dimensional conformation of a protein directly from its primary sequence.

## PROTEIN FOLDING

### The Native Conformation of a Protein Is Thermodynamically Favored

The number of distinct combinations of phi and psi angles specifying potential conformations of even a relatively small—15-kDa—polypeptide is unbelievably vast. Proteins are guided through this vast labyrinth of possibilities by thermodynamics. Since the biologically relevant—or native—conformation of a protein generally is that which is most energetically favored, knowledge of the native conformation is specified in the primary sequence. However, if one were to wait for a polypeptide to find its native conformation by random exploration of all possible conformations, the process would require billions of years to complete. Clearly, protein folding in cells takes place in a more orderly and guided fashion.

### Folding Is Modular

Protein folding generally occurs via a stepwise process. In the first stage, the newly synthesized polypeptide emerges from ribosomes, and short segments fold into secondary structural units that provide local regions of organized structure. Folding is now reduced to the selection of an appropriate arrangement of this relatively small number of secondary structural elements. In the second stage, the forces that drive hydrophobic regions into the interior of the protein away from solvent drive the partially folded polypeptide into a "molten globule" in which the modules of secondary structure rearrange to arrive at the mature conformation of the protein. This process is orderly but not rigid. Considerable flexibility exists in the ways and in the order in which elements of secondary structure can be rearranged. In general, each element of secondary or supersecondary structure facilitates proper folding by directing the folding process toward the native conformation and away from unproductive alternatives. For oligomeric proteins, individual protomers tend to fold before they associate with other subunits.

### Auxiliary Proteins Assist Folding

Under appropriate conditions, many proteins will spontaneously refold after being previously **denatured** (ie, unfolded) by treatment with acid or base, chaotropic agents, or detergents. However, unlike the folding process in vivo, refolding under laboratory conditions is a far slower process. Moreover, some proteins fail to spontaneously refold in vitro, often forming insoluble **aggregates,** disordered complexes of unfolded or partially folded polypeptides held together by hydrophobic interactions. Aggregates represent unproductive dead ends in the folding process. Cells employ auxiliary proteins to speed the process of folding and to guide it toward a productive conclusion.

### Chaperones

**Chaperone** proteins participate in the folding of over half of mammalian proteins. The hsp70 (70-kDa heat shock protein) family of chaperones binds short sequences of hydrophobic amino acids in newly synthesized polypeptides, shielding them from solvent. Chaperones prevent aggregation, thus providing an opportunity for the formation of appropriate secondary structural elements and their subsequent coalescence into a molten globule. The hsp60 family of chaperones, sometimes called **chaperonins,** differ in sequence and structure from hsp70 and its homologs. Hsp60 acts later in the folding process, often together with an hsp70 chaperone. The central cavity of the donut-shaped hsp60 chaperone provides a sheltered environment in which a polypeptide can fold until all hydrophobic regions are buried in its interior, eliminating aggregation. Chaperone proteins can also "rescue" proteins that have become thermodynamically trapped in a misfolded dead end by unfolding hydrophobic regions and providing a second chance to fold productively.

### Protein Disulfide Isomerase

Disulfide bonds between and within polypeptides stabilize tertiary and quaternary structure. However, disulfide bond formation is nonspecific. Under oxidizing conditions, a given cysteine can form a disulfide bond with the —SH of any accessible cysteinyl residue. By catalyzing disulfide exchange, the rupture of an S—S bond and its reformation with a different partner cysteine, protein disulfide isomerase facilitates the formation of disulfide bonds that stabilize their native conformation.

### Proline-*cis,trans*-Isomerase

All X-Pro peptide bonds—where X represents any residue—are synthesized in the trans configuration. However, of the X-Pro bonds of mature proteins, approximately 6% are cis. The cis configuration is particularly common in β-turns. Isomerization from trans to cis is catalyzed by the enzyme proline-*cis,trans*-isomerase (Figure 5–9).

## SEVERAL NEUROLOGIC DISEASES RESULT FROM ALTERED PROTEIN CONFORMATION

### Prions

The transmissible spongiform encephalopathies, or **prion diseases,** are fatal neurodegenerative diseases characterized by spongiform changes, astrocytic gliomas, and neuronal loss resulting from the deposition of insoluble protein aggregates in neural cells. They include Creutzfeldt-Jakob disease in humans, scrapie in sheep, and bovine spongiform encephalopathy (mad cow disease) in cattle. Prion diseases may manifest themselves as infectious, genetic, or sporadic disorders. Because no viral or bacterial gene encoding the pathologic prion protein could be identified, the source and mechanism of transmission of prion disease long remained elusive. Today it is believed that prion diseases are protein conformation diseases transmitted by altering the conformation, and hence the physical properties, of proteins endogenous to the host. Human prion-related protein, PrP, a glycoprotein encoded on the short arm of chromosome 20, normally is monomeric and rich in α helix. Pathologic prion proteins serve as the templates for the conformational transformation of normal PrP, known as PrPc, into PrPsc. PrPsc is rich in β sheet with many hydrophobic aminoacyl side chains exposed to solvent. PrPsc molecules therefore associate strongly with one other, forming insoluble protease-resistant aggregates. Since one pathologic prion or prion-related protein can serve as template for the conformational transformation of many times its number of PrPc molecules, prion diseases can be transmitted by the protein alone without involvement of DNA or RNA.

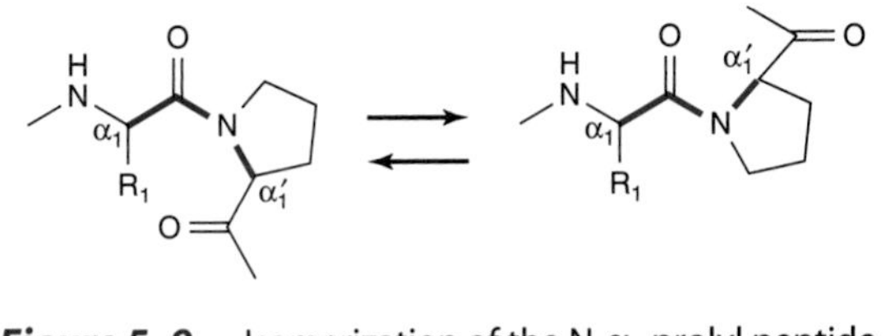

### Alzheimer's Disease

Refolding or misfolding of another protein endogenous to human brain tissue, β-amyloid, is also a prominent feature of Alzheimer's disease. While the root cause of Alzheimer's disease remains elusive, the characteristic senile plaques and neurofibrillary bundles contain aggregates of the protein β-amyloid, a 4.3-kDa polypeptide produced by proteolytic cleavage of a larger protein known as amyloid precursor protein. In Alzheimer's disease patients, levels of β-amyloid become elevated, and this protein undergoes a conformational transformation from a soluble α helix–rich state to a state rich in β sheet and prone to self-aggregation. Apolipoprotein E has been implicated as a potential mediator of this conformational transformation.

**Figure 5–9.** Isomerization of the N-$\alpha_1$ prolyl peptide bond from a cis to a trans configuration relative to the backbone of the polypeptide.

## COLLAGEN ILLUSTRATES THE ROLE OF POSTTRANSLATIONAL PROCESSING IN PROTEIN MATURATION

### Protein Maturation Often Involves Making & Breaking Covalent Bonds

The maturation of proteins into their final structural state often involves the cleavage or formation (or both) of covalent bonds, a process termed **posttranslational modification.** Many polypeptides are initially synthesized as larger precursors, called **proproteins.** The "extra" polypeptide segments in these proproteins often serve as leader sequences that target a polypeptide to a particular organelle or facilitate its passage through a membrane. Others ensure that the potentially harmful activity of a protein such as the proteases trypsin and chymotrypsin remains inhibited until these proteins reach their final destination. However, once these transient requirements are fulfilled, the now superfluous peptide regions are removed by selective proteolysis. Other covalent modifications may take place that add new chemical functionalities to a protein. The maturation of collagen illustrates both of these processes.

## Collagen Is a Fibrous Protein

Collagen is the most abundant of the fibrous proteins that constitute more than 25% of the protein mass in the human body. Other prominent fibrous proteins include keratin and myosin. These proteins represent a primary source of structural strength for cells (ie, the cytoskeleton) and tissues. Skin derives its strength and flexibility from a crisscrossed mesh of collagen and keratin fibers, while bones and teeth are buttressed by an underlying network of collagen fibers analogous to the steel strands in reinforced concrete. Collagen also is present in connective tissues such as ligaments and tendons. The high degree of tensile strength required to fulfill these structural roles requires elongated proteins characterized by repetitive amino acid sequences and a regular secondary structure.

## Collagen Forms a Unique Triple Helix

Tropocollagen consists of three fibers, each containing about 1000 amino acids, bundled together in a unique conformation, the collagen triple helix (Figure 5–10). A mature collagen fiber forms an elongated rod with an axial ratio of about 200. Three intertwined polypeptide strands, which twist to the left, wrap around one another in a right-handed fashion to form the collagen triple helix. The opposing handedness of this superhelix and its component polypeptides makes the collagen triple helix highly resistant to unwinding—the same principle used in the steel cables of suspension bridges. A collagen triple helix has 3.3 residues per turn and a rise per residue nearly twice that of an α helix. The R groups of each polypeptide strand of the triple helix pack so closely that in order to fit, one must be glycine. Thus, every third amino acid residue in collagen is a glycine residue. Staggering of the three strands provides appropriate positioning of the requisite glycines throughout the helix. Collagen is also rich in proline and hydroxyproline, yielding a repetitive Gly-X-Y pattern (Figure 5–10) in which Y generally is proline or hydroxyproline.

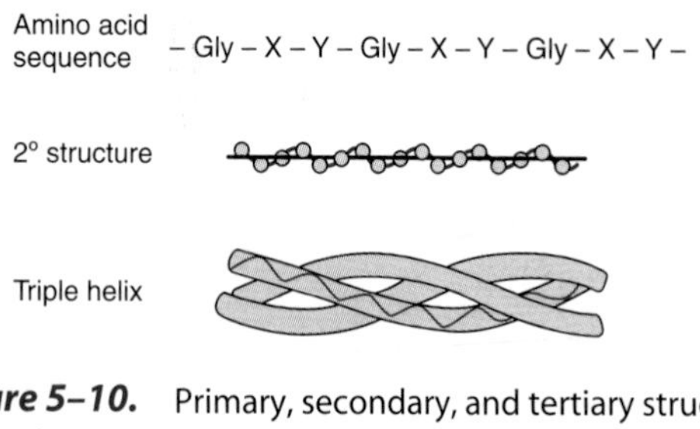

Amino acid sequence – Gly – X – Y – Gly – X – Y – Gly – X – Y –

2° structure

Triple helix

***Figure 5–10.*** Primary, secondary, and tertiary structures of collagen.

Collagen triple helices are stabilized by hydrogen bonds between residues in *different* polypeptide chains. The hydroxyl groups of hydroxyprolyl residues also participate in interchain hydrogen bonding. Additional stability is provided by covalent cross-links formed between modified lysyl residues both within and between polypeptide chains.

## Collagen Is Synthesized as a Larger Precursor

Collagen is initially synthesized as a larger precursor polypeptide, procollagen. Numerous prolyl and lysyl residues of procollagen are hydroxylated by prolyl hydroxylase and lysyl hydroxylase, enzymes that require ascorbic acid (vitamin C). Hydroxyprolyl and hydroxylysyl residues provide additional hydrogen bonding capability that stabilizes the mature protein. In addition, glucosyl and galactosyl transferases attach glucosyl or galactosyl residues to the hydroxyl groups of specific hydroxylysyl residues.

The central portion of the precursor polypeptide then associates with other molecules to form the characteristic triple helix. This process is accompanied by the removal of the globular amino terminal and carboxyl terminal extensions of the precursor polypeptide by selective proteolysis. Certain lysyl residues are modified by lysyl oxidase, a copper-containing protein that converts ε-amino groups to aldehydes. The aldehydes can either undergo an aldol condensation to form a C=C double bond or to form a Schiff base (eneimine) with the ε-amino group of an unmodified lysyl residue, which is subsequently reduced to form a C—N single bond. These covalent bonds cross-link the individual polypeptides and imbue the fiber with exceptional strength and rigidity.

## Nutritional & Genetic Disorders Can Impair Collagen Maturation

The complex series of events in collagen maturation provide a model that illustrates the biologic consequences of incomplete polypeptide maturation. The best-known defect in collagen biosynthesis is scurvy, a result of a dietary deficiency of vitamin C required by prolyl and lysyl hydroxylases. The resulting deficit in the number of hydroxyproline and hydroxylysine residues undermines the conformational stability of collagen fibers, leading to bleeding gums, swelling joints, poor wound healing, and ultimately to death. Menkes' syndrome, characterized by kinky hair and growth retardation, reflects a dietary deficiency of the copper required by lysyl oxidase, which catalyzes a key step in formation of the covalent cross-links that strengthen collagen fibers.

Genetic disorders of collagen biosynthesis include several forms of osteogenesis imperfecta, characterized by fragile bones. In Ehlers-Dahlos syndrome, a group of connective tissue disorders that involve impaired integrity of supporting structures, defects in the genes that encode $\alpha$ collagen-1, procollagen *N*-peptidase, or lysyl hydroxylase result in mobile joints and skin abnormalities.

## SUMMARY

- Proteins may be classified on the basis of the solubility, shape, or function or of the presence of a prosthetic group such as heme. Proteins perform complex physical and catalytic functions by positioning specific chemical groups in a precise three-dimensional arrangement that is both functionally efficient and physically strong.
- The gene-encoded primary structure of a polypeptide is the sequence of its amino acids. Its secondary structure results from folding of polypeptides into hydrogen-bonded motifs such as the $\alpha$ helix, the $\beta$-pleated sheet, $\beta$ bends, and loops. Combinations of these motifs can form supersecondary motifs.
- Tertiary structure concerns the relationships between secondary structural domains. Quaternary structure of proteins with two or more polypeptides (oligomeric proteins) is a feature based on the spatial relationships between various types of polypeptides.
- Primary structures are stabilized by covalent peptide bonds. Higher orders of structure are stabilized by weak forces—multiple hydrogen bonds, salt (electrostatic) bonds, and association of hydrophobic R groups.
- The phi ($\Phi$) angle of a polypeptide is the angle about the $C_\alpha$—N bond; the psi ($\Psi$) angle is that about the $C_\alpha$-$C_o$ bond. Most combinations of phi-psi angles are disallowed due to steric hindrance. The phi-psi angles that form the $\alpha$ helix and the $\beta$ sheet fall within the lower and upper left-hand quadrants of a Ramachandran plot, respectively.
- Protein folding is a poorly understood process. Broadly speaking, short segments of newly synthesized polypeptide fold into secondary structural units. Forces that bury hydrophobic regions from solvent then drive the partially folded polypeptide into a "molten globule" in which the modules of secondary structure are rearranged to give the native conformation of the protein.
- Proteins that assist folding include protein disulfide isomerase, proline-*cis,trans*,-isomerase, and the chaperones that participate in the folding of over half of mammalian proteins. Chaperones shield newly synthesized polypeptides from solvent and provide an environment for elements of secondary structure to emerge and coalesce into molten globules.
- Techniques for study of higher orders of protein structure include x-ray crystallography, NMR spectroscopy, analytical ultracentrifugation, gel filtration, and gel electrophoresis.
- Silk fibroin and collagen illustrate the close linkage of protein structure and biologic function. Diseases of collagen maturation include Ehlers-Danlos syndrome and the vitamin C deficiency disease scurvy.
- Prions—protein particles that lack nucleic acid—cause fatal transmissible spongiform encephalopathies such as Creutzfeldt-Jakob disease, scrapie, and bovine spongiform encephalopathy. Prion diseases involve an altered secondary-tertiary structure of a naturally occurring protein, PrPc. When PrPc interacts with its pathologic isoform PrPSc, its conformation is transformed from a predominantly $\alpha$-helical structure to the $\beta$-sheet structure characteristic of PrPSc.

## REFERENCES


Branden C, Tooze J: *Introduction to Protein Structure.* Garland, 1991.

Burkhard P, Stetefeld J, Strelkov SV: Coiled coils: A highly versatile protein folding motif. Trends Cell Biol 2001;11:82.

Collinge J: Prion diseases of humans and animals: Their causes and molecular basis. Annu Rev Neurosci 2001;24:519.

Frydman J: Folding of newly translated proteins in vivo: The role of molecular chaperones. Annu Rev Biochem 2001;70:603.

Radord S: Protein folding: Progress made and promises ahead. Trends Biochem Sci 2000;25:611.

Schmid FX: Proly isomerase: Enzymatic catalysis of slow protein folding reactions. Ann Rev Biophys Biomol Struct 1993;22: 123.

Segrest MP et al: The amphipathic alpha-helix: A multifunctional structural motif in plasma lipoproteins. Adv Protein Chem 1995;45:1.

Soto C: Alzheimer's and prion disease as disorders of protein conformation: Implications for the design of novel therapeutic approaches. J Mol Med 1999;77:412.

# Proteins: Myoglobin & Hemoglobin 6

*Victor W. Rodwell, PhD, & Peter J. Kennelly, PhD*

## BIOMEDICAL IMPORTANCE

The heme proteins myoglobin and hemoglobin maintain a supply of oxygen essential for oxidative metabolism. Myoglobin, a monomeric protein of red muscle, stores oxygen as a reserve against oxygen deprivation. Hemoglobin, a tetrameric protein of erythrocytes, transports $O_2$ to the tissues and returns $CO_2$ and protons to the lungs. Cyanide and carbon monoxide kill because they disrupt the physiologic function of the heme proteins cytochrome oxidase and hemoglobin, respectively. The secondary-tertiary structure of the subunits of hemoglobin resembles myoglobin. However, the tetrameric structure of hemoglobin permits cooperative interactions that are central to its function. For example, 2,3-bisphosphoglycerate (BPG) promotes the efficient release of $O_2$ by stabilizing the quaternary structure of deoxyhemoglobin. Hemoglobin and myoglobin illustrate both protein structure-function relationships and the molecular basis of genetic diseases such as sickle cell disease and the thalassemias.

## HEME & FERROUS IRON CONFER THE ABILITY TO STORE & TO TRANSPORT OXYGEN

Myoglobin and hemoglobin contain **heme,** a cyclic tetrapyrrole consisting of four molecules of pyrrole linked by α-methylene bridges. This planar network of conjugated double bonds absorbs visible light and colors heme deep red. The substituents at the β-positions of heme are methyl (M), vinyl (V), and propionate (Pr) groups arranged in the order M, V, M, V, M, Pr, Pr, M (Figure 6–1). One atom of ferrous iron ($Fe_2^+$) resides at the center of the planar tetrapyrrole. Other proteins with metal-containing tetrapyrrole prosthetic groups include the cytochromes (Fe and Cu) and chlorophyll (Mg) (see Chapter 12). Oxidation and reduction of the Fe and Cu atoms of cytochromes is essential to their biologic function as carriers of electrons. By contrast, oxidation of the $Fe_2^+$ of myoglobin or hemoglobin to $Fe_3^+$ destroys their biologic activity.

### Myoglobin Is Rich in α Helix

Oxygen stored in red muscle myoglobin is released during $O_2$ deprivation (eg, severe exercise) for use in muscle mitochondria for aerobic synthesis of ATP (see Chapter 12). A 153-aminoacyl residue polypeptide (MW 17,000), myoglobin folds into a compact shape that measures 4.5 × 3.5 × 2.5 nm (Figure 6–2). Unusually high proportions, about 75%, of the residues are present in eight right-handed, 7–20 residue α helices. Starting at the amino terminal, these are termed helices A–H. Typical of globular proteins, the surface of myoglobin is polar, while—with only two exceptions—the interior contains only nonpolar residues such as Leu, Val, Phe, and Met. The exceptions are His E7 and His F8, the seventh and eighth residues in helices E and F, which lie close to the heme iron where they function in $O_2$ binding.

### Histidines F8 & E7 Perform Unique Roles in Oxygen Binding

The heme of myoglobin lies in a crevice between helices E and F oriented with its polar propionate groups facing the surface of the globin (Figure 6–2). The remainder resides in the nonpolar interior. The fifth coordination position of the iron is linked to a ring nitrogen of the **proximal histidine,** His F8. The **distal histidine,** His E7, lies on the side of the heme ring opposite to His F8.

### The Iron Moves Toward the Plane of the Heme When Oxygen Is Bound

The iron of unoxygenated myoglobin lies 0.03 nm (0.3 Å) outside the plane of the heme ring, toward His F8. The heme therefore "puckers" slightly. When $O_2$ occupies the sixth coordination position, the iron moves to within 0.01 nm (0.1 Å) of the plane of the heme ring. Oxygenation of myoglobin thus is accompanied by motion of the iron, of His F8, and of residues linked to His F8.

***Figure 6–1.*** Heme. The pyrrole rings and methylene bridge carbons are coplanar, and the iron atom ($Fe_2^+$) resides in almost the same plane. The fifth and sixth coordination positions of $Fe_2^+$ are directed perpendicular to—and directly above and below—the plane of the heme ring. Observe the nature of the substituent groups on the β carbons of the pyrrole rings, the central iron atom, and the location of the polar side of the heme ring (at about 7 o'clock) that faces the surface of the myoglobin molecule.

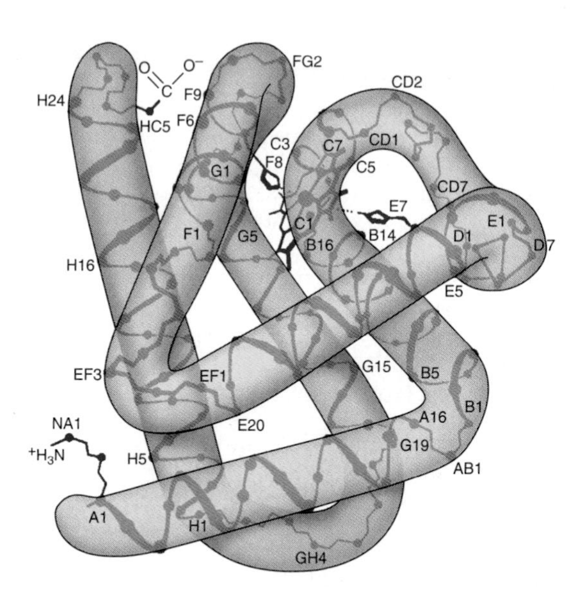


***Figure 6–2.*** A model of myoglobin at low resolution. Only the α-carbon atoms are shown. The α-helical regions are named A through H. (Based on Dickerson RE in: The Proteins, 2nd ed. Vol 2. Neurath H [editor]. Academic Press, 1964. Reproduced with permission.)

## Apomyoglobin Provides a Hindered Environment for Heme Iron

When $O_2$ binds to myoglobin, the bond between the first oxygen atom and the $Fe_2^+$ is perpendicular to the plane of the heme ring. The bond linking the first and second oxygen atoms lies at an angle of 121 degrees to the plane of the heme, orienting the second oxygen away from the distal histidine (Figure 6–3, left). Isolated heme binds carbon monoxide (CO) 25,000 times more strongly than oxygen. Since CO is present in small quantities in the atmosphere and arises in cells from the catabolism of heme, why is it that CO does not completely displace $O_2$ from heme iron? The accepted explanation is that the apoproteins of myoglobin and hemoglobin create a **hindered environment.** While CO can bind to isolated heme in its preferred orientation, ie, with all three atoms (Fe, C, and O) perpendicular to the plane of the heme, in myoglobin and hemoglobin the distal histidine sterically precludes this orientation. Binding at a less favored angle reduces the strength of the heme-CO bond to about 200 times that of the heme-$O_2$ bond (Figure 6–3, right) at which level the great excess of $O_2$ over CO normally present dominates. Nevertheless, about 1% of myoglobin typically is present combined with carbon monoxide.

## THE OXYGEN DISSOCIATION CURVES FOR MYOGLOBIN & HEMOGLOBIN SUIT THEIR PHYSIOLOGIC ROLES

Why is myoglobin unsuitable as an $O_2$ transport protein but well suited for $O_2$ storage? The relationship between the concentration, or partial pressure, of $O_2$ ($PO_2$) and the quantity of $O_2$ bound is expressed as an $O_2$ saturation isotherm (Figure 6–4). The oxygen-

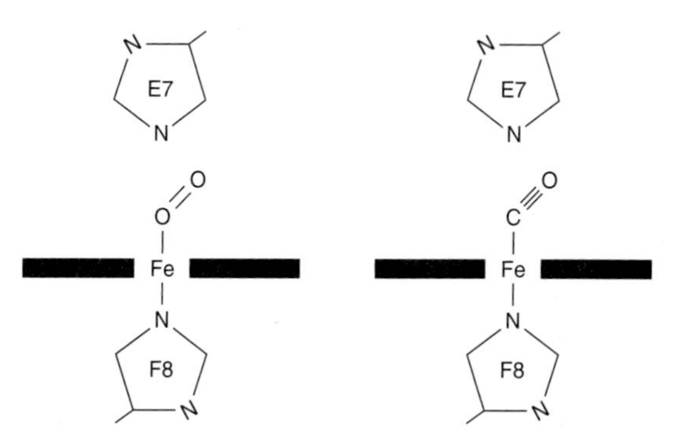


***Figure 6–3.*** Angles for bonding of oxygen and carbon monoxide to the heme iron of myoglobin. The distal E7 histidine hinders bonding of CO at the preferred (180 degree) angle to the plane of the heme ring.

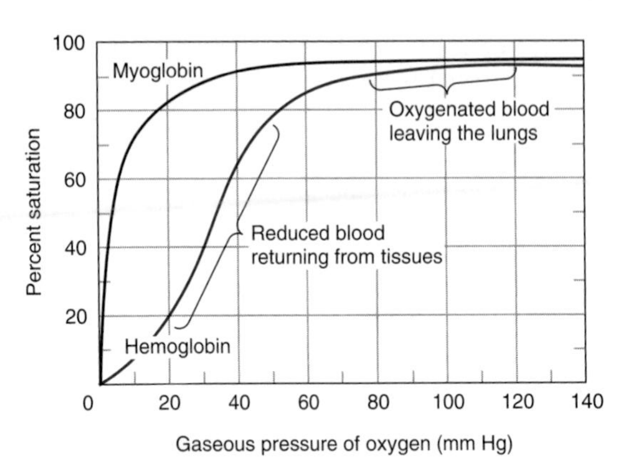


***Figure 6–4.*** Oxygen-binding curves of both hemoglobin and myoglobin. Arterial oxygen tension is about 100 mm Hg; mixed venous oxygen tension is about 40 mm Hg; capillary (active muscle) oxygen tension is about 20 mm Hg; and the minimum oxygen tension required for cytochrome oxidase is about 5 mm Hg. Association of chains into a tetrameric structure (hemoglobin) results in much greater oxygen delivery than would be possible with single chains. (Modified, with permission, from Scriver CR et al [editors]: The Molecular and Metabolic Bases of Inherited Disease, 7th ed. McGraw-Hill, 1995.)

binding curve for myoglobin is hyperbolic. Myoglobin therefore loads $O_2$ readily at the $PO_2$ of the lung capillary bed (100 mm Hg). However, since myoglobin releases only a small fraction of its bound $O_2$ at the $PO_2$ values typically encountered in active muscle (20 mm Hg) or other tissues (40 mm Hg), it represents an ineffective vehicle for delivery of $O_2$. However, when strenuous exercise lowers the $PO_2$ of muscle tissue to about 5 mm Hg, myoglobin releases $O_2$ for mitochondrial synthesis of ATP, permitting continued muscular activity.

## THE ALLOSTERIC PROPERTIES OF HEMOGLOBINS RESULT FROM THEIR QUATERNARY STRUCTURES

The properties of individual hemoglobins are consequences of their quaternary as well as of their secondary and tertiary structures. The quaternary structure of hemoglobin confers striking additional properties, absent from monomeric myoglobin, which adapts it to its unique biologic roles. The **allosteric** (Gk *allos* "other," *steros* "space") properties of hemoglobin provide, in addition, a model for understanding other allosteric proteins (see Chapter 11).

### Hemoglobin Is Tetrameric

Hemoglobins are tetramers comprised of pairs of two different polypeptide subunits. Greek letters are used to designate each subunit type. The subunit composition of the principal hemoglobins are $\alpha_2\beta_2$ (HbA; normal adult hemoglobin), $\alpha_2\gamma_2$ (HbF; fetal hemoglobin), $\alpha_2S_2$ (HbS; sickle cell hemoglobin), and $\alpha_2\delta_2$ ($HbA_2$; a minor adult hemoglobin). The primary structures of the β, γ, and δ chains of human hemoglobin are highly conserved.

### Myoglobin & the β Subunits of Hemoglobin Share Almost Identical Secondary and Tertiary Structures

Despite differences in the kind and number of amino acids present, myoglobin and the β polypeptide of hemoglobin A have almost identical secondary and tertiary structures. Similarities include the location of the heme and the eight helical regions and the presence of amino acids with similar properties at comparable locations. Although it possesses seven rather than eight helical regions, the α polypeptide of hemoglobin also closely resembles myoglobin.

### Oxygenation of Hemoglobin Triggers Conformational Changes in the Apoprotein

Hemoglobins bind four molecules of $O_2$ per tetramer, one per heme. A molecule of $O_2$ binds to a hemoglobin tetramer more readily if other $O_2$ molecules are already bound (Figure 6–4). Termed **cooperative binding,** this phenomenon permits hemoglobin to maximize both the quantity of $O_2$ loaded at the $PO_2$ of the lungs and the quantity of $O_2$ released at the $PO_2$ of the peripheral tissues. Cooperative interactions, an exclusive property of multimeric proteins, are critically important to aerobic life.

### $P_{50}$ Expresses the Relative Affinities of Different Hemoglobins for Oxygen

The quantity $P_{50}$, a measure of $O_2$ concentration, is the partial pressure of $O_2$ that half-saturates a given hemoglobin. Depending on the organism, $P_{50}$ can vary widely, but in all instances it will exceed the $PO_2$ of the peripheral tissues. For example, values of $P_{50}$ for HbA and fetal HbF are 26 and 20 mm Hg, respectively. In the placenta, this difference enables HbF to extract oxygen from the HbA in the mother's blood. However, HbF is suboptimal postpartum since its high affinity for $O_2$ dictates that it can deliver less $O_2$ to the tissues.

The subunit composition of hemoglobin tetramers undergoes complex changes during development. The human fetus initially synthesizes a $\zeta_2\varepsilon_2$ tetramer. By the end of the first trimester, $\zeta$ and $\gamma$ subunits have been replaced by $\alpha$ and $\varepsilon$ subunits, forming HbF ($\alpha_2\gamma_2$), the hemoglobin of late fetal life. While synthesis of $\beta$ subunits begins in the third trimester, $\beta$ subunits do not completely replace $\gamma$ subunits to yield adult HbA ($\alpha_2\beta_2$) until some weeks postpartum (Figure 6–5).

## Oxygenation of Hemoglobin Is Accompanied by Large Conformational Changes

The binding of the first $O_2$ molecule to deoxyHb shifts the heme iron towards the plane of the heme ring from a position about 0.6 nm beyond it (Figure 6–6). This motion is transmitted to the proximal (F8) histidine and to the residues attached thereto, which in turn causes the rupture of salt bridges between the carboxyl terminal residues of all four subunits. As a consequence, one pair of $\alpha/\beta$ subunits rotates 15 degrees with respect to the other, compacting the tetramer (Figure 6–7). Profound changes in secondary, tertiary, and quaternary structure accompany the high-affinity $O_2$-induced transition of hemoglobin from the low-affinity **T (taut) state** to the **R (relaxed) state.** These changes significantly increase the affinity of the remaining unoxygenated hemes for $O_2$, as subsequent binding events require the rupture of fewer salt bridges (Figure 6–8). The terms T and R also are used to refer to the low-affinity and high-affinity conformations of allosteric enzymes, respectively.

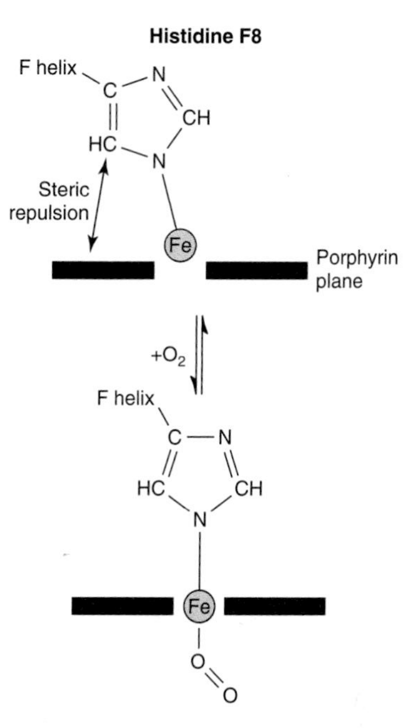


***Figure 6–6.*** The iron atom moves into the plane of the heme on oxygenation. Histidine F8 and its associated residues are pulled along with the iron atom. (Slightly modified and reproduced, with permission, from Stryer L: *Biochemistry,* 4th ed. Freeman, 1995.)

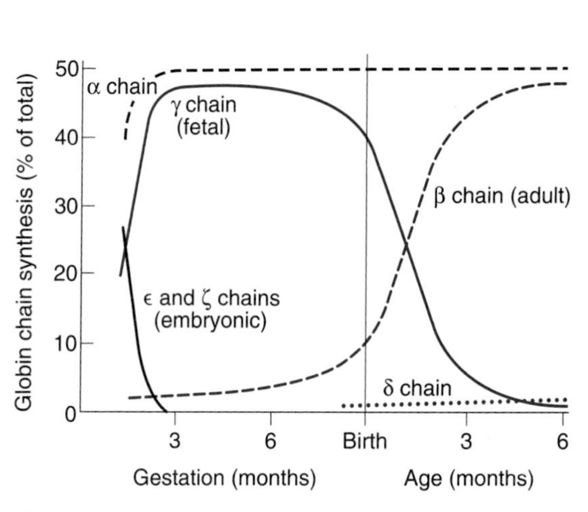


***Figure 6–5.*** Developmental pattern of the quaternary structure of fetal and newborn hemoglobins. (Reproduced, with permission, from Ganong WF: *Review of Medical Physiology,* 20th ed. McGraw-Hill, 2001.)

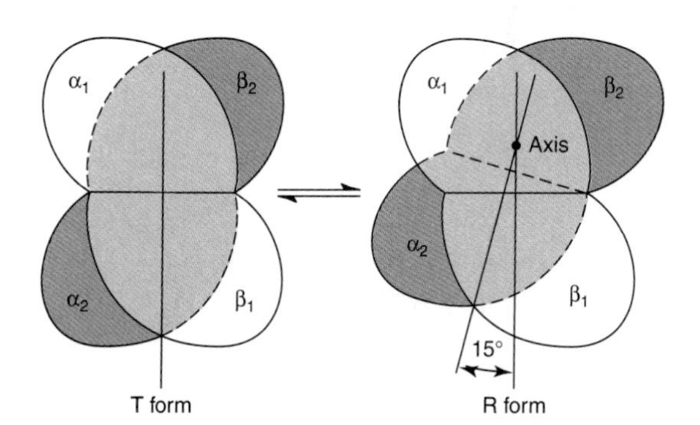


***Figure 6–7.*** During transition of the T form to the R form of hemoglobin, one pair of subunits ($\alpha_2/\beta_2$) rotates through 15 degrees relative to the other pair ($\alpha_1/\beta_1$). The axis of rotation is eccentric, and the $\alpha_2/\beta_2$ pair also shifts toward the axis somewhat. In the diagram, the unshaded $\alpha_1/\beta_1$ pair is shown fixed while the colored $\alpha_2/\beta_2$ pair both shifts and rotates.

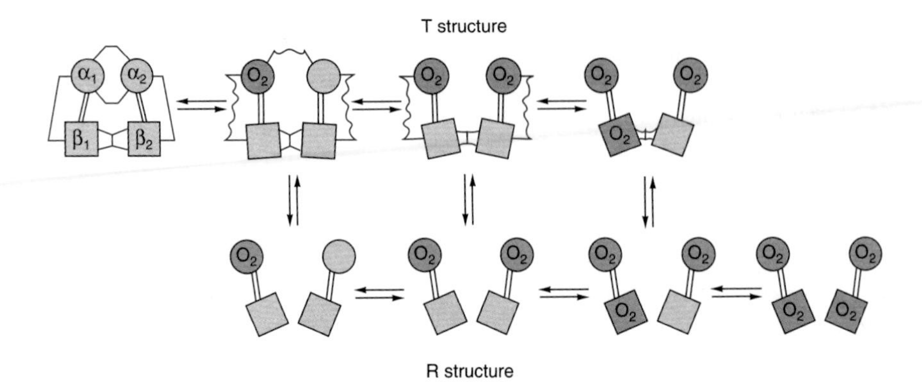


***Figure 6–8.*** Transition from the T structure to the R structure. In this model, salt bridges (thin lines) linking the subunits in the T structure break progressively as oxygen is added, and even those salt bridges that have not yet ruptured are progressively weakened (wavy lines). The transition from T to R does not take place after a fixed number of oxygen molecules have been bound but becomes more probable as each successive oxygen binds. The transition between the two structures is influenced by protons, carbon dioxide, chloride, and BPG; the higher their concentration, the more oxygen must be bound to trigger the transition. Fully oxygenated molecules in the T structure and fully deoxygenated molecules in the R structure are not shown because they are unstable. (Modified and redrawn, with permission, from Perutz MF: Hemoglobin structure and respiratory transport. Sci Am [Dec] 1978;239:92.)

## After Releasing $O_2$ at the Tissues, Hemoglobin Transports $CO_2$ & Protons to the Lungs

In addition to transporting $O_2$ from the lungs to peripheral tissues, hemoglobin transports $CO_2$, the byproduct of respiration, and protons from peripheral tissues to the lungs. Hemoglobin carries $CO_2$ as carbamates formed with the amino terminal nitrogens of the polypeptide chains.

$$CO_2 + Hb{-}NH_3^{+} = 2H^{+} + Hb{-}\underset{H}{N}{-}\overset{O}{\overset{\|}{C}}{-}O^{-}$$

Carbamates change the charge on amino terminals from positive to negative, favoring salt bond formation between the α and β chains.

Hemoglobin carbamates account for about 15% of the $CO_2$ in venous blood. Much of the remaining $CO_2$ is carried as bicarbonate, which is formed in erythrocytes by the hydration of $CO_2$ to carbonic acid ($H_2CO_3$), a process catalyzed by carbonic anhydrase. At the pH of venous blood, $H_2CO_3$ dissociates into bicarbonate and a proton.

$$CO_2 + H_2O \underset{}{\overset{\text{CARBONIC ANHYDRASE}}{\rightleftharpoons}} \underset{\text{Carbonic acid}}{H_2CO_3} \overset{\text{(Spontaneous)}}{\rightleftharpoons} HCO_3^{-} + H^{+}$$

Deoxyhemoglobin binds one proton for every two $O_2$ molecules released, contributing significantly to the buffering capacity of blood. The somewhat lower pH of peripheral tissues, aided by carbamation, stabilizes the T state and thus enhances the delivery of $O_2$. In the lungs, the process reverses. As $O_2$ binds to deoxyhemoglobin, protons are released and combine with bicarbonate to form carbonic acid. Dehydration of $H_2CO_3$, catalyzed by carbonic anhydrase, forms $CO_2$, which is exhaled. Binding of oxygen thus drives the exhalation of $CO_2$ (Figure 6–9). This reciprocal coupling of proton and $O_2$ binding is termed the **Bohr effect.** The Bohr effect is dependent upon **cooperative interactions between the hemes of the hemoglobin tetramer.** Myoglobin, a monomer, exhibits no Bohr effect.

## Protons Arise From Rupture of Salt Bonds When $O_2$ Binds

Protons responsible for the Bohr effect arise from rupture of salt bridges during the binding of $O_2$ to T state

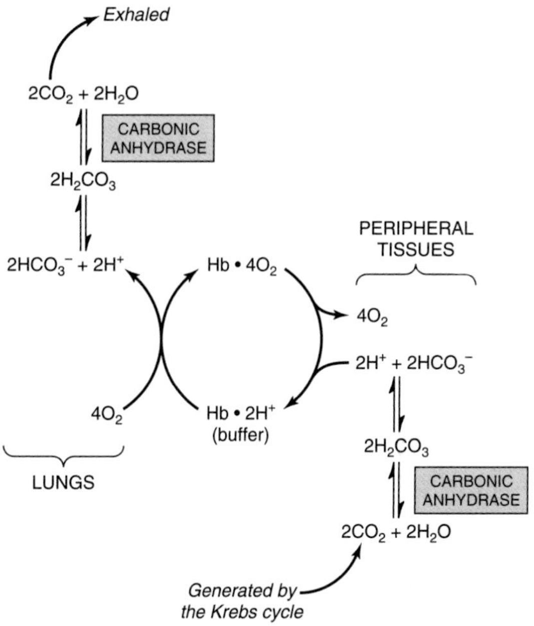

***Figure 6–9.*** The Bohr effect. Carbon dioxide generated in peripheral tissues combines with water to form carbonic acid, which dissociates into protons and bicarbonate ions. Deoxyhemoglobin acts as a buffer by binding protons and delivering them to the lungs. In the lungs, the uptake of oxygen by hemoglobin releases protons that combine with bicarbonate ion, forming carbonic acid, which when dehydrated by carbonic anhydrase becomes carbon dioxide, which then is exhaled.

hemoglobin. Conversion to the oxygenated R state breaks salt bridges involving β-chain residue His 146. The subsequent dissociation of protons from His 146 drives the conversion of bicarbonate to carbonic acid (Figure 6–9). Upon the release of $O_2$, the T structure and its salt bridges re-form. This conformational change increases the p$K_a$ of the β-chain His 146 residues, which bind protons. By facilitating the re-formation of salt bridges, an increase in proton concentration enhances the release of $O_2$ from oxygenated (R state) hemoglobin. Conversely, an increase in $PO_2$ promotes proton release.

## 2,3-Bisphosphoglycerate (BPG) Stabilizes the T Structure of Hemoglobin

A low $PO_2$ in peripheral tissues promotes the synthesis in erythrocytes of 2,3-bisphosphoglycerate (BPG) from the glycolytic intermediate 1,3-bisphosphoglycerate.

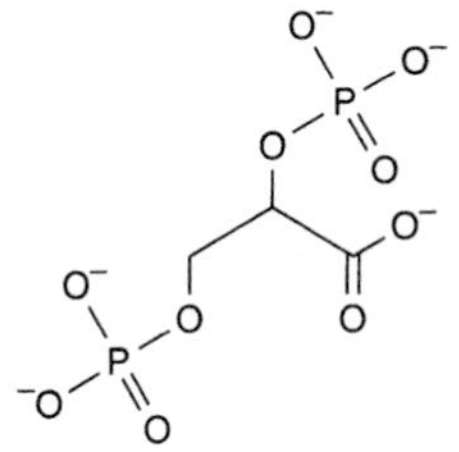

The hemoglobin tetramer binds one molecule of BPG in the central cavity formed by its four subunits. However, the space between the H helices of the β chains lining the cavity is sufficiently wide to accommodate BPG only when hemoglobin is in the T state. BPG forms salt bridges with the terminal amino groups of both β chains via Val NA1 and with Lys EF6 and His H21 (Figure 6–10). BPG therefore stabilizes deoxygenated (T state) hemoglobin by forming additional salt bridges that must be broken prior to conversion to the R state.

Residue H21 of the γ subunit of fetal hemoglobin (HbF) is Ser rather than His. Since Ser cannot form a salt bridge, BPG binds more weakly to HbF than to HbA. The lower stabilization afforded to the T state by BPG accounts for HbF having a higher affinity for $O_2$ than HbA.

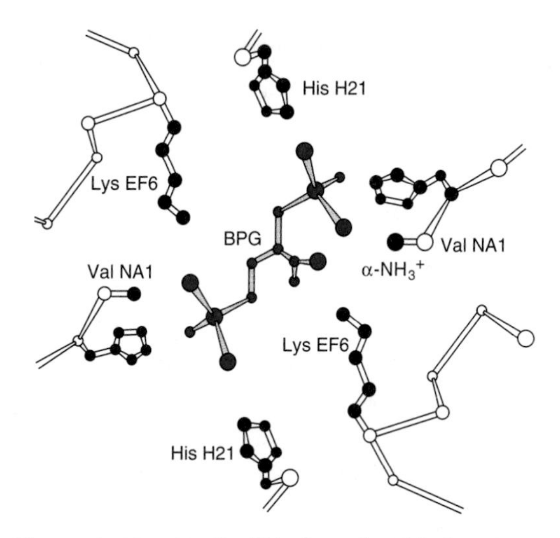


***Figure 6–10.*** Mode of binding of 2,3-bisphosphoglycerate to human deoxyhemoglobin. BPG interacts with three positively charged groups on each β chain. (Based on Arnone A: X-ray diffraction study of binding of 2,3-diphosphoglycerate to human deoxyhemoglobin. Nature 1972;237:146. Reproduced with permission.)

### Adaptation to High Altitude

Physiologic changes that accompany prolonged exposure to high altitude include an increase in the number of erythrocytes and in their concentrations of hemoglobin and of BPG. Elevated BPG lowers the affinity of HbA for $O_2$ (decreases $P_{50}$), which enhances release of $O_2$ at the tissues.

## NUMEROUS MUTANT HUMAN HEMOGLOBINS HAVE BEEN IDENTIFIED

Mutations in the genes that encode the α or β subunits of hemoglobin potentially can affect its biologic function. However, almost all of the over 800 known mutant human hemoglobins are both extremely rare and benign, presenting no clinical abnormalities. When a mutation does compromise biologic function, the condition is termed a **hemoglobinopathy.** The URL http://globin.cse.psu.edu/ (Globin Gene Server) provides information about—and links for—normal and mutant hemoglobins.

### Methemoglobin & Hemoglobin M

In methemoglobinemia, the heme iron is ferric rather than ferrous. Methemoglobin thus can neither bind nor transport $O_2$. Normally, the enzyme methemoglobin reductase reduces the $Fe_3^+$ of methemoglobin to $Fe_2^+$. Methemoglobin can arise by oxidation of $Fe_2^+$ to $Fe_3^+$ as a side effect of agents such as sulfonamides, from hereditary hemoglobin M, or consequent to reduced activity of the enzyme methemoglobin reductase.

In hemoglobin M, histidine F8 (His F8) has been replaced by tyrosine. The iron of HbM forms a tight ionic complex with the phenolate anion of tyrosine that stabilizes the $Fe_3^+$ form. In α-chain hemoglobin M variants, the R-T equilibrium favors the T state. Oxygen affinity is reduced, and the Bohr effect is absent. β-Chain hemoglobin M variants exhibit R-T switching, and the Bohr effect is therefore present.

Mutations (eg, hemoglobin Chesapeake) that favor the R state increase $O_2$ affinity. These hemoglobins therefore fail to deliver adequate $O_2$ to peripheral tissues. The resulting tissue hypoxia leads to **polycythemia,** an increased concentration of erythrocytes.

### Hemoglobin S

In HbS, the nonpolar amino acid valine has replaced the polar surface residue Glu6 of the β subunit, generating a hydrophobic **"sticky patch"** on the surface of the β subunit of both oxyHbS and deoxyHbS. Both HbA and HbS contain a complementary sticky patch on their surfaces that is exposed only in the deoxygenated, R state. Thus, at low $Po_2$, deoxyHbS can polymerize to form long, insoluble fibers. Binding of deoxyHbA terminates fiber polymerization, since HbA lacks the second sticky patch necessary to bind another Hb molecule (Figure 6–11). These twisted helical fibers distort the erythrocyte into a characteristic sickle shape, rendering it vulnerable to lysis in the interstices of the splenic sinusoids. They also cause multiple secondary clinical effects. A low $Po_2$ such as that at high altitudes exacerbates the tendency to polymerize.

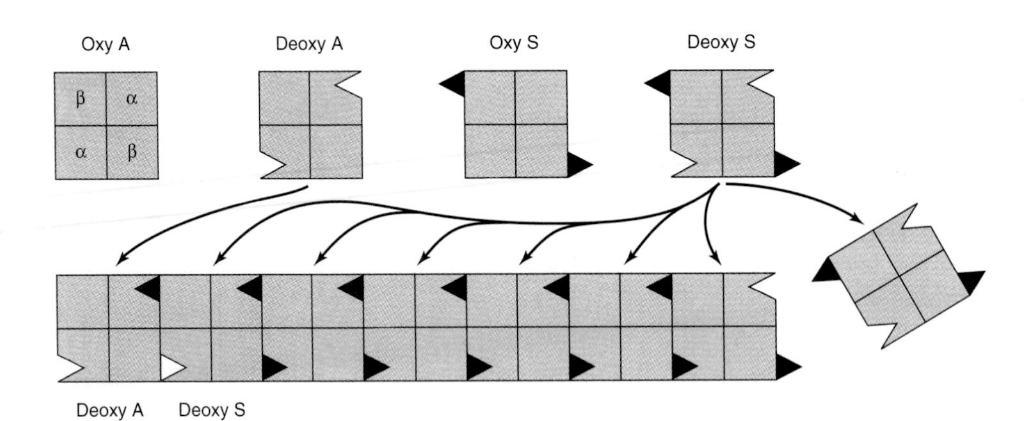


***Figure 6–11.*** Representation of the sticky patch (▲) on hemoglobin S and its "receptor" (△) on deoxyhemoglobin A and deoxyhemoglobin S. The complementary surfaces allow deoxyhemoglobin S to polymerize into a fibrous structure, but the presence of deoxyhemoglobin A will terminate the polymerization by failing to provide sticky patches. (Modified and reproduced, with permission, from Stryer L: *Biochemistry*, 4th ed. Freeman, 1995.)

## BIOMEDICAL IMPLICATIONS

### Myoglobinuria

Following massive crush injury, myoglobin released from damaged muscle fibers colors the urine dark red. Myoglobin can be detected in plasma following a myocardial infarction, but assay of serum enzymes (see Chapter 7) provides a more sensitive index of myocardial injury.

### Anemias

Anemias, reductions in the number of red blood cells or of hemoglobin in the blood, can reflect impaired synthesis of hemoglobin (eg, in iron deficiency; Chapter 51) or impaired production of erythrocytes (eg, in folic acid or vitamin $B_{12}$ deficiency; Chapter 45). Diagnosis of anemias begins with spectroscopic measurement of blood hemoglobin levels.

### Thalassemias

The genetic defects known as thalassemias result from the partial or total absence of one or more $\alpha$ or $\beta$ chains of hemoglobin. Over 750 different mutations have been identified, but only three are common. Either the $\alpha$ chain (alpha thalassemias) or $\beta$ chain (beta thalassemias) can be affected. A superscript indicates whether a subunit is completely absent ($\alpha^0$ or $\beta^0$) or whether its synthesis is reduced ($\alpha^+$ or $\beta^+$). Apart from marrow transplantation, treatment is symptomatic.

Certain mutant hemoglobins are common in many populations, and a patient may inherit more than one type. Hemoglobin disorders thus present a complex pattern of clinical phenotypes. The use of DNA probes for their diagnosis is considered in Chapter 40.

### Glycosylated Hemoglobin ($HbA_{1c}$)

When blood glucose enters the erythrocytes it glycosylates the $\varepsilon$-amino group of lysine residues and the amino terminals of hemoglobin. The fraction of hemoglobin glycosylated, normally about 5%, is proportionate to blood glucose concentration. Since the half-life of an erythrocyte is typically 60 days, the level of glycosylated hemoglobin ($HbA_{1c}$) reflects the mean blood glucose concentration over the preceding 6–8 weeks. Measurement of $HbA_{1c}$ therefore provides valuable information for management of diabetes mellitus.

## SUMMARY


- Myoglobin is monomeric; hemoglobin is a tetramer of two subunit types ($\alpha_2\beta_2$ in HbA). Despite having different primary structures, myoglobin and the subunits of hemoglobin have nearly identical secondary and tertiary structures.
- Heme, an essentially planar, slightly puckered, cyclic tetrapyrrole, has a central $Fe_2^+$ linked to all four nitrogen atoms of the heme, to histidine F8, and, in oxyMb and oxyHb, also to $O_2$.
- The $O_2$-binding curve for myoglobin is hyperbolic, but for hemoglobin it is sigmoidal, a consequence of cooperative interactions in the tetramer. Cooperativity maximizes the ability of hemoglobin both to load $O_2$ at the $PO_2$ of the lungs and to deliver $O_2$ at the $PO_2$ of the tissues.
- Relative affinities of different hemoglobins for oxygen are expressed as $P_{50}$, the $PO_2$ that half-saturates them with $O_2$. Hemoglobins saturate at the partial pressures of their respective respiratory organ, eg, the lung or placenta.
- On oxygenation of hemoglobin, the iron, histidine F8, and linked residues move toward the heme ring. Conformational changes that accompany oxygenation include rupture of salt bonds and loosening of quaternary structure, facilitating binding of additional $O_2$.
- 2,3-Bisphosphoglycerate (BPG) in the central cavity of deoxyHb forms salt bonds with the $\beta$ subunits that stabilize deoxyHb. On oxygenation, the central cavity contracts, BPG is extruded, and the quaternary structure loosens.
- Hemoglobin also functions in $CO_2$ and proton transport from tissues to lungs. Release of $O_2$ from oxyHb at the tissues is accompanied by uptake of protons due to lowering of the $pK_a$ of histidine residues.
- In sickle cell hemoglobin (HbS), Val replaces the $\beta$6 Glu of HbA, creating a "sticky patch" that has a complement on deoxyHb (but not on oxyHb). DeoxyHbS polymerizes at low $O_2$ concentrations, forming fibers that distort erythrocytes into sickle shapes.
- Alpha and beta thalassemias are anemias that result from reduced production of $\alpha$ and $\beta$ subunits of HbA, respectively.


## REFERENCES


Bettati S et al: Allosteric mechanism of haemoglobin: Rupture of salt-bridges raises the oxygen affinity of the T-structure. J Mol Biol 1998;281:581.

Bunn HF: Pathogenesis and treatment of sickle cell disease. N Engl J Med 1997;337:762.

Faustino P et al: Dominantly transmitted beta-thalassemia arising from the production of several aberrant mRNA species and one abnormal peptide. Blood 1998;91:685.

Manning JM et al: Normal and abnormal protein subunit interactions in hemoglobins. J Biol Chem 1998;273:19359.

Mario N, Baudin B, Giboudeau J: Qualitative and quantitative analysis of hemoglobin variants by capillary isoelectric focusing. J Chromatogr B Biomed Sci Appl 1998;706:123.

Reed W, Vichinsky EP: New considerations in the treatment of sickle cell disease. Annu Rev Med 1998;49:461.

Unzai S et al: Rate constants for $O_2$ and CO binding to the alpha and beta subunits within the R and T states of human hemoglobin. J Biol Chem 1998;273:23150.

Weatherall DJ et al: The hemoglobinopathies. Chapter 181 in *The Metabolic and Molecular Bases of Inherited Disease,* 8th ed. Scriver CR et al (editors). McGraw-Hill, 2000.

# Enzymes: Mechanism of Action 7

Victor W. Rodwell, PhD, & Peter J. Kennelly, PhD

## BIOMEDICAL IMPORTANCE

Enzymes are biologic polymers that catalyze the chemical reactions which make life as we know it possible. The presence and maintenance of a complete and balanced set of enzymes is essential for the breakdown of nutrients to supply energy and chemical building blocks; the assembly of those building blocks into proteins, DNA, membranes, cells, and tissues; and the harnessing of energy to power cell motility and muscle contraction. With the exception of a few catalytic RNA molecules, or ribozymes, the vast majority of enzymes are proteins. Deficiencies in the quantity or catalytic activity of key enzymes can result from genetic defects, nutritional deficits, or toxins. Defective enzymes can result from genetic mutations or infection by viral or bacterial pathogens (eg, *Vibrio cholerae*). Medical scientists address imbalances in enzyme activity by using pharmacologic agents to inhibit specific enzymes and are investigating gene therapy as a means to remedy deficits in enzyme level or function.

## ENZYMES ARE EFFECTIVE & HIGHLY SPECIFIC CATALYSTS

The enzymes that catalyze the conversion of one or more compounds **(substrates)** into one or more different compounds **(products)** enhance the rates of the corresponding noncatalyzed reaction by factors of at least $10^6$. Like all catalysts, enzymes are neither consumed nor permanently altered as a consequence of their participation in a reaction.

In addition to being highly efficient, enzymes are also extremely selective catalysts. Unlike most catalysts used in synthetic chemistry, enzymes are specific both for the type of reaction catalyzed and for a single substrate or a small set of closely related substrates. Enzymes are also stereospecific catalysts and typically catalyze reactions only of specific stereoisomers of a given compound—for example, D- but not L-sugars, L- but not D-amino acids. Since they bind substrates through at least "three points of attachment," enzymes can even convert nonchiral substrates to chiral products. Figure 7–1 illustrates why the enzyme-catalyzed reduction of the nonchiral substrate pyruvate produces L-lactate rather a racemic mixture of D- and L-lactate. The exquisite specificity of enzyme catalysts imbues living cells with the ability to simultaneously conduct and independently control a broad spectrum of chemical processes.

## ENZYMES ARE CLASSIFIED BY REACTION TYPE & MECHANISM

A system of enzyme nomenclature that is comprehensive, consistent, and at the same time easy to use has proved elusive. The common names for most enzymes derive from their most distinctive characteristic: their ability to catalyze a specific chemical reaction. In general, an enzyme's name consists of a term that identifies the type of reaction catalyzed followed by the suffix *-ase*. For example, dehydrogen*ases* remove hydrogen atoms, prote*ases* hydrolyze proteins, and isomer*ases* catalyze rearrangements in configuration. One or more modifiers usually precede this name. Unfortunately, while many modifiers name the specific substrate involved (xanthine oxidase), others identify the source of the enzyme (pancreatic ribonuclease), specify its mode of regulation (hormone-sensitive lipase), or name a distinguishing characteristic of its mechanism (a cysteine protease). When it was discovered that multiple forms of some enzymes existed, alphanumeric designators were added to distinguish between them (eg, RNA polymerase III; protein kinase Cβ). To address the ambiguity and confusion arising from these inconsistencies in nomenclature and the continuing discovery of new enzymes, the International Union of Biochemists (IUB) developed a complex but unambiguous system of enzyme nomenclature. In the IUB system, each enzyme has a unique name and code number that reflect the type of reaction catalyzed and the substrates involved. Enzymes are grouped into six classes, each with several subclasses. For example, the enzyme commonly called "hexokinase" is designated "ATP:D-hexose-6-phosphotransferase E.C. 2.7.1.1." This identifies hexokinase as a member of class 2 (transferases), subclass 7 (transfer of a phosphoryl group), sub-subclass 1 (alcohol is the phosphoryl acceptor). Finally, the term "hexose-6" indicates that the alcohol phosphorylated is that of carbon six of a hexose. Listed below are the six IUB classes of enzymes and the reactions they catalyze.

1. **Oxidoreductases** catalyze oxidations and reductions.

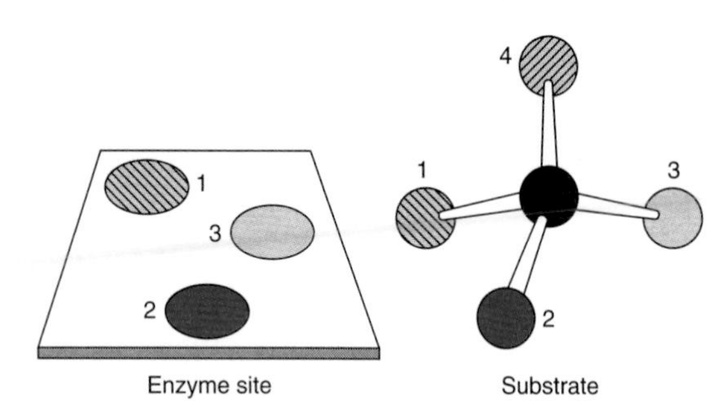


***Figure 7–1.*** Planar representation of the "three-point attachment" of a substrate to the active site of an enzyme. Although atoms 1 and 4 are identical, once atoms 2 and 3 are bound to their complementary sites on the enzyme, only atom 1 can bind. Once bound to an enzyme, apparently identical atoms thus may be distinguishable, permitting a stereospecific chemical change.

2. **Transferases** catalyze transfer of groups such as methyl or glycosyl groups from a donor molecule to an acceptor molecule.
3. **Hydrolases** catalyze the hydrolytic cleavage of C—C, C—O, C—N, P—O, and certain other bonds, including acid anhydride bonds.
4. **Lyases** catalyze cleavage of C—C, C—O, C—N, and other bonds by elimination, leaving double bonds, and also add groups to double bonds.
5. **Isomerases** catalyze geometric or structural changes within a single molecule.
6. **Ligases** catalyze the joining together of two molecules, coupled to the hydrolysis of a pyrophosphoryl group in ATP or a similar nucleoside triphosphate.

Despite the many advantages of the IUB system, texts tend to refer to most enzymes by their older and shorter, albeit sometimes ambiguous names.

## PROSTHETIC GROUPS, COFACTORS, & COENZYMES PLAY IMPORTANT ROLES IN CATALYSIS

Many enzymes contain small nonprotein molecules and metal ions that participate directly in substrate binding or catalysis. Termed **prosthetic groups, cofactors,** and **coenzymes,** these extend the repertoire of catalytic capabilities beyond those afforded by the limited number of functional groups present on the aminoacyl side chains of peptides.

### Prosthetic Groups Are Tightly Integrated Into an Enzyme's Structure

Prosthetic groups are distinguished by their tight, stable incorporation into a protein's structure by covalent or noncovalent forces. Examples include pyridoxal phosphate, flavin mononucleotide (FMN), flavin dinucleotide (FAD), thiamin pyrophosphate, biotin, and the metal ions of Co, Cu, Mg, Mn, Se, and Zn. Metals are the most common prosthetic groups. The roughly one-third of all enzymes that contain tightly bound metal ions are termed **metalloenzymes.** Metal ions that participate in redox reactions generally are complexed to prosthetic groups such as heme (Chapter 6) or iron-sulfur clusters (Chapter 12). Metals also may facilitate the binding and orientation of substrates, the formation of covalent bonds with reaction intermediates ($Co^{2+}$ in coenzyme $B_{12}$), or interaction with substrates to render them more **electrophilic** (electron-poor) or **nucleophilic** (electron-rich).

### Cofactors Associate Reversibly With Enzymes or Substrates

**Cofactors** serve functions similar to those of prosthetic groups but bind in a transient, dissociable manner either to the enzyme or to a substrate such as ATP. Unlike the stably associated prosthetic groups, cofactors therefore must be present in the medium surrounding the enzyme for catalysis to occur. The most common cofactors also are metal ions. Enzymes that require a metal ion cofactor are termed **metal-activated enzymes** to distinguish them from the **metalloenzymes** for which metal ions serve as prosthetic groups.

### Coenzymes Serve as Substrate Shuttles

**Coenzymes** serve as recyclable shuttles—or group transfer reagents—that transport many substrates from their point of generation to their point of utilization. Association with the coenzyme also stabilizes substrates such as hydrogen atoms or hydride ions that are unstable in the aqueous environment of the cell. Other chemical moieties transported by coenzymes include methyl groups (folates), acyl groups (coenzyme A), and oligosaccharides (dolichol).

### Many Coenzymes, Cofactors, & Prosthetic Groups Are Derivatives of B Vitamins

The water-soluble B vitamins supply important components of numerous coenzymes. Many coenzymes contain, in addition, the adenine, ribose, and phosphoryl moieties of AMP or ADP (Figure 7–2). **Nicotinamide** and **riboflavin** are components of the redox coenzymes

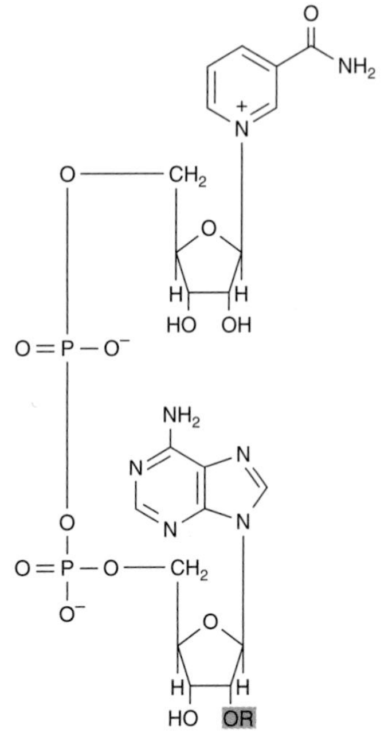

***Figure 7–2.*** Structure of $NAD^+$ and $NADP^+$. For $NAD^+$, R = H. For $NADP^+$, R = $PO_3^{2-}$.

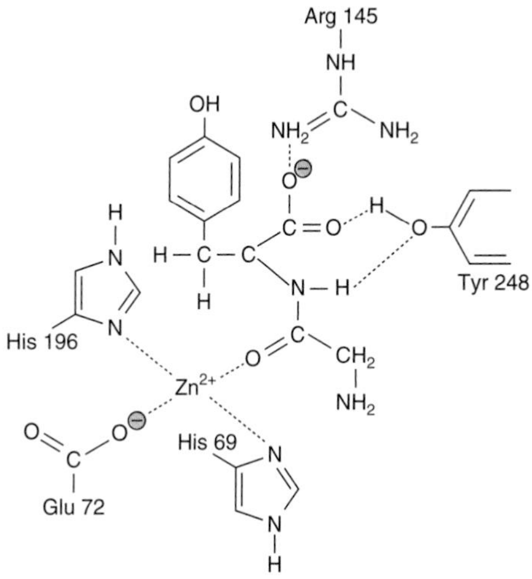

***Figure 7–3.*** Two-dimensional representation of a dipeptide substrate, glycyl-tyrosine, bound within the active site of carboxypeptidase A.

NAD and NADP and FMN and FAD, respectively. **Pantothenic acid** is a component of the acyl group carrier coenzyme A. As its pyrophosphate, **thiamin** participates in decarboxylation of α-keto acids and **folic acid** and **cobamide** coenzymes function in one-carbon metabolism.

## CATALYSIS OCCURS AT THE ACTIVE SITE

The extreme substrate specificity and high catalytic efficiency of enzymes reflect the existence of an environment that is exquisitely tailored to a single reaction. Termed the **active site,** this environment generally takes the form of a cleft or pocket. The active sites of multimeric enzymes often are located at the interface between subunits and recruit residues from more than one monomer. The three-dimensional active site both shields substrates from solvent and facilitates catalysis. Substrates bind to the active site at a region complementary to a portion of the substrate that will *not* undergo chemical change during the course of the reaction. This simultaneously aligns portions of the substrate that *will* undergo change with the chemical functional groups of peptidyl aminoacyl residues. The active site also binds and orients cofactors or prosthetic groups. Many amino acyl residues drawn from diverse portions of the polypeptide chain (Figure 7–3) contribute to the extensive size and three-dimensional character of the active site.

## ENZYMES EMPLOY MULTIPLE MECHANISMS TO FACILITATE CATALYSIS

Four general mechanisms account for the ability of enzymes to achieve dramatic catalytic enhancement of the rates of chemical reactions.

### Catalysis by Proximity

For molecules to react, they must come within bond-forming distance of one another. The higher their concentration, the more frequently they will encounter one another and the greater will be the rate of their reaction. When an enzyme binds substrate molecules in its active site, it creates a region of high local substrate concentration. This environment also orients the substrate molecules spatially in a position ideal for them to interact, resulting in rate enhancements of at least a thousandfold.

### Acid-Base Catalysis

The ionizable functional groups of aminoacyl side chains and (where present) of prosthetic groups contribute to catalysis by acting as acids or bases. Acid-base catalysis can be either specific or general. By "specific" we mean only protons ($H_3O^+$) or $OH^-$ ions. In **specific acid** or **specific base catalysis,** the rate of reaction is sensitive to changes in the concentration of protons but independent of the concentrations of other acids (proton donors) or bases (proton acceptors) present in solution or at the active site. Reactions whose rates are responsive to *all* the acids or bases present are said to be subject to **general acid** or **general base catalysis.**

### Catalysis by Strain

Enzymes that catalyze lytic reactions which involve breaking a covalent bond typically bind their substrates in a conformation slightly unfavorable for the bond that will undergo cleavage. The resulting strain stretches or distorts the targeted bond, weakening it and making it more vulnerable to cleavage.

### Covalent Catalysis

The process of **covalent catalysis** involves the formation of a covalent bond between the enzyme and one or more substrates. The modified enzyme then becomes a reactant. Covalent catalysis introduces a new reaction pathway that is energetically more favorable—and therefore faster—than the reaction pathway in homogeneous solution. The chemical modification of the enzyme is, however, transient. On completion of the reaction, the enzyme returns to its original unmodified state. Its role thus remains catalytic. Covalent catalysis is particularly common among enzymes that catalyze group transfer reactions. Residues on the enzyme that participate in covalent catalysis generally are cysteine or serine and occasionally histidine. Covalent catalysis often follows a "ping-pong" mechanism—one in which the first substrate is bound and its product released prior to the binding of the second substrate (Figure 7–4).

## SUBSTRATES INDUCE CONFORMATIONAL CHANGES IN ENZYMES

Early in the last century, Emil Fischer compared the highly specific fit between enzymes and their substrates to that of a lock and its key. While the "lock and key model" accounted for the exquisite specificity of enzyme-substrate interactions, the implied rigidity of the enzyme's active site failed to account for the dynamic changes that accompany catalysis. This drawback was addressed by Daniel Koshland's **induced fit** model, which states that when substrates approach and bind to an enzyme they induce a conformational change, a change analogous to placing a hand (substrate) into a glove (enzyme) (Figure 7–5). A corollary is that the enzyme induces reciprocal changes in its substrates, harnessing the energy of binding to facilitate the transformation of substrates into products. The induced fit model has been amply confirmed by biophysical studies of enzyme motion during substrate binding.

## HIV PROTEASE ILLUSTRATES ACID-BASE CATALYSIS

Enzymes of the **aspartic protease family,** which includes the digestive enzyme pepsin, the lysosomal cathepsins, and the protease produced by the human immunodeficiency virus (HIV), share a common catalytic mechanism. Catalysis involves two conserved aspartyl residues which act as acid-base catalysts. In the first stage of the reaction, an aspartate functioning as a general base (Asp X, Figure 7–6) extracts a proton from a water molecule, making it more nucleophilic. This resulting nucleophile then attacks the electrophilic carbonyl carbon of the peptide bond targeted for hydrolysis, forming a **tetrahedral transition state intermediate.** A second aspartate (Asp Y, Figure 7–6) then facilitates the decomposition of this tetrahedral intermediate by donating a proton to the amino group produced by rupture of the peptide bond. Two different active site aspartates thus can act simultaneously as a general base or as a general acid. This is possible because their immediate environment favors ionization of one but not the other.

## CHYMOTRYPSIN & FRUCTOSE-2,6-BISPHOSPHATASE ILLUSTRATE COVALENT CATALYSIS

### Chymotrypsin

While catalysis by aspartic proteases involves the direct hydrolytic attack of water on a peptide bond, catalysis

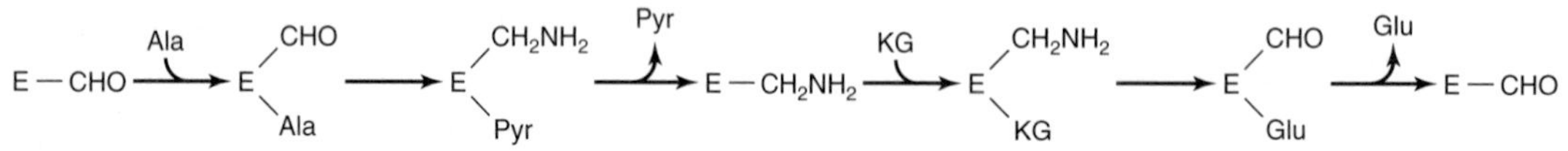

$$E{-}CHO \xrightarrow{Ala} E\begin{smallmatrix}CHO\\Ala\end{smallmatrix} \longrightarrow E\begin{smallmatrix}CH_2NH_2\\Pyr\end{smallmatrix} \xrightarrow{Pyr} E{-}CH_2NH_2 \xrightarrow{KG} E\begin{smallmatrix}CH_2NH_2\\KG\end{smallmatrix} \longrightarrow E\begin{smallmatrix}CHO\\Glu\end{smallmatrix} \xrightarrow{Glu} E{-}CHO$$

***Figure 7–4.*** Ping-pong mechanism for transamination. E—CHO and E—$CH_2NH_2$ represent the enzyme-pyridoxal phosphate and enzyme-pyridoxamine complexes, respectively. (Ala, alanine; Pyr, pyruvate; KG, α-ketoglutarate; Glu, glutamate).

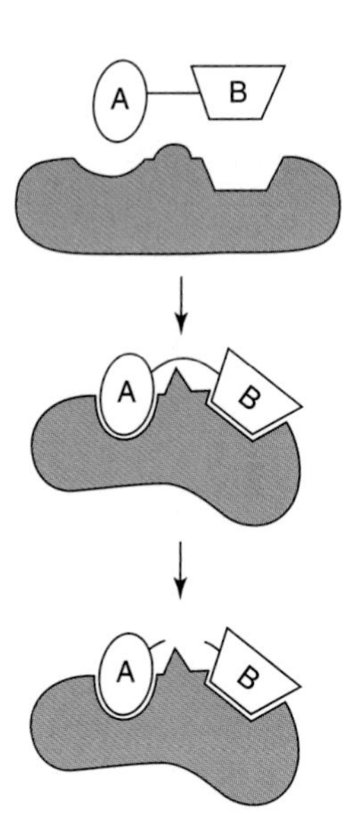


**Figure 7–5.** Two-dimensional representation of Koshland's induced fit model of the active site of a lyase. Binding of the substrate A—B induces conformational changes in the enzyme that aligns catalytic residues which participate in catalysis and strains the bond between A and B, facilitating its cleavage.

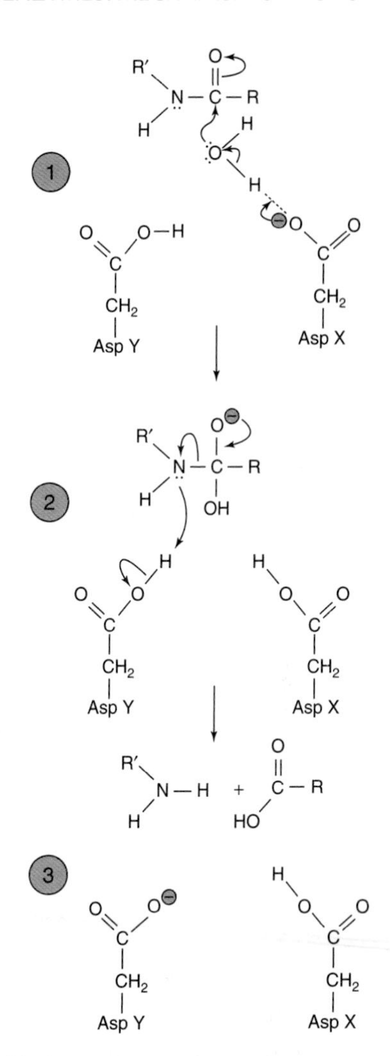


**Figure 7–6.** Mechanism for catalysis by an aspartic protease such as HIV protease. Curved arrows indicate directions of electron movement. ① Aspartate X acts as a base to activate a water molecule by abstracting a proton. ② The activated water molecule attacks the peptide bond, forming a transient tetrahedral intermediate. ③ Aspartate Y acts as an acid to facilitate breakdown of the tetrahedral intermediate and release of the split products by donating a proton to the newly formed amino group. Subsequent shuttling of the proton on Asp X to Asp Y restores the protease to its initial state.

by the **serine protease** chymotrypsin involves prior formation of a covalent acyl enzyme intermediate. A highly reactive seryl residue, serine 195, participates in a charge-relay network with histidine 57 and aspartate 102. Far apart in primary structure, in the active site these residues are within bond-forming distance of one another. Aligned in the order Asp 102-His 57-Ser 195, they constitute a "charge-relay network" that functions as a "proton shuttle."

Binding of substrate initiates proton shifts that in effect transfer the hydroxyl proton of Ser 195 to Asp 102 (Figure 7–7). The enhanced nucleophilicity of the seryl oxygen facilitates its attack on the carbonyl carbon of the peptide bond of the substrate, forming a covalent acyl-enzyme intermediate. The hydrogen on Asp 102 then shuttles through His 57 to the amino group liberated when the peptide bond is cleaved. The portion of the original peptide with a free amino group then leaves the active site and is replaced by a water molecule. The charge-relay network now activates the water molecule by withdrawing a proton through His 57 to Asp 102. The resulting hydroxide ion attacks the acyl-enzyme in-

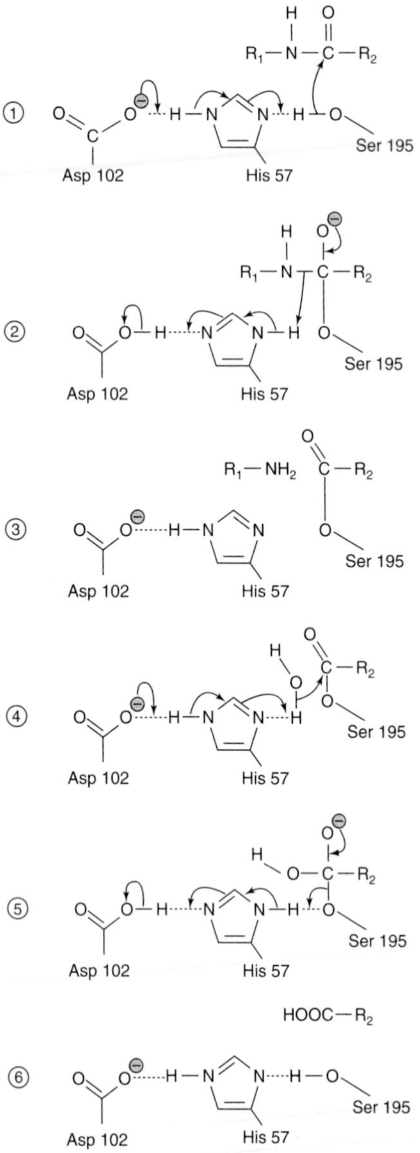

***Figure 7–7.*** Catalysis by chymotrypsin. ① The charge-relay system removes a proton from Ser 195, making it a stronger nucleophile. ② Activated Ser 195 attacks the peptide bond, forming a transient tetrahedral intermediate. ③ Release of the amino terminal peptide is facilitated by donation of a proton to the newly formed amino group by His 57 of the charge-relay system, yielding an acyl-Ser 195 intermediate. ④ His 57 and Asp 102 collaborate to activate a water molecule, which attacks the acyl-Ser 195, forming a second tetrahedral intermediate. ⑤ The charge-relay system donates a proton to Ser 195, facilitating breakdown of tetrahedral intermediate to release the carboxyl terminal peptide ⑥.

termediate and a reverse proton shuttle returns a proton to Ser 195, restoring its original state. While modified during the process of catalysis, chymotrypsin emerges unchanged on completion of the reaction. Trypsin and elastase employ a similar catalytic mechanism, but the numbers of the residues in their Ser-His-Asp proton shuttles differ.

### Fructose-2,6-Bisphosphatase

Fructose-2,6-bisphosphatase, a regulatory enzyme of gluconeogenesis (Chapter 19), catalyzes the hydrolytic release of the phosphate on carbon 2 of fructose 2,6-bisphosphate. Figure 7–8 illustrates the roles of seven active site residues. Catalysis involves a "catalytic triad" of one Glu and two His residues and a covalent phosphohistidyl intermediate.

## CATALYTIC RESIDUES ARE HIGHLY CONSERVED

Members of an enzyme family such as the aspartic or serine proteases employ a similar mechanism to catalyze a common reaction type but act on different substrates. Enzyme families appear to arise through gene duplication events that create a second copy of the gene which encodes a particular enzyme. The proteins encoded by the two genes can then evolve independently to recognize different substrates—resulting, for example, in chymotrypsin, which cleaves peptide bonds on the carboxyl terminal side of large hydrophobic amino acids; and trypsin, which cleaves peptide bonds on the carboxyl terminal side of basic amino acids. The common ancestry of enzymes can be inferred from the presence of specific amino acids in the same position in each family member. These residues are said to be **conserved residues.** Proteins that share a large number of conserved residues are said to be **homologous** to one another. Table 7–1 illustrates the primary structural conservation of two components of the charge-relay network for several serine proteases. Among the most highly conserved residues are those that participate directly in catalysis.

## ISOZYMES ARE DISTINCT ENZYME FORMS THAT CATALYZE THE SAME REACTION

Higher organisms often elaborate several physically distinct versions of a given enzyme, each of which catalyzes the same reaction. Like the members of other protein families, these protein catalysts or **isozymes** arise through gene duplication. Isozymes may exhibit subtle differences in properties such as sensitivity to

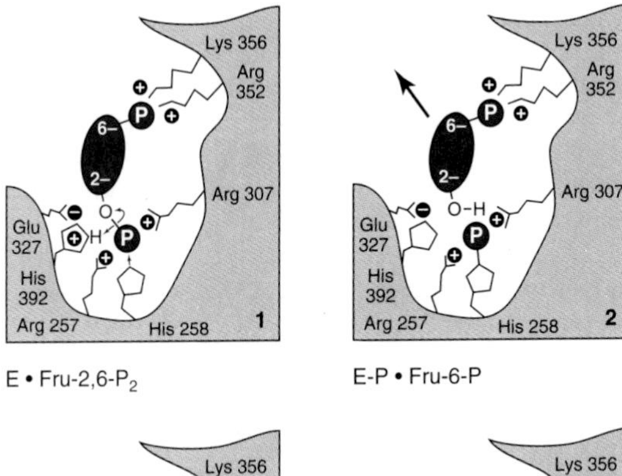


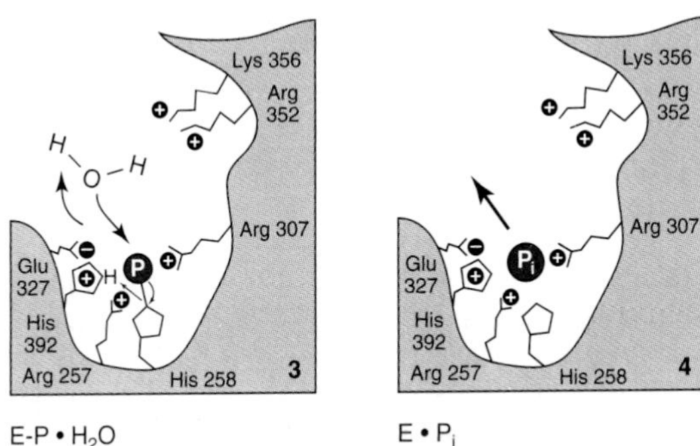


***Figure 7–8.*** Catalysis by fructose-2,6-bisphosphatase. (1) Lys 356 and Arg 257, 307, and 352 stabilize the quadruple negative charge of the substrate by charge-charge interactions. Glu 327 stabilizes the positive charge on His 392. (2) The nucleophile His 392 attacks the C-2 phosphoryl group and transfers it to His 258, forming a phosphoryl-enzyme intermediate. Fructose 6-phosphate leaves the enzyme. (3) Nucleophilic attack by a water molecule, possibly assisted by Glu 327 acting as a base, forms inorganic phosphate. (4) Inorganic orthophosphate is released from Arg 257 and Arg 307. (Reproduced, with permission, from Pilkis SJ et al: 6-Phosphofructo-2-kinase/fructose-2,6-bisphosphatase: A metabolic signaling enzyme. Annu Rev Biochem 1995;64:799.)

particular regulatory factors (Chapter 9) or substrate affinity (eg, hexokinase and glucokinase) that adapt them to specific tissues or circumstances. Some isozymes may also enhance survival by providing a "back-up" copy of an essential enzyme.

## THE CATALYTIC ACTIVITY OF ENZYMES FACILITATES THEIR DETECTION

The minute quantities of enzymes present in cells complicate determination of their presence and concentration. However, the ability to rapidly transform thousands of molecules of a specific substrate into products imbues each enzyme with the ability to reveal its presence. Assays of the catalytic activity of enzymes are frequently used in research and clinical laboratories. Under appropriate conditions (see Chapter 8), the rate of the catalytic reaction being monitored is proportionate to the amount of enzyme present, which allows its concentration to be inferred.

### Enzyme-Linked Immunoassays

The sensitivity of enzyme assays can also be exploited to detect proteins that lack catalytic activity. **Enzyme-linked immunoassays** (ELISAs) use antibodies covalently linked to a "reporter enzyme" such as alkaline phosphatase or horseradish peroxidase, enzymes whose products are readily detected. When serum or other samples to be tested are placed in a plastic microtiter plate, the proteins adhere to the plastic surface and are immobilized. Any remaining absorbing areas of the well are then "blocked" by adding a nonantigenic protein such as bovine serum albumin. A solution of antibody covalently linked to a reporter enzyme is then added. The antibodies adhere to the immobilized antigen and these are themselves immobilized. Excess free antibody molecules are then removed by washing. The presence and quantity of bound antibody are then determined by adding the substrate for the reporter enzyme.

***Table 7–1.*** Amino acid sequences in the neighborhood of the catalytic sites of several bovine proteases. Regions shown are those on either side of the catalytic site seryl (S) and histidyl (H) residues.

| Enzyme | Sequence Around Serine Ⓢ | Sequence Around Histidine Ⓗ |
|---|---|---|
| Trypsin | D S C Q D G Ⓢ G G P V V C S G K | V V S A A Ⓗ C Y K S G |
| Chymotrypsin A | S S C M G D Ⓢ G G P L V C K K N | V V T A A Ⓗ G G V T T |
| Chymotrypsin B | S S C M G D Ⓢ G G P L V C Q K N | V V T A A Ⓗ C G V T T |
| Thrombin | D A C E G D Ⓢ G G P F V M K S P | V L T A A Ⓗ C L L Y P |

### NAD(P)$^+$-Dependent Dehydrogenases Are Assayed Spectrophotometrically

The physicochemical properties of the reactants in an enzyme-catalyzed reaction dictate the options for the assay of enzyme activity. Spectrophotometric assays exploit the ability of a substrate or product to absorb light. The reduced coenzymes NADH and NADPH, written as NAD(P)H, absorb light at a wavelength of 340 nm, whereas their oxidized forms NAD(P)$^+$ do not (Figure 7–9). When NAD(P)$^+$ is reduced, the absorbance at 340 nm therefore increases in proportion to—and at a rate determined by—the quantity of NAD(P)H produced. Conversely, for a dehydrogenase that catalyzes the oxidation of NAD(P)H, a decrease in absorbance at 340 nm will be observed. In each case, the rate of change in optical density at 340 nm will be proportionate to the quantity of enzyme present.

### Many Enzymes Are Assayed by Coupling to a Dehydrogenase

The assay of enzymes whose reactions are not accompanied by a change in absorbance or fluorescence is generally more difficult. In some instances, the product or remaining substrate can be transformed into a more readily detected compound. In other instances, the reaction product may have to be separated from unreacted substrate prior to measurement—a process facilitated by the use of radioactive substrates. An alternative strategy is to devise a synthetic substrate whose product absorbs light. For example, *p*-nitrophenyl phosphate is an artificial substrate for certain phosphatases and for chymotrypsin that does not absorb visible light. However, following hydrolysis, the resulting *p*-nitrophenylate anion absorbs light at 419 nm.

Another quite general approach is to employ a "coupled" assay (Figure 7–10). Typically, a dehydrogenase whose substrate is the product of the enzyme of interest is added in catalytic excess. The rate of appearance or disappearance of NAD(P)H then depends on the rate of the enzyme reaction to which the dehydrogenase has been coupled.

## THE ANALYSIS OF CERTAIN ENZYMES AIDS DIAGNOSIS

Of the thousands of different enzymes present in the human body, those that fulfill functions indispensable to cell vitality are present throughout the body tissues. Other enzymes or isozymes are expressed only in specific cell types, during certain periods of development, or in response to specific physiologic or pathophysiologic changes. Analysis of the presence and distribution of enzymes and isozymes—whose expression is normally tissue-, time-, or circumstance-specific—often aids diagnosis.

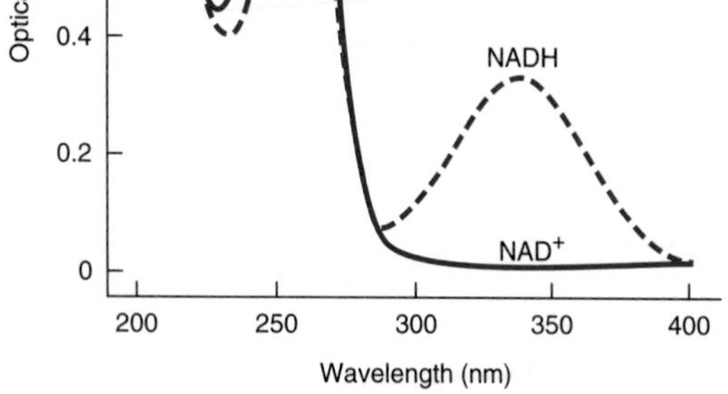

***Figure 7–9.*** Absorption spectra of NAD$^+$ and NADH. Densities are for a 44 mg/L solution in a cell with a 1 cm light path. NADP$^+$ and NADPH have spectrums analogous to NAD$^+$ and NADH, respectively.

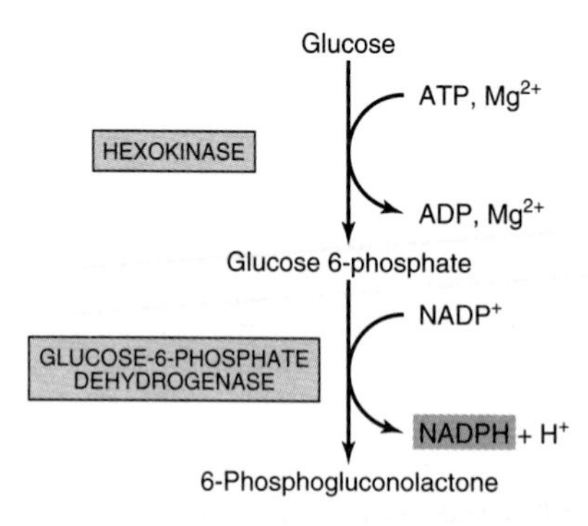


***Figure 7–10.*** Coupled enzyme assay for hexokinase activity. The production of glucose 6-phosphate by hexokinase is coupled to the oxidation of this product by glucose-6-phosphate dehydrogenase in the presence of added enzyme and NADP$^+$. When an excess of glucose-6-phosphate dehydrogenase is present, the rate of formation of NADPH, which can be measured at 340 nm, is governed by the rate of formation of glucose 6-phosphate by hexokinase.

## Nonfunctional Plasma Enzymes Aid Diagnosis & Prognosis

Certain enzymes, proenzymes, and their substrates are present at all times in the circulation of normal individuals and perform a physiologic function in the blood. Examples of these **functional plasma enzymes** include lipoprotein lipase, pseudocholinesterase, and the proenzymes of blood coagulation and blood clot dissolution (Chapters 9 and 51). The majority of these enzymes are synthesized in and secreted by the liver.

Plasma also contains numerous other enzymes that perform no known physiologic function in blood. These apparently **nonfunctional plasma enzymes** arise from the routine normal destruction of erythrocytes, leukocytes, and other cells. Tissue damage or necrosis resulting from injury or disease is generally accompanied by increases in the levels of several nonfunctional plasma enzymes. Table 7–2 lists several enzymes used in diagnostic enzymology.

## Isozymes of Lactate Dehydrogenase Are Used to Detect Myocardial Infarctions

L-Lactate dehydrogenase is a tetrameric enzyme whose four subunits occur in two isoforms, designated H (for heart) and M (for muscle). The subunits can combine as shown below to yield catalytically active isozymes of L-lactate dehydrogenase:

| Lactate Dehydrogenase Isozyme | Subunits |
|---|---|
| $I_1$ | HHHH |
| $I_2$ | HHHM |
| $I_3$ | HHMM |
| $I_4$ | HMMM |
| $I_5$ | MMMM |

Distinct genes whose expression is differentially regulated in various tissues encode the H and M subunits. Since heart expresses the H subunit almost exclusively, isozyme $I_1$ predominates in this tissue. By contrast, isozyme $I_5$ predominates in liver. Small quantities of lactate dehydrogenase are normally present in plasma. Following a myocardial infarction or in liver disease, the damaged tissues release characteristic lactate dehydrogenase isoforms into the blood. The resulting elevation in the levels of the $I_1$ or $I_5$ isozymes is detected by separating the different oligomers of lactate dehydrogenase by electrophoresis and assaying their catalytic activity (Figure 7–11).

***Table 7–2.*** Principal serum enzymes used in clinical diagnosis. Many of the enzymes are not specific for the disease listed.

| Serum Enzyme | Major Diagnostic Use |
|---|---|
| Aminotransferases | |
| Aspartate aminotransferase (AST, or SGOT) | Myocardial infarction |
| Alanine aminotransferase (ALT, or SGPT) | Viral hepatitis |
| Amylase | Acute pancreatitis |
| Ceruloplasmin | Hepatolenticular degeneration (Wilson's disease) |
| Creatine kinase | Muscle disorders and myocardial infarction |
| γ-Glutamyl transpeptidase | Various liver diseases |
| Lactate dehydrogenase (isozymes) | Myocardial infarction |
| Lipase | Acute pancreatitis |
| Phosphatase, acid | Metastatic carcinoma of the prostate |
| Phosphatase, alkaline (isozymes) | Various bone disorders, obstructive liver diseases |

## ENZYMES FACILITATE DIAGNOSIS OF GENETIC DISEASES

While many diseases have long been known to result from alterations in an individual's DNA, tools for the detection of genetic mutations have only recently become widely available. These techniques rely upon the catalytic efficiency and specificity of enzyme catalysts. For example, the **polymerase chain reaction** (PCR) relies upon the ability of enzymes to serve as catalytic amplifiers to analyze the DNA present in biologic and forensic samples. In the PCR technique, a thermostable DNA polymerase, directed by appropriate oligonucleotide primers, produces thousands of copies of a sample of DNA that was present initially at levels too low for direct detection.

The detection of **restriction fragment length polymorphisms** (RFLPs) facilitates prenatal detection of hereditary disorders such as sickle cell trait, beta-thalassemia, infant phenylketonuria, and Huntington's disease. Detection of RFLPs involves cleavage of double-stranded DNA by restriction endonucleases, which can detect subtle alterations in DNA that affect their recognized sites. Chapter 40 provides further details concerning the use of PCR and restriction enzymes for diagnosis.

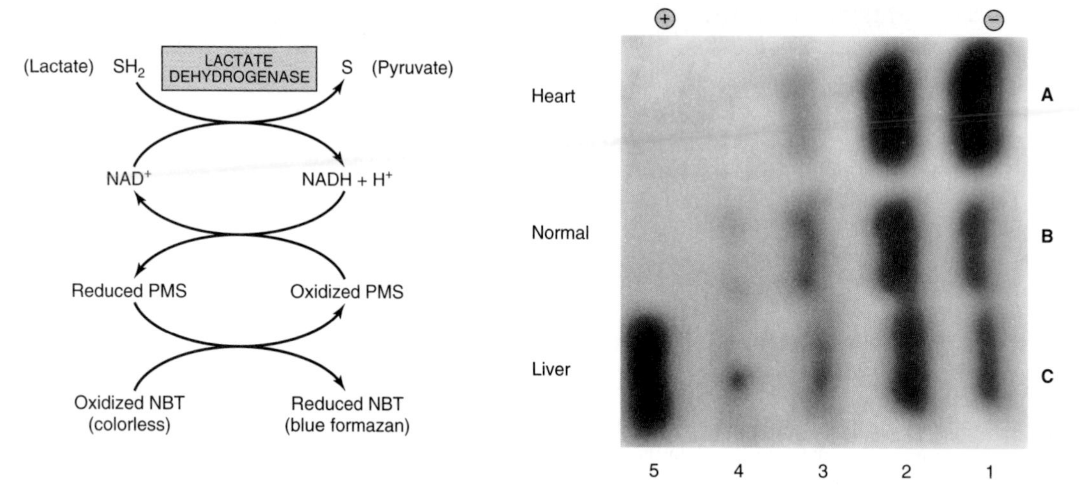


***Figure 7–11.*** Normal and pathologic patterns of lactate dehydrogenase (LDH) isozymes in human serum. LDH isozymes of serum were separated by electrophoresis and visualized using the coupled reaction scheme shown on the left. (NBT, nitroblue tetrazolium; PMS, phenazine methylsulfate). At right is shown the stained electropherogram. Pattern A is serum from a patient with a myocardial infarct; B is normal serum; and C is serum from a patient with liver disease. Arabic numerals denote specific LDH isozymes.

## RECOMBINANT DNA PROVIDES AN IMPORTANT TOOL FOR STUDYING ENZYMES

Recombinant DNA technology has emerged as an important asset in the study of enzymes. Highly purified samples of enzymes are necessary for the study of their structure and function. The isolation of an individual enzyme, particularly one present in low concentration, from among the thousands of proteins present in a cell can be extremely difficult. If the gene for the enzyme of interest has been cloned, it generally is possible to produce large quantities of its encoded protein in *Escherichia coli* or yeast. However, not all animal proteins can be expressed in active form in microbial cells, nor do microbes perform certain posttranslational processing tasks. For these reasons, a gene may be expressed in cultured animal cell systems employing the baculovirus expression vector to transform cultured insect cells. For more details concerning recombinant DNA techniques, see Chapter 40.

### Recombinant Fusion Proteins Are Purified by Affinity Chromatography

Recombinant DNA technology can also be used to create modified proteins that are readily purified by affinity chromatography. The gene of interest is linked to an oligonucleotide sequence that encodes a carboxyl or amino terminal extension to the encoded protein. The resulting modified protein, termed a **fusion protein,** contains a domain tailored to interact with a specific affinity support. One popular approach is to attach an oligonucleotide that encodes six consecutive histidine residues. The expressed "His tag" protein binds to chromatographic supports that contain an immobilized divalent metal ion such as $Ni^{2+}$. Alternatively, the substrate-binding domain of glutathione S-transferase (GST) can serve as a "GST tag." Figure 7–12 illustrates the purification of a GST-fusion protein using an affinity support containing bound glutathione. Fusion proteins also often encode a cleavage site for a highly specific protease such as thrombin in the region that links the two portions of the protein. This permits removal of the added fusion domain following affinity purification.

### Site-Directed Mutagenesis Provides Mechanistic Insights

Once the ability to express a protein from its cloned gene has been established, it is possible to employ **site-directed mutagenesis** to change specific aminoacyl residues by altering their codons. Used in combination with kinetic analyses and x-ray crystallography, this approach facilitates identification of the specific roles of given aminoacyl residues in substrate binding and catalysis. For example, the inference that a particular aminoacyl residue functions as a general acid can be tested by replacing it with an aminoacyl residue incapable of donating a proton.

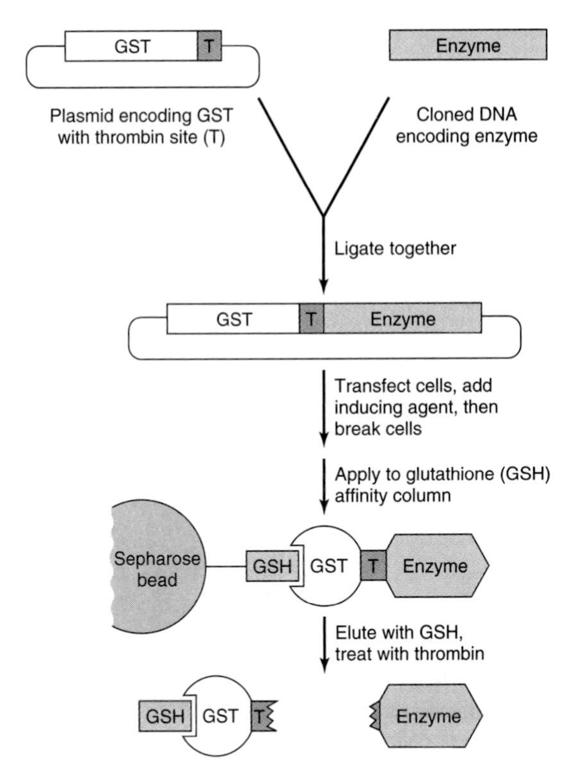


***Figure 7–12.*** Use of glutathione S-transferase (GST) fusion proteins to purify recombinant proteins. (GSH, glutathione.)

## SUMMARY

- Enzymes are highly effective and extremely specific catalysts.
- Organic and inorganic prosthetic groups, cofactors, and coenzymes play important roles in catalysis. Coenzymes, many of which are derivatives of B vitamins, serve as "shuttles."
- Catalytic mechanisms employed by enzymes include the introduction of strain, approximation of reactants, acid-base catalysis, and covalent catalysis.
- Aminoacyl residues that participate in catalysis are highly conserved among all classes of a given enzyme activity.
- Substrates and enzymes induce mutual conformational changes in one another that facilitate substrate recognition and catalysis.
- The catalytic activity of enzymes reveals their presence, facilitates their detection, and provides the basis for enzyme-linked immunoassays.
- Many enzymes can be assayed spectrophotometrically by coupling them to an $NAD(P)^+$-dependent dehydrogenase.
- Assay of plasma enzymes aids diagnosis and prognosis. For example, a myocardial infarction elevates serum levels of lactate dehydrogenase isozyme $I_1$.
- Restriction endonucleases facilitate diagnosis of genetic diseases by revealing restriction fragment length polymorphisms.
- Site-directed mutagenesis, used to change residues suspected of being important in catalysis or substrate binding, provides insights into the mechanisms of enzyme action.
- Recombinant fusion proteins such as His-tagged or GST fusion enzymes are readily purified by affinity chromatography.

## REFERENCES

Conyers GB et al: Metal requirements of a diadenosine pyrophosphatase from *Bartonella bacilliformis.* Magnetic resonance and kinetic studies of the role of $Mn^{2+}$. Biochemistry 2000; 39:2347.

Fersht A: *Structure and Mechanism in Protein Science: A Guide to Enzyme Catalysis and Protein Folding.* Freeman, 1999.

Suckling CJ: *Enzyme Chemistry.* Chapman & Hall, 1990.

Walsh CT: *Enzymatic Reaction Mechanisms.* Freeman, 1979.

# Enzymes: Kinetics

# 8

*Victor W. Rodwell, PhD, & Peter J. Kennelly, PhD*

## BIOMEDICAL IMPORTANCE

**Enzyme kinetics** is the field of biochemistry concerned with the quantitative measurement of the rates of enzyme-catalyzed reactions and the systematic study of factors that affect these rates. Kinetic analyses permit scientists to reconstruct the **number** and **order** of the individual steps by which enzymes transform substrates into products. The study of enzyme kinetics also represents the principal way to identify potential therapeutic agents that selectively enhance or inhibit the rates of specific enzyme-catalyzed processes. Together with site-directed mutagenesis and other techniques that probe protein structure, kinetic analysis can also reveal details of the catalytic mechanism. A complete, balanced set of enzyme activities is of fundamental importance for maintaining homeostasis. An understanding of enzyme kinetics thus is important for understanding how physiologic stresses such as anoxia, metabolic acidosis or alkalosis, toxins, and pharmacologic agents affect that balance.

## CHEMICAL REACTIONS ARE DESCRIBED USING BALANCED EQUATIONS

A **balanced chemical equation** lists the initial chemical species (substrates) present and the new chemical species (products) formed for a particular chemical reaction, all in their correct proportions or **stoichiometry.** For example, balanced equation (1) below describes the reaction of one molecule each of substrates A and B to form one molecule each of products P and Q.

$$A + B \rightleftarrows P + Q \quad \textbf{(1)}$$

The double arrows indicate reversibility, an intrinsic property of all chemical reactions. Thus, for reaction (1), if A and B can form P and Q, then P and Q can also form A and B. Designation of a particular reactant as a "substrate" or "product" is therefore somewhat arbitrary since the products for a reaction written in one direction are the substrates for the reverse reaction. The term "products" is, however, often used to designate the reactants whose formation is thermodynamically favored. Reactions for which thermodynamic factors strongly favor formation of the products to which the arrow points often are represented with a single arrow as if they were "irreversible":

$$A + B \rightarrow P + Q \quad \textbf{(2)}$$

Unidirectional arrows are also used to describe reactions in living cells where the products of reaction (2) are immediately consumed by a subsequent enzyme-catalyzed reaction. The rapid removal of product P or Q therefore precludes occurrence of the reverse reaction, rendering equation (2) **functionally irreversible under physiologic conditions.**

## CHANGES IN FREE ENERGY DETERMINE THE DIRECTION & EQUILIBRIUM STATE OF CHEMICAL REACTIONS

The Gibbs free energy change $\Delta G$ (also called either the free energy or Gibbs energy) describes both the *direction* in which a chemical reaction will tend to proceed and the concentrations of reactants and products that will be present at equilibrium. $\Delta G$ for a chemical reaction equals the sum of the free energies of formation of the reaction products $\Delta G_P$ minus the sum of the free energies of formation of the substrates $\Delta G_S$. $\Delta G^0$ denotes the change in free energy that accompanies transition from the standard state, one-molar concentrations of substrates and products, to equilibrium. A more useful biochemical term is $\Delta G^{0'}$, which defines $\Delta G^0$ at a standard state of $10^{-7}$ M protons, pH 7.0 (Chapter 10). If the free energy of the products is lower than that of the substrates, the signs of $\Delta G^0$ and $\Delta G^{0'}$ will be negative, indicating that the reaction as written is favored in the direction left to right. Such reactions are referred to as **spontaneous.** The **sign** and the **magnitude** of the free energy change determine how far the reaction will proceed. Equation (3)—

$$\Delta G^0 = -RT \ln K_{eq} \quad \textbf{(3)}$$

—illustrates the relationship between the equilibrium constant $K_{eq}$ and $\Delta G^0$, where R is the gas constant (1.98 cal/mol/°K or 8.31 J/mol/°K) and T is the absolute temperature in degrees Kelvin. $K_{eq}$ is equal to the product of the concentrations of the reaction products, each raised to the power of their stoichiometry, divided by the product of the substrates, each raised to the power of their stoichiometry.

For the reaction A + B → P + Q—

$$K_{eq} = \frac{[P][Q]}{[A][B]} \quad (4)$$

and for reaction (5)

$$A + A \rightleftarrows P \quad (5)$$

$$K_{eq} = \frac{[P]}{[A]^2} \quad (6)$$

—$\Delta G^0$ may be calculated from equation (3) if the concentrations of substrates and products present at equilibrium are known. If $\Delta G^0$ is a negative number, $K_{eq}$ will be greater than unity and the concentration of products at equilibrium will exceed that of substrates. If $\Delta G^0$ is positive, $K_{eq}$ will be less than unity and the formation of substrates will be favored.

Notice that, since $\Delta G^0$ is a function exclusively of the initial and final states of the reacting species, it can provide information only about the *direction* and *equilibrium state* of the reaction. **$\Delta G^0$** is independent of the mechanism of the reaction and therefore provides no information concerning ***rates*** of reactions. Consequently—and as explained below—although a reaction may have a large negative $\Delta G^0$ or $\Delta G^{0'}$, it may nevertheless take place at a negligible rate.

## THE RATES OF REACTIONS ARE DETERMINED BY THEIR ACTIVATION ENERGY

### Reactions Proceed via Transition States

The concept of the **transition state** is fundamental to understanding the chemical and thermodynamic basis of catalysis. Equation (7) depicts a displacement reaction in which an entering group E displaces a leaving group L, attached initially to R.

$$E + R{-}L \rightleftarrows E{-}R + L \quad (7)$$

Midway through the displacement, the bond between R and L has weakened but has not yet been completely severed, and the new bond between E and R is as yet incompletely formed. This transient intermediate—in which neither free substrate nor product exists—is termed the **transition state, E···R···L.** Dotted lines represent the "partial" bonds that are undergoing formation and rupture.

Reaction (7) can be thought of as consisting of two "partial reactions," the first corresponding to the formation (F) and the second to the subsequent decay (D) of the transition state intermediate. As for all reactions, characteristic changes in free energy, $\Delta G_F$, and $\Delta G_D$ are associated with each partial reaction.

$$E + R{-}L \rightleftarrows E\cdots R\cdots L \qquad \Delta G_F \quad (8)$$

$$E\cdots R\cdots L \rightleftarrows E{-}R + L \qquad \Delta G_D \quad (9)$$

$$E + R{-}L \rightleftarrows E{-}R + L \quad \Delta G = \Delta G_F + \Delta G_D \quad (8\text{-}10)$$

For the overall reaction (10), $\Delta G$ is the sum of $\Delta G_F$ and $\Delta G_D$. As for any equation of two terms, it is not possible to infer from $\Delta G$ either the sign or the magnitude of $\Delta G_F$ or $\Delta G_D$.

Many reactions involve multiple transition states, each with an associated change in free energy. For these reactions, the overall $\Delta G$ represents the sum of *all* of the free energy changes associated with the formation and decay of *all* of the transition states. **Therefore, it is not possible to infer from the overall $\Delta G$ the number or type of transition states through which the reaction proceeds.** Stated another way: overall thermodynamics tells us nothing about kinetics.

### $\Delta G_F$ Defines the Activation Energy

Regardless of the sign or magnitude of $\Delta G$, $\Delta G_F$ for the overwhelming majority of chemical reactions has a positive sign. The formation of transition state intermediates therefore requires surmounting of energy barriers. For this reason, $\Delta G_F$ is often termed the **activation energy,** $E_{act}$, the energy required to surmount a given energy barrier. The ease—and hence the frequency—with which this barrier is overcome is inversely related to $E_{act}$. The thermodynamic parameters that determine how *fast* a reaction proceeds thus are the $\Delta G_F$ values for formation of the transition states through which the reaction proceeds. For a simple reaction, where ∝ means "proportionate to,"

$$\text{Rate} \propto e^{\frac{-E_{act}}{RT}} \quad (11)$$

The activation energy for the reaction proceeding in the opposite direction to that drawn is equal to $-\Delta G_D$.

## NUMEROUS FACTORS AFFECT THE REACTION RATE

The **kinetic theory**—also called the **collision theory**—of chemical kinetics states that for two molecules to react they must (1) approach within bond-forming distance of one another, or "collide"; and (2) must possess sufficient kinetic energy to overcome the energy barrier for reaching the transition state. It therefore follows that anything which increases the *frequency* or *energy* of collision between substrates will increase the rate of the reaction in which they participate.

## Temperature

Raising the temperature increases the kinetic energy of molecules. As illustrated in Figure 8–1, the total number of molecules whose kinetic energy exceeds the energy barrier $E_{act}$ (vertical bar) for formation of products increases from low (A), through intermediate (B), to high (C) temperatures. Increasing the kinetic energy of molecules also increases their motion and therefore the frequency with which they collide. This combination of more frequent and more highly energetic and productive collisions increases the reaction rate.

## Reactant Concentration

The frequency with which molecules collide is directly proportionate to their concentrations. For two different molecules A and B, the frequency with which they collide will double if the concentration of either A or B is doubled. If the concentrations of both A and B are doubled, the probability of collision will increase fourfold.

For a chemical reaction proceeding at constant temperature that involves one molecule each of A and B,

$$A + B \rightarrow P \quad (12)$$

the number of molecules that possess kinetic energy sufficient to overcome the activation energy barrier will be a constant. The number of collisions with sufficient energy to produce product P therefore will be directly proportionate to the number of collisions between A and B and thus to their molar concentrations, denoted by square brackets.

$$\text{Rate} \propto [A][B] \quad (13)$$

Similarly, for the reaction represented by

$$A + 2B \rightarrow P \quad (14)$$

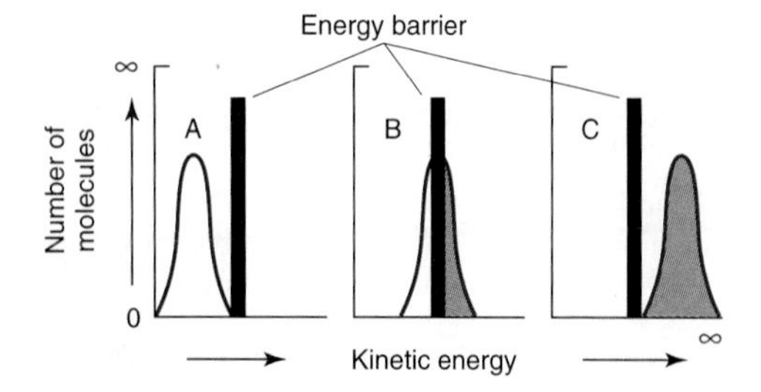


***Figure 8–1.*** The energy barrier for chemical reactions.

which can also be written as

$$A + B + B \rightarrow P \quad (15)$$

the corresponding rate expression is

$$\text{Rate} \propto [A][B][B] \quad (16)$$

or

$$\text{Rate} \propto [A][B]^2 \quad (17)$$

For the general case when n molecules of A react with m molecules of B,

$$nA + mB \rightarrow P \quad (18)$$

the rate expression is

$$\text{Rate} \propto [A]^n[B]^m \quad (19)$$

Replacing the proportionality constant with an equal sign by introducing a proportionality or **rate constant k** characteristic of the reaction under study gives equations (20) and (21), in which the subscripts 1 and −1 refer to the rate constants for the forward and reverse reactions, respectively.

$$\text{Rate}_1 = k_1[A]^n[B]^m \quad (20)$$

$$\text{Rate}_{-1} = k_{-1}[P] \quad (21)$$

## $K_{eq}$ Is a Ratio of Rate Constants

While all chemical reactions are to some extent reversible, at equilibrium the *overall* concentrations of reactants and products remain constant. At equilibrium, the rate of conversion of substrates to products therefore equals the rate at which products are converted to substrates.

$$\text{Rate}_1 = \text{Rate}_{-1} \quad (22)$$

Therefore,

$$k_1[A]^n[B]^m = k_{-1}[P] \quad (23)$$

and

$$\frac{k_1}{k_{-1}} = \frac{[P]}{[A]^n[B]^m} \quad (24)$$

**The ratio of $k_1$** to $k_{-1}$ is termed the equilibrium constant, $K_{eq}$. The following important properties of a system at equilibrium must be kept in mind:

**(1)** The equilibrium constant is a ratio of the reaction *rate constants* (not the reaction *rates*).

(2) At equilibrium, the reaction *rates* (not the *rate constants*) of the forward and back reactions are equal.
(3) Equilibrium is a *dynamic* state. Although there is no *net* change in the concentration of substrates or products, individual substrate and product molecules are continually being interconverted.
(4) The numeric value of the equilibrium constant $K_{eq}$ can be calculated either from the concentrations of substrates and products at equilibrium or from the ratio $k_1/k_{-1}$.

## THE KINETICS OF ENZYMATIC CATALYSIS

### Enzymes Lower the Activation Energy Barrier for a Reaction

All enzymes accelerate reaction rates by providing transition states with a lowered $\Delta G_F$ for formation of the transition states. However, they may differ in the way this is achieved. Where the mechanism or the sequence of chemical steps at the active site is essentially the same as those for the same reaction proceeding in the absence of a catalyst, **the environment of the active site lowers $\Delta G_F$** by stabilizing the transition state intermediates. As discussed in Chapter 7, stabilization can involve (1) acid-base groups suitably positioned to transfer protons to or from the developing transition state intermediate, (2) suitably positioned charged groups or metal ions that stabilize developing charges, or (3) the imposition of steric strain on substrates so that their geometry approaches that of the transition state. HIV protease (Figure 7–6) illustrates catalysis by an enzyme that lowers the activation barrier by stabilizing a transition state intermediate.

Catalysis by enzymes that proceeds via a *unique* reaction mechanism typically occurs when the transition state intermediate forms a covalent bond with the enzyme **(covalent catalysis).** The catalytic mechanism of the serine protease chymotrypsin (Figure 7–7) illustrates how an enzyme utilizes covalent catalysis to provide a unique reaction pathway.

## ENZYMES DO NOT AFFECT $K_{eq}$

Enzymes accelerate reaction rates by lowering the activation barrier $\Delta G_F$. While they may undergo transient modification during the process of catalysis, enzymes emerge unchanged at the completion of the reaction. The presence of an enzyme therefore has no effect on $\Delta G^0$ for the *overall* reaction, which is a function solely of the *initial and final states* of the reactants. Equation (25) shows the relationship between the equilibrium constant for a reaction and the standard free energy change for that reaction:

$$\Delta G^0 = -RT \ln K_{eq} \quad (25)$$

If we include the presence of the enzyme (E) in the calculation of the equilibrium constant for a reaction,

$$A + B + Enz \rightleftarrows P + Q + Enz \quad (26)$$

the expression for the equilibrium constant,

$$K_{eq} = \frac{[P][Q][Enz]}{[A][B][Enz]} \quad (27)$$

reduces to one identical to that for the reaction in the absence of the enzyme:

$$K_{eq} = \frac{[P][Q]}{[A][B]} \quad (28)$$

**Enzymes therefore have no effect on $K_{eq}$.**

## MULTIPLE FACTORS AFFECT THE RATES OF ENZYME-CATALYZED REACTIONS

### Temperature

Raising the temperature increases the rate of both uncatalyzed and enzyme-catalyzed reactions by increasing the kinetic energy and the collision frequency of the reacting molecules. However, heat energy can also increase the kinetic energy of the enzyme to a point that exceeds the energy barrier for disrupting the noncovalent interactions that maintain the enzyme's three-dimensional structure. The polypeptide chain then begins to unfold, or **denature,** with an accompanying rapid loss of catalytic activity. The temperature range over which an enzyme maintains a stable, catalytically competent conformation depends upon—and typically moderately exceeds—the normal temperature of the cells in which it resides. Enzymes from humans generally exhibit stability at temperatures up to 45–55 °C. By contrast, enzymes from the thermophilic microorganisms that reside in volcanic hot springs or undersea hydrothermal vents may be stable up to or above 100 °C.

The **$Q_{10}$, or temperature coefficient,** is the factor by which the rate of a biologic process increases for a 10 °C increase in temperature. For the temperatures over which enzymes are stable, the rates of most biologic processes typically double for a 10 °C rise in temperature ($Q_{10}$ = 2). Changes in the rates of enzyme-catalyzed reactions that accompany a rise or fall in body temperature constitute a prominent survival feature for "cold-blooded" life forms such as lizards or fish, whose body temperatures are dictated by the external environment. However, for mammals and other homeothermic organisms, changes in enzyme reaction rates with temperature assume physiologic importance only in circumstances such as fever or hypothermia.

### Hydrogen Ion Concentration

The rate of almost all enzyme-catalyzed reactions exhibits a significant dependence on hydrogen ion concentration. Most intracellular enzymes exhibit optimal activity at pH values between 5 and 9. The relationship of activity to hydrogen ion concentration (Figure 8–2) reflects the balance between enzyme denaturation at high or low pH and effects on the charged state of the enzyme, the substrates, or both. For enzymes whose mechanism involves acid-base catalysis, the residues involved must be in the appropriate state of protonation for the reaction to proceed. The binding and recognition of substrate molecules with dissociable groups also typically involves the formation of salt bridges with the enzyme. The most common charged groups are the negative carboxylate groups and the positively charged groups of protonated amines. Gain or loss of critical charged groups thus will adversely affect substrate binding and thus will retard or abolish catalysis.

## ASSAYS OF ENZYME-CATALYZED REACTIONS TYPICALLY MEASURE THE INITIAL VELOCITY

Most measurements of the rates of enzyme-catalyzed reactions employ relatively short time periods, conditions that approximate **initial rate conditions.** Under these conditions, only traces of product accumulate, hence the rate of the reverse reaction is negligible. The **initial velocity ($v_i$)** of the reaction thus is essentially that of the rate of the forward reaction. Assays of enzyme activity almost always employ a large ($10^3$–$10^7$) molar excess of substrate over enzyme. Under these conditions, $v_i$ is proportionate to the concentration of enzyme. Measuring the initial velocity therefore permits one to estimate the quantity of enzyme present in a biologic sample.

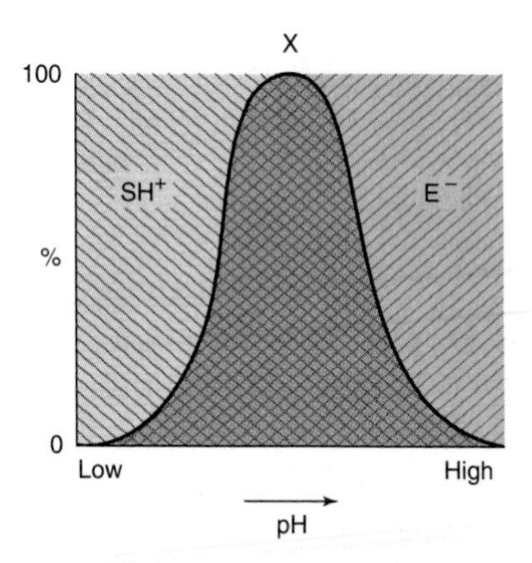


*Figure 8–2.* Effect of pH on enzyme activity. Consider, for example, a negatively charged enzyme ($EH^-$) that binds a positively charged substrate ($SH^+$). Shown is the proportion (%) of $SH^+$ [\\\] and of $EH^-$ [///] as a function of pH. Only in the cross-hatched area do both the enzyme and the substrate bear an appropriate charge.

## SUBSTRATE CONCENTRATION AFFECTS REACTION RATE

In what follows, enzyme reactions are treated as if they had only a single substrate and a single product. While most enzymes have more than one substrate, the principles discussed below apply with equal validity to enzymes with multiple substrates.

For a typical enzyme, as substrate concentration is increased, $v_i$ increases until it reaches a maximum value $V_{max}$ (Figure 8–3). When further increases in substrate concentration do not further increase $v_i$, the enzyme is said to be "saturated" with substrate. Note that the shape of the curve that relates activity to substrate concentration (Figure 8–3) is hyperbolic. At any given instant, only substrate molecules that are combined with the enzyme as an ES complex can be transformed into product. Second, the equilibrium constant for the formation of the enzyme-substrate complex is not infinitely large. Therefore, even when the substrate is present in excess (points A and B of Figure 8–4), only a fraction of the enzyme may be present as an ES complex. At points A or B, increasing or decreasing [S] therefore will increase or decrease the number of ES complexes with a corresponding change in $v_i$. At point C (Figure 8–4), essentially all the enzyme is present as the ES complex. Since no free enzyme remains available for forming ES, further increases in [S] cannot increase the rate of the reaction. Under these saturating conditions, $v_i$ depends solely on—and thus is limited by—the rapidity with which free enzyme is released to combine with more substrate.

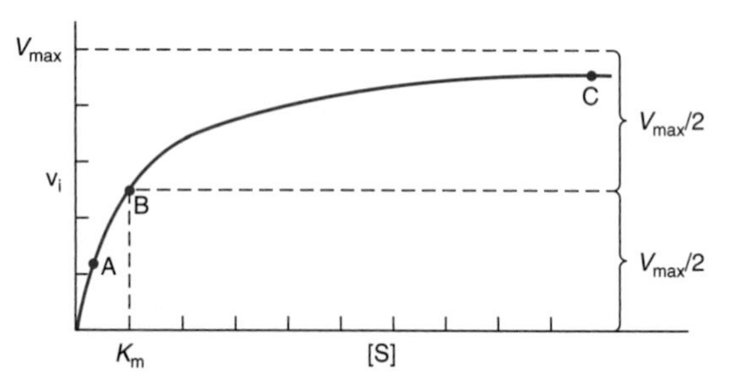


*Figure 8–3.* Effect of substrate concentration on the initial velocity of an enzyme-catalyzed reaction.

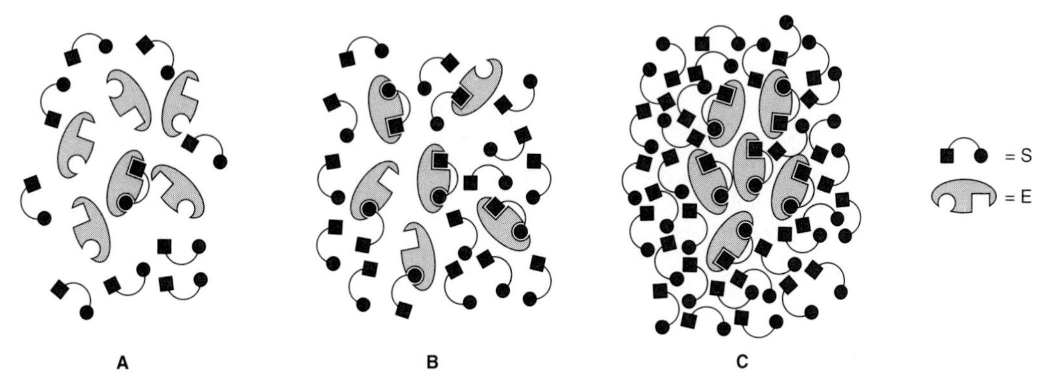


***Figure 8–4.*** Representation of an enzyme at low **(A)**, at high **(C)**, and at a substrate concentration equal to $K_m$ **(B)**. Points A, B, and C correspond to those points in Figure 8–3.

## THE MICHAELIS-MENTEN & HILL EQUATIONS MODEL THE EFFECTS OF SUBSTRATE CONCENTRATION

### The Michaelis-Menten Equation

The Michaelis-Menten equation (29) illustrates in mathematical terms the relationship between initial reaction velocity $v_i$ and substrate concentration [S], shown graphically in Figure 8–3.

$$v_i = \frac{V_{max}[S]}{K_m + [S]} \quad \mathbf{(29)}$$

**The Michaelis constant $K_m$ is the substrate concentration at which $v_i$ is half the maximal velocity ($V_{max}/2$) attainable at a particular concentration of enzyme.** $K_m$ thus has the dimensions of substrate concentration. The dependence of initial reaction velocity on [S] and $K_m$ may be illustrated by evaluating the Michaelis-Menten equation under three conditions.

**(1)** When [S] is much less than $K_m$ (point A in Figures 8–3 and 8–4), the term $K_m$ + [S] is essentially equal to $K_m$. Replacing $K_m$ + [S] with $K_m$ reduces equation (29) to

$$v_1 = \frac{V_{max}[S]}{K_m + [S]} \quad v_1 \approx \frac{V_{max}[S]}{K_m} \approx \left(\frac{V_{max}}{K_m}\right)[S] \quad \mathbf{(30)}$$

where ≈ means "approximately equal to." Since $V_{max}$ and $K_m$ are both constants, their ratio is a constant. In other words, when [S] is considerably below $K_m$, $v_i \propto k[S]$. The initial reaction velocity therefore is directly proportionate to [S].

**(2)** When [S] is much greater than $K_m$ (point C in Figures 8–3 and 8–4), the term $K_m$ + [S] is essentially equal to [S]. Replacing $K_m$ + [S] with [S] reduces equation (29) to

$$v_i = \frac{V_{max}[S]}{K_m + [S]} \quad v_i \approx \frac{V_{max}[S]}{[S]} \approx V_{max} \quad \mathbf{(31)}$$

Thus, when [S] greatly exceeds $K_m$, the reaction velocity is maximal ($V_{max}$) and unaffected by further increases in substrate concentration.

**(3)** When [S] = $K_m$ (point B in Figures 8–3 and 8–4).

$$v_i = \frac{V_{max}[S]}{K_m + [S]} = \frac{V_{max}[S]}{2[S]} = \frac{V_{max}}{2} \quad \mathbf{(32)}$$

Equation (32) states that when [S] equals $K_m$, the initial velocity is half-maximal. Equation (32) also reveals that $K_m$ is—and may be determined experimentally from—the substrate concentration at which the initial velocity is half-maximal.

### A Linear Form of the Michaelis-Menten Equation Is Used to Determine $K_m$ & $V_{max}$

The direct measurement of the numeric value of $V_{max}$ and therefore the calculation of $K_m$ often requires impractically high concentrations of substrate to achieve saturating conditions. A linear form of the Michaelis-Menten equation circumvents this difficulty and permits $V_{max}$ and $K_m$ to be extrapolated from initial velocity data obtained at less than saturating concentrations of substrate. Starting with equation (29),

$$v_i = \frac{V_{max}[S]}{K_m + [S]} \quad \mathbf{(29)}$$

invert

$$\frac{1}{v_1} = \frac{K_m + [S]}{V_{max}[S]} \quad (33)$$

factor

$$\frac{1}{v_i} = \frac{K_m}{V_{max}[S]} + \frac{[S]}{V_{max}[S]} \quad (34)$$

and simplify

$$\frac{1}{v_i} = \left(\frac{K_m}{V_{max}}\right)\frac{1}{[S]} + \frac{1}{V_{max}} \quad (35)$$

Equation (35) is the equation for a straight line, $y = ax + b$, where $y = 1/v_i$ and $x = 1/[S]$. A plot of $1/v_i$ as $y$ as a function of $1/[S]$ as $x$ therefore gives a straight line whose $y$ intercept is $1/V_{max}$ and whose slope is $K_m/V_{max}$. Such a plot is called a **double reciprocal** or **Lineweaver-Burk plot** (Figure 8–5). Setting the $y$ term of equation (36) equal to zero and solving for $x$ reveals that the $x$ intercept is $-1/K_m$.

$$0 = ax + b;\ \text{therefore, } x = \frac{-b}{a} = \frac{-1}{K_m} \quad (36)$$

$K_m$ is thus most easily calculated from the $x$ intercept.

## $K_m$ May Approximate a Binding Constant

The affinity of an enzyme for its substrate is the inverse of the dissociation constant $K_d$ for dissociation of the enzyme substrate complex ES.

$$E + S \underset{k_{-1}}{\overset{k_1}{\rightleftarrows}} ES \quad (37)$$

$$K_d = \frac{k_{-1}}{k_1} \quad (38)$$

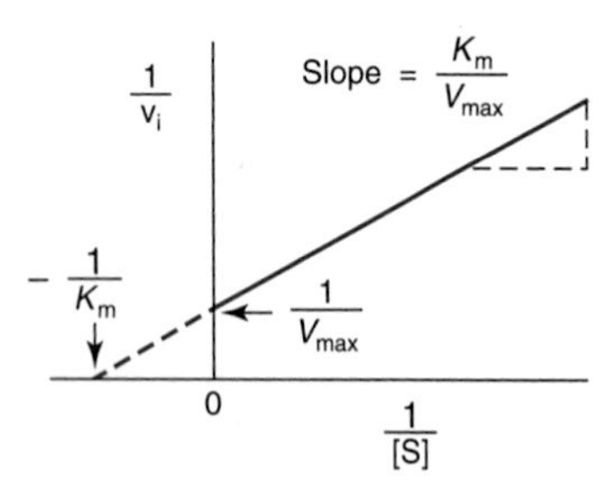


***Figure 8–5.*** Double reciprocal or Lineweaver-Burk plot of $1/v_i$ versus $1/[S]$ used to evaluate $K_m$ and $V_{max}$.

Stated another way, the smaller the tendency of the enzyme and its substrate to *dissociate,* the *greater* the affinity of the enzyme for its substrate. While the Michaelis constant $K_m$ often approximates the dissociation constant $K_d$, this is by no means always the case. For a typical enzyme-catalyzed reaction,

$$E + S \underset{k_{-1}}{\overset{k_1}{\rightleftarrows}} ES \overset{k_2}{\rightarrow} E + P \quad (39)$$

the value of [S] that gives $v_i = V_{max}/2$ is

$$[S] = \frac{k_{-1} + k_2}{k_1} = K_m \quad (40)$$

When $k_{-1} \gg k_2$, then

$$k_{-1} + k_2 \approx k_{-1} \quad (41)$$

and

$$[S] \approx \frac{k_1}{k_{-1}} \approx K_d \quad (42)$$

Hence, $1/K_m$ only approximates $1/K_d$ under conditions where the association and dissociation of the ES complex is rapid relative to the rate-limiting step in catalysis. For the many enzyme-catalyzed reactions for which $k_{-1} + k_2$ is *not* approximately equal to $k_{-1}$, $1/K_m$ will underestimate $1/K_d$.

## The Hill Equation Describes the Behavior of Enzymes That Exhibit Cooperative Binding of Substrate

While most enzymes display the simple **saturation kinetics** depicted in Figure 8–3 and are adequately described by the Michaelis-Menten expression, some enzymes bind their substrates in a cooperative fashion analogous to the binding of oxygen by hemoglobin (Chapter 6). Cooperative behavior may be encountered for multimeric enzymes that bind substrate at multiple sites. For enzymes that display positive cooperativity in binding substrate, the shape of the curve that relates changes in $v_i$ to changes in [S] is sigmoidal (Figure 8–6). Neither the Michaelis-Menten expression nor its derived double-reciprocal plots can be used to evaluate cooperative saturation kinetics. Enzymologists therefore employ a graphic representation of the **Hill equation** originally derived to describe the cooperative binding of $O_2$ by hemoglobin. Equation (43) represents the Hill equation arranged in a form that predicts a straight line, where k′ is a complex constant.

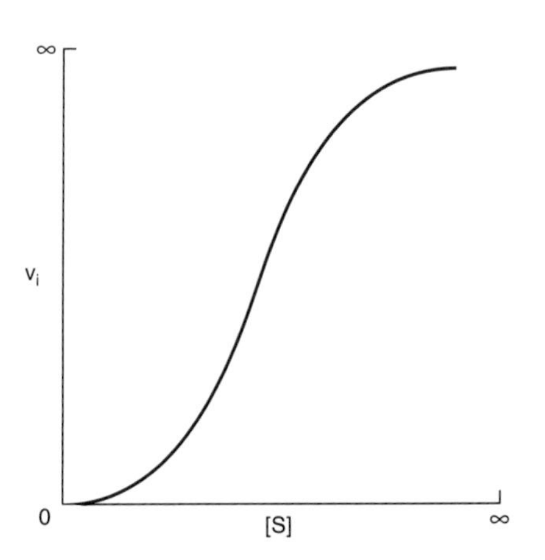


***Figure 8–6.*** Representation of sigmoid substrate saturation kinetics.

$$\frac{\log v_1}{V_{max} - v_1} = n \log[S] - \log k' \qquad \textbf{(43)}$$

Equation (43) states that when [S] is low relative to k′, the initial reaction velocity increases as the nth power of [S].

A graph of log $v_i/(V_{max} - v_i)$ versus log[S] gives a straight line (Figure 8–7), where the slope of the line **n** is the **Hill coefficient,** an empirical parameter whose value is a function of the number, kind, and strength of the interactions of the multiple substrate-binding sites on the enzyme. When n = 1, all binding sites behave independently, and simple Michaelis-Menten kinetic behavior is observed. If n is greater than 1, the enzyme is said to exhibit positive cooperativity. Binding of the first substrate molecule then enhances the affinity of the enzyme for binding additional substrate. The greater the value for n, the higher the degree of cooperativity and the more sigmoidal will be the plot of $v_i$ versus [S]. A perpendicular dropped from the point where the *y* term log $v_i/(V_{max} - v_i)$ is zero intersects the *x* axis at a substrate concentration termed $\mathbf{S_{50}}$, the substrate concentration that results in half-maximal velocity. $S_{50}$ thus is analogous to the $P_{50}$ for oxygen binding to hemoglobin (Chapter 6).

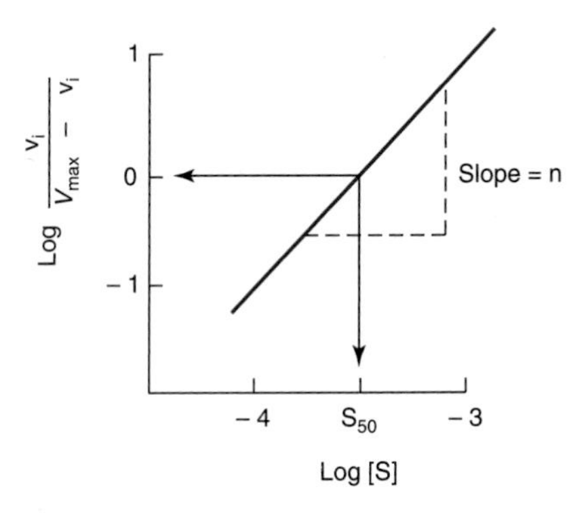


***Figure 8–7.*** A graphic representation of a linear form of the Hill equation is used to evaluate $S_{50}$, the substrate concentration that produces half-maximal velocity, and the degree of cooperativity **n.**

## KINETIC ANALYSIS DISTINGUISHES COMPETITIVE FROM NONCOMPETITIVE INHIBITION

Inhibitors of the catalytic activities of enzymes provide both pharmacologic agents and research tools for study of the mechanism of enzyme action. Inhibitors can be classified based upon their site of action on the enzyme, on whether or not they chemically modify the enzyme, or on the kinetic parameters they influence. Kinetically, we distinguish two classes of inhibitors based upon whether raising the substrate concentration does or does not overcome the inhibition.

### Competitive Inhibitors Typically Resemble Substrates

The effects of competitive inhibitors can be overcome by raising the concentration of the substrate. Most frequently, in competitive inhibition the inhibitor, **I,** binds to the substrate-binding portion of the active site and blocks access by the substrate. The structures of most classic competitive inhibitors therefore tend to resemble the structures of a substrate and thus are termed **substrate analogs.** Inhibition of the enzyme succinate dehydrogenase by malonate illustrates competitive inhibition by a substrate analog. Succinate dehydrogenase catalyzes the removal of one hydrogen atom from each of the two methylene carbons of succinate (Figure 8–8). Both succinate and its structural analog malonate ($^-OOC—CH_2—COO^-$) can bind to the active site of succinate dehydrogenase, forming an ES or an EI complex, respectively. However, since malonate contains

H
|
H — C — COO⁻
|
⁻OOC — C — H
|
H

**Succinate**

−2H → SUCCINATE DEHYDROGENASE

H — C — COO⁻
||
⁻OOC — C — H

**Fumarate**

***Figure 8–8.*** The succinate dehydrogenase reaction.

only one methylene carbon, it cannot undergo dehydrogenation. The formation and dissociation of the EI complex is a dynamic process described by

$$\text{EnzI} \underset{k_{-1}}{\overset{k_1}{\rightleftarrows}} \text{Enz} + \text{I} \qquad \textbf{(44)}$$

for which the equilibrium constant $K_i$ is

$$K_1 = \frac{[\text{Enz}][\text{I}]}{[\text{EnzI}]} = \frac{k_1}{k_{-1}} \qquad \textbf{(45)}$$

In effect, **a competitive inhibitor acts by decreasing the number of free enzyme molecules available to bind substrate, ie, to form ES, and thus eventually to form product,** as described below:

$$\text{E} \underset{}{\overset{\pm\text{I}}{\rightleftarrows}} \text{E-I}$$
$$\text{E} \overset{\pm\text{S}}{\rightleftarrows} \text{E-S} \rightarrow \text{E} + \text{P} \qquad \textbf{(46)}$$

A competitive inhibitor and substrate exert reciprocal effects on the concentration of the EI and ES complexes. Since binding substrate removes free enzyme available to combine with inhibitor, increasing the [S] decreases the concentration of the EI complex and raises the reaction velocity. The extent to which [S] must be increased to completely overcome the inhibition depends upon the concentration of inhibitor present, its affinity for the enzyme $K_i$, and the $K_m$ of the enzyme for its substrate.

## Double Reciprocal Plots Facilitate the Evaluation of Inhibitors

Double reciprocal plots distinguish between competitive and noncompetitive inhibitors and simplify evaluation of inhibition constants $K_i$. $v_i$ is determined at several substrate concentrations both in the presence and in the absence of inhibitor. For classic competitive inhibition, the lines that connect the experimental data points meet at the $y$ axis (Figure 8–9). Since the $y$ intercept is equal to $1/V_{max}$, this pattern indicates that **when 1/[S] approaches 0, $v_i$ is independent of the presence of inhibitor.** Note, however, that the intercept on the $x$ axis does vary with *inhibitor* concentration—and that since $-1/K_m'$ is smaller than $1/K_m$, $K_m'$ (the "apparent $K_m$") becomes larger in the presence of increasing concentrations of inhibitor. Thus, **a competitive inhibitor has no effect on $V_{max}$ but raises $K'_m$, the apparent $K_m$ for the substrate.**

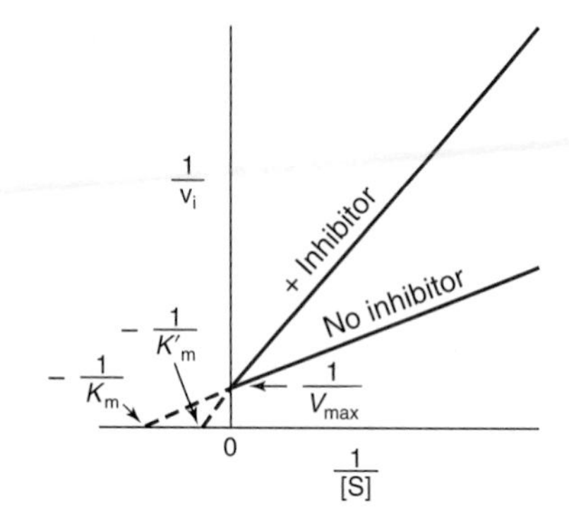


***Figure 8–9.*** Lineweaver-Burk plot of competitive inhibition. Note the complete relief of inhibition at high [S] (ie, low 1/[S]).

For simple competitive inhibition, the intercept on the $x$ axis is

$$x = \frac{-1}{K_m}\left(1 + \frac{[\text{I}]}{K_i}\right) \qquad \textbf{(47)}$$

Once $K_m$ has been determined in the absence of inhibitor, $K_i$ can be calculated from equation (47). $K_i$ values are used to compare different inhibitors of the same enzyme. The lower the value for $K_i$, the more effective the inhibitor. For example, the statin drugs that act as competitive inhibitors of HMG-CoA reductase (Chapter 26) have $K_i$ values several orders of magnitude lower than the $K_m$ for the substrate HMG-CoA.

## Simple Noncompetitive Inhibitors Lower $V_{max}$ but Do Not Affect $K_m$

In noncompetitive inhibition, binding of the inhibitor does not affect binding of substrate. Formation of both EI and EIS complexes is therefore possible. However, while the enzyme-inhibitor complex can still bind substrate, its efficiency at transforming substrate to product, reflected by $V_{max}$, is decreased. Noncompetitive inhibitors bind enzymes at sites distinct from the substrate-binding site and generally bear little or no structural resemblance to the substrate.

For simple noncompetitive inhibition, E and EI possess identical affinity for substrate, and the EIS complex generates product at a negligible rate (Figure 8–10). More complex noncompetitive inhibition occurs when binding of the inhibitor *does* affect the apparent affinity of the enzyme for substrate, causing the lines to intercept in either the third or fourth quadrants of a double reciprocal plot (not shown).

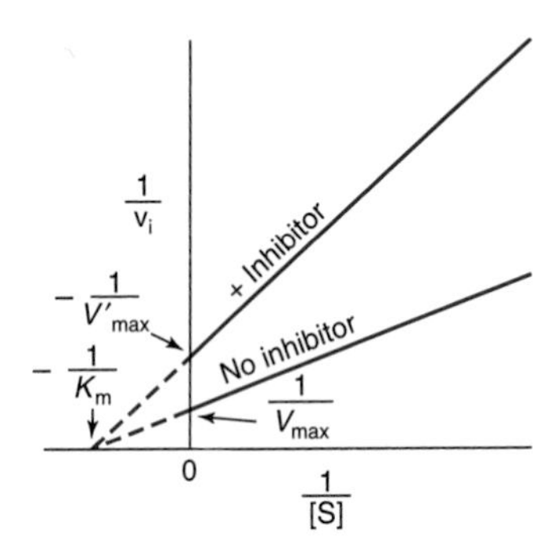


***Figure 8–10.*** Lineweaver-Burk plot for simple noncompetitive inhibition.

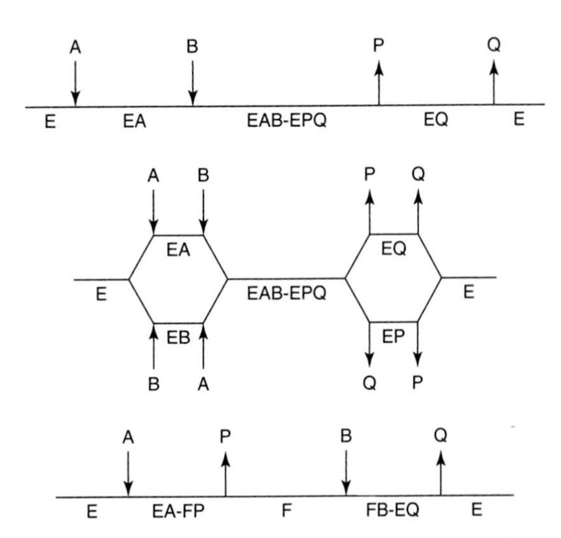


***Figure 8–11.*** Representations of three classes of Bi-Bi reaction mechanisms. Horizontal lines represent the enzyme. Arrows indicate the addition of substrates and departure of products. **Top:** An ordered Bi-Bi reaction, characteristic of many NAD(P)H-dependent oxidoreductases. **Center:** A random Bi-Bi reaction, characteristic of many kinases and some dehydrogenases. **Bottom:** A ping-pong reaction, characteristic of aminotransferases and **serine proteases.**

### Irreversible Inhibitors "Poison" Enzymes

In the above examples, the inhibitors form a dissociable, dynamic complex with the enzyme. Fully active enzyme can therefore be recovered simply by removing the inhibitor from the surrounding medium. However, a variety of other inhibitors act irreversibly by chemically modifying the enzyme. These modifications generally involve making or breaking covalent bonds with aminoacyl residues essential for substrate binding, catalysis, or maintenance of the enzyme's functional conformation. Since these covalent changes are relatively stable, an enzyme that has been "poisoned" by an irreversible inhibitor remains inhibited even after removal of the remaining inhibitor from the surrounding medium.

## MOST ENZYME-CATALYZED REACTIONS INVOLVE TWO OR MORE SUBSTRATES

While many enzymes have a single substrate, many others have two—and sometimes more than two—substrates and products. The fundamental principles discussed above, while illustrated for single-substrate enzymes, apply also to multisubstrate enzymes. The mathematical expressions used to evaluate multisubstrate reactions are, however, complex. While detailed kinetic analysis of multisubstrate reactions exceeds the scope of this chapter, two-substrate, two-product reactions (termed "Bi-Bi" reactions) are considered below.

### Sequential or Single Displacement Reactions

In **sequential reactions,** both substrates must combine with the enzyme to form a ternary complex before catalysis can proceed (Figure 8–11, top). Sequential reactions are sometimes referred to as single displacement reactions because the group undergoing transfer is usually passed directly, in a single step, from one substrate to the other. Sequential Bi-Bi reactions can be further distinguished based on whether the two substrates add in a **random** or in a **compulsory** order. For random-order reactions, either substrate A or substrate B may combine first with the enzyme to form an EA or an EB complex (Figure 8–11, center). For compulsory-order reactions, A must first combine with E before B can combine with the EA complex. One explanation for a compulsory-order mechanism is that the addition of A induces a conformational change in the enzyme that aligns residues which recognize and bind B.

### Ping-Pong Reactions

The term **"ping-pong"** applies to mechanisms in which one or more products are released from the enzyme before all the substrates have been added. Ping-pong reactions involve covalent catalysis and a transient, modified form of the enzyme (Figure 7–4). Ping-pong Bi-Bi reactions are **double displacement reactions.** The group undergoing transfer is first displaced from substrate A by the enzyme to form product

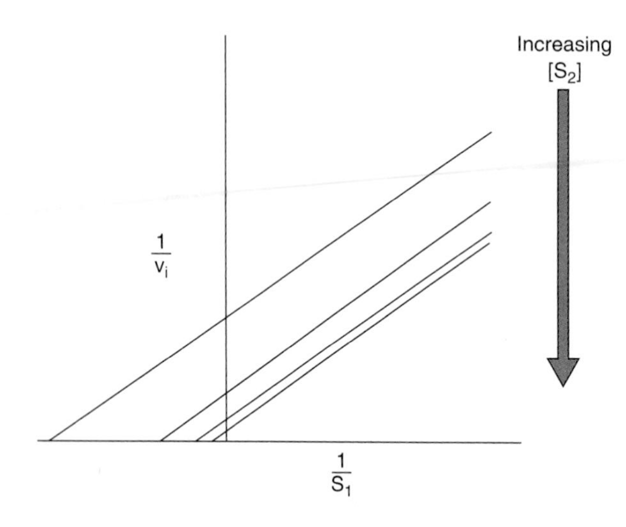


***Figure 8–12.*** Lineweaver-Burk plot for a two-substrate ping-pong reaction. An increase in concentration of one substrate ($S_1$) while that of the other substrate ($S_2$) is maintained constant changes both the *x* and *y* intercepts, but not the slope.

P and a modified form of the enzyme (F). The subsequent group transfer from F to the second substrate B, forming product Q and regenerating E, constitutes the second displacement (Figure 8–11, bottom).

## Most Bi-Bi Reactions Conform to Michaelis-Menten Kinetics

Most Bi-Bi reactions conform to a somewhat more complex form of Michaelis-Menten kinetics in which $V_{max}$ refers to the reaction rate attained when both substrates are present at saturating levels. Each substrate has its own characteristic $K_m$ value which corresponds to the concentration that yields half-maximal velocity when the second substrate is present at saturating levels. As for single-substrate reactions, double-reciprocal plots can be used to determine $V_{max}$ and $K_m$. $v_i$ is measured as a function of the concentration of one substrate (the variable substrate) while the concentration of the other substrate (the fixed substrate) is maintained constant. If the lines obtained for several fixed-substrate concentrations are plotted on the same graph, it is possible to distinguish between a ping-pong enzyme, which yields parallel lines, and a sequential mechanism, which yields a pattern of intersecting lines (Figure 8–12).

Product inhibition studies are used to complement kinetic analyses and to distinguish between ordered and random Bi-Bi reactions. For example, in a random-order Bi-Bi reaction, each product will be a competitive inhibitor regardless of which substrate is designated the variable substrate. However, for a sequential mechanism (Figure 8–11, bottom), only product Q will give the pattern indicative of competitive inhibition when A is the variable substrate, while only product P will produce this pattern with B as the variable substrate. The other combinations of product inhibitor and variable substrate will produce forms of complex noncompetitive inhibition.

## SUMMARY

- The study of enzyme kinetics—the factors that affect the rates of enzyme-catalyzed reactions—reveals the individual steps by which enzymes transform substrates into products.
- $\Delta G$, the overall change in free energy for a reaction, is independent of reaction mechanism and provides no information concerning *rates* of reactions.
- Enzymes do not affect $K_{eq}$. $K_{eq}$, a ratio of reaction *rate constants,* may be calculated from the concentrations of substrates and products at equilibrium or from the ratio $k_1/k_{-1}$.
- Reactions proceed via transition states in which $\Delta G_F$ is the activation energy. Temperature, hydrogen ion concentration, enzyme concentration, substrate concentration, and inhibitors all affect the rates of enzyme-catalyzed reactions.
- A measurement of the rate of an enzyme-catalyzed reaction generally employs initial rate conditions, for which the essential absence of product precludes the reverse reaction.
- A linear form of the Michaelis-Menten equation simplifies determination of $K_m$ and $V_{max}$.
- A linear form of the Hill equation is used to evaluate the cooperative substrate-binding kinetics exhibited by some multimeric enzymes. The slope **n,** the Hill coefficient, reflects the number, nature, and strength of the interactions of the substrate-binding sites. A

value of **n** greater than 1 indicates positive cooperativity.

- The effects of competitive inhibitors, which typically resemble substrates, are overcome by raising the concentration of the substrate. Noncompetitive inhibitors lower $V_{max}$ but do not affect $K_m$.
- Substrates may add in a random order (either substrate may combine first with the enzyme) or in a compulsory order (substrate A must bind before substrate B).
- In ping-pong reactions, one or more products are released from the enzyme before all the substrates have added.

## REFERENCES

Fersht A: *Structure and Mechanism in Protein Science: A Guide to Enzyme Catalysis and Protein Folding.* Freeman, 1999.

Schultz AR: *Enzyme Kinetics: From Diastase to Multi-enzyme Systems.* Cambridge Univ Press, 1994.

Segel IH: *Enzyme Kinetics.* Wiley Interscience, 1975.

# Enzymes: Regulation of Activities 9

Victor W. Rodwell, PhD, & Peter J. Kennelly, PhD

## BIOMEDICAL IMPORTANCE

The 19th-century physiologist Claude Bernard enunciated the conceptual basis for metabolic regulation. He observed that living organisms respond in ways that are both quantitatively and temporally appropriate to permit them to survive the multiple challenges posed by changes in their external and internal environments. Walter Cannon subsequently coined the term "homeostasis" to describe the ability of animals to maintain a constant intracellular environment despite changes in their external environment. We now know that organisms respond to changes in their external and internal environment by balanced, coordinated changes in the rates of specific metabolic reactions. Many human diseases, including cancer, diabetes, cystic fibrosis, and Alzheimer's disease, are characterized by regulatory dysfunctions triggered by pathogenic agents or genetic mutations. For example, many oncogenic viruses elaborate protein-tyrosine kinases that modify the regulatory events which control patterns of gene expression, contributing to the initiation and progression of cancer. The toxin from *Vibrio cholerae,* the causative agent of cholera, disables sensor-response pathways in intestinal epithelial cells by ADP-ribosylating the GTP-binding proteins (G-proteins) that link cell surface receptors to adenylyl cyclase. The consequent activation of the cyclase triggers the flow of water into the intestines, resulting in massive diarrhea and dehydration. *Yersinia pestis,* the causative agent of plague, elaborates a protein-tyrosine phosphatase that hydrolyzes phosphoryl groups on key cytoskeletal proteins. Knowledge of factors that control the rates of enzyme-catalyzed reactions thus is essential to an understanding of the molecular basis of disease. This chapter outlines the patterns by which metabolic processes are controlled and provides illustrative examples. Subsequent chapters provide additional examples.

## REGULATION OF METABOLITE FLOW CAN BE ACTIVE OR PASSIVE

Enzymes that operate at their maximal rate cannot respond to an increase in substrate concentration, and can respond only to a precipitous decrease in substrate concentration. For most enzymes, therefore, the average intracellular concentration of their substrate tends to be close to the $K_m$ value, so that changes in substrate concentration generate corresponding changes in metabolite flux (Figure 9–1). Responses to changes in substrate level represent an important but *passive* means for coordinating metabolite flow and maintaining homeostasis in quiescent cells. However, they offer limited scope for responding to changes in environmental variables. The mechanisms that regulate enzyme activity in an *active* manner in response to internal and external signals are discussed below.

### Metabolite Flow Tends to Be Unidirectional

Despite the existence of short-term oscillations in metabolite concentrations and enzyme levels, living cells exist in a dynamic steady state in which the mean concentrations of metabolic intermediates remain relatively constant over time (Figure 9–2). While all chemical reactions are to some extent reversible, in living cells the reaction products serve as substrates for—and are removed by—other enzyme-catalyzed reactions. Many nominally reversible reactions thus occur unidirectionally. This succession of coupled metabolic reactions is accompanied by an overall change in free energy that favors unidirectional metabolite flow (Chapter 10). The unidirectional flow of metabolites through a pathway with a large overall negative change in free energy is analogous to the flow of water through a pipe in which one end is lower than the other. Bends or kinks in the pipe simulate individual enzyme-catalyzed steps with a small negative or positive change in free energy. Flow of water through the pipe nevertheless remains unidirectional due to the overall change in height, which corresponds to the overall change in free energy in a pathway (Figure 9–3).

## COMPARTMENTATION ENSURES METABOLIC EFFICIENCY & SIMPLIFIES REGULATION

In eukaryotes, anabolic and catabolic pathways that interconvert common products may take place in specific subcellular compartments. For example, many of the enzymes that degrade proteins and polysaccharides reside inside organelles called lysosomes. Similarly, fatty acid biosynthesis occurs in the cytosol, whereas fatty

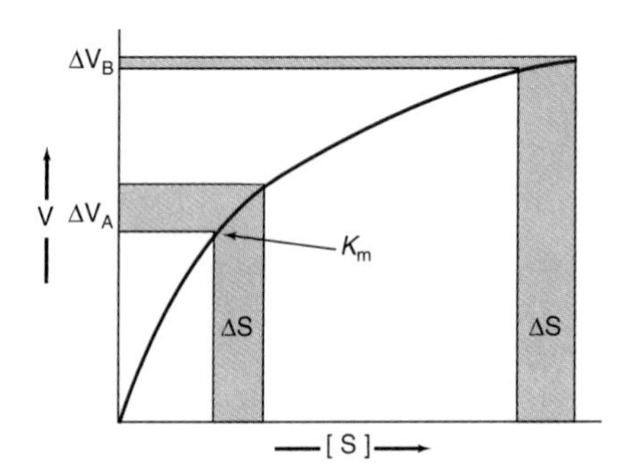


***Figure 9–1.*** Differential response of the rate of an enzyme-catalyzed reaction, ΔV, to the same incremental change in substrate concentration at a substrate concentration of $K_m$ ($\Delta V_A$) or far above $K_m$ ($\Delta V_B$).

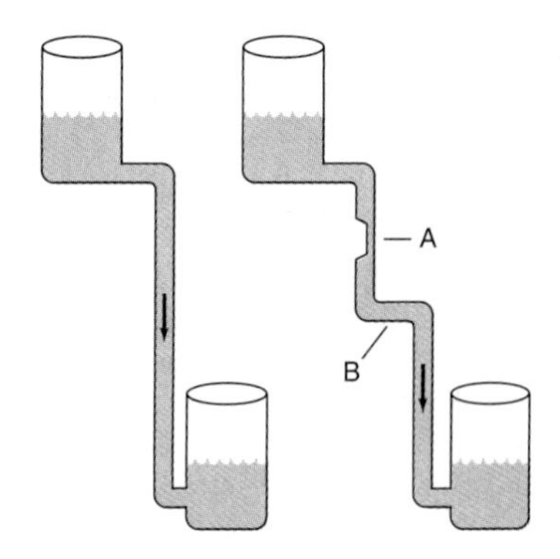


***Figure 9–3.*** Hydrostatic analogy for a pathway with a rate-limiting step **(A)** and a step with a ΔG value near zero **(B).**

acid oxidation takes place within mitochondria (Chapters 21 and 22). Segregation of certain metabolic pathways within specialized cell types can provide further physical compartmentation. Alternatively, possession of one or more *unique intermediates* can permit apparently opposing pathways to coexist even in the absence of physical barriers. For example, despite many shared intermediates and enzymes, both glycolysis and gluconeogenesis are favored energetically. This cannot be true if *all* the reactions were the same. If one pathway was favored energetically, the other would be accompanied by a change in free energy G equal in magnitude but opposite in sign. Simultaneous spontaneity of both pathways results from substitution of one or more reactions by different reactions favored thermodynamically in the opposite direction. The glycolytic enzyme phosphofructokinase (Chapter 17) is replaced by the gluconeogenic enzyme fructose-1,6-bisphosphatase (Chapter 19). The ability of enzymes to discriminate between the structurally similar coenzymes $NAD^+$ and $NADP^+$ also results in a form of compartmentation, since it segregates the electrons of NADH that are destined for ATP generation from those of NADPH that participate in the reductive steps in many biosynthetic pathways.

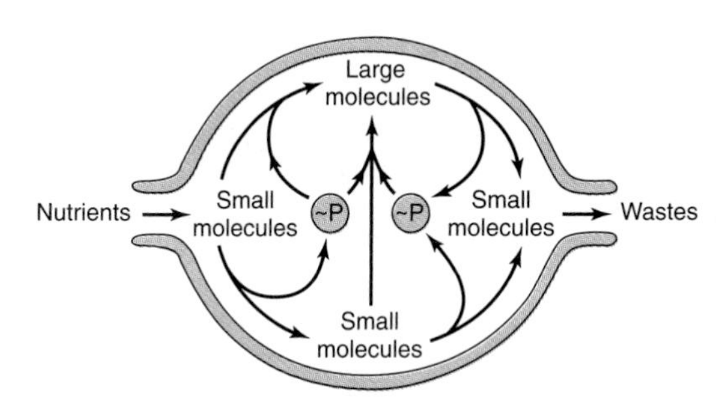


***Figure 9–2.*** An idealized cell in steady state. Note that metabolite flow is unidirectional.

## Controlling an Enzyme That Catalyzes a Rate-Limiting Reaction Regulates an Entire Metabolic Pathway

While the flux of metabolites through metabolic pathways involves catalysis by numerous enzymes, active control of homeostasis is achieved by regulation of only a small number of enzymes. The ideal enzyme for regulatory intervention is one whose quantity or catalytic efficiency dictates that the reaction it catalyzes is slow relative to all others in the pathway. Decreasing the catalytic efficiency or the quantity of the catalyst for the "bottleneck" or **rate-limiting reaction** immediately reduces metabolite flux through the entire pathway. Conversely, an increase in either its quantity or catalytic efficiency enhances flux through the pathway as a whole. For example, acetyl-CoA carboxylase catalyzes the synthesis of malonyl-CoA, the first committed reaction of fatty acid biosynthesis (Chapter 21). When synthesis of malonyl-CoA is inhibited, subsequent reactions of fatty acid synthesis cease due to lack of substrates. Enzymes that catalyze rate-limiting steps serve as natural "governors" of metabolic flux. Thus, they constitute efficient targets for regulatory intervention by drugs. For example, inhibition by "statin" drugs of HMG-CoA reductase, which catalyzes the rate-limiting reaction of cholesterogenesis, curtails synthesis of cholesterol.

## REGULATION OF ENZYME QUANTITY

The catalytic capacity of the rate-limiting reaction in a metabolic pathway is the product of the concentration of enzyme molecules and their intrinsic catalytic efficiency. It therefore follows that catalytic capacity can be influenced both by changing the quantity of enzyme present and by altering its intrinsic catalytic efficiency.

### Control of Enzyme Synthesis

Enzymes whose concentrations remain essentially constant over time are termed **constitutive enzymes.** By contrast, the concentrations of many other enzymes depend upon the presence of **inducers,** typically substrates or structurally related compounds, that initiate their synthesis. *Escherichia coli* grown on glucose will, for example, only catabolize lactose after addition of a β-galactoside, an inducer that initiates synthesis of a β-galactosidase and a galactoside permease (Figure 39–3). Inducible enzymes of humans include tryptophan pyrrolase, threonine dehydrase, tyrosine-α-ketoglutarate aminotransferase, enzymes of the urea cycle, HMG-CoA reductase, and cytochrome P450. Conversely, an excess of a metabolite may curtail synthesis of its cognate enzyme via **repression.** Both induction and repression involve cis elements, specific DNA sequences located upstream of regulated genes, and trans-acting regulatory proteins. The molecular mechanisms of induction and repression are discussed in Chapter 39.

### Control of Enzyme Degradation

The absolute quantity of an enzyme reflects the net balance between enzyme synthesis and enzyme degradation, where $k_s$ and $k_{deg}$ represent the rate constants for the overall processes of synthesis and degradation, respectively. Changes in both the $k_s$ and $k_{deg}$ of specific enzymes occur in human subjects.

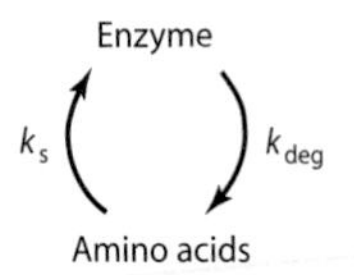


**Protein turnover** represents the net result of enzyme synthesis and degradation. By measuring the rates of incorporation of $^{15}N$-labeled amino acids into protein and the rates of loss of $^{15}N$ from protein, Schoenheimer deduced that body proteins are in a state of "dynamic equilibrium" in which they are continuously synthesized and degraded. Mammalian proteins are degraded both by ATP and ubiquitin-dependent pathways and by ATP-independent pathways (Chapter 29). Susceptibility to proteolytic degradation can be influenced by the presence of ligands such as substrates, coenzymes, or metal ions that alter protein conformation. Intracellular ligands thus can influence the rates at which specific enzymes are degraded.

Enzyme levels in mammalian tissues respond to a wide range of physiologic, hormonal, or dietary factors. For example, glucocorticoids increase the concentration of tyrosine aminotransferase by stimulating $k_s$, and glucagon—despite its antagonistic physiologic effects—increases $k_s$ fourfold to fivefold. Regulation of liver arginase can involve changes either in $k_s$ or in $k_{deg}$. After a protein-rich meal, liver arginase levels rise and arginine synthesis decreases (Chapter 29). Arginase levels also rise in starvation, but here arginase degradation decreases, whereas $k_s$ remains unchanged. Similarly, injection of glucocorticoids and ingestion of tryptophan both elevate levels of tryptophan oxygenase. While the hormone raises $k_s$ for oxygenase synthesis, tryptophan specifically lowers $k_{deg}$ by stabilizing the oxygenase against proteolytic digestion.

## MULTIPLE OPTIONS ARE AVAILABLE FOR REGULATING CATALYTIC ACTIVITY

In humans, the induction of protein synthesis is a complex multistep process that typically requires hours to produce significant changes in overall enzyme level. By contrast, changes in intrinsic catalytic efficiency effected by binding of dissociable ligands **(allosteric regulation)** or by **covalent modification** achieve regulation of enzymic activity within seconds. Changes in protein level serve long-term adaptive requirements, whereas changes in catalytic efficiency are best suited for rapid and transient alterations in metabolite flux.

## ALLOSTERIC EFFECTORS REGULATE CERTAIN ENZYMES

Feedback inhibition refers to inhibition of an enzyme in a biosynthetic pathway by an end product of that pathway. For example, for the biosynthesis of D from A catalyzed by enzymes $Enz_1$ through $Enz_3$,

$$A \xrightarrow{Enz_1} B \xrightarrow{Enz_2} C \xrightarrow{Enz_3} D$$

high concentrations of D inhibit conversion of A to B. Inhibition results not from the "backing up" of intermediates but from the ability of D to bind to and inhibit $Enz_1$. Typically, D binds at an **allosteric site** spatially distinct from the catalytic site of the target enzyme. Feedback inhibitors thus are allosteric effectors and typically bear little or no structural similarity to the substrates of the enzymes they inhibit. In this example, the feedback inhibitor D acts as a **negative allosteric effector** of $Enz_1$.

In a branched biosynthetic pathway, the initial reactions participate in the synthesis of several products. Figure 9–4 shows a hypothetical branched biosynthetic pathway in which curved arrows lead from feedback inhibitors to the enzymes whose activity they inhibit. The sequences $S_3 \rightarrow A$, $S_4 \rightarrow B$, $S_4 \rightarrow C$, and $S_3 \rightarrow \rightarrow D$ each represent linear reaction sequences that are feedback-inhibited by their end products. The pathways of nucleotide biosynthesis (Chapter 34) provide specific examples.

The kinetics of feedback inhibition may be competitive, noncompetitive, partially competitive, or mixed. Feedback inhibitors, which frequently are the small molecule building blocks of macromolecules (eg, amino acids for proteins, nucleotides for nucleic acids), typically inhibit the first committed step in a particular biosynthetic sequence. A much-studied example is inhibition of bacterial aspartate transcarbamoylase by CTP (see below and Chapter 34).

Multiple feedback loops can provide additional fine control. For example, as shown in Figure 9–5, the presence of excess product B decreases the requirement for substrate $S_2$. However, $S_2$ is also required for synthesis of A, C, and D. Excess B should therefore also curtail synthesis of all four end products. To circumvent this potential difficulty, each end product typically only partially inhibits catalytic activity. The effect of an excess of two or more end products may be strictly additive or, alternatively, may be greater than their individual effect (cooperative feedback inhibition).

## Aspartate Transcarbamoylase Is a Model Allosteric Enzyme

Aspartate transcarbamoylase (ATCase), the catalyst for the first reaction unique to pyrimidine biosynthesis (Figure 34–7), is feedback-inhibited by cytidine triphosphate (CTP). Following treatment with mercurials, ATCase loses its sensitivity to inhibition by CTP but retains its full activity for synthesis of carbamoyl aspartate. This suggests that CTP is bound at a different (allosteric) site from either substrate. ATCase consists of multiple catalytic and regulatory subunits. Each catalytic subunit contains four aspartate (substrate) sites and each regulatory subunit at least two CTP (regulatory) sites (Chapter 34).

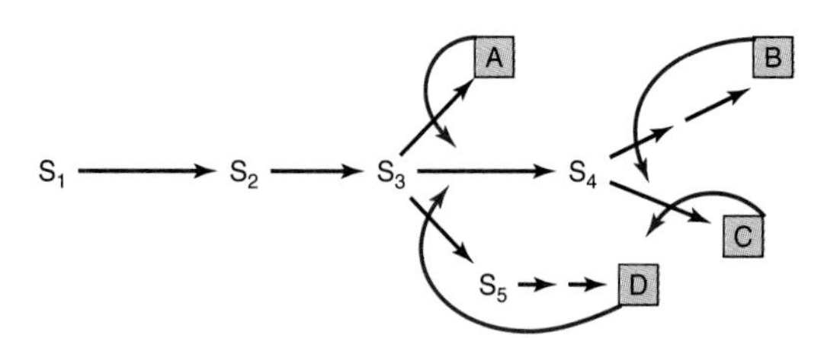


***Figure 9–4.*** Sites of feedback inhibition in a branched biosynthetic pathway. $S_1$–$S_5$ are intermediates in the biosynthesis of end products A–D. Straight arrows represent enzymes catalyzing the indicated conversions. Curved arrows represent feedback loops and indicate sites of feedback inhibition by specific end products.

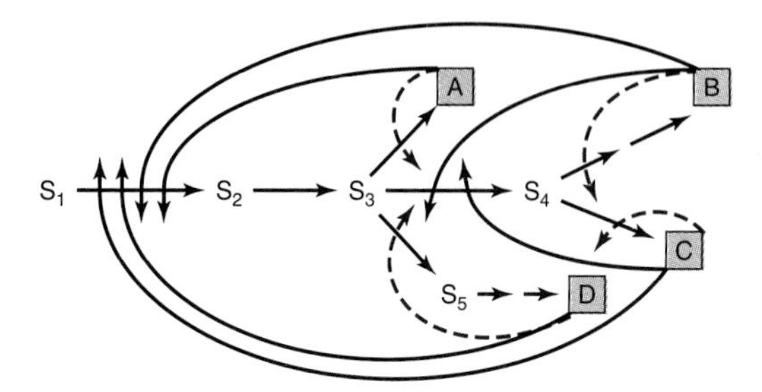


***Figure 9–5.*** Multiple feedback inhibition in a branched biosynthetic pathway. Superimposed on simple feedback loops (dashed, curved arrows) are multiple feedback loops (solid, curved arrows) that regulate enzymes common to biosynthesis of several end products.

## Allosteric & Catalytic Sites Are Spatially Distinct

The lack of structural similarity between a feedback inhibitor and the substrate for the enzyme whose activity it regulates suggests that these effectors are not **isosteric** with a substrate but **allosteric** ("occupy another space"). Jacques Monod therefore proposed the existence of allosteric sites that are physically distinct from the catalytic site. **Allosteric enzymes** thus are those whose activity at the active site may be modulated by the presence of effectors at an allosteric site. This hypothesis has been confirmed by many lines of evidence, including x-ray crystallography and site-directed mutagenesis, demonstrating the existence of spatially distinct active and allosteric sites on a variety of enzymes.

## Allosteric Effects May Be on $K_m$ or on $V_{max}$

To refer to the kinetics of allosteric inhibition as "competitive" or "noncompetitive" with substrate carries misleading mechanistic implications. We refer instead to two classes of regulated enzymes: K-series and V-series enzymes. For K-series allosteric enzymes, the substrate saturation kinetics are competitive in the sense that $K_m$ is raised without an effect on $V_{max}$. For V-series allosteric enzymes, the allosteric inhibitor lowers $V_{max}$

without affecting the $K_m$. Alterations in $K_m$ or $V_{max}$ probably result from conformational changes at the catalytic site induced by binding of the allosteric effector at the allosteric site. For a K-series allosteric enzyme, this conformational change may weaken the bonds between substrate and substrate-binding residues. For a V-series allosteric enzyme, the primary effect may be to alter the orientation or charge of catalytic residues, lowering $V_{max}$. Intermediate effects on $K_m$ and $V_{max}$, however, may be observed consequent to these conformational changes.

## FEEDBACK REGULATION IS NOT SYNONYMOUS WITH FEEDBACK INHIBITION

In both mammalian and bacterial cells, end products "feed back" and control their own synthesis, in many instances by feedback inhibition of an early biosynthetic enzyme. We must, however, distinguish between **feedback regulation,** a phenomenologic term devoid of mechanistic implications, and **feedback inhibition,** a mechanism for regulation of enzyme activity. For example, while dietary cholesterol decreases hepatic synthesis of cholesterol, this feedback **regulation** does not involve feedback **inhibition.** HMG-CoA reductase, the rate-limiting enzyme of cholesterologenesis, is affected, but cholesterol does not feedback-inhibit its activity. Regulation in response to dietary cholesterol involves curtailment by cholesterol or a cholesterol metabolite of the expression of the gene that encodes HMG-CoA reductase (enzyme repression) (Chapter 26).

## MANY HORMONES ACT THROUGH ALLOSTERIC SECOND MESSENGERS

Nerve impulses—and binding of hormones to cell surface receptors—elicit changes in the rate of enzyme-catalyzed reactions within target cells by inducing the release or synthesis of specialized allosteric effectors called **second messengers.** The primary or "first" messenger is the hormone molecule or nerve impulse. Second messengers include 3′,5′-cAMP, synthesized from ATP by the enzyme adenylyl cyclase in response to the hormone epinephrine, and $Ca^{2+}$, which is stored inside the endoplasmic reticulum of most cells. Membrane depolarization resulting from a nerve impulse opens a membrane channel that releases calcium ion into the cytoplasm, where it binds to and activates enzymes involved in the regulation of contraction and the mobilization of stored glucose from glycogen. Glucose then supplies the increased energy demands of muscle contraction. Other second messengers include 3′,5′-cGMP and polyphosphoinositols, produced by the hydrolysis of inositol phospholipids by hormone-regulated phospholipases.

## REGULATORY COVALENT MODIFICATIONS CAN BE REVERSIBLE OR IRREVERSIBLE

In mammalian cells, the two most common forms of covalent modification are **partial proteolysis** and **phosphorylation.** Because cells lack the ability to reunite the two portions of a protein produced by hydrolysis of a peptide bond, proteolysis constitutes an irreversible modification. By contrast, phosphorylation is a reversible modification process. The phosphorylation of proteins on seryl, threonyl, or tyrosyl residues, catalyzed by protein kinases, is thermodynamically spontaneous. Equally spontaneous is the hydrolytic removal of these phosphoryl groups by enzymes called protein phosphatases.

## PROTEASES MAY BE SECRETED AS CATALYTICALLY INACTIVE PROENZYMES

Certain proteins are synthesized and secreted as inactive precursor proteins known as **proproteins.** The proproteins of enzymes are termed **proenzymes** or **zymogens.** Selective proteolysis converts a proprotein by one or more successive proteolytic "clips" to a form that exhibits the characteristic activity of the mature protein, eg, its enzymatic activity. Proteins synthesized as proproteins include the hormone insulin (proprotein = proinsulin), the digestive enzymes pepsin, trypsin, and chymotrypsin (proproteins = pepsinogen, trypsinogen, and chymotrypsinogen, respectively), several factors of the blood clotting and blood clot dissolution cascades (see Chapter 51), and the connective tissue protein collagen (proprotein = procollagen).

### Proenzymes Facilitate Rapid Mobilization of an Activity in Response to Physiologic Demand

The synthesis and secretion of proteases as catalytically inactive proenzymes protects the tissue of origin (eg, the pancreas) from autodigestion, such as can occur in pancreatitis. Certain physiologic processes such as digestion are intermittent but fairly regular and predictable. Others such as blood clot formation, clot dissolution, and tissue repair are brought "on line" only in response to pressing physiologic or pathophysiologic need. The processes of blood clot formation and dissolution clearly must be temporally coordinated to achieve homeostasis. Enzymes needed intermittently but rapidly often are secreted in an initially inactive form since the secretion process or new synthesis of the required proteins might be insufficiently rapid for response to a pressing pathophysiologic demand such as the loss of blood.

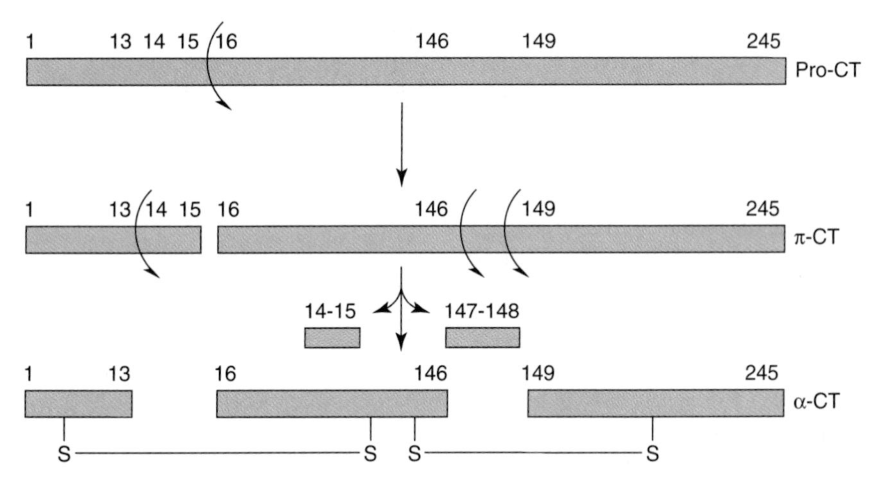


***Figure 9–6.*** Selective proteolysis and associated conformational changes form the active site of chymotrypsin, which includes the Asp102-His57-Ser195 catalytic triad. Successive proteolysis forms prochymotrypsin (pro-CT), π-chymotrypsin (π-CT), and ultimately α-chymotrypsin (α-CT), an active protease whose three peptides remain associated by covalent inter-chain disulfide bonds.

### Activation of Prochymotrypsin Requires Selective Proteolysis

Selective proteolysis involves one or more highly specific proteolytic clips that may or may not be accompanied by separation of the resulting peptides. Most importantly, selective proteolysis often results in conformational changes that "create" the catalytic site of an enzyme. Note that while His 57 and Asp 102 reside on the B peptide of α-chymotrypsin, Ser 195 resides on the C peptide (Figure 9–6). The conformational changes that accompany selective proteolysis of prochymotrypsin (chymotrypsinogen) align the three residues of the charge-relay network, creating the catalytic site. Note also that contact and catalytic residues can be located on different peptide chains but still be within bond-forming distance of bound substrate.

## REVERSIBLE COVALENT MODIFICATION REGULATES KEY MAMMALIAN ENZYMES

Mammalian proteins are the targets of a wide range of covalent modification processes. Modifications such as glycosylation, hydroxylation, and fatty acid acylation introduce new structural features into newly synthesized proteins that tend to persist for the lifetime of the protein. Among the covalent modifications that regulate protein function (eg, methylation, adenylylation), the most common by far is phosphorylation-dephosphorylation. **Protein kinases** phosphorylate proteins by catalyzing transfer of the terminal phosphoryl group of ATP to the hydroxyl groups of seryl, threonyl, or tyrosyl residues, forming *O*-phosphoseryl, *O*-phosphothreonyl, or *O*-phosphotyrosyl residues, respectively (Figure 9–7). Some protein kinases target the side chains of histidyl, lysyl, arginyl, and aspartyl residues. The unmodified form of the protein can be regenerated by hydrolytic removal of phosphoryl groups, catalyzed by **protein phosphatases.**

A typical mammalian cell possesses over 1000 phosphorylated proteins and several hundred protein kinases and protein phosphatases that catalyze their interconversion. The ease of interconversion of enzymes between their phospho- and dephospho- forms in part

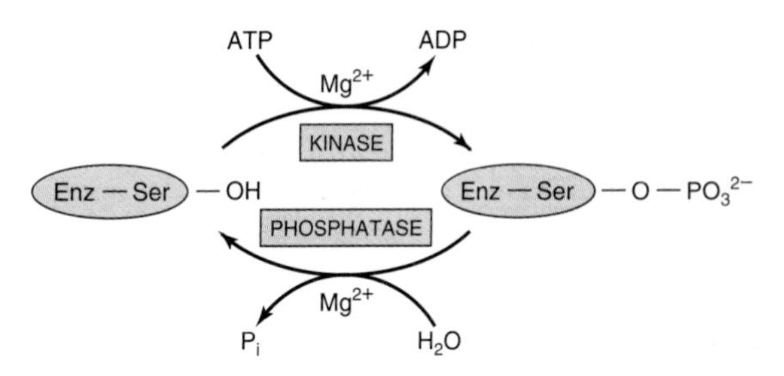


***Figure 9–7.*** Covalent modification of a regulated enzyme by phosphorylation-dephosphorylation of a seryl residue.

accounts for the frequency of phosphorylation-dephosphorylation as a mechanism for regulatory control. Phosphorylation-dephosphorylation permits the functional properties of the affected enzyme to be altered only for as long as it serves a specific need. Once the need has passed, the enzyme can be converted back to its original form, poised to respond to the next stimulatory event. A second factor underlying the widespread use of protein phosphorylation-dephosphorylation lies in the chemical properties of the phosphoryl group itself. In order to alter an enzyme's functional properties, any modification of its chemical structure must influence the protein's three-dimensional configuration. The high charge density of protein-bound phosphoryl groups—generally −2 at physiologic pH—and their propensity to form salt bridges with arginyl residues make them potent agents for modifying protein structure and function. Phosphorylation generally targets amino acids distant from the catalytic site itself. Consequent conformational changes then influence an enzyme's intrinsic catalytic efficiency or other properties. In this sense, the sites of phosphorylation and other covalent modifications can be considered another form of allosteric site. However, in this case the "allosteric ligand" binds covalently to the protein.

## PROTEIN PHOSPHORYLATION IS EXTREMELY VERSATILE

Protein phosphorylation-dephosphorylation is a highly versatile and selective process. Not all proteins are subject to phosphorylation, and of the many hydroxyl groups on a protein's surface, only one or a small subset are targeted. While the most common enzyme function affected is the protein's catalytic efficiency, phosphorylation can also alter the affinity for substrates, location within the cell, or responsiveness to regulation by allosteric ligands. Phosphorylation can increase an enzyme's catalytic efficiency, converting it to its active form in one protein, while phosphorylation of another converts it into an intrinsically inefficient, or inactive, form (Table 9–1).

Many proteins can be phosphorylated at multiple sites or are subject to regulation both by phosphorylation-dephosphorylation and by the binding of allosteric ligands. Phosphorylation-dephosphorylation at any one site can be catalyzed by multiple protein kinases or protein phosphatases. Many protein kinases and most protein phosphatases act on more than one protein and are themselves interconverted between active and inactive forms by the binding of second messengers or by covalent modification by phosphorylation-dephosphorylation.

The interplay between protein kinases and protein phosphatases, between the functional consequences of phosphorylation at different sites, or between phosphorylation sites and allosteric sites provides the basis for regulatory networks that integrate multiple environmental input signals to evoke an appropriate coordinated cellular response. In these sophisticated regulatory networks, individual enzymes respond to different environmental signals. For example, if an enzyme can be phosphorylated at a single site by more than one protein kinase, it can be converted from a catalytically efficient to an inefficient (inactive) form, or vice versa, in response to any one of several signals. If the protein kinase is activated in response to a signal different from the signal that activates the protein phosphatase, the phosphoprotein becomes a decision node. The functional output, generally catalytic activity, reflects the phosphorylation state. This state or degree of phosphorylation is determined by the relative activities of the protein kinase and protein phosphatase, a reflection of the presence and relative strength of the environmental signals that act through each. The ability of many protein kinases and protein phosphatases to target more than one protein provides a means for an environmental signal to coordinately regulate multiple metabolic processes. For example, the enzymes 3-hydroxy-3-methylglutaryl-CoA reductase and acetyl-CoA carboxylase—the rate-controlling enzymes for cholesterol and fatty acid biosynthesis, respectively—are phosphorylated and inactivated by the AMP-activated protein kinase. When this protein kinase is activated either through phosphorylation by yet another protein kinase or in response to the binding of its allosteric activator 5′-AMP, the two major pathways responsible for the synthesis of lipids from acetyl-CoA both are inhibited. Interconvertible enzymes and the enzymes responsible for their interconversion do not act as mere on and off switches working independently of one another.

***Table 9–1.*** Examples of mammalian enzymes whose catalytic activity is altered by covalent phosphorylation-dephosphorylation.

| Enzyme | Activity State[1] | |
|---|---|---|
| | Low | High |
| Acetyl-CoA carboxylase | EP | E |
| Glycogen synthase | EP | E |
| Pyruvate dehydrogenase | EP | E |
| HMG-CoA reductase | EP | E |
| Glycogen phosphorylase | E | EP |
| Citrate lyase | E | EP |
| Phosphorylase b kinase | E | EP |
| HMG-CoA reductase kinase | E | EP |

[1]E, dephosphoenzyme; EP, phosphoenzyme.

They form the building blocks of biomolecular computers that maintain homeostasis in cells that carry out a complex array of metabolic processes that must be regulated in response to a broad spectrum of environmental factors.

### Covalent Modification Regulates Metabolite Flow

Regulation of enzyme activity by phosphorylation-dephosphorylation has analogies to regulation by feedback inhibition. Both provide for short-term, readily reversible regulation of metabolite flow in response to specific physiologic signals. Both act without altering gene expression. Both act on early enzymes of a protracted, often biosynthetic metabolic sequence, and both act at allosteric rather than catalytic sites. Feedback inhibition, however, involves a single protein and lacks hormonal and neural features. By contrast, regulation of mammalian enzymes by phosphorylation-dephosphorylation involves several proteins and ATP and is under direct neural and hormonal control.

## SUMMARY

- Homeostasis involves maintaining a relatively constant intracellular and intra-organ environment despite wide fluctuations in the external environment via appropriate changes in the rates of biochemical reactions in response to physiologic need.
- The substrates for most enzymes are usually present at a concentration close to $K_m$. This facilitates passive control of the rates of product formation response to changes in levels of metabolic intermediates.
- Active control of metabolite flux involves changes in the concentration, catalytic activity, or both of an enzyme that catalyzes a committed, rate-limiting reaction.
- Selective proteolysis of catalytically inactive proenzymes initiates conformational changes that form the active site. Secretion as an inactive proenzyme facilitates rapid mobilization of activity in response to injury or physiologic need and may protect the tissue of origin (eg, autodigestion by proteases).
- Binding of metabolites and second messengers to sites distinct from the catalytic site of enzymes triggers conformational changes that alter $V_{max}$ or the $K_m$.
- Phosphorylation by protein kinases of specific seryl, threonyl, or tyrosyl residues—and subsequent dephosphorylation by protein phosphatases—regulates the activity of many human enzymes. The protein kinases and phosphatases that participate in regulatory cascades which respond to hormonal or second messenger signals constitute a "bio-organic computer" that can process and integrate complex environmental information to produce an appropriate and comprehensive cellular response.

## REFERENCES


Bray D: Protein molecules as computational elements in living cells. Nature 1995;376:307.

Graves DJ, Martin BL, Wang JH: *Co- and Post-translational Modification of Proteins: Chemical Principles and Biological Effects.* Oxford Univ Press, 1994.

Johnson LN, Barford D: The effect of phosphorylation on the structure and function of proteins. Annu Rev Biophys Biomol Struct 1993;22:199.

Marks F (editor): *Protein Phosphorylation.* VCH Publishers, 1996.

Pilkis SJ et al: 6-Phosphofructo-2-kinase/fructose-2,6-bisphosphatase: A metabolic signaling enzyme. Annu Rev Biochem 1995;64:799.

Scriver CR et al (editors): *The Metabolic and Molecular Bases of Inherited Disease,* 8th ed. McGraw-Hill, 2000.

Sitaramayya A (editor): *Introduction to Cellular Signal Transduction.* Birkhauser, 1999.

Stadtman ER, Chock PB (editors): *Current Topics in Cellular Regulation.* Academic Press, 1969 to the present.

Weber G (editor): *Advances in Enzyme Regulation.* Pergamon Press, 1963 to the present.

# SECTION II

# Bioenergetics & the Metabolism of Carbohydrates & Lipids

## Bioenergetics: The Role of ATP — 10

*Peter A. Mayes, PhD, DSc, & Kathleen M. Botham, PhD, DSc*

### BIOMEDICAL IMPORTANCE

Bioenergetics, or biochemical thermodynamics, is the study of the energy changes accompanying biochemical reactions. Biologic systems are essentially **isothermic** and use chemical energy to power living processes. How an animal obtains suitable fuel from its food to provide this energy is basic to the understanding of normal nutrition and metabolism. Death from **starvation** occurs when available energy reserves are depleted, and certain forms of malnutrition are associated with energy imbalance **(marasmus).** Thyroid hormones control the rate of energy release (metabolic rate), and disease results when they malfunction. Excess storage of surplus energy causes **obesity,** one of the most common diseases of Western society.

### FREE ENERGY IS THE USEFUL ENERGY IN A SYSTEM

Gibbs change in free energy ($\Delta G$) is that portion of the total energy change in a system that is available for doing work—ie, the useful energy, also known as the chemical potential.

#### Biologic Systems Conform to the General Laws of Thermodynamics

The first law of thermodynamics states that **the total energy of a system, including its surroundings, remains constant.** It implies that within the total system, energy is neither lost nor gained during any change. However, energy may be transferred from one part of the system to another or may be transformed into another form of energy. In living systems, chemical energy may be transformed into heat or into electrical, radiant, or mechanical energy.

The second law of thermodynamics states that **the total entropy of a system must increase if a process is to occur spontaneously. Entropy** is the extent of disorder or randomness of the system and becomes maximum as equilibrium is approached. Under conditions of constant temperature and pressure, the relationship between the free energy change ($\Delta G$) of a reacting system and the change in entropy ($\Delta S$) is expressed by the following equation, which combines the two laws of thermodynamics:

$$\Delta G = \Delta H - T\Delta S$$

where $\Delta H$ is the change in **enthalpy** (heat) and T is the absolute temperature.

In biochemical reactions, because $\Delta H$ is approximately equal to $\Delta E$, the total change in internal energy of the reaction, the above relationship may be expressed in the following way:

$$\Delta G = \Delta E - T\Delta S$$

If $\Delta G$ is negative, the reaction proceeds spontaneously with loss of free energy; ie, it is **exergonic.** If, in addition, $\Delta G$ is of great magnitude, the reaction goes virtually to completion and is essentially irreversible. On the other hand, if $\Delta G$ is positive, the reaction proceeds only if free energy can be gained; ie, it is **endergonic.** If, in addition, the magnitude of $\Delta G$ is great, the system is stable, with little or no tendency for a reaction to occur. If ΔG is zero, the system is at equilibrium and no net change takes place.

When the reactants are present in concentrations of 1.0 mol/L, $\Delta G^0$ is the standard free energy change. For biochemical reactions, a standard state is defined as having a pH of 7.0. The standard free energy change at this standard state is denoted by $\Delta G^{0'}$.

The standard free energy change can be calculated from the equilibrium constant $K_{eq}$.

$$\Delta G^{0'} = -RT \ln K'_{eq}$$

where R is the gas constant and T is the absolute temperature (Chapter 8). It is important to note that the actual ΔG may be larger or smaller than $\Delta G^{0'}$ depending on the concentrations of the various reactants, including the solvent, various ions, and proteins.

In a biochemical system, an enzyme only speeds up the attainment of equilibrium; it never alters the final concentrations of the reactants at equilibrium.

## ENDERGONIC PROCESSES PROCEED BY COUPLING TO EXERGONIC PROCESSES

The vital processes—eg, synthetic reactions, muscular contraction, nerve impulse conduction, and active transport—obtain energy by chemical linkage, or **coupling,** to oxidative reactions. In its simplest form, this type of coupling may be represented as shown in Figure 10–1. The conversion of metabolite A to metabolite B occurs with release of free energy. It is coupled to another reaction, in which free energy is required to convert metabolite C to metabolite D. The terms **exergonic** and **endergonic** rather than the normal chemical terms "exothermic" and "endothermic" are used to indicate that a process is accompanied by loss or gain, respectively, of free energy in any form, not necessarily as heat. In practice, an endergonic process cannot exist independently but must be a component of a coupled exergonic-endergonic system where the overall net change is exergonic. The exergonic reactions are termed **catabolism** (generally, the breakdown or oxidation of fuel molecules), whereas the synthetic reactions that build up substances are termed **anabolism.** The combined catabolic and anabolic processes constitute **metabolism.**

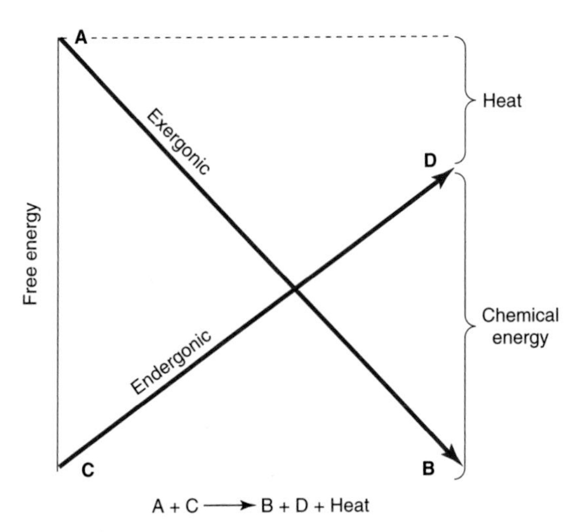


***Figure 10–1.*** Coupling of an exergonic to an endergonic reaction.

If the reaction shown in Figure 10–1 is to go from left to right, then the overall process must be accompanied by loss of free energy as heat. One possible mechanism of coupling could be envisaged if a common obligatory intermediate (I) took part in both reactions, ie,

$$A + C \rightarrow I \rightarrow B + D$$

Some exergonic and endergonic reactions in biologic systems are coupled in this way. This type of system has a built-in mechanism for biologic control of the rate of oxidative processes since the common obligatory intermediate allows the rate of utilization of the product of the synthetic path (D) to determine by mass action the rate at which A is oxidized. Indeed, these relationships supply a basis for the concept of **respiratory control,** the process that prevents an organism from burning out of control. An extension of the coupling concept is provided by dehydrogenation reactions, which are coupled to hydrogenations by an intermediate carrier (Figure 10–2).

An alternative method of coupling an exergonic to an endergonic process is to synthesize a compound of high-energy potential in the exergonic reaction and to incorporate this new compound into the endergonic reaction, thus effecting a transference of free energy from the exergonic to the endergonic pathway (Figure 10–3). The biologic advantage of this mechanism is that the compound of high potential energy, ~Ⓔ, unlike I

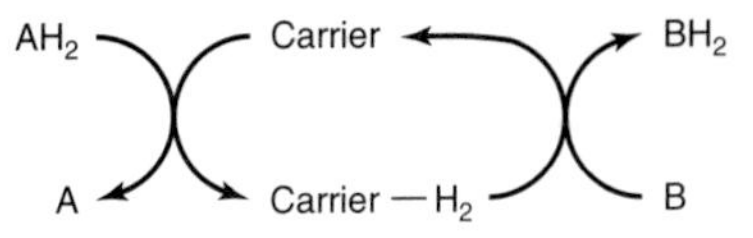

$AH_2$ — Carrier — $BH_2$
A — Carrier — $H_2$ — B

***Figure 10–2.*** Coupling of dehydrogenation and hydrogenation reactions by an intermediate carrier.

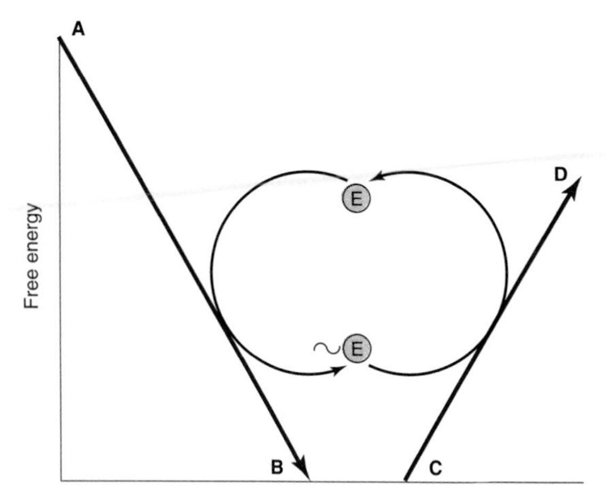


***Figure 10–3.*** Transfer of free energy from an exergonic to an endergonic reaction via a high-energy intermediate compound (~Ⓔ).

in the previous system, need not be structurally related to A, B, C, or D, allowing Ⓔ to serve as a transducer of energy from a wide range of exergonic reactions to an equally wide range of endergonic reactions or processes, such as biosyntheses, muscular contraction, nervous excitation, and active transport. In the living cell, the principal high-energy intermediate or carrier compound (designated ~Ⓔ in Figure 10–3) is **adenosine triphosphate (ATP).**

## HIGH-ENERGY PHOSPHATES PLAY A CENTRAL ROLE IN ENERGY CAPTURE AND TRANSFER

In order to maintain living processes, all organisms must obtain supplies of free energy from their environment. **Autotrophic** organisms utilize simple exergonic processes; eg, the energy of sunlight (green plants), the reaction $Fe^{2+} \rightarrow Fe^{3+}$ (some bacteria). On the other hand, **heterotrophic** organisms obtain free energy by coupling their metabolism to the breakdown of complex organic molecules in their environment. In all these organisms, ATP plays a central role in the transference of free energy from the exergonic to the endergonic processes (Figure 10–3). ATP is a nucleoside triphosphate containing adenine, ribose, and three phosphate groups. In its reactions in the cell, it functions as the $Mg^{2+}$ complex (Figure 10–4).

The importance of phosphates in intermediary metabolism became evident with the discovery of the role of ATP, adenosine diphosphate (ADP), and inorganic phosphate ($P_i$) in glycolysis (Chapter 17).

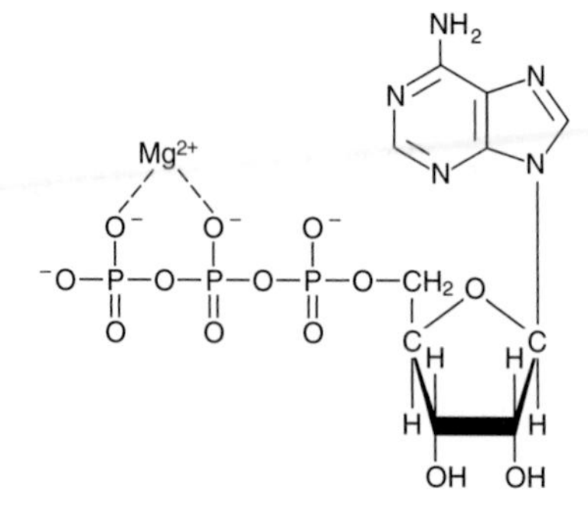


***Figure 10–4.*** Adenosine triphosphate (ATP) shown as the magnesium complex. ADP forms a similar complex with $Mg^{2+}$.

### The Intermediate Value for the Free Energy of Hydrolysis of ATP Has Important Bioenergetic Significance

The standard free energy of hydrolysis of a number of biochemically important phosphates is shown in Table 10–1. An estimate of the comparative tendency of each of the phosphate groups to transfer to a suitable acceptor may be obtained from the $\Delta G^{0'}$ of hydrolysis at 37 °C. The value for the hydrolysis of the terminal

***Table 10–1.*** Standard free energy of hydrolysis of some organophosphates of biochemical importance.[1,2]

| Compound | $\Delta G^{0'}$ | |
|---|---|---|
| | **kJ/mol** | **kcal/mol** |
| Phosphoenolpyruvate | –61.9 | –14.8 |
| Carbamoyl phosphate | –51.4 | –12.3 |
| 1,3-Bisphosphoglycerate (to 3-phosphoglycerate) | –49.3 | –11.8 |
| Creatine phosphate | –43.1 | –10.3 |
| $ATP \rightarrow ADP + P_i$ | –30.5 | –7.3 |
| $ADP \rightarrow AMP + P_i$ | –27.6 | –6.6 |
| Pyrophosphate | –27.6 | –6.6 |
| Glucose 1-phosphate | –20.9 | –5.0 |
| Fructose 6-phosphate | –15.9 | –3.8 |
| AMP | –14.2 | –3.4 |
| Glucose 6-phosphate | –13.8 | –3.3 |
| Glycerol 3-phosphate | –9.2 | –2.2 |

[1] $P_i$, inorganic orthophosphate.
[2] Values for ATP and most others taken from Krebs and Kornberg (1957). They differ between investigators depending on the precise conditions under which the measurements are made.

phosphate of ATP divides the list into two groups. **Low-energy phosphates,** exemplified by the ester phosphates found in the intermediates of glycolysis, have $\Delta G^{0'}$ values smaller than that of ATP, while in **high-energy phosphates** the value is higher than that of ATP. The components of this latter group, including ATP, are usually anhydrides (eg, the 1-phosphate of 1,3-bisphosphoglycerate), enolphosphates (eg, phosphoenolpyruvate), and phosphoguanidines (eg, creatine phosphate, arginine phosphate). The intermediate position of ATP allows it to play an important role in energy transfer. The high free energy change on hydrolysis of ATP is due to relief of charge repulsion of adjacent negatively charged oxygen atoms and to stabilization of the reaction products, especially phosphate, as resonance hybrids. Other "high-energy compounds" are thiol esters involving coenzyme A (eg, acetyl-CoA), acyl carrier protein, amino acid esters involved in protein synthesis, *S*-adenosylmethionine (active methionine), UDPGlc (uridine diphosphate glucose), and PRPP (5-phosphoribosyl-1-pyrophosphate).

### High-Energy Phosphates Are Designated by ~ Ⓟ

The symbol ~Ⓟ indicates that the group attached to the bond, on transfer to an appropriate acceptor, results in transfer of the larger quantity of free energy. For this reason, the term **group transfer potential** is preferred by some to "high-energy bond." Thus, ATP contains two high-energy phosphate groups and ADP contains one, whereas the phosphate in AMP (adenosine monophosphate) is of the low-energy type, since it is a normal ester link (Figure 10–5).

## HIGH-ENERGY PHOSPHATES ACT AS THE "ENERGY CURRENCY" OF THE CELL

ATP is able to act as a donor of high-energy phosphate to form those compounds below it in Table 10–1. Likewise, with the necessary enzymes, ADP can accept high-energy phosphate to form ATP from those compounds above ATP in the table. In effect, an **ATP/ADP cycle** connects those processes that generate ~Ⓟ to those processes that utilize ~Ⓟ (Figure 10–6), continuously consuming and regenerating ATP. This occurs at a very rapid rate, since the total ATP/ADP pool is extremely small and sufficient to maintain an active tissue for only a few seconds.

There are three major sources of ~Ⓟ taking part in **energy conservation** or **energy capture:**

**(1) Oxidative phosphorylation:** The greatest quantitative source of ~Ⓟ in aerobic organisms. Free energy

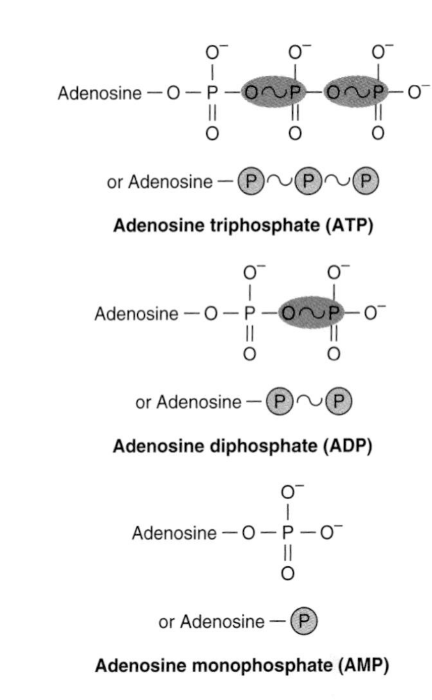


***Figure 10–5.*** Structure of ATP, ADP, and AMP showing the position and the number of high-energy phosphates (~Ⓟ).

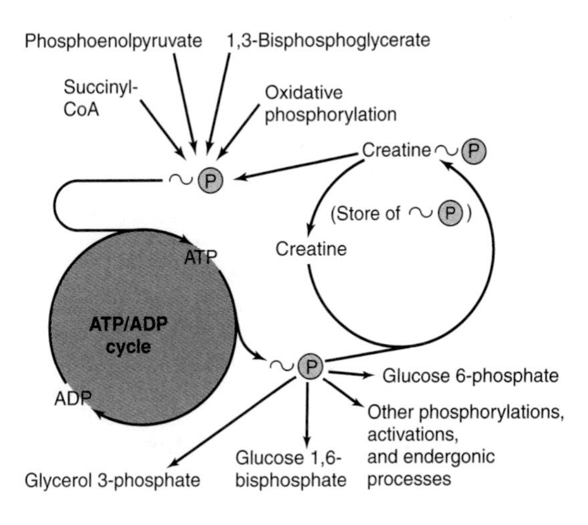


***Figure 10–6.*** Role of ATP/ADP cycle in transfer of high-energy phosphate.

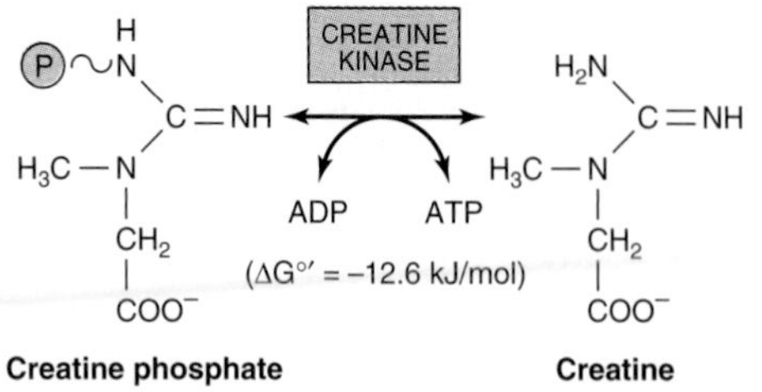

CREATINE KINASE

Creatine phosphate ($H_3C-N(CH_2COO^-)-C(=NH)-NH\sim P$) + ADP ⇌ Creatine ($H_3C-N(CH_2COO^-)-C(=NH)-NH_2$) + ATP

($\Delta G^{0'} = -12.6$ kJ/mol)

**Creatine phosphate** **Creatine**

***Figure 10–7.*** Transfer of high-energy phosphate between ATP and creatine.

comes from respiratory chain oxidation using molecular $O_2$ within mitochondria (Chapter 11).

**(2) Glycolysis:** A net formation of two ~Ⓟ results from the formation of lactate from one molecule of glucose, generated in two reactions catalyzed by phosphoglycerate kinase and pyruvate kinase, respectively (Figure 17–2).

**(3) The citric acid cycle:** One ~Ⓟ is generated directly in the cycle at the succinyl thiokinase step (Figure 16–3).

**Phosphagens** act as storage forms of high-energy phosphate and include creatine phosphate, occurring in vertebrate skeletal muscle, heart, spermatozoa, and brain; and arginine phosphate, occurring in invertebrate muscle. When ATP is rapidly being utilized as a source of energy for muscular contraction, phosphagens permit its concentrations to be maintained, but when the ATP/ADP ratio is high, their concentration can increase to act as a store of high-energy phosphate (Figure 10–7).

When ATP acts as a phosphate donor to form those compounds of lower free energy of hydrolysis (Table 10–1), the phosphate group is invariably converted to one of low energy, eg,

GLYCEROL KINASE

Glycerol + Adenosine — Ⓟ ~ Ⓟ ~ Ⓟ → Glycerol— Ⓟ + Adenosine— Ⓟ ~ Ⓟ

## ATP Allows the Coupling of Thermodynamically Unfavorable Reactions to Favorable Ones

The phosphorylation of glucose to glucose 6-phosphate, the first reaction of glycolysis (Figure 17–2), is highly endergonic and cannot proceed under physiologic conditions.

$$(1)\ \text{Glucose} + P_i \rightarrow \text{Glucose 6-phosphate} + H_2O \quad (\Delta G^{0'} = +13.8\ \text{kJ/mol})$$

To take place, the reaction must be coupled with another—more exergonic—reaction such as the hydrolysis of the terminal phosphate of ATP.

$$(2)\ \text{ATP} \rightarrow \text{ADP} + P_i \quad (\Delta G^{0'} = -30.5\ \text{kJ/mol})$$

When (1) and (2) are coupled in a reaction catalyzed by hexokinase, phosphorylation of glucose readily proceeds in a highly exergonic reaction that under physiologic conditions is irreversible. Many "activation" reactions follow this pattern.

## Adenylyl Kinase (Myokinase) Interconverts Adenine Nucleotides

This enzyme is present in most cells. It catalyzes the following reaction:

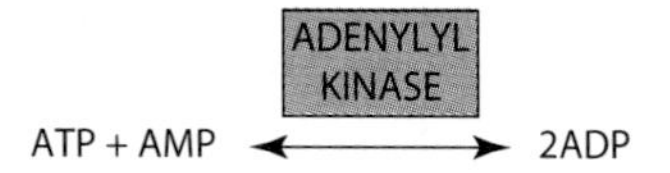


This allows:

**(1)** High-energy phosphate in ADP to be used in the synthesis of ATP.

**(2)** AMP, formed as a consequence of several activating reactions involving ATP, to be recovered by rephosphorylation to ADP.

**(3)** AMP to increase in concentration when ATP becomes depleted and act as a metabolic (allosteric) signal to increase the rate of catabolic reactions, which in turn lead to the generation of more ATP (Chapter 19).

## When ATP Forms AMP, Inorganic Pyrophosphate ($PP_i$) Is Produced

This occurs, for example, in the activation of long-chain fatty acids (Chapter 22):

ACYL-CoA SYNTHETASE

ATP + CoA • SH + R • COOH → AMP + $PP_i$ + R • CO — SCoA

This reaction is accompanied by loss of free energy as heat, which ensures that the activation reaction will go to the right; and is further aided by the hydrolytic splitting of $PP_i$, catalyzed by **inorganic pyrophosphatase,** a reaction that itself has a large $\Delta G^{0'}$ of −27.6 kJ/

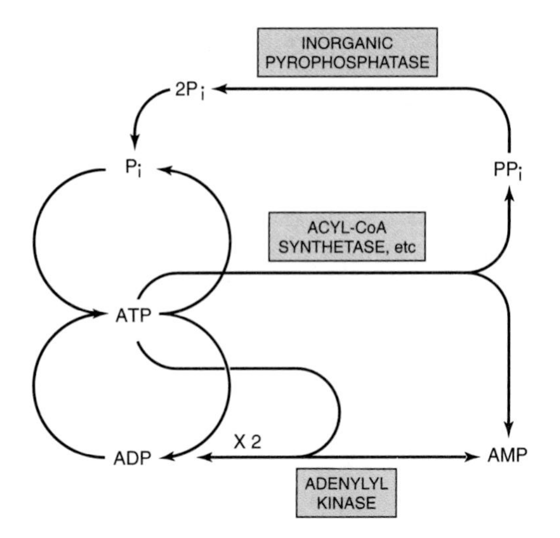


***Figure 10–8.*** Phosphate cycles and interchange of adenine nucleotides.

mol. Note that activations via the pyrophosphate pathway result in the loss of two ~Ⓟ rather than one ~Ⓟ as occurs when ADP and $P_i$ are formed.

INORGANIC PYROPHOSPHATASE

$$PP_i + H_2O \longrightarrow 2\,P_i$$

A combination of the above reactions makes it possible for phosphate to be recycled and the adenine nucleotides to interchange (Figure 10–8).

### Other Nucleoside Triphosphates Participate in the Transfer of High-Energy Phosphate

By means of the enzyme **nucleoside diphosphate kinase,** UTP, GTP, and CTP can be synthesized from their diphosphates, eg,

NUCLEOSIDE DIPHOSPHATE KINASE

$$\text{ATP} + \text{UDP} \longleftrightarrow \text{ADP} + \text{UTP}$$

(uridine triphosphate)

All of these triphosphates take part in phosphorylations in the cell. Similarly, specific nucleoside monophosphate kinases catalyze the formation of nucleoside diphosphates from the corresponding monophosphates.

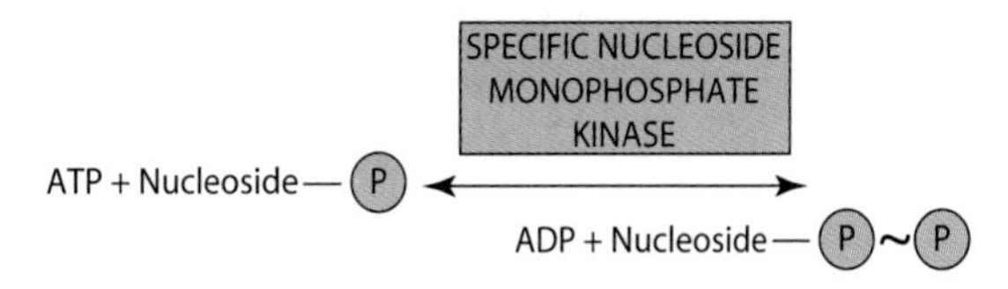


Thus, adenylyl kinase is a specialized monophosphate kinase.

## SUMMARY

- Biologic systems use chemical energy to power the living processes.
- Exergonic reactions take place spontaneously with loss of free energy ($\Delta G$ is negative). Endergonic reactions require the gain of free energy ($\Delta G$ is positive) and only occur when coupled to exergonic reactions.
- ATP acts as the "energy currency" of the cell, transferring free energy derived from substances of higher energy potential to those of lower energy potential.

## REFERENCES

de Meis L: The concept of energy-rich phosphate compounds: Water, transport ATPases, and entropy energy. Arch Biochem Biophys 1993;306:287.

Ernster L (editor): *Bioenergetics.* Elsevier, 1984.

Harold FM: *The Vital Force: A Study of Bioenergetics.* Freeman, 1986.

Klotz IM: *Introduction to Biomolecular Energetics.* Academic Press, 1986.

Krebs HA, Kornberg HL: *Energy Transformations in Living Matter.* Springer, 1957.

# Biologic Oxidation

11

*Peter A. Mayes, PhD, DSc, & Kathleen M. Botham, PhD, DSc*

## BIOMEDICAL IMPORTANCE

Chemically, **oxidation** is defined as the removal of electrons and **reduction** as the gain of electrons. Thus, oxidation is always accompanied by reduction of an electron acceptor. This principle of oxidation-reduction applies equally to biochemical systems and is an important concept underlying understanding of the nature of biologic oxidation. Note that many biologic oxidations can take place without the participation of molecular oxygen, eg, dehydrogenations. The life of higher animals is absolutely dependent upon a supply of oxygen for **respiration,** the process by which cells derive energy in the form of ATP from the controlled reaction of hydrogen with oxygen to form water. In addition, molecular oxygen is incorporated into a variety of substrates by enzymes designated as **oxygenases;** many drugs, pollutants, and chemical carcinogens (xenobiotics) are metabolized by enzymes of this class, known as the **cytochrome P450 system.** Administration of oxygen can be lifesaving in the treatment of patients with respiratory or circulatory failure.

## FREE ENERGY CHANGES CAN BE EXPRESSED IN TERMS OF REDOX POTENTIAL

In reactions involving oxidation and reduction, the free energy change is proportionate to the tendency of reactants to donate or accept electrons. Thus, in addition to expressing free energy change in terms of $\Delta G^{0'}$ (Chapter 10), it is possible, in an analogous manner, to express it numerically as an **oxidation-reduction** or **redox potential** ($E'_0$). The redox potential of a system ($E_0$) is usually compared with the potential of the hydrogen electrode (0.0 volts at pH 0.0). However, for biologic systems, the redox potential ($E'_0$)is normally expressed at pH 7.0, at which pH the electrode potential of the hydrogen electrode is −0.42 volts. The redox potentials of some redox systems of special interest in mammalian biochemistry are shown in Table 11–1. The relative positions of redox systems in the table allows prediction of the direction of flow of electrons from one redox couple to another.

Enzymes involved in oxidation and reduction are called **oxidoreductases** and are classified into four groups: **oxidases, dehydrogenases, hydroperoxidases,** and **oxygenases.**

## OXIDASES USE OXYGEN AS A HYDROGEN ACCEPTOR

Oxidases catalyze the removal of hydrogen from a substrate using oxygen as a hydrogen acceptor.* They form water or hydrogen peroxide as a reaction product (Figure 11–1).

### Some Oxidases Contain Copper

**Cytochrome oxidase** is a hemoprotein widely distributed in many tissues, having the typical heme prosthetic group present in myoglobin, hemoglobin, and other cytochromes (Chapter 6). It is the terminal component of the chain of respiratory carriers found in mitochondria and transfers electrons resulting from the oxidation of substrate molecules by dehydrogenases to their final acceptor, oxygen. The enzyme is poisoned by carbon monoxide, cyanide, and hydrogen sulfide. It has also been termed cytochrome $a_3$. It is now known that cytochromes $a$ and $a_3$ are combined in a single protein, and the complex is known as **cytochrome $aa_3$.** It contains two molecules of heme, each having one Fe atom that oscillates between $Fe^{3+}$ and $Fe^{2+}$ during oxidation and reduction. Furthermore, two atoms of Cu are present, each associated with a heme unit.

### Other Oxidases Are Flavoproteins

Flavoprotein enzymes contain **flavin mononucleotide (FMN)** or **flavin adenine dinucleotide (FAD)** as prosthetic groups. FMN and FAD are formed in the body from the vitamin **riboflavin** (Chapter 45). FMN and FAD are usually tightly—but not covalently—bound to their respective apoenzyme proteins. Metalloflavoproteins contain one or more metals as essential cofactors.

Examples of flavoprotein enzymes include **L-amino acid oxidase,** an FMN-linked enzyme found in kidney with general specificity for the oxidative deamination of

* The term "oxidase" is sometimes used collectively to denote all enzymes that catalyze reactions involving molecular oxygen.

***Table 11–1.*** Some redox potentials of special interest in mammalian oxidation systems.

| System | $E'_0$ Volts |
|---|---|
| $H^+/H_2$ | −0.42 |
| $NAD^+$/NADH | −0.32 |
| Lipoate; ox/red | −0.29 |
| Acetoacetate/3-hydroxybutyrate | −0.27 |
| Pyruvate/lactate | −0.19 |
| Oxaloacetate/malate | −0.17 |
| Fumarate/succinate | +0.03 |
| Cytochrome *b*; $Fe^{3+}/Fe^{2+}$ | +0.08 |
| Ubiquinone; ox/red | +0.10 |
| Cytochrome $c_1$; $Fe^{3+}/Fe^{2+}$ | +0.22 |
| Cytochrome *a*; $Fe^{3+}/Fe^{2+}$ | +0.29 |
| Oxygen/water | +0.82 |

the naturally occurring L-amino acids; **xanthine oxidase,** which contains molybdenum and plays an important role in the conversion of purine bases to uric acid (Chapter 34), and is of particular significance in uricotelic animals (Chapter 29); and **aldehyde dehydrogenase,** an FAD-linked enzyme present in mammalian livers, which contains molybdenum and nonheme iron and acts upon aldehydes and N-heterocyclic substrates. The mechanisms of oxidation and reduction of these enzymes are complex. Evidence suggests a two-step reaction as shown in Figure 11–2.

## DEHYDROGENASES CANNOT USE OXYGEN AS A HYDROGEN ACCEPTOR

There are a large number of enzymes in this class. They perform two main functions:

**(1)** Transfer of hydrogen from one substrate to another in a coupled oxidation-reduction reaction (Figure 11–3). These dehydrogenases are specific for their substrates but often utilize common coenzymes or hydrogen carriers, eg, $NAD^+$. Since the reactions are reversible, these properties enable reducing equivalents to be freely transferred within the cell. This type of reaction, which enables one substrate to be oxidized at the expense of another, is particularly useful in enabling oxidative processes to occur in the absence of oxygen, such as during the anaerobic phase of glycolysis (Figure 17–2).

**(2)** As components in the **respiratory chain** of electron transport from substrate to oxygen (Figure 12–3).

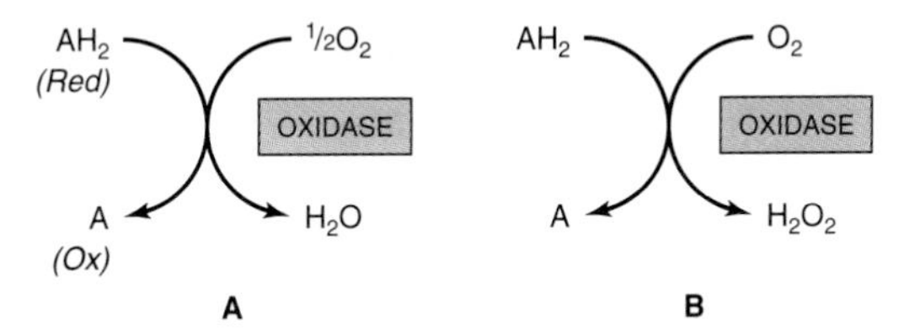


***Figure 11–1.*** Oxidation of a metabolite catalyzed by an oxidase **(A)** forming $H_2O$, **(B)** forming $H_2O_2$.

### Many Dehydrogenases Depend on Nicotinamide Coenzymes

These dehydrogenases use **nicotinamide adenine dinucleotide ($NAD^+$)** or **nicotinamide adenine dinucleotide phosphate ($NADP^+$)**—or both—are formed in the body from the vitamin **niacin** (Chapter 45). The coenzymes are reduced by the specific substrate of the dehydrogenase and reoxidized by a suitable electron acceptor (Figure 11–4).They may freely and reversibly dissociate from their respective apoenzymes.

Generally, NAD-linked dehydrogenases catalyze oxidoreduction reactions in the oxidative pathways of metabolism, particularly in glycolysis, in the citric acid cycle, and in the respiratory chain of mitochondria. NADP-linked dehydrogenases are found characteristically in reductive syntheses, as in the extramitochondrial pathway of fatty acid synthesis and steroid synthesis—and also in the pentose phosphate pathway.

### Other Dehydrogenases Depend on Riboflavin

The flavin groups associated with these dehydrogenases are similar to FMN and FAD occurring in oxidases. They are generally more tightly bound to their apoenzymes than are the nicotinamide coenzymes. Most of the riboflavin-linked dehydrogenases are concerned with electron transport in (or to) the respiratory chain (Chapter 12). **NADH dehydrogenase** acts as a carrier of electrons between NADH and the components of higher redox potential (Figure 12–3). Other dehydrogenases such as **succinate dehydrogenase, acyl-CoA dehydrogenase,** and **mitochondrial glycerol-3-phosphate dehydrogenase** transfer reducing equivalents directly from the substrate to the respiratory chain (Figure 12–4). Another role of the flavin-dependent dehydrogenases is in the dehydrogenation (by **dihydrolipoyl dehydrogenase**) of reduced lipoate, an intermediate in the oxidative decarboxylation of pyruvate and α-ketoglutarate (Figures 12–4 and 17–5). The **electron-transferring flavoprotein** is an intermediary carrier of electrons between acyl-CoA dehydrogenase and the respiratory chain (Figure 12–4).

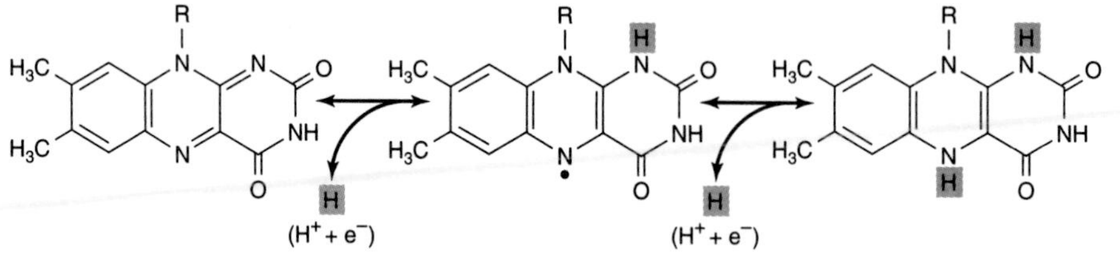

***Figure 11–2.*** Oxidoreduction of isoalloxazine ring in flavin nucleotides via a semiquinone (free radical) intermediate (center).

## Cytochromes May Also Be Regarded as Dehydrogenases

The cytochromes are iron-containing hemoproteins in which the iron atom oscillates between $Fe^{3+}$ and $Fe^{2+}$ during oxidation and reduction. Except for cytochrome oxidase (previously described), they are classified as dehydrogenases. In the respiratory chain, they are involved as carriers of electrons from flavoproteins on the one hand to cytochrome oxidase on the other (Figure 12–4). Several identifiable cytochromes occur in the respiratory chain, ie, cytochromes $b$, $c_1$, $c$, $a$, and $a_3$ (cytochrome oxidase). Cytochromes are also found in other locations, eg, the endoplasmic reticulum (cytochromes P450 and $b_5$), and in plant cells, bacteria, and yeasts.

# HYDROPEROXIDASES USE HYDROGEN PEROXIDE OR AN ORGANIC PEROXIDE AS SUBSTRATE

Two type of enzymes found both in animals and plants fall into this category: **peroxidases** and **catalase.**

Hydroperoxidases protect the body against harmful peroxides. Accumulation of peroxides can lead to generation of free radicals, which in turn can disrupt membranes and perhaps cause cancer and atherosclerosis. (See Chapters 14 and 45.)

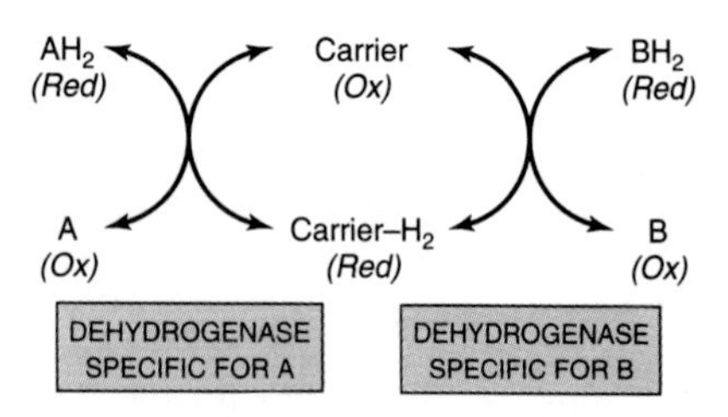


***Figure 11–3.*** Oxidation of a metabolite catalyzed by coupled dehydrogenases.

## Peroxidases Reduce Peroxides Using Various Electron Acceptors

Peroxidases are found in milk and in leukocytes, platelets, and other tissues involved in eicosanoid metabolism (Chapter 23). The prosthetic group is protoheme. In the reaction catalyzed by peroxidase, hydrogen peroxide is reduced at the expense of several substances that will act as electron acceptors, such as ascorbate, quinones, and cytochrome $c$. The reaction catalyzed by peroxidase is complex, but the overall reaction is as follows:

$$H_2O_2 + AH_2 \xrightarrow{\text{PEROXIDASE}} 2H_2O + A$$

In erythrocytes and other tissues, the enzyme **glutathione peroxidase,** containing **selenium** as a prosthetic group, catalyzes the destruction of $H_2O_2$ and lipid hydroperoxides by reduced glutathione, protecting membrane lipids and hemoglobin against oxidation by peroxides (Chapter 20).

## Catalase Uses Hydrogen Peroxide as Electron Donor & Electron Acceptor

Catalase is a hemoprotein containing four heme groups. In addition to possessing peroxidase activity, it is able to use one molecule of $H_2O_2$ as a substrate electron donor and another molecule of $H_2O_2$ as an oxidant or electron acceptor.

$$2H_2O_2 \xrightarrow{\text{CATALASE}} 2H_2O + O_2$$

Under most conditions in vivo, the peroxidase activity of catalase seems to be favored. Catalase is found in blood, bone marrow, mucous membranes, kidney, and liver. Its function is assumed to be the destruction of hydrogen peroxide formed by the action of oxidases.

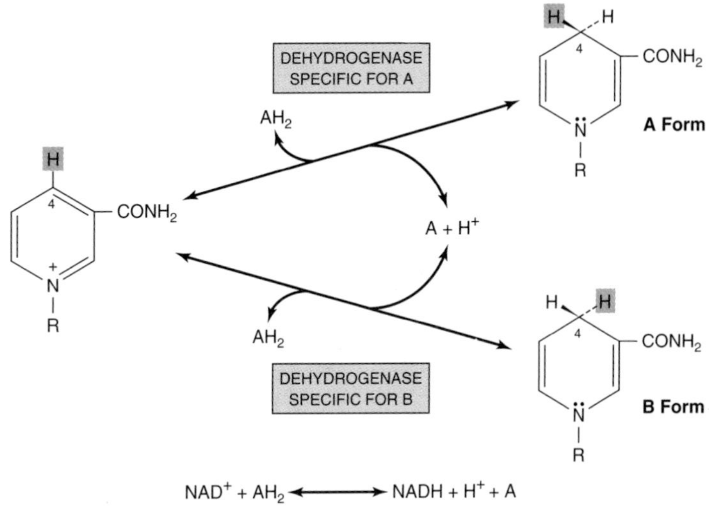

***Figure 11–4.*** Mechanism of oxidation and reduction of nicotinamide coenzymes. There is stereospecificity about position 4 of nicotinamide when it is reduced by a substrate $AH_2$. One of the hydrogen atoms is removed from the substrate as a hydrogen nucleus with two electrons (hydride ion, $H^-$) and is transferred to the 4 position, where it may be attached in either the A or the B position according to the specificity determined by the particular dehydrogenase catalyzing the reaction. The remaining hydrogen of the hydrogen pair removed from the substrate remains free as a hydrogen ion.

**Peroxisomes** are found in many tissues, including liver. They are rich in oxidases and in catalase, Thus, the enzymes that produce $H_2O_2$ are grouped with the enzyme that destroys it. However, mitochondrial and microsomal electron transport systems as well as xanthine oxidase must be considered as additional sources of $H_2O_2$.

## OXYGENASES CATALYZE THE DIRECT TRANSFER & INCORPORATION OF OXYGEN INTO A SUBSTRATE MOLECULE

Oxygenases are concerned with the synthesis or degradation of many different types of metabolites. They catalyze the incorporation of oxygen into a substrate molecule in two steps: (1) oxygen is bound to the enzyme at the active site, and (2) the bound oxygen is reduced or transferred to the substrate. Oxygenases may be divided into two subgroups, as follows.

### Dioxygenases Incorporate Both Atoms of Molecular Oxygen Into the Substrate

The basic reaction is shown below:

$$A + O_2 \rightarrow AO_2$$

Examples include the liver enzymes, **homogentisate dioxygenase** (oxidase) and **3-hydroxyanthranilate dioxygenase** (oxidase), that contain iron; and **L-tryptophan dioxygenase** (tryptophan pyrrolase) (Chapter 30), that utilizes heme.

### Monooxygenases (Mixed-Function Oxidases, Hydroxylases) Incorporate Only One Atom of Molecular Oxygen Into the Substrate

The other oxygen atom is reduced to water, an additional electron donor or cosubstrate (Z) being necessary for this purpose.

$$A—H + O_2 + ZH_2 \rightarrow A—OH + H_2O + Z$$

### Cytochromes P450 Are Monooxygenases Important for the Detoxification of Many Drugs & for the Hydroxylation of Steroids

Cytochromes P450 are an important superfamily of heme-containing monooxgenases, and more than 1000 such enzymes are known. Both NADH and NADPH donate reducing equivalents for the reduction of these cytochromes (Figure 11–5), which in turn are oxidized by substrates in a series of enzymatic reactions collectively known as the **hydroxylase cycle** (Figure 11–6). In liver microsomes, cytochromes P450 are found together with **cytochrome $b_5$** and have an important role in detoxification. Benzpyrene, aminopyrine, aniline, morphine, and benzphetamine are hydroxylated, increasing their solubility and aiding their excretion. Many drugs such as phenobarbital have the ability to induce the formation of microsomal enzymes and of cytochromes P450.

Mitochondrial cytochrome P450 systems are found in steroidogenic tissues such as adrenal cortex, testis, ovary, and placenta and are concerned with the biosyn-

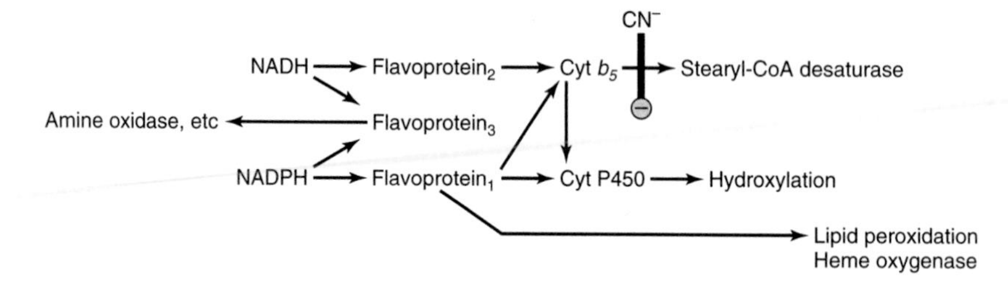


***Figure 11–5.*** Electron transport chain in microsomes. Cyanide ($CN^-$) inhibits the indicated step.

thesis of steroid hormones from cholesterol (hydroxylation at $C_{22}$ and $C_{20}$ in side-chain cleavage and at the 11β and 18 positions). In addition, renal systems catalyzing 1α- and 24-hydroxylations of 25-hydroxycholecalciferol in vitamin D metabolism—and cholesterol 7α-hydroxylase and sterol 27-hydroxylase involved in bile acid biosynthesis in the liver (Chapter 26)—are P450 enzymes.

## SUPEROXIDE DISMUTASE PROTECTS AEROBIC ORGANISMS AGAINST OXYGEN TOXICITY

Transfer of a single electron to $O_2$ generates the potentially damaging **superoxide anion free radical** ($O_2^{\bar{\cdot}}$), the destructive effects of which are amplified by its giving rise to free radical chain reactions (Chapter 14). The ease with which superoxide can be formed from oxygen in tissues and the occurrence of **superoxide dismutase,** the enzyme responsible for its removal in all aerobic organisms (although not in obligate anaerobes) indicate that the potential toxicity of oxygen is due to its conversion to superoxide.

Superoxide is formed when reduced flavins—present, for example, in xanthine oxidase—are reoxidized univalently by molecular oxygen.

$$\text{Enz}-\text{Flavin}-\text{H}_2+\text{O}_2 \rightarrow \text{Enz}-\text{Flavin}-\text{H}+\text{O}_2^{\bar{\cdot}}+\text{H}^+$$

Superoxide can reduce oxidized cytochrome *c*

$$\text{O}_2^{\bar{\cdot}}+\text{Cyt}\ c\ (\text{Fe}^{3+}) \rightarrow \text{O}_2+\text{Cyt}\ c\ (\text{Fe}^{2+})$$

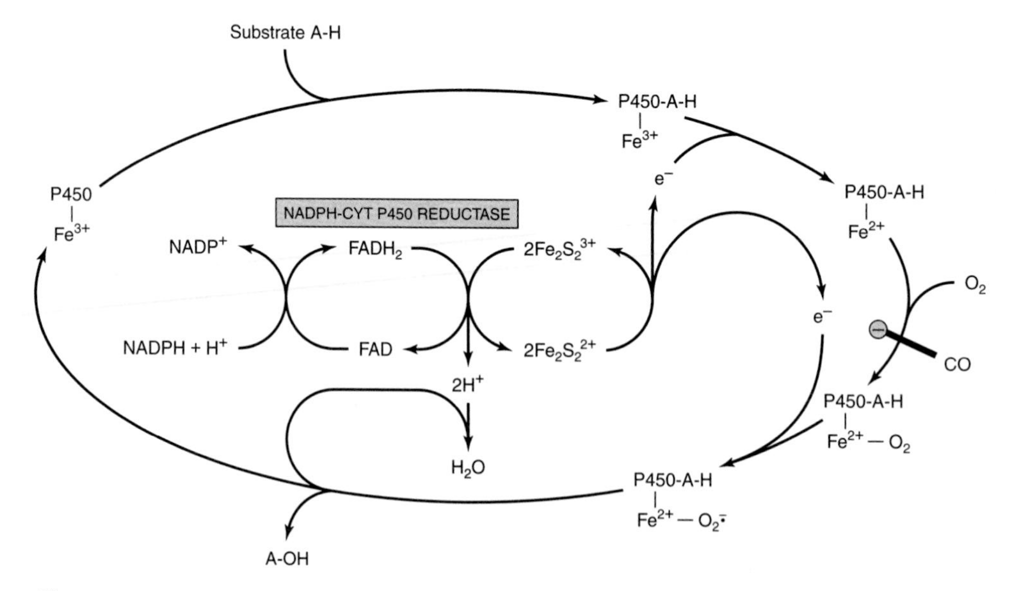


***Figure 11–6.*** Cytochrome P450 hydroxylase cycle in microsomes. The system shown is typical of steroid hydroxylases of the adrenal cortex. Liver microsomal cytochrome P450 hydroxylase does not require the iron-sulfur protein $Fe_2S_2$. Carbon monoxide (CO) inhibits the indicated step.

or be removed by superoxide dismutase.

$$O_2^{-\cdot} + O_2^{-\cdot} + 2H^+ \longrightarrow H_2O_2 + O_2$$

In this reaction, superoxide acts as both oxidant and reductant. Thus, superoxide dismutase protects aerobic organisms against the potential deleterious effects of superoxide. The enzyme occurs in all major aerobic tissues in the mitochondria and the cytosol. Although exposure of animals to an atmosphere of 100% oxygen causes an adaptive increase in superoxide dismutase, particularly in the lungs, prolonged exposure leads to lung damage and death. Antioxidants, eg, α-tocopherol (vitamin E), act as scavengers of free radicals and reduce the toxicity of oxygen (Chapter 45).

## SUMMARY

- In biologic systems, as in chemical systems, oxidation (loss of electrons) is always accompanied by reduction of an electron acceptor.
- Oxidoreductases have a variety of functions in metabolism; oxidases and dehydrogenases play major roles in respiration; hydroperoxidases protect the body against damage by free radicals; and oxygenases mediate the hydroxylation of drugs and steroids.
- Tissues are protected from oxygen toxicity caused by the superoxide free radical by the specific enzyme superoxide dismutase.

## REFERENCES


Babcock GT, Wikstrom M: Oxygen activation and the conservation of energy in cell respiration. Nature 1992;356:301.

Coon MJ et al: Cytochrome P450: Progress and predictions. FASEB J 1992;6:669.

Ernster L (editor): *Bioenergetics.* Elsevier, 1984.

Mammaerts GP, Van Veldhoven PP: Role of peroxisomes in mammalian metabolism. Cell Biochem Funct 1992;10:141.

Nicholls DG: *Cytochromes and Cell Respiration.* Carolina Biological Supply Company, 1984.

Raha S, Robinson BH: Mitochondria, oxygen free radicals, disease and aging. Trends Biochem Sci 2000;25:502.

Tyler DD: *The Mitochondrion in Health and Disease.* VCH Publishers, 1992.

Tyler DD, Sutton CM: Respiratory enzyme systems in mitochondrial membranes. In: *Membrane Structure and Function,* vol 5. Bittar EE (editor). Wiley, 1984.

Yang CS, Brady JF, Hong JY: Dietary effects on cytochromes P450, xenobiotic metabolism, and toxicity. FASEB J 1992; 6:737.

# The Respiratory Chain & Oxidative Phosphorylation

12

*Peter A. Mayes, PhD, DSc, & Kathleen M. Botham, PhD, DSc*

## BIOMEDICAL IMPORTANCE

Aerobic organisms are able to capture a far greater proportion of the available free energy of respiratory substrates than anaerobic organisms. Most of this takes place inside mitochondria, which have been termed the "powerhouses" of the cell. Respiration is coupled to the generation of the high-energy intermediate, ATP, by **oxidative phosphorylation,** and the **chemiosmotic theory** offers insight into how this is accomplished. A number of drugs (eg, **amobarbital**) and poisons (eg, **cyanide, carbon monoxide**) inhibit oxidative phosphorylation, usually with fatal consequences. Several inherited defects of mitochondria involving components of the respiratory chain and oxidative phosphorylation have been reported. Patients present with **myopathy** and **encephalopathy** and often have **lactic acidosis.**

## SPECIFIC ENZYMES ACT AS MARKERS OF COMPARTMENTS SEPARATED BY THE MITOCHONDRIAL MEMBRANES

Mitochondria have an **outer membrane** that is permeable to most metabolites, an **inner membrane** that is selectively permeable, and a **matrix** within (Figure 12–1). The outer membrane is characterized by the presence of various enzymes, including acyl-CoA synthetase and glycerolphosphate acyltransferase. Adenylyl kinase and creatine kinase are found in the intermembrane space. The phospholipid cardiolipin is concentrated in the inner membrane together with the enzymes of the respiratory chain.

## THE RESPIRATORY CHAIN COLLECTS & OXIDIZES REDUCING EQUIVALENTS

Most of the energy liberated during the oxidation of carbohydrate, fatty acids, and amino acids is made available within mitochondria as reducing equivalents (—H or electrons) (Figure 12–2). Mitochondria contain the **respiratory chain,** which collects and transports reducing equivalents directing them to their final reaction with oxygen to form water, the machinery for trapping the liberated free energy as high-energy phosphate, and the enzymes of β-oxidation and of the citric acid cycle (Chapters 22 and 16) that produce most of the reducing equivalents.

### Components of the Respiratory Chain Are Arranged in Order of Increasing Redox Potential

Hydrogen and electrons flow through the respiratory chain (Figure 12–3) through a redox span of 1.1 V from $NAD^+$/NADH to $O_2/2H_2O$ (Table 11–1). The respiratory chain consists of a number of redox carriers that proceed from the NAD-linked dehydrogenase systems, through flavoproteins and cytochromes, to molecular oxygen. Not all substrates are linked to the respiratory chain through NAD-specific dehydrogenases; some, because their redox potentials are more positive (eg, fumarate/succinate; Table 11–1), are linked directly to flavoprotein dehydrogenases, which in turn are linked to the cytochromes of the respiratory chain (Figure 12–4).

**Ubiquinone or Q (coenzyme Q)** (Figure 12–5) links the flavoproteins to cytochrome *b*, the member of the cytochrome chain of lowest redox potential. Q exists in the oxidized quinone or reduced quinol form under aerobic or anaerobic conditions, respectively. The structure of Q is very similar to that of vitamin K and vitamin E (Chapter 45) and of plastoquinone, found in chloroplasts. Q acts as a mobile component of the respiratory chain that collects reducing equivalents from the more fixed flavoprotein complexes and passes them on to the cytochromes.

An additional component is the **iron-sulfur protein (FeS;** nonheme iron) (Figure 12–6). It is associated with the flavoproteins (metalloflavoproteins) and with cytochrome *b*. The sulfur and iron are thought to take part in the oxidoreduction mechanism between flavin and Q, which involves only a single $e^-$ change, the iron atom undergoing oxidoreduction between $Fe^{2+}$ and $Fe^{3+}$.

Pyruvate and α-ketoglutarate dehydrogenase have complex systems involving lipoate and FAD prior to the passage of electrons to NAD, while electron trans-

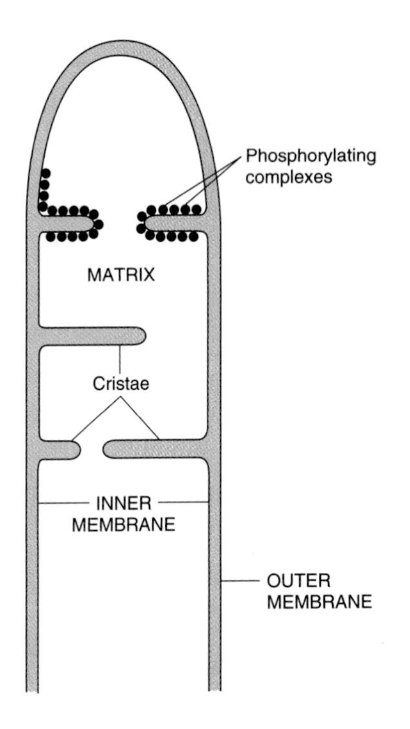


***Figure 12–1.*** Structure of the mitochondrial membranes. Note that the inner membrane contains many folds, or cristae.

fers from other dehydrogenases, eg, L(+)-3-hydroxyacyl-CoA dehydrogenase, couple directly with NAD.

The reduced NADH of the respiratory chain is in turn oxidized by a metalloflavoprotein enzyme—**NADH dehydrogenase.** This enzyme contains FeS and FMN, is tightly bound to the respiratory chain, and passes reducing equivalents on to Q.

Electrons flow from Q through the series of cytochromes in order of increasing redox potential to molecular oxygen (Figure 12–4). The terminal cytochrome $aa_3$ (cytochrome oxidase), responsible for the final combination of reducing equivalents with molecular oxygen, has a very high affinity for oxygen, allowing the respiratory chain to function at maximum rate until the tissue has become depleted of $O_2$. Since this is an irreversible reaction (the only one in the chain), it gives direction to the movement of reducing equivalents and to the production of ATP, to which it is coupled.

Functionally and structurally, the components of the respiratory chain are present in the inner mitochondrial membrane as four **protein-lipid respiratory chain complexes** that span the membrane. Cytochrome *c* is the only soluble cytochrome and, together with Q, seems to be a more mobile component of the respiratory chain connecting the fixed complexes (Figures 12–7 and 12–8).

## THE RESPIRATORY CHAIN PROVIDES MOST OF THE ENERGY CAPTURED DURING CATABOLISM

ADP captures, in the form of high-energy phosphate, a significant proportion of the free energy released by catabolic processes. The resulting ATP has been called the energy "currency" of the cell because it passes on this free energy to drive those processes requiring energy (Figure 10–6).

There is a net direct capture of two high-energy phosphate groups in the glycolytic reactions (Table 17–1), equivalent to approximately 103.2 kJ/mol of glucose. (In vivo, $\Delta G$ for the synthesis of ATP from ADP has been calculated as approximately 51.6 kJ/mol. (It is greater than $\Delta G^{0'}$ for the hydrolysis of ATP as given in Table 10–1, which is obtained under standard

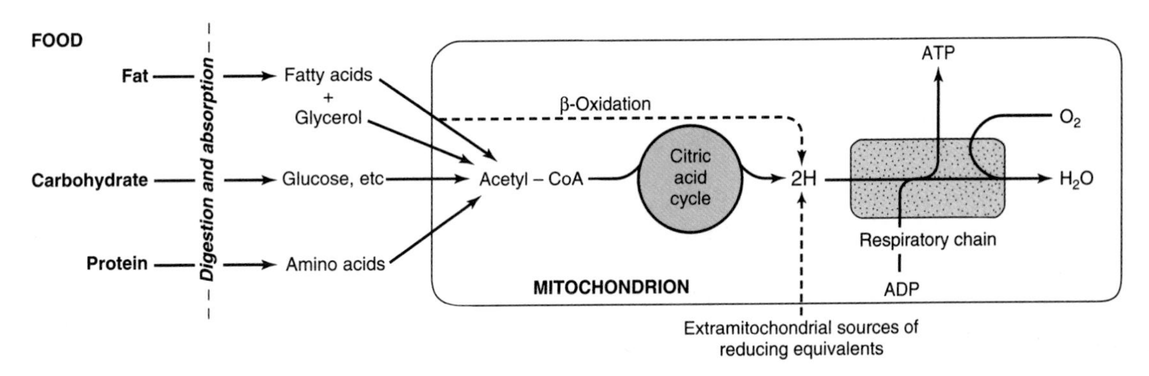


***Figure 12–2.*** Role of the respiratory chain of mitochondria in the conversion of food energy to ATP. Oxidation of the major foodstuffs leads to the generation of reducing equivalents (2H) that are collected by the respiratory chain for oxidation and coupled generation of ATP.

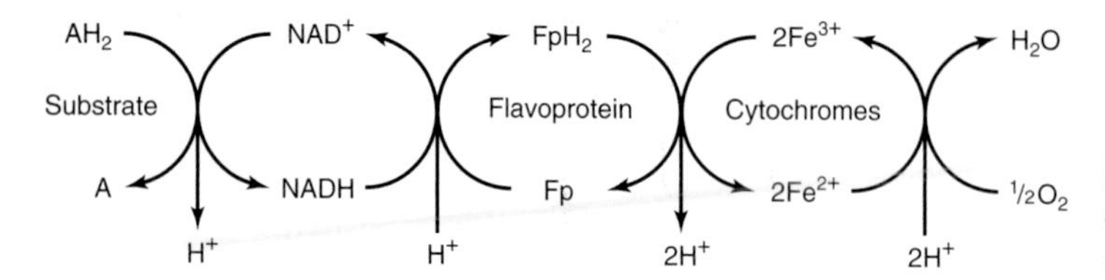


***Figure 12–3.*** Transport of reducing equivalents through the respiratory chain.

concentrations of 1.0 mol/L.) Since 1 mol of glucose yields approximately 2870 kJ on complete combustion, the energy captured by phosphorylation in glycolysis is small. Two more high-energy phosphates per mole of glucose are captured in the citric acid cycle during the conversion of succinyl CoA to succinate. All of these phosphorylations occur at the **substrate level.** When substrates are oxidized via an NAD-linked dehydrogenase and the respiratory chain, approximately 3 mol of inorganic phosphate are incorporated into 3 mol of ADP to form 3 mol of ATP per half mol of $O_2$ consumed; ie, the P:O ratio = 3 (Figure 12–7). On the other hand, when a substrate is oxidized via a flavoprotein-linked dehydrogenase, only 2 mol of ATP are formed; ie, P:O = 2. These reactions are known as **oxidative phosphorylation at the respiratory chain level.** Such dehydrogenations plus phosphorylations at the substrate level can now account for 68% of the free energy resulting from the combustion of glucose, captured in the form of high-energy phosphate. It is evident that the respiratory chain is responsible for a large proportion of total ATP formation.

## Respiratory Control Ensures a Constant Supply of ATP

The rate of respiration of mitochondria can be controlled by the availability of ADP. This is because oxidation and phosphorylation are tightly coupled; ie, oxidation cannot proceed via the respiratory chain without concomitant phosphorylation of ADP. Table 12–1 shows the five conditions controlling the rate of respiration in mitochondria. Most cells in the resting state are in state 4, and respiration is controlled by the availability of ADP. When work is performed, ATP is converted to ADP, allowing more respiration to occur, which in turn replenishes the store of ATP. Under certain conditions, the concentration of inorganic phosphate can also affect the rate of functioning of the respiratory chain. As respiration increases (as in exercise),

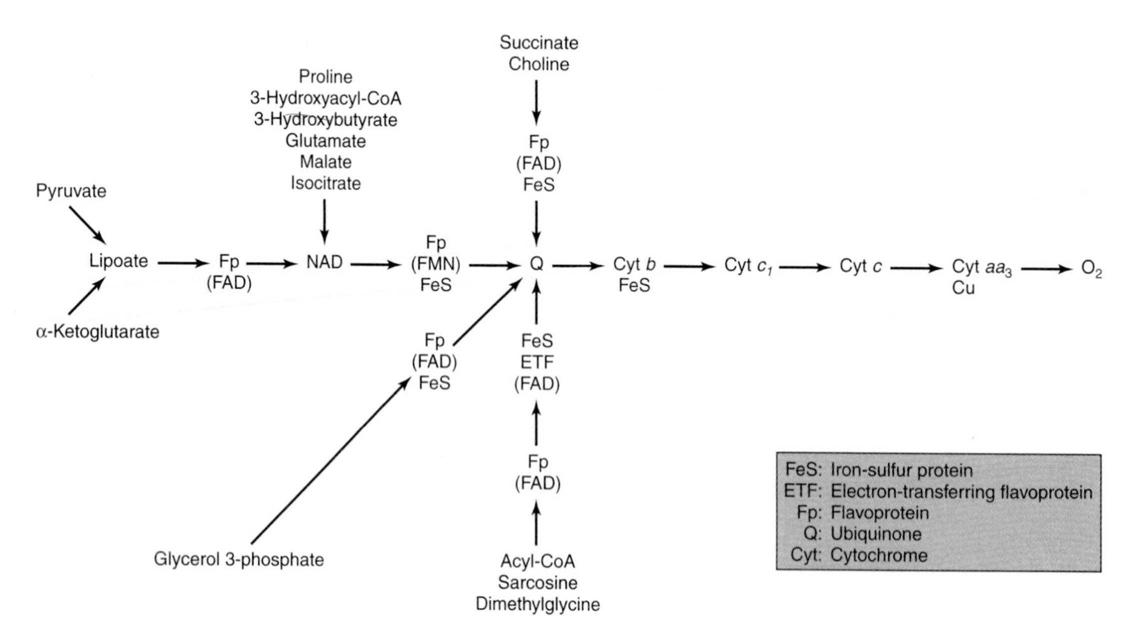


***Figure 12–4.*** Components of the respiratory chain in mitochondria, showing the collecting points for reducing equivalents from important substrates. FeS occurs in the sequences on the $O_2$ side of Fp or Cyt *b*.

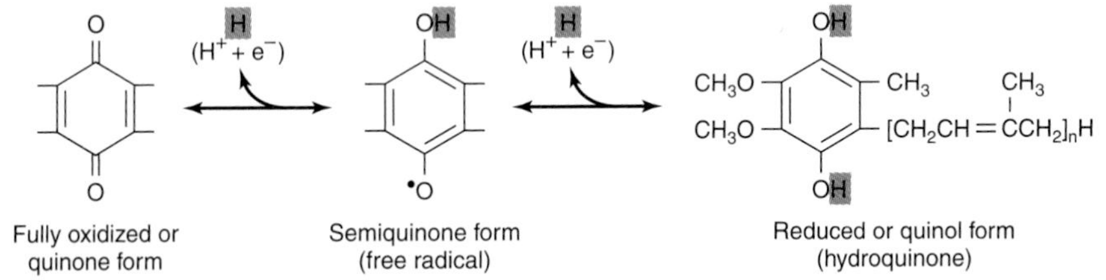

Fully oxidized or quinone form

Semiquinone form (free radical)

Reduced or quinol form (hydroquinone)

***Figure 12–5.*** Structure of ubiquinone (Q). n = Number of isoprenoid units, which is 10 in higher animals, ie, $Q_{10}$.

the cell approaches state 3 or state 5 when either the capacity of the respiratory chain becomes saturated or the $PO_2$ decreases below the $K_m$ for cytochrome $a_3$. There is also the possibility that the ADP/ATP transporter (Figure 12–9), which facilitates entry of cytosolic ADP into and ATP out of the mitochondrion, becomes rate-limiting.

Thus, the manner in which biologic oxidative processes allow the free energy resulting from the oxidation of foodstuffs to become available and to be captured is stepwise, efficient (approximately 68%), and controlled—rather than explosive, inefficient, and uncontrolled, as in many nonbiologic processes. The remaining free energy that is not captured as high-energy phosphate is liberated as **heat.** This need not be considered "wasted," since it ensures that the respiratory system as a whole is sufficiently exergonic to be removed from equilibrium, allowing continuous unidirectional flow and constant provision of ATP. It also contributes to maintenance of body temperature.

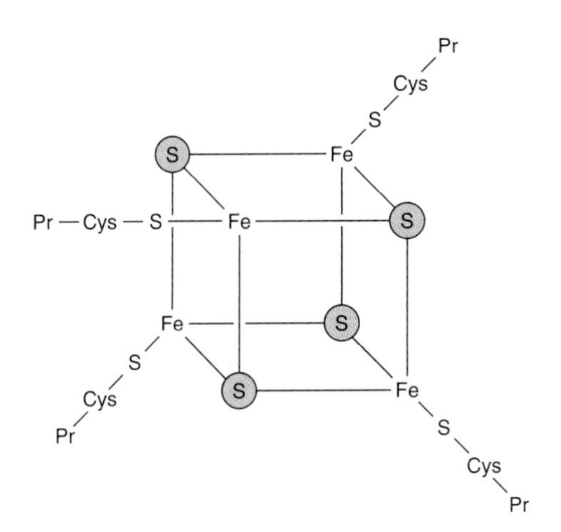


***Figure 12–6.*** Iron-sulfur-protein complex ($Fe_4S_4$). Ⓢ, acid-labile sulfur; Pr, apoprotein; Cys, cysteine residue. Some iron-sulfur proteins contain two iron atoms and two sulfur atoms ($Fe_2S_2$).

## MANY POISONS INHIBIT THE RESPIRATORY CHAIN

Much information about the respiratory chain has been obtained by the use of inhibitors, and, conversely, this has provided knowledge about the mechanism of action of several poisons (Figure 12–7). They may be classified as inhibitors of the respiratory chain, inhibitors of oxidative phosphorylation, and uncouplers of oxidative phosphorylation.

**Barbiturates** such as amobarbital inhibit NAD-linked dehydrogenases by blocking the transfer from FeS to Q. At sufficient dosage, they are fatal in vivo. **Antimycin A** and **dimercaprol** inhibit the respiratory chain between cytochrome *b* and cytochrome *c*. The classic poisons $\mathbf{H_2S}$, **carbon monoxide,** and **cyanide** inhibit cytochrome oxidase and can therefore totally arrest respiration. **Malonate** is a competitive inhibitor of succinate dehydrogenase.

**Atractyloside** inhibits oxidative phosphorylation by inhibiting the transporter of ADP into and ATP out of the mitochondrion (Figure 12–10).

The action of **uncouplers** is to dissociate oxidation in the respiratory chain from phosphorylation. These compounds are toxic in vivo, causing respiration to become uncontrolled, since the rate is no longer limited by the concentration of ADP or $P_i$. The uncoupler that has been used most frequently is **2,4-dinitrophenol,** but other compounds act in a similar manner. The antibiotic **oligomycin** completely blocks oxidation and phosphorylation by acting on a step in phosphorylation (Figures 12–7 and 12–8).

## THE CHEMIOSMOTIC THEORY EXPLAINS THE MECHANISM OF OXIDATIVE PHOSPHORYLATION

**Mitchell's chemiosmotic theory** postulates that the energy from oxidation of components in the respiratory chain is coupled to the translocation of hydrogen ions (protons, $H^+$) from the inside to the outside of the inner mitochondrial membrane. The electrochemical potential difference resulting from the asymmetric dis-

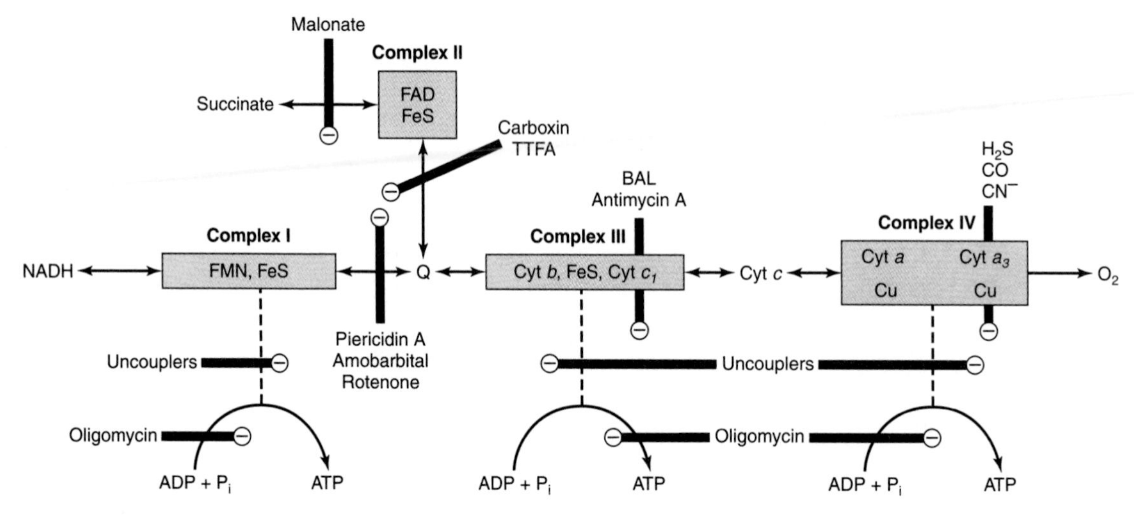


***Figure 12–7.*** Proposed sites of inhibition (⊖) of the respiratory chain by specific drugs, chemicals, and antibiotics. The sites that appear to support phosphorylation are indicated. BAL, dimercaprol. TTFA, an Fe-chelating agent. Complex I, NADH:ubiquinone oxidoreductase; complex II, succinate:ubiquinone oxidoreductase; complex III, ubiquinol:ferricytochrome c oxidoreductase; complex IV, ferrocytochrome c:oxygen oxidoreductase. Other abbreviations as in Figure 12–4.

tribution of the hydrogen ions is used to drive the mechanism responsible for the formation of ATP (Figure 12–8).

## The Respiratory Chain Is a Proton Pump

Each of the respiratory chain complexes I, III, and IV (Figures 12–7 and 12–8) acts as a **proton pump.** The inner membrane is impermeable to ions in general but particularly to protons, which accumulate outside the membrane, creating an **electrochemical potential difference** across the membrane ($\Delta\mu_{H^+}$).This consists of a chemical potential (difference in pH) and an electrical potential.

## A Membrane-Located ATP Synthase Functions as a Rotary Motor to Form ATP

The electrochemical potential difference is used to drive a membrane-located **ATP synthase** which in the presence of $P_i$ + ADP forms ATP (Figure 12–8). Scattered over the surface of the inner membrane are the phosphorylating complexes, ATP synthase, responsible for the production of ATP (Figure 12–1). These consist of several protein subunits, collectively known as $F_1$, which project into the matrix and which contain the phosphorylation mechanism (Figure 12–8). These subunits are attached to a membrane protein complex known as $F_0$, which also consists of several protein subunits. $F_0$ spans the membrane and forms the proton channel. The flow of protons through $F_0$ causes it to rotate, driving the production of ATP in the $F_1$ complex (Figure 12–9). Estimates suggest that for each NADH oxidized, complex I translocates four protons and complexes III and IV translocate 6 between them. As four protons are taken into the mitochondrion for each ATP exported, the P:O ratio would not necessarily be a complete integer, ie, 3, but possibly 2.5. However, for simplicity, a value of 3 for the oxidation of NADH + $H^+$ and 2 for the oxidation of $FADH_2$ will continue to be used throughout this text.

## Experimental Findings Support the Chemiosmotic Theory

(1) Addition of protons (acid) to the external medium of intact mitochondria leads to the generation of ATP.

(2) Oxidative phosphorylation does not occur in soluble systems where there is no possibility of a vectorial ATP synthase. A closed membrane must be present in order to achieve oxidative phosphorylation (Figure 12–8).

(3) The respiratory chain contains components organized in a sided manner (transverse asymmetry) as required by the chemiosmotic theory.

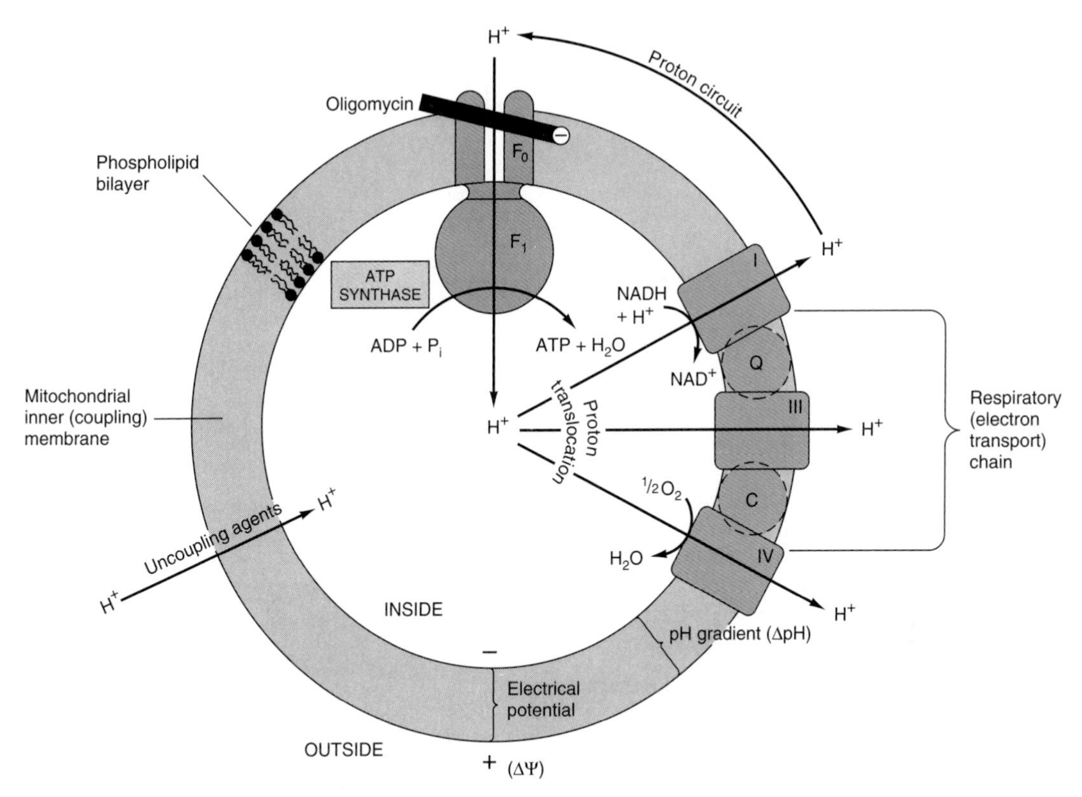


***Figure 12–8.*** Principles of the chemiosmotic theory of oxidative phosphorylation. The main proton circuit is created by the coupling of oxidation in the respiratory chain to proton translocation from the inside to the outside of the membrane, driven by the respiratory chain complexes I, III, and IV, each of which acts as a *proton pump.* Q, ubiquinone; C, cytochrome c; $F_1$, $F_0$, protein subunits which utilize energy from the proton gradient to promote phosphorylation. Uncoupling agents such as dinitrophenol allow leakage of $H^+$ across the membrane, thus collapsing the electrochemical proton gradient. Oligomycin specifically blocks conduction of $H^+$ through $F_0$.

***Table 12–1.*** States of respiratory control.

| | Conditions Limiting the Rate of Respiration |
|---|---|
| State 1 | Availability of ADP and substrate |
| State 2 | Availability of substrate only |
| State 3 | The capacity of the respiratory chain itself, when all substrates and components are present in saturating amounts |
| State 4 | Availability of ADP only |
| State 5 | Availability of oxygen only |

## The Chemiosmotic Theory Can Account for Respiratory Control and the Action of Uncouplers

The electrochemical potential difference across the membrane, once established as a result of proton translocation, inhibits further transport of reducing equivalents through the respiratory chain unless discharged by back-translocation of protons across the membrane through the vectorial ATP synthase. This in turn depends on availability of ADP and $P_i$.

Uncouplers (eg, dinitrophenol) are amphipathic (Chapter 14) and increase the permeability of the lipoid inner mitochondrial membrane to protons (Figure 12–8), thus reducing the electrochemical potential and short-circuiting the ATP synthase. In this way, oxidation can proceed without phosphorylation.

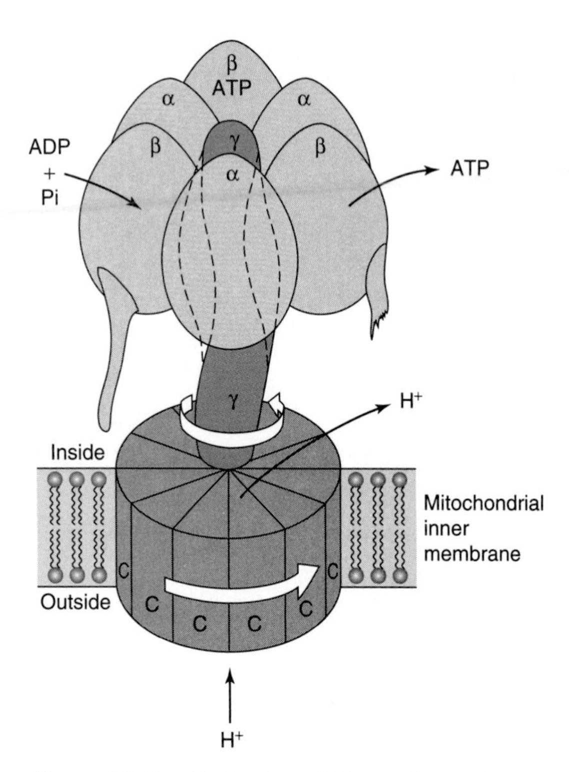


***Figure 12–9.*** Mechanism of ATP production by ATP synthase. The enzyme complex consists of an $F_0$ subcomplex which is a disk of "C" protein subunits. Attached is a γ-subunit in the form of a "bent axle." Protons passing through the disk of "C" units cause it and the attached γ-subunit to rotate. The γ-subunit fits inside the $F_1$ subcomplex of three α- and three β-subunits, which are fixed to the membrane and do not rotate. ADP and $P_i$ are taken up sequentially by the β-subunits to form ATP, which is expelled as the rotating γ-subunit squeezes each β-subunit in turn. Thus, three ATP molecules are generated per revolution. For clarity, not all the subunits that have been identified are shown—eg, the "axle" also contains an ε-subunit.

## THE RELATIVE IMPERMEABILITY OF THE INNER MITOCHONDRIAL MEMBRANE NECESSITATES EXCHANGE TRANSPORTERS

Exchange diffusion systems are present in the membrane for exchange of anions against $OH^-$ ions and cations against $H^+$ ions. Such systems are necessary for uptake and output of ionized metabolites while preserving electrical and osmotic equilibrium. The inner bilipoid mitochondrial membrane is freely permeable to uncharged small molecules, such as oxygen, water, $CO_2$, and $NH_3$, and to monocarboxylic acids, such as 3-hydroxybutyric, acetoacetic, and acetic. Long-chain fatty acids are transported into mitochondria via the carnitine system (Figure 22–1), and there is also a special carrier for pyruvate involving a symport that utilizes the $H^+$ gradient from outside to inside the mitochondrion (Figure 12–10). However, dicarboxylate and tri-

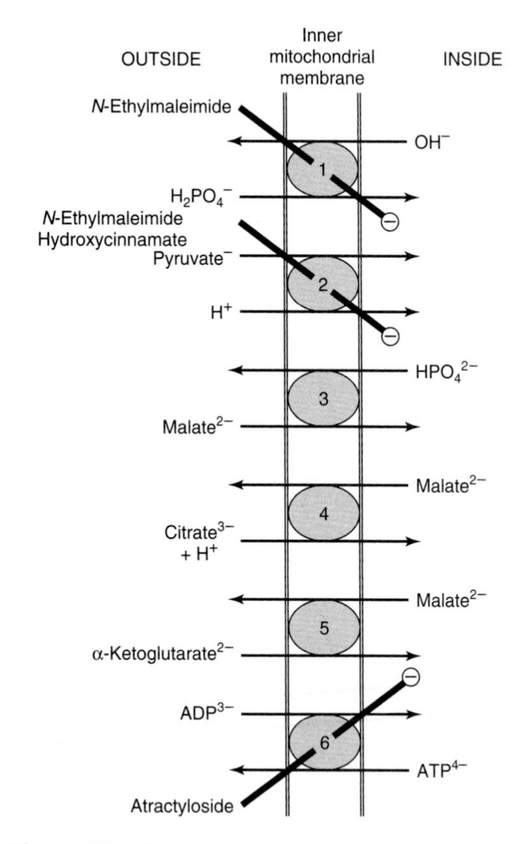


***Figure 12–10.*** Transporter systems in the inner mitochondrial membrane. ①, phosphate transporter; ②, pyruvate symport; ③, dicarboxylate transporter; ④, tricarboxylate transporter; ⑤, α-ketoglutarate transporter; ⑥, adenine nucleotide transporter. *N*-Ethylmaleimide, hydroxycinnamate, and atractyloside inhibit (⊖) the indicated systems. Also present (but not shown) are transporter systems for glutamate/aspartate (Figure 12–13), glutamine, ornithine, neutral amino acids, and carnitine (Figure 22–1).

carboxylate anions and amino acids require specific transporter or carrier systems to facilitate their passage across the membrane. Monocarboxylic acids penetrate more readily in their undissociated and more lipid-soluble form.

The transport of di- and tricarboxylate anions is closely linked to that of inorganic phosphate, which penetrates readily as the $H_2PO_4^-$ ion in exchange for $OH^-$. The net uptake of malate by the dicarboxylate transporter requires inorganic phosphate for exchange in the opposite direction. The net uptake of citrate, isocitrate, or *cis*-aconitate by the tricarboxylate transporter requires malate in exchange. α-Ketoglutarate transport also requires an exchange with malate. The adenine nucleotide transporter allows the exchange of ATP and ADP but not AMP. It is vital in allowing ATP exit from mitochondria to the sites of extramitochondrial utilization and in allowing the return of ADP for ATP production within the mitochondrion (Figure 12–11). $Na^+$ can be exchanged for $H^+$, driven by the proton gradient. It is believed that active uptake of $Ca^{2+}$ by mitochondria occurs with a net charge transfer of 1 ($Ca^+$ uniport), possibly through a $Ca^{2+}/H^+$ antiport. Calcium release from mitochondria is facilitated by exchange with $Na^+$.

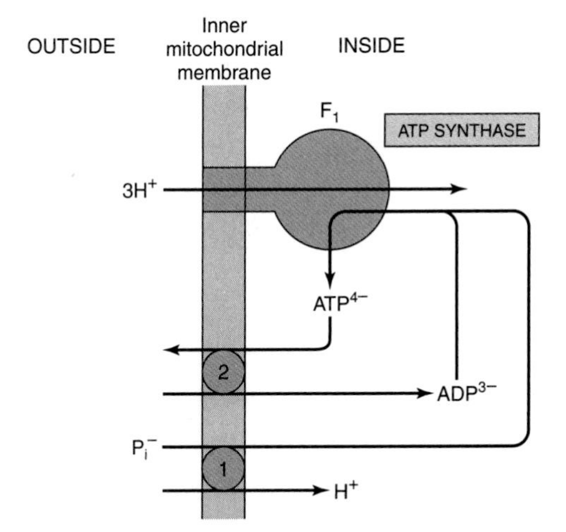


***Figure 12–11.*** Combination of phosphate transporter (①) with the adenine nucleotide transporter (②) in ATP synthesis. The $H^+/P_i$ symport shown is equivalent to the $P_i/OH^-$ antiport shown in Figure 12–10. Four protons are taken into the mitochondrion for each ATP exported. However, one less proton would be taken in when ATP is used inside the mitochondrion.

## Ionophores Permit Specific Cations to Penetrate Membranes

Ionophores are lipophilic molecules that complex specific cations and facilitate their transport through biologic membranes, eg, **valinomycin** ($K^+$). The classic uncouplers such as dinitrophenol are, in fact, proton ionophores.

## A Proton-Translocating Transhydrogenase Is a Source of Intramitochondrial NADPH

Energy-linked transhydrogenase, a protein in the inner mitochondrial membrane, couples the passage of protons down the electrochemical gradient from outside to inside the mitochondrion with the transfer of H from intramitochondrial NADH to NADPH for intramitochondrial enzymes such as glutamate dehydrogenase and hydroxylases involved in steroid synthesis.

## Oxidation of Extramitochondrial NADH Is Mediated by Substrate Shuttles

NADH cannot penetrate the mitochondrial membrane, but it is produced continuously in the cytosol by 3-phosphoglyceraldehyde dehydrogenase, an enzyme in the glycolysis sequence (Figure 17–2). However, under aerobic conditions, extramitochondrial NADH does not accumulate and is presumed to be oxidized by the respiratory chain in mitochondria. The transfer of reducing equivalents through the mitochondrial membrane requires substrate pairs, linked by suitable dehydrogenases on each side of the mitochondrial membrane. The mechanism of transfer using the **glycerophosphate shuttle** is shown in Figure 12–12). Since the mitochondrial enzyme is linked to the respiratory chain via a flavoprotein rather than NAD, only 2 mol rather than 3 mol of ATP are formed per atom of oxygen consumed. Although this shuttle is present in some tissues (eg, brain, white muscle), in others (eg, heart muscle) it is deficient. It is therefore believed that the **malate shuttle** system (Figure 12–13) is of more universal utility. The complexity of this system is due to the impermeability of the mitochondrial membrane to oxaloacetate, which must react with glutamate and transaminate to aspartate and α-ketoglutarate before transport through the mitochondrial membrane and reconstitution to oxaloacetate in the cytosol.

## Ion Transport in Mitochondria Is Energy-Linked

Mitochondria maintain or accumulate cations such as $K^+$, $Na^+$, $Ca^{2+}$, and $Mg^{2+}$, and $P_i$. It is assumed that a primary proton pump drives cation exchange.

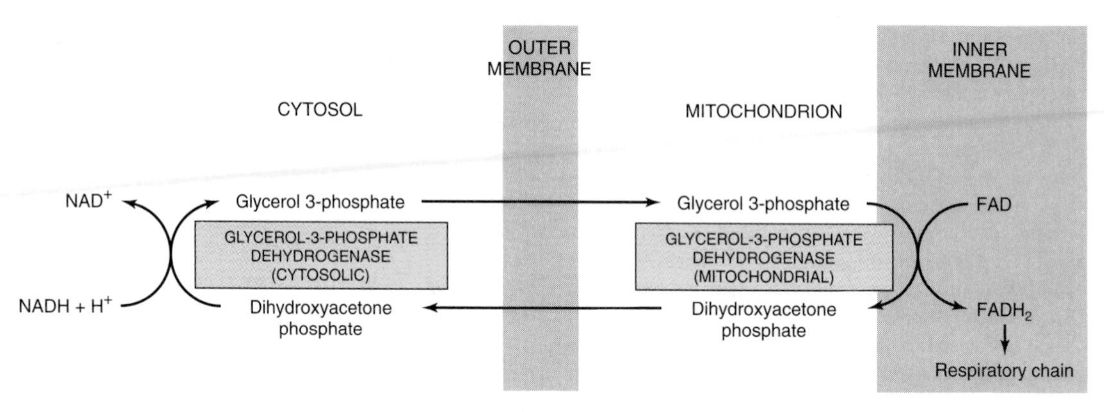


***Figure 12–12.*** Glycerophosphate shuttle for transfer of reducing equivalents from the cytosol into the mitochondrion.

## The Creatine Phosphate Shuttle Facilitates Transport of High-Energy Phosphate From Mitochondria

This shuttle (Figure 12–14) augments the functions of creatine phosphate as an energy buffer by acting as a dynamic system for transfer of high-energy phosphate from mitochondria in active tissues such as heart and skeletal muscle. An isoenzyme of creatine kinase ($CK_m$) is found in the mitochondrial intermembrane space, catalyzing the transfer of high-energy phosphate to creatine from ATP emerging from the adenine nucleotide transporter. In turn, the creatine phosphate is transported into the cytosol via protein pores in the outer mitochondrial membrane, becoming available for generation of extramitochondrial ATP.

# CLINICAL ASPECTS

The condition known as **fatal infantile mitochondrial myopathy and renal dysfunction** involves severe diminution or absence of most oxidoreductases of the respiratory chain. **MELAS** (mitochondrial encephalopathy, lactic acidosis, and stroke) is an inherited condition due to NADH:ubiquinone oxidoreductase (complex I) or cytochrome oxidase deficiency. It is caused by a muta-

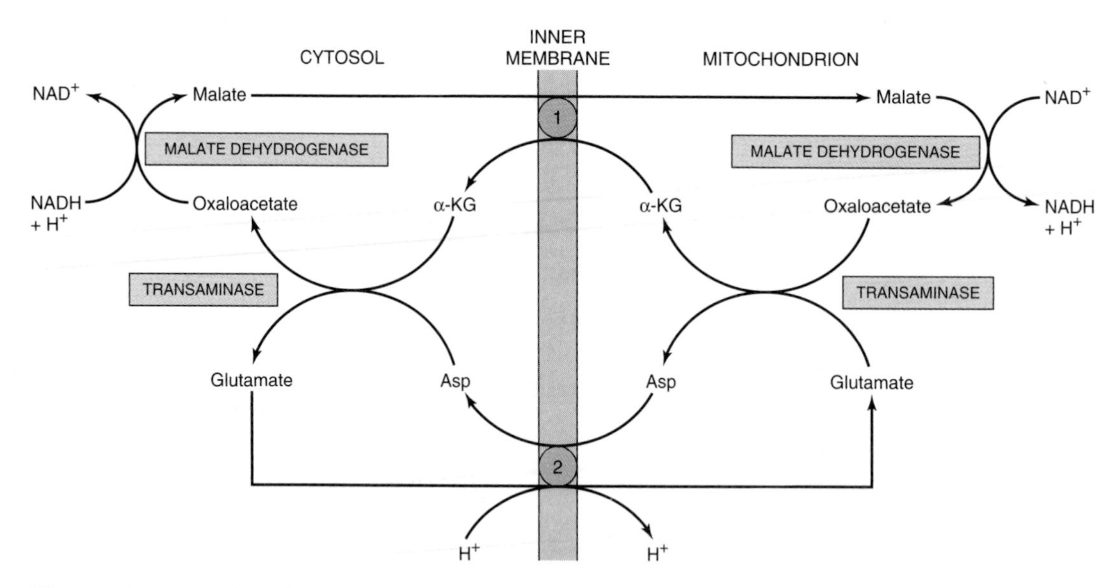


***Figure 12–13.*** Malate shuttle for transfer of reducing equivalents from the cytosol into the mitochondrion. ① Ketoglutarate transporter; ②, glutamate/aspartate transporter (note the proton symport with glutamate).

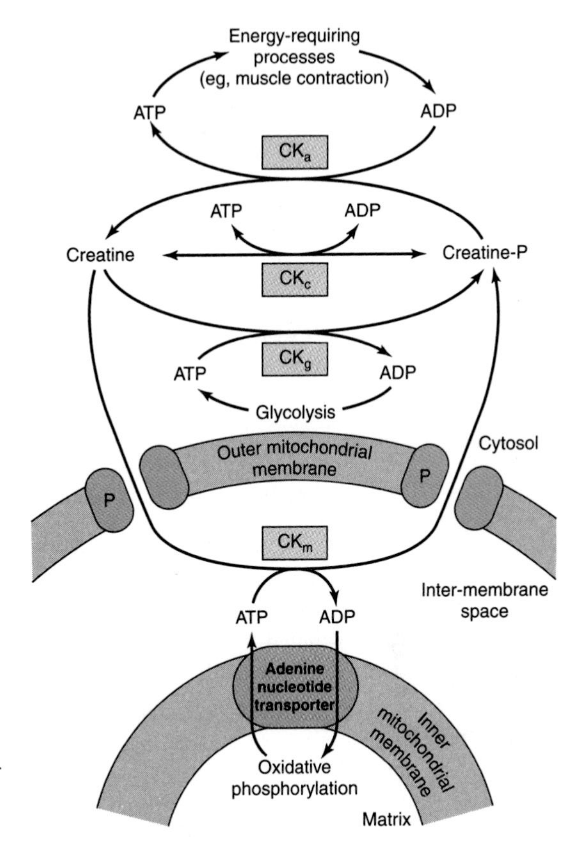


***Figure 12–14.*** The creatine phosphate shuttle of heart and skeletal muscle. The shuttle allows rapid transport of high-energy phosphate from the mitochondrial matrix into the cytosol. $CK_a$, creatine kinase concerned with large requirements for ATP, eg, muscular contraction; $CK_c$, creatine kinase for maintaining equilibrium between creatine and creatine phosphate and ATP/ADP; $CK_g$, creatine kinase coupling glycolysis to creatine phosphate synthesis; $CK_m$, mitochondrial creatine kinase mediating creatine phosphate production from ATP formed in oxidative phosphorylation; P, pore protein in outer mitochondrial membrane.

tion in mitochondrial DNA and may be involved in Alzheimer's disease and diabetes mellitus. A number of drugs and poisons act by inhibition of oxidative phosphorylation (see above).

## SUMMARY

- Virtually all energy released from the oxidation of carbohydrate, fat, and protein is made available in mitochondria as reducing equivalents (—H or $e^-$). These are funneled into the respiratory chain, where they are passed down a redox gradient of carriers to their final reaction with oxygen to form water.
- The redox carriers are grouped into respiratory chain complexes in the inner mitochondrial membrane. These use the energy released in the redox gradient to pump protons to the outside of the membrane, creating an electrochemical potential across the membrane.
- Spanning the membrane are ATP synthase complexes that use the potential energy of the proton gradient to synthesize ATP from ADP and $P_i$. In this way, oxidation is closely coupled to phosphorylation to meet the energy needs of the cell.
- Because the inner mitochondrial membrane is impermeable to protons and other ions, special exchange transporters span the membrane to allow passage of ions such as $OH^-$, $P_i^-$, $ATP^{4-}$, $ADP^{3-}$, and metabolites, without discharging the electrochemical gradient across the membrane.
- Many well-known poisons such as cyanide arrest respiration by inhibition of the respiratory chain.

## REFERENCES

Balaban RS: Regulation of oxidative phosphorylation in the mammalian cell. Am J Physiol 1990;258:C377.

Hinkle PC et al: Mechanistic stoichiometry of mitochondrial oxidative phosphorylation. Biochemistry 1991;30:3576.

Mitchell P: Keilin's respiratory chain concept and its chemiosmotic consequences. Science 1979;206:1148.

Smeitink J et al: The genetics and pathology of oxidative phosphorylation. Nat Rev Genet 2001;2:342.

Tyler DD: *The Mitochondrion in Health and Disease.* VCH Publishers, 1992.

Wallace DC: Mitochondrial DNA in aging and disease. Sci Am 1997;277(2):22.

Yoshida M et al: ATP synthase—a marvellous rotary engine of the cell. Nat Rev Mol Cell Biol 2001;2:669.

# Carbohydrates of Physiologic Significance

# 13

Peter A. Mayes, PhD, DSc, & David A. Bender, PhD

## BIOMEDICAL IMPORTANCE

Carbohydrates are widely distributed in plants and animals; they have important structural and metabolic roles. In plants, glucose is synthesized from carbon dioxide and water by photosynthesis and stored as starch or used to synthesize cellulose of the plant framework. Animals can synthesize carbohydrate from lipid glycerol and amino acids, but most animal carbohydrate is derived ultimately from plants. **Glucose** is the most important carbohydrate; most dietary carbohydrate is absorbed into the bloodstream as glucose, and other sugars are converted into glucose in the liver. Glucose is the major metabolic fuel of mammals (except ruminants) and a universal fuel of the fetus. It is the precursor for synthesis of all the other carbohydrates in the body, including **glycogen** for storage; **ribose** and **deoxyribose** in nucleic acids; and **galactose** in lactose of milk, in glycolipids, and in combination with protein in glycoproteins and proteoglycans. Diseases associated with carbohydrate metabolism include **diabetes mellitus, galactosemia, glycogen storage diseases,** and **lactose intolerance.**

## CARBOHYDRATES ARE ALDEHYDE OR KETONE DERIVATIVES OF POLYHYDRIC ALCOHOLS

(1) **Monosaccharides** are those carbohydrates that cannot be hydrolyzed into simpler carbohydrates: They may be classified as **trioses, tetroses, pentoses, hexoses,** or **heptoses,** depending upon the number of carbon atoms; and as **aldoses** or **ketoses** depending upon whether they have an aldehyde or ketone group. Examples are listed in Table 13–1.

(2) **Disaccharides** are condensation products of two monosaccharide units. Examples are maltose and sucrose.

(3) **Oligosaccharides** are condensation products of two to ten monosaccharides; maltotriose* is an example.

*Note that this is not a true triose but a trisaccharide containing three α-glucose residues.

(4) **Polysaccharides** are condensation products of more than ten monosaccharide units; examples are the starches and dextrins, which may be linear or branched polymers. Polysaccharides are sometimes classified as hexosans or pentosans, depending upon the identity of the constituent monosaccharides.

## BIOMEDICALLY, GLUCOSE IS THE MOST IMPORTANT MONOSACCHARIDE

### The Structure of Glucose Can Be Represented in Three Ways

The straight-chain structural formula (aldohexose; Figure 13–1A) can account for some of the properties of glucose, but a cyclic structure is favored on thermodynamic grounds and accounts for the remainder of its chemical properties. For most purposes, the structural formula is represented as a simple ring in perspective as proposed by Haworth (Figure 13–1B). In this representation, the molecule is viewed from the side and above the plane of the ring. By convention, the bonds nearest to the viewer are bold and thickened. The six-membered ring containing one oxygen atom is in the form of a chair (Figure 13–1C).

### Sugars Exhibit Various Forms of Isomerism

Glucose, with four asymmetric carbon atoms, can form 16 isomers. The more important types of isomerism found with glucose are as follows.

(1) **D and L isomerism:** The designation of a sugar isomer as the D form or of its mirror image as the L form

***Table 13–1.*** Classification of important sugars.

| | Aldoses | Ketoses |
|---|---|---|
| Trioses ($C_3H_6O_3$) | Glycerose | Dihydroxyacetone |
| Tetroses ($C_4H_8O_4$) | Erythrose | Erythrulose |
| Pentoses ($C_5H_{10}O_5$) | Ribose | Ribulose |
| Hexoses ($C_6H_{12}O_6$) | Glucose | Fructose |

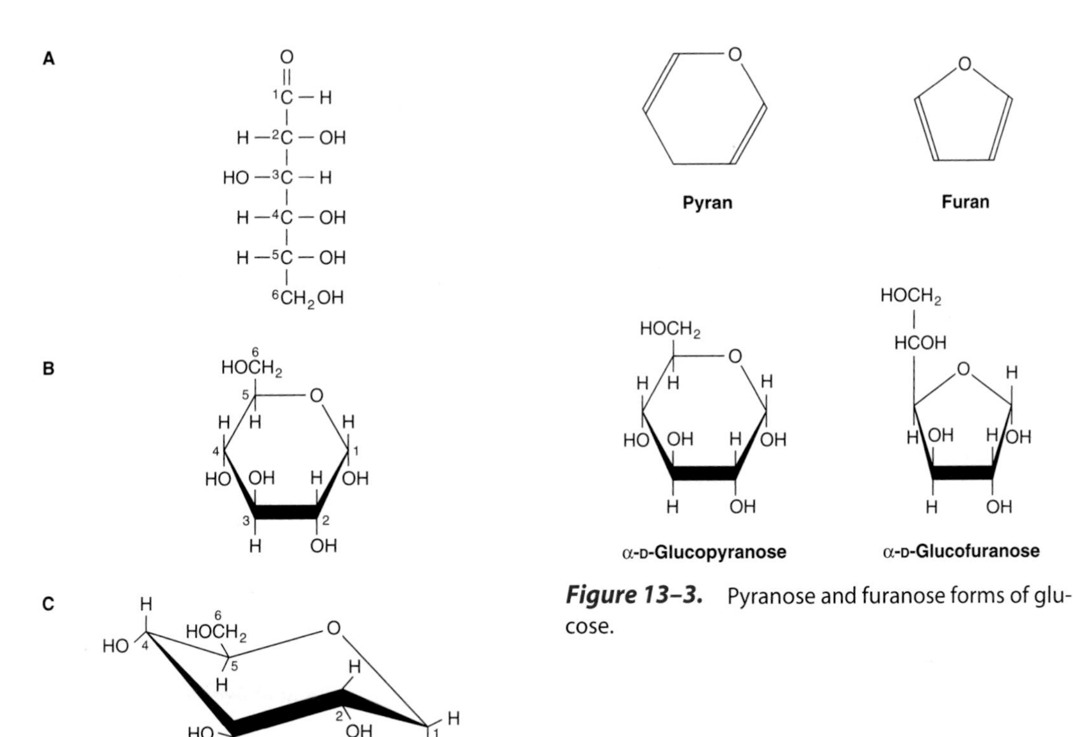


***Figure 13–1.*** D-Glucose. A: straight chain form. B: α-D-glucose; Haworth projection. C: α-D-glucose; chair form.

***Figure 13–3.*** Pyranose and furanose forms of glucose.

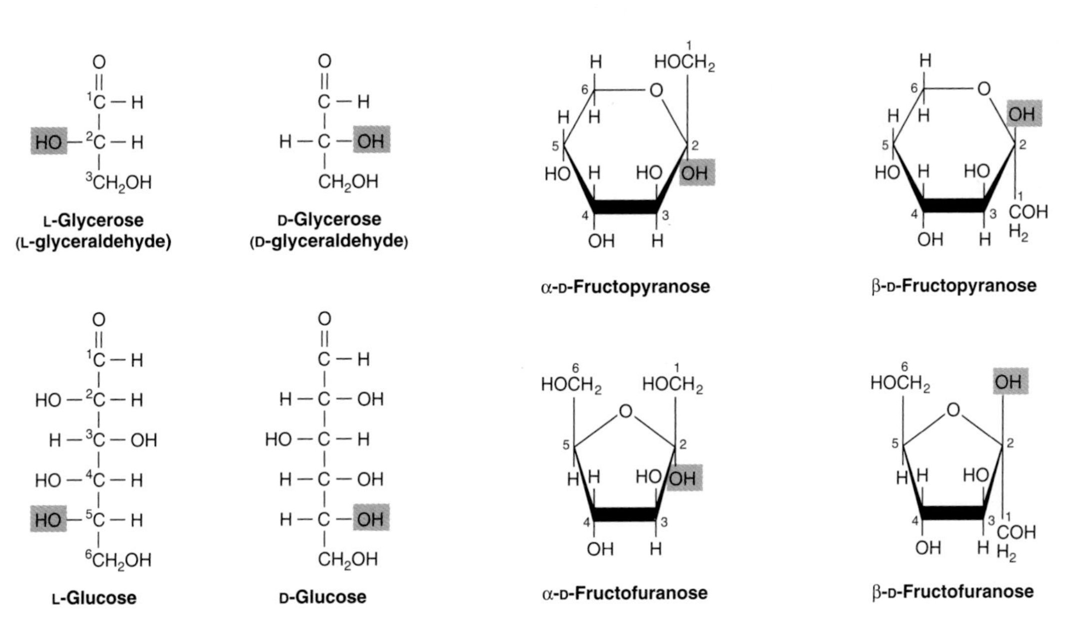


***Figure 13–2.*** D- and L-isomerism of glycerose and glucose.

***Figure 13–4.*** Pyranose and furanose forms of fructose.

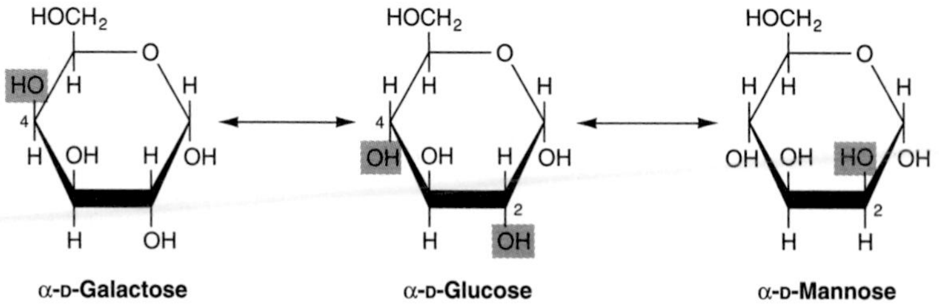

α-D-Galactose ↔ α-D-Glucose ↔ α-D-Mannose

***Figure 13–5.*** Epimerization of glucose.

is determined by its spatial relationship to the parent compound of the carbohydrates, the three-carbon sugar glycerose (glyceraldehyde). The L and D forms of this sugar, and of glucose, are shown in Figure 13–2. The orientation of the —H and —OH groups around the carbon atom adjacent to the terminal primary alcohol carbon (carbon 5 in glucose) determines whether the sugar belongs to the D or L series. When the —OH group on this carbon is on the right (as seen in Figure 13–2), the sugar is the D-isomer; when it is on the left, it is the L-isomer. Most of the monosaccharides occurring in mammals are D sugars, and the enzymes responsible for their metabolism are specific for this configuration. In solution, glucose is dextrorotatory—hence the alternative name **dextrose,** often used in clinical practice.

The presence of asymmetric carbon atoms also confers **optical activity** on the compound. When a beam of plane-polarized light is passed through a solution of an **optical isomer,** it will be rotated either to the right, dextrorotatory (+); or to the left, levorotatory (−). The direction of rotation is independent of the stereochemistry of the sugar, so it may be designated D(−), D(+), L(−), or L(+). For example, the naturally occurring form of fructose is the D(−) isomer.

(2) **Pyranose and furanose ring structures:** The stable ring structures of monosaccharides are similar to the ring structures of either pyran (a six-membered ring) or furan (a five-membered ring) (Figures 13–3 and 13–4). For glucose in solution, more than 99% is in the pyranose form.

(3) **Alpha and beta anomers:** The ring structure of an aldose is a hemiacetal, since it is formed by combination of an aldehyde and an alcohol group. Similarly, the ring structure of a ketose is a hemiketal. Crystalline glucose is α-D-glucopyranose. The cyclic structure is retained in solution, but isomerism occurs about position 1, the carbonyl or **anomeric carbon atom,** to give a mixture of α-glucopyranose (38%) and β-glucopyranose (62%). Less than 0.3% is represented by α and β anomers of glucofuranose.

(4) **Epimers:** Isomers differing as a result of variations in configuration of the —OH and —H on carbon atoms 2, 3, and 4 of glucose are known as epimers. Biologically, the most important epimers of glucose are mannose and galactose, formed by epimerization at carbons 2 and 4, respectively (Figure 13–5).

(5) **Aldose-ketose isomerism:** Fructose has the same molecular formula as glucose but differs in its structural formula, since there is a potential keto group in position 2, the anomeric carbon of fructose (Figures 13–4 and 13–7), whereas there is a potential aldehyde group in position 1, the anomeric carbon of glucose (Figures 13–2 and 13–6).

## Many Monosaccharides Are Physiologically Important

Derivatives of trioses, tetroses, and pentoses and of a seven-carbon sugar (sedoheptulose) are formed as metabolic intermediates in glycolysis and the pentose phosphate pathway. Pentoses are important in nucleotides,

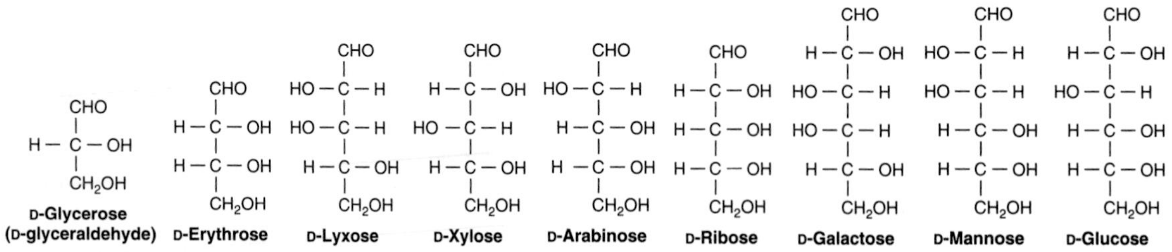

D-Glycerose (D-glyceraldehyde), D-Erythrose, D-Lyxose, D-Xylose, D-Arabinose, D-Ribose, D-Galactose, D-Mannose, D-Glucose

***Figure 13–6.*** Examples of aldoses of physiologic significance.

***Table 13–2.*** Pentoses of physiologic importance.

| Sugar | Where Found | Biochemical Importance | Clinical Significance |
|---|---|---|---|
| D-Ribose | Nucleic acids. | Structural elements of nucleic acids and coenzymes, eg, ATP, NAD, NADP, flavoproteins. Ribose phosphates are intermediates in pentose phosphate pathway. | |
| D-Ribulose | Formed in metabolic processes. | Ribulose phosphate is an intermediate in pentose phosphate pathway. | |
| D-Arabinose | Gum arabic. Plum and cherry gums. | Constituent of glycoproteins. | |
| D-Xylose | Wood gums, proteoglycans, glycosaminoglycans. | Constituent of glycoproteins. | |
| D-Lyxose | Heart muscle. | A constituent of a lyxoflavin isolated from human heart muscle. | |
| L-Xylulose | Intermediate in uronic acid pathway. | | Found in urine in essential pentosuria. |

nucleic acids, and several coenzymes (Table 13–2). Glucose, galactose, fructose, and mannose are physiologically the most important hexoses (Table 13–3). The biochemically important aldoses are shown in Figure 13–6, and important ketoses in Figure 13–7.

In addition, carboxylic acid derivatives of glucose are important, including D-glucuronate (for glucuronide formation and in glycosaminoglycans) and its metabolic derivative, L-iduronate (in glycosaminoglycans) (Figure 13–8) and L-gulonate (an intermediate in the uronic acid pathway; see Figure 20–4).

## Sugars Form Glycosides With Other Compounds & With Each Other

**Glycosides** are formed by condensation between the hydroxyl group of the anomeric carbon of a monosaccharide, or monosaccharide residue, and a second compound that may—or may not (in the case of an **aglycone**)—be another monosaccharide. If the second group is a hydroxyl, the O-glycosidic bond is an **acetal** link because it results from a reaction between a hemiacetal group (formed from an aldehyde and an —OH group) and an-

***Table 13–3.*** Hexoses of physiologic importance.

| Sugar | Source | Importance | Clinical Significance |
|---|---|---|---|
| D-Glucose | Fruit juices. Hydrolysis of starch, cane sugar, maltose, and lactose. | The "sugar" of the body. The sugar carried by the blood, and the principal one used by the tissues. | Present in the urine (glycosuria) in diabetes mellitus owing to raised blood glucose (hyperglycemia). |
| D-Fructose | Fruit juices. Honey. Hydrolysis of cane sugar and of inulin (from the Jerusalem artichoke). | Can be changed to glucose in the liver and so used in the body. | Hereditary fructose intolerance leads to fructose accumulation and hypoglycemia. |
| D-Galactose | Hydrolysis of lactose. | Can be changed to glucose in the liver and metabolized. Synthesized in the mammary gland to make the lactose of milk. A constituent of glycolipids and glycoproteins. | Failure to metabolize leads to galactosemia and cataract. |
| D-Mannose | Hydrolysis of plant mannans and gums. | A constituent of many glycoproteins. | |

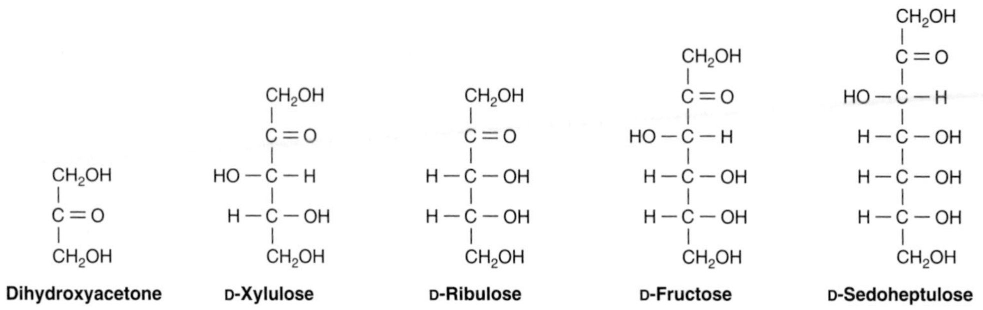

Dihydroxyacetone D-Xylulose D-Ribulose D-Fructose D-Sedoheptulose

***Figure 13–7.*** Examples of ketoses of physiologic significance.

other —OH group. If the hemiacetal portion is glucose, the resulting compound is a **glucoside;** if galactose, a **galactoside;** and so on. If the second group is an amine, an N-glycosidic bond is formed, eg, between adenine and ribose in nucleotides such as ATP (Figure 10–4).

Glycosides are widely distributed in nature; the aglycone may be methanol, glycerol, a sterol, a phenol, or a base such as adenine. The glycosides that are important in medicine because of their action on the heart **(cardiac glycosides)** all contain steroids as the aglycone. These include derivatives of digitalis and strophanthus such as **ouabain,** an inhibitor of the $Na^+$-$K^+$ ATPase of cell membranes. Other glycosides include antibiotics such as **streptomycin.**

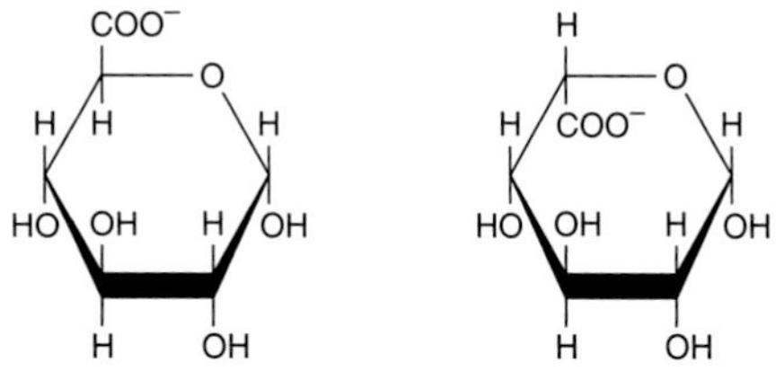

***Figure 13–8.*** α-D-Glucuronate (left) and β-L-iduronate (right).

## Deoxy Sugars Lack an Oxygen Atom

Deoxy sugars are those in which a hydroxyl group has been replaced by hydrogen. An example is **deoxyribose** (Figure 13–9) in DNA. The deoxy sugar L-fucose (Figure 13–15) occurs in glycoproteins; 2-deoxyglucose is used experimentally as an inhibitor of glucose metabolism.

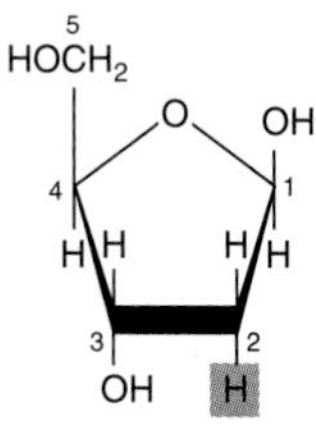


***Figure 13–9.*** 2-Deoxy-D-ribofuranose (β form).

## Amino Sugars (Hexosamines) Are Components of Glycoproteins, Gangliosides, & Glycosaminoglycans

The amino sugars include D-glucosamine, a constituent of hyaluronic acid (Figure 13–10), D-galactosamine (chondrosamine), a constituent of chondroitin; and D-mannosamine. Several **antibiotics** (eg, erythromycin) contain amino sugars believed to be important for their antibiotic activity.

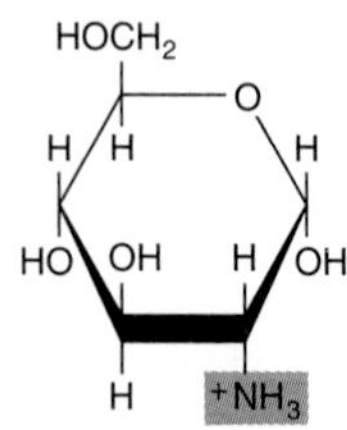


***Figure 13–10.*** Glucosamine (2-amino-D-glucopyranose) (α form). Galactosamine is 2-amino-D-galactopyranose. Both glucosamine and galactosamine occur as N-acetyl derivatives in more complex carbohydrates, eg, glycoproteins.

# MALTOSE, SUCROSE, & LACTOSE ARE IMPORTANT DISACCHARIDES

The physiologically important disaccharides are maltose, sucrose, and lactose (Table 13–4; Figure 13–11). Hydrolysis of sucrose yields a mixture of glucose and

***Table 13–4.*** Disaccharides.

| Sugar | Source | Clinical Significance |
|---|---|---|
| Maltose | Digestion by amylase or hydrolysis of starch. Germinating cereals and malt. | |
| Lactose | Milk. May occur in urine during pregnancy. | In lactase deficiency, malabsorption leads to diarrhea and flatulence. |
| Sucrose | Cane and beet sugar. Sorghum. Pineapple. Carrot roots. | In sucrase deficiency, malabsorption leads to diarrhea and flatulence. |
| Trehalose[1] | Fungi and yeasts. The major sugar of insect hemolymph. | |

[1] *O*-α-D-Glucopyranosyl-(1 → 1)-α-D-glucopyranoside.

fructose which is called "invert sugar" because the strongly levorotatory fructose changes (inverts) the previous dextrorotatory action of sucrose.

## POLYSACCHARIDES SERVE STORAGE & STRUCTURAL FUNCTIONS

Polysaccharides include the following physiologically important carbohydrates.

**Starch** is a homopolymer of glucose forming an α-glucosidic chain, called a **glucosan** or **glucan.** It is the most abundant dietary carbohydrate in cereals, potatoes, legumes, and other vegetables. The two main constituents are **amylose** (15–20%), which has a nonbranching helical structure (Figure 13–12); and **amylopectin** (80–85%), which consists of branched chains composed of 24–30 glucose residues united by 1 → 4 linkages in the chains and by 1 → 6 linkages at the branch points.

**Glycogen** (Figure 13–13) is the storage polysaccharide in animals. It is a more highly branched structure than amylopectin, with chains of 12–14 α-D-glucopyranose residues (in α[1 → 4]-glucosidic linkage), with branching by means of α(1 → 6)-glucosidic bonds.

**Maltose**

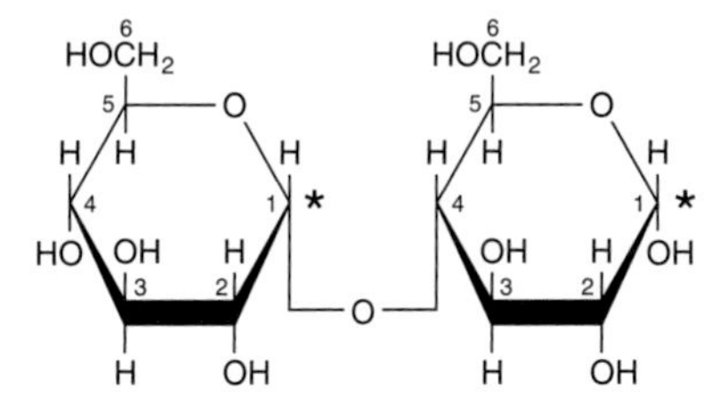


*O*-α-D-Glucopyranosyl-(1 → 4)-α-D-glucopyranose

**Lactose**

*O*-β-D-Galactopyranosyl-(1 → 4)-β-D-glucopyranose

**Sucrose**

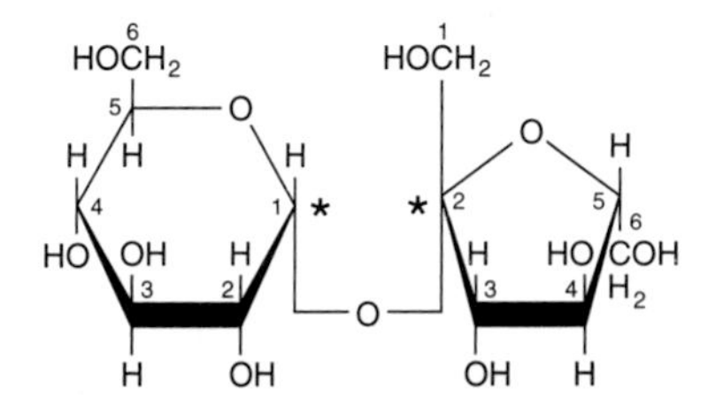


*O*-α-D-Glucopyranosyl-(1 → 2)-β-D-fructofuranoside

***Figure 13–11.*** Structures of important disaccharides. The α and β refer to the configuration at the anomeric carbon atom (asterisk). When the anomeric carbon of the second residue takes part in the formation of the glycosidic bond, as in sucrose, the residue becomes a glycoside known as a furanoside or pyranoside. As the disaccharide no longer has an anomeric carbon with a free potential aldehyde or ketone group, it no longer exhibits reducing properties. The configuration of the β-fructofuranose residue in sucrose results from turning the β-fructofuranose molecule depicted in Figure 13–4 through 180 degrees and inverting it.

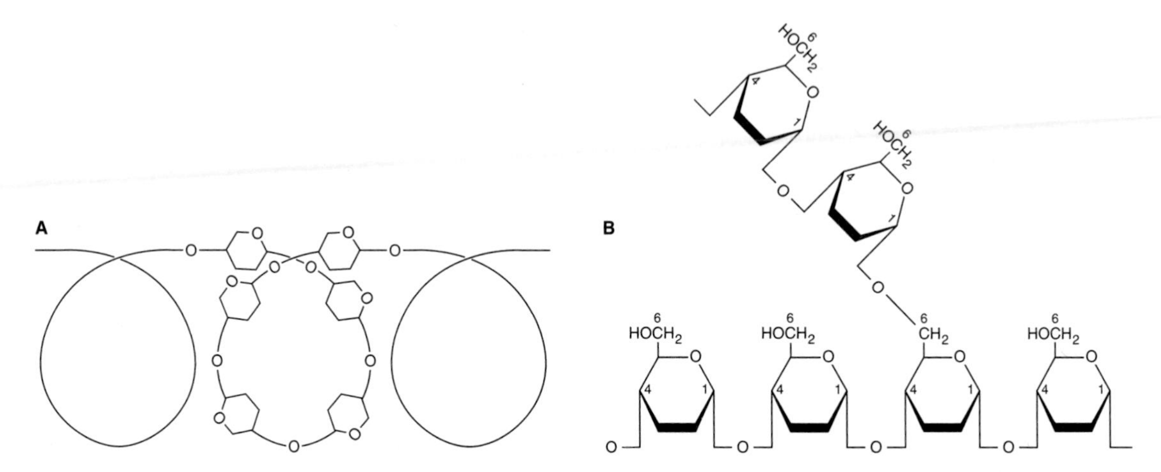


***Figure 13–12.*** Structure of starch. **A:** Amylose, showing helical coil structure. **B:** Amylopectin, showing 1 → 6 branch point.

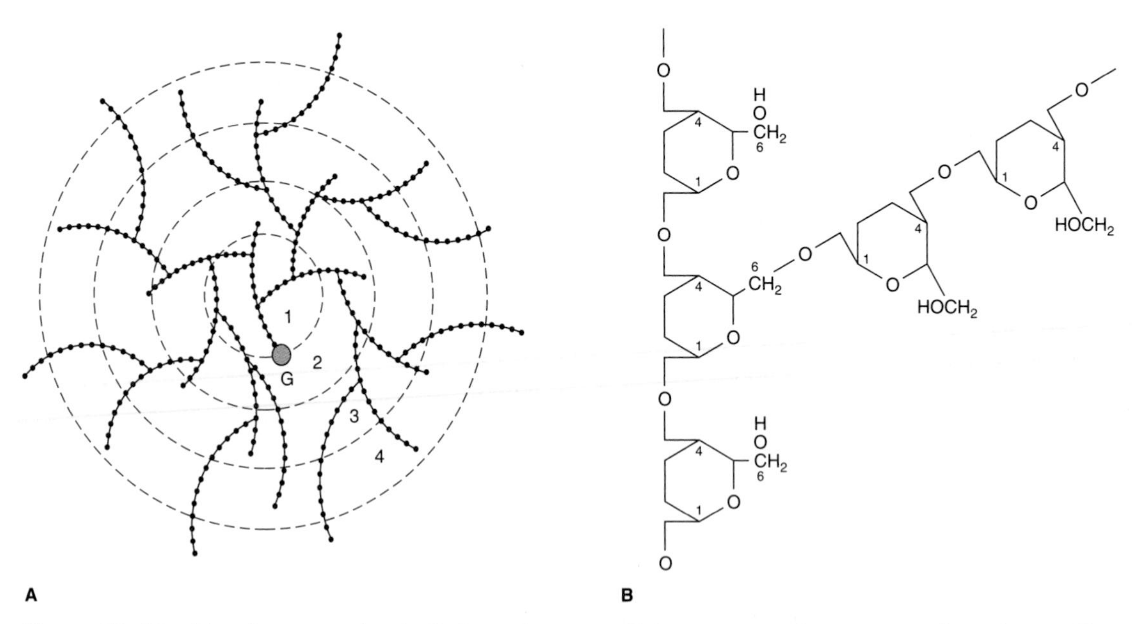


***Figure 13–13.*** The glycogen molecule. **A:** General structure. **B:** Enlargement of structure at a branch point. The molecule is a sphere approximately 21 nm in diameter that can be visualized in electron micrographs. It has a molecular mass of $10^7$ Da and consists of polysaccharide chains each containing about 13 glucose residues. The chains are either branched or unbranched and are arranged in 12 concentric layers (only four are shown in the figure). The branched chains (each has two branches) are found in the inner layers and the unbranched chains in the outer layer. (G, glycogenin, the primer molecule for glycogen synthesis.)

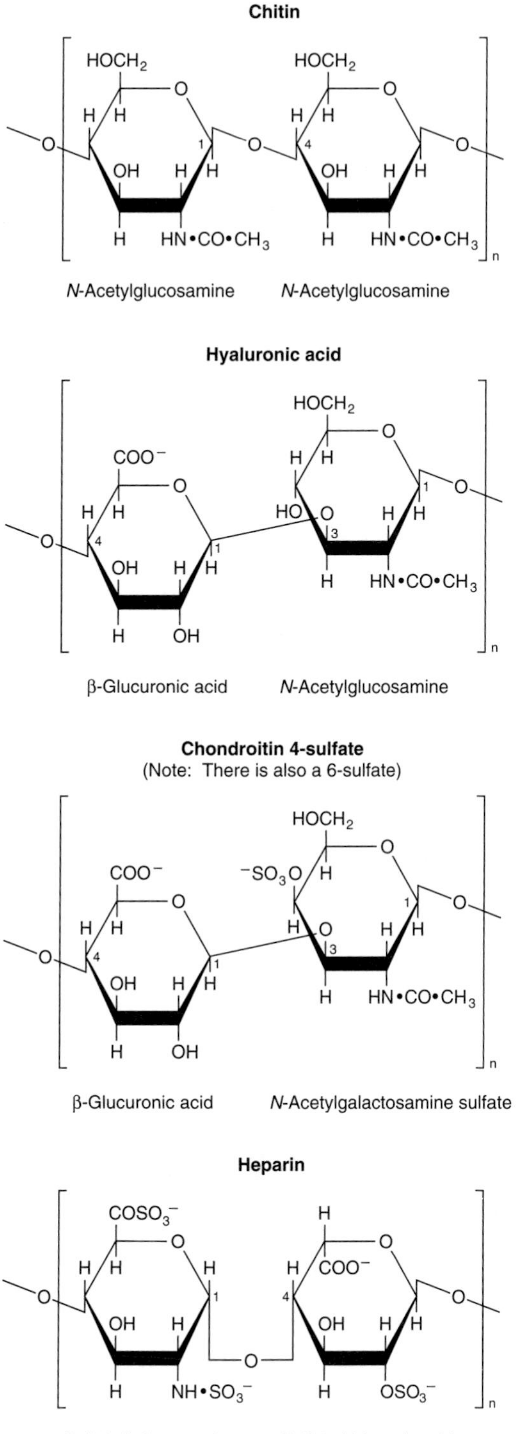

**Inulin** is a polysaccharide of fructose (and hence a fructosan) found in tubers and roots of dahlias, artichokes, and dandelions. It is readily soluble in water and is used to determine the glomerular filtration rate. **Dextrins** are intermediates in the hydrolysis of starch. **Cellulose** is the chief constituent of the framework of plants. It is insoluble and consists of β-D-glucopyranose units linked by β(1 → 4) bonds to form long, straight chains strengthened by cross-linked hydrogen bonds. Cellulose cannot be digested by mammals because of the absence of an enzyme that hydrolyzes the β linkage. It is an important source of "bulk" in the diet. Microorganisms in the gut of ruminants and other herbivores can hydrolyze the β linkage and ferment the products to short-chain fatty acids as a major energy source. There is limited bacterial metabolism of cellulose in the human colon. **Chitin** is a structural polysaccharide in the exoskeleton of crustaceans and insects and also in mushrooms. It consists of *N*-acetyl-D-glucosamine units joined by β (1 → 4)-glycosidic linkages (Figure 13–14).

**Glycosaminoglycans** (mucopolysaccharides) are complex carbohydrates characterized by their content of **amino sugars** and **uronic acids.** When these chains are attached to a protein molecule, the result is a **proteoglycan.** Proteoglycans provide the ground or packing substance of connective tissues. Their property of holding large quantities of water and occupying space, thus cushioning or lubricating other structures, is due to the large number of —OH groups and negative charges on the molecules, which, by repulsion, keep the carbohydrate chains apart. Examples are **hyaluronic acid, chondroitin sulfate,** and **heparin** (Figure 13–14).

**Glycoproteins** (mucoproteins) occur in many different situations in fluids and tissues, including the cell membranes (Chapters 41 and 47). They are proteins

***Figure 13–14.*** Structure of some complex polysaccharides and glycosaminoglycans.

***Table 13–5.*** Carbohydrates found in glycoproteins.

| | |
|---|---|
| Hexoses | Mannose (Man)<br>Galactose (Gal) |
| Acetyl hexosamines | *N*-Acetylglucosamine (GlcNAc)<br>*N*-Acetylgalactosamine (GalNAc) |
| Pentoses | Arabinose (Ara)<br>Xylose (Xyl) |
| Methyl pentose | L-Fucose (Fuc; see Figure 13–15) |
| Sialic acids | N-Acyl derivatives of neuraminic acid, eg, *N*-acetylneuraminic acid (NeuAc; see Figure 13–16), the predominant sialic acid. |

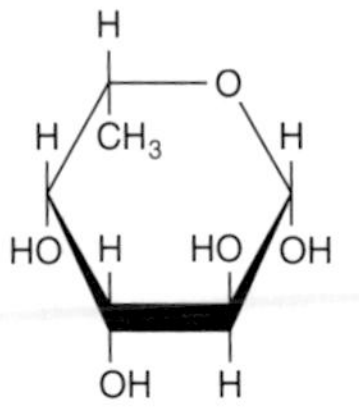

***Figure 13–15.*** β-L-Fucose (6-deoxy-β-L-galactose).

containing branched or unbranched oligosaccharide chains (see Table 13–5). The **sialic acids** are N- or O-acyl derivatives of neuraminic acid (Figure 13–16). **Neuraminic acid** is a nine-carbon sugar derived from mannosamine (an epimer of glucosamine) and pyruvate. Sialic acids are constituents of both **glycoproteins** and **gangliosides** (Chapters 14 and 47).

## CARBOHYDRATES OCCUR IN CELL MEMBRANES & IN LIPOPROTEINS

In addition to the lipid of cell membranes (see Chapters 14 and 41), approximately 5% is carbohydrate in glycoproteins and glycolipids. Carbohydrates are also present in apo B of lipoproteins. Their presence on the outer surface of the plasma membrane (the **glycocalyx**) has been shown with the use of plant **lectins,** protein agglutinins that bind with specific glycosyl residues. For example, **concanavalin A** binds α-glucosyl and α-mannosyl residues. **Glycophorin** is a major integral membrane glycoprotein of human erythrocytes and spans the lipid membrane, having free polypeptide portions outside both the external and internal (cytoplasmic) surfaces. Carbohydrate chains are only attached to the amino terminal portion outside the external surface (Chapter 41).

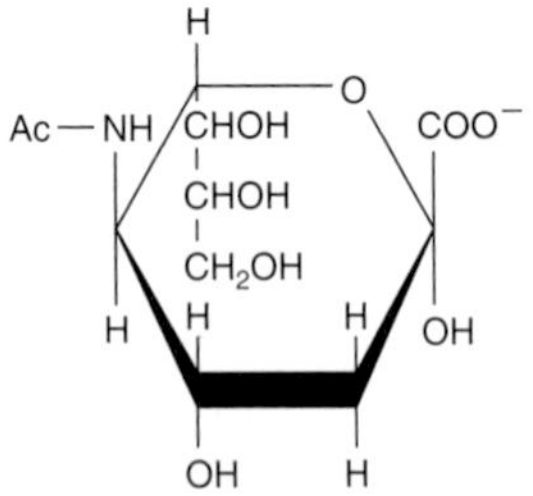

***Figure 13–16.*** Structure of *N*-acetylneuraminic acid, a sialic acid (Ac = $CH_3$—CO—).

## SUMMARY

- Carbohydrates are major constituents of animal food and animal tissues. They are characterized by the type and number of monosaccharide residues in their molecules.
- Glucose is the most important carbohydrate in mammalian biochemistry because nearly all carbohydrate in food is converted to glucose for metabolism.
- Sugars have large numbers of stereoisomers because they contain several asymmetric carbon atoms.
- The monosaccharides include glucose, the "blood sugar"; and ribose, an important constituent of nucleotides and nucleic acids.
- The disaccharides include maltose (glucosyl glucose), an intermediate in the digestion of starch; sucrose (glucosyl fructose), important as a dietary constituent containing fructose; and lactose (galactosyl glucose), in milk.
- Starch and glycogen are storage polymers of glucose in plants and animals, respectively. Starch is the major source of energy in the diet.
- Complex carbohydrates contain other sugar derivatives such as amino sugars, uronic acids, and sialic acids. They include proteoglycans and glycosaminoglycans, associated with structural elements of the tissues; and glycoproteins, proteins containing attached oligosaccharide chains. They are found in many situations including the cell membrane.

## REFERENCES


Binkley RW: *Modern Carbohydrate Chemistry.* Marcel Dekker, 1988.

Collins PM (editor): *Carbohydrates.* Chapman & Hall, 1988.

El-Khadem HS: *Carbohydrate Chemistry: Monosaccharides and Their Oligomers.* Academic Press, 1988.

Lehman J (editor) (translated by Haines A.): *Carbohydrates: Structure and Biology.* Thieme, 1998.

Lindahl U, Höök M: Glycosaminoglycans and their binding to biological macromolecules. Annu Rev Biochem 1978;47:385.

Melendes-Hevia E, Waddell TG, Shelton ED: Optimization of molecular design in the evolution of metabolism: the glycogen molecule. Biochem J 1993;295:477.

# Lipids of Physiologic Significance 14

Peter A. Mayes, PhD, DSc, & Kathleen M. Botham, PhD, DS

## BIOMEDICAL IMPORTANCE

The lipids are a heterogeneous group of compounds, including fats, oils, steroids, waxes, and related compounds, which are related more by their physical than by their chemical properties. They have the common property of being (1) relatively **insoluble in water** and (2) **soluble in nonpolar solvents** such as ether and chloroform. They are important dietary constituents not only because of their high energy value but also because of the fat-soluble vitamins and the essential fatty acids contained in the fat of natural foods. Fat is stored in **adipose tissue,** where it also serves as a thermal insulator in the subcutaneous tissues and around certain organs. Nonpolar lipids act as **electrical insulators,** allowing rapid propagation of depolarization waves along **myelinated nerves.** Combinations of lipid and protein (lipoproteins) are important cellular constituents, occurring both in the cell **membrane** and in the mitochondria, and serving also as the means of **transporting lipids** in the blood. Knowledge of lipid biochemistry is necessary in understanding many important biomedical areas, eg, **obesity, diabetes mellitus, atherosclerosis,** and the role of various **polyunsaturated fatty acids** in nutrition and health.

## LIPIDS ARE CLASSIFIED AS SIMPLE OR COMPLEX

1. **Simple lipids:** Esters of fatty acids with various alcohols.
   a. **Fats:** Esters of fatty acids with glycerol. **Oils** are fats in the liquid state.
   b. **Waxes:** Esters of fatty acids with higher molecular weight monohydric alcohols.
2. **Complex lipids:** Esters of fatty acids containing groups in addition to an alcohol and a fatty acid.
   a. **Phospholipids:** Lipids containing, in addition to fatty acids and an alcohol, a phosphoric acid residue. They frequently have nitrogen-containing bases and other substituents, eg, in **glycerophospholipids** the alcohol is glycerol and in **sphingophospholipids** the alcohol is sphingosine.
   b. **Glycolipids (glycosphingolipids):** Lipids containing a fatty acid, sphingosine, and carbohydrate.
   c. **Other complex lipids:** Lipids such as sulfolipids and aminolipids. Lipoproteins may also be placed in this category.
3. **Precursor and derived lipids:** These include fatty acids, glycerol, steroids, other alcohols, fatty aldehydes, and ketone bodies (Chapter 22), hydrocarbons, lipid-soluble vitamins, and hormones.

Because they are uncharged, acylglycerols (glycerides), cholesterol, and cholesteryl esters are termed **neutral lipids.**

## FATTY ACIDS ARE ALIPHATIC CARBOXYLIC ACIDS

Fatty acids occur mainly as esters in natural fats and oils but do occur in the unesterified form as **free fatty acids,** a transport form found in the plasma. Fatty acids that occur in natural fats are usually straight-chain derivatives containing an even number of carbon atoms. The chain may be **saturated** (containing no double bonds) or **unsaturated** (containing one or more double bonds).

### Fatty Acids Are Named After Corresponding Hydrocarbons

The most frequently used systematic nomenclature names the fatty acid after the hydrocarbon with the same number and arrangement of carbon atoms, with **-oic** being substituted for the final **-e** (Genevan system). Thus, saturated acids end in **-anoic,** eg, octanoic acid, and unsaturated acids with double bonds end in **-enoic,** eg, octadecenoic acid (oleic acid).

Carbon atoms are numbered from the carboxyl carbon (carbon No. 1). The carbon atoms adjacent to the carboxyl carbon (Nos. 2, 3, and 4) are also known as the $\alpha$, $\beta$, and $\gamma$ carbons, respectively, and the terminal methyl carbon is known as the $\omega$ or n-carbon.

Various conventions use $\Delta$ for indicating the number and position of the double bonds (Figure 14–1); eg, $\Delta^9$ indicates a double bond between carbons 9 and 10 of the fatty acid; $\omega 9$ indicates a double bond on the ninth carbon counting from the $\omega$- carbon. In animals, additional double bonds are introduced only between the existing double bond (eg, $\omega 9$, $\omega 6$, or $\omega 3$) and the

18:1;9 *or* $\Delta^9$ 18:1

$$\overset{18}{C}H_3(CH_2)_7\overset{10}{C}H{=}\overset{9}{C}H(CH_2)_7\overset{1}{C}OOH$$

*or*

ω9,C18:1 *or* n–9, 18:1

$$\underset{n\ 17}{\overset{\omega\ \ 2\ \ 3\ \ 4\ \ 5\ \ 6\ \ 7\ \ 8\ \ 9}{CH_3CH_2CH_2CH_2CH_2CH_2CH_2CH_2}}\underset{10}{\overset{10}{C}H}{=}\underset{9}{C}H(CH_2)_7\underset{1}{\overset{18}{C}OOH}$$

***Figure 14–1.*** Oleic acid. n – 9 (n minus 9) is equivalent to ω9.

carboxyl carbon, leading to three series of fatty acids known as the ω9, ω6, and ω3 families, respectively.

## Saturated Fatty Acids Contain No Double Bonds

Saturated fatty acids may be envisaged as based on acetic acid ($CH_3$—COOH) as the first member of the series in which —$CH_2$— is progressively added between the terminal $CH_3$— and —COOH groups. Examples are shown in Table 14–1. Other higher members of the series are known to occur, particularly in waxes. A few branched-chain fatty acids have also been isolated from both plant and animal sources.

***Table 14–1.*** Saturated fatty acids.

| Common Name | Number of C Atoms | |
|---|---|---|
| Acetic | 2 | Major end product of carbohydrate fermentation by rumen organisms[1] |
| Propionic | 3 | An end product of carbohydrate fermentation by rumen organisms[1] |
| Butyric | 4 | In certain fats in small amounts (especially butter). An end product of carbohydrate fermentation by rumen organisms[1] |
| Valeric | 5 | |
| Caproic | 6 | |
| Lauric | 12 | Spermaceti, cinnamon, palm kernel, coconut oils, laurels, butter |
| Myristic | 14 | Nutmeg, palm kernel, coconut oils, myrtles, butter |
| Palmitic | 16 | Common in all animal and plant fats |
| Stearic | 18 | |

[1]Also formed in the cecum of herbivores and to a lesser extent in the colon of humans.

## Unsaturated Fatty Acids Contain One or More Double Bonds (Table 14–2)

Fatty acids may be further subdivided as follows:

(1) **Monounsaturated** (monoethenoid, monoenoic) acids, containing one double bond.

(2) **Polyunsaturated** (polyethenoid, polyenoic) acids, containing two or more double bonds.

(3) **Eicosanoids:** These compounds, derived from eicosa- (20-carbon) polyenoic fatty acids, comprise the **prostanoids, leukotrienes (LTs),** and **lipoxins (LXs).** Prostanoids include **prostaglandins (PGs), prostacyclins (PGIs),** and **thromboxanes (TXs).**

**Prostaglandins** exist in virtually every mammalian tissue, acting as local hormones; they have important physiologic and pharmacologic activities. They are synthesized in vivo by cyclization of the center of the carbon chain of 20-carbon (eicosanoic) polyunsaturated fatty acids (eg, arachidonic acid) to form a cyclopentane ring (Figure 14–2). A related series of compounds, the **thromboxanes,** have the cyclopentane ring interrupted with an oxygen atom (oxane ring) (Figure 14–3). Three different eicosanoic fatty acids give rise to three groups of eicosanoids characterized by the number of double bonds in the side chains, eg, $PG_1$, $PG_2$, $PG_3$. Different substituent groups attached to the rings give rise to series of prostaglandins and thromboxanes, labeled A, B, etc—eg, the "E" type of prostaglandin (as in $PGE_2$) has a keto group in position 9, whereas the "F" type has a hydroxyl group in this position. The **leukotrienes** and **lipoxins** are a third group of eicosanoid derivatives formed via the lipoxygenase pathway (Figure 14–4). They are characterized by the presence of three or four conjugated double bonds, respectively. Leukotrienes cause bronchoconstriction as well as being potent proinflammatory agents and play a part in **asthma.**

## Most Naturally Occurring Unsaturated Fatty Acids Have *cis* Double Bonds

The carbon chains of saturated fatty acids form a zigzag pattern when extended, as at low temperatures. At higher temperatures, some bonds rotate, causing chain shortening, which explains why biomembranes become thinner with increases in temperature. A type of **geometric isomerism** occurs in unsaturated fatty acids, depending on the orientation of atoms or groups around the axes of double bonds, which do not allow rotation. If the acyl chains are on the same side of the bond, it is *cis-*, as in oleic acid; if on opposite sides, it is *trans-*, as in elaidic acid, the *trans* isomer of oleic acid (Fig-

***Table 14–2.*** Unsaturated fatty acids of physiologic and nutritional significance.

| Number of C Atoms and Number and Position of Double Bonds | Family | Common Name | Systematic Name | Occurrence |
|---|---|---|---|---|
| **Monoenoic acids (one double bond)** | | | | |
| 16:1;9 | ω7 | Palmitoleic | *cis*-9-Hexadecenoic | In nearly all fats. |
| 18:1;9 | ω9 | Oleic | *cis*-9-Octadecenoic | Possibly the most common fatty acid in natural fats. |
| 18:1;9 | ω9 | Elaidic | *trans*-9-Octadecenoic | Hydrogenated and ruminant fats. |
| **Dienoic acids (two double bonds)** | | | | |
| 18:2;9,12 | ω6 | Linoleic | all-*cis*-9,12-Octadecadienoic | Corn, peanut, cottonseed, soybean, and many plant oils. |
| **Trienoic acids (three double bonds)** | | | | |
| 18:3;6,9,12 | ω6 | γ-Linolenic | all-*cis*-6,9,12-Octadecatrienoic | Some plants, eg, oil of evening primrose, borage oil; minor fatty acid in animals. |
| 18:3;9,12,15 | ω3 | α-Linolenic | all-*cis*-9,12,15-Octadecatrienoic | Frequently found with linoleic acid but particularly in linseed oil. |
| **Tetraenoic acids (four double bonds)** | | | | |
| 20:4;5,8,11,14 | ω6 | Arachidonic | all-*cis*-5,8,11,14-Eicosatetraenoic | Found in animal fats and in peanut oil; important component of phospholipids in animals. |
| **Pentaenoic acids (five double bonds)** | | | | |
| 20:5;5,8,11,14,17 | ω3 | Timnodonic | all-*cis*-5,8,11,14,17-Eicosapentaenoic | Important component of fish oils, eg, cod liver, mackerel, menhaden, salmon oils. |
| **Hexaenoic acids (six double bonds)** | | | | |
| 22:6;4,7,10,13,16,19 | ω3 | Cervonic | all-*cis*-4,7,10,13,16,19-Docosahexaenoic | Fish oils, phospholipids in brain. |

ure 14–5). Naturally occurring unsaturated long-chain fatty acids are nearly all of the *cis* configuration, the molecules being "bent" 120 degrees at the double bond. Thus, oleic acid has an L shape, whereas elaidic acid remains "straight." Increase in the number of *cis* double bonds in a fatty acid leads to a variety of possible spatial configurations of the molecule—eg, arachidonic acid, with four *cis* double bonds, has "kinks" or a U shape. This has profound significance on molecular packing in membranes and on the positions occupied by fatty acids in more complex molecules such as phospholipids. *Trans* double bonds alter these spatial relationships. *Trans* fatty acids are present in certain foods, arising as a by-product of the saturation of fatty acids during hydrogenation, or "hardening," of natural oils in the manufacture of margarine. An additional small

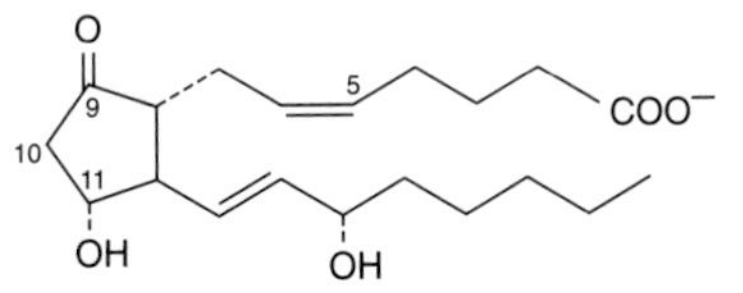

***Figure 14–2.*** Prostaglandin $E_2$ ($PGE_2$).

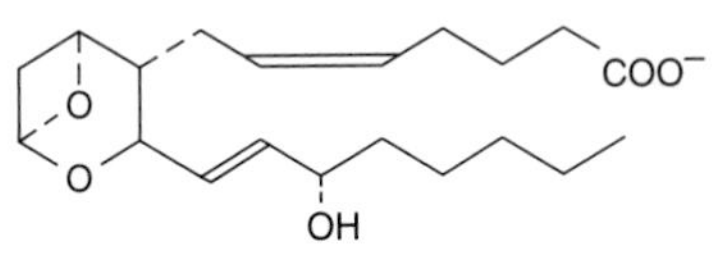

***Figure 14–3.*** Thromboxane $A_2$ ($TXA_2$).

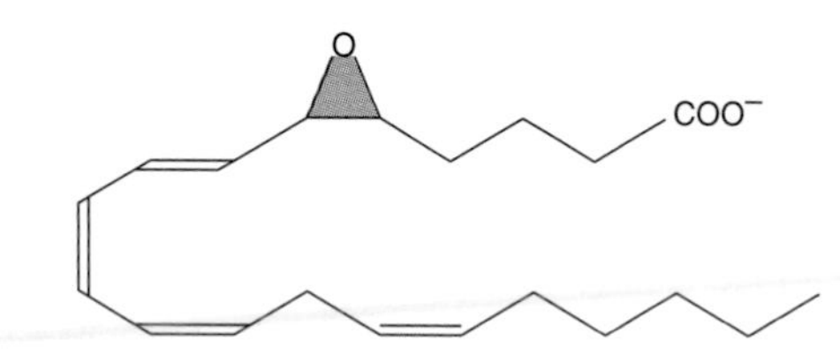


***Figure 14–4.*** Leukotriene $A_4$ ($LTA_4$).

contribution comes from the ingestion of ruminant fat that contains *trans* fatty acids arising from the action of microorganisms in the rumen.

### Physical and Physiologic Properties of Fatty Acids Reflect Chain Length and Degree of Unsaturation

The melting points of even-numbered-carbon fatty acids increase with chain length and decrease according to unsaturation. A triacylglycerol containing three saturated fatty acids of 12 carbons or more is solid at body temperature, whereas if the fatty acid residues are 18:2, it is liquid to below 0 °C. In practice, natural acylglycerols contain a mixture of fatty acids tailored to suit their functional roles. The membrane lipids, which must be fluid at all environmental temperatures, are more unsaturated than storage lipids. Lipids in tissues that are subject to cooling, eg, in hibernators or in the extremities of animals, are more unsaturated.

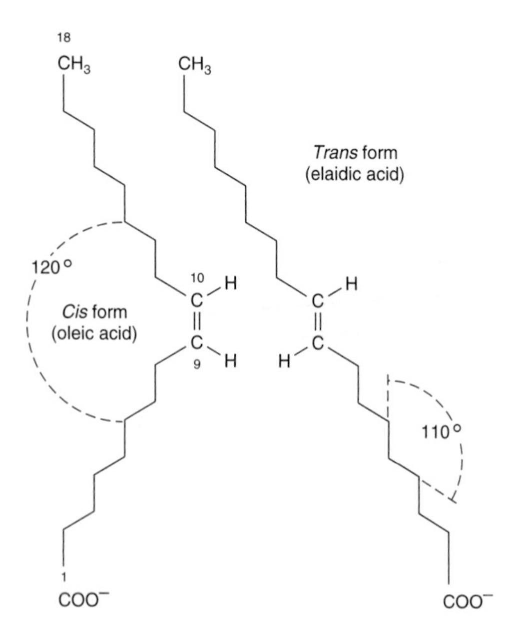


***Figure 14–5.*** Geometric isomerism of $\Delta^9$, 18:1 fatty acids (oleic and elaidic acids).

## TRIACYLGLYCEROLS (TRIGLYCERIDES)* ARE THE MAIN STORAGE FORMS OF FATTY ACIDS

The triacylglycerols (Figure 14–6) are esters of the trihydric alcohol glycerol and fatty acids. Mono- and diacylglycerols wherein one or two fatty acids are esterified with glycerol are also found in the tissues. These are of particular significance in the synthesis and hydrolysis of triacylglycerols.

### Carbons 1 & 3 of Glycerol Are Not Identical

To number the carbon atoms of glycerol unambiguously, the *-sn-* (stereochemical numbering) system is used. It is important to realize that carbons 1 and 3 of glycerol are not identical when viewed in three dimensions (shown as a projection formula in Figure 14–7). Enzymes readily distinguish between them and are nearly always specific for one or the other carbon; eg, glycerol is always phosphorylated on *sn*-3 by glycerol kinase to give glycerol 3-phosphate and not glycerol 1-phosphate.

## PHOSPHOLIPIDS ARE THE MAIN LIPID CONSTITUENTS OF MEMBRANES

Phospholipids may be regarded as derivatives of **phosphatidic acid** (Figure 14–8), in which the phosphate is esterified with the —OH of a suitable alcohol. Phosphatidic acid is important as an intermediate in the synthesis of triacylglycerols as well as phosphoglycerols but is not found in any great quantity in tissues.

### Phosphatidylcholines (Lecithins) Occur in Cell Membranes

Phosphoacylglycerols containing choline (Figure 14–8) are the most abundant phospholipids of the cell mem-

* According to the standardized terminology of the International Union of Pure and Applied Chemistry (IUPAC) and the International Union of Biochemistry (IUB), the monoglycerides, diglycerides, and triglycerides should be designated monoacylglycerols, diacylglycerols, and triacylglycerols, respectively. However, the older terminology is still widely used, particularly in clinical medicine.

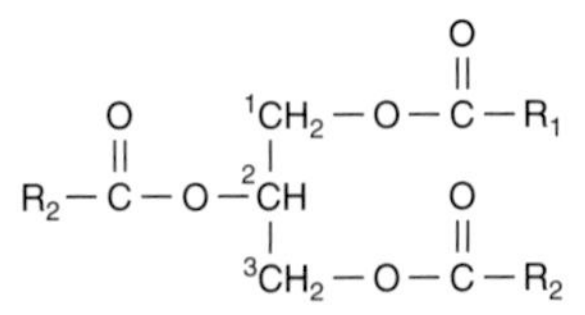

***Figure 14–6.*** Triacylglycerol.

brane and represent a large proportion of the body's store of choline. Choline is important in nervous transmission, as acetylcholine, and as a store of labile methyl groups. **Dipalmitoyl lecithin** is a very effective surface-active agent and a major constituent of the **surfactant** preventing adherence, due to surface tension, of the inner surfaces of the lungs. Its absence from the lungs of premature infants causes **respiratory distress syndrome.** Most phospholipids have a saturated acyl radical in the *sn*-1 position but an unsaturated radical in the *sn*-2 position of glycerol.

**Phosphatidylethanolamine (cephalin)** and **phosphatidylserine** (found in most tissues) differ from phosphatidylcholine only in that ethanolamine or serine, respectively, replaces choline (Figure 14–8).

## Phosphatidylinositol Is a Precursor of Second Messengers

The inositol is present in **phosphatidylinositol** as the stereoisomer, myoinositol (Figure 14–8). **Phosphatidylinositol 4,5-bisphosphate** is an important constituent of cell membrane phospholipids; upon stimulation by a suitable hormone agonist, it is cleaved into **diacylglycerol** and **inositol trisphosphate,** both of which act as internal signals or second messengers.

## Cardiolipin Is a Major Lipid of Mitochondrial Membranes

Phosphatidic acid is a precursor of **phosphatidylglycerol** which, in turn, gives rise to **cardiolipin** (Figure 14–8).

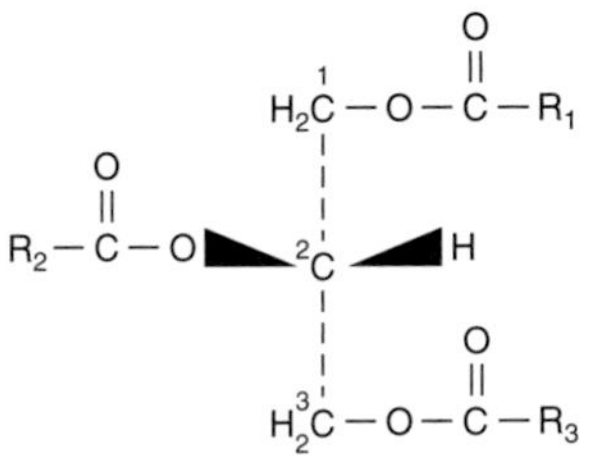

***Figure 14–7.*** Triacyl-*sn*-glycerol.

**Phosphatidic acid**

A **Choline**

B **Ethanolamine**

C **Serine**

D **Myoinositol**

E

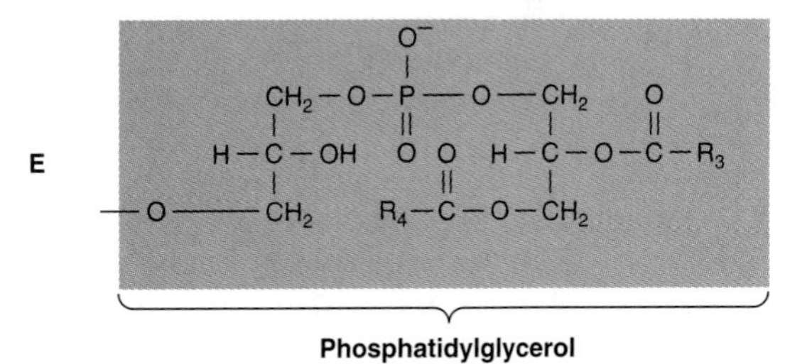


**Phosphatidylglycerol**

***Figure 14–8.*** Phosphatidic acid and its derivatives. The $O^-$ shown shaded in phosphatidic acid is substituted by the substituents shown to form in **(A)** 3-phosphatidylcholine, **(B)** 3-phosphatidylethanolamine, **(C)** 3-phosphatidylserine, **(D)** 3-phosphatidylinositol, and **(E)** cardiolipin (diphosphatidylglycerol).

## Lysophospholipids Are Intermediates in the Metabolism of Phosphoglycerols

These are phosphoacylglycerols containing only one acyl radical, eg, **lysophosphatidylcholine (lysolecithin),** important in the metabolism and interconversion of phospholipids (Figure 14–9).It is also found in oxidized lipoproteins and has been implicated in some of their effects in promoting **atherosclerosis.**

## Plasmalogens Occur in Brain & Muscle

These compounds constitute as much as 10% of the phospholipids of brain and muscle. Structurally, the plasmalogens resemble phosphatidylethanolamine but possess an ether link on the *sn*-1 carbon instead of the ester link found in acylglycerols. Typically, the alkyl radical is an unsaturated alcohol (Figure 14–10). In some instances, choline, serine, or inositol may be substituted for ethanolamine.

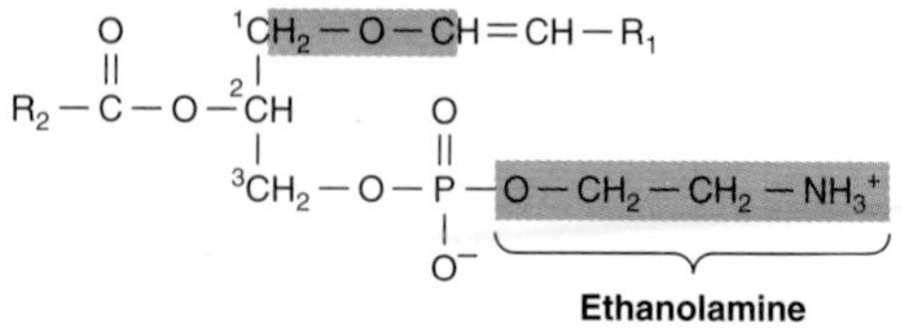

## Sphingomyelins Are Found in the Nervous System

Sphingomyelins are found in large quantities in brain and nerve tissue. On hydrolysis, the sphingomyelins yield a fatty acid, phosphoric acid, choline, and a complex amino alcohol, **sphingosine** (Figure 14–11). No glycerol is present. The combination of sphingosine plus fatty acid is known as **ceramide,** a structure also found in the glycosphingolipids (see below).

# GLYCOLIPIDS (GLYCOSPHINGOLIPIDS) ARE IMPORTANT IN NERVE TISSUES & IN THE CELL MEMBRANE

Glycolipids are widely distributed in every tissue of the body, particularly in nervous tissue such as brain. They occur particularly in the outer leaflet of the plasma membrane, where they contribute to **cell surface carbohydrates.**

The major glycolipids found in animal tissues are glycosphingolipids. They contain ceramide and one or more sugars. **Galactosylceramide** is a major glycosphingolipid of brain and other nervous tissue, found in relatively low amounts elsewhere. It contains a number of characteristic $C_{24}$ fatty acids, eg, cerebronic acid. Galactosylceramide (Figure 14–12) can be converted to sulfogalactosylceramide **(sulfatide),** present in high amounts in myelin. Glucosylceramide is the predominant simple glycosphingolipid of extraneural tissues, also occurring in the brain in small amounts. **Gangliosides** are complex glycosphingolipids derived from glucosylceramide that contain in addition one or more molecules of a **sialic acid.** Neuraminic acid (NeuAc; see Chapter 13) is the principal sialic acid found in human tissues. Gangliosides are also present in nervous tissues in high concentration. They appear to have receptor and other functions. The simplest ganglioside found in tissues is $G_{M3}$, which contains ceramide, one molecule of glucose, one molecule of galactose, and one molecule of NeuAc. In the shorthand nomenclature used, G represents ganglioside; M is a monosialo-containing species; and the subscript 3 is a number assigned on the basis of chromatographic migration. $G_{M1}$ (Figure 14–13), a more complex ganglioside derived from $G_{M3}$, is of considerable biologic interest, as it is known to be the receptor in human intestine for cholera toxin. Other gangliosides can contain anywhere from one to five molecules of sialic acid, giving rise to di-, trisialogangliosides, etc.

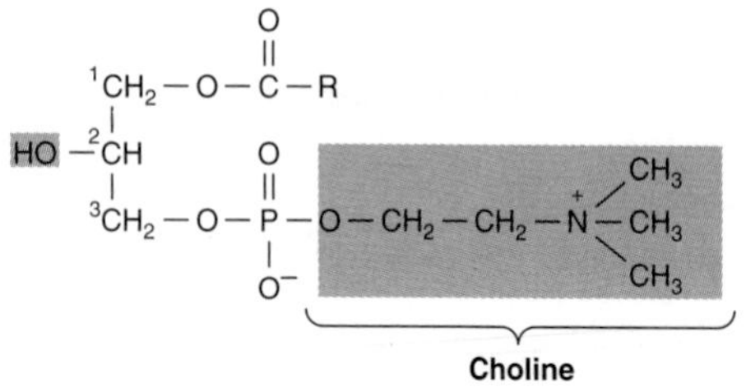

$^{1}CH_2-O-C(=O)-R$
$HO-{}^{2}CH$
$^{3}CH_2-O-P(=O)(O^-)-O-CH_2-CH_2-N^+(CH_3)_3$

Choline

**Figure 14–9.** Lysophosphatidylcholine (lysolecithin).

$^{1}CH_2-O-CH=CH-R_1$
$R_2-C(=O)-O-{}^{2}CH$
$^{3}CH_2-O-P(=O)(O^-)-O-CH_2-CH_2-NH_3^+$

Ethanolamine

**Figure 14–10.** Plasmalogen.

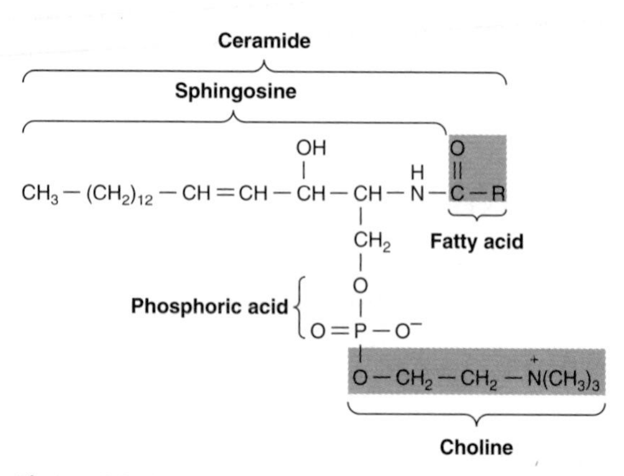


**Figure 14–11.** A sphingomyelin.

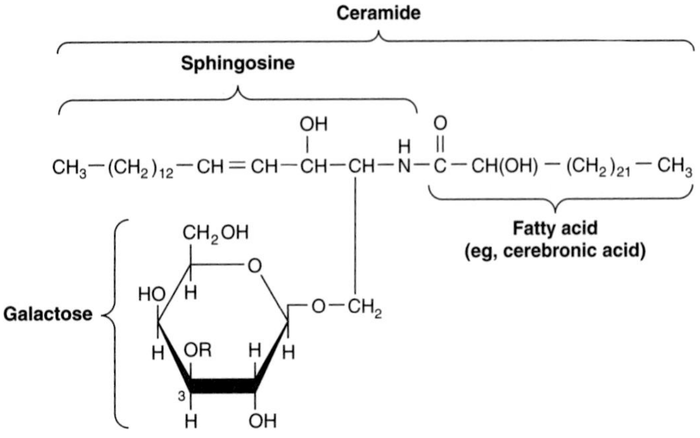

***Figure 14–12.*** Structure of galactosylceramide (galactocerebroside, R = H), and sulfogalactosylceramide (a sulfatide, R = $SO_4^{2-}$).

## STEROIDS PLAY MANY PHYSIOLOGICALLY IMPORTANT ROLES

**Cholesterol** is probably the best known steroid because of its association with **atherosclerosis.** However, biochemically it is also of significance because it is the precursor of a large number of equally important steroids that include the bile acids, adrenocortical hormones, sex hormones, D vitamins, cardiac glycosides, sitosterols of the plant kingdom, and some alkaloids.

All of the steroids have a similar cyclic nucleus resembling phenanthrene (rings A, B, and C) to which a cyclopentane ring (D) is attached. The carbon positions on the steroid nucleus are numbered as shown in Figure 14–14. It is important to realize that in structural formulas of steroids, a simple hexagonal ring denotes a completely saturated six-carbon ring with all valences satisfied by hydrogen bonds unless shown otherwise; ie, it is not a benzene ring. All double bonds are shown as such. Methyl side chains are shown as single bonds unattached at the farther (methyl) end. These occur typically at positions 10 and 13 (constituting C atoms 19 and 18). A side chain at position 17 is usual (as in cholesterol). If the compound has one or more hydroxyl groups and no carbonyl or carboxyl groups, it is a **sterol,** and the name terminates in -ol.

### Because of Asymmetry in the Steroid Molecule, Many Stereoisomers Are Possible

Each of the six-carbon rings of the steroid nucleus is capable of existing in the three-dimensional conformation either of a "chair" or a "boat" (Figure 14–15). In naturally occurring steroids, virtually all the rings are in the "chair" form, which is the more stable conformation. With respect to each other, the rings can be either *cis* or *trans* (Figure 14–16). The junction between the A and B rings can be *cis* or *trans* in naturally occurring steroids. That between B and C is *trans,* as is usually the C/D junction. Bonds attaching substituent groups above the plane of the rings (β bonds) are shown with bold solid lines, whereas those bonds attaching groups below (α bonds) are indicated with broken lines. The A ring of a 5α steroid is always *trans* to the B ring, whereas it is *cis* in a 5β steroid. The methyl groups attached to $C_{10}$ and $C_{13}$ are invariably in the β configuration.

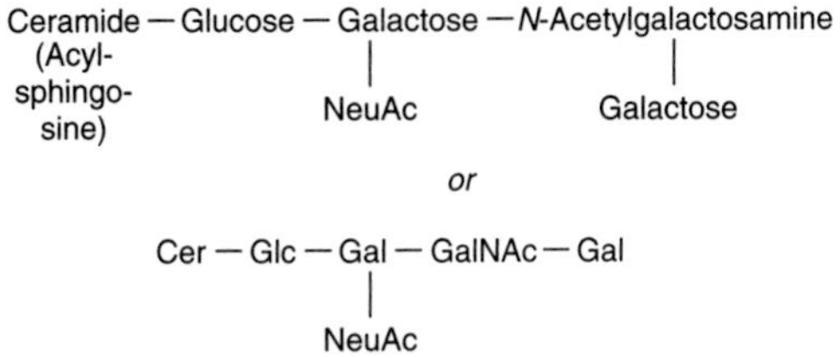

Ceramide — Glucose — Galactose — *N*-Acetylgalactosamine
(Acylsphingosine) | NeuAc | Galactose

*or*

Cer — Glc — Gal — GalNAc — Gal
| NeuAc

***Figure 14–13.*** $G_{M1}$ ganglioside, a monosialoganglioside, the receptor in human intestine for cholera toxin.

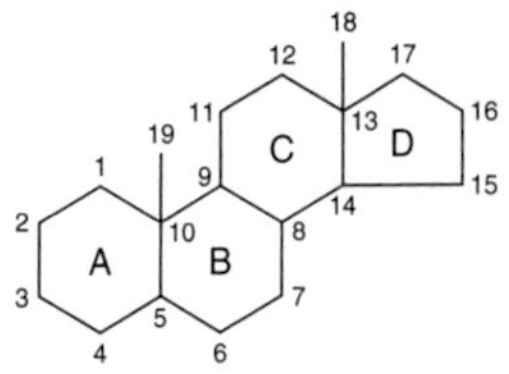

***Figure 14–14.*** The steroid nucleus.

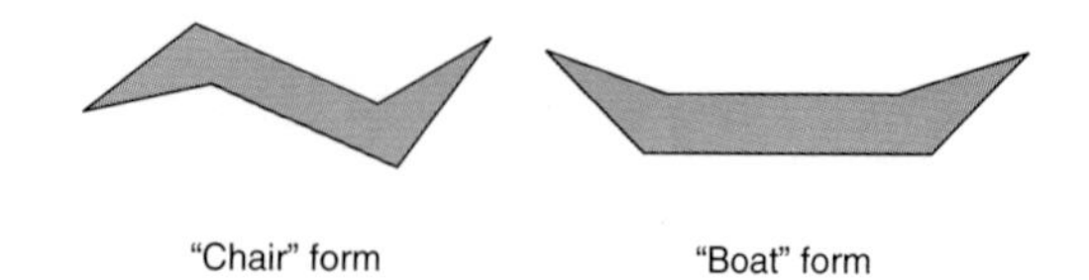


***Figure 14–15.*** Conformations of stereoisomers of the steroid nucleus.

### Cholesterol Is a Significant Constituent of Many Tissues

Cholesterol (Figure 14–17) is widely distributed in all cells of the body but particularly in nervous tissue. It is a major constituent of the plasma membrane and of plasma lipoproteins. It is often found as **cholesteryl ester,** where the hydroxyl group on position 3 is esterified with a long-chain fatty acid. It occurs in animals but not in plants.

### Ergosterol Is a Precursor of Vitamin D

Ergosterol occurs in plants and yeast and is important as a precursor of vitamin D (Figure 14–18). When irradiated with ultraviolet light, it acquires antirachitic properties consequent to the opening of ring B.

### Polyprenoids Share the Same Parent Compound as Cholesterol

Although not steroids, these compounds are related because they are synthesized, like cholesterol (Figure 26–2), from five-carbon isoprene units (Figure 14–19). They include **ubiquinone** (Chapter 12), a member of the respiratory chain in mitochondria, and the long-chain alcohol **dolichol** (Figure 14–20), which takes part in glycoprotein synthesis by transferring carbohydrate residues to asparagine residues of the polypeptide (Chapter 47). Plant-derived isoprenoid compounds include rubber, camphor, the fat-soluble vitamins A, D, E, and K, and β-carotene (provitamin A).

## LIPID PEROXIDATION IS A SOURCE OF FREE RADICALS

Peroxidation **(auto-oxidation)** of lipids exposed to oxygen is responsible not only for deterioration of foods **(rancidity)** but also for damage to tissues in vivo, where it may be a cause of cancer, inflammatory diseases, atherosclerosis, and aging. The deleterious effects are considered to be caused by free radicals ($ROO^•$, $RO^•$, $OH^•$) produced during peroxide formation from fatty acids containing methylene-interrupted double bonds, ie, those found in the naturally occurring polyunsaturated fatty acids (Figure 14–21). Lipid peroxidation is a chain reaction providing a continuous supply of free radicals that initiate further peroxidation. The whole process can be depicted as follows:

**(1)** Initiation:

$$ROOH + Metal^{(n)+} \rightarrow ROO^{\bullet} + Metal^{(n-1)+} + H^+$$
$$X^{\bullet} + RH \rightarrow R^{\bullet} + XH$$

**(2)** Propagation:

$$R^{\bullet} + O_2 \rightarrow ROO^{\bullet}$$
$$ROO^{\bullet} + RH \rightarrow ROOH + R^{\bullet}, \text{ etc}$$

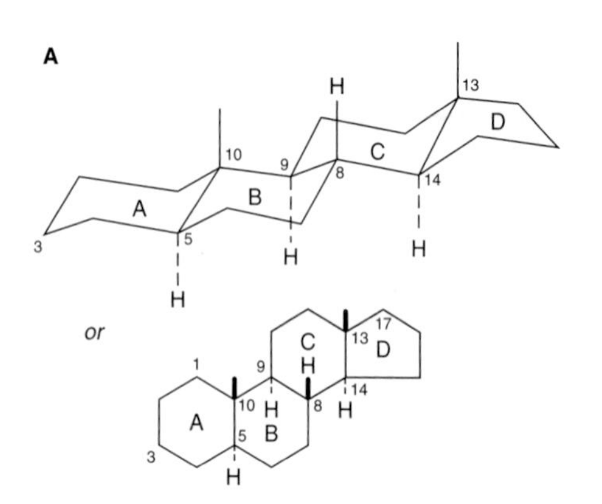


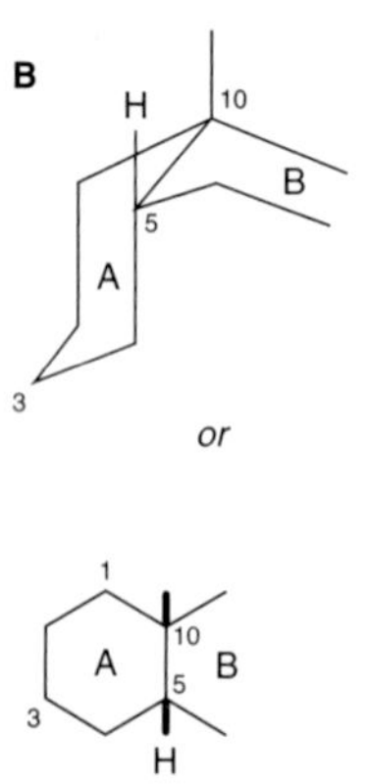


***Figure 14–16.*** Generalized steroid nucleus, showing **(A)** an all-*trans* configuration between adjacent rings and **(B)** a *cis* configuration between rings A and B.

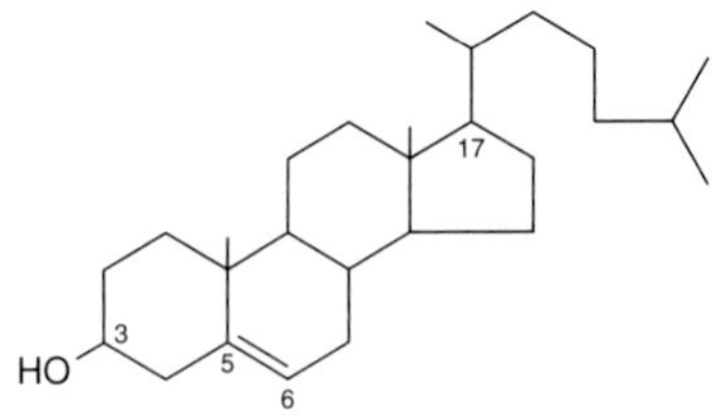

***Figure 14–17.*** Cholesterol, 3-hydroxy-5,6-cholestene.

(3) Termination:

$$ROO^{\bullet} + ROO^{\bullet} \rightarrow ROOR + O_2$$
$$ROO^{\bullet} + R^{\bullet} \rightarrow ROOR$$
$$R^{\bullet} + R^{\bullet} \rightarrow RR$$

Since the molecular precursor for the initiation process is generally the hydroperoxide product ROOH, lipid peroxidation is a chain reaction with potentially devastating effects. To control and reduce lipid peroxidation, both humans in their activities and nature invoke the use of **antioxidants.** Propyl gallate, butylated hydroxyanisole (BHA), and butylated hydroxytoluene (BHT) are antioxidants used as food additives. Naturally occurring antioxidants include vitamin E (tocopherol), which is lipid-soluble, and urate and vitamin C, which are water-soluble. Beta-carotene is an antioxidant at low $PO_2$. Antioxidants fall into two classes: (1) preventive antioxidants, which reduce the rate of chain initiation; and (2) chain-breaking antioxidants, which interfere with chain propagation. Preventive antioxidants include catalase and other peroxidases that react with ROOH and chelators of metal ions such as EDTA (ethylenediaminetetraacetate) and DTPA (diethylenetriaminepentaacetate). In vivo, the principal chain-breaking antioxidants are superoxide dismutase, which acts in the aqueous phase to trap superoxide free radicals ($O_2^{\bullet-}$); perhaps urate; and vitamin E, which acts in the lipid phase to trap $ROO^{\bullet}$ radicals (Figure 45–6).

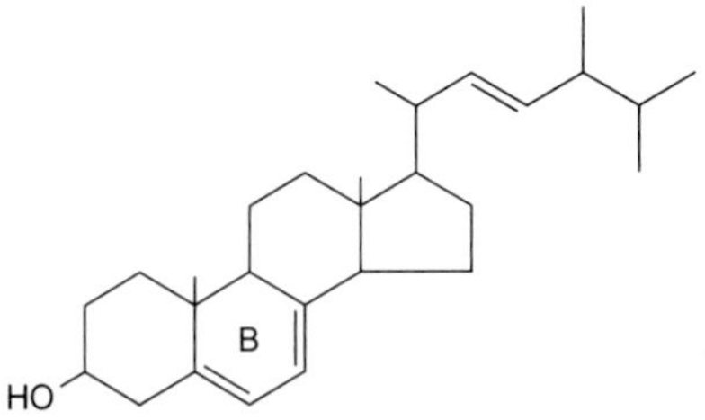

***Figure 14–18.*** Ergosterol.

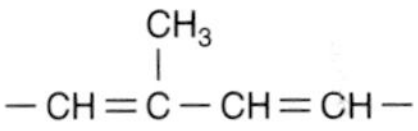

$$\begin{array}{c} CH_3 \\ | \\ -CH{=}C-CH{=}CH- \end{array}$$

***Figure 14–19.*** Isoprene unit.

Peroxidation is also catalyzed in vivo by heme compounds and by **lipoxygenases** found in platelets and leukocytes. Other products of auto-oxidation or enzymic oxidation of physiologic significance include **oxysterols** (formed from cholesterol) and **isoprostanes** (prostanoids).

## AMPHIPATHIC LIPIDS SELF-ORIENT AT OIL:WATER INTERFACES

### They Form Membranes, Micelles, Liposomes, & Emulsions

In general, lipids are insoluble in water since they contain a predominance of nonpolar (hydrocarbon) groups. However, fatty acids, phospholipids, sphingolipids, bile salts, and, to a lesser extent, cholesterol contain polar groups. Therefore, part of the molecule is **hydrophobic,** or water-insoluble; and part is **hydrophilic,** or water-soluble. Such molecules are described as **amphipathic** (Figure 14–22). They become oriented at oil:water interfaces with the polar group in the water phase and the nonpolar group in the oil phase. A bilayer of such amphipathic lipids has been regarded as a basic structure in biologic membranes (Chapter 41). When a critical concentration of these lipids is present in an aqueous medium, they form **micelles.** Aggregations of bile salts into micelles and liposomes and the formation of mixed micelles with the products of fat digestion are important in facilitating absorption of lipids from the intestine. Liposomes may be formed by sonicating an amphipathic lipid in an aqueous medium. They consist of spheres of lipid bilayers that enclose part of the aqueous medium. They are of potential clinical use—particularly when combined with tissue-specific antibodies—as carriers of drugs in the circulation, targeted to specific organs, eg, in cancer therapy. In addition, they are being used for gene transfer into vascular cells and as carriers for topical and transdermal

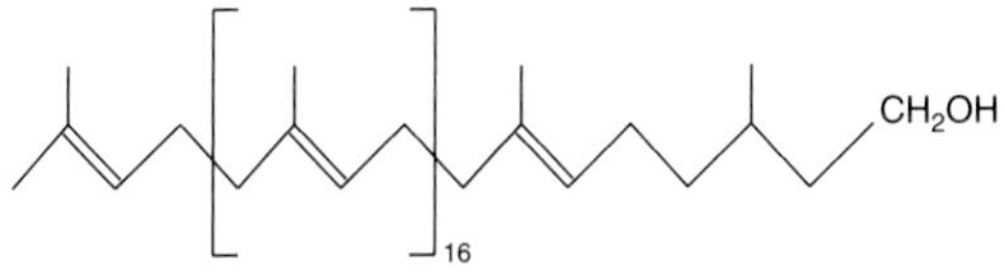

***Figure 14–20.*** Dolichol—a $C_{95}$ alcohol.

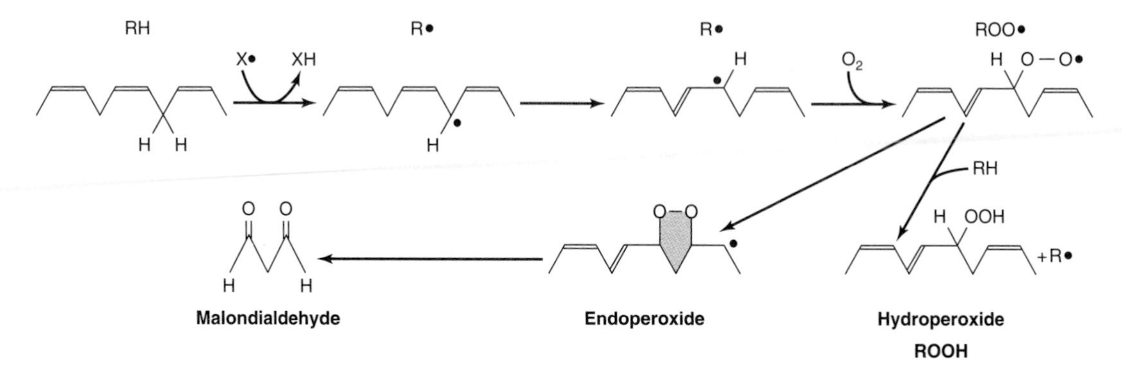


***Figure 14–21.*** Lipid peroxidation. The reaction is initiated by an existing free radical (X˙), by light, or by metal ions. Malondialdehyde is only formed by fatty acids with three or more double bonds and is used as a measure of lipid peroxidation together with ethane from the terminal two carbons of ω3 fatty acids and pentane from the terminal five carbons of ω6 fatty acids.

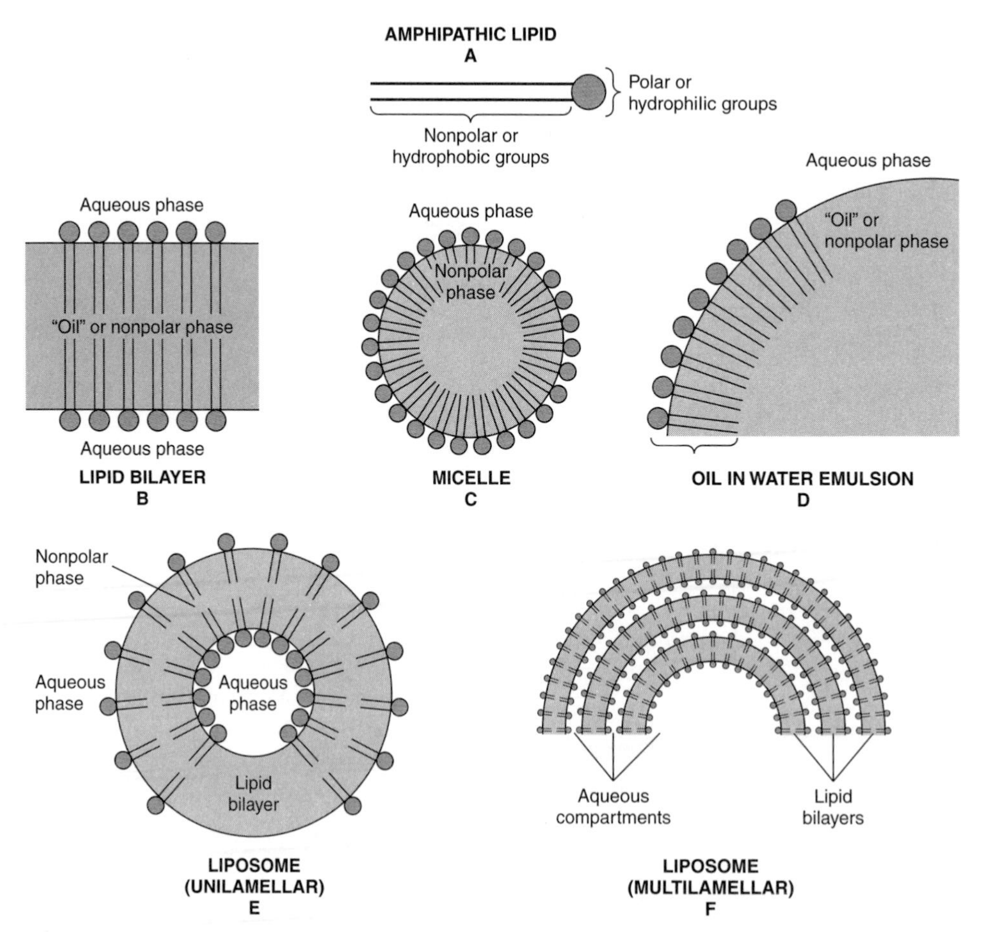


***Figure 14–22.*** Formation of lipid membranes, micelles, emulsions, and liposomes from amphipathic lipids, eg, phospholipids.

delivery of drugs and cosmetics. **Emulsions** are much larger particles, formed usually by nonpolar lipids in an aqueous medium. These are stabilized by emulsifying agents such as amphipathic lipids (eg, lecithin), which form a surface layer separating the main bulk of the nonpolar material from the aqueous phase (Figure 14–22).

## SUMMARY

- Lipids have the common property of being relatively insoluble in water (hydrophobic) but soluble in nonpolar solvents. Amphipathic lipids also contain one or more polar groups, making them suitable as constituents of membranes at lipid:water interfaces.
- The lipids of major physiologic significance are fatty acids and their esters, together with cholesterol and other steroids.
- Long-chain fatty acids may be saturated, monounsaturated, or polyunsaturated, according to the number of double bonds present. Their fluidity decreases with chain length and increases according to degree of unsaturation.
- Eicosanoids are formed from 20-carbon polyunsaturated fatty acids and make up an important group of physiologically and pharmacologically active compounds known as prostaglandins, thromboxanes, leukotrienes, and lipoxins.
- The esters of glycerol are quantitatively the most significant lipids, represented by triacylglycerol ("fat"), a major constituent of lipoproteins and the storage form of lipid in adipose tissue. Phosphoacylglycerols are amphipathic lipids and have important roles—as major constituents of membranes and the outer layer of lipoproteins, as surfactant in the lung, as precursors of second messengers, and as constituents of nervous tissue.
- Glycolipids are also important constituents of nervous tissue such as brain and the outer leaflet of the cell membrane, where they contribute to the carbohydrates on the cell surface.
- Cholesterol, an amphipathic lipid, is an important component of membranes. It is the parent molecule from which all other steroids in the body, including major hormones such as the adrenocortical and sex hormones, D vitamins, and bile acids, are synthesized.
- Peroxidation of lipids containing polyunsaturated fatty acids leads to generation of free radicals that may damage tissues and cause disease.

## REFERENCES


Benzie IFF: Lipid peroxidation: a review of causes, consequences, measurement and dietary influences. Int J Food Sci Nutr 1996;47:233.

Christie WW: *Lipid Analysis,* 2nd ed. Pergamon Press, 1982.

Cullis PR, Fenske DB, Hope MJ: Physical properties and functional roles of lipids in membranes. In: *Biochemistry of Lipids, Lipoproteins and Membranes.* Vance DE, Vance JE (editors). Elsevier, 1996.

Gunstone FD, Harwood JL, Padley FB: *The Lipid Handbook.* Chapman & Hall, 1986.

Gurr MI, Harwood JL: *Lipid Biochemistry: An Introduction,* 4th ed. Chapman & Hall, 1991.

# Overview of Metabolism

# 15

Peter A. Mayes, PhD, DSc, & David A. Bender, PhD

## BIOMEDICAL IMPORTANCE

The fate of dietary components after digestion and absorption constitutes metabolism—the metabolic pathways taken by individual molecules, their interrelationships, and the mechanisms that regulate the flow of metabolites through the pathways. Metabolic pathways fall into three categories: (1) **Anabolic pathways** are those involved in the synthesis of compounds. Protein synthesis is such a pathway, as is the synthesis of fuel reserves of triacylglycerol and glycogen. Anabolic pathways are endergonic. (2) **Catabolic pathways** are involved in the breakdown of larger molecules, commonly involving oxidative reactions; they are exergonic, producing reducing equivalents and, mainly via the respiratory chain, ATP. (3) **Amphibolic pathways** occur at the "crossroads" of metabolism, acting as links between the anabolic and catabolic pathways, eg, the citric acid cycle.

A knowledge of normal metabolism is essential for an understanding of abnormalities underlying disease. Normal metabolism includes adaptation to periods of starvation, exercise, pregnancy, and lactation. Abnormal metabolism may result from nutritional deficiency, enzyme deficiency, abnormal secretion of hormones, or the actions of drugs and toxins. An important example of a metabolic disease is **diabetes mellitus.**

## PATHWAYS THAT PROCESS THE MAJOR PRODUCTS OF DIGESTION

The nature of the diet sets the basic pattern of metabolism. There is a need to process the products of digestion of dietary carbohydrate, lipid, and protein. These are mainly glucose, fatty acids and glycerol, and amino acids, respectively. In ruminants (and to a lesser extent in other herbivores), dietary cellulose is fermented by symbiotic microorganisms to short-chain fatty acids (acetic, propionic, butyric), and metabolism in these animals is adapted to use these fatty acids as major substrates. All the products of digestion are metabolized to a **common product, acetyl-CoA,** which is then oxidized by the **citric acid cycle** (Figure 15–1).

### Carbohydrate Metabolism Is Centered on the Provision & Fate of Glucose (Figure 15–2)

Glucose is metabolized to pyruvate by the pathway of **glycolysis,** which can occur anaerobically (in the absence of oxygen), when the end product is lactate. Aerobic tissues metabolize pyruvate to **acetyl-CoA,** which can enter the **citric acid cycle** for complete oxidation to $CO_2$ and $H_2O$, linked to the formation of ATP in the process of **oxidative phosphorylation** (Figure 16–2). Glucose is the major fuel of most tissues.

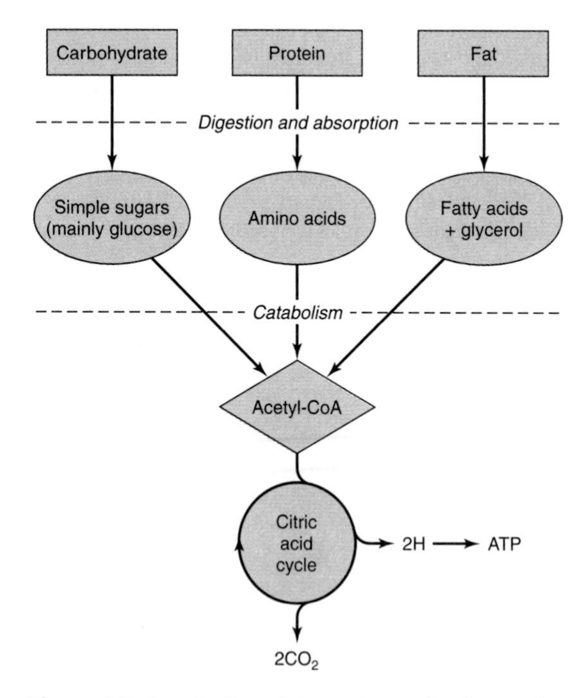


***Figure 15–1.*** Outline of the pathways for the catabolism of dietary carbohydrate, protein, and fat. All the pathways lead to the production of acetyl-CoA, which is oxidized in the citric acid cycle, ultimately yielding ATP in the process of oxidative phosphorylation.

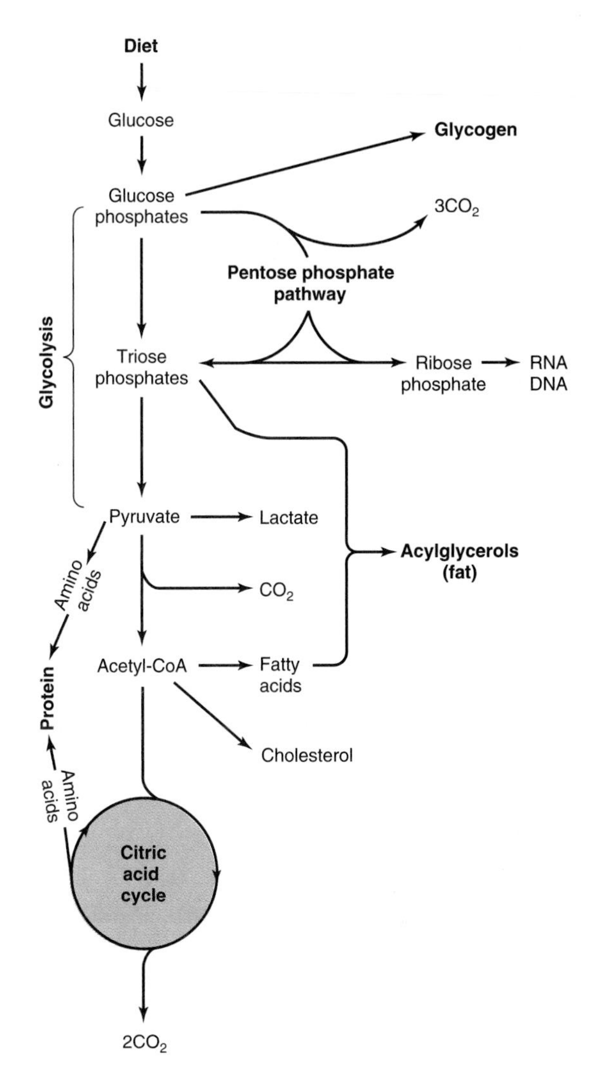


***Figure 15–2.*** Overview of carbohydrate metabolism showing the major pathways and end products. Gluconeogenesis is not shown.

Glucose and its metabolites also take part in other processes. Examples: (1) Conversion to the storage polymer **glycogen** in skeletal muscle and liver. (2) The **pentose phosphate pathway,** an alternative to part of the pathway of glycolysis, is a source of reducing equivalents (NADPH) for biosynthesis and the source of **ribose** for nucleotide and nucleic acid synthesis. (3) Triose phosphate gives rise to the **glycerol moiety** of triacylglycerols. (4) Pyruvate and intermediates of the citric acid cycle provide the carbon skeletons for the synthesis of **amino acids;** and acetyl-CoA, the precursor of **fatty acids** and **cholesterol** (and hence of all steroids synthesized in the body). **Gluconeogenesis** is the process of forming glucose from noncarbohydrate precursors, eg, lactate, amino acids, and glycerol.

## Lipid Metabolism Is Concerned Mainly With Fatty Acids & Cholesterol (Figure 15–3)

The source of long-chain fatty acids is either dietary lipid or de novo synthesis from acetyl-CoA derived from carbohydrate. Fatty acids may be oxidized to **acetyl-CoA** (β-oxidation) or esterified with glycerol, forming **triacylglycerol** (fat) as the body's main fuel reserve.

Acetyl-CoA formed by β-oxidation may undergo several fates:

**(1)** As with acetyl-CoA arising from glycolysis, it is **oxidized** to $CO_2$ + $H_2O$ via the **citric acid cycle.**

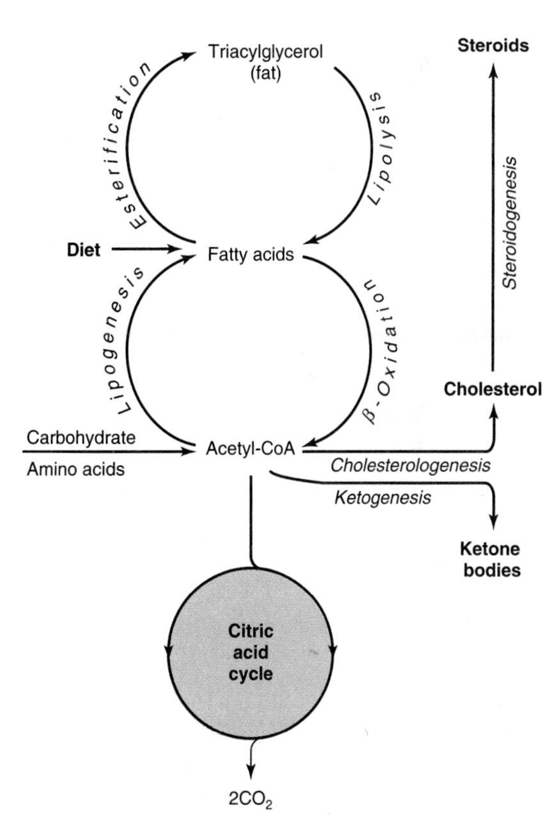


***Figure 15–3.*** Overview of fatty acid metabolism showing the major pathways and end products. Ketone bodies comprise the substances acetoacetate, 3-hydroxybutyrate, and acetone.

(2) It is the precursor for synthesis of **cholesterol** and other **steroids.**

(3) In the liver, it forms **ketone bodies** (acetone, acetoacetate, and 3-hydroxybutyrate) that are important fuels in prolonged starvation.

### Much of Amino Acid Metabolism Involves Transamination (Figure 15–4)

The amino acids are required for protein synthesis. Some must be supplied in the diet (the **essential amino acids**) since they cannot be synthesized in the body. The remainder are **nonessential amino acids** that are supplied in the diet but can be formed from metabolic intermediates by **transamination,** using the amino nitrogen from other amino acids. After **deamination,** amino nitrogen is excreted as **urea,** and the carbon skeletons that remain after transamination (1) are oxidized to $CO_2$ via the citric acid cycle, (2) form glucose (gluconeogenesis), or (3) form ketone bodies.

Several amino acids are also the precursors of other compounds, eg, purines, pyrimidines, hormones such as epinephrine and thyroxine, and neurotransmitters.

## METABOLIC PATHWAYS MAY BE STUDIED AT DIFFERENT LEVELS OF ORGANIZATION

In addition to studies in the whole organism, the location and integration of metabolic pathways is revealed by studies at several levels of organization. At the **tissue and organ level,** the nature of the substrates entering and metabolites leaving tissues and organs is defined. At the **subcellular level,** each cell organelle (eg, the mitochondrion) or compartment (eg, the cytosol) has specific roles that form part of a subcellular pattern of metabolic pathways.

### At the Tissue and Organ Level, the Blood Circulation Integrates Metabolism

**Amino acids** resulting from the digestion of dietary protein and glucose resulting from the digestion of carbohydrate are absorbed and directed to the liver via the **hepatic portal vein.** The liver has the role of regulating the blood concentration of most water-soluble metabolites (Figure 15–5). In the case of glucose, this is achieved by taking up glucose in excess of immediate requirements and converting it to glycogen **(glycogene-**

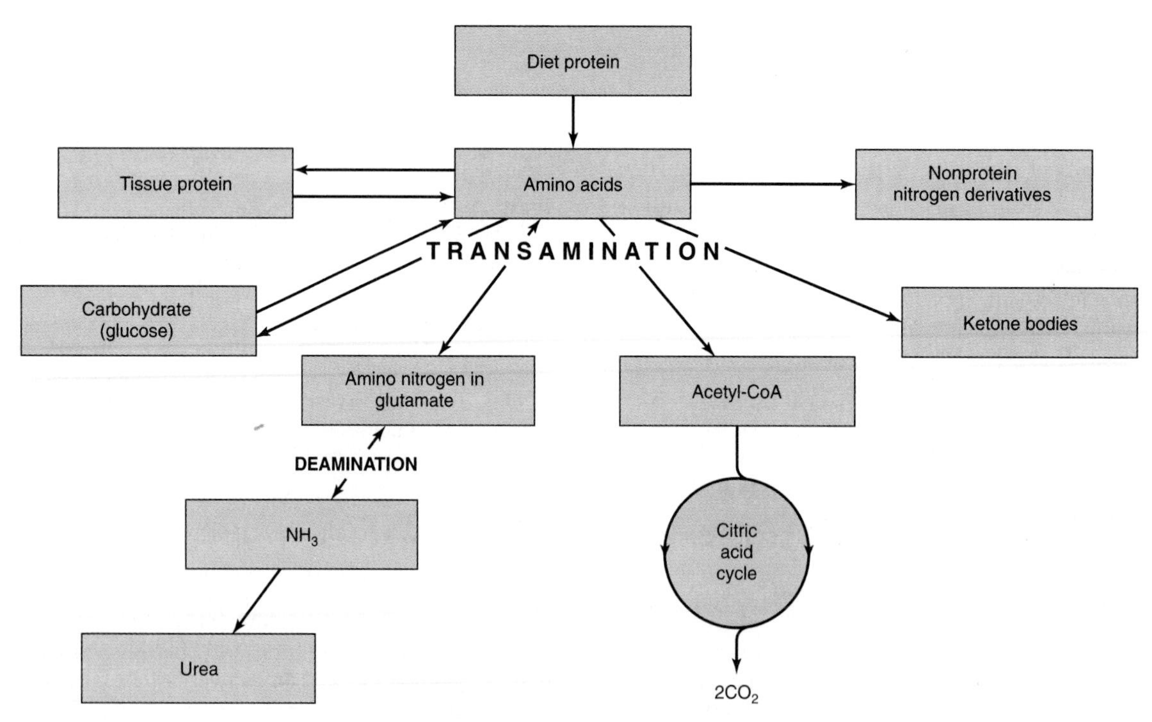


***Figure 15–4.*** Overview of amino acid metabolism showing the major pathways and end products.

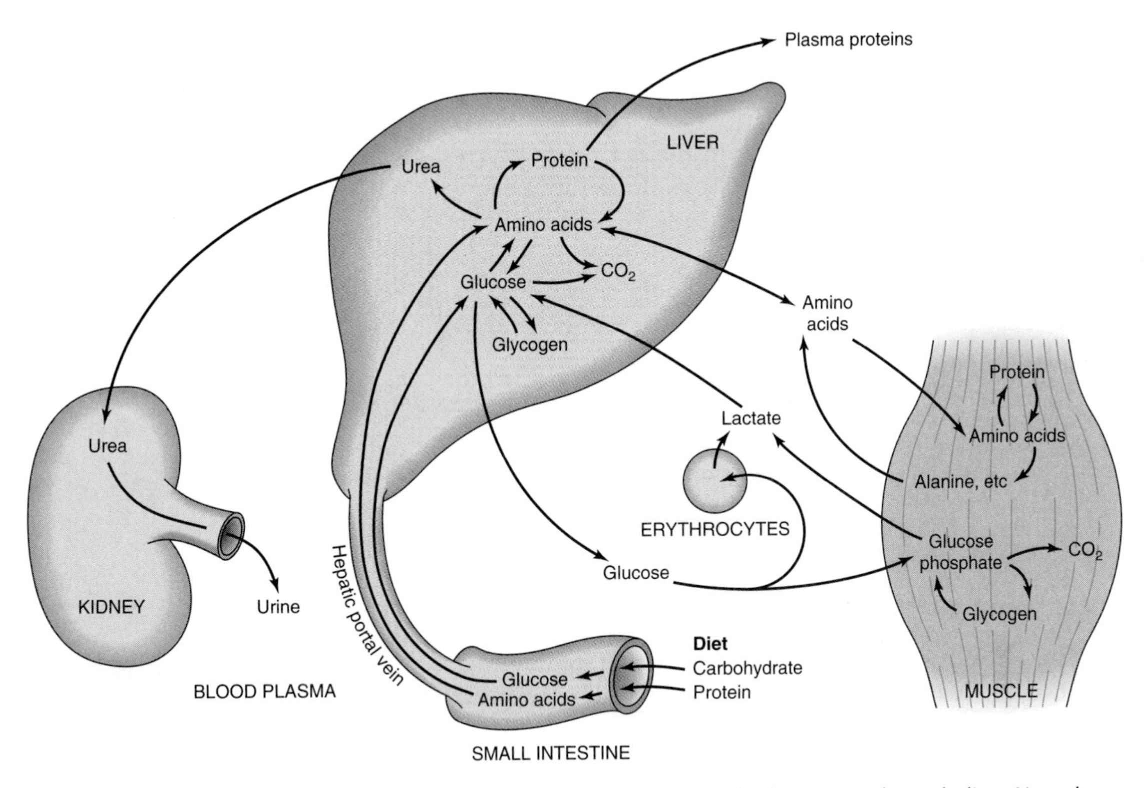


***Figure 15–5.*** Transport and fate of major carbohydrate and amino acid substrates and metabolites. Note that there is little free glucose in muscle, since it is rapidly phosphorylated upon entry.

**sis)** or to fat **(lipogenesis).** Between meals, the liver acts to maintain the blood glucose concentration from glycogen **(glycogenolysis)** and, together with the kidney, by converting noncarbohydrate metabolites such as lactate, glycerol, and amino acids to glucose **(gluconeogenesis).** Maintenance of an adequate concentration of blood glucose is vital for those tissues in which it is the major fuel (the brain) or the only fuel (the erythrocytes). The liver also **synthesizes the major plasma proteins** (eg, albumin) and **deaminates amino acids** that are in excess of requirements, forming urea, which is transported to the kidney and excreted.

**Skeletal muscle** utilizes glucose as a fuel, forming both lactate and $CO_2$. It stores glycogen as a fuel for its use in muscular contraction and synthesizes muscle protein from plasma amino acids. Muscle accounts for approximately 50% of body mass and consequently represents a considerable store of protein that can be drawn upon to supply amino acids for gluconeogenesis in starvation.

**Lipids** in the diet (Figure 15–6) are mainly triacylglycerol and are hydrolyzed to monoacylglycerols and fatty acids in the gut, then reesterified in the intestinal mucosa. Here they are packaged with protein and secreted into the lymphatic system and thence into the blood stream as **chylomicrons,** the largest of the plasma **lipoproteins.** Chylomicrons also contain other lipid-soluble nutrients, eg, vitamins. Unlike glucose and amino acids, chylomicron triacylglycerol is not taken up directly by the liver. It is first metabolized by tissues that have **lipoprotein lipase,** which hydrolyzes the triacylglycerol, releasing fatty acids that are incorporated into tissue lipids or oxidized as fuel. The other major source of long-chain fatty acid is synthesis **(lipogenesis)** from carbohydrate, mainly in adipose tissue and the liver.

Adipose tissue triacylglycerol is the main fuel reserve of the body. On hydrolysis **(lipolysis)** free fatty acids are released into the circulation. These are taken up by most tissues (but not brain or erythrocytes) and esterified to acylglycerols or oxidized as a fuel. In the liver, triacylglycerol arising from lipogenesis, free fatty acids, and chylomicron remnants (see Figures 25–3 and 25–4) is secreted into the circulation as **very low density lipoprotein** (VLDL). This triacylglycerol undergoes a fate similar to that of chylomicrons. Partial oxidation of fatty acids in the liver leads to **ketone body** production (keto-

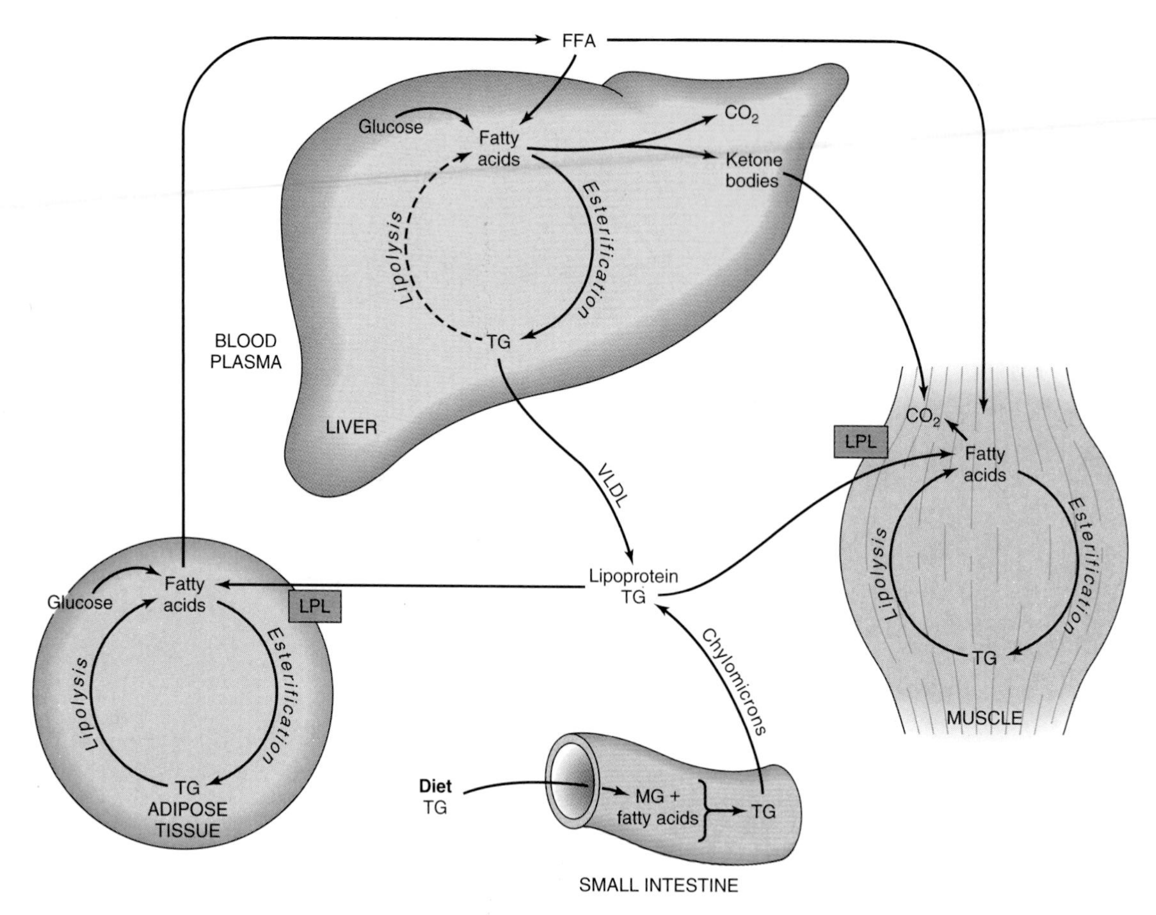


***Figure 15–6.*** Transport and fate of major lipid substrates and metabolites. (FFA, free fatty acids; LPL, lipoprotein lipase; MG, monoacylglycerol; TG, triacylglycerol; VLDL, very low density lipoprotein.)

genesis). Ketone bodies are transported to extrahepatic tissues, where they act as a fuel source in starvation.

### At the Subcellular Level, Glycolysis Occurs in the Cytosol & the Citric Acid Cycle in the Mitochondria

Compartmentation of pathways in separate subcellular compartments or organelles permits integration and regulation of metabolism. Not all pathways are of equal importance in all cells. Figure 15–7 depicts the subcellular compartmentation of metabolic pathways in a hepatic parenchymal cell.

The central role of the **mitochondrion** is immediately apparent, since it acts as the focus of carbohydrate, lipid, and amino acid metabolism. It contains the enzymes of the citric acid cycle, β-oxidation of fatty acids, and ketogenesis, as well as the respiratory chain and ATP synthase.

Glycolysis, the pentose phosphate pathway, and fatty acid synthesis are all found in the cytosol. In gluconeogenesis, substrates such as lactate and pyruvate, which are formed in the cytosol, enter the mitochondrion to yield **oxaloacetate** before formation of glucose.

The membranes of the **endoplasmic reticulum** contain the enzyme system for **acylglycerol synthesis,** and the **ribosomes** are responsible for **protein synthesis.**

## THE FLUX OF METABOLITES IN METABOLIC PATHWAYS MUST BE REGULATED IN A CONCERTED MANNER

Regulation of the overall flux through a pathway is important to ensure an appropriate supply, when required, of the products of that pathway. Regulation is achieved by control of one or more key reactions in the pathway, catalyzed by **"regulatory enzymes."** The physicochemical factors that control the rate of an

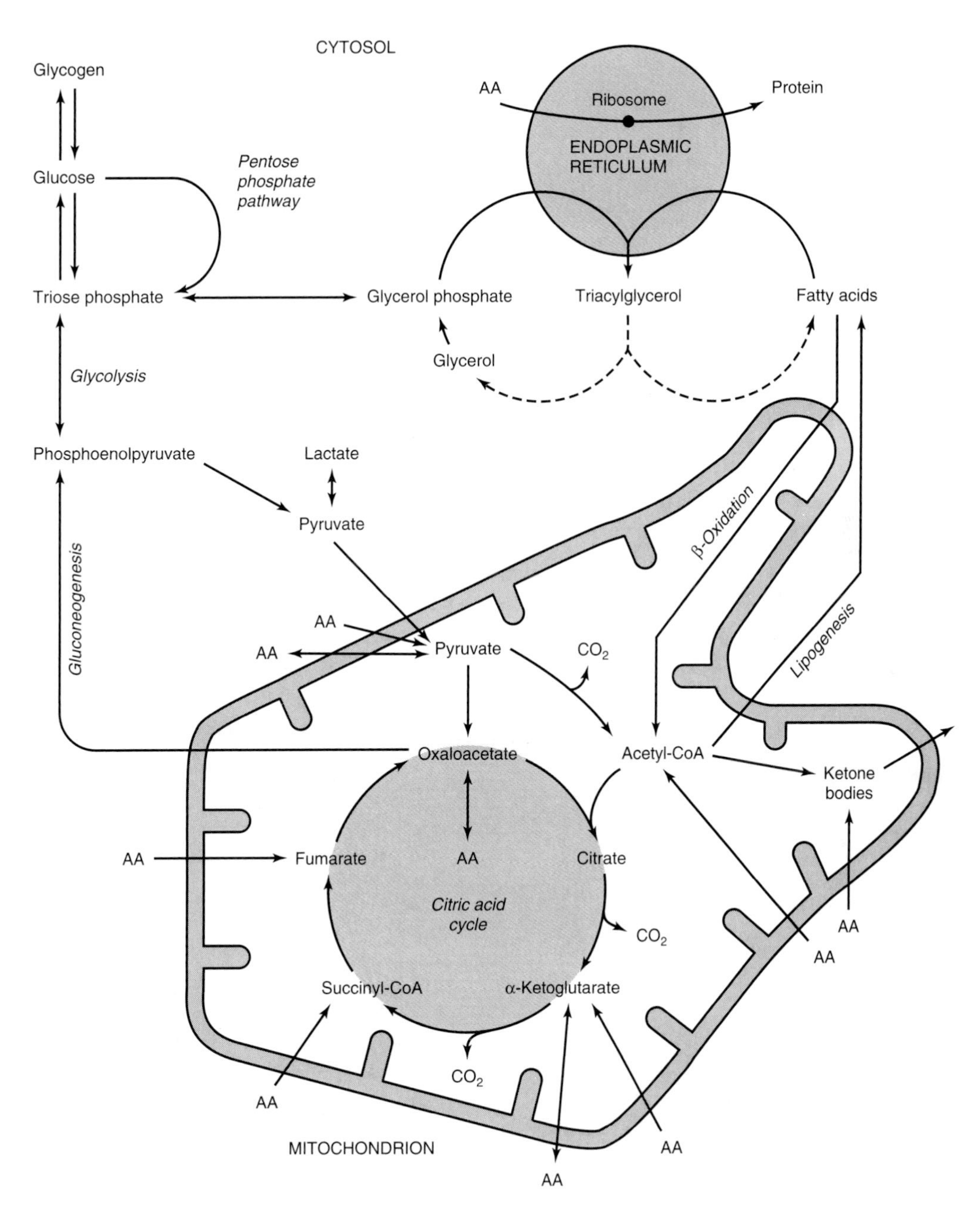


***Figure 15–7.*** Intracellular location and overview of major metabolic pathways in a liver parenchymal cell. (AA →, metabolism of one or more essential amino acids; AA ↔, metabolism of one or more nonessential amino acids.)

enzyme-catalyzed reaction, eg, substrate concentration, are of primary importance in the control of the overall rate of a metabolic pathway (Chapter 9).

## "Nonequilibrium" Reactions Are Potential Control Points

In a reaction at equilibrium, the forward and reverse reactions occur at equal rates, and there is therefore no net flux in either direction:

$$A \leftrightarrow B \leftrightarrow C \leftrightarrow D$$

In vivo, under "steady-state" conditions, there is a net flux from left to right because there is a continuous supply of A and removal of D. In practice, there are invariably one or more **nonequilibrium reactions** in a metabolic pathway, where the reactants are present in concentrations that are far from equilibrium. In attempting to reach equilibrium, large losses of free energy occur as heat, making this type of reaction essentially irreversible, eg,

$$A \leftrightarrow B \overset{\text{Heat}}{\not\leftrightarrow} C \leftrightarrow D$$

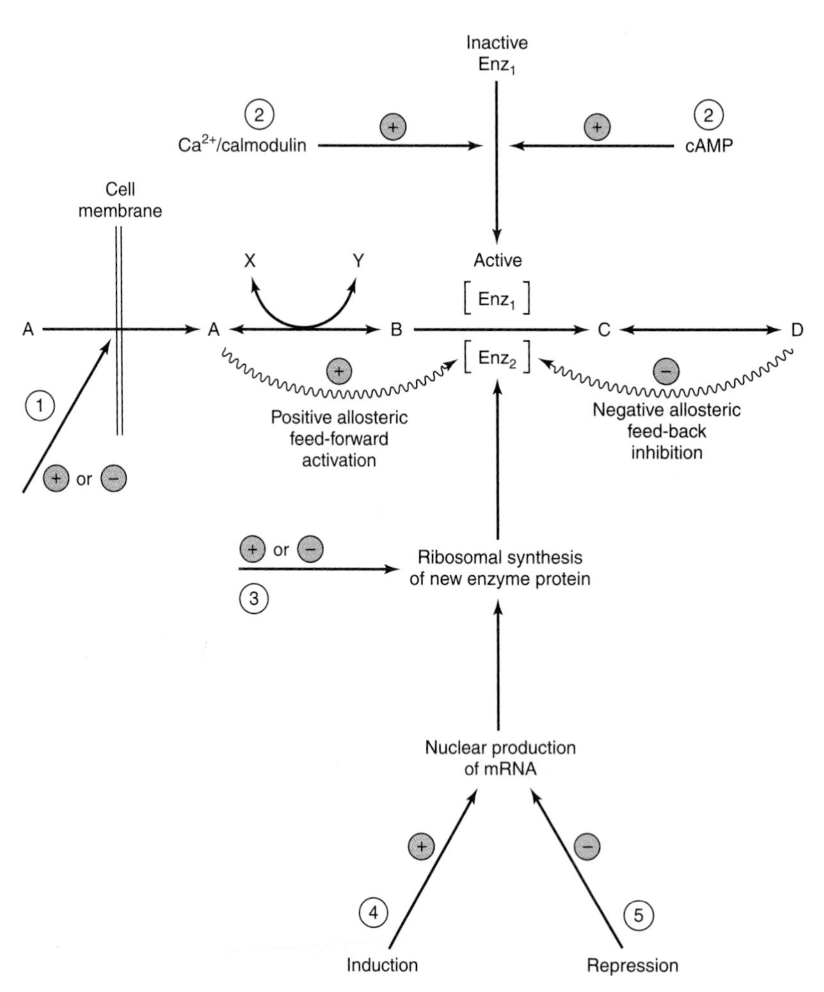


***Figure 15–8.*** Mechanisms of control of an enzyme-catalyzed reaction. Circled numbers indicate possible sites of action of hormones. ①, Alteration of membrane permeability; ②, conversion of an inactive to an active enzyme, usually involving phosphorylation/dephosphorylation reactions; ③, alteration of the rate of translation of mRNA at the ribosomal level; ④, induction of new mRNA formation; and ⑤, repression of mRNA formation. ① and ② are rapid, whereas ③–⑤ are slower ways of regulating enzyme activity.

Such a pathway has both flow and direction. The enzymes catalyzing nonequilibrium reactions are usually present in low concentrations and are subject to a variety of regulatory mechanisms. However, many of the reactions in metabolic pathways cannot be classified as equilibrium or nonequilibrium but fall somewhere between the two extremes.

### The Flux-Generating Reaction Is the First Reaction in a Pathway That Is Saturated With Substrate

It may be identified as a nonequilibrium reaction in which the $K_m$ of the enzyme is considerably lower than the normal substrate concentration. The first reaction in glycolysis, catalyzed by **hexokinase** (Figure 17–2), is such a flux-generating step because its $K_m$ for glucose of 0.05 mmol/L is well below the normal blood glucose concentration of 5 mmol/L.

## ALLOSTERIC & HORMONAL MECHANISMS ARE IMPORTANT IN THE METABOLIC CONTROL OF ENZYME-CATALYZED REACTIONS

A hypothetical metabolic pathway is shown in Figure 15–8, in which reactions A ↔ B and C ↔ D are equilibrium reactions and B → C is a nonequilibrium reaction. The flux through such a pathway can be regulated by the availability of substrate A. This depends on its supply from the blood, which in turn depends on either food intake or key reactions that maintain and release substrates from tissue reserves to the blood, eg, the glycogen phosphorylase in liver (Figure 18–1) and hormone-sensitive lipase in adipose tissue (Figure 25–7). The flux also depends on the transport of substrate A across the cell membrane. Flux is also determined by the removal of the end product D and the availability of cosubstrate or cofactors represented by X and Y. Enzymes catalyzing nonequilibrium reactions are often allosteric proteins subject to the rapid actions of "feed-back" or "feed-forward" control by **allosteric modifiers** in immediate response to the needs of the cell (Chapter 9). Frequently, the product of a biosynthetic pathway will inhibit the enzyme catalyzing the first reaction in the pathway. Other control mechanisms depend on the action of **hormones** responding to the needs of the body as a whole; they may act rapidly, by altering the activity of existing enzyme molecules, or slowly, by altering the rate of enzyme synthesis.

## SUMMARY


- The products of digestion provide the tissues with the building blocks for the biosynthesis of complex molecules and also with the fuel to power the living processes.
- Nearly all products of digestion of carbohydrate, fat, and protein are metabolized to a common metabolite, acetyl-CoA, before final oxidation to $CO_2$ in the citric acid cycle.
- Acetyl-CoA is also used as the precursor for biosynthesis of long-chain fatty acids; steroids, including cholesterol; and ketone bodies.
- Glucose provides carbon skeletons for the glycerol moiety of fat and of several nonessential amino acids.
- Water-soluble products of digestion are transported directly to the liver via the hepatic portal vein. The liver regulates the blood concentrations of glucose and amino acids.
- Pathways are compartmentalized within the cell. Glycolysis, glycogenesis, glycogenolysis, the pentose phosphate pathway, and lipogenesis occur in the cytosol. The mitochondrion contains the enzymes of the citric acid cycle, β-oxidation of fatty acids, and of oxidative phosphorylation. The endoplasmic reticulum also contains the enzymes for many other processes, including protein synthesis, glycerolipid formation, and drug metabolism.
- Metabolic pathways are regulated by rapid mechanisms affecting the activity of existing enzymes, eg, allosteric and covalent modification (often in response to hormone action); and slow mechanisms affecting the synthesis of enzymes.


## REFERENCES


Cohen P: *Control of Enzyme Activity,* 2nd ed. Chapman & Hall, 1983.

Fell D: *Understanding the Control of Metabolism.* Portland Press, 1997.

Frayn KN: *Metabolic Regulation—A Human Perspective.* Portland Press, 1996.

Newsholme EA, Crabtree B: Flux-generating and regulatory steps in metabolic control. Trends Biochem Sci 1981;6:53.

# The Citric Acid Cycle: The Catabolism of Acetyl-CoA

# 16

*Peter A. Mayes, PhD, DSc, & David A. Bender, PhD*

## BIOMEDICAL IMPORTANCE

The citric acid cycle (Krebs cycle, tricarboxylic acid cycle) is a series of reactions in mitochondria that oxidize acetyl residues (as acetyl-CoA) and reduce coenzymes that upon reoxidation are linked to the formation of ATP.

The citric acid cycle is the final common pathway for the aerobic oxidation of carbohydrate, lipid, and protein because glucose, fatty acids, and most amino acids are metabolized to acetyl-CoA or intermediates of the cycle. It also has a central role in gluconeogenesis, lipogenesis, and interconversion of amino acids. Many of these processes occur in most tissues, but the liver is the only tissue in which all occur to a significant extent. The repercussions are therefore profound when, for example, large numbers of hepatic cells are damaged as in acute **hepatitis** or replaced by connective tissue (as in **cirrhosis**). Very few, if any, genetic abnormalities of citric acid cycle enzymes have been reported; such abnormalities would be incompatible with life or normal development.

## THE CITRIC ACID CYCLE PROVIDES SUBSTRATE FOR THE RESPIRATORY CHAIN

The cycle starts with reaction between the acetyl moiety of acetyl-CoA and the four-carbon dicarboxylic acid oxaloacetate, forming a six-carbon tricarboxylic acid, citrate. In the subsequent reactions, two molecules of $CO_2$ are released and oxaloacetate is regenerated (Figure 16–1). Only a small quantity of oxaloacetate is needed for the oxidation of a large quantity of acetyl-CoA; oxaloacetate may be considered to play a **catalytic role.**

The citric acid cycle is an integral part of the process by which much of the free energy liberated during the oxidation of fuels is made available. During oxidation of acetyl-CoA, coenzymes are reduced and subsequently reoxidized in the respiratory chain, linked to the formation of ATP (oxidative phosphorylation; see Figure 16–2 and also Chapter 12). This process is **aerobic,** requiring oxygen as the final oxidant of the reduced coenzymes. The enzymes of the citric acid cycle are located in the **mitochondrial matrix,** either free or attached to the inner mitochondrial membrane, where the enzymes of the respiratory chain are also found.

## REACTIONS OF THE CITRIC ACID CYCLE LIBERATE REDUCING EQUIVALENTS & $CO_2$ (Figure 16–3)*

The initial reaction between acetyl-CoA and oxaloacetate to form citrate is catalyzed by **citrate synthase** which forms a carbon-carbon bond between the methyl carbon of acetyl-CoA and the carbonyl carbon of oxaloacetate. The thioester bond of the resultant citryl-CoA is hydrolyzed, releasing citrate and CoASH—an exergonic reaction.

Citrate is isomerized to isocitrate by the enzyme **aconitase** (aconitate hydratase); the reaction occurs in two steps: dehydration to *cis*-aconitate, some of which remains bound to the enzyme; and rehydration to isocitrate. Although citrate is a symmetric molecule, aconitase reacts with citrate asymmetrically, so that the two carbon atoms that are lost in subsequent reactions of the cycle are not those that were added from acetyl-CoA. This asymmetric behavior is due to **channeling**—transfer of the product of citrate synthase directly onto the active site of aconitase without entering free solution. This provides integration of citric acid cycle activity and the provision of citrate in the cytosol as a source of acetyl-CoA for fatty acid synthesis. The poison **fluoroacetate** is toxic because fluoroacetyl-CoA condenses with oxaloacetate to form fluorocitrate, which inhibits aconitase, causing citrate to accumulate.

Isocitrate undergoes dehydrogenation catalyzed by **isocitrate dehydrogenase** to form, initially, oxalosuccinate, which remains enzyme-bound and undergoes decarboxylation to α-ketoglutarate. The decarboxylation

*From Circular No. 200 of the Committee of Editors of Biochemical Journals Recommendations (1975): "According to standard biochemical convention, the ending *ate* in, eg, palmitate, denotes any mixture of free acid and the ionized form(s) (according to pH) in which the cations are not specified." The same convention is adopted in this text for all carboxylic acids.

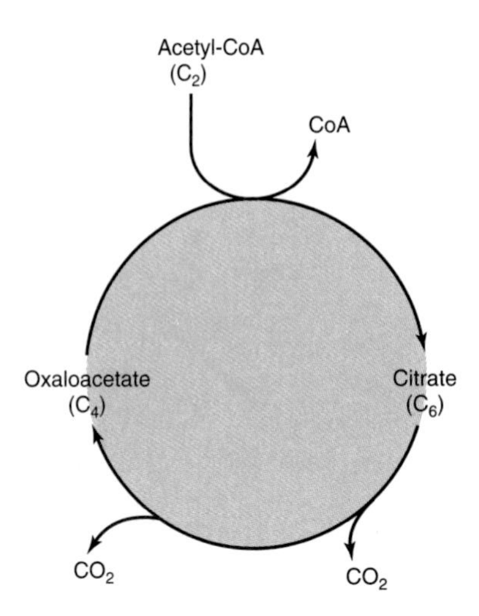


***Figure 16–1.*** Citric acid cycle, illustrating the catalytic role of oxaloacetate.

requires $Mg^{2+}$ or $Mn^{2+}$ ions. There are three isoenzymes of isocitrate dehydrogenase. One, which uses $NAD^+$, is found only in mitochondria. The other two use $NADP^+$ and are found in mitochondria and the cytosol. Respiratory chain-linked oxidation of isocitrate proceeds almost completely through the $NAD^+$-dependent enzyme.

α-Ketoglutarate undergoes **oxidative decarboxylation** in a reaction catalyzed by a multi-enzyme complex similar to that involved in the oxidative decarboxylation of pyruvate (Figure 17–5). The **α-ketoglutarate dehydrogenase complex** requires the same cofactors as the pyruvate dehydrogenase complex—thiamin diphosphate, lipoate, $NAD^+$, FAD, and CoA—and results in the formation of succinyl-CoA. The equilibrium of this reaction is so much in favor of succinyl-CoA formation that it must be considered physiologically unidirectional. As in the case of pyruvate oxidation (Chapter 17), arsenite inhibits the reaction, causing the substrate, **α-ketoglutarate,** to accumulate.

Succinyl-CoA is converted to succinate by the enzyme **succinate thiokinase (succinyl-CoA synthetase).** This is the only example in the citric acid cycle of substrate-level phosphorylation. Tissues in which gluconeogenesis occurs (the liver and kidney) contain two isoenzymes of succinate thiokinase, one specific for GDP and the other for ADP. The GTP formed is used for the decarboxylation of oxaloacetate to phosphoenolpyruvate in gluconeogenesis and provides a regulatory link between citric acid cycle activity and the withdrawal of oxaloacetate for gluconeogenesis. Nongluconeogenic tissues have only the isoenzyme that uses ADP.

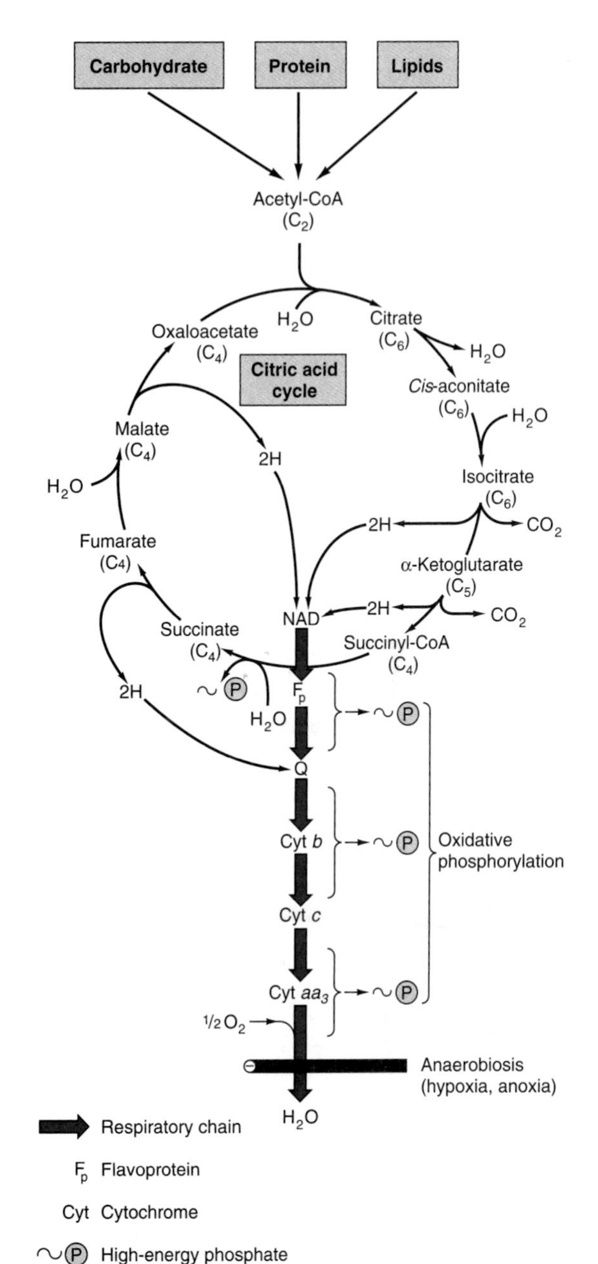


***Figure 16–2.*** The citric acid cycle: the major catabolic pathway for acetyl-CoA in aerobic organisms. Acetyl-CoA, the product of carbohydrate, protein, and lipid catabolism, is taken into the cycle, together with $H_2O$, and oxidized to $CO_2$ with the release of reducing equivalents (2H). Subsequent oxidation of 2H in the respiratory chain leads to coupled phosphorylation of ADP to ATP. For one turn of the cycle, 11~Ⓟ are generated via oxidative phosphorylation and one ~Ⓟ arises at substrate level from the conversion of succinyl-CoA to succinate.

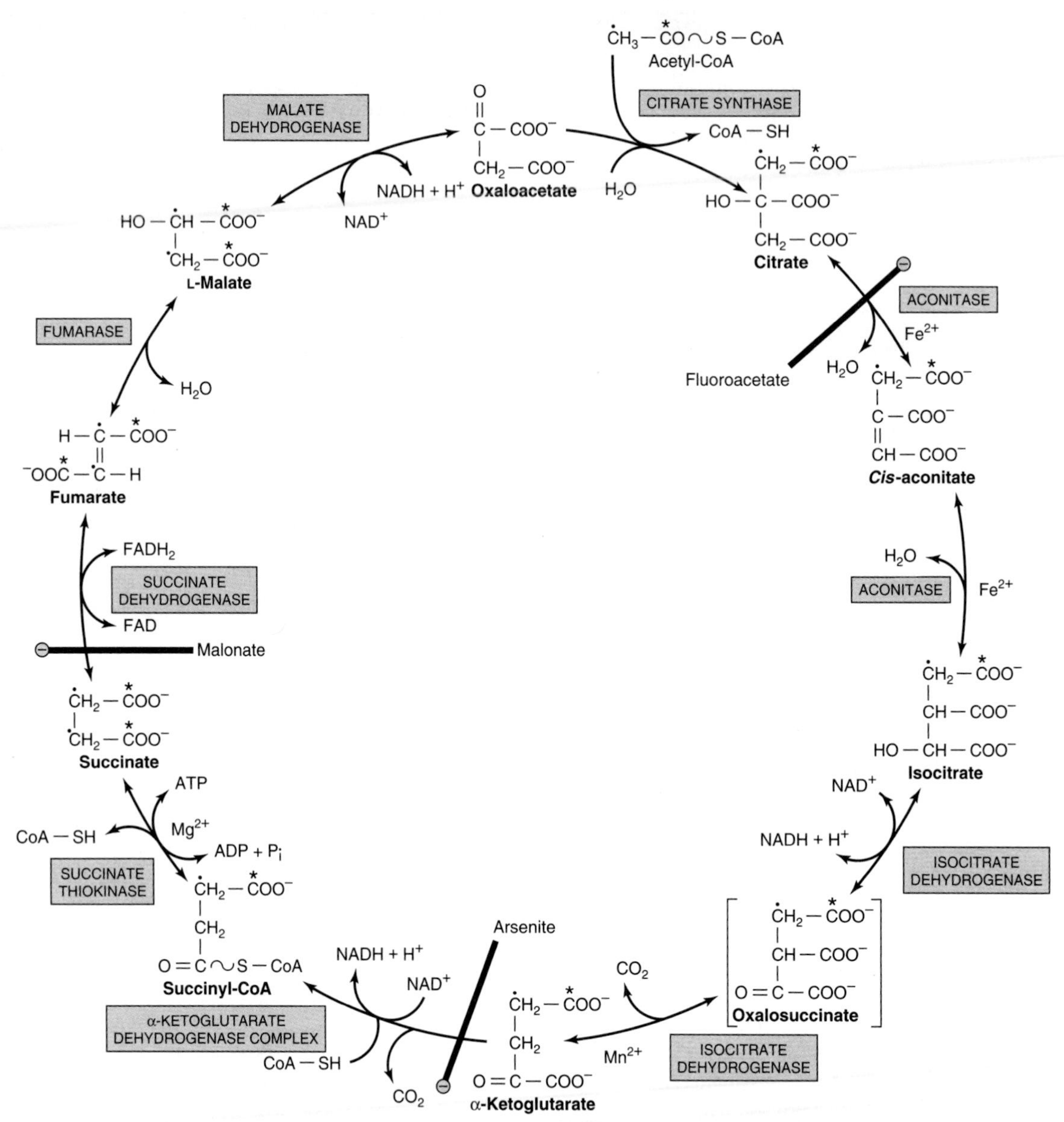


***Figure 16–3.*** Reactions of the citric acid (Krebs) cycle. Oxidation of NADH and $FADH_2$ in the respiratory chain leads to the generation of ATP via oxidative phosphorylation. In order to follow the passage of acetyl-CoA through the cycle, the two carbon atoms of the acetyl radical are shown labeled on the carboxyl carbon (designated by asterisk) and on the methyl carbon (using the designation •). Although two carbon atoms are lost as $CO_2$ in one revolution of the cycle, these atoms are not derived from the acetyl-CoA that has immediately entered the cycle but from that portion of the citrate molecule that was derived from oxaloacetate. However, on completion of a single turn of the cycle, the oxaloacetate that is regenerated is now labeled, which leads to labeled $CO_2$ being evolved during the second turn of the cycle. Because succinate is a symmetric compound and because succinate dehydrogenase does not differentiate between its two carboxyl groups, "randomization" of label occurs at this step such that all four carbon atoms of oxaloacetate appear to be labeled after one turn of the cycle. During gluconeogenesis, some of the label in oxaloacetate is incorporated into glucose and glycogen (Figure 19–1). For a discussion of the stereochemical aspects of the citric acid cycle, see Greville (1968). The sites of inhibition (⊖) by fluoroacetate, malonate, and arsenite are indicated.

When ketone bodies are being metabolized in extrahepatic tissues there is an alternative reaction catalyzed by **succinyl-CoA–acetoacetate-CoA transferase (thiophorase)**—involving transfer of CoA from succinyl-CoA to acetoacetate, forming acetoacetyl-CoA (Chapter 22).

The onward metabolism of succinate, leading to the regeneration of oxaloacetate, is the same sequence of chemical reactions as occurs in the β-oxidation of fatty acids: dehydrogenation to form a carbon-carbon double bond, addition of water to form a hydroxyl group, and a further dehydrogenation to yield the oxo- group of oxaloacetate.

The first dehydrogenation reaction, forming fumarate, is catalyzed by **succinate dehydrogenase,** which is bound to the inner surface of the inner mitochondrial membrane. The enzyme contains FAD and iron-sulfur (Fe:S) protein and directly reduces ubiquinone in the respiratory chain. **Fumarase (fumarate hydratase)** catalyzes the addition of water across the double bond of fumarate, yielding malate. Malate is converted to oxaloacetate by **malate dehydrogenase,** a reaction requiring $NAD^+$. Although the equilibrium of this reaction strongly favors malate, the net flux is toward the direction of oxaloacetate because of the continual removal of oxaloacetate (either to form citrate, as a substrate for gluconeogenesis, or to undergo transamination to aspartate) and also because of the continual reoxidation of NADH.

## TWELVE ATP ARE FORMED PER TURN OF THE CITRIC ACID CYCLE

As a result of oxidations catalyzed by the dehydrogenases of the citric acid cycle, three molecules of NADH and one of $FADH_2$ are produced for each molecule of acetyl-CoA catabolized in one turn of the cycle. These reducing equivalents are transferred to the respiratory chain (Figure 16–2), where reoxidation of each NADH results in formation of 3 ATP and reoxidation of $FADH_2$ in formation of 2 ATP. In addition, 1 ATP (or GTP) is formed by substrate-level phosphorylation catalyzed by succinate thiokinase.

## VITAMINS PLAY KEY ROLES IN THE CITRIC ACID CYCLE

Four of the B vitamins are essential in the citric acid cycle and therefore in energy-yielding metabolism: (1) **riboflavin,** in the form of flavin adenine dinucleotide (FAD), a cofactor in the α-ketoglutarate dehydrogenase complex and in succinate dehydrogenase; (2) **niacin,** in the form of nicotinamide adenine dinucleotide (NAD), the coenzyme for three dehydrogenases in the cycle—isocitrate dehydrogenase, α-ketoglutarate dehydrogenase, and malate dehydrogenase; (3) **thiamin (vitamin $B_1$),** as thiamin diphosphate, the coenzyme for decarboxylation in the α-ketoglutarate dehydrogenase reaction; and (4) **pantothenic acid,** as part of coenzyme A, the cofactor attached to "active" carboxylic acid residues such as acetyl-CoA and succinyl-CoA.

## THE CITRIC ACID CYCLE PLAYS A PIVOTAL ROLE IN METABOLISM

The citric acid cycle is not only a pathway for oxidation of two-carbon units—it is also a major pathway for interconversion of metabolites arising from **transamination** and **deamination** of amino acids. It also provides the substrates for **amino acid synthesis** by transamination, as well as for **gluconeogenesis** and **fatty acid synthesis**. Because it functions in both oxidative and synthetic processes, it is **amphibolic** (Figure 16–4).

### The Citric Acid Cycle Takes Part in Gluconeogenesis, Transamination, & Deamination

All the intermediates of the cycle are potentially glucogenic, since they can give rise to oxaloacetate and thus net production of glucose (in the liver and kidney, the organs that carry out gluconeogenesis; see Chapter 19). The key enzyme that catalyzes net transfer out of the cycle into gluconeogenesis is **phosphoenolpyruvate carboxykinase,** which decarboxylates oxaloacetate to phosphoenolpyruvate, with GTP acting as the donor phosphate (Figure 16–4).

Net transfer into the cycle occurs as a result of several different reactions. Among the most important of such **anaplerotic reactions** is the formation of oxaloacetate by the carboxylation of pyruvate, catalyzed by **pyruvate carboxylase.** This reaction is important in maintaining an adequate concentration of oxaloacetate for the condensation reaction with acetyl-CoA. If acetyl-CoA accumulates, it acts both as an allosteric activator of pyruvate carboxylase and as an inhibitor of pyruvate dehydrogenase, thereby ensuring a supply of oxaloacetate. Lactate, an important substrate for gluconeogenesis, enters the cycle via oxidation to pyruvate and then carboxylation to oxaloacetate.

**Aminotransferase** (transaminase) reactions form pyruvate from alanine, oxaloacetate from aspartate, and α-ketoglutarate from glutamate. Because these reactions are reversible, the cycle also serves as a source of carbon skeletons for the synthesis of these amino acids. Other amino acids contribute to gluconeogenesis because their carbon skeletons give rise to citric acid cycle

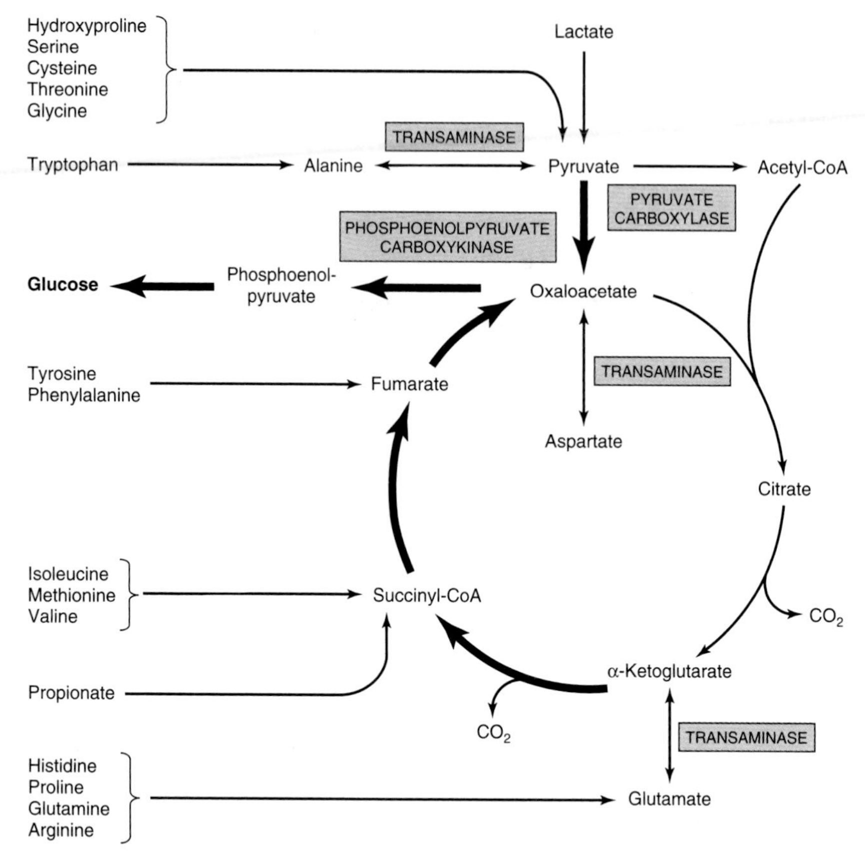


***Figure 16–4.*** Involvement of the citric acid cycle in transamination and gluconeogenesis. The bold arrows indicate the main pathway of gluconeogenesis.

intermediates. Alanine, cysteine, glycine, hydroxyproline, serine, threonine, and tryptophan yield pyruvate; arginine, histidine, glutamine, and proline yield α-ketoglutarate; isoleucine, methionine, and valine yield succinyl-CoA; and tyrosine and phenylalanine yield fumarate (Figure 16–4).

In ruminants, whose main metabolic fuel is short-chain fatty acids formed by bacterial fermentation, the conversion of propionate, the major glucogenic product of rumen fermentation, to succinyl-CoA via the methylmalonyl-CoA pathway (Figure 19–2) is especially important.

## The Citric Acid Cycle Takes Part in Fatty Acid Synthesis (Figure 16–5)

Acetyl-CoA, formed from pyruvate by the action of pyruvate dehydrogenase, is the major building block for long-chain fatty acid synthesis in nonruminants. (In ruminants, acetyl-CoA is derived directly from acetate.) Pyruvate dehydrogenase is a mitochondrial enzyme, and fatty acid synthesis is a cytosolic pathway, but the mitochondrial membrane is impermeable to acetyl-CoA. Acetyl-CoA is made available in the cytosol from citrate synthesized in the mitochondrion, transported into the cytosol and cleaved in a reaction catalyzed by **ATP-citrate lyase.**

## Regulation of the Citric Acid Cycle Depends Primarily on a Supply of Oxidized Cofactors

In most tissues, where the primary role of the citric acid cycle is in energy-yielding metabolism, **respiratory control** via the respiratory chain and oxidative phosphorylation regulates citric acid cycle activity (Chapter 14). Thus, activity is immediately dependent on the supply of $NAD^+$, which in turn, because of the tight coupling between oxidation and phosphorylation, is dependent on the availability of ADP and hence, ulti-

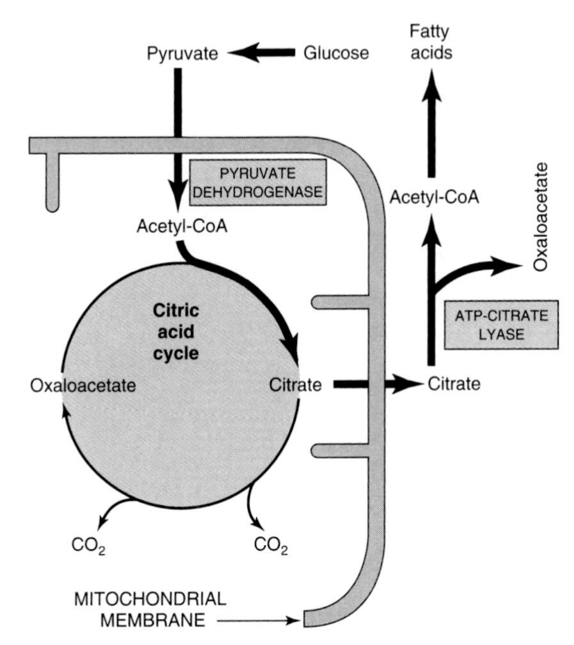


***Figure 16–5.*** Participation of the citric acid cycle in fatty acid synthesis from glucose. See also Figure 21–5.

mately, on the rate of utilization of ATP in chemical and physical work. In addition, individual enzymes of the cycle are regulated. The most likely sites for regulation are the nonequilibrium reactions catalyzed by pyruvate dehydrogenase, citrate synthase, isocitrate dehydrogenase, and α-ketoglutarate dehydrogenase. The dehydrogenases are activated by $Ca^{2+}$, which increases in concentration during muscular contraction and secretion, when there is increased energy demand. In a tissue such as brain, which is largely dependent on carbohydrate to supply acetyl-CoA, control of the citric acid cycle may occur at pyruvate dehydrogenase. Several enzymes are responsive to the energy status, as shown by the [ATP]/[ADP] and [NADH]/[$NAD^+$] ratios. Thus, there is allosteric inhibition of citrate synthase by ATP and long-chain fatty acyl-CoA. Allosteric activation of mitochondrial NAD-dependent isocitrate dehydrogenase by ADP is counteracted by ATP and NADH. The α-ketoglutarate dehydrogenase complex is regulated in the same way as is pyruvate dehydrogenase (Figure 17–6). Succinate dehydrogenase is inhibited by oxaloacetate, and the availability of oxaloacetate, as controlled by malate dehydrogenase, depends on the [NADH]/[$NAD^+$] ratio. Since the $K_m$ for oxaloacetate of citrate synthase is of the same order of magnitude as the intramitochondrial concentration, it is likely that the concentration of oxaloacetate controls the rate of citrate formation. Which of these mechanisms are important in vivo has still to be resolved.

## SUMMARY

- The citric acid cycle is the final pathway for the oxidation of carbohydrate, lipid, and protein whose common end-metabolite, acetyl-CoA, reacts with oxaloacetate to form citrate. By a series of dehydrogenations and decarboxylations, citrate is degraded, releasing reduced coenzymes and $2CO_2$ and regenerating oxaloacetate.
- The reduced coenzymes are oxidized by the respiratory chain linked to formation of ATP. Thus, the cycle is the major route for the generation of ATP and is located in the matrix of mitochondria adjacent to the enzymes of the respiratory chain and oxidative phosphorylation.
- The citric acid cycle is amphibolic, since in addition to oxidation it is important in the provision of carbon skeletons for gluconeogenesis, fatty acid synthesis, and interconversion of amino acids.

## REFERENCES

Baldwin JE, Krebs HA: The evolution of metabolic cycles. Nature 1981;291:381.

Goodwin TW (editor): *The Metabolic Roles of Citrate.* Academic Press, 1968.

Greville GD: Vol 1, p 297, in: *Carbohydrate Metabolism and Its Disorders.* Dickens F, Randle PJ, Whelan WJ (editors). Academic Press, 1968.

Kay J, Weitzman PDJ (editors): *Krebs' Citric Acid Cycle—Half a Century and Still Turning.* Biochemical Society, London, 1987.

Srere PA: The enzymology of the formation and breakdown of citrate. Adv Enzymol 1975;43:57.

Tyler DD: *The Mitochondrion in Health and Disease.* VCH Publishers, 1992.

# Glycolysis & the Oxidation of Pyruvate

# 17

*Peter A. Mayes, PhD, DSc, & David A. Bender, PhD*

## BIOMEDICAL IMPORTANCE

Most tissues have at least some requirement for glucose. In brain, the requirement is substantial. Glycolysis, the major pathway for glucose metabolism, occurs in the cytosol of all cells. It is unique in that it can function either aerobically or anaerobically. Erythrocytes, which lack mitochondria, are completely reliant on glucose as their metabolic fuel and metabolize it by anaerobic glycolysis. However, to oxidize glucose beyond pyruvate (the end product of glycolysis) requires both oxygen and mitochondrial enzyme systems such as the pyruvate dehydrogenase complex, the citric acid cycle, and the respiratory chain.

Glycolysis is both the principal route for glucose metabolism and the main pathway for the metabolism of fructose, galactose, and other carbohydrates derived from the diet. The ability of glycolysis to provide ATP in the absence of oxygen is especially important because it allows skeletal muscle to perform at very high levels when oxygen supply is insufficient and because it allows tissues to survive anoxic episodes. However, heart muscle, which is adapted for aerobic performance, has relatively low glycolytic activity and poor survival under conditions of **ischemia.** Diseases in which enzymes of glycolysis (eg, pyruvate kinase) are deficient are mainly seen as **hemolytic anemias** or, if the defect affects skeletal muscle (eg, phosphofructokinase), as **fatigue.** In fast-growing cancer cells, glycolysis proceeds at a higher rate than is required by the citric acid cycle, forming large amounts of pyruvate, which is reduced to lactate and exported. This produces a relatively acidic local environment in the tumor which may have implications for cancer therapy. The lactate is used for gluconeogenesis in the liver, an energy-expensive process responsible for much of the **hypermetabolism** seen in **cancer cachexia. Lactic acidosis** results from several causes, including impaired activity of pyruvate dehydrogenase.

## GLYCOLYSIS CAN FUNCTION UNDER ANAEROBIC CONDITIONS

When a muscle contracts in an anaerobic medium, ie, one from which oxygen is excluded, **glycogen disappears** and **lactate appears** as the principal end product. When oxygen is admitted, aerobic recovery takes place and lactate disappears. However, if contraction occurs under aerobic conditions, lactate does not accumulate and pyruvate is the major end product of glycolysis. Pyruvate is oxidized further to $CO_2$ and water (Figure 17–1). When oxygen is in short supply, mitochondrial reoxidation of NADH formed from $NAD^+$ during glycolysis is impaired, and NADH is reoxidized by reducing pyruvate to lactate, so permitting glycolysis to proceed (Figure 17–1). While glycolysis can occur under anaerobic conditions, this has a price, for it limits the amount of ATP formed per mole of glucose oxidized, so that much more glucose must be metabolized under anaerobic than under aerobic conditions.

## THE REACTIONS OF GLYCOLYSIS CONSTITUTE THE MAIN PATHWAY OF GLUCOSE UTILIZATION

The overall equation for glycolysis from glucose to lactate is as follows:

$$\text{Glucose} + 2\text{ADP} + 2\text{P}_i \rightarrow 2\text{L}(+)-\text{Lactate} + 2\text{ATP} + 2\text{H}_2\text{O}$$

All of the enzymes of glycolysis (Figure 17–2) are found in the cytosol. Glucose enters glycolysis by phosphorylation to glucose 6-phosphate, catalyzed by **hexokinase,** using ATP as the phosphate donor. Under physiologic conditions, the phosphorylation of glucose to glucose 6-phosphate can be regarded as irreversible. Hexokinase is inhibited allosterically by its product, glucose 6-phosphate. In tissues other than the liver and pancreatic B islet cells, the availability of glucose for

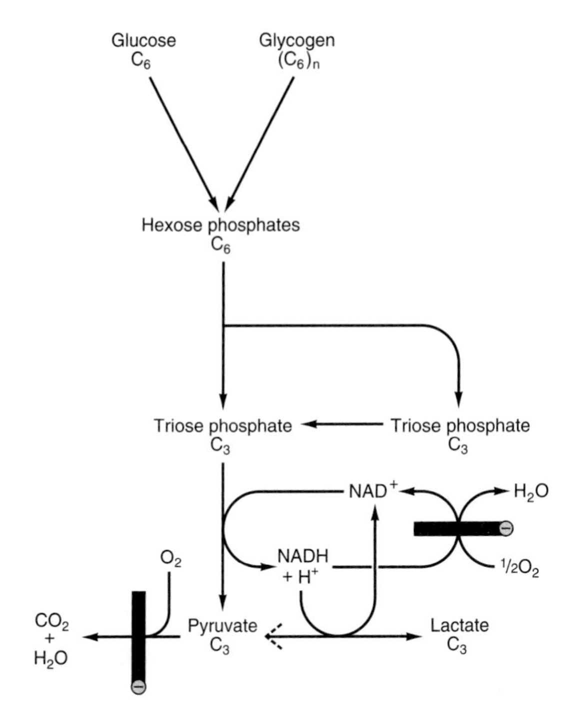


***Figure 17–1.*** Summary of glycolysis. ⊖, blocked by anaerobic conditions or by absence of mitochondria containing key respiratory enzymes, eg, as in erythrocytes.

glycolysis (or glycogen synthesis in muscle and lipogenesis in adipose tissue) is controlled by transport into the cell, which in turn is regulated by **insulin.** Hexokinase has a high affinity (low $K_m$) for its substrate, glucose, and in the liver and pancreatic B islet cells is saturated under all normal conditions and so acts at a constant rate to provide glucose 6-phosphate to meet the cell's need. Liver and pancreatic B islet cells also contain an isoenzyme of hexokinase, **glucokinase,** which has a $K_m$ very much higher than the normal intracellular concentration of glucose. The function of glucokinase in the liver is to remove glucose from the blood following a meal, providing glucose 6-phosphate in excess of requirements for glycolysis, which will be used for glycogen synthesis and lipogenesis. In the pancreas, the glucose 6-phosphate formed by glucokinase signals increased glucose availability and leads to the secretion of insulin.

Glucose 6-phosphate is an important compound at the junction of several metabolic pathways (glycolysis, gluconeogenesis, the pentose phosphate pathway, glycogenesis, and glycogenolysis). In glycolysis, it is converted to fructose 6-phosphate by **phosphohexose-isomerase,** which involves an aldose-ketose isomerization. This reaction is followed by another phosphorylation with ATP catalyzed by the enzyme **phosphofructokinase (phosphofructokinase-1),** forming fructose 1,6-bisphosphate. The phosphofructokinase reaction may be considered to be functionally irreversible under physiologic conditions; it is both inducible and subject to allosteric regulation and has a major role in regulating the rate of glycolysis. Fructose 1,6-bisphosphate is cleaved by **aldolase** (fructose 1,6-bisphosphate aldolase) into two triose phosphates, glyceraldehyde 3-phosphate and dihydroxyacetone phosphate. Glyceraldehyde 3-phosphate and dihydroxyacetone phosphate are interconverted by the enzyme **phosphotriose isomerase.**

Glycolysis continues with the oxidation of glyceraldehyde 3-phosphate to 1,3-bisphosphoglycerate. The enzyme catalyzing this oxidation, **glyceraldehyde 3-phosphate dehydrogenase,** is NAD-dependent. Structurally, it consists of four identical polypeptides (monomers) forming a tetramer. —SH groups are present on each polypeptide, derived from cysteine residues within the polypeptide chain. One of the —SH groups at the active site of the enzyme (Figure 17–3) combines with the substrate forming a thiohemiacetal that is oxidized to a thiol ester; the hydrogens removed in this oxidation are transferred to $NAD^+$. The thiol ester then undergoes phosphorolysis; inorganic phosphate ($P_i$) is added, forming 1,3-bisphosphoglycerate, and the —SH group is reconstituted.

In the next reaction, catalyzed by **phosphoglycerate kinase,** phosphate is transferred from 1,3-bisphosphoglycerate onto ADP, forming ATP (substrate-level phosphorylation) and 3-phosphoglycerate. Since two molecules of triose phosphate are formed per molecule of glucose, two molecules of ATP are generated at this stage per molecule of glucose undergoing glycolysis. The toxicity of arsenic is due to competition of arsenate with inorganic phosphate ($P_i$) in the above reactions to give 1-arseno-3-phosphoglycerate, which hydrolyzes spontaneously to give 3-phosphoglycerate plus heat, without generating ATP. 3-Phosphoglycerate is isomerized to 2-phosphoglycerate by **phosphoglycerate mutase.** It is likely that 2,3-bisphosphoglycerate (diphosphoglycerate; DPG) is an intermediate in this reaction.

The subsequent step is catalyzed by **enolase** and involves a dehydration, forming phosphoenolpyruvate. Enolase is inhibited by **fluoride.** To prevent glycolysis in the estimation of glucose, blood is collected in tubes containing fluoride. The enzyme is also dependent on the presence of either $Mg^{2+}$ or $Mn^{2+}$. The phosphate of phosphoenolpyruvate is transferred to ADP by **pyruvate kinase** to generate, at this stage, two molecules of ATP per molecule of glucose oxidized. The product of the enzyme-catalyzed reaction, enolpyruvate, undergoes spontaneous (nonenzymic) isomerization to pyruvate and so is not available to

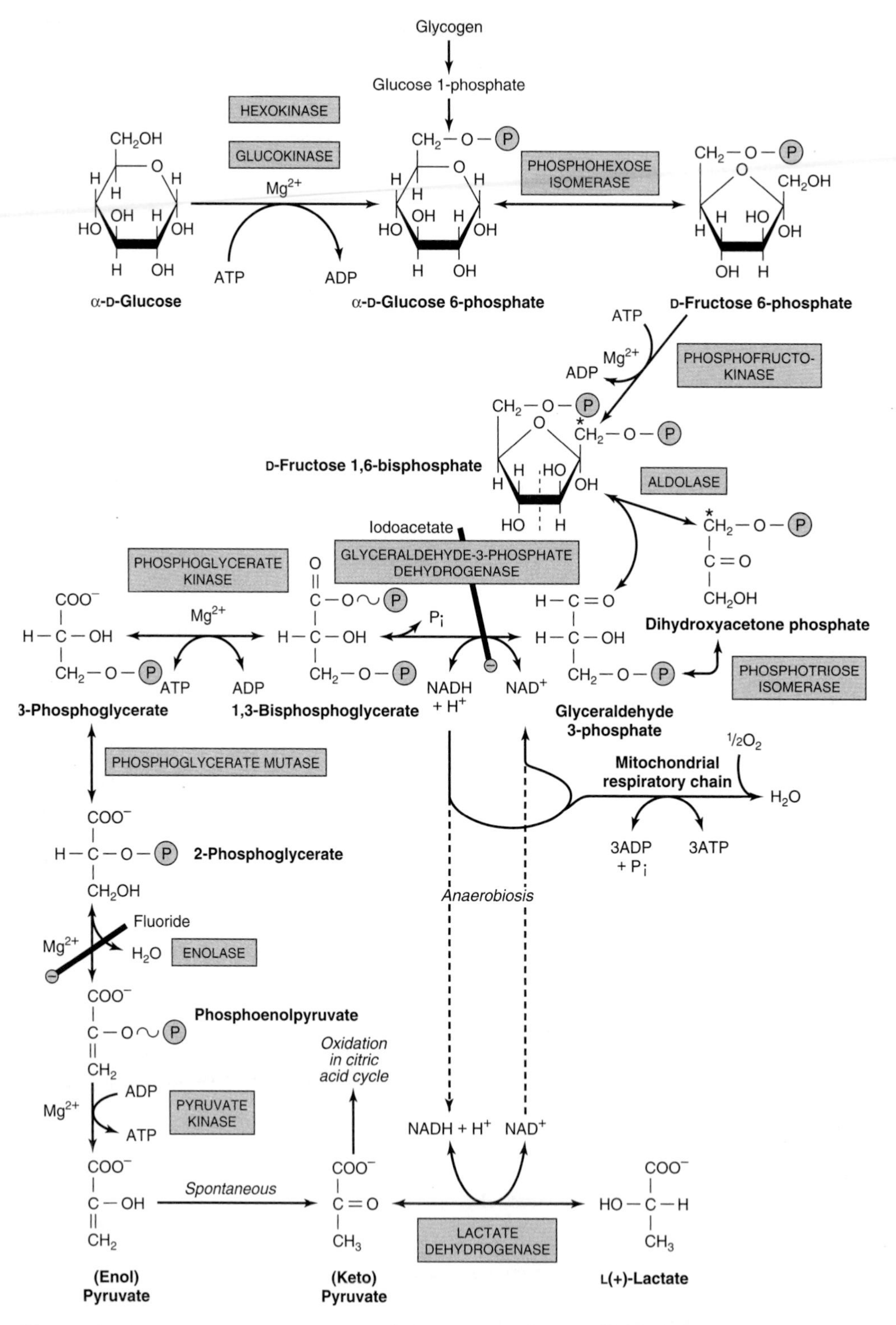


***Figure 17–2.*** The pathway of glycolysis. (Ⓟ, $—PO_3^{2-}$; $P_i$, $HOPO_3^{2-}$; ⊖, inhibition.) At asterisk: Carbon atoms 1–3 of fructose bisphosphate form dihydroxyacetone phosphate, whereas carbons 4–6 form glyceraldehyde 3-phosphate. The term "bis-," as in bisphosphate, indicates that the phosphate groups are separated, whereas diphosphate, as in adenosine diphosphate, indicates that they are joined.

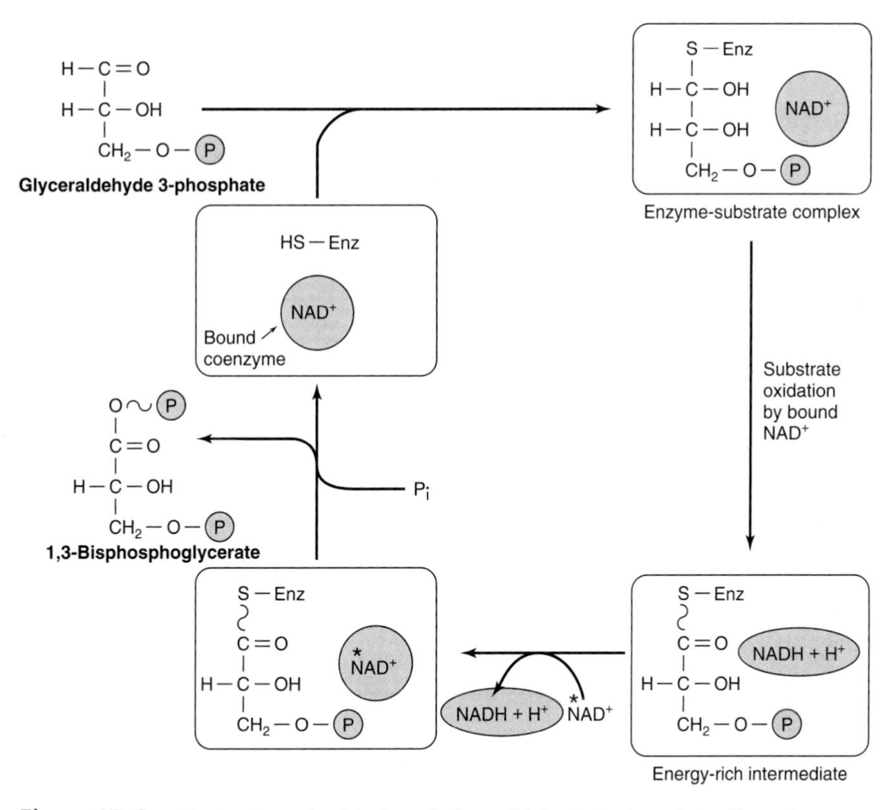


***Figure 17–3.*** Mechanism of oxidation of glyceraldehyde 3-phosphate. (Enz, glyceraldehyde-3-phosphate dehydrogenase.) The enzyme is inhibited by the —SH poison iodoacetate, which is thus able to inhibit glycolysis. The NADH produced on the enzyme is not as firmly bound to the enzyme as is $NAD^+$. Consequently, NADH is easily displaced by another molecule of $NAD^+$.

undergo the reverse reaction. The pyruvate kinase reaction is thus also irreversible under physiologic conditions.

The redox state of the tissue now determines which of two pathways is followed. Under **anaerobic** conditions, the reoxidation of NADH through the respiratory chain to oxygen is prevented. Pyruvate is reduced by the NADH to lactate, the reaction being catalyzed by **lactate dehydrogenase.** Several tissue-specific isoenzymes of this enzyme have been described and have clinical significance (Chapter 7). The reoxidation of NADH via lactate formation allows glycolysis to proceed in the absence of oxygen by regenerating sufficient $NAD^+$ for another cycle of the reaction catalyzed by glyceraldehyde-3-phosphate dehydrogenase. Under **aerobic conditions,** pyruvate is taken up into mitochondria and after conversion to acetyl-CoA is oxidized to $CO_2$ by the citric acid cycle. The reducing equivalents from the NADH + $H^+$ formed in glycolysis are taken up into mitochondria for oxidation via one of the two shuttles described in Chapter 12.

## Tissues That Function Under Hypoxic Circumstances Tend to Produce Lactate (Figure 17–2)

This is true of skeletal muscle, particularly the white fibers, where the rate of work output—and therefore the need for ATP formation—may exceed the rate at which oxygen can be taken up and utilized. Glycolysis in erythrocytes, even under aerobic conditions, always terminates in lactate, because the subsequent reactions of pyruvate are mitochondrial, and erythrocytes lack mitochondria. Other tissues that normally derive much of their energy from glycolysis and produce lactate include brain, gastrointestinal tract, renal medulla, retina, and skin. The liver, kidneys, and heart usually take up lactate and oxidize it but will produce it under hypoxic conditions.

### Glycolysis Is Regulated at Three Steps Involving Nonequilibrium Reactions

Although most of the reactions of glycolysis are reversible, three are markedly exergonic and must therefore be considered physiologically irreversible. These reactions, catalyzed by **hexokinase** (and glucokinase), **phosphofructokinase,** and **pyruvate kinase,** are the major sites of regulation of glycolysis. Cells that are capable of reversing the glycolytic pathway **(gluconeogenesis)** have different enzymes that catalyze reactions which effectively reverse these irreversible reactions. The importance of these steps in the regulation of glycolysis and gluconeogenesis is discussed in Chapter 19.

### In Erythrocytes, the First Site in Glycolysis for ATP Generation May Be Bypassed

In the erythrocytes of many mammals, the reaction catalyzed by **phosphoglycerate kinase** may be bypassed by a process that effectively dissipates as heat the free energy associated with the high-energy phosphate of 1,3-bisphosphoglycerate (Figure 17–4). **Bisphosphoglycerate mutase** catalyzes the conversion of 1,3-bisphosphoglycerate to 2,3-bisphosphoglycerate, which is converted to 3-phosphoglycerate by **2,3-bisphosphoglycerate phosphatase** (and possibly also phosphoglycerate mutase). This alternative pathway involves no net yield of ATP from glycolysis. However, it does serve to provide 2,3-bisphosphoglycerate, which binds to hemoglobin, decreasing its affinity for oxygen and so making oxygen more readily available to tissues (see Chapter 6).

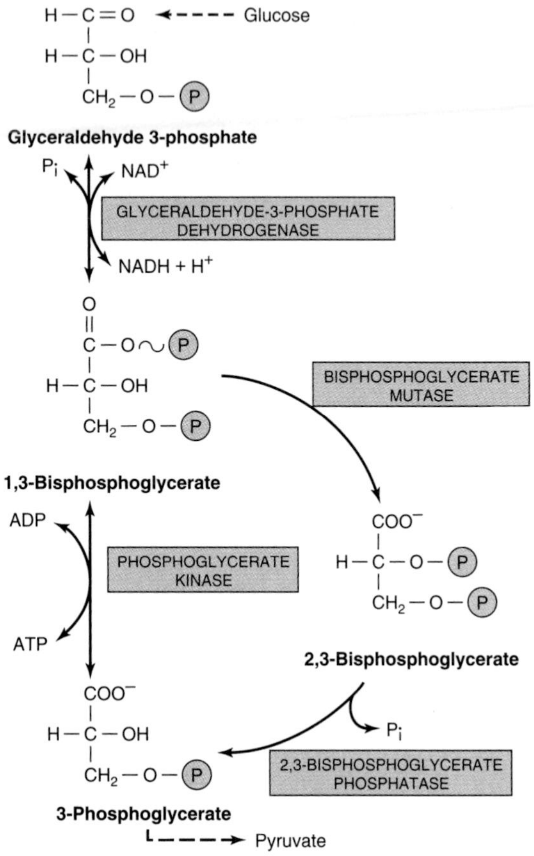

***Figure 17–4.*** 2,3-Bisphosphoglycerate pathway in erythrocytes.

## THE OXIDATION OF PYRUVATE TO ACETYL-CoA IS THE IRREVERSIBLE ROUTE FROM GLYCOLYSIS TO THE CITRIC ACID CYCLE

Pyruvate, formed in the cytosol, is transported into the mitochondrion by a proton symporter (Figure 12–10). Inside the mitochondrion, pyruvate is oxidatively decarboxylated to acetyl-CoA by a multienzyme complex that is associated with the inner mitochondrial membrane. This **pyruvate dehydrogenase complex** is analogous to the α-ketoglutarate dehydrogenase complex of the citric acid cycle (Figure 16–3). Pyruvate is decarboxylated by the **pyruvate dehydrogenase** component of the enzyme complex to a hydroxyethyl derivative of the thiazole ring of enzyme-bound **thiamin diphosphate,** which in turn reacts with oxidized lipoamide, the prosthetic group of **dihydrolipoyl transacetylase,** to form acetyl lipoamide (Figure 17–5). Thiamin is vitamin $B_1$ (Chapter 45), and in thiamin deficiency glucose metabolism is impaired and there is significant (and potentially life-threatening) lactic and pyruvic acidosis. Acetyl lipoamide reacts with coenzyme A to form acetyl-CoA and reduced lipoamide. The cycle of reaction is completed when the reduced lipoamide is reoxidized by a flavoprotein, **dihydrolipoyl dehydrogenase,** containing FAD. Finally, the reduced flavoprotein is oxidized by $NAD^+$, which in turn transfers reducing equivalents to the respiratory chain.

$$\text{Pyruvate} + NAD^+ + \text{CoA} \rightarrow \text{Acetyl-CoA} + NADH + H^+ + CO_2$$

The pyruvate dehydrogenase complex consists of a number of polypeptide chains of each of the three component enzymes, all organized in a regular spatial configuration. Movement of the individual enzymes appears to be restricted, and the metabolic intermediates do not dissociate freely but remain bound to the enzymes. Such a complex of enzymes, in which the sub-

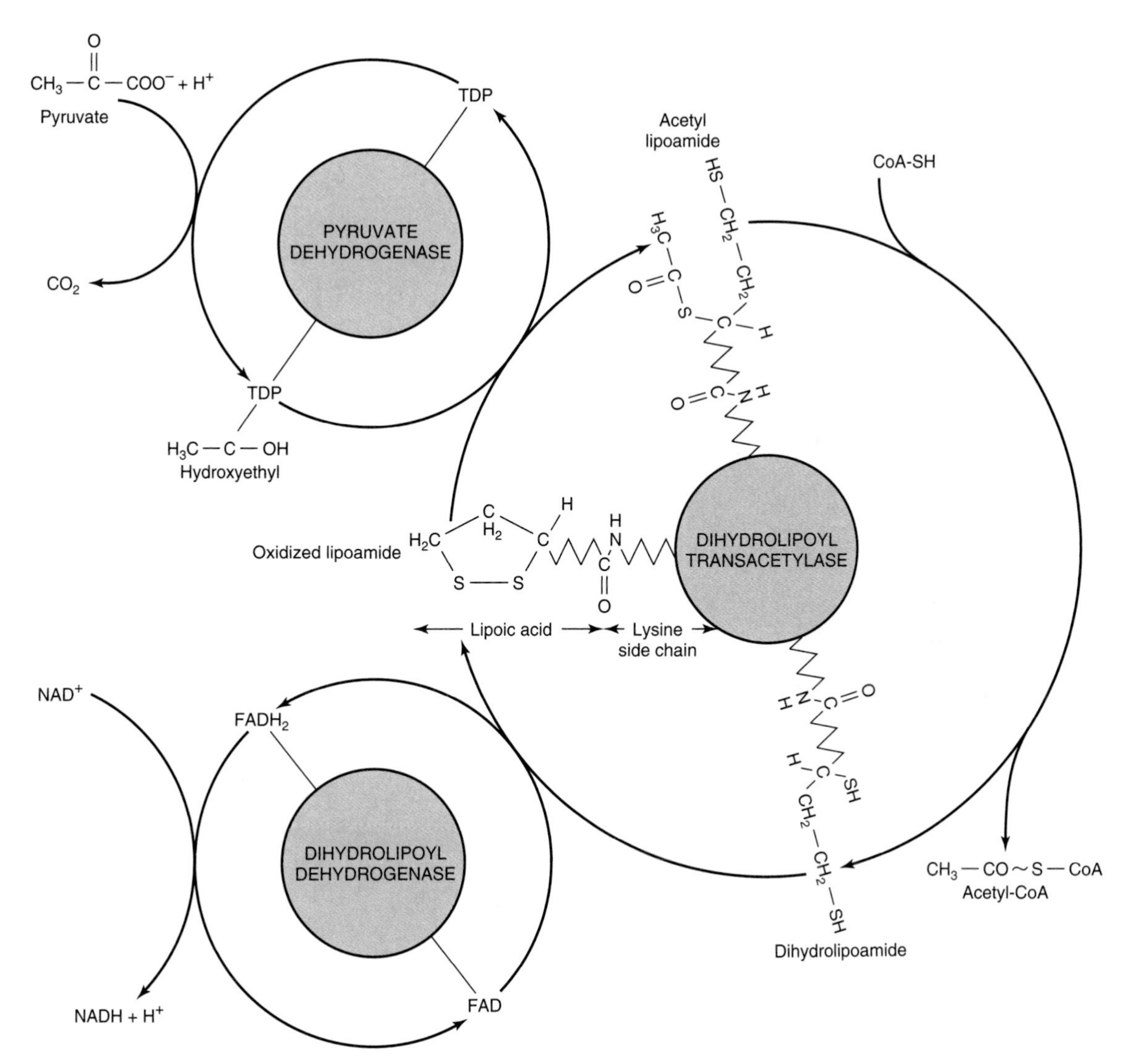


***Figure 17–5.*** Oxidative decarboxylation of pyruvate by the pyruvate dehydrogenase complex. Lipoic acid is joined by an amide link to a lysine residue of the transacetylase component of the enzyme complex. It forms a long flexible arm, allowing the lipoic acid prosthetic group to rotate sequentially between the active sites of each of the enzymes of the complex. ($NAD^+$, nicotinamide adenine dinucleotide; FAD, flavin adenine dinucleotide; TDP, thiamin diphosphate.)

strates are handed on from one enzyme to the next, increases the reaction rate and eliminates side reactions, increasing overall efficiency.

## Pyruvate Dehydrogenase Is Regulated by End-Product Inhibition & Covalent Modification

Pyruvate dehydrogenase is inhibited by its products, acetyl-CoA and NADH (Figure 17–6). It is also regulated by phosphorylation by a kinase of three serine residues on the pyruvate dehydrogenase component of the multienzyme complex, resulting in decreased activity, and by dephosphorylation by a phosphatase that causes an increase in activity. The kinase is activated by increases in the [ATP]/[ADP], [acetyl-CoA]/[CoA], and [NADH]/[$NAD^+$] ratios. Thus, pyruvate dehydrogenase—and therefore glycolysis—is inhibited not only by a high-energy potential but also when fatty acids are being oxidized. Thus, in starvation, when free fatty acid

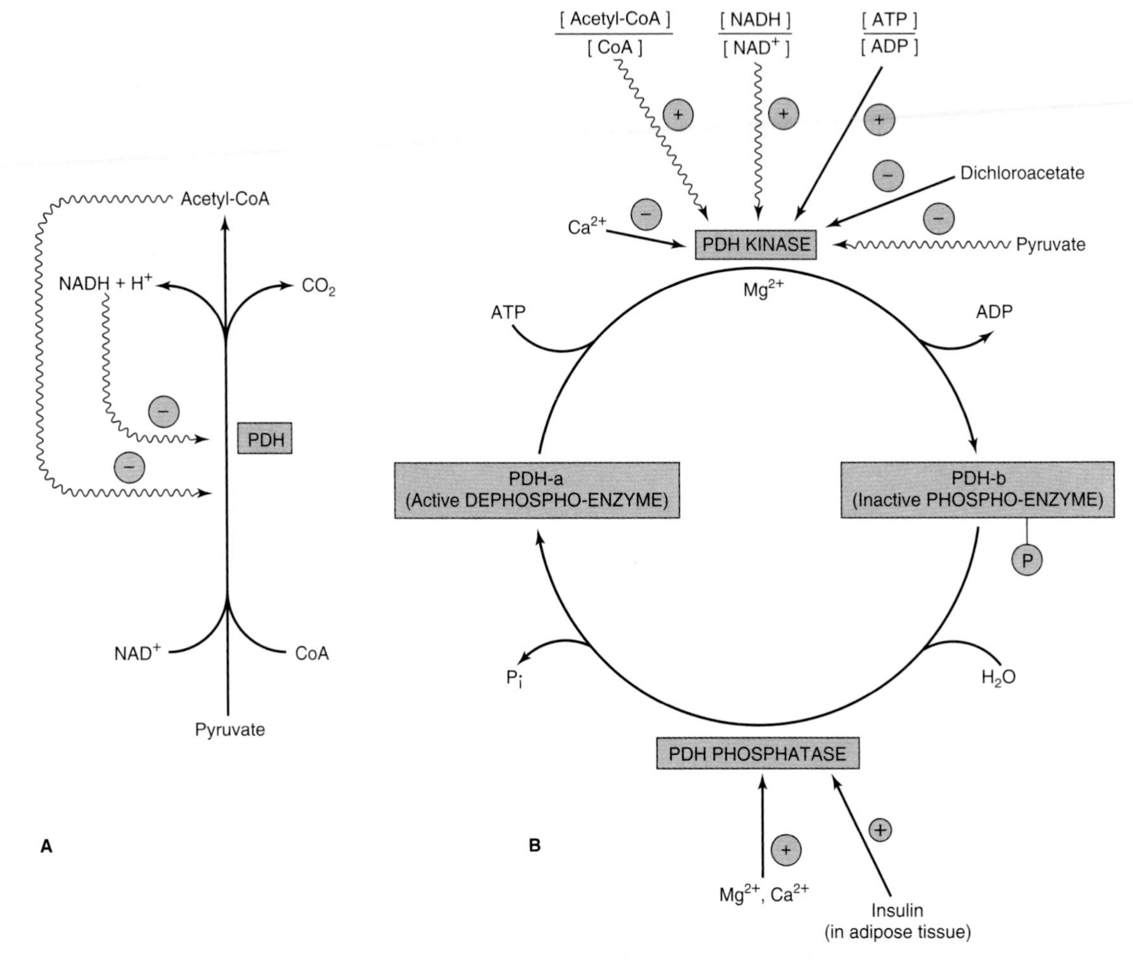


***Figure 17–6.*** Regulation of pyruvate dehydrogenase (PDH). Arrows with wavy shafts indicate allosteric effects. **A:** Regulation by end-product inhibition. **B:** Regulation by interconversion of active and inactive forms.

concentrations increase, there is a decrease in the proportion of the enzyme in the active form, leading to a sparing of carbohydrate. In adipose tissue, where glucose provides acetyl CoA for lipogenesis, the enzyme is activated in response to insulin.

## Oxidation of Glucose Yields Up to 38 Mol of ATP Under Aerobic Conditions But Only 2 Mol When $O_2$ Is Absent

When 1 mol of glucose is combusted in a calorimeter to $CO_2$ and water, approximately 2870 kJ are liberated as heat. When oxidation occurs in the tissues, approximately 38 mol of ATP are generated per molecule of glucose oxidized to $CO_2$ and water. In vivo, $\Delta G$ for the ATP synthase reaction has been calculated as approximately 51.6 kJ. It follows that the total energy captured in ATP per mole of glucose oxidized is 1961 kJ, or approximately 68% of the energy of combustion. Most of the ATP is formed by oxidative phosphorylation resulting from the reoxidation of reduced coenzymes by the respiratory chain. The remainder is formed by substrate-level phosphorylation (Table 17–1).

# CLINICAL ASPECTS

## Inhibition of Pyruvate Metabolism Leads to Lactic Acidosis

Arsenite and mercuric ions react with the —SH groups of lipoic acid and inhibit pyruvate dehydrogenase, as

***Table 17–1.*** Generation of high-energy phosphate in the catabolism of glucose.

| Pathway | Reaction Catalyzed by | Method of ~Ⓟ Production | Number of ~Ⓟ Formed per Mole of Glucose |
|---|---|---|---|
| Glycolysis | Glyceraldehyde-3-phosphate dehydrogenase | Respiratory chain oxidation of 2 NADH | 6* |
| | Phosphoglycerate kinase | Phosphorylation at substrate level | 2 |
| | Pyruvate kinase | Phosphorylation at substrate level | 2 |
| | | | 10 |
| Allow for consumption of ATP by reactions catalyzed by hexokinase and phosphofructokinase | | | −2 |
| | | | Net 8 |
| Citric acid cycle | Pyruvate dehydrogenase | Respiratory chain oxidation of 2 NADH | 6 |
| | Isocitrate dehydrogenase | Respiratory chain oxidation of 2 NADH | 6 |
| | α-Ketoglutarate dehydrogenase | Respiratory chain oxidation of 2 NADH | 6 |
| | Succinate thiokinase | Phosphorylation at substrate level | 2 |
| | Succinate dehydrogenase | Respiratory chain oxidation of 2 $FADH_2$ | 4 |
| | Malate dehydrogenase | Respiratory chain oxidation of 2 NADH | 6 |
| | | | Net 30 |
| Total per mole of glucose under aerobic conditions | | | 38 |
| Total per mole of glucose under anaerobic conditions | | | 2 |

*It is assumed that NADH formed in glycolysis is transported into mitochondria via the malate shuttle (see Figure 12–13). If the glycerophosphate shuttle is used, only 2 ~Ⓟ would be formed per mole of NADH, the total net production being 26 instead of 38. The calculation ignores the small loss of ATP due to a transport of $H^+$ into the mitochondrion with pyruvate and a similar transport of $H^+$ in the operation of the malate shuttle, totaling about 1 mol of ATP. Note that there is a substantial benefit under anaerobic conditions if glycogen is the starting point, since the net production of high-energy phosphate in glycolysis is increased from 2 to 3, as ATP is no longer required by the hexokinase reaction.

does a **dietary deficiency of thiamin,** allowing pyruvate to accumulate. Nutritionally deprived alcoholics are thiamin-deficient and may develop potentially fatal pyruvic and lactic acidosis. Patients with **inherited pyruvate dehydrogenase deficiency,** which can be due to defects in one or more of the components of the enzyme complex, also present with lactic acidosis, particularly after a glucose load. Because of its dependence on glucose as a fuel, brain is a prominent tissue where these metabolic defects manifest themselves in neurologic disturbances.

Inherited aldolase A deficiency and pyruvate kinase deficiency in erythrocytes cause **hemolytic anemia.** The exercise capacity of patients with **muscle phosphofructokinase deficiency** is low, particularly on high-carbohydrate diets. By providing an alternative lipid fuel, eg, during starvation, when blood free fatty acids and ketone bodies are increased, work capacity is improved.

## SUMMARY


- Glycolysis is the cytosolic pathway of all mammalian cells for the metabolism of glucose (or glycogen) to pyruvate and lactate.
- It can function anaerobically by regenerating oxidized $NAD^+$ (required in the glyceraldehyde-3-phosphate dehydrogenase reaction) by reducing pyruvate to lactate.
- Lactate is the end product of glycolysis under anaerobic conditions (eg, in exercising muscle) or when the metabolic machinery is absent for the further oxidation of pyruvate (eg, in erythrocytes).
- Glycolysis is regulated by three enzymes catalyzing nonequilibrium reactions: hexokinase, phosphofructokinase, and pyruvate kinase.
- In erythrocytes, the first site in glycolysis for generation of ATP may be bypassed, leading to the formation of 2,3-bisphosphoglycerate, which is important in decreasing the affinity of hemoglobin for $O_2$.
- Pyruvate is oxidized to acetyl-CoA by a multienzyme complex, pyruvate dehydrogenase, that is dependent on the vitamin cofactor thiamin diphosphate.
- Conditions that involve an inability to metabolize pyruvate frequently lead to lactic acidosis.


## REFERENCES


Behal RH et al: Regulation of the pyruvate dehydrogenase multienzyme complex. Annu Rev Nutr 1993;13:497.

Boiteux A, Hess B: Design of glycolysis. Phil Trans R Soc London B 1981;293:5.

Fothergill-Gilmore LA: The evolution of the glycolytic pathway. Trends Biochem Sci 1986;11:47.

Scriver CR et al (editors): *The Metabolic and Molecular Bases of Inherited Disease,* 8th ed. McGraw-Hill, 2001.

Sols A: Multimodulation of enzyme activity. Curr Top Cell Reg 1981;19:77.

Srere PA: Complexes of sequential metabolic enzymes. Annu Rev Biochem 1987;56:89.

# Metabolism of Glycogen 18

*Peter A. Mayes, PhD, DSc, & David A. Bender, PhD*

## BIOMEDICAL IMPORTANCE

Glycogen is the major storage carbohydrate in animals, corresponding to starch in plants; it is a branched polymer of α-D-glucose. It occurs mainly in liver (up to 6%) and muscle, where it rarely exceeds 1%. However, because of its greater mass, muscle contains about three to four times as much glycogen as does liver (Table 18–1).

Muscle glycogen is a readily available source of glucose for glycolysis within the muscle itself. Liver glycogen functions to store and export glucose to maintain **blood glucose** between meals. After 12–18 hours of fasting, the liver glycogen is almost totally depleted. **Glycogen storage diseases** are a group of inherited disorders characterized by deficient mobilization of glycogen or deposition of abnormal forms of glycogen, leading to muscular weakness or even death.

## GLYCOGENESIS OCCURS MAINLY IN MUSCLE & LIVER

### The Pathway of Glycogen Biosynthesis Involves a Special Nucleotide of Glucose (Figure 18–1)

As in glycolysis, glucose is phosphorylated to glucose 6-phosphate, catalyzed by **hexokinase** in muscle and **glucokinase** in liver. Glucose 6-phosphate is isomerized to glucose 1-phosphate by **phosphoglucomutase.** The enzyme itself is phosphorylated, and the phospho-group takes part in a reversible reaction in which glucose 1,6-bisphosphate is an intermediate. Next, glucose 1-phosphate reacts with uridine triphosphate (UTP) to form the active nucleotide **uridine diphosphate glucose (UDPGlc)*** and pyrophosphate (Figure 18–2), catalyzed by **UDPGlc pyrophosphorylase. Pyrophosphatase** catalyzes hydrolysis of pyrophosphate to 2 mol of inorganic phosphate, shifting the equilibrium of the main reaction by removing one of its products.

**Glycogen synthase** catalyzes the formation of a glycoside bond between $C_1$ of the activated glucose of UDPGlc and $C_4$ of a terminal glucose residue of glycogen, liberating uridine diphosphate (UDP). A preexisting glycogen molecule, or "glycogen primer," must be present to initiate this reaction. The glycogen primer may in turn be formed on a primer known as **glycogenin,** which is a 37-kDa protein that is glycosylated on a specific tyrosine residue by UDPGlc. Further glucose residues are attached in the 1→4 position to make a short chain that is a substrate for glycogen synthase. In skeletal muscle, glycogenin remains attached in the center of the glycogen molecule (Figure 13–15), whereas in liver the number of glycogen molecules is greater than the number of glycogenin molecules.

### Branching Involves Detachment of Existing Glycogen Chains

The addition of a glucose residue to a preexisting glycogen chain, or "primer," occurs at the nonreducing, outer end of the molecule so that the "branches" of the glycogen "tree" become elongated as successive 1→4 linkages are formed (Figure 18–3). When the chain has been lengthened to at least 11 glucose residues, **branching enzyme** transfers a part of the 1→4 chain (at least six glucose residues) to a neighboring chain to form a 1→6 linkage, establishing a **branch point.** The branches grow by further additions of 1→4-glucosyl units and further branching.

## GLYCOGENOLYSIS IS NOT THE REVERSE OF GLYCOGENESIS BUT IS A SEPARATE PATHWAY (Figure 18–1)

**Glycogen phosphorylase** catalyzes the rate-limiting step in glycogenolysis by promoting the phosphorylytic cleavage by inorganic phosphate (phosphorylysis; cf hy-

* Other nucleoside diphosphate sugar compounds are known, eg, UDPGal. In addition, the same sugar may be linked to different nucleotides. For example, glucose may be linked to uridine (as shown above) as well as to guanosine, thymidine, adenosine, or cytidine nucleotides.

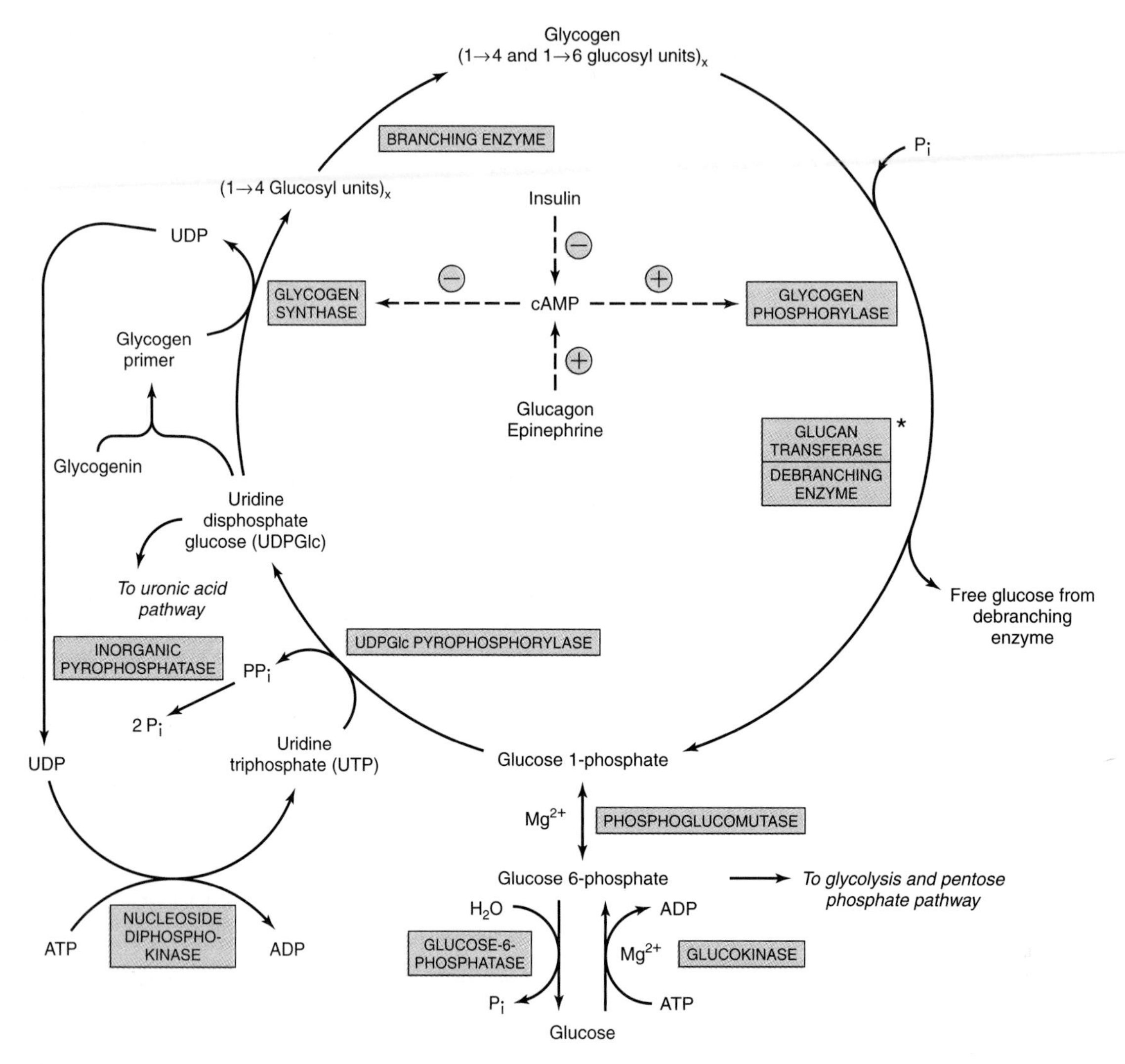


***Figure 18–1.*** Pathway of glycogenesis and of glycogenolysis in the liver. Two high-energy phosphates are used in the incorporation of 1 mol of glucose into glycogen. ⊕, stimulation; ⊖, inhibition. Insulin decreases the level of cAMP only after it has been raised by glucagon or epinephrine—ie, it antagonizes their action. Glucagon is active in heart muscle but not in skeletal muscle. At asterisk: Glucan transferase and debranching enzyme appear to be two separate activities of the same enzyme.

***Table 18–1.*** Storage of carbohydrate in postabsorptive normal adult humans (70 kg).

| | | | |
|---|---|---|---|
| Liver glycogen | 4.0% | = | 72 g[1] |
| Muscle glycogen | 0.7% | = | 245 g[2] |
| Extracellular glucose | 0.1% | = | 10 g[3] |
| | | | 327 g |

[1]Liver weight 1800 g.
[2]Muscle mass 35 kg.
[3]Total volume 10 L.

drolysis) of the 1→4 linkages of glycogen to yield glucose 1-phosphate. The terminal glucosyl residues from the outermost chains of the glycogen molecule are removed sequentially until approximately four glucose residues remain on either side of a 1→6 branch (Figure 18–4). Another enzyme **(α-[1→4]→α-[1→4] glucan transferase)** transfers a trisaccharide unit from one branch to the other, exposing the 1→6 branch point. **Hydrolysis** of the 1→6 linkages requires the **debranching enzyme.** Further phosphorylase action can

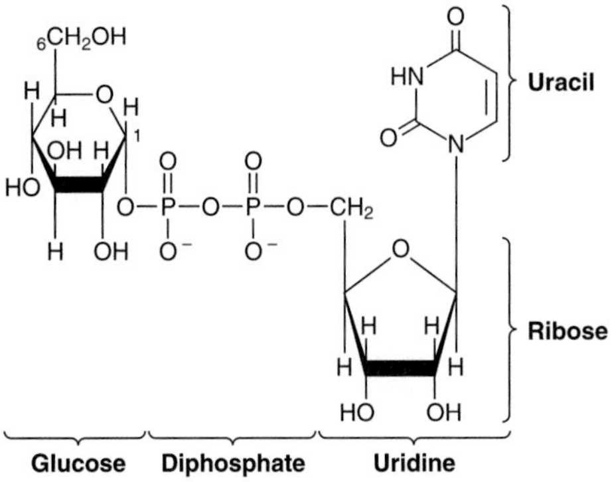

**Figure 18–2.** Uridine diphosphate glucose (UDPGlc).

then proceed. The combined action of phosphorylase and these other enzymes leads to the complete breakdown of glycogen. The reaction catalyzed by phosphoglucomutase is reversible, so that glucose 6-phosphate can be formed from glucose 1-phosphate. In **liver** (and **kidney**), but not in muscle, there is a specific enzyme, **glucose-6-phosphatase,** that hydrolyzes glucose 6-phosphate, yielding glucose that is exported, leading to an increase in the blood glucose concentration.

## CYCLIC AMP INTEGRATES THE REGULATION OF GLYCOGENOLYSIS & GLYCOGENESIS

The principal enzymes controlling glycogen metabolism—glycogen phosphorylase and glycogen synthase—are regulated by allosteric mechanisms and covalent modifications due to reversible phosphorylation and dephosphorylation of enzyme protein in response to hormone action (Chapter 9).

Cyclic AMP (cAMP) (Figure 18–5) is formed from ATP by **adenylyl cyclase** at the inner surface of cell membranes and acts as an intracellular **second messenger** in response to hormones such as **epinephrine, norepinephrine,** and **glucagon.** cAMP is hydrolyzed by **phosphodiesterase,** so terminating hormone action. In liver, insulin increases the activity of phosphodiesterase.

### Phosphorylase Differs Between Liver & Muscle

In liver, one of the serine hydroxyl groups of active **phosphorylase a** is phosphorylated. It is inactivated by hydrolytic removal of the phosphate by **protein phosphatase-1** to form **phosphorylase b.** Reactivation requires rephosphorylation catalyzed by **phosphorylase kinase.**

Muscle phosphorylase is distinct from that of liver. It is a dimer, each monomer containing 1 mol of pyridoxal phosphate (vitamin $B_6$). It is present in two forms: **phosphorylase a,** which is phosphorylated and active in either the presence or absence of 5′-AMP (its allosteric modifier); and **phosphorylase b,** which is dephosphorylated and active only in the presence of 5′-AMP. This occurs during exercise when the level of 5′-AMP rises, providing, by this mechanism, fuel for the muscle. Phosphorylase a is the normal physiologically active form of the enzyme.

### cAMP Activates Muscle Phosphorylase

Phosphorylase in muscle is activated in response to epinephrine (Figure 18–6) acting via cAMP. Increasing the concentration of cAMP activates **cAMP-dependent**

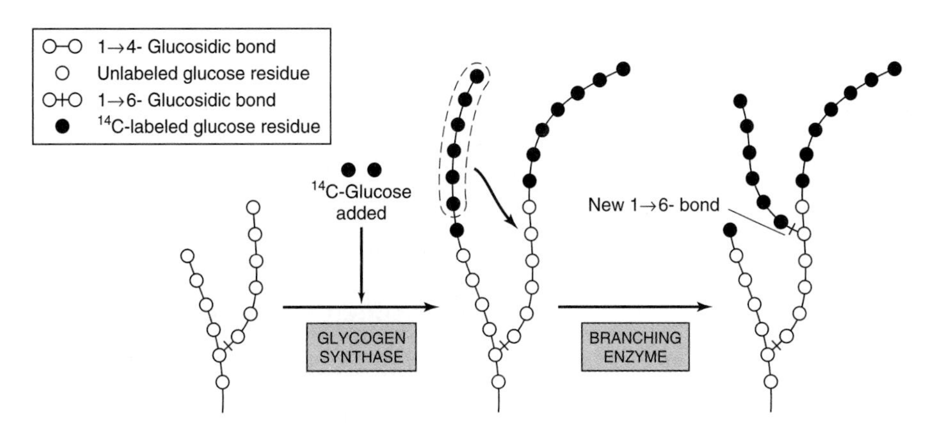


**Figure 18–3.** The biosynthesis of glycogen. The mechanism of branching as revealed by adding $^{14}$C-labeled glucose to the diet in the living animal and examining the liver glycogen at further intervals.

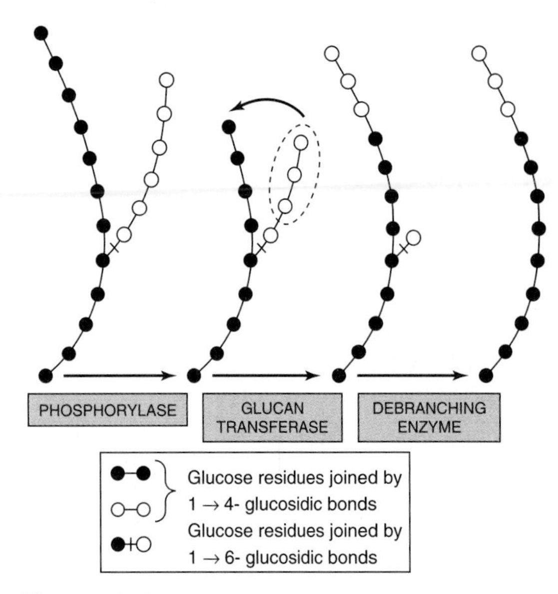


***Figure 18–4.*** Steps in glycogenolysis.

**protein kinase,** which catalyzes the phosphorylation by ATP of inactive **phosphorylase kinase b** to active **phosphorylase kinase a,** which in turn, by means of a further phosphorylation, activates phosphorylase b to phosphorylase a.

## $Ca^{2+}$ Synchronizes the Activation of Phosphorylase With Muscle Contraction

Glycogenolysis increases in muscle several hundred-fold immediately after the onset of contraction. This involves the rapid activation of phosphorylase by activation of phosphorylase kinase by $Ca^{2+}$, the same signal as that which initiates contraction in response to nerve stimulation. Muscle phosphorylase kinase has four types of subunits—α, β, γ, and δ—in a structure represented as $(\alpha\beta\gamma\delta)_4$. The α and β subunits contain serine residues that are phosphorylated by cAMP-dependent protein kinase. The δ subunit binds four $Ca^{2+}$ and is identical to the $Ca^{2+}$-binding protein **calmodulin** (Chapter 43). The binding of $Ca^{2+}$ activates the catalytic site of the γ subunit while the molecule remains in the dephosphorylated b configuration. However, the phosphorylated a form is only fully activated in the presence of $Ca^{2+}$. A second molecule of calmodulin, or TpC (the structurally similar $Ca^{2+}$-binding protein in muscle), can interact with phosphorylase kinase, causing further activation. Thus, activation of muscle contraction and glycogenolysis are carried out by the same $Ca^{2+}$-binding protein, ensuring their synchronization.

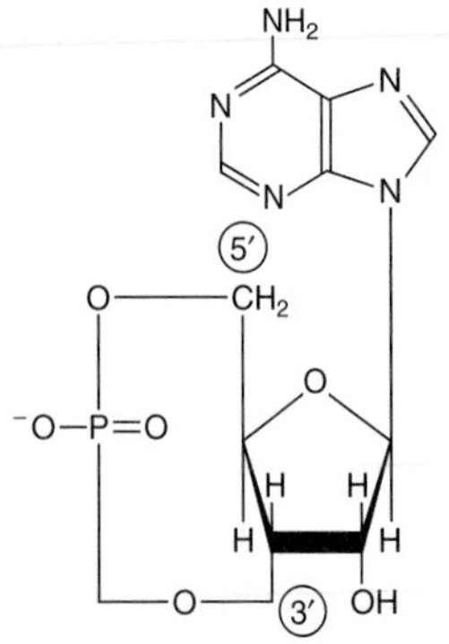

***Figure 18–5.*** 3′,5′-Adenylic acid (cyclic AMP; cAMP).

## Glycogenolysis in Liver Can Be cAMP-Independent

In addition to the action of **glucagon** in causing formation of cAMP and activation of phosphorylase in liver, **$\alpha_1$-adrenergic** receptors mediate stimulation of glycogenolysis by epinephrine and norepinephrine. This involves a **cAMP-independent** mobilization of $Ca^{2+}$ from mitochondria into the cytosol, followed by the stimulation of a **$Ca^{2+}$**/calmodulin-sensitive phosphorylase kinase. cAMP-independent glycogenolysis is also caused by vasopressin, oxytocin, and angiotensin II acting through calcium or the phosphatidylinositol bisphosphate pathway (Figure 43–7).

## Protein Phosphatase-1 Inactivates Phosphorylase

Both phosphorylase a and phosphorylase kinase a are dephosphorylated and inactivated by **protein phosphatase-1.** Protein phosphatase-1 is inhibited by a protein, **inhibitor-1,** which is active only after it has been phosphorylated by cAMP-dependent protein kinase. Thus, cAMP controls both the activation and inactivation of phosphorylase (Figure 18–6). **Insulin** reinforces this effect by inhibiting the activation of phosphorylase b. It does this indirectly by increasing uptake of glucose, leading to increased formation of glucose 6-phosphate, which is an inhibitor of phosphorylase kinase.

## Glycogen Synthase & Phosphorylase Activity Are Reciprocally Regulated (Figure 18–7)

Like phosphorylase, glycogen synthase exists in either a phosphorylated or nonphosphorylated state. However, unlike phosphorylase, the active form is dephosphorylated **(glycogen synthase a)** and may be inactivated to

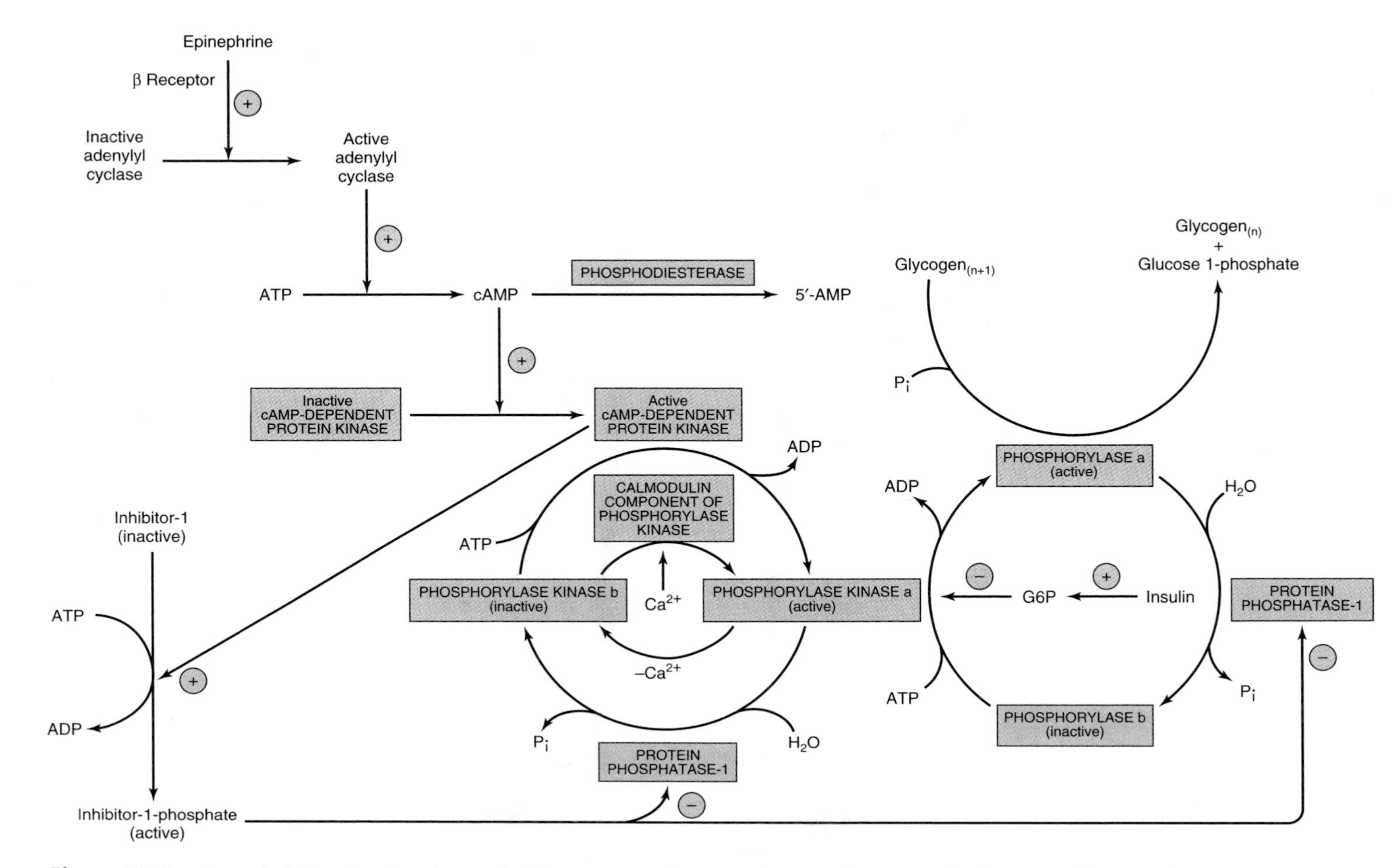


***Figure 18–6.*** Control of phosphorylase in muscle. The sequence of reactions arranged as a cascade allows amplification of the hormonal signal at each step. (n = number of glucose residues; G6P, glucose 6-phosphate.)

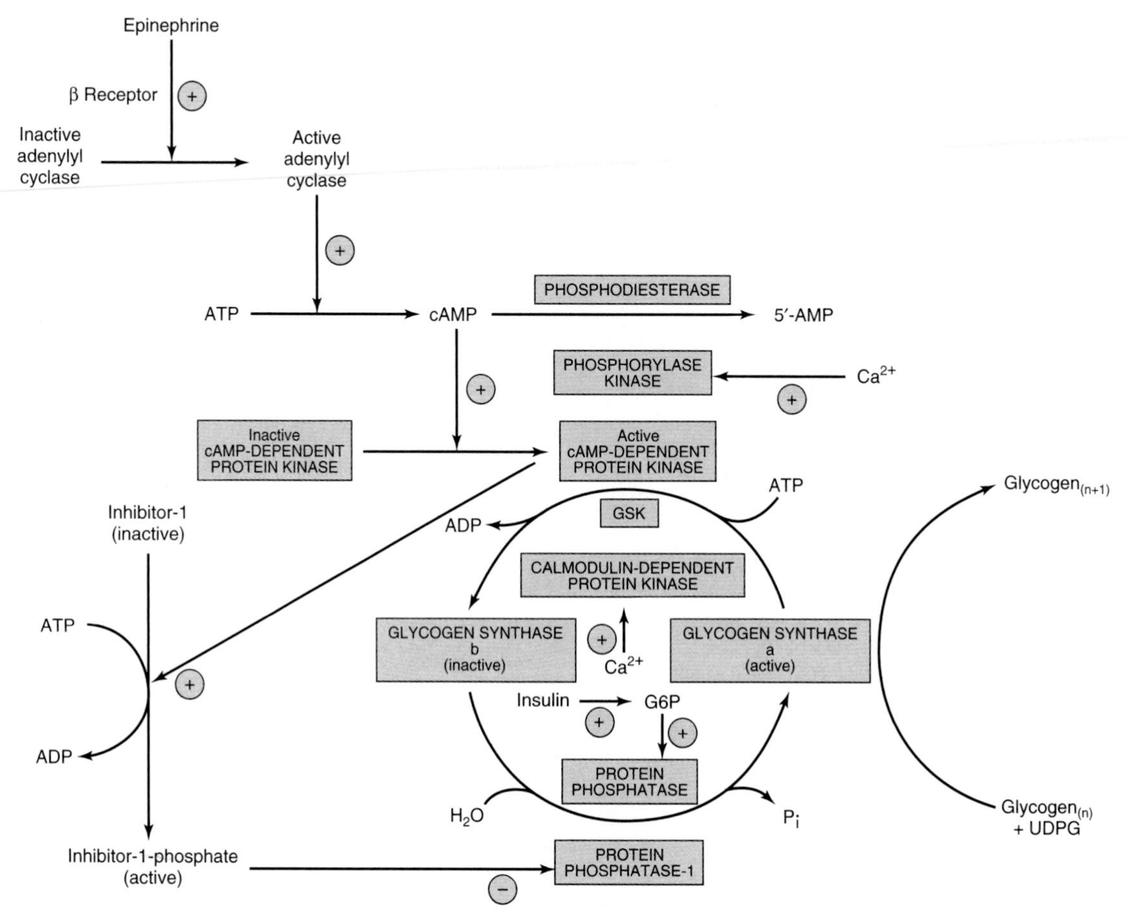


***Figure 18–7.*** Control of glycogen synthase in muscle (n = number of glucose residues). The sequence of reactions arranged in a cascade causes amplification at each step, allowing only nanomole quantities of hormone to cause major changes in glycogen concentration. (GSK, glycogen synthase kinase-3, -4, and -5; G6P, glucose 6-phosphate.)

**glycogen synthase b** by phosphorylation on serine residues by no fewer than six different protein kinases. Two of the protein kinases are $Ca^{2+}$/calmodulin-dependent (one of these is phosphorylase kinase). Another kinase is cAMP-dependent protein kinase, which allows cAMP-mediated hormonal action to inhibit glycogen synthesis synchronously with the activation of glycogenolysis. Insulin also promotes glycogenesis in muscle at the same time as inhibiting glycogenolysis by raising glucose 6-phosphate concentrations, which stimulates the dephosphorylation and activation of glycogen synthase. Dephosphorylation of glycogen synthase b is carried out by protein phosphatase-1, which is under the control of cAMP-dependent protein kinase.

## REGULATION OF GLYCOGEN METABOLISM IS EFFECTED BY A BALANCE IN ACTIVITIES BETWEEN GLYCOGEN SYNTHASE & PHOSPHORYLASE (Figure 18–8)

Not only is phosphorylase activated by a rise in concentration of cAMP (via phosphorylase kinase), but glycogen synthase is at the same time converted to the inactive form; both effects are mediated via **cAMP-dependent protein kinase.** Thus, inhibition of glycogenolysis enhances net glycogenesis, and inhibition of glycogenesis enhances net glycogenolysis. Furthermore,

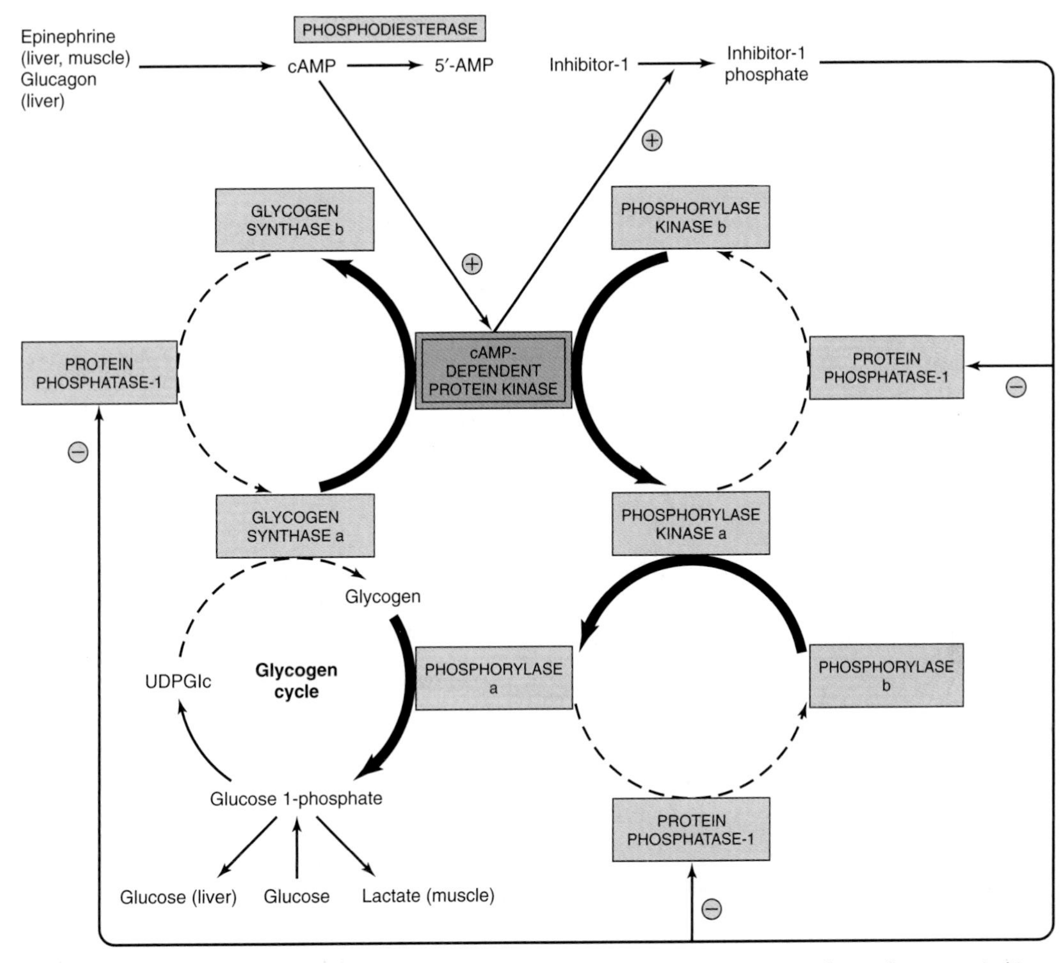


***Figure 18–8.*** Coordinated control of glycogenolysis and glycogenesis by cAMP-dependent protein kinase. The reactions that lead to glycogenolysis as a result of an increase in cAMP concentrations are shown with bold arrows, and those that are inhibited by activation of protein phosphatase-1 are shown as broken arrows. The reverse occurs when cAMP concentrations decrease as a result of phosphodiesterase activity, leading to glycogenesis.

the dephosphorylation of phosphorylase a, phosphorylase kinase a, and glycogen synthase b is catalyzed by a single enzyme of wide specificity—**protein phosphatase-1.** In turn, protein phosphatase-1 is inhibited by cAMP-dependent protein kinase via inhibitor-1. Thus, glycogenolysis can be terminated and glycogenesis can be stimulated synchronously, or vice versa, because both processes are keyed to the activity of cAMP-dependent protein kinase. Both phosphorylase kinase and glycogen synthase may be reversibly phosphorylated in more than one site by separate kinases and phosphatases. These secondary phosphorylations modify the sensitivity of the primary sites to phosphorylation and dephosphorylation **(multisite phosphorylation).** What is more, they allow insulin, via glucose 6-phosphate elevation, to have effects that act reciprocally to those of cAMP (Figures 18–6 and 18–7).

## CLINICAL ASPECTS

### Glycogen Storage Diseases Are Inherited

"Glycogen storage disease" is a generic term to describe a group of inherited disorders characterized by deposition of an abnormal type or quantity of glycogen in the tissues. The principal glycogenoses are summarized in Table 18–2. Deficiencies of **adenylyl kinase** and **cAMP-dependent protein kinase** have also been re-

***Table 18–2.*** Glycogen storage diseases.

| Glycogenosis | Name | Cause of Disorder | Characteristics |
|---|---|---|---|
| Type I | Von Gierke's disease | Deficiency of glucose-6-phosphatase | Liver cells and renal tubule cells loaded with glycogen. Hypoglycemia, lacticacidemia, ketosis, hyperlipemia. |
| Type II | Pompe's disease | Deficiency of lysosomal $\alpha$-1→4- and 1→6-glucosidase (acid maltase) | Fatal, accumulation of glycogen in lysosomes, heart failure. |
| Type III | Limit dextrinosis, Forbes' or Cori's disease | Absence of debranching enzyme | Accumulation of a characteristic branched polysaccharide. |
| Type IV | Amylopectinosis, Andersen's disease | Absence of branching enzyme | Accumulation of a polysaccharide having few branch points. Death due to cardiac or liver failure in first year of life. |
| Type V | Myophosphorylase deficiency, McArdle's syndrome | Absence of muscle phosphorylase | Diminished exercise tolerance; muscles have abnormally high glycogen content (2.5–4.1%). Little or no lactate in blood after exercise. |
| Type VI | Hers' disease | Deficiency of liver phosphorylase | High glycogen content in liver, tendency toward hypoglycemia. |
| Type VII | Tarui's disease | Deficiency of phosphofructokinase in muscle and erythrocytes | As for type V but also possibility of hemolytic anemia. |
| Type VIII | | Deficiency of liver phosphorylase kinase | As for type VI. |

ported. Some of the conditions described have benefited from liver transplantation.

## SUMMARY

- Glycogen represents the principal storage form of carbohydrate in the mammalian body, mainly in the liver and muscle.
- In the liver, its major function is to provide glucose for extrahepatic tissues. In muscle, it serves mainly as a ready source of metabolic fuel for use in muscle.
- Glycogen is synthesized from glucose by the pathway of glycogenesis. It is broken down by a separate pathway known as glycogenolysis. Glycogenolysis leads to glucose formation in liver and lactate formation in muscle owing to the respective presence or absence of glucose-6-phosphatase.
- Cyclic AMP integrates the regulation of glycogenolysis and glycogenesis by promoting the simultaneous activation of phosphorylase and inhibition of glycogen synthase. Insulin acts reciprocally by inhibiting glycogenolysis and stimulating glycogenesis.
- Inherited deficiencies in specific enzymes of glycogen metabolism in both liver and muscle are the causes of glycogen storage diseases.

## REFERENCES


Bollen M, Keppens S, Stalmans W: Specific features of glycogen metabolism in the liver. Biochem J 1998;336:19.

Cohen P: The role of protein phosphorylation in the hormonal control of enzyme activity. Eur J Biochem 1985;151:439.

Ercan N, Gannon MC, Nuttall FQ: Incorporation of glycogenin into a hepatic proteoglycogen after oral glucose administration. J Biol Chem 1994;269:22328.

Geddes R: Glycogen: a metabolic viewpoint. Bioscience Rep 1986;6:415.

McGarry JD et al: From dietary glucose to liver glycogen: the full circle round. Annu Rev Nutr 1987;7:51.

Meléndez-Hevia E, Waddell TG, Shelton ED: Optimization of molecular design in the evolution of metabolism: the glycogen molecule. Biochem J 1993;295:477.

Raz I, Katz A, Spencer MK: Epinephrine inhibits insulin-mediated glycogenesis but enhances glycolysis in human skeletal muscle. Am J Physiol 1991;260:E430.

Scriver CR et al (editors): *The Metabolic and Molecular Bases of Inherited Disease,* 8th ed. McGraw-Hill, 2001.

Shulman GI, Landau BR: Pathways of glycogen repletion. Physiol Rev 1992;72:1019.

Villar-Palasi C: On the mechanism of inactivation of muscle glycogen phosphorylase by insulin. Biochim Biophys Acta 1994;1224:384.

# Gluconeogenesis & Control of the Blood Glucose

# 19

*Peter A. Mayes, PhD, DSc, & David A. Bender, PhD*

## BIOMEDICAL IMPORTANCE

Gluconeogenesis is the term used to include all pathways responsible for converting noncarbohydrate precursors to glucose or glycogen. The major substrates are the glucogenic amino acids and lactate, glycerol, and propionate. Liver and kidney are the major gluconeogenic tissues. Gluconeogenesis meets the needs of the body for glucose when carbohydrate is not available in sufficient amounts from the diet or from glycogen reserves. A supply of glucose is necessary especially for the nervous system and erythrocytes. Failure of gluconeogenesis is usually fatal. **Hypoglycemia** causes brain dysfunction, which can lead to coma and death. Glucose is also important in maintaining the level of intermediates of the citric acid cycle even when fatty acids are the main source of acetyl-CoA in the tissues. In addition, gluconeogenesis clears lactate produced by muscle and erythrocytes and glycerol produced by adipose tissue. Propionate, the principal glucogenic fatty acid produced in the digestion of carbohydrates by ruminants, is a major substrate for gluconeogenesis in these species.

## GLUCONEOGENESIS INVOLVES GLYCOLYSIS, THE CITRIC ACID CYCLE, & SOME SPECIAL REACTIONS (Figure 19–1)

### Thermodynamic Barriers Prevent a Simple Reversal of Glycolysis

Three nonequilibrium reactions catalyzed by hexokinase, phosphofructokinase, and pyruvate kinase prevent simple reversal of glycolysis for glucose synthesis (Chapter 17). They are circumvented as follows:

#### A. Pyruvate & Phosphoenolpyruvate

Mitochondrial **pyruvate carboxylase** catalyzes the carboxylation of pyruvate to oxaloacetate, an ATP-requiring reaction in which the vitamin biotin is the coenzyme. Biotin binds $CO_2$ from bicarbonate as carboxybiotin prior to the addition of the $CO_2$ to pyruvate (Figure 45–17). A second enzyme, **phosphoenolpyruvate carboxykinase,** catalyzes the decarboxylation and phosphorylation of oxaloacetate to phosphoenolpyruvate using GTP (or ITP) as the phosphate donor. Thus, reversal of the reaction catalyzed by pyruvate kinase in glycolysis involves two endergonic reactions.

In pigeon, chicken, and rabbit liver, phosphoenolpyruvate carboxykinase is a mitochondrial enzyme, and phosphoenolpyruvate is transported into the cytosol for gluconeogenesis. In the rat and the mouse, the enzyme is cytosolic. Oxaloacetate does not cross the mitochondrial inner membrane; it is converted to malate, which is transported into the cytosol, and converted back to oxaloacetate by cytosolic malate dehydrogenase. In humans, the guinea pig, and the cow, the enzyme is equally distributed between mitochondria and cytosol.

The main source of GTP for phosphoenolpyruvate carboxykinase inside the mitochondrion is the reaction of succinyl-CoA synthetase (Chapter 16). This provides a link and limit between citric acid cycle activity and the extent of withdrawal of oxaloacetate for gluconeogenesis.

#### B. Fructose 1,6-Bisphosphate & Fructose 6-Phosphate

The conversion of fructose 1,6-bisphosphate to fructose 6-phosphate, to achieve a reversal of glycolysis, is catalyzed by **fructose-1,6-bisphosphatase.** Its presence determines whether or not a tissue is capable of synthesizing glycogen not only from pyruvate but also from triosephosphates. It is present in liver, kidney, and skeletal muscle but is probably absent from heart and smooth muscle.

#### C. Glucose 6-Phosphate & Glucose

The conversion of glucose 6-phosphate to glucose is catalyzed by **glucose-6-phosphatase.** It is present in liver and kidney but absent from muscle and adipose tissue, which, therefore, cannot export glucose into the bloodstream.

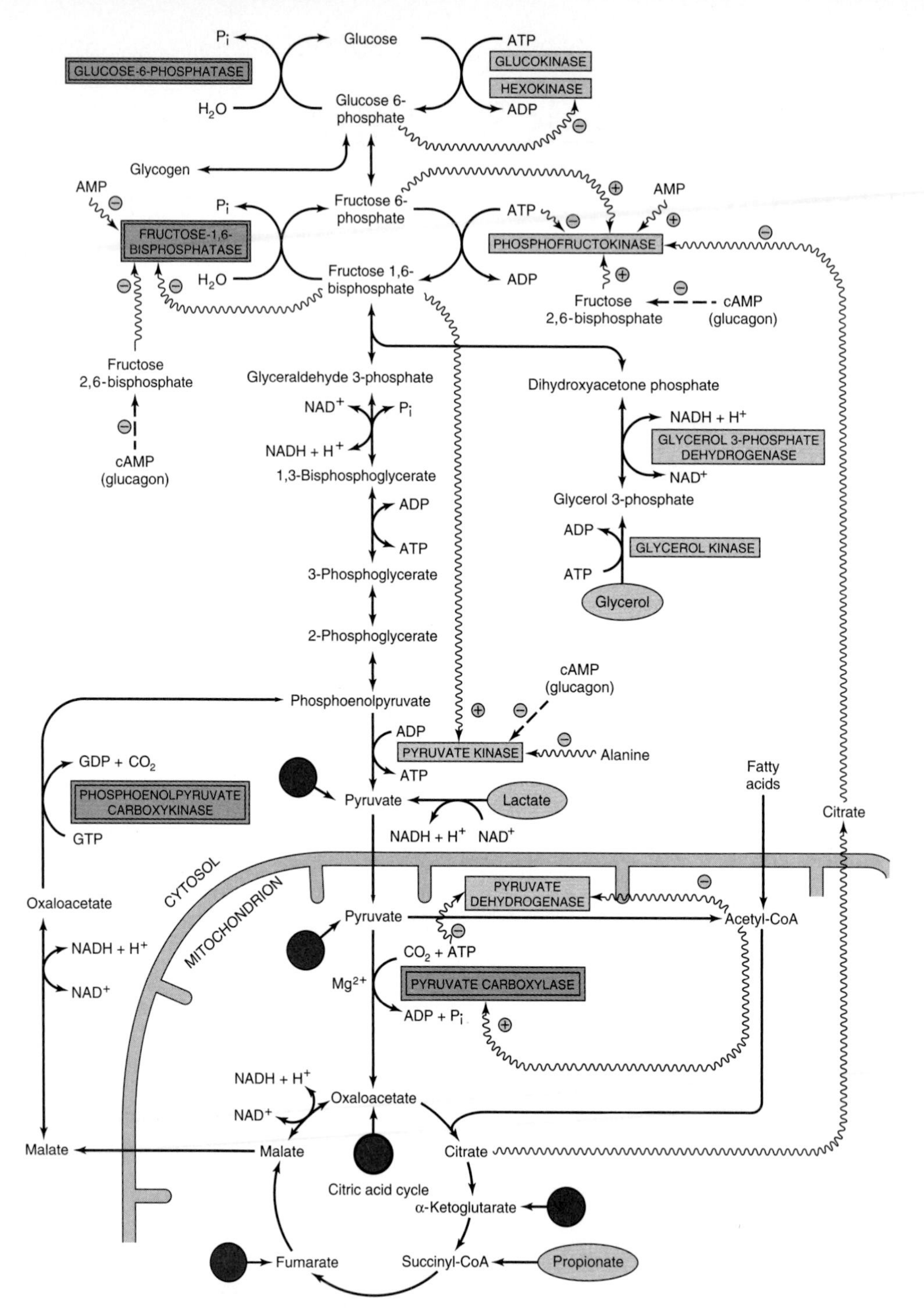


***Figure 19–1.*** Major pathways and regulation of gluconeogenesis and glycolysis in the liver. Entry points of glucogenic amino acids after transamination are indicated by arrows extended from circles. (See also Figure 16–4.) The key gluconeogenic enzymes are enclosed in double-bordered boxes. The ATP required for gluconeogenesis is supplied by the oxidation of long-chain fatty acids. Propionate is of quantitative importance only in ruminants. Arrows with wavy shafts signify allosteric effects; dash-shafted arrows, covalent modification by reversible phosphorylation. High concentrations of alanine act as a "gluconeogenic signal" by inhibiting glycolysis at the pyruvate kinase step.

### D. Glucose 1-Phosphate & Glycogen

The breakdown of glycogen to glucose 1-phosphate is catalyzed by phosphorylase. Glycogen synthesis involves a different pathway via uridine diphosphate glucose and **glycogen synthase** (Figure 18–1).

The relationships between gluconeogenesis and the glycolytic pathway are shown in Figure 19–1. After transamination or deamination, glucogenic amino acids yield either pyruvate or intermediates of the citric acid cycle. Therefore, the reactions described above can account for the conversion of both glucogenic amino acids and lactate to glucose or glycogen. Propionate is a major source of glucose in ruminants and enters gluconeogenesis via the citric acid cycle. Propionate is esterified with CoA, then propionyl-CoA, is carboxylated to D-methylmalonyl-CoA, catalyzed by **propionyl-CoA carboxylase,** a biotin-dependent enzyme (Figure 19–2). **Methylmalonyl-CoA racemase** catalyzes the conversion of D-methylmalonyl-CoA to L-methylmalonyl-CoA, which then undergoes isomerization to succinyl-CoA catalyzed by **methylmalonyl-CoA isomerase.** This enzyme requires vitamin $B_{12}$ as a coenzyme, and deficiency of this vitamin results in the excretion of methylmalonate **(methylmalonic aciduria).**

$C_{15}$ and $C_{17}$ fatty acids are found particularly in the lipids of ruminants. Dietary odd-carbon fatty acids upon oxidation yield propionate (Chapter 22), which is a substrate for gluconeogenesis in human liver.

Glycerol is released from adipose tissue as a result of lipolysis, and only tissues such as liver and kidney that possess **glycerol kinase,** which catalyzes the conversion of glycerol to glycerol 3-phosphate, can utilize it. Glycerol 3-phosphate may be oxidized to dihydroxyacetone phosphate by $NAD^+$ catalyzed by **glycerol-3-phosphate dehydrogenase.**

## SINCE GLYCOLYSIS & GLUCONEOGENESIS SHARE THE SAME PATHWAY BUT IN OPPOSITE DIRECTIONS, THEY MUST BE REGULATED RECIPROCALLY

Changes in the availability of substrates are responsible for most changes in metabolism either directly or indirectly acting via changes in hormone secretion. Three mechanisms are responsible for regulating the activity of enzymes in carbohydrate metabolism: (1) changes in the rate of enzyme synthesis, (2) covalent modification by reversible phosphorylation, and (3) allosteric effects.

### Induction & Repression of Key Enzyme Synthesis Requires Several Hours

The changes in enzyme activity in the liver that occur under various metabolic conditions are listed in Table 19–1. The enzymes involved catalyze nonequilibrium (physiologically irreversible) reactions. The effects are generally reinforced because the activity of the enzymes catalyzing the changes in the opposite direction varies reciprocally (Figure 19–1). The enzymes involved in the utilization of glucose (ie, those of glycolysis and lipogenesis) all become more active when there is a superfluity of glucose, and under these conditions the enzymes responsible for gluconeogenesis all have low activity. The secretion of insulin, in response to increased blood glucose, enhances the synthesis of the key

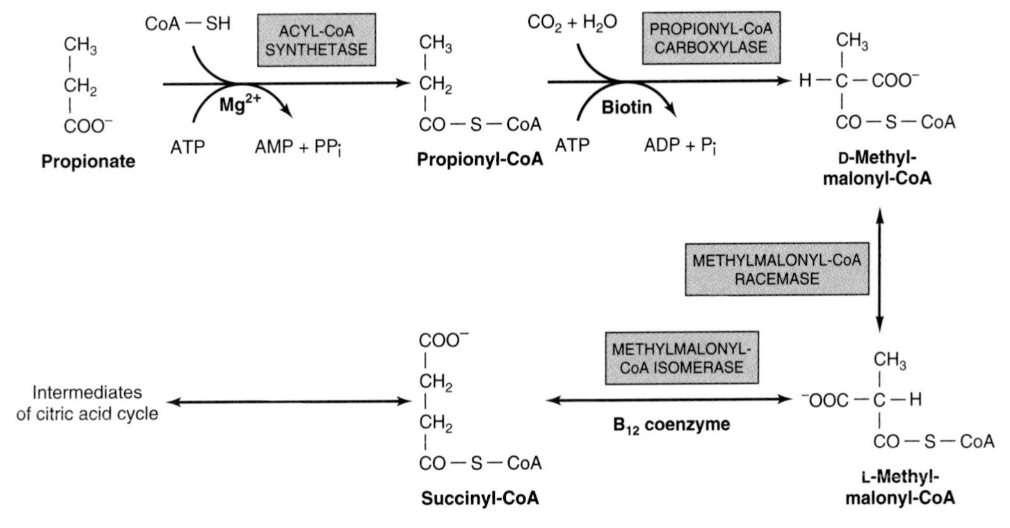


*Figure 19–2.* Metabolism of propionate.

***Table 19–1.*** Regulatory and adaptive enzymes of the rat (mainly liver).

| | Activity In | | | | | |
|---|---|---|---|---|---|---|
| | **Carbohydrate Feeding** | **Starvation and Diabetes** | **Inducer** | **Repressor** | **Activator** | **Inhibitor** |
| **Enzymes of glycogenesis, glycolysis, and pyruvate oxidation** | | | | | | |
| Glycogen synthase system | ↑ | ↓ | Insulin | | Insulin Glucose 6-phosphate[1] | Glucagon (cAMP) phosphorylase, glycogen |
| Hexokinase | | | | | | Glucose 6-phosphate[1] |
| Glucokinase | ↑ | ↓ | Insulin | Glucagon (cAMP) | | |
| Phosphofructokinase-1 | ↑ | ↓ | Insulin | Glucagon (cAMP) | AMP, fructose 6-phosphate, $P_i$, fructose 2,6-bisphosphate[1] | Citrate (fatty acids, ketone bodies),[1] ATP,[1] glucagon (cAMP) |
| Pyruvate kinase | ↑ | ↓ | Insulin, fructose | Glucagon (cAMP) | Fructose 1,6-bisphosphate[1], insulin | ATP, alanine, glucagon (cAMP), epinephrine |
| Pyruvate dehydrogenase | ↑ | ↓ | | | CoA, $NAD^+$, insulin,[2] ADP, pyruvate | Acetyl-CoA, NADH, ATP (fatty acids, ketone bodies) |
| **Enzymes of gluconeogenesis** | | | | | | |
| Pyruvate carboxylase | ↓ | ↑ | Glucocorticoids, glucagon, epinephrine (cAMP) | Insulin | Acetyl-CoA[1] | ADP[1] |
| Phosphoenolpyruvate carboxykinase | ↓ | ↑ | Glucocorticoids, glucagon, epinephrine (cAMP) | Insulin | Glucagon? | |
| Fructose-1,6-bisphosphatase | ↓ | ↑ | Glucocorticoids, glucagon, epinephrine (cAMP) | Insulin | Glucagon (cAMP) | Fructose 1,6- bisphosphate, AMP, fructose 2,6-bisphosphate[1] |
| Glucose-6-phosphatase | ↓ | ↑ | Glucocorticoids, glucagon, epinephrine (cAMP) | Insulin | | |
| **Enzymes of the pentose phosphate pathway and lipogenesis** | | | | | | |
| Glucose-6-phosphate dehydrogenase | ↑ | ↓ | Insulin | | | |
| 6-Phosphogluconate dehydrogenase | ↑ | ↓ | Insulin | | | |
| "Malic enzyme" | ↑ | ↓ | Insulin | | | |
| ATP-citrate lyase | ↑ | ↓ | Insulin | | | |
| Acetyl-CoA carboxylase | ↑ | ↓ | Insulin? | | Citrate,[1] insulin | Long-chain acyl-CoA, cAMP, glucagon |
| Fatty acid synthase | ↑ | ↓ | Insulin? | | | |

[1]Allosteric.
[2]In adipose tissue but not in liver.

enzymes in glycolysis. Likewise, it antagonizes the effect of the glucocorticoids and glucagon-stimulated cAMP, which induce synthesis of the key enzymes responsible for gluconeogenesis.

Both dehydrogenases of the pentose phosphate pathway can be classified as adaptive enzymes, since they increase in activity in the well-fed animal and when insulin is given to a diabetic animal. Activity is low in diabetes or starvation. "Malic enzyme" and ATP-citrate lyase behave similarly, indicating that these two enzymes are involved in lipogenesis rather than gluconeogenesis (Chapter 21).

### Covalent Modification by Reversible Phosphorylation Is Rapid

**Glucagon,** and to a lesser extent **epinephrine,** hormones that are responsive to decreases in blood glucose, inhibit glycolysis and stimulate gluconeogenesis in the liver by increasing the concentration of cAMP. This in turn activates cAMP-dependent protein kinase, leading to the phosphorylation and inactivation of **pyruvate kinase.** They also affect the concentration of fructose 2,6-bisphosphate and therefore glycolysis and gluconeogenesis, as explained below.

### Allosteric Modification Is Instantaneous

In gluconeogenesis, pyruvate carboxylase, which catalyzes the synthesis of oxaloacetate from pyruvate, requires acetyl-CoA as an **allosteric activator.** The presence of acetyl-CoA results in a change in the tertiary structure of the protein, lowering the $K_m$ value for bicarbonate. This means that as acetyl-CoA is formed from pyruvate, it automatically ensures the provision of oxaloacetate and, therefore, its further oxidation in the citric acid cycle. The activation of pyruvate carboxylase and the reciprocal inhibition of pyruvate dehydrogenase by acetyl-CoA derived from the oxidation of fatty acids explains the action of fatty acid oxidation in sparing the oxidation of pyruvate and in stimulating gluconeogenesis. The reciprocal relationship between these two enzymes in both liver and kidney alters the metabolic fate of pyruvate as the tissue changes from carbohydrate oxidation, via glycolysis, to gluconeogenesis during transition from a fed to a starved state (Figure 19–1). A major role of fatty acid oxidation in promoting gluconeogenesis is to supply the requirement for ATP. **Phosphofructokinase (phosphofructokinase-1)** occupies a key position in regulating glycolysis and is also subject to feedback control. It is inhibited by citrate and by ATP and is activated by 5′-AMP. 5′-AMP acts as an indicator of the energy status of the cell. The presence of **adenylyl kinase** in liver and many other tissues allows rapid equilibration of the reaction:

$$ATP + AMP \leftrightarrow 2ADP$$

Thus, when ATP is used in energy-requiring processes resulting in formation of ADP, [AMP] increases. As [ATP] may be 50 times [AMP] at equilibrium, a small fractional decrease in [ATP] will cause a severalfold increase in [AMP]. Thus, a large change in [AMP] acts as a metabolic amplifier of a small change in [ATP]. This mechanism allows the activity of phosphofructokinase-1 to be highly sensitive to even small changes in energy status of the cell and to control the quantity of carbohydrate undergoing glycolysis prior to its entry into the citric acid cycle. The increase in [AMP] can also explain why glycolysis is increased during hypoxia when [ATP] decreases. Simultaneously, AMP activates phosphorylase, increasing glycogenolysis. The inhibition of phosphofructokinase-1 by citrate and ATP is another explanation of the sparing action of fatty acid oxidation on glucose oxidation and also of the **Pasteur effect,** whereby aerobic oxidation (via the citric acid cycle) inhibits the anaerobic degradation of glucose. A consequence of the inhibition of phosphofructokinase-1 is an accumulation of glucose 6-phosphate that, in turn, inhibits further uptake of glucose in extrahepatic tissues by allosteric inhibition of hexokinase.

### Fructose 2,6-Bisphosphate Plays a Unique Role in the Regulation of Glycolysis & Gluconeogenesis in Liver

The most potent positive allosteric effector of phosphofructokinase-1 and inhibitor of fructose-1,6-bisphosphatase in liver is **fructose 2,6-bisphosphate.** It relieves inhibition of phosphofructokinase-1 by ATP and increases affinity for fructose 6-phosphate. It inhibits fructose-1,6-bisphosphatase by increasing the $K_m$ for fructose 1,6-bisphosphate. Its concentration is under both substrate (allosteric) and hormonal control (covalent modification) (Figure 19–3).

Fructose 2,6-bisphosphate is formed by phosphorylation of fructose 6-phosphate by **phosphofructokinase-2.** The same enzyme protein is also responsible for its breakdown, since it has **fructose-2,6-bisphosphatase** activity. This **bifunctional enzyme** is under the allosteric control of fructose 6-phosphate, which stimulates the kinase and inhibits the phosphatase. Hence, when glucose is abundant, the concentration of fructose 2,6-bisphosphate increases, stimulating glycolysis by activating phosphofructokinase-1 and inhibiting

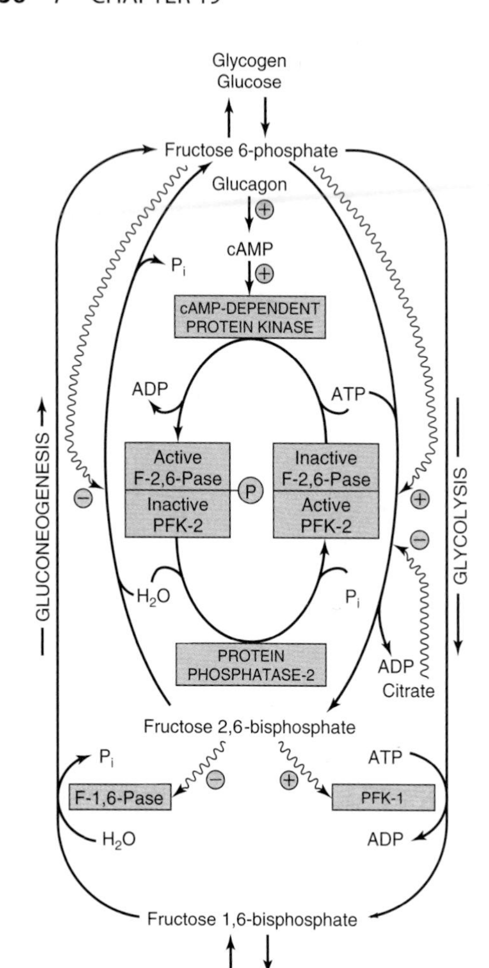


***Figure 19–3.*** Control of glycolysis and gluconeogenesis in the liver by fructose 2,6-bisphosphate and the bifunctional enzyme PFK-2/F-2,6-Pase (6-phosphofructo-2-kinase/fructose-2,6-bisphosphatase). (PFK-1, phosphofructokinase-1 [6-phosphofructo-1-kinase]; F-1,6-Pase, fructose-1,6-bisphosphatase. Arrows with wavy shafts indicate allosteric effects.)

fructose-1,6-bisphosphatase. When glucose is short, glucagon stimulates the production of cAMP, activating cAMP-dependent protein kinase, which in turn inactivates phosphofructokinase-2 and activates fructose 2,6-bisphosphatase by phosphorylation. Therefore, gluconeogenesis is stimulated by a decrease in the concentration of fructose 2,6-bisphosphate, which deactivates phosphofructokinase-1 and deinhibits fructose-1,6-bisphosphatase. This mechanism also ensures that glucagon stimulation of glycogenolysis in liver results in glucose release rather than glycolysis.

### Substrate (Futile) Cycles Allow Fine Tuning

It will be apparent that the control points in glycolysis and glycogen metabolism involve a cycle of phosphorylation and dephosphorylation catalyzed by: glucokinase and glucose-6-phosphatase; phosphofructokinase-1 and fructose-1,6-bisphosphatase; pyruvate kinase, pyruvate carboxylase, and phosphoenolypyruvate carboxykinase; and glycogen synthase and phosphorylase. If these were allowed to cycle unchecked, they would amount to futile cycles whose net result was hydrolysis of ATP. This does not occur extensively due to the various control mechanisms, which ensure that one reaction is inhibited as the other is stimulated. However, there is a physiologic advantage in allowing some cycling. The rate of net glycolysis may increase several thousand-fold in response to stimulation, and this is more readily achieved by both increasing the activity of phosphofructokinase and decreasing that of fructose bisphosphatase if both are active, than by switching one enzyme "on" and the other "off" completely. This "fine tuning" of metabolic control occurs at the expense of some loss of ATP.

## THE CONCENTRATION OF BLOOD GLUCOSE IS REGULATED WITHIN NARROW LIMITS

In the postabsorptive state, the concentration of blood glucose in most mammals is maintained between 4.5 and 5.5 mmol/L. After the ingestion of a carbohydrate meal, it may rise to 6.5–7.2 mmol/L, and in starvation, it may fall to 3.3–3.9 mmol/L. A sudden decrease in blood glucose will cause convulsions, as in insulin overdose, owing to the immediate dependence of the brain on a supply of glucose. However, much lower concentrations can be tolerated, provided progressive adaptation is allowed. The blood glucose level in birds is considerably higher (14.0 mmol/L) and in ruminants considerably lower (approximately 2.2 mmol/L in sheep and 3.3 mmol/L in cattle). These lower normal levels appear to be associated with the fact that ruminants ferment virtually all dietary carbohydrate to lower (volatile) fatty acids, and these largely replace glucose as the main metabolic fuel of the tissues in the fed condition.

## BLOOD GLUCOSE IS DERIVED FROM THE DIET, GLUCONEOGENESIS, & GLYCOGENOLYSIS

The digestible dietary carbohydrates yield glucose, galactose, and fructose that are transported via the **hepatic portal vein** to the liver where galactose and fructose are readily converted to glucose (Chapter 20).

Glucose is formed from two groups of compounds that undergo gluconeogenesis (Figures 16–4 and 19–1): (1) those which involve a direct net conversion to glucose without significant recycling, such as some **amino acids** and **propionate;** and (2) those which are the products of the metabolism of glucose in tissues. Thus, **lactate,** formed by glycolysis in skeletal muscle and erythrocytes, is transported to the liver and kidney where it re-forms glucose, which again becomes available via the circulation for oxidation in the tissues. This process is known as the **Cori cycle,** or **lactic acid cycle** (Figure 19–4). Triacylglycerol glycerol in adipose tissue is derived from blood glucose. This triacylglycerol is continuously undergoing hydrolysis to form free **glycerol,** which cannot be utilized by adipose tissue and is converted back to glucose by gluconeogenic mechanisms in the liver and kidney (Figure 19–1).

Of the amino acids transported from muscle to the liver during starvation, alanine predominates. The **glucose-alanine cycle** (Figure 19–4) transports glucose from liver to muscle with formation of pyruvate, followed by transamination to alanine, then transports alanine to the liver, followed by gluconeogenesis back to glucose. A net transfer of amino nitrogen from muscle to liver and of free energy from liver to muscle is effected. The energy required for the hepatic synthesis of glucose from pyruvate is derived from the oxidation of fatty acids.

Glucose is also formed from liver glycogen by glycogenolysis (Chapter 18).

## Metabolic & Hormonal Mechanisms Regulate the Concentration of the Blood Glucose

The maintenance of stable levels of glucose in the blood is one of the most finely regulated of all homeostatic mechanisms, involving the liver, extrahepatic tissues, and several hormones. Liver cells are freely permeable to glucose (via the GLUT 2 transporter), whereas cells of extrahepatic tissues (apart from pancreatic B islets) are relatively impermeable, and their glucose transporters are regulated by insulin. As a result, uptake from the bloodstream is the rate-limiting step in the utilization of glucose in extrahepatic tissues. The role of various glucose transporter proteins found in cell membranes, each having 12 transmembrane domains, is shown in Table 19–2.

## Glucokinase Is Important in Regulating Blood Glucose After a Meal

Hexokinase has a low $K_m$ for glucose and in the liver is saturated and acting at a constant rate under all normal conditions. Glucokinase has a considerably higher $K_m$ (lower affinity) for glucose, so that its activity increases over the physiologic range of glucose concentrations (Figure 19–5). It promotes hepatic uptake of large amounts of glucose at the high concentrations found in the hepatic portal vein after a carbohydrate meal. It is absent from the liver of ruminants, which have little

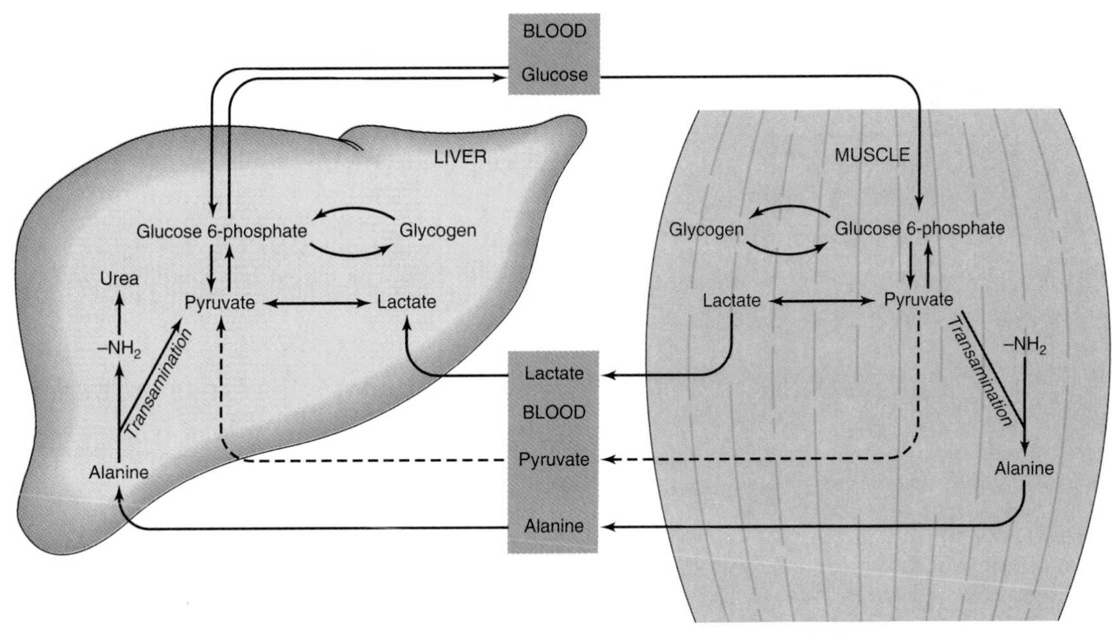


***Figure 19–4.*** The lactic acid (Cori) cycle and glucose-alanine cycle.

***Table 19–2.*** Glucose transporters.

| | Tissue Location | Functions |
|---|---|---|
| **Facilitative bidirectional transporters** | | |
| GLUT 1 | Brain, kidney, colon, placenta, erythrocyte | Uptake of glucose |
| GLUT 2 | Liver, pancreatic B cell, small intestine, kidney | Rapid uptake and release of glucose |
| GLUT 3 | Brain, kidney, placenta | Uptake of glucose |
| GLUT 4 | Heart and skeletal muscle, adipose tissue | Insulin-stimulated uptake of glucose |
| GLUT 5 | Small intestine | Absorption of glucose |
| **Sodium-dependent unidirectional transporter** | | |
| SGLT 1 | Small intestine and kidney | Active uptake of glucose from lumen of intestine and reabsorption of glucose in proximal tubule of kidney against a concentration gradient |

glucose entering the portal circulation from the intestines.

At normal systemic-blood glucose concentrations (4.5–5.5 mmol/L), the liver is a net producer of glucose. However, as the glucose level rises, the output of glucose ceases, and there is a net uptake.

## Insulin Plays a Central Role in Regulating Blood Glucose

In addition to the direct effects of hyperglycemia in enhancing the uptake of glucose into the liver, the hormone insulin plays a central role in regulating blood glucose. It is produced by the B cells of the islets of Langerhans in the pancreas in response to hyperglycemia. The B islet cells are freely permeable to glucose via the GLUT 2 transporter, and the glucose is phosphorylated by glucokinase. Therefore, increasing blood glucose increases metabolic flux through glycolysis, the citric acid cycle, and the generation of ATP. Increase in [ATP] inhibits ATP-sensitive $K^+$ channels, causing depolarization of the B cell membrane, which increases $Ca^{2+}$ influx via voltage-sensitive $Ca^{2+}$ channels, stimulating exocytosis of insulin. Thus, the concentration of insulin in the blood parallels that of the blood glucose. Other substances causing release of insulin from the pancreas include amino acids, free fatty acids, ketone bodies, glucagon, secretin, and the sulfonylurea drugs tolbutamide and glyburide. These drugs are used to stimulate insulin secretion in type 2 diabetes mellitus (NIDDM, non-insulin-dependent diabetes mellitus); they act by inhibiting the ATP-sensitive $K^+$ channels. Epinephrine and norepinephrine block the release of insulin. Insulin lowers blood glucose immediately by enhancing glucose transport into adipose tissue and muscle by recruitment of glucose transporters (GLUT 4) from the interior of the cell to the plasma membrane. Although it does not affect glucose uptake into the liver directly, insulin does enhance long-term uptake as a result of its actions on the enzymes controlling glycolysis, glycogenesis, and gluconeogenesis (Chapter 18).

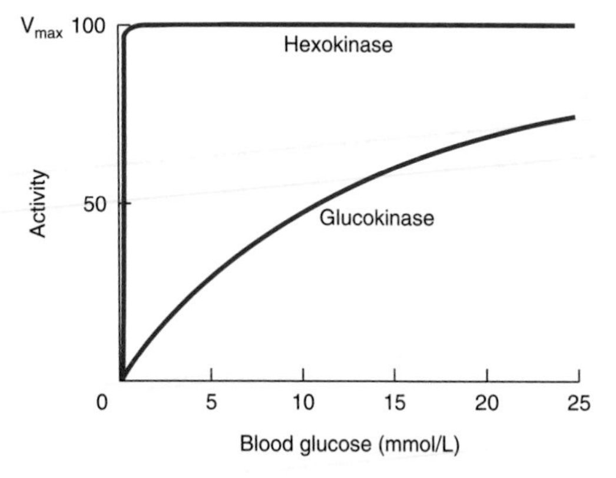


***Figure 19–5.*** Variation in glucose phosphorylating activity of hexokinase and glucokinase with increase of blood glucose concentration. The $K_m$ for glucose of hexokinase is 0.05 mmol/L and of glucokinase is 10 mmol/L.

## Glucagon Opposes the Actions of Insulin

Glucagon is the hormone produced by the A cells of the pancreatic islets. Its secretion is stimulated by hypoglycemia. In the liver, it stimulates glycogenolysis by activating phosphorylase. Unlike epinephrine, glucagon does not have an effect on muscle phosphorylase. Glucagon also enhances gluconeogenesis from amino acids and lactate. In all these actions, glucagon acts via generation of cAMP (Table 19–1). Both hepatic glycogenolysis and gluconeogenesis contribute to the

**hyperglycemic effect** of glucagon, whose actions oppose those of insulin. Most of the endogenous glucagon (and insulin) is cleared from the circulation by the liver.

### Other Hormones Affect Blood Glucose

The **anterior pituitary gland** secretes hormones that tend to elevate the blood glucose and therefore antagonize the action of insulin. These are growth hormone, ACTH (corticotropin), and possibly other "diabetogenic" hormones. Growth hormone secretion is stimulated by hypoglycemia; it decreases glucose uptake in muscle. Some of this effect may not be direct, since it stimulates mobilization of free fatty acids from adipose tissue, which themselves inhibit glucose utilization. The **glucocorticoids** (11-oxysteroids) are secreted by the adrenal cortex and increase gluconeogenesis. This is a result of enhanced hepatic uptake of amino acids and increased activity of aminotransferases and key enzymes of gluconeogenesis. In addition, glucocorticoids inhibit the utilization of glucose in extrahepatic tissues. In all these actions, glucocorticoids act in a manner antagonistic to insulin.

**Epinephrine** is secreted by the adrenal medulla as a result of stressful stimuli (fear, excitement, hemorrhage, hypoxia, hypoglycemia, etc) and leads to glycogenolysis in liver and muscle owing to stimulation of phosphorylase via generation of cAMP. In muscle, glycogenolysis results in increased glycolysis, whereas in liver glucose is the main product leading to increase in blood glucose.

## FURTHER CLINICAL ASPECTS

### Glucosuria Occurs When the Renal Threshold for Glucose Is Exceeded

When the blood glucose rises to relatively high levels, the kidney also exerts a regulatory effect. Glucose is continuously filtered by the glomeruli but is normally completely reabsorbed in the renal tubules by active transport. The capacity of the tubular system to reabsorb glucose is limited to a rate of about 350 mg/min, and in hyperglycemia (as occurs in poorly controlled diabetes mellitus) the glomerular filtrate may contain more glucose than can be reabsorbed, resulting in **glucosuria.** Glucosuria occurs when the venous blood glucose concentration exceeds 9.5–10.0 mmol/L; this is termed the **renal threshold** for glucose.

### Hypoglycemia May Occur During Pregnancy & in the Neonate

During pregnancy, fetal glucose consumption increases and there is a risk of maternal and possibly fetal hypoglycemia, particularly if there are long intervals between meals or at night. Furthermore, premature and low-birth-weight babies are more susceptible to hypoglycemia, since they have little adipose tissue to generate alternative fuels such as free fatty acids or ketone bodies during the transition from fetal dependency to the free-living state. The enzymes of gluconeogenesis may not be completely functional at this time, and the process is dependent on a supply of free fatty acids for energy. Glycerol, which would normally be released from adipose tissue, is less available for gluconeogenesis.

### The Body's Ability to Utilize Glucose May Be Ascertained by Measuring Its Glucose Tolerance

Glucose tolerance is the ability to regulate the blood glucose concentration after the administration of a test dose of glucose (normally 1 g/kg body weight) (Figure 19–6). **Diabetes mellitus** (type 1, or insulin-dependent diabetes mellitus; IDDM) is characterized by decreased glucose tolerance due to decreased secretion of insulin in response to the glucose challenge. Glucose tolerance is also impaired in type 2 diabetes mellitus (NIDDM), which is often associated with obesity and raised levels of plasma free fatty acids and in conditions where the liver is damaged; in some infections; and in response to some drugs. Poor glucose tolerance can also be expected

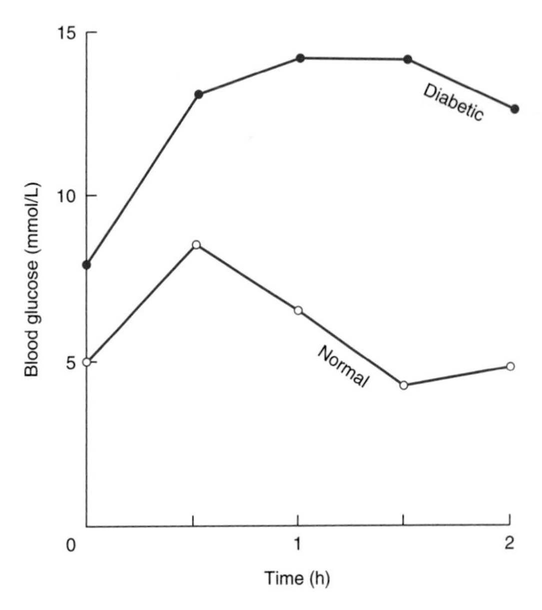


***Figure 19–6.*** Glucose tolerance test. Blood glucose curves of a normal and a diabetic individual after oral administration of 50 g of glucose. Note the initial raised concentration in the diabetic. A criterion of normality is the return of the curve to the initial value within 2 hours.

due to hyperactivity of the pituitary or adrenal cortex because of the antagonism of the hormones secreted by these glands to the action of insulin.

Administration of insulin (as in the treatment of diabetes mellitus type 1) lowers the blood glucose and increases its utilization and storage in the liver and muscle as glycogen. An excess of insulin may cause **hypoglycemia,** resulting in convulsions and even in death unless glucose is administered promptly. Increased tolerance to glucose is observed in pituitary or adrenocortical insufficiency—attributable to a decrease in the antagonism to insulin by the hormones normally secreted by these glands.

## SUMMARY


- Gluconeogenesis is the process of converting noncarbohydrates to glucose or glycogen. It is of particular importance when carbohydrate is not available from the diet. Significant substrates are amino acids, lactate, glycerol, and propionate.
- The pathway of gluconeogenesis in the liver and kidney utilizes those reactions in glycolysis which are reversible plus four additional reactions that circumvent the irreversible nonequilibrium reactions.
- Since glycolysis and gluconeogenesis share the same pathway but operate in opposite directions, their activities are regulated reciprocally.
- The liver regulates the blood glucose after a meal because it contains the high-$K_m$ glucokinase that promotes increased hepatic utilization of glucose.
- Insulin is secreted as a direct response to hyperglycemia; it stimulates the liver to store glucose as glycogen and facilitates uptake of glucose into extrahepatic tissues.
- Glucagon is secreted as a response to hypoglycemia and activates both glycogenolysis and gluconeogenesis in the liver, causing release of glucose into the blood.


## REFERENCES


Burant CF et al: Mammalian glucose transporters: structure and molecular regulation. Recent Prog Horm Res 1991;47:349.

Krebs HA: Gluconeogenesis. Proc R Soc London (Biol) 1964; 159:545.

Lenzen S: Hexose recognition mechanisms in pancreatic B-cells. Biochem Soc Trans 1990;18:105.

Newgard CB, McGarry JD: Metabolic coupling factors in pancreatic beta-cell signal transduction. Annu Rev Biochem 1995; 64:689.

Newsholme EA, Start C: *Regulation in Metabolism.* Wiley, 1973.

Nordlie RC, Foster JD, Lange AJ: Regulation of glucose production by the liver. Annu Rev Nutr 1999;19:379.

Pilkis SJ, El-Maghrabi MR, Claus TH: Hormonal regulation of hepatic gluconeogenesis and glycolysis. Annu Rev Biochem 1988;57:755.

Pilkis SJ, Granner DK: Molecular physiology of the regulation of hepatic gluconeogenesis and glycolysis. Annu Rev Physiol 1992;54:885.

Yki-Jarvinen H: Action of insulin on glucose metabolism in vivo. Baillieres Clin Endocrinol Metab 1993;7:903.

# The Pentose Phosphate Pathway & Other Pathways of Hexose Metabolism

20

*Peter A. Mayes, PhD, DSc, & David A. Bender, PhD*

## BIOMEDICAL IMPORTANCE

The pentose phosphate pathway is an alternative route for the metabolism of glucose. It does not generate ATP but has two major functions: (1) The formation of **NADPH** for synthesis of fatty acids and steroids and (2) the synthesis of **ribose** for nucleotide and nucleic acid formation. Glucose, fructose, and galactose are the main hexoses absorbed from the gastrointestinal tract, derived principally from dietary starch, sucrose, and lactose, respectively. Fructose and galactose are converted to glucose, mainly in the liver.

Genetic deficiency of glucose 6-phosphate dehydrogenase, the first enzyme of the pentose phosphate pathway, is a major cause of hemolysis of red blood cells, resulting in **hemolytic anemia** and affecting approximately 100 million people worldwide. Glucuronic acid is synthesized from glucose via the **uronic acid pathway,** of major significance for the excretion of metabolites and foreign chemicals (xenobiotics) as **glucuronides.** A deficiency in the pathway leads to **essential pentosuria.** The lack of one enzyme of the pathway (gulonolactone oxidase) in primates and some other animals explains why **ascorbic acid** (vitamin C) is a dietary requirement for humans but not most other mammals. Deficiencies in the enzymes of fructose and galactose metabolism lead to **essential fructosuria** and the **galactosemias.**

## THE PENTOSE PHOSPHATE PATHWAY GENERATES NADPH & RIBOSE PHOSPHATE (Figure 20–1)

The pentose phosphate pathway (hexose monophosphate shunt) is a more complex pathway than glycolysis. Three molecules of glucose 6-phosphate give rise to three molecules of $CO_2$ and three five-carbon sugars. These are rearranged to regenerate two molecules of glucose 6-phosphate and one molecule of the glycolytic intermediate, glyceraldehyde 3-phosphate. Since two molecules of glyceraldehyde 3-phosphate can regenerate glucose 6-phosphate, the pathway can account for the complete oxidation of glucose.

## REACTIONS OF THE PENTOSE PHOSPHATE PATHWAY OCCUR IN THE CYTOSOL

The enzymes of the pentose phosphate pathway, as of glycolysis, are cytosolic. As in glycolysis, oxidation is achieved by dehydrogenation; but **$NADP^+$** and not **$NAD^+$** is the hydrogen acceptor. The sequence of reactions of the pathway may be divided into two phases: an **oxidative nonreversible phase** and a **nonoxidative reversible phase.** In the first phase, glucose 6-phosphate undergoes dehydrogenation and decarboxylation to yield a pentose, ribulose 5-phosphate. In the second phase, ribulose 5-phosphate is converted back to glucose 6-phosphate by a series of reactions involving mainly two enzymes: **transketolase** and **transaldolase** (Figure 20–1).

### The Oxidative Phase Generates NADPH (Figures 20–1 and 20–2)

Dehydrogenation of glucose 6-phosphate to 6-phosphogluconate occurs via the formation of 6-phosphogluconolactone, catalyzed by **glucose-6-phosphate dehydrogenase,** an NADP-dependent enzyme. The hydrolysis of 6-phosphogluconolactone is accomplished by the enzyme **gluconolactone hydrolase.** A second oxidative step is catalyzed by **6-phosphogluconate dehydrogenase,** which also requires $NADP^+$ as hydrogen acceptor and involves decarboxylation followed by formation of the ketopentose, ribulose 5-phosphate.

### The Nonoxidative Phase Generates Ribose Precursors

Ribulose 5-phosphate is the substrate for two enzymes. **Ribulose 5-phosphate 3-epimerase** alters the configuration about carbon 3, forming another ketopentose, xylulose 5-phosphate. **Ribose 5-phosphate ketoisomerase** converts ribulose 5-phosphate to the corresponding aldopentose, ribose 5-phosphate, which is the precursor of the ribose required for nucleotide and nucleic acid synthesis. **Transketolase** transfers the two-carbon

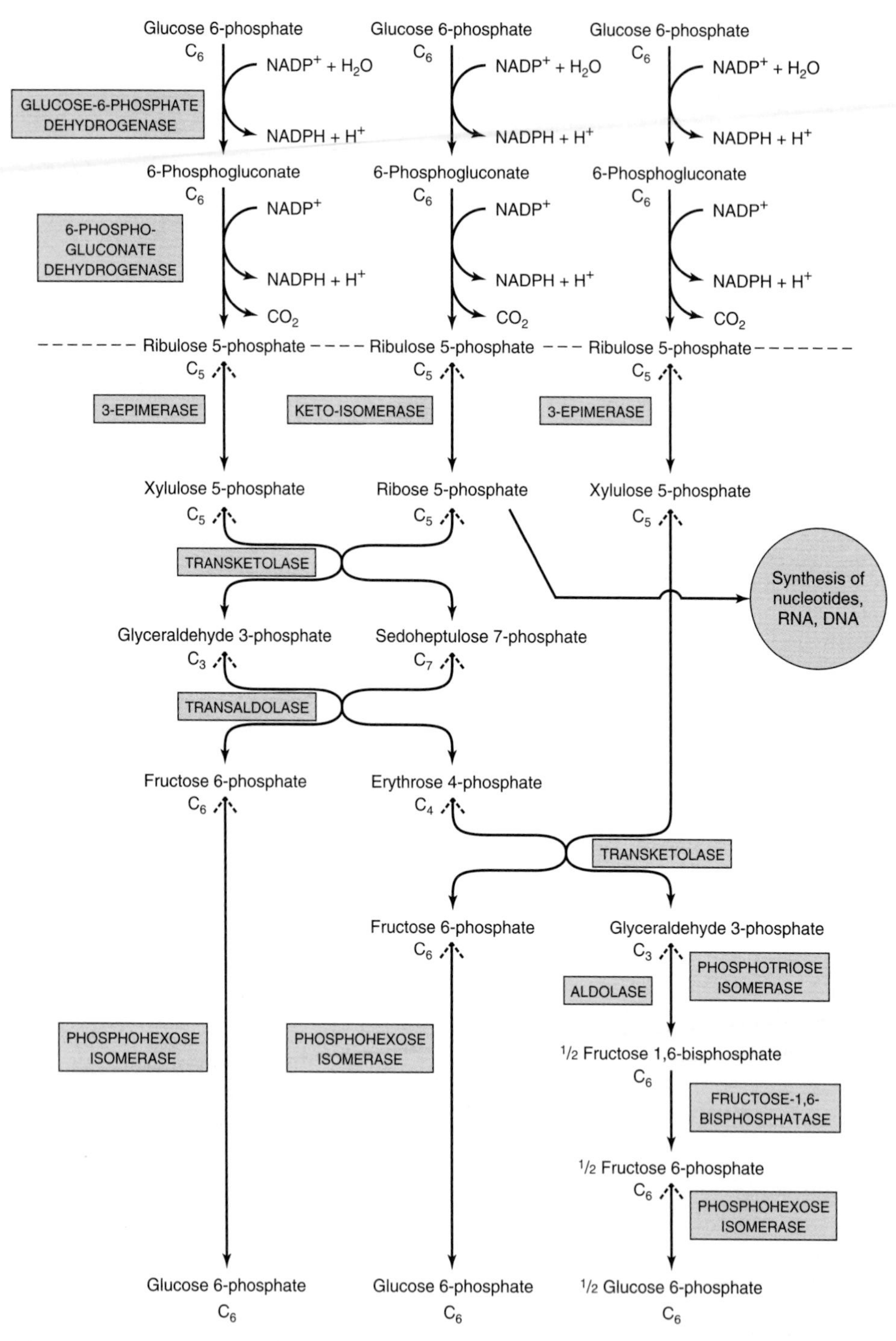


***Figure 20–1.*** Flow chart of pentose phosphate pathway and its connections with the pathway of glycolysis. The full pathway, as indicated, consists of three interconnected cycles in which glucose 6-phosphate is both substrate and end product. The reactions above the broken line are nonreversible, whereas all reactions under that line are freely reversible apart from that catalyzed by fructose-1,6-bisphosphatase.

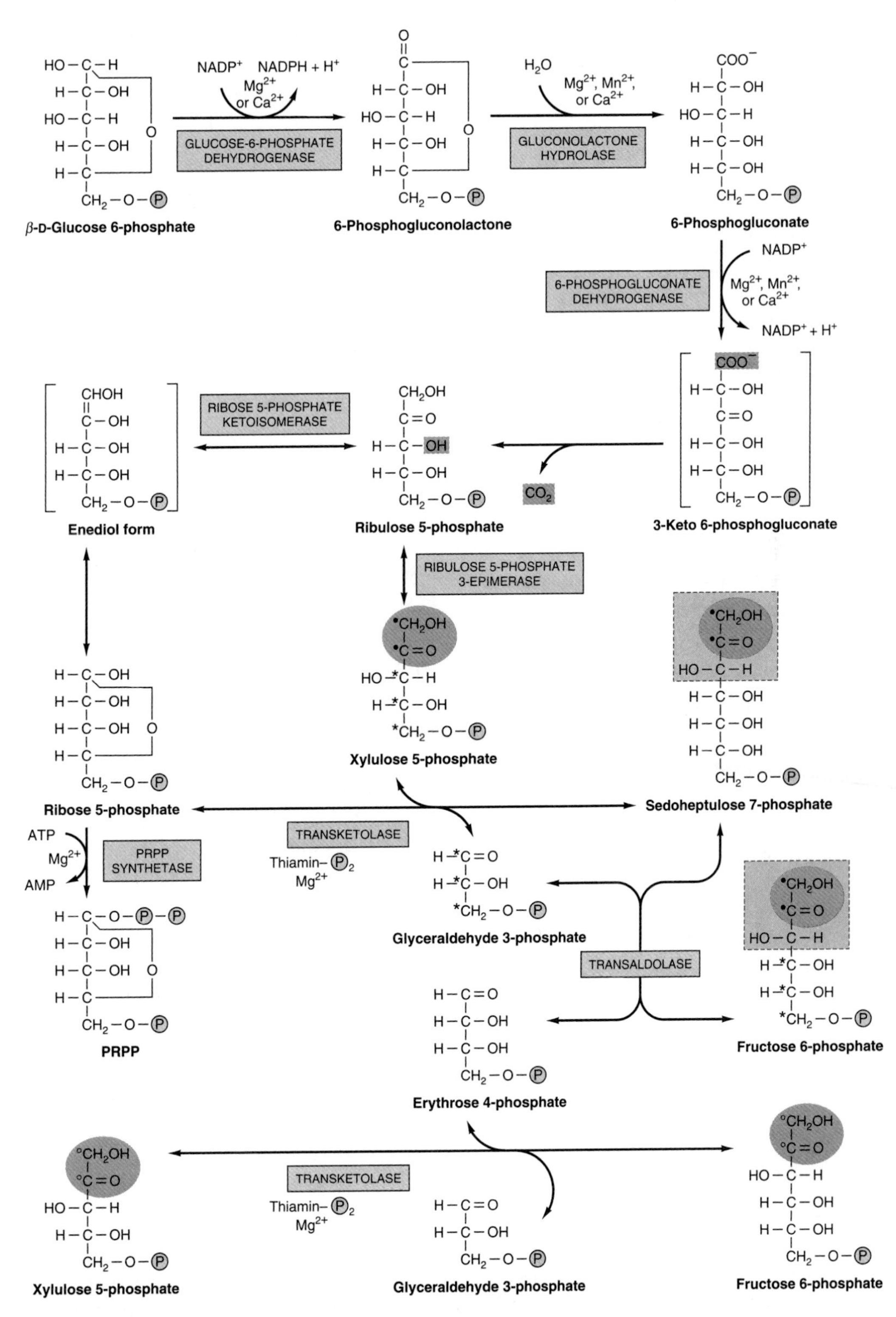


***Figure 20–2.*** The pentose phosphate pathway. (Ⓟ, —$PO_3^{2-}$; PRPP, 5-phosphoribosyl 1-pyrophosphate.)

unit comprising carbons 1 and 2 of a ketose onto the aldehyde carbon of an aldose sugar. It therefore effects the conversion of a ketose sugar into an aldose with two carbons less and simultaneously converts an aldose sugar into a ketose with two carbons more. The reaction requires $Mg^{2+}$ and **thiamin diphosphate** (vitamin $B_1$) as coenzyme. Thus, transketolase catalyzes the transfer of the two-carbon unit from xylulose 5-phosphate to ribose 5-phosphate, producing the seven-carbon ketose sedoheptulose 7-phosphate and the aldose glyceraldehyde 3-phosphate. **Transaldolase** allows the transfer of a three-carbon dihydroxyacetone moiety (carbons 1–3) from the ketose sedoheptulose 7-phosphate onto the aldose glyceraldehyde 3-phosphate to form the ketose fructose 6-phosphate and the four-carbon aldose erythrose 4-phosphate. In a further reaction catalyzed by **transketolase,** xylulose 5-phosphate donates a two-carbon unit to erythrose 4-phosphate to form fructose 6-phosphate and glyceraldehyde 3-phosphate.

In order to oxidize glucose completely to $CO_2$ via the pentose phosphate pathway, there must be enzymes present in the tissue to convert glyceraldehyde 3-phosphate to glucose 6-phosphate. This involves reversal of glycolysis and the gluconeogenic enzyme **fructose 1,6-bisphosphatase.** In tissues that lack this enzyme, glyceraldehyde 3-phosphate follows the normal pathway of glycolysis to pyruvate.

### The Two Major Pathways for the Catabolism of Glucose Have Little in Common

Although glucose 6-phosphate is common to both pathways, the pentose phosphate pathway is markedly different from glycolysis. Oxidation utilizes NADP rather than NAD, and $CO_2$, which is not produced in glycolysis, is a characteristic product. No ATP is generated in the pentose phosphate pathway, whereas ATP is a major product of glycolysis.

### Reducing Equivalents Are Generated in Those Tissues Specializing in Reductive Syntheses

The pentose phosphate pathway is active in liver, adipose tissue, adrenal cortex, thyroid, erythrocytes, testis, and lactating mammary gland. Its activity is low in nonlactating mammary gland and skeletal muscle. Those tissues in which the pathway is active use NADPH in reductive syntheses, eg, of fatty acids, steroids, amino acids via glutamate dehydrogenase, and reduced glutathione. The synthesis of glucose-6-phosphate dehydrogenase and 6-phosphogluconate dehydrogenase may also be induced by insulin during conditions associated with the "fed state" (Table 19–1), when lipogenesis increases.

### Ribose Can Be Synthesized in Virtually All Tissues

Little or no ribose circulates in the bloodstream, so tissues must synthesize the ribose required for nucleotide and nucleic acid synthesis (Figure 20–2). The source of ribose 5-phosphate is the pentose phosphate pathway (Chapter 34). Muscle has only low activity of glucose-6-phosphate dehydrogenase and 6-phosphogluconate dehydrogenase. Nevertheless, like most other tissues, it is capable of synthesizing ribose 5-phosphate by reversal of the nonoxidative phase of the pentose phosphate pathway utilizing fructose 6-phosphate. It is not necessary to have a completely functioning pentose phosphate pathway for a tissue to synthesize ribose phosphates.

## THE PENTOSE PHOSPHATE PATHWAY & GLUTATHIONE PEROXIDASE PROTECT ERYTHROCYTES AGAINST HEMOLYSIS

In erythrocytes, the pentose phosphate pathway provides NADPH for the reduction of oxidized glutathione catalyzed by **glutathione reductase,** a flavoprotein containing FAD. Reduced glutathione removes $H_2O_2$ in a reaction catalyzed by **glutathione peroxidase,** an enzyme that contains the **selenium** analogue of cysteine (selenocysteine) at the active site (Figure 20–3). This reaction is important, since accumulation of $H_2O_2$ may decrease the life span of the erythrocyte by causing oxidative damage to the cell membrane, leading to hemolysis.

## GLUCURONATE, A PRECURSOR OF PROTEOGLYCANS & CONJUGATED GLUCURONIDES, IS A PRODUCT OF THE URONIC ACID PATHWAY

In liver, the **uronic acid pathway** catalyzes the conversion of glucose to glucuronic acid, ascorbic acid, and pentoses (Figure 20–4). It is also an alternative oxidative pathway for glucose, but—like the pentose phosphate pathway—it does not lead to the generation of ATP. Glucose 6-phosphate is isomerized to glucose 1-phosphate, which then reacts with uridine triphosphate (UTP) to form uridine diphosphate glucose (UDPGlc) in a reaction catalyzed by **UDPGlc pyrophosphorylase,** as occurs in glycogen synthesis (Chapter 18). UDPGlc is oxidized at carbon 6 by NAD-dependent **UDPGlc dehydrogenase** in a two-step reaction to yield UDP-glucuronate. UDP-glucuronate is the "active" form of glucuronate for reactions involving incorporation of glucuronic acid into proteoglycans or for reactions in which substrates such as steroid hormones, bilirubin, and a number of drugs are conjugated with glucuronate for excretion in urine or bile (Figure 32–13).

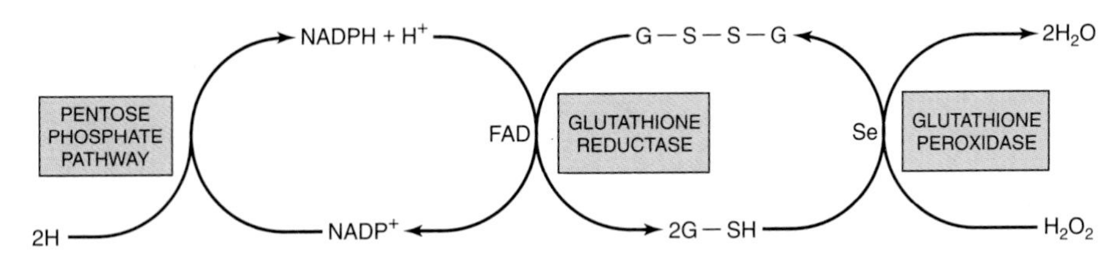


***Figure 20–3.*** Role of the pentose phosphate pathway in the glutathione peroxidase reaction of erythrocytes. (G-S-S-G, oxidized glutathione; G-SH, reduced glutathione; Se, selenium cofactor.)

Glucuronate is reduced to L-gulonate in an NADPH-dependent reaction; L-gulonate is the direct precursor of **ascorbate** in those animals capable of synthesizing this vitamin. In humans and other primates as well as guinea pigs, ascorbic acid cannot be synthesized because of the absence of **L-gulonolactone oxidase.** L-Gulonate is metabolized ultimately to D-xylulose 5-phosphate, a constituent of the pentose phosphate pathway.

## INGESTION OF LARGE QUANTITIES OF FRUCTOSE HAS PROFOUND METABOLIC CONSEQUENCES

Diets high in sucrose or in high-fructose syrups used in manufactured foods and beverages lead to large amounts of fructose (and glucose) entering the hepatic portal vein. Fructose undergoes more rapid glycolysis in the liver than does glucose because it bypasses the regulatory step catalyzed by phosphofructokinase (Figure 20–6). This allows fructose to flood the pathways in the liver, leading to enhanced fatty acid synthesis, increased esterification of fatty acids, and increased VLDL secretion, which may raise serum triacylglycerols and ultimately raise LDL cholesterol concentrations (Figure 25–6). A specific kinase, **fructokinase,** in liver (and kidney and intestine) catalyzes the phosphorylation of fructose to fructose 1-phosphate. This enzyme does not act on glucose, and, unlike glucokinase, its activity is not affected by fasting or by insulin, which may explain why fructose is cleared from the blood of diabetic patients at a normal rate. Fructose 1-phosphate is cleaved to D-glyceraldehyde and dihydroxyacetone phosphate by **aldolase B,** an enzyme found in the liver, which also functions in glycolysis by cleaving fructose 1,6-bisphosphate. D-Glyceraldehyde enters glycolysis via phosphorylation to glyceraldehyde 3-phosphate, catalyzed by **triokinase.** The two triose phosphates, dihydroxyacetone phosphate and glyceraldehyde 3-phosphate, may be degraded by glycolysis or may be substrates for aldolase and hence gluconeogenesis, which is the fate of much of the fructose metabolized in the liver.

In extrahepatic tissues, hexokinase catalyzes the phosphorylation of most hexose sugars, including fructose. However, glucose inhibits the phosphorylation of fructose since it is a better substrate for hexokinase. Nevertheless, some fructose can be metabolized in adipose tissue and muscle. Fructose, a potential fuel, is found in seminal plasma and in the fetal circulation of ungulates and whales. **Aldose reductase** is found in the placenta of the ewe and is responsible for the secretion of sorbitol into the fetal blood. The presence of **sorbitol dehydrogenase** in the liver, including the fetal liver, is responsible for the conversion of sorbitol into fructose. This pathway is also responsible for the occurrence of fructose in seminal fluid.

## GALACTOSE IS NEEDED FOR THE SYNTHESIS OF LACTOSE, GLYCOLIPIDS, PROTEOGLYCANS, & GLYCOPROTEINS

Galactose is derived from intestinal hydrolysis of the disaccharide **lactose,** the sugar of milk. It is readily converted in the liver to glucose. **Galactokinase** catalyzes the phosphorylation of galactose, using ATP as phosphate donor (Figure 20–6A). Galactose 1-phosphate reacts with uridine diphosphate glucose (UDPGlc) to form uridine diphosphate galactose (UDPGal) and glucose 1-phosphate, in a reaction catalyzed by **galactose 1-phosphate uridyl transferase.** The conversion of UDPGal to UDPGlc is catalyzed by **UDPGal 4-epimerase.** Epimerization involves an oxidation and reduction at carbon 4 with $NAD^+$ as coenzyme. Finally, glucose is liberated from UDPGlc after conversion to glucose 1-phosphate, probably via incorporation into glycogen followed by phosphorolysis (Chapter 18).

Since the epimerase reaction is freely reversible, glucose can be converted to galactose, so that galactose is not a dietary essential. Galactose is required in the body not only in the formation of lactose but also as a constituent of glycolipids (cerebrosides), proteoglycans, and glycoproteins. In the synthesis of lactose in the mammary gland, UDPGal condenses with glucose to yield lactose, catalyzed by **lactose synthase** (Figure 20–6B).

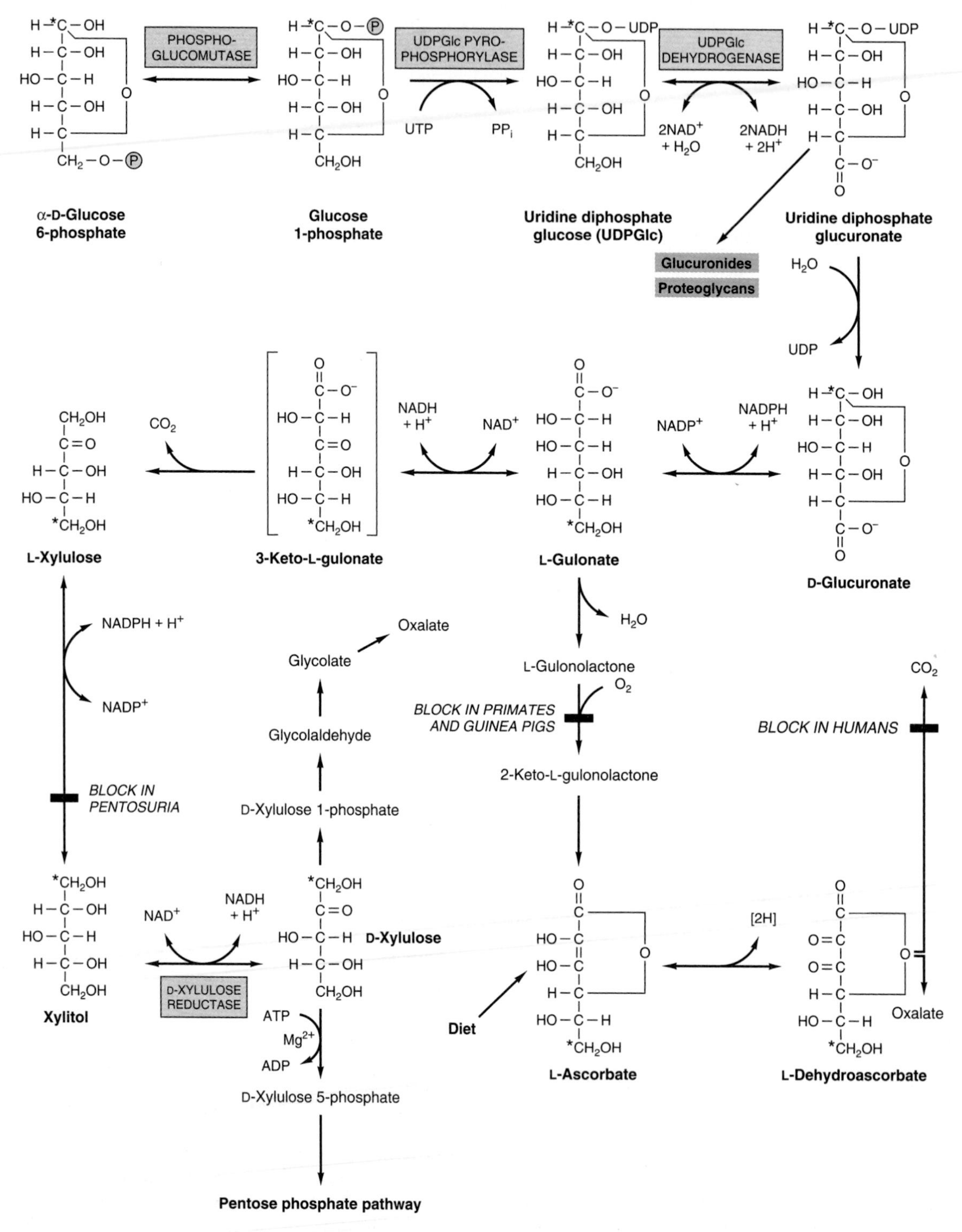


***Figure 20–4.*** Uronic acid pathway. (Asterisk indicates the fate of carbon 1 of glucose; Ⓟ, —$PO_3^{2-}$.)

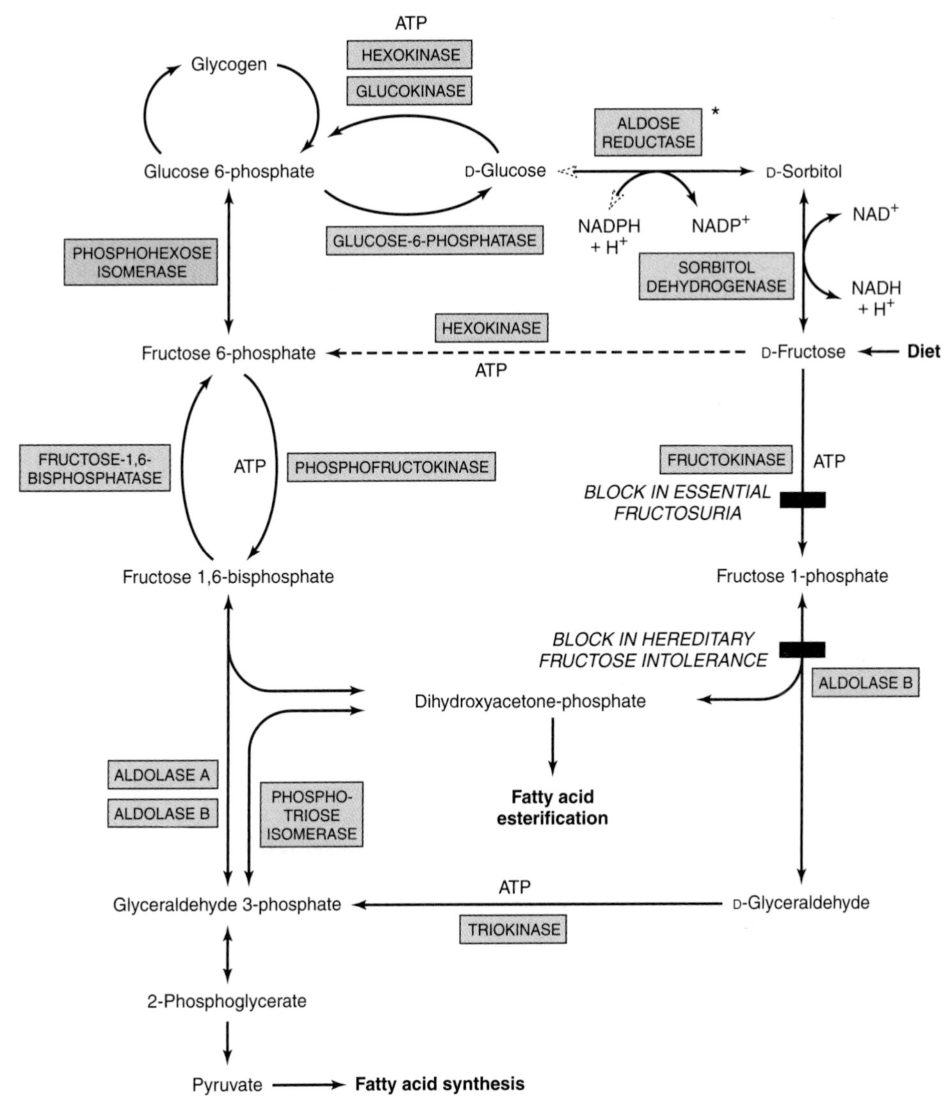


***Figure 20–5.*** Metabolism of fructose. Aldolase A is found in all tissues, whereas aldolase B is the predominant form in liver. ($A^{+*}$, not found in liver.)

## Glucose Is the Precursor of All Amino Sugars (Hexosamines)

Amino sugars are important components of **glycoproteins** (Chapter 47), of certain **glycosphingolipids** (eg, gangliosides) (Chapter 14), and of **glycosaminoglycans** (Chapter 48). The major amino sugars are **glucosamine, galactosamine,** and **mannosamine** and the nine-carbon compound **sialic acid.** The principal sialic acid found in human tissues is ***N*-acetylneuraminic acid** (NeuAc). A summary of the metabolic interrelationships among the amino sugars is shown in Figure 20–7.

# CLINICAL ASPECTS

## Impairment of the Pentose Phosphate Pathway Leads to Erythrocyte Hemolysis

Genetic deficiency of glucose-6-phosphate dehydrogenase, with consequent impairment of the generation of NADPH, is common in populations of Mediterranean and Afro-Caribbean origin. The defect is manifested as red cell hemolysis **(hemolytic anemia)** when susceptible individuals are subjected to oxidants, such as the antimalarial primaquine, aspirin, or sulfonamides or when

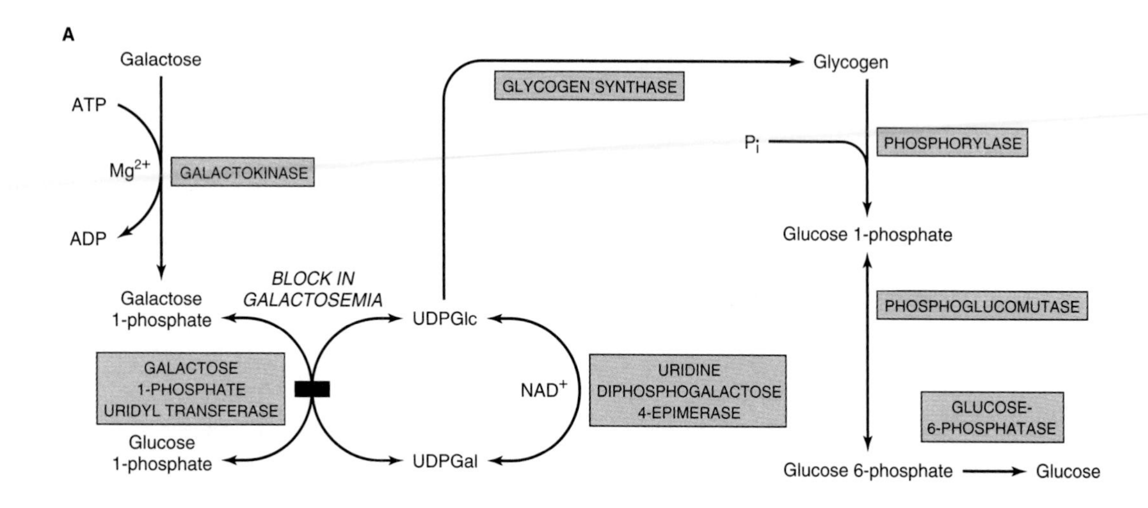


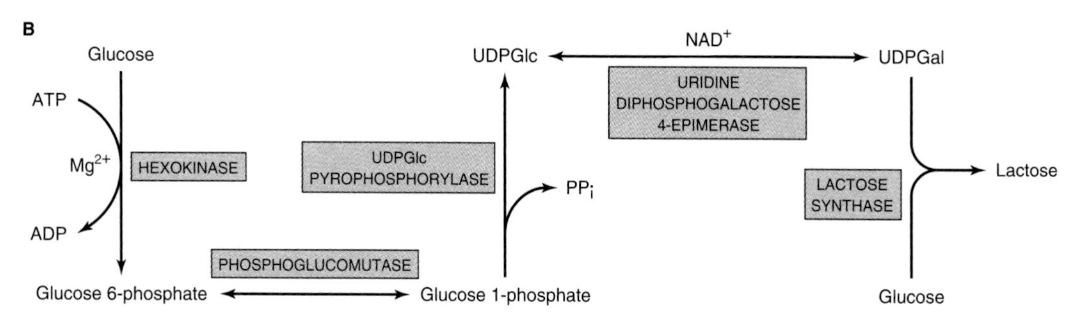


***Figure 20–6.*** Pathway of conversion of **(A)** galactose to glucose in the liver and **(B)** glucose to lactose in the lactating mammary gland.

they have eaten fava beans (*Vicia fava*—hence the term **favism**). Glutathione peroxidase is dependent upon a supply of NADPH, which in erythrocytes can be formed only via the pentose phosphate pathway. It reduces organic peroxides and $H_2O_2$ as part of the body's defense against lipid peroxidation (Figure 14–21). Measurement of erythrocyte **transketolase** and its activation by thiamin diphosphate is used to assess thiamin nutritional status (Chapter 45).

## Disruption of the Uronic Acid Pathway Is Caused by Enzyme Defects & Some Drugs

In the rare hereditary disease **essential pentosuria,** considerable quantities of **L-xylulose** appear in the urine because of absence of the enzyme necessary to reduce L-xylulose to xylitol. Parenteral administration of xylitol may lead to **oxalosis,** involving calcium oxalate deposition in brain and kidneys (Figure 20–4). Various drugs markedly increase the rate at which glucose enters the uronic acid pathway. For example, administration of barbital or of chlorobutanol to rats results in a significant increase in the conversion of glucose to glucuronate, L-gulonate, and ascorbate.

## Loading of the Liver With Fructose May Potentiate Hyperlipidemia & Hyperuricemia

In the liver, fructose increases triacylglycerol synthesis and VLDL secretion, leading to hypertriacylglycerolemia—and increased LDL cholesterol—which can be regarded as potentially atherogenic (Chapter 26). In addition, acute loading of the liver with fructose, as can occur with intravenous infusion or following very high fructose intakes, causes sequestration of inorganic phosphate in fructose 1-phosphate and diminished ATP synthesis. As a result there is less inhibition of de novo purine synthesis by ATP and uric acid formation is in-

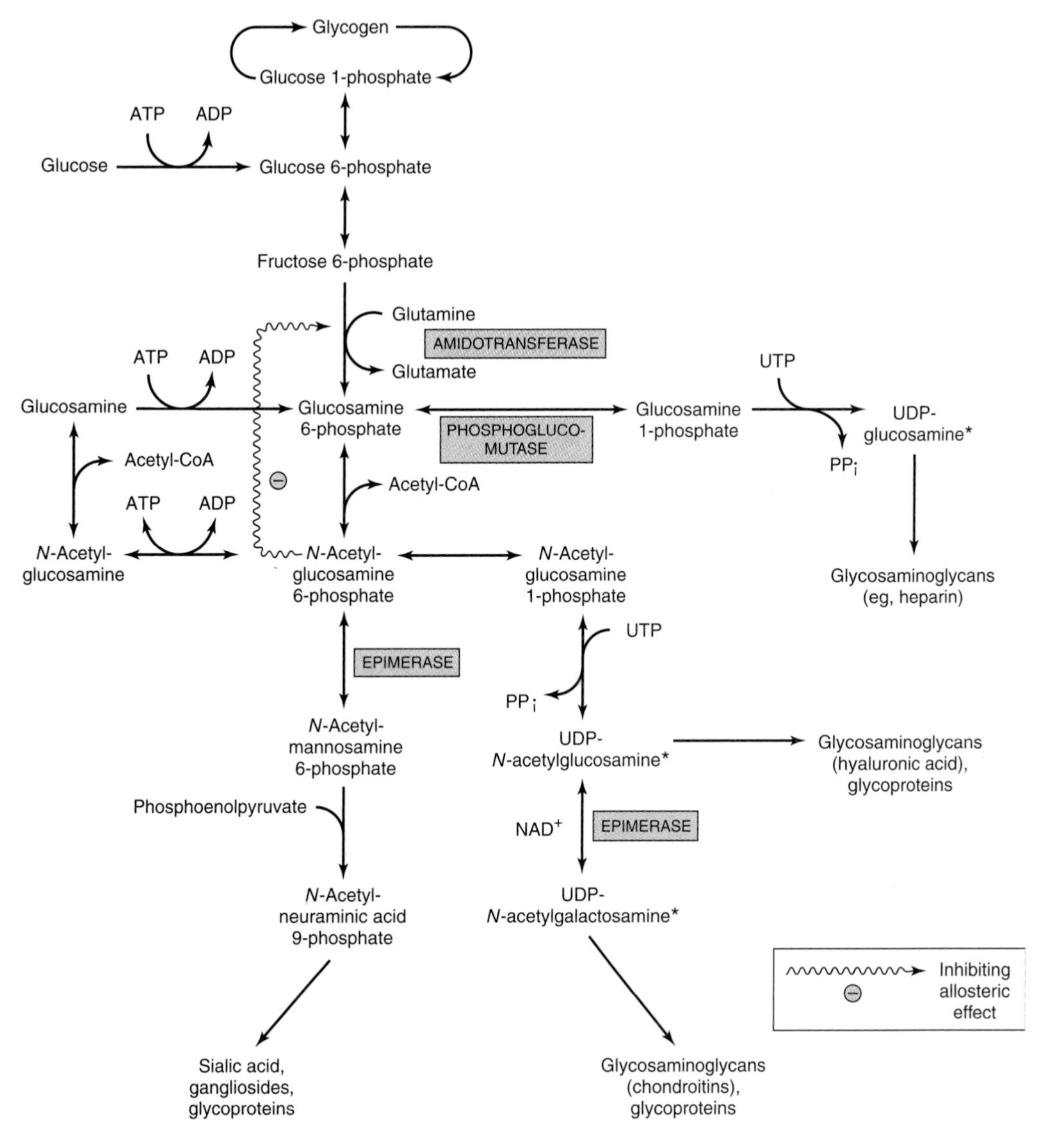


***Figure 20–7.*** Summary of the interrelationships in metabolism of amino sugars. (At asterisk: Analogous to UDPGlc.) Other purine or pyrimidine nucleotides may be similarly linked to sugars or amino sugars. Examples are thymidine diphosphate (TDP)-glucosamine and TDP-*N*-acetylglucosamine.

creased, causing hyperuricemia, which is a cause of gout (Chapter 34).

## Defects in Fructose Metabolism Cause Disease (Figure 20–5)

Lack of hepatic fructokinase causes **essential fructosuria,** and absence of hepatic aldolase B, which cleaves fructose 1-phosphate, leads to **hereditary fructose intolerance.** Diets low in fructose, sorbitol, and sucrose are beneficial for both conditions. One consequence of hereditary fructose intolerance and of another condition due to **fructose-1,6-bisphosphatase deficiency** is fructose-induced **hypoglycemia** despite the presence of high glycogen reserves. The accumulation of fructose 1-phosphate and fructose 1,6-bisphosphate allosterically inhibits the activity of liver phosphorylase. The sequestration of inorganic phosphate also leads to depletion of ATP and hyperuricemia.

### Fructose & Sorbitol in the Lens Are Associated With Diabetic Cataract

Both fructose and sorbitol are found in the lens of the eye in increased concentrations in diabetes mellitus and may be involved in the pathogenesis of **diabetic cataract.** The **sorbitol (polyol) pathway** (not found in liver) is responsible for fructose formation from glucose (Figure 20–5) and increases in activity as the glucose concentration rises in diabetes in those tissues that are not insulin-sensitive, ie, the lens, peripheral nerves, and renal glomeruli. Glucose is reduced to sorbitol by **aldose reductase,** followed by oxidation of sorbitol to fructose in the presence of $NAD^+$ and sorbitol dehydrogenase (polyol dehydrogenase). Sorbitol does not diffuse through cell membranes easily and accumulates, causing osmotic damage. Simultaneously, myoinositol levels fall. Sorbitol accumulation, myoinositol depletion, and diabetic cataract can be prevented by aldose reductase inhibitors in diabetic rats, and promising results have been obtained in clinical trials.

When sorbitol is administered intravenously, it is converted to fructose rather than to glucose. It is poorly absorbed in the small intestine, and much is fermented by colonic bacteria to short-chain fatty acids, $CO_2$, and $H_2$, leading to abdominal pain and diarrhea **(sorbitol intolerance).**

### Enzyme Deficiencies in the Galactose Pathway Cause Galactosemia

Inability to metabolize galactose occurs in the **galactosemias,** which may be caused by inherited defects in galactokinase, uridyl transferase, or 4-epimerase (Figure 20–6A), though a deficiency in **uridyl transferase** is the best known cause. The galactose concentration in the blood and in the eye is reduced by aldose reductase to galactitol, which accumulates, causing cataract. In uridyl transferase deficiency, galactose 1-phosphate accumulates and depletes the liver of inorganic phosphate. Ultimately, liver failure and mental deterioration result. As the epimerase is present in adequate amounts, the galactosemic individual can still form UDPGal from glucose, and normal growth and development can occur regardless of the galactose-free diets used to control the symptoms of the disease.

## SUMMARY

- The pentose phosphate pathway, present in the cytosol, can account for the complete oxidation of glucose, producing NADPH and $CO_2$ but not ATP.
- The pathway has an oxidative phase, which is irreversible and generates NADPH; and a nonoxidative phase, which is reversible and provides ribose precursors for nucleotide synthesis. The complete pathway is present only in those tissues having a requirement for NADPH for reductive syntheses, eg, lipogenesis or steroidogenesis, whereas the nonoxidative phase is present in all cells requiring ribose.
- In erythrocytes, the pathway has a major function in preventing hemolysis by providing NADPH to maintain glutathione in the reduced state as the substrate for glutathione peroxidase.
- The uronic acid pathway is the source of glucuronic acid for conjugation of many endogenous and exogenous substances before excretion as glucuronides in urine and bile.
- Fructose bypasses the main regulatory step in glycolysis, catalyzed by phosphofructokinase, and stimulates fatty acid synthesis and hepatic triacylglycerol secretion.
- Galactose is synthesized from glucose in the lactating mammary gland and in other tissues where it is required for the synthesis of glycolipids, proteoglycans, and glycoproteins.

## REFERENCES

Couet C, Jan P, Debry G: Lactose and cataract in humans: a review. J Am Coll Nutr 1991;10:79.

Cox TM: Aldolase B and fructose intolerance. FASEB J 1994;8:62.

Cross NCP, Cox TM: Hereditary fructose intolerance. Int J Biochem 1990;22:685.

Kador PF: The role of aldose reductase in the development of diabetic complications. Med Res Rev 1988;8:325.

Kaufman FR, Devgan S: Classical galactosemia: a review. Endocrinologist 1995;5:189.

Macdonald I, Vrana A (editors): *Metabolic Effects of Dietary Carbohydrates.* Karger, 1986.

Mayes PA: Intermediary metabolism of fructose. Am J Clin Nutr 1993(5 Suppl);58:754S.

Van den Berghe G: Inborn errors of fructose metabolism. Annu Rev Nutr 1994;14:41.

Wood T: Physiological functions of the pentose phosphate pathway. Cell Biol Funct 1986;4:241.

# Biosynthesis of Fatty Acids

# 21

*Peter A. Mayes, PhD, DSc, & Kathleen M. Botham, PhD, DSc*

## BIOMEDICAL IMPORTANCE

Fatty acids are synthesized by an **extramitochondrial system**, which is responsible for the complete synthesis of palmitate from acetyl-CoA in the cytosol. In the rat, the pathway is well represented in adipose tissue and liver, whereas in humans adipose tissue may not be an important site, and liver has only low activity. In birds, lipogenesis is confined to the liver, where it is particularly important in providing lipids for egg formation. In most mammals, glucose is the primary substrate for lipogenesis, but in ruminants it is acetate, the main fuel molecule produced by the diet. Critical diseases of the pathway have not been reported in humans. However, inhibition of lipogenesis occurs in type 1 (insulin-dependent) **diabetes mellitus,** and variations in its activity may affect the nature and extent of **obesity**.

## THE MAIN PATHWAY FOR DE NOVO SYNTHESIS OF FATTY ACIDS (LIPOGENESIS) OCCURS IN THE CYTOSOL

This system is present in many tissues, including liver, kidney, brain, lung, mammary gland, and adipose tissue. Its cofactor requirements include NADPH, ATP, $Mn^{2+}$, biotin, and $HCO_3^-$ (as a source of $CO_2$). **Acetyl-CoA** is the immediate substrate, and **free palmitate** is the end product.

### Production of Malonyl-CoA Is the Initial & Controlling Step in Fatty Acid Synthesis

Bicarbonate as a source of $CO_2$ is required in the initial reaction for the carboxylation of acetyl-CoA to **malonyl-CoA** in the presence of ATP and **acetyl-CoA carboxylase.** Acetyl-CoA carboxylase has a requirement for the vitamin **biotin** (Figure 21–1). The enzyme is a **multienzyme protein** containing a variable number of identical subunits, each containing biotin, biotin carboxylase, biotin carboxyl carrier protein, and transcarboxylase, as well as a regulatory allosteric site. The reaction takes place in two steps: (1) carboxylation of biotin involving ATP and (2) transfer of the carboxyl to acetyl-CoA to form malonyl-CoA.

### The Fatty Acid Synthase Complex Is a Polypeptide Containing Seven Enzyme Activities

In bacteria and plants, the individual enzymes of the **fatty acid synthase** system are separate, and the acyl radicals are found in combination with a protein called the **acyl carrier protein (ACP).** However, in yeast, mammals, and birds, the synthase system is a multienzyme polypeptide complex that incorporates ACP, which takes over the role of CoA. It contains the vitamin **pantothenic acid** in the form of 4′-phosphopantetheine (Figure 45–18). The use of one multienzyme functional unit has the advantages of achieving the effect of compartmentalization of the process within the cell without the erection of permeability barriers, and synthesis of all enzymes in the complex is coordinated since it is encoded by a single gene.

In mammals, the fatty acid synthase complex is a dimer comprising two identical monomers, each containing all seven enzyme activities of fatty acid synthase on one polypeptide chain (Figure 21–2). Initially, a priming molecule of acetyl-CoA combines with a cysteine —SH group catalyzed by **acetyl transacylase** (Figure 21–3, reaction 1a). Malonyl-CoA combines with the adjacent —SH on the 4′-phosphopantetheine of ACP of the other monomer, catalyzed by **malonyl transacylase** (reaction 1b), to form **acetyl (acyl)-malonyl enzyme.** The acetyl group attacks the methylene group of the malonyl residue, catalyzed by **3-ketoacyl synthase,** and liberates $CO_2$, forming 3-ketoacyl enzyme (acetoacetyl enzyme) (reaction 2), freeing the cysteine —SH group. Decarboxylation allows the reaction to go to completion, pulling the whole sequence of reactions in the forward direction. The 3-ketoacyl group is reduced, dehydrated, and reduced again (reactions 3, 4, 5) to form the corresponding saturated acyl-S-enzyme. A new malonyl-CoA molecule combines with the —SH of 4′-phosphopantetheine, displacing the saturated acyl residue onto the free cysteine —SH group. The sequence of reactions is repeated six more times until a saturated 16-carbon acyl radical (palmityl) has been assembled. It is liberated from the enzyme complex by the activity of a seventh enzyme in the complex, **thioesterase** (deacylase). The free palmitate must be activated to acyl-CoA before it can proceed via any other

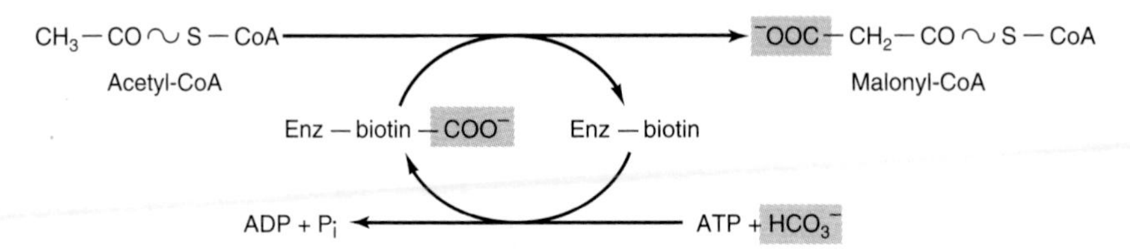


***Figure 21–1.*** Biosynthesis of malonyl-CoA. (Enz, acetyl-CoA carboxylase.)

metabolic pathway. Its usual fate is esterification into acylglycerols, chain elongation or desaturation, or esterification to cholesteryl ester. In mammary gland, there is a separate thioesterase specific for acyl residues of $C_8$, $C_{10}$, or $C_{12}$, which are subsequently found in milk lipids.

The equation for the overall synthesis of palmitate from acetyl-CoA and malonyl-CoA is:

$$CH_2CO\cdot S\cdot CoA + 7HOOC\cdot CH_2CO\cdot S\cdot CoA + 14NADPH + 14H^+$$
$$\rightarrow CH_3(CH_2)_{14}COOH + 7CO_2 + 6H_2O + 8CoA\cdot SH + 14NADP^+$$

The acetyl-CoA used as a primer forms carbon atoms 15 and 16 of palmitate. The addition of all the subsequent $C_2$ units is via malonyl-CoA. Propionyl-CoA acts as primer for the synthesis of long-chain fatty

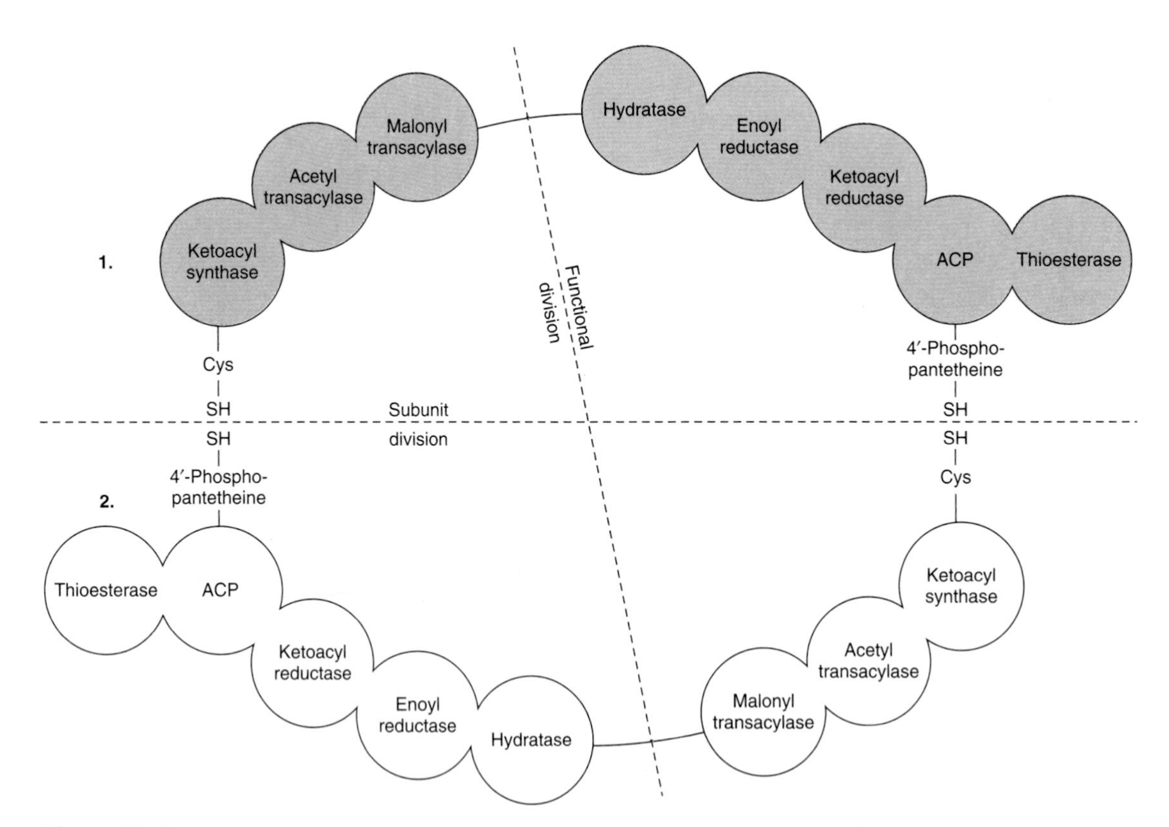


***Figure 21–2.*** Fatty acid synthase multienzyme complex. The complex is a dimer of two identical polypeptide monomers, 1 and 2, each consisting of seven enzyme activities and the acyl carrier protein (ACP). (Cys—SH, cysteine thiol.) The —SH of the 4′-phosphopantetheine of one monomer is in close proximity to the —SH of the cysteine residue of the ketoacyl synthase of the other monomer, suggesting a "head-to-tail" arrangement of the two monomers. Though each monomer contains all the partial activities of the reaction sequence, the actual functional unit consists of one-half of one monomer interacting with the complementary half of the other. Thus, two acyl chains are produced simultaneously. The sequence of the enzymes in each monomer is based on Wakil.

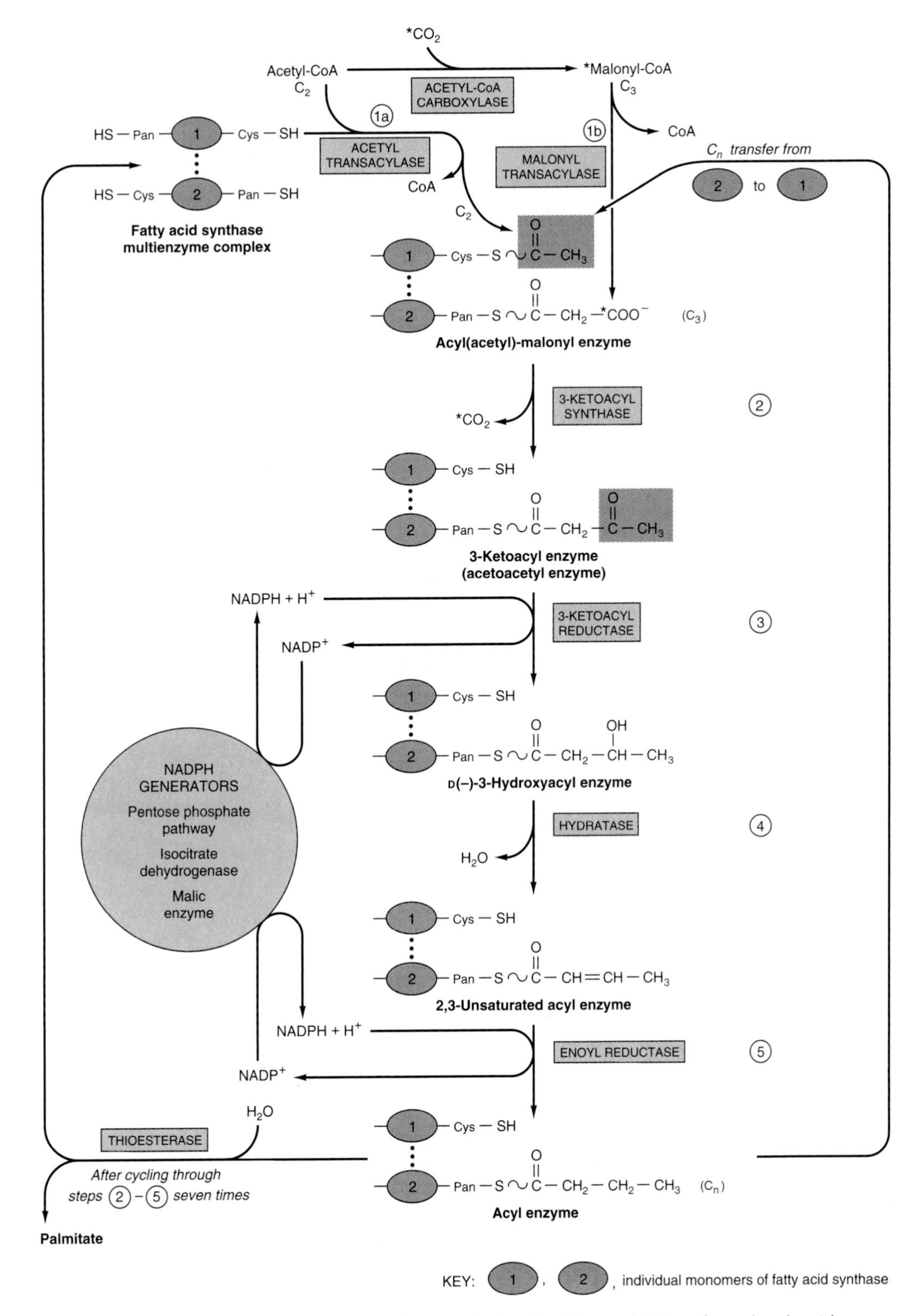


***Figure 21–3.*** Biosynthesis of long-chain fatty acids. Details of how addition of a malonyl residue causes the acyl chain to grow by two carbon atoms. (Cys, cysteine residue; Pan, 4′-phosphopantetheine.) The blocks shown in dark blue contain initially a $C_2$ unit derived from acetyl-CoA (as illustrated) and subsequently the $C_n$ unit formed in reaction 5.

acids having an odd number of carbon atoms, found particularly in ruminant fat and milk.

## The Main Source of NADPH for Lipogenesis Is the Pentose Phosphate Pathway

NADPH is involved as donor of reducing equivalents in both the reduction of the 3-ketoacyl and of the 2,3-unsaturated acyl derivatives (Figure 21–3, reactions 3 and 5). The oxidative reactions of the pentose phosphate pathway (see Chapter 20) are the chief source of the hydrogen required for the reductive synthesis of fatty acids. Significantly, tissues specializing in active lipogenesis—ie, liver, adipose tissue, and the lactating mammary gland—also possess an active pentose phosphate pathway. Moreover, both metabolic pathways are found in the cytosol of the cell, so there are no membranes or permeability barriers against the transfer of NADPH. Other sources of NADPH include the reaction that converts malate to pyruvate catalyzed by the **"malic enzyme"** (NADP malate dehydrogenase) (Figure 21–4) and the extramitochondrial **isocitrate dehydrogenase** reaction (probably not a substantial source, except in ruminants).

## Acetyl-CoA Is the Principal Building Block of Fatty Acids

Acetyl-CoA is formed from glucose via the oxidation of pyruvate within the mitochondria. However, it does not diffuse readily into the extramitochondrial cytosol,

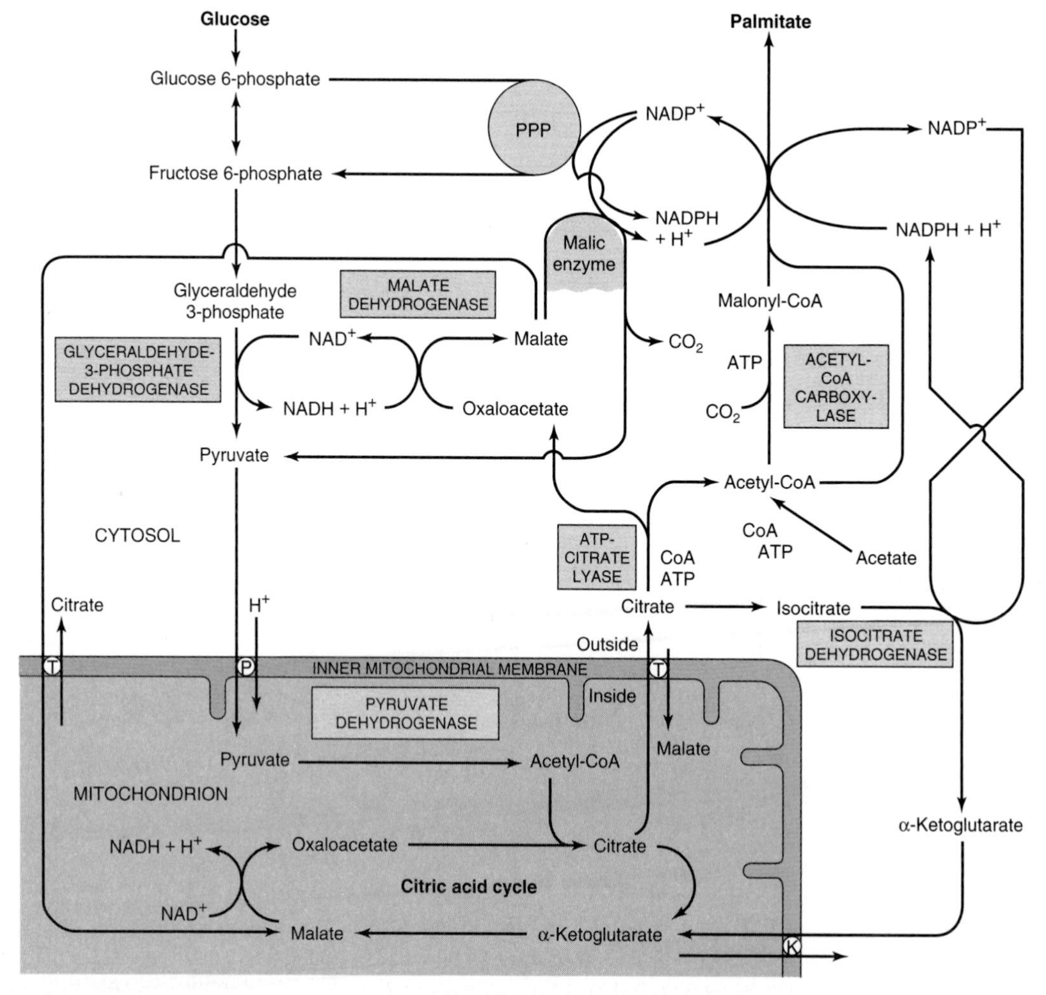


***Figure 21–4.*** The provision of acetyl-CoA and NADPH for lipogenesis. (PPP, pentose phosphate pathway; T, tricarboxylate transporter; K, α-ketoglutarate transporter; P, pyruvate transporter.)

the principal site of fatty acid synthesis. Citrate, formed after condensation of acetyl-CoA with oxaloacetate in the citric acid cycle within mitochondria, is translocated into the extramitochondrial compartment via the tricarboxylate transporter, where in the presence of CoA and ATP it undergoes cleavage to acetyl-CoA and oxaloacetate catalyzed by **ATP-citrate lyase,** which increases in activity in the well-fed state. The acetyl-CoA is then available for malonyl-CoA formation and synthesis to palmitate (Figure 21–4). The resulting oxaloacetate can form malate via NADH-linked malate dehydrogenase, followed by the generation of NADPH via the malic enzyme. The NADPH becomes available for lipogenesis, and the pyruvate can be used to regenerate acetyl-CoA after transport into the mitochondrion. This pathway is a means of transferring reducing equivalents from extramitochondrial NADH to NADP. Alternatively, malate itself can be transported into the mitochondrion, where it is able to re-form oxaloacetate. Note that the citrate (tricarboxylate) transporter in the mitochondrial membrane requires malate to exchange with citrate (see Figure 12-10). There is little ATP-citrate lyase or malic enzyme in ruminants, probably because in these species acetate (derived from the rumen and activated to acetyl CoA extramitochondrially) is the main source of acetyl-CoA.

## Elongation of Fatty Acid Chains Occurs in the Endoplasmic Reticulum

This pathway (the "microsomal system") elongates saturated and unsaturated fatty acyl-CoAs (from $C_{10}$ upward) by two carbons, using malonyl-CoA as acetyl donor and NADPH as reductant, and is catalyzed by the microsomal **fatty acid elongase** system of enzymes (Figure 21–5). Elongation of stearyl-CoA in brain increases rapidly during myelination in order to provide $C_{22}$ and $C_{24}$ fatty acids for sphingolipids.

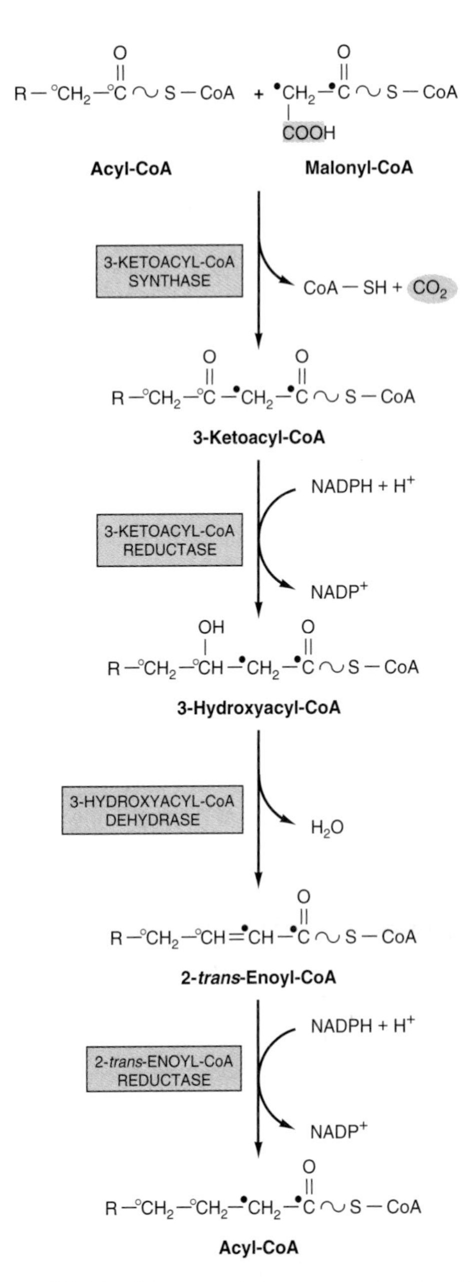


***Figure 21–5.*** Microsomal elongase system for fatty acid chain elongation. NADH is also used by the reductases, but NADPH is preferred.

## THE NUTRITIONAL STATE REGULATES LIPOGENESIS

Excess carbohydrate is stored as fat in many animals in anticipation of periods of caloric deficiency such as starvation, hibernation, etc, and to provide energy for use between meals in animals, including humans, that take their food at spaced intervals. Lipogenesis converts surplus glucose and intermediates such as pyruvate, lactate, and acetyl-CoA to fat, assisting the anabolic phase of this feeding cycle. The nutritional state of the organism is the main factor regulating the rate of lipogenesis. Thus, the rate is high in the well-fed animal whose diet contains a high proportion of carbohydrate. It is depressed under conditions of restricted caloric intake, on a fat diet, or when there is a deficiency of insulin, as in diabetes mellitus. These latter conditions are associated with increased concentrations of plasma free fatty acids, and an inverse relationship has been demonstrated between hepatic lipogenesis and the concentration of serum-free fatty acids. Lipogenesis is increased when sucrose is fed instead of glucose because fructose bypasses the phosphofructokinase control point in glycolysis and floods the lipogenic pathway (Figure 20–5).

## SHORT- & LONG-TERM MECHANISMS REGULATE LIPOGENESIS

Long-chain fatty acid synthesis is controlled in the short term by allosteric and covalent modification of enzymes and in the long term by changes in gene expression governing rates of synthesis of enzymes.

### Acetyl-CoA Carboxylase Is the Most Important Enzyme in the Regulation of Lipogenesis

Acetyl-CoA carboxylase is an allosteric enzyme and is activated by **citrate,** which increases in concentration in the well-fed state and is an indicator of a plentiful supply of acetyl-CoA. Citrate converts the enzyme from an inactive dimer to an active polymeric form, having a molecular mass of several million. Inactivation is promoted by phosphorylation of the enzyme and by long-chain acyl-CoA molecules, an example of negative feedback inhibition by a product of a reaction. Thus, if acyl-CoA accumulates because it is not esterified quickly enough or because of increased lipolysis or an influx of free fatty acids into the tissue, it will automatically reduce the synthesis of new fatty acid. Acyl-CoA may also inhibit the mitochondrial **tricarboxylate transporter,** thus preventing activation of the enzyme by egress of citrate from the mitochondria into the cytosol.

Acetyl-CoA carboxylase is also regulated by hormones such as **glucagon, epinephrine,** and **insulin** via changes in its phosphorylation state (details in Figure 21–6).

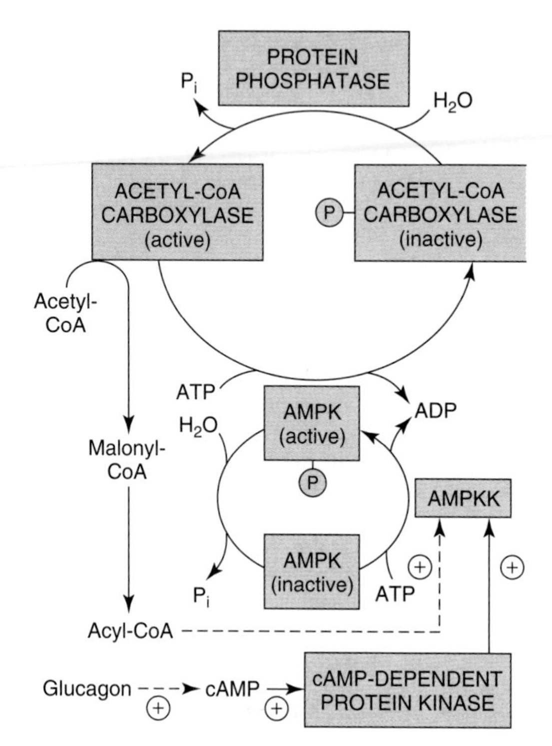


***Figure 21–6.*** Regulation of acetyl-CoA carboxylase by phosphorylation/dephosphorylation. The enzyme is inactivated by phosphorylation by AMP-activated protein kinase (AMPK), which in turn is phosphorylated and activated by AMP-activated protein kinase kinase (AMPKK). Glucagon (and epinephrine), after increasing cAMP, activate this latter enzyme via cAMP-dependent protein kinase. The kinase kinase enzyme is also believed to be activated by acyl-CoA. Insulin activates acetyl-CoA carboxylase, probably through an "activator" protein and an insulin-stimulated protein kinase.

### Pyruvate Dehydrogenase Is Also Regulated by Acyl-CoA

Acyl-CoA causes an inhibition of pyruvate dehydrogenase by inhibiting the ATP-ADP exchange transporter of the inner mitochondrial membrane, which leads to increased intramitochondrial [ATP]/[ADP] ratios and therefore to conversion of active to inactive pyruvate dehydrogenase (see Figure 17–6), thus regulating the availability of acetyl-CoA for lipogenesis. Furthermore, oxidation of acyl-CoA due to increased levels of free fatty acids may increase the ratios of [acetyl-CoA]/[CoA] and [NADH]/[$NAD^+$] in mitochondria, inhibiting pyruvate dehydrogenase.

### Insulin Also Regulates Lipogenesis by Other Mechanisms

**Insulin** stimulates lipogenesis by several other mechanisms as well as by increasing acetyl-CoA carboxylase activity. It increases the transport of glucose into the cell (eg, in adipose tissue), increasing the availability of both pyruvate for fatty acid synthesis and glycerol 3-phosphate for esterification of the newly formed fatty acids, and also converts the inactive form of pyruvate dehydrogenase to the active form in adipose tissue but not in liver. Insulin also—by its ability to depress the level of intracellular cAMP—**inhibits lipolysis** in adipose tissue and thereby reduces the concentration of plasma free fatty acids and therefore long-chain acyl-CoA, an inhibitor of lipogenesis.

### The Fatty Acid Synthase Complex & Acetyl-CoA Carboxylase Are Adaptive Enzymes

These enzymes adapt to the body's physiologic needs by increasing in total amount in the fed state and by decreasing in starvation, feeding of fat, and in diabetes. **Insulin** is an important hormone causing gene expression and induction of enzyme biosynthesis, and glucagon (via cAMP) antagonizes this effect. Feeding fats containing polyunsaturated fatty acids coordinately regulates the inhibition of expression of key enzymes of glycolysis and lipogenesis. These mechanisms for longer-term regulation of lipogenesis take several days to become fully manifested and augment the direct and immediate effect of free fatty acids and hormones such as insulin and glucagon.

## SUMMARY

- The synthesis of long-chain fatty acids (lipogenesis) is carried out by two enzyme systems: acetyl-CoA carboxylase and fatty acid synthase.
- The pathway converts acetyl-CoA to palmitate and requires NADPH, ATP, $Mn^{2+}$, biotin, pantothenic acid, and $HCO_3^-$ as cofactors.
- Acetyl-CoA carboxylase is required to convert acetyl-CoA to malonyl-CoA. In turn, fatty acid synthase, a multienzyme complex of one polypeptide chain with seven separate enzymatic activities, catalyzes the assembly of palmitate from one acetyl-CoA and seven malonyl-CoA molecules.
- Lipogenesis is regulated at the acetyl-CoA carboxylase step by allosteric modifiers, phosphorylation/dephosphorylation, and induction and repression of enzyme synthesis. Citrate activates the enzyme, and long-chain acyl-CoA inhibits its activity. Insulin activates acetyl-CoA carboxylase whereas glucagon and epinephrine have opposite actions.

## REFERENCES

Hudgins LC et al: Human fatty acid synthesis is stimulated by a eucaloric low fat, high carbohydrate diet. J Clin Invest 1996;97:2081.

Jump DB et al: Coordinate regulation of glycolytic and lipogenic gene expression by polyunsaturated fatty acids. J Lipid Res 1994;35:1076.

Kim KH: Regulation of mammalian acetyl-coenzyme A carboxylase. Annu Rev Nutr 1997;17:77.

Salati LM, Goodridge AG: Fatty acid synthesis in eukaryotes. In: *Biochemistry of Lipids, Lipoproteins and Membranes.* Vance DE, Vance JE (editors). Elsevier, 1996.

Wakil SJ: Fatty acid synthase, a proficient multifunctional enzyme. Biochemistry 1989;28:4523.

# Oxidation of Fatty Acids: Ketogenesis

22

*Peter A. Mayes, PhD, DSc, & Kathleen M. Botham, PhD, DSc*

## BIOMEDICAL IMPORTANCE

Although fatty acids are both oxidized to acetyl-CoA and synthesized from acetyl-CoA, fatty acid oxidation is not the simple reverse of fatty acid biosynthesis but an entirely different process taking place in a separate compartment of the cell. The separation of fatty acid oxidation in mitochondria from biosynthesis in the cytosol allows each process to be individually controlled and integrated with tissue requirements. Each step in fatty acid oxidation involves acyl-CoA derivatives catalyzed by separate enzymes, utilizes $NAD^+$ and FAD as coenzymes, and generates ATP. It is an aerobic process, requiring the presence of oxygen.

Increased fatty acid oxidation is a characteristic of starvation and of diabetes mellitus, leading to **ketone body** production by the liver **(ketosis).** Ketone bodies are acidic and when produced in excess over long periods, as in diabetes, cause **ketoacidosis,** which is ultimately fatal. Because gluconeogenesis is dependent upon fatty acid oxidation, any impairment in fatty acid oxidation leads to **hypoglycemia.** This occurs in various states of **carnitine deficiency** or deficiency of essential enzymes in fatty acid oxidation, eg, **carnitine palmitoyltransferase,** or inhibition of fatty acid oxidation by poisons, eg, **hypoglycin.**

## OXIDATION OF FATTY ACIDS OCCURS IN MITOCHONDRIA

### Fatty Acids Are Transported in the Blood as Free Fatty Acids (FFA)

Free fatty acids—also called unesterified (UFA) or nonesterified (NEFA) fatty acids—are fatty acids that are in the **unesterified state.** In plasma, longer-chain FFA are combined with **albumin,** and in the cell they are attached to a **fatty acid-binding protein,** so that in fact they are never really "free." Shorter-chain fatty acids are more water-soluble and exist as the un-ionized acid or as a fatty acid anion.

### Fatty Acids Are Activated Before Being Catabolized

Fatty acids must first be converted to an active intermediate before they can be catabolized. This is the only step in the complete degradation of a fatty acid that requires energy from ATP. In the presence of ATP and coenzyme A, the enzyme **acyl-CoA synthetase (thiokinase)** catalyzes the conversion of a fatty acid (or free fatty acid) to an "active fatty acid" or acyl-CoA, which uses one high-energy phosphate with the formation of AMP and $PP_i$ (Figure 22–1). The $PP_i$ is hydrolyzed by **inorganic pyrophosphatase** with the loss of a further high-energy phosphate, ensuring that the overall reaction goes to completion. Acyl-CoA synthetases are found in the endoplasmic reticulum, peroxisomes, and inside and on the outer membrane of mitochondria.

### Long-Chain Fatty Acids Penetrate the Inner Mitochondrial Membrane as Carnitine Derivatives

Carnitine (β-hydroxy-γ-trimethylammonium butyrate), $(CH_3)_3N^+—CH_2—CH(OH)—CH_2—COO^-$, is widely distributed and is particularly abundant in muscle. Long-chain acyl-CoA (or FFA) will not penetrate the inner membrane of mitochondria. However, **carnitine palmitoyltransferase-I,** present in the outer mitochondrial membrane, converts long-chain acyl-CoA to acylcarnitine, which is able to penetrate the inner membrane and gain access to the β-oxidation system of enzymes (Figure 22–1). **Carnitine-acylcarnitine translocase** acts as an inner membrane exchange transporter. Acylcarnitine is transported in, coupled with the transport out of one molecule of carnitine. The acylcarnitine then reacts with CoA, cat-

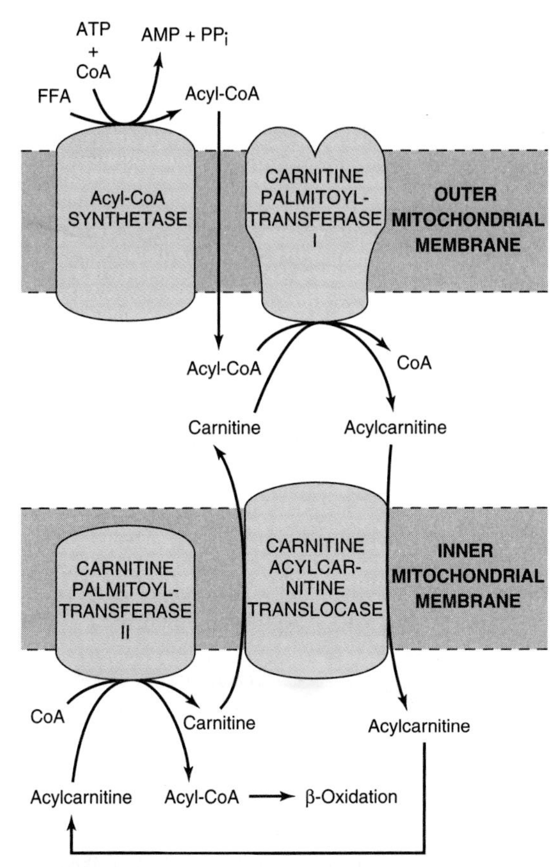


***Figure 22–1.*** Role of carnitine in the transport of long-chain fatty acids through the inner mitochondrial membrane. Long-chain acyl-CoA cannot pass through the inner mitochondrial membrane, but its metabolic product, acylcarnitine, can.

alyzed by **carnitine palmitoyltransferase-II,** located on the inside of the inner membrane. Acyl-CoA is re-formed in the mitochondrial matrix, and carnitine is liberated.

## β-OXIDATION OF FATTY ACIDS INVOLVES SUCCESSIVE CLEAVAGE WITH RELEASE OF ACETYL-CoA

In β-oxidation (Figure 22–2), two carbons at a time are cleaved from acyl-CoA molecules, starting at the carboxyl end. The chain is broken between the α(2)- and β(3)-carbon atoms—hence the name β-oxidation. The two-carbon units formed are acetyl-CoA; thus, palmitoyl-CoA forms eight acetyl-CoA molecules.

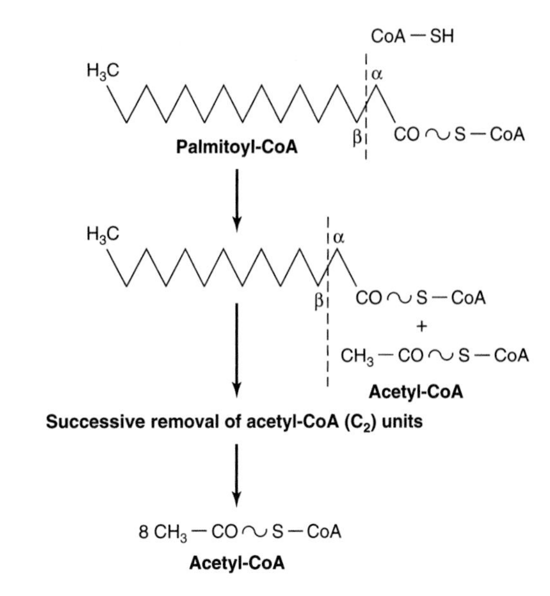


***Figure 22–2.*** Overview of β-oxidation of fatty acids.

### The Cyclic Reaction Sequence Generates $FADH_2$ & NADH

Several enzymes, known collectively as "fatty acid oxidase," are found in the mitochondrial matrix or inner membrane adjacent to the respiratory chain. These catalyze the oxidation of acyl-CoA to acetyl-CoA, the system being coupled with the phosphorylation of ADP to ATP (Figure 22–3).

The first step is the removal of two hydrogen atoms from the 2(α)- and 3(β)-carbon atoms, catalyzed by **acyl-CoA dehydrogenase** and requiring FAD. This results in the formation of $\Delta^2$-*trans*-enoyl-CoA and $FADH_2$. The reoxidation of $FADH_2$ by the respiratory chain requires the mediation of another flavoprotein, termed **electron-transferring flavoprotein** (Chapter 11). Water is added to saturate the double bond and form 3-hydroxyacyl-CoA, catalyzed by **$\Delta^2$-enoyl-CoA hydratase.** The 3-hydroxy derivative undergoes further dehydrogenation on the 3-carbon catalyzed by **L(+)-3-hydroxyacyl-CoA dehydrogenase** to form the corresponding 3-ketoacyl-CoA compound. In this case, $NAD^+$ is the coenzyme involved. Finally, 3-ketoacyl-CoA is split at the 2,3- position by **thiolase** (3-ketoacyl-CoA-thiolase), forming acetyl-CoA and a new acyl-CoA two carbons shorter than the original acyl-CoA molecule. The acyl-CoA formed in the cleavage reaction reenters the oxidative pathway at reaction 2 (Figure 22–3). In this way, a long-chain fatty acid may be degraded completely to acetyl-CoA ($C_2$ units). Since acetyl-CoA can be oxidized to $CO_2$ and water via the

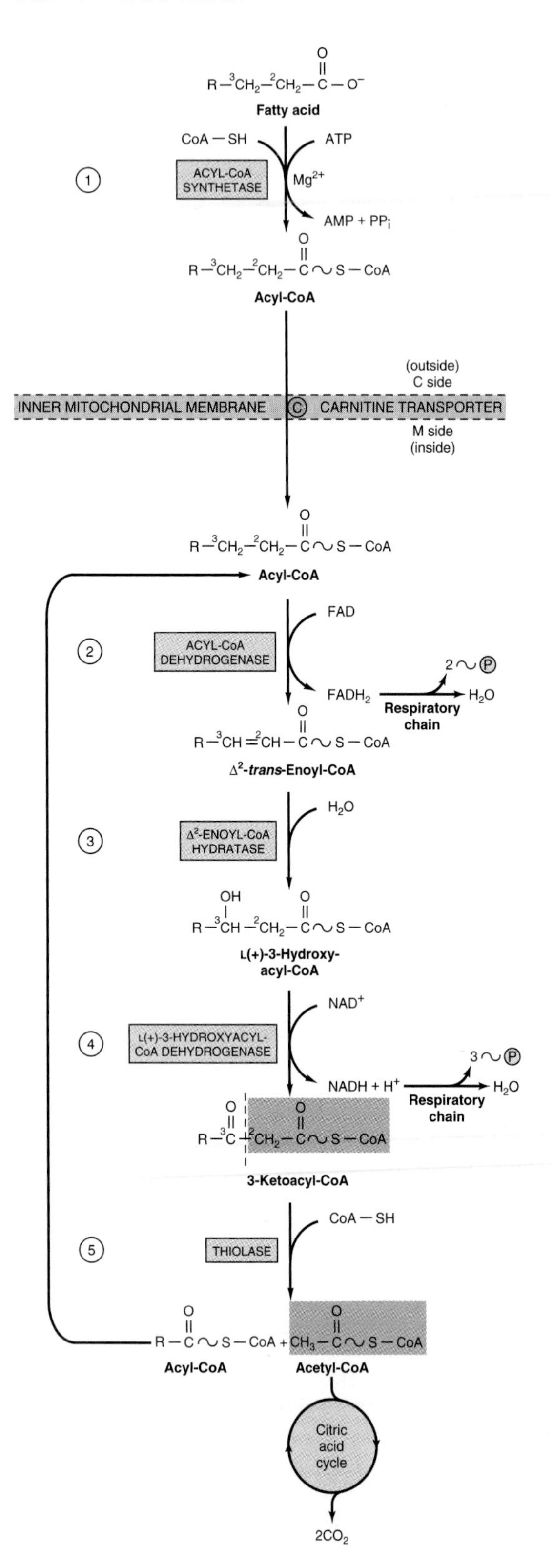


***Figure 22–3.*** β-Oxidation of fatty acids. Long-chain acyl-CoA is cycled through reactions 2–5, acetyl-CoA being split off, each cycle, by thiolase (reaction 5). When the acyl radical is only four carbon atoms in length, two acetyl-CoA molecules are formed in reaction 5.

citric acid cycle (which is also found within the mitochondria), the complete oxidation of fatty acids is achieved.

## Oxidation of a Fatty Acid With an Odd Number of Carbon Atoms Yields Acetyl-CoA Plus a Molecule of Propionyl-CoA

Fatty acids with an odd number of carbon atoms are oxidized by the pathway of β-oxidation, producing acetyl-CoA, until a three-carbon (propionyl-CoA) residue remains. This compound is converted to succinyl-CoA, a constituent of the citric acid cycle (Figure 19–2). Hence, **the propionyl residue from an odd-chain fatty acid is the only part of a fatty acid that is glucogenic.**

## Oxidation of Fatty Acids Produces a Large Quantity of ATP

Transport in the respiratory chain of electrons from $FADH_2$ and NADH will lead to the synthesis of five high-energy phosphates (Chapter 12) for each of the first seven acetyl-CoA molecules formed by β-oxidation of palmitate ($7 \times 5 = 35$). A total of 8 mol of acetyl-CoA is formed, and each will give rise to 12 mol of ATP on oxidation in the citric acid cycle, making $8 \times 12 = 96$ mol. Two must be subtracted for the initial activation of the fatty acid, yielding a net gain of 129 mol of ATP per mole of palmitate, or $129 \times 51.6^* = 6656$ kJ. This represents 68% of the free energy of combustion of palmitic acid.

## Peroxisomes Oxidize Very Long Chain Fatty Acids

A modified form of β-oxidation is found in peroxisomes and leads to the formation of acetyl-CoA and $H_2O_2$ (from the flavoprotein-linked dehydrogenase step), which is broken down by catalase. Thus, this dehydrogenation in peroxisomes is not linked directly to phosphorylation and the generation of ATP. The system facilitates the oxidation of very long chain fatty acids (eg, $C_{20}$, $C_{22}$). These enzymes are induced by

* ΔG for the ATP reaction, as explained in Chapter 17.

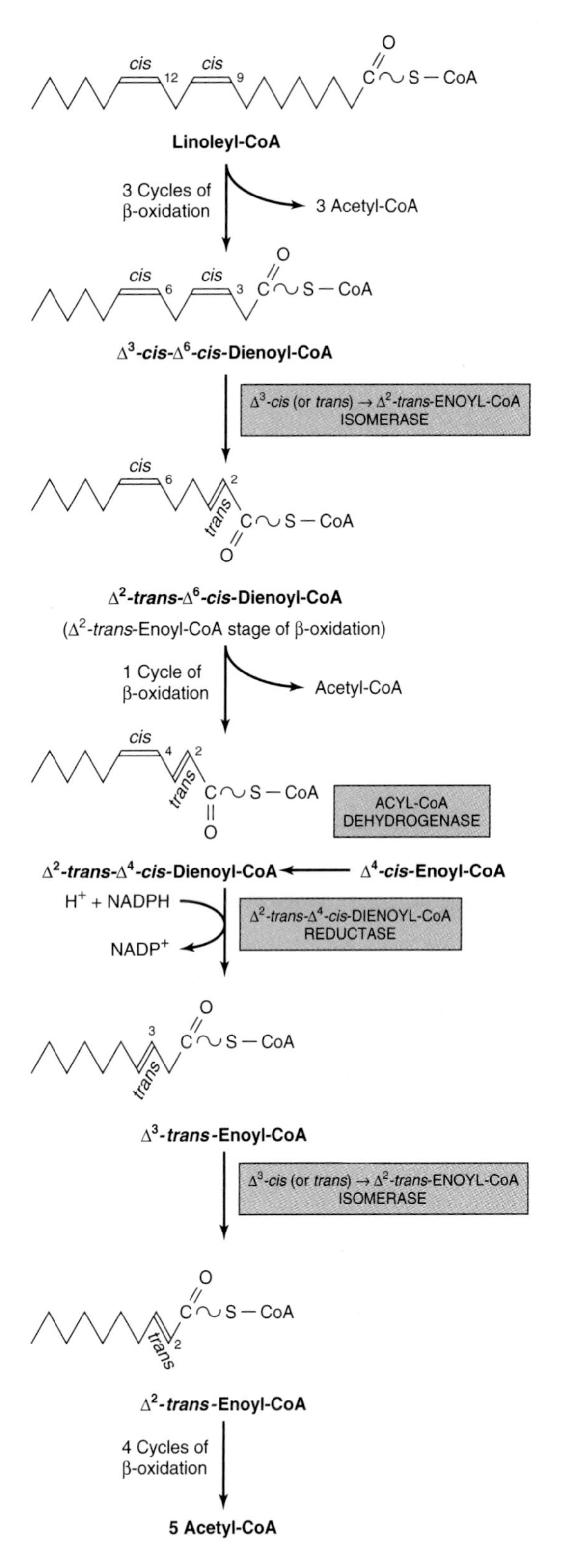


***Figure 22–4.*** Sequence of reactions in the oxidation of unsaturated fatty acids, eg, linoleic acid. $\Delta^4$-*cis*-fatty acids or fatty acids forming $\Delta^4$-*cis*-enoyl-CoA enter the pathway at the position shown. NADPH for the dienoyl-CoA reductase step is supplied by intramitochondrial sources such as glutamate dehydrogenase, isocitrate dehydrogenase, and NAD(P)H transhydrogenase.

high-fat diets and in some species by hypolipidemic drugs such as clofibrate.

The enzymes in peroxisomes do not attack shorter-chain fatty acids; the β-oxidation sequence ends at octanoyl-CoA. Octanoyl and acetyl groups are both further oxidized in mitochondria. Another role of peroxisomal β-oxidation is to shorten the side chain of cholesterol in bile acid formation (Chapter 26). Peroxisomes also take part in the synthesis of ether glycerolipids (Chapter 24), cholesterol, and dolichol (Figure 26–2).

## OXIDATION OF UNSATURATED FATTY ACIDS OCCURS BY A MODIFIED β-OXIDATION PATHWAY

The CoA esters of these acids are degraded by the enzymes normally responsible for β-oxidation until either a $\Delta^3$-*cis*-acyl-CoA compound or a $\Delta^4$-*cis*-acyl-CoA compound is formed, depending upon the position of the double bonds (Figure 22–4). The former compound is isomerized **($\Delta^3$*cis* → $\Delta^2$-*trans*-enoyl-CoA isomerase)** to the corresponding $\Delta^2$-*trans*-CoA stage of β-oxidation for subsequent hydration and oxidation. Any $\Delta^4$-*cis*-acyl-CoA either remaining, as in the case of linoleic acid, or entering the pathway at this point after conversion by acyl-CoA dehydrogenase to $\Delta^2$-*trans*-$\Delta^4$-*cis*-dienoyl-CoA, is then metabolized as indicated in Figure 22–4.

## KETOGENESIS OCCURS WHEN THERE IS A HIGH RATE OF FATTY ACID OXIDATION IN THE LIVER

Under metabolic conditions associated with a high rate of fatty acid oxidation, the liver produces considerable quantities of **acetoacetate** and **D(−)-3-hydroxybutyrate** (β-hydroxybutyrate). Acetoacetate continually undergoes spontaneous decarboxylation to yield **acetone.** These three substances are collectively known as the **ketone bodies** (also called acetone bodies or [incorrectly*] "ketones") (Figure 22–5). Acetoacetate and 3-hydroxybu-

* The term "ketones" should not be used because 3-hydroxybutyrate is not a ketone and there are ketones in blood that are not ketone bodies, eg, pyruvate, fructose.

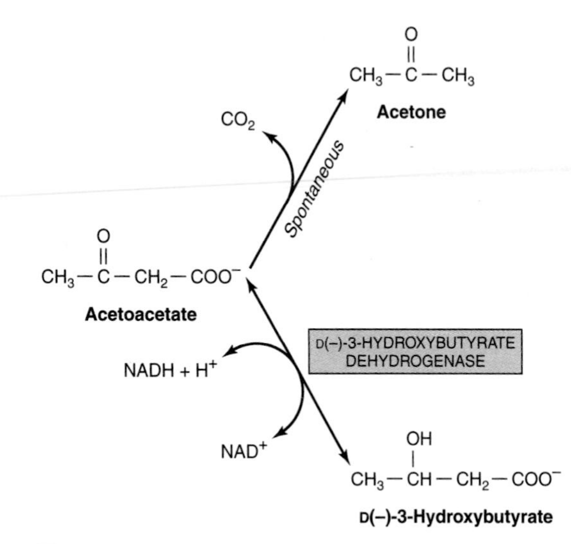


***Figure 22–5.*** Interrelationships of the ketone bodies. D(−)-3-hydroxybutyrate dehydrogenase is a mitochondrial enzyme.

tyrate are interconverted by the mitochondrial enzyme **D(−)-3-hydroxybutyrate dehydrogenase;** the equilibrium is controlled by the mitochondrial [$NAD^+$]/[NADH] ratio, ie, the **redox state.** The concentration of total ketone bodies in the blood of well-fed mammals does not normally exceed 0.2 mmol/L except in ruminants, where 3-hydroxybutyrate is formed continuously from butyric acid (a product of ruminal fermentation) in the rumen wall. In vivo, the liver appears to be the only organ in nonruminants to add significant quantities of ketone bodies to the blood. Extrahepatic tissues utilize them as respiratory substrates. The net flow of ketone bodies from the liver to the extrahepatic tissues results from active hepatic synthesis coupled with very low utilization. The reverse situation occurs in extrahepatic tissues (Figure 22–6).

## 3-Hydroxy-3-Methylglutaryl-CoA (HMG-CoA) Is an Intermediate in the Pathway of Ketogenesis

Enzymes responsible for ketone body formation are associated mainly with the mitochondria. Two acetyl-CoA molecules formed in β-oxidation condense with one another to form acetoacetyl-CoA by a reversal of the **thiolase** reaction. Acetoacetyl-CoA, which is the

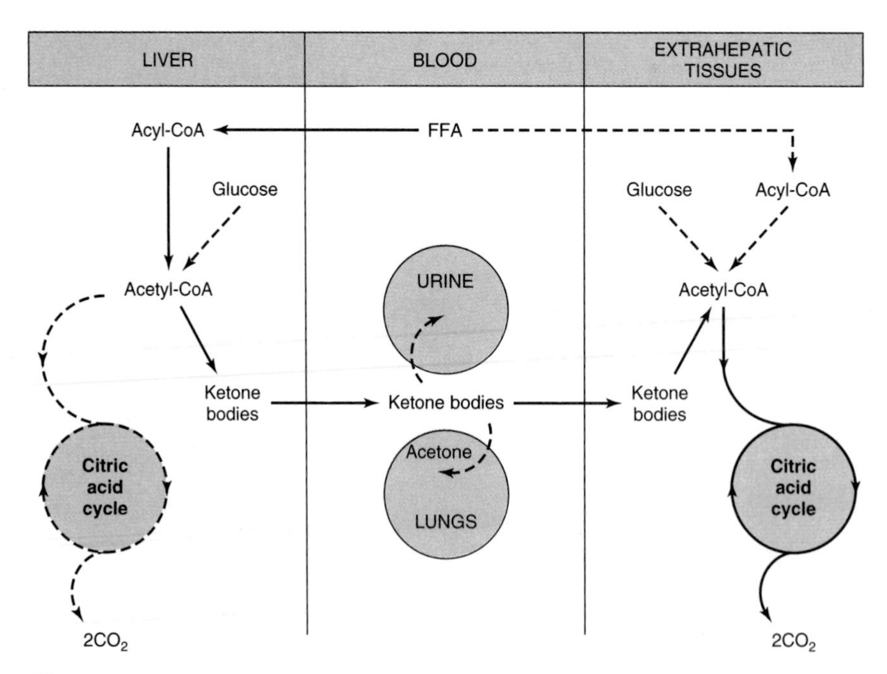


***Figure 22–6.*** Formation, utilization, and excretion of ketone bodies. (The main pathway is indicated by the solid arrows.)

starting material for ketogenesis, also arises directly from the terminal four carbons of a fatty acid during β-oxidation (Figure 22–7). Condensation of acetoacetyl-CoA with another molecule of acetyl-CoA by **3-hydroxy-3-methylglutaryl-CoA synthase** forms **HMG-CoA. 3-Hydroxy-3-methylglutaryl-CoA lyase** then causes acetyl-CoA to split off from the HMG-CoA, leaving free acetoacetate. The carbon atoms split off in the acetyl-CoA molecule are derived from the original acetoacetyl-CoA molecule. **Both enzymes must be present in mitochondria for ketogenesis to take place.** This occurs solely in liver and rumen epithelium. D(−)-3-Hydroxybutyrate is quantitatively the predominant ketone body present in the blood and urine in ketosis.

## Ketone Bodies Serve as a Fuel for Extrahepatic Tissues

While an active enzymatic mechanism produces acetoacetate from acetoacetyl-CoA in the liver, acetoacetate once formed cannot be reactivated directly except in the cytosol, where it is used in a much less active pathway as a precursor in cholesterol synthesis. This accounts for the net production of ketone bodies by the liver.

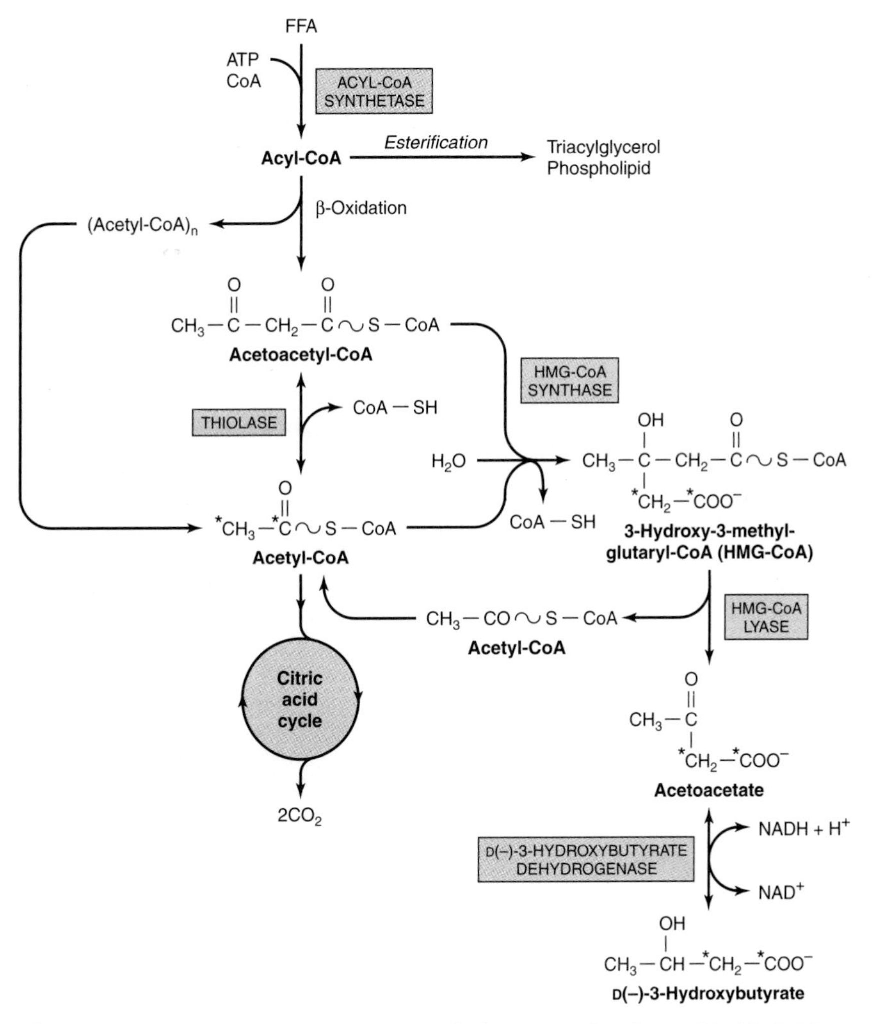


***Figure 22–7.*** Pathways of ketogenesis in the liver. (FFA, free fatty acids; HMG, 3-hydroxy-3-methylglutaryl.)

In extrahepatic tissues, acetoacetate is activated to acetoacetyl-CoA by **succinyl-CoA-acetoacetate CoA transferase.** CoA is transferred from succinyl-CoA to form acetoacetyl-CoA (Figure 22–8). The acetoacetyl-CoA is split to acetyl-CoA by thiolase and oxidized in the citric acid cycle. If the blood level is raised, oxidation of ketone bodies increases until, at a concentration of approximately 12 mmol/L, they saturate the oxidative machinery. When this occurs, a large proportion of the oxygen consumption may be accounted for by the oxidation of ketone bodies.

In most cases, **ketonemia is due to increased production of ketone bodies** by the liver rather than to a deficiency in their utilization by extrahepatic tissues. While acetoacetate and D(−)-3-hydroxybutyrate are readily oxidized by extrahepatic tissues, acetone is difficult to oxidize in vivo and to a large extent is volatilized in the lungs.

In moderate ketonemia, the loss of ketone bodies via the urine is only a few percent of the total ketone body production and utilization. Since there are renal threshold-like effects (there is not a true threshold) that vary between species and individuals, measurement of the ketonemia, not the ketonuria, is the preferred method of assessing the severity of ketosis.

## KETOGENESIS IS REGULATED AT THREE CRUCIAL STEPS

**(1)** Ketosis does not occur in vivo unless there is an increase in the level of circulating free fatty acids that arise from lipolysis of triacylglycerol in adipose tissue. **Free fatty acids are the precursors of ketone bodies in the liver.** The liver, both in fed and in fasting conditions, extracts about 30% of the free fatty acids passing through it, so that at high concentrations the flux passing into the liver is substantial. **Therefore, the factors regulating mobilization of free fatty acids from adipose tissue are important in controlling ketogenesis** (Figures 22–9 and 25–8).

**(2)** After uptake by the liver, free fatty acids are either **β-oxidized** to $CO_2$ or ketone bodies or **esterified** to triacylglycerol and phospholipid. There is regulation of entry of fatty acids into the oxidative pathway by **carnitine palmitoyltransferase-I** (CPT-I), and the remainder of the fatty acid uptake is esterified. CPT-I activity is

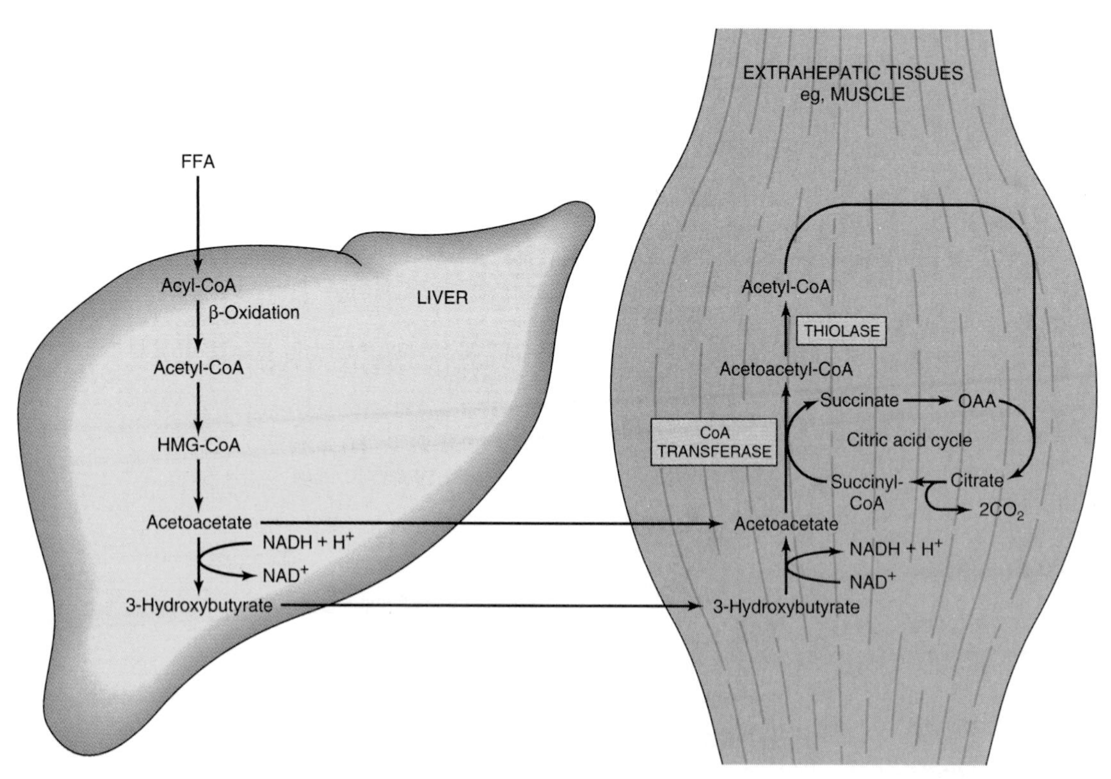


***Figure 22–8.*** Transport of ketone bodies from the liver and pathways of utilization and oxidation in extrahepatic tissues.

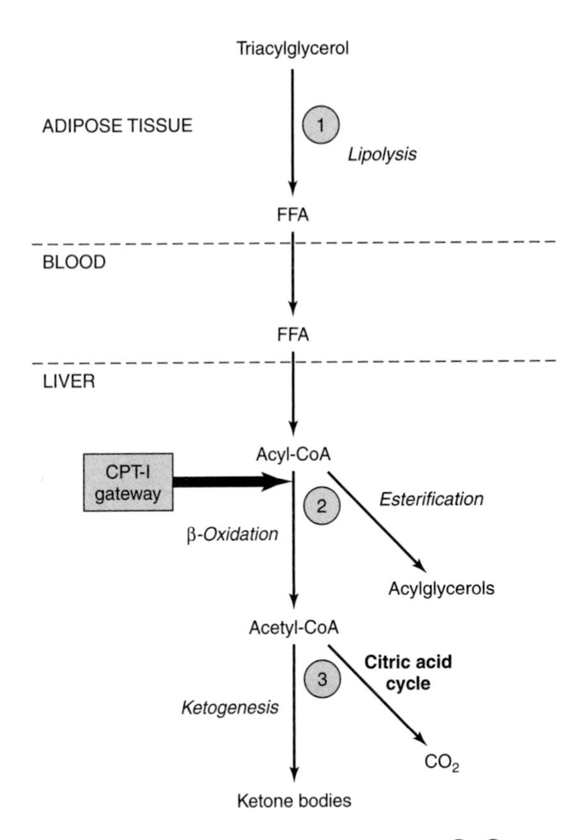


***Figure 22–9.*** Regulation of ketogenesis. ①–③ show three crucial steps in the pathway of metabolism of free fatty acids (FFA) that determine the magnitude of ketogenesis. (CPT-I, carnitine palmitoyltransferase-I.)

low in the fed state, leading to depression of fatty acid oxidation, and high in starvation, allowing fatty acid oxidation to increase. Malonyl-CoA, the initial intermediate in fatty acid biosynthesis (Figure 21–1), formed by acetyl-CoA carboxylase in the fed state, is a potent inhibitor of CPT-I (Figure 22–10). Under these conditions, free fatty acids enter the liver cell in low concentrations and are nearly all esterified to acylglycerols and transported out of the liver in very low density lipoproteins (VLDL). However, as the concentration of free fatty acids increases with the onset of starvation, acetyl-CoA carboxylase is inhibited directly by acyl-CoA, and [malonyl-CoA] decreases, releasing the inhibition of CPT-I and allowing more acyl-CoA to be β-oxidized. These events are reinforced in starvation by decrease in the **[insulin]/[glucagon] ratio.** Thus, β-oxidation from free fatty acids is controlled by the CPT-I gateway into the mitochondria, and the balance of the free fatty acid uptake not oxidized is esterified.

(**3**) In turn, the acetyl-CoA formed in β-oxidation is oxidized in the citric acid cycle, or it enters the pathway of ketogenesis to form ketone bodies. As the level of serum free fatty acids is raised, proportionately more free fatty acid is converted to ketone bodies and less is oxidized via the citric acid cycle to $CO_2$. The partition of acetyl-CoA between the ketogenic pathway and the pathway of oxidation to $CO_2$ is so regulated that the total free energy captured in ATP which results from the oxidation of free fatty acids remains constant. This may be appreciated when it is realized that complete oxidation of 1 mol of palmitate involves a net production of 129 mol of ATP via β-oxidation and $CO_2$ production in the citric acid cycle (see above), whereas only 33 mol of ATP are produced when acetoacetate is the end product and only 21 mol when 3-hydroxybutyrate is the end product. Thus, ketogenesis may be regarded as a mechanism that allows the liver to oxidize increasing quantities of fatty acids within the constraints of a tightly coupled system of oxidative phosphorylation—without increasing its total energy expenditure.

Theoretically, a fall in concentration of oxaloacetate, particularly within the mitochondria, could impair the ability of the citric acid cycle to metabolize acetyl-CoA and divert fatty acid oxidation toward ketogenesis. Such a fall may occur because of an increase in the $[NADH]/[NAD^+]$ ratio caused by increased β-oxidation affecting the equilibrium between oxaloacetate and malate and decreasing the concentration of oxaloacetate. However, pyruvate carboxylase, which catalyzes the conversion of pyruvate to oxaloacetate, is activated by acetyl-CoA. Consequently, when there are significant amounts of acetyl-CoA, there should be sufficient oxaloacetate to initiate the condensing reaction of the citric acid cycle.

## CLINICAL ASPECTS

### Impaired Oxidation of Fatty Acids Gives Rise to Diseases Often Associated With Hypoglycemia

**Carnitine deficiency** can occur particularly in the newborn—and especially in preterm infants—owing to inadequate biosynthesis or renal leakage. Losses can also occur in hemodialysis. This suggests a vitamin-like dietary requirement for carnitine in some individuals. Symptoms of deficiency include hypoglycemia, which is a consequence of impaired fatty acid oxidation and lipid accumulation with muscular weakness. Treatment is by oral supplementation with carnitine.

Inherited **CPT-I deficiency** affects only the liver, resulting in reduced fatty acid oxidation and ketogenesis, with hypoglycemia. **CPT-II deficiency** affects pri-

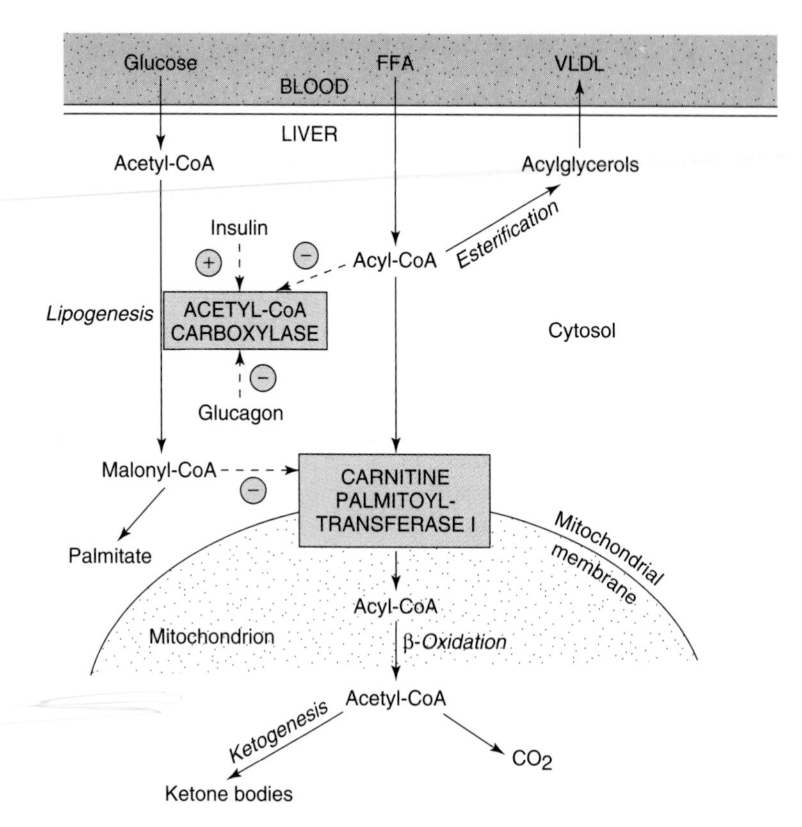


***Figure 22–10.*** Regulation of long-chain fatty acid oxidation in the liver. (FFA, free fatty acids; VLDL, very low density lipoprotein.) Positive (⊕) and negative (⊖) regulatory effects are represented by broken arrows and substrate flow by solid arrows.

marily skeletal muscle and, when severe, the liver. The sulfonylurea drugs **(glyburide [glibenclamide]** and **tolbutamide),** used in the treatment of type 2 diabetes mellitus, reduce fatty acid oxidation and, therefore, hyperglycemia by inhibiting CPT-I.

Inherited defects in the enzymes of β-oxidation and ketogenesis also lead to nonketotic hypoglycemia, coma, and fatty liver. Defects are known in long- and short-chain 3-hydroxyacyl-CoA dehydrogenase (deficiency of the long-chain enzyme may be a cause of **acute fatty liver of pregnancy). 3-Ketoacyl-CoA thiolase** and **HMG-CoA lyase deficiency** also affect the degradation of leucine, a ketogenic amino acid (Chapter 30).

**Jamaican vomiting sickness** is caused by eating the unripe fruit of the akee tree, which contains a toxin, **hypoglycin,** that inactivates medium- and short-chain acyl-CoA dehydrogenase, inhibiting β-oxidation and causing hypoglycemia. **Dicarboxylic aciduria** is characterized by the excretion of $C_6$–$C_{10}$ ω-dicarboxylic acids and by nonketotic hypoglycemia. It is caused by a lack of mitochondrial **medium-chain acyl-CoA dehydrogenase. Refsum's disease** is a rare neurologic disorder due to a defect that causes the accumulation of phytanic acid, which is found in plant foodstuffs and blocks β-oxidation. **Zellweger's (cerebrohepatorenal) syndrome** occurs in individuals with a rare inherited absence of peroxisomes in all tissues. They accumulate $C_{26}$–$C_{38}$ polyenoic acids in brain tissue and also exhibit a generalized loss of peroxisomal functions, eg, impaired bile acid and ether lipid synthesis.

## Ketoacidosis Results From Prolonged Ketosis

Higher than normal quantities of ketone bodies present in the blood or urine constitute **ketonemia** (hyperketonemia) or **ketonuria,** respectively. The overall condition is called **ketosis.** Acetoacetic and 3-hydroxybutyric acids are both moderately strong acids and are buffered when present in blood or other tissues. However, their continual excretion in quantity progressively depletes the alkali reserve, causing **ketoacidosis.** This may be fatal in uncontrolled **diabetes mellitus.**

The basic form of ketosis occurs in **starvation** and involves depletion of available carbohydrate coupled with mobilization of free fatty acids. This general pattern of metabolism is exaggerated to produce the pathologic states found in **diabetes mellitus, twin lamb disease,** and **ketosis in lactating cattle.** Nonpathologic forms of ketosis are found under conditions of high-fat feeding and after severe exercise in the postabsorptive state.

## SUMMARY

- Fatty acid oxidation in mitochondria leads to the generation of large quantities of ATP by a process called β-oxidation that cleaves acetyl-CoA units sequentially from fatty acyl chains. The acetyl-CoA is oxidized in the citric acid cycle, generating further ATP.
- The ketone bodies (acetoacetate, 3-hydroxybutyrate, and acetone) are formed in hepatic mitochondria when there is a high rate of fatty acid oxidation. The pathway of ketogenesis involves synthesis and breakdown of 3-hydroxy-3-methylglutaryl-CoA (HMG-CoA) by two key enzymes, HMG-CoA synthase and HMG-CoA lyase.
- Ketone bodies are important fuels in extrahepatic tissues.
- Ketogenesis is regulated at three crucial steps: (1) control of free fatty acid mobilization from adipose tissue; (2) the activity of carnitine palmitoyltransferase-I in liver, which determines the proportion of the fatty acid flux that is oxidized rather than esterified; and (3) partition of acetyl-CoA between the pathway of ketogenesis and the citric acid cycle.
- Diseases associated with impairment of fatty acid oxidation lead to hypoglycemia, fatty infiltration of organs, and hypoketonemia.
- Ketosis is mild in starvation but severe in diabetes mellitus and ruminant ketosis.

## REFERENCES

Eaton S, Bartlett K, Pourfarzam M: Mammalian mitochondrial β-oxidation. Biochem J 1996;320:345.

Mayes PA, Laker ME: Regulation of ketogenesis in the liver. Biochem Soc Trans 1981;9:339.

McGarry JD, Foster DW: Regulation of hepatic fatty acid oxidation and ketone body production. Annu Rev Biochem 1980;49:395.

Osmundsen H, Hovik R: β-Oxidation of polyunsaturated fatty acids. Biochem Soc Trans 1988;16:420.

Reddy JK, Mannaerts GP: Peroxisomal lipid metabolism. Annu Rev Nutr 1994;14:343.

Scriver CR et al (editors): *The Metabolic and Molecular Bases of Inherited Disease,* 8th ed. McGraw-Hill, 2001.

Treem WR et al: Acute fatty liver of pregnancy and long-chain 3-hydroxyacyl-coenzyme A dehydrogenase deficiency. Hepatology 1994;19:339.

Wood PA: Defects in mitochondrial beta-oxidation of fatty acids. Curr Opin Lipidol 1999;10:107.

# Metabolism of Unsaturated Fatty Acids & Eicosanoids

# 23

Peter A. Mayes, PhD, DSc, & Kathleen M. Botham, PhD, DSc

## BIOMEDICAL IMPORTANCE

Unsaturated fatty acids in phospholipids of the cell membrane are important in maintaining membrane fluidity. A high ratio of polyunsaturated fatty acids to saturated fatty acids (P:S ratio) in the diet is a major factor in lowering plasma cholesterol concentrations and is considered to be beneficial in preventing coronary heart disease. Animal tissues have limited capacity for desaturating fatty acids, and that process requires certain dietary polyunsaturated fatty acids derived from plants. These **essential fatty acids** are used to form eicosanoic ($C_{20}$) fatty acids, which in turn give rise to the prostaglandins and thromboxanes and to leukotrienes and lipoxins—known collectively as **eicosanoids.** The prostaglandins and thromboxanes are local hormones that are synthesized rapidly when required. Prostaglandins mediate **inflammation,** produce **pain,** and induce **sleep** as well as being involved in the regulation of **blood coagulation** and **reproduction.** Nonsteroidal anti-inflammatory drugs such as **aspirin** act by inhibiting prostaglandin synthesis. Leukotrienes have muscle contractant and chemotactic properties and are important in allergic reactions and inflammation.

## SOME POLYUNSATURATED FATTY ACIDS CANNOT BE SYNTHESIZED BY MAMMALS & ARE NUTRITIONALLY ESSENTIAL

Certain long-chain unsaturated fatty acids of metabolic significance in mammals are shown in Figure 23–1. Other $C_{20}$, $C_{22}$, and $C_{24}$ polyenoic fatty acids may be derived from oleic, linoleic, and α-linolenic acids by chain elongation. Palmitoleic and oleic acids are not essential in the diet because the tissues can introduce a double bond at the $\Delta^9$ position of a saturated fatty acid. **Linoleic and α-linolenic acids** are the only fatty acids known to be essential for the complete nutrition of many species of animals, including humans, and are known as the **nutritionally essential fatty acids.** In most mammals, **arachidonic acid** can be formed from linoleic acid (Figure 23–4). Double bonds can be introduced at the $\Delta^4$, $\Delta^5$, $\Delta^6$, and $\Delta^9$ positions (see Chapter 14) in most animals, but never beyond the $\Delta^9$ position. In contrast, plants are able to synthesize the nutritionally essential fatty acids by introducing double bonds at the $\Delta^{12}$ and $\Delta^{15}$ positions.

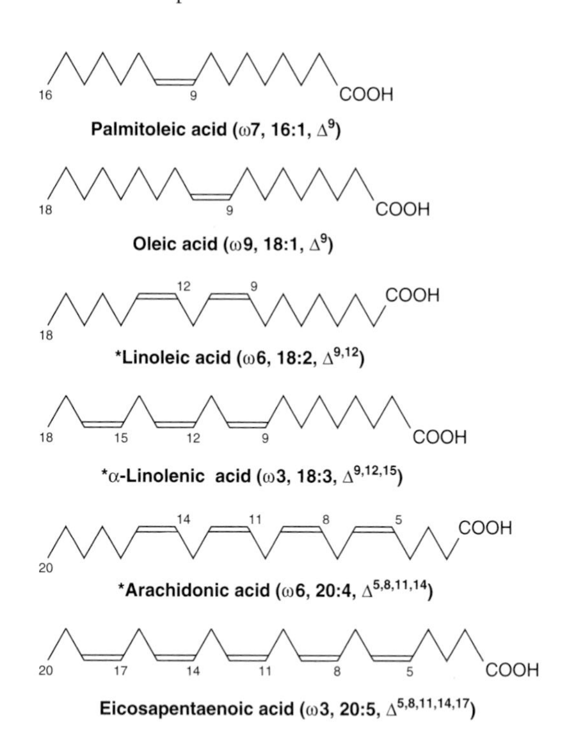


***Figure 23–1.*** Structure of some unsaturated fatty acids. Although the carbon atoms in the molecules are conventionally numbered—ie, numbered from the carboxyl terminal—the ω numbers (eg, ω7 in palmitoleic acid) are calculated from the reverse end (the methyl terminal) of the molecules. The information in parentheses shows, for instance, that α-linolenic acid contains double bonds starting at the third carbon from the methyl terminal, has 18 carbons and 3 double bonds, and has these double bonds at the 9th, 12th, and 15th carbons from the carboxyl terminal. (Asterisks: Classified as "essential fatty acids.")

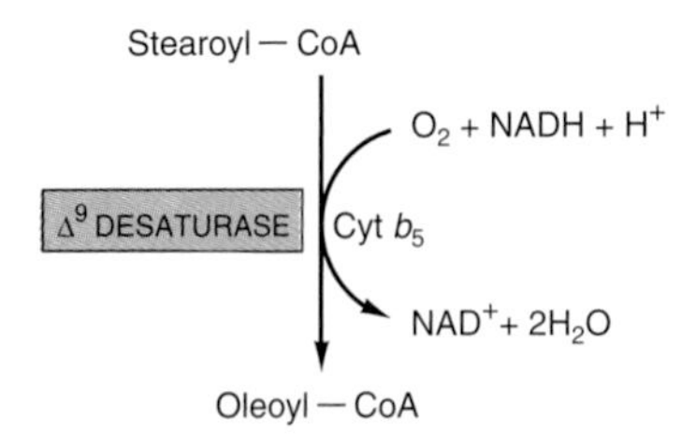


***Figure 23–2.*** Microsomal $\Delta^9$ desaturase.

## MONOUNSATURATED FATTY ACIDS ARE SYNTHESIZED BY A $\Delta^9$ DESATURASE SYSTEM

Several tissues including the liver are considered to be responsible for the formation of nonessential monounsaturated fatty acids from saturated fatty acids. The first double bond introduced into a saturated fatty acid is nearly always in the $\Delta^9$ position. An enzyme system—**$\Delta^9$ desaturase** (Figure 23–2)—in the endoplasmic reticulum will catalyze the conversion of palmitoyl-CoA or stearoyl-CoA to palmitoleoyl-CoA or oleoyl-CoA, respectively. Oxygen and either NADH or NADPH are necessary for the reaction. The enzymes appear to be similar to a monooxygenase system involving cytochrome $b_5$ (Chapter 11).

## SYNTHESIS OF POLYUNSATURATED FATTY ACIDS INVOLVES DESATURASE & ELONGASE ENZYME SYSTEMS

Additional double bonds introduced into existing monounsaturated fatty acids are always separated from each other by a methylene group (methylene interrupted) except in bacteria. Since animals have a $\Delta^9$ desaturase, they are able to synthesize the ω9 (oleic acid) family of unsaturated fatty acids completely by a combination of chain elongation and desaturation (Figure 23–3). However, as indicated above, linoleic (ω6) or α-linolenic (ω3) acids required for the synthesis of the other members of the ω6 or ω3 families must be supplied in the diet. Linoleate may be converted to arachidonate via **γ-linolenate** by the pathway shown in Figure 23–4. The nutritional requirement for arachidonate may thus be dispensed with if there is adequate linoleate in the diet. The desaturation and chain elongation system is greatly diminished in the starving state, in response to glucagon and epinephrine administration, and in the absence of insulin as in type 1 diabetes mellitus.

## DEFICIENCY SYMPTOMS ARE PRODUCED WHEN THE ESSENTIAL FATTY ACIDS (EFA) ARE ABSENT FROM THE DIET

Rats fed a purified nonlipid diet containing vitamins A and D exhibit a reduced growth rate and reproductive deficiency which may be cured by the addition of **linoleic, α-linolenic,** and **arachidonic acids** to the diet. These fatty acids are found in high concentrations in vegetable oils (Table 14–2) and in small amounts in animal carcasses. These essential fatty acids are required for prostaglandin, thromboxane, leukotriene, and lipoxin formation (see below), and they also have various other functions which are less well defined. Essential fatty acids are found in the structural lipids of the cell, often in the 2 position of phospholipids, and are concerned with the structural integrity of the mitochondrial membrane.

Arachidonic acid is present in membranes and accounts for 5–15% of the fatty acids in phospholipids. Docosahexaenoic acid (DHA; ω3, 22:6), which is syn-

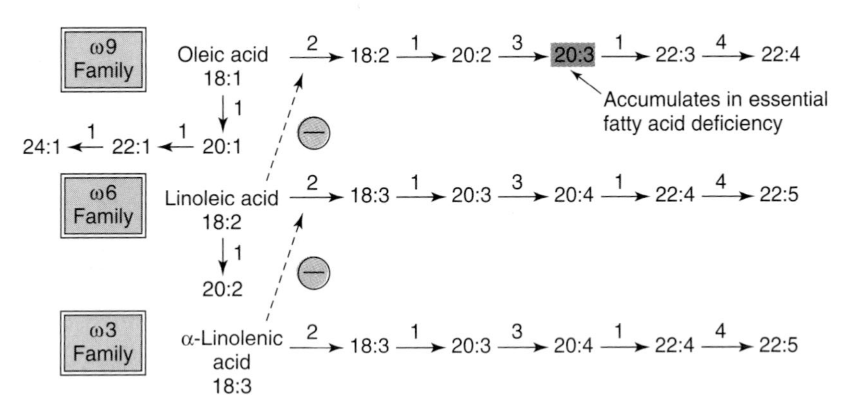


***Figure 23–3.*** Biosynthesis of the ω9, ω6, and ω3 families of polyunsaturated fatty acids. Each step is catalyzed by the microsomal chain elongation or desaturase system: 1, elongase; 2, $\Delta^6$ desaturase; 3, $\Delta^5$ desaturase; 4, $\Delta^4$ desaturase. (⊖, Inhibition.)

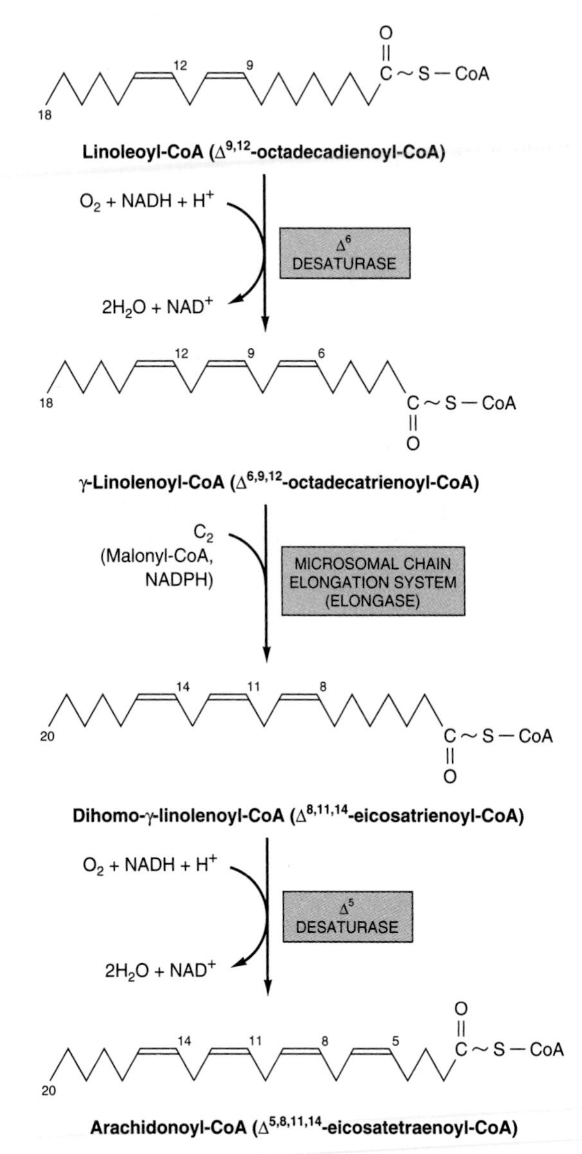


***Figure 23–4.*** Conversion of linoleate to arachidonate. Cats cannot carry out this conversion owing to absence of $\Delta^6$ desaturase and must obtain arachidonate in their diet.

thesized from α-linolenic acid or obtained directly from fish oils, is present in high concentrations in retina, cerebral cortex, testis, and sperm. DHA is particularly needed for development of the brain and retina and is supplied via the placenta and milk. Patients with **retinitis pigmentosa** are reported to have low blood levels of DHA. In **essential fatty acid deficiency,** nonessential polyenoic acids of the ω9 family replace the essential fatty acids in phospholipids, other complex lipids, and membranes, particularly with $\Delta^{5,8,11}$-eicosatrienoic acid (ω9 20:3) (Figure 23–3). The triene:tetraene ratio in plasma lipids can be used to diagnose the extent of essential fatty acid deficiency.

### Trans Fatty Acids Are Implicated in Various Disorders

Small amounts of trans-unsaturated fatty acids are found in ruminant fat (eg, butter fat has 2–7%), where they arise from the action of microorganisms in the rumen, but the main source in the human diet is from partially hydrogenated vegetable oils (eg, margarine). Trans fatty acids compete with essential fatty acids and may exacerbate essential fatty acid deficiency. Moreover, they are structurally similar to saturated fatty acids (Chapter 14) and have comparable effects in the promotion of hypercholesterolemia and atherosclerosis (Chapter 26).

## EICOSANOIDS ARE FORMED FROM $C_{20}$ POLYUNSATURATED FATTY ACIDS

Arachidonate and some other $C_{20}$ polyunsaturated fatty acids give rise to **eicosanoids,** physiologically and pharmacologically active compounds known as **prostaglandins (PG), thromboxanes (TX), leukotrienes (LT),** and **lipoxins (LX)** (Chapter 14). Physiologically, they are considered to act as local hormones functioning through G-protein-linked receptors to elicit their biochemical effects.

There are three groups of eicosanoids that are synthesized from $C_{20}$ eicosanoic acids derived from the essential fatty acids **linoleate** and **α-linolenate,** or directly from dietary arachidonate and eicosapentaenoate (Figure 23–5). Arachidonate, usually derived from the 2 position of phospholipids in the plasma membrane by the action of phospholipase $A_2$ (Figure 24–6)—but also from the diet—is the substrate for the synthesis of the $PG_2$, $TX_2$ series **(prostanoids)** by the **cyclooxygenase pathway,** or the $LT_4$ and $LX_4$ series by the **lipoxygenase pathway,** with the two pathways competing for the arachidonate substrate (Figure 23–5).

## THE CYCLOOXYGENASE PATHWAY IS RESPONSIBLE FOR PROSTANOID SYNTHESIS

Prostanoid synthesis (Figure 23–6) involves the consumption of two molecules of $O_2$ catalyzed by **prostaglandin H synthase (PGHS),** which consists of two enzymes, **cyclooxygenase** and **peroxidase.** PGHS is present as two isoenzymes, PGHS-1 and PGHS-2. The product, an endoperoxide (PGH), is converted to prostaglandins D, E, and F as well as to a thromboxane

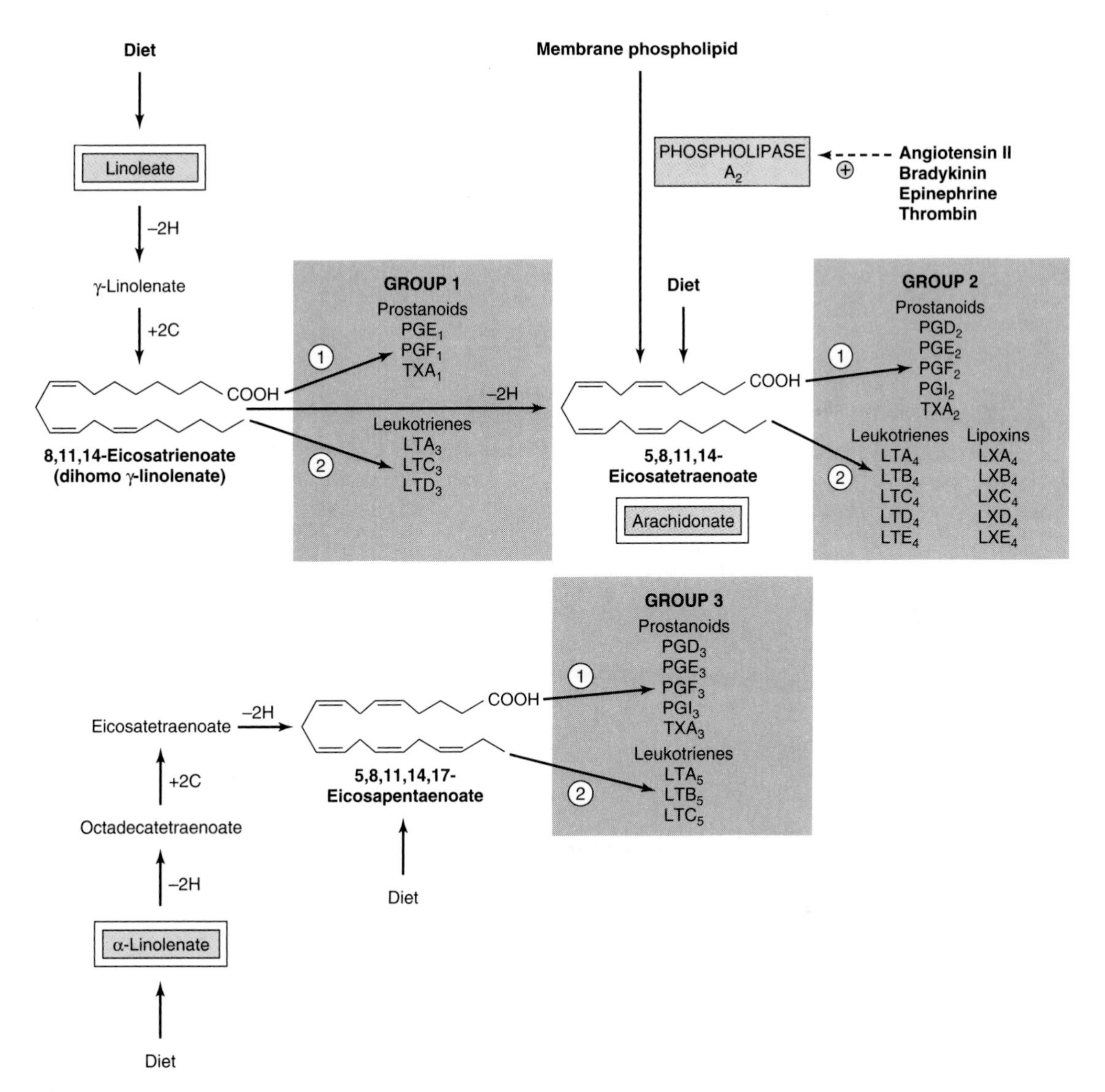


***Figure 23–5.*** The three groups of eicosanoids and their biosynthetic origins. (PG, prostaglandin; PGI, prostacyclin; TX, thromboxane; LT, leukotriene; LX, lipoxin; ①, cyclooxygenase pathway; ②, lipoxygenase pathway.) The subscript denotes the total number of double bonds in the molecule and the series to which the compound belongs.

($TXA_2$) and prostacyclin ($PGI_2$). Each cell type produces only one type of prostanoid. **Aspirin,** a nonsteroidal anti-inflammatory drug (NSAID), inhibits cyclooxygenase of both PGHS-1 and PGHS-2 by acetylation. Most other NSAIDs, such as indomethacin and ibuprofen, inhibit cyclooxygenases by competing with arachidonate. Transcription of PGHS-2—but not of PGHS-1—is completely inhibited by **anti-inflammatory corticosteroids.**

## Essential Fatty Acids Do Not Exert All Their Physiologic Effects Via Prostaglandin Synthesis

The role of essential fatty acids in membrane formation is unrelated to prostaglandin formation. Prostaglandins do not relieve symptoms of essential fatty acid deficiency, and an essential fatty acid deficiency is not caused by inhibition of prostaglandin synthesis.

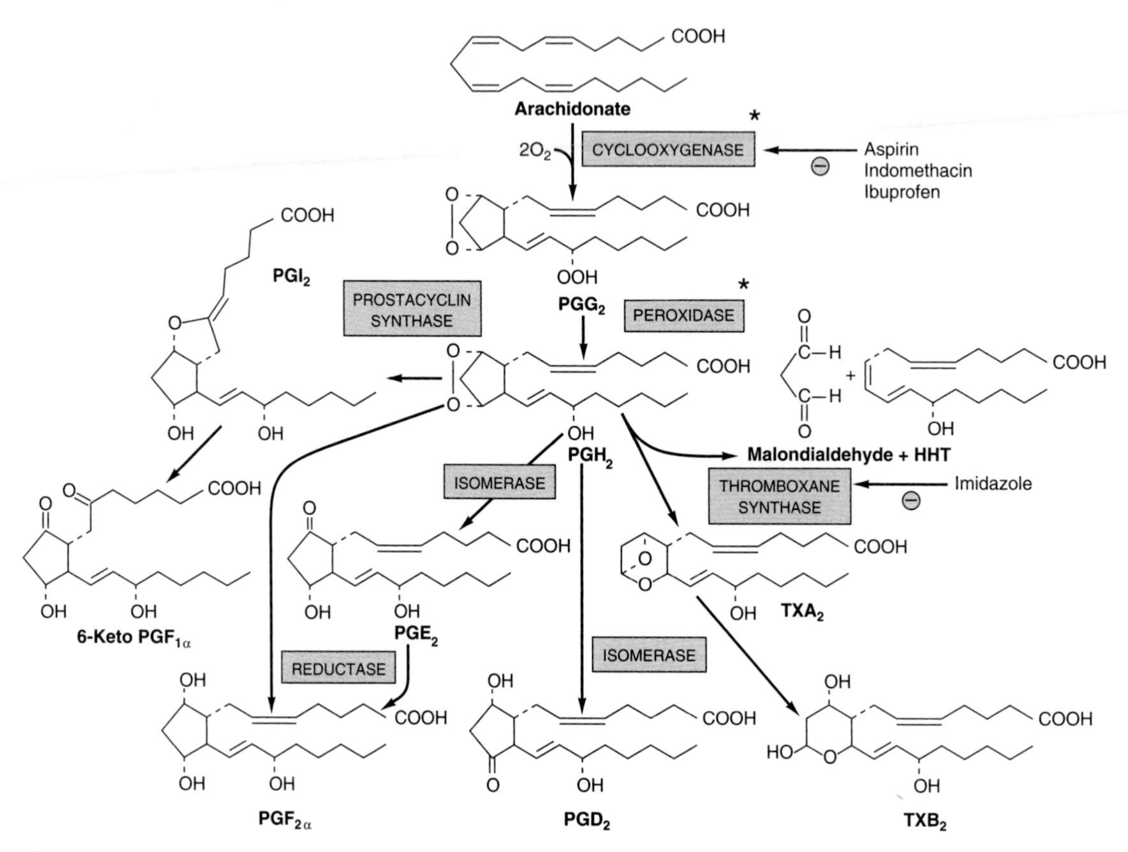


***Figure 23–6.*** Conversion of arachidonic acid to prostaglandins and thromboxanes of series 2. (PG, prostaglandin; TX, thromboxane; PGI, prostacyclin; HHT, hydroxyheptadecatrienoate.) (Asterisk: Both of these starred activities are attributed to one enzyme: prostaglandin H synthase. Similar conversions occur in prostaglandins and thromboxanes of series 1 and 3.)

## Cyclooxygenase Is a "Suicide Enzyme"

"Switching off" of prostaglandin activity is partly achieved by a remarkable property of cyclooxygenase—that of self-catalyzed destruction; ie, it is a "suicide enzyme." Furthermore, the inactivation of prostaglandins by **15-hydroxyprostaglandin dehydrogenase** is rapid. Blocking the action of this enzyme with sulfasalazine or indomethacin can prolong the half-life of prostaglandins in the body.

# LEUKOTRIENES & LIPOXINS ARE FORMED BY THE LIPOXYGENASE PATHWAY

The leukotrienes are a family of conjugated trienes formed from eicosanoic acids in leukocytes, mastocytoma cells, platelets, and macrophages by the **lipoxygenase pathway** in response to both immunologic and nonimmunologic stimuli. Three different lipoxygenases (dioxygenases) insert oxygen into the 5, 12, and 15 positions of arachidonic acid, giving rise to hydroperoxides (HPETE). Only **5-lipoxygenase** forms leukotrienes (details in Figure 23–7). Lipoxins are a family of conjugated tetraenes also arising in leukocytes. They are formed by the combined action of more than one lipoxygenase (Figure 23–7).

# CLINICAL ASPECTS

## Symptoms of Essential Fatty Acid Deficiency in Humans Include Skin Lesions & Impairment of Lipid Transport

In adults subsisting on ordinary diets, no signs of essential fatty acid deficiencies have been reported. How-

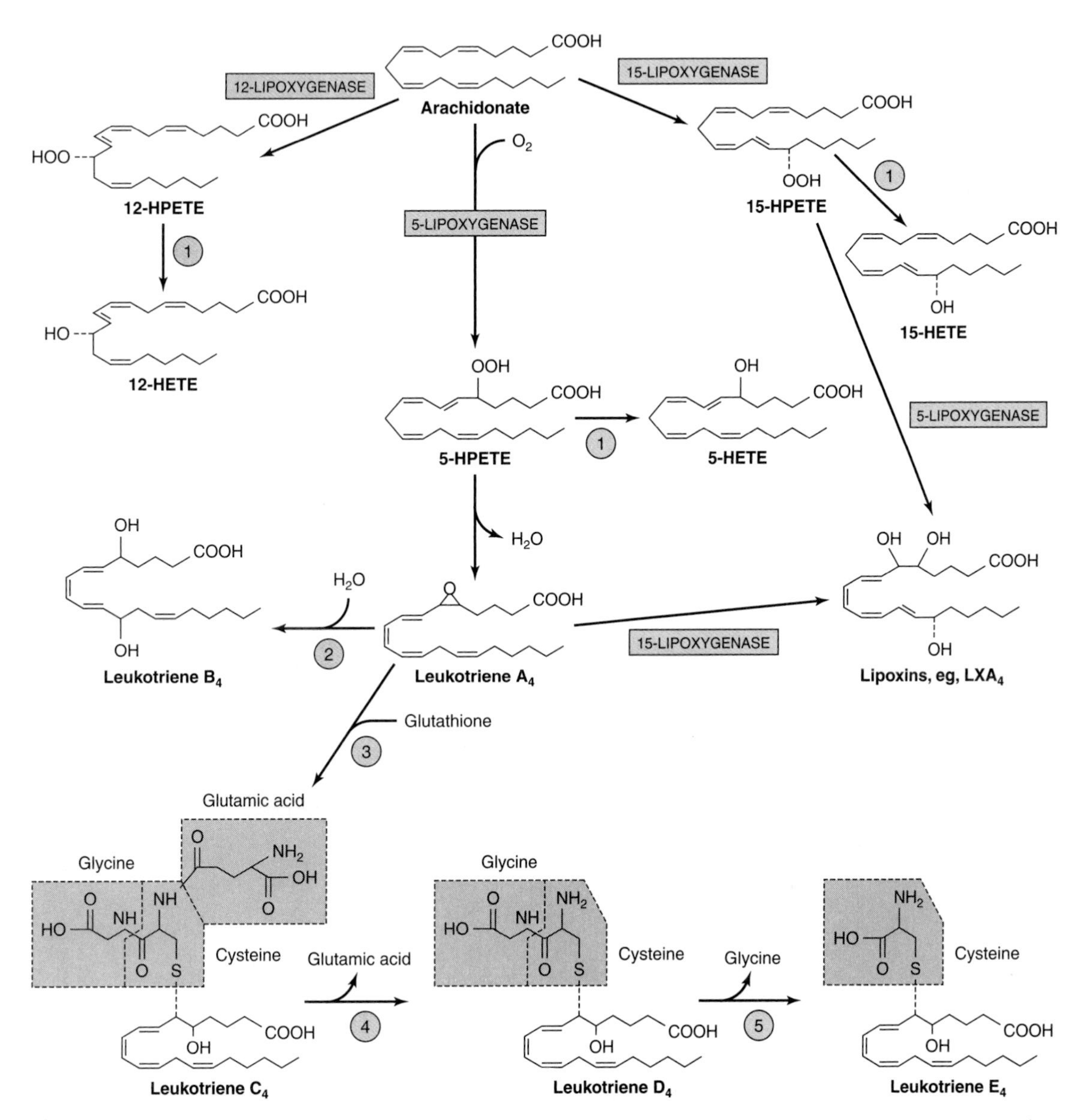


***Figure 23–7.*** Conversion of arachidonic acid to leukotrienes and lipoxins of series 4 via the lipoxygenase pathway. Some similar conversions occur in series 3 and 5 leukotrienes. (HPETE, hydroperoxyeicosatetraenoate; HETE, hydroxyeicosatetraenoate; ①, peroxidase; ②, leukotriene $A_4$ epoxide hydrolase; ③, glutathione *S*-transferase; ④, γ-glutamyltranspeptidase; ⑤, cysteinyl-glycine dipeptidase.)

ever, infants receiving formula diets low in fat and patients maintained for long periods exclusively by intravenous nutrition low in essential fatty acids show deficiency symptoms that can be prevented by an essential fatty acid intake of 1–2% of the total caloric requirement.

## Abnormal Metabolism of Essential Fatty Acids Occurs in Several Diseases

Abnormal metabolism of essential fatty acids, which may be connected with dietary insufficiency, has been noted in cystic fibrosis, acrodermatitis enteropathica, hepatorenal syndrome, Sjögren-Larsson syndrome, multisystem neuronal degeneration, Crohn's disease, cirrhosis and alcoholism, and Reye's syndrome. Elevated levels of very long chain polyenoic acids have been found in the brains of patients with Zellweger's syndrome (Chapter 22). Diets with a high P:S (polyunsaturated:saturated fatty acid) ratio reduce serum cholesterol levels and are considered to be beneficial in terms of the risk of development of coronary heart disease.

### Prostanoids Are Potent Biologically Active Substances

**Thromboxanes** are synthesized in platelets and upon release cause vasoconstriction and platelet aggregation. Their synthesis is specifically inhibited by low-dose aspirin. **Prostacyclins ($PGI_2$)** are produced by blood vessel walls and are potent inhibitors of platelet aggregation. Thus, thromboxanes and prostacyclins are antagonistic. $PG_3$ and $TX_3$, formed from eicosapentaenoic acid (EPA) in fish oils, inhibit the release of arachidonate from phospholipids and the formation of $PG_2$ and $TX_2$. $PGI_3$ is as potent an antiaggregator of platelets as $PGI_2$, but $TXA_3$ is a weaker aggregator than $TXA_2$, changing the balance of activity and favoring longer clotting times. As little as 1 ng/mL of plasma prostaglandins causes contraction of smooth muscle in animals. Potential therapeutic uses include prevention of conception, induction of labor at term, termination of pregnancy, prevention or alleviation of gastric ulcers, control of inflammation and of blood pressure, and relief of asthma and nasal congestion. In addition, $PGD_2$ is a potent sleep-promoting substance. Prostaglandins increase cAMP in platelets, thyroid, corpus luteum, fetal bone, adenohypophysis, and lung but reduce cAMP in renal tubule cells and adipose tissue (Chapter 25).

### Leukotrienes & Lipoxins Are Potent Regulators of Many Disease Processes

Slow-reacting substance of anaphylaxis **(SRS-A)** is a mixture of leukotrienes $C_4$, $D_4$, and $E_4$. This mixture of leukotrienes is a potent constrictor of the bronchial airway musculature. These leukotrienes together with leukotriene $B_4$ also cause vascular permeability and attraction and activation of leukocytes and are important regulators in many diseases involving inflammatory or immediate hypersensitivity reactions, such as asthma. Leukotrienes are vasoactive, and 5-lipoxygenase has been found in arterial walls. Evidence supports a role for lipoxins in vasoactive and immunoregulatory function, eg, as counterregulatory compounds (chalones) of the immune response.

## SUMMARY

- Biosynthesis of unsaturated long-chain fatty acids is achieved by desaturase and elongase enzymes, which introduce double bonds and lengthen existing acyl chains, respectively.
- Higher animals have $\Delta^4$, $\Delta^5$, $\Delta^6$, and $\Delta^9$ desaturases but cannot insert new double bonds beyond the 9 position of fatty acids. Thus, the essential fatty acids linoleic (ω6) and α-linolenic (ω3) must be obtained from the diet.
- Eicosanoids are derived from $C_{20}$ (eicosanoic) fatty acids synthesized from the essential fatty acids and comprise important groups of physiologically and pharmacologically active compounds, including the prostaglandins, thromboxanes, leukotrienes, and lipoxins.

## REFERENCES


Connor WE: The beneficial effects of omega-3 fatty acids: cardiovascular disease and neurodevelopment. Curr Opin Lipidol 1997;8:1.

Fischer S: Dietary polyunsaturated fatty acids and eicosanoid formation in humans. Adv Lipid Res 1989;23:169.

Lagarde M, Gualde N, Rigaud M: Metabolic interactions between eicosanoids in blood and vascular cells. Biochem J 1989; 257:313.

Neuringer M, Anderson GJ, Connor WE: The essentiality of n-3 fatty acids for the development and function of the retina and brain. Annu Rev Nutr 1988;8:517.

Serhan CN: Lipoxin biosynthesis and its impact in inflammatory and vascular events. Biochim Biophys Acta 1994;1212:1.

Smith WL, Fitzpatrick FA: The eicosanoids: Cyclooxygenase, lipoxygenase, and epoxygenase pathways. In: *Biochemistry of Lipids, Lipoproteins and Membranes.* Vance DE, Vance JE (editors). Elsevier, 1996.

Tocher DR, Leaver MJ, Hodgson PA: Recent advances in the biochemistry and molecular biology of fatty acyl desaturases. Prog Lipid Res 1998;37:73.

Valenzuela A, Morgado N: Trans fatty acid isomers in human health and the food industry. Biol Res 1999;32:273.

# Metabolism of Acylglycerols & Sphingolipids

# 24

*Peter A. Mayes, PhD, DSc, & Kathleen M. Botham, PhD, DSc*

## BIOMEDICAL IMPORTANCE

Acylglycerols constitute the majority of lipids in the body. Triacylglycerols are the major lipids in fat deposits and in food, and their roles in lipid transport and storage and in various diseases such as obesity, diabetes, and hyperlipoproteinemia will be described in subsequent chapters. The amphipathic nature of phospholipids and sphingolipids makes them ideally suitable as the main lipid component of cell membranes. Phospholipids also take part in the metabolism of many other lipids. Some phospholipids have specialized functions; eg, dipalmitoyl lecithin is a major component of **lung surfactant,** which is lacking in respiratory distress syndrome of the newborn. Inositol phospholipids in the cell membrane act as precursors of **hormone second messengers,** and **platelet-activating factor** is an alkylphospholipid. Glycosphingolipids, containing sphingosine and sugar residues as well as fatty acid and found in the outer leaflet of the plasma membrane with their oligosaccharide chains facing outward, form part of the glycocalyx of the cell surface and are important (1) in cell adhesion and cell recognition; (2) as receptors for bacterial toxins (eg, the toxin that causes cholera); and (3) as ABO blood group substances. A dozen or so **glycolipid storage diseases** have been described (eg, Gaucher's disease, Tay-Sachs disease), each due to a genetic defect in the pathway for glycolipid degradation in the lysosomes.

## HYDROLYSIS INITIATES CATABOLISM OF TRIACYLGLYCEROLS

Triacylglycerols must be hydrolyzed by a **lipase** to their constituent fatty acids and glycerol before further catabolism can proceed. Much of this hydrolysis (lipolysis) occurs in adipose tissue with release of free fatty acids into the plasma, where they are found combined with serum albumin. This is followed by free fatty acid uptake into tissues (including liver, heart, kidney, muscle, lung, testis, and adipose tissue, but not readily by brain), where they are oxidized or reesterified. The utilization of glycerol depends upon whether such tissues possess **glycerol kinase,** found in significant amounts in liver, kidney, intestine, brown adipose tissue, and lactating mammary gland.

## TRIACYLGLYCEROLS & PHOSPHOGLYCEROLS ARE FORMED BY ACYLATION OF TRIOSE PHOSPHATES

The major pathways of triacylglycerol and phosphoglycerol biosynthesis are outlined in Figure 24–1. Important substances such as triacylglycerols, phosphatidylcholine, phosphatidylethanolamine, phosphatidylinositol, and cardiolipin, a constituent of mitochondrial membranes, are formed from glycerol-3-phosphate. Significant branch points in the pathway occur at the phosphatidate and diacylglycerol steps. From dihydroxyacetone phosphate are derived phosphoglycerols containing an ether link (—C—O—C—), the best-known of which are plasmalogens and platelet-activating factor (PAF). Glycerol 3-phosphate and dihydroxyacetone phosphate are intermediates in glycolysis, making a very important connection between carbohydrate and lipid metabolism.

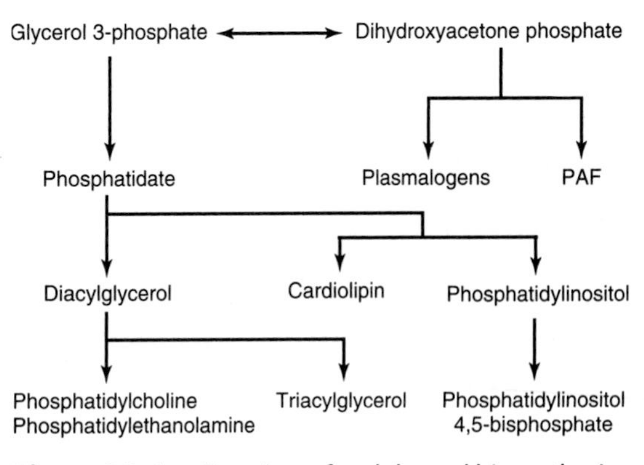


***Figure 24–1.*** Overview of acylglycerol biosynthesis. (PAF, platelet-activating factor.)

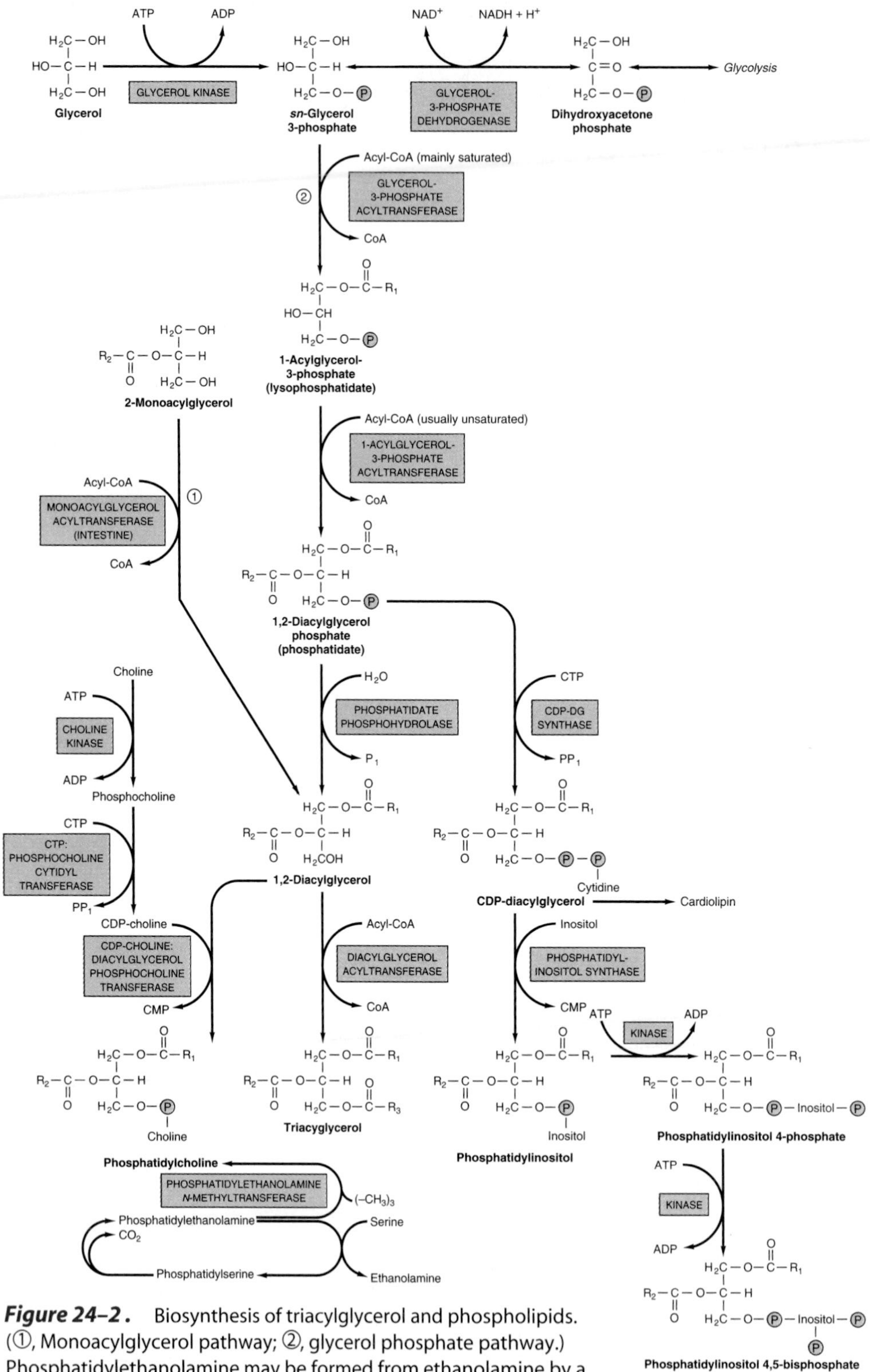


***Figure 24–2.*** Biosynthesis of triacylglycerol and phospholipids. (①, Monoacylglycerol pathway; ②, glycerol phosphate pathway.) Phosphatidylethanolamine may be formed from ethanolamine by a pathway similar to that shown for the formation of phosphatidylcholine from choline.

### Phosphatidate Is the Common Precursor in the Biosynthesis of Triacylglycerols, Many Phosphoglycerols, & Cardiolipin

Both glycerol and fatty acids must be activated by ATP before they can be incorporated into acylglycerols. Glycerol kinase catalyzes the activation of glycerol to *sn*-glycerol 3-phosphate. If the activity of this enzyme is absent or low, as in muscle or adipose tissue, most of the glycerol 3-phosphate is formed from dihydroxyacetone phosphate by **glycerol-3-phosphate dehydrogenase** (Figure 24–2).

#### A. Biosynthesis of Triacylglycerols

Two molecules of acyl-CoA, formed by the activation of fatty acids by **acyl-CoA synthetase** (Chapter 22), combine with glycerol 3-phosphate to form **phosphatidate** (1,2-diacylglycerol phosphate). This takes place in two stages, catalyzed by **glycerol-3-phosphate acyltransferase** and **1-acylglycerol-3-phosphate acyltransferase.** Phosphatidate is converted by **phosphatidate phosphohydrolase** and **diacylglycerol acyltransferase** to 1,2-diacylglycerol and then triacylglycerol. In intestinal mucosa, **monoacylglycerol acyltransferase** converts **monoacylglycerol** to 1,2-diacylglycerol in the **monoacylglycerol pathway.** Most of the activity of these enzymes resides in the endoplasmic reticulum of the cell, but some is found in mitochondria. Phosphatidate phosphohydrolase is found mainly in the cytosol, but the active form of the enzyme is membrane-bound.

In the biosynthesis of phosphatidylcholine and phosphatidylethanolamine (Figure 24–2), choline or ethanolamine must first be activated by phosphorylation by ATP followed by linkage to CTP. The resulting CDP-choline or CDP-ethanolamine reacts with 1,2-diacylglycerol to form either phosphatidylcholine or phosphatidylethanolamine, respectively. Phosphatidylserine is formed from phosphatidylethanolamine directly by reaction with serine (Figure 24–2). Phosphatidylserine may re-form phosphatidylethanolamine by decarboxylation. An alternative pathway in liver enables phosphatidylethanolamine to give rise directly to phosphatidylcholine by progressive methylation of the ethanolamine residue. In spite of these sources of choline, it is considered to be an essential nutrient in many mammalian species, but this has not been established in humans.

The regulation of triacylglycerol, phosphatidylcholine, and phosphatidylethanolamine biosynthesis is driven by the availability of free fatty acids. Those that escape oxidation are preferentially converted to phospholipids, and when this requirement is satisfied they are used for triacylglycerol synthesis.

A phospholipid present in mitochondria is **cardiolipin** (diphosphatidylglycerol; Figure 14–8). It is formed from phosphatidylglycerol, which in turn is synthesized from CDP-diacylglycerol (Figure 24–2) and glycerol 3-phosphate according to the scheme shown in Figure 24–3. Cardiolipin, found in the inner membrane of mitochondria, is specifically required for the functioning of the phosphate transporter and for cytochrome oxidase activity.

#### B. Biosynthesis of Glycerol Ether Phospholipids

This pathway is located in peroxisomes. Dihydroxyacetone phosphate is the precursor of the glycerol moiety of glycerol ether phospholipids (Figure 24–4). This compound combines with acyl-CoA to give 1-acyldihydroxyacetone phosphate. The ether link is formed in the next reaction, producing 1-alkyldihydroxyacetone phosphate, which is then converted to 1-alkylglycerol 3-phosphate. After further acylation in the 2 position, the resulting 1-alkyl-2-acylglycerol 3-phosphate (analogous to phosphatidate in Figure 24–2) is hydrolyzed to give the free glycerol derivative. **Plasmalogens,** which comprise much of the phospholipid in mitochondria, are formed by desaturation of the analogous 3-phosphoethanolamine derivative (Figure 24–4). **Platelet-activating factor (PAF)** (1-alkyl-2-acetyl-*sn*-glycerol-3-phosphocholine) is synthesized from the corresponding 3-phosphocholine derivative. It is formed by many blood cells and other tissues and aggregates platelets at concentrations as low as $10^{-11}$ mol/L. It also has hypotensive and ulcerogenic properties and is involved in a variety of biologic responses, including inflammation, chemotaxis, and protein phosphorylation.

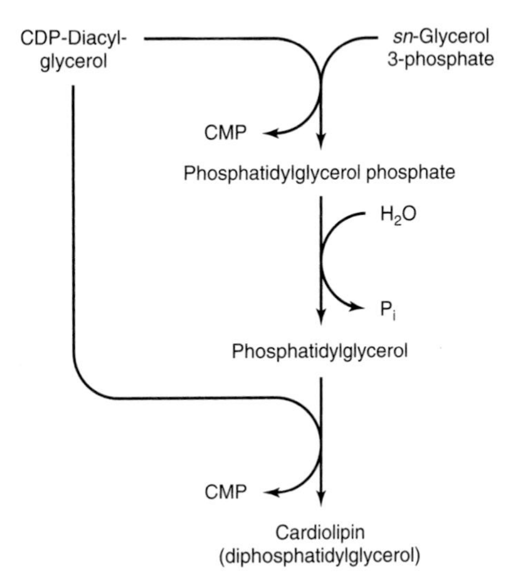


***Figure 24–3.*** Biosynthesis of cardiolipin.

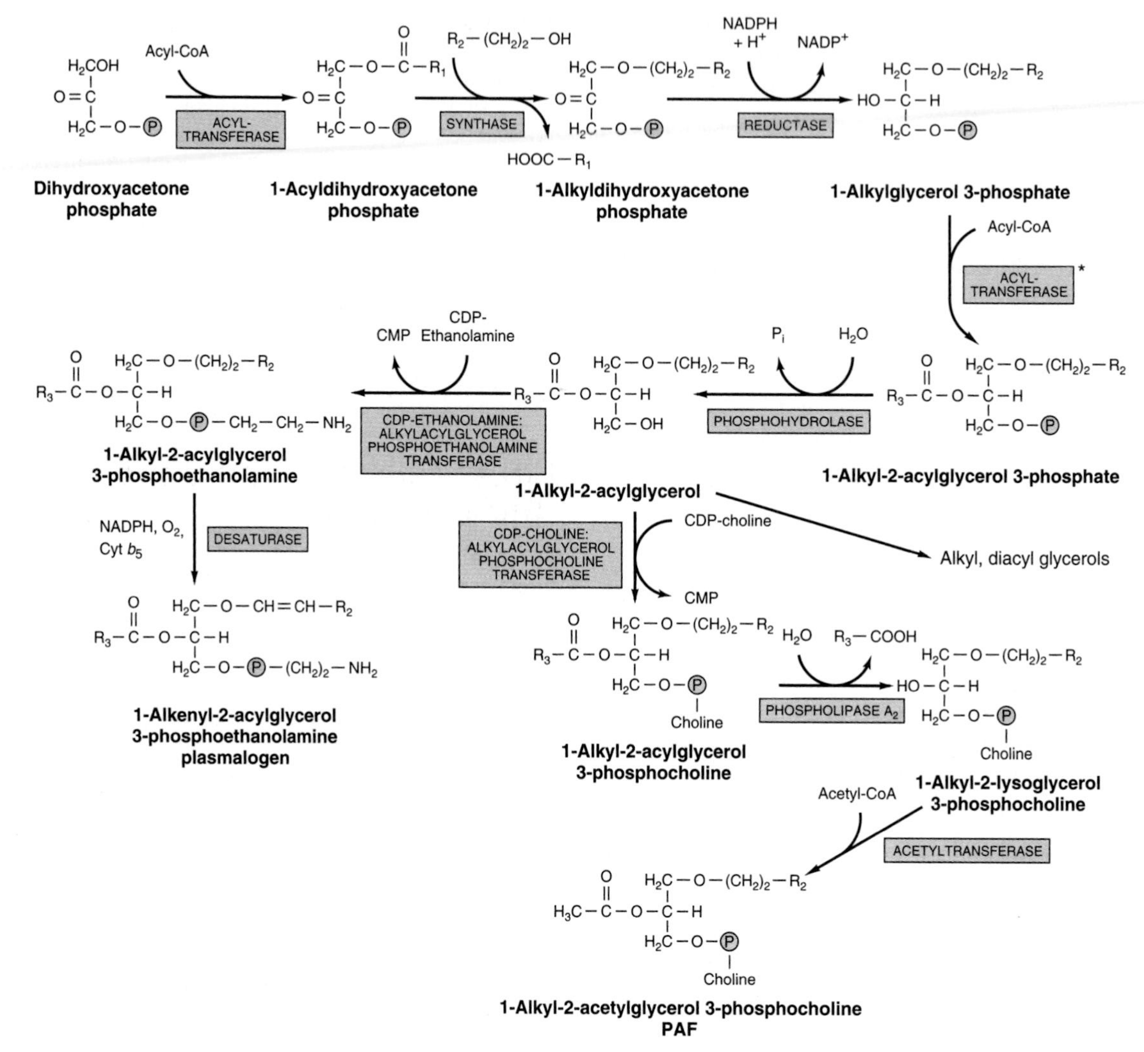


***Figure 24–4.*** Biosynthesis of ether lipids, including plasmalogens, and platelet-activating factor (PAF). In the de novo pathway for PAF synthesis, acetyl-CoA is incorporated at stage *, avoiding the last two steps in the pathway shown here.

## Phospholipases Allow Degradation & Remodeling of Phosphoglycerols

Although phospholipids are actively degraded, each portion of the molecule turns over at a different rate—eg, the turnover time of the phosphate group is different from that of the 1-acyl group. This is due to the presence of enzymes that allow partial degradation followed by resynthesis (Figure 24–5). **Phospholipase $A_2$** catalyzes the hydrolysis of glycerophospholipids to form a free fatty acid and lysophospholipid, which in turn may be reacylated by acyl-CoA in the presence of an acyltransferase. Alternatively, lysophospholipid (eg, lysolecithin) is attacked by **lysophospholipase**, forming the corresponding glyceryl phosphoryl base, which in turn may be split by a hydrolase liberating glycerol 3-phosphate plus base. **Phospholipases $A_1$, $A_2$, B, C, and D** attack the bonds indicated in Figure 24–6. **Phospholipase $A_2$** is found in pancreatic fluid and snake venom as well as in many types of cells; **phospholipase C** is one of the major toxins secreted by bacteria; and **phospholipase D** is known to be involved in mammalian signal transduction.

**Lysolecithin (lysophosphatidylcholine)** may be formed by an alternative route that involves **lecithin: cholesterol acyltransferase (LCAT).** This enzyme,

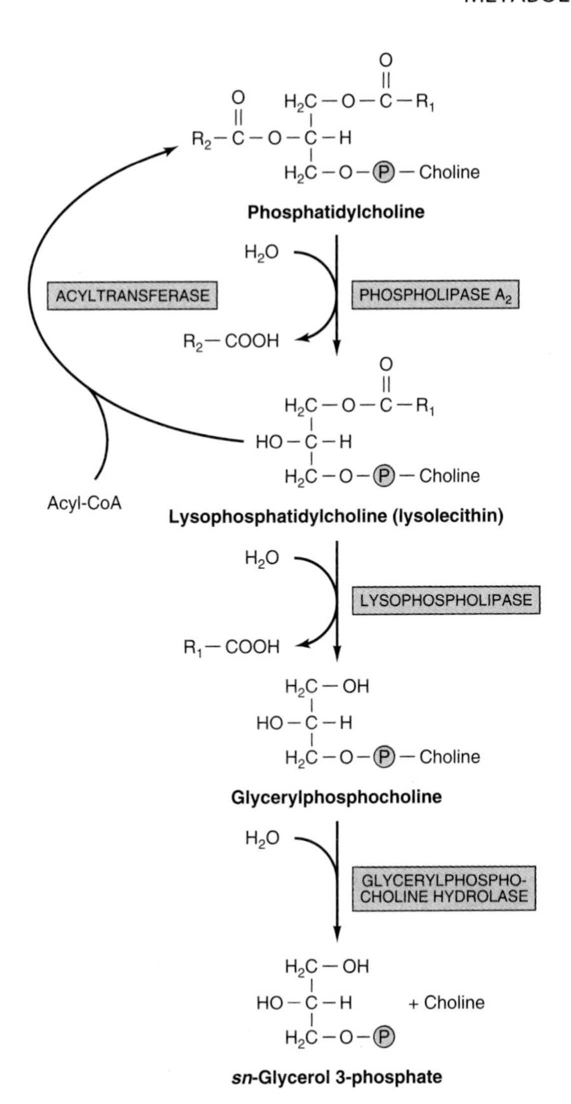


***Figure 24–5.*** Metabolism of phosphatidylcholine (lecithin).

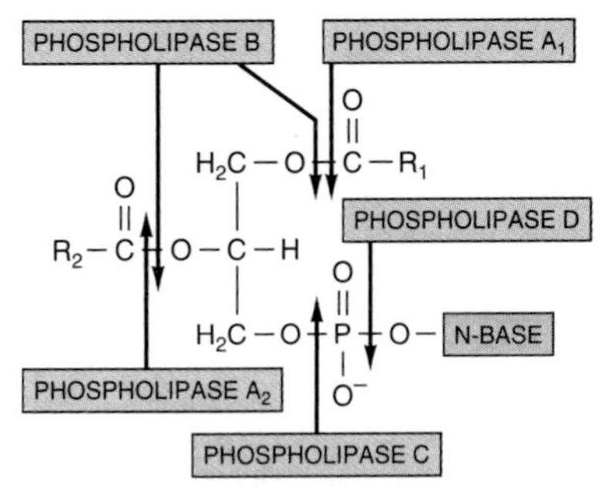


***Figure 24–6.*** Sites of the hydrolytic activity of phospholipases on a phospholipid substrate.

found in plasma, catalyzes the transfer of a fatty acid residue from the 2 position of lecithin to cholesterol to form cholesteryl ester and lysolecithin and is considered to be responsible for much of the cholesteryl ester in plasma lipoproteins. Long-chain saturated fatty acids are found predominantly in the 1 position of phospholipids, whereas the polyunsaturated acids (eg, the precursors of prostaglandins) are incorporated more into the 2 position. The incorporation of fatty acids into lecithin occurs by complete synthesis of the phospholipid, by transacylation between cholesteryl ester and lysolecithin, and by direct acylation of lysolecithin by acyl-CoA. Thus, a continuous exchange of the fatty acids is possible, particularly with regard to introducing essential fatty acids into phospholipid molecules.

## ALL SPHINGOLIPIDS ARE FORMED FROM CERAMIDE

**Ceramide** is synthesized in the endoplasmic reticulum from the amino acid serine according to Figure 24–7. Ceramide is an important signaling molecule (second messenger) regulating pathways including **apoptosis** (processes leading to cell death), cell senescence, and differentiation, and opposes some of the actions of diacylglycerol.

**Sphingomyelins** (Figure 14–11) are phospholipids and are formed when ceramide reacts with phosphatidylcholine to form sphingomyelin plus diacylglycerol (Figure 24–8A). This occurs mainly in the Golgi apparatus and to a lesser extent in the plasma membrane.

### Glycosphingolipids Are a Combination of Ceramide With One or More Sugar Residues

The simplest glycosphingolipids **(cerebrosides)** are **galactosylceramide (GalCer)** and **glucosylceramide (GlcCer).** GalCer is a major lipid of myelin, whereas GlcCer is the major glycosphingolipid of extraneural tissues and a precursor of most of the more complex glycosphingolipids. Galactosylceramide (Figure 24–8B) is formed in a reaction between ceramide and UDPGal (formed by epimerization from UDPGlc—Figure 20–6). **Sulfogalactosylceramide** and other sulfolipids such as the **sulfo(galacto)-glycerolipids** and the **steroid sulfates** are formed after further reactions involving 3′-phosphoadenosine-5′-phosphosulfate (PAPS; "active sulfate"). **Gangliosides** are synthesized from ceramide by the stepwise addition of activated sugars (eg, UDPGlc and UDPGal) and a **sialic acid,** usually ***N*-acetylneuraminic acid** (Figure 24–9). A large number of gangliosides of increasing molecular weight may be formed. Most of the enzymes transferring sugars from

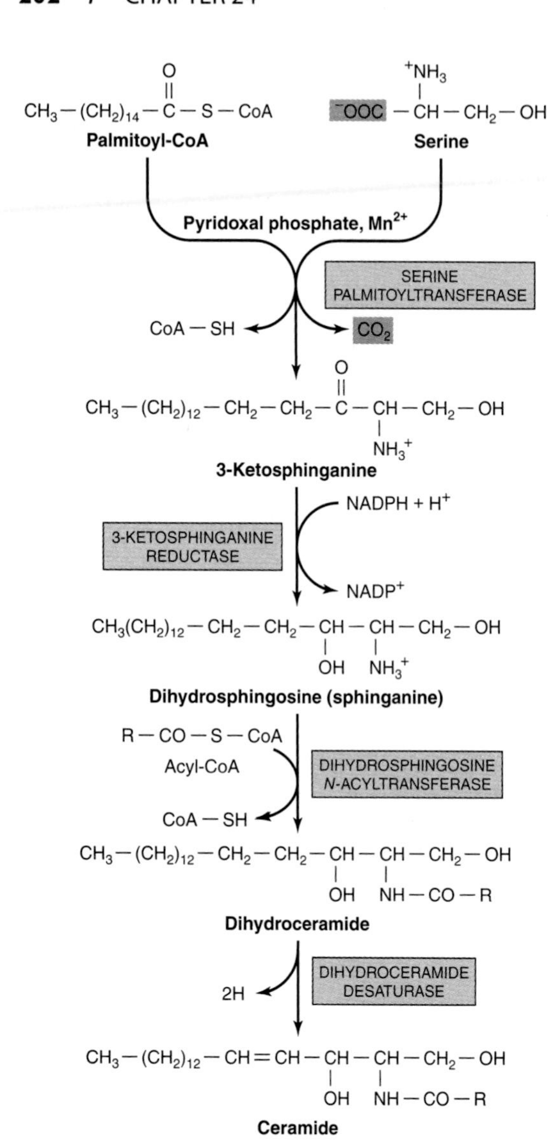


***Figure 24–7.*** Biosynthesis of ceramide.

nucleotide sugars (glycosyl transferases) are found in the Golgi apparatus.

**Glycosphingolipids** are constituents of the outer leaflet of plasma membranes and are important in **cell adhesion** and **cell recognition.** Some are antigens, eg, ABO blood group substances. Certain gangliosides function as receptors for bacterial toxins (eg, for cholera toxin, which subsequently activates adenylyl cyclase).

## CLINICAL ASPECTS

### Deficiency of Lung Surfactant Causes Respiratory Distress Syndrome

**Lung surfactant** is composed mainly of lipid with some proteins and carbohydrate and prevents the alveoli from collapsing. Surfactant activity is largely attributed to **dipalmitoylphosphatidylcholine,** which is synthesized shortly before parturition in full-term infants. Deficiency of lung surfactant in the lungs of many preterm newborns gives rise to **respiratory distress syndrome.** Administration of either natural or artificial surfactant has been of therapeutic benefit.

### Phospholipids & Sphingolipids Are Involved in Multiple Sclerosis and Lipidoses

Certain diseases are characterized by abnormal quantities of these lipids in the tissues, often in the nervous system. They may be classified into two groups: (1) true demyelinating diseases and (2) sphingolipidoses.

In **multiple sclerosis,** which is a demyelinating disease, there is loss of both phospholipids (particularly ethanolamine plasmalogen) and of sphingolipids from white matter. Thus, the lipid composition of white matter resembles that of gray matter. The cerebrospinal fluid shows raised phospholipid levels.

The **sphingolipidoses (lipid storage diseases)** are a group of inherited diseases that are often manifested in childhood. These diseases are part of a larger group of lysosomal disorders and exhibit several constant features: (1) Complex lipids containing ceramide accumulate in cells, particularly neurons, causing neurodegen-

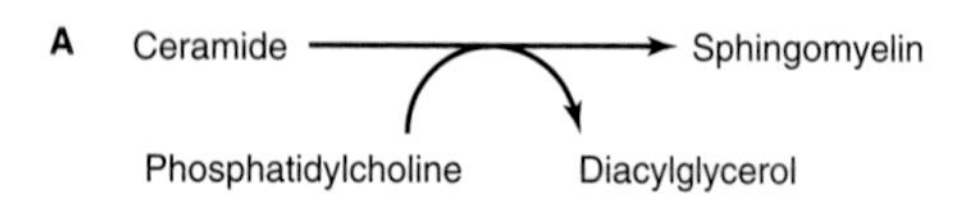


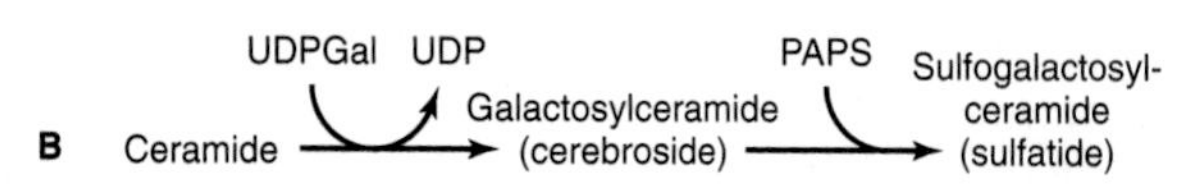


***Figure 24–8.*** Biosynthesis of sphingomyelin (A), galactosylceramide and its sulfo derivative (B). (PAPS, "active sulfate," adenosine 3′-phosphate-5′-phosphosulfate.)

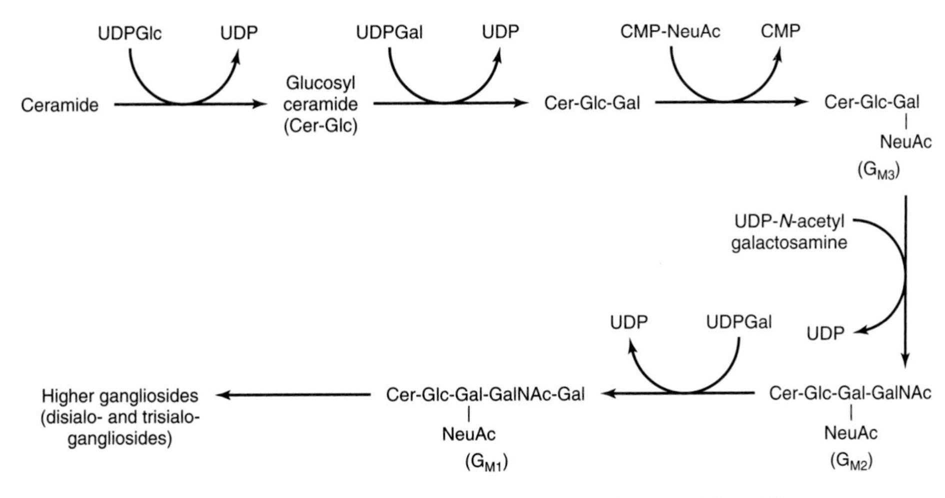


***Figure 24–9.*** Biosynthesis of gangliosides. (NeuAc, *N*-acetylneuraminic acid.)

eration and shortening the life span. (2) The rate of **synthesis** of the stored lipid is normal. (3) The enzymatic defect is in the **lysosomal degradation pathway** of sphingolipids. (4) The extent to which the activity of the affected enzyme is decreased is similar in all tissues. There is no effective treatment for many of the diseases, though some success has been achieved with enzymes that have been chemically modified to ensure binding to receptors of target cells, eg, to macrophages in the liver in order to deliver β-glucosidase (glucocerebrosidase) in the treatment of Gaucher's disease. A recent promising approach is substrate reduction therapy to inhibit the synthesis of sphingolipids, and gene therapy for lysosomal disorders is currently under investigation. Some examples of the more important lipid storage diseases are shown in Table 24–1.

**Multiple sulfatase deficiency** results in accumulation of sulfogalactosylceramide, steroid sulfates, and proteoglycans owing to a combined deficiency of arylsulfatases A, B, and C and steroid sulfatase.

***Table 24–1.*** Examples of sphingolipidoses.

| Disease | Enzyme Deficiency | Lipid Accumulating[1] | Clinical Symptoms |
|---|---|---|---|
| Tay-Sachs disease | Hexosaminidase A | Cer—Glc—Gal(NeuAc)-⁝-GalNAc<br>$G_{M2}$ Ganglioside | Mental retardation, blindness, muscular weakness. |
| Fabry's disease | α-Galactosidase | Cer—Glc—Gal-⁝-Gal<br>Globotriaosylceramide | Skin rash, kidney failure (full symptoms only in males; X-linked recessive). |
| Metachromatic leukodystrophy | Arylsulfatase A | Cer—Gal-⁝-$OSO_3$<br>3-Sulfogalactosylceramide | Mental retardation and psychologic disturbances in adults; demyelination. |
| Krabbe's disease | β-Galactosidase | Cer-⁝-Gal<br>Galactosylceramide | Mental retardation; myelin almost absent. |
| Gaucher's disease | β-Glucosidase | Cer-⁝-Glc<br>Glucosylceramide | Enlarged liver and spleen, erosion of long bones, mental retardation in infants. |
| Niemann-Pick disease | Sphingomyelinase | Cer-⁝-P—choline<br>Sphingomyelin | Enlarged liver and spleen, mental retardation; fatal in early life. |
| Farber's disease | Ceramidase | Acyl-⁝-Sphingosine<br>Ceramide | Hoarseness, dermatitis, skeletal deformation, mental retardation; fatal in early life. |

[1]NeuAc, *N*-acetylneuraminic acid; Cer, ceramide; Glc, glucose; Gal, galactose. -⁝-, site of deficient enzyme reaction.

## SUMMARY


- Triacylglycerols are the major energy-storing lipids, whereas phosphoglycerols, sphingomyelin, and glycosphingolipids are amphipathic and have structural functions in cell membranes as well as other specialized roles.
- Triacylglycerols and some phosphoglycerols are synthesized by progressive acylation of glycerol 3-phosphate. The pathway bifurcates at phosphatidate, forming inositol phospholipids and cardiolipin on the one hand and triacylglycerol and choline and ethanolamine phospholipids on the other.
- Plasmalogens and platelet-activating factor (PAF) are ether phospholipids formed from dihydroxyacetone phosphate.
- Sphingolipids are formed from ceramide (*N*-acylsphingosine). Sphingomyelin is present in membranes of organelles involved in secretory processes (eg, Golgi apparatus). The simplest glycosphingolipids are a combination of ceramide plus a sugar residue (eg, GalCer in myelin). Gangliosides are more complex glycosphingolipids containing more sugar residues plus sialic acid. They are present in the outer layer of the plasma membrane, where they contribute to the glycocalyx and are important as antigens and cell receptors.
- Phospholipids and sphingolipids are involved in several disease processes, including respiratory distress syndrome (lack of lung surfactant), multiple sclerosis (demyelination), and sphingolipidoses (inability to break down sphingolipids in lysosomes due to inherited defects in hydrolase enzymes).


## REFERENCES


Griese M: Pulmonary surfactant in health and human lung diseases: state of the art. Eur Respir J 1999;13:1455.

Merrill AH, Sweeley CC: Sphingolipids: metabolism and cell signaling. In: *Biochemistry of Lipids, Lipoproteins and Membranes.* Vance DE, Vance JE (editors). Elsevier, 1996.

Prescott SM et al: Platelet-activating factor and related lipid mediators. Annu Rev Biochem 2000;69:419.

Ruvolo PP: Ceramide regulates cellular homeostasis via diverse stress signaling pathways. Leukemia 2001;15:1153.

Schuette CG et al: The glycosphingolipidoses—from disease to basic principles of metabolism. Biol Chem 1999;380:759.

Scriver CR et al (editors): *The Metabolic and Molecular Bases of Inherited Disease,* 8th ed. McGraw-Hill, 2001.

Tijburg LBM, Geelen MJH, van Golde LMG: Regulation of the biosynthesis of triacylglycerol, phosphatidylcholine and phosphatidylethanolamine in the liver. Biochim Biophys Acta 1989;1004:1.

Vance DE: Glycerolipid biosynthesis in eukaryotes. In: *Biochemistry of Lipids, Lipoproteins and Membranes.* Vance DE, Vance JE (editors). Elsevier, 1996.

van Echten G, Sandhoff K: Ganglioside metabolism. Enzymology, topology, and regulation. J Biol Chem 1993;268:5341.

Waite M: Phospholipases. In: *Biochemistry of Lipids, Lipoproteins and Membranes.* Vance DE, Vance JE (editors). Elsevier, 1996.

# Lipid Transport & Storage

# 25

*Peter A. Mayes, PhD, DSc, & Kathleen M. Botham, PhD, DSc*

## BIOMEDICAL IMPORTANCE

Fat absorbed from the diet and lipids synthesized by the liver and adipose tissue must be transported between the various tissues and organs for utilization and storage. Since lipids are insoluble in water, the problem of how to transport them in the aqueous blood plasma is solved by associating nonpolar lipids (triacylglycerol and cholesteryl esters) with amphipathic lipids (phospholipids and cholesterol) and proteins to make water-miscible lipoproteins.

In a meal-eating omnivore such as the human, excess calories are ingested in the anabolic phase of the feeding cycle, followed by a period of negative caloric balance when the organism draws upon its carbohydrate and fat stores. Lipoproteins mediate this cycle by transporting lipids from the intestines as chylomicrons—and from the liver as very low density lipoproteins (VLDL)—to most tissues for oxidation and to adipose tissue for storage. Lipid is mobilized from adipose tissue as free fatty acids (FFA) attached to serum albumin. Abnormalities of lipoprotein metabolism cause various **hypo-** or **hyperlipoproteinemias.** The most common of these is **diabetes mellitus,** where insulin deficiency causes excessive mobilization of FFA and underutilization of chylomicrons and VLDL, leading to **hypertriacylglycerolemia.** Most other pathologic conditions affecting lipid transport are due primarily to inherited defects, some of which cause **hypercholesterolemia,** and premature **atherosclerosis. Obesity**—particularly abdominal obesity—is a risk factor for increased mortality, hypertension, type 2 diabetes mellitus, hyperlipidemia, hyperglycemia, and various endocrine dysfunctions.

## LIPIDS ARE TRANSPORTED IN THE PLASMA AS LIPOPROTEINS

### Four Major Lipid Classes Are Present in Lipoproteins

Plasma lipids consist of **triacylglycerols** (16%), **phospholipids** (30%), **cholesterol** (14%), and **cholesteryl esters** (36%) and a much smaller fraction of unesterified long-chain fatty acids (free fatty acids) (4%). This latter fraction, the **free fatty acids (FFA),** is metabolically the most active of the plasma lipids.

### Four Major Groups of Plasma Lipoproteins Have Been Identified

Because fat is less dense than water, the density of a lipoprotein decreases as the proportion of lipid to protein increases (Table 25–1). In addition to FFA, four major groups of lipoproteins have been identified that are important physiologically and in clinical diagnosis. These are (1) **chylomicrons,** derived from intestinal absorption of triacylglycerol and other lipids; (2) **very low density lipoproteins** (VLDL, or pre-β-lipoproteins), derived from the liver for the export of triacylglycerol; (3) **low-density lipoproteins** (LDL, or β-lipoproteins), representing a final stage in the catabolism of VLDL; and (4) **high-density lipoproteins** (HDL, or α-lipoproteins), involved in VLDL and chylomicron metabolism and also in cholesterol transport. Triacylglycerol is the predominant lipid in chylomicrons and VLDL, whereas cholesterol and phospholipid are the predominant lipids in LDL and HDL, respectively (Table 25–1). Lipoproteins may be separated according to their electrophoretic properties into **α-, β-,** and **pre-β-lipoproteins.**

### Lipoproteins Consist of a Nonpolar Core & a Single Surface Layer of Amphipathic Lipids

The **nonpolar lipid core** consists of mainly **triacylglycerol** and **cholesteryl ester** and is surrounded by a **single surface layer** of **amphipathic phospholipid** and **cholesterol** molecules (Figure 25–1). These are oriented so that their polar groups face outward to the aqueous medium, as in the cell membrane (Chapter 14). The protein moiety of a lipoprotein is known as an **apolipoprotein** or **apoprotein,** constituting nearly 70% of some HDL and as little as 1% of chylomicrons. Some apolipoproteins are integral and cannot be removed, whereas others are free to transfer to other lipoproteins.

### The Distribution of Apolipoproteins Characterizes the Lipoprotein

One or more apolipoproteins (proteins or polypeptides) are present in each lipoprotein. The major apolipoproteins of HDL (α-lipoprotein) are designated A (Table

***Table 25–1.*** Composition of the lipoproteins in plasma of humans.

| Lipoprotein | Source | Diameter (nm) | Density (g/mL) | Composition | | Main Lipid Components | Apolipoproteins |
|---|---|---|---|---|---|---|---|
| | | | | Protein (%) | Lipid (%) | | |
| Chylomicrons | Intestine | 90–1000 | < 0.95 | 1–2 | 98–99 | Triacylglycerol | A-I, A-II, A-IV,[1] B-48, C-I, C-II, C-III, E |
| Chylomicron remnants | Chylomicrons | 45–150 | < 1.006 | 6–8 | 92–94 | Triacylglycerol, phospholipids, cholesterol | B-48, E |
| VLDL | Liver (intestine) | 30–90 | 0.95–1.006 | 7–10 | 90–93 | Triacylglycerol | B-100, C-I, C-II, C-III |
| IDL | VLDL | 25–35 | 1.006–1.019 | 11 | 89 | Triacylglycerol, cholesterol | B-100, E |
| LDL | VLDL | 20–25 | 1.019–1.063 | 21 | 79 | Cholesterol | B-100 |
| HDL | Liver, intestine, VLDL, chylomicrons | | | | | Phospholipids, cholesterol | A-I, A-II, A-IV, C-I, C-II, C-III, D,[2] E |
| HDL$_1$ | | 20–25 | 1.019–1.063 | 32 | 68 | | |
| HDL$_2$ | | 10–20 | 1.063–1.125 | 33 | 67 | | |
| HDL$_3$ | | 5–10 | 1.125–1.210 | 57 | 43 | | |
| Preβ-HDL[3] | | < 5 | > 1.210 | | | | A-I |
| Albumin/free fatty acids | Adipose tissue | | > 1.281 | 99 | 1 | Free fatty acids | |

**Abbreviations:** HDL, high-density lipoproteins; IDL, intermediate-density lipoproteins; LDL, low-density lipoproteins; VLDL, very low density lipoproteins.
[1]Secreted with chylomicrons but transfers to HDL.
[2]Associated with HDL$_2$ and HDL$_3$ subfractions.
[3]Part of a minor fraction known as very high density lipoproteins (VHDL).

25–1). The main apolipoprotein of LDL (β-lipoprotein) is apolipoprotein B (B-100) and is found also in VLDL. Chylomicrons contain a truncated form of apo B (B-48) that is synthesized in the intestine, while B-100 is synthesized in the liver. Apo B-100 is one of the longest single polypeptide chains known, having 4536 amino acids and a molecular mass of 550,000 Da. Apo B-48 (48% of B-100) is formed from the same mRNA as apo B-100 after the introduction of a stop signal by an RNA editing enzyme. Apo C-I, C-II, and C-III are smaller polypeptides (molecular mass 7000–9000 Da) freely transferable between several different lipoproteins. Apo E is found in VLDL, HDL, chylomicrons, and chylomicron remnants; it accounts for 5–10% of total VLDL apolipoproteins in normal subjects.

Apolipoproteins carry out several roles: (1) they can form part of the structure of the lipoprotein, eg, apo B; (2) they are enzyme cofactors, eg, C-II for lipoprotein lipase, A-I for lecithin:cholesterol acyltransferase, or enzyme inhibitors, eg, apo A-II and apo C-III for lipoprotein lipase, apo C-I for cholesteryl ester transfer protein; and (3) they act as ligands for interaction with lipoprotein receptors in tissues, eg, apo B-100 and apo E for the LDL receptor, apo E for the LDL receptor-related protein (LRP), which has been identified as the remnant receptor, and apo A-I for the HDL receptor. The functions of apo A-IV and apo D, however, are not yet clearly defined.

## FREE FATTY ACIDS ARE RAPIDLY METABOLIZED

The free fatty acids (FFA, nonesterified fatty acids, unesterified fatty acids) arise in the plasma from lipolysis of triacylglycerol in adipose tissue or as a result of the action of lipoprotein lipase during uptake of plasma triacylglycerols into tissues. They are found **in combination with albumin,** a very effective solubilizer, in concentrations varying between 0.1 and 2.0 μeq/mL of plasma. Levels are low in the fully fed condition and rise to 0.7–0.8 μeq/mL in the starved state. In uncontrolled **diabetes mellitus,** the level may rise to as much as 2 μeq/mL.

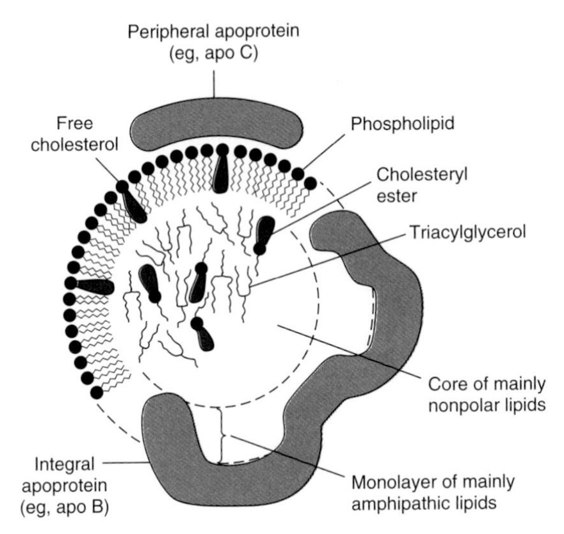


***Figure 25–1.*** Generalized structure of a plasma lipoprotein. The similarities with the structure of the plasma membrane are to be noted. Small amounts of cholesteryl ester and triacylglycerol are to be found in the surface layer and a little free cholesterol in the core.

Free fatty acids are removed from the blood extremely rapidly and oxidized (fulfilling 25–50% of energy requirements in starvation) or esterified to form triacylglycerol in the tissues. In starvation, esterified lipids from the circulation or in the tissues are oxidized as well, particularly in heart and skeletal muscle cells, where considerable stores of lipid are to be found.

The free fatty acid uptake by tissues is related directly to the plasma free fatty acid concentration, which in turn is determined by the rate of lipolysis in adipose tissue. After dissociation of the fatty acid-albumin complex at the plasma membrane, fatty acids bind to a **membrane fatty acid transport protein** that acts as a transmembrane cotransporter with $Na^+$. On entering the cytosol, free fatty acids are bound by intracellular **fatty acid-binding proteins.** The role of these proteins in intracellular transport is thought to be similar to that of serum albumin in extracellular transport of long-chain fatty acids.

## TRIACYLGLYCEROL IS TRANSPORTED FROM THE INTESTINES IN CHYLOMICRONS & FROM THE LIVER IN VERY LOW DENSITY LIPOPROTEINS

By definition, **chylomicrons** are found in **chyle** formed only by the lymphatic system **draining the intestine.** They are responsible for the transport of all dietary lipids into the circulation. Small quantities of VLDL are also to be found in chyle; however, most of the plasma **VLDL** are of hepatic origin. **They are the vehicles of transport of triacylglycerol from the liver to the extrahepatic tissues.**

There are striking similarities in the mechanisms of formation of chylomicrons by intestinal cells and of VLDL by hepatic parenchymal cells (Figure 25–2), perhaps because—apart from the mammary gland—the intestine and liver are the only tissues from which particulate lipid is secreted. Newly secreted or "nascent" chylomicrons and VLDL contain only a small amount of apolipoproteins C and E, and the full complement is acquired from HDL in the circulation (Figures 25–3 and 25–4). Apo B is essential for chylomicron and VLDL formation. In **abetalipoproteinemia** (a rare disease), lipoproteins containing apo B are not formed and lipid droplets accumulate in the intestine and liver.

A more detailed account of the factors controlling hepatic VLDL secretion is given below.

## CHYLOMICRONS & VERY LOW DENSITY LIPOPROTEINS ARE RAPIDLY CATABOLIZED

The clearance of labeled chylomicrons from the blood is rapid, the half-time of disappearance being under 1 hour in humans. Larger particles are catabolized more quickly than smaller ones. Fatty acids originating from chylomicron triacylglycerol are delivered mainly to adipose tissue, heart, and muscle (80%), while about 20% goes to the liver. However, **the liver does not metabolize native chylomicrons or VLDL significantly;** thus, the fatty acids in the liver must be secondary to their metabolism in extrahepatic tissues.

### Triacylglycerols of Chylomicrons & VLDL Are Hydrolyzed by Lipoprotein Lipase

**Lipoprotein lipase** is located on the walls of blood capillaries, anchored to the endothelium by negatively charged proteoglycan chains of heparan sulfate. It has been found in heart, adipose tissue, spleen, lung, renal medulla, aorta, diaphragm, and lactating mammary gland, though it is not active in adult liver. It is not normally found in blood; however, following injection of **heparin,** lipoprotein lipase is released from its heparan sulfate binding into the circulation. **Hepatic lipase** is bound to the sinusoidal surface of liver cells and is released by heparin. This enzyme, however, does not react readily with chylomicrons or VLDL but is concerned with chylomicron remnant and HDL metabolism.

Both **phospholipids** and **apo C-II** are required as cofactors for lipoprotein lipase activity, while apo A-II

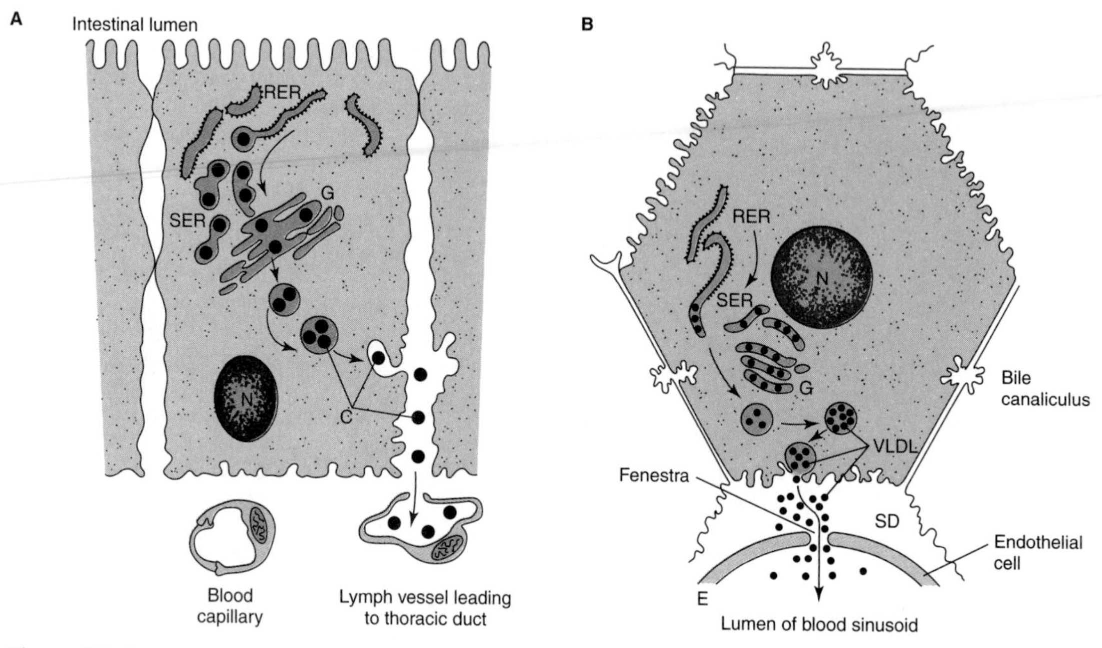


***Figure 25–2.*** The formation and secretion of **(A)** chylomicrons by an intestinal cell and **(B)** very low density lipoproteins by a hepatic cell. (RER, rough endoplasmic reticulum; SER, smooth endoplasmic reticulum; G, Golgi apparatus; N, nucleus; C, chylomicrons; VLDL, very low density lipoproteins; E, endothelium; SD, space of Disse, containing blood plasma.) Apolipoprotein B, synthesized in the RER, is incorporated into lipoproteins in the SER, the main site of synthesis of triacylglycerol. After addition of carbohydrate residues in G, they are released from the cell by reverse pinocytosis. Chylomicrons pass into the lymphatic system. VLDL are secreted into the space of Disse and then into the hepatic sinusoids through fenestrae in the endothelial lining.

and apo C-III act as inhibitors. Hydrolysis takes place while the lipoproteins are attached to the enzyme on the endothelium. Triacylglycerol is hydrolyzed progressively through a diacylglycerol to a monoacylglycerol that is finally hydrolyzed to free fatty acid plus glycerol. Some of the released free fatty acids return to the circulation, attached to albumin, but the bulk is transported into the tissue (Figures 25–3 and 25–4). Heart lipoprotein lipase has a low $K_m$ for triacylglycerol, about one-tenth of that for the enzyme in adipose tissue. This enables the delivery of fatty acids from triacylglycerol to be **redirected from adipose tissue to the heart in the starved state** when the plasma triacylglycerol decreases. A similar redirection to the mammary gland occurs during lactation, allowing uptake of lipoprotein triacylglycerol fatty acid for milk fat synthesis. The **VLDL receptor** plays an important part in the delivery of fatty acids from VLDL triacylglycerol to adipocytes by binding VLDL and bringing it into close contact with lipoprotein lipase. In adipose tissue, insulin enhances lipoprotein lipase synthesis in adipocytes and its translocation to the luminal surface of the capillary endothelium.

## The Action of Lipoprotein Lipase Forms Remnant Lipoproteins

Reaction with lipoprotein lipase results in the loss of approximately 90% of the triacylglycerol of chylomicrons and in the loss of apo C (which returns to HDL) but not apo E, which is retained. The resulting **chylomicron remnant** is about half the diameter of the parent chylomicron and is relatively enriched in cholesterol and cholesteryl esters because of the loss of triacylglycerol (Figure 25–3). Similar changes occur to VLDL, with the formation of VLDL remnants or IDL (intermediate-density lipoprotein) (Figure 25–4).

## The Liver Is Responsible for the Uptake of Remnant Lipoproteins

Chylomicron remnants are taken up by the liver by receptor-mediated endocytosis, and the cholesteryl esters and triacylglycerols are hydrolyzed and metabolized. Uptake is mediated by a **receptor specific for apo E** (Figure 25–3), and both the **LDL (apo B-100, E) receptor** and the **LRP (LDL receptor-related protein)**

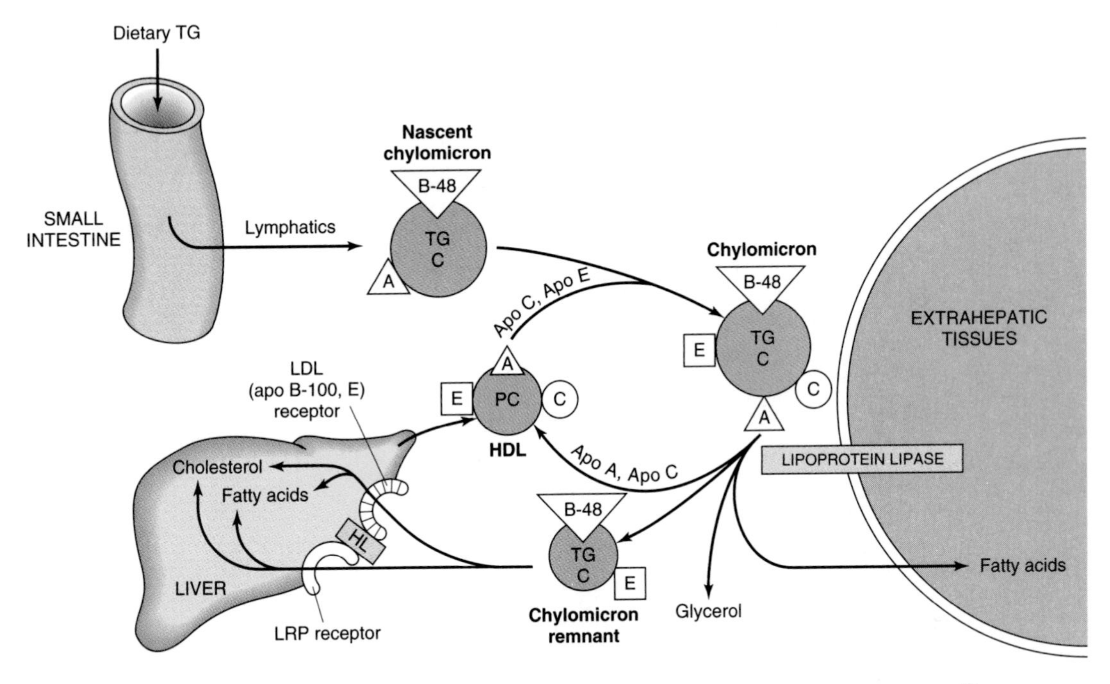


***Figure 25–3.*** Metabolic fate of chylomicrons. (A, apolipoprotein A; B-48, apolipoprotein B-48; ©, apolipoprotein C; E, apolipoprotein E; HDL, high-density lipoprotein; TG, triacylglycerol; C, cholesterol and cholesteryl ester; P, phospholipid; HL, hepatic lipase; LRP, LDL receptor-related protein.) Only the predominant lipids are shown.

are believed to take part. Hepatic lipase has a dual role: (1) in acting as a ligand to the lipoprotein and (2) in hydrolyzing its triacylglycerol and phospholipid.

VLDL is the precursor of IDL, which is then converted to LDL. Only one molecule of apo B-100 is present in each of these lipoprotein particles, and this is conserved during the transformations. Thus, each LDL particle is derived from only one VLDL particle (Figure 25–4). Two possible fates await IDL. It can be taken up by the liver directly via the LDL (apo B-100, E) receptor, or it is converted to LDL. In humans, a relatively large proportion forms LDL, accounting for the increased concentrations of LDL in humans compared with many other mammals.

## LDL IS METABOLIZED VIA THE LDL RECEPTOR

The liver and many extrahepatic tissues express the **LDL (B-100, E) receptor.** It is so designated because it is specific for apo B-100 but not B-48, which lacks the carboxyl terminal domain of B-100 containing the LDL receptor ligand, and it also takes up lipoproteins rich in apo E. This receptor is defective in **familial hypercholesterolemia.** Approximately 30% of LDL is degraded in extrahepatic tissues and 70% in the liver. A positive correlation exists between the incidence of **coronary atherosclerosis** and the plasma concentration of LDL cholesterol. For further discussion of the regulation of the LDL receptor, see Chapter 26.

## HDL TAKES PART IN BOTH LIPOPROTEIN TRIACYLGLYCEROL & CHOLESTEROL METABOLISM

HDL is synthesized and secreted from both liver and intestine (Figure 25–5). However, apo C and apo E are synthesized in the liver and transferred from liver HDL to intestinal HDL when the latter enters the plasma. A major function of HDL is to act as a repository for the apo C and apo E required in the metabolism of chylomicrons and VLDL. Nascent HDL consists of discoid phospholipid bilayers containing apo A and free cholesterol. These lipoproteins are similar to the particles found in the plasma of patients with a deficiency of the plasma enzyme **lecithin:cholesterol acyltransferase (LCAT)** and in the plasma of patients with obstructive jaundice. LCAT—and the LCAT activator apo A-I—bind to the disk, and the surface phospholipid and free cholesterol are converted into cholesteryl esters and

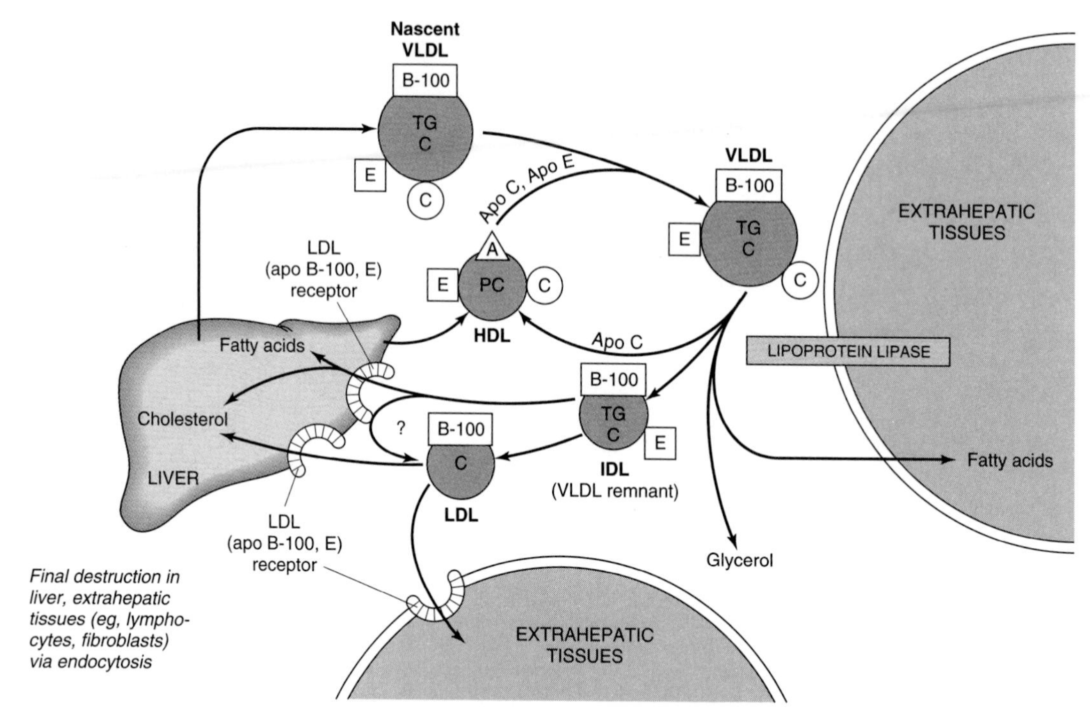


***Figure 25–4.*** Metabolic fate of very low density lipoproteins (VLDL) and production of low-density lipoproteins (LDL). (A, apolipoprotein A; B-100, apolipoprotein B-100; ©, apolipoprotein C; E, apolipoprotein E; HDL, high-density lipoprotein; TG, triacylglycerol; IDL, intermediate-density lipoprotein; C, cholesterol and cholesteryl ester; P, phospholipid.) Only the predominant lipids are shown. It is possible that some IDL is also metabolized via the LRP.

lysolecithin (Chapter 24). The nonpolar cholesteryl esters move into the hydrophobic interior of the bilayer, whereas lysolecithin is transferred to plasma albumin. Thus, a nonpolar core is generated, forming a spherical, pseudomicellar HDL covered by a surface film of polar lipids and apolipoproteins. In this way, the LCAT system is involved in the removal of excess unesterified cholesterol from lipoproteins and tissues. The **class B scavenger receptor B1 (SR-B1)** has recently been identified as an **HDL receptor** in the liver and in steroidogenic tissues. HDL binds to the receptor via apo A-I and cholesteryl ester is selectively delivered to the cells, but the particle itself, including apo A-I, is not taken up. The transport of cholesterol from the tissues to the liver is known as **reverse cholesterol transport** and is mediated by an HDL cycle (Figure 25–5). The smaller $HDL_3$ accepts cholesterol from the tissues via the **ATP-binding cassette transporter-1 (ABC-1).** ABC-1 is a member of a family of transporter proteins that couple the hydrolysis of ATP to the binding of a substrate, enabling it to be transported across the membrane. After being accepted by $HDL_3$, the cholesterol is then esterified by LCAT, increasing the size of the particles to form the less dense $HDL_2$. The cycle is completed by the re-formation of $HDL_3$, either after selective delivery of cholesteryl ester to the liver via the SR-B1 or by hydrolysis of $HDL_2$ phospholipid and triacylglycerol by hepatic lipase. In addition, free apo A-I is released by these processes and forms **preβ-HDL** after associating with a minimum amount of phospholipid and cholesterol. Preβ-HDL is the most potent form of HDL in inducing cholesterol efflux from the tissues to form discoidal HDL. Surplus apo A-I is destroyed in the kidney.

HDL concentrations vary reciprocally with plasma triacylglycerol concentrations and directly with the activity of lipoprotein lipase. This may be due to surplus surface constituents, eg, phospholipid and apo A-I being released during hydrolysis of chylomicrons and VLDL and contributing toward the formation of preβ-HDL and discoidal HDL. $HDL_2$ concentrations are **inversely related to the incidence of coronary atherosclerosis,** possibly because they reflect the efficiency of reverse cholesterol transport. $HDL_c$ ($HDL_1$) is found in

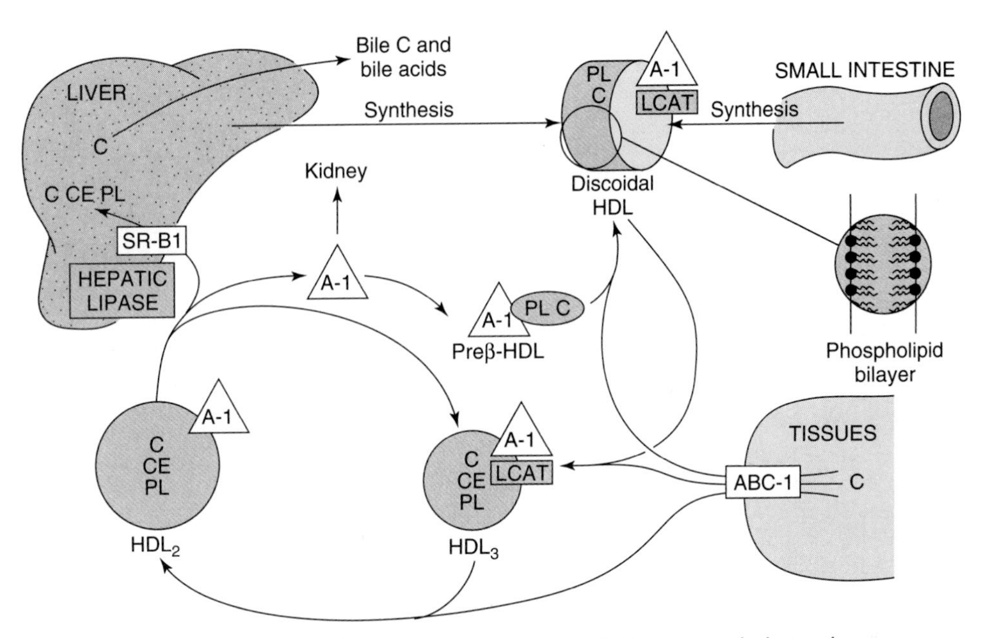


***Figure 25–5.*** Metabolism of high-density lipoprotein (HDL) in reverse cholesterol transport. (LCAT, lecithin:cholesterol acyltransferase; C, cholesterol; CE, cholesteryl ester; PL, phospholipid; A-I, apolipoprotein A-I; SR-B1, scavenger receptor B1; ABC-1, ATP binding cassette transporter 1.) Preβ-HDL, $HDL_2$, $HDL_3$—see Table 25–1. Surplus surface constituents from the action of lipoprotein lipase on chylomicrons and VLDL are another source of preβ-HDL. Hepatic lipase activity is increased by androgens and decreased by estrogens, which may account for higher concentrations of plasma $HDL_2$ in women.

the blood of diet-induced hypercholesterolemic animals. It is rich in cholesterol, and its sole apolipoprotein is apo E. It appears that all plasma lipoproteins are interrelated components of one or more metabolic cycles that together are responsible for the complex process of plasma lipid transport.

## THE LIVER PLAYS A CENTRAL ROLE IN LIPID TRANSPORT & METABOLISM

The liver carries out the following major functions in lipid metabolism: (1) It facilitates the digestion and absorption of lipids by the production of bile, which contains cholesterol and bile salts synthesized within the liver de novo or from uptake of lipoprotein cholesterol (Chapter 26). (2) The liver has active enzyme systems for synthesizing and oxidizing fatty acids (Chapters 21 and 22) and for synthesizing triacylglycerols and phospholipids (Chapter 24). (3) It converts fatty acids to ketone bodies (ketogenesis) (Chapter 22). (4) It plays an integral part in the synthesis and metabolism of plasma lipoproteins (this chapter).

### Hepatic VLDL Secretion Is Related to Dietary & Hormonal Status

The cellular events involved in VLDL formation and secretion have been described above. Hepatic triacylglycerol synthesis provides the immediate stimulus for the formation and secretion of VLDL. The fatty acids used are derived from two possible sources: (1) synthesis within the liver from acetyl-CoA derived mainly from carbohydrate (perhaps not so important in humans) and (2) uptake of free fatty acids from the circulation. The first source is predominant in the well-fed condition, when fatty acid synthesis is high and the level of circulating free fatty acids is low. As triacylglycerol does not normally accumulate in the liver under this condition, it must be inferred that it is transported from the liver in VLDL as rapidly as it is synthesized and that the synthesis of apo B-100 is not rate-limiting. Free fatty acids from the circulation are the main source during starvation, the feeding of high-fat diets, or in diabetes mellitus, when hepatic lipogenesis is inhibited. Factors that enhance both the synthesis of triacylglycerol and the secretion of VLDL by the liver include (1) the fed state rather than the starved state; (2) the feeding of diets high in carbohydrate (particularly if they contain sucrose or fructose), leading to high rates of lipogenesis and esterification of fatty acids; (3) high levels of circulating free fatty acids; (4) ingestion of ethanol; and (5) the presence of high concentrations of insulin and low concentrations of glucagon, which enhance fatty acid synthesis and esterification and inhibit their oxidation (Figure 25–6).

## CLINICAL ASPECTS

### Imbalance in the Rate of Triacylglycerol Formation & Export Causes Fatty Liver

For a variety of reasons, lipid—mainly as triacylglycerol—can accumulate in the liver (Figure 25–6). Extensive accumulation is regarded as a pathologic condition. When accumulation of lipid in the liver becomes chronic, fibrotic changes occur in the cells that progress to **cirrhosis** and impaired liver function.

Fatty livers fall into two main categories. The first type is associated with **raised levels of plasma free fatty acids** resulting from mobilization of fat from adipose tissue or from the hydrolysis of lipoprotein triacylglycerol by lipoprotein lipase in extrahepatic tissues. The production of VLDL does not keep pace with the increasing influx and esterification of free fatty acids, allowing triacylglycerol to accumulate, causing a fatty liver. This occurs during **starvation** and the feeding of **high-fat diets.** The ability to secrete VLDL may also be impaired (eg, in starvation). In uncontrolled **diabetes mellitus, twin lamb disease,** and **ketosis in cattle,** fatty infiltration is sufficiently severe to cause visible pallor (fatty appearance) and enlargement of the liver with possible liver dysfunction.

The second type of fatty liver is usually due to a **metabolic block in the production of plasma lipoproteins,** thus allowing triacylglycerol to accumulate. Theoretically, the lesion may be due to (1) a block in apolipoprotein synthesis, (2) a block in the synthesis of the lipoprotein from lipid and apolipoprotein, (3) a failure in provision of phospholipids that are found in lipoproteins, or (4) a failure in the secretory mechanism itself.

One type of fatty liver that has been studied extensively in rats is due to a deficiency of **choline,** which has therefore been called a **lipotropic factor.** The antibiotic puromycin, ethionine ($\alpha$-amino-$\gamma$-mercaptobutyric acid), carbon tetrachloride, chloroform, phosphorus, lead, and arsenic all cause fatty liver and a marked reduction in concentration of VLDL in rats. Choline will not protect the organism against these agents but appears to aid in recovery. The action of carbon tetrachloride probably involves formation of free radicals causing lipid peroxidation. Some protection against this is provided by the antioxidant action of vitamin E-supplemented diets. The action of ethionine is thought to be due to a reduction in availability of ATP due to its replacing methionine in *S*-adenosylmethionine, trapping available adenine and preventing synthesis of ATP. Orotic acid also causes fatty liver; it is believed to interfere with glycosylation of the lipoprotein, thus inhibiting release, and may also impair the recruitment of triacylglycerol to the particles. A deficiency of vitamin E enhances the hepatic necrosis of the choline deficiency type of fatty liver. Added vitamin E or a source of selenium has a protective effect by combating lipid peroxidation. In addition to protein deficiency, essential fatty acid and vitamin deficiencies (eg, linoleic acid, pyridoxine, and pantothenic acid) can cause fatty infiltration of the liver.

### Ethanol Also Causes Fatty Liver

**Alcoholism** leads to fat accumulation in the liver, hyperlipidemia, and ultimately **cirrhosis.** The exact mechanism of action of ethanol in the long term is still uncertain. Ethanol consumption over a long period leads to the accumulation of fatty acids in the liver that are derived from endogenous synthesis rather than from increased mobilization from adipose tissue. There is no impairment of hepatic synthesis of protein after ethanol ingestion. Oxidation of ethanol by **alcohol dehydrogenase** leads to excess production of NADH.

$$\underset{\textbf{Ethanol}}{CH_3-CH_2-OH} \xrightarrow[NAD^+ \quad NADH + H^+]{\text{ALCOHOL DEHYDROGENASE}} \underset{\textbf{Acetaldehyde}}{CH_3-CHO}$$

The NADH generated competes with reducing equivalents from other substrates, including fatty acids, for the respiratory chain, inhibiting their oxidation, and decreasing activity of the citric acid cycle. The net effect of inhibiting fatty acid oxidation is to cause increased esterification of fatty acids in triacylglycerol, resulting in the fatty liver. Oxidation of ethanol leads to the formation of acetaldehyde, which is oxidized by **aldehyde dehydrogenase,** producing acetate. Other effects of ethanol may include increased lipogenesis and cholesterol synthesis from acetyl-CoA, and lipid peroxidation. The increased [NADH]/[$NAD^+$] ratio also causes increased [lactate]/[pyruvate], resulting in **hyperlactic-acidemia,** which decreases excretion of uric acid, aggravating gout. Some metabolism of ethanol takes place via a cytochrome P450-dependent microsomal ethanol oxidizing system (MEOS) involving NADPH and $O_2$. This system increases in activity in **chronic alcoholism**

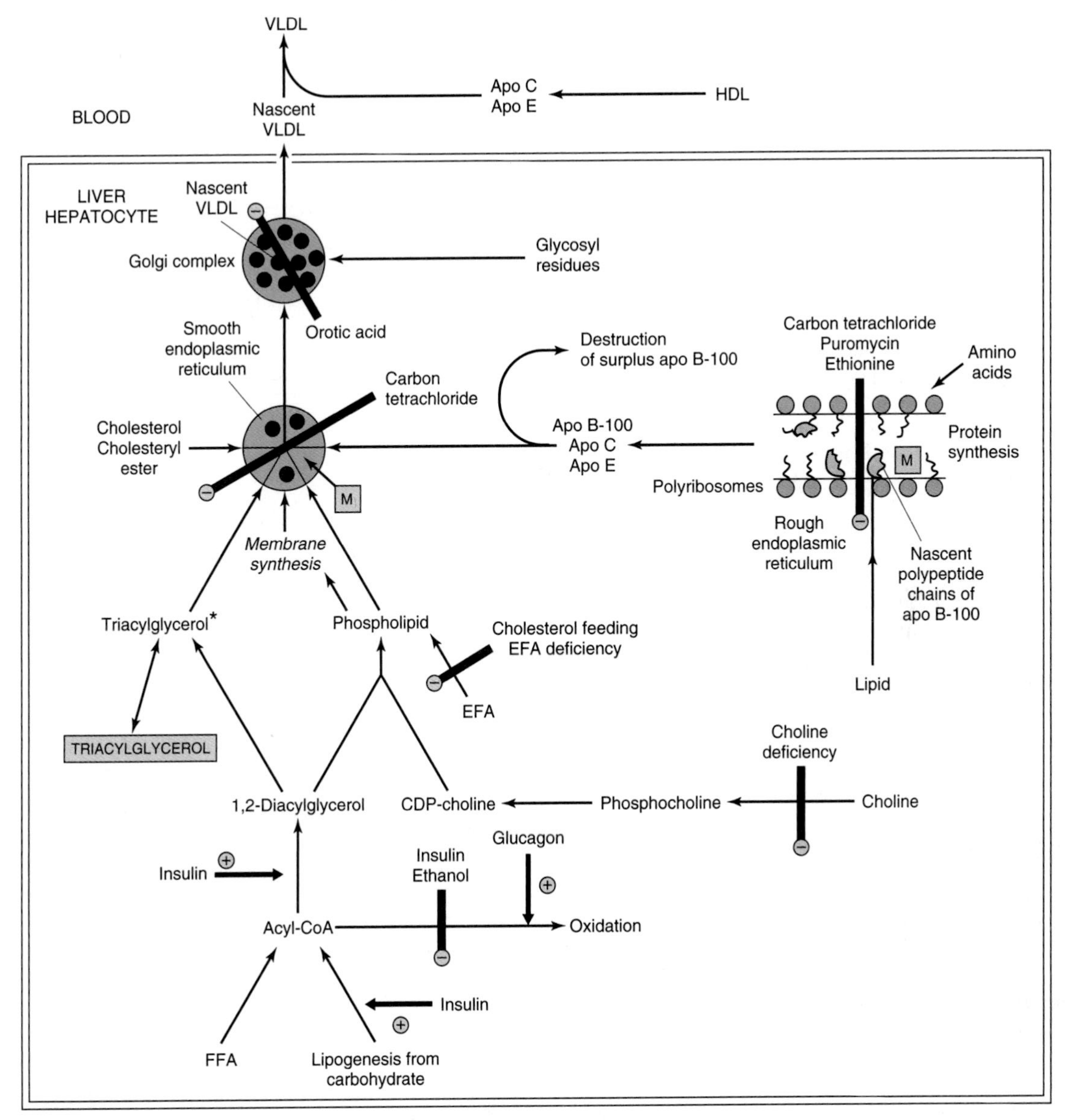


***Figure 25–6.*** The synthesis of very low density lipoprotein (VLDL) in the liver and the possible loci of action of factors causing accumulation of triacylglycerol and a fatty liver. (EFA, essential fatty acids; FFA, free fatty acids; HDL, high-density lipoproteins; Apo, apolipoprotein; M, microsomal triacylglycerol transfer protein.) The pathways indicated form a basis for events depicted in Figure 25–2. The main triacylglycerol pool in liver is not on the direct pathway of VLDL synthesis from acyl-CoA. Thus, FFA, insulin, and glucagon have immediate effects on VLDL secretion as their effects impinge directly on the small triacylglycerol* precursor pool. In the fully fed state, apo B-100 is synthesized in excess of requirements for VLDL secretion and the surplus is destroyed in the liver. During translation of apo B-100, microsomal transfer protein-mediated lipid transport enables lipid to become associated with the nascent polypeptide chain. After release from the ribosomes, these particles fuse with more lipids from the smooth endoplasmic reticulum, producing nascent VLDL.

and may account for the increased metabolic clearance in this condition. Ethanol will also inhibit the metabolism of some drugs, eg, barbiturates, by competing for cytochrome P450-dependent enzymes.

$$\underset{\textbf{Ethanol}}{CH_3-CH_2-OH} + NADPH + H^+ + O_2 \xrightarrow{\text{MEOS}} \underset{\textbf{Acetaldehyde}}{CH_3-CHO} + NADP^+ + 2H_2O$$

In some Asian populations and Native Americans, alcohol consumption results in increased adverse reactions to acetaldehyde owing to a genetic defect of mitochondrial aldehyde dehydrogenase.

## ADIPOSE TISSUE IS THE MAIN STORE OF TRIACYLGLYCEROL IN THE BODY

The triacylglycerol stores in adipose tissue are continually undergoing lipolysis (hydrolysis) and reesterification (Figure 25–7). These two processes are entirely different pathways involving different reactants and enzymes. This allows the processes of esterification or lipolysis to be regulated separately by many nutritional, metabolic, and hormonal factors. The resultant of these two processes determines the magnitude of the free fatty acid pool in adipose tissue, which in turn determines the level of free fatty acids circulating in the plasma. Since the latter has most profound effects upon the metabolism of other tissues, particularly liver and muscle, the factors operating in adipose tissue that regulate the outflow of free fatty acids exert an influence far beyond the tissue itself.

### The Provision of Glycerol 3-Phosphate Regulates Esterification: Lipolysis Is Controlled by Hormone-Sensitive Lipase (Figure 25–7)

Triacylglycerol is synthesized from acyl-CoA and glycerol 3-phosphate (Figure 24–2). Because the enzyme **glycerol kinase** is not expressed in adipose tissue, glycerol cannot be utilized for the provision of glycerol 3-phosphate, which must be supplied by glucose via glycolysis.

Triacylglycerol undergoes hydrolysis by a **hormone-sensitive lipase** to form free fatty acids and glycerol. This lipase is distinct from lipoprotein lipase that catalyzes lipoprotein triacylglycerol hydrolysis before its uptake into extrahepatic tissues (see above). Since glycerol cannot be utilized, it diffuses into the blood, whence it is utilized by tissues such as those of the liver and kidney, which possess an active glycerol kinase.

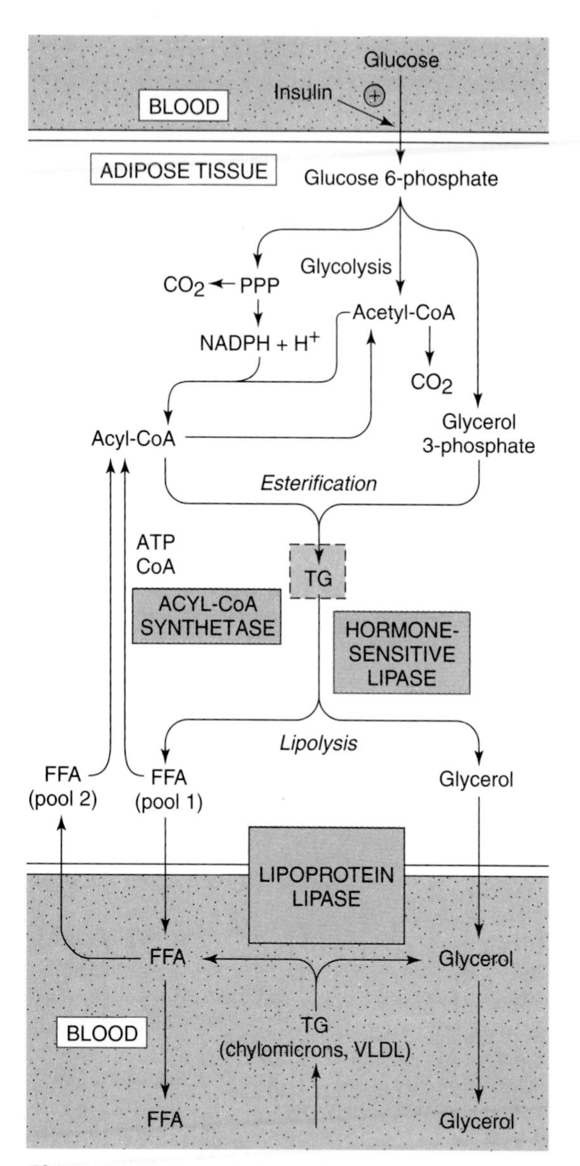


***Figure 25–7.*** Metabolism of adipose tissue. Hormone-sensitive lipase is activated by ACTH, TSH, glucagon, epinephrine, norepinephrine, and vasopressin and inhibited by insulin, prostaglandin $E_1$, and nicotinic acid. Details of the formation of glycerol 3-phosphate from intermediates of glycolysis are shown in Figure 24–2. (PPP, pentose phosphate pathway; TG, triacylglycerol; FFA, free fatty acids; VLDL, very low density lipoprotein.)

The free fatty acids formed by lipolysis can be reconverted in the tissue to acyl-CoA by **acyl-CoA synthetase** and reesterified with glycerol 3-phosphate to form triacylglycerol. Thus, **there is a continuous cycle of lipolysis and reesterification within the tissue.** However, when the rate of reesterification is not sufficient to match the rate of lipolysis, free fatty acids accumulate and diffuse into the plasma, where they bind to albumin and raise the concentration of plasma free fatty acids.

### Increased Glucose Metabolism Reduces the Output of Free Fatty Acids

When the utilization of glucose by adipose tissue is increased, the free fatty acid outflow decreases. However, the release of glycerol continues, demonstrating that the effect of glucose is not mediated by reducing the rate of lipolysis. The effect is due to the provision of glycerol 3-phosphate, which enhances esterification of free fatty acids. Glucose can take several pathways in adipose tissue, including oxidation to $CO_2$ via the citric acid cycle, oxidation in the pentose phosphate pathway, conversion to long-chain fatty acids, and formation of acylglycerol via glycerol 3-phosphate (Figure 25–7). When glucose utilization is high, a larger proportion of the uptake is oxidized to $CO_2$ and converted to fatty acids. However, as total glucose utilization decreases, the greater proportion of the glucose is directed to the formation of glycerol 3-phosphate for the esterification of acyl-CoA, which helps to minimize the efflux of free fatty acids.

## HORMONES REGULATE FAT MOBILIZATION

### Insulin Reduces the Output of Free Fatty Acids

The rate of release of free fatty acids from adipose tissue is affected by many hormones that influence either the rate of esterification or the rate of lipolysis. Insulin inhibits the release of free fatty acids from adipose tissue, which is followed by a fall in circulating plasma free fatty acids. It enhances lipogenesis and the synthesis of acylglycerol and increases the oxidation of glucose to $CO_2$ via the pentose phosphate pathway. All of these effects are dependent on the presence of glucose and can be explained, to a large extent, on the basis of the ability of insulin to enhance the uptake of glucose into adipose cells via the GLUT 4 transporter. Insulin also increases the activity of pyruvate dehydrogenase, acetyl-CoA carboxylase, and glycerol phosphate acyltransferase, reinforcing the effects of increased glucose uptake on the enhancement of fatty acid and acylglycerol synthesis. These three enzymes are now known to be regulated in a coordinate manner by phosphorylation-dephosphorylation mechanisms.

A principal action of insulin in adipose tissue is to inhibit the activity of **hormone-sensitive lipase,** reducing the release not only of free fatty acids but of glycerol as well. Adipose tissue is much more sensitive to insulin than are many other tissues, which points to adipose tissue as a major site of insulin action in vivo.

### Several Hormones Promote Lipolysis

Other hormones accelerate the release of free fatty acids from adipose tissue and raise the plasma free fatty acid concentration by increasing the rate of lipolysis of the triacylglycerol stores (Figure 25–8). These include epinephrine, norepinephrine, glucagon, adrenocorticotropic hormone (ACTH), α- and β-melanocyte-stimulating hormones (MSH), thyroid-stimulating hormone (TSH), growth hormone (GH), and vasopressin. Many of these activate the hormone-sensitive lipase. For an optimal effect, most of these lipolytic processes require the presence of **glucocorticoids** and **thyroid hormones.** These hormones act in a **facilitatory** or **permissive** capacity with respect to other lipolytic endocrine factors.

The hormones that act rapidly in promoting lipolysis, ie, catecholamines, do so by stimulating the activity of **adenylyl cyclase,** the enzyme that converts ATP to cAMP. The mechanism is analogous to that responsible for hormonal stimulation of glycogenolysis (Chapter 18). cAMP, by stimulating **cAMP-dependent protein kinase,** activates hormone-sensitive lipase. Thus, processes which destroy or preserve cAMP influence lipolysis. cAMP is degraded to 5′-AMP by the enzyme **cyclic 3′,5′-nucleotide phosphodiesterase.** This enzyme is inhibited by methylxanthines such as **caffeine** and **theophylline. Insulin** antagonizes the effect of the lipolytic hormones. Lipolysis appears to be more sensitive to changes in concentration of insulin than are glucose utilization and esterification. The antilipolytic effects of insulin, nicotinic acid, and prostaglandin $E_1$ are accounted for by inhibition of the synthesis of cAMP at the adenylyl cyclase site, acting through a $G_i$ protein. Insulin also stimulates phosphodiesterase and the lipase phosphatase that inactivates hormone-sensitive lipase. The effect of growth hormone in promoting lipolysis is dependent on synthesis of proteins involved in the formation of cAMP. Glucocorticoids promote lipolysis via synthesis of new lipase protein by a cAMP-independent pathway, which may be inhibited by insulin, and also by promoting transcription of genes involved in the cAMP signal cascade. These findings help to explain the role of the pituitary gland and the adrenal cortex in enhancing fat mobilization. The recently discovered body weight regulatory hormone, **leptin,** stimulates

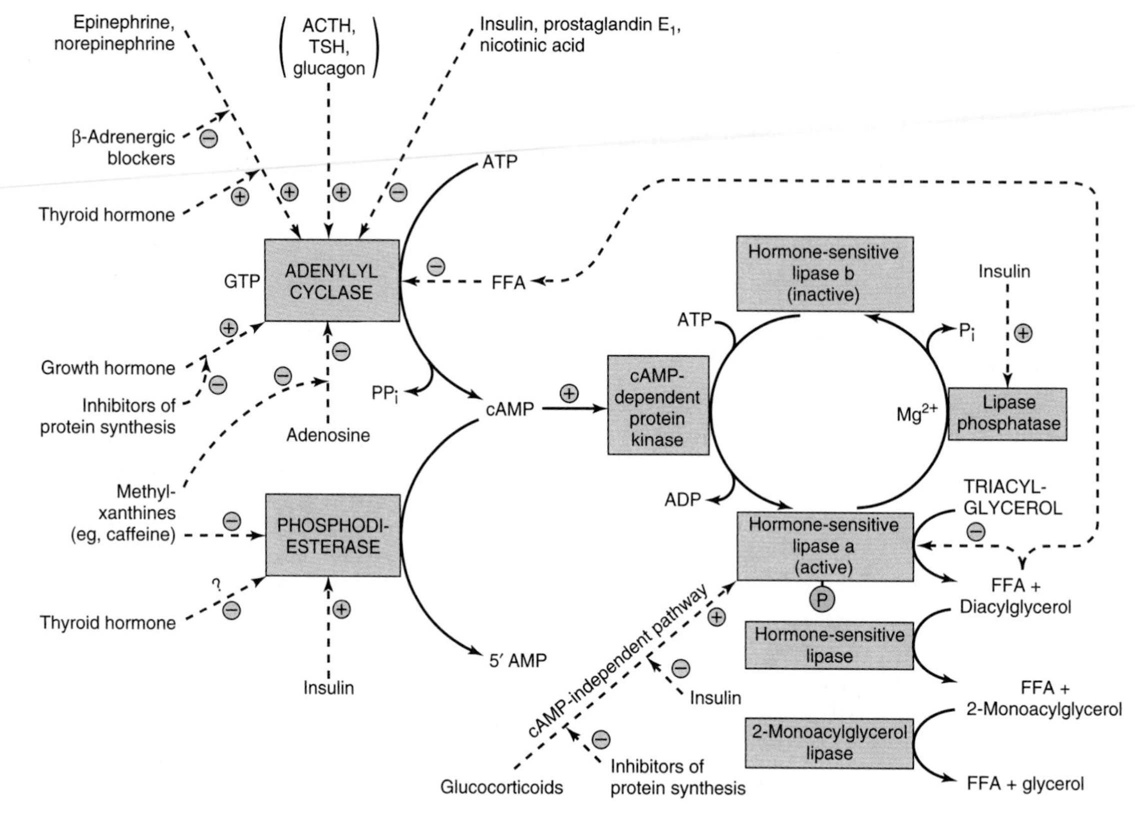


***Figure 25–8.*** Control of adipose tissue lipolysis. (TSH, thyroid-stimulating hormone; FFA, free fatty acids.) Note the cascade sequence of reactions affording amplification at each step. The lipolytic stimulus is "switched off" by removal of the stimulating hormone; the action of lipase phosphatase; the inhibition of the lipase and adenylyl cyclase by high concentrations of FFA; the inhibition of adenylyl cyclase by adenosine; and the removal of cAMP by the action of phosphodiesterase. ACTH, TSH, and glucagon may not activate adenylyl cyclase in vivo, since the concentration of each hormone required in vitro is much higher than is found in the circulation. Positive (⊕) and negative (⊖) regulatory effects are represented by broken lines and substrate flow by solid lines.

lipolysis and inhibits lipogenesis by influencing the activity of the enzymes in the pathways for the breakdown and synthesis of fatty acids.

The sympathetic nervous system, through liberation of norepinephrine in adipose tissue, plays a central role in the mobilization of free fatty acids. Thus, the increased lipolysis caused by many of the factors described above can be reduced or abolished by denervation of adipose tissue or by ganglionic blockade.

## A Variety of Mechanisms Have Evolved for Fine Control of Adipose Tissue Metabolism

Human adipose tissue may not be an important site of lipogenesis. There is no significant incorporation of glucose or pyruvate into long-chain fatty acids; ATP-citrate lyase, a key enzyme in lipogenesis, does not appear to be present, and other lipogenic enzymes—eg, glucose-6-phosphate dehydrogenase and the malic enzyme—do not undergo adaptive changes. Indeed, it has been suggested that in humans there is a "carbohydrate excess syndrome" due to a unique limitation in ability to dispose of excess carbohydrate by lipogenesis. In birds, lipogenesis is confined to the liver, where it is particularly important in providing lipids for egg formation, stimulated by estrogens. Human adipose tissue is unresponsive to most of the lipolytic hormones apart from the catecholamines.

On consideration of the profound derangement of metabolism in **diabetes mellitus** (due in large part to increased release of free fatty acids from the depots) and the fact that insulin to a large extent corrects the condi-

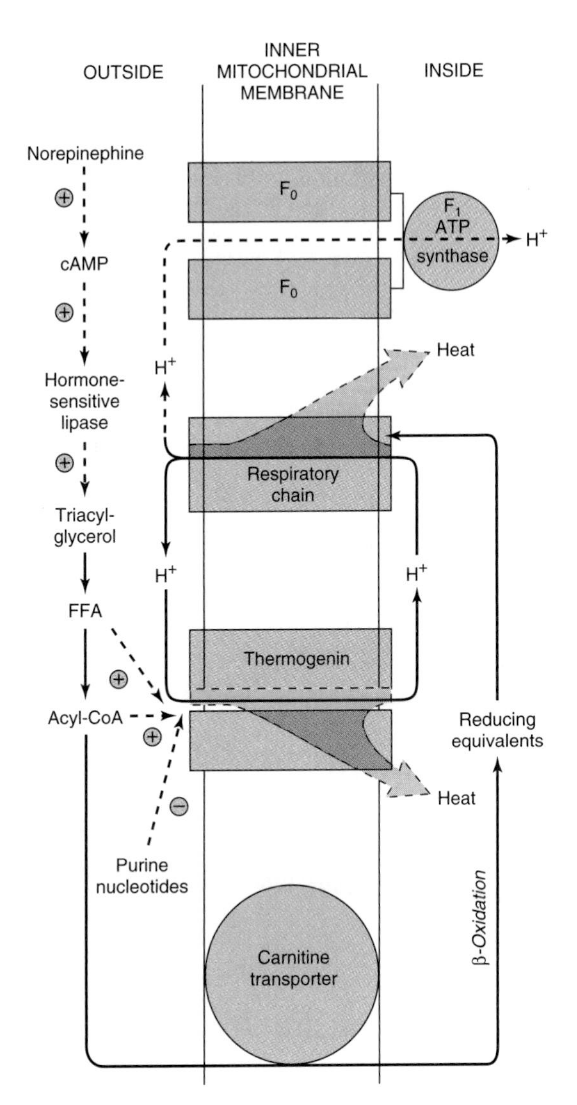


***Figure 25–9.*** Thermogenesis in brown adipose tissue. Activity of the respiratory chain produces heat in addition to translocating protons (Chapter 12). These protons dissipate more heat when returned to the inner mitochondrial compartment via thermogenin instead of generating ATP when returning via the $F_1$ ATP synthase. The passage of $H^+$ via thermogenin is inhibited by purine nucleotides when brown adipose tissue is unstimulated. Under the influence of norepinephrine, the inhibition is removed by the production of free fatty acids (FFA) and acyl-CoA. Note the dual role of acyl-CoA in both facilitating the action of thermogenin and supplying reducing equivalents for the respiratory chain. ⊕ and ⊖ signify positive or negative regulatory effects.

tion, it must be concluded that insulin plays a prominent role in the regulation of adipose tissue metabolism.

## BROWN ADIPOSE TISSUE PROMOTES THERMOGENESIS

Brown adipose tissue is involved in metabolism particularly at times when heat generation is necessary. Thus, the tissue is extremely active in some species in arousal from hibernation, in animals exposed to cold (nonshivering thermogenesis), and in heat production in the newborn animal. Though not a prominent tissue in humans, it is present in normal individuals, where it could be responsible for **"diet-induced thermogenesis."** It is noteworthy that brown adipose tissue is reduced or absent in obese persons. The tissue is characterized by a well-developed blood supply and a high content of mitochondria and cytochromes but low activity of ATP synthase. Metabolic emphasis is placed on oxidation of both glucose and fatty acids. **Norepinephrine** liberated from sympathetic nerve endings is important in increasing lipolysis in the tissue and increasing synthesis of lipoprotein lipase to enhance utilization of triacylglycerol-rich lipoproteins from the circulation. Oxidation and phosphorylation are not coupled in mitochondria of this tissue, and the phosphorylation that does occur is at the substrate level, eg, at the succinate thiokinase step and in glycolysis. Thus, **oxidation produces much heat, and little free energy is trapped in ATP.** A thermogenic uncoupling protein, **thermogenin,** acts as a proton conductance pathway dissipating the electrochemical potential across the mitochondrial membrane (Figure 25–9).

## SUMMARY

- Since nonpolar lipids are insoluble in water, for transport between the tissues in the aqueous blood plasma they are combined with amphipathic lipids and proteins to make water-miscible lipoproteins.
- Four major groups of lipoproteins are recognized: Chylomicrons transport lipids resulting from digestion and absorption. Very low density lipoproteins (VLDL) transport triacylglycerol from the liver. Low-density lipoproteins (LDL) deliver cholesterol to the tissues, and high-density lipoproteins (HDL) remove cholesterol from the tissues in the process known as reverse cholesterol transport.
- Chylomicrons and VLDL are metabolized by hydrolysis of their triacylglycerol, and lipoprotein remnants are left in the circulation. These are taken up by liver, but some of the remnants (IDL) resulting from VLDL form LDL which is taken up by the liver and other tissues via the LDL receptor.

- Apolipoproteins constitute the protein moiety of lipoproteins. They act as enzyme activators (eg, apo C-II and apo A-I) or as ligands for cell receptors (eg, apo A-I, apo E, and apo B-100).
- Triacylglycerol is the main storage lipid in adipose tissue. Upon mobilization, free fatty acids and glycerol are released. Free fatty acids are an important fuel source.
- Brown adipose tissue is the site of "nonshivering thermogenesis." It is found in hibernating and newborn animals and is present in small quantity in humans. Thermogenesis results from the presence of an uncoupling protein, thermogenin, in the inner mitochondrial membrane.

## REFERENCES

Chappell DA, Medh JD: Receptor-mediated mechanisms of lipoprotein remnant catabolism. Prog Lipid Res 1998;37:393.

Eaton S et al: Multiple biochemical effects in the pathogenesis of fatty liver. Eur J Clin Invest 1997;27:719.

Goldberg IJ, Merkel M: Lipoprotein lipase: physiology, biochemistry and molecular biology. Front Biosci 2001;6:D388.

Holm C et al: Molecular mechanisms regulating hormone sensitive lipase and lipolysis. Annu Rev Nutr 2000;20:365.

Kaikans RM, Bass NM, Ockner RK: Functions of fatty acid binding proteins. Experientia 1990;46:617.

Lardy H, Shrago E: Biochemical aspects of obesity. Annu Rev Biochem 1990;59:689.

Rye K-A et al: Overview of plasma lipid transport. In: *Plasma Lipids and Their Role in Disease.* Barter PJ, Rye K-A (editors). Harwood Academic Publishers, 1999.

Shelness GS, Sellers JA: Very-low-density lipoprotein assembly and secretion. Curr Opin Lipidol 2001;12:151.

Various authors: *Biochemistry of Lipids, Lipoproteins and Membranes.* Vance DE, Vance JE (editors). Elsevier, 1996.

Various authors: Brown adipose tissue—role in nutritional energetics. (Symposium.) Proc Nutr Soc 1989;48:165.

# Cholesterol Synthesis, Transport, & Excretion

# 26

Peter A. Mayes, PhD, DSc, & Kathleen M. Botham, PhD, DSc

## BIOMEDICAL IMPORTANCE

Cholesterol is present in tissues and in plasma either as free cholesterol or as a storage form, combined with a long-chain fatty acid as cholesteryl ester. In plasma, both forms are transported in lipoproteins (Chapter 25). Cholesterol is an amphipathic lipid and as such is an essential structural component of membranes and of the outer layer of plasma lipoproteins. It is synthesized in many tissues from acetyl-CoA and is the precursor of all other steroids in the body such as corticosteroids, sex hormones, bile acids, and vitamin D. As a typical product of animal metabolism, cholesterol occurs in foods of animal origin such as egg yolk, meat, liver, and brain. Plasma low-density lipoprotein (LDL) is the vehicle of uptake of cholesterol and cholesteryl ester into many tissues. Free cholesterol is removed from tissues by plasma high-density lipoprotein (HDL) and transported to the liver, where it is eliminated from the body either unchanged or after conversion to bile acids in the process known as **reverse cholesterol transport.** Cholesterol is a major constituent of **gallstones.** However, its chief role in pathologic processes is as a factor in the genesis of **atherosclerosis** of vital arteries, causing cerebrovascular, coronary, and peripheral vascular disease.

## CHOLESTEROL IS DERIVED ABOUT EQUALLY FROM THE DIET & FROM BIOSYNTHESIS

A little more than half the cholesterol of the body arises by synthesis (about 700 mg/d), and the remainder is provided by the average diet. The liver and intestine account for approximately 10% each of total synthesis in humans. Virtually all tissues containing nucleated cells are capable of cholesterol synthesis, which occurs in the endoplasmic reticulum and the cytosol.

### Acetyl-CoA Is the Source of All Carbon Atoms in Cholesterol

The biosynthesis of cholesterol may be divided into five steps: (1) Synthesis of mevalonate occurs from acetyl-CoA (Figure 26–1). (2) Isoprenoid units are formed from mevalonate by loss of $CO_2$ (Figure 26–2). (3) Six isoprenoid units condense to form squalene. (4) Squalene cyclizes to give rise to the parent steroid, lanosterol. (5) Cholesterol is formed from lanosterol (Figure 26–3).

**Step 1—Biosynthesis of Mevalonate:** HMG-CoA (3-hydroxy-3-methylglutaryl-CoA) is formed by the reactions used in mitochondria to synthesize ketone bodies (Figure 22–7). However, since cholesterol synthesis is extramitochondrial, the two pathways are distinct. Initially, two molecules of acetyl-CoA condense to form acetoacetyl-CoA catalyzed by cytosolic **thiolase.** Acetoacetyl-CoA condenses with a further molecule of acetyl-CoA catalyzed by **HMG-CoA synthase** to form HMG-CoA, which is reduced to **mevalonate** by NADPH catalyzed by **HMG-CoA reductase.** This is the principal regulatory step in the pathway of cholesterol synthesis and is the site of action of the most effective class of cholesterol-lowering drugs, the HMG-CoA reductase inhibitors (statins) (Figure 26–1).

**Step 2—Formation of Isoprenoid Units:** Mevalonate is phosphorylated sequentially by ATP by three kinases, and after decarboxylation (Figure 26–2) the active isoprenoid unit, **isopentenyl diphosphate,** is formed.

**Step 3—Six Isoprenoid Units Form Squalene:** Isopentenyl diphosphate is isomerized by a shift of the double bond to form **dimethylallyl diphosphate,** then condensed with another molecule of isopentenyl diphosphate to form the ten-carbon intermediate **geranyl diphosphate** (Figure 26–2). A further condensation with isopentenyl diphosphate forms **farnesyl diphosphate.** Two molecules of farnesyl diphosphate condense at the diphosphate end to form **squalene.** Initially, inorganic pyrophosphate is eliminated, forming presqualene diphosphate, which is then reduced by NADPH with elimination of a further inorganic pyrophosphate molecule.

**Step 4—Formation of Lanosterol:** Squalene can fold into a structure that closely resembles the steroid nucleus (Figure 26–3). Before ring closure occurs, squalene is converted to squalene 2,3-epoxide by a mixed-

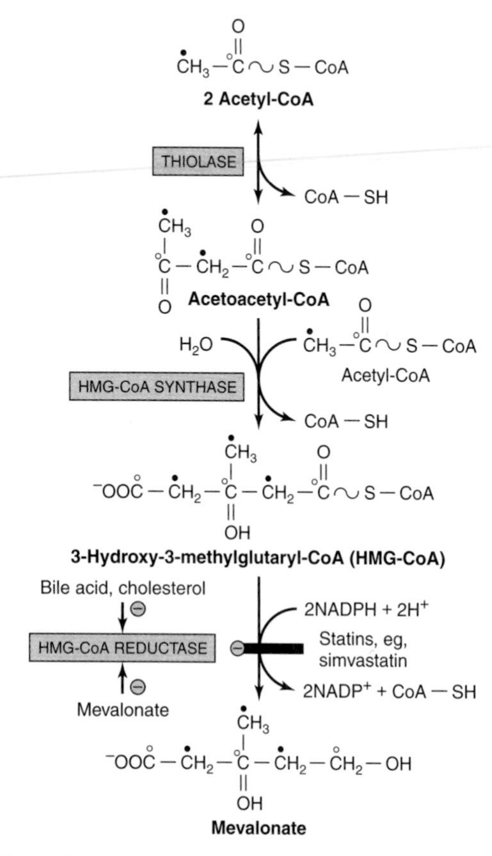


***Figure 26–1.*** Biosynthesis of mevalonate. HMG-CoA reductase is inhibited by atorvastatin, pravastatin, and simvastatin. The open and solid circles indicate the fate of each of the carbons in the acetyl moiety of acetyl-CoA.

function oxidase in the endoplasmic reticulum, **squalene epoxidase.** The methyl group on $C_{14}$ is transferred to $C_{13}$ and that on $C_8$ to $C_{14}$ as cyclization occurs, catalyzed by **oxidosqualene:lanosterol cyclase.**

**Step 5—Formation of Cholesterol:** The formation of cholesterol from **lanosterol** takes place in the membranes of the endoplasmic reticulum and involves changes in the steroid nucleus and side chain (Figure 26–3). The methyl groups on $C_{14}$ and $C_4$ are removed to form 14-desmethyl lanosterol and then zymosterol. The double bond at $C_8$–$C_9$ is subsequently moved to $C_5$–$C_6$ in two steps, forming **desmosterol.** Finally, the double bond of the side chain is reduced, producing cholesterol. The exact order in which the steps described actually take place is not known with certainty.

## Farnesyl Diphosphate Gives Rise to Dolichol & Ubiquinone

The polyisoprenoids **dolichol** (Figure 14–20 and Chapter 47) and **ubiquinone** (Figure 12–5) are formed from farnesyl diphosphate by the further addition of up to 16 (dolichol) or 3–7 (ubiquinone) isopentenyl diphosphate residues, respectively. Some GTP-binding proteins in the cell membrane are prenylated with farnesyl or geranylgeranyl (20 carbon) residues. Protein prenylation is believed to facilitate the anchoring of proteins into lipoid membranes and may also be involved in protein-protein interactions and membrane-associated protein trafficking.

# CHOLESTEROL SYNTHESIS IS CONTROLLED BY REGULATION OF HMG-CoA REDUCTASE

Regulation of cholesterol synthesis is exerted near the beginning of the pathway, at the HMG-CoA reductase step. The reduced synthesis of cholesterol in starving animals is accompanied by a decrease in the activity of the enzyme. However, it is only hepatic synthesis that is inhibited by dietary cholesterol. HMG-CoA reductase in liver is inhibited by mevalonate, the immediate product of the pathway, and by cholesterol, the main product. Cholesterol (or a metabolite, eg, oxygenated sterol) represses transcription of the HMG-CoA reductase gene and is also believed to influence translation. A **diurnal variation** occurs in both cholesterol synthesis and reductase activity. In addition to these mechanisms regulating the rate of protein synthesis, the enzyme activity is also modulated more rapidly by posttranslational modification (Figure 26–4). Insulin or thyroid hormone increases HMG-CoA reductase activity, whereas glucagon or glucocorticoids decrease it. Activity is reversibly modified by phosphorylation-dephosphorylation mechanisms, some of which may be cAMP-dependent and therefore immediately responsive to glucagon. Attempts to lower plasma cholesterol in humans by reducing the amount of cholesterol in the diet produce variable results. Generally, a decrease of 100 mg in dietary cholesterol causes a decrease of approximately 0.13 mmol/L of serum.

# MANY FACTORS INFLUENCE THE CHOLESTEROL BALANCE IN TISSUES

In tissues, cholesterol balance is regulated as follows (Figure 26–5): Cell cholesterol increase is due to uptake of cholesterol-containing lipoproteins by receptors, eg, the LDL receptor or the scavenger receptor; uptake of free cholesterol from cholesterol-rich lipoproteins to the cell

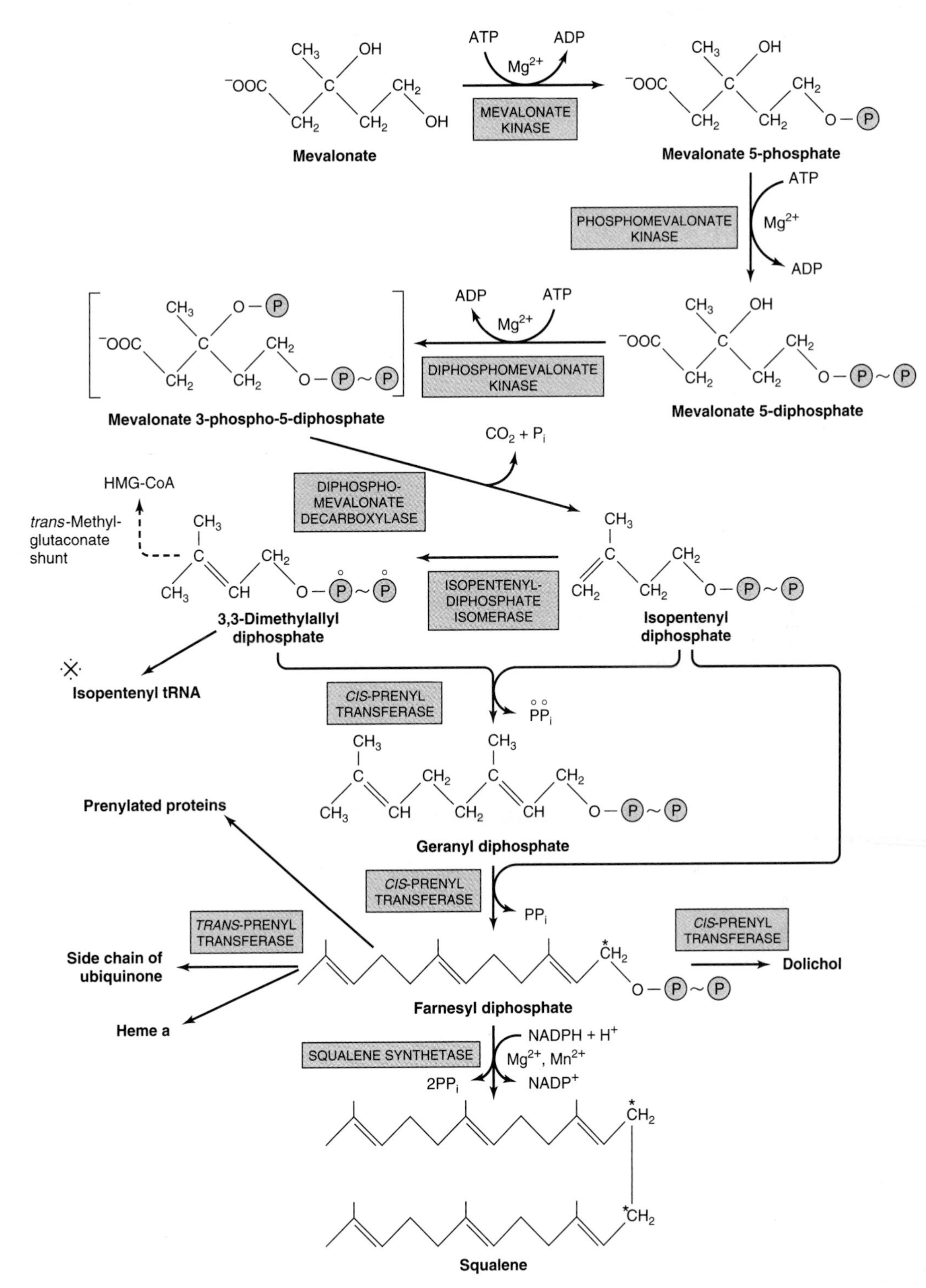


***Figure 26–2.*** Biosynthesis of squalene, ubiquinone, dolichol, and other polyisoprene derivatives. (HMG, 3-hydroxy-3-methylglutaryl; ⋇, cytokinin.) A farnesyl residue is present in heme a of cytochrome oxidase. The carbon marked with asterisk becomes $C_{11}$ or $C_{12}$ in squalene. Squalene synthetase is a microsomal enzyme; all other enzymes indicated are soluble cytosolic proteins, and some are found in peroxisomes.

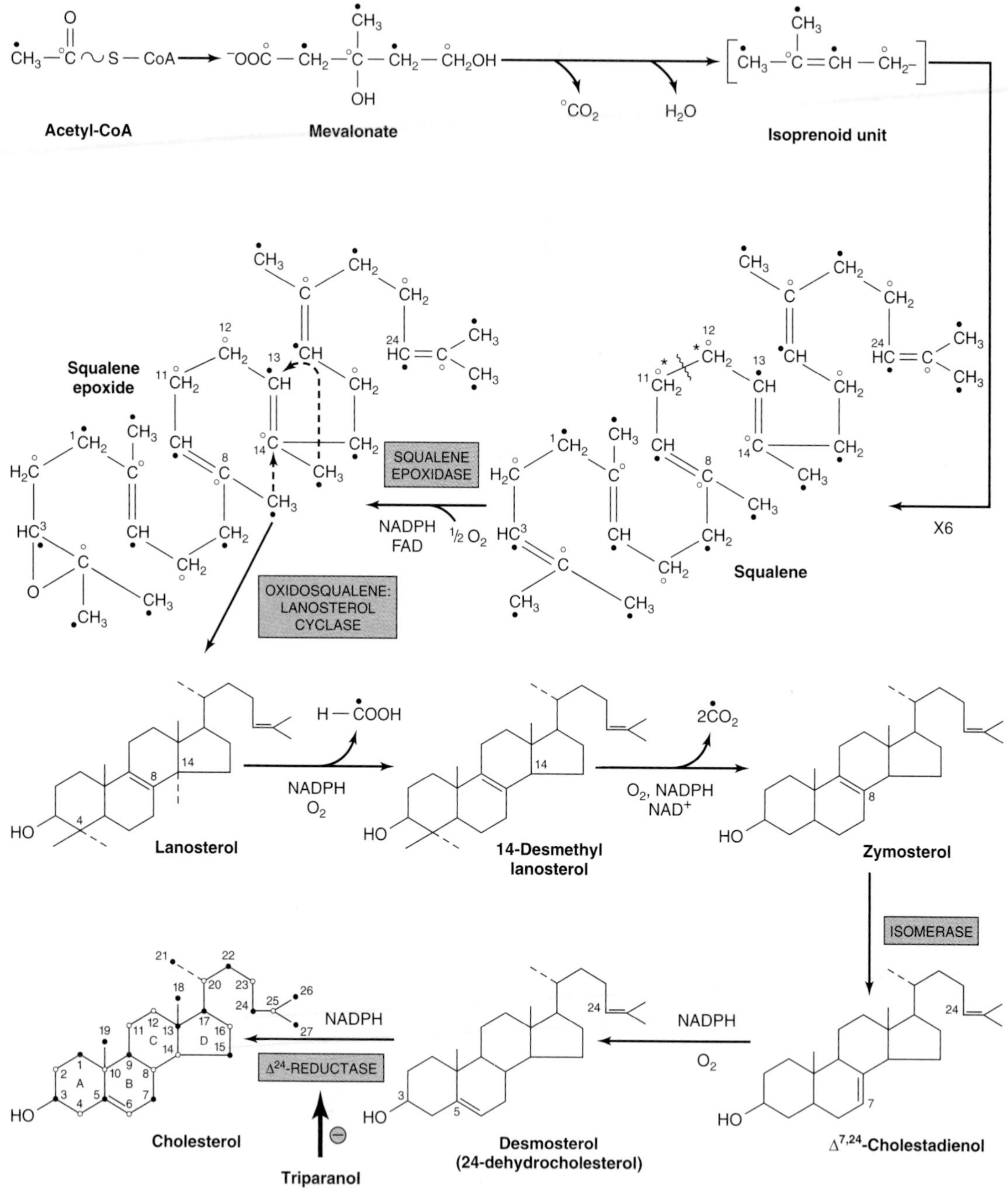

Acetyl-CoA
Mevalonate
$CO_2$
$H_2O$
Isoprenoid unit
X6
Squalene
SQUALENE EPOXIDASE
NADPH
FAD
½ $O_2$
Squalene epoxide
OXIDOSQUALENE: LANOSTEROL CYCLASE
Lanosterol
H—COOH
NADPH
$O_2$
14-Desmethyl lanosterol
$2CO_2$
$O_2$, NADPH
$NAD^+$
Zymosterol
ISOMERASE
$\Delta^{7,24}$-Cholestadienol
NADPH
$O_2$
Desmosterol (24-dehydrocholesterol)
NADPH
$\Delta^{24}$-REDUCTASE
Triparanol
Cholesterol

***Figure 26–3.*** Biosynthesis of cholesterol. The numbered positions are those of the steroid nucleus and the open and solid circles indicate the fate of each of the carbons in the acetyl moiety of acetyl-CoA. Asterisks: Refer to labeling of squalene in Figure 26–2.

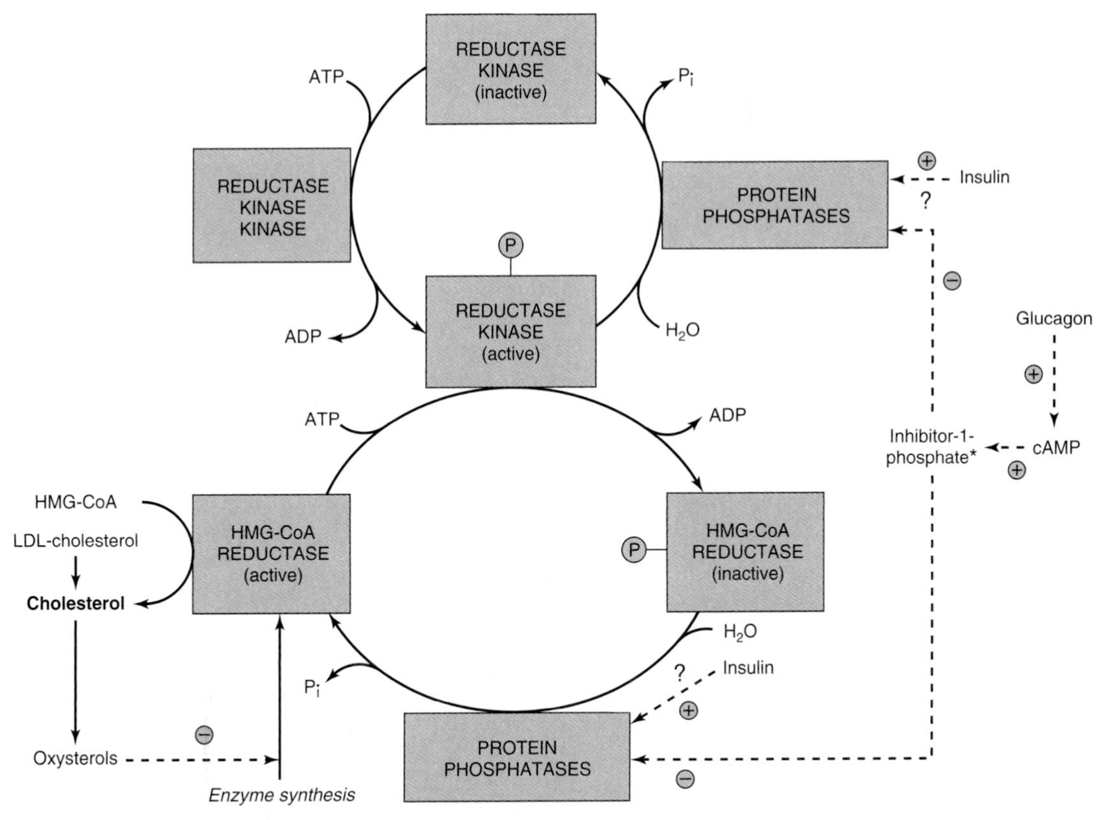


***Figure 26–4.*** Possible mechanisms in the regulation of cholesterol synthesis by HMG-CoA reductase. Insulin has a dominant role compared with glucagon. Asterisk: See Figure 18–6.

membrane; cholesterol synthesis; and hydrolysis of cholesteryl esters by the enzyme **cholesteryl ester hydrolase.** Decrease is due to efflux of cholesterol from the membrane to HDL, promoted by **LCAT** (lecithin:cholesterol acyltransferase) (Chapter 25); esterification of cholesterol by **ACAT** (acyl-CoA:cholesterol acyltransferase); and utilization of cholesterol for synthesis of other steroids, such as hormones, or bile acids in the liver.

## The LDL Receptor Is Highly Regulated

LDL (apo B-100, E) receptors occur on the cell surface in pits that are coated on the cytosolic side of the cell membrane with a protein called clathrin. The glycoprotein receptor spans the membrane, the B-100 binding region being at the exposed amino terminal end. After binding, LDL is taken up intact by endocytosis. The apoprotein and cholesteryl ester are then hydrolyzed in the lysosomes, and cholesterol is translocated into the cell. The receptors are recycled to the cell surface. This influx of cholesterol inhibits in a coordinated manner HMG-CoA synthase, HMG-CoA reductase, and, therefore, cholesterol synthesis; stimulates ACAT activity; and down-regulates synthesis of the LDL receptor. Thus, the number of LDL receptors on the cell surface is regulated by the cholesterol requirement for membranes, steroid hormones, or bile acid synthesis (Figure 26–5). The apo B-100, E receptor is a "high-affinity" LDL receptor, which may be saturated under most circumstances. Other "low-affinity" LDL receptors also appear to be present in addition to a scavenger pathway, which is not regulated.

# CHOLESTEROL IS TRANSPORTED BETWEEN TISSUES IN PLASMA LIPOPROTEINS (Figure 26–6)

In Western countries, the total plasma cholesterol in humans is about 5.2 mmol/L, rising with age, though there are wide variations between individuals. The greater part is found in the esterified form. It is transported in lipoproteins of the plasma, and the highest proportion of cholesterol is found in the LDL. Dietary cholesterol equilibrates with plasma cholesterol in days

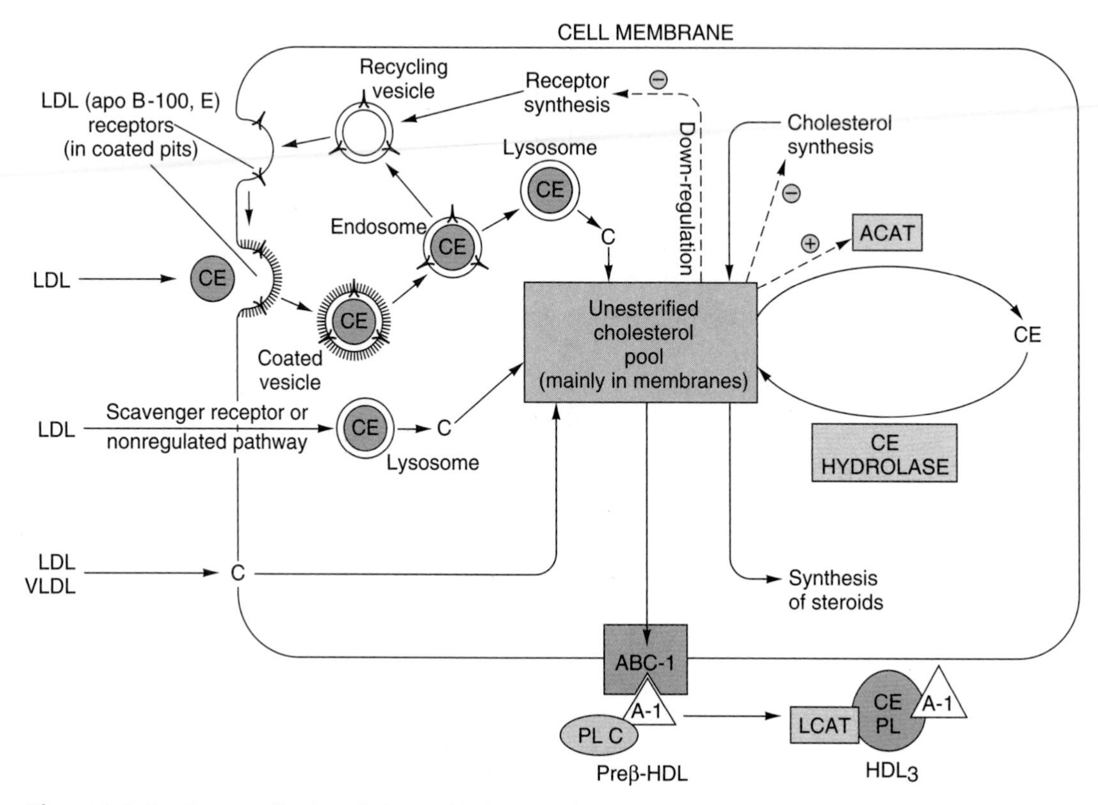


***Figure 26–5.*** Factors affecting cholesterol balance at the cellular level. Reverse cholesterol transport may be initiated by preβ HDL binding to the ABC-1 transporter protein via apo A-I. Cholesterol is then moved out of the cell via the transporter, lipidating the HDL, and the larger particles then dissociate from the ABC-1 molecule. (C, cholesterol; CE, cholesteryl ester; PL, phospholipid; ACAT, acyl-CoA:cholesterol acyltransferase; LCAT, lecithin:cholesterol acyltransferase; A-I, apolipoprotein A-I; LDL, low-density lipoprotein; VLDL, very low density lipoprotein.) LDL and HDL are not shown to scale.

and with tissue cholesterol in weeks. Cholesteryl ester in the diet is hydrolyzed to cholesterol, which is then absorbed by the intestine together with dietary unesterified cholesterol and other lipids. With cholesterol synthesized in the intestines, it is then incorporated into chylomicrons. Of the cholesterol absorbed, 80–90% is esterified with long-chain fatty acids in the intestinal mucosa. Ninety-five percent of the chylomicron cholesterol is delivered to the liver in chylomicron remnants, and most of the cholesterol secreted by the liver in VLDL is retained during the formation of IDL and ultimately LDL, which is taken up by the LDL receptor in liver and extrahepatic tissues (Chapter 25).

## Plasma LCAT Is Responsible for Virtually All Plasma Cholesteryl Ester in Humans

LCAT activity is associated with HDL containing apo A-I. As cholesterol in HDL becomes esterified, it creates a concentration gradient and draws in cholesterol from tissues and from other lipoproteins (Figures 26–5 and 26–6), thus enabling HDL to function in **reverse cholesterol transport** (Figure 25–5).

## Cholesteryl Ester Transfer Protein Facilitates Transfer of Cholesteryl Ester From HDL to Other Lipoproteins

This protein is found in plasma of humans and many other species, associated with HDL. It facilitates transfer of cholesteryl ester from HDL to VLDL, IDL, and LDL in exchange for triacylglycerol, relieving product inhibition of LCAT activity in HDL. Thus, in humans, much of the cholesteryl ester formed by LCAT finds its way to the liver via VLDL remnants (IDL) or LDL (Figure 26–6). The triacylglycerol-enriched $HDL_2$ delivers its cholesterol to the liver in the HDL cycle (Figure 25–6).

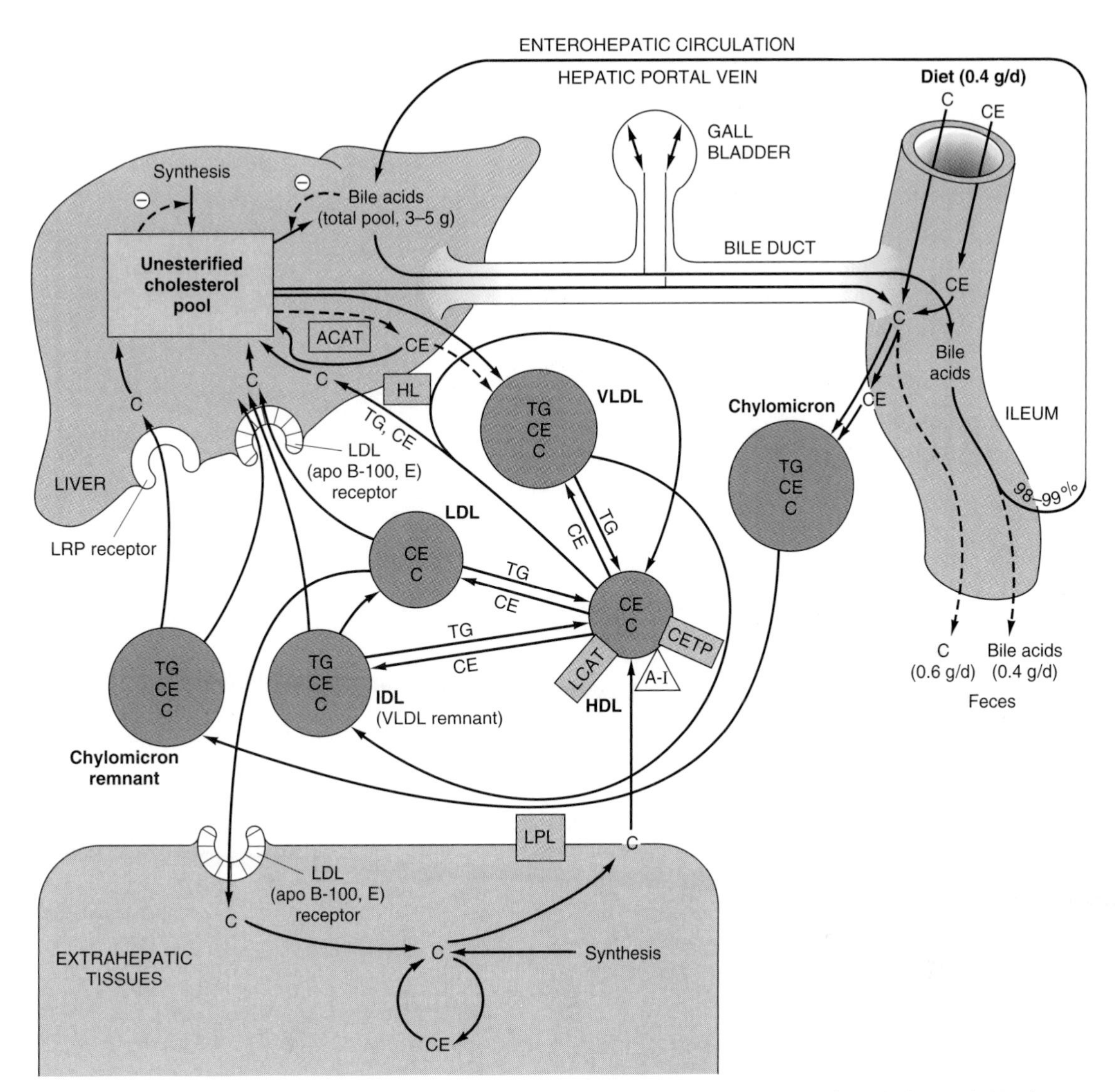


***Figure 26–6.*** Transport of cholesterol between the tissues in humans. (C, unesterified cholesterol; CE, cholesteryl ester; TG, triacylglycerol; VLDL, very low density lipoprotein; IDL, intermediate-density lipoprotein; LDL, low-density lipoprotein; HDL, high-density lipoprotein; ACAT, acyl-CoA:cholesterol acyltransferase; LCAT, lecithin:cholesterol acyltransferase; A-I, apolipoprotein A-I; CETP, cholesteryl ester transfer protein; LPL, lipoprotein lipase; HL, hepatic lipase.)

## CHOLESTEROL IS EXCRETED FROM THE BODY IN THE BILE AS CHOLESTEROL OR BILE ACIDS (SALTS)

About 1 g of cholesterol is eliminated from the body per day. Approximately half is excreted in the feces after conversion to bile acids. The remainder is excreted as cholesterol. **Coprostanol** is the principal sterol in the feces; it is formed from cholesterol by the bacteria in the lower intestine.

### Bile Acids Are Formed From Cholesterol

The **primary bile acids** are synthesized in the liver from cholesterol. These are **cholic acid** (found in the largest amount) and **chenodeoxycholic acid** (Figure 26–7).

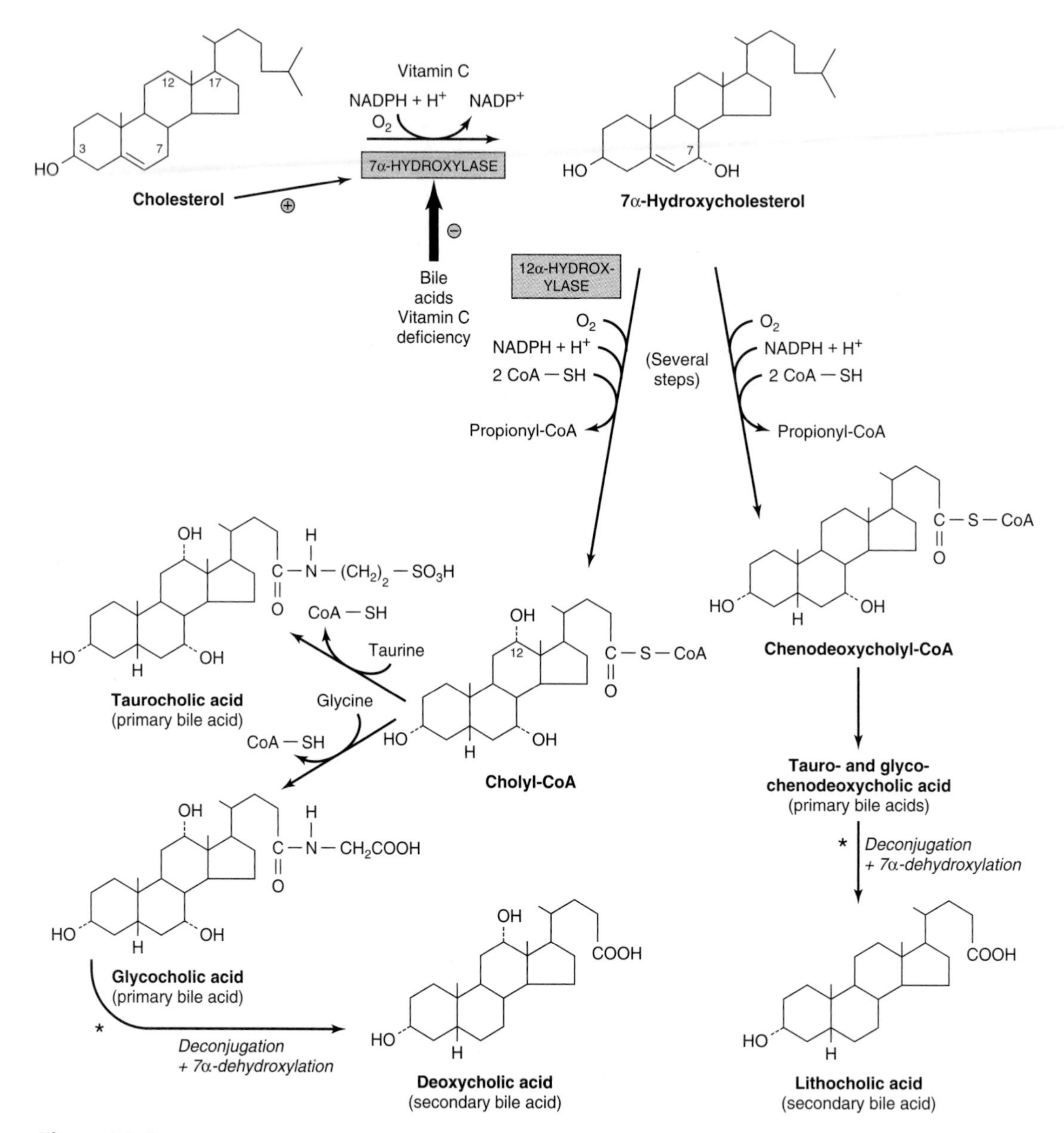


***Figure 26–7.*** Biosynthesis and degradation of bile acids. A second pathway in mitochondria involves hydroxylation of cholesterol by sterol 27-hydroxylase. Asterisk: Catalyzed by microbial enzymes.

The 7α-hydroxylation of cholesterol is the first and principal regulatory step in the biosynthesis of bile acids catalyzed by **7α-hydroxylase,** a microsomal enzyme. A typical monooxygenase, it requires oxygen, NADPH, and cytochrome P450. Subsequent hydroxylation steps are also catalyzed by monooxygenases. The pathway of bile acid biosynthesis divides early into one subpathway leading to **cholyl-CoA,** characterized by an extra α-OH group on position 12, and another pathway leading to **chenodeoxycholyl-CoA** (Figure 26–7). A second pathway in mitochondria involving the 27-hydroxylation of cholesterol by **sterol 27-hydroxylase** as the first step is responsible for a significant proportion of the primary bile acids synthesized. The primary bile acids (Figure 26–7) enter the bile as glycine or taurine conjugates. Conjugation takes place in peroxisomes. In humans, the ratio of the glycine to the taurine conjugates is normally 3:1. In the alkaline bile, the bile acids and their conjugates are assumed to be in a salt form—hence the term "bile salts."

A portion of the primary bile acids in the intestine is subjected to further changes by the activity of the intestinal bacteria. These include deconjugation and 7α-dehydroxylation, which produce the **secondary bile acids,** deoxycholic acid and lithocholic acid.

### Most Bile Acids Return to the Liver in the Enterohepatic Circulation

Although products of fat digestion, including cholesterol, are absorbed in the first 100 cm of small intestine, the primary and secondary bile acids are absorbed almost exclusively in the ileum, and 98–99% are returned to the liver via the portal circulation. This is known as the **enterohepatic circulation.** However, lithocholic acid, because of its insolubility, is not reabsorbed to any significant extent. Only a small fraction of the bile salts escapes absorption and is therefore eliminated in the feces. Nonetheless, this represents a major pathway for the elimination of cholesterol. Each day the small pool of bile acids (about 3–5 g) is cycled through the intestine six to ten times and an amount of bile acid equivalent to that lost in the feces is synthesized from cholesterol, so that a pool of bile acids of constant size is maintained. This is accomplished by a system of feedback controls.

### Bile Acid Synthesis Is Regulated at the 7α-Hydroxylase Step

The principal rate-limiting step in the biosynthesis of bile acids is at the **cholesterol 7α-hydroxylase reaction** (Figure 26–7). The activity of the enzyme is feedback-regulated via the nuclear bile acid-binding receptor **farnesoid X receptor (FXR).** When the size of the bile acid pool in the enterohepatic circulation increases, FXR is activated and transcription of the cholesterol 7α-hydroxylase gene is suppressed. Chenodeoxycholic acid is particularly important in activating FXR. Cholesterol 7α-hydroxylase activity is also enhanced by cholesterol of endogenous and dietary origin and regulated by insulin, glucagon, glucocorticoids, and thyroid hormone.

## CLINICAL ASPECTS

### The Serum Cholesterol Is Correlated With the Incidence of Atherosclerosis & Coronary Heart Disease

While cholesterol is believed to be chiefly concerned in the relationship, other serum lipids such as triacylglycerols may also play a role. Atherosclerosis is characterized by the deposition of cholesterol and cholesteryl ester from the plasma lipoproteins into the artery wall. Diseases in which prolonged elevated levels of VLDL, IDL, chylomicron remnants, or LDL occur in the blood (eg, diabetes mellitus, lipid nephrosis, hypothyroidism, and other conditions of hyperlipidemia) are often accompanied by premature or more severe atherosclerosis. There is also an inverse relationship between HDL ($HDL_2$) concentrations and coronary heart disease, and some consider that the most predictive relationship is the **LDL:HDL cholesterol ratio.** This is consistent with the function of HDL in reverse cholesterol transport. Susceptibility to atherosclerosis varies widely among species, and humans are one of the few in which the disease can be induced by diets high in cholesterol.

### Diet Can Play an Important Role in Reducing Serum Cholesterol

Hereditary factors play the greatest role in determining individual serum cholesterol concentrations; however, dietary and environmental factors also play a part, and the most beneficial of these is the substitution in the diet of **polyunsaturated and monounsaturated fatty acids** for saturated fatty acids. Plant oils such as corn oil and sunflower seed oil contain a high proportion of polyunsaturated fatty acids, while olive oil contains a high concentration of monounsaturated fatty acids. On the other hand, butterfat, beef fat, and palm oil contain a high proportion of saturated fatty acids. Sucrose and fructose have a greater effect in raising blood lipids, particularly triacylglycerols, than do other carbohydrates.

The reason for the cholesterol-lowering effect of polyunsaturated fatty acids is still not fully understood. It is clear, however, that one of the mechanisms involved is the up-regulation of LDL receptors by poly- and monounsaturated as compared with saturated fatty acids, causing an increase in the catabolic rate of LDL, the main atherogenic lipoprotein. In addition, saturated fatty acids cause the formation of smaller VLDL particles that contain relatively more cholesterol, and they are utilized by extrahepatic tissues at a slower rate than are larger particles—tendencies that may be regarded as atherogenic.

### Lifestyle Affects the Serum Cholesterol Level

Additional factors considered to play a part in coronary heart disease include high blood pressure, smoking, male gender, obesity (particularly abdominal obesity), lack of exercise, and drinking soft as opposed to hard water. Factors associated with elevation of plasma FFA followed by increased output of triacylglycerol and cholesterol into the circulation in VLDL include emotional stress and coffee drinking. Premenopausal women appear to be protected against many of these deleterious factors, and this is thought to be related to the beneficial effects of estrogen. There is an association between moderate alcohol consumption and a lower incidence of coronary heart disease. This may be due to elevation of HDL concentrations resulting from increased synthesis of apo A-I and changes in activity of cholesteryl ester transfer protein. It has been claimed that red wine is particularly beneficial, perhaps because of its content of antioxidants. Regular exercise lowers plasma LDL

***Table 26–1.*** Primary disorders of plasma lipoproteins (dyslipoproteinemias).

| Name | Defect | Remarks |
|---|---|---|
| **Hypolipoproteinemias**<br>Abetalipoproteinemia | No chylomicrons, VLDL, or LDL are formed because of defect in the loading of apo B with lipid. | Rare; blood acylglycerols low; intestine and liver accumulate acylglycerols. Intestinal malabsorption. Early death avoidable by administration of large doses of fat-soluble vitamins, particularly vitamin E. |
| Familial alpha-lipoprotein deficiency<br>Tangier disease<br>Fish-eye disease<br>Apo-A-I deficiencies | All have low or near absence of HDL. | Tendency toward hypertriacylglycerolemia as a result of absence of apo C-II, causing inactive LPL. Low LDL levels. Atherosclerosis in the elderly. |
| **Hyperlipoproteinemias**<br>Familial lipoprotein lipase deficiency (type I) | Hypertriacylglycerolemia due to deficiency of LPL, abnormal LPL, or apo C-II deficiency causing inactive LPL. | Slow clearance of chylomicrons and VLDL. Low levels of LDL and HDL. No increased risk of coronary disease. |
| Familial hypercholesterolemia (type IIa) | Defective LDL receptors or mutation in ligand region of apo B-100. | Elevated LDL levels and hypercholesterolemia, resulting in atherosclerosis and coronary disease. |
| Familial type III hyperlipoproteinemia (broad beta disease, remnant removal disease, familial dysbetalipoproteinemia) | Deficiency in remnant clearance by the liver is due to abnormality in apo E. Patients lack isoforms E3 and E4 and have only E2, which does not react with the E receptor.[1] | Increase in chylomicron and VLDL remnants of density < 1.019 (β-VLDL). Causes hypercholesterolemia, xanthomas, and atherosclerosis. |
| Familial hypertriacylglycerolemia (type IV) | Overproduction of VLDL often associated with glucose intolerance and hyperinsulinemia. | Cholesterol levels rise with the VLDL concentration. LDL and HDL tend to be subnormal. This type of pattern is commonly associated with coronary heart disease, type II diabetes mellitus, obesity, alcoholism, and administration of progestational hormones. |
| Familial hyperalphalipoproteinemia | Increased concentrations of HDL. | A rare condition apparently beneficial to health and longevity. |
| Hepatic lipase deficiency | Deficiency of the enzyme leads to accumulation of large triacylglycerol-rich HDL and VLDL remnants. | Patients have xanthomas and coronary heart disease. |
| Familial lecithin:cholesterol acyltransferase (LCAT) deficiency | Absence of LCAT leads to block in reverse cholesterol transport. HDL remains as nascent disks incapable of taking up and esterifying cholesterol. | Plasma concentrations of cholesteryl esters and lysolecithin are low. Present is an abnormal LDL fraction, lipoprotein X, found also in patients with cholestasis. VLDL is abnormal (β-VLDL). |
| Familial lipoprotein(a) excess | Lp(a) consists of 1 mol of LDL attached to 1 mol of apo(a). Apo(a) shows structural homologies to plasminogen. | Premature coronary heart disease due to atherosclerosis, plus thrombosis due to inhibition of fibrinolysis. |

[1]There is an association between patients possessing the apo E4 allele and the incidence of Alzheimer's disease. Apparently, apo E4 binds more avidly to β-amyloid found in neuritic plaques.

but raises HDL. Triacylglycerol concentrations are also reduced, due most likely to increased insulin sensitivity, which enhances expression of lipoprotein lipase.

## When Diet Changes Fail, Hypolipidemic Drugs Will Reduce Serum Cholesterol & Triacylglycerol

Significant reductions of plasma cholesterol can be effected medically by the use of **cholestyramine resin** or surgically by the ileal exclusion operations. Both procedures block the reabsorption of bile acids, causing increased bile acid synthesis in the liver. This increases cholesterol excretion and up-regulates LDL receptors, lowering plasma cholesterol. **Sitosterol** is a hypocholesterolemic agent that acts by blocking the absorption of cholesterol from the gastrointestinal tract.

Several drugs are known to block the formation of cholesterol at various stages in the biosynthetic pathway. The **statins** inhibit HMG-CoA reductase, thus up-regulating LDL receptors. Statins currently in use include **atorvastatin, simvastatin,** and **pravastatin.** Fibrates such as **clofibrate** and **gemfibrozil** act mainly to lower plasma triacylglycerols by decreasing the secretion of triacylglycerol and cholesterol-containing VLDL by the liver. In addition, they stimulate hydrolysis of VLDL triacylglycerols by lipoprotein lipase. **Probucol** appears to increase LDL catabolism via receptor-independent pathways, but its antioxidant properties may be more important in preventing accumulation of oxidized LDL, which has enhanced atherogenic properties, in arterial walls. **Nicotinic acid** reduces the flux of FFA by inhibiting adipose tissue lipolysis, thereby inhibiting VLDL production by the liver.

## Primary Disorders of the Plasma Lipoproteins (Dyslipoproteinemias) Are Inherited

Inherited defects in lipoprotein metabolism lead to the primary condition of either **hypo-** or **hyperlipoproteinemia** (Table 26–1). In addition, diseases such as diabetes mellitus, hypothyroidism, kidney disease (nephrotic syndrome), and atherosclerosis are associated with secondary abnormal lipoprotein patterns that are very similar to one or another of the primary inherited conditions. Virtually all of the primary conditions are due to a defect at a stage in lipoprotein formation, transport, or destruction (see Figures 25–4, 26–5, and 26–6). Not all of the abnormalities are harmful.

## SUMMARY


- Cholesterol is the precursor of all other steroids in the body, eg, corticosteroids, sex hormones, bile acids, and vitamin D. It also plays an important structural role in membranes and in the outer layer of lipoproteins.
- Cholesterol is synthesized in the body entirely from acetyl-CoA. Three molecules of acetyl-CoA form mevalonate via the important regulatory reaction for the pathway, catalyzed by HMG-CoA reductase. Next, a five-carbon isoprenoid unit is formed, and six of these condense to form squalene. Squalene undergoes cyclization to form the parent steroid lanosterol, which, after the loss of three methyl groups, forms cholesterol.
- Cholesterol synthesis in the liver is regulated partly by cholesterol in the diet. In tissues, cholesterol balance is maintained between the factors causing gain of cholesterol (eg, synthesis, uptake via the LDL or scavenger receptors) and the factors causing loss of cholesterol (eg, steroid synthesis, cholesteryl ester formation, excretion). The activity of the LDL receptor is modulated by cellular cholesterol levels to achieve this balance. In reverse cholesterol transport, HDL (preβ-HDL, discoidal, or $HD_3$) takes up cholesterol from the tissues and LCAT esterifies it and deposits it in the core of HDL, which is converted to $HDL_2$. The cholesteryl ester in $HDL_2$ is taken up by the liver, either directly or after transfer to VLDL, IDL, or LDL via the cholesteryl ester transfer protein.
- Excess cholesterol is excreted from the liver in the bile as cholesterol or bile salts. A large proportion of bile salts is absorbed into the portal circulation and returned to the liver as part of the enterohepatic circulation.
- Elevated levels of cholesterol present in VLDL, IDL, or LDL are associated with atherosclerosis, whereas high levels of HDL have a protective effect.
- Inherited defects in lipoprotein metabolism lead to a primary condition of hypo- or hyperlipoproteinemia. Conditions such as diabetes mellitus, hypothyroidism, kidney disease, and atherosclerosis exhibit secondary abnormal lipoprotein patterns that resemble certain primary conditions.


## REFERENCES


Illingworth DR: Management of hypercholesterolemia. Med Clin North Am 2000;84:23.

Ness GC, Chambers CM: Feedback and hormonal regulation of hepatic 3-hydroxy-3-methylglutaryl coenzyme A reductase: the concept of cholesterol buffering capacity. Proc Soc Exp Biol Med 2000;224:8.

Parks DJ et al: Bile acids: natural ligands for a nuclear orphan receptor. Science 1999;284:1365.

Princen HMG: Regulation of bile acid synthesis. Curr Pharm Design 1997;3:59.

Russell DW: Cholesterol biosynthesis and metabolism. Cardiovascular Drugs Therap 1992;6:103.

Spady DK, Woollett LA, Dietschy JM: Regulation of plasma LDL-cholesterol levels by dietary cholesterol and fatty acids. Annu Rev Nutr 1993;13:355.

Tall A: Plasma lipid transfer proteins. Annu Rev Biochem 1995; 64:235.

Various authors: *Biochemistry of Lipids, Lipoproteins and Membranes.* Vance DE, Vance JE (editors). Elsevier, 1996.

Various authors: The cholesterol facts. A summary of the evidence relating dietary fats, serum cholesterol, and coronary heart disease. Circulation 1990;81:1721.

Zhang FL, Casey PJ: Protein prenylation: Molecular mechanisms and functional consequences. Annu Rev Biochem 1996; 65:241.

# Integration of Metabolism—The Provision of Metabolic Fuels

# 27

*David A Bender, PhD, & Peter A. Mayes, PhD, DSc*

## BIOMEDICAL IMPORTANCE

An adult human weighing 70 kg requires about 10–12 MJ (2400–2900 kcal) from metabolic fuels each day. This requirement is met from carbohydrates (40–60%), lipids (mainly triacylglycerol, 30–40%), protein (10–15%), and alcohol if consumed. The mix being oxidized varies depending on whether the subject is in the fed or starving state and on the intensity of physical work. The requirement for metabolic fuels is relatively constant throughout the day, since average physical activity only increases metabolic rate by about 40–50% over the basal metabolic rate. However, most people consume their daily intake of metabolic fuels in two or three meals, so there is a need to form reserves of carbohydrate (glycogen in liver and muscle) and lipid (triacylglycerol in adipose tissue) for use between meals.

If the intake of fuels is consistently greater than energy expenditure, the surplus is stored, largely as fat, leading to the development of **obesity** and its associated health hazards. If the intake of fuels is consistently lower than energy expenditure, there will be negligible fat and carbohydrate reserves, and amino acids arising from protein turnover will be used for energy rather than replacement protein synthesis, leading to emaciation and eventually death.

After a normal meal there is an ample supply of carbohydrate, and the fuel for most tissues is glucose. In the starving state, glucose must be spared for use by the central nervous system (which is largely dependent on glucose) and the erythrocytes (which are wholly reliant on glucose). Other tissues can utilize alternative fuels such as fatty acids and ketone bodies. As glycogen reserves become depleted, so amino acids arising from protein turnover and glycerol arising from lipolysis are used for gluconeogenesis. These events are largely controlled by the hormones insulin and glucagon. In **diabetes mellitus** there is either impaired synthesis and secretion of insulin (type 1 diabetes mellitus) or impaired sensitivity of tissues to insulin action (type 2 diabetes mellitus), leading to severe metabolic derangement. In cattle the demands of heavy lactation can lead to ketosis, as can the demands of twin pregnancy in sheep.

## MANY METABOLIC FUELS ARE INTERCONVERTIBLE

Carbohydrate in excess of immediate requirements as fuel or for synthesis of glycogen in muscle and liver may be used for lipogenesis (Chapter 21) and hence triacylglycerol synthesis in both adipose tissue and liver (whence it is exported in very low density lipoprotein). The importance of lipogenesis in human beings is unclear; in Western countries, dietary fat provides 35–45% of energy intake, while in less developed countries where carbohydrate may provide 60–75% of energy intake the total intake of food may be so low that there is little surplus for lipogenesis. A high intake of fat inhibits lipogenesis.

Fatty acids (and ketone bodies formed from them) cannot be used for the synthesis of glucose. The reaction of pyruvate dehydrogenase, forming acetyl-CoA, is irreversible, and for every two-carbon unit from acetyl-CoA that enters the citric acid cycle there is a loss of two carbon atoms as carbon dioxide before only one molecule of oxaloacetate is re-formed—ie, there is no net increase. This means that acetyl-CoA (and therefore any substrates that yield acetyl-CoA) can never be used for gluconeogenesis (Chapter 19). The (relatively rare) fatty acids with an odd number of carbon atoms yield propionyl-CoA as the product of the final cycle of β-oxidation (Chapter 22), and this can be a substrate for gluconeogenesis, as can the glycerol released by lipolysis of adipose tissue triacylglycerol reserves. Most of the amino acids in excess of requirements for protein synthesis (arising from the diet or from tissue protein turnover) yield pyruvate, or five- and four-carbon intermediates of the citric acid cycle. Pyruvate can be carboxylated to oxaloacetate, which is the primary substrate for gluconeogenesis, and the five- and four-carbon intermediates also result in a net increase in the formation of oxaloacetate, which is then available for gluconeogenesis. These amino acids are classified as **glucogenic.** Lysine and leucine yield only acetyl-CoA on oxidation and thus cannot be used for gluconeogenesis, while phenylalanine, tyrosine, tryptophan, and isoleucine give rise to both acetyl-CoA and to intermediates of the citric acid cycle that can be used for gluconeogenesis. Those amino acids that give rise to acetyl-CoA are classified as **ketogenic** because in the starving state much of the acetyl-CoA will be used for synthesis of ketone bodies in the liver.

## A SUPPLY OF METABOLIC FUELS IS PROVIDED IN BOTH THE FED & STARVING STATES (Figure 27–1)

### Glucose Is Always Required by the Central Nervous System & Erythrocytes

Erythrocytes lack mitochondria and hence are wholly reliant on glycolysis and the pentose phosphate pathway. The brain can metabolize ketone bodies to meet about 20% of its energy requirements; the remainder must be supplied by glucose. The metabolic changes that occur in starvation are the consequences of the need to preserve glucose and the limited reserves of glycogen in liver for use by the brain and erythrocytes and to ensure the provision of alternative fuels for other tissues. The fetus and synthesis of lactose in milk also require a significant amount of glucose.

### In the Fed State, Metabolic Fuel Reserves Are Laid Down

For several hours after a meal, while the products of digestion are being absorbed, there is an abundant supply of metabolic fuels. Under these conditions, glucose is the major fuel for oxidation in most tissues; this is observed as an increase in the respiratory quotient (the ratio of carbon dioxide produced to oxygen consumed) from about 0.8 in the starved state to near 1 (Table 27–1).

Glucose uptake into muscle and adipose tissue is controlled by insulin, which is secreted by the B islet cells of the pancreas in response to an increased concentration of glucose in the portal blood. An early response to insulin in muscle and adipose tissue is the migration of glucose transporter vesicles to the cell surface, exposing active glucose transporters (GLUT 4). These insulin-sensitive tissues will only take up glucose from the blood stream to any significant extent in the presence of the hormone. As insulin secretion falls in the starved state, so the transporters are internalized again, reducing glucose uptake.

The uptake of glucose into the liver is independent of insulin, but liver has an isoenzyme of hexokinase (glucokinase) with a high $K_m$, so that as the concentration of glucose entering the liver increases, so does the rate of synthesis of glucose 6-phosphate. This is in excess of the liver's requirement for energy and is used mainly for synthesis of glycogen. In both liver and skeletal muscle, insulin acts to stimulate glycogen synthase and inhibit glycogen phosphorylase. Some of the glucose entering the liver may also be used for lipogenesis and synthesis of triacylglycerol. In adipose tissue, insulin stimulates glucose uptake, its conversion to fatty acids, and their esterification; and inhibits intracellular lipolysis and the release of free fatty acids.

The products of lipid digestion enter the circulation as triacylglycerol-rich chylomicrons (Chapter 25). In adipose tissue and skeletal muscle, lipoprotein lipase is activated in response to insulin; the resultant free fatty acids are largely taken up to form triacylglycerol reserves, while the glycerol remains in the blood stream and is taken up by the liver and used for glycogen synthesis or lipogenesis. Free fatty acids remaining in the blood stream are taken up by the liver and reesterified. The lipid-depleted chylomicron remnants are also cleared by the liver, and surplus liver triacylglycerol—including that from lipogenesis—is exported in very low density lipoprotein.

Under normal feeding patterns the rate of tissue protein catabolism is more or less constant throughout the day; it is only in cachexia that there is an increased rate of protein catabolism. There is net protein catabolism in the postabsorptive phase of the feeding cycle and net protein synthesis in the absorptive phase, when the rate of synthesis increases by about 20–25%. The increased rate of protein synthesis is, again, a response to insulin action. Protein synthesis is an energy-expensive process, accounting for up to almost 20% of energy expenditure in the fed state, when there is an ample supply of amino acids from the diet, but under 9% in the starved state.

### Metabolic Fuel Reserves Are Mobilized in the Starving State

There is a small fall in plasma glucose upon starvation, then little change as starvation progresses (Table 27–2; Figure 27–2). Plasma free fatty acids increase with onset of starvation but then plateau. There is an initial delay in ketone body production, but as starvation progresses the plasma concentration of ketone bodies increases markedly.

In the postabsorptive state, as the concentration of glucose in the portal blood falls, so insulin secretion decreases, resulting in skeletal muscle and adipose tissue taking up less glucose. The increase in secretion of glucagon from the A cells of the pancreas inhibits glycogen synthase and activates glycogen phosphorylase in liver. The resulting glucose 6-phosphate in liver is

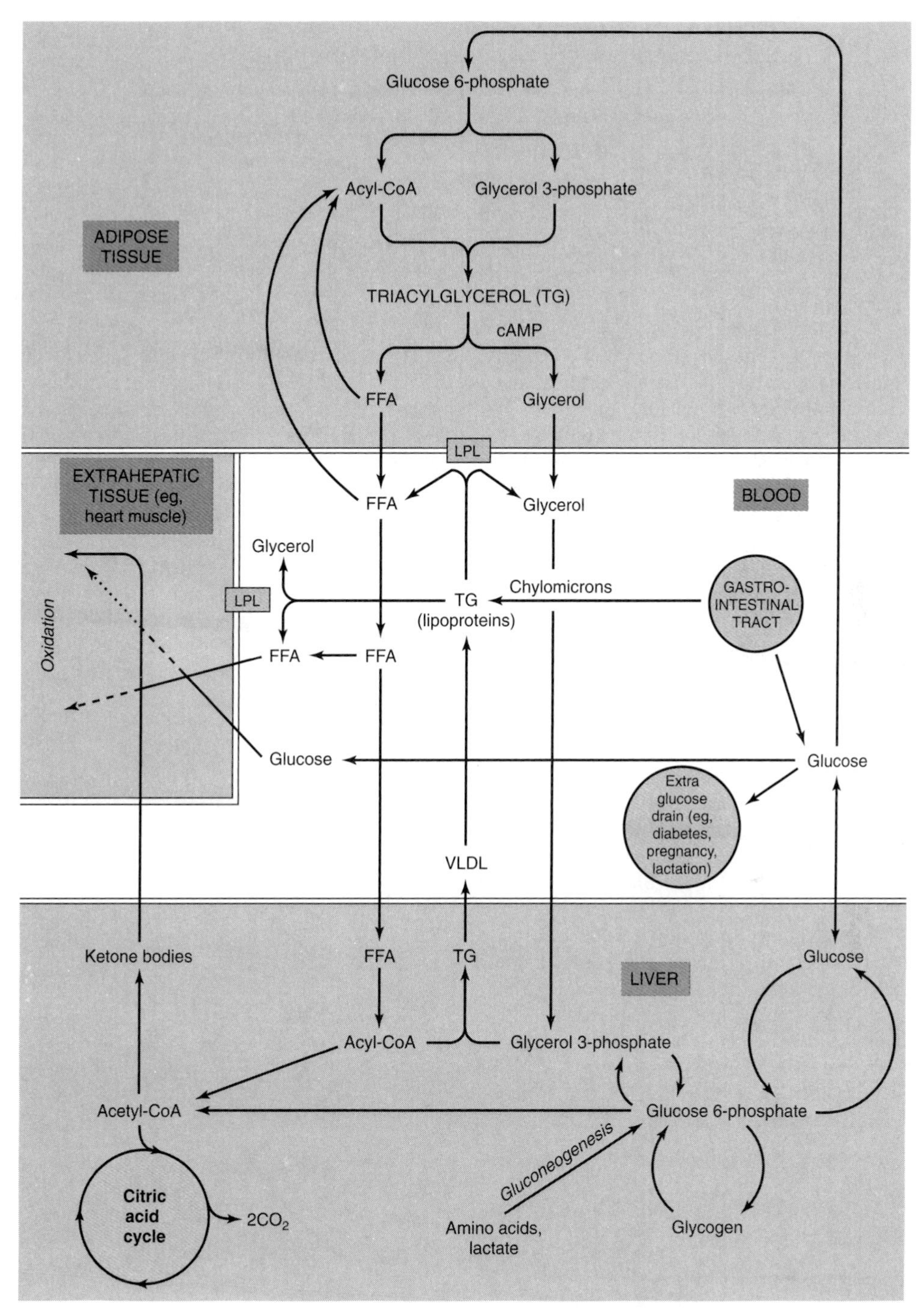


***Figure 27–1.*** Metabolic interrelationships between adipose tissue, the liver, and extrahepatic tissues. In extrahepatic tissues such as heart, metabolic fuels are oxidized in the following order of preference: (1) ketone bodies, (2) fatty acids, (3) glucose. (LPL, lipoprotein lipase; FFA, free fatty acids; VLDL, very low density lipoproteins.)

***Table 27–1.*** Energy yields, oxygen consumption, and carbon dioxide production in the oxidation of metabolic fuels.

| | Energy Yield (kJ/g) | $O_2$ Consumed (L/g) | $CO_2$ Produced (L/g) | RQ | Oxygen (kJ/L) |
|---|---|---|---|---|---|
| Carbohydrate | 16 | 0.829 | 0.829 | 1.00 | 20 |
| Protein | 17 | 0.966 | 0.782 | 0.81 | 20 |
| Fat | 37 | 2.016 | 1.427 | 0.71 | 20 |

hydrolyzed by glucose-6-phosphatase, and glucose is released into the blood stream for use by other tissues, particularly the brain and erythrocytes.

Muscle glycogen cannot contribute directly to plasma glucose, since muscle lacks glucose-6-phosphatase, and the primary purpose of muscle glycogen is to provide a source of glucose 6-phosphate for energy-yielding metabolism in the muscle itself. However, acetyl-CoA formed by oxidation of fatty acids in muscle inhibits pyruvate dehydrogenase and leads to citrate accumulation, which in turn inhibits phosphofructokinase and therefore glycolysis, thus sparing glucose. Any accumulated pyruvate is transaminated to alanine at the expense of amino acids arising from breakdown of protein reserves. The alanine—and much of the keto acids resulting from this transamination—are exported from muscle and taken up by the liver, where the alanine is transaminated to yield pyruvate. The resultant amino acids are largely exported back to muscle to provide amino groups for formation of more alanine, while the pyruvate is a major substrate for gluconeogenesis in the liver.

In adipose tissue, the effect of the decrease in insulin and increase in glucagon results in inhibition of lipogenesis, inactivation of lipoprotein lipase, and activation of hormone-sensitive lipase (Chapter 25). This leads to release of increased amounts of glycerol (a substrate for gluconeogenesis in the liver) and free fatty acids, which are used by skeletal muscle and liver as their preferred metabolic fuels, so sparing glucose.

***Table 27–2.*** Plasma concentrations of metabolic fuels (mmol/L) in the fed and starving states.

| | Fed | 40 Hours Starvation | 7 Days Starvation |
|---|---|---|---|
| Glucose | 5.5 | 3.6 | 3.5 |
| Free fatty acids | 0.30 | 1.15 | 1.19 |
| Ketone bodies | Negligible | 2.9 | 4.5 |

Although muscle takes up and preferentially oxidizes free fatty acids in the starving state, it cannot meet all of its energy requirements by β-oxidation. By contrast, the liver has a greater capacity for β-oxidation than it requires to meet its own energy needs and forms more acetyl-CoA than can be oxidized. This acetyl-CoA is used to synthesize ketone bodies (Chapter 22), which are major metabolic fuels for skeletal and heart muscle and can meet some of the brain's energy needs. In prolonged starvation, glucose may represent less than 10% of whole body energy-yielding metabolism. Furthermore, as a result of protein catabolism, an increasing number of amino acids are released and utilized in the liver and kidneys for gluconeogenesis.

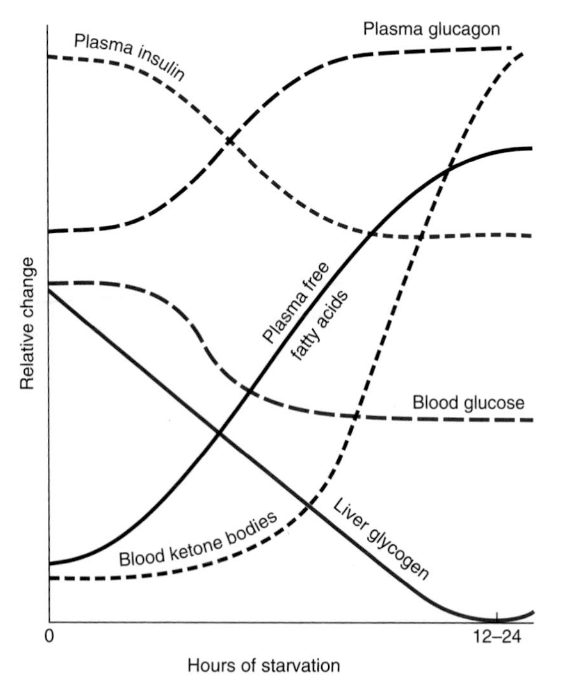


***Figure 27–2.*** Relative changes in metabolic parameters during the onset of starvation.

***Table 27–3.*** Summary of the major and unique features of metabolism of the principal organs.

| Organ | Major Function | Major Pathways | Main Substrates | Major Products | Specialist Enzymes |
|---|---|---|---|---|---|
| Liver | Service for the other organs and tissues | Most represented, including gluconeogenesis; β-oxidation; ketogenesis; lipoprotein formation; urea, uric acid, and bile acid formation; cholesterol synthesis; lipogenesis[1] | Free fatty acids, glucose (well fed), lactate, glycerol, fructose, amino acids<br>(Ethanol) | Glucose, VLDL (triacylglycerol), HDL, ketone bodies, urea, uric acid, bile acids, plasma proteins<br>(Acetate) | Glucokinase, glucose-6-phosphatase, glycerol kinase, phosphoenolpyruvate carboxykinase, fructokinase, arginase, HMG-CoA synthase and lyase, 7α-hydroxylase<br>(Alcohol dehydrogenase) |
| Brain | Coordination of the nervous system | Glycolysis, amino acid metabolism | Glucose, amino acid, ketone bodies (in starvation)<br>Polyunsaturated fatty acids in neonate | Lactate | |
| Heart | Pumping of blood | Aerobic pathways, eg, β-oxidation and citric acid cycle | Free fatty acids, lactate, ketone bodies, VLDL and chylomicron triacylglycerol, some glucose | | Lipoprotein lipase. Respiratory chain well developed. |
| Adipose tissue | Storage and breakdown of triacylglycerol | Esterification of fatty acids and lipolysis; lipogenesis[1] | Glucose, lipoprotein triacylglycerol | Free fatty acids, glycerol | Lipoprotein lipase, hormone-sensitive lipase |
| Muscle<br>Fast twitch<br>Slow twitch | <br>Rapid movement<br>Sustained movement | <br>Glycolysis<br>Aerobic pathways, eg, β-oxidation and citric acid cycle | <br>Glucose<br>Ketone bodies, triacylglycerol in VLDL and chylomicrons, free fatty acids | <br>Lactate | <br><br>Lipoprotein lipase.<br>Respiratory chain well developed. |
| Kidney | Excretion and gluconeogenesis | Gluconeogenesis | Free fatty acids, lactate, glycerol | Glucose | Glycerol kinase, phosphoenolpyruvate carboxykinase |
| Erythrocytes | Transport of $O_2$ | Glycolysis, pentose phosphate pathway. No mitochondria and therefore no β-oxidation or citric acid cycle. | Glucose | Lactate | (Hemoglobin) |

[1]In many species but not very active in humans.

## CLINICAL ASPECTS

In prolonged starvation, as adipose tissue reserves are depleted there is a very considerable increase in the net rate of protein catabolism to provide amino acids not only as substrates for gluconeogenesis but also as the main metabolic fuel of the tissues. Death results when essential tissue proteins are catabolized beyond the point at which they can sustain this metabolic drain. In patients with cachexia as a result of release of cytokines in response to tumors and a number of other pathologic conditions, there is an increase in the rate of tissue protein catabolism as well as a considerably increased metabolic rate, resulting in a state of advanced starvation. Again, death results when essential tissue proteins have been catabolized.

The high demand for glucose by the fetus and for synthesis of lactose in lactation can lead to ketosis. This may be seen as mild ketosis with hypoglycemia in women, but in lactating cattle and in ewes carrying twins there may be very pronounced ketosis and profound hypoglycemia.

In poorly controlled type 1 diabetes mellitus, patients may become hyperglycemic, partly as a result of lack of insulin to stimulate uptake and utilization of glucose and partly because of increased gluconeogenesis from amino acids in the liver. At the same time, the lack of insulin results in increased lipolysis in adipose tissue, and the resultant free fatty acids are substrates for ketogenesis in the liver. It is possible that in very severe diabetes utilization of ketone bodies in muscle (and other tissues) is impaired because of lack of oxaloacetate (most tissues have a requirement for some glucose metabolism to maintain an adequate amount of oxaloacetate for citric acid cycle activity). In uncontrolled diabetes, the magnitude of ketosis may be such as to result in severe acidosis (ketoacidosis) since acetoacetic acid and 3-hydroxybutyric acid are relatively strong acids. Coma results from both the acidosis and the considerably increased osmolarity of extracellular fluid (mainly due to the hyperglycemia).

A summary of the major and unique metabolic features of the principal tissues is presented in Table 27–3.

## SUMMARY

- The body can interconvert the majority of foodstuffs. However, there is no net conversion of most fatty acids (or other acetyl-CoA-forming substances) to glucose. Most amino acids, arising from the diet or from tissue protein, can be used for gluconeogenesis, as can the glycerol from triacylglycerol.
- In starvation, glucose must be provided for the brain and erythrocytes; initially, this is supplied from liver glycogen reserves. To spare glucose, muscle and other tissues reduce glucose uptake in response to lowered insulin secretion; they also oxidize fatty acids and ketone bodies preferentially to glucose.
- Adipose tissue releases free fatty acids in starvation, and these are used by many tissues as fuel. Furthermore, in the liver they are the substrate for synthesis of ketone bodies.
- Ketosis, a metabolic adaptation to starvation, is exacerbated in pathologic conditions such as diabetes mellitus and ruminant ketosis.

## REFERENCES

Bender DA: *Introduction to Nutrition and Metabolism,* 3rd edition. Taylor & Francis, 2002.

Caprio S et al: Oxidative fuel metabolism during mild hypoglycemia: critical role of free fatty acids. Am J Physiol 1989;256:E413.

Fell D: *Understanding the Control of Metabolism.* Portland Press, 1997.

Frayn KN: *Metabolic Regulation—A Human Perspective.* Portland Press, 1996.

McNamara JP: Role and regulation of metabolism in adipose tissue during lactation. J Nutr Biochem 1995;6:120.

Randle PJ: The glucose-fatty acid cycle—biochemical aspects. Atherosclerosis Rev 1991;22:183.

# SECTION III
## Metabolism of Proteins & Amino Acids

# Biosynthesis of the Nutritionally Nonessential Amino Acids — 28

*Victor W. Rodwell, PhD*

## BIOMEDICAL IMPORTANCE

All 20 of the amino acids present in proteins are essential for health. While comparatively rare in the Western world, amino acid deficiency states are endemic in certain regions of West Africa where the diet relies heavily on grains that are poor sources of amino acids such as tryptophan and lysine. These disorders include kwashiorkor, which results when a child is weaned onto a starchy diet poor in protein; and marasmus, in which both caloric intake and specific amino acids are deficient.

Humans can synthesize 12 of the 20 common amino acids from the amphibolic intermediates of glycolysis and of the citric acid cycle (Table 28–1). While *nutritionally* nonessential, these 12 amino acids are not "nonessential." All 20 amino acids are *biologically* essential. Of the 12 nutritionally nonessential amino acids, nine are formed from amphibolic intermediates and three (cysteine, tyrosine and hydroxylysine) from nutritionally essential amino acids. Identification of the twelve amino acids that humans can synthesize rested primarily on data derived from feeding diets in which purified amino acids replaced protein. This chapter considers only the biosynthesis of the twelve amino acids that are synthesized in human tissues, not the other eight that are synthesized by plants.

## NUTRITIONALLY NONESSENTIAL AMINO ACIDS HAVE SHORT BIOSYNTHETIC PATHWAYS

The enzymes glutamate dehydrogenase, glutamine synthetase, and aminotransferases occupy central positions in amino acid biosynthesis. The combined effect of those three enzymes is to transform ammonium ion into the α-amino nitrogen of various amino acids.

**Glutamate and Glutamine.** Reductive amination of α-ketoglutarate is catalyzed by glutamate dehydrogenase (Figure 28–1). Amination of glutamate to glutamine is catalyzed by glutamine synthetase (Figure 28–2).

**Alanine.** Transamination of pyruvate forms alanine (Figure 28–3).

**Aspartate and Asparagine.** Transamination of oxaloacetate forms aspartate. The conversion of aspartate

***Table 28–1.*** Amino acid requirements of humans.

| Nutritionally Essential | Nutritionally Nonessential |
|---|---|
| Arginine[1] | Alanine |
| Histidine | Asparagine |
| Isoleucine | Aspartate |
| Leucine | Cysteine |
| Lysine | Glutamate |
| Methionine | Glutamine |
| Phenylalanine | Glycine |
| Threonine | Hydroxyproline[2] |
| Tryptophan | Hydroxylysine[2] |
| Valine | Proline |
| | Serine |
| | Tyrosine |

[1]"Nutritionally semiessential." Synthesized at rates inadequate to support growth of children.
[2]Not necessary for protein synthesis but formed during post-translational processing of collagen.

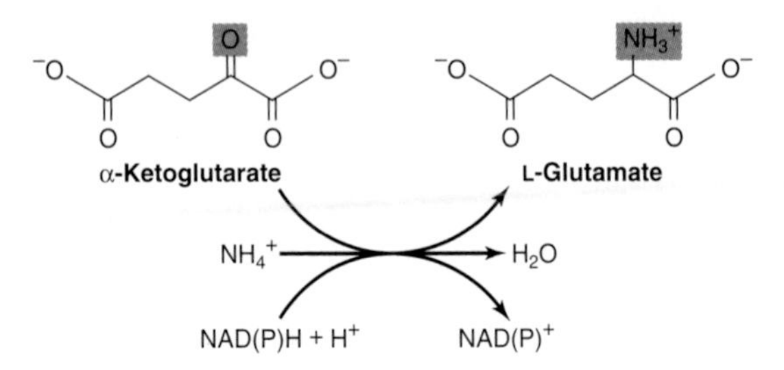


***Figure 28–1.*** The glutamate dehydrogenase reaction.

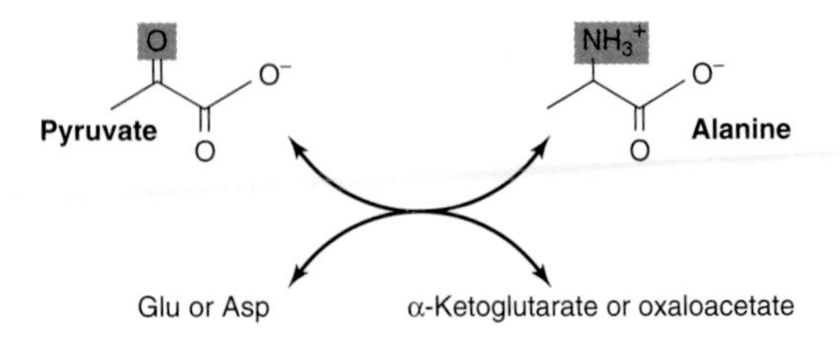


***Figure 28–3.*** Formation of alanine by transamination of pyruvate. The amino donor may be glutamate or aspartate. The other product thus is α-ketoglutarate or oxaloacetate.

to asparagine is catalyzed by asparagine synthetase (Figure 28–4), which resembles glutamine synthetase (Figure 28–2) except that glutamine, not ammonium ion, provides the nitrogen. Bacterial asparagine synthetases can, however, also use ammonium ion. Coupled hydrolysis of $PP_i$ to $P_i$ by pyrophosphatase ensures that the reaction is strongly favored.

**Serine.** Oxidation of the α-hydroxyl group of the glycolytic intermediate 3-phosphoglycerate converts it to an oxo acid, whose subsequent transamination and dephosphorylation leads to serine (Figure 28–5).

**Glycine.** Glycine aminotransferases can catalyze the synthesis of glycine from glyoxylate and glutamate or alanine. Unlike most aminotransferase reactions, these strongly favor glycine synthesis. Additional important mammalian routes for glycine formation are from choline (Figure 28–6) and from serine (Figure 28–7).

**Proline.** Proline is formed from glutamate by reversal of the reactions of proline catabolism (Figure 28–8).

**Cysteine.** Cysteine, while not nutritionally essential, is formed from methionine, which is nutritionally essential. Following conversion of methionine to ho-

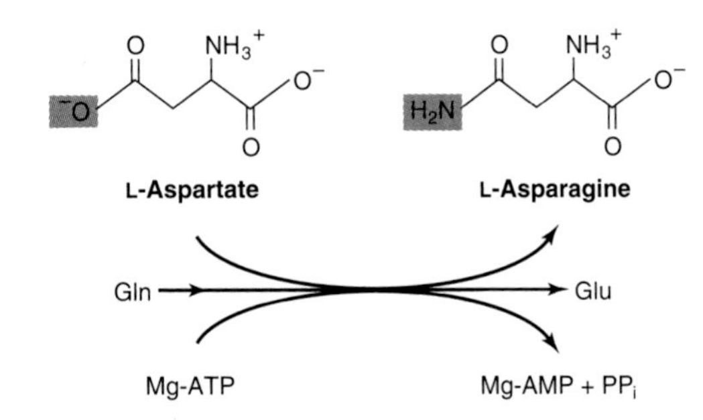


***Figure 28–4.*** The asparagine synthetase reaction. Note similarities to and differences from the glutamine synthetase reaction (Figure 28–2).

***Figure 28–5.*** Serine biosynthesis. (α-AA, α-amino acids; α-KA, α-keto acids.)

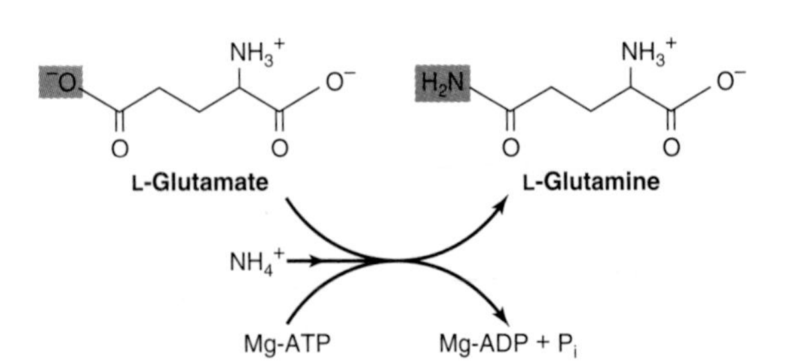


***Figure 28–2.*** The glutamine synthetase reaction.

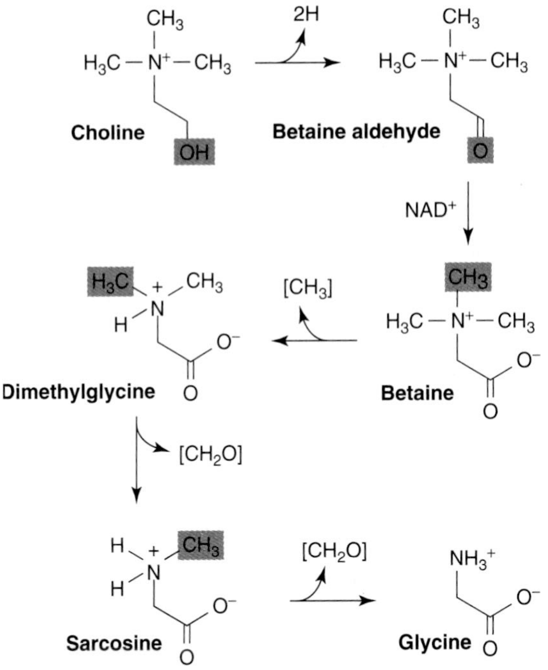

***Figure 28–6.*** Formation of glycine from choline.

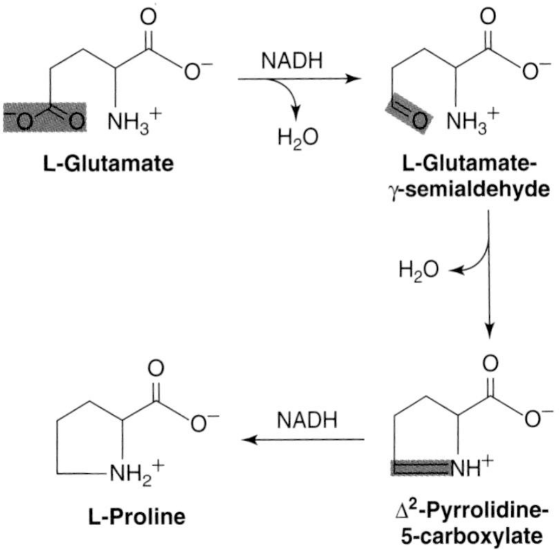

***Figure 28–8.*** Biosynthesis of proline from glutamate by reversal of reactions of proline catabolism.

mocysteine (see Chapter 30), homocysteine and serine form cysteine and homoserine (Figure 28–9).

**Tyrosine.** Phenylalanine hydroxylase converts phenylalanine to tyrosine (Figure 28–10). Provided that the diet contains adequate nutritionally essential phenylalanine, tyrosine is nutritionally nonessential. But since the reaction is irreversible, dietary tyrosine cannot replace phenylalanine. Catalysis by this mixed-function oxygenase incorporates one atom of $O_2$ into phenylalanine and reduces the other atom to water. Reducing power, provided as tetrahydrobiopterin, derives ultimately from NADPH.

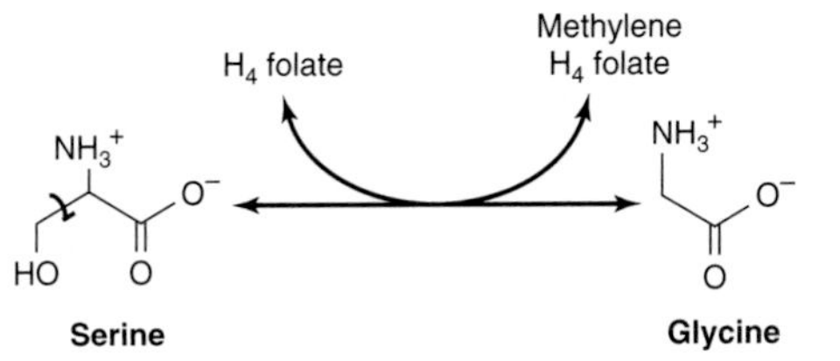

***Figure 28–7.*** The serine hydroxymethyltransferase reaction. The reaction is freely reversible. ($H_4$ folate, tetrahydrofolate.)

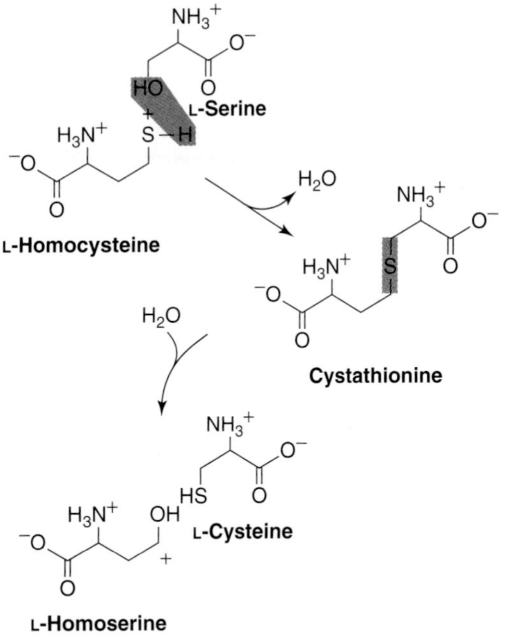

***Figure 28–9.*** Conversion of homocysteine and serine to homoserine and cysteine. The sulfur of cysteine derives from methionine and the carbon skeleton from serine.

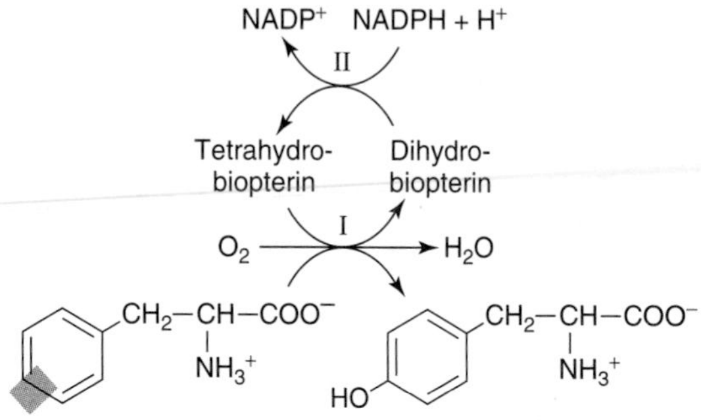

***Figure 28–10.*** The phenylalanine hydroxylase reaction. Two distinct enzymatic activities are involved. Activity II catalyzes reduction of dihydrobiopterin by NADPH, and activity I the reduction of $O_2$ to $H_2O$ and of phenylalanine to tyrosine. This reaction is associated with several defects of phenylalanine metabolism discussed in Chapter 30.

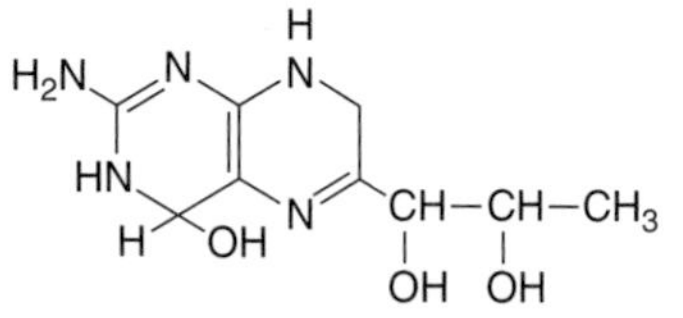

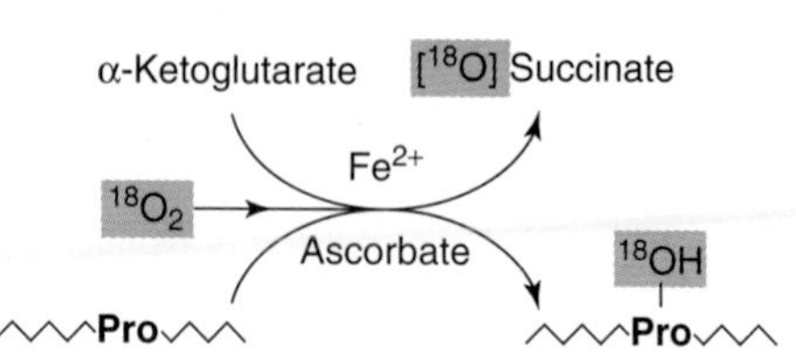


***Figure 28–11.*** The prolyl hydroxylase reaction. The substrate is a proline-rich peptide. During the course of the reaction, molecular oxygen is incorporated into both succinate and proline. Lysyl hydroxylase catalyzes an analogous reaction.

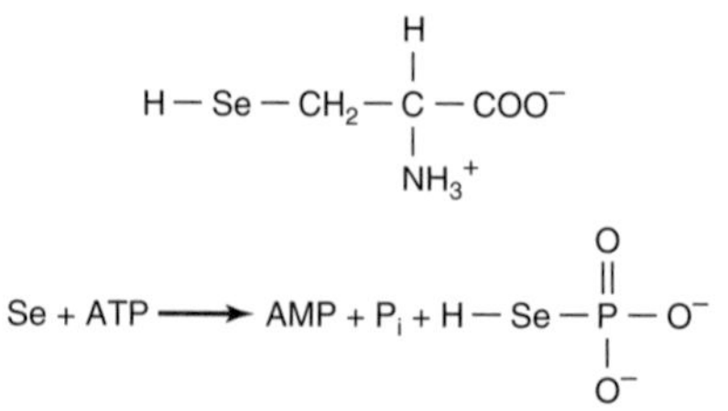

**Hydroxyproline and Hydroxylysine.** Hydroxyproline and hydroxylysine are present principally in collagen. Since there is no tRNA for either hydroxylated amino acid, neither dietary hydroxyproline nor hydroxylysine is incorporated into protein. Both are completely degraded (see Chapter 30). Hydroxyproline and hydroxylysine arise from proline and lysine, but only after these amino acids have been incorporated into peptides. Hydroxylation of peptide-bound prolyl and lysyl residues is catalyzed by prolyl hydroxylase and lysyl hydroxylase of tissues, including skin and skeletal muscle, and of granulating wounds (Figure 28–11). The hydroxylases are mixed-function oxygenases that require substrate, molecular $O_2$, ascorbate, $Fe^{2+}$, and α-ketoglutarate. For every mole of proline or lysine hydroxylated, one mole of α-ketoglutarate is decarboxylated to succinate. One atom of $O_2$ is incorporated into proline or lysine, the other into succinate (Figure 28–11). A deficiency of the vitamin C required for these hydroxylases results in scurvy.

**Valine, Leucine, and Isoleucine.** While leucine, valine, and isoleucine are all nutritionally essential amino acids, tissue aminotransferases reversibly interconvert all three amino acids and their corresponding α-keto acids. These α-keto acids thus can replace their amino acids in the diet.

**Selenocysteine.** While not normally considered an amino acid present in proteins, selenocysteine occurs at the active sites of several enzymes. Examples include the human enzymes thioredoxin reductase, glutathione peroxidase, and the deiodinase that converts thyroxine to triiodothyronine. Unlike hydroxyproline or hydroxylysine, selenocysteine arises co-translationally during its incorporation into peptides. The UGA anticodon of the unusual tRNA designated $tRNA^{Sec}$ normally signals STOP. The ability of the protein synthetic apparatus to identify a selenocysteine-specific UGA codon involves the selenocysteine insertion element, a stem-loop structure in the untranslated region of the mRNA. Selenocysteine-$tRNA^{Sec}$ is first charged with serine by the ligase that charges $tRNA^{Ser}$. Subsequent replacement of the serine oxygen by selenium involves selenophosphate formed by selenophosphate synthase (Figure 28–12).

***Figure 28–12.*** Selenocysteine *(top)* and the reaction catalyzed by selenophosphate synthetase *(bottom).*

## SUMMARY


- All vertebrates can form certain amino acids from amphibolic intermediates or from other dietary amino acids. The intermediates and the amino acids to which they give rise are α-ketoglutarate (Glu, Gln, Pro, Hyp), oxaloacetate (Asp, Asn) and 3-phosphoglycerate (Ser, Gly).
- Cysteine, tyrosine, and hydroxylysine are formed from nutritionally essential amino acids. Serine provides the carbon skeleton and homocysteine the sulfur for cysteine biosynthesis. Phenylalanine hydroxylase converts phenylalanine to tyrosine.
- Neither dietary hydroxyproline nor hydroxylysine is incorporated into proteins because no codon or tRNA dictates their insertion into peptides.
- Peptidyl hydroxyproline and hydroxylysine are formed by hydroxylation of peptidyl proline or lysine in reactions catalyzed by mixed-function oxidases that require vitamin C as cofactor. The nutritional disease scurvy reflects impaired hydroxylation due to a deficiency of vitamin C.
- Selenocysteine, an essential active site residue in several mammalian enzymes, arises by co-translational insertion of a previously modified tRNA.


## REFERENCES


Brown KM, Arthur JR: Selenium, selenoproteins and human health: a review. Public Health Nutr 2001;4:593.

Combs GF, Gray WP: Chemopreventive agents—selenium. Pharmacol Ther 1998;79:179.

Mercer LP, Dodds SJ, Smith DI: Dispensable, indispensable, and conditionally indispensable amino acid ratios in the diet. In: *Absorption and Utilization of Amino Acids.* Friedman M (editor). CRC Press, 1989.

Nordberg J et al: Mammalian thioredoxin reductase is irreversibly inhibited by dinitrohalobenzenes by alkylation of both the redox active selenocysteine and its neighboring cysteine residue. J Biol Chem 1998;273:10835.

Scriver CR et al (editors): *The Metabolic and Molecular Bases of Inherited Disease,* 8th ed. McGraw-Hill, 2001.

St Germain DL, Galton VA: The deiodinase family of selenoproteins. Thyroid 1997;7:655.

# Catabolism of Proteins & of Amino Acid Nitrogen

# 29

*Victor W. Rodwell, PhD*

## BIOMEDICAL IMPORTANCE

This chapter describes how the nitrogen of amino acids is converted to urea and the rare disorders that accompany defects in urea biosynthesis. In normal adults, nitrogen intake matches nitrogen excreted. Positive nitrogen balance, an excess of ingested over excreted nitrogen, accompanies growth and pregnancy. Negative nitrogen balance, where output exceeds intake, may follow surgery, advanced cancer, and kwashiorkor or marasmus.

While ammonia, derived mainly from the α-amino nitrogen of amino acids, is highly toxic, tissues convert ammonia to the amide nitrogen of nontoxic glutamine. Subsequent deamination of glutamine in the liver releases ammonia, which is then converted to nontoxic urea. If liver function is compromised, as in cirrhosis or hepatitis, elevated blood ammonia levels generate clinical signs and symptoms. Rare metabolic disorders involve each of the five urea cycle enzymes.

## PROTEIN TURNOVER OCCURS IN ALL FORMS OF LIFE

The continuous degradation and synthesis of cellular proteins occur in all forms of life. Each day humans turn over 1–2% of their total body protein, principally muscle protein. High rates of protein degradation occur in tissues undergoing structural rearrangement—eg, uterine tissue during pregnancy, tadpole tail tissue during metamorphosis, or skeletal muscle in starvation. Of the liberated amino acids, approximately 75% are reutilized. The excess nitrogen forms urea. Since excess amino acids are not stored, those not immediately incorporated into new protein are rapidly degraded.

## PROTEASES & PEPTIDASES DEGRADE PROTEINS TO AMINO ACIDS

The susceptibility of a protein to degradation is expressed as its half-life ($t_{1/2}$), the time required to lower its concentration to half the initial value. Half-lives of liver proteins range from under 30 minutes to over 150 hours. Typical "housekeeping" enzymes have $t_{1/2}$ values of over 100 hours. By contrast, many key regulatory enzymes have a $t_{1/2}$ of 0.5–2 hours. PEST sequences, regions rich in proline (P), glutamate (E), serine (S), and threonine (T), target some proteins for rapid degradation. Intracellular proteases hydrolyze internal peptide bonds. The resulting peptides are then degraded to amino acids by endopeptidases that cleave internal bonds and by aminopeptidases and carboxypeptidases that remove amino acids sequentially from the amino and carboxyl terminals, respectively. Degradation of circulating peptides such as hormones follows loss of a sialic acid moiety from the nonreducing ends of their oligosaccharide chains. Asialoglycoproteins are internalized by liver cell asialoglycoprotein receptors and degraded by lysosomal proteases termed cathepsins.

Extracellular, membrane-associated, and long-lived intracellular proteins are degraded in lysosomes by ATP-independent processes. By contrast, degradation of abnormal and other short-lived proteins occurs in the cytosol and requires ATP and ubiquitin. Ubiquitin, so named because it is present in all eukaryotic cells, is a small (8.5 kDa) protein that targets many intracellular proteins for degradation. The primary structure of ubiquitin is highly conserved. Only 3 of 76 residues differ between yeast and human ubiquitin. Several molecules of ubiquitin are attached by non-α-peptide bonds formed between the carboxyl terminal of ubiquitin and the ε-amino groups of lysyl residues in the target protein (Figure 29–1). The residue present at its amino terminal affects whether a protein is ubiquitinated. Amino terminal Met or Ser retards whereas Asp or Arg accelerates ubiquitination. Degradation occurs in a multicatalytic complex of proteases known as the proteasome.

## ANIMALS CONVERT α-AMINO NITROGEN TO VARIED END PRODUCTS

Different animals excrete excess nitrogen as ammonia, uric acid, or urea. The aqueous environment of teleostean fish, which are ammonotelic (excrete ammonia), compels them to excrete water continuously, facilitating excretion of highly toxic ammonia. Birds, which must conserve water and maintain low weight, are uricotelic and excrete uric acid as semisolid guano. Many

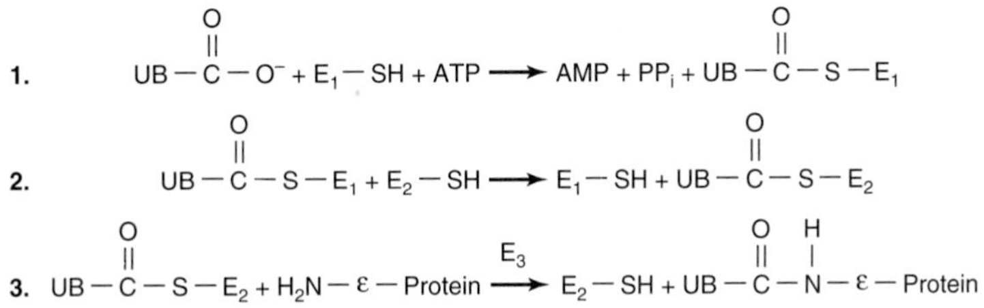

$$1.\quad UB-\overset{O}{\overset{\|}{C}}-O^- + E_1-SH + ATP \longrightarrow AMP + PP_i + UB-\overset{O}{\overset{\|}{C}}-S-E_1$$

$$2.\quad UB-\overset{O}{\overset{\|}{C}}-S-E_1 + E_2-SH \longrightarrow E_1-SH + UB-\overset{O}{\overset{\|}{C}}-S-E_2$$

$$3.\quad UB-\overset{O}{\overset{\|}{C}}-S-E_2 + H_2N-\varepsilon-\text{Protein} \xrightarrow{E_3} E_2-SH + UB-\overset{O}{\overset{\|}{C}}-\overset{H}{\overset{|}{N}}-\varepsilon-\text{Protein}$$

***Figure 29–1.*** Partial reactions in the attachment of ubiquitin (UB) to proteins. (1) The terminal COOH of ubiquitin forms a thioester bond with an -SH of $E_1$ in a reaction driven by conversion of ATP to AMP and $PP_i$. Subsequent hydrolysis of $PP_i$ by pyrophosphatase ensures that reaction 1 will proceed readily. (2) A thioester exchange reaction transfers activated ubiquitin to $E_2$. (3) $E_3$ catalyzes transfer of ubiquitin to ε-amino groups of lysyl residues of target proteins.

land animals, including humans, are ureotelic and excrete nontoxic, water-soluble urea. High blood urea levels in renal disease are a consequence—not a cause—of impaired renal function.

## BIOSYNTHESIS OF UREA

Urea biosynthesis occurs in four stages: (1) transamination, (2) oxidative deamination of glutamate, (3) ammonia transport, and (4) reactions of the urea cycle (Figure 29–2).

### Transamination Transfers α-Amino Nitrogen to α-Ketoglutarate, Forming Glutamate

Transamination interconverts pairs of α-amino acids and α-keto acids (Figure 29–3). All the protein amino acids except lysine, threonine, proline, and hydroxyproline participate in transamination. Transamination is readily reversible, and aminotransferases also function in amino acid biosynthesis. The coenzyme pyridoxal phosphate (PLP) is present at the catalytic site of aminotransferases and of many other enzymes that act on amino acids. PLP, a derivative of vitamin $B_6$, forms an enzyme-bound Schiff base intermediate that can rearrange in various ways. During transamination, bound PLP serves as a carrier of amino groups. Rearrangement forms an α-keto acid and enzyme-bound pyridoxamine phosphate, which forms a Schiff base with a second keto acid. Following removal of α-amino nitrogen by transamination, the remaining carbon "skeleton" is degraded by pathways discussed in Chapter 30.

Alanine-pyruvate aminotransferase (alanine aminotransferase) and glutamate-α-ketoglutarate aminotransferase (glutamate aminotransferase) catalyze the transfer

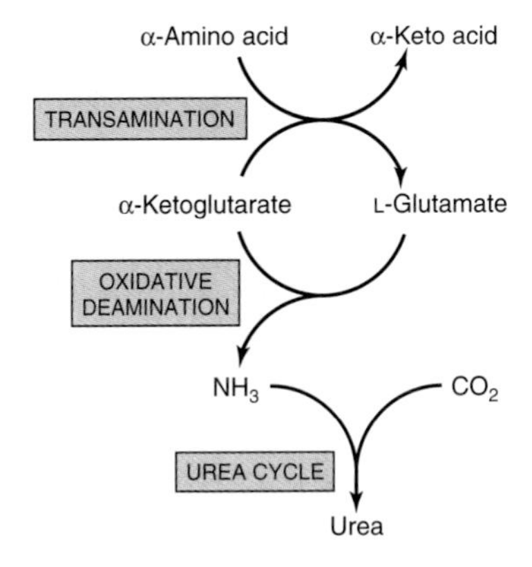


***Figure 29–2.*** Overall flow of nitrogen in amino acid catabolism.

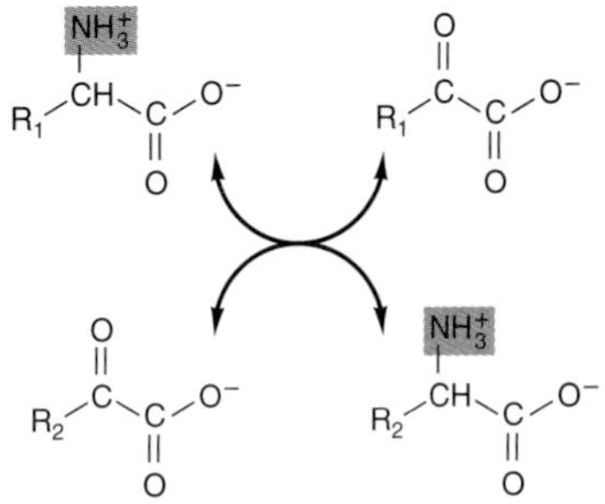

***Figure 29–3.*** Transamination. The reaction is freely reversible with an equilibrium constant close to unity.

of amino groups to pyruvate (forming alanine) or to α-ketoglutarate (forming glutamate) (Figure 29–4). Each aminotransferase is specific for one pair of substrates but nonspecific for the other pair. Since alanine is also a substrate for glutamate aminotransferase, all the amino nitrogen from amino acids that undergo transamination can be concentrated in glutamate. This is important because L-glutamate is the only amino acid that undergoes oxidative deamination at an appreciable rate in mammalian tissues. The formation of ammonia from α-amino groups thus occurs mainly via the α-amino nitrogen of L-glutamate.

Transamination is not restricted to α-amino groups. The δ-amino group of ornithine—but not the ε-amino group of lysine—readily undergoes transamination. Serum levels of aminotransferases are elevated in some disease states (see Figure 7–11).

## L-GLUTAMATE DEHYDROGENASE OCCUPIES A CENTRAL POSITION IN NITROGEN METABOLISM

Transfer of amino nitrogen to α-ketoglutarate forms L-glutamate. Release of this nitrogen as ammonia is then catalyzed by hepatic **L-glutamate dehydrogenase (GDH),** which can use either $NAD^+$ or $NADP^+$ (Figure 29–5). Conversion of α-amino nitrogen to ammonia by the concerted action of glutamate aminotransferase and GDH is often termed "transdeamination." Liver GDH activity is allosterically inhibited by ATP, GTP, and NADH and activated by ADP. The reaction catalyzed by GDH is freely reversible and functions also in amino acid biosynthesis (see Figure 28–1).

### Amino Acid Oxidases Also Remove Nitrogen as Ammonia

While their physiologic role is uncertain, L-amino acid oxidases of liver and kidney convert amino acids to an α-imino acid that decomposes to an α-keto acid with release of ammonium ion (Figure 29–6). The reduced flavin is reoxidized by molecular oxygen, forming hydrogen peroxide ($H_2O_2$), which then is split to $O_2$ and $H_2O$ by **catalase.**

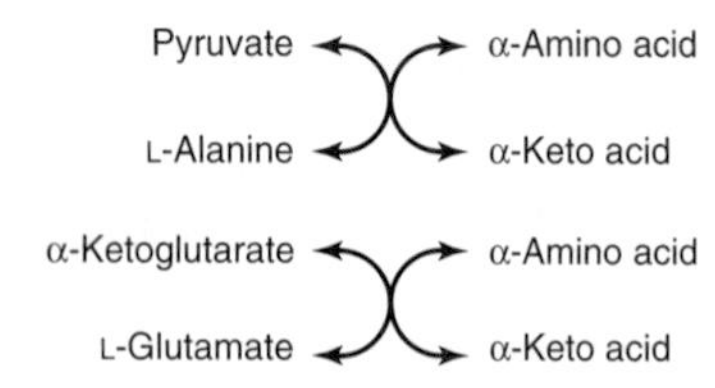


***Figure 29–4.*** Alanine aminotransferase *(top)* and glutamate aminotransferase *(bottom).*

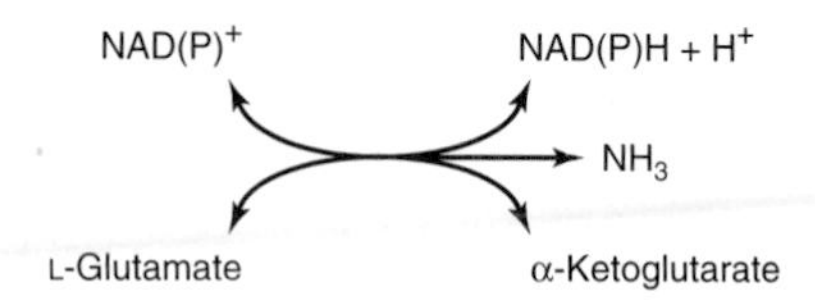


***Figure 29–5.*** The L-glutamate dehydrogenase reaction. $NAD(P)^+$ means that either $NAD^+$ or $NADP^+$ can serve as co-substrate. The reaction is reversible but favors glutamate formation.

### Ammonia Intoxication Is Life-Threatening

The ammonia produced by enteric bacteria and absorbed into portal venous blood and the ammonia produced by tissues are rapidly removed from circulation by the liver and converted to urea. Only traces (10–20 μg/dL) thus normally are present in peripheral blood. This is essential, since ammonia is toxic to the central nervous system. Should portal blood bypass the liver, systemic blood ammonia levels may rise to toxic levels. This occurs in severely impaired hepatic function or the development of collateral links between the portal and systemic veins in cirrhosis. Symptoms of **ammonia intoxication** include tremor, slurred speech, blurred vision, coma, and ultimately death. Ammonia may be toxic to the brain in part because it reacts with α-ketoglutarate to form glutamate. The resulting depleted levels of α-ketoglutarate then impair function of the tricarboxylic acid (TCA) cycle in neurons.

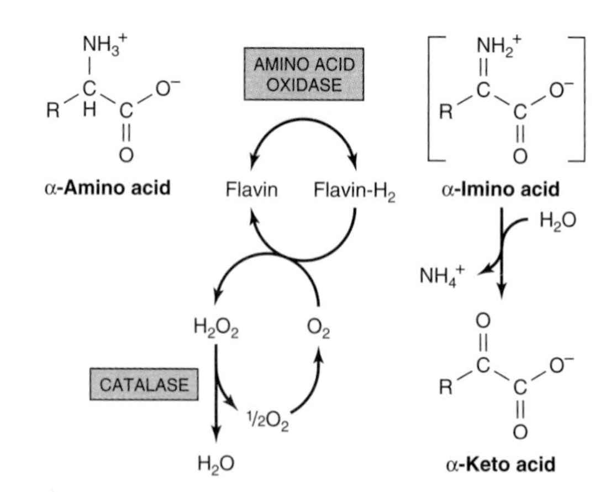


***Figure 29–6.*** Oxidative deamination catalyzed by L-amino acid oxidase (L-α-amino acid:$O_2$ oxidoreductase). The α-imino acid, shown in brackets, is not a stable intermediate.

## Glutamine Synthase Fixes Ammonia as Glutamine

Formation of glutamine is catalyzed by mitochondrial **glutamine synthase** (Figure 29–7). Since amide bond synthesis is coupled to the hydrolysis of ATP to ADP and $P_i$, the reaction strongly favors glutamine synthesis. One function of glutamine is to sequester ammonia in a nontoxic form.

## Glutaminase & Asparaginase Deamidate Glutamine & Asparagine

Hydrolytic release of the amide nitrogen of glutamine as ammonia, catalyzed by **glutaminase** (Figure 29–8), strongly favors glutamate formation. The concerted action of glutamine synthase and glutaminase thus catalyzes the interconversion of free ammonium ion and glutamine. An analogous reaction is catalyzed by L-asparaginase.

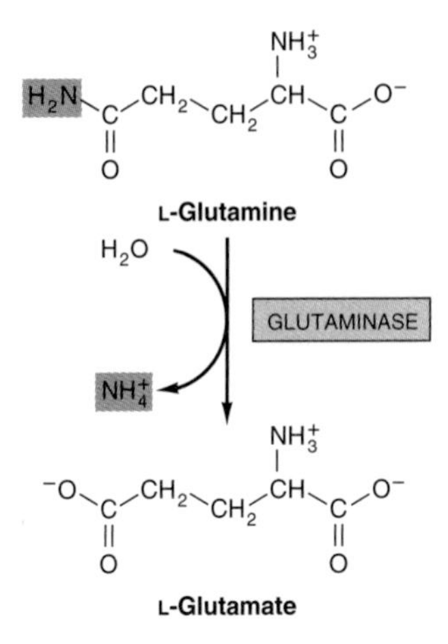


***Figure 29–8.*** The glutaminase reaction proceeds essentially irreversibly in the direction of glutamate and $NH_4^+$ formation. Note that the *amide* nitrogen, not the α-amino nitrogen, is removed.

## Formation & Secretion of Ammonia Maintains Acid-Base Balance

Excretion into urine of ammonia produced by renal tubular cells facilitates cation conservation and regulation of acid-base balance. Ammonia production from intracellular renal amino acids, especially glutamine, increases in **metabolic acidosis** and decreases in **metabolic alkalosis.**

# UREA IS THE MAJOR END PRODUCT OF NITROGEN CATABOLISM IN HUMANS

Synthesis of 1 mol of urea requires 3 mol of ATP plus 1 mol each of ammonium ion and of the α-amino nitrogen of aspartate. Five enzymes catalyze the numbered reactions of Figure 29–9. Of the six participating amino acids, *N*-acetylglutamate functions solely as an enzyme activator. The others serve as carriers of the atoms that ultimately become urea. The major metabolic role of **ornithine, citrulline,** and **argininosuccinate** in mammals is urea synthesis. Urea synthesis is a cyclic process. Since the ornithine consumed in reaction 2 is regenerated in reaction 5, there is no net loss or gain of ornithine, citrulline, argininosuccinate, or arginine. Ammonium ion, $CO_2$, ATP, and aspartate are, however, consumed. Some reactions of urea synthesis occur in the matrix of the mitochondrion, other reactions in the cytosol (Figure 29–9).

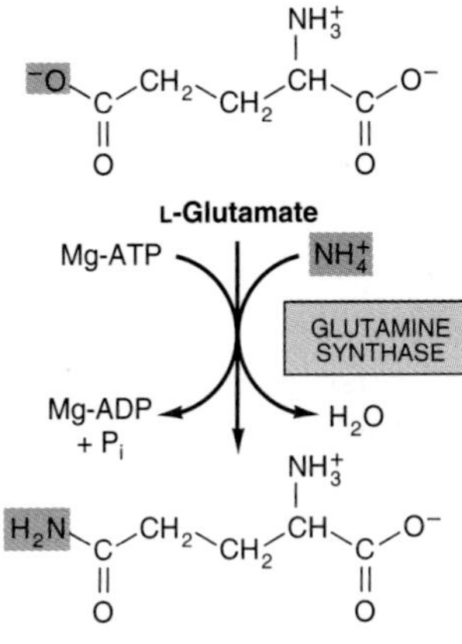

L-Glutamate

Mg-ATP $NH_4^+$

GLUTAMINE SYNTHASE

Mg-ADP + $P_i$ $H_2O$

L-Glutamine

***Figure 29–7.*** The glutamine synthase reaction strongly favors glutamine synthesis.

## Carbamoyl Phosphate Synthase I Initiates Urea Biosynthesis

Condensation of $CO_2$, ammonia, and ATP to form **carbamoyl phosphate** is catalyzed by mitochondrial **carbamoyl phosphate synthase I** (reaction 1, Figure 29–9). A cytosolic form of this enzyme, carbamoyl phosphate synthase II, uses glutamine rather than ammonia as the nitrogen donor and functions in pyrimidine biosynthesis (see Chapter 34). Carbamoyl phosphate synthase I, the rate-limiting enzyme of the urea cycle, is active only in the presence of its allosteric activator ***N*-acetylglutamate,** which enhances the affinity of the synthase for ATP. Formation of carbamoyl phosphate requires 2 mol of ATP, one of which serves as a phosphate donor. Conversion of the second ATP to AMP and pyrophosphate, coupled to the hydrolysis of pyrophosphate to orthophosphate, provides the driving

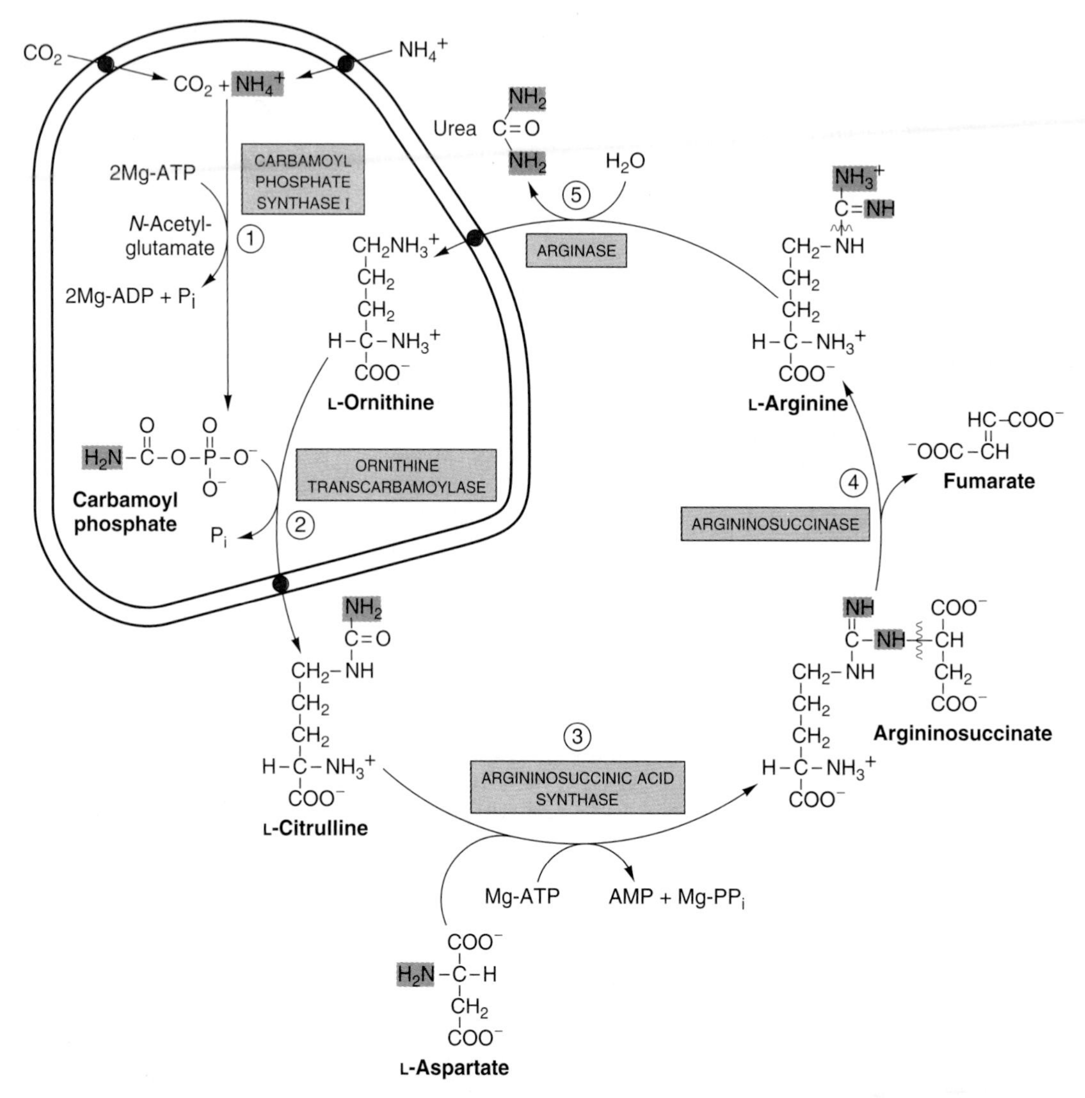


***Figure 29–9.*** Reactions and intermediates of urea biosynthesis. The nitrogen-containing groups that contribute to the formation of urea are shaded. Reactions ① and ② occur in the matrix of liver mitochondria and reactions ③, ④, and ⑤ in liver cytosol. $CO_2$ (as bicarbonate), ammonium ion, ornithine, and citrulline enter the mitochondrial matrix via specific carriers (see heavy dots) present in the inner membrane of liver mitochondria.

force for synthesis of the amide bond and the mixed acid anhydride bond of carbamoyl phosphate. The concerted action of GDH and carbamoyl phosphate synthase I thus shuttles nitrogen into carbamoyl phosphate, a compound with high group transfer potential. The reaction proceeds stepwise. Reaction of bicarbonate with ATP forms carbonyl phosphate and ADP. Ammonia then displaces ADP, forming carbamate and orthophosphate. Phosphorylation of carbamate by the second ATP then forms carbamoyl phosphate.

## Carbamoyl Phosphate Plus Ornithine Forms Citrulline

**L-Ornithine transcarbamoylase** catalyzes transfer of the carbamoyl group of carbamoyl phosphate to ornithine, forming citrulline and orthophosphate (reaction 2, Figure 29–9). While the reaction occurs in the mitochondrial matrix, both the formation of ornithine and the subsequent metabolism of citrulline take place in the cytosol. Entry of ornithine into mitochondria and exodus of citrulline from mitochondria therefore involve mitochondrial inner membrane transport systems (Figure 29–9).

### Citrulline Plus Aspartate Forms Argininosuccinate

**Argininosuccinate synthase** links aspartate and citrulline via the amino group of aspartate (reaction 3, Figure 29–9) and provides the second nitrogen of urea. The reaction requires ATP and involves intermediate formation of citrullyl-AMP. Subsequent displacement of AMP by aspartate then forms citrulline.

### Cleavage of Argininosuccinate Forms Arginine & Fumarate

Cleavage of argininosuccinate, catalyzed by **argininosuccinase,** proceeds with retention of nitrogen in arginine and release of the aspartate skeleton as fumarate (reaction 4, Figure 29–9). Addition of water to fumarate forms L-malate, and subsequent $NAD^+$-dependent oxidation of malate forms oxaloacetate. These two reactions are analogous to reactions of the citric acid cycle (see Figure 16–3) but are catalyzed by *cytosolic* fumarase and malate dehydrogenase. Transamination of oxaloacetate by glutamate aminotransferase then re-forms aspartate. The carbon skeleton of aspartate-fumarate thus acts as a carrier of the nitrogen of glutamate into a precursor of urea.

### Cleavage of Arginine Releases Urea & Re-forms Ornithine

Hydrolytic cleavage of the guanidino group of arginine, catalyzed by liver **arginase,** releases urea (reaction 5, Figure 29–9). The other product, ornithine, reenters liver mitochondria for additional rounds of urea synthesis. Ornithine and lysine are potent inhibitors of arginase, competitive with arginine. Arginine also serves as the precursor of the potent muscle relaxant nitric oxide (NO) in a $Ca^{2+}$-dependent reaction catalyzed by NO synthase (see Figure 49–15).

### Carbamoyl Phosphate Synthase I Is the Pacemaker Enzyme of the Urea Cycle

The activity of carbamoyl phosphate synthase I is determined by *N*-acetylglutamate, whose steady-state level is dictated by its rate of synthesis from acetyl-CoA and glutamate and its rate of hydrolysis to acetate and glutamate. These reactions are catalyzed by *N*-acetylglutamate synthase and *N*-acetylglutamate hydrolase, respectively. Major changes in diet can increase the concentrations of individual urea cycle enzymes 10-fold to 20-fold. Starvation, for example, elevates enzyme levels, presumably to cope with the increased production of ammonia that accompanies enhanced protein degradation.

## METABOLIC DISORDERS ARE ASSOCIATED WITH EACH REACTION OF THE UREA CYCLE

Metabolic disorders of urea biosynthesis, while extremely rare, illustrate four important principles: (1) Defects in any of several enzymes of a metabolic pathway enzyme can result in similar clinical signs and symptoms. (2) The identification of intermediates and of ancillary products that accumulate prior to a metabolic block provides insight into the reaction that is impaired. (3) Precise diagnosis requires quantitative assay of the activity of the enzyme thought to be defective. (4) Rational therapy must be based on an understanding of the underlying biochemical reactions in normal and impaired individuals.

All defects in urea synthesis result in ammonia intoxication. Intoxication is more severe when the metabolic block occurs at reactions 1 or 2 since some covalent linking of ammonia to carbon has already occurred if citrulline can be synthesized. Clinical symptoms common to all urea cycle disorders include vomiting, avoidance of high-protein foods, intermittent ataxia, irritability, lethargy, and mental retardation. The clinical features and treatment of all five disorders discussed below are similar. Significant improvement and minimization of brain damage accompany a low-protein diet ingested as frequent small meals to avoid sudden increases in blood ammonia levels.

**Hyperammonemia Type 1.** A consequence of **carbamoyl phosphate synthase I** deficiency (reaction 1, Figure 29–9), this relatively infrequent condition (estimated frequency 1:62,000) probably is familial.

**Hyperammonemia Type 2.** A deficiency of **ornithine transcarbamoylase** (reaction 2, Figure 29–9) produces this X chromosome–linked deficiency. The mothers also exhibit hyperammonemia and an aversion to high-protein foods. Levels of glutamine are elevated in blood, cerebrospinal fluid, and urine, probably due to enhanced glutamine synthesis in response to elevated levels of tissue ammonia.

**Citrullinemia.** In this rare disorder, plasma and cerebrospinal fluid citrulline levels are elevated and 1–2 g of citrulline are excreted daily. One patient lacked detectable **argininosuccinate synthase** activity (reaction 3, Figure 29–9). In another, the $K_m$ for citrulline was 25 times higher than normal. Citrulline and argininosuccinate, which contain nitrogen destined for urea synthesis, serve as alternative carriers of excess nitrogen. Feeding arginine enhanced excretion of citrulline in these patients. Similarly, feeding benzoate diverts ammonia nitrogen to hippurate via glycine (see Figure 31–1).

**Argininosuccinicaciduria.** A rare disease characterized by elevated levels of argininosuccinate in blood, cerebrospinal fluid, and urine is associated with friable, tufted hair (trichorrhexis nodosa). Both early-onset and late-onset types are known. The metabolic defect is the absence of **argininosuccinase** (reaction 4, Figure 29–9). Diagnosis by measurement of erythrocyte argininosuccinase activity can be performed on umbilical cord blood or amniotic fluid cells. As for citrullinemia, feeding arginine and benzoate promotes nitrogen excretion.

**Hyperargininemia.** This defect is characterized by elevated blood and cerebrospinal fluid arginine levels, low erythrocyte levels of arginase (reaction 5, Figure 29–9), and a urinary amino acid pattern resembling that of lysine-cystinuria. This pattern may reflect competition by arginine with lysine and cystine for reabsorption in the renal tubule. A low-protein diet lowers plasma ammonia levels and abolishes lysine-cystinuria.

## Gene Therapy Offers Promise for Correcting Defects in Urea Biosynthesis

Gene therapy for rectification of defects in the enzymes of the urea cycle is an area of active investigation. Encouraging preliminary results have been obtained, for example, in animal models using an adenoviral vector to treat citrullinemia.

## SUMMARY

- Human subjects degrade 1–2% of their body protein daily at rates that vary widely between proteins and with physiologic state. Key regulatory enzymes often have short half-lives.
- Proteins are degraded by both ATP-dependent and ATP-independent pathways. Ubiquitin targets many intracellular proteins for degradation. Liver cell surface receptors bind and internalize circulating asialoglycoproteins destined for lysosomal degradation.
- Ammonia is highly toxic. Fish excrete $NH_3$ directly; birds convert $NH_3$ to uric acid. Higher vertebrates convert $NH_3$ to urea.
- Transamination channels α-amino acid nitrogen into glutamate. L-Glutamate dehydrogenase (GDH) occupies a central position in nitrogen metabolism.
- Glutamine synthase converts $NH_3$ to nontoxic glutamine. Glutaminase releases $NH_3$ for use in urea synthesis.
- $NH_3$, $CO_2$, and the amide nitrogen of aspartate provide the atoms of urea.
- Hepatic urea synthesis takes place in part in the mitochondrial matrix and in part in the cytosol. Inborn errors of metabolism are associated with each reaction of the urea cycle.
- Changes in enzyme levels and allosteric regulation of carbamoyl phosphate synthase by *N*-acetylglutamate regulate urea biosynthesis.

## REFERENCES

Brooks P et al: Subcellular localization of proteasomes and their regulatory complexes in mammalian cells. Biochem J 2000;346:155.

Curthoys NP, Watford M: Regulation of glutaminase activity and glutamine metabolism. Annu Rev Nutr 1995;15:133.

Hershko A, Ciechanover A: The ubiquitin system. Annu Rev Biochem 1998;67:425.

Iyer R et al: The human arginases and arginase deficiency. J Inherit Metab Dis 1998;21:86.

Lim CB et al: Reduction in the rates of protein and amino acid catabolism to slow down the accumulation of endogenous ammonia: a strategy potentially adopted by mudskippers during aerial exposure in constant darkness. J Exp Biol 2001;204:1605.

Patejunas G et al: Evaluation of gene therapy for citrullinaemia using murine and bovine models. J Inherit Metab Dis 1998;21:138.

Pickart CM. Mechanisms underlying ubiquitination. Annu Rev Biochem 2001;70:503.

Scriver CR et al (editors): *The Metabolic and Molecular Bases of Inherited Disease,* 8th ed. McGraw-Hill, 2001.

Tuchman M et al: The biochemical and molecular spectrum of ornithine transcarbamoylase deficiency. J Inherit Metab Dis 1998;21:40.

Turner MA et al: Human argininosuccinate lyase: a structural basis for intragenic complementation. Proc Natl Acad Sci U S A 1997;94:9063.

# Catabolism of the Carbon Skeletons of Amino Acids

# 30

*Victor W. Rodwell, PhD*

## BIOMEDICAL IMPORTANCE

This chapter considers conversion of the carbon skeletons of the common L-amino acids to amphibolic intermediates and the metabolic diseases or "inborn errors of metabolism" associated with these processes. Left untreated, they can result in irreversible brain damage and early mortality. Prenatal or early postnatal detection and timely initiation of treatment thus are essential. Many of the enzymes concerned can be detected in cultured amniotic fluid cells, which facilitates early diagnosis by amniocentesis. Treatment consists primarily of feeding diets low in the amino acids whose catabolism is impaired. While many changes in the primary structure of enzymes have no adverse effects, others modify the three-dimensional structure of catalytic or regulatory sites, lower catalytic efficiency (lower $V_{max}$ or elevate $K_m$), or alter the affinity for an allosteric regulator of activity. A variety of mutations thus may give rise to the same clinical signs and symptoms.

## TRANSAMINATION TYPICALLY INITIATES AMINO ACID CATABOLISM

Removal of α-amino nitrogen by transamination (see Figure 28–3) is the first catabolic reaction of amino acids except in the case of proline, hydroxyproline, threonine, and lysine. The residual hydrocarbon skeleton is then degraded to amphibolic intermediates as outlined in Figure 30–1.

**Asparagine, Aspartate, Glutamine, and Glutamate.** All four carbons of asparagine and aspartate form **oxaloacetate** (Figure 30–2, top). Analogous reactions convert glutamine and glutamate to **α-ketoglutarate** (Figure 30–2, bottom). Since the enzymes also fulfill anabolic functions, no metabolic defects are associated with the catabolism of these four amino acids.

**Proline.** Proline forms dehydroproline, glutamate-γ-semialdehyde, glutamate, and, ultimately, **α-ketoglutarate** (Figure 30–3, top). The metabolic block in **type I hyperprolinemia** is at **proline dehydrogenase.**

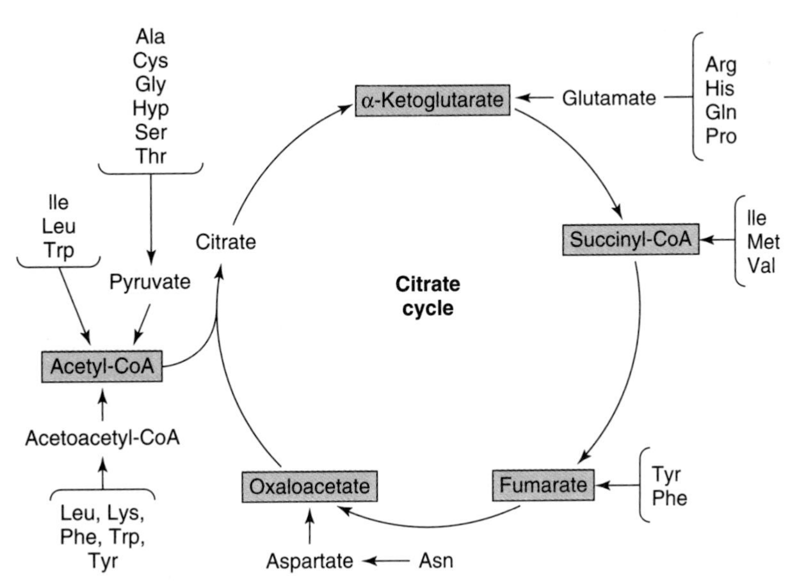


***Figure 30–1.*** Amphibolic intermediates formed from the carbon skeletons of amino acids.

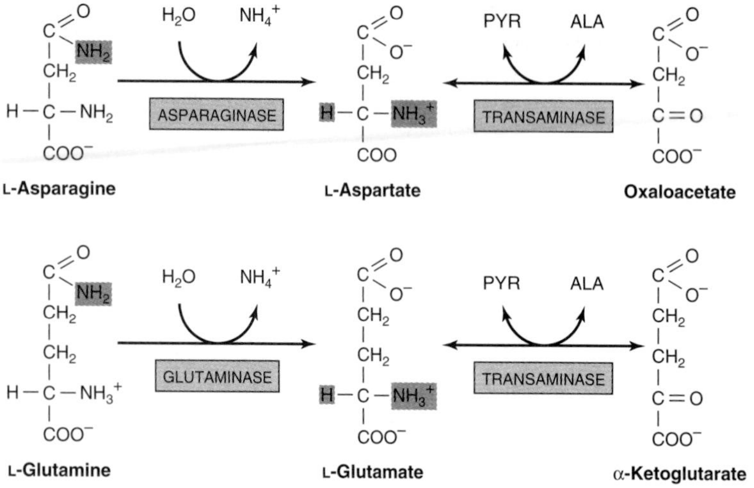

***Figure 30–2.*** Catabolism of L-asparagine (*top*) and of L-glutamine (*bottom*) to amphibolic intermediates. (PYR, pyruvate; ALA, L-alanine.) In this and subsequent figures, color highlights portions of the molecules undergoing chemical change.

There is no associated impairment of hydroxyproline catabolism. The metabolic block in **type II hyperprolinemia** is at **glutamate-γ-semialdehyde dehydrogenase,** which also functions in hydroxyproline catabolism. Both proline and hydroxyproline catabolism thus are affected and $\Delta^1$-pyrroline-3-hydroxy-5-carboxylate (see Figure 30–10) is excreted.

**Arginine and Ornithine.** Arginine is converted to ornithine, glutamate γ-semialdehyde, and then **α-ketoglutarate** (Figure 30–3, bottom). Mutations in **ornithine δ-aminotransferase** elevate plasma and urinary ornithine and cause **gyrate atrophy of the retina.** Treatment involves restricting dietary arginine. In **hyperornithinemia-hyperammonemia syndrome,** a defective mitochondrial **ornithine-citrulline antiporter** (see Figure 29–9) impairs transport of ornithine into mitochondria for use in urea synthesis.

**Histidine.** Catabolism of histidine proceeds via urocanate, 4-imidazolone-5-propionate, and *N*-formiminoglutamate (Figlu). Formimino group transfer to tetrahydrofolate forms glutamate, then **α-ketoglutarate** (Figure 30–4). In folic acid deficiency, group transfer is impaired and Figlu is excreted. Excretion of Figlu following a dose of histidine thus has been used to detect folic acid deficiency. Benign disorders of histidine catabolism include **histidinemia** and **urocanic aciduria** associated with impaired **histidase.**

## SIX AMINO ACIDS FORM PYRUVATE

All of the carbons of glycine, serine, alanine, and cysteine and two carbons of threonine form pyruvate and subsequently acetyl-CoA.

**Glycine.** The **glycine synthase complex** of liver mitochondria splits glycine to $CO_2$ and $NH_4^+$ and forms $N^5,N^{10}$-methylene tetrahydrofolate (Figure 30–5).

**Glycinuria** results from a defect in renal tubular reabsorption. The defect in **primary hyperoxaluria** is the failure to catabolize glyoxylate formed by deamination of glycine. Subsequent oxidation of glyoxylate to oxalate results in urolithiasis, nephrocalcinosis, and early mortality from renal failure or hypertension.

**Serine.** Following conversion to glycine, catalyzed by **serine hydroxymethyltransferase** (Figure 30–5), serine catabolism merges with that of glycine (Figure 30–6).

**Alanine.** Transamination of alanine forms pyruvate. Perhaps for the reason advanced under glutamate and aspartate catabolism, there is no known metabolic defect of alanine catabolism. **Cysteine.** Cystine is first reduced to cysteine by **cystine reductase** (Figure 30–7). Two different pathways then convert cysteine to pyruvate (Figure 30–8).

There are numerous abnormalities of cysteine metabolism. Cystine, lysine, arginine, and ornithine are excreted in **cystine-lysinuria (cystinuria),** a defect in renal reabsorption. Apart from cystine calculi, cystinuria is benign. The mixed disulfide of L-cysteine and L-homocysteine (Figure 30–9) excreted by cystinuric patients is more soluble than cystine and reduces formation of cystine calculi. Several metabolic defects result in vitamin $B_6$-responsive or -unresponsive **homocystinurias.** Defective carrier-mediated transport of cystine results in **cystinosis (cystine storage disease)** with deposition of cystine crystals in tissues and early mortality from acute renal failure. Despite

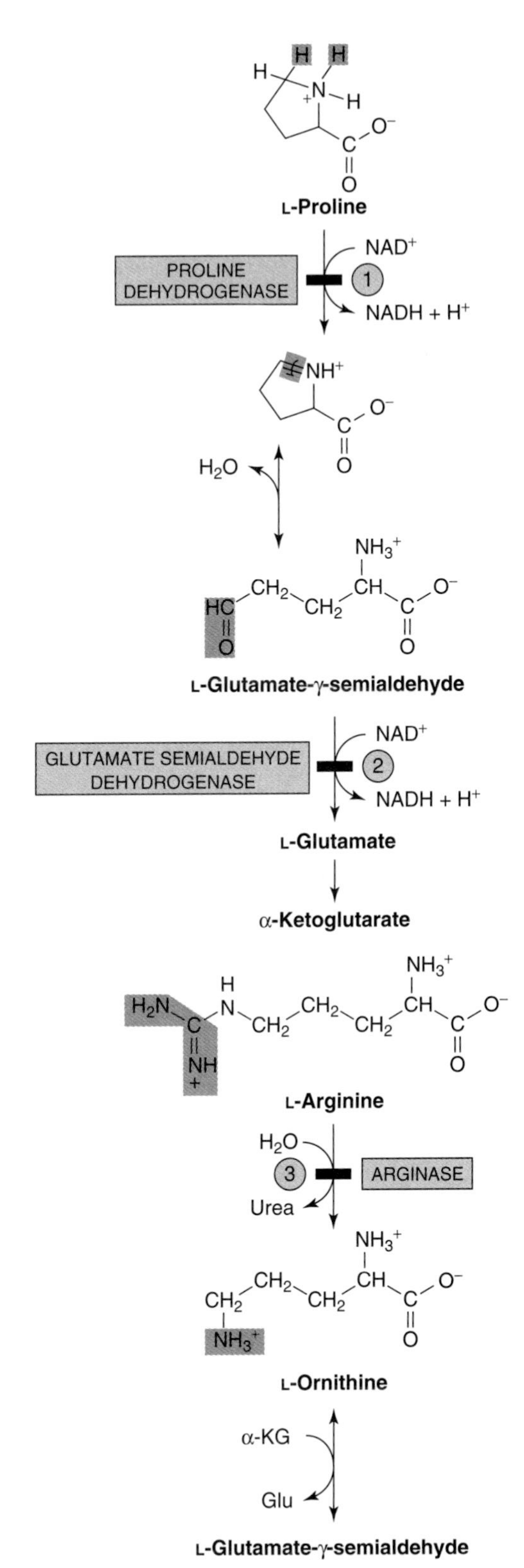


***Figure 30–3.*** **Top:** Catabolism of proline. Numerals indicate sites of the metabolic defects in ① type I and ② type II hyperprolinemias. **Bottom:** Catabolism of arginine. Glutamate-γ-semialdehyde forms α-ketoglutarate as shown above. ③, site of the metabolic defect in hyperargininemia.

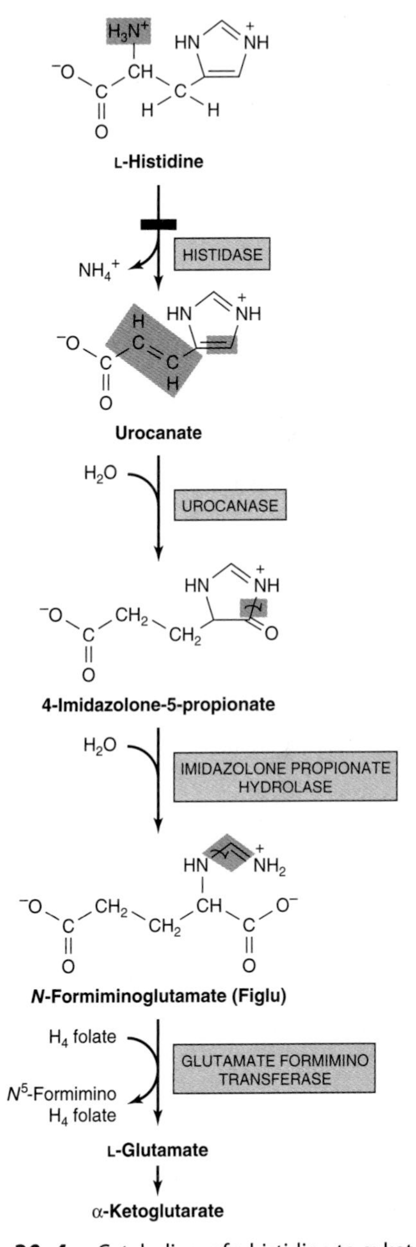


***Figure 30–4.*** Catabolism of L-histidine to α-ketoglutarate. ($H_4$ folate, tetrahydrofolate.) Histidase is the probable site of the metabolic defect in histidinemia.

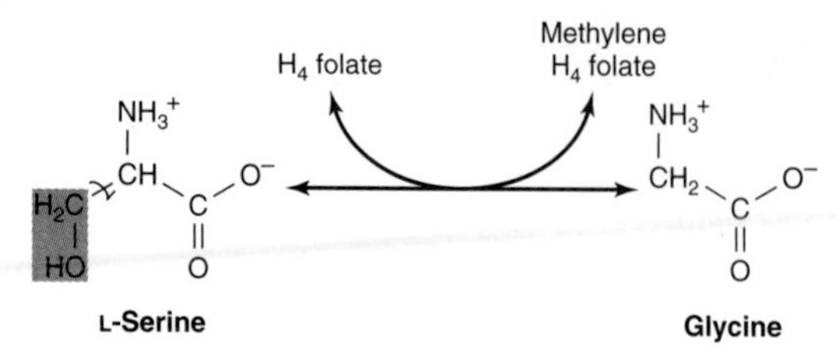


**Figure 30–5.** Interconversion of serine and glycine catalyzed by serine hydroxymethyltransferase. ($H_4$ folate, tetrahydrofolate.)

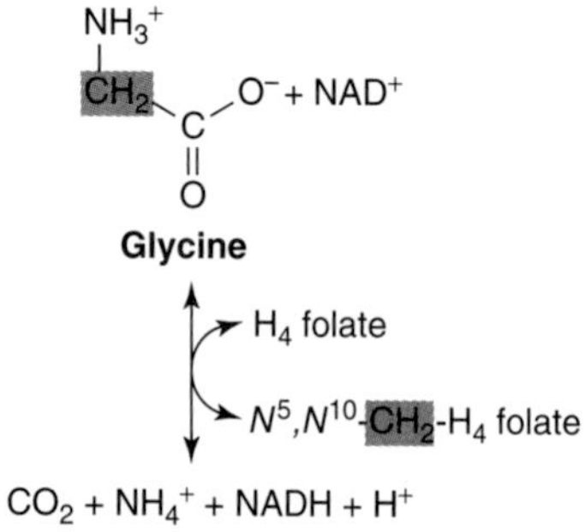

$NH_3^+$–$CH_2$–$COO^-$ + $NAD^+$

**Glycine**

$H_4$ folate

$N^5,N^{10}$-$CH_2$-$H_4$ folate

$CO_2 + NH_4^+ + NADH + H^+$

**Figure 30–6.** Reversible cleavage of glycine by the mitochondrial glycine synthase complex. (PLP, pyridoxal phosphate.)

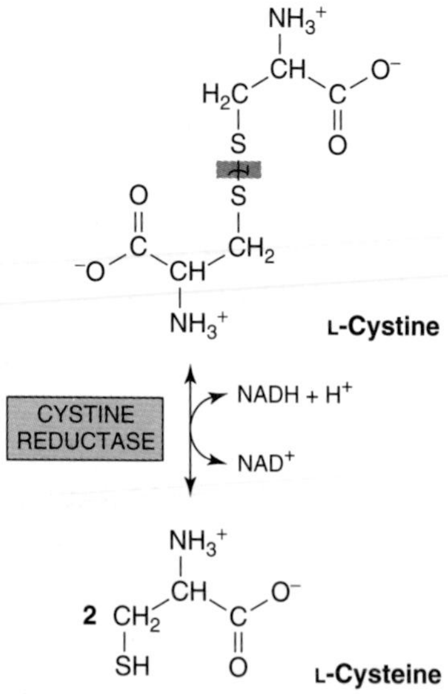

**L-Cystine**

CYSTINE REDUCTASE

$NADH + H^+$

$NAD^+$

2 **L-Cysteine**

**Figure 30–7.** The cystine reductase reaction.

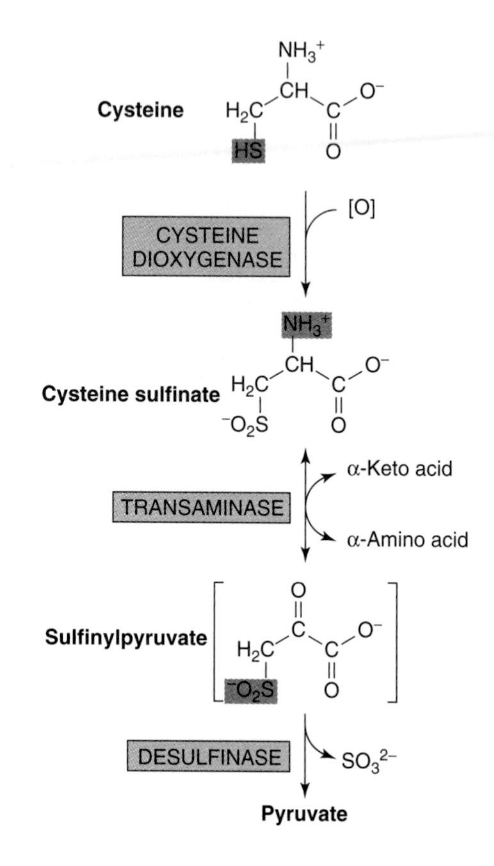


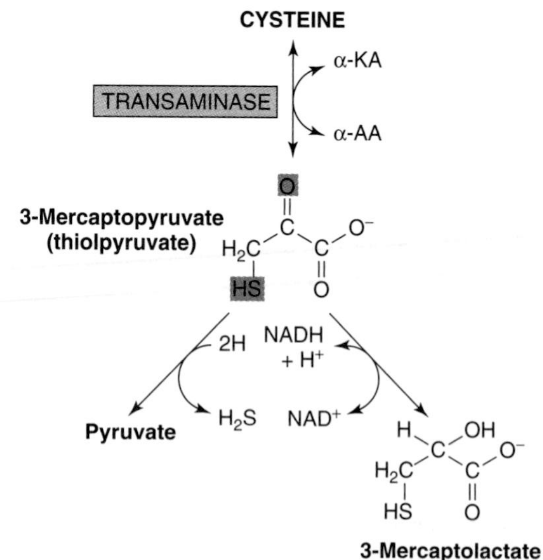


**Figure 30–8.** Catabolism of L-cysteine via the cysteine sulfinate pathway *(top)* and by the 3-mercaptopyruvate pathway *(bottom)*.

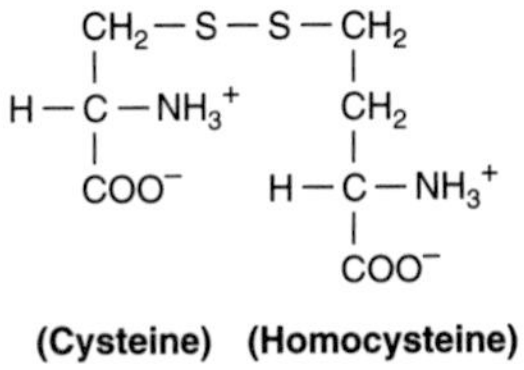

(Cysteine) (Homocysteine)

***Figure 30–9.*** Mixed disulfide of cysteine and homocysteine.

L-Threonine
THREONINE ALDOLASE
Glycine
Acetaldehyde
ALDEHYDE DEHYDROGENASE
$H_2O$
$NAD^+$
$NADH+H^+$
Acetate
CoASH
Mg-ATP
ACETATE THIOKINASE
$H_2O$
Mg-ADP
Acetyl-CoA

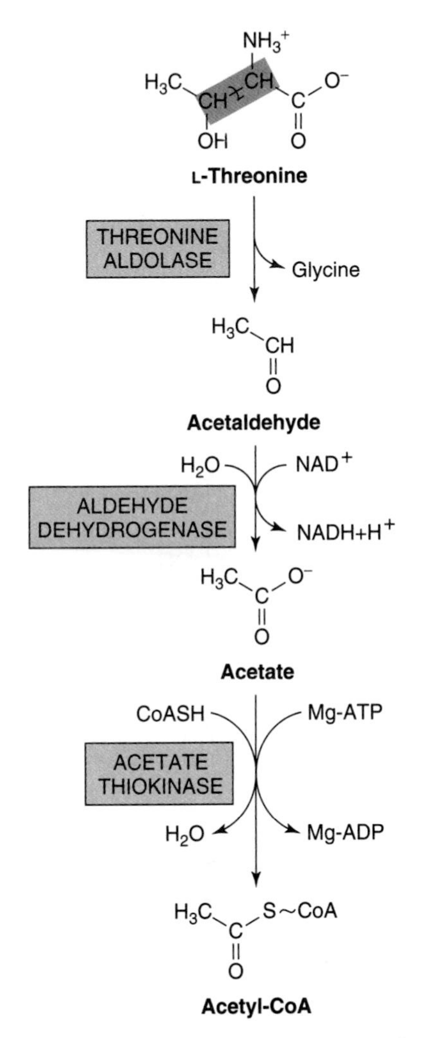

***Figure 30–10.*** Conversion of threonine to glycine (see Figure 30–6) and acetyl-CoA.

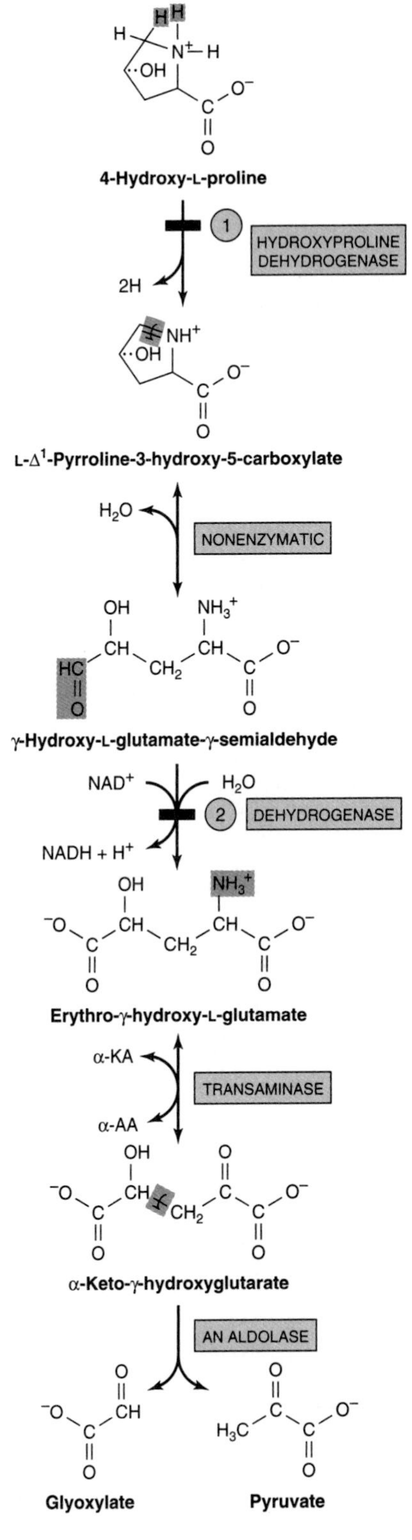

4-Hydroxy-L-proline
1
HYDROXYPROLINE DEHYDROGENASE
2H
L-Δ1-Pyrroline-3-hydroxy-5-carboxylate
$H_2O$
NONENZYMATIC
γ-Hydroxy-L-glutamate-γ-semialdehyde
$NAD^+$
$H_2O$
2
DEHYDROGENASE
$NADH + H^+$
Erythro-γ-hydroxy-L-glutamate
α-KA
TRANSAMINASE
α-AA
α-Keto-γ-hydroxyglutarate
AN ALDOLASE
Glyoxylate
Pyruvate

***Figure 30–11.*** Intermediates in L-hydroxyproline catabolism. (α-KA, α-keto acid; α-AA, α-amino acid.) Numerals identify sites of metabolic defects in ① hyperhydroxyprolinemia and ② type II hyperprolinemia.

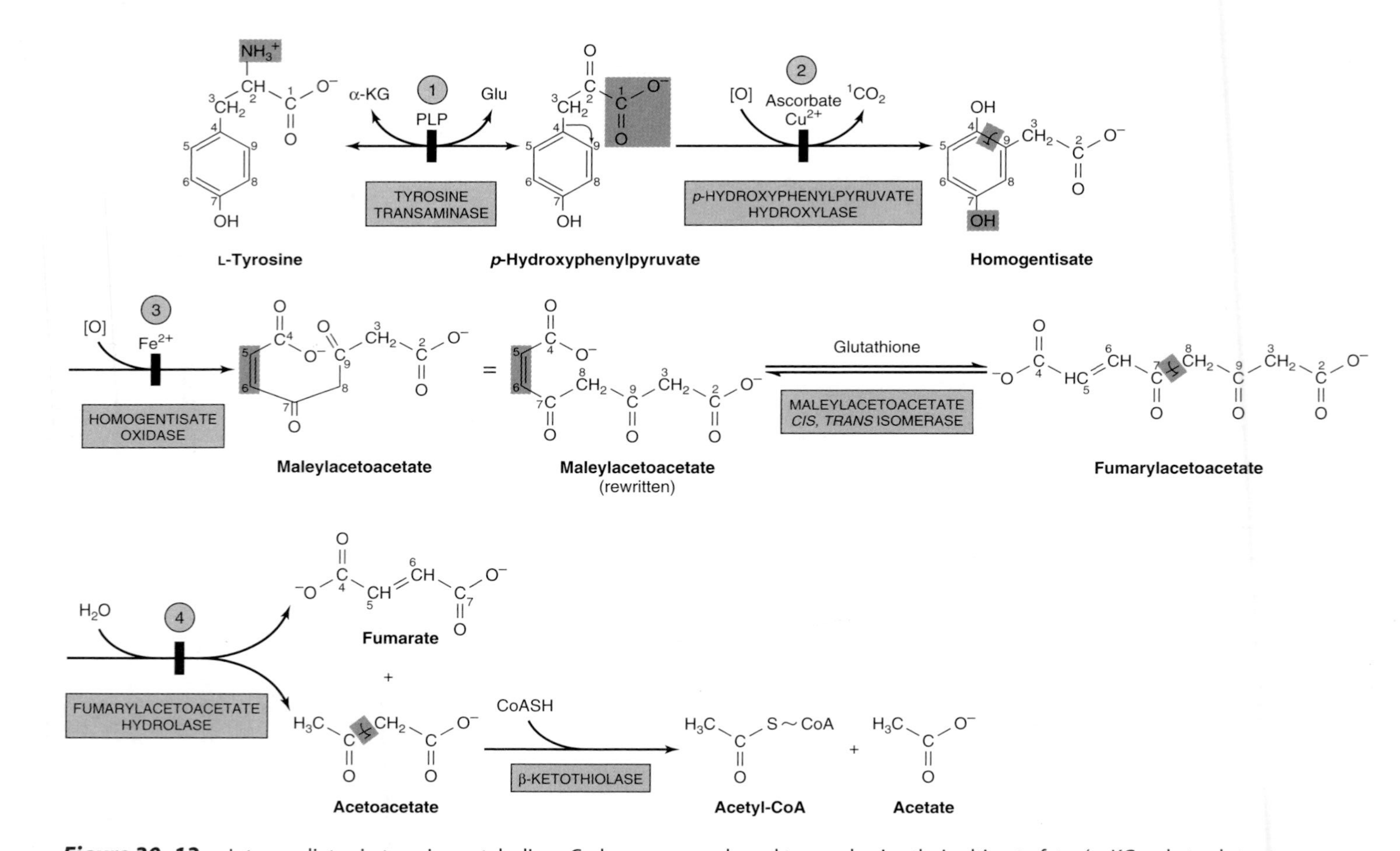


***Figure 30–12.*** Intermediates in tyrosine catabolism. Carbons are numbered to emphasize their ultimate fate. (α-KG, α-ketoglutarate; Glu, glutamate; PLP, pyridoxal phosphate.) Circled numerals represent the probable sites of the metabolic defects in ① type II tyrosinemia; ② neonatal tyrosinemia; ③ alkaptonuria; and ④ type I tyrosinemia, or tyrosinosis.

epidemiologic data suggesting a relationship between plasma homocysteine and cardiovascular disease, whether homocysteine represents a causal cardiovascular risk factor remains controversial.

**Threonine.** Threonine is cleaved to acetaldehyde and glycine. Oxidation of acetaldehyde to acetate is followed by formation of acetyl-CoA (Figure 30–10). Catabolism of glycine is discussed above.

**4-Hydroxyproline.** Catabolism of 4-hydroxy-L-proline forms, successively, L-$\Delta^1$-pyrroline-3-hydroxy-5-carboxylate, $\gamma$-hydroxy-L-glutamate-$\gamma$-semialdehyde, erythro-$\gamma$-hydroxy-L-glutamate, and $\alpha$-keto-$\gamma$-hydroxyglutarate. An aldol-type cleavage then forms glyoxylate plus pyruvate (Figure 30–11). A defect in **4-hydroxyproline dehydrogenase** results in **hyperhydroxyprolinemia,** which is benign. There is no associated impairment of proline catabolism.

## TWELVE AMINO ACIDS FORM ACETYL-CoA

**Tyrosine.** Figure 30–12 diagrams the conversion of tyrosine to amphibolic intermediates. Since ascorbate is the reductant for conversion of *p*-hydroxyphenylpyruvate to homogentisate, scorbutic patients excrete incompletely oxidized products of tyrosine catabolism. Subsequent catabolism forms maleylacetoacetate, fumarylacetoacetate, fumarate, acetoacetate, and ultimately **acetyl-CoA.**

The probable metabolic defect in **type I tyrosinemia (tyrosinosis)** is at **fumarylacetoacetate hydrolase** (reaction 4, Figure 30–12). Therapy employs a diet low in tyrosine and phenylalanine. Untreated acute and chronic tyrosinosis leads to death from liver failure. Alternate metabolites of tyrosine are also excreted in **type II tyrosinemia (Richner-Hanhart syndrome),** a defect in **tyrosine aminotransferase** (reaction 1, Figure 30–12), and in **neonatal tyrosinemia,** due to lowered *p*-hydroxyphenylpyruvate hydroxylase activity (reaction 2, Figure 30–12). Therapy employs a diet low in protein.

**Alkaptonuria** was first described in the 16th century. Characterized in 1859, it provided the basis for Garrod's classic ideas concerning heritable metabolic disorders. The defect is lack of **homogentisate oxidase** (reaction 3, Figure 30–12). The urine darkens on exposure to air due to oxidation of excreted homogentisate. Late in the disease, there is arthritis and connective tissue pigmentation (ochronosis) due to oxidation of homogentisate to benzoquinone acetate, which polymerizes and binds to connective tissue.

**Phenylalanine.** Phenylalanine is first converted to tyrosine (see Figure 28–10). Subsequent reactions are those of tyrosine (Figure 30–12). **Hyperphenylalaninemias** arise from defects in phenylalanine hydroxylase itself (**type I, classic phenylketonuria** or **PKU**), in dihydrobiopterin reductase (**types II and III**), or in dihydrobiopterin biosynthesis (**types IV and V**) (Figure 28–10). Alternative catabolites are excreted (Figure 30–13). DNA probes facilitate prenatal diagnosis of defects in phenylalanine hydroxylase or dihydrobiopterin reductase. A diet low in phenylalanine can prevent the mental retardation of PKU (frequency 1:10,000

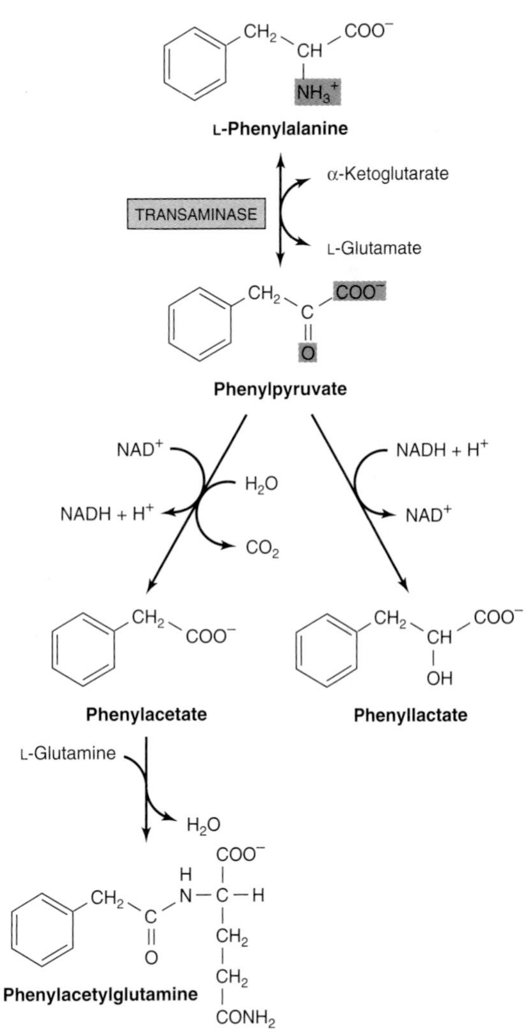

L-Phenylalanine

TRANSAMINASE

α-Ketoglutarate

L-Glutamate

Phenylpyruvate

$NAD^+$

$NADH + H^+$

$H_2O$

$CO_2$

$NADH + H^+$

$NAD^+$

Phenylacetate

Phenyllactate

L-Glutamine

$H_2O$

Phenylacetylglutamine

***Figure 30–13.*** Alternative pathways of phenylalanine catabolism in phenylketonuria. The reactions also occur in normal liver tissue but are of minor significance.

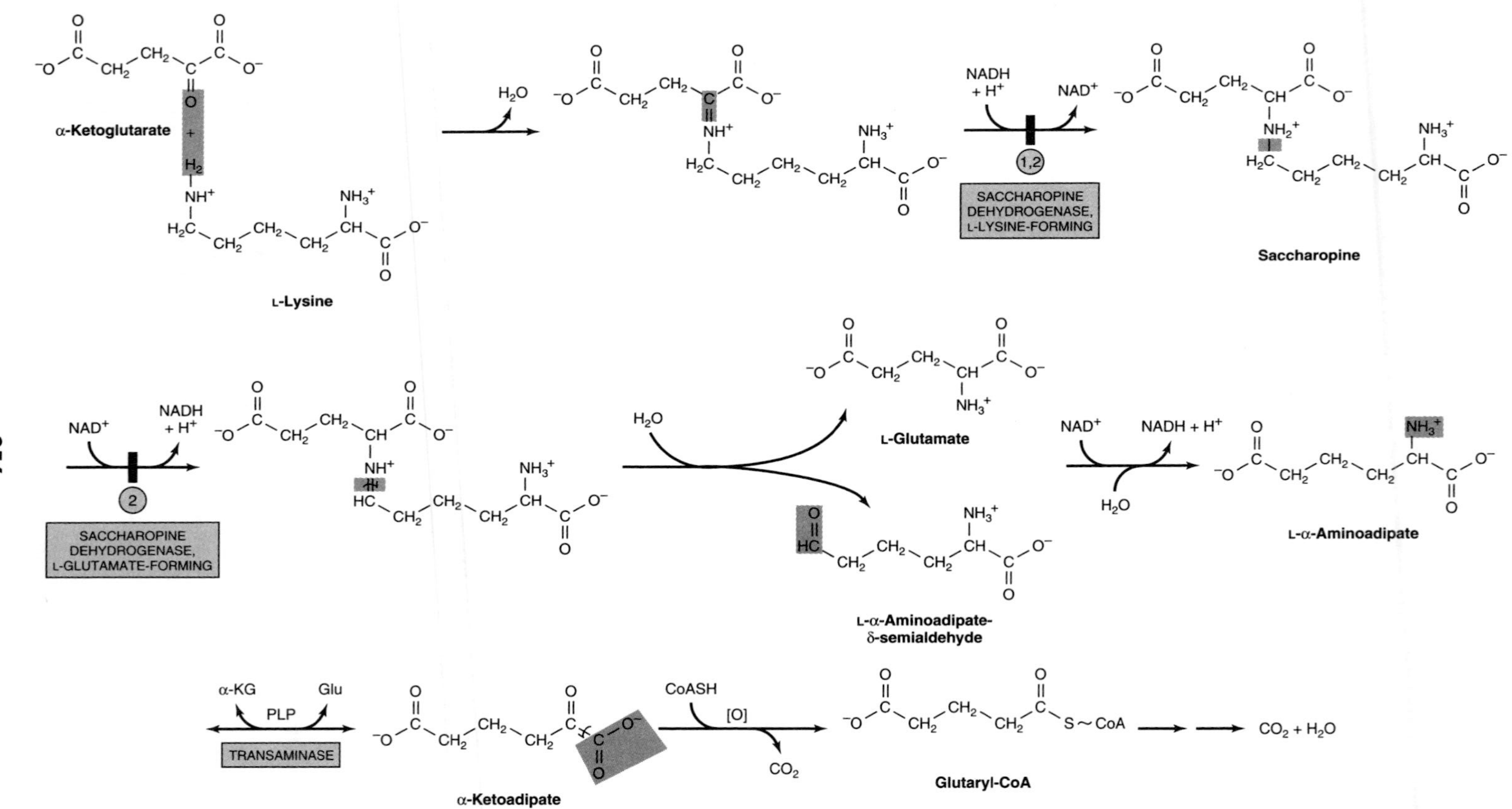


***Figure 30–14.*** Catabolism of L-lysine. (α-KG, α-ketoglutarate; Glu, glutamate; PLP, pyridoxal phosphate.) Circled numerals indicate the probable sites of the metabolic defects in ① periodic hyperlysinemia with associated hyperammonemia; and ② persistent hyperlysinemia without associated hyperammonemia.

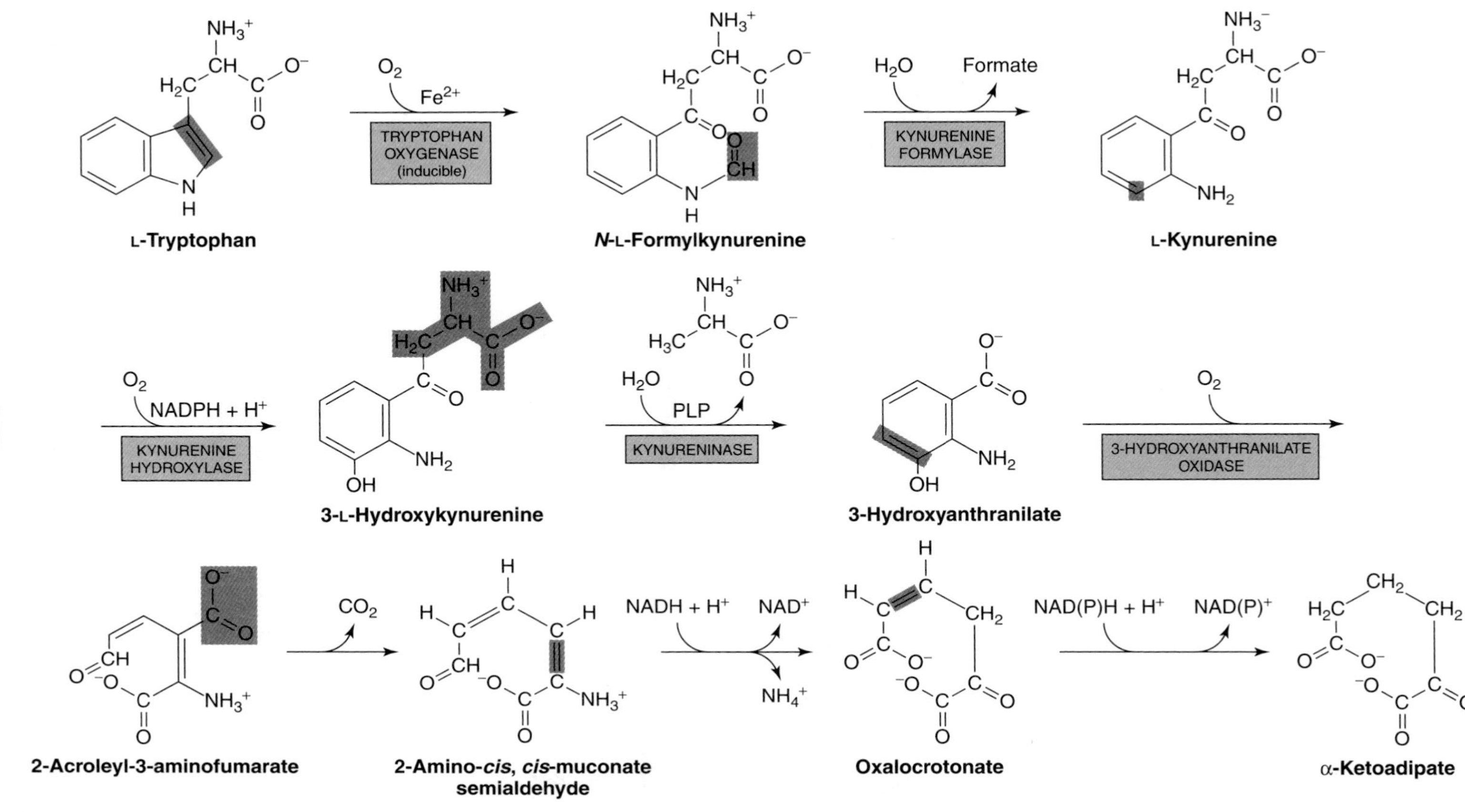


**Figure 30–15.** Catabolism of L-tryptophan. (PLP, pyridoxal phosphate.)

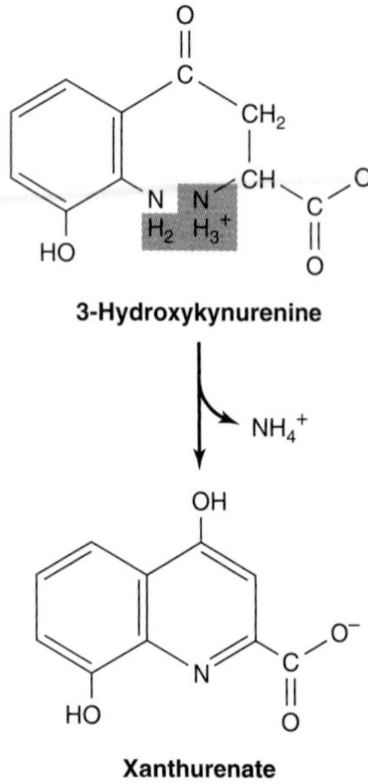

***Figure 30–16.*** Formation of xanthurenate in vitamin $B_6$ deficiency. Conversion of the tryptophan metabolite 3-hydroxykynurenine to 3-hydroxyanthranilate is impaired (see Figure 30–15). A large portion is therefore converted to xanthurenate.

births). Elevated blood phenylalanine may not be detectable until 3–4 days postpartum. False-positives in premature infants may reflect delayed maturation of enzymes of phenylalanine catabolism. A less reliable screening test employs $FeCl_3$ to detect urinary phenylpyruvate. $FeCl_3$ screening for PKU of the urine of newborn infants is compulsory in the United States and many other countries.

**Lysine.** Figure 30–14 summarizes the catabolism of lysine. Lysine first forms a Schiff base with α-ketoglutarate, which is reduced to **saccharopine.** In one form of **periodic hyperlysinemia**, elevated lysine competitively inhibits liver **arginase** (see Figure 29–9), causing hyperammonemia. Restricting dietary lysine relieves the ammonemia, whereas ingestion of a lysine load precipitates severe crises and coma. In a different **periodic hyperlysinemia,** lysine catabolites accumulate, but even a lysine load does not trigger hyperammonemia. In addition to impaired synthesis of saccharopine, some patients cannot cleave saccharopine.

**Tryptophan.** Tryptophan is degraded to amphibolic intermediates via the kynurenine-anthranilate pathway (Figure 30–15). **Tryptophan oxygenase (tryptophan pyrrolase)** opens the indole ring, incorporates molecular oxygen, and forms *N*-formylkynurenine. An iron porphyrin metalloprotein that is inducible in liver by adrenal corticosteroids and by tryptophan, tryptophan oxygenase is feedback-inhibited by nicotinic acid derivatives, including NADPH. Hydrolytic removal of the formyl group of *N*-formylkynurenine, catalyzed by **kynurenine formylase,** produces kynurenine. Since **kynureninase** requires pyridoxal phosphate, excretion of xanthurenate (Figure 30–16) in response to a tryptophan load is diagnostic of vitamin $B_6$ deficiency. **Hartnup disease** reflects impaired intestinal and renal transport of tryptophan and other neutral amino acids. Indole derivatives of unabsorbed tryptophan formed by intestinal bacteria are excreted. The defect limits tryptophan availability for niacin biosynthesis and accounts for the pellagra-like signs and symptoms.

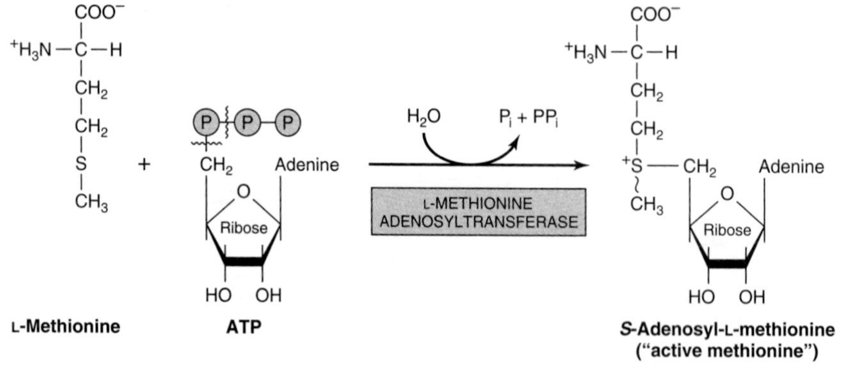

***Figure 30–17.*** Formation of *S*-adenosylmethionine. ~$CH_3$ represents the high group transfer potential of "active methionine."

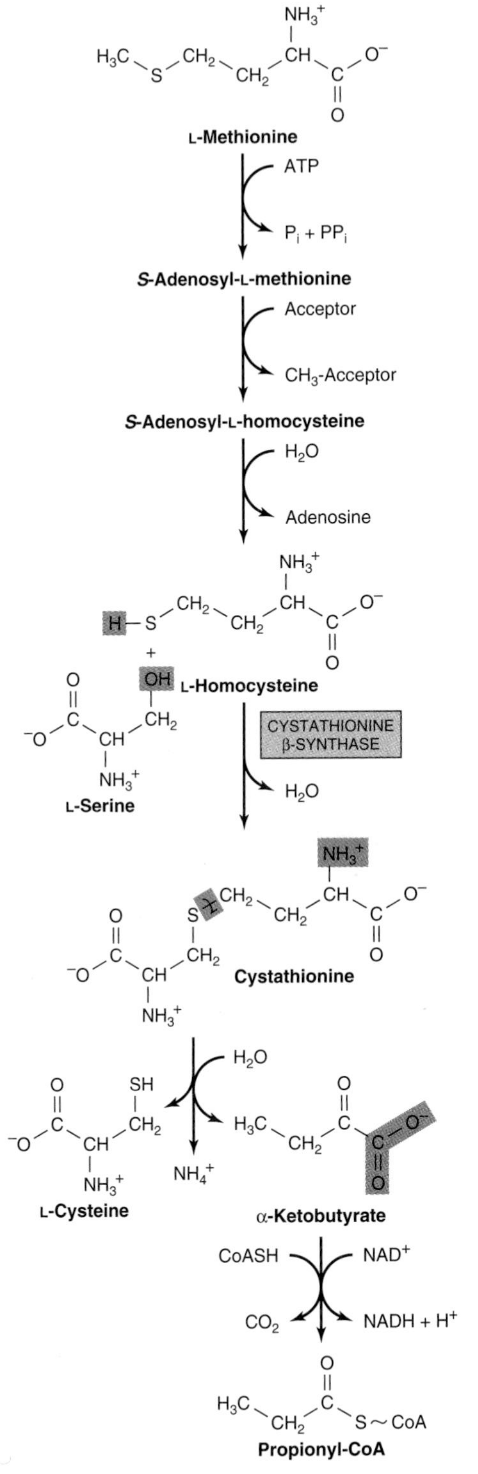

**Figure 30–18.** Conversion of methionine to propionyl-CoA.

**Methionine.** Methionine reacts with ATP forming *S*-adenosylmethionine, "active methionine" (Figure 30–17). Subsequent reactions form propionyl-CoA (Figure 30–18) and ultimately succinyl-CoA (see Figure 19–2).

## THE INITIAL REACTIONS ARE COMMON TO ALL THREE BRANCHED-CHAIN AMINO ACIDS

Reactions 1–3 of Figure 30–19 are analogous to those of fatty acid catabolism. Following transamination, all three α-keto acids undergo oxidative decarboxylation catalyzed by mitochondrial **branched-chain α-keto acid dehydrogenase.** This multimeric enzyme complex of a decarboxylase, a transacylase, and a dihydrolipoyl dehydrogenase closely resembles pyruvate dehydrogenase (see Figure 17–5). Its regulation also parallels that of pyruvate dehydrogenase, being inactivated by phosphorylation and reactivated by dephosphorylation (see Figure 17–6).

Reaction 3 is analogous to the dehydrogenation of fatty acyl-CoA thioesters (see Figure 22–3). In **isovaleric acidemia,** ingestion of protein-rich foods elevates isovalerate, the deacylation product of isovaleryl-CoA. Figures 30–20, 30–21, and 30–22 illustrate the subsequent reactions unique to each amino acid skeleton.

## METABOLIC DISORDERS OF BRANCHED-CHAIN AMINO ACID CATABOLISM

As the name implies, the odor of urine in **maple syrup urine disease (branched-chain ketonuria)** suggests maple syrup or burnt sugar. The biochemical defect involves the **α-keto acid decarboxylase complex** (reaction 2, Figure 30–19). Plasma and urinary levels of leucine, isoleucine, valine, α-keto acids, and α-hydroxy acids (reduced α-keto acids) are elevated. The mechanism of toxicity is unknown. Early diagnosis, especially prior to 1 week of age, employs enzymatic analysis. Prompt replacement of dietary protein by an amino acid mixture that lacks leucine, isoleucine, and valine averts brain damage and early mortality.

Mutation of the dihydrolipoate reductase component impairs decarboxylation of branched-chain α-keto acids, of pyruvate, and of α-ketoglutarate. In **intermittent branched-chain ketonuria,** the α-keto acid decarboxylase retains some activity, and symptoms occur later in life. The impaired enzyme in **isovaleric acidemia** is **isovaleryl-CoA dehydrogenase** (reaction 3, Figure 30–19). Vomiting, acidosis, and coma follow ingestion of excess protein. Accumulated

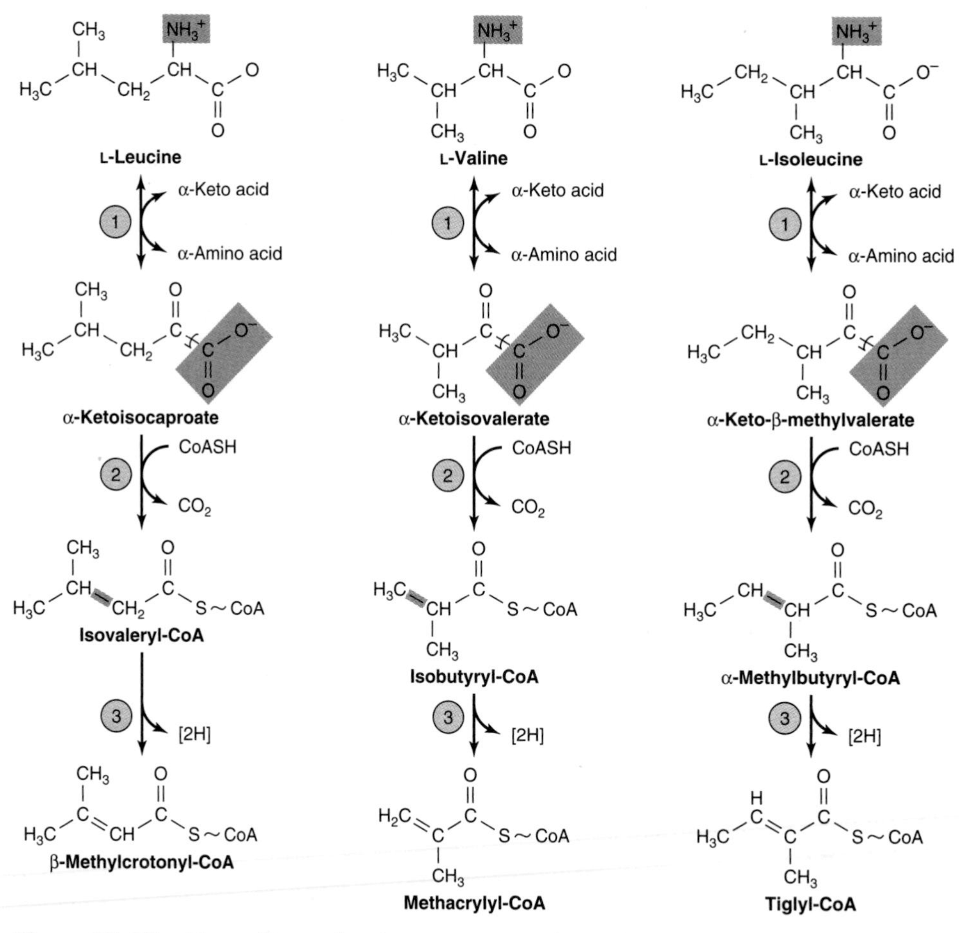


***Figure 30–19.*** The analogous first three reactions in the catabolism of leucine, valine, and isoleucine. Note also the analogy of reactions ② and ③ to reactions of the catabolism of fatty acids (see Figure 22–3). The analogy to fatty acid catabolism continues, as shown in subsequent figures.

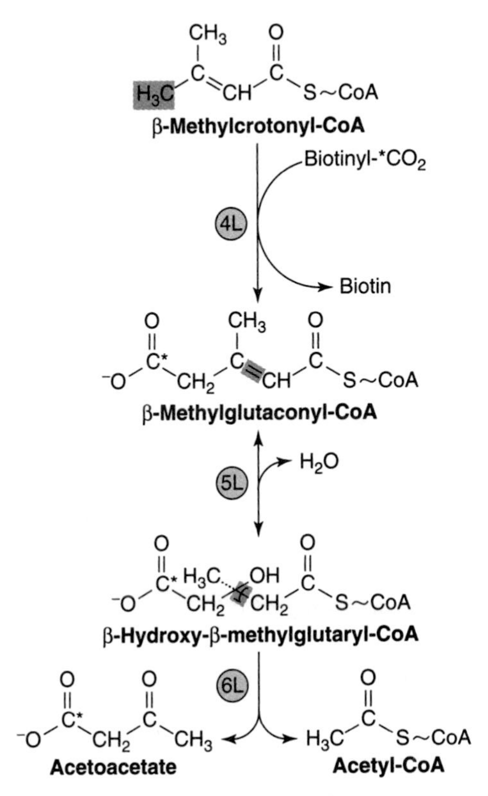


***Figure 30–20.*** Catabolism of the β-methylcrotonyl-CoA formed from L-leucine. Asterisks indicate carbon atoms derived from $CO_2$.

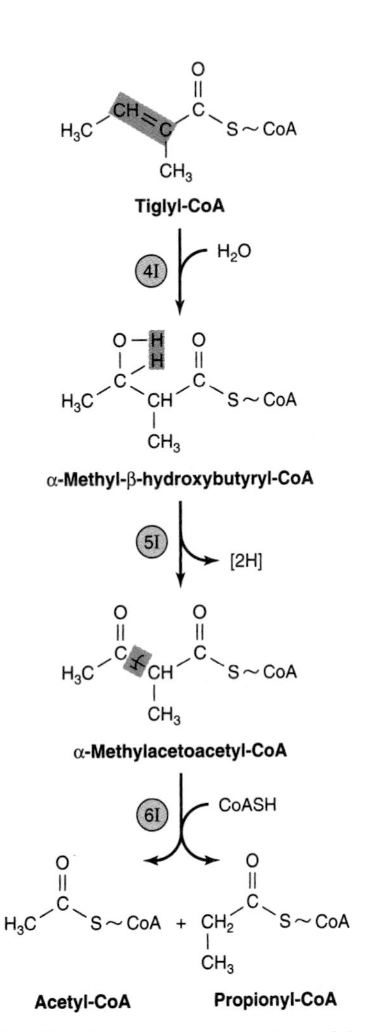


***Figure 30–21.*** Subsequent catabolism of the tiglyl-CoA formed from L-isoleucine.

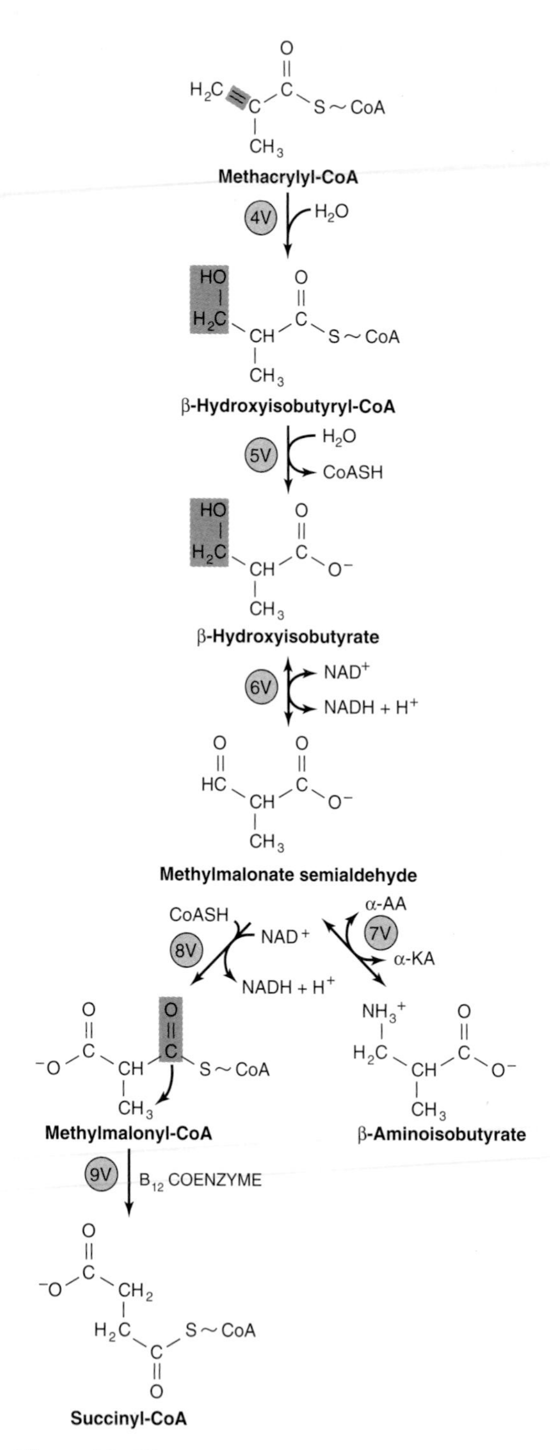


***Figure 30–22.*** Subsequent catabolism of the methacrylyl-CoA formed from L-valine (see Figure 30–19). (α-KA, α-keto acid; α-AA, α-amino acid.)

isovaleryl-CoA is hydrolyzed to isovalerate and excreted.

## SUMMARY


- Excess amino acids are catabolized to amphibolic intermediates used as sources of energy or for carbohydrate and lipid biosynthesis.
- Transamination is the most common initial reaction of amino acid catabolism. Subsequent reactions remove any additional nitrogen and restructure the hydrocarbon skeleton for conversion to oxaloacetate, α-ketoglutarate, pyruvate, and acetyl-CoA.
- Metabolic diseases associated with glycine catabolism include glycinuria and primary hyperoxaluria.
- Two distinct pathways convert cysteine to pyruvate. Metabolic disorders of cysteine catabolism include cystine-lysinuria, cystine storage disease, and the homocystinurias.
- Threonine catabolism merges with that of glycine after threonine aldolase cleaves threonine to glycine and acetaldehyde.
- Following transamination, the carbon skeleton of tyrosine is degraded to fumarate and acetoacetate. Metabolic diseases of tyrosine catabolism include tyrosinosis, Richner-Hanhart syndrome, neonatal tyrosinemia, and alkaptonuria.
- Metabolic disorders of phenylalanine catabolism include phenylketonuria (PKU) and several hyperphenylalaninemias.
- Neither nitrogen of lysine undergoes transamination. Metabolic diseases of lysine catabolism include periodic and persistent forms of hyperlysinemia-ammonemia.
- The catabolism of leucine, valine, and isoleucine presents many analogies to fatty acid catabolism. Metabolic disorders of branched-chain amino acid catabolism include hypervalinemia, maple syrup urine disease, intermittent branched-chain ketonuria, isovaleric acidemia, and methylmalonic aciduria.


## REFERENCES


Blacher J, Safar ME: Homocysteine, folic acid, B vitamins and cardiovascular risk. J Nutr Health Aging 2001;5:196.

Cooper AJL: Biochemistry of the sulfur-containing amino acids. Annu Rev Biochem 1983;52:187.

Gjetting T et al: A phenylalanine hydroxylase amino acid polymorphism with implications for molecular diagnostics. Mol Genet Metab 2001;73:280.

Harris RA et al: Molecular cloning of the branched-chain α-ketoacid dehydrogenase kinase and the CoA-dependent methylmalonate semialdehyde dehydrogenase. Adv Enzyme Regul 1993;33:255.

Scriver CR: Garrod's foresight; our hindsight. J Inherit Metab Dis 2001;24:93.

Scriver CR et al (editors): *The Metabolic and Molecular Bases of Inherited Disease,* 8th ed. McGraw-Hill, 2001.

Waters PJ, Scriver CR, Parniak MA: Homomeric and heteromeric interactions between wild-type and mutant phenylalanine hydroxylase subunits: evaluation of two-hybrid approaches for functional analysis of mutations causing hyperphenylalaninemia. Mol Genet Metab 2001;73:230.

# Conversion of Amino Acids to Specialized Products

# 31

*Victor W. Rodwell, PhD*

## BIOMEDICAL IMPORTANCE

Important products derived from amino acids include heme, purines, pyrimidines, hormones, neurotransmitters, and biologically active peptides. In addition, many proteins contain amino acids that have been modified for a specific function such as binding calcium or as intermediates that serve to stabilize proteins—generally structural proteins—by subsequent covalent cross-linking. The amino acid residues in those proteins serve as precursors for these modified residues. Small peptides or peptide-like molecules not synthesized on ribosomes fulfill specific functions in cells. Histamine plays a central role in many allergic reactions. Neurotransmitters derived from amino acids include γ-aminobutyrate, 5-hydroxytryptamine (serotonin), dopamine, norepinephrine, and epinephrine. Many drugs used to treat neurologic and psychiatric conditions affect the metabolism of these neurotransmitters.

### Glycine

Metabolites and pharmaceuticals excreted as water-soluble glycine conjugates include glycocholic acid (Chapter 24) and hippuric acid formed from the food additive benzoate (Figure 31–1). Many drugs, drug metabolites, and other compounds with carboxyl groups are excreted in the urine as glycine conjugates. Glycine is incorporated into creatine (see Figure 31–6), the nitrogen and α-carbon of glycine are incorporated into the pyrrole rings and the methylene bridge carbons of heme (Chapter 32), and the entire glycine molecule becomes atoms 4, 5, and 7 of purines (Figure 34–1).

### β-Alanine

β-Alanine, a metabolite of cysteine (Figure 34–9), is present in coenzyme A and as β-alanyl dipeptides, principally carnosine (see below). Mammalian tissues form β-alanine from cytosine (Figure 34–9), carnosine, and anserine (Figure 31–2). Mammalian tissues transaminate β-alanine, forming malonate semialdehyde. Body fluid and tissue levels of β-alanine, taurine, and β-aminoisobutyrate are elevated in the rare metabolic disorder hyperbeta-alaninemia.

### β-Alanyl Dipeptides

The β-alanyl dipeptides carnosine and anserine (*N*-methylcarnosine) (Figure 31–2) activate myosin ATPase, chelate copper, and enhance copper uptake. β-Alanyl-imidazole buffers the pH of anaerobically contracting skeletal muscle. Biosynthesis of carnosine is catalyzed by carnosine synthetase in a two-stage reaction that involves initial formation of an enzyme-bound acyl-adenylate of β-alanine and subsequent transfer of the β-alanyl moiety to L-histidine.

$$\text{ATP} + \beta\text{-Alanine} \rightarrow \beta\text{-Alanyl} - \text{AMP} \rightarrow + \text{PP}_i$$
$$\beta\text{-Alanyl} - \text{AMP} + \text{L-Histidine} \rightarrow \text{Carnosine} + \text{AMP}$$

Hydrolysis of carnosine to β-alanine and L-histidine is catalyzed by carnosinase. The heritable disorder carnosinase deficiency is characterized by carnosinuria.

Homocarnosine (Figure 31–2), present in human brain at higher levels than carnosine, is synthesized in brain tissue by carnosine synthetase. Serum carnosinase does not hydrolyze homocarnosine. Homocarnosinosis, a rare genetic disorder, is associated with progressive spastic paraplegia and mental retardation.

### Phosphorylated Serine, Threonine, & Tyrosine

The phosphorylation and dephosphorylation of seryl, threonyl, and tyrosyl residues regulate the activity of certain enzymes of lipid and carbohydrate metabolism and the properties of proteins that participate in signal transduction cascades.

### Methionine

*S*-Adenosylmethionine, the principal source of methyl groups in the body, also contributes its carbon skeleton for the biosynthesis of the 3-diaminopropane portions of the polyamines spermine and spermidine (Figure 31–4).

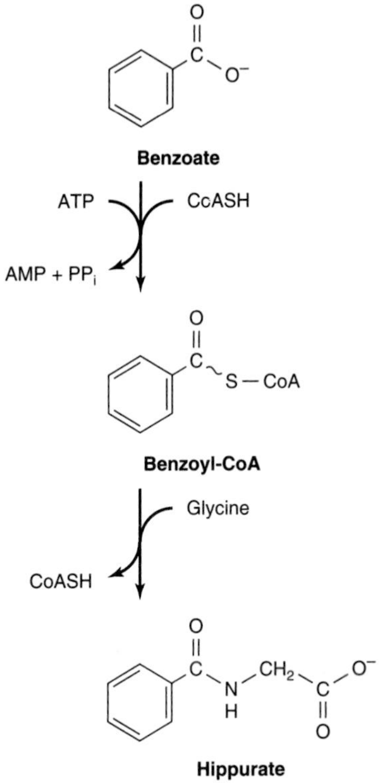

Benzoate

ATP → CoASH

AMP + $PP_i$

Benzoyl-CoA

Glycine

CoASH

Hippurate

***Figure 31–1.*** Biosynthesis of hippurate. Analogous reactions occur with many acidic drugs and catabolites.

Ergothioneine

Carnosine

Anserine

Homocarnosine

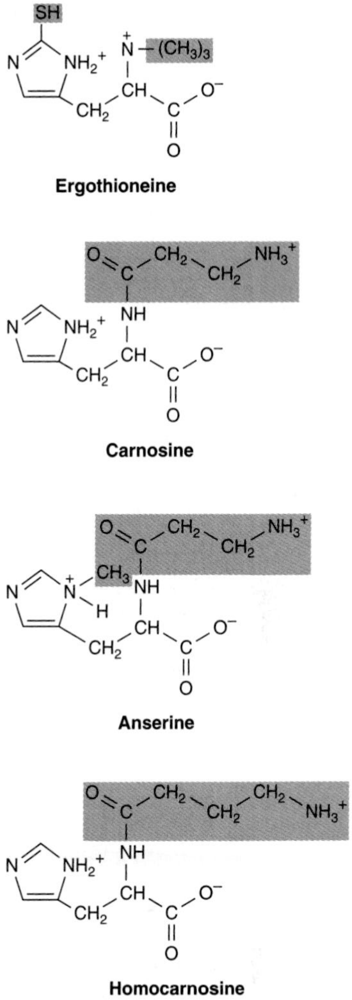

***Figure 31–2.*** Compounds related to histidine. The boxes surround the components not derived from histidine. The SH group of ergothioneine derives from cysteine.

## Cysteine

L-Cysteine is a precursor of the thioethanolamine portion of coenzyme A and of the taurine that conjugates with bile acids such as taurocholic acid (Chapter 26).

## Histidine

Decarboxylation of histidine to histamine is catalyzed by a broad-specificity aromatic L-amino acid decarboxylase that also catalyzes the decarboxylation of dopa, 5-hydroxytryptophan, phenylalanine, tyrosine, and tryptophan. α-Methyl amino acids, which inhibit decarboxylase activity, find application as antihypertensive agents. Histidine compounds present in the human body include ergothioneine, carnosine, and dietary anserine (Figure 31–2). Urinary levels of 3-methylhistidine are unusually low in patients with Wilson's disease.

## Ornithine & Arginine

Arginine is the formamidine donor for creatine synthesis (Figure 31–6) and via ornithine to putrescine, spermine, and spermidine (Figure 31–3) Arginine is also the precursor of the intercellular signaling molecule nitric oxide (NO) that serves as a neurotransmitter, smooth muscle relaxant, and vasodilator. Synthesis of NO, catalyzed by NO synthase, involves the NADPH-dependent reaction of L-arginine with $O_2$ to yield L-citrulline and NO.

## Polyamines

The polyamines spermidine and spermine (Figure 31–4) function in cell proliferation and growth, are growth factors for cultured mammalian cells, and stabilize intact cells, subcellular organelles, and membranes. Pharmacologic doses of polyamines are hypothermic

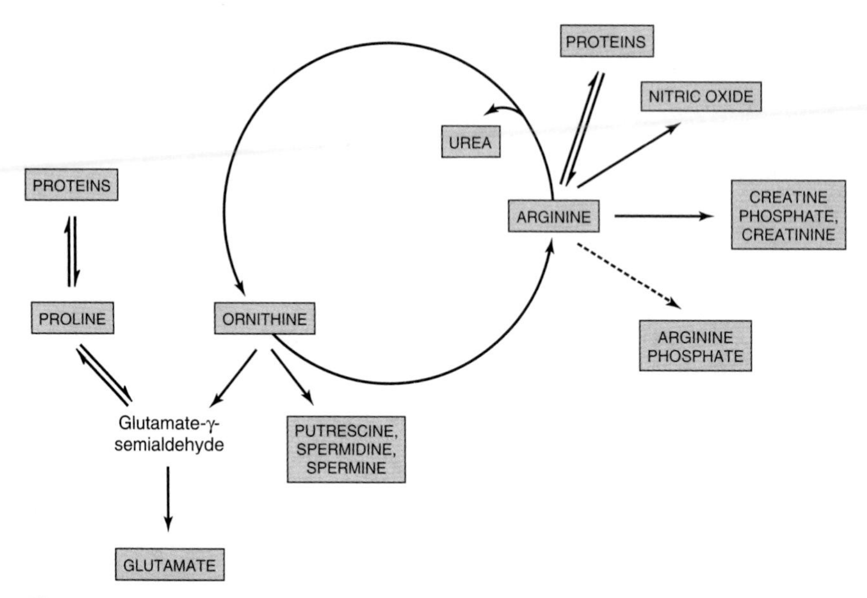


***Figure 31–3.*** Arginine, ornithine, and proline metabolism. Reactions with solid arrows all occur in mammalian tissues. Putrescine and spermine synthesis occurs in both mammals and bacteria. Arginine phosphate of invertebrate muscle functions as a phosphagen analogous to creatine phosphate of mammalian muscle (see Figure 31–6).

and hypotensive. Since they bear multiple positive charges, polyamines associate readily with DNA and RNA. Figure 31–4 summarizes polyamine biosynthesis.

## Tryptophan

Following hydroxylation of tryptophan to 5-hydroxytryptophan by liver tyrosine hydroxylase, subsequent decarboxylation forms serotonin (5-hydroxytryptamine), a potent vasoconstrictor and stimulator of smooth muscle contraction. Catabolism of serotonin is initiated by monoamine oxidase-catalyzed oxidative deamination to 5-hydroxyindoleacetate. The psychic stimulation that follows administration of iproniazid results from its ability to prolong the action of serotonin by inhibiting monoamine oxidase. In carcinoid (argentaffinoma), tumor cells overproduce serotonin. Urinary metabolites of serotonin in patients with carci-

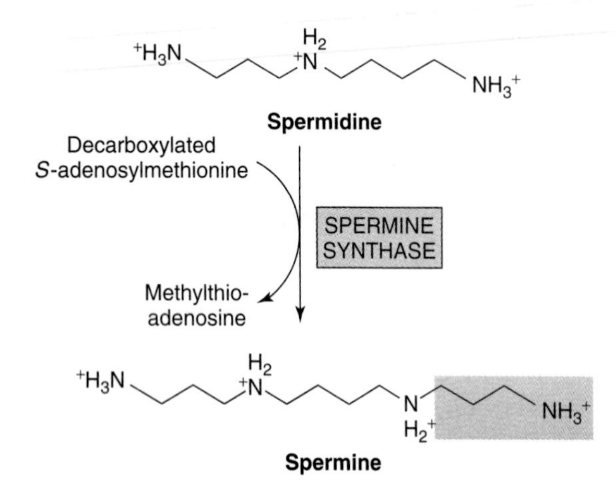


***Figure 31–4.*** Conversion of spermidine to spermine. Spermidine formed from putrescine (decarboxylated L-ornithine) by transfer of a propylamine moiety from decarboxylated S-adenosylmethionine accepts a second propylamine moiety to form spermidine.

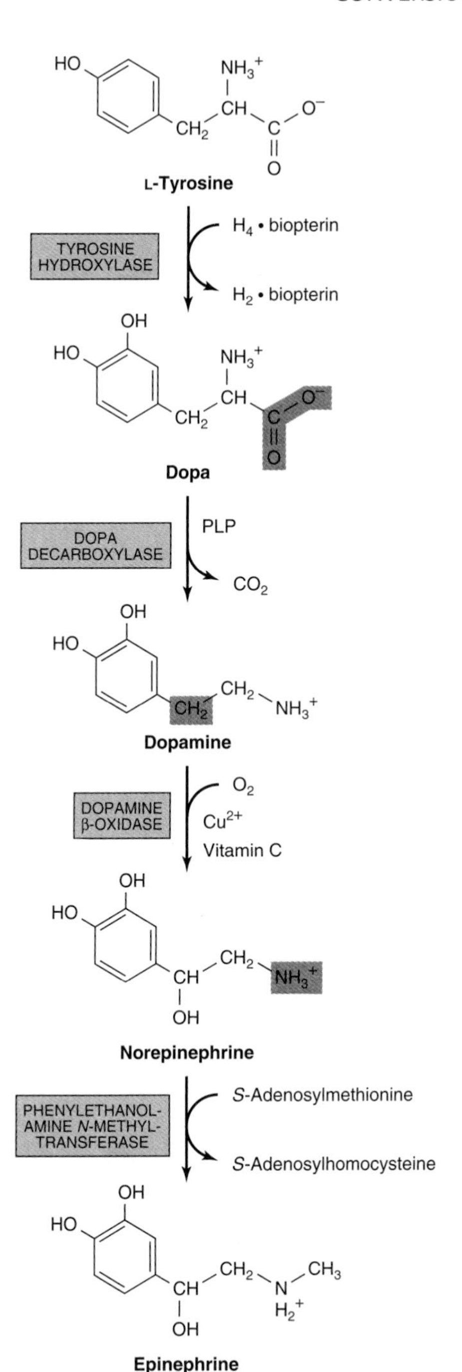


***Figure 31–5.*** Conversion of tyrosine to epinephrine and norepinephrine in neuronal and adrenal cells. (PLP, pyridoxal phosphate.)

noid include *N*-acetylserotonin glucuronide and the glycine conjugate of 5-hydroxyindoleacetate. Serotonin and 5-methoxytryptamine are metabolized to the corresponding acids by monoamine oxidase. *N*-Acetylation of serotonin, followed by *O*-methylation in the pineal body, forms melatonin. Circulating melatonin is taken up by all tissues, including brain, but is rapidly metabolized by hydroxylation followed by conjugation with sulfate or with glucuronic acid.

Kidney tissue, liver tissue, and fecal bacteria all convert tryptophan to tryptamine, then to indole 3-acetate. The principal normal urinary catabolites of tryptophan are 5-hydroxyindoleacetate and indole 3-acetate.

## Tyrosine

Neural cells convert tyrosine to epinephrine and norepinephrine (Figure 31–5). While dopa is also an intermediate in the formation of melanin, different enzymes hydroxylate tyrosine in melanocytes. Dopa decarboxylase, a pyridoxal phosphate-dependent enzyme, forms dopamine. Subsequent hydroxylation by dopamine β-oxidase then forms norepinephrine. In the adrenal medulla, phenylethanolamine-*N*-methyltransferase utilizes *S*-adenosylmethionine to methylate the primary amine of norepinephrine, forming epinephrine (Figure 31–5). Tyrosine is also a precursor of triiodothyronine and thyroxine (Chapter 42).

## Creatinine

Creatinine is formed in muscle from creatine phosphate by irreversible, nonenzymatic dehydration and loss of phosphate (Figure 31–6). The 24-hour urinary excretion of creatinine is proportionate to muscle mass. Glycine, arginine, and methionine all participate in creatine biosynthesis. Synthesis of creatine is completed by methylation of guanidoacetate by *S*-adenosylmethionine (Figure 31–6).

## γ-Aminobutyrate

γ-Aminobutyrate (GABA) functions in brain tissue as an inhibitory neurotransmitter by altering transmembrane potential differences. It is formed by decarboxylation of L-glutamate, a reaction catalyzed by L-glutamate decarboxylase (Figure 31–7). Transamination of γ-aminobutyrate forms succinate semialdehyde (Figure 31–7), which may then undergo reduction to γ-hydroxybutyrate, a reaction catalyzed by L-lactate dehydrogenase, or oxidation to succinate and thence via the citric acid cycle to $CO_2$ and $H_2O$. A rare genetic disorder of GABA metabolism involves a defective GABA aminotransferase, an enzyme that participates in the catabolism of GABA subsequent to its postsynaptic release in brain tissue.

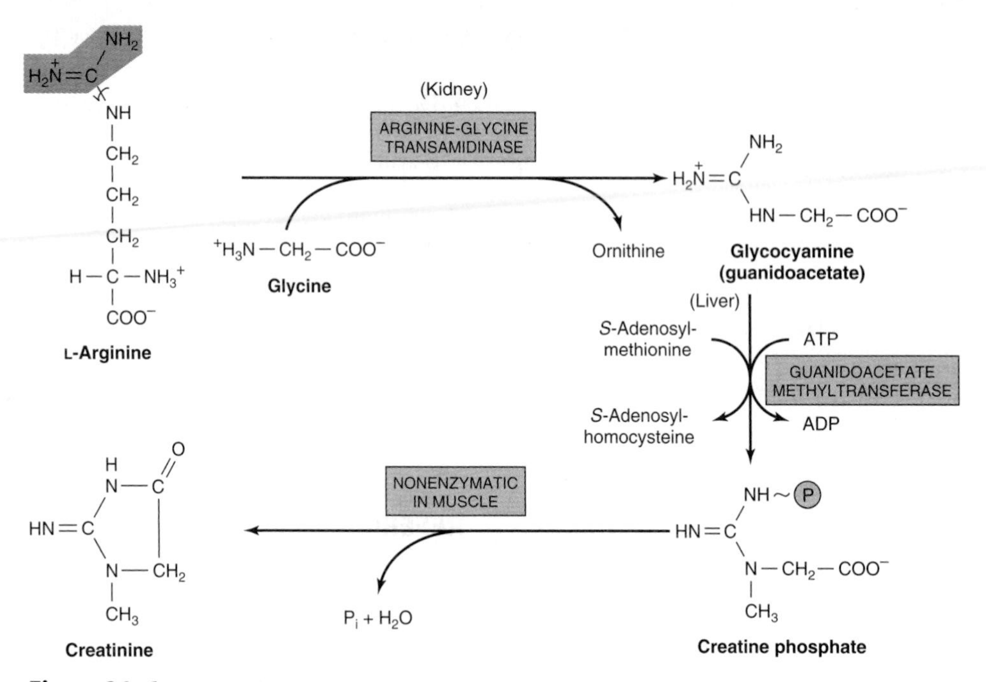


***Figure 31–6.*** Biosynthesis and metabolism of creatine and creatinine.

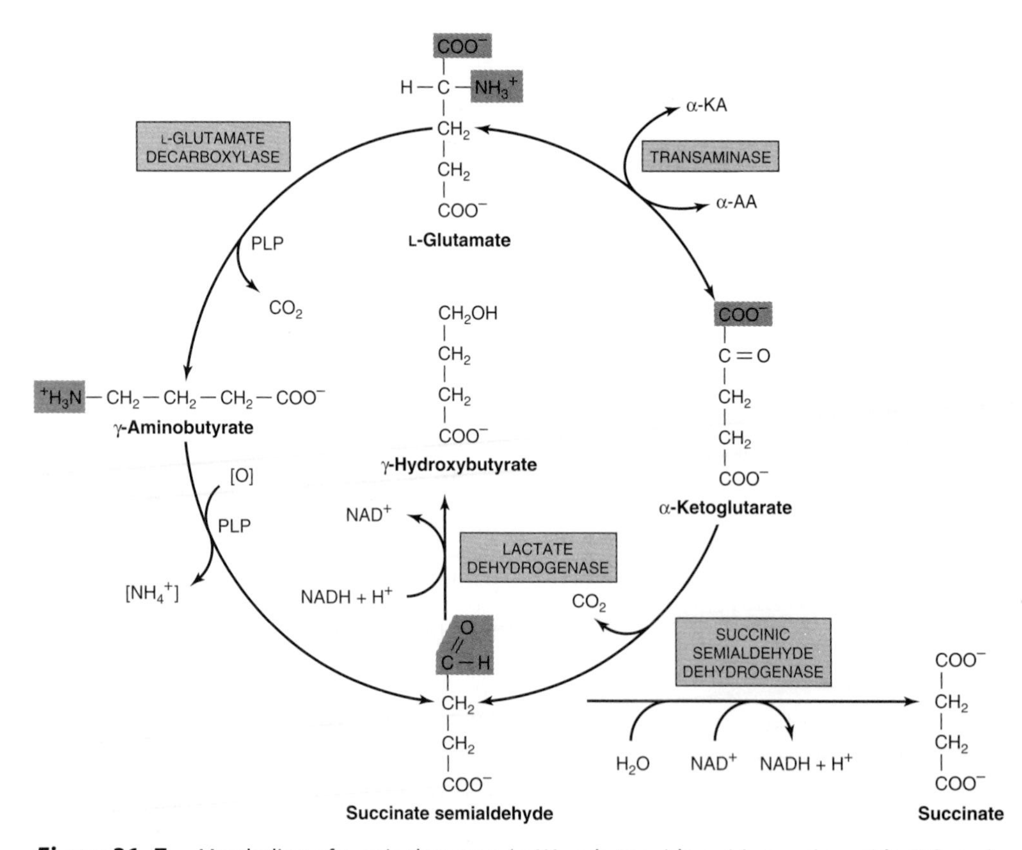


***Figure 31–7.*** Metabolism of γ-aminobutyrate. (α-KA, α-keto acids; α-AA, α-amino acids; PLP, pyridoxal phosphate.)

## SUMMARY


- In addition to their roles in proteins and polypeptides, amino acids participate in a wide variety of additional biosynthetic processes.
- Glycine participates in the biosynthesis of heme, purines, and creatine and is conjugated to bile acids and to the urinary metabolites of many drugs.
- In addition to its roles in phospholipid and sphingosine biosynthesis, serine provides carbons 2 and 8 of purines and the methyl group of thymine.
- *S*-Adenosylmethionine, the methyl group donor for many biosynthetic processes, also participates directly in spermine and spermidine biosynthesis.
- Glutamate and ornithine form the neurotransmitter γ-aminobutyrate (GABA).
- The thioethanolamine of coenzyme A and the taurine of taurocholic acid arise from cysteine.
- Decarboxylation of histidine forms histamine, and several dipeptides are derived from histidine and β-alanine.
- Arginine serves as the formamidine donor for creatine biosynthesis, participates in polyamine biosynthesis, and provides the nitrogen of nitric oxide (NO).
- Important tryptophan metabolites include serotonin, melanin, and melatonin.
- Tyrosine forms both epinephrine and norepinephrine, and its iodination forms thyroid hormone.


## REFERENCE


Scriver CR et al (editors): *The Metabolic and Molecular Bases of Inherited Disease,* 8th ed. McGraw-Hill, 2001.

# Porphyrins & Bile Pigments

32

Robert K. Murray, MD, PhD

## BIOMEDICAL IMPORTANCE

The biochemistry of the porphyrins and of the bile pigments is presented in this chapter. These topics are closely related, because heme is synthesized from porphyrins and iron, and the products of degradation of heme are the bile pigments and iron.

Knowledge of the biochemistry of the porphyrins and of heme is basic to understanding the varied functions of hemoproteins (see below) in the body. The **porphyrias** are a group of diseases caused by abnormalities in the pathway of biosynthesis of the various porphyrins. Although porphyrias are not very prevalent, physicians must be aware of them. A much more prevalent clinical condition is **jaundice,** due to elevation of bilirubin in the plasma. This elevation is due to overproduction of bilirubin or to failure of its excretion and is seen in numerous diseases ranging from hemolytic anemias to viral hepatitis and to cancer of the pancreas.

## METALLOPORPHYRINS & HEMOPROTEINS ARE IMPORTANT IN NATURE

Porphyrins are cyclic compounds formed by the linkage of four pyrrole rings through —HC= methenyl bridges (Figure 32–1). A characteristic property of the porphyrins is the formation of complexes with metal ions bound to the nitrogen atom of the pyrrole rings. Examples are the **iron porphyrins** such as **heme** of hemoglobin and the **magnesium**-containing porphyrin **chlorophyll,** the photosynthetic pigment of plants.

Proteins that contain heme (hemoproteins) are widely distributed in nature. Examples of their importance in humans and animals are listed in Table 32–1.

### Natural Porphyrins Have Substituent Side Chains on the Porphin Nucleus

The porphyrins found in nature are compounds in which various **side chains** are substituted for the eight hydrogen atoms numbered in the porphin nucleus shown in Figure 32–1. As a simple means of showing these substitutions, Fischer proposed a shorthand formula in which the methenyl bridges are omitted and each pyrrole ring is shown as indicated with the eight substituent positions numbered as shown in Figure 32–2. Various porphyrins are represented in Figures 32–2, 32–3, and 32–4.

The arrangement of the acetate (A) and propionate (P) substituents in the uroporphyrin shown in Figure 32–2 is asymmetric (in ring IV, the expected order of the A and P substituents is reversed). A porphyrin with this type of **asymmetric substitution** is classified as a type III porphyrin. A porphyrin with a completely symmetric arrangement of the substituents is classified as a type I porphyrin. Only types I and III are found in nature, and the type III series is far more abundant (Figure 32–3)—and more important because it includes heme.

Heme and its immediate precursor, protoporphyrin IX (Figure 32–4), are both type III porphyrins (ie, the methyl groups are asymmetrically distributed, as in type III coproporphyrin). However, they are sometimes identified as belonging to series IX, because they were designated ninth in a series of isomers postulated by Hans Fischer, the pioneer worker in the field of porphyrin chemistry.

## HEME IS SYNTHESIZED FROM SUCCINYL-COA & GLYCINE

Heme is synthesized in living cells by a pathway that has been much studied. The two starting materials are **succinyl-CoA,** derived from the citric acid cycle in mitochondria, and the amino acid **glycine.** Pyridoxal phosphate is also necessary in this reaction to "activate" glycine. The product of the condensation reaction between succinyl-CoA and glycine is α-amino-β-ketoadipic acid, which is rapidly decarboxylated to form α-aminolevulinate (ALA) (Figure 32–5). This reaction sequence is catalyzed by **ALA synthase,** the rate-controlling enzyme in porphyrin biosynthesis in mammalian liver. Synthesis of ALA occurs in **mitochondria.** In the cytosol, two molecules of ALA are condensed by the enzyme **ALA dehydratase** to form two molecules of water and one of **porphobilinogen** (PBG) (Figure 32–5). ALA dehydratase is a zinc-containing enzyme and is sensitive to inhibition by **lead,** as can occur in lead poisoning.

The formation of a cyclic tetrapyrrole—ie, a porphyrin—occurs by condensation of four molecules of PBG (Figure 32–6). These four molecules condense in a head-to-tail manner to form a linear tetrapyrrole, hy-

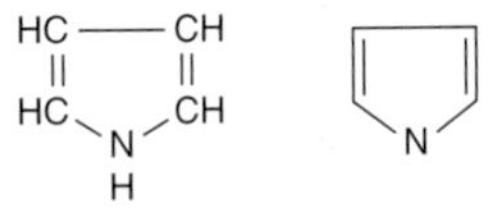

Pyrrole

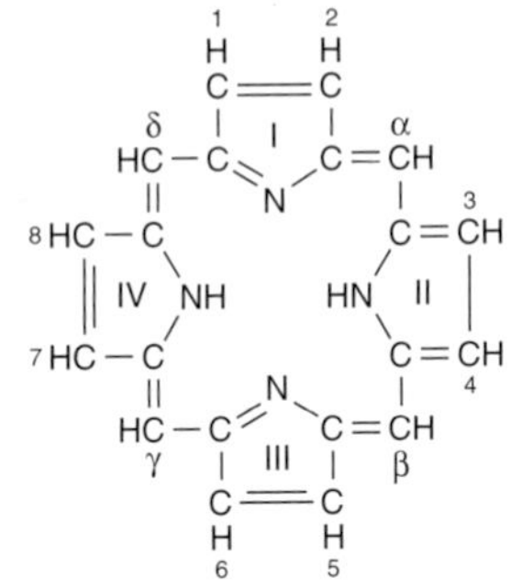

Porphin
($C_{20}H_{14}N_4$)

***Figure 32–1.*** The porphin molecule. Rings are labeled I, II, III, and IV. Substituent positions on the rings are labeled 1, 2, 3, 4, 5, 6, 7, and 8. The methenyl bridges (—HC═) are labeled α, β, γ, and δ.

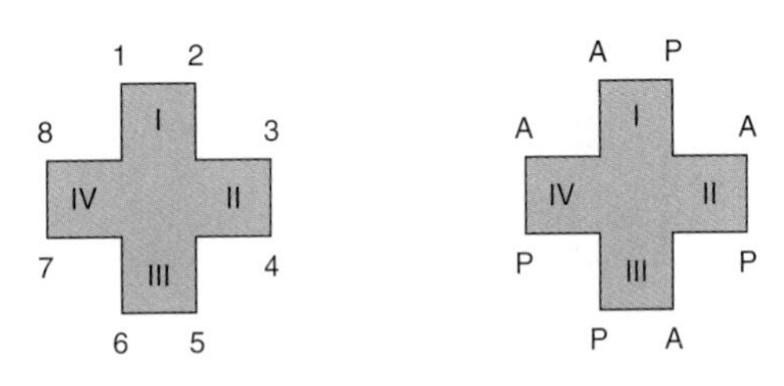


***Figure 32–2.*** Uroporphyrin III. A (acetate) = —$CH_2COOH$; P (propionate) = —$CH_2CH_2COOH$.

droxymethylbilane (HMB). The reaction is catalyzed by uroporphyrinogen I synthase, also named PBG deaminase or HMB synthase. HMB cyclizes spontaneously to form **uroporphyrinogen I** (left-hand side of Figure 32–6) or is converted to **uroporphyrinogen III** by the action of uroporphyrinogen III synthase (right-hand side of Figure 32–6). Under normal conditions, the uroporphyrinogen formed is almost exclusively the III isomer, but in certain of the porphyrias (discussed below), the type I isomers of porphyrinogens are formed in excess.

Note that both of these uroporphyrinogens have the pyrrole rings connected by methylene bridges (—$CH_2$—), which do not form a conjugated ring system. Thus, these compounds are colorless (as are all porphyrinogens). However, the porphyrinogens are readily auto-oxidized to their respective colored porphyrins. These oxidations are catalyzed by light and by the porphyrins that are formed.

Uroporphyrinogen III is converted to coproporphyrinogen III by decarboxylation of all of the acetate (A) groups, which changes them to methyl (M) substituents. The reaction is catalyzed by **uroporphyrinogen decarboxylase,** which is also capable of converting uroporphyrinogen I to coproporphyrinogen I (Figure 32–7). Coproporphyrinogen III then enters the mitochondria, where it is converted to **protoporphyrinogen III** and then to **protoporphyrin III.** Several steps are involved in this conversion. The mitochondrial enzyme **coproporphyrinogen oxidase** catalyzes the decarboxylation and oxidation of two propionic side chains to form protoporphyrinogen. This enzyme is able to act only on type III coproporphyrinogen, which would explain why type I protoporphyrins do not generally occur in nature. The oxidation of protoporphyrinogen to protoporphyrin is catalyzed by another mitochondrial enzyme, **protoporphyrinogen oxidase.** In mammalian liver, the conversion of coproporphyrinogen to protoporphyrin requires molecular oxygen.

## Formation of Heme Involves Incorporation of Iron Into Protoporphyrin

The final step in heme synthesis involves the incorporation of ferrous iron into protoporphyrin in a reaction catalyzed by **ferrochelatase (heme synthase),** another mitochondrial enzyme (Figure 32–4).

A summary of the steps in the biosynthesis of the porphyrin derivatives from PBG is given in Figure 32–8. The last three enzymes in the pathway and ALA synthase are located in the mitochondrion, whereas the other enzymes are cytosolic. Both erythroid and nonerythroid ("housekeeping") forms of the first four enzymes are found. Heme biosynthesis occurs in most mammalian cells with the exception of mature erythrocytes, which do not contain mitochondria. However,

***Table 32–1.*** Examples of some important human and animal hemoproteins.[1]

| Protein | Function |
|---|---|
| Hemoglobin | Transport of oxygen in blood |
| Myoglobin | Storage of oxygen in muscle |
| Cytochrome *c* | Involvement in electron transport chain |
| Cytochrome P450 | Hydroxylation of xenobiotics |
| Catalase | Degradation of hydrogen peroxide |
| Tryptophan pyrrolase | Oxidation of trypotophan |

[1]The functions of the above proteins are described in various chapters of this text.

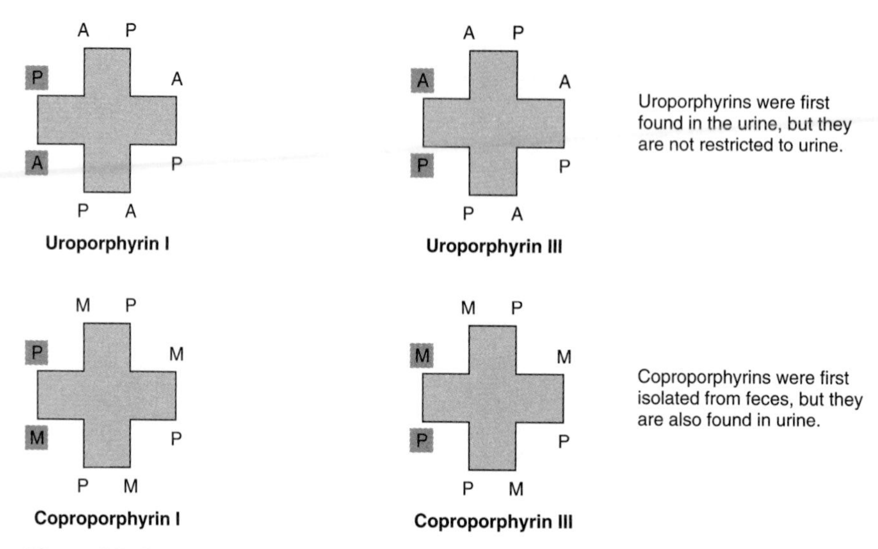


***Figure 32–3.*** Uroporphyrins and coproporphyrins. A (acetate); P (propionate); M (methyl) = $-CH_3$; V (vinyl) = $-CH-CH_2$.

approximately 85% of heme synthesis occurs in erythroid precursor cells in the **bone marrow** and the majority of the remainder in **hepatocytes.**

The porphyrinogens described above are colorless, containing six extra hydrogen atoms as compared with the corresponding colored porphyrins. These **reduced porphyrins** (the porphyrinogens) and not the corresponding porphyrins are the actual intermediates in the biosynthesis of protoporphyrin and of heme.

## ALA Synthase Is the Key Regulatory Enzyme in Hepatic Biosynthesis of Heme

ALA synthase occurs in both hepatic (ALAS1) and erythroid (ALAS2) forms. The rate-limiting reaction in the synthesis of heme in liver is that catalyzed by ALAS1 (Figure 32–5), a regulatory enzyme. It appears that heme, probably acting through an aporepressor molecule, acts as a negative regulator of the synthesis of ALAS1. This repression-derepression mechanism is depicted diagrammatically in Figure 32–9. Thus, the rate of synthesis of ALAS1 increases greatly in the absence of heme and is diminished in its presence. The turnover rate of ALAS1 in rat liver is normally rapid (half-life about 1 hour), a common feature of an enzyme catalyzing a rate-limiting reaction. Heme also affects translation of the enzyme and its transfer from the cytosol to the mitochondrion.

Many drugs when administered to humans can result in a marked increase in ALAS1. Most of these drugs are metabolized by a system in the liver that utilizes a specific hemoprotein, **cytochrome P450** (see Chapter 53). During their metabolism, the utilization of heme by cytochrome P450 is greatly increased, which in turn diminishes the intracellular heme concentration. This latter event effects a derepression of ALAS1 with a corresponding increased rate of heme synthesis to meet the needs of the cells.

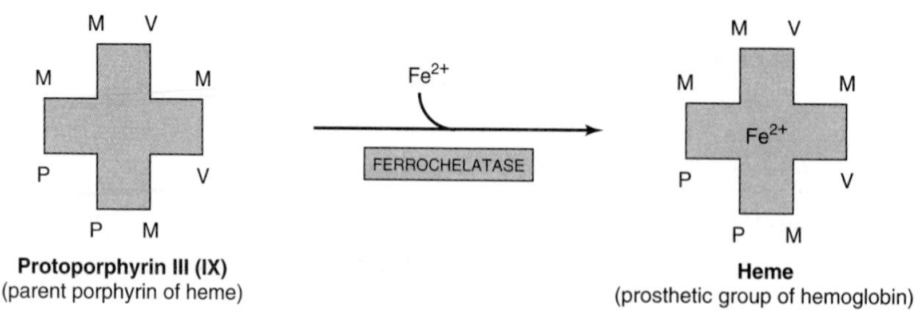


***Figure 32–4.*** Addition of iron to protoporphyrin to form heme.

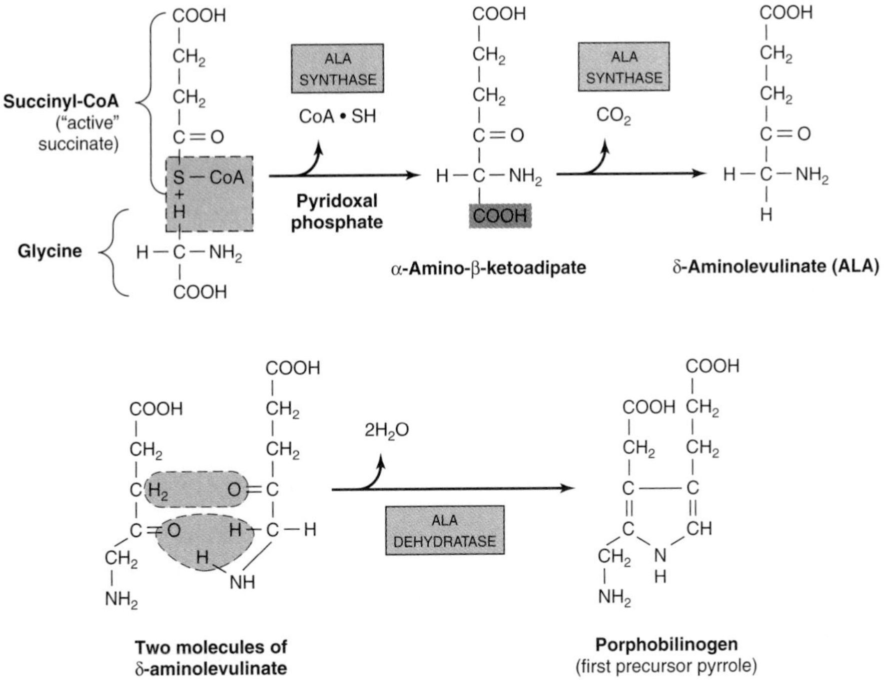

***Figure 32–5.*** Biosynthesis of porphobilinogen. ALA synthase occurs in the mitochondria, whereas ALA dehydratase is present in the cytosol.

Several factors affect drug-mediated derepression of ALAS1 in liver—eg, the administration of glucose can prevent it, as can the administration of hematin (an oxidized form of heme).

The importance of some of these regulatory mechanisms is further discussed below when the porphyrias are described.

Regulation of the **erythroid** form of ALAS (ALAS2) differs from that of ALAS1. For instance, it is not induced by the drugs that affect ALAS1, and it does not undergo feedback regulation by heme.

## PORPHYRINS ARE COLORED & FLUORESCE

The various porphyrinogens are colorless, whereas the various porphyrins are all colored. In the study of porphyrins or porphyrin derivatives, the characteristic absorption spectrum that each exhibits—in both the visible and the ultraviolet regions of the spectrum—is of great value. An example is the absorption curve for a solution of porphyrin in 5% hydrochloric acid (Figure 32–10). Note particularly the sharp absorption band near 400 nm. This is a distinguishing feature of the porphin ring and is characteristic of all porphyrins regardless of the side chains present. This band is termed the **Soret band** after its discoverer, the French physicist Charles Soret.

When porphyrins dissolved in strong mineral acids or in organic solvents are illuminated by ultraviolet light, they emit a strong red **fluorescence.** This fluorescence is so characteristic that it is often used to detect small amounts of free porphyrins. The double bonds joining the pyrrole rings in the porphyrins are responsible for the characteristic absorption and fluorescence of these compounds; these double bonds are absent in the porphyrinogens.

An interesting application of the photodynamic properties of porphyrins is their possible use in the treatment of certain types of cancer, a procedure called **cancer phototherapy.** Tumors often take up more porphyrins than do normal tissues. Thus, hematoporphyrin or other related compounds are administered to a patient with an appropriate tumor. The tumor is then exposed to an argon laser, which excites the porphyrins, producing cytotoxic effects.

### Spectrophotometry Is Used to Test for Porphyrins & Their Precursors

Coproporphyrins and uroporphyrins are of clinical interest because they are excreted in increased amounts in

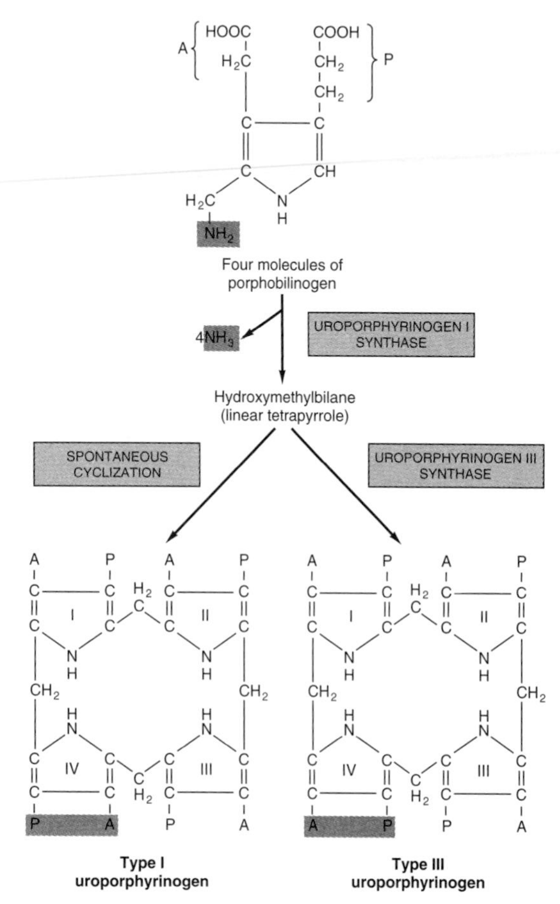


***Figure 32–6.*** Conversion of porphobilinogen to uroporphyrinogens. Uroporphyrinogen synthase I is also called porphobilinogen (PBG) deaminase or hydroxymethylbilane (HMB) synthase.

the porphyrias. These compounds, when present in urine or feces, can be separated from each other by extraction with appropriate solvent mixtures. They can then be identified and quantified using spectrophotometric methods.

ALA and PBG can also be measured in urine by appropriate colorimetric tests.

## THE PORPHYRIAS ARE GENETIC DISORDERS OF HEME METABOLISM

The **porphyrias** are a group of disorders due to abnormalities in the pathway of biosynthesis of heme; they can be genetic or acquired. They are not prevalent, but it is important to consider them in certain circumstances (eg, in the differential diagnosis of abdominal pain and of a variety of neuropsychiatric findings); otherwise, patients will be subjected to inappropriate treatments. It has been speculated that King George III had a type of porphyria, which may account for his periodic confinements in Windsor Castle and perhaps for some of his views regarding American colonists. Also, the **photosensitivity** (favoring nocturnal activities) and severe **disfigurement** exhibited by some victims of congenital erythropoietic porphyria have led to the suggestion that these individuals may have been the prototypes of so-called werewolves. No evidence to support this notion has been adduced.

### Biochemistry Underlies the Causes, Diagnoses, & Treatments of the Porphyrias

Six major types of porphyria have been described, resulting from depressions in the activities of enzymes 3 through 8 shown in Figure 32–9 (see also Table 32–2). Assay of the activity of one or more of these enzymes using an appropriate source (eg, red blood cells) is thus important in making a definitive diagnosis in a suspected case of porphyria. Individuals with low activities of enzyme 1 (ALAS2) develop anemia, not porphyria (see Table 32–2). Patients with low activities of enzyme 2 (ALA dehydratase) have been reported, but very rarely; the resulting condition is called ALA dehydratase-deficient porphyria.

In general, the porphyrias described are inherited in an autosomal dominant manner, with the exception of congenital erythropoietic porphyria, which is inherited in a recessive mode. The precise abnormalities in the genes directing synthesis of the enzymes involved in heme biosynthesis have been determined in some instances. Thus, the use of appropriate gene probes has made possible the prenatal diagnosis of some of the porphyrias.

As is true of most inborn errors, the signs and symptoms of porphyria result from either a deficiency of metabolic products beyond the enzymatic block or from an accumulation of metabolites behind the block.

If the enzyme lesion occurs early in the pathway prior to the formation of porphyrinogens (eg, enzyme 3 of Figure 32–9, which is affected in acute intermittent porphyria), ALA and PBG will accumulate in body tissues and fluids (Figure 32–11). Clinically, patients complain of abdominal pain and neuropsychiatric symptoms. The precise biochemical cause of these symptoms has not been determined but may relate to elevated levels of ALA or PBG or to a deficiency of heme.

On the other hand, enzyme blocks later in the pathway result in the accumulation of the porphyrinogens

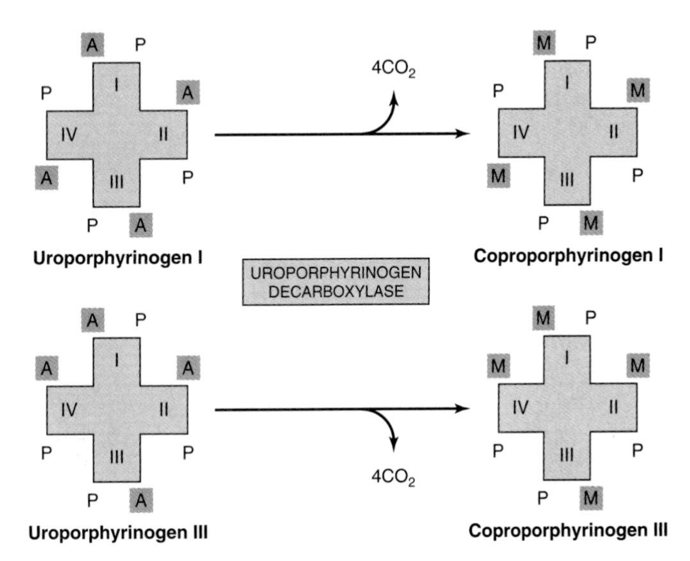


***Figure 32–7.*** Decarboxylation of uroporphyrinogens to coproporphyrinogens in cytosol. (A, acetyl; M, methyl; P, propionyl.)

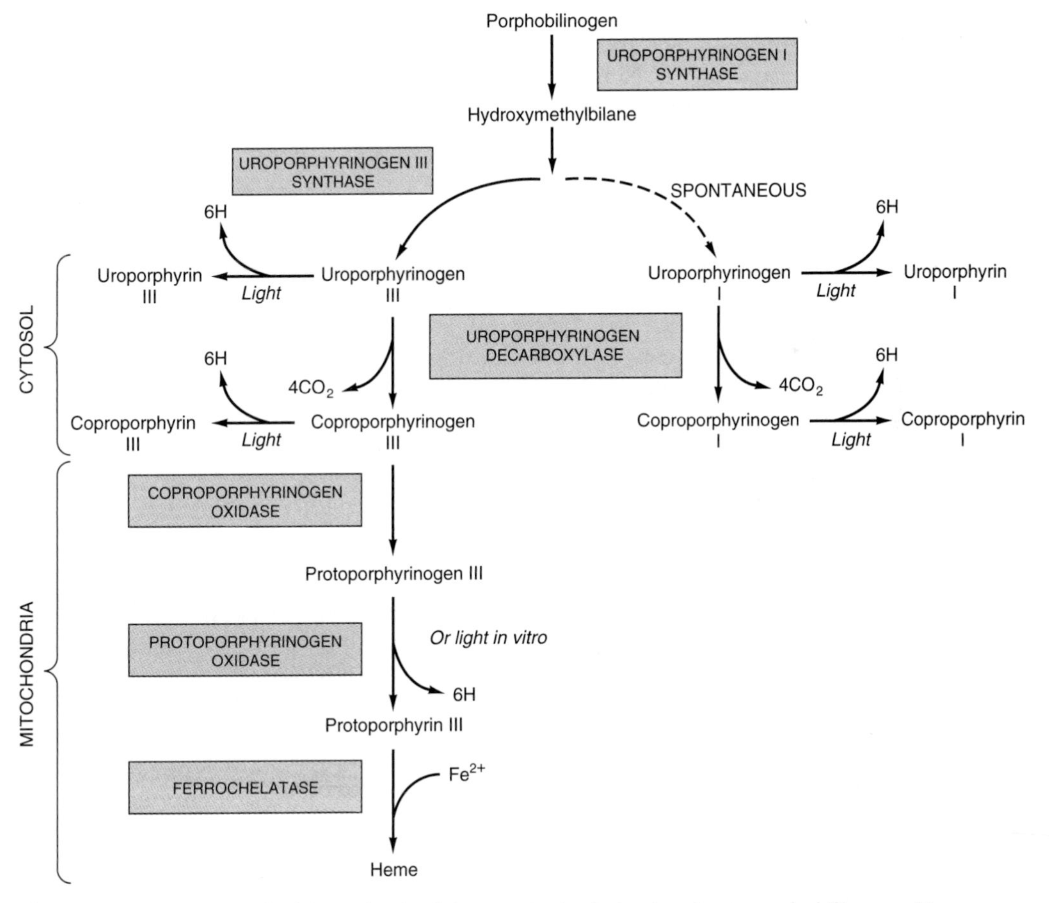


***Figure 32–8.*** Steps in the biosynthesis of the porphyrin derivatives from porphobilinogen. Uroporphyrinogen I synthase is also called porphobilinogen deaminase or hydroxymethylbilane synthase.

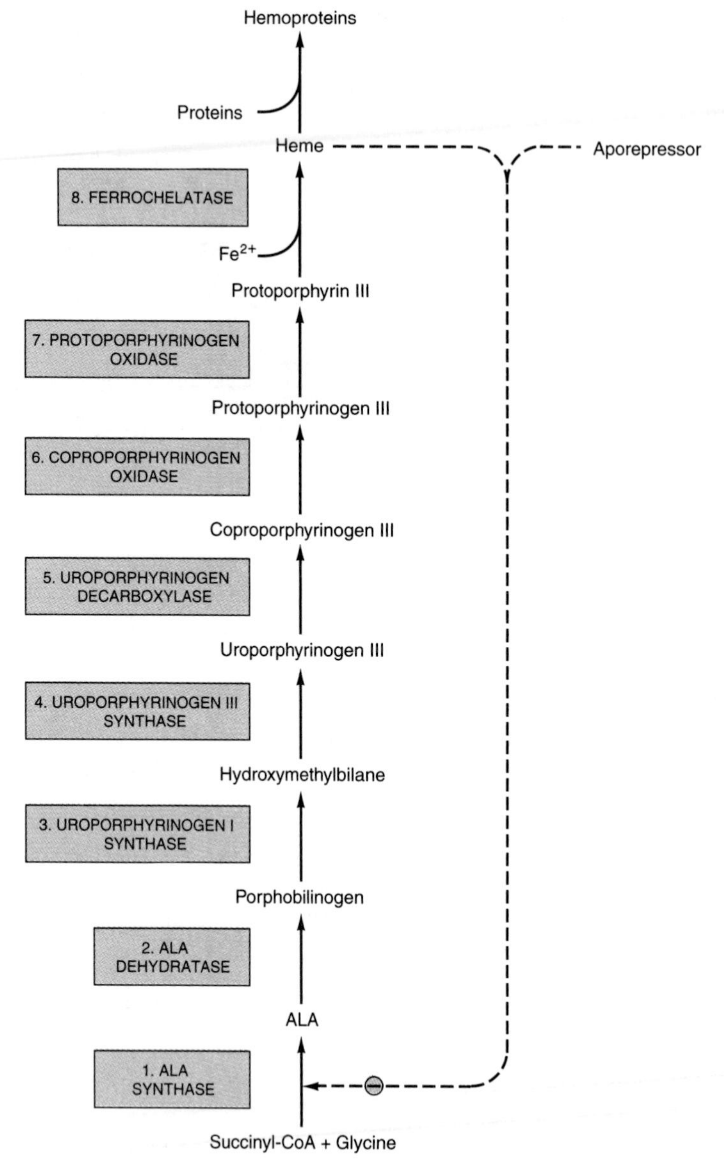


***Figure 32–9.*** Intermediates, enzymes, and regulation of heme synthesis. The enzyme numbers are those referred to in column 1 of Table 32–2. Enzymes 1, 6, 7, and 8 are located in mitochondria, the others in the cytosol. Mutations in the gene encoding enzyme 1 causes X-linked sideroblastic anemia. Mutations in the genes encoding enzymes 2–8 cause the porphyrias, though only a few cases due to deficiency of enzyme 2 have been reported. Regulation of hepatic heme synthesis occurs at ALA synthase (ALAS1) by a repression-derepression mechanism mediated by heme and its hypothetical aporepressor. The dotted lines indicate the negative (⊖) regulation by repression. Enzyme 3 is also called porphobilinogen deaminase or hydroxymethylbilane synthase.

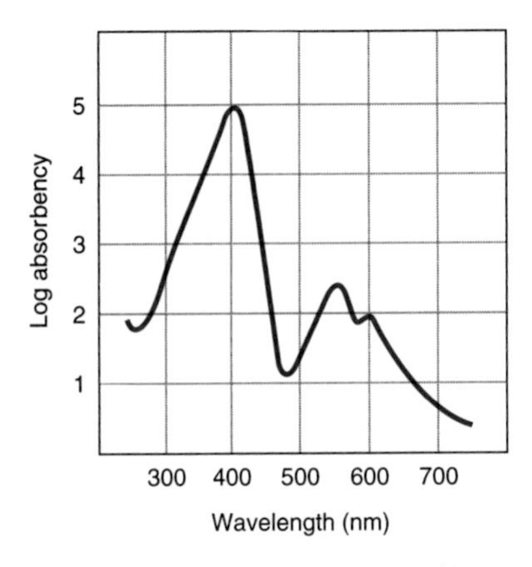


***Figure 32–10.*** Absorption spectrum of hematoporphyrin (0.01% solution in 5% HCl).

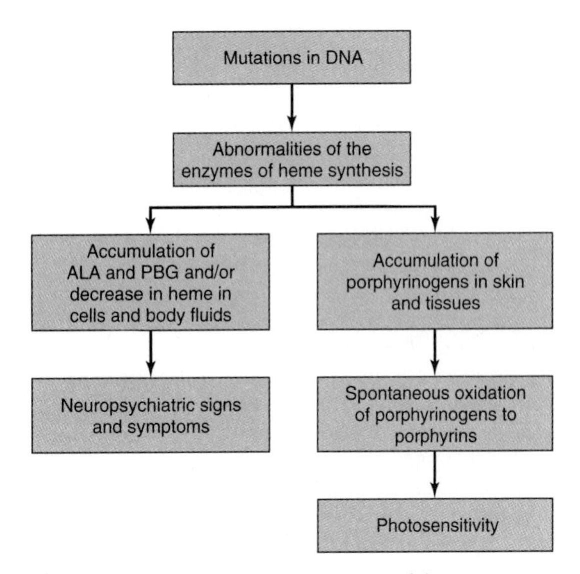


***Figure 32–11.*** Biochemical causes of the major signs and symptoms of the porphyrias.

***Table 32–2.*** Summary of major findings in the porphyrias.[1]

| Enzyme Involved[2] | Type, Class, and MIM Number | Major Signs and Symptoms | Results of Laboratory Tests |
|---|---|---|---|
| 1. ALA synthase (erythroid form) | X-linked sideroblastic anemia[3] (erythropoietic) (MIM 201300) | Anemia | Red cell counts and hemoglobin decreased |
| 2. ALA dehydratase | ALA dehydratase deficiency (hepatic) (MIM 125270) | Abdominal pain, neuropsychiatric symptoms | Urinary δ-aminolevulinic acid |
| 3. Uroporphyrinogen I synthase[4] | Acute intermittent porphyria (hepatic) (MIM 176000) | Abdominal pain, neuropsychiatric symptoms | Urinary porphobilinogen positive, uroporphyrin positive |
| 4. Uroporphyrinogen III synthase | Congenital erythropoietic (erythropoietic) (MIM 263700) | No photosensitivity | Uroporphyrin positive, porphobilinogen negative |
| 5. Uroporphyrinogen decarboxylase | Porphyria cutanea tarda (hepatic) (MIM 176100) | Photosensitivity | Uroporphyrin positive, porphobilinogen negative |
| 6. Coproporphyrinogen oxidase | Hereditary coproporphyria (hepatic) (MIM 121300) | Photosensitivity, abdominal pain, neuropsychiatric symptoms | Urinary porphobilinogen positive, urinary uroporphyrin positive, fecal protoporphyrin positive |
| 7. Protoporphyrinogen oxidase | Variegate porphyria (hepatic) (MIM 176200) | Photosensitivity, abdominal pain, neuropsychiatric symptoms | Urinary porphobilinogen positive, fecal protoporphyrin positive |
| 8. Ferrochelatase | Protoporphyria (erythropoietic) (MIM 177000) | Photosensitivity | Fecal protoporphyrin positive, red cell protoporphyrin positive |

[1]Only the biochemical findings in the active stages of these diseases are listed. Certain biochemical abnormalities are detectable in the latent stages of some of the above conditions. Conditions 3, 5, and 8 are generally the most prevalent porphyrias.
[2]The numbering of the enzymes in this table corresponds to that used in Figure 32-9.
[3]X-linked sideroblastic anemia is not a porphyria but is included here because δ–aminolevulinic acid synthase is involved.
[4]This enzyme is also called porphobilinogen deaminase or hydroxymethylbilane synthase.

indicated in Figures 32–9 and 32–11. Their oxidation products, the corresponding porphyrin derivatives, cause photosensitivity, a reaction to visible light of about 400 nm. The porphyrins, when exposed to light of this wavelength, are thought to become "excited" and then react with molecular oxygen to form oxygen radicals. These latter species injure lysosomes and other organelles. Damaged lysosomes release their degradative enzymes, causing variable degrees of skin damage, including scarring.

The porphyrias can be **classified** on the basis of the organs or cells that are most affected. These are generally organs or cells in which synthesis of heme is particularly active. The bone marrow synthesizes considerable hemoglobin, and the liver is active in the synthesis of another hemoprotein, cytochrome P450. Thus, one classification of the porphyrias is to designate them as predominantly either **erythropoietic** or **hepatic;** the types of porphyrias that fall into these two classes are so characterized in Table 32–2. Porphyrias can also be classified as acute or cutaneous on the basis of their clinical features. Why do specific types of porphyria affect certain organs more markedly than others? A partial answer is that the levels of metabolites that cause damage (eg, ALA, PBG, specific porphyrins, or lack of heme) can vary markedly in different organs or cells depending upon the differing activities of their heme-forming enzymes.

As described above, **ALAS1** is the key regulatory enzyme of the heme biosynthetic pathway in liver. A large number of **drugs** (eg, barbiturates, griseofulvin) induce the enzyme. Most of these drugs do so by inducing cytochrome P450 (see Chapter 53), which uses up heme and thus derepresses (induces) ALAS1. In patients with porphyria, increased activities of ALAS1 result in increased levels of potentially harmful heme precursors prior to the metabolic block. Thus, taking drugs that cause induction of cytochrome P450 (so-called microsomal inducers) can precipitate attacks of porphyria.

The **diagnosis** of a specific type of porphyria can generally be established by consideration of the clinical and family history, the physical examination, and appropriate laboratory tests. The major findings in the six principal types of porphyria are listed in Table 32–2.

High levels of **lead** can affect heme metabolism by combining with SH groups in enzymes such as ferrochelatase and ALA dehydratase. This affects porphyrin metabolism. Elevated levels of protoporphyrin are found in red blood cells, and elevated levels of ALA and of coproporphyrin are found in urine.

It is hoped that **treatment** of the porphyrias at the gene level will become possible. In the meantime, treatment is essentially symptomatic. It is important for patients to avoid drugs that cause induction of cytochrome P450. Ingestion of large amounts of carbohydrates (glucose loading) or administration of hematin (a hydroxide of heme) may repress ALAS1, resulting in diminished production of harmful heme precursors. Patients exhibiting photosensitivity may benefit from administration of β-carotene; this compound appears to lessen production of free radicals, thus diminishing photosensitivity. Sunscreens that filter out visible light can also be helpful to such patients.

## CATABOLISM OF HEME PRODUCES BILIRUBIN

Under physiologic conditions in the human adult, 1–2 $\times 10^8$ erythrocytes are destroyed per hour. Thus, in 1 day, a 70-kg human turns over approximately 6 g of hemoglobin. When hemoglobin is destroyed in the body, **globin** is degraded to its constituent amino acids, which are reused, and the **iron** of heme enters the iron pool, also for reuse. The iron-free **porphyrin** portion of heme is also degraded, mainly in the reticuloendothelial cells of the liver, spleen, and bone marrow.

The catabolism of heme from all of the heme proteins appears to be carried out in the microsomal fractions of cells by a complex enzyme system called **heme oxygenase.** By the time the heme derived from heme proteins reaches the oxygenase system, the iron has usually been oxidized to the ferric form, constituting **hemin.** The heme oxygenase system is substrate-inducible. As depicted in Figure 32–12, the hemin is reduced to heme with NADPH, and, with the aid of more NADPH, oxygen is added to the α-methenyl bridge between pyrroles I and II of the porphyrin. The ferrous iron is again oxidized to the ferric form. With the further addition of oxygen, **ferric ion** is released, **carbon monoxide** is produced, and an equimolar quantity of **biliverdin** results from the splitting of the tetrapyrrole ring.

In birds and amphibia, the green biliverdin IX is excreted; in mammals, a soluble enzyme called **biliverdin reductase** reduces the methenyl bridge between pyrrole III and pyrrole IV to a methylene group to produce **bilirubin,** a yellow pigment (Figure 32–12).

It is estimated that 1 g of hemoglobin yields 35 mg of bilirubin. The daily bilirubin formation in human adults is approximately 250–350 mg, deriving mainly from hemoglobin but also from ineffective erythropoiesis and from various other heme proteins such as cytochrome P450.

The chemical conversion of heme to bilirubin by reticuloendothelial cells can be observed in vivo as the purple color of the heme in a hematoma is slowly converted to the yellow pigment of bilirubin.

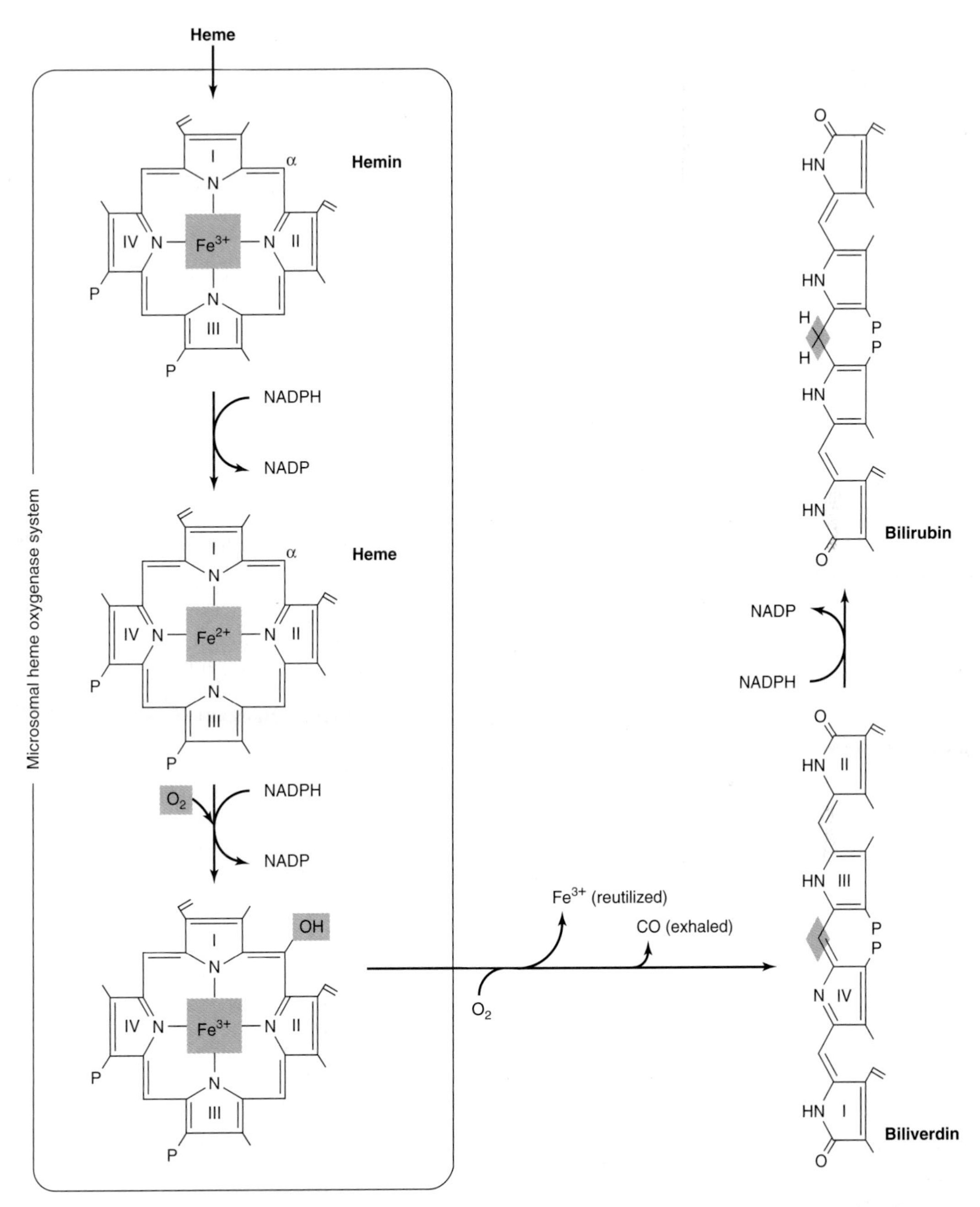


***Figure 32–12.*** Schematic representation of the microsomal heme oxygenase system. (Modified from Schmid R, McDonough AF in: *The Porphyrins.* Dolphin D [editor]. Academic Press, 1978.)

Bilirubin formed in peripheral tissues is transported to the liver by plasma albumin. The further metabolism of bilirubin occurs primarily in the liver. It can be divided into three processes: (1) uptake of bilirubin by liver parenchymal cells, (2) conjugation of bilirubin with glucuronate in the endoplasmic reticulum, and (3) secretion of conjugated bilirubin into the bile. Each of these processes will be considered separately.

## THE LIVER TAKES UP BILIRUBIN

Bilirubin is only sparingly soluble in water, but its solubility in plasma is increased by noncovalent binding to albumin. Each molecule of albumin appears to have one high-affinity site and one low-affinity site for bilirubin. In 100 mL of plasma, approximately 25 mg of bilirubin can be tightly bound to albumin at its high-affinity site. Bilirubin in excess of this quantity can be bound only loosely and thus can easily be detached and diffuse into tissues. A number of compounds such as antibiotics and other drugs compete with bilirubin for the high-affinity binding site on albumin. Thus, these compounds can displace bilirubin from albumin and have significant clinical effects.

In the liver, the bilirubin is removed from albumin and taken up at the sinusoidal surface of the hepatocytes by a carrier-mediated saturable system. This **facilitated transport system** has a very large capacity, so that even under pathologic conditions the system does not appear to be rate-limiting in the metabolism of bilirubin.

Since this facilitated transport system allows the equilibrium of bilirubin across the sinusoidal membrane of the hepatocyte, the net uptake of bilirubin will be dependent upon the removal of bilirubin via subsequent metabolic pathways.

Once bilirubin enters the hepatocytes, it can bind to certain cytosolic proteins, which help to keep it solubilized prior to conjugation. Ligandin (a family of glutathione S-transferases) and protein Y are the involved proteins. They may also help to prevent efflux of bilirubin back into the blood stream.

### Conjugation of Bilirubin With Glucuronic Acid Occurs in the Liver

Bilirubin is nonpolar and would persist in cells (eg, bound to lipids) if not rendered water-soluble. Hepatocytes convert bilirubin to a polar form, which is readily excreted in the bile, by adding glucuronic acid molecules to it. This process is called conjugation and can employ polar molecules other than glucuronic acid (eg, sulfate). Many steroid hormones and drugs are also converted to water-soluble derivatives by conjugation in preparation for excretion (see Chapter 53).

The conjugation of bilirubin is catalyzed by a specific glucuronosyltransferase. The enzyme is mainly located in the endoplasmic reticulum, uses UDP-glucuronic acid as the glucuronosyl donor, and is referred to as bilirubin-UGT. Bilirubin monoglucuronide is an intermediate and is subsequently converted to the diglucuronide (Figures 32–13 and 32–14). Most of the bilirubin excreted in the bile of mammals is in the form of bilirubin diglucuronide. However, when bilirubin conjugates exist abnormally in human plasma (eg, in obstructive jaundice), they are predominantly monoglucuronides. Bilirubin-UGT activity can be **induced** by a number of clinically useful drugs, including phenobarbital. More information about glucuronosylation is presented below in the discussion of inherited disorders of bilirubin conjugation.

### Bilirubin Is Secreted Into Bile

Secretion of conjugated bilirubin into the bile occurs by an active transport mechanism, which is probably rate-limiting for the entire process of hepatic bilirubin metabolism. The protein involved is MRP-2 (multidrug resistance-like protein 2), also called multispecific organic anion transporter (MOAT). It is located in the plasma membrane of the bile canalicular membrane and handles a number of organic anions. It is a member of the family of ATP-binding cassette (ABC) transporters. The hepatic transport of conjugated bilirubin into the bile is inducible by those same drugs that are capable of inducing the conjugation of bilirubin. Thus, the conjugation and excretion systems for bilirubin behave as a coordinated functional unit.

Figure 32–15 summarizes the three major processes involved in the transfer of bilirubin from blood to bile. Sites that are affected in a number of conditions causing jaundice (see below) are also indicated.

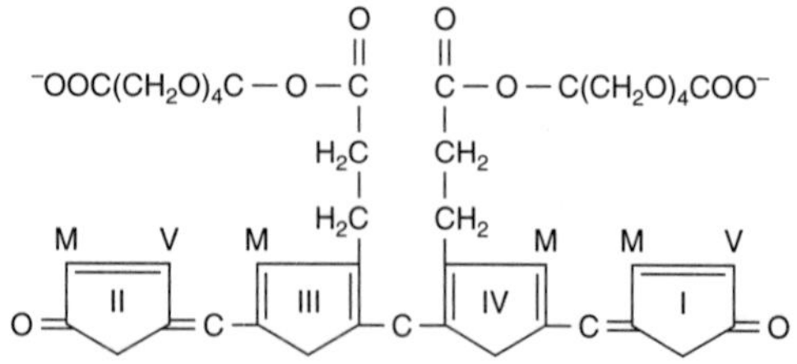

***Figure 32–13.*** Structure of bilirubin diglucuronide (conjugated, "direct-reacting" bilirubin). Glucuronic acid is attached via ester linkage to the two propionic acid groups of bilirubin to form an acylglucuronide.

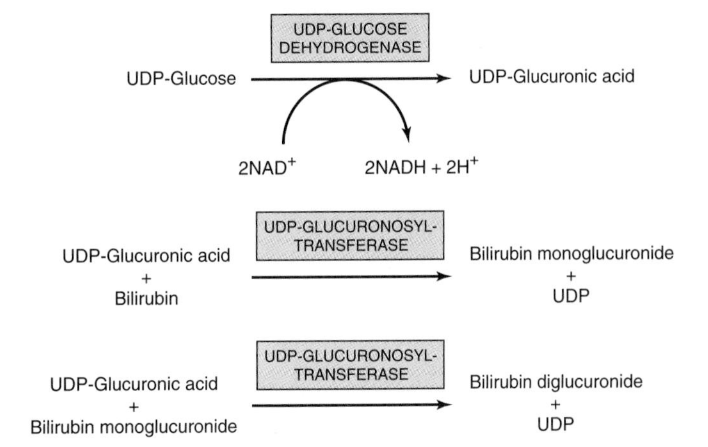


***Figure 32–14.*** Conjugation of bilirubin with glucuronic acid. The glucuronate donor, UDP-glucuronic acid, is formed from UDP-glucose as depicted. The UDP-glucuronosyl-transferase is also called bilirubin-UGT.

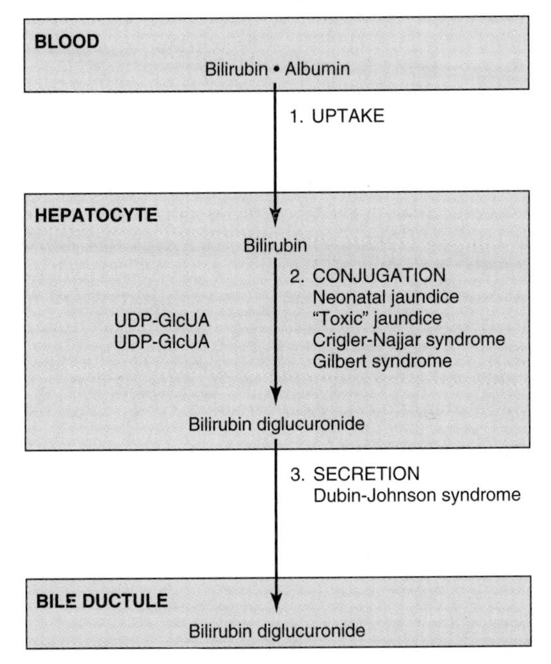


***Figure 32–15.*** Diagrammatic representation of the three major processes (uptake, conjugation, and secretion) involved in the transfer of bilirubin from blood to bile. Certain proteins of hepatocytes, such as ligandin (a family of glutathione *S*-transferase) and Y protein, bind intracellular bilirubin and may prevent its efflux into the blood stream. The process affected in a number of conditions causing jaundice is also shown.

## Conjugated Bilirubin Is Reduced to Urobilinogen by Intestinal Bacteria

As the conjugated bilirubin reaches the terminal ileum and the large intestine, the glucuronides are removed by specific bacterial enzymes **(β-glucuronidases),** and the pigment is subsequently reduced by the fecal flora to a group of colorless tetrapyrrolic compounds called **urobilinogens** (Figure 32–16). In the terminal ileum and large intestine, a small fraction of the urobilinogens is reabsorbed and reexcreted through the liver to constitute the **enterohepatic urobilinogen cycle.** Under abnormal conditions, particularly when excessive bile pigment is formed or liver disease interferes with this intrahepatic cycle, urobilinogen may also be excreted in the urine.

Normally, most of the colorless urobilinogens formed in the colon by the fecal flora are oxidized there to urobilins (colored compounds) and are excreted in the feces (Figure 32–16). Darkening of feces upon standing in air is due to the oxidation of residual urobilinogens to urobilins.

# HYPERBILIRUBINEMIA CAUSES JAUNDICE

When bilirubin in the blood exceeds 1 mg/dL (17.1 μmol/L), hyperbilirubinemia exists. Hyperbilirubinemia may be due to the production of more bilirubin than the normal liver can excrete, or it may result from the failure of a damaged liver to excrete bilirubin produced in normal amounts. In the absence of hepatic damage, obstruction of the excretory ducts of the liver—by preventing the excretion of bilirubin—will also cause hyperbilirubinemia. In all these situations, bilirubin accumulates in the blood, and when it reaches a certain concentration (approximately 2–2.5 mg/dL),

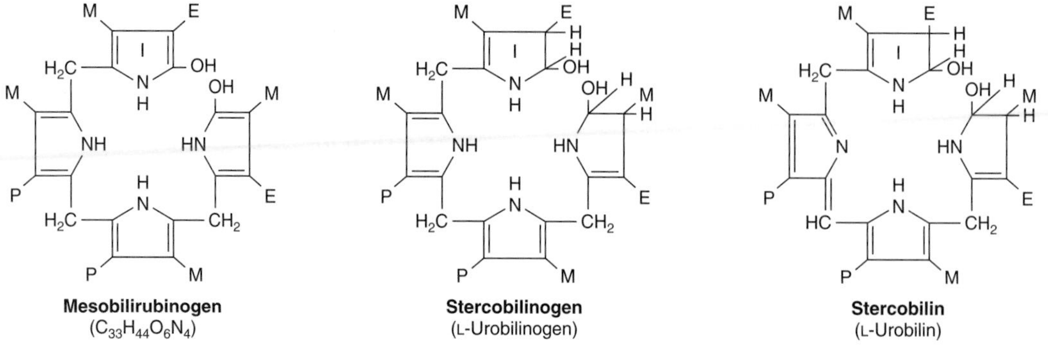

**Mesobilirubinogen** ($C_{33}H_{44}O_6N_4$)

**Stercobilinogen** (L-Urobilinogen)

**Stercobilin** (L-Urobilin)

***Figure 32–16.*** Structure of some bile pigments.

it diffuses into the tissues, which then become yellow. That condition is called **jaundice** or **icterus.**

In clinical studies of jaundice, measurement of bilirubin in the serum is of great value. A method for quantitatively assaying the bilirubin content of the serum was first devised by van den Bergh by application of Ehrlich's test for bilirubin in urine. The Ehrlich reaction is based on the coupling of diazotized sulfanilic acid (Ehrlich's diazo reagent) and bilirubin to produce a reddish-purple azo compound. In the original procedure as described by Ehrlich, methanol was used to provide a solution in which both bilirubin and the diazo regent were soluble. Van den Bergh inadvertently omitted the methanol on an occasion when assay of bile pigment in human bile was being attempted. To his surprise, normal development of the color occurred "directly." This form of bilirubin that would react without the addition of methanol was thus termed **"direct-reacting."** It was then found that this same direct reaction would also occur in serum from cases of jaundice due to biliary obstruction. However, it was still necessary to add methanol to detect bilirubin in normal serum or that which was present in excess in serum from cases of hemolytic jaundice where no evidence of obstruction was to be found. To that form of bilirubin which could be measured only after the addition of methanol, the term **"indirect-reacting"** was applied.

It was subsequently discovered that the indirect bilirubin is "free" (unconjugated) bilirubin en route to the liver from the reticuloendothelial tissues, where the bilirubin was originally produced by the breakdown of heme porphyrins. Since this bilirubin is not water-soluble, it requires methanol to initiate coupling with the diazo reagent. In the liver, the free bilirubin becomes conjugated with glucuronic acid, and the conjugate, bilirubin glucuronide, can then be excreted into the bile. Furthermore, conjugated bilirubin, being water-soluble, can react directly with the diazo reagent, so that the "direct bilirubin" of van den Bergh is actually a bilirubin conjugate (bilirubin glucuronide).

Depending on the type of bilirubin present in plasma—ie, unconjugated or conjugated—hyperbilirubinemia may be classified as **retention hyperbilirubinemia,** due to overproduction, or **regurgitation hyperbilirubinemia,** due to reflux into the bloodstream because of biliary obstruction.

Because of its hydrophobicity, only unconjugated bilirubin can cross the blood-brain barrier into the central nervous system; thus, encephalopathy due to hyperbilirubinemia **(kernicterus)** can occur only in connection with unconjugated bilirubin, as found in retention hyperbilirubinemia. On the other hand, because of its water-solubility, only conjugated bilirubin can appear in urine. Accordingly, **choluric jaundice** (choluria is the presence of bile pigments in the urine) occurs only in regurgitation hyperbilirubinemia, and **acholuric jaundice** occurs only in the presence of an excess of unconjugated bilirubin.

## Elevated Amounts of Unconjugated Bilirubin in Blood Occur in a Number of Conditions

### A. Hemolytic Anemias

Hemolytic anemias are important causes of unconjugated hyperbilirubinemia, though unconjugated hyperbilirubinemia is usually only slight (< 4 mg/dL; < 68.4 μmol/L) even in the event of extensive hemolysis because of the healthy liver's large capacity for handling bilirubin.

### B. Neonatal "Physiologic Jaundice"

This transient condition is the most common cause of unconjugated hyperbilirubinemia. It results from an accelerated hemolysis around the time of birth and an immature hepatic system for the uptake, conjugation, and secretion of bilirubin. Not only is the bilirubin-UGT activity reduced, but there probably is reduced synthesis of the substrate for that enzyme, UDP-glucuronic acid. Since the increased amount of bilirubin is unconjugated, it is capable of penetrating the blood-brain barrier when its concentration in plasma exceeds that which can be tightly bound by albumin (20–25 mg/dL). This can result in a hyperbilirubinemic toxic encephalopathy, or **kernicterus,** which can cause mental retardation. Because of the recognized inducibility of this bilirubin-metabolizing system, phenobarbital has been administered to jaundiced neonates and is effective in this disorder. In addition, exposure to blue light (phototherapy) promotes the hepatic excretion of unconjugated bilirubin by converting some of the bilirubin to other derivatives such as maleimide fragments and geometric isomers that are excreted in the bile.

#### C. Crigler-Najjar Syndrome, Type I; Congenital Nonhemolytic Jaundice

Type I Crigler-Najjar syndrome is a rare autosomal recessive disorder. It is characterized by severe congenital jaundice (serum bilirubin usually exceeds 20 mg/dL) due to mutations in the gene encoding bilirubin-UGT activity in hepatic tissues. The disease is often fatal within the first 15 months of life. Children with this condition have been treated with phototherapy, resulting in some reduction in plasma bilirubin levels. Phenobarbital has no effect on the formation of bilirubin glucuronides in patients with type I Crigler-Najjar syndrome. A liver transplant may be curative.

#### D. Crigler-Najjar Syndrome, Type II

This rare inherited disorder also results from mutations in the gene encoding bilirubin-UGT, but some activity of the enzyme is retained and the condition has a more benign course than type I. Serum bilirubin concentrations usually do not exceed 20 mg/dL. Patients with this condition can respond to treatment with large doses of phenobarbital.

#### E. Gilbert Syndrome

Again, this is caused by mutations in the gene encoding bilirubin-UGT, but approximately 30% of the enzyme's activity is preserved and the condition is entirely harmless.

#### F. Toxic Hyperbilirubinemia

Unconjugated hyperbilirubinemia can result from toxin-induced liver dysfunction such as that caused by chloroform, arsphenamines, carbon tetrachloride, acetaminophen, hepatitis virus, cirrhosis, and *Amanita* mushroom poisoning. These acquired disorders are due to hepatic parenchymal cell damage, which impairs conjugation.

### Obstruction in the Biliary Tree Is the Commonest Cause of Conjugated Hyperbilirubinemia

#### A. Obstruction of the Biliary Tree

Conjugated hyperbilirubinemia commonly results from blockage of the hepatic or common bile ducts, most often due to a gallstone or to cancer of the head of the pancreas. Because of the obstruction, bilirubin diglucuronide cannot be excreted. It thus regurgitates into the hepatic veins and lymphatics, and conjugated bilirubin appears in the blood and urine (choluric jaundice).

The term **cholestatic jaundice** is used to include all cases of extrahepatic obstructive jaundice. It also covers those cases of jaundice that exhibit conjugated hyperbilirubinemia due to micro-obstruction of intrahepatic biliary ductules by swollen, damaged hepatocytes (eg, as may occur in infectious hepatitis).

#### B. Dubin-Johnson Syndrome

This benign autosomal recessive disorder consists of conjugated hyperbilirubinemia in childhood or during adult life. The hyperbilirubinemia is caused by mutations in the gene encoding MRP-2 (see above), the protein involved in the secretion of conjugated bilirubin into bile. The centrilobular hepatocytes contain an abnormal black pigment that may be derived from epinephrine.

#### C. Rotor Syndrome

This is a rare benign condition characterized by chronic conjugated hyperbilirubinemia and normal liver histology. Its precise cause has not been identified, but it is thought to be due to an abnormality in hepatic storage.

### Some Conjugated Bilirubin Can Bind Covalently to Albumin

When levels of conjugated bilirubin remain high in plasma, a fraction can bind covalently to albumin (delta bilirubin). Because it is bound covalently to albumin, this fraction has a longer half-life in plasma than does conventional conjugated bilirubin. Thus, it remains elevated during the recovery phase of obstructive jaundice after the remainder of the conjugated bilirubin has declined to normal levels; this explains why some patients continue to appear jaundiced after conjugated bilirubin levels have returned to normal.

***Table 32–3.*** Laboratory results in normal patients and patients with three different causes of jaundice.

| Condition | Serum Bilirubin | Urine Urobilinogen | Urine Bilirubin | Fecal Urobilinogen |
|---|---|---|---|---|
| Normal | Direct: 0.1–0.4 mg/dL<br>Indirect: 0.2–0.7 mg/dL | 0–4 mg/24 h | Absent | 40–280 mg/24 h |
| Hemolytic anemia | ↑ Indirect | Increased | Absent | Increased |
| Hepatitis | ↑ Direct and indirect | Decreased if micro-obstruction is present | Present if micro-obstruction occurs | Decreased |
| Obstructive jaundice[1] | ↑ Direct | Absent | Present | Trace to absent |

[1]The commonest causes of obstructive (posthepatic) jaundice are cancer of the head of the pancreas and a gallstone lodged in the common bile duct. The presence of bilirubin in the urine is sometimes referred to as choluria—therefore, hepatitis and obstruction of the common bile duct cause choluric jaundice, whereas the jaundice of hemolytic anemia is referred to as acholuric. The laboratory results in patients with hepatitis are variable, depending on the extent of damage to parenchymal cells and the extent of micro-obstruction to bile ductules. Serum levels of ALT and AST are usually markedly elevated in hepatitis, whereas serum levels of alkaline phosphatase are elevated in obstructive liver disease.

## Urobilinogen & Bilirubin in Urine Are Clinical Indicators

Normally, there are mere traces of urobilinogen in the urine. In **complete obstruction of the bile duct,** no urobilinogen is found in the urine, since bilirubin has no access to the intestine, where it can be converted to urobilinogen. In this case, the presence of bilirubin (conjugated) in the urine without urobilinogen suggests obstructive jaundice, either intrahepatic or posthepatic.

In **jaundice secondary to hemolysis,** the increased production of bilirubin leads to increased production of urobilinogen, which appears in the urine in large amounts. Bilirubin is not usually found in the urine in hemolytic jaundice (because unconjugated bilirubin does not pass into the urine), so that the combination of increased urobilinogen and absence of bilirubin is suggestive of hemolytic jaundice. Increased blood destruction from any cause brings about an increase in urine urobilinogen.

Table 32–3 summarizes laboratory results obtained on patients with three different causes of jaundice—hemolytic anemia (a prehepatic cause), hepatitis (a hepatic cause), and obstruction of the common bile duct (a posthepatic cause). Laboratory tests on blood (evaluation of the possibility of a hemolytic anemia and measurement of prothrombin time) and on serum (eg, electrophoresis of proteins; activities of the enzymes ALT, AST, and alkaline phosphatase) are also important in helping to distinguish between prehepatic, hepatic, and posthepatic causes of jaundice.

## SUMMARY

- Hemoproteins, such as hemoglobin and the cytochromes, contain heme. Heme is an iron-porphyrin compound ($Fe^{2+}$-protoporphyrin IX) in which four pyrrole rings are joined by methenyl bridges. The eight side groups (methyl, vinyl, and propionyl substituents) on the four pyrrole rings of heme are arranged in a specific sequence.
- Biosynthesis of the heme ring occurs in mitochondria and cytosol via eight enzymatic steps. It commences with formation of δ-aminolevulinate (ALA) from succinyl-CoA and glycine in a reaction catalyzed by ALA synthase, the regulatory enzyme of the pathway.
- Genetically determined abnormalities of seven of the eight enzymes involved in heme biosynthesis result in the inherited porphyrias. Red blood cells and liver are the major sites of metabolic expression of the porphyrias. Photosensitivity and neurologic problems are common complaints. Intake of certain compounds (such as lead) can cause acquired porphyrias. Increased amounts of porphyrins or their precursors can be detected in blood and urine, facilitating diagnosis.
- Catabolism of the heme ring is initiated by the enzyme heme oxygenase, producing a linear tetrapyrrole.
- Biliverdin is an early product of catabolism and on reduction yields bilirubin. The latter is transported by albumin from peripheral tissues to the liver, where it is taken up by hepatocytes. The iron of heme and the amino acids of globin are conserved and reutilized.
- In the liver, bilirubin is made water-soluble by conjugation with two molecules of glucuronic acid and is secreted into the bile. The action of bacterial enzymes in the gut produces urobilinogen and urobilin, which are excreted in the feces and urine.
- Jaundice is due to elevation of the level of bilirubin in the blood. The causes of jaundice can be classified

as prehepatic (eg, hemolytic anemias), hepatic (eg, hepatitis), and posthepatic (eg, obstruction of the common bile duct). Measurements of plasma total and nonconjugated bilirubin, of urinary urobilinogen and bilirubin, and of certain serum enzymes as well as inspection of stool samples help distinguish between these causes.

## REFERENCES

Anderson KE et al: Disorders of heme biosynthesis: X-linked sideroblastic anemia and the porphyrias. In: *The Metabolic and Molecular Bases of Inherited Disease,* 8th ed. Scriver CR et al (editors). McGraw-Hill, 2001.

Berk PD, Wolkoff AW: Bilirubin metabolism and the hyperbilirubinemias. In: *Harrison's Principles of Internal Medicine,* 15th ed. Braunwald E et al (editors). McGraw-Hill, 2001.

Chowdhury JR et al: Hereditary jaundice and disorders of bilirubin metabolism. In: *The Metabolic and Molecular Bases of Inherited Disease,* 8th ed. Scriver CR et al (editors). McGraw-Hill, 2001.

Desnick RJ: The porphyrias. In: *Harrison's Principles of Internal Medicine,* 15th ed. Braunwald E et al (editors). McGraw-Hill, 2001.

Elder GH: Haem synthesis and the porphyrias. In: *Scientific Foundations of Biochemistry in Clinical Practice,* 2nd ed. Williams DL, Marks V (editors). Butterworth-Heinemann, 1994.

# SECTION IV

## Structure, Function, & Replication of Informational Macromolecules

# Nucleotides 33

*Victor W. Rodwell, PhD*

## BIOMEDICAL IMPORTANCE

Nucleotides—the monomer units or building blocks of nucleic acids—serve multiple additional functions. They form a part of many coenzymes and serve as donors of phosphoryl groups (eg, ATP or GTP), of sugars (eg, UDP- or GDP-sugars), or of lipid (eg, CDP-acylglycerol). Regulatory nucleotides include the second messengers cAMP and cGMP, the control by ADP of oxidative phosphorylation, and allosteric regulation of enzyme activity by ATP, AMP, and CTP. Synthetic purine and pyrimidine analogs that contain halogens, thiols, or additional nitrogen are employed for chemotherapy of cancer and AIDS and as suppressors of the immune response during organ transplantation.

## PURINES, PYRIMIDINES, NUCLEOSIDES, & NUCLEOTIDES

Purines and pyrimidines are nitrogen-containing heterocycles, cyclic compounds whose rings contain both carbon and other elements (hetero atoms). Note that the smaller pyrimidine has the *longer* name and the larger purine the *shorter* name and that their six-atom rings are numbered in opposite directions (Figure 33–1). The planar character of purines and pyrimidines facilitates their close association, or "stacking," which stabilizes double-stranded DNA (Chapter 36). The oxo and amino groups of purines and pyrimidines exhibit keto-enol and amine-imine tautomerism (Figure 33–2), but physiologic conditions strongly favor the amino and oxo forms.

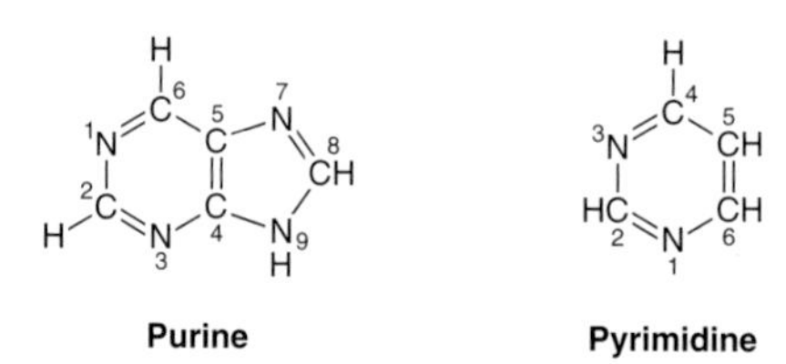


***Figure 33–1.*** Purine and pyrimidine. The atoms are numbered according to the international system.

### Nucleosides & Nucleotides

Nucleo**sides** are derivatives of purines and pyrimidines that have a sugar linked to a ring nitrogen. Numerals with a prime (eg, 2′ or 3′) distinguish atoms of the sugar from those of the heterocyclic base. The sugar in **ribonucleosides** is D-ribose, and in **deoxyribonucleosides** it is 2-deoxy-D-ribose. The sugar is linked to the heterocyclic base via a **β-N-glycosidic bond,** almost always to N-1 of a pyrimidine or to N-9 of a purine (Figure 33–3).

***Figure 33–2.*** Tautomerism of the oxo and amino functional groups of purines and pyrimidines.

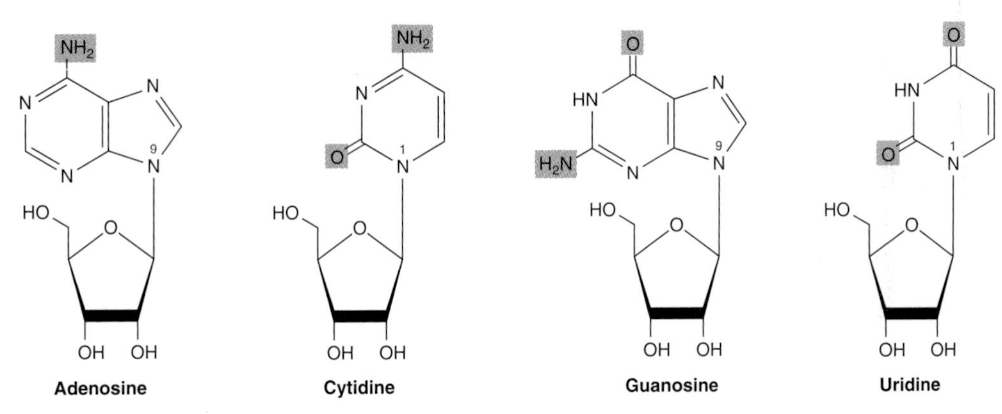


***Figure 33–3.*** Ribonucleosides, drawn as the syn conformers.

Mononucleo**tides** are nucleo**sides** with a phosphoryl group esterified to a hydroxyl group of the sugar. The 3′- and 5′-nucleo**tides** are nucleo**sides** with a phosphoryl group on the 3′- or 5′-hydroxyl group of the sugar, respectively. Since most nucleotides are 5′-, the prefix "5′-" is usually omitted when naming them. UMP and dAMP thus represent nucleotides with a phosphoryl group on C-5 of the pentose. Additional phosphoryl groups linked by **acid anhydride bonds** to the phosphoryl group of a mononucleotide form nucleoside **diphosphates** and **triphosphates** (Figure 33–4).

Steric hindrance by the base restricts rotation about the β-N-glycosidic bond of nucleosides and nucleotides. Both therefore exist as syn or anti conformers (Figure 33–5). While both conformers occur in nature, anti conformers predominate. Table 33–1 lists the major purines and pyrimidines and their nucleoside and nucleotide derivatives. Single-letter abbreviations are used to identify adenine (A), guanine (G), cytosine (C), thymine (T), and uracil (U), whether free or present in nucleosides or nucleotides. The prefix "d" (deoxy) indicates that the sugar is 2′-deoxy-D-ribose (eg, dGTP) (Figure 33–6).

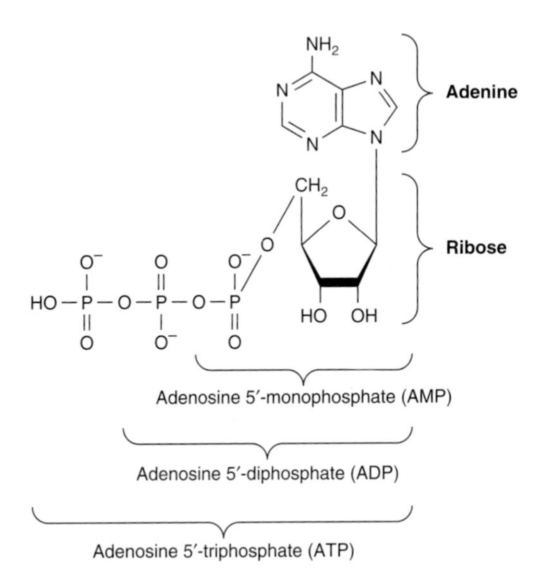


***Figure 33–4.*** ATP, its diphosphate, and its monophosphate.

## Nucleic Acids Also Contain Additional Bases

Small quantities of additional purines and pyrimidines occur in DNA and RNAs. Examples include 5-methylcytosine of bacterial and human DNA, 5-hydroxymethylcytosine of bacterial and viral nucleic acids, and mono- and di-N-methylated adenine and guanine of

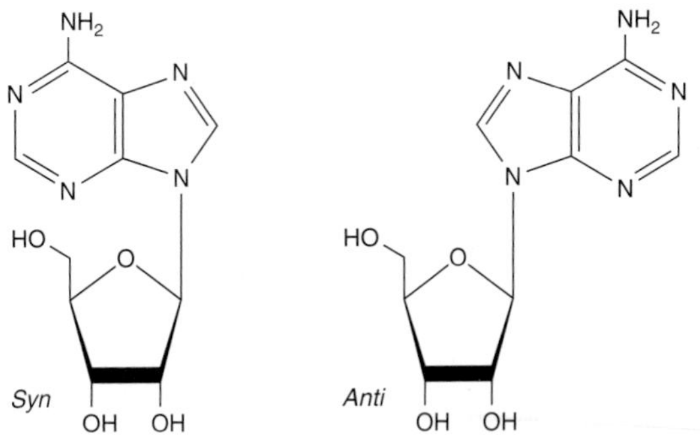

*Syn*

*Anti*

***Figure 33–5.*** The syn and anti conformers of adenosine differ with respect to orientation about the N-glycosidic bond.

***Table 33–1.*** Bases, nucleosides, and nucleotides.

| Base Formula | Base X = H | Nucleoside X = Ribose or Deoxyribose | Nucleotide, Where X = Ribose Phosphate |
|---|---|---|---|
| | Adenine<br>A | Adenosine<br>A | Adenosine monophosphate<br>AMP |
| | Guanine<br>G | Guanosine<br>G | Guanosine monophosphate<br>GMP |
| | Cytosine<br>C | Cytidine<br>C | Cytidine monophosphate<br>CMP |
| | Uracil<br>U | Uridine<br>U | Uridine monophosphate<br>UMP |
| | Thymine<br>T | Thymidine<br>T | Thymidine monophosphate<br>TMP |

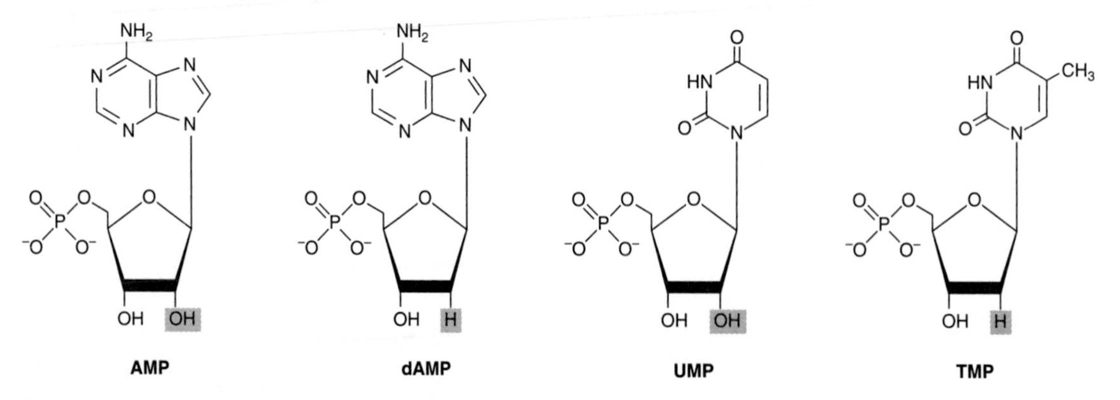


***Figure 33–6.*** AMP, dAMP, UMP, and TMP.

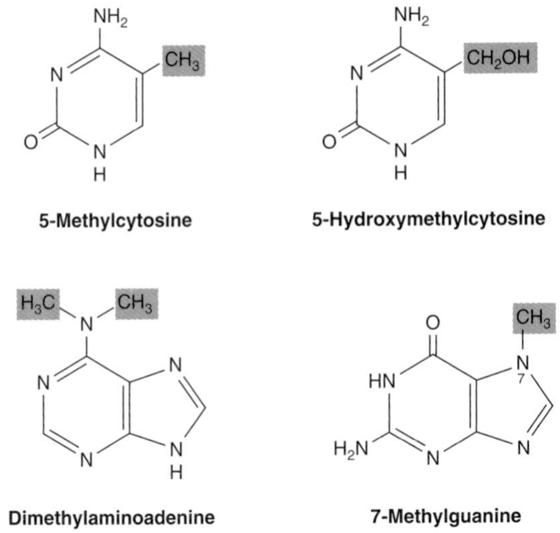

5-Methylcytosine

5-Hydroxymethylcytosine

Dimethylaminoadenine

7-Methylguanine

***Figure 33–7.*** Four uncommon naturally occurring pyrimidines and purines.

mammalian messenger RNAs (Figure 33–7). These atypical bases function in oligonucleotide recognition and in regulating the half-lives of RNAs. Free nucleotides include hypoxanthine, xanthine, and uric acid (see Figure 34–8), intermediates in the catabolism of adenine and guanine. Methylated heterocyclic bases of plants include the xanthine derivatives caffeine of coffee, theophylline of tea, and theobromine of cocoa (Figure 33–8).

Posttranslational modification of preformed polynucleotides can generate additional bases such as pseudouridine, in which D-ribose is linked to C-5 of uracil by a carbon-to-carbon bond rather than by a β-N-glycosidic bond. The nucleotide pseudouridylic acid Ψ arises by rearrangement of UMP of a preformed tRNA. Similarly, methylation by *S*-adenosylmethionine of a UMP of preformed tRNA forms TMP (thymidine monophosphate), which contains ribose rather than deoxyribose.

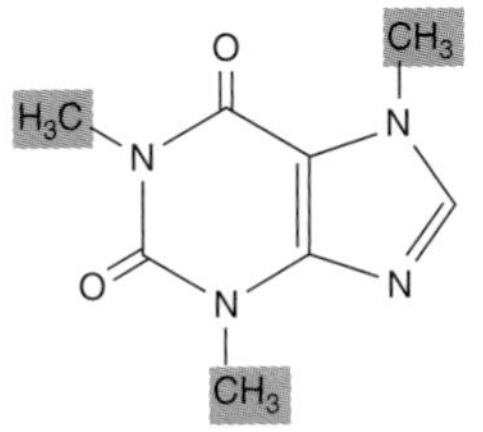

***Figure 33–8.*** Caffeine, a trimethylxanthine. The dimethylxanthines theobromine and theophylline are similar but lack the methyl group at N-1 and at N-7, respectively.

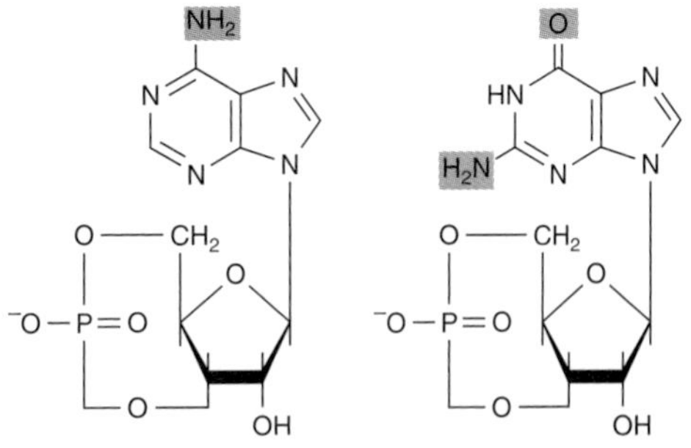

***Figure 33–9.*** cAMP, 3′,5′-cyclic AMP, and cGMP.

## Nucleotides Serve Diverse Physiologic Functions

Nucleotides participate in reactions that fulfill physiologic functions as diverse as protein synthesis, nucleic acid synthesis, regulatory cascades, and signal transduction pathways.

## Nucleoside Triphosphates Have High Group Transfer Potential

Acid anhydrides, unlike phosphate esters, have high group transfer potential. $\Delta^{0\prime}$ for the hydrolysis of each of the terminal phosphates of nucleoside triphosphates is about −7 kcal/mol (−30 kJ/mol). The high group transfer potential of purine and pyrimidine nucleoside triphosphates permits them to function as group transfer reagents. Cleavage of an acid anhydride bond typically is coupled with a highly endergonic process such as covalent bond synthesis—eg, polymerization of nucleoside triphosphates to form a nucleic acid.

In addition to their roles as precursors of nucleic acids, ATP, GTP, UTP, CTP, and their derivatives each serve unique physiologic functions discussed in other chapters. Selected examples include the role of ATP as the principal biologic transducer of free energy; the second messenger cAMP (Figure 33–9); adenosine 3′-phosphate-5′-phosphosulfate (Figure 33–10), the sulfate donor for sulfated proteoglycans (Chapter 48) and for sulfate conjugates of drugs; and the methyl group donor *S*-adenosylmethionine (Figure 33–11).

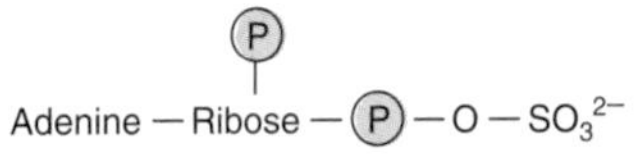

Adenine — Ribose — Ⓟ — O — $SO_3^{2-}$ (with Ⓟ attached to Ribose)

***Figure 33–10.*** Adenosine 3′-phosphate-5′-phosphosulfate.

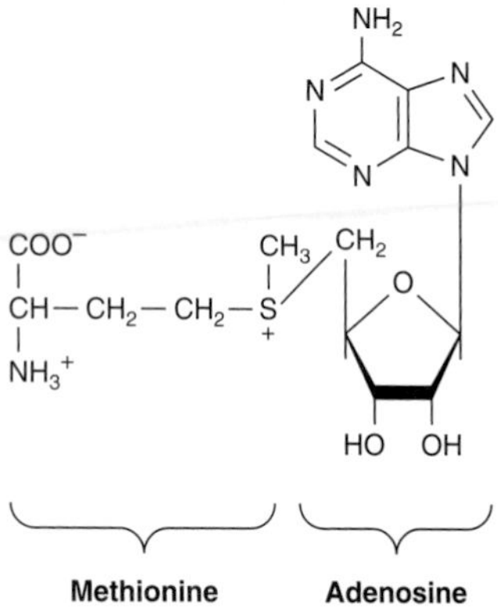

***Figure 33–11.*** *S*-Adenosylmethionine.

GTP serves as an allosteric regulator and as an energy source for protein synthesis, and cGMP (Figure 33–9) serves as a second messenger in response to nitric oxide (NO) during relaxation of smooth muscle (Chapter 48). UDP-sugar derivatives participate in sugar epimerizations and in biosynthesis of glycogen, glucosyl disaccharides, and the oligosaccharides of glycoproteins and proteoglycans (Chapters 47 and 48). UDP-glucuronic acid forms the urinary glucuronide conjugates of bilirubin (Chapter 32) and of drugs such as aspirin. CTP participates in biosynthesis of phosphoglycerides, sphingomyelin, and other substituted sphingosines (Chapter 24). Finally, many coenzymes incorporate nucleotides as well as structures similar to purine and pyrimidine nucleotides (see Table 33–2).

## Nucleotides Are Polyfunctional Acids

Nucleosides or free purine or pyrimidine bases are uncharged at physiologic pH. By contrast, the primary phosphoryl groups (p*K* about 1.0) and secondary phosphoryl groups (p*K* about 6.2) of nucleotides ensure that they bear a negative charge at physiologic pH. Nucleotides can, however, act as proton donors or acceptors at pH values two or more units above or below neutrality.

## Nucleotides Absorb Ultraviolet Light

The conjugated double bonds of purine and pyrimidine derivatives absorb ultraviolet light. The mutagenic effect of ultraviolet light results from its absorption by nucleotides in DNA with accompanying chemical changes. While spectra are pH-dependent, at pH 7.0 all the common nucleotides absorb light at a wavelength close to 260 nm. The concentration of nucleotides and nucleic acids thus often is expressed in terms of "absorbance at 260 nm."

***Table 33–2.*** Many coenzymes and related compounds are derivatives of adenosine monophosphate.

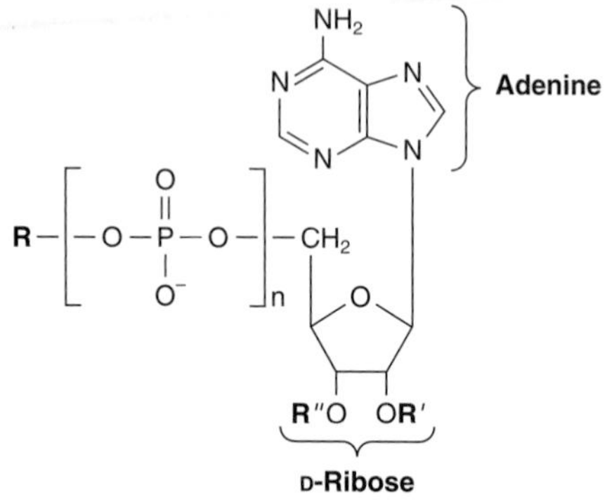

| Coenzyme | R | R′ | R″ | n |
|---|---|---|---|---|
| Active methionine | Methionine* | H | H | 0 |
| Amino acid adenylates | Amino acid | H | H | 1 |
| Active sulfate | $SO_3^{2-}$ | H | $PO_3^{2-}$ | 1 |
| 3′,5′-Cyclic AMP | | H | $PO_3^{2-}$ | 1 |
| NAD* | † | H | H | 2 |
| NADP* | † | $PO_3^{2-}$ | H | 2 |
| FAD | † | H | H | 2 |
| CoASH | † | H | $PO_3^{2-}$ | 2 |

*Replaces phosphoryl group.
†R is a B vitamin derivative.

## SYNTHETIC NUCLEOTIDE ANALOGS ARE USED IN CHEMOTHERAPY

Synthetic analogs of purines, pyrimidines, nucleosides, and nucleotides altered in either the heterocyclic ring or the sugar moiety have numerous applications in clinical medicine. Their toxic effects reflect either inhibition of enzymes essential for nucleic acid synthesis or their incorporation into nucleic acids with resulting disruption of base-pairing. Oncologists employ 5-fluoro- or 5-iodouracil, 3-deoxyuridine, 6-thioguanine and 6-mercaptopurine, 5- or 6-azauridine, 5- or 6-azacytidine, and 8-azaguanine (Figure 33–12), which are incorporated into DNA prior to cell division. The purine analog allopurinol, used in treatment of hyperuricemia and gout, inhibits purine biosynthesis and xanthine oxidase activity. Cytarabine is used in chemotherapy of cancer. Finally, azathioprine, which is catabolized to 6-mercaptopurine, is employed during organ transplantation to suppress immunologic rejection.

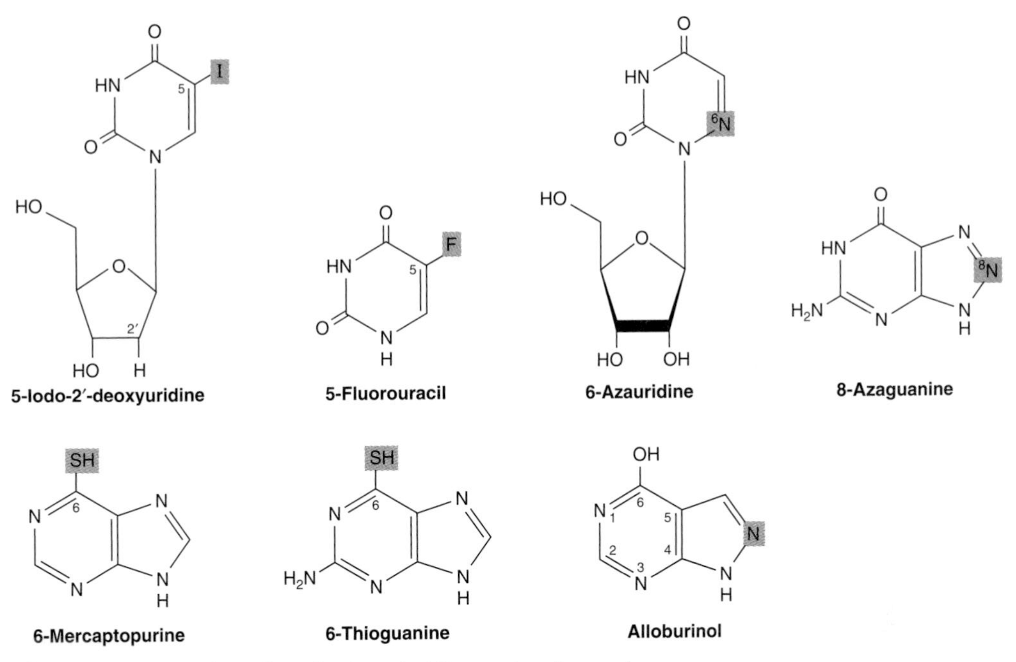


***Figure 33–12.*** Selected synthetic pyrimidine and purine analogs.

## Nonhydrolyzable Nucleoside Triphosphate Analogs Serve as Research Tools

Synthetic nonhydrolyzable analogs of nucleoside triphosphates (Figure 33–13) allow investigators to distinguish the effects of nucleotides due to phosphoryl transfer from effects mediated by occupancy of allosteric nucleotide-binding sites on regulated enzymes.

# POLYNUCLEOTIDES

The 5′-phosphoryl group of a mononucleotide can esterify a second —OH group, forming a **phosphodiester.** Most commonly, this second —OH group is the 3′-OH of the pentose of a second nucleotide. This forms a **dinucleotide** in which the pentose moieties are linked by a 3′ → 5′ phosphodiester bond to form the "backbone" of RNA and DNA.

While formation of a dinucleotide may be represented as the elimination of water between two monomers, the reaction in fact strongly favors phosphodiester hydrolysis. **Phosphodiesterases** rapidly catalyze the hydrolysis of phosphodiester bonds whose spontaneous hydrolysis is an extremely slow process. Consequently, DNA persists for considerable periods and has been detected even in fossils. RNAs are far less stable than DNA since the 2′-hydroxyl group of RNA (absent from DNA) functions as a nucleophile during hydrolysis of the 3′,5′-phosphodiester bond.

## Polynucleotides Are Directional Macromolecules

Phosphodiester bonds link the 3′- and 5′-carbons of adjacent monomers. Each end of a nucleotide polymer thus is distinct. We therefore refer to the "5′- end" or the "3′- end" of polynucleotides, the 5′- end being the one with a free or phosphorylated 5′-hydroxyl.

## Polynucleotides Have Primary Structure

The base sequence or **primary structure** of a polynucleotide can be represented as shown below. The phosphodiester bond is represented by P or p, bases by a single letter, and pentoses by a vertical line.

A T C A

P P P P OH

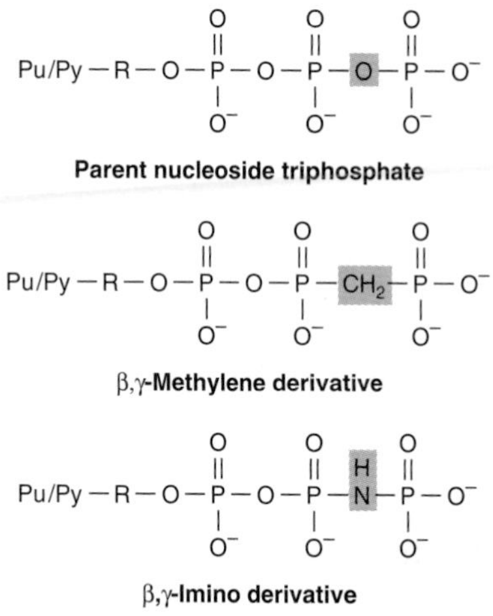

Pu/Py — R — O — P(=O)(O⁻) — O — P(=O)(O⁻) — O — P(=O)(O⁻) — O⁻

**Parent nucleoside triphosphate**

Pu/Py — R — O — P(=O)(O⁻) — O — P(=O)(O⁻) — $CH_2$ — P(=O)(O⁻) — O⁻

**β,γ-Methylene derivative**

Pu/Py — R — O — P(=O)(O⁻) — O — P(=O)(O⁻) — NH — P(=O)(O⁻) — O⁻

**β,γ-Imino derivative**

***Figure 33–13.*** Synthetic derivatives of nucleoside triphosphates incapable of undergoing hydrolytic release of the terminal phosphoryl group. (Pu/Py, a purine or pyrimidine base; R, ribose or deoxyribose.) Shown are the parent (hydrolyzable) nucleoside triphosphate **(top)** and the unhydrolyzable β-methylene **(center)** and γ-imino derivatives **(bottom).**

Where all the phosphodiester bonds are 5′ → 3′, a more compact notation is possible:

pGpGpApTpCpA

This representation indicates that the 5′-hydroxyl—but not the 3′-hydroxyl—is phosphorylated.

The most compact representation shows only the base sequence with the 5′- end on the left and the 3′- end on the right. The phosphoryl groups are assumed but not shown:

GGATCA

## SUMMARY

- Under physiologic conditions, the amino and oxo tautomers of purines, pyrimidines, and their derivatives predominate.
- Nucleic acids contain, in addition to A, G, C, T, and U, traces of 5-methylcytosine, 5-hydroxymethylcytosine, pseudouridine (Ψ), or N-methylated bases.
- Most nucleosides contain D-ribose or 2-deoxy-D-ribose linked to N-1 of a pyrimidine or to N-9 of a purine by a β-glycosidic bond whose syn conformers predominate.
- A primed numeral locates the position of the phosphate on the sugars of mononucleotides (eg, 3′-GMP, 5′-dCMP). Additional phosphoryl groups linked to the first by acid anhydride bonds form nucleoside diphosphates and triphosphates.
- Nucleoside triphosphates have high group transfer potential and participate in covalent bond syntheses. The cyclic phosphodiesters cAMP and cGMP function as intracellular second messengers.
- Mononucleotides linked by 3′ → 5′-phosphodiester bonds form polynucleotides, directional macromolecules with distinct 3′- and 5′- ends. For pTpGpTp or TGCATCA, the 5′- end is at the left, and all phosphodiester bonds are 3′ → 5′.
- Synthetic analogs of purine and pyrimidine bases and their derivatives serve as anticancer drugs either by inhibiting an enzyme of nucleotide biosynthesis or by being incorporated into DNA or RNA.

## REFERENCES

Adams RLP, Knowler JT, Leader DP: *The Biochemistry of the Nucleic Acids,* 11th ed. Chapman & Hall, 1992.

Blackburn GM, Gait MJ: *Nucleic Acids in Chemistry & Biology.* IRL Press, 1990.

Bugg CE, Carson WM, Montgomery JA: Drugs by design. Sci Am 1992;269(6):92.

# Metabolism of Purine & Pyrimidine Nucleotides

# 34

Victor W. Rodwell, PhD

## BIOMEDICAL IMPORTANCE

The biosynthesis of purines and pyrimidines is stringently regulated and coordinated by feedback mechanisms that ensure their production in quantities and at times appropriate to varying physiologic demand. Genetic diseases of purine metabolism include gout, Lesch-Nyhan syndrome, adenosine deaminase deficiency, and purine nucleoside phosphorylase deficiency. By contrast, apart from the orotic acidurias, there are few clinically significant disorders of pyrimidine catabolism.

## PURINES & PYRIMIDINES ARE DIETARILY NONESSENTIAL

Human tissues can synthesize purines and pyrimidines from amphibolic intermediates. Ingested nucleic acids and nucleotides, which therefore are dietarily nonessential, are degraded in the intestinal tract to mononucleotides, which may be absorbed or converted to purine and pyrimidine bases. The purine bases are then oxidized to uric acid, which may be absorbed and excreted in the urine. While little or no dietary purine or pyrimidine is incorporated into tissue nucleic acids, injected compounds are incorporated. The incorporation of injected [$^3$H]thymidine into newly synthesized DNA thus is used to measure the rate of DNA synthesis.

## BIOSYNTHESIS OF PURINE NUCLEOTIDES

Purine and pyrimidine nucleotides are synthesized in vivo at rates consistent with physiologic need. Intracellular mechanisms sense and regulate the pool sizes of nucleotide triphosphates (NTPs), which rise during growth or tissue regeneration when cells are rapidly dividing. Early investigations of nucleotide biosynthesis employed birds, and later ones used *Escherichia coli.* Isotopic precursors fed to pigeons established the source of each atom of a purine base (Figure 34–1) and initiated study of the intermediates of purine biosynthesis.

Three processes contribute to purine nucleotide biosynthesis. These are, in order of decreasing importance: (1) synthesis from amphibolic intermediates (synthesis de novo), (2) phosphoribosylation of purines, and (3) phosphorylation of purine nucleosides.

## INOSINE MONOPHOSPHATE (IMP) IS SYNTHESIZED FROM AMPHIBOLIC INTERMEDIATES

Figure 34–2 illustrates the intermediates and reactions for conversion of α-D-ribose 5-phosphate to inosine monophosphate (IMP). Separate branches then lead to AMP and GMP (Figure 34–3). Subsequent phosphoryl transfer from ATP converts AMP and GMP to ADP and GDP. Conversion of GDP to GTP involves a second phosphoryl transfer from ATP, whereas conversion of ADP to ATP is achieved primarily by oxidative phosphorylation (see Chapter 12).

### Multifunctional Catalysts Participate in Purine Nucleotide Biosynthesis

In prokaryotes, each reaction of Figure 34–2 is catalyzed by a different polypeptide. By contrast, in eukaryotes, the enzymes are polypeptides with multiple catalytic activities whose adjacent catalytic sites facilitate channeling of intermediates between sites. Three distinct multifunctional enzymes catalyze reactions 3, 4, and 6, reactions 7 and 8, and reactions 10 and 11 of Figure 34–2.

### Antifolate Drugs or Glutamine Analogs Block Purine Nucleotide Biosynthesis

The carbons added in reactions 4 and 5 of Figure 34–2 are contributed by derivatives of tetrahydrofolate. Purine deficiency states, which are rare in humans, generally reflect a deficiency of folic acid. Compounds that inhibit formation of tetrahydrofolates and therefore block purine synthesis have been used in cancer chemotherapy. Inhibitory compounds and the reactions they inhibit include azaserine (reaction 5, Figure 34–2), diazanorleucine (reaction 2), 6-mercaptopurine (reactions 13 and 14), and mycophenolic acid (reaction 14).

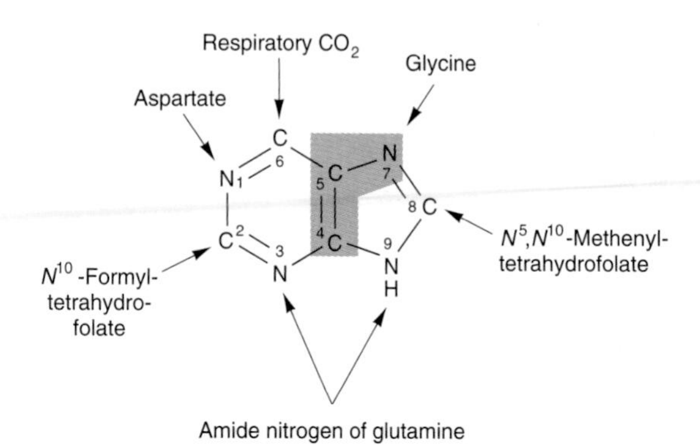


***Figure 34–1.*** Sources of the nitrogen and carbon atoms of the purine ring. Atoms 4, 5, and 7 (shaded) derive from glycine.

## "SALVAGE REACTIONS" CONVERT PURINES & THEIR NUCLEOSIDES TO MONONUCLEOTIDES

Conversion of purines, their ribonucleo**sides**, and their deoxyribonucleo**sides** to mononucleo**tides** involves so-called "salvage reactions" that require far less energy than de novo synthesis. The more important mechanism involves phosphoribosylation by PRPP (structure II, Figure 34–2) of a free purine (Pu) to form a purine 5′-mononucleotide (Pu-RP).

$$\text{Pu} + \text{PR} - \text{PP} \rightarrow \text{PRP} + \text{PP}_i$$

Two phosphoribosyl transferases then convert adenine to AMP and hypoxanthine and guanine to IMP or GMP (Figure 34–4). A second salvage mechanism involves phosphoryl transfer from ATP to a purine ribonucleo**side** (PuR):

$$\text{PuR} + \text{ATP} \rightarrow \text{PuR} - \text{P} + \text{ADP}$$

Adenosine kinase catalyzes phosphorylation of adenosine and deoxyadenosine to AMP and dAMP, and deoxycytidine kinase phosphorylates deoxycytidine and 2′-deoxyguanosine to dCMP and dGMP.

Liver, the major site of purine nucleotide biosynthesis, provides purines and purine nucleosides for salvage and utilization by tissues incapable of their biosynthesis. For example, human brain has a low level of PRPP amidotransferase (reaction 2, Figure 34–2) and hence depends in part on exogenous purines. Erythrocytes and polymorphonuclear leukocytes cannot synthesize 5-phosphoribosylamine (structure III, Figure 34–2) and therefore utilize exogenous purines to form nucleotides.

### AMP & GMP Feedback-Regulate PRPP Glutamyl Amidotransferase

Since biosynthesis of IMP consumes glycine, glutamine, tetrahydrofolate derivatives, aspartate, and ATP, it is advantageous to regulate purine biosynthesis. The major determinant of the rate of de novo purine nucleotide biosynthesis is the concentration of PRPP, whose pool size depends on its rates of synthesis, utilization, and degradation. The rate of PRPP synthesis depends on the availability of ribose 5-phosphate and on the activity of PRPP synthase, an enzyme sensitive to feedback inhibition by AMP, ADP, GMP, and GDP.

### AMP & GMP Feedback-Regulate Their Formation From IMP

Two mechanisms regulate conversion of IMP to GMP and AMP. AMP and GMP feedback-inhibit adenylosuccinate synthase and IMP dehydrogenase (reactions 12 and 14, Figure 34–3), respectively. Furthermore, conversion of IMP to adenylosuccinate en route to AMP requires GTP, and conversion of xanthinylate (XMP) to GMP requires ATP. This cross-regulation between the pathways of IMP metabolism thus serves to decrease synthesis of one purine nucleotide when there is a deficiency of the other nucleotide. AMP and GMP also inhibit hypoxanthine-guanine phosphoribosyltransferase, which converts hypoxanthine and guanine to IMP and GMP (Figure 34–4), and GMP feedback-inhibits PRPP glutamyl amidotransferase (reaction 2, Figure 34–2).

## REDUCTION OF RIBONUCLEOSIDE DIPHOSPHATES FORMS DEOXYRIBONUCLEOSIDE DIPHOSPHATES

Reduction of the 2′-hydroxyl of purine and pyrimidine ribonucleotides, catalyzed by the **ribonucleotide reductase complex** (Figure 34–5), forms deoxyribonucleoside diphosphates (dNDPs). The enzyme complex is active only when cells are actively synthesizing DNA. Reduction requires thioredoxin, thioredoxin reductase, and NADPH. The immediate reductant, reduced thioredoxin, is produced by NADPH:thioredoxin reductase (Figure 34–5). Reduction of ribonucleoside diphosphates (NDPs) to deoxyribonucleoside diphosphates (dNDPs) is subject to complex regulatory controls that achieve balanced production of deoxyribonucleotides for synthesis of DNA (Figure 34–6).

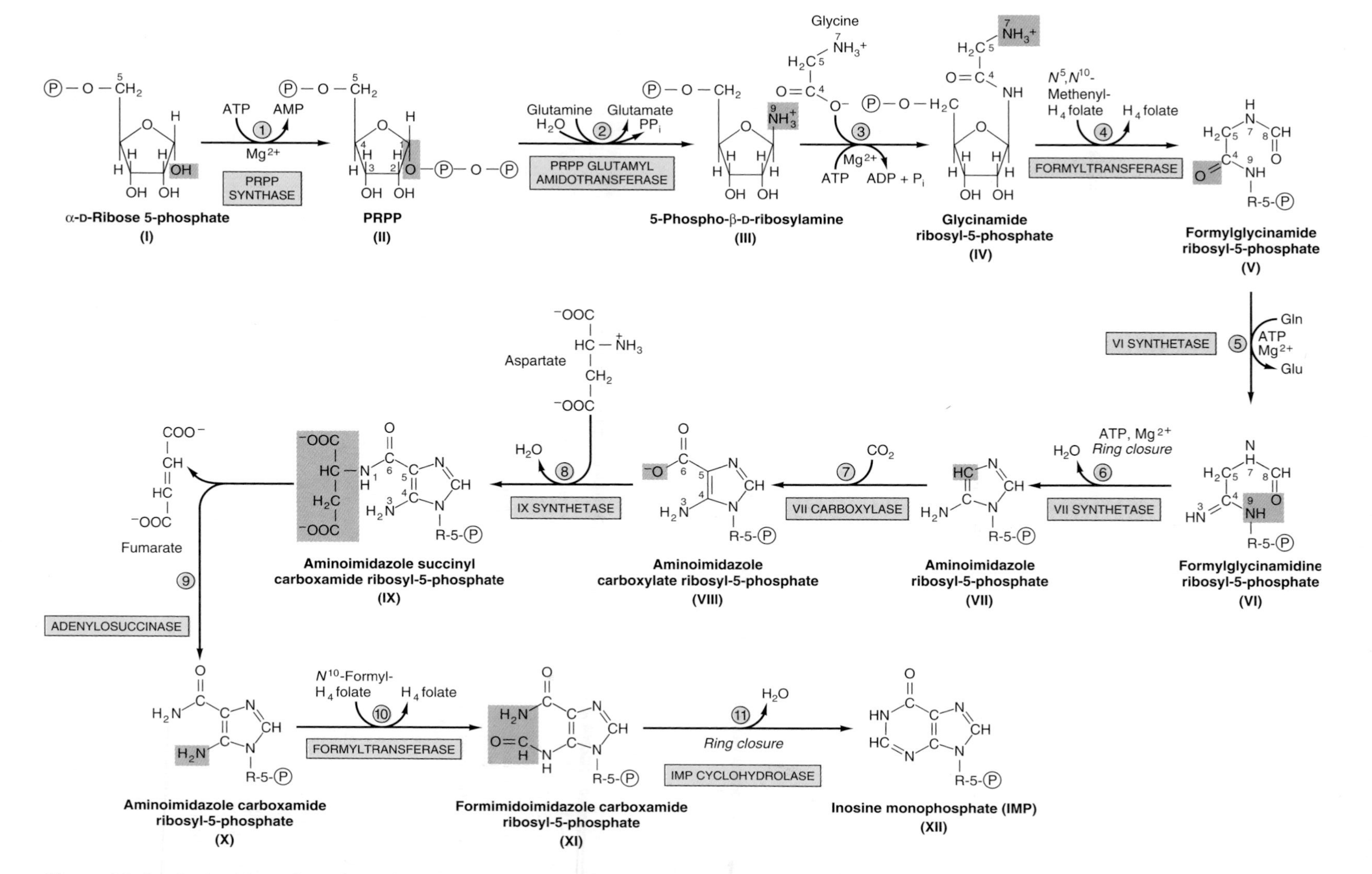


***Figure 34–2.*** Purine biosynthesis from ribose 5-phosphate and ATP. See text for explanations. (Ⓟ, $PO_3^{2-}$ or $PO_2^-$.)

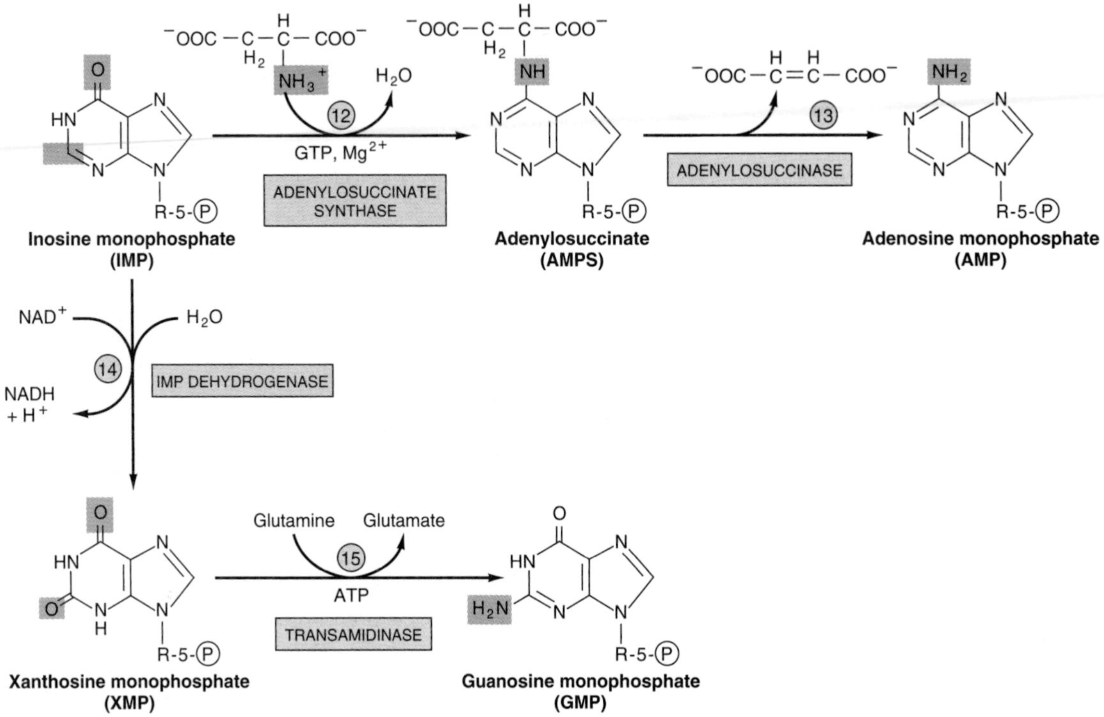

***Figure 34–3.*** Conversion of IMP to AMP and GMP.

## BIOSYNTHESIS OF PYRIMIDINE NUCLEOTIDES

Figure 34–7 summarizes the roles of the intermediates and enzymes of pyrimidine nucleotide biosynthesis. The catalyst for the initial reaction is *cytosolic* carbamoyl phosphate synthase II, a different enzyme from the *mitochondrial* carbamoyl phosphate synthase I of urea synthesis (Figure 29–9). Compartmentation thus provides two independent pools of carbamoyl phosphate. PRPP, an early participant in purine nucleotide synthesis (Figure 34–2), is a much later participant in pyrimidine biosynthesis.

### Multifunctional Proteins Catalyze the Early Reactions of Pyrimidine Biosynthesis

Five of the first six enzyme activities of pyrimidine biosynthesis reside on multifunctional polypeptides. One such polypeptide catalyzes the first three reactions of Figure 34–2 and ensures efficient channeling of carbamoyl phosphate to pyrimidine biosynthesis. A second bifunctional enzyme catalyzes reactions 5 and 6.

## THE DEOXYRIBONUCLEOSIDES OF URACIL & CYTOSINE ARE SALVAGED

While mammalian cells reutilize few free pyrimidines, "salvage reactions" convert the ribonucleosides uridine and cytidine and the deoxyribonucleosides thymidine and deoxycytidine to their respective nucleotides. ATP-dependent phosphoryltransferases (kinases) catalyze the phosphorylation of the nucleoside diphosphates 2′-deoxycytidine, 2′-deoxyguanosine, and 2′-deoxyadenosine to their corresponding nucleoside triphosphates. In addition, orotate phosphoribosyltransferase (reaction 5, Figure 34–7), an enzyme of pyrimidine nucleotide synthesis, salvages orotic acid by converting it to orotidine monophosphate (OMP).

### Methotrexate Blocks Reduction of Dihydrofolate

Reaction 12 of Figure 34–7 is the only reaction of pyrimidine nucleotide biosynthesis that requires a tetrahydrofolate derivative. The methylene group of $N^5$,$N^{10}$-methylene-tetrahydrofolate is reduced to the methyl group that is transferred, and tetrahydrofolate is oxidized to dihydro-

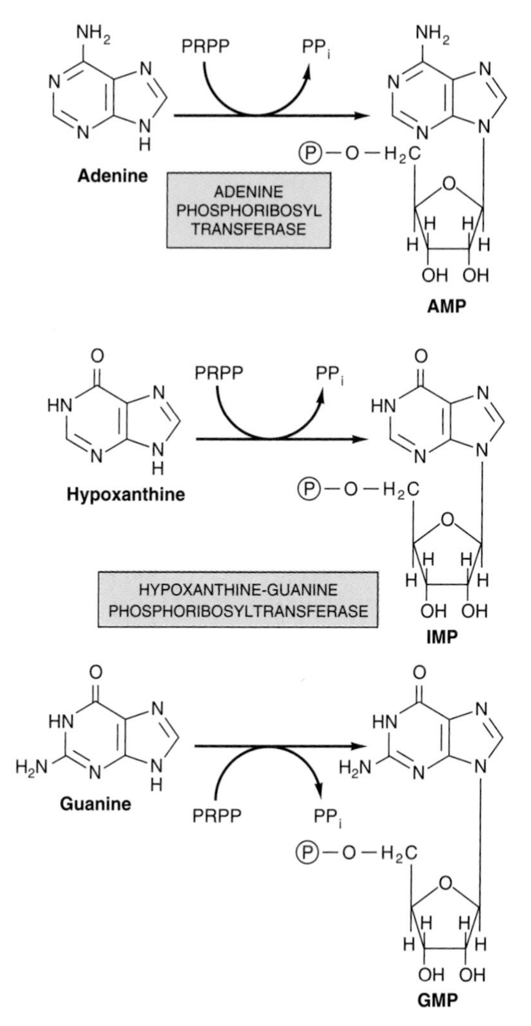


***Figure 34–4.*** Phosphoribosylation of adenine, hypoxanthine, and guanine to form AMP, IMP, and GMP, respectively.

folate. For further pyrimidine synthesis to occur, dihydrofolate must be reduced back to tetrahydrofolate, a reaction catalyzed by dihydrofolate reductase. Dividing cells, which must generate TMP and dihydrofolate, thus are especially sensitive to inhibitors of dihydrofolate reductase such as the anticancer drug **methotrexate.**

## Certain Pyrimidine Analogs Are Substrates for Enzymes of Pyrimidine Nucleotide Biosynthesis

Orotate phosphoribosyltransferase (reaction 5, Figure 34–7) converts the drug **allopurinol** (Figure 33–12) to

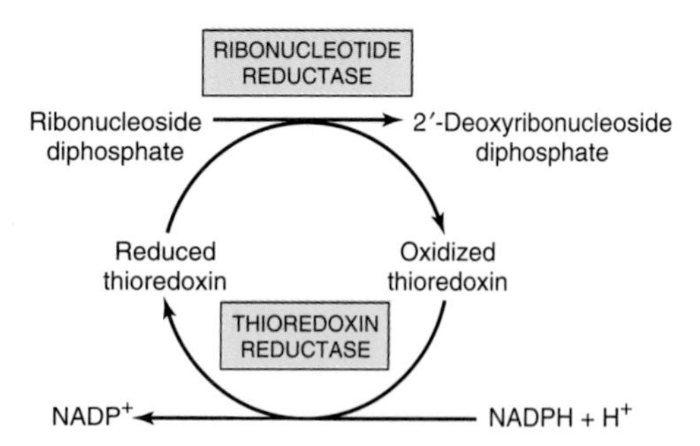


***Figure 34–5.*** Reduction of ribonucleoside diphosphates to 2′-deoxyribonucleoside diphosphates.

a nucleotide in which the ribosyl phosphate is attached to N-1 of the pyrimidine ring. The anticancer drug **5-fluorouracil** (Figure 33–12) is also phosphoribosylated by orotate phosphoribosyl transferase.

# REGULATION OF PYRIMIDINE NUCLEOTIDE BIOSYNTHESIS

## Gene Expression & Enzyme Activity Both Are Regulated

The activities of the first and second enzymes of pyrimidine nucleotide biosynthesis are controlled by allosteric

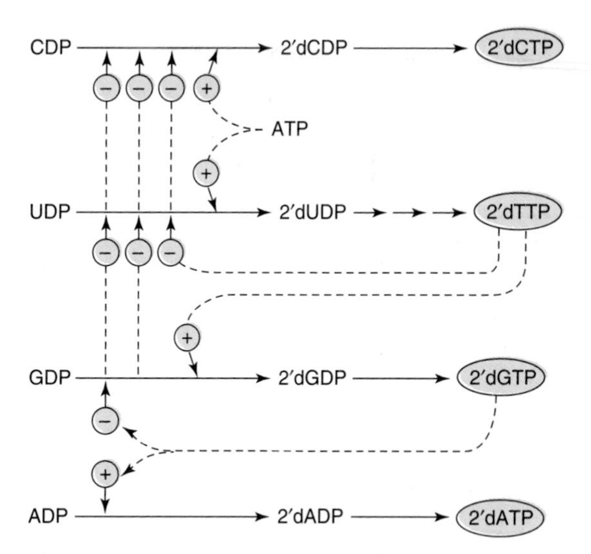


***Figure 34–6.*** Regulation of the reduction of purine and pyrimidine ribonucleotides to their respective 2′-deoxyribonucleotides. Solid lines represent chemical flow. Broken lines show negative (⊖) or positive (⊕) feedback regulation.

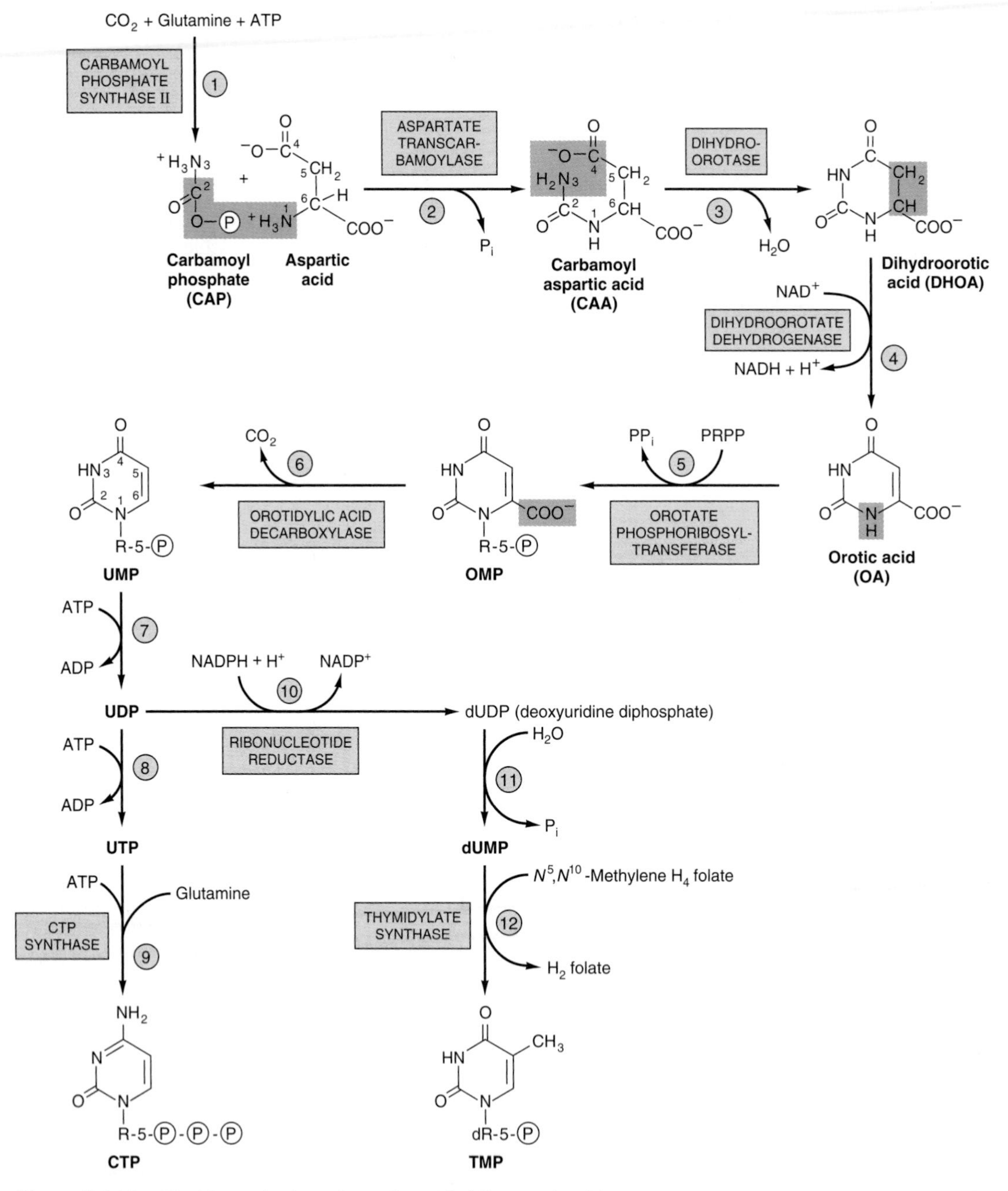


***Figure 34–7.*** The biosynthetic pathway for pyrimidine nucleotides.

regulation. Carbamoyl phosphate synthase II (reaction 1, Figure 34–7) is inhibited by UTP and purine nucleotides but activated by PRPP. Aspartate transcarbamoylase (reaction 2, Figure 34–7) is inhibited by CTP but activated by ATP. In addition, the first three and the last two enzymes of the pathway are regulated by coordinate repression and derepression.

### Purine & Pyrimidine Nucleotide Biosynthesis Are Coordinately Regulated

Purine and pyrimidine biosynthesis parallel one another mole for mole, suggesting coordinated control of their biosynthesis. Several sites of cross-regulation characterize purine and pyrimidine nucleotide biosynthesis. The PRPP synthase reaction (reaction 1, Figure 34–2), which forms a precursor essential for both processes, is feedback-inhibited by both purine and pyrimidine nucleotides.

## HUMANS CATABOLIZE PURINES TO URIC ACID

Humans convert adenosine and guanosine to uric acid (Figure 34–8). Adenosine is first converted to inosine by adenosine deaminase. In mammals other than higher primates, uricase converts uric acid to the water-soluble product allantoin. However, since humans lack uricase, the end product of purine catabolism in humans is uric acid.

## GOUT IS A METABOLIC DISORDER OF PURINE CATABOLISM

Various genetic defects in PRPP synthetase (reaction 1, Figure 34–2) present clinically as gout. Each defect—eg, an elevated $V_{max}$, increased affinity for ribose 5-phosphate, or resistance to feedback inhibition—results in overproduction and overexcretion of purine catabolites. When serum urate levels exceed the solubility limit, sodium urate crystalizes in soft tissues and joints and causes an inflammatory reaction, **gouty arthritis.** However, most cases of gout reflect abnormalities in renal handling of uric acid.

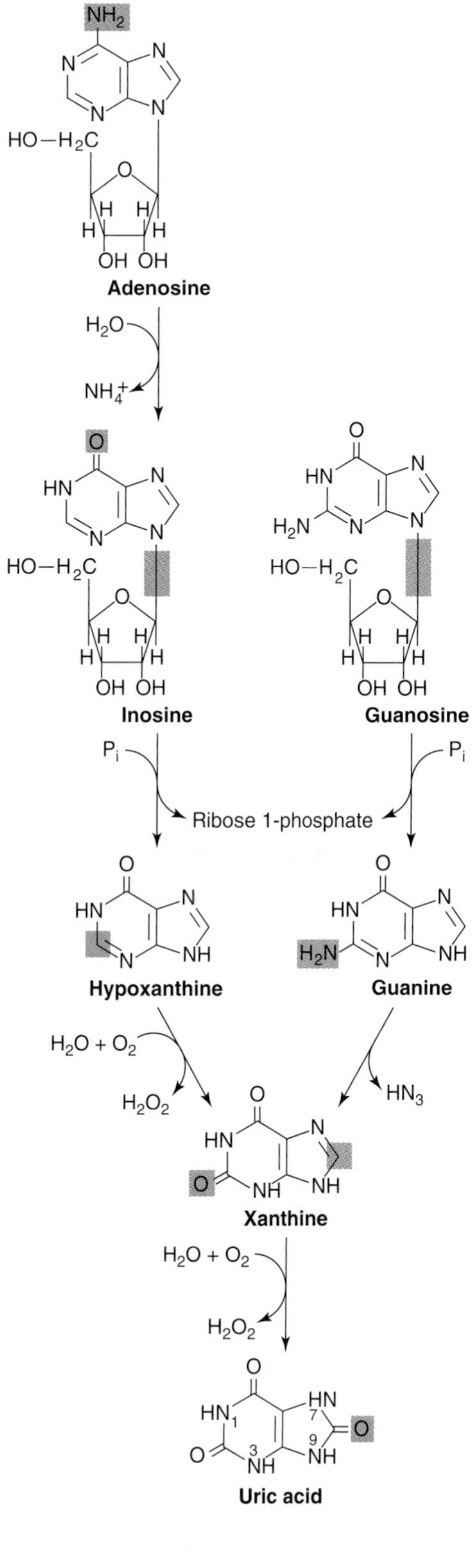

***Figure 34–8.*** Formation of uric acid from purine nucleosides by way of the purine bases hypoxanthine, xanthine, and guanine. Purine deoxyribonucleosides are degraded by the same catabolic pathway and enzymes, all of which exist in the mucosa of the mammalian gastrointestinal tract.

## OTHER DISORDERS OF PURINE CATABOLISM

While purine deficiency states are rare in human subjects, there are numerous genetic disorders of purine catabolism. **Hyperuricemias** may be differentiated based on whether patients excrete normal or excessive quantities of total urates. Some hyperuricemias reflect specific enzyme defects. Others are secondary to diseases such as cancer or psoriasis that enhance tissue turnover.

### Lesch-Nyhan Syndrome

Lesch-Nyhan syndrome, an overproduction hyperuricemia characterized by frequent episodes of uric acid lithiasis and a bizarre syndrome of self-mutilation, reflects a defect in **hypoxanthine-guanine phosphoribosyl transferase,** an enzyme of purine salvage (Figure 34–4). The accompanying rise in intracellular PRPP results in purine overproduction. Mutations that decrease or abolish hypoxanthine-guanine phosphoribosyltransferase activity include deletions, frameshift mutations, base substitutions, and aberrant mRNA splicing.

### Von Gierke's Disease

Purine overproduction and hyperuricemia in von Gierke's disease **(glucose-6-phosphatase deficiency)** occurs secondary to enhanced generation of the PRPP precursor ribose 5-phosphate. An associated lactic acidosis elevates the renal threshold for urate, elevating total body urates.

### Hypouricemia

Hypouricemia and increased excretion of hypoxanthine and xanthine are associated with **xanthine oxidase deficiency** due to a genetic defect or to severe liver damage. Patients with a severe enzyme deficiency may exhibit xanthinuria and xanthine lithiasis.

### Adenosine Deaminase & Purine Nucleoside Phosphorylase Deficiency

**Adenosine deaminase deficiency** is associated with an immunodeficiency disease in which both thymus-derived lymphocytes (T cells) and bone marrow-derived lymphocytes (B cells) are sparse and dysfunctional. **Purine nucleoside phosphorylase deficiency** is associated with a severe deficiency of T cells but apparently normal B cell function. Immune dysfunctions appear to result from accumulation of dGTP and dATP, which inhibit ribonucleotide reductase and thereby deplete cells of DNA precursors.

## CATABOLISM OF PYRIMIDINES PRODUCES WATER-SOLUBLE METABOLITES

Unlike the end products of purine catabolism, those of pyrimidine catabolism are highly water-soluble: $CO_2$, $NH_3$, β-alanine, and β-aminoisobutyrate (Figure 34–9). Excretion of β-aminoisobutyrate increases in leukemia and severe x-ray radiation exposure due to increased destruction of DNA. However, many persons of Chinese or Japanese ancestry routinely excrete β-aminoisobutyrate. Humans probably transaminate β-aminoisobutyrate to methylmalonate semialdehyde, which then forms succinyl-CoA (Figure 19–2).

### Pseudouridine Is Excreted Unchanged

Since no human enzyme catalyzes hydrolysis or phosphorolysis of pseudouridine, this unusual nucleoside is excreted unchanged in the urine of normal subjects.

## OVERPRODUCTION OF PYRIMIDINE CATABOLITES IS ONLY RARELY ASSOCIATED WITH CLINICALLY SIGNIFICANT ABNORMALITIES

Since the end products of pyrimidine catabolism are highly water-soluble, pyrimidine overproduction results in few clinical signs or symptoms. In hyperuricemia associated with severe overproduction of PRPP, there is overproduction of pyrimidine nucleotides and increased excretion of β-alanine. Since $N^5,N^{10}$-methylene-tetrahydrofolate is required for thymidylate synthesis, disorders of folate and vitamin $B_{12}$ metabolism result in deficiencies of TMP.

### Orotic Acidurias

The orotic aciduria that accompanies **Reye's syndrome** probably is a consequence of the inability of severely damaged mitochondria to utilize carbamoyl phosphate, which then becomes available for cytosolic overproduction of orotic acid. **Type I orotic aciduria** reflects a deficiency of both orotate phosphoribosyltransferase and orotidylate decarboxylase (reactions 5 and 6, Figure 34–7); the rarer **type II orotic aciduria** is due to a deficiency only of orotidylate decarboxylase (reaction 6, Figure 34–7).

### Deficiency of a Urea Cycle Enzyme Results in Excretion of Pyrimidine Precursors

Increased excretion of orotic acid, uracil, and uridine accompanies a deficiency in liver mitochondrial ornithine transcarbamoylase (reaction 2, Figure 29–9).

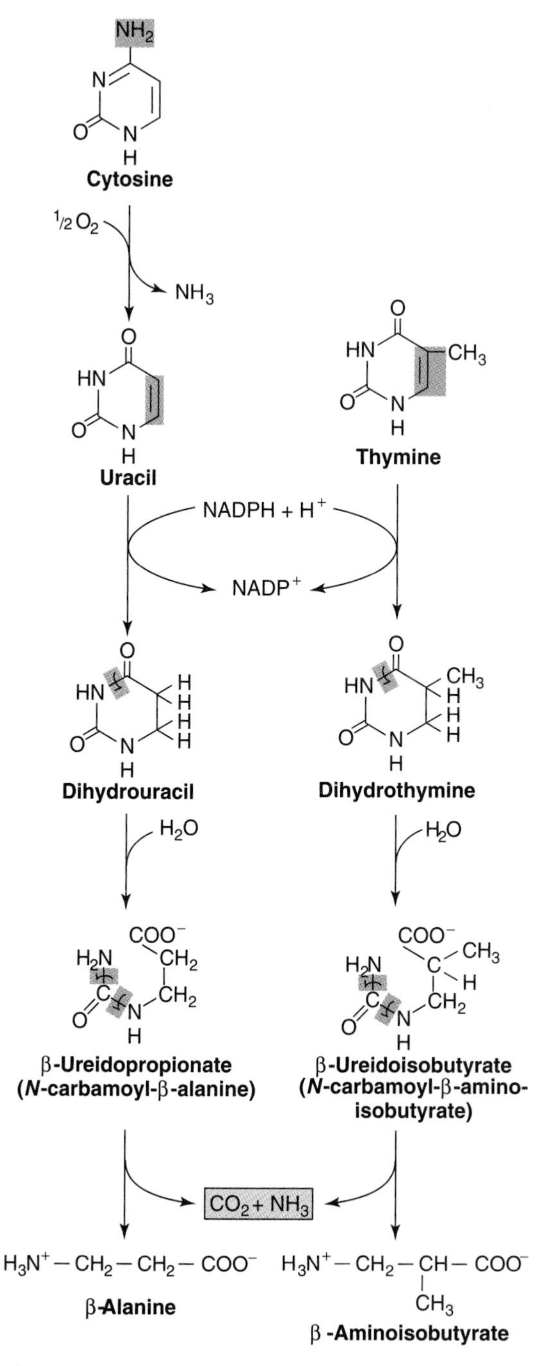


***Figure 34–9.*** Catabolism of pyrimidines.

Excess carbamoyl phosphate exits to the cytosol, where it stimulates pyrimidine nucleotide biosynthesis. The resulting mild **orotic aciduria** is increased by high-nitrogen foods.

## Drugs May Precipitate Orotic Aciduria

Allopurinol (Figure 33–12), an alternative substrate for orotate phosphoribosyltransferase (reaction 5, Figure 34–7), competes with orotic acid. The resulting nucleotide product also inhibits orotidylate decarboxylase (reaction 6, Figure 34–7), resulting in **orotic aciduria** and **orotidinuria.** 6-Azauridine, following conversion to 6-azauridylate, also competitively inhibits orotidylate decarboxylase (reaction 6, Figure 34–7), enhancing excretion of orotic acid and orotidine.

## SUMMARY


- Ingested nucleic acids are degraded to purines and pyrimidines. New purines and pyrimidines are formed from amphibolic intermediates and thus are dietarily nonessential.
- Several reactions of IMP biosynthesis require folate derivatives and glutamine. Consequently, antifolate drugs and glutamine analogs inhibit purine biosynthesis.
- Oxidation and amination of IMP forms AMP and GMP, and subsequent phosphoryl transfer from ATP forms ADP and GDP. Further phosphoryl transfer from ATP to GDP forms GTP. ADP is converted to ATP by oxidative phosphorylation. Reduction of NDPs forms dNDPs.
- Hepatic purine nucleotide biosynthesis is stringently regulated by the pool size of PRPP and by feedback inhibition of PRPP-glutamyl amidotransferase by AMP and GMP.
- Coordinated regulation of purine and pyrimidine nucleotide biosynthesis ensures their presence in proportions appropriate for nucleic acid biosynthesis and other metabolic needs.
- Humans catabolize purines to uric acid (p$K_a$ 5.8), present as the relatively insoluble acid at acidic pH or as its more soluble sodium urate salt at a pH near neutrality. Urate crystals are diagnostic of gout. Other disorders of purine catabolism include Lesch-Nyhan syndrome, von Gierke's disease, and hypouricemias.
- Since pyrimidine catabolites are water-soluble, their overproduction does not result in clinical abnormalities. Excretion of pyrimidine precursors can, however, result from a deficiency of ornithine transcarbamoylase because excess carbamoyl phosphate is available for pyrimidine biosynthesis.

## REFERENCES

Benkovic SJ: The transformylase enzymes in de novo purine biosynthesis. Trends Biochem Sci 1994;9:320.

Brooks EM et al: Molecular description of three macro-deletions and an Alu-Alu recombination-mediated duplication in the HPRT gene in four patients with Lesch-Nyhan disease. Mutat Res 2001;476:43.

Curto R, Voit EO, Cascante M: Analysis of abnormalities in purine metabolism leading to gout and to neurological dysfunctions in man. Biochem J 1998;329:477.

Harris MD, Siegel LB, Alloway JA: Gout and hyperuricemia. Am Family Physician 1999;59:925.

Lipkowitz MS et al: Functional reconstitution, membrane targeting, genomic structure, and chromosomal localization of a human urate transporter. J Clin Invest 2001;107:1103.

Martinez J et al: Human genetic disorders, a phylogenetic perspective. J Mol Biol 2001;308:587.

Puig JG et al: Gout: new questions for an ancient disease. Adv Exp Med Biol 1998;431:1.

Scriver CR et al (editors): *The Metabolic and Molecular Bases of Inherited Disease,* 8th ed. McGraw-Hill, 2001.

Tvrdik T et al: Molecular characterization of two deletion events involving Alu-sequences, one novel base substitution and two tentative hotspot mutations in the hypoxanthine phosphoribosyltransferase gene in five patients with Lesch-Nyhan-syndrome. Hum Genet 1998;103:311.

Zalkin H, Dixon JE: De novo purine nucleotide synthesis. Prog Nucleic Acid Res Mol Biol 1992;42:259.

# Nucleic Acid Structure & Function 35

Daryl K. Granner, MD

## BIOMEDICAL IMPORTANCE

The discovery that genetic information is coded along the length of a polymeric molecule composed of only four types of monomeric units was one of the major scientific achievements of the twentieth century. This polymeric molecule, **DNA,** is the chemical basis of heredity and is organized into genes, the fundamental units of genetic information. The basic information pathway—ie, DNA directs the synthesis of RNA, which in turn directs protein synthesis—has been elucidated. Genes do not function autonomously; their replication and function are controlled by various gene products, often in collaboration with components of various signal transduction pathways. Knowledge of the structure and function of nucleic acids is essential in understanding genetics and many aspects of pathophysiology as well as the genetic basis of disease.

## DNA CONTAINS THE GENETIC INFORMATION

The demonstration that DNA contained the genetic information was first made in 1944 in a series of experiments by Avery, MacLeod, and McCarty. They showed that the genetic determination of the character (type) of the capsule of a specific pneumococcus could be transmitted to another of a different capsular type by introducing purified DNA from the former coccus into the latter. These authors referred to the agent (later shown to be DNA) accomplishing the change as "transforming factor." Subsequently, this type of genetic manipulation has become commonplace. Similar experiments have recently been performed utilizing yeast, cultured mammalian cells, and insect and mammalian embryos as recipients and cloned DNA as the donor of genetic information.

### DNA Contains Four Deoxynucleotides

The chemical nature of the monomeric deoxynucleotide units of DNA—**deoxyadenylate, deoxyguanylate, deoxycytidylate,** and **thymidylate**—is described in Chapter 33. These monomeric units of DNA are held in polymeric form by 3′,5′-phosphodiester bridges constituting a single strand, as depicted in Figure 35–1. The informational content of DNA (the genetic code) resides in the sequence in which these monomers—purine and pyrimidine deoxyribonucleotides—are ordered. The polymer as depicted possesses a polarity; one end has a 5′-hydroxyl or phosphate terminal while the other has a 3′-phosphate or hydroxyl terminal. The importance of this polarity will become evident. Since the genetic information resides in the order of the monomeric units within the polymers, there must exist a mechanism of reproducing or replicating this specific information with a high degree of fidelity. That requirement, together with x-ray diffraction data from the DNA molecule and the observation of Chargaff that in DNA molecules the concentration of deoxyadenosine (A) nucleotides equals that of thymidine (T) nucleotides (A = T), while the concentration of deoxyguanosine (G) nucleotides equals that of deoxycytidine (C) nucleotides (G = C), led Watson, Crick, and Wilkins to propose in the early 1950s a model of a double-stranded DNA molecule. The model they proposed is depicted in Figure 35–2. The two strands of this double-stranded helix are held in register by **hydrogen bonds** between the purine and pyrimidine bases of the respective linear molecules. The pairings between the purine and pyrimidine nucleotides on the opposite strands are very specific and are dependent upon hydrogen bonding of **A with T** and **G with C** (Figure 35–3).

This common form of DNA is said to be right-handed because as one looks down the double helix the base residues form a spiral in a clockwise direction. In the double-stranded molecule, restrictions imposed by the rotation about the phosphodiester bond, the favored anti configuration of the glycosidic bond (Figure 33–8), and the predominant tautomers (see Figure 33–3) of the four bases (A, G, T, and C) allow A to pair only with T and G only with C, as depicted in Figure 35–3. This base-pairing restriction explains the earlier observation that in a double-stranded DNA molecule the content of A equals that of T and the content of G equals that of C. The two strands of the double-helical molecule, each of which possesses a polarity, are **antiparallel;** ie, one strand runs in the 5′ to 3′ direction and the other in the 3′ to 5′ direction. This is analogous to two parallel streets, each running one way but carrying traffic in opposite directions. In the double-stranded DNA molecules, the genetic information re-

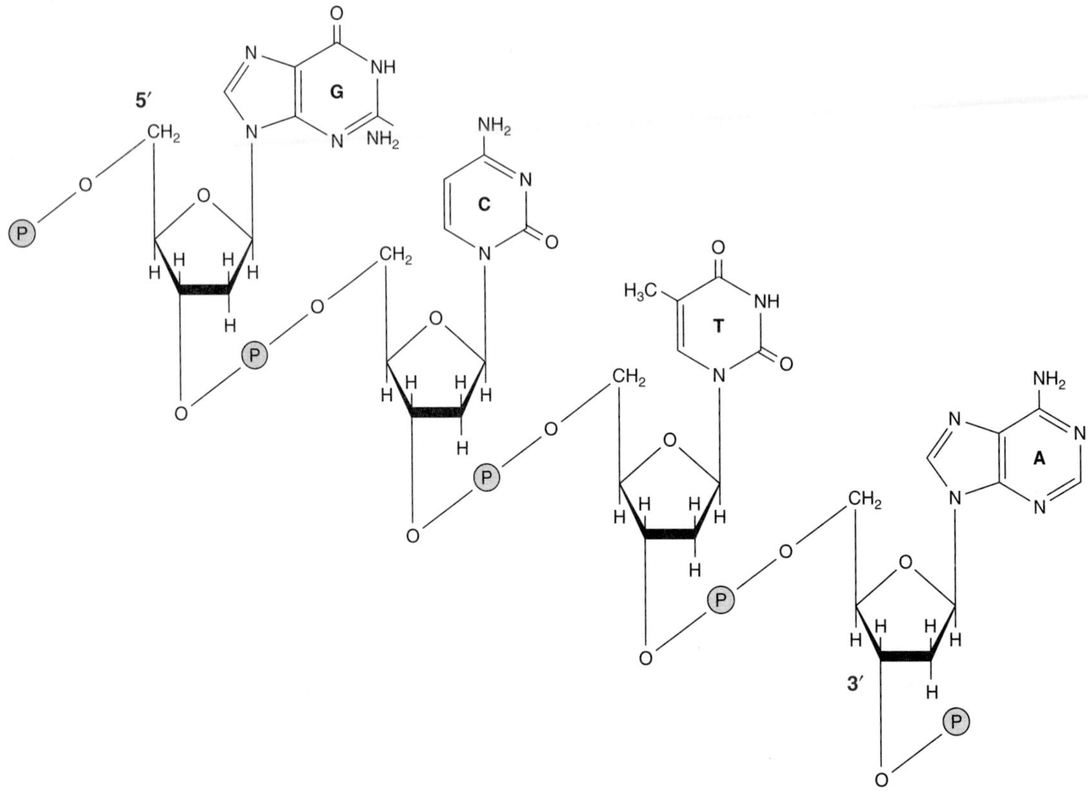

***Figure 35–1.*** A segment of one strand of a DNA molecule in which the purine and pyrimidine bases guanine (G), cytosine (C), thymine (T), and adenine (A) are held together by a phosphodiester backbone between 2′-deoxyribosyl moieties attached to the nucleobases by an *N*-glycosidic bond. Note that the backbone has a polarity (ie, a direction). Convention dictates that a single-stranded DNA sequence is written in the 5′ to 3′ direction (ie, pGpCpTpA, where G, C, T, and A represent the four bases and p represents the interconnecting phosphates).

sides in the sequence of nucleotides on one strand, the **template strand.** This is the strand of DNA that is copied during nucleic acid synthesis. It is sometimes referred to as the **noncoding strand.** The opposite strand is considered the **coding strand** because it matches the RNA transcript that encodes the protein.

The two strands, in which opposing bases are held together by hydrogen bonds, wind around a central axis in the form of a **double helix.** Double-stranded DNA exists in at least six forms (A–E and Z). The B form is usually found under physiologic conditions (low salt, high degree of hydration). A single turn of B-DNA about the axis of the molecule contains ten base pairs. The distance spanned by one turn of B-DNA is 3.4 nm. The width (helical diameter) of the double helix in B-DNA is 2 nm.

As depicted in Figure 35–3, three hydrogen bonds hold the deoxyguanosine nucleotide to the deoxycytidine nucleotide, whereas the other pair, the A–T pair, is held together by two hydrogen bonds. Thus, the G–C bonds are much more resistant to denaturation, or "melting," than A–T-rich regions.

## The Denaturation (Melting) of DNA Is Used to Analyze Its Structure

The double-stranded structure of DNA can be separated into two component strands (melted) in solution by increasing the temperature or decreasing the salt concentration. Not only do the two stacks of bases pull apart but the bases themselves unstack while still connected in the polymer by the phosphodiester backbone. Concomitant with this denaturation of the DNA molecule is an increase in the optical absorbance of the purine and pyrimidine bases—a phenomenon referred to as **hyperchromicity** of denaturation. Because of the

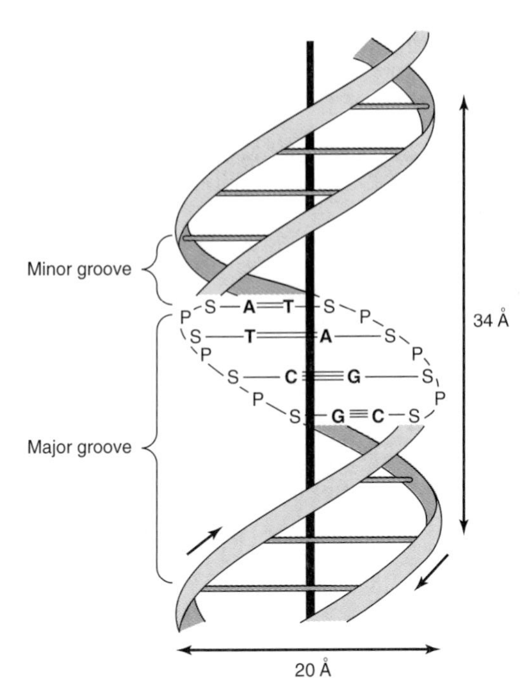


***Figure 35–2.*** A diagrammatic representation of the Watson and Crick model of the double-helical structure of the B form of DNA. The horizontal arrow indicates the width of the double helix (20 Å), and the vertical arrow indicates the distance spanned by one complete turn of the double helix (34 Å). One turn of B-DNA includes ten base pairs (bp), so the rise is 3.4 Å per bp. The central axis of the double helix is indicated by the vertical rod. The short arrows designate the polarity of the antiparallel strands. The major and minor grooves are depicted. (A, adenine; C, cytosine; G, guanine; T, thymine; P, phosphate; S, sugar [deoxyribose].)

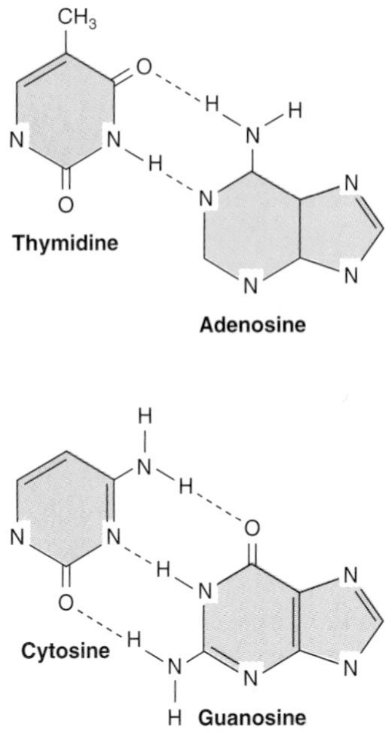

stacking of the bases and the hydrogen bonding between the stacks, the double-stranded DNA molecule exhibits properties of a rigid rod and in solution is a viscous material that loses its viscosity upon denaturation.

The strands of a given molecule of DNA separate over a temperature range. The midpoint is called the **melting temperature, or $T_m$**. The $T_m$ is influenced by the base composition of the DNA and by the salt concentration of the solution. DNA rich in G–C pairs, which have three hydrogen bonds, melts at a higher temperature than that rich in A–T pairs, which have two hydrogen bonds. A tenfold increase of monovalent cation concentration increases the $T_m$ by 16.6 °C. Formamide, which is commonly used in recombinant DNA experiments, destabilizes hydrogen bonding between bases, thereby lowering the $T_m$. This allows the strands of DNA

***Figure 35–3.*** Base pairing between deoxyadenosine and thymidine involves the formation of two hydrogen bonds. Three such bonds form between deoxycytidine and deoxyguanosine. The broken lines represent hydrogen bonds.

or DNA-RNA hybrids to be separated at much lower temperatures and minimizes the phosphodiester bond breakage that occurs at high temperatures.

## Renaturation of DNA Requires Base Pair Matching

Separated strands of DNA will renature or reassociate when appropriate physiologic temperature and salt conditions are achieved. The rate of reassociation depends upon the concentration of the complementary strands. Reassociation of the two complementary DNA strands of a chromosome after DNA replication is a physiologic example of renaturation (see below). At a given temperature and salt concentration, a particular nucleic acid strand will associate tightly only with a complementary strand. Hybrid molecules will also form under appropriate conditions. For example, DNA will form a hybrid with a complementary DNA (cDNA) or with a cognate messenger RNA (mRNA; see below). When combined with gel electrophoresis techniques that separate hybrid molecules by size and radioactive labeling to provide a detectable signal, the resulting analytic techniques are called **Southern (DNA/cDNA)** and **Northern blotting (DNA/RNA)**, respectively. These procedures allow for very specific identification of hybrids from mixtures of DNA or RNA (see Chapter 40).

### There Are Grooves in the DNA Molecule

Careful examination of the model depicted in Figure 35–2 reveals a **major groove** and a **minor groove** winding along the molecule parallel to the phosphodiester backbones. In these grooves, proteins can interact specifically with exposed atoms of the nucleotides (usually by H bonding) and thereby recognize and bind to specific nucleotide sequences without disrupting the base pairing of the double-helical DNA molecule. As discussed in Chapters 37 and 39, regulatory proteins control the expression of specific genes via such interactions.

### DNA Exists in Relaxed & Supercoiled Forms

In some organisms such as bacteria, bacteriophages, and many DNA-containing animal viruses, the ends of the DNA molecules are joined to create a closed circle with no covalently free ends. This of course does not destroy the polarity of the molecules, but it eliminates all free 3′ and 5′ hydroxyl and phosphoryl groups. Closed circles exist in relaxed or supercoiled forms. Supercoils are introduced when a closed circle is twisted around its own axis or when a linear piece of duplex DNA, whose ends are fixed, is twisted. This energy-requiring process puts the molecule under stress, and the greater the number of supercoils, the greater the stress or torsion (test this by twisting a rubber band). **Negative supercoils** are formed when the molecule is twisted in the direction opposite from the clockwise turns of the right-handed double helix found in B-DNA. Such DNA is said to be underwound. The energy required to achieve this state is, in a sense, stored in the supercoils. The transition to another form that requires energy is thereby facilitated by the underwinding. One such transition is strand separation, which is a prerequisite for DNA replication and transcription. Supercoiled DNA is therefore a preferred form in biologic systems. Enzymes that catalyze topologic changes of DNA are called **topoisomerases.** Topoisomerases can relax or insert supercoils. The best-characterized example is **bacterial gyrase,** which induces negative supercoiling in DNA using ATP as energy source. Homologs of this enzyme exist in all organisms and are important targets for cancer chemotherapy.

## DNA PROVIDES A TEMPLATE FOR REPLICATION & TRANSCRIPTION

The genetic information stored in the nucleotide sequence of DNA serves two purposes. It is the source of information for the synthesis of all protein molecules of the cell and organism, and it provides the information inherited by daughter cells or offspring. Both of these functions require that the DNA molecule serve as a template—in the first case for the transcription of the information into RNA and in the second case for the replication of the information into daughter DNA molecules.

The complementarity of the Watson and Crick double-stranded model of DNA strongly suggests that replication of the DNA molecule occurs in a semiconservative manner. Thus, when each strand of the double-stranded parental DNA molecule separates from its complement during replication, each serves as a template on which a new complementary strand is synthesized (Figure 35–4). The two newly formed double-stranded daughter DNA molecules, each containing one strand (but complementary rather than identical) from the parent double-stranded DNA molecule, are then sorted between the two daughter cells (Figure 35–5). Each daughter cell contains DNA molecules with information identical to that which the parent possessed; yet in each daughter cell the DNA molecule of the parent cell has been only semiconserved.

## THE CHEMICAL NATURE OF RNA DIFFERS FROM THAT OF DNA

Ribonucleic acid (RNA) is a polymer of purine and pyrimidine ribonucleotides linked together by 3′,5′-phosphodiester bridges analogous to those in DNA (Figure 35–6). Although sharing many features with DNA, RNA possesses several specific differences:

**(1)** In RNA, the sugar moiety to which the phosphates and purine and pyrimidine bases are attached is ribose rather than the 2′-deoxyribose of DNA.

**(2)** The pyrimidine components of RNA differ from those of DNA. Although RNA contains the ribonucleotides of adenine, guanine, and cytosine, it does not possess thymine except in the rare case mentioned below. Instead of thymine, RNA contains the ribonucleotide of uracil.

**(3)** RNA exists as a single strand, whereas DNA exists as a double-stranded helical molecule. However, given the proper complementary base sequence with opposite polarity, the single strand of RNA—as demonstrated in Figure 35–7—is capable of folding back on itself like a hairpin and thus acquiring double-stranded characteristics.

**(4)** Since the RNA molecule is a single strand complementary to only one of the two strands of a gene, its guanine content does not necessarily equal its cytosine content, nor does its adenine content necessarily equal its uracil content.

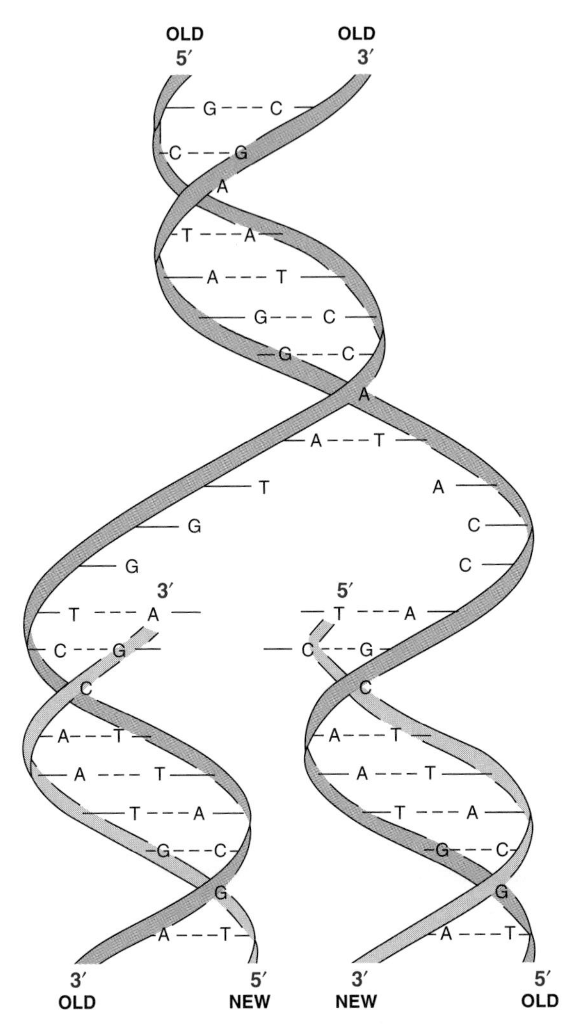


***Figure 35–4.*** The double-stranded structure of DNA and the template function of each old strand (dark shading) on which a new (light shading) complementary strand is synthesized.

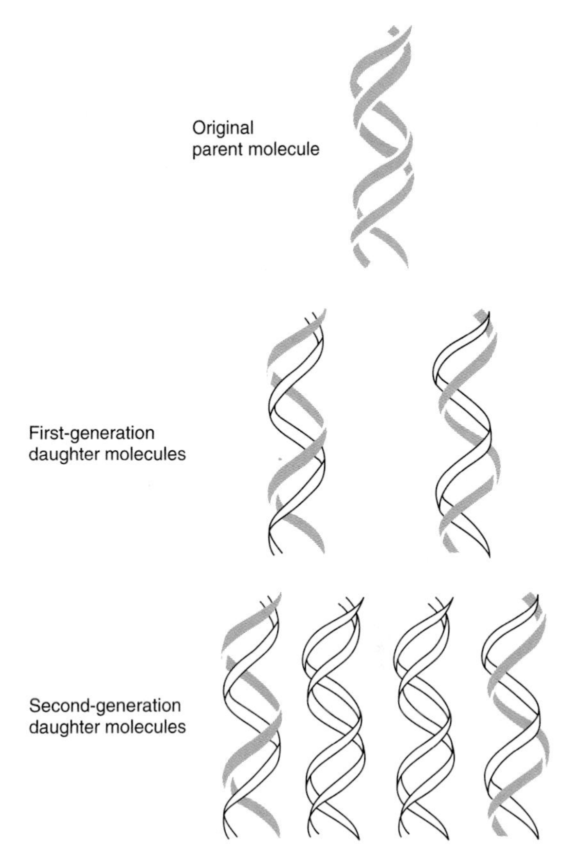


***Figure 35–5.*** DNA replication is semiconservative. During a round of replication, each of the two strands of DNA is used as a template for synthesis of a new, complementary strand.

(5) RNA can be hydrolyzed by alkali to 2′,3′ cyclic diesters of the mononucleotides, compounds that cannot be formed from alkali-treated DNA because of the absence of a 2′-hydroxyl group. The alkali lability of RNA is useful both diagnostically and analytically.

Information within the single strand of RNA is contained in its sequence ("primary structure") of purine and pyrimidine nucleotides within the polymer. The sequence is complementary to the template strand of the gene from which it was transcribed. Because of this complementarity, an RNA molecule can bind specifically via the base-pairing rules to its template DNA strand; it will not bind ("hybridize") with the other (coding) strand of its gene. The sequence of the RNA molecule (except for U replacing T) is the same as that of the coding strand of the gene (Figure 35–8).

## Nearly All of the Several Species of RNA Are Involved in Some Aspect of Protein Synthesis

Those cytoplasmic RNA molecules that serve as templates for protein synthesis (ie, that transfer genetic information from DNA to the protein-synthesizing machinery) are designated **messenger RNAs,** or **mRNAs.** Many other cytoplasmic RNA molecules (**ribosomal RNAs; rRNAs**) have structural roles wherein they con-

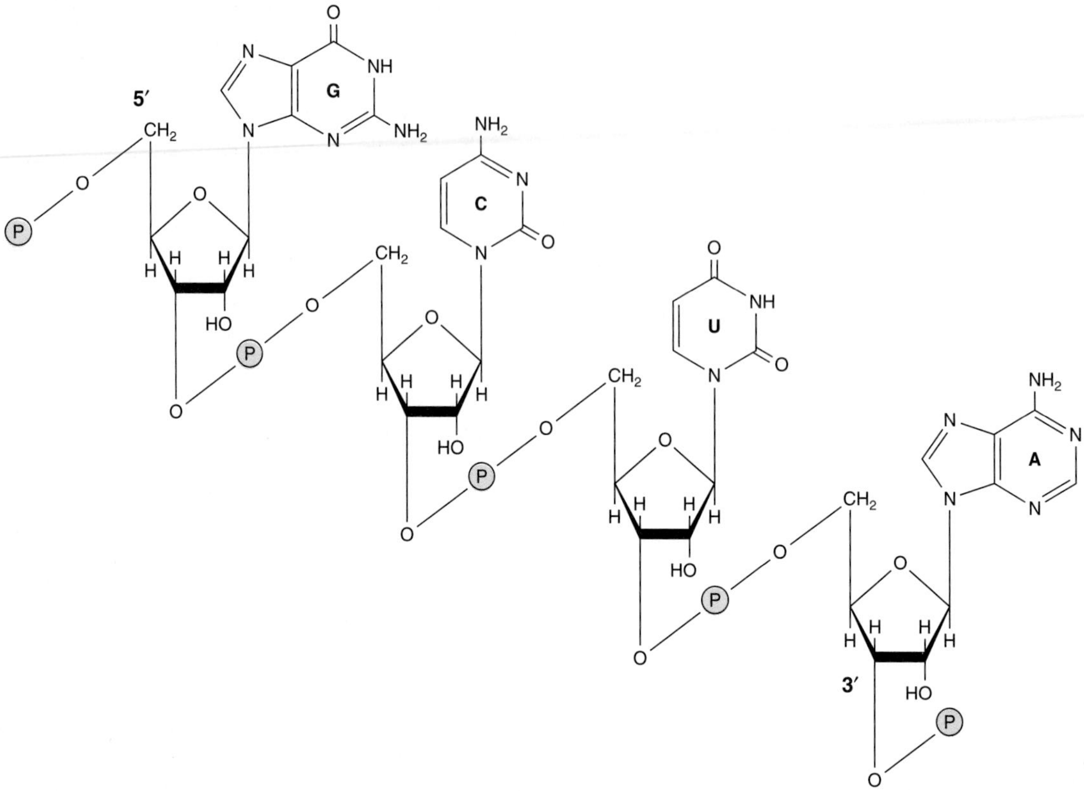

***Figure 35–6.*** A segment of a ribonucleic acid (RNA) molecule in which the purine and pyrimidine bases—guanine (G), cytosine (C), uracil (U), and adenine (A)—are held together by phosphodiester bonds between ribosyl moieties attached to the nucleobases by *N*-glycosidic bonds. Note that the polymer has a polarity as indicated by the labeled 3′- and 5′-attached phosphates.

tribute to the formation and function of ribosomes (the organellar machinery for protein synthesis) or serve as adapter molecules **(transfer RNAs; tRNAs)** for the translation of RNA information into specific sequences of polymerized amino acids.

Some RNA molecules have intrinsic catalytic activity. The activity of these **ribozymes** often involves the cleavage of a nucleic acid. An example is the role of RNA in catalyzing the processing of the primary transcript of a gene into mature messenger RNA.

Much of the RNA synthesized from DNA templates in eukaryotic cells, including mammalian cells, is degraded within the nucleus, and it never serves as either a structural or an informational entity within the cellular cytoplasm.

In all eukaryotic cells there are **small nuclear RNA (snRNA)** species that are not directly involved in protein synthesis but play pivotal roles in RNA processing. These relatively small molecules vary in size from 90 to about 300 nucleotides (Table 35–1).

The genetic material for some animal and plant viruses is RNA rather than DNA. Although some RNA viruses never have their information transcribed into a DNA molecule, many animal RNA viruses—specifically, the retroviruses (the HIV virus, for example)—are transcribed by an RNA-dependent DNA polymerase, the so-called **reverse transcriptase,** to produce a double-stranded DNA copy of their RNA genome. In many cases, the resulting double-stranded DNA transcript is integrated into the host genome and subsequently serves as a template for gene expression and from which new viral RNA genomes can be transcribed.

## RNA Is Organized in Several Unique Structures

In all prokaryotic and eukaryotic organisms, three main classes of RNA molecules exist: messenger RNA (mRNA), transfer RNA (tRNA), and ribosomal RNA

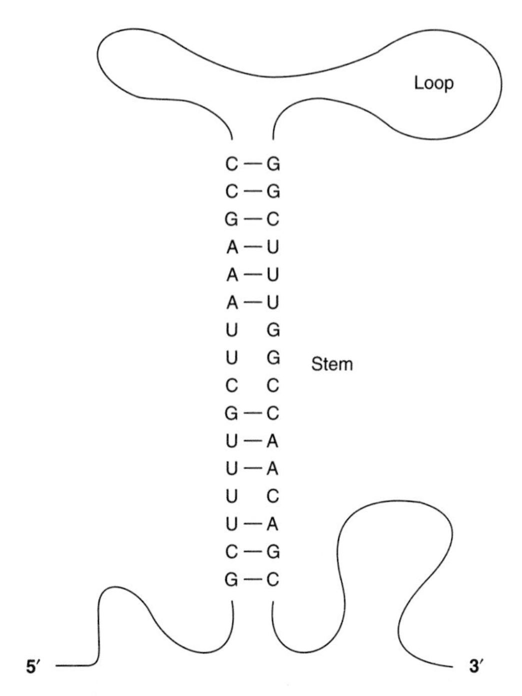


***Figure 35–7.*** Diagrammatic representation of the secondary structure of a single-stranded RNA molecule in which a stem loop, or "hairpin," has been formed and is dependent upon the intramolecular base pairing. Note that A forms hydrogen bonds with U in RNA.

(rRNA). Each differs from the others by size, function, and general stability.

## A. Messenger RNA (mRNA)

This class is the most heterogeneous in size and stability. All members of the class function as messengers conveying the information in a gene to the protein-synthesizing machinery, where each serves as a template on which a specific sequence of amino acids is polymerized to form a specific protein molecule, the ultimate gene product (Figure 35–9).

***Table 35–1.*** Some of the species of small stable RNAs found in mammalian cells.

| Name | Length (nucleotides) | Molecules per Cell | Localization |
|---|---|---|---|
| U1 | 165 | $1 \times 10^6$ | Nucleoplasm/hnRNA |
| U2 | 188 | $5 \times 10^5$ | Nucleoplasm |
| U3 | 216 | $3 \times 10^5$ | Nucleolus |
| U4 | 139 | $1 \times 10^5$ | Nucleoplasm |
| U5 | 118 | $2 \times 10^5$ | Nucleoplasm |
| U6 | 106 | $3 \times 10^5$ | Perichromatin granules |
| 4.5S | 91–95 | $3 \times 10^5$ | Nucleus and cytoplasm |
| 7S | 280 | $5 \times 10^5$ | Nucleus and cytoplasm |
| 7-2 | 290 | $1 \times 10^5$ | Nucleus and cytoplasm |
| 7-3 | 300 | $2 \times 10^5$ | Nucleus |

Messenger RNAs, particularly in eukaryotes, have some unique chemical characteristics. The 5′ terminal of mRNA is "capped" by a 7-methylguanosine triphosphate that is linked to an adjacent 2′-*O*-methyl ribonucleoside at its 5′-hydroxyl through the three phosphates (Figure 35–10). The mRNA molecules frequently contain internal 6-methyladenylates and other 2′-*O*-ribose methylated nucleotides. The cap is involved in the recognition of mRNA by the translating machinery, and it probably helps stabilize the mRNA by preventing the attack of 5′-exonucleases. The protein-synthesizing machinery begins translating the mRNA into proteins beginning downstream of the 5′ or capped terminal. The other end of most mRNA molecules, the 3′-hydroxyl terminal, has an attached polymer of adenylate residues 20–250 nucleotides in length. The specific function of the **poly(A) "tail"** at the 3′-hydroxyl terminal of mRNAs is not fully understood, but it seems that it maintains the intracellular stability of the specific mRNA by preventing the attack of 3′-exonucleases. Some mRNAs, including those for some histones, do not contain poly(A). The poly(A) tail, because it will form a base pair with oligodeoxythymidine polymers attached to a solid substrate like cellulose, can be used to separate mRNA from other species of RNA, including mRNA molecules that lack this tail.

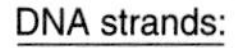

***Figure 35–8.*** The relationship between the sequences of an RNA transcript and its gene, in which the coding and template strands are shown with their polarities. The RNA transcript with a 5′ to 3′ polarity is complementary to the template strand with its 3′ to 5′ polarity. Note that the sequence in the RNA transcript and its polarity is the same as that in the coding strand, except that the U of the transcript replaces the T of the gene.

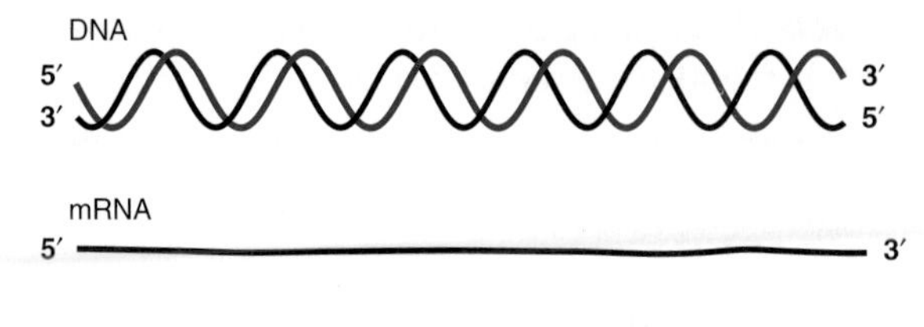


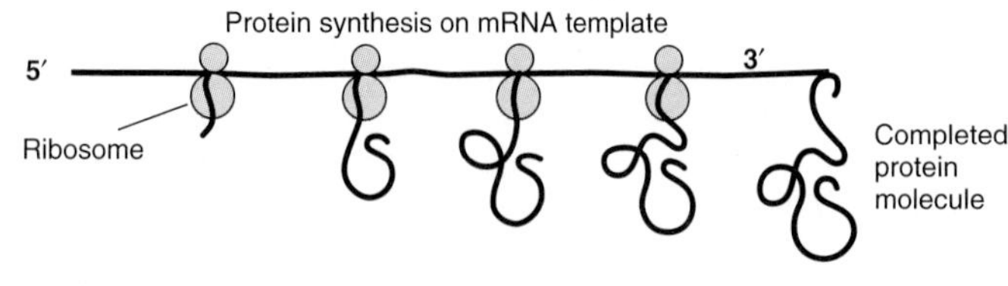


***Figure 35–9.*** The expression of genetic information in DNA into the form of an mRNA transcript. This is subsequently translated by ribosomes into a specific protein molecule.

In mammalian cells, including cells of humans, the mRNA molecules present in the cytoplasm are not the RNA products immediately synthesized from the DNA template but must be formed by processing from a precursor molecule before entering the cytoplasm. Thus, in mammalian nuclei, the immediate products of gene transcription constitute a fourth class of RNA molecules. These nuclear RNA molecules are very heterogeneous in size and are quite large. The **heterogeneous nuclear RNA (hnRNA)** molecules may have a molecular weight in excess of $10^7$, whereas the molecular weight of mRNA molecules is generally less than $2 \times 10^6$. As discussed in Chapter 37, hnRNA molecules are processed to generate the mRNA molecules which then enter the cytoplasm to serve as templates for protein synthesis.

### B. Transfer RNA (tRNA)

tRNA molecules vary in length from 74 to 95 nucleotides. They also are generated by nuclear processing of a precursor molecule (Chapter 37). The tRNA molecules serve as adapters for the translation of the information in the sequence of nucleotides of the mRNA into specific amino acids. There are at least 20 species of tRNA molecules in every cell, at least one (and often several) corresponding to each of the 20 amino acids required for protein synthesis. Although each specific tRNA differs from the others in its sequence of nucleotides, the tRNA molecules as a class have many features in common. The primary structure—ie, the nucleotide sequence—of all tRNA molecules allows extensive folding and intrastrand complementarity to generate a secondary structure that appears like a cloverleaf (Figure 35–11).

All tRNA molecules contain four main arms. The **acceptor arm** terminates in the nucleotides CpCpAOH. These three nucleotides are added posttranscriptionally. The tRNA-appropriate amino acid is attached to the 3′-OH group of the A moiety of the acceptor arm. The **D, TΨC,** and **extra arms** help define a specific tRNA.

Although tRNAs are quite stable in prokaryotes, they are somewhat less stable in eukaryotes. The opposite is true for mRNAs, which are quite unstable in prokaryotes but generally stable in eukaryotic organisms.

### C. Ribosomal RNA (rRNA)

A ribosome is a cytoplasmic nucleoprotein structure that acts as the machinery for the synthesis of proteins from the mRNA templates. On the ribosomes, the mRNA and tRNA molecules interact to translate into a specific protein molecule information transcribed from the gene. In active protein synthesis, many ribosomes are associated with an mRNA molecule in an assembly called the **polysome.**

The components of the mammalian ribosome, which has a molecular weight of about $4.2 \times 10^6$ and a sedimentation velocity of 80S (Svedberg units), are shown in Table 35–2. The mammalian ribosome contains two major nucleoprotein subunits—a larger one with a molecular weight of $2.8 \times 10^6$ (60S) and a smaller subunit with a molecular weight of $1.4 \times 10^6$ (40S). The 60S subunit contains a 5S ribosomal RNA (rRNA), a 5.8S rRNA, and a 28S rRNA; there are also probably more than 50 specific polypeptides. The 40S subunit is smaller and contains a single 18S rRNA and approximately 30 distinct polypeptide chains. All of the ribosomal RNA molecules except the 5S rRNA are processed from a single 45S precursor RNA molecule in the nucleolus (Chapter 37). 5S rRNA is independently transcribed. The highly methylated ribosomal RNA molecules are packaged in the nucleolus with the specific ribosomal proteins. In the cytoplasm, the ribosomes remain quite stable and capable of many translation cycles. The functions of the ribosomal RNA molecules in the ribosomal particle are not fully understood, but they are necessary for ribosomal assembly and seem to play key roles in the binding of mRNA to

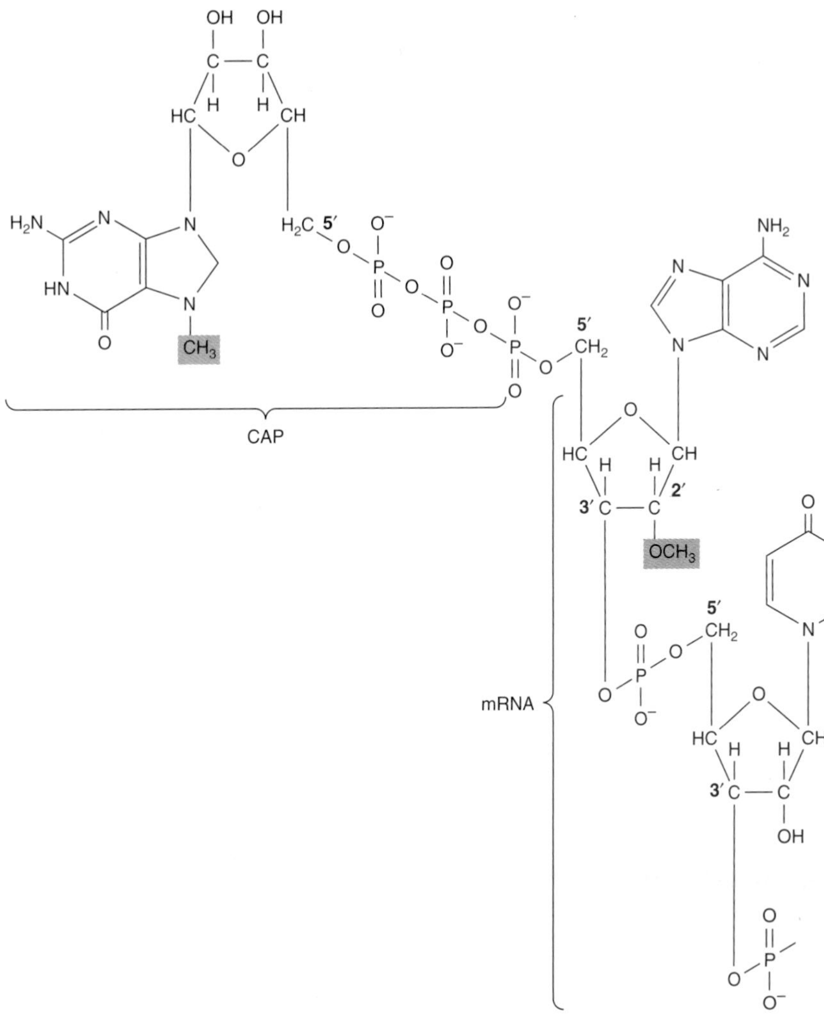

***Figure 35–10.*** The cap structure attached to the 5′ terminal of most eukaryotic messenger RNA molecules. A 7-methylguanosine triphosphate (black) is attached at the 5′ terminal of the mRNA (shown in blue), which usually contains a 2′-*O*-methylpurine nucleotide. These modifications (the cap and methyl group) are added after the mRNA is transcribed from DNA.

ribosomes and its translation. Recent studies suggest that an rRNA component performs the peptidyl transferase activity and thus is an enzyme (a ribozyme).

### D. Small Stable RNA

A large number of discrete, highly conserved, and small stable RNA species are found in eukaryotic cells. The majority of these molecules are complexed with proteins to form ribonucleoproteins and are distributed in the nucleus, in the cytoplasm, or in both. They range in size from 90 to 300 nucleotides and are present in 100,000–1,000,000 copies per cell.

Small nuclear RNAs (snRNAs), a subset of these RNAs, are significantly involved in mRNA processing and gene regulation. Of the several snRNAs, U1, U2, U4, U5, and U6 are involved in intron removal and the processing of hnRNA into mRNA (Chapter 37). The U7 snRNA may be involved in production of the correct 3′ ends of histone mRNA—which lacks a poly(A) tail. The U4 and U6 snRNAs may also be required for poly(A) processing.

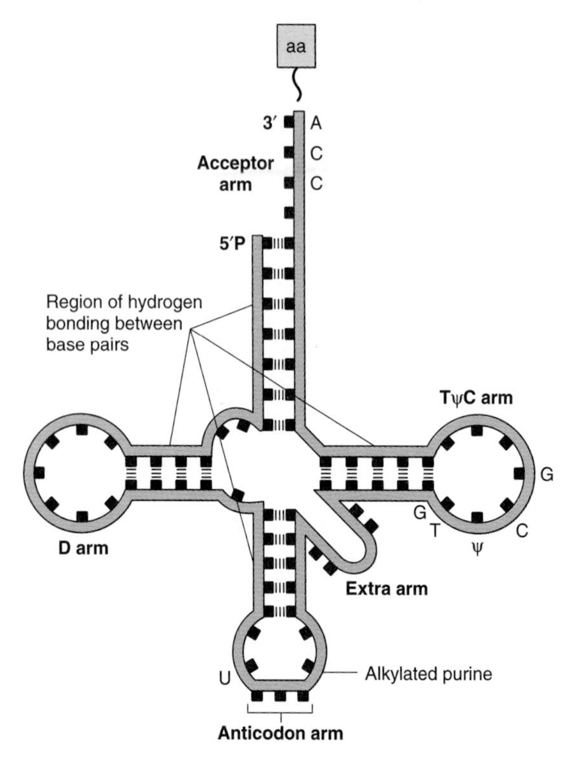


***Figure 35–11.*** Typical aminoacyl tRNA in which the amino acid (aa) is attached to the 3′ CCA terminal. The anticodon, TΨC, and dihydrouracil (D) arms are indicated, as are the positions of the intramolecular hydrogen bonding between these base pairs. (From Watson JD: *Molecular Biology of the Gene,* 3rd ed. Copyright © 1976, 1970, 1965, by W.A. Benjamin, Inc., Menlo Park, California.)

## SPECIFIC NUCLEASES DIGEST NUCLEIC ACIDS

Enzymes capable of degrading nucleic acids have been recognized for many years. These nucleases can be classified in several ways. Those which exhibit specificity for deoxyribonucleic acid are referred to as **deoxyribonucleases.** Those which specifically hydrolyze ribonucleic acids are **ribonucleases.** Within both of these classes are enzymes capable of cleaving internal phosphodiester bonds to produce either 3′-hydroxyl and 5′-phosphoryl terminals or 5′-hydroxyl and 3′-phosphoryl terminals. These are referred to as **endonucleases.** Some are capable of hydrolyzing both strands of a **double-stranded** molecule, whereas others can only cleave **single strands** of nucleic acids. Some nucleases can hydrolyze only unpaired single strands, while others are capable of hydrolyzing single strands participating in the formation of a double-stranded molecule. There exist classes of endonucleases that recognize specific sequences in DNA; the majority of these are the **restriction endonucleases,** which have in recent years become important tools in molecular genetics and medical sciences. A list of some currently recognized restriction endonucleases is presented in Table 40–2.

Some nucleases are capable of hydrolyzing a nucleotide only when it is present at a terminal of a molecule; these are referred to as **exonucleases.** Exonucleases act in one direction (3′ → 5′ or 5′ → 3′) only. In bacteria, a 3′ → 5′ exonuclease is an integral part of the DNA replication machinery and there serves to edit—or proofread—the most recently added deoxynucleotide for base-pairing errors.

***Table 35–2.*** Components of mammalian ribosomes.[1]

| Component | Mass (mw) | Protein | | RNA | | |
|---|---|---|---|---|---|---|
| | | Number | Mass | Size | Mass | Bases |
| 40S subunit | $1.4 \times 10^6$ | ~35 | $7 \times 10^5$ | 18S | $7 \times 10^5$ | 1900 |
| 60S subunit | $2.8 \times 10^6$ | ~50 | $1 \times 10^6$ | 5S | 35,000 | 120 |
| | | | | 5.8S | 45,000 | 160 |
| | | | | 28S | $1.6 \times 10^6$ | 4700 |

[1]The ribosomal subunits are defined according to their sedimentation velocity in Svedberg units (40S or 60S). This table illustrates the total mass (MW) of each. The number of unique proteins and their total mass (MW) and the RNA components of each subunit in size (Svedberg units), mass, and number of bases are listed.

## SUMMARY

- DNA consists of four bases—A, G, C, and T—which are held in linear array by phosphodiester bonds through the 3′ and 5′ positions of adjacent deoxyribose moieties.
- DNA is organized into two strands by the pairing of bases A to T and G to C on complementary strands. These strands form a double helix around a central axis.
- The $3 \times 10^9$ base pairs of DNA in humans are organized into the haploid complement of 23 chromosomes. The exact sequence of these 3 billion nucleotides defines the uniqueness of each individual.
- DNA provides a template for its own replication and thus maintenance of the genotype and for the transcription of the 30,000–50,000 genes into a variety of RNA molecules.
- RNA exists in several different single-stranded structures, most of which are involved in protein synthesis. The linear array of nucleotides in RNA consists of A, G, C, and U, and the sugar moiety is ribose.
- The major forms of RNA include messenger RNA (mRNA), ribosomal RNA (rRNA), and transfer RNA (tRNA). Certain RNA molecules act as catalysts (ribozymes).

## REFERENCES

Green R, Noller HF: Ribosomes and translation. Annu Rev Biochem 1997;66:689.

Guthrie C, Patterson B: Spliceosomal snRNAs. Ann Rev Genet 1988;22:387.

Hunt T: *DNA Makes RNA Makes Protein.* Elsevier, 1983.

Watson JD, Crick FHC: Molecular structure of nucleic acids. Nature 1953;171:737.

Watson JD: *The Double Helix.* Atheneum, 1968.

Watson JD et al: *Molecular Biology of the Gene,* 5th ed. Benjamin-Cummings, 2000.

# DNA Organization, Replication, & Repair

# 36

*Daryl K. Granner, MD, & P. Anthony Weil, PhD*

## BIOMEDICAL IMPORTANCE*

The genetic information in the DNA of a chromosome can be transmitted by exact replication or it can be exchanged by a number of processes, including crossing over, recombination, transposition, and conversion. These provide a means of ensuring adaptability and diversity for the organism but, when these processes go awry, can also result in disease. A number of enzyme systems are involved in DNA replication, alteration, and repair. Mutations are due to a change in the base sequence of DNA and may result from the faulty replication, movement, or repair of DNA and occur with a frequency of about one in every $10^6$ cell divisions. Abnormalities in gene products (either in protein function or amount) can be the result of mutations that occur in coding or regulatory-region DNA. A mutation in a germ cell will be transmitted to offspring (so-called vertical transmission of hereditary disease). A number of factors, including viruses, chemicals, ultraviolet light, and ionizing radiation, increase the rate of mutation. Mutations often affect somatic cells and so are passed on to successive generations of cells, but only within an organism. It is becoming apparent that a number of diseases—and perhaps most cancers—are due to the combined effects of vertical transmission of mutations as well as horizontal transmission of induced mutations.

## CHROMATIN IS THE CHROMOSOMAL MATERIAL EXTRACTED FROM NUCLEI OF CELLS OF EUKARYOTIC ORGANISMS

Chromatin consists of very long double-stranded **DNA molecules** and a nearly equal mass of rather small basic proteins termed **histones** as well as a smaller amount of **nonhistone proteins** (most of which are acidic and larger than histones) and a small quantity of **RNA.** The nonhistone proteins include enzymes involved in DNA replication, such as DNA topoisomerases. Also included are proteins involved in transcription, such as the RNA polymerase complex. The double-stranded DNA helix in each chromosome has a length that is thousands of times the diameter of the cell nucleus. One purpose of the molecules that comprise chromatin, particularly the histones, is to condense the DNA. Electron microscopic studies of chromatin have demonstrated dense spherical particles called **nucleosomes,** which are approximately 10 nm in diameter and connected by DNA filaments (Figure 36–1). Nucleosomes are composed of DNA wound around a collection of histone molecules.

### Histones Are the Most Abundant Chromatin Proteins

The histones are a small family of closely related basic proteins. **H1 histones** are the ones least tightly bound to chromatin Figure 36–1) and are, therefore, easily removed with a salt solution, after which chromatin becomes soluble. The organizational unit of this soluble chromatin is the nucleosome. Nucleosomes contain four classes of histones: **H2A, H2B, H3,** and **H4.** The structures of all four histones—H2A, H2B, H3, and H4, the so-called core histones forming the nucleosome—have been highly conserved between species. This extreme conservation implies that the function of histones is identical in all eukaryotes and that the entire molecule is involved quite specifically in carrying out this function. The carboxyl terminal two-thirds of the molecules have a typical random amino acid composition, while their amino terminal thirds are particularly rich in basic amino acids. **These four core histones are subject to at least five types of covalent modification:** acetylation, methylation, phosphorylation, ADP-ribosylation, and covalent linkage (H2A only) to ubiquitin. These histone modifications probably play an important role in chromatin structure and function as illustrated in Table 36–1.

The histones interact with each other in very specific ways. **H3 and H4 form a tetramer** containing two mol-

*So far as is possible, the discussion in this chapter and in Chapters 37, 38, and 39 will pertain to mammalian organisms, which are, of course, among the higher eukaryotes. At times it will be necessary to refer to observations in prokaryotic organisms such as bacteria and viruses, but in such cases the information will be of a kind that can be extrapolated to mammalian organisms.

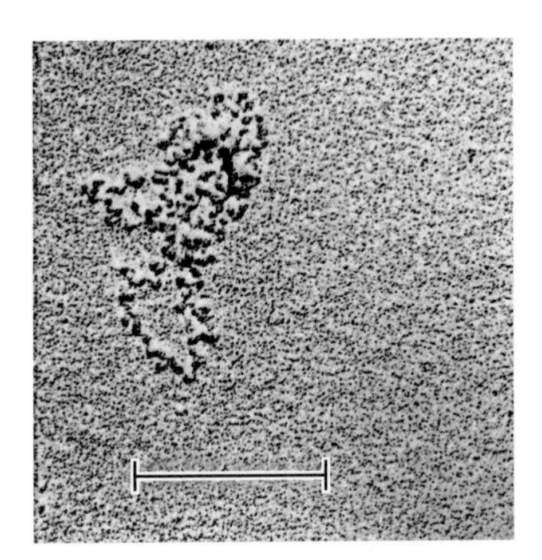

***Figure 36–1.*** Electron micrograph of nucleosomes attached by strands of nucleic acid. (The bar represents 2.5 μm.) (Reproduced, with permission, from Oudet P, Gross-Bellard M, Chambon P: Electron microscopic and biochemical evidence that chromatin structure is a repeating unit. Cell 1975;4:281.)

ecules of each $(H3/H4)_2$, while **H2A and H2B form dimers** (H2A-H2B). Under physiologic conditions, these histone oligomers associate to form the **histone octamer** of the composition $(H3/H4)_2$-$(H2A\text{-}H2B)_2$.

## The Nucleosome Contains Histone & DNA

When the histone octamer is mixed with purified, double-stranded DNA, the same x-ray diffraction pattern is formed as that observed in freshly isolated chromatin. Electron microscopic studies confirm the existence of reconstituted nucleosomes. Furthermore, the reconstitution of nucleosomes from DNA and histones H2A, H2B, H3, and H4 is independent of the organismal or cellular origin of the various components. The histone H1 and the nonhistone proteins are not necessary for the reconstitution of the nucleosome core.

***Table 36–1.*** Possible roles of modified histones.

1. Acetylation of histones H3 and H4 is associated with the activation or inactivation of gene transcription (Chapter 37).
2. Acetylation of core histones is associated with chromosomal assembly during DNA replication.
3. Phosphorylation of histone H1 is associated with the condensation of chromosomes during the replication cycle.
4. ADP-ribosylation of histones is associated with DNA repair.
5. Methylation of histones is correlated with activation and repression of gene transcription.

In the nucleosome, the DNA is supercoiled in a left-handed helix over the surface of the disk-shaped histone octamer (Figure 36–2). The majority of core histone proteins interact with the DNA on the inside of the supercoil without protruding, though the amino terminal tails of all the histones probably protrude outside of this structure and are available for regulatory covalent modifications (see Table 36–1).

The $(H3/H4)_2$ tetramer itself can confer nucleosome-like properties on DNA and thus has a central role in the formation of the nucleosome. The addition of two H2A-H2B dimers stabilizes the primary particle and firmly binds two additional half-turns of DNA previously bound only loosely to the $(H3/H4)_2$. Thus, 1.75 superhelical turns of DNA are wrapped around the surface of the histone octamer, protecting 146 base pairs of DNA and forming the nucleosome core particle (Figure 36–2). The core particles are separated by an about 30-bp linker region of DNA. Most of the DNA is in a repeating series of these structures, giving the so-called "beads-on-a-string" appearance when examined by electron microscopy (see Figure 36–1).

The assembly of nucleosomes is mediated by one of several chromatin assembly factors facilitated by histone chaperones, proteins such as the anionic nuclear protein **nucleoplasmin.** As the nucleosome is assembled, histones are released from the histone chaperones. Nucleosomes appear to exhibit preference for certain regions on specific DNA molecules, but the basis for this nonrandom distribution, termed **phasing,** is not completely

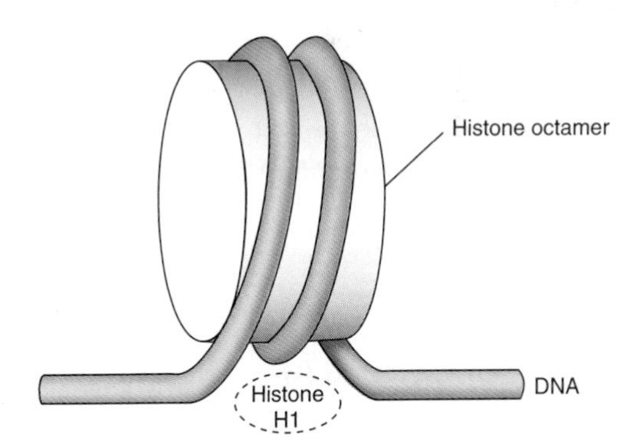


***Figure 36–2.*** Model for the structure of the nucleosome, in which DNA is wrapped around the surface of a flat protein cylinder consisting of two each of histones H2A, H2B, H3, and H4 that form the histone octamer. The 146 base pairs of DNA, consisting of 1.75 superhelical turns, are in contact with the histone octamer. This protects the DNA from digestion by a nuclease. The position of histone H1, when it is present, is indicated by the dashed outline at the bottom of the figure.

understood. It is probably related to the relative physical flexibility of certain nucleotide sequences that are able to accommodate the regions of kinking within the supercoil as well as the presence of other DNA-bound factors that limit the sites of nucleosome deposition.

The super-packing of nucleosomes in nuclei is seemingly dependent upon the interaction of the H1 histones with adjacent nucleosomes.

## HIGHER-ORDER STRUCTURES PROVIDE FOR THE COMPACTION OF CHROMATIN

Electron microscopy of chromatin reveals two higher orders of structure—the 10-nm fibril and the 30-nm chromatin fiber—beyond that of the nucleosome itself. The disk-like nucleosome structure has a 10-nm diameter and a height of 5 nm. The **10-nm fibril** consists of nucleosomes arranged with their edges separated by a small distance (30 bp of DNA) with their flat faces parallel with the fibril axis (Figure 36–3). The 10-nm fibril is probably further supercoiled with six or seven nucleosomes per turn to form the **30-nm chromatin fiber** (Figure 36–3). Each turn of the supercoil is relatively flat, and the faces of the nucleosomes of successive turns would be nearly parallel to each other. H1 histones appear to stabilize the 30-nm fiber, but their position and that of the variable length spacer DNA are not clear. It is probable that nucleosomes can form a variety of packed structures. In order to form a mitotic chromosome, the 30-nm fiber must be compacted in length another 100-fold (see below).

In **interphase chromosomes,** chromatin fibers appear to be organized into 30,000–100,000 bp **loops or domains** anchored in a scaffolding (or supporting matrix) within the nucleus. Within these domains, some DNA sequences may be located nonrandomly. It has been suggested that each looped domain of chromatin corresponds to one or more separate genetic functions, containing both coding and noncoding regions of the cognate gene or genes.

## SOME REGIONS OF CHROMATIN ARE "ACTIVE" & OTHERS ARE "INACTIVE"

Generally, every cell of an individual metazoan organism contains the same genetic information. Thus, the differences between cell types within an organism must be explained by differential expression of the common genetic information. Chromatin containing active genes (ie, transcriptionally active chromatin) has been shown to differ in several ways from that of inactive regions. The nucleosome structure of active chromatin appears to be altered, sometimes quite extensively, in highly active regions. DNA in active chromatin contains large regions (about 100,000 bases long) that are **sensitive to digestion by a nuclease** such as DNase I. DNase I makes single-strand cuts in any segment of DNA (no sequence specificity). It will digest DNA not protected by protein into its component deoxynucleotides. The sensitivity to DNase I of chromatin regions being actively transcribed reflects only a potential for transcription rather than transcription itself and in several systems can be correlated with a relative lack of 5-methyldeoxycytidine in the DNA and particular histone covalent modifications (phosphorylation, acetylation, etc; see Table 36–1).

Within the large regions of active chromatin there exist shorter stretches of 100–300 nucleotides that exhibit an even greater (another tenfold) sensitivity to DNase I. These **hypersensitive sites** probably result from a structural conformation that favors access of the nuclease to the DNA. These regions are often located immediately upstream from the active gene and are the location of interrupted nucleosomal structure caused by the binding of nonhistone regulatory transcription factor proteins. (See Chapters 37 and 39.) In many cases, it seems that if a gene is capable of being transcribed, it very often has a DNase-hypersensitive site(s) in the chromatin immediately upstream. As noted above, nonhistone regulatory proteins involved in transcription control and those involved in maintaining access to the template strand lead to the formation of hypersensitive sites. Hypersensitive sites often provide the first clue about the presence and location of a transcription control element.

Transcriptionally inactive chromatin is densely packed during interphase as observed by electron microscopic studies and is referred to as **heterochromatin;** transcriptionally active chromatin stains less densely and is referred to as **euchromatin.** Generally, euchromatin is replicated earlier than heterochromatin in the mammalian cell cycle (see below).

There are two types of heterochromatin: constitutive and facultative. **Constitutive heterochromatin** is always condensed and thus inactive. It is found in the regions near the chromosomal centromere and at chromosomal ends (telomeres). **Facultative heterochromatin** is at times condensed, but at other times it is actively transcribed and, thus, uncondensed and appears as euchromatin. Of the two members of the X chromosome pair in mammalian females, one X chromosome is almost completely inactive transcriptionally and is heterochromatic. However, the heterochromatic X chromosome decondenses during gametogenesis and becomes transcriptionally active during early embryogenesis—thus, it is facultative heterochromatin.

Certain cells of insects, eg, *Chironomus,* contain giant chromosomes that have been replicated for ten cycles without separation of daughter chromatids. These copies of DNA line up side by side in precise register and produce a banded chromosome containing regions of condensed chromatin and lighter bands of

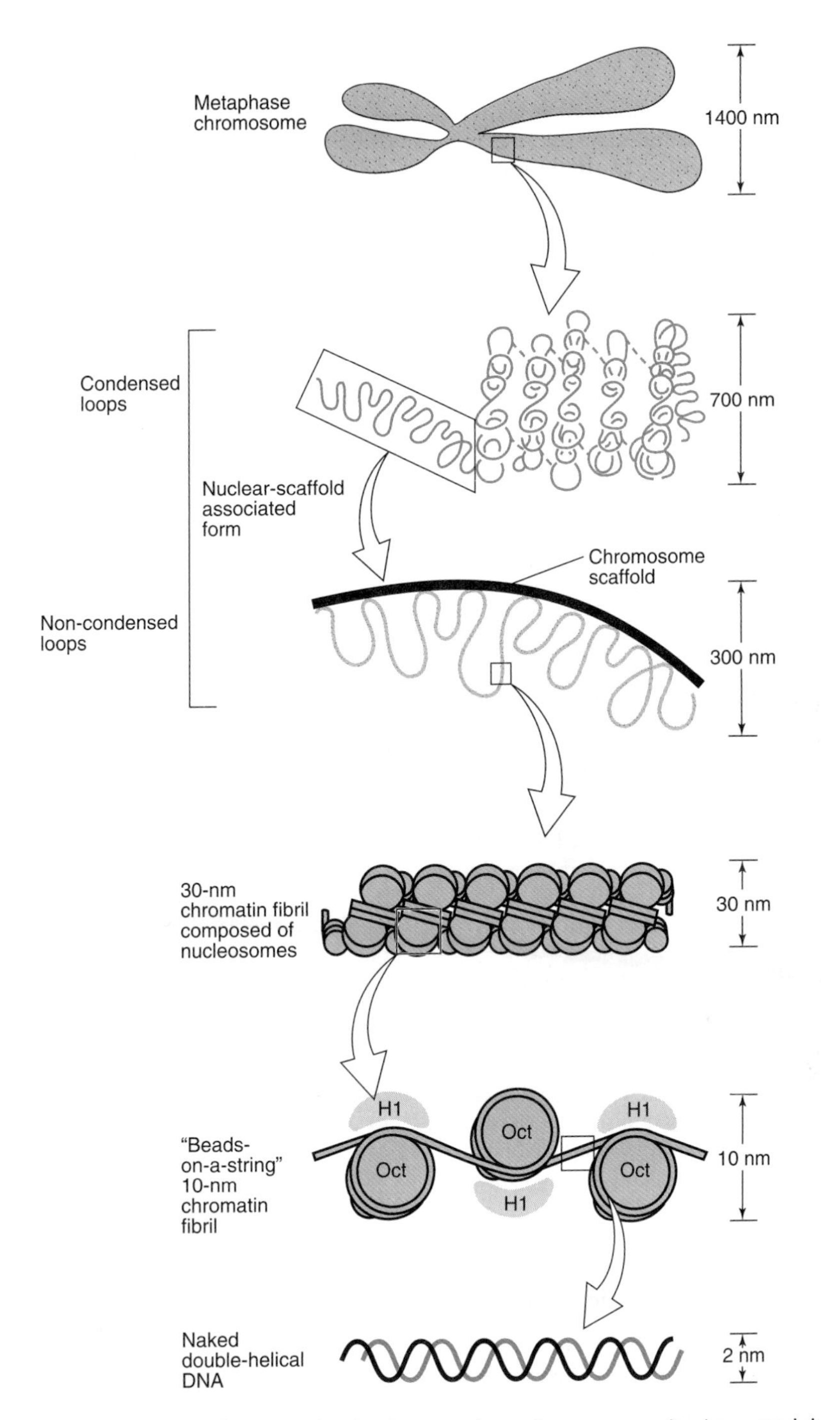


***Figure 36–3.*** Shown is the extent of DNA packaging in metaphase chromosomes (*top*) to noted duplex DNA (*bottom*). Chromosomal DNA is packaged and organized at several levels as shown (see Table 36–2). Each phase of condensation or compaction and organization (*bottom to top*) decreases overall DNA accessibility to an extent that the DNA sequences in metaphase chromosomes are almost totally transcriptionally inert. In toto, these five levels of DNA compaction result in nearly a $10^4$-fold linear decrease in end-to-end DNA length. Complete condensation and decondensation of the linear DNA in chromosomes occur in the space of hours during the normal replicative cell cycle (see Figure 36–20).

more extended chromatin. Transcriptionally active regions of these **polytene chromosomes** are especially decondensed into **"puffs"** that can be shown to contain the enzymes responsible for transcription and to be the sites of RNA synthesis (Figure 36–4).

## DNA IS ORGANIZED INTO CHROMOSOMES

At metaphase, mammalian **chromosomes** possess a twofold symmetry, with the identical duplicated **sister chromatids** connected at a **centromere,** the relative position of which is characteristic for a given chromosome (Figure 36–5). The centromere is an adenine-thymine (A–T) rich region ranging in size from $10^2$ (brewers' yeast) to $10^6$ (mammals) base pairs. It binds several proteins with high affinity. This complex, called the **kinetochore,** provides the anchor for the mitotic spindle. It thus is an essential structure for chromosomal segregation during mitosis.

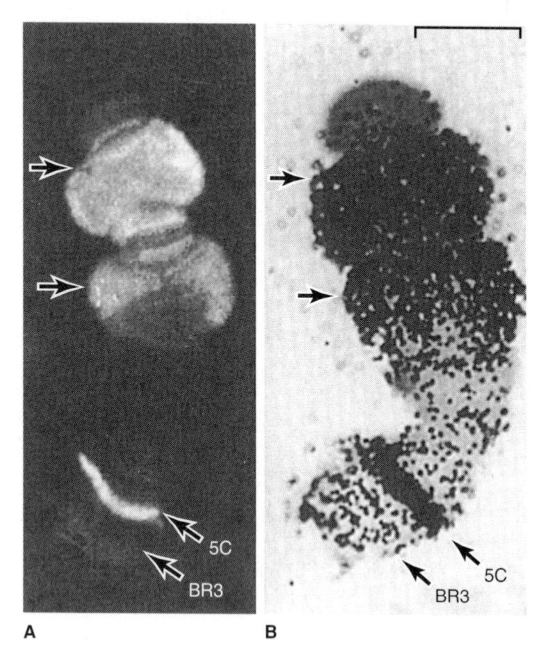


***Figure 36–4.*** Illustration of the tight correlation between the presence of RNA polymerase II and RNA synthesis. A number of genes are activated when *Chironomus tentans* larvae are subjected to heat shock (39 °C for 30 minutes). **A:** Distribution of RNA polymerase II (also called type B) in isolated chromosome IV from the salivary gland (at arrows). The enzyme was detected by immunofluorescence using an antibody directed against the polymerase. The 5C and BR3 are specific bands of chromosome IV, and the arrows indicate puffs. **B:** Autoradiogram of a chromosome IV that was incubated in $^3$H-uridine to label the RNA. Note the correspondence of the immunofluorescence and presence of the radioactive RNA (black dots). Bar = 7 μm. (Reproduced, with permission, from Sass H: RNA polymerase B in polytene chromosomes. Cell 1982;28:274. Copyright © 1982 by the Massachusetts Institute of Technology.)

The ends of each chromosome contain structures called **telomeres.** Telomeres consist of short, repeat TG-rich sequences. Human telomeres have a variable number of repeats of the sequence 5′-TTAGGG-3′, which can extend for several kilobases. **Telomerase,** a multisubunit RNA-containing complex related to viral RNA-dependent DNA polymerases (reverse transcriptases), is the enzyme responsible for telomere synthesis and thus for maintaining the length of the telomere. Since telomere shortening has been associated with both malignant transformation and aging, telomerase has become an attractive target for cancer chemotherapy and drug development. Each sister chromatid contains one double-stranded DNA molecule. During interphase, the packing of the DNA molecule is less dense than it is in the condensed chromosome during metaphase. Metaphase chromosomes are nearly completely transcriptionally inactive.

The human haploid genome consists of about $3 \times 10^9$ bp and about $1.7 \times 10^7$ nucleosomes. Thus, each of the 23 chromatids in the human haploid genome would contain on the average $1.3 \times 10^8$ nucleotides in one double-stranded DNA molecule. The length of each DNA molecule must be compressed about 8000-fold to generate the structure of a condensed metaphase chromosome! In metaphase chromosomes, the 30-nm chromatin fibers are also folded into a series of **looped domains,** the proximal portions of which are anchored to a nonhistone proteinaceous scaffolding within the nucleus (Figure 36–3). The packing ratios of each of the orders of DNA structure are summarized in Table 36–2.

The packaging of nucleoproteins within chromatids is not random, as evidenced by the characteristic patterns observed when chromosomes are stained with specific dyes such as quinacrine or Giemsa's stain (Figure 36–6).

From individual to individual within a single species, the pattern of staining (banding) of the entire chromosome complement is highly reproducible; nonetheless, it differs significantly from other species, even those closely related. Thus, the packaging of the nucleoproteins in chromosomes of higher eukaryotes must in some way be dependent upon species-specific characteristics of the DNA molecules.

A combination of specialized staining techniques and high-resolution microscopy has allowed geneticists

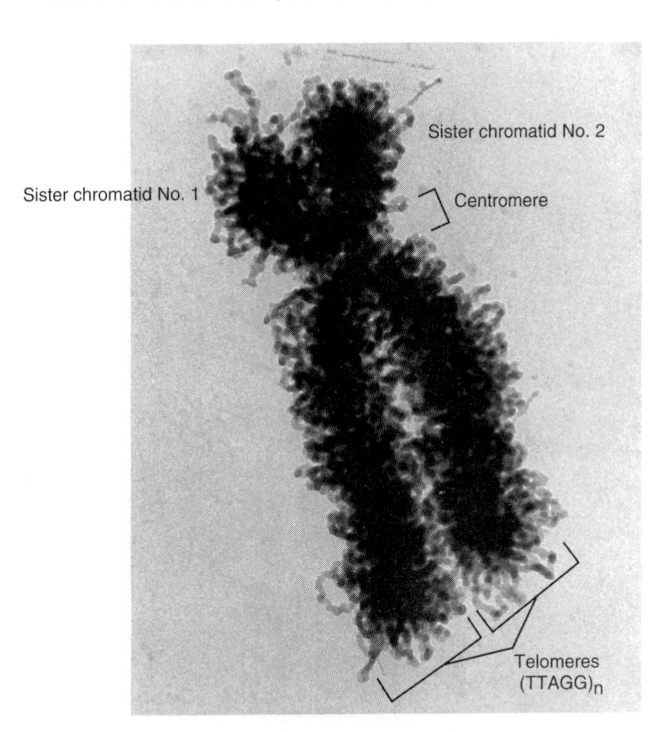


***Figure 36–5.*** The two sister chromatids of human chromosome 12 (× 27,850). The location of the A+T-rich centromeric region connecting sister chromatids is indicated, as are two of the four telomeres residing at the very ends of the chromatids that are attached one to the other at the centromere. (Modified and reproduced, with permission, from DuPraw EJ: *DNA and Chromosomes.* Holt, Rinehart, and Winston, 1970.)

to quite precisely map thousands of genes to specific regions of mouse and human chromosomes. With the recent elucidation of the human and mouse genome sequences, it has become clear that many of these visual mapping methods were remarkably accurate.

## Coding Regions Are Often Interrupted by Intervening Sequences

The **protein coding regions of DNA,** the transcripts of which ultimately appear in the cytoplasm as single mRNA molecules, are usually **interrupted in the eukaryotic genome by large intervening sequences of nonprotein coding DNA.** Accordingly, the primary transcripts of DNA (mRNA precursors, originally termed hnRNA because this species of RNA was quite heterogeneous in size [length] and mostly restricted to the nucleus), contain noncoding intervening sequences of RNA that must be removed in a process which also joins together the appropriate coding segments to form the mature mRNA. Most coding sequences for a single mRNA are interrupted in the genome (and thus in the primary transcript) by at least one—and in some cases as many as 50—noncoding intervening sequences **(introns).** In most cases, the introns are much longer than the continuous coding regions **(exons).** The processing of the primary transcript, which involves removal of introns and splicing of adjacent exons, is described in detail in Chapter 37.

The function of the intervening sequences, or introns, is not clear. They may serve to separate functional domains (exons) of coding information in a form that permits genetic rearrangement by recombination to occur more rapidly than if all coding regions for a given genetic function were contiguous. Such an enhanced rate of genetic rearrangement of functional domains might allow more rapid evolution of biologic function. The relationships among chromosomal DNA, gene clusters on the chromosome, the exon-intron structure of genes, and the final mRNA product are illustrated in Figure 36–7.

***Table 36–2.*** The packing ratios of each of the orders of DNA structure.

| Chromatin Form | Packing Ratio |
|---|---|
| Naked double-helical DNA | ~1.0 |
| 10-nm fibril of nucleosomes | 7–10 |
| 25- to 30-nm chromatin fiber of superhelical nucleosomes | 40–60 |
| Condensed metaphase chromosome of loops | 8000 |

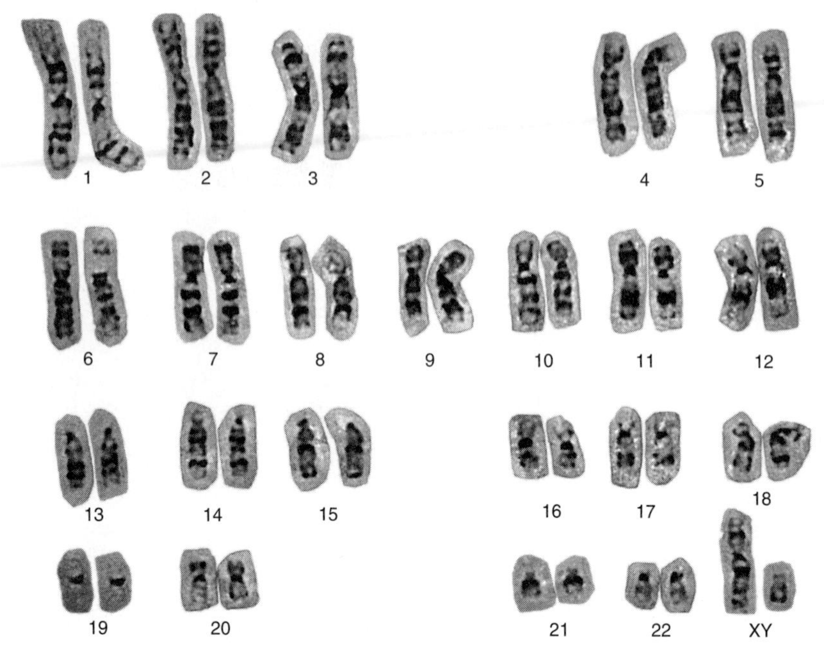


***Figure 36–6.*** A human karyotype (of a man with a normal 46,XY constitution), in which the metaphase chromosomes have been stained by the Giemsa method and aligned according to the Paris Convention. (Courtesy of H Lawce and F Conte.)

## MUCH OF THE MAMMALIAN GENOME IS REDUNDANT & MUCH IS NOT TRANSCRIBED

The haploid genome of each human cell consists of $3 \times 10^9$ base pairs of DNA subdivided into 23 chromosomes. The entire haploid genome contains sufficient DNA to code for nearly 1.5 million average-sized genes. However, studies of mutation rates and of the complexities of the genomes of higher organisms strongly suggest that humans have < 100,000 proteins encoded by the ~1.1% of the human genome that is composed of exonic DNA. This implies that most of the DNA is noncoding—ie, its information is never translated into an amino acid sequence of a protein molecule. Certainly, some of the excess DNA sequences serve to regulate the expression of genes during development, differentiation, and adaptation to the environment. Some excess clearly makes up the intervening sequences or introns (24% of the total human genome) that split the coding regions of genes, but much of the excess appears to be composed of many families of repeated sequences for which no functions have been clearly defined. A summary of the salient features of the human genome is presented in Chapter 40.

The DNA in a eukaryotic genome can be divided into different "sequence classes." These are **unique-sequence, or nonrepetitive, DNA** and **repetitive-sequence DNA.** In the haploid genome, unique-sequence DNA generally includes the single copy genes that code for proteins. The repetitive DNA in the haploid genome includes sequences that vary in copy number from two to as many as $10^7$ copies per cell.

### More Than Half the DNA in Eukaryotic Organisms Is in Unique or Nonrepetitive Sequences

This estimation (and the distribution of repetitive-sequence DNA) is based on a variety of DNA-RNA hybridization techniques and, more recently, on direct DNA sequencing. Similar techniques are used to estimate the number of active genes in a population of unique-sequence DNA. In brewers' yeast (*Saccharomyces cerevisiae,* a lower eukaryote), about two thirds of its 6200 genes are expressed. In typical tissues in a higher eukaryote (eg, mammalian liver and kidney), between 10,000 and 15,000 genes are expressed. Different combinations of genes are expressed in each tissue,

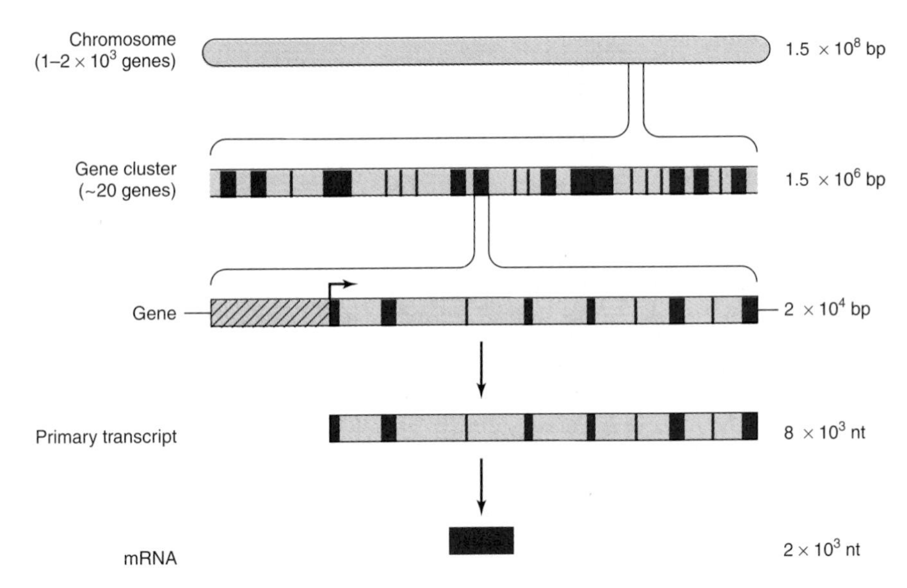


***Figure 36–7.*** The relationship between chromosomal DNA and mRNA. The human haploid DNA complement of $3 \times 10^9$ base pairs (bp) is distributed between 23 chromosomes. Genes are clustered on these chromosomes. An average gene is 27,000 bp in length, including the regulatory region (hatched area), which is usually located at the 5′ end of the gene. The regulatory region is shown here as being adjacent to the transcription initiation site (arrow). Most eukaryotic genes have alternating exons and introns. In this example, there are nine exons (dark blue areas) and eight introns (light blue areas). The introns are removed from the primary transcript by the processing reaction, and the exons are ligated together in sequence to form the mature mRNA. (nt, nucleotides.)

of course, and how this is accomplished is one of the major unanswered questions in biology.

## In Human DNA, at Least 30% of the Genome Consists of Repetitive Sequences

Repetitive-sequence DNA can be broadly classified as moderately repetitive or as highly repetitive. The highly repetitive sequences consist of 5–500 base pair lengths repeated many times in tandem. These sequences are usually clustered in centromeres and telomeres of the chromosome and are present in about 1–10 million copies per haploid genome. These sequences are transcriptionally inactive and may play a structural role in the chromosome (see Chapter 40).

The moderately repetitive sequences, which are defined as being present in numbers of less than $10^6$ copies per haploid genome, are not clustered but are interspersed with unique sequences. In many cases, these long interspersed repeats are transcribed by RNA polymerase II and contain caps indistinguishable from those on mRNA.

Depending on their length, moderately repetitive sequences are classified as **long interspersed repeat sequences (LINEs)** or **short interspersed repeat sequences (SINEs).** Both types appear to be **retroposons,** ie, they arose from movement from one location to another **(transposition)** through an RNA intermediate by the action of reverse transcriptase that transcribes an RNA template into DNA. Mammalian genomes contain 20–50 thousand copies of the 6–7 kb LINEs. These represent species-specific families of repeat elements. SINEs are shorter (70–300 bp), and there may be more than 100,000 copies per genome. Of the SINEs in the human genome, one family, the **Alu family,** is present in about 500,000 copies per haploid genome and accounts for at least 5–6% of the human genome. Members of the human Alu family and their closely related analogs in other animals are transcribed as integral components of hnRNA or as discrete RNA molecules, including the well-studied 4.5S RNA and 7S RNA. These particular family members are highly conserved within a species as well as between mammalian species. Components of the short interspersed repeats, including the members of the Alu family, may be mobile elements, capable of jumping into and out of various sites within the genome (see below). This can have disastrous results, as exemplified by the insertion of Alu sequences into a gene, which, when so mutated, causes neurofibromatosis.

## Microsatellite Repeat Sequences

One category of repeat sequences exists as both dispersed and grouped tandem arrays. The sequences consist of 2–6 bp repeated up to 50 times. These **microsatellite sequences** most commonly are found as dinucleotide repeats of AC on one strand and TG on the opposite strand, but several other forms occur, including CG, AT, and CA. The AC repeat sequences are estimated to occur at 50,000–100,000 locations in the genome. At any locus, the number of these repeats may vary on the two chromosomes, thus providing heterozygosity of the number of copies of a particular microsatellite number in an individual. This is a heritable trait, and, because of their number and the ease of detecting them using the polymerase chain reaction (PCR) (Chapter 40), AC repeats are very useful in constructing genetic linkage maps. Most genes are associated with one or more microsatellite markers, so the relative position of genes on chromosomes can be assessed, as can the association of a gene with a disease. Using PCR, a large number of family members can be rapidly screened for a certain **microsatellite polymorphism.** The association of a specific polymorphism with a gene in affected family members—and the lack of this association in unaffected members—may be the first clue about the genetic basis of a disease.

Trinucleotide sequences that increase in number (microsatellite instability) can cause disease. The unstable $p(CGG)_n$ repeat sequence is associated with the fragile X syndrome. Other trinucleotide repeats that undergo dynamic mutation (usually an increase) are associated with Huntington's chorea (CAG), myotonic dystrophy (CTG), spinobulbar muscular atrophy (CAG), and Kennedy's disease (CAG).

## ONE PERCENT OF CELLULAR DNA IS IN MITOCHONDRIA

The majority of the peptides in mitochondria (about 54 out of 67) are coded by nuclear genes. The rest are coded by genes found in mitochondrial (mt) DNA. Human mitochondria contain two to ten copies of a small circular double-stranded DNA molecule that makes up approximately 1% of total cellular DNA. This mtDNA codes for mt ribosomal and transfer RNAs and for 13 proteins that play key roles in the respiratory chain. The linearized structural map of the human mitochondrial genes is shown in Figure 36–8. Some of the features of mtDNA are shown in Table 36–3.

An important feature of human mitochondrial mtDNA is that—because all mitochondria are contributed by the ovum during zygote formation—it is transmitted by maternal nonmendelian inheritance.

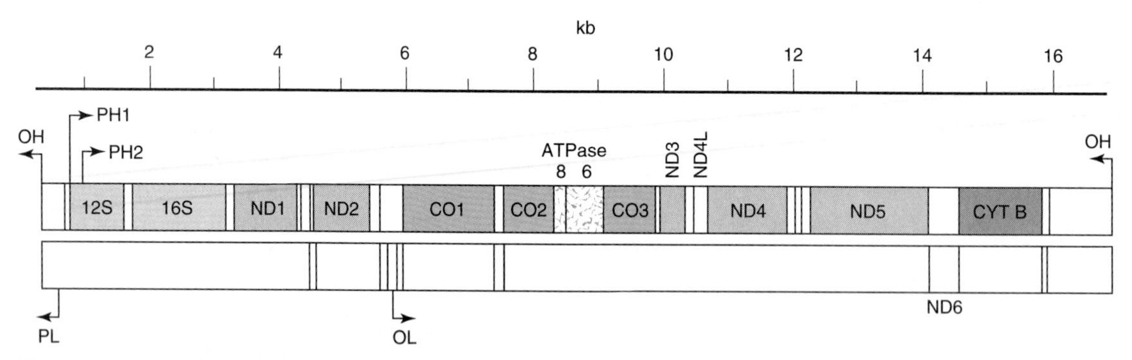


***Figure 36–8.*** Maps of human mitochondrial genes. The maps represent the heavy (upper strand) and light (lower map) strands of linearized mitochondrial (mt) DNA, showing the genes for the subunits of NADH-coenzyme Q oxidoreductase (ND1 through ND6), cytochrome *c* oxidase (CO1 through CO3), cytochrome *b* (CYT B), and ATP synthase (ATPase 8 and 6) and for the 12S and 16S ribosomal mt rRNAs. The transfer RNAs are denoted by small open boxes. The origin of heavy-strand (OH) and light-strand (OL) replication and the promoters for the initiation of heavy-strand (PH1 and PH2) and light-strand (PL) transcription are indicated by arrows. (Reproduced, with permission, from Moraes CT et al: Mitochondrial DNA deletions in progressive external ophthalmoplegia and Kearns-Sayre syndrome. N Engl J Med 1989;320:1293.)

***Table 36–3.*** Some major features of the structure and function of human mitochondrial DNA.[1]

- Is circular, double-stranded, and composed of heavy (H) and a light (L) chains or strands.
- Contains 16,569 bp.
- Encodes 13 protein subunits of the respiratory chain (of a total of about 67):
  - Seven subunits of NADH dehydrogenase (complex I)
  - Cytochrome *b* of complex III
  - Three subunits of cytochrome oxidase (complex IV)
  - Two subunits of ATP synthase
- Encodes large (16s) and small (12s) mt ribosomal RNAs.
- Encodes 22 mt tRNA molecules.
- Genetic code differs slightly from the standard code:
  - UGA (standard stop codon) is read as Trp.
  - AGA and AGG (standard codons for Arg) are read as stop codons.
- Contains very few untranslated sequences.
- High mutation rate (five to ten times that of nuclear DNA).
- Comparisons of mtDNA sequences provide evidence about evolutionary origins of primates and other species.

[1]Adapted from Harding AE: Neurological disease and mitochondrial genes. Trends Neurol Sci 1991;14:132.

Thus, in diseases resulting from mutations of mtDNA, an affected mother would in theory pass the disease to all of her children but only her daughters would transmit the trait. However, in some cases, deletions in mtDNA occur during oogenesis and thus are not inherited from the mother. A number of diseases have now been shown to be due to mutations of mtDNA. These include a variety of myopathies, neurologic disorders, and some cases of diabetes mellitus.

## GENETIC MATERIAL CAN BE ALTERED & REARRANGED

An alteration in the sequence of purine and pyrimidine bases in a gene due to a change—a removal or an insertion—of one or more bases may result in an altered gene product. Such alteration in the genetic material results in a **mutation** whose consequences are discussed in detail in Chapter 38.

### Chromosomal Recombination Is One Way of Rearranging Genetic Material

Genetic information can be exchanged between similar or homologous chromosomes. The exchange or **recombination** event occurs primarily during meiosis in mammalian cells and requires alignment of homologous metaphase chromosomes, an alignment that almost always occurs with great exactness. A process of crossing over occurs as shown in Figure 36–9. This usually results in an equal and reciprocal exchange of genetic information between homologous chromosomes. If the homologous chromosomes possess different alleles of the same genes, the crossover may produce noticeable and heritable genetic linkage differences. In the rare case where the alignment of homologous chromosomes is not exact, the crossing over or recombination event may result in an unequal exchange of information. One chromosome may receive less genetic material and thus a deletion, while the other partner of the chromosome pair receives more genetic material and thus an insertion or duplication (Figure 36–9). Unequal crossing over does occur in humans, as evidenced by the existence of hemoglobins designated Lepore and anti-Lepore (Figure 36–10). The farther apart two sequences are on an individual chromosome, the greater the likelihood of a crossover recombination

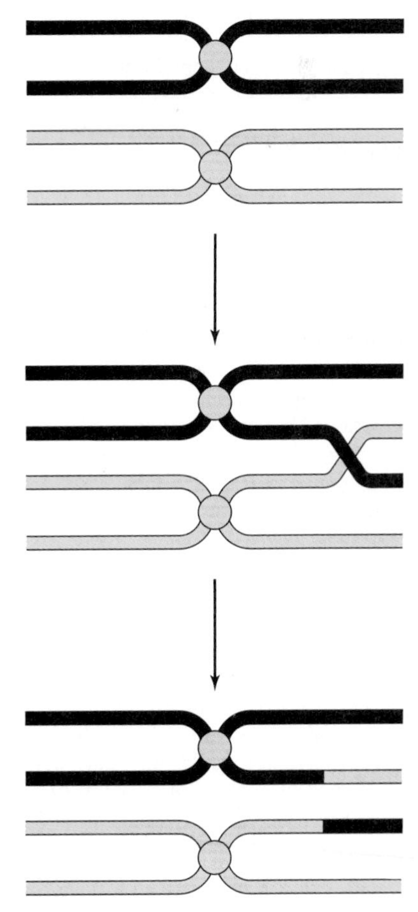

***Figure 36–9.*** The process of crossing-over between homologous metaphase chromosomes to generate recombinant chromosomes. See also Figure 36–12.

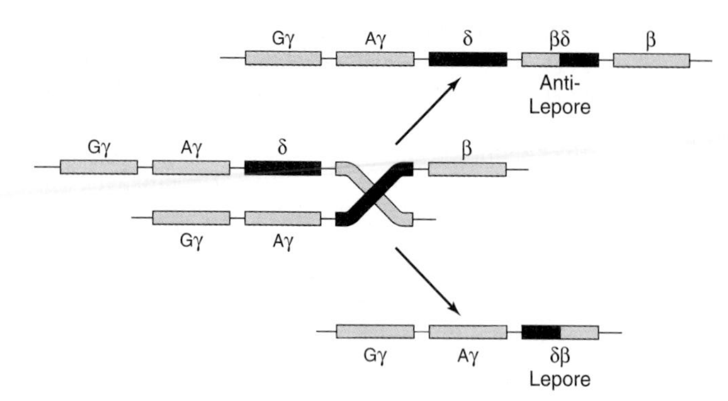


***Figure 36–10.*** The process of unequal crossover in the region of the mammalian genome that harbors the structural genes encoding hemoglobins and the generation of the unequal recombinant products hemoglobin delta-beta Lepore and beta-delta anti-Lepore. The examples given show the locations of the crossover regions between amino acid residues. (Redrawn and reproduced, with permission, from Clegg JB, Weatherall DJ: $\beta^0$ Thalassemia: Time for a reappraisal? Lancet 1974;2:133.)

event. This is the basis for genetic mapping methods. **Unequal crossover** affects tandem arrays of repeated DNAs whether they are related globin genes, as in Figure 36–10, or more abundant repetitive DNA. Unequal crossover through slippage in the pairing can result in expansion or contraction in the copy number of the repeat family and may contribute to the expansion and fixation of variant members throughout the array.

## Chromosomal Integration Occurs With Some Viruses

Some bacterial viruses (bacteriophages) are capable of recombining with the DNA of a bacterial host in such a way that the genetic information of the bacteriophage is incorporated in a linear fashion into the genetic information of the host. This integration, which is a form of recombination, occurs by the mechanism illustrated in Figure 36–11. The backbone of the circular bacteriophage genome is broken, as is that of the DNA molecule of the host; the appropriate ends are resealed with the proper polarity. The bacteriophage DNA is figuratively straightened out ("linearized") as it is integrated into the bacterial DNA molecule—frequently a closed circle as well. The site at which the bacteriophage genome integrates or recombines with the bacterial genome is chosen by one of two mechanisms. If the bacteriophage contains a DNA sequence **homologous** to a sequence in the host DNA molecule, then a recombination event analogous to that occurring between homologous chromosomes can occur. However, some bacteriophages synthesize proteins that bind specific sites on bacterial chromosomes to a **nonhomologous** site characteristic of the bacteriophage DNA molecule. Integration occurs at the site and is said to be **"site-specific."**

Many animal viruses, particularly the oncogenic viruses—either directly or, in the case of RNA viruses such as HIV that causes AIDS, their DNA transcripts generated by the action of the viral RNA-dependent DNA polymerase, or reverse transcriptase—can be integrated into chromosomes of the mammalian cell. The integration of the animal virus DNA into the animal genome generally is not "site-specific" but does display site preferences.

## Transposition Can Produce Processed Genes

In eukaryotic cells, small DNA elements that clearly are not viruses are capable of transposing themselves in and

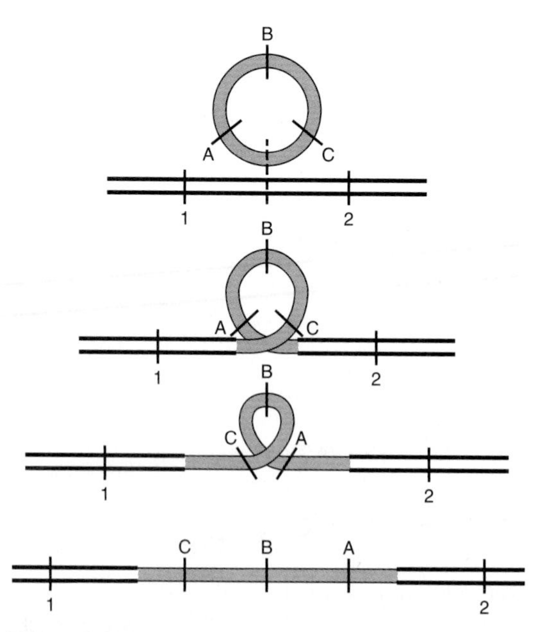


***Figure 36–11.*** The integration of a circular genome from a virus (with genes A, B, and C) into the DNA molecule of a host (with genes 1 and 2) and the consequent ordering of the genes.

out of the host genome in ways that affect the function of neighboring DNA sequences. These mobile elements, sometimes called "jumping DNA," can carry flanking regions of DNA and, therefore, profoundly affect evolution. As mentioned above, the Alu family of moderately repeated DNA sequences has structural characteristics similar to the termini of retroviruses, which would account for the ability of the latter to move into and out of the mammalian genome.

Direct evidence for the transposition of other small DNA elements into the human genome has been provided by the discovery of "processed genes" for immunoglobulin molecules, α-globin molecules, and several others. These **processed genes** consist of DNA sequences identical or nearly identical to those of the messenger RNA for the appropriate gene product. That is, the 5′ nontranscribed region, the coding region without intron representation, and the 3′ poly(A) tail are all present contiguously. This particular DNA sequence arrangement must have resulted from the reverse transcription of an appropriately processed messenger RNA molecule from which the intron regions had been removed and the poly(A) tail added. The only recognized mechanism this reverse transcript could have used to integrate into the genome would have been a transposition event. In fact, these "processed genes" have short terminal repeats at each end, as do known transposed sequences in lower organisms. In the absence of their transcription and thus genetic selection for function, many of the processed genes have been randomly altered through evolution so that they now contain nonsense codons which preclude their ability to encode a functional, intact protein (see Chapter 38). Thus, they are referred to as **"pseudogenes."**

## Gene Conversion Produces Rearrangements

Besides unequal crossover and transposition, a third mechanism can effect rapid changes in the genetic material. Similar sequences on homologous or nonhomologous chromosomes may occasionally pair up and eliminate any mismatched sequences between them. This may lead to the accidental fixation of one variant or another throughout a family of repeated sequences and thereby homogenize the sequences of the members of repetitive DNA families. This latter process is referred to as **gene conversion.**

## Sister Chromatids Exchange

In diploid eukaryotic organisms such as humans, after cells progress through the S phase they contain a tetraploid content of DNA. This is in the form of sister chromatids of chromosome pairs. Each of these sister chromatids contains identical genetic information since each is a product of the semiconservative replication of the original parent DNA molecule of that chromosome. Crossing over occurs between these genetically identical sister chromatids. Of course, these **sister chromatid exchanges** (Figure 36–12) have no genetic consequence as long as the exchange is the result of an equal crossover.

## Immunoglobulin Genes Rearrange

In mammalian cells, some interesting gene rearrangements occur normally during development and differentiation. For example, in mice the $V_L$ and $C_L$ genes for a single immunoglobulin molecule (see Chapter 39) are widely separated in the germ line DNA. In the DNA of a differentiated immunoglobulin-producing (plasma) cell, the same $V_L$ and $C_L$ genes have been moved physically closer together in the genome and into the same transcription unit. However, even then, this rearrangement of DNA during differentiation does not bring the $V_L$ and $C_L$ genes into contiguity in the DNA. Instead, the DNA

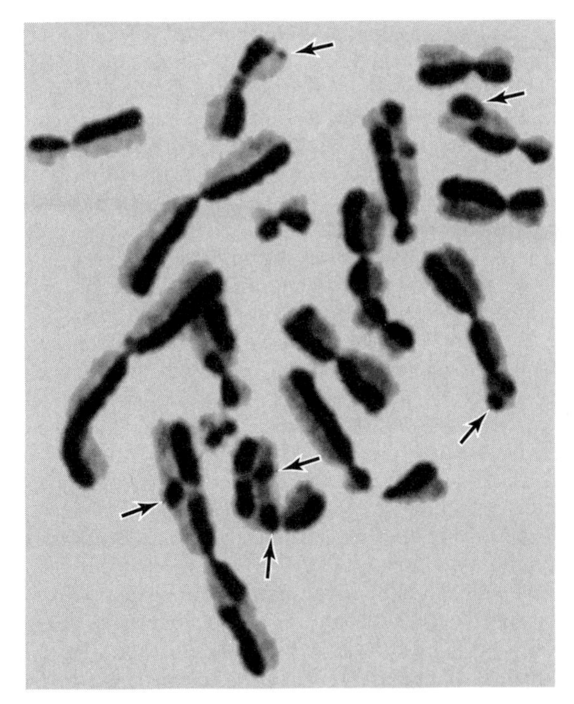

***Figure 36–12.*** Sister chromatid exchanges between human chromosomes. These are detectable by Giemsa staining of the chromosomes of cells replicated for two cycles in the presence of bromodeoxyuridine. The arrows indicate some regions of exchange. (Courtesy of S Wolff and J Bodycote.)

contains an interspersed or interruption sequence of about 1200 base pairs at or near the junction of the V and C regions. The interspersed sequence is transcribed into RNA along with the $V_L$ and $C_L$ genes, and the interspersed information is removed from the RNA during its nuclear processing (Chapters 37 and 39).

## DNA SYNTHESIS & REPLICATION ARE RIGIDLY CONTROLLED

The primary function of DNA replication is understood to be the provision of progeny with the genetic information possessed by the parent. Thus, the replication of DNA must be complete and carried out in such a way as to maintain genetic stability within the organism and the species. The process of DNA replication is complex and involves many cellular functions and several verification procedures to ensure fidelity in replication. About 30 proteins are involved in the replication of the *E coli* chromosome, and this process is almost certainly more complex in eukaryotic organisms. The first enzymologic observations on DNA replication were made in *E coli* by Kornberg, who described in that organism the existence of an enzyme now called DNA polymerase I. This enzyme has multiple catalytic activities, a complex structure, and a requirement for the triphosphates of the four deoxyribonucleosides of adenine, guanine, cytosine, and thymine. The polymerization reaction catalyzed by DNA polymerase I of *E coli* has served as a prototype for all DNA polymerases of both prokaryotes and eukaryotes, even though it is now recognized that the major role of this polymerase is to complete replication on the lagging strand.

In all cells, replication can occur only from a single-stranded DNA (ssDNA) template. Mechanisms must exist to target the site of initiation of replication and to unwind the double-stranded DNA (dsDNA) in that region. The replication complex must then form. After replication is complete in an area, the parent and daughter strands must re-form dsDNA. In eukaryotic cells, an additional step must occur. The dsDNA must precisely re-form the chromatin structure, including nucleosomes, that existed prior to the onset of replication. Although this entire process is not well understood in eukaryotic cells, replication has been quite precisely described in prokaryotic cells, and the general principles are thought to be the same in both. The major steps are listed in Table 36–4, illustrated in Figure 36–13, and discussed, in sequence, below. A number of proteins, most with specific enzymatic action, are involved in this process (Table 36–5).

**Table 36–4.** Steps involved in DNA replication in eukaryotes.

1. Identification of the origins of replication.
2. Unwinding (denaturation) of dsDNA to provide an ssDNA template.
3. Formation of the replication fork.
4. Initiation of DNA synthesis and elongation.
5. Formation of replication bubbles with ligation of the newly synthesized DNA segments.
6. Reconstitution of chromatin structure.

### The Origin of Replication

At the **origin of replication (ori)**, there is an association of sequence-specific dsDNA-binding proteins with a series of direct repeat DNA sequences. In bacteriophage λ, the oriλ is bound by the λ-encoded O protein to four adjacent sites. In *E coli,* the oriC is bound by the protein dnaA. In both cases, a complex is formed consisting of 150–250 bp of DNA and multimers of the DNA-binding protein. This leads to the local denaturation and unwinding of an adjacent A+T-rich region of DNA. Functionally similar **autonomously replicating sequences (ARS)** have been identified in yeast cells. The ARS contains a somewhat degenerate 11-bp sequence called the **origin replication element (ORE).** The ORE binds a set of proteins, analogous to the dnaA protein of *E coli,* which is collectively called the **origin recognition complex (ORC).** The ORE is located adjacent to an approximately 80-bp A+T-rich sequence that is easy to unwind. This is called the **DNA unwinding element (DUE).** The DUE is the origin of replication in yeast.

Consensus sequences similar to ori or ARS in structure or function have not been precisely defined in mammalian cells, though several of the proteins that participate in ori recognition and function have been identified and appear quite similar to their yeast counterparts in both amino acid sequence and function.

### Unwinding of DNA

The interaction of proteins with ori defines the start site of replication and provides a short region of ssDNA essential for initiation of synthesis of the nascent DNA strand. This process requires the formation of a number of protein-protein and protein-DNA interactions. A critical step is provided by a DNA helicase that allows for processive unwinding of DNA. In uninfected *E coli,* this function is provided by a complex of dnaB helicase and the dnaC protein. Single-stranded DNA-binding proteins (SSBs) stabilize this complex. In λ phage-infected *E coli,* the phage protein P binds to dnaB and the P/dnaB complex binds to oriλ by interacting with the O protein. dnaB is not an active helicase when in the P/dnaB/O complex. Three *E coli* heat shock proteins (dnaK, dnaJ, and GrpE) cooperate to remove the P

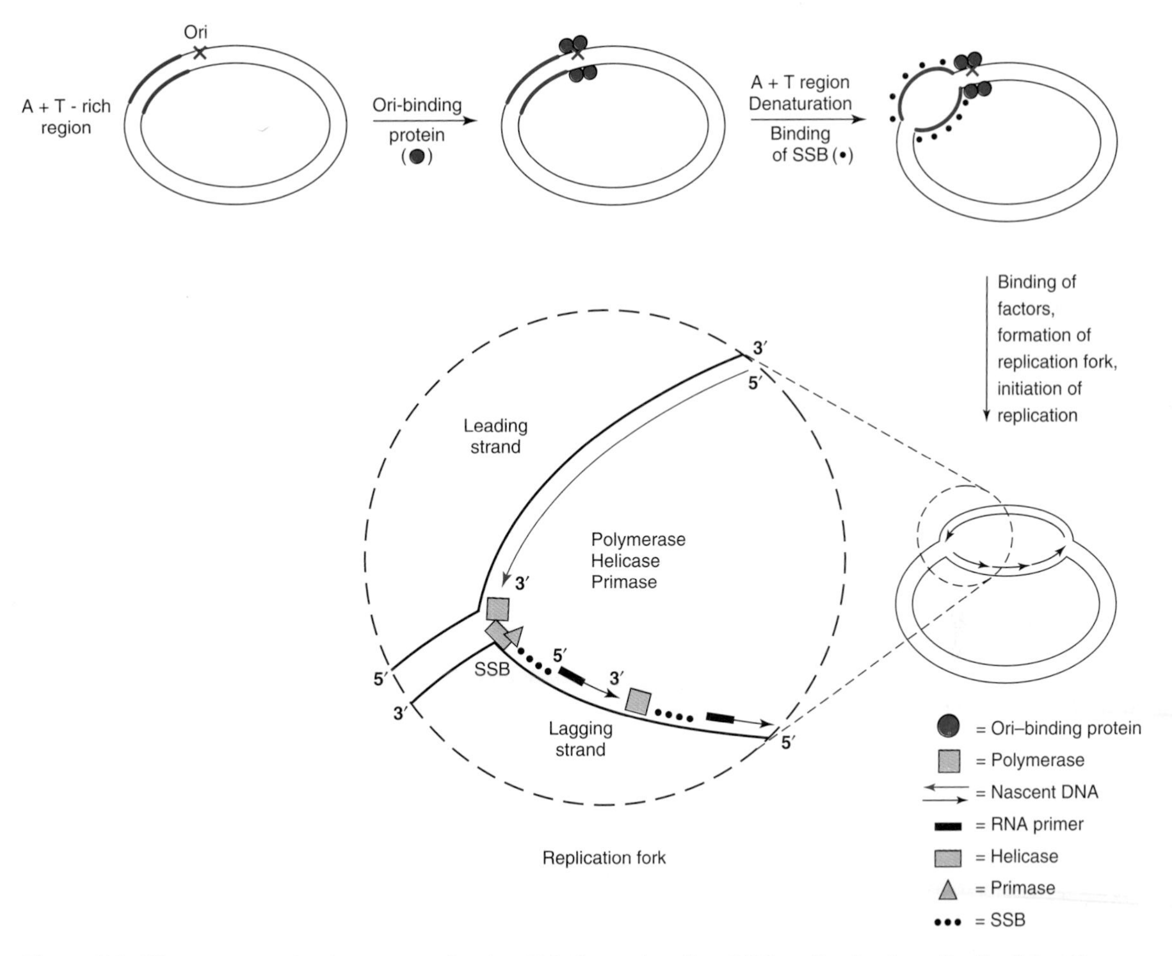


***Figure 36–13.*** Steps involved in DNA replication. This figure describes DNA replication in an *E coli* cell, but the general steps are similar in eukaryotes. A specific interaction of a protein (the O protein) to the origin of replication (ori) results in local unwinding of DNA at an adjacent A+T-rich region. The DNA in this area is maintained in the single-strand conformation (ssDNA) by single-strand-binding proteins (SSBs). This allows a variety of proteins, including helicase, primase, and DNA polymerase, to bind and to initiate DNA synthesis. The replication fork proceeds as DNA synthesis occurs continuously (long arrow) on the leading strand and discontinuously (short arrows) on the lagging strand. The nascent DNA is always synthesized in the 5′ to 3′ direction, as DNA polymerases can add a nucleotide only to the 3′ end of a DNA strand.

protein and activate the dnaB helicase. In cooperation with SSB, this leads to DNA unwinding and active replication. In this way, the replication of the λ phage is accomplished at the expense of replication of the host *E coli* cell.

## Formation of the Replication Fork

A replication fork consists of four components that form in the following sequence: (1) the DNA helicase unwinds a short segment of the parental duplex DNA; (2) a primase initiates synthesis of an RNA molecule that is essential for priming DNA synthesis; (3) the DNA polymerase initiates nascent, daughter strand synthesis; and (4) SSBs bind to ssDNA and prevent premature reannealing of ssDNA to dsDNA. These reactions are illustrated in Figure 36–13.

The polymerase III holoenzyme (the *dnaE* gene product in *E coli*) binds to template DNA as part of a multiprotein complex that consists of several polymerase accessory factors (β, γ, δ, δ′, and τ). DNA polymerases only synthesize DNA in the 5′ to 3′ direction,

***Table 36–5.*** Classes of proteins involved in replication.

| Protein | Function |
|---|---|
| DNA polymerases | Deoxynucleotide polymerization |
| Helicases | Processive unwinding of DNA |
| Topoisomerases | Relieve torsional strain that results from helicase-induced unwinding |
| DNA primase | Initiates synthesis of RNA primers |
| Single-strand binding proteins | Prevent premature reannealing of dsDNA |
| DNA ligase | Seals the single strand nick between the nascent chain and Okazaki fragments on lagging strand |

and only one of the several different types of polymerases is involved at the replication fork. Because the DNA strands are antiparallel (Chapter 35), the polymerase functions asymmetrically. On the **leading (forward) strand,** the DNA is synthesized continuously. On the **lagging (retrograde) strand,** the DNA is synthesized in short (1–5 kb) fragments, the so-called **Okazaki fragments.** Several Okazaki fragments (up to 250) must be synthesized, in sequence, for each replication fork. To ensure that this happens, the helicase acts on the lagging strand to unwind dsDNA in a 5′ to 3′ direction. The helicase associates with the primase to afford the latter proper access to the template. This allows the RNA primer to be made and, in turn, the polymerase to begin replicating the DNA. This is an important reaction sequence since DNA polymerases cannot initiate DNA synthesis de novo. The mobile complex between helicase and primase has been called a **primosome.** As the synthesis of an Okazaki fragment is completed and the polymerase is released, a new primer has been synthesized. The same polymerase molecule remains associated with the replication fork and proceeds to synthesize the next Okazaki fragment.

## The DNA Polymerase Complex

A number of different DNA polymerase molecules engage in DNA replication. These share three important properties: (1) **chain elongation,** (2) **processivity,** and (3) **proofreading.** Chain elongation accounts for the rate (in nucleotides per second) at which polymerization occurs. Processivity is an expression of the number of nucleotides added to the nascent chain before the polymerase disengages from the template. The proofreading function identifies copying errors and corrects them. In *E coli,* polymerase III (pol III) functions at the replication fork. Of all polymerases, it catalyzes the highest rate of chain elongation and is the most processive. It is capable of polymerizing 0.5 Mb of DNA during one cycle on the leading strand. Pol III is a large (> 1 MDa), ten-subunit protein complex in *E coli.* The two identical β subunits of pol III encircle the DNA template in a sliding "clamp," which accounts for the stability of the complex and for the high degree of processivity the enzyme exhibits.

Polymerase II (pol II) is mostly involved in proofreading and DNA repair. Polymerase I (pol I) completes chain synthesis between Okazaki fragments on the lagging strand. Eukaryotic cells have counterparts for each of these enzymes plus some additional ones. A comparison is shown in Table 36–6.

In mammalian cells, the polymerase is capable of polymerizing about 100 nucleotides per second, a rate at least tenfold slower than the rate of polymerization of deoxynucleotides by the bacterial DNA polymerase complex. This reduced rate may result from interference by nucleosomes. It is not known how the replication complex negotiates nucleosomes.

## Initiation & Elongation of DNA Synthesis

The initiation of DNA synthesis (Figure 36–14) requires **priming by a short length of RNA,** about 10–200 nucleotides long. This priming process involves the nucleophilic attack by the 3′-hydroxyl group of the RNA primer on the α phosphate of the first entering deoxynucleoside triphosphate (N in Figure 36–14) with the splitting off of pyrophosphate. The 3′-hydroxyl group of the recently attached deoxyribonucleoside monophosphate is then free to carry out a **nucleophilic attack** on the next entering deoxyribonucleoside triphosphate (N + 1 in Figure 36–14), again at its α phosphate moiety, with the splitting off of pyrophosphate. Of course, selection of the proper deoxyribonucleotide whose terminal 3′-hydroxyl group is to be attacked is dependent upon **proper base pairing**

***Table 36–6.*** A comparison of prokaryotic and eukaryotic DNA polymerases.

| *E coli* | Mammalian | Function |
|---|---|---|
| I | α | Gap filling and synthesis of lagging strand |
| II | ε | DNA proofreading and repair |
| | β | DNA repair |
| | γ | Mitochondrial DNA synthesis |
| III | δ | Processive, leading strand synthesis |

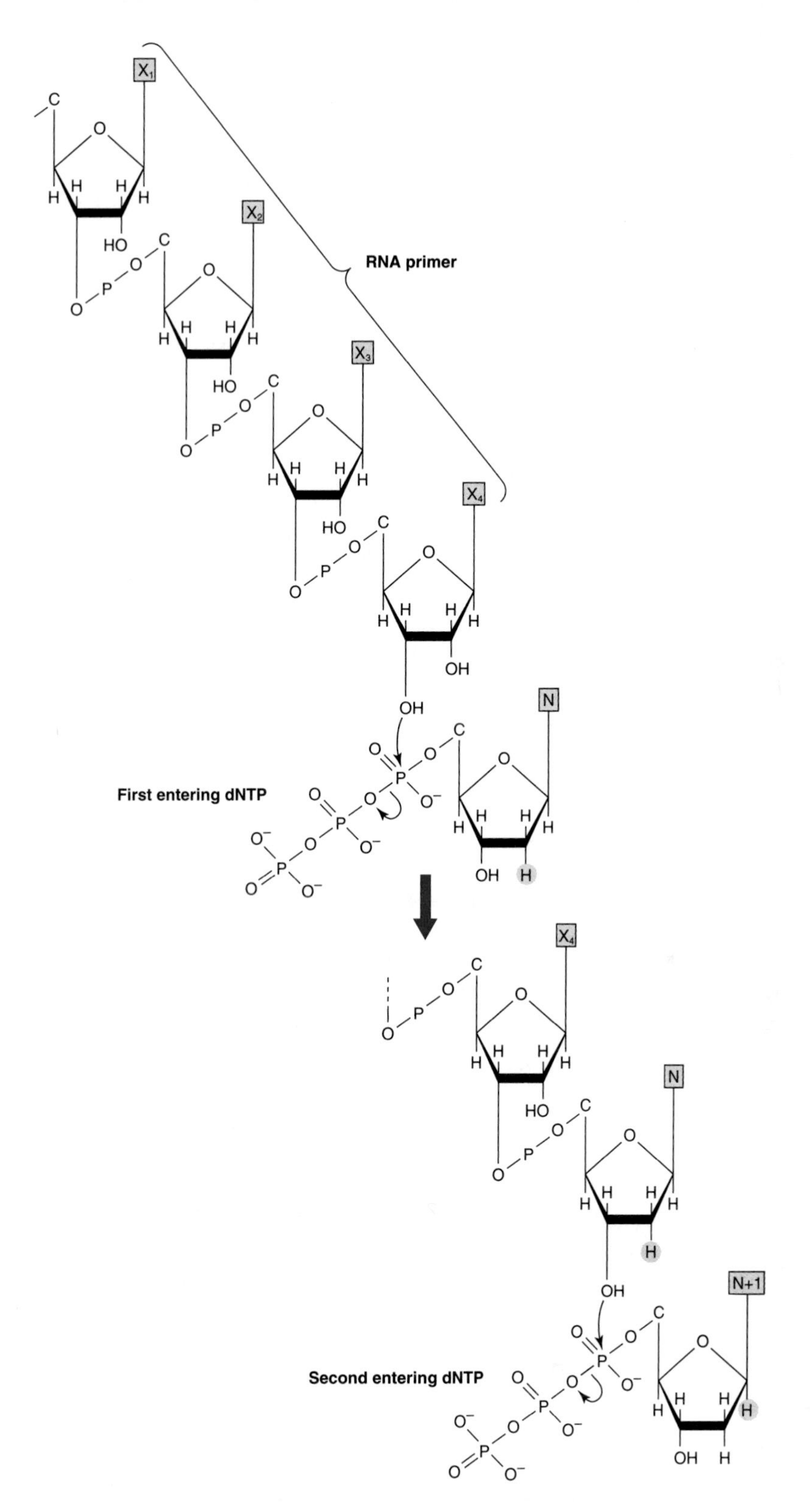


***Figure 36–14.*** The initiation of DNA synthesis upon a primer of RNA and the subsequent attachment of the second deoxyribonucleoside triphosphate.

**with the other strand** of the DNA molecule according to the rules proposed originally by Watson and Crick (Figure 36–15). When an adenine deoxyribonucleoside monophosphoryl moiety is in the template position, a thymidine triphosphate will enter and its α phosphate will be attacked by the 3′-hydroxyl group of the deoxyribonucleoside monophosphoryl most recently added to the polymer. By this stepwise process, the template dictates which deoxyribonucleoside triphosphate is complementary and by hydrogen bonding holds it in place while the 3′-hydroxyl group of the growing strand attacks and incorporates the new nucleotide into the polymer. These segments of DNA attached to an RNA initiator component are the **Okazaki fragments** (Figure 36–16). In mammals, after many Okazaki fragments are generated, the replication complex begins to remove the RNA primers, to fill in the gaps left by their removal with the proper base-paired deoxynucleotide, and then to seal the fragments of newly synthesized DNA by enzymes referred to as **DNA ligases.**

## Replication Exhibits Polarity

As has already been noted, DNA molecules are double-stranded and the two strands are antiparallel, ie, running in opposite directions. The replication of DNA in prokaryotes and eukaryotes occurs on both strands simultaneously. However, an enzyme capable of polymerizing DNA in the 3′ to 5′ direction does not exist in any organism, so that both of the newly replicated DNA strands cannot grow in the same direction simultaneously. Nevertheless, the same enzyme does replicate both strands at the same time. The single enzyme replicates one strand ("leading strand") in a continuous manner in the 5′ to 3′ direction, with the same overall forward direction. It replicates the other strand ("lagging strand") discontinuously while polymerizing the

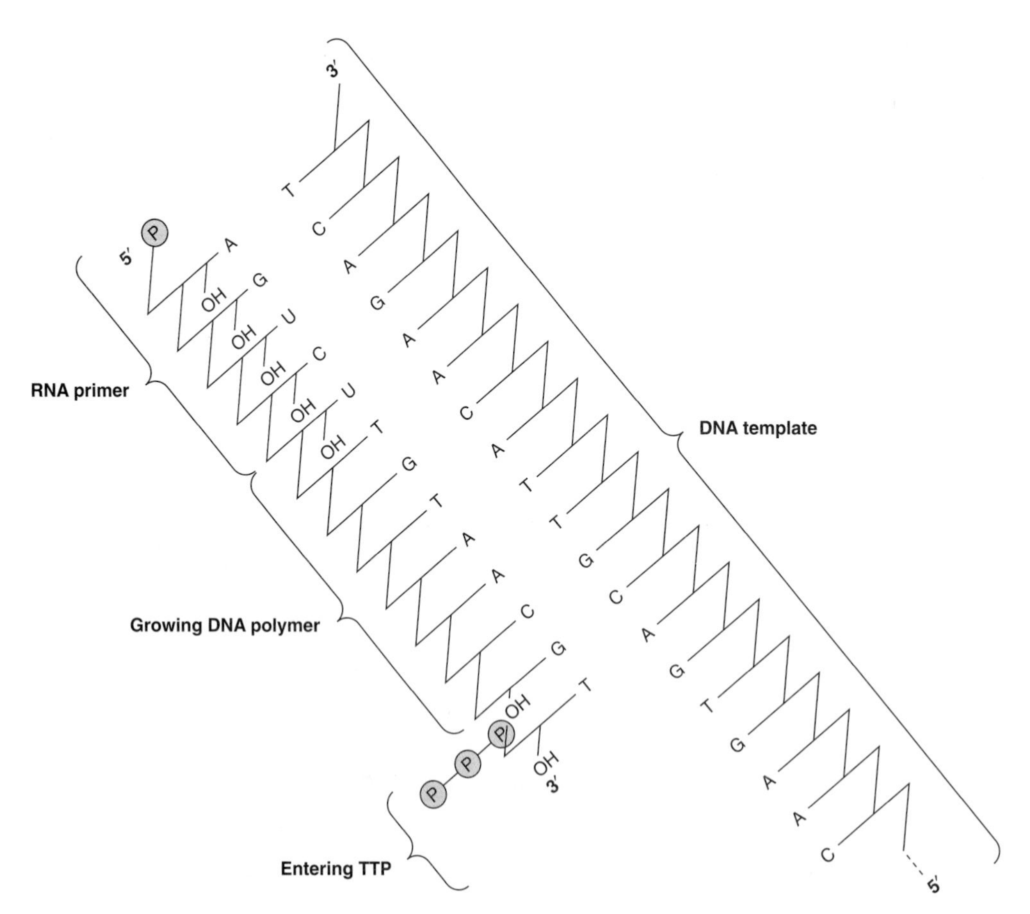


***Figure 36–15.*** The RNA-primed synthesis of DNA demonstrating the template function of the complementary strand of parental DNA.

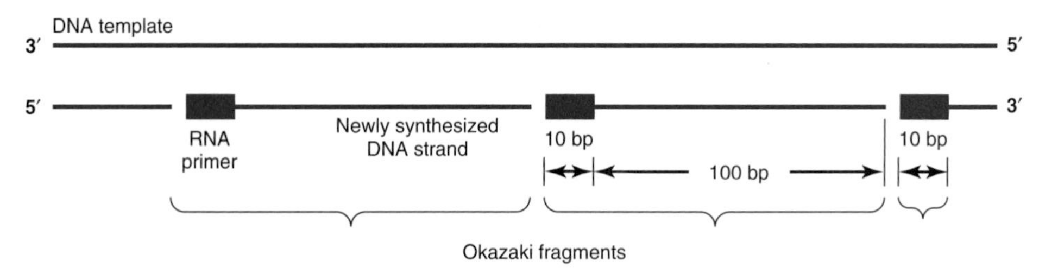


***Figure 36–16.*** The discontinuous polymerization of deoxyribonucleotides on the lagging strand; formation of Okazaki fragments during lagging strand DNA synthesis is illustrated.

nucleotides in short spurts of 150–250 nucleotides, again in the 5′ to 3′ direction, but at the same time it faces toward the back end of the preceding RNA primer rather than toward the unreplicated portion. This process of **semidiscontinuous DNA synthesis** is shown diagrammatically in Figures 36–13 and 36–16.

In the mammalian nuclear genome, most of the RNA primers are eventually removed as part of the replication process, whereas after replication of the mitochondrial genome the small piece of RNA remains as an integral part of the closed circular DNA structure.

## Formation of Replication Bubbles

Replication proceeds from a single ori in the circular bacterial chromosome, composed of roughly $6 \times 10^6$ bp of DNA. This process is completed in about 30 minutes, a replication rate of $3 \times 10^5$ bp/min. The entire mammalian genome replicates in approximately 9 hours, the average period required for formation of a tetraploid genome from a diploid genome in a replicating cell. If a mammalian genome ($3 \times 10^9$ bp) replicated at the same rate as bacteria (ie, $3 \times 10^5$ bp/min) from but a single ori, replication would take over 150 hours! Metazoan organisms get around this problem using two strategies. First, replication is bidirectional. Second, replication proceeds from multiple origins in each chromosome (a total of as many as 100 in humans). Thus, replication occurs in both directions along all of the chromosomes, and both strands are replicated simultaneously. This replication process generates **"replication bubbles"** (Figure 36–17).

The multiple sites that serve as origins for DNA replication in eukaryotes are poorly defined except in a few animal viruses and in yeast. However, it is clear that initiation is regulated both spatially and temporally, since clusters of adjacent sites initiate replication synchronously. There are suggestions that functional domains of chromatin replicate as intact units, implying that the origins of replication are specifically located with respect to transcription units.

During the replication of DNA, there must be a separation of the two strands to allow each to serve as a template by hydrogen bonding its nucleotide bases to the incoming deoxynucleoside triphosphate. The separation of the DNA double helix is promoted by SSBs, specific protein molecules that stabilize the single-stranded structure as the replication fork progresses. These stabi-

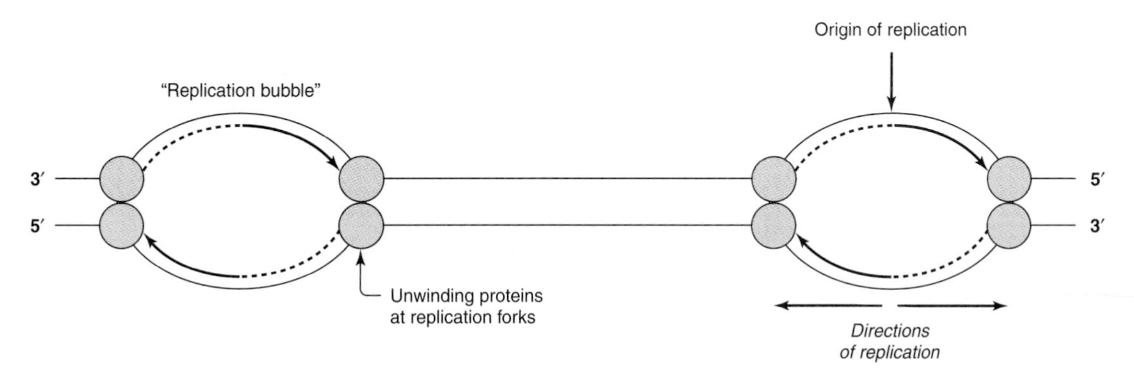


***Figure 36–17.*** The generation of "replication bubbles" during the process of DNA synthesis. The bidirectional replication and the proposed positions of unwinding proteins at the replication forks are depicted.

lizing proteins bind cooperatively and stoichiometrically to the single strands without interfering with the abilities of the nucleotides to serve as templates (Figure 36–13). In addition to separating the two strands of the double helix, there must be an unwinding of the molecule (once every 10 nucleotide pairs) to allow strand separation. This must happen in segments, given the time during which DNA replication occurs. There are multiple "swivels" interspersed in the DNA molecules of all organisms. The swivel function is provided by specific enzymes that introduce **"nicks" in one strand of the unwinding double helix,** thereby allowing the unwinding process to proceed. The nicks are quickly resealed without requiring energy input, because of the formation of a high-energy covalent bond between the nicked phosphodiester backbone and the nicking-sealing enzyme. The nicking-resealing enzymes are called **DNA topoisomerases.** This process is depicted diagrammatically in Figure 36–18 and there compared with the ATP-dependent resealing carried out by the DNA ligases. Topoisomerases are also capable of unwinding supercoiled DNA. Supercoiled DNA is a higher-ordered structure occurring in circular DNA molecules wrapped around a core, as depicted in Figure 36–19.

There exists in one species of animal viruses (retroviruses) a class of enzymes capable of synthesizing a sin-

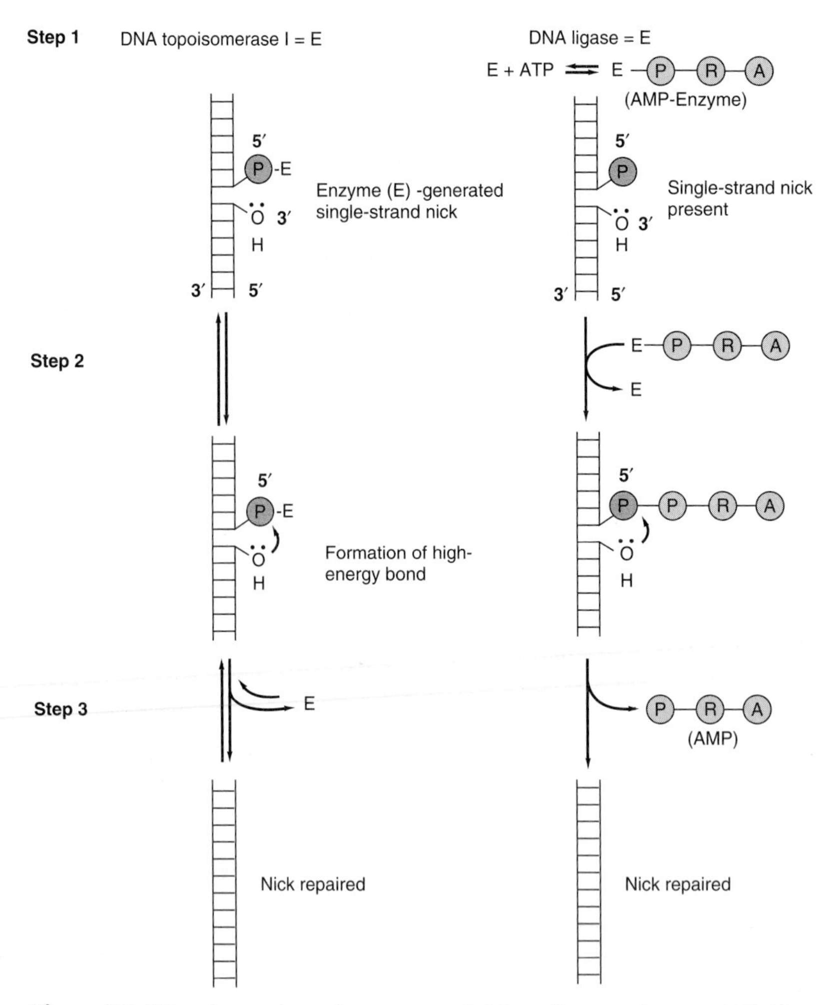


***Figure 36–18.*** Comparison of two types of nick-sealing reactions on DNA. The series of reactions at left is catalyzed by DNA topoisomerase I, that at right by DNA ligase. (Slightly modified and reproduced, with permission, from Lehninger AL: *Biochemistry,* 2nd ed. Worth, 1975.)

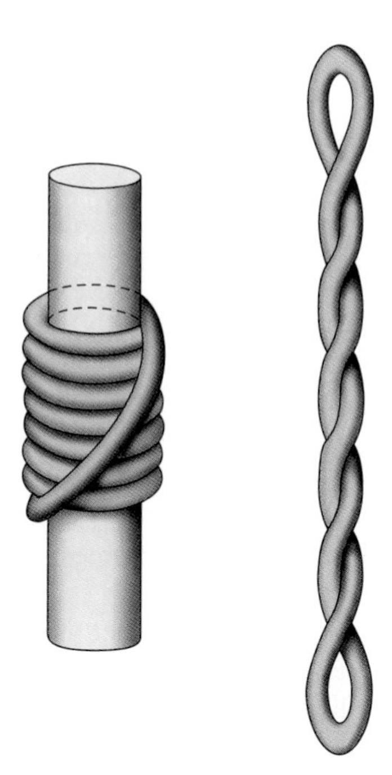

***Figure 36–19.*** Supercoiling of DNA. A left-handed toroidal (solenoidal) supercoil, at left, will convert to a right-handed interwound supercoil, at right, when the cylindric core is removed. Such a transition is analogous to that which occurs when nucleosomes are disrupted by the high salt extraction of histones from chromatin.

gle-stranded and then a double-stranded DNA molecule from a single-stranded RNA template. This polymerase, RNA-dependent DNA polymerase, or **"reverse transcriptase,"** first synthesizes a DNA-RNA hybrid molecule utilizing the RNA genome as a template. A specific nuclease, RNase H, degrades the RNA strand, and the remaining DNA strand in turn serves as a template to form a double-stranded DNA molecule containing the information originally present in the RNA genome of the animal virus.

## Reconstitution of Chromatin Structure

There is evidence that nuclear organization and chromatin structure are involved in determining the regulation and initiation of DNA synthesis. As noted above, the rate of polymerization in eukaryotic cells, which have chromatin and nucleosomes, is tenfold slower than that in prokaryotic cells, which have naked DNA. It is also clear that chromatin structure must be re-formed after replication. Newly replicated DNA is rapidly assembled into nucleosomes, and the preexisting and newly assembled histone octamers are randomly distributed to each arm of the replication fork.

## DNA Synthesis Occurs During the S Phase of the Cell Cycle

In animal cells, including human cells, the replication of the DNA genome occurs only at a specified time during the life span of the cell. This period is referred to as the synthetic or S phase. This is usually temporally separated from the mitotic phase by nonsynthetic periods referred to as gap 1 (G1) and gap 2 (G2), occurring before and after the S phase, respectively (Figure 36–20). Among other things, the cell prepares for DNA synthesis in G1 and for mitosis in G2. The cell regulates its DNA synthesis grossly by allowing it to occur only at specific times and mostly in cells preparing to divide by a mitotic process.

It appears that all eukaryotic cells have gene products that govern the transition from one phase of the cell cycle to another. The **cyclins** are a family of proteins whose concentration increases and decreases throughout the cell cycle—thus their name. The cyclins turn on, at the appropriate time, different **cyclin-dependent protein kinases (CDKs)** that phosphorylate substrates essential for progression through the cell cycle (Figure 36–21). For example, cyclin D levels rise in late G1 phase and allow progression beyond the **start (yeast)** or **restriction point (mammals)**, the point beyond which cells irrevocably proceed into the S or DNA synthesis phase.

The D cyclins activate CDK4 and CDK6. These two kinases are also synthesized during G1 in cells undergoing active division. The D cyclins and CDK4 and CDK6 are nuclear proteins that assemble as a complex in late G1 phase. The complex is an active serine-threonine protein kinase. One substrate for this kinase is the retinoblastoma (Rb) protein. Rb is a cell cycle regulator because it binds to and inactivates a transcription factor (E2F) necessary for the transcription of certain genes (histone genes, DNA replication proteins, etc) needed for progression from G1 to S phase. The phosphorylation of Rb by CDK4 or CDK6 results in the release of E2F from Rb-mediated transcription repression—thus, gene activation ensues and cell cycle progression takes place.

Other cyclins and CDKs are involved in different aspects of cell cycle progression (Table 36–7). Cyclin E and CDK2 form a complex in late G1. Cyclin E is rapidly degraded, and the released CDK2 then forms a complex with cyclin A. This sequence is necessary for the initiation of DNA synthesis in S phase. A complex between cyclin B and CDK1 is rate-limiting for the G2/M transition in eukaryotic cells.

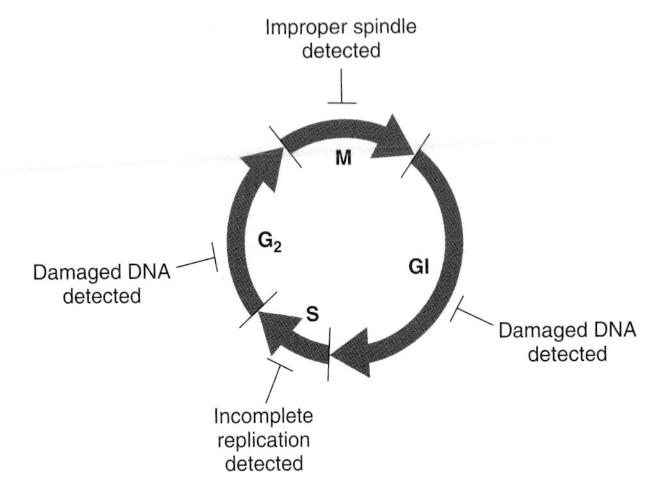


***Figure 36–20.*** Mammalian cell cycle and cell cycle checkpoints. DNA, chromosome, and chromosome segregation integrity is continuously monitored throughout the cell cycle. If DNA damage is detected in either the G1 or the G2 phase of the cell cycle, if the genome is incompletely replicated, or if normal chromosome segregation machinery is incomplete (ie, a defective spindle), cells will not progress through the phase of the cycle in which defects are detected. In some cases, if the damage cannot be repaired, such cells undergo programmed cell death (apoptosis).

Many of the cancer-causing viruses (oncoviruses) and cancer-inducing genes (oncogenes) are capable of alleviating or disrupting the apparent restriction that normally controls the entry of mammalian cells from G1 into the S phase. From the foregoing, one might have surmised that excessive production of a cyclin—or production at an inappropriate time—might result in abnormal or unrestrained cell division. In this context it is noteworthy that the *bcl* oncogene associated with B cell lymphoma appears to be the cyclin D1 gene. Similarly, the oncoproteins (or transforming proteins) produced by several DNA viruses target the Rb transcription repressor for inactivation, inducing cell division inappropriately.

During the S phase, mammalian cells contain greater quantities of DNA polymerase than during the nonsynthetic phases of the cell cycle. Furthermore, those enzymes responsible for formation of the substrates for DNA synthesis—ie, deoxyribonucleoside triphosphates—are also increased in activity, and their activity will diminish following the synthetic phase until the reappearance of the signal for renewed DNA

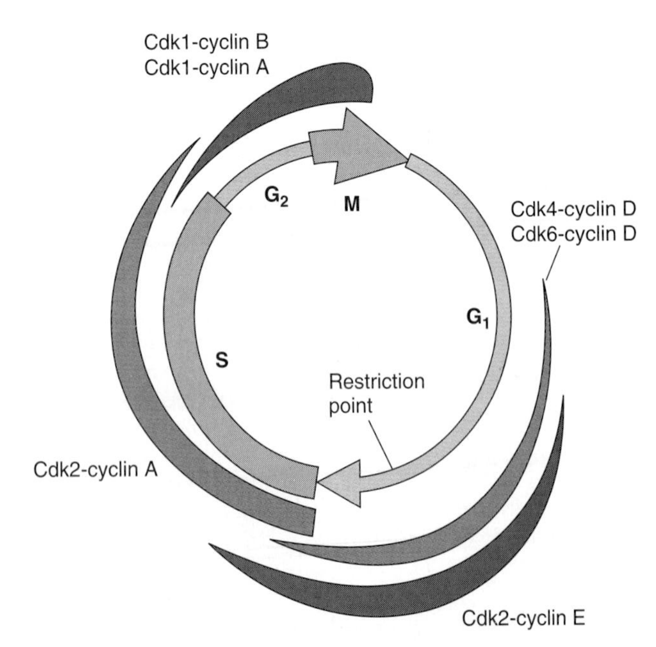


***Figure 36–21.*** Schematic illustration of the points during the mammalian cell cycle during which the indicated cyclins and cyclin-dependent kinases are activated. The thickness of the various colored lines is indicative of the extent of activity.

***Table 36–7.*** Cyclins and cyclin-dependent kinases involved in cell cycle progression.

| Cyclin | Kinase | Function |
|---|---|---|
| D | CDK4, CDK6 | Progression past restriction point at G1/S boundary |
| E, A | CDK2 | Initiation of DNA synthesis in early S phase |
| B | CDK1 | Transition from G2 to M |

synthesis. During the S phase, the nuclear DNA is **completely replicated once and only once.** It seems that once chromatin has been replicated, it is marked so as to prevent its further replication until it again passes through mitosis. The molecular mechanisms for this phenomenon have yet to be elucidated.

In general, a given pair of chromosomes will replicate simultaneously and within a fixed portion of the S phase upon every replication. On a chromosome, clusters of replication units replicate coordinately. The nature of the signals that regulate DNA synthesis at these levels is unknown, but the regulation does appear to be an intrinsic property of each individual chromosome.

## Enzymes Repair Damaged DNA

The maintenance of the integrity of the information in DNA molecules is of utmost importance to the survival of a particular organism as well as to survival of the species. Thus, it can be concluded that surviving species have evolved mechanisms for repairing DNA damage occurring as a result of either replication errors or environmental insults.

As described in Chapter 35, the major responsibility for the fidelity of replication resides in the specific pairing of nucleotide bases. Proper pairing is dependent upon the presence of the favored tautomers of the purine and pyrimidine nucleotides, but the equilibrium whereby one tautomer is more stable than another is only about $10^4$ or $10^5$ in favor of that with the greater stability. Although this is not favorable enough to ensure the high fidelity that is necessary, favoring of the preferred tautomers—and thus of the proper base pairing—could be ensured by monitoring the base pairing twice. Such double monitoring does appear to occur in both bacterial and mammalian systems: once at the time of insertion of the deoxyribonucleoside triphosphates, and later by a follow-up energy-requiring mechanism that removes all improper bases which may occur in the newly formed strand. This "proofreading" prevents tautomer-induced misincorporation from occurring more frequently than once every $10^8$–$10^{10}$ base pairs of DNA synthesized. The mechanisms responsible for this monitoring mechanism in *E coli* include the 3′ to 5′ exonuclease activities of one of the subunits of the pol III complex and of the pol I molecule. The analogous mammalian enzymes ($\delta$ and $\alpha$) do not seem to possess such a nuclease proofreading function. Other enzymes provide this repair function.

Replication errors, even with a very efficient repair system, lead to the accumulation of mutations. A human has $10^{14}$ nucleated cells each with $3 \times 10^9$ base pairs of DNA. If about $10^{16}$ cell divisions occur in a lifetime and $10^{-10}$ mutations per base pair per cell generation escape repair, there may eventually be as many as one mutation per $10^6$ bp in the genome. Fortunately, most of these will probably occur in DNA that does not encode proteins or will not affect the function of encoded proteins and so are of no consequence. In addition, spontaneous and chemically induced damage to DNA must be repaired.

Damage to DNA by environmental, physical, and chemical agents may be classified into four types (Table 36–8). Abnormal regions of DNA, either from copying errors or DNA damage, are replaced by four mechanisms: (1) mismatch repair, (2) base excision-repair, (3) nucleotide excision-repair, and (4) double-strand break repair (Table 36–9). These mechanisms exploit the redundancy of information inherent in the double helical DNA structure. The defective region in one strand can be returned to its original form by relying on the complementary information stored in the unaffected strand.

***Table 36–8.*** Types of damage to DNA.

I. Single-base alteration
   A. Depurination
   B. Deamination of cytosine to uracil
   C. Deamination of adenine to hypoxanthine
   D. Alkylation of base
   E. Insertion or deletion of nucleotide
   F. Base-analog incorporation
II. Two-base alteration
   A. UV light–induced thymine-thymine (pyrimidine) dimer
   B. Bifunctional alkylating agent cross-linkage
III. Chain breaks
   A. Ionizing radiation
   B. Radioactive disintegration of backbone element
   C. Oxidative free radical formation
IV. Cross-linkage
   A. Between bases in same or opposite strands
   B. Between DNA and protein molecules (eg, histones)

***Table 36–9.*** Mechanism of DNA repair

| Mechanism | Problem | Solution |
|---|---|---|
| Mismatch repair | Copying errors (single base or two- to five-base unpaired loops) | Methyl-directed strand cutting, exonuclease digestion, and replacement |
| Base excision-repair | Spontaneous, chemical, or radiation damage to a single base | Base removal by *N*-glycosylase, abasic sugar removal, replacement |
| Nucleotide excision-repair | Spontaneous, chemical, or radiation damage to a DNA segment | Removal of an approximately 30-nucleotide oligomer and replacement |
| Double-strand break repair | Ionizing radiation, chemotherapy, oxidative free radicals | Synapsis, unwinding, alignment, ligation |

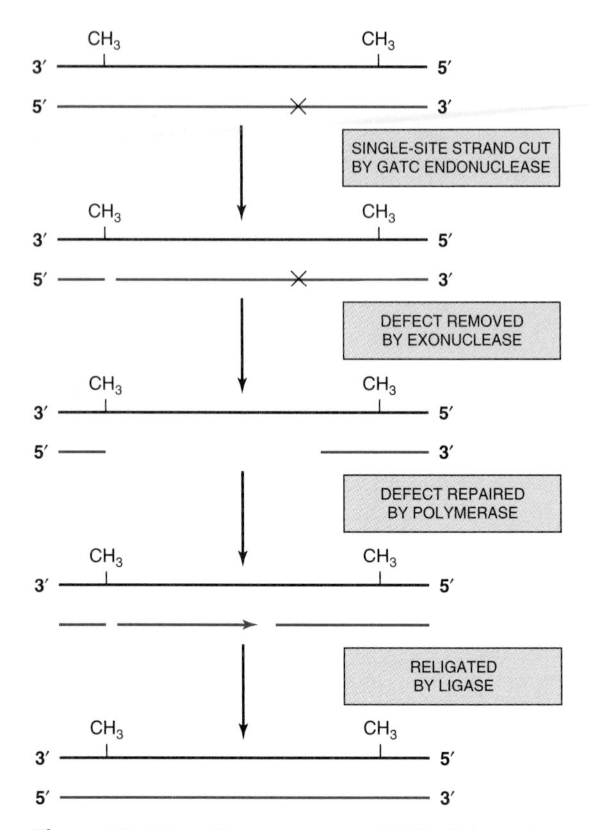


***Figure 36–22.*** Mismatch repair of DNA. This mechanism corrects a single mismatch base pair (eg, C to A rather than T to A) or a short region of unpaired DNA. The defective region is recognized by an endonuclease that makes a single-strand cut at an adjacent methylated GATC sequence. The DNA strand is removed through the mutation, replaced, and religated.

## Mismatch Repair

Mismatch repair corrects errors made when DNA is copied. For example, a C could be inserted opposite an A, or the polymerase could slip or stutter and insert two to five extra unpaired bases. Specific proteins scan the newly synthesized DNA, using adenine methylation within a GATC sequence as the point of reference (Figure 36–22). The template strand is methylated, and the newly synthesized strand is not. This difference allows the repair enzymes to identify the strand that contains the errant nucleotide which requires replacement. If a mismatch or small loop is found, a GATC endonuclease cuts the strand bearing the mutation at a site corresponding to the GATC. An exonuclease then digests this strand from the GATC through the mutation, thus removing the faulty DNA. This can occur from either end if the defect is bracketed by two GATC sites. This defect is then filled in by normal cellular enzymes according to base pairing rules. In *E coli,* three proteins (Mut S, Mut C, and Mut H) are required for recognition of the mutation and nicking of the strand. Other cellular enzymes, including ligase, polymerase, and SSBs, remove and replace the strand. The process is somewhat more complicated in mammalian cells, as about six proteins are involved in the first steps.

Faulty mismatch repair has been linked to hereditary nonpolyposis colon cancer (HNPCC), one of the most common inherited cancers. Genetic studies linked HNPCC in some families to a region of chromosome 2. The gene located, designated *hMSH2,* was subsequently shown to encode the human analog of the *E coli* MutS protein that is involved in mismatch repair (see above). Mutations of *hMSH2* account for 50–60% of HNPCC cases. Another gene, *hMLH1,* is associated with most of the other cases. *hMLH1* is the human analog of the bacterial mismatch repair gene *MutL.* How does faulty mismatch repair result in colon cancer? The human genes were localized because microsatellite instability was detected. That is, the cancer cells had a microsatellite of a length different from that found in the normal cells of the individual. It appears that the affected cells, which harbor a mutated *hMSH2* or *hMLH1* mismatch repair enzyme, are unable to remove small loops of unpaired DNA, and the microsatellite thus increases in size. Ultimately, microsatellite DNA expansion must affect either the expression or the function of a protein critical in surveillance of the cell cycle in these colon cells.

## Base Excision-Repair

The **depurination of DNA,** which happens spontaneously owing to the thermal lability of the purine N-glycosidic bond, occurs at a rate of 5000–10,000/cell/d at 37 °C. Specific enzymes recognize a depurinated site and replace the appropriate purine directly, without interruption of the phosphodiester backbone.

Cytosine, adenine, and guanine bases in DNA spontaneously form uracil, hypoxanthine, or xanthine, respectively. Since none of these normally exist in DNA, it is not surprising that specific **N-glycosylases** can recognize these abnormal bases and remove the base itself from the DNA. This removal marks the site of the defect and allows an **apurinic or apyrimidinic endonuclease** to excise the abasic sugar. The proper base is then replaced by a repair DNA polymerase, and a **ligase** returns the DNA to its original state (Figure 36–23). This series of events is called **base excision-repair.** By a similar series of steps involving initially the recognition of the defect, alkylated bases and base analogs can be removed from DNA and the DNA returned to its original informational content. This mechanism is suitable for replacement of a single base but is not effective at replacing regions of damaged DNA.

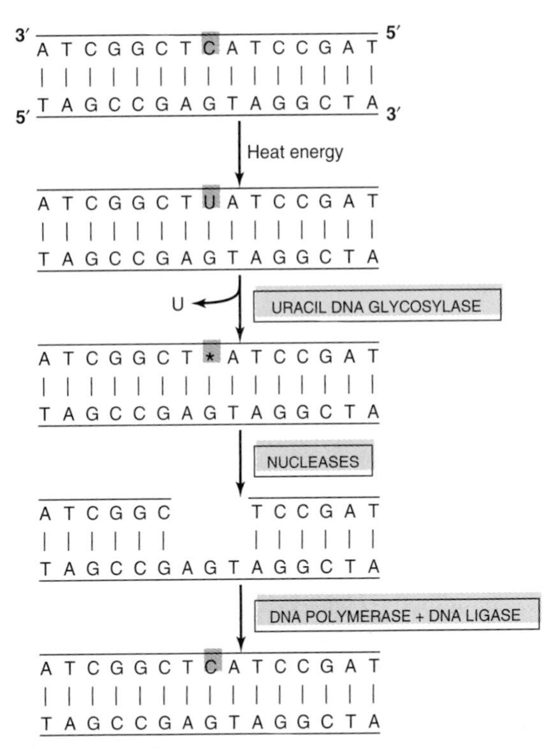


***Figure 36–23.*** Base excision-repair of DNA. The enzyme uracil DNA glycosylase removes the uracil created by spontaneous deamination of cytosine in the DNA. An endonuclease cuts the backbone near the defect; then, after an endonuclease removes a few bases, the defect is filled in by the action of a repair polymerase and the strand is rejoined by a ligase. (Courtesy of B Alberts.)

## Nucleotide Excision-Repair

This mechanism is used to replace regions of damaged DNA up to 30 bases in length. Common examples of DNA damage include ultraviolet (UV) light, which induces the formation of cyclobutane pyrimidine-pyrimidine dimers, and smoking, which causes formation of benzo[*a*]pyrene-guanine adducts. Ionizing radiation, cancer chemotherapeutic agents, and a variety of chemicals found in the environment cause base modification, strand breaks, cross-linkage between bases on opposite strands or between DNA and protein, and numerous other defects. These are repaired by a process called nucleotide excision-repair (Figure 36–24). This complex process, which involves more gene products than the two other types of repair, essentially involves the hydrolysis of two phosphodiester bonds on the strand containing the defect. A special excision nuclease (exinuclease), consisting of at least three subunits in *E coli* and 16 polypeptides in humans, accomplishes this task. In eukaryotic cells the enzymes cut between the third to fifth phosphodiester bond 3′ from the lesion, and on the 5′ side the cut is somewhere between the twenty-first and twenty-fifth bonds. Thus, a fragment of DNA 27–29 nucleotides long is excised. After the strand is removed it is replaced, again by exact base pairing, through the action of yet another polymerase (δ/ε in humans), and the ends are joined to the existing strands by DNA ligase.

**Xeroderma pigmentosum (XP)** is an autosomal recessive genetic disease. The clinical syndrome includes marked sensitivity to sunlight (ultraviolet) with subsequent formation of multiple skin cancers and premature death. The risk of developing skin cancer is increased 1000- to 2000-fold. The inherited defect seems to involve the repair of damaged DNA, particularly thymine dimers. Cells cultured from patients with xeroderma pigmentosum exhibit low activity for the nucleotide excision-repair process. Seven complementation groups have been identified using hybrid cell analyses, so at least seven gene products (XPA–XPG) are involved. Two of these (XPA and XPC) are involved in recognition and excision. XPB and XPD are helicases and, interestingly, are subunits of the transcription factor TFIIH (see Chapter 37).

## Double-Strand Break Repair

The repair of double-strand breaks is part of the physiologic process of immunoglobulin gene rearrangement. It

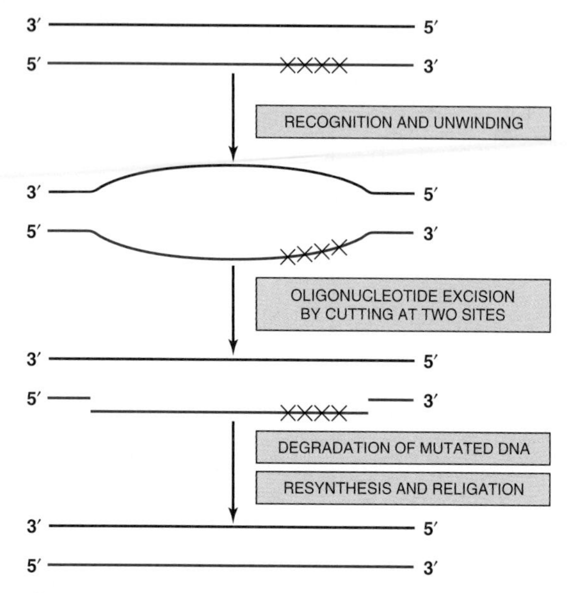


***Figure 36–24.*** Nucleotide excision-repair. This mechanism is employed to correct larger defects in DNA and generally involves more proteins than either mismatch or base excision-repair. After defect recognition (indicated by XXXX) and unwinding of the DNA encompassing the defect, an excision nuclease (exinuclease) cuts the DNA upstream and downstream of the defective region. This gap is then filled in by a polymerase (δ/ε in humans) and religated.

is also an important mechanism for repairing damaged DNA, such as occurs as a result of ionizing radiation or oxidative free radical generation. Some chemotherapeutic agents destroy cells by causing ds breaks or preventing their repair.

Two proteins are initially involved in the nonhomologous rejoining of a ds break. **Ku**, a heterodimer of 70 kDa and 86 kDa subunits, binds to free DNA ends and has latent ATP-dependent helicase activity. The DNA-bound Ku heterodimer recruits a unique protein kinase, **DNA-dependent protein kinase (DNA-PK).** DNA-PK has a binding site for DNA free ends and another for dsDNA just inside these ends. It therefore allows for the approximation of the two separated ends. The free end DNA-Ku-DNA-PK complex activates the kinase activity in the latter. DNA-PK reciprocally phosphorylates Ku and the other DNA-PK molecule, on the opposing strand, in trans. DNA-PK then dissociates from the DNA and Ku, resulting in activation of the Ku helicase. This results in unwinding of the two ends. The unwound, approximated DNA forms base pairs; the extra nucleotide tails are removed by an exonuclease; and the gaps are filled and closed by DNA ligase. This repair mechanism is illustrated in Figure 36–25.

## Some Repair Enzymes Are Multifunctional

Somewhat surprising is the recent observation that DNA repair proteins can serve other purposes. For example, some repair enzymes are also found as components of the large TFIIH complex that plays a central role in gene transcription (Chapter 37). Another component of TFIIH is involved in cell cycle regulation. Thus, three critical cellular processes may be linked through use of common proteins. There is also good evidence that some repair enzymes are involved in gene rearrangements that occur normally.

In patients with **ataxia-telangiectasia,** an autosomal recessive disease in humans resulting in the development of cerebellar ataxia and lymphoreticular neoplasms, there appears to exist an increased sensitivity to damage by x-ray. Patients with **Fanconi's anemia,** an autosomal recessive anemia characterized also by an increased frequency of cancer and by chromosomal instability, probably have defective repair of cross-linking damage.

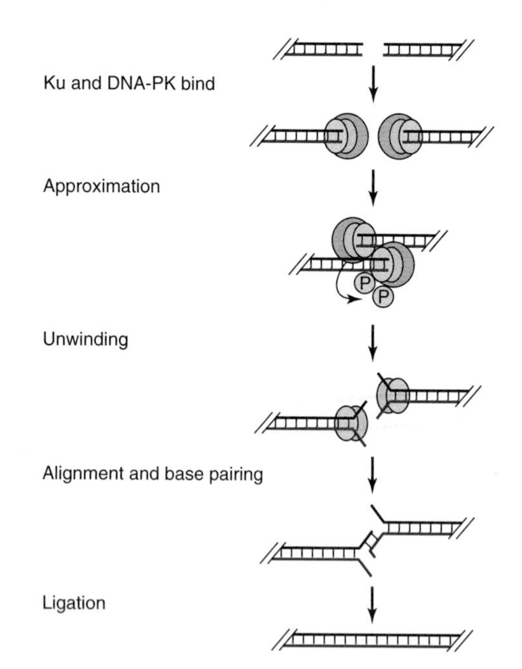


***Figure 36–25.*** Double-strand break repair of DNA. The proteins Ku and DNA-dependent protein kinase combine to approximate the two strands and unwind them. The aligned fragments form base pairs; the extra ends are removed, probably by a DNA-PK-associated endo- or exonuclease, and the gaps are filled in; and continuity is restored by ligation.

All three of these clinical syndromes are associated with an increased frequency of cancer. It is likely that other human diseases resulting from disordered DNA repair capabilities will be found in the future.

### DNA & Chromosome Integrity Is Monitored Throughout the Cell Cycle

Given the importance of normal DNA and chromosome function to survival, it is not surprising that eukaryotic cells have developed elaborate mechanisms to monitor the integrity of the genetic material. As detailed above, a number of complex multi-subunit enzyme systems have evolved to repair damaged DNA at the nucleotide sequence level. Similarly, DNA mishaps at the chromosome level are also monitored and repaired. As shown in Figure 36–19, DNA integrity and chromosomal integrity are continuously monitored throughout the cell cycle. The four specific steps at which this monitoring occurs have been termed **checkpoint controls.** If problems are detected at any of these checkpoints, progression through the cycle is interrupted and transit through the cell cycle is halted until the damage is repaired. The molecular mechanisms underlying detection of DNA damage during the G1 and G2 phases of the cycle are understood better than those operative during S and M phases.

The **tumor suppressor** p53, a protein of MW 53 kDa, plays a key role in both G1 and G2 checkpoint control. Normally a very unstable protein, p53 is a DNA binding transcription factor, one of a family of related proteins, that is somehow stabilized in response to DNA damage, perhaps by direct p53-DNA interactions. Increased levels of p53 activate transcription of an ensemble of genes that collectively serve to delay transit through the cycle. One of these induced proteins, $p21^{CIP}$, is a potent CDK-cyclin inhibitor (CKI) that is capable of efficiently inhibiting the action of all CDKs. Clearly, inhibition of CDKs will halt progression through the cell cycle (see Figures 36–19 and 36–20). If DNA damage is too extensive to repair, the affected cells undergo **apoptosis** (programmed cell death) in a p53-dependent fashion. In this case, p53 induces the activation of a collection of genes that induce apoptosis. Cells lacking functional p53 fail to undergo apoptosis in response to high levels of radiation or DNA-active chemotherapeutic agents. It may come as no surprise, then, that *p53* is one of the most frequently mutated genes in human cancers. Additional research into the mechanisms of checkpoint control will prove invaluable for the development of effective anticancer therapeutic options.

## SUMMARY

- DNA in eukaryotic cells is associated with a variety of proteins, resulting in a structure called chromatin.
- Much of the DNA is associated with histone proteins to form a structure called the nucleosome. Nucleosomes are composed of an octamer of histones and 150 bp of DNA.
- Nucleosomes and higher-order structures formed from them serve to compact the DNA.
- As much as 90% of DNA may be transcriptionally inactive as a result of being nuclease-resistant, highly compacted, and nucleosome-associated.
- DNA in transcriptionally active regions is sensitive to nuclease attack; some regions are exceptionally sensitive and are often found to contain transcription control sites.
- Transcriptionally active DNA (the genes) is often clustered in regions of each chromosome. Within these regions, genes may be separated by inactive DNA in nucleosomal structures. The transcription unit—that portion of a gene that is copied by RNA polymerase—consists of coding regions of DNA (exons) interrupted by intervening sequences of noncoding DNA (introns).
- After transcription, during RNA processing, introns are removed and the exons are ligated together to form the mature mRNA that appears in the cytoplasm.
- DNA in each chromosome is exactly replicated according to the rules of base pairing during the S phase of the cell cycle.
- Each strand of the double helix is replicated simultaneously but by somewhat different mechanisms. A complex of proteins, including DNA polymerase, replicates the leading strand continuously in the 5′ to 3′ direction. The lagging strand is replicated discontinuously, in short pieces of 150–250 nucleotides, in the 3′ to 5′ direction.
- DNA replication occurs at several sites—called replication bubbles—in each chromosome. The entire process takes about 9 hours in a typical cell.
- A variety of mechanisms employing different enzymes repair damaged DNA, as after exposure to chemical mutagens or ultraviolet radiation.

## REFERENCES

DePamphilis ML: Origins of DNA replication in metazoan chromosomes. J Biol Chem 1993;268:1.

Hartwell LH, Kastan MB: Cell cycle control and cancer. Science 1994;266:1821.

Jenuwein T, Allis CD: Translating the histone code. Science 2001; 293:1074.

Lander ES et al: Initial sequencing and analysis of the human genome. Nature 2001;409:860.

Luger L et al: Crystal structure of the nucleosome core particle at 2.8 Å resolution. Nature 1997;398:251.

Marians KJ: Prokaryotic DNA replication. Annu Rev Biochem 1992;61:673.

Michelson RJ, Weinart T: Closing the gaps among a web of DNA repair disorders. Bioessays J 2002;22:966.

Moll UM, Erster S, Zaika A: p53, p63 and p73—solos, alliances and feuds among family members. Biochim Biophys Acta 2001;1552:47.

Mouse Genome Sequencing Consortium: Initial sequencing and comparative analysis of the mouse genome. Nature 2002; 420:520.

Narlikar GJ et al: Cooperation between complexes that regulate chromatin structure and transcription. Cell 2002;108:475.

Sullivan et al: Determining centromere identity: cyclical stories and forking paths. Nat Rev Genet 2001;2:584.

van Holde K, Zlatanova J: Chromatin higher order: chasing a mirage? J Biol Chem 1995;270:8373.

Venter JC et al: The sequence of the human genome. Science 2002;291:1304.

Wallace DC: Mitochondrial DNA in aging and disease. Sci Am 1997 Aug;277:40.

Wood RD: Nucleotide excision repair in mammalian cells. J Biol Chem 1997;272:23465.

# RNA Synthesis, Processing, & Modification

37

Daryl K. Granner, MD, & P. Anthony Weil, PhD

## BIOMEDICAL IMPORTANCE

The synthesis of an RNA molecule from DNA is a complex process involving one of the group of RNA polymerase enzymes and a number of associated proteins. The general steps required to synthesize the primary transcript are initiation, elongation, and termination. Most is known about initiation. A number of DNA regions (generally located upstream from the initiation site) and protein factors that bind to these sequences to regulate the initiation of transcription have been identified. Certain RNAs—mRNAs in particular—have very different life spans in a cell. It is important to understand the basic principles of messenger RNA synthesis and metabolism, for modulation of this process results in altered rates of protein synthesis and thus a variety of metabolic changes. This is how all organisms adapt to changes of environment. It is also how differentiated cell structures and functions are established and maintained. The RNA molecules synthesized in mammalian cells are made as precursor molecules that have to be processed into mature, active RNA. Errors or changes in synthesis, processing, and splicing of mRNA transcripts are a cause of disease.

## RNA EXISTS IN FOUR MAJOR CLASSES

All eukaryotic cells have four major classes of RNA: ribosomal RNA (rRNA), messenger RNA (mRNA), transfer RNA (tRNA), and small nuclear RNA (snRNA). The first three are involved in protein synthesis, and snRNA is involved in mRNA splicing. As shown in Table 37–1, these various classes of RNA are different in their diversity, stability, and abundance in cells.

## RNA IS SYNTHESIZED FROM A DNA TEMPLATE BY AN RNA POLYMERASE

The processes of DNA and RNA synthesis are similar in that they involve (1) the general steps of initiation, elongation, and termination with 5′ to 3′ polarity; (2) large, multicomponent initiation complexes; and (3) adherence to Watson-Crick base-pairing rules. These processes differ in several important ways, including the following: (1) ribonucleotides are used in RNA synthesis rather than deoxyribonucleotides; (2) U replaces T as the complementary base pair for A in RNA; (3) a primer is not involved in RNA synthesis; (4) only a very small portion of the genome is transcribed or copied into RNA, whereas the entire genome must be copied during DNA replication; and (5) there is no proofreading function during RNA transcription.

The process of synthesizing RNA from a DNA template has been characterized best in prokaryotes. Although in mammalian cells the regulation of RNA synthesis and the processing of the RNA transcripts are different from those in prokaryotes, the process of RNA synthesis per se is quite similar in these two classes of organisms. Therefore, the description of RNA synthesis in prokaryotes, where it is better understood, is applicable to eukaryotes even though the enzymes involved and the regulatory signals are different.

### The Template Strand of DNA Is Transcribed

The sequence of ribonucleotides in an RNA molecule is complementary to the sequence of deoxyribonucleotides in one strand of the double-stranded DNA molecule (Figure 35–8). The strand that is transcribed or copied into an RNA molecule is referred to as the **template strand** of the DNA. The other DNA strand is frequently referred to as the **coding strand** of that gene. It is called this because, with the exception of T for U changes, it corresponds exactly to the sequence of the primary transcript, which encodes the protein product of the gene. In the case of a double-stranded DNA molecule containing many genes, the template strand for each gene will not necessarily be the same strand of the DNA double helix (Figure 37–1). Thus, a given strand of a double-stranded DNA molecule will serve as the template strand for some genes and the coding strand of other genes. Note that the nucleotide sequence of an RNA transcript will be the same (except for U replacing T) as that of the coding strand. The information in the template strand is read out in the 3′ to 5′ direction.

***Table 37–1.*** Classes of eukaryotic RNA.

| RNA | Types | Abundance | Stability |
|---|---|---|---|
| Ribosomal (rRNA) | 28S, 18S, 5.8S, 5S | 80% of total | Very stable |
| Messenger (mRNA) | ~$10^5$ different species | 2–5% of total | Unstable to very stable |
| Transfer (tRNA) | ~60 different species | ~15% of total | Very stable |
| Small nuclear (snRNA) | ~30 different species | ≤ 1% of total | Very stable |

## DNA-Dependent RNA Polymerase Initiates Transcription at a Distinct Site, the Promoter

DNA-dependent RNA polymerase is the enzyme responsible for the polymerization of ribonucleotides into a sequence complementary to the template strand of the gene (see Figures 37–2 and 37–3). The enzyme attaches at a specific site—the promoter—on the template strand. This is followed by initiation of RNA synthesis at the starting point, and the process continues until a termination sequence is reached (Figure 37–3). A **transcription unit** is defined as that region of DNA that includes the signals for transcription initiation, elongation, and termination. The RNA product, which is synthesized in the 5′ to 3′ direction, is the **primary transcript.** In prokaryotes, this can represent the product of several contiguous genes; in mammalian cells, it usually represents the product of a single gene. The 5′ terminals of the primary RNA transcript and the mature cytoplasmic RNA are identical. **Thus, the starting point of transcription corresponds to the 5′ nucleotide of the mRNA.** This is designated position +1, as is the corresponding nucleotide in the DNA. The

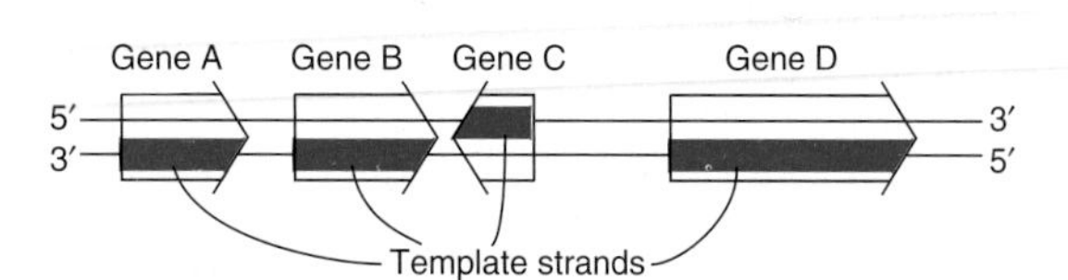


***Figure 37–1.*** This figure illustrates that genes can be transcribed off both strands of DNA. The arrowheads indicate the direction of transcription (polarity). Note that the template strand is always read in the 3′ to 5′ direction. The opposite strand is called the coding strand because it is identical (except for T for U changes) to the mRNA transcript (the primary transcript in eukaryotic cells) that encodes the protein product of the gene.

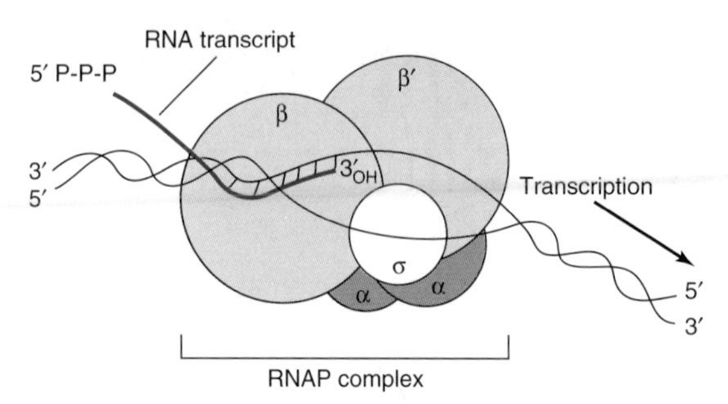


***Figure 37–2.*** RNA polymerase (RNAP) catalyzes the polymerization of ribonucleotides into an RNA sequence that is complementary to the template strand of the gene. The RNA transcript has the same polarity (5′ to 3′) as the coding strand but contains U rather than T. *E coli* RNAP consists of a core complex of two α subunits and two β subunits (β and β′). The holoenzyme contains the σ subunit bound to the $\alpha_2\beta\beta'$ core assembly. The ω subunit is not shown. The transcription "bubble" is an approximately 20-bp area of melted DNA, and the entire complex covers 30–75 bp, depending on the conformation of RNAP.

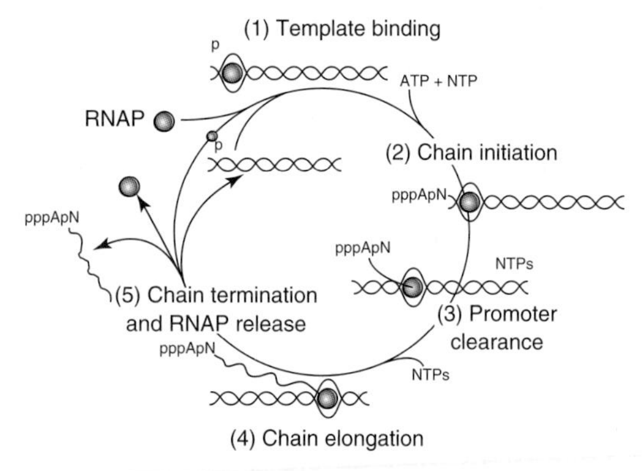


***Figure 37–3.*** The transcription cycle in bacteria. Bacterial RNA transcription is described in four steps: **(1) Template binding:** RNA polymerase (RNAP) binds to DNA and locates a promoter (P) melts the two DNA strands to form a preinitiation complex (PIC). **(2) Chain initiation:** RNAP holoenzyme (core + one of multiple sigma factors) catalyzes the coupling of the first base (usually ATP or GTP) to a second ribonucleoside triphosphate to form a dinucleotide. **(3) Chain elongation:** Successive residues are added to the 3′-OH terminus of the nascent RNA molecule. **(4) Chain termination and release:** The completed RNA chain and RNAP are released from the template. The RNAP holoenzyme re-forms, finds a promoter, and the cycle is repeated.

numbers increase as the sequence proceeds *downstream.* This convention makes it easy to locate particular regions, such as intron and exon boundaries. The nucleotide in the promoter adjacent to the transcription initiation site is designated −1, and these negative numbers increase as the sequence proceeds *upstream,* away from the initiation site. This provides a conventional way of defining the location of regulatory elements in the promoter.

The primary transcripts generated by RNA polymerase II—one of three distinct nuclear DNA-dependent RNA polymerases in eukaryotes—are promptly capped by 7-methylguanosine triphosphate caps (Figure 35–10) that persist and eventually appear on the 5′ end of mature cytoplasmic mRNA. These caps are necessary for the subsequent processing of the primary transcript to mRNA, for the translation of the mRNA, and for protection of the mRNA against exonucleolytic attack.

### Bacterial DNA-Dependent RNA Polymerase Is a Multisubunit Enzyme

The DNA-dependent RNA polymerase (RNAP) of the bacterium *Escherichia coli* exists as an approximately 400 kDa core complex consisting of two identical α subunits, similar but not identical β and β′ subunits, and an ω subunit. Beta is thought to be the catalytic subunit (Figure 37–2). RNAP, a metalloenzyme, also contains two zinc molecules. The core RNA polymerase associates with a specific protein factor (the sigma [σ] factor) that helps the core enzyme recognize and bind to the specific deoxynucleotide sequence of the promoter region (Figure 37–5) to form the preinitiation complex (PIC). Sigma factors have a dual role in the process of promoter recognition; σ association with core RNA polymerase decreases its affinity for nonpromoter DNA while simultaneously increasing holoenzyme affinity for promoter DNA. Bacteria contain multiple σ factors, each of which acts as a regulatory protein that modifies the **promoter recognition specificity** of the RNA polymerase. The appearance of different σ factors can be correlated temporally with various programs of gene expression in prokaryotic systems such as bacteriophage development, sporulation, and the response to heat shock.

### Mammalian Cells Possess Three Distinct Nuclear DNA-Dependent RNA Polymerases

The properties of mammalian polymerases are described in Table 37–2. Each of these DNA-dependent RNA polymerases is responsible for transcription of different sets of genes. The sizes of the RNA polymerases range from MW 500,000 to MW 600,000. These enzymes are much more complex than prokaryotic RNA polymerases. They all have two large subunits and a number of smaller subunits—as many as 14 in the case of RNA pol III. The eukaryotic RNA polymerases have extensive amino acid homologies with prokaryotic RNA polymerases. This homology has been shown recently to extend to the level of three-dimensional structures. The functions of each of the subunits are not yet fully understood. Many could have regulatory functions, such as serving to assist the polymerase in the recognition of specific sequences like promoters and termination signals.

***Table 37–2.*** Nomenclature and properties of mammalian nuclear DNA-dependent RNA polymerases.

| Form of RNA Polymerase | Sensitivity to α-Amanitin | Major Products |
|---|---|---|
| I (A) | Insensitive | rRNA |
| II (B) | High sensitivity | mRNA |
| III (C) | Intermediate sensitivity | tRNA/5S rRNA |

One peptide toxin from the mushroom *Amanita phalloides,* α-amanitin, is a specific differential inhibitor of the eukaryotic nuclear DNA-dependent RNA polymerases and as such has proved to be a powerful research tool (Table 37–2). α-Amanitin blocks the translocation of RNA polymerase during transcription.

## RNA SYNTHESIS IS A CYCLICAL PROCESS & INVOLVES INITIATION, ELONGATION, & TERMINATION

The process of RNA synthesis in bacteria—depicted in Figure 37–3—involves first the binding of the RNA holopolymerase molecule to the template at the promoter site to form a PIC. Binding is followed by a conformational change of the RNAP, and the first nucleotide (almost always a purine) then associates with the initiation site on the β subunit of the enzyme. In the presence of the appropriate nucleotide, the RNAP catalyzes the formation of a phosphodiester bond, and the nascent chain is now attached to the polymerization site on the β subunit of RNAP. (The analogy to the A and P sites on the ribosome should be noted; see Figure 38–9.)

**Initiation** of formation of the RNA molecule at its 5′ end then follows, while elongation of the RNA molecule from the 5′ to its 3′ end continues cyclically, antiparallel to its template. The enzyme polymerizes the ribonucleotides in a specific sequence dictated by the template strand and interpreted by Watson-Crick base-pairing rules. Pyrophosphate is released in the polymerization reaction. This pyrophosphate ($PP_i$) is rapidly degraded to 2 mol of inorganic phosphate ($P_i$) by ubiquitous pyrophosphatases, thereby providing irreversibility on the overall synthetic reaction. In both prokaryotes and eukaryotes, a purine ribonucleotide is usually the first to be polymerized into the RNA molecule. As with eukaryotes, 5′ triphosphate of this first nucleotide is maintained in prokaryotic mRNA.

As the **elongation** complex containing the core RNA polymerase progresses along the DNA molecule, **DNA unwinding** must occur in order to provide access for the appropriate base pairing to the nucleotides of the coding strand. The extent of this transcription bubble (ie, DNA unwinding) is constant throughout transcription and has been estimated to be about 20 base pairs per polymerase molecule. Thus, it appears that the size of the unwound DNA region is dictated by the polymerase and is independent of the DNA sequence in the complex. This suggests that RNA polymerase has associated with it an "unwindase" activity that opens the DNA helix. The fact that the DNA double helix must unwind and the strands part at least transiently for transcription implies some disruption of the nucleosome structure of eukaryotic cells. Topoisomerase both precedes and follows the progressing RNAP to prevent the formation of superhelical complexes.

**Termination** of the synthesis of the RNA molecule in bacteria is signaled by a sequence in the template strand of the DNA molecule—a signal that is recognized by a termination protein, the rho (ρ) factor. Rho is an ATP-dependent RNA-stimulated helicase that disrupts the nascent RNA-DNA complex. After termination of synthesis of the RNA molecule, the enzyme separates from the DNA template and probably dissociates to free core enzyme and free σ factor. With the assistance of another σ factor, the core enzyme then recognizes a promoter at which the synthesis of a new RNA molecule commences. In eukaryotic cells, termination is less well defined. It appears to be somehow linked both to initiation and to addition of the 3′ polyA tail of mRNA and could involve destabilization of the RNA-DNA complex at a region of A–U base pairs. More than one RNA polymerase molecule may transcribe the same template strand of a gene simultaneously, but the process is phased and spaced in such a way that at any one moment each is transcribing a different portion of the DNA sequence. An electron micrograph of extremely active RNA synthesis is shown in Figure 37–4.

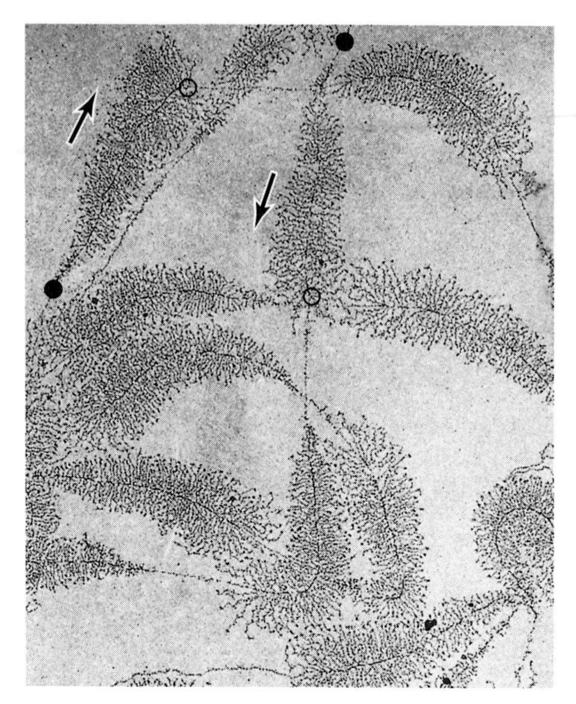

***Figure 37–4.*** Electron photomicrograph of multiple copies of amphibian ribosomal RNA genes in the process of being transcribed. The magnification is about 6000 ×. Note that the length of the transcripts increases as the RNA polymerase molecules progress along the individual ribosomal RNA genes; transcription start sites (filled circles) to transcription termination sites (open circles). RNA polymerase I (not visualized here) is at the base of the nascent rRNA transcripts. Thus, the proximal end of the transcribed gene has short transcripts attached to it, while much longer transcripts are attached to the distal end of the gene. The arrows indicate the direction (5′ to 3′) of transcription. (Reproduced with permission, from Miller OL Jr, Beatty BR: Portrait of a gene. J Cell Physiol 1969;74[Suppl 1]:225.)

## THE FIDELITY & FREQUENCY OF TRANSCRIPTION IS CONTROLLED BY PROTEINS BOUND TO CERTAIN DNA SEQUENCES

The DNA sequence analysis of specific genes has allowed the recognition of a number of sequences important in gene transcription. From the large number of bacterial genes studied it is possible to construct consensus models of transcription initiation and termination signals.

The question, "How does RNAP find the correct site to initiate transcription?" is not trivial when the complexity of the genome is considered. *E coli* has $4 \times 10^3$ transcription initiation sites in $4 \times 10^6$ base pairs (bp) of DNA. The situation is even more complex in humans, where perhaps $10^5$ transcription initiation sites are distributed throughout in $3 \times 10^9$ bp of DNA. RNAP can bind to many regions of DNA, but it scans the DNA sequence—at a rate of $\geq 10^3$ bp/s—until it recognizes certain specific regions of DNA to which it binds with higher affinity. This region is called the promoter, and it is the association of RNAP with the promoter that ensures accurate initiation of transcription. The promoter recognition-utilization process is the target for regulation in both bacteria and humans.

## Bacterial Promoters Are Relatively Simple

Bacterial promoters are approximately 40 nucleotides (40 bp or four turns of the DNA double helix) in length, a region small enough to be covered by an *E coli* RNA holopolymerase molecule. In this consensus promoter region are two short, conserved sequence elements. Approximately 35 bp upstream of the transcription start site there is a consensus sequence of eight nucleotide pairs (5′-TGTTGACA-3′) to which the RNAP binds to form the so-called **closed complex.** More proximal to the transcription start site—about ten nucleotides upstream—is a six-nucleotide-pair A+T-rich sequence (5′-TATAAT-3′). These conserved sequence elements comprising the promoter are shown schematically in Figure 37–5. The latter sequence has a low melting temperature because of its deficiency of GC nucleotide pairs. Thus, the **TATA box** is thought to ease the dissociation between the two DNA strands so that RNA polymerase bound to the promoter region can have access to the nucleotide sequence of its immediately downstream template strand. Once this process occurs, the combination of RNA polymerase plus promoter is called the **open complex.** Other bacteria have slightly different consensus sequences in their promoters, but all generally have two components to the promoter; these tend to be in the same position relative to the transcription start site, and in all cases the sequences between the boxes have no similarity but still provide critical spacing functions facilitating recognition of −35 and −10 sequence by RNA polymerase holoenzyme. Within a bacterial cell, different sets of genes are often

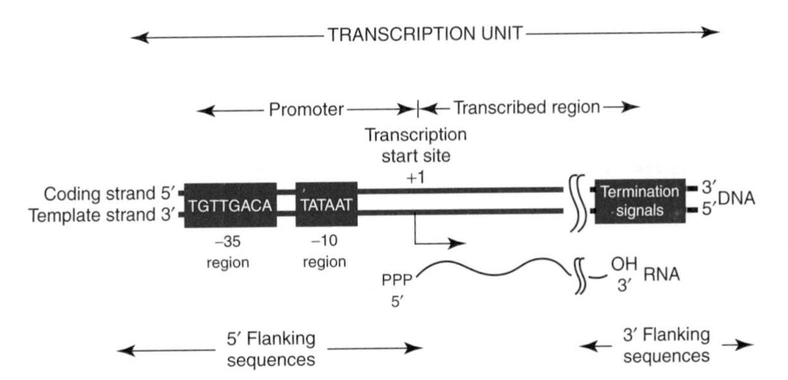


***Figure 37–5.*** Bacterial promoters, such as that from *E coli* shown here, share two regions of highly conserved nucleotide sequence. These regions are located 35 and 10 bp upstream (in the 5′ direction of the coding strand) from the start site of transcription, which is indicated as +1. By convention, all nucleotides upstream of the transcription initiation site (at +1) are numbered in a negative sense and are referred to as 5′-flanking sequences. Also by convention, the DNA regulatory sequence elements (TATA box, etc) are described in the 5′ to 3′ direction and as being on the coding strand. These elements function only in double-stranded DNA, however. Note that the transcript produced from this transcription unit has the same polarity or "sense" (ie, 5′ to 3′ orientation) as the coding strand. Termination *cis*-elements reside at the end of the transcription unit (see Figure 37–6 for more detail). By convention the sequences downstream of the site at which transcription termination occurs are termed 3′-flanking sequences.

coordinately regulated. One important way that this is accomplished is through the fact that these co-regulated genes share unique −35 and −10 promoter sequences. These unique promoters are recognized by different σ factors bound to core RNA polymerase.

Rho-dependent transcription **termination signals** in *E coli* also appear to have a distinct consensus sequence, as shown in Figure 37–6. The conserved consensus sequence, which is about 40 nucleotide pairs in length, can be seen to contain a hyphenated or interrupted inverted repeat followed by a series of AT base pairs. As transcription proceeds through the hyphenated, inverted repeat, the generated transcript can form the intramolecular hairpin structure, also depicted in Figure 37–6.

Transcription continues into the AT region, and with the aid of the ρ termination protein the RNA polymerase stops, dissociates from the DNA template, and releases the nascent transcript.

## Eukaryotic Promoters Are More Complex

It is clear that the signals in DNA which control transcription in eukaryotic cells are of several types. Two types of sequence elements are promoter-proximal. One of these defines **where transcription is to commence** along the DNA, and the other contributes to the mechanisms that control **how frequently** this event is to occur. For example, in the thymidine kinase gene of the herpes simplex virus, which utilizes transcription factors of its mammalian host for gene expression, there is a single unique transcription start site, and accurate transcription from this start site depends upon a nucleotide sequence located 32 nucleotides upstream from the start site (ie, at −32) (Figure 37–7). This region has the sequence of **TATAAAAG** and bears remarkable similarity to the functionally related TATA box that is located about 10 bp upstream from the prokaryotic mRNA start site (Figure 37–5). Mutation or inactivation of the TATA box markedly reduces transcription of this and many other genes that contain this consensus *cis* element (see Figures 37–7, 37–8). Most mammalian genes have a TATA box that is usually located 25–30 bp upstream from the transcription start site. The consensus sequence for a TATA box is TATAAA, though numerous variations have been characterized. The TATA box is bound by 34 kDa **TATA binding protein (TBP),** which in turn binds several other proteins called **TBP-associated factors (TAFs).** This complex of TBP and TAFs is referred to as TFIID. Binding of TFIID to the TATA box sequence is thought to represent the first step in the formation of the transcription complex on the promoter.

A small number of genes lack a TATA box. In such instances, two additional *cis* elements, an **initiator sequence (Inr)** and the so-called **downstream promoter element (DPE),** direct RNA polymerase II to the promoter and in so doing provide basal transcription starting from the correct site. The Inr element spans the start

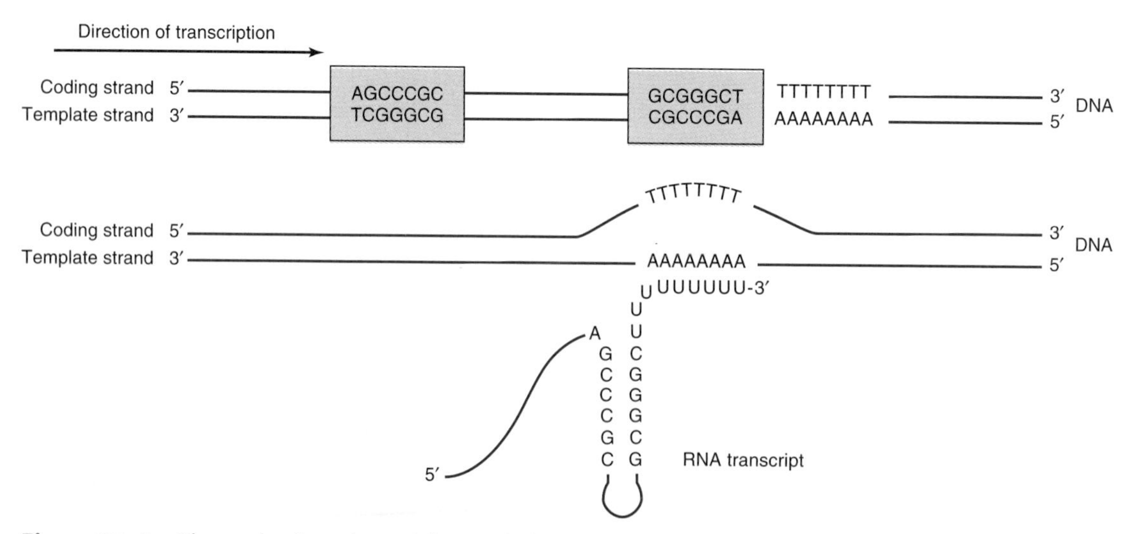


***Figure 37–6.*** The predominant bacterial transcription termination signal contains an inverted, hyphenated repeat (the two boxed areas) followed by a stretch of AT base pairs (top figure). The inverted repeat, when transcribed into RNA, can generate the secondary structure in the RNA transcript shown at the bottom of the figure. Formation of this RNA hairpin causes RNA polymerase to pause and subsequently the ρ termination factor interacts with the paused polymerase and somehow induces chain termination.

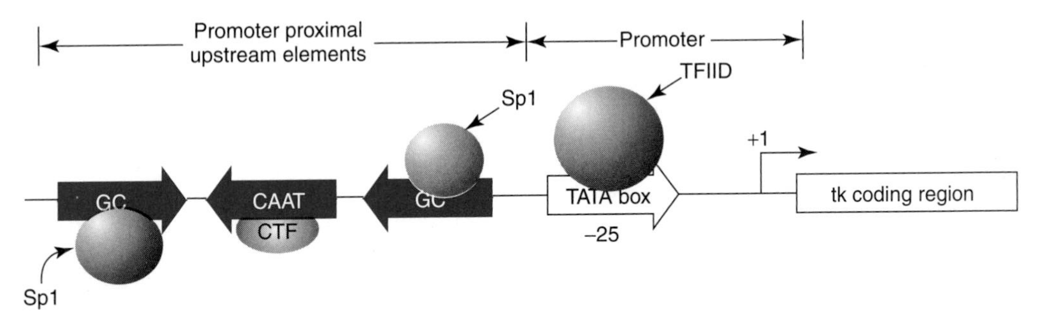


***Figure 37–7.*** Transcription elements and binding factors in the herpes simplex virus thymidine kinase *(tk)* gene. DNA-dependent RNA polymerase II binds to the region of the TATA box (which is bound by transcription factor TFIID) to form a multicomponent preinitiation complex capable of initiate transcription at a single nucleotide (+1). The frequency of this event is increased by the presence of upstream *cis*-acting elements (the GC and CAAT boxes). These elements bind *trans*-acting transcription factors, in this example Sp1 and CTF (also called C/EBP, NF1, NFY). These *cis* elements can function independently of orientation *(arrows).*

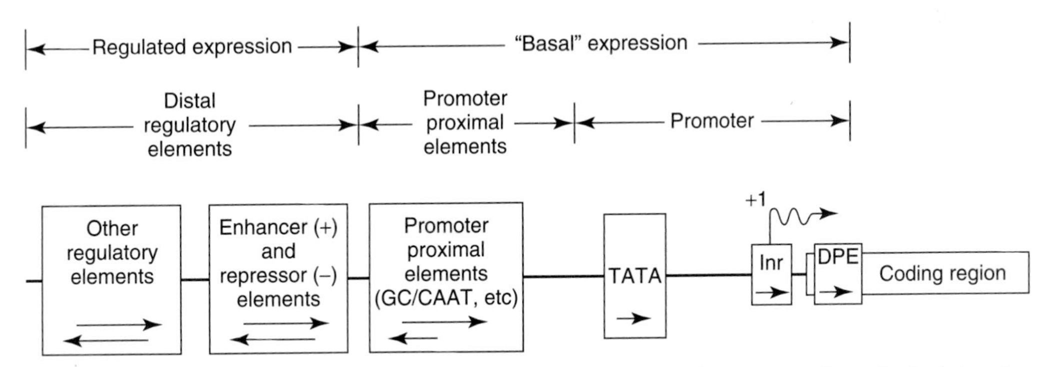


***Figure 37–8.*** Schematic diagram showing the transcription control regions in a hypothetical class II (mRNA-producing) eukaryotic gene. Such a gene can be divided into its coding and regulatory regions, as defined by the transcription start site *(arrow;* +1*).* The coding region contains the DNA sequence that is transcribed into mRNA, which is ultimately translated into protein. The regulatory region consists of two classes of elements. One class is responsible for ensuring basal expression. These elements generally have two components. The proximal component, generally the TATA box, or Inr or DPE elements direct RNA polymerase II to the correct site (fidelity). In TATA-less promoters, an initiator (Inr) element that spans the initiation site (+1) may direct the polymerase to this site. Another component, the upstream elements, specifies the frequency of initiation. Among the best studied of these is the CAAT box, but several other elements (Sp1, NF1, AP1, etc) may be used in various genes. A second class of regulatory *cis*-acting elements is responsible for regulated expression. This class consists of elements that enhance or repress expression and of others that mediate the response to various signals, including hormones, heat shock, heavy metals, and chemicals. Tissue-specific expression also involves specific sequences of this sort. The orientation dependence of all the elements is indicated by the arrows within the boxes. For example, the proximal element (the TATA box) must be in the 5′ to 3′ orientation. The upstream elements work best in the 5′ to 3′ orientation, but some of them can be reversed. The locations of some elements are not fixed with respect to the transcription start site. Indeed, some elements responsible for regulated expression can be located either interspersed with the upstream elements, or they can be located downstream from the start site.

site (from −3 to +5) and consists of the general consensus sequence TCA$_{+1}$ G/T T T/C which is similar to the initiation site sequence per se. (A+1 indicates the first nucleotide transcribed.) The proteins that bind to Inr in order to direct pol II binding include TFIID. Promoters that have both a TATA box and an Inr may be stronger than those that have just one of these elements. The DPE has the consensus sequence A/GGA/T CGTG and is localized about 25 bp downstream of the +1 start site. Like the Inr, DPE sequences are also bound by the TAF subunits of TFIID. In a survey of over 200 eukaryotic genes, roughly 30% contained a TATA box and Inr, 25% contained Inr and DPE, 15% contained all three elements, while ~30% contained just the Inr.

Sequences farther upstream from the start site determine how frequently the transcription event occurs. Mutations in these regions reduce the frequency of transcriptional starts tenfold to twentyfold. Typical of these DNA elements are the GC and CAAT boxes, so named because of the DNA sequences involved. As illustrated in Figure 37–7, each of these boxes binds a protein, Sp1 in the case of the GC box and CTF (or C/EPB,NF1,NFY) by the CAAT box; both bind through their distinct DNA binding domains (DBDs). The frequency of transcription initiation is a consequence of these protein-DNA interactions and complex interactions between particular domains of the transcription factors (distinct from the DBD domains—so-called activation domains; ADs) of these proteins and the rest of the transcription machinery (RNA polymerase II and the basal factors TFIIA, B, D, E, F). (See below and Figures 37–9 and 37–10). The protein-DNA interaction at the TATA box involving RNA polymerase II and other components of the basal transcription machinery ensures the fidelity of initiation.

Together, then, the promoter and promoter-proximal *cis*-active upstream elements confer fidelity and frequency of initiation upon a gene. The TATA box has a particularly rigid requirement for both position and orientation. Single-base changes in any of these *cis* elements have dramatic effects on function by reducing the binding affinity of the cognate *trans* factors (either TFIID/TBP or Sp1, CTF, and similar factors). The spacing of these elements with respect to the transcription start site can also be critical. This is particularly true for the TATA box Inr and DPE.

A third class of sequence elements can either increase or decrease the rate of transcription initiation of eukaryotic genes. These elements are called either **enhancers** or **repressors (or silencers),** depending on which effect they have. They have been found in a variety of locations both upstream and downstream of the transcription start site and even within the transcribed portions of some genes. In contrast to proximal and upstream promoter elements, enhancers and silencers can exert their effects when located hundreds or even thousands of bases away from transcription units located on the same chromosome. Surprisingly, enhancers and silencers can function in an orientation-independent fashion. Literally hundreds of these elements have been described. In some cases, the sequence requirements for binding are rigidly constrained; in others, considerable sequence variation is

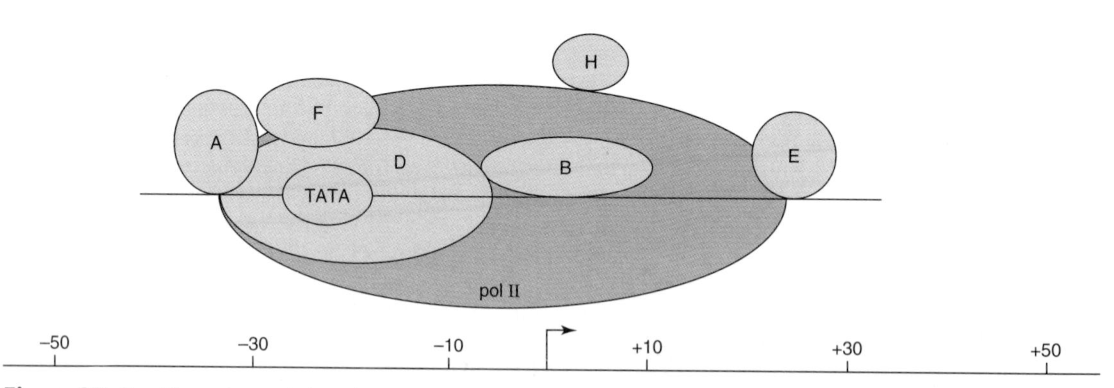


***Figure 37–9.*** The eukaryotic basal transcription complex. Formation of the basal transcription complex begins when TFIID binds to the TATA box. It directs the assembly of several other components by protein-DNA and protein-protein interactions. The entire complex spans DNA from position −30 to +30 relative to the initiation site (+1, marked by bent arrow). The atomic level, x-ray-derived structures of RNA polymerase II alone and of TBP bound to TATA promoter DNA in the presence of either TFIIB or TFIIA have all been solved at 3 Å resolution. The structure of TFIID complexes have been determined by electron microscopy at 30 Å resolution. Thus, the molecular structures of the transcription machinery are beginning to be elucidated. Much of this structural information is consistent with the models presented here.

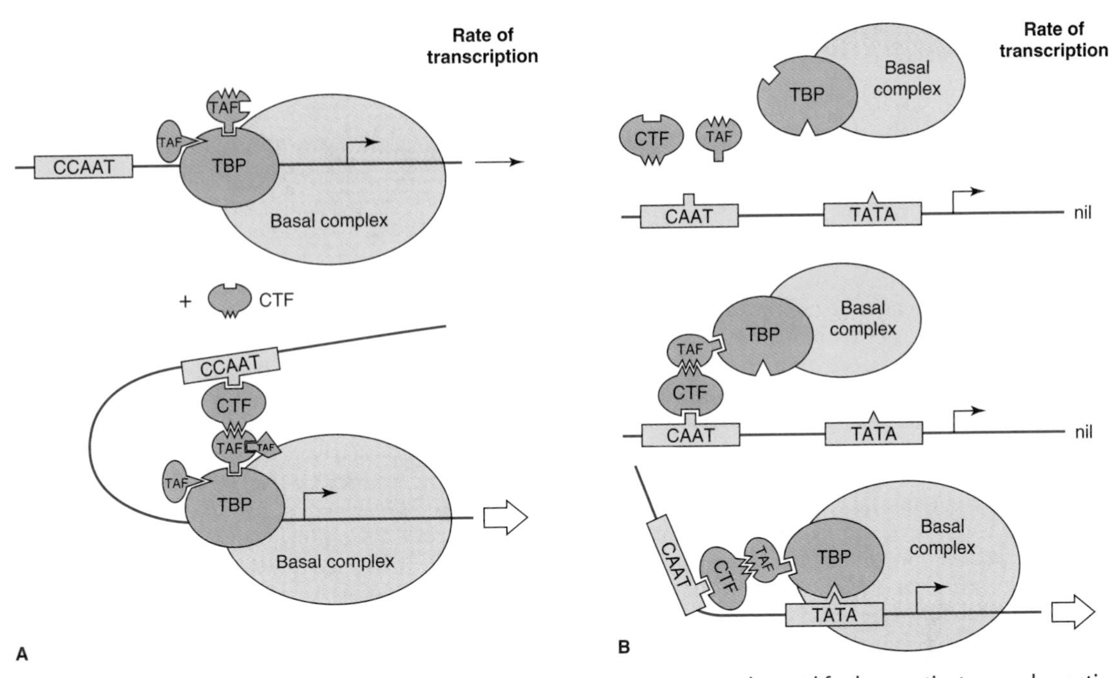


***Figure 37–10.*** Two models for assembly of the active transcription complex and for how activators and coactivators might enhance transcription. Shown here as a small oval is TBP, which contains TFIID, a large oval that contains all the components of the basal transcription complex illustrated in Figure 37–9 (ie, RNAP II and TFIIA, TFIIB, TFIIE, TFIIF, and TFIIH). **Panel A:** The basal transcription complex is assembled on the promoter after the TBP subunit of TFIID is bound to the TATA box. Several TAFs (coactivators) are associated with TBP. In this example, a transcription activator, CTF, is shown bound to the CAAT box, forming a loop complex by interacting with a TAF bound to TBP. **Panel B:** The recruitment model. The transcription activator CTF binds to the CAAT box and interacts with a coactivator (TAF in this case). This allows for an interaction with the preformed TBP-basal transcription complex. TBP can now bind to the TATA box, and the assembled complex is fully active.

allowed. Some sequences bind only a single protein, but the majority bind several different proteins. Similarly, a single protein can bind to more than one element.

**Hormone response elements** (for steroids, $T_3$, retinoic acid, peptides, etc) act as—or in conjunction with—enhancers or silencers (Chapter 43). Other processes that enhance or silence gene expression—such as the response to heat shock, heavy metals ($Cd^{2+}$ and $Zn^{2+}$), and some toxic chemicals (eg, dioxin)—are mediated through specific regulatory elements. Tissue-specific expression of genes (eg, the albumin gene in liver, the hemoglobin gene in reticulocytes) is also mediated by specific DNA sequences.

## Specific Signals Regulate Transcription Termination

The **signals for the termination of transcription** by eukaryotic RNA polymerase II are very poorly understood. However, it appears that the termination signals exist far downstream of the coding sequence of eukaryotic genes. For example, the transcription termination signal for mouse β-globin occurs at several positions 1000–2000 bases beyond the site at which the poly(A) tail will eventually be added. Little is known about the termination process or whether specific termination factors similar to the bacterial ρ factor are involved. However, it is known that the mRNA 3′ terminal is generated posttranscriptionally, is somehow coupled to events or structures formed at the time and site of initiation, depends on a special structure in one of the subunits of RNA polymerase II (the CTD; see below), and appears to involve at least two steps. After RNA polymerase II has traversed the region of the transcription unit encoding the 3′ end of the transcript, an RNA endonuclease cleaves the primary transcript at a position about 15 bases 3′ of the consensus sequence **AAUAAA** that serves in eukaryotic transcripts as a cleavage signal.

Finally, this newly formed 3′ terminal is polyadenylated in the nucleoplasm, as described below.

## THE EUKARYOTIC TRANSCRIPTION COMPLEX

A complex apparatus consisting of as many as 50 unique proteins provides accurate and regulatable transcription of eukaryotic genes. The RNA polymerase enzymes (pol I, pol II, and pol III for class I, II, and III genes, respectively) transcribe information contained in the template strand of DNA into RNA. These polymerases must recognize a specific site in the promoter in order to initiate transcription at the proper nucleotide. In contrast to the situation in prokaryotes, eukaryotic RNA polymerases alone are not able to discriminate between promoter sequences and other regions of DNA; thus, other proteins known as general transcription factors or GTFs facilitate promoter-specific binding of these enzymes and formation of the preinitiation complex (PIC). This combination of components can catalyze basal or (non)-unregulated transcription in vitro. Another set of proteins—coactivators—help regulate the rate of transcription initiation by interacting with transcription activators that bind to upstream DNA elements (see below).

### Formation of the Basal Transcription Complex

In bacteria, a σ factor–polymerase complex selectively binds to DNA in the promoter forming the PIC. The situation is more complex in eukaryotic genes. Class II genes—those transcribed by pol II to make mRNA—are described as an example. In class II genes, the function of σ factors is assumed by a number of proteins. **Basal transcription requires, in addition to pol II, a number of GTFs called TFIIA, TFIIB, TFIID, TFIIE, TFIIF, and TFIIH.** These GTFs serve to promote RNA polymerase II transcription on essentially all genes. Some of these GTFs are composed of multiple subunits. **TFIID, which binds to the TATA box promoter element, is the only one of these factors capable of binding to specific sequences of DNA.** As described above, TFIID consists of TATA binding protein (TBP) and 14 TBP-associated factors (TAFs).

TBP binds to the TATA box in the minor groove of DNA (most transcription factors bind in the major groove) and causes an approximately 100-degree bend or kink of the DNA helix. This bending is thought to facilitate the interaction of TBP-associated factors with other components of the transcription initiation complex and possibly with factors bound to upstream elements. Although defined as a component of class II gene promoters, TBP, by virtue of its association with distinct, polymerase-specific sets of TAFs, is also an important component of class I and class III initiation complexes even if they do not contain TATA boxes.

The binding of TBP marks a specific promoter for transcription and is the only step in the assembly process that is entirely dependent on specific, high-affinity protein-DNA interaction. Of several subsequent in vitro steps, the first is the binding of TFIIB to the TFIID-promoter complex. This results in a stable ternary complex which is then more precisely located and more tightly bound at the transcription initiation site. This complex then attracts and tethers the pol II-TFIIF complex to the promoter. TFIIF is structurally and functionally similar to the bacterial σ factor and is required for the delivery of pol II to the promoter. TFIIA binds to this assembly and may allow the complex to respond to activators, perhaps by the displacement of repressors. Addition of TFIIE and TFIIH is the final step in the assembly of the PIC. TFIIE appears to join the complex with pol II-TFIIF, and TFIIH is then recruited. Each of these binding events extends the size of the complex so that finally about 60 bp (from −30 to +30 relative to +1, the nucleotide from which transcription commences) are covered (Figure 37–9). The PIC is now complete and capable of basal transcription initiated from the correct nucleotide. In genes that lack a TATA box, the same factors, including TBP, are required. In such cases, an Inr or the DPEs (see Figure 37–8) position the complex for accurate initiation of transcription.

### Phosphorylation Activates Pol II

Eukaryotic pol II consists of 12 subunits. The two largest subunits, both about 200 kDa, are homologous to the bacterial β and β′ subunits. In addition to the increased number of subunits, eukaryotic pol II differs from its prokaryotic counterpart in that it has a series of heptad repeats with consensus sequence Tyr-Ser-Pro-Thr-Ser-Pro-Ser at the carboxyl terminal of the largest pol II subunit. This **carboxyl terminal repeat domain (CTD)** has 26 repeated units in brewers' yeast and 52 units in mammalian cells. The CTD is both a substrate for several kinases, including the kinase component of TFIIH, and a binding site for a wide array of proteins. The CTD has been shown to interact with RNA processing enzymes; such binding may be involved with RNA polyadenylation. The association of the factors with the CTD of RNA polymerase II (and other components of the basal machinery) somehow serves to couple initiation with mRNA 3′ end formation. Pol II is activated when phosphorylated on the Ser and Thr residues and displays reduced activity when the CTD is dephosphorylated. Pol II lacking the CTD tail is incapable of activating transcription, which underscores the importance of this domain.

Pol II associates with other proteins to form a holoenzyme complex. In yeast, at least nine gene products—called Srb (for suppressor of RNA polymerase B)—bind to the CTD. The Srb proteins—or mediators, as they are also called—are essential for pol II transcription, though their exact role in this process has not been defined. Related proteins comprising even more complex forms of RNA polymerase II have been described in human cells.

## The Role of Transcription Activators & Coactivators

TFIID was originally considered to be a single protein. However, several pieces of evidence led to the important discovery that TFIID is actually a complex consisting of TBP and the 14 TAFs. The first evidence that TFIID was more complex than just the TBP molecules came from the observation that TBP binds to a 10-bp segment of DNA, immediately over the TATA box of the gene, whereas native holo-TFIID covers a 35 bp or larger region (Figure 37–9). Second, TBP has a molecular mass of 20–40 kDa (depending on the species), whereas the TFIID complex has a mass of about 1000 kDa. Finally, and perhaps most importantly, TBP supports basal transcription but not the augmented transcription provided by certain activators, eg, Sp1 bound to the GC box. TFIID, on the other hand, supports both basal and enhanced transcription by Sp1, Oct1, AP1, CTF, ATF, etc. (Table 37–3). The TAFs are essential for this activator-enhanced transcription. It is not yet clear whether there are one or several forms of TFIID that might differ slightly in their complement of TAFs. It is conceivable that different combinations of TAFs with TBP—or one of several recently discovered TBP-like factors (TLFs)—may bind to different promoters, and recent reports suggest that this may account for selective activation noted in various promoters and for the different strengths of certain promoters. **TAFs, since they are required for the action of activators, are often called coactivators.** There are thus three classes of transcription factors involved in the regulation of class II genes: basal factors, coactivators, and activator-repressors (Table 37–4). How these classes of proteins interact to govern both the site and frequency of transcription is a question of central importance.

***Table 37–3.*** Some of the transcription control elements, their consensus sequences, and the factors that bind to them which are found in mammalian genes transcribed by RNA polymerase II. A complete list would include dozens of examples. The asterisks mean that there are several members of this family.

| Element | Consensus Sequence | Factor |
|---|---|---|
| TATA box | TATAAA | TBP |
| CAAT box | CCAATC | C/EBP*, NF-Y* |
| GC box | GGGCGG | Sp1* |
| | CAACTGAC | Myo D |
| | T/CGGA/CN$_5$GCCAA | NF1* |
| Ig octamer | ATGCAAAT | Oct1, 2, 4, 6* |
| AP1 | TGAG/CTC/AA | Jun, Fos, ATF* |
| Serum response | GATGCCCATA | SRF |
| Heat shock | (NGAAN)$_3$ | HSF |

## Two Models Explain the Assembly of the Preinitiation Complex

The formation of the PIC described above is based on the sequential addition of purified components in in vitro experiments. An essential feature of this model is that the assembly takes place on the DNA template. Accordingly, transcription activators, which have autonomous DNA binding and activation domains (see Chapter 39), are thought to function by stimulating either PIC formation or PIC function. The TAF coactivators are viewed as bridging factors that communicate between the upstream activators, the proteins associated with pol II, or the many other components of TFIID. This view, which assumes that there is **stepwise assembly** of the PIC—promoted by various interactions between activators, coactivators, and PIC components—is illustrated in panel A of Figure 37–10. This model was supported by observations that many of these proteins could indeed bind to one another in vitro.

Recent evidence suggests that there is another possible mechanism of PIC formation and transcription regulation. First, large preassembled complexes of GTFs and pol II are found in cell extracts, and this complex can associate with a promoter in a single step. Second, the rate of transcription achieved when activators are added to limiting concentrations of pol II holoenzyme can be matched by increasing the concentration of the pol II holoenzyme in the absence of activators. Thus,

***Table 37–4.*** Three classes of transcription factors in class II genes.

| General Mechanisms | Specific Components |
|---|---|
| Basal components | TBP, TFIIA, B, E, F, and H |
| Coactivators | TAFs (TBP + TAFs) = TFIID; Srbs |
| Activators | SP1, ATF, CTF, AP1, etc |

activators are not in themselves absolutely essential for PIC formation. These observations led to the **"recruitment" hypothesis,** which has now been tested experimentally. Simply stated, the role of activators and coactivators may be solely to recruit a preformed holoenzyme-GTF complex to the promoter. The requirement for an activation domain is circumvented when either a component of TFIID or the pol II holoenzyme is artificially tethered, using recombinant DNA techniques, to the DNA binding domain (DBD) of an activator. This anchoring, through the DBD component of the activator molecule, leads to a transcriptionally competent structure, and there is no further requirement for the activation domain of the activator. In this view, the role of activation domains and TAFs is to form an assembly that directs the preformed holoenzyme-GTF complex to the promoter; they do not assist in PIC assembly (see panel B, Figure 37–10). The efficiency of this recruitment process determines the rate of transcription at a given promoter.

Hormones—and other effectors that serve to transmit information related to the extracellular environment—modulate gene expression by influencing the assembly and activity of the activator and coactivator complexes and the subsequent formation of the PIC at the promoter of target genes (see Chapter 43). The numerous components involved provide for an abundance of possible combinations and therefore a range of transcriptional activity of a given gene. It is important to note that the two models are not mutually exclusive—stepwise versus holoenzyme-mediated PIC formation. Indeed, one can envision various more complex models invoking elements of both models operating on a gene.

## RNA MOLECULES ARE USUALLY PROCESSED BEFORE THEY BECOME FUNCTIONAL

In prokaryotic organisms, the primary transcripts of mRNA-encoding genes begin to serve as translation templates even before their transcription has been completed. This is because the site of transcription is not compartmentalized into a nucleus as it is in eukaryotic organisms. Thus, transcription and translation are coupled in prokaryotic cells. Consequently, prokaryotic mRNAs are subjected to little processing prior to carrying out their intended function in protein synthesis. Indeed, appropriate regulation of some genes (eg, the *Trp* operon) relies upon this coupling of transcription and translation. Prokaryotic rRNA and tRNA molecules are transcribed in units considerably longer than the ultimate molecule. In fact, many of the tRNA transcription units contain more than one molecule. Thus, in prokaryotes the processing of these rRNA and tRNA precursor molecules is required for the generation of the mature functional molecules.

Nearly all eukaryotic RNA primary transcripts undergo extensive processing between the time they are synthesized and the time at which they serve their ultimate function, whether it be as mRNA or as a component of the translation machinery such as rRNA, 5S RNA, or tRNA or RNA processing machinery, snRNAs. Processing occurs primarily within the nucleus and includes nucleolytic cleavage to smaller molecules and coupled **nucleolytic and ligation reactions (splicing of exons).** In mammalian cells, 50–75% of the nuclear RNA does not contribute to the cytoplasmic mRNA. This nuclear RNA loss is significantly greater than can be reasonably accounted for by the loss of intervening sequences alone (see below). Thus, the exact function of the seemingly excessive transcripts in the nucleus of a mammalian cell is not known.

### The Coding Portions (Exons) of Most Eukaryotic Genes Are Interrupted by Introns

Interspersed within the amino acid-coding portions **(exons)** of many genes are long sequences of DNA that do not contribute to the genetic information ultimately translated into the amino acid sequence of a protein molecule (see Chapter 36). In fact, these sequences actually interrupt the coding region of structural genes. These **intervening sequences (introns)** exist within most but not all mRNA encoding genes of higher eukaryotes. The primary transcripts of the structural genes contain RNA complementary to the interspersed sequences. However, the intron RNA sequences are cleaved out of the transcript, and the exons of the transcript are appropriately spliced together in the nucleus before the resulting mRNA molecule appears in the cytoplasm for translation (Figures 37–11 and 37–12). One speculation is that exons, which often encode an activity domain of a protein, represent a convenient means of shuffling genetic information, permitting organisms to quickly test the results of combining novel protein functional domains.

### Introns Are Removed & Exons Are Spliced Together

The mechanisms whereby introns are removed from the primary transcript in the nucleus, exons are ligated to form the mRNA molecule, and the mRNA molecule is transported to the cytoplasm are being elucidated. Four different splicing reaction mechanisms have been described. The one most frequently used in eukaryotic cells is described below. Although the sequences of nu-

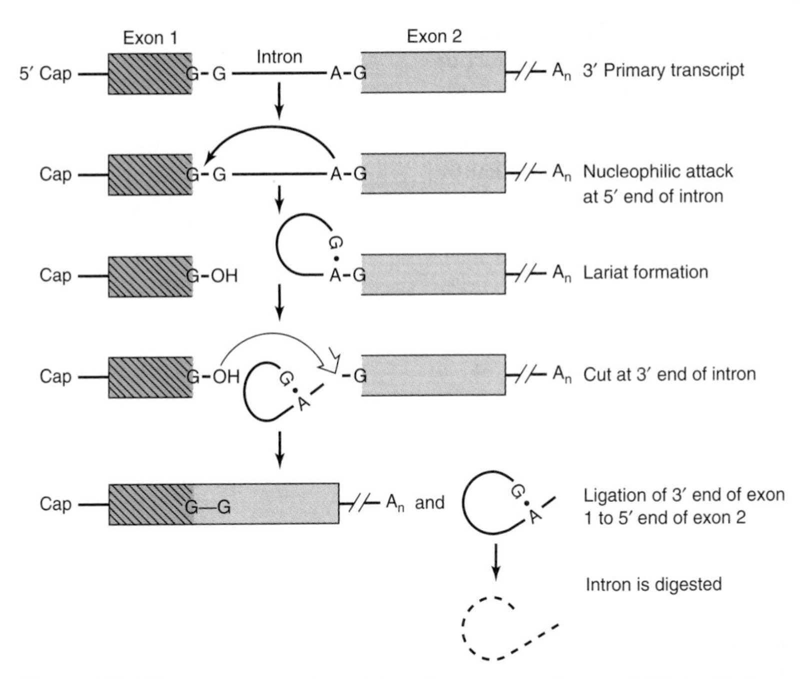


***Figure 37–11.*** The processing of the primary transcript to mRNA. In this hypothetical transcript, the 5′ (left) end of the intron is cut (↓) and a lariat forms between the G at the 5′ end of the intron and an A near the 3′ end, in the consensus sequence UACUAAC. This sequence is called the branch site, and it is the 3′ most A that forms the 5′–2′ bond with the G. The 3′ (right) end of the intron is then cut (⇓). This releases the lariat, which is digested, and exon 1 is joined to exon 2 at G residues.

cleotides in the introns of the various eukaryotic transcripts—and even those within a single transcript—are quite heterogeneous, there are reasonably conserved sequences at each of the two exon-intron (splice) junctions and at the branch site, which is located 20–40 nucleotides upstream from the 3′ splice site (see consensus sequences in Figure 37–12). A special structure, the **spliceosome,** is involved in converting the primary transcript into mRNA. Spliceosomes consist of the primary transcript, five small nuclear RNAs (U1, U2, U5, U4, and U6) and more than 60 proteins. Collectively, these form a **small nucleoprotein (snRNP) complex,** sometimes called a **"snurp."** It is likely that this penta-snRNP spliceosome forms prior to interaction with mRNA precursors. Snurps are thought to position the RNA segments for the necessary splicing reactions. The splicing reaction starts with a cut at the junction of the 5′ exon (donor or left) and intron (Figure 37–11). This

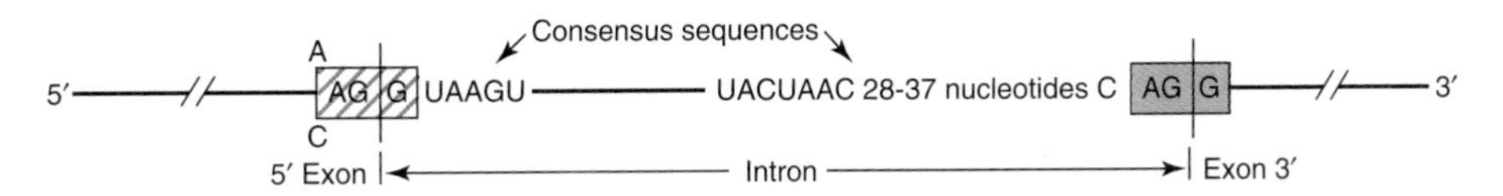


***Figure 37–12.*** Consensus sequences at splice junctions. The 5′ (donor or left) and 3′ (acceptor or right) sequences are shown. Also shown is the yeast consensus sequence (UACUAAC) for the branch site. In mammalian cells, this consensus sequence is PyNPyPyPuAPy, where Py is a pyrimidine, Pu is a purine, and N is any nucleotide. The branch site is located 20–40 nucleotides upstream from the 3′ site.

is accomplished by a nucleophilic attack by an adenylyl residue in the branch point sequence located just upstream from the 3′ end of this intron. The free 5′ terminal then forms a loop or lariat structure that is linked by an unusual 5′–2′ phosphodiester bond to the reactive A in the PyNPyPyPuAPy branch site sequence (Figure 37–12). This adenylyl residue is typically located 28–37 nucleotides upstream from the 3′ end of the intron being removed. The branch site identifies the 3′ splice site. A second cut is made at the junction of the intron with the 3′ exon (donor on right). In this second transesterification reaction, the 3′ hydroxyl of the upstream exon attacks the 5′ phosphate at the downstream exon-intron boundary, and the lariat structure containing the intron is released and hydrolyzed. The 5′ and 3′ exons are ligated to form a continuous sequence.

The snRNAs and associated proteins are required for formation of the various structures and intermediates. U1 within the snRNP complex binds first by base pairing to the 5′ exon-intron boundary. U2 within the snRNP complex then binds by base pairing to the branch site, and this exposes the nucleophilic A residue. U5/U4/U6 within the snRNP complex mediates an ATP-dependent protein-mediated unwinding that results in disruption of the base-paired U4-U6 complex with the release of U4. U6 is then able to interact first with U2, then with U1. These interactions serve to approximate the 5′ splice site, the branch point with its reactive A, and the 3′ splice site. This alignment is enhanced by U5. This process also results in the formation of the loop or lariat structure. **The two ends are cleaved, probably by the U2-U6 within the snRNP complex.** U6 is certainly essential, since yeasts deficient in this snRNA are not viable. It is important to note that RNA serves as the catalytic agent. This sequence is then repeated in genes containing multiple introns. In such cases, a definite pattern is followed for each gene, and the introns are not necessarily removed in sequence—1, then 2, then 3, etc.

The relationship between hnRNA and the corresponding mature mRNA in eukaryotic cells is now apparent. The hnRNA molecules are the primary transcripts plus their early processed products, which, after the addition of caps and poly(A) tails and removal of the portion corresponding to the introns, are transported to the cytoplasm as mature mRNA molecules.

## Alternative Splicing Provides for Different mRNAs

**The processing of hnRNA molecules is a site for regulation of gene expression.** Alternative patterns of RNA splicing result from tissue-specific adaptive and developmental control mechanisms. As mentioned above, the sequence of exon-intron splicing events generally follows a hierarchical order for a given gene. The fact that very complex RNA structures are formed during splicing—and that a number of snRNAs and proteins are involved—affords numerous possibilities for a change of this order and for the generation of different mRNAs. Similarly, the use of alternative termination-cleavage-polyadenylation sites also results in mRNA heterogeneity. Some schematic examples of these processes, all of which occur in nature, are shown in Figure 37–13.

**Faulty splicing can cause disease.** At least one form of β-thalassemia, a disease in which the β-globin gene of hemoglobin is severely underexpressed, appears to result from a nucleotide change at an exon-intron junction, precluding removal of the intron and therefore leading to diminished or absent synthesis of the β-chain protein. This is a consequence of the fact that the normal translation reading frame of the mRNA is disrupted—a defect in this fundamental process (splicing) that underscores the accuracy which the process of RNA-RNA splicing must achieve.

## Alternative Promoter Utilization Provides a Form of Regulation

Tissue-specific regulation of gene expression can be provided by control elements in the promoter or by the

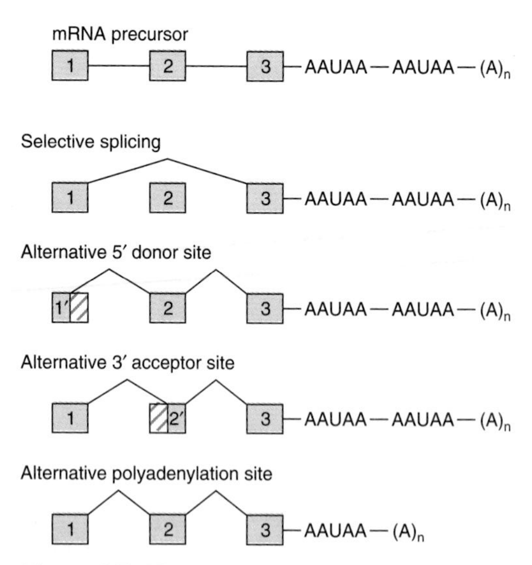


***Figure 37–13.*** Mechanisms of alternative processing of mRNA precursors. This form of RNA processing involves the selective inclusion or exclusion of exons, the use of alternative 5′ donor or 3′ acceptor sites, and the use of different polyadenylation sites.

use of alternative promoters. The glucokinase *(GK)* gene consists of ten exons interrupted by nine introns. The sequence of exons 2–10 is identical in liver and pancreatic B cells, the primary tissues in which GK protein is expressed. Expression of the *GK* gene is regulated very differently—by two different promoters—in these two tissues. The liver promoter and exon 1L are located near exons 2–10; exon 1L is ligated directly to exon 2. In contrast, the pancreatic B cell promoter is located about 30 kbp upstream. In this case, the 3′ boundary of exon 1B is ligated to the 5′ boundary of exon 2. The liver promoter and exon 1L are excluded and removed during the splicing reaction (see Figure 37–14). The existence of multiple distinct promoters allows for cell- and tissue-specific expression patterns of a particular gene (mRNA).

### Both Ribosomal RNAs & Most Transfer RNAs Are Processed From Larger Precursors

In mammalian cells, the three rRNA molecules are transcribed as part of a single large precursor molecule. **The precursor is subsequently processed in the nucleolus** to provide the RNA component for the ribosome subunits found in the cytoplasm. The rRNA genes are located in the nucleoli of mammalian cells. Hundreds of copies of these genes are present in every cell. This large number of genes is required to synthesize sufficient copies of each type of rRNA to form the $10^7$ ribosomes required for each cell replication. Whereas a single mRNA molecule may be copied into $10^5$ protein molecules, providing a large amplification, the rRNAs are end products. This lack of amplification requires a large number of genes. Similarly, transfer RNAs are often synthesized as precursors, with extra sequences both 5′ and 3′ of the sequences comprising the mature tRNA. A small fraction of tRNAs even contain introns.

## RNAS CAN BE EXTENSIVELY MODIFIED

Essentially all RNAs are covalently modified after transcription. It is clear that at least some of these modifications are regulatory.

### Messenger RNA (mRNA) Is Modified at the 5′ & 3′ Ends

As mentioned above, mammalian mRNA molecules contain a 7-methylguanosine cap structure at their 5′ terminal, and most have a poly(A) tail at the 3′ terminal. The cap structure is added to the 5′ end of the newly transcribed mRNA precursor in the nucleus prior to transport of the mRNA molecule to the cytoplasm. The **5′ cap** of the RNA transcript is required both for efficient translation initiation and protection of the 5′ end of mRNA from attack by 5′ → 3′ exonucleases. The secondary methylations of mRNA molecules, those on the 2′-hydroxy and the $N^6$ of adenylyl residues, occur after the mRNA molecule has appeared in the cytoplasm.

Poly(A) tails are added to the 3′ end of mRNA molecules in a posttranscriptional processing step. The mRNA is first cleaved about 20 nucleotides downstream from an AAUAA recognition sequence. Another enzyme, poly(A) polymerase, adds a poly(A) tail which is subsequently extended to as many as 200 A residues. The **poly(A) tail** appears to protect the 3′ end of mRNA from 3′ → 5′ exonuclease attack. The presence or absence of the poly(A) tail does not determine whether a precursor molecule in the nucleus appears in the cytoplasm, because all poly(A)-tailed hnRNA molecules do not contribute to cytoplasmic mRNA, nor do all cytoplasmic mRNA molecules contain poly(A) tails

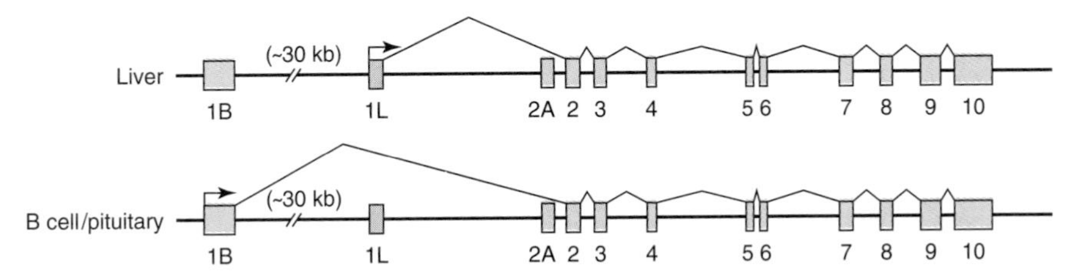


***Figure 37–14.*** Alternative promoter use in the liver and pancreatic B cell glucokinase genes. Differential regulation of the glucokinase (*GK*) gene is accomplished by the use of tissue-specific promoters. The B cell *GK* gene promoter and exon 1B are located about 30 kbp upstream from the liver promoter and exon 1L. Each promoter has a unique structure and is regulated differently. Exons 2–10 are identical in the two genes, and the GK proteins encoded by the liver and B cell mRNAs have identical kinetic properties.

(the histones are most notable in this regard). Cytoplasmic enzymes in mammalian cells can both add and remove adenylyl residues from the poly(A) tails; this process has been associated with an alteration of mRNA stability and translatability.

The size of some cytoplasmic mRNA molecules, even after the poly(A) tail is removed, is still considerably greater than the size required to code for the specific protein for which it is a template, often by a factor of 2 or 3. **The extra nucleotides occur in untranslated (non-protein coding) regions** both 5′ and 3′ of the coding region; the longest untranslated sequences are usually at the 3′ end. The exact function of these sequences is unknown, but they have been implicated in RNA processing, transport, degradation, and translation; each of these reactions potentially contributes additional levels of control of gene expression.

### RNA Editing Changes mRNA After Transcription

The central dogma states that for a given gene and gene product there is a linear relationship between the coding sequence in DNA, the mRNA sequence, and the protein sequence (Figure 36–7). Changes in the DNA sequence should be reflected in a change in the mRNA sequence and, depending on codon usage, in protein sequence. However, exceptions to this dogma have been recently documented. Coding information can be changed at the mRNA level by **RNA editing.** In such cases, the coding sequence of the mRNA differs from that in the cognate DNA. An example is the apolipoprotein B *(apoB)* gene and mRNA. In liver, the single *apoB* gene is transcribed into an mRNA that directs the synthesis of a 100-kDa protein, apoB100. In the intestine, the same gene directs the synthesis of the primary transcript; however, a cytidine deaminase converts a CAA codon in the mRNA to UAA at a single specific site. Rather than encoding glutamine, this codon becomes a termination signal, and a 48-kDa protein (apoB48) is the result. ApoB100 and apoB48 have different functions in the two organs. A growing number of other examples include a glutamine to arginine change in the glutamate receptor and several changes in trypanosome mitochondrial mRNAs, generally involving the addition or deletion of uridine. The exact extent of RNA editing is unknown, but current estimates suggest that $< 0.01\%$ of mRNAs are edited in this fashion.

### Transfer RNA (tRNA) Is Extensively Processed & Modified

As described in Chapters 35 and 38, the tRNA molecules serve as adapter molecules for the translation of mRNA into protein sequences. The tRNAs contain many modifications of the standard bases A, U, G, and C, including methylation, reduction, deamination, and rearranged glycosidic bonds. Further modification of the tRNA molecules includes nucleotide alkylations and the attachment of the characteristic CpCpAOH terminal at the 3′ end of the molecule by the enzyme nucleotidyl transferase. The 3′ OH of the A ribose is the point of attachment for the specific amino acid that is to enter into the polymerization reaction of protein synthesis. The methylation of mammalian tRNA precursors probably occurs in the nucleus, whereas the cleavage and attachment of CpCpAOH are cytoplasmic functions, since the terminals turn over more rapidly than do the tRNA molecules themselves. Enzymes within the cytoplasm of mammalian cells are required for the attachment of amino acids to the CpCpAOH residues. (See Chapter 38.)

## RNA CAN ACT AS A CATALYST

In addition to the catalytic action served by the snRNAs in the formation of mRNA, several other enzymatic functions have been attributed to RNA. **Ribozymes** are RNA molecules with catalytic activity. These generally involve transesterification reactions, and most are concerned with RNA metabolism (splicing and endoribonuclease). Recently, a ribosomal RNA component was noted to hydrolyze an aminoacyl ester and thus to play a central role in peptide bond function (peptidyl transferases; see Chapter 38). These observations, made in organelles from plants, yeast, viruses, and higher eukaryotic cells, show that RNA can act as an enzyme. This has revolutionized thinking about enzyme action and the origin of life itself.

## SUMMARY

- RNA is synthesized from a DNA template by the enzyme RNA polymerase.
- There are three distinct nuclear DNA-dependent RNA polymerases in mammals: RNA polymerases I, II, and III. These enzymes control the transcriptional function—the transcription of rRNA, mRNA, and small RNA (tRNA/5S rRNA, snRNA) genes, respectively.
- RNA polymerases interact with unique *cis*-active regions of genes, termed promoters, in order to form preinitiation complexes (PICs) capable of initiation. In eukaryotes the process of PIC formation is facilitated by multiple general transcription factors (GTFs), TFIIA, B, D, E, F, and H.
- Eukaryotic PIC formation can occur either stepwise—by the sequential, ordered interactions of

GTFs and RNA polymerase with promoters—or in one step by the recognition of the promoter by a preformed GTF-RNA polymerase holoenzyme complex.

- Transcription exhibits three phases: initiation, elongation, and termination. All are dependent upon distinct DNA *cis*-elements and can be modulated by distinct *trans*-acting protein factors.
- Most eukaryotic RNAs are synthesized as precursors that contain excess sequences which are removed prior to the generation of mature, functional RNA.
- Eukaryotic mRNA synthesis results in a pre-mRNA precursor that contains extensive amounts of excess RNA (introns) that must be precisely removed by RNA splicing to generate functional, translatable mRNA composed of exonic coding and noncoding sequences.
- All steps—from changes in DNA template, sequence, and accessibility in chromatin to RNA stability—are subject to modulation and hence are potential control sites for eukaryotic gene regulation.

## REFERENCES

Busby S, Ebright RH: Promoter structure, promoter recognition, and transcription activation in prokaryotes. Cell 1994;79: 743.

Cramer P, Bushnell DA, Kornberg R: Structural basis of transcription: RNA polymerase II at 2.8 angstrom resolution. Science 2001;292:1863.

Fedor MJ: Ribozymes. Curr Biol 1998;8:R441.

Gott JM, Emeson RB: Functions and mechanisms of RNA editing. Ann Rev Genet 2000;34:499.

Hirose Y, Manley JL: RNA polymerase II and the integration of nuclear events. Genes Dev 2000;14:1415.

Keaveney M, Struhl K: Activator-mediated recruitment of the RNA polymerase machinery is the predominant mechanism for transcriptional activation in yeast. Mol Cell 1998;1:917.

Lemon B, Tjian R: Orchestrated response: a symphony of transcription factors for gene control. Genes Dev 2000;14:2551.

Maniatis T, Reed R: An extensive network of coupling among gene expression machines. Nature 2002;416:499.

Orphanides G, Reinberg D: A unified theory of gene expression. Cell 2002;108:439.

Shatkin AJ, Manley JL: The ends of the affair: capping and polyadenylation. Nat Struct Biol 2000;7:838.

Stevens SW et al: Composition and functional characterization of the yeast spliceosomal penta-snRNP. Mol Cell 2002;9:31.

Tucker M, Parker R: Mechanisms and control of mRNA decapping in *Saccharomyces cerevisiae.* Ann Rev Biochem 2000;69: 571.

Woychik NA, Hampsey M: The RNA polymerase II machinery: structure illuminates function. Cell 2002;108:453.

# Protein Synthesis & the Genetic Code

38

*Daryl K. Granner, MD*

## BIOMEDICAL IMPORTANCE

The letters A, G, T, and C correspond to the nucleotides found in DNA. They are organized into three-letter code words called **codons,** and the collection of these codons makes up the **genetic code.** It was impossible to understand protein synthesis—or to explain mutations—before the genetic code was elucidated. The code provides a foundation for explaining the way in which protein defects may cause genetic disease and for the diagnosis and perhaps the treatment of these disorders. In addition, the pathophysiology of many viral infections is related to the ability of these agents to disrupt host cell protein synthesis. Many antibacterial agents are effective because they selectively disrupt protein synthesis in the invading bacterial cell but do not affect protein synthesis in eukaryotic cells.

## GENETIC INFORMATION FLOWS FROM DNA TO RNA TO PROTEIN

The genetic information within the nucleotide sequence of DNA is transcribed in the nucleus into the specific nucleotide sequence of an RNA molecule. The sequence of nucleotides in the RNA transcript is complementary to the nucleotide sequence of the template strand of its gene in accordance with the base-pairing rules. Several different classes of RNA combine to direct the synthesis of proteins.

In prokaryotes there is a linear correspondence between the gene, the **messenger RNA (mRNA)** transcribed from the gene, and the polypeptide product. The situation is more complicated in higher eukaryotic cells, in which the primary transcript is much larger than the mature mRNA. The large mRNA precursors contain coding regions **(exons)** that will form the mature mRNA and long intervening sequences **(introns)** that separate the exons. The hnRNA is processed within the nucleus, and the introns, which often make up much more of this RNA than the exons, are removed. Exons are spliced together to form mature mRNA, which is transported to the cytoplasm, where it is translated into protein.

The cell must possess the machinery necessary to translate information accurately and efficiently from the nucleotide sequence of an mRNA into the sequence of amino acids of the corresponding specific protein. Clarification of our understanding of this process, which is termed **translation,** awaited deciphering of the genetic code. It was realized early that mRNA molecules themselves have no affinity for amino acids and, therefore, that the translation of the information in the mRNA nucleotide sequence into the amino acid sequence of a protein requires an intermediate adapter molecule. This adapter molecule must recognize a specific nucleotide sequence on the one hand as well as a specific amino acid on the other. With such an adapter molecule, the cell can direct a specific amino acid into the proper sequential position of a protein during its synthesis as dictated by the nucleotide sequence of the specific mRNA. In fact, the functional groups of the amino acids do not themselves actually come into contact with the mRNA template.

## THE NUCLEOTIDE SEQUENCE OF AN mRNA MOLECULE CONSISTS OF A SERIES OF CODONS THAT SPECIFY THE AMINO ACID SEQUENCE OF THE ENCODED PROTEIN

Twenty different amino acids are required for the synthesis of the cellular complement of proteins; thus, there must be at least 20 distinct codons that make up the genetic code. Since there are only four different nucleotides in mRNA, each codon must consist of more than a single purine or pyrimidine nucleotide. Codons consisting of two nucleotides each could provide for only 16 ($4^2$) specific codons, whereas codons of three nucleotides could provide 64 ($4^3$) specific codons.

It is now known that each codon consists of a sequence of three nucleotides; ie, **it is a triplet code** (see Table 38–1). The deciphering of the genetic code depended heavily on the chemical synthesis of nucleotide polymers, particularly triplets in repeated sequence.

***Table 38–1.*** The genetic code (codon assignments in mammalian messenger RNA).[1]

| First Nucleotide | Second Nucleotide | | | | Third Nucleotide |
|---|---|---|---|---|---|
| | U | C | A | G | |
| U | Phe | Ser | Tyr | Cys | U |
| | Phe | Ser | Tyr | Cys | C |
| | Leu | Ser | Term | Term[2] | A |
| | Leu | Ser | Term | Trp | G |
| C | Leu | Pro | His | Arg | U |
| | Leu | Pro | His | Arg | C |
| | Leu | Pro | Gln | Arg | A |
| | Leu | Pro | Gln | Arg | G |
| A | Ile | Thr | Asn | Ser | U |
| | Ile | Thr | Asn | Ser | C |
| | Ile[2] | Thr | Lys | Arg[2] | A |
| | Met | Thr | Lys | Arg[2] | G |
| G | Val | Ala | Asp | Gly | U |
| | Val | Ala | Asp | Gly | C |
| | Val | Ala | Glu | Gly | A |
| | Val | Ala | Glu | Gly | G |

[1]The terms first, second, and third nucleotide refer to the individual nucleotides of a triplet codon. U, uridine nucleotide; C, cytosine nucleotide; A, adenine nucleotide; G, guanine nucleotide; Term, chain terminator codon. AUG, which codes for Met, serves as the initiator codon in mammalian cells and encodes for internal methionines in a protein. (Abbreviations of amino acids are explained in Chapter 3.)
[2]In mammalian mitochondria, AUA codes for Met and UGA for Trp, and AGA and AGG serve as chain terminators.

## THE GENETIC CODE IS DEGENERATE, UNAMBIGUOUS, NONOVERLAPPING, WITHOUT PUNCTUATION, & UNIVERSAL

Three of the 64 possible codons do not code for specific amino acids; these have been termed **nonsense codons.** These nonsense codons are utilized in the cell as **termination signals;** they specify where the polymerization of amino acids into a protein molecule is to stop. The remaining 61 codons code for 20 amino acids (Table 38–1). Thus, there must be **"degeneracy"** in the genetic code—ie, multiple codons must decode the same amino acid. Some amino acids are encoded by several codons; for example, six different codons specify serine. Other amino acids, such as methionine and tryptophan, have a single codon. In general, the third nucleotide in a codon is less important than the first two in determining the specific amino acid to be incorporated, and this accounts for most of the degeneracy of the code. However, for any specific codon, only a single amino acid is indicated; with rare exceptions, the genetic code is **unambiguous**—ie, given a specific codon, only a single amino acid is indicated. **The distinction between ambiguity and degeneracy is an important concept.**

The unambiguous but degenerate code can be explained in molecular terms. The recognition of specific codons in the mRNA by the tRNA adapter molecules is dependent upon their **anticodon region** and specific base-pairing rules. Each tRNA molecule contains a specific sequence, complementary to a codon, which is termed its anticodon. For a given codon in the mRNA, only a single species of tRNA molecule possesses the proper anticodon. Since each tRNA molecule can be charged with only one specific amino acid, each codon therefore specifies only one amino acid. However, some tRNA molecules can utilize the anticodon to recognize more than one codon. **With few exceptions, given a specific codon, only a specific amino acid will be incorporated—although, given a specific amino acid, more than one codon may be used.**

As discussed below, the reading of the genetic code during the process of protein synthesis does not involve any overlap of codons. **Thus, the genetic code is nonoverlapping.** Furthermore, once the reading is commenced at a specific codon, there is **no punctuation** between codons, and the message is read in a continuing sequence of nucleotide triplets until a translation stop codon is reached.

Until recently, the genetic code was thought to be universal. It has now been shown that the set of tRNA molecules in mitochondria (which contain their own separate and distinct set of translation machinery) from lower and higher eukaryotes, including humans, reads four codons differently from the tRNA molecules in the cytoplasm of even the same cells. As noted in Table 38–1, the codon AUA is read as Met, and UGA codes for Trp in mammalian mitochondria. In addition, in mitochondria, the codons AGA and AGG are read as stop or chain terminator codons rather than as Arg. As a result, mitochondria require only 22 tRNA molecules to read their genetic code, whereas the cytoplasmic translation system possesses a full complement of 31 tRNA species. These exceptions noted, **the genetic code is universal.** The frequency of use of each amino acid codon varies considerably between species and among different tissues within a species. The specific tRNA levels generally mirror these codon usage biases. Thus, a particular abundantly used codon is decoded by a similarly abundant specific tRNA which recognizes that particular codon. Tables of **codon usage** are becoming more accurate as more genes are sequenced. This is of considerable importance because investigators

***Table 38–2.*** Features of the genetic code.

| |
|---|
| • Degenerate |
| • Unambiguous |
| • Nonoverlapping |
| • Not punctuated |
| • Universal |

often need to deduce mRNA structure from the primary sequence of a portion of protein in order to synthesize an oligonucleotide probe and initiate a recombinant DNA cloning project. The main features of the genetic code are listed in Table 38–2.

## AT LEAST ONE SPECIES OF TRANSFER RNA (tRNA) EXISTS FOR EACH OF THE 20 AMINO ACIDS

tRNA molecules have extraordinarily similar functions and three-dimensional structures. The adapter function of the tRNA molecules requires the charging of each specific tRNA with its specific amino acid. Since there is no affinity of nucleic acids for specific functional groups of amino acids, this recognition must be carried out by a protein molecule capable of recognizing both a specific tRNA molecule and a specific amino acid. At least 20 specific enzymes are required for these specific recognition functions and for the proper attachment of the 20 amino acids to specific tRNA molecules. The process of **recognition and attachment (charging)** proceeds in two steps by one enzyme for each of the 20 amino acids. These enzymes are termed **aminoacyl-tRNA synthetases.** They form an activated intermediate of aminoacyl-AMP-enzyme complex (Figure 38–1). The specific aminoacyl-AMP-enzyme complex then recognizes a specific tRNA to which it attaches the aminoacyl moiety at the 3′-hydroxyl adenosyl terminal. The charging reactions have an error rate of less than $10^{-4}$ and so are extremely accurate. The amino acid remains attached to its specific tRNA in an ester linkage until it is polymerized at a specific position in the fabrication of a polypeptide precursor of a protein molecule.

The regions of the tRNA molecule referred to in Chapter 35 (and illustrated in Figure 35–11) now become important. The thymidine-pseudouridine-cytidine (TΨC) arm is involved in binding of the aminoacyl-tRNA to the ribosomal surface at the site of protein synthesis. The D arm is one of the sites important for the proper recognition of a given tRNA species by its proper aminoacyl-tRNA synthetase. The acceptor arm, located at the 3′-hydroxyl adenosyl terminal, is the site of attachment of the specific amino acid.

The anticodon region consists of seven nucleotides, and it recognizes the three-letter codon in mRNA (Figure 38–2). The sequence read from the 3′ to 5′ direction in that anticodon loop consists of a variable base–modified purine–XYZ–pyrimidine–pyrimidine-5′. Note that this direction of reading the anticodon is 3′ to 5′, whereas the genetic code in Table 38–1 is read 5′ to 3′, since the codon and the anticodon loop of the mRNA and tRNA molecules, respectively, are **antiparallel** in their complementarity just like all other intermolecular interactions between nucleic acid strands.

The degeneracy of the genetic code resides mostly in the last nucleotide of the codon triplet, suggesting that the base pairing between this last nucleotide and the corresponding nucleotide of the anticodon is not strictly

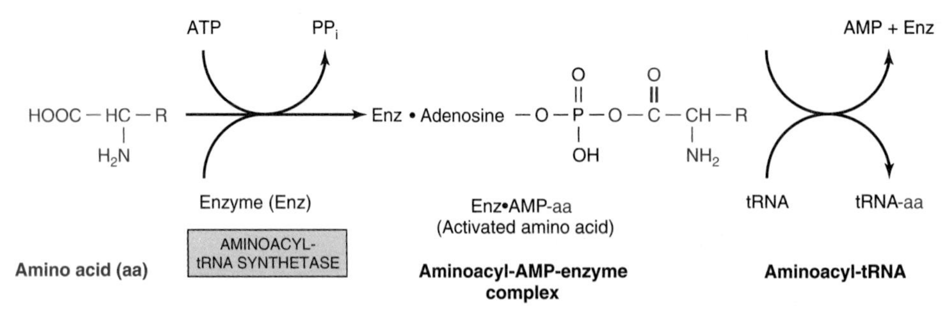


***Figure 38–1.*** Formation of aminoacyl-tRNA. A two-step reaction, involving the enzyme aminoacyl-tRNA synthetase, results in the formation of aminoacyl-tRNA. The first reaction involves the formation of an AMP-amino acid-enzyme complex. This activated amino acid is next transferred to the corresponding tRNA molecule. The AMP and enzyme are released, and the latter can be reutilized. The charging reactions have an error rate of less than $10^{-4}$ and so are extremely accurate.

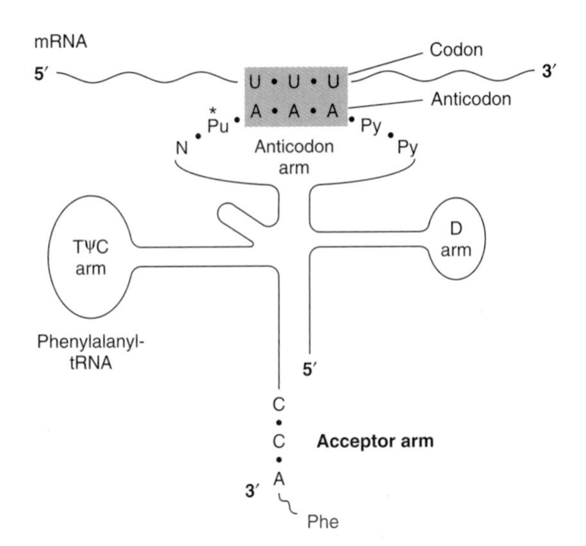


***Figure 38–2.*** Recognition of the codon by the anticodon. One of the codons for phenylalanine is UUU. tRNA charged with phenylalanine (Phe) has the complementary sequence AAA; hence, it forms a base-pair complex with the codon. The anticodon region typically consists of a sequence of seven nucleotides: variable (N), modified purine ((Pu*), X, Y, Z, and two pyrimidines (Py) in the 3′ to 5′ direction.

by the Watson-Crick rule. This is called **wobble;** the pairing of the codon and anticodon can "wobble" at this specific nucleotide-to-nucleotide pairing site. For example, the two codons for arginine, AGA and AGG, can bind to the same anticodon having a uracil at its 5′ end (UCU). Similarly, three codons for glycine—GGU, GGC, and GGA—can form a base pair from one anticodon, CCI. I is an inosine nucleotide, another of the peculiar bases appearing in tRNA molecules.

## MUTATIONS RESULT WHEN CHANGES OCCUR IN THE NUCLEOTIDE SEQUENCE

Although the initial change may not occur in the template strand of the double-stranded DNA molecule for that gene, after replication, daughter DNA molecules with mutations in the template strand will segregate and appear in the population of organisms.

### Some Mutations Occur by Base Substitution

Single-base changes **(point mutations)** may be **transitions** or **transversions.** In the former, a given pyrimidine is changed to the other pyrimidine or a given purine is changed to the other purine. Transversions are changes from a purine to either of the two pyrimidines or the change of a pyrimidine into either of the two purines, as shown in Figure 38–3.

If the nucleotide sequence of the gene containing the mutation is transcribed into an RNA molecule, then the RNA molecule will possess a complementary base change at this corresponding locus.

Single-base changes in the mRNA molecules may have one of several effects when translated into protein:

(1) There may be no detectable effect because of the degeneracy of the code. This would be more likely if the changed base in the mRNA molecule were to be at the third nucleotide of a codon; such mutations are often referred to as **silent mutations.** Because of wobble, the translation of a codon is least sensitive to a change at the third position.

(2) A **missense effect** will occur when a different amino acid is incorporated at the corresponding site in the protein molecule. This mistaken amino acid—or missense, depending upon its location in the specific protein—might be acceptable, partially acceptable, or unacceptable to the function of that protein molecule. From a careful examination of the genetic code, one can conclude that most single-base changes would result in the replacement of one amino acid by another with rather similar functional groups. This is an effective mechanism to avoid drastic change in the physical properties of a protein molecule. If an acceptable missense effect occurs, the resulting protein molecule may not be distinguishable from the normal one. A partially acceptable missense will result in a protein molecule with partial but abnormal function. If an unacceptable missense effect occurs, then the protein molecule will not be capable of functioning in its assigned role.

(3) A **nonsense** codon may appear that would then result in the **premature termination** of amino acid incorporation into a peptide chain and the production of only a fragment of the intended protein molecule. The probability is high that a prematurely terminated protein molecule or peptide fragment will not function in its assigned role.

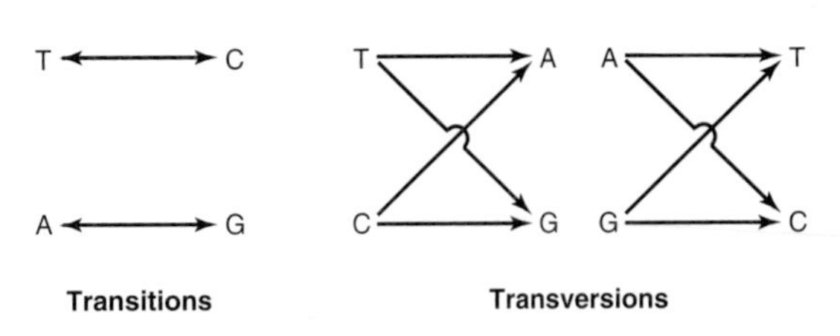


***Figure 38–3.*** Diagrammatic representation of transition mutations and transversion mutations.

## Hemoglobin Illustrates the Effects of Single-Base Changes in Structural Genes

**Some mutations have no apparent effect.** The gene system that encodes hemoglobin is one of the best-studied in humans. The lack of effect of a single-base change is demonstrable only by sequencing the nucleotides in the mRNA molecules or structural genes. The sequencing of a large number of hemoglobin mRNAs and genes from many individuals has shown that the codon for valine at position 67 of the β chain of hemoglobin is not identical in all persons who possess a normally functional β chain of hemoglobin. Hemoglobin Milwaukee has at position 67 a glutamic acid; hemoglobin Bristol contains aspartic acid at position 67. In order to account for the amino acid change by the change of a single nucleotide residue in the codon for amino acid 67, one must infer that the mRNA encoding hemoglobin Bristol possessed a GUU or GUC codon prior to a later change to GAU or GAC, both codons for aspartic acid. However, the mRNA encoding hemoglobin Milwaukee would have to possess at position 67 a codon GUA or GUG in order that a single nucleotide change could provide for the appearance of the glutamic acid codons GAA or GAG. Hemoglobin Sydney, which contains an alanine at position 67, could have arisen by the change of a single nucleotide in any of the four codons for valine (GUU, GUC, GUA, or GUG) to the alanine codons (GCU, GCC, GCA, or GCG, respectively).

## Substitution of Amino Acids Causes Missense Mutations

### A. Acceptable Missense Mutations

An example of an acceptable missense mutation (Figure 38–4, top) in the structural gene for the β chain of hemoglobin could be detected by the presence of an electrophoretically altered hemoglobin in the red cells of an apparently healthy individual. Hemoglobin Hikari has been found in at least two families of Japanese people. This hemoglobin has asparagine substituted for lysine at the 61 position in the β chain. The corresponding

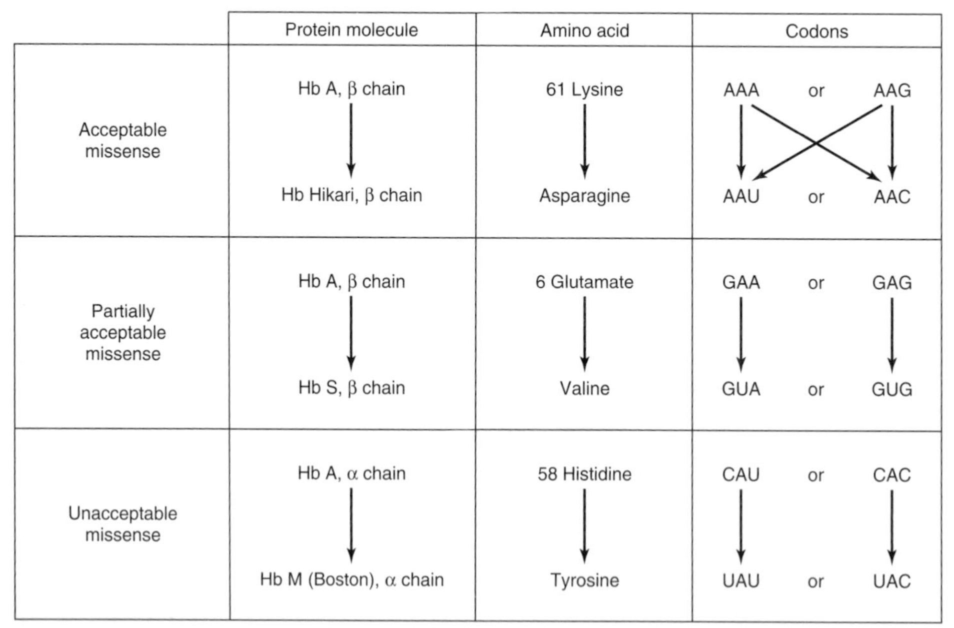


***Figure 38–4.*** Examples of three types of missense mutations resulting in abnormal hemoglobin chains. The amino acid alterations and possible alterations in the respective codons are indicated. The hemoglobin Hikari β-chain mutation has apparently normal physiologic properties but is electrophoretically altered. Hemoglobin S has a β-chain mutation and partial function; hemoglobin S binds oxygen but precipitates when deoxygenated. Hemoglobin M Boston, an α-chain mutation, permits the oxidation of the heme ferrous iron to the ferric state and so will not bind oxygen at all.

transversion might be either AAA or AAG changed to either AAU or AAC. The replacement of the specific lysine with asparagine apparently does not alter the normal function of the β chain in these individuals.

### B. Partially Acceptable Missense Mutations

A partially acceptable missense mutation (Figure 38–4, center) is best exemplified by **hemoglobin S,** which is found in sickle cell anemia. Here glutamic acid, the normal amino acid in position 6 of the β chain, has been replaced by valine. The corresponding single nucleotide change within the codon would be GAA or GAG of glutamic acid to GUA or GUG of valine. Clearly, this missense mutation hinders normal function and results in sickle cell anemia when the mutant gene is present in the homozygous state. The glutamate-to-valine change may be considered to be partially acceptable because hemoglobin S does bind and release oxygen, although abnormally.

### C. Unacceptable Missense Mutations

An unacceptable missense mutation (Figure 38–4, bottom) in a hemoglobin gene generates a nonfunctioning hemoglobin molecule. For example, the hemoglobin M mutations generate molecules that allow the $Fe^{2+}$ of the heme moiety to be oxidized to $Fe^{3+}$, producing methemoglobin. Methemoglobin cannot transport oxygen (see Chapter 6).

## Frameshift Mutations Result From Deletion or Insertion of Nucleotides in DNA That Generates Altered mRNAs

The deletion of a single nucleotide from the coding strand of a gene results in an altered reading frame in the mRNA. The machinery translating the mRNA does not recognize that a base was missing, since there is no punctuation in the reading of codons. Thus, a major alteration in the sequence of polymerized amino acids, as depicted in example 1, Figure 38–5, results. Altering the reading frame results in a garbled translation of the mRNA distal to the single nucleotide deletion. Not only is the sequence of amino acids distal to this deletion garbled, but reading of the message can also result in the appearance of a nonsense codon and thus the production of a polypeptide both garbled and prematurely terminated (example 3, Figure 38–5).

If three nucleotides or a multiple of three are deleted from a coding region, the corresponding mRNA when translated will provide a protein from which is missing the corresponding number of amino acids (example 2, Figure 38–5). Because the reading frame is a triplet, the reading phase will not be disturbed for those codons distal to the deletion. If, however, deletion of one or two nucleotides occurs just prior to or within the normal termination codon (nonsense codon), the reading of the normal termination signal is disturbed. Such a deletion might result in reading through a termination signal until another nonsense codon is encountered (example 1, Figure 38–5). Examples of this phenomenon are described in discussions of hemoglobinopathies.

Insertions of one or two or nonmultiples of three nucleotides into a gene result in an mRNA in which the reading frame is distorted upon translation, and the same effects that occur with deletions are reflected in the mRNA translation. This may result in garbled amino acid sequences distal to the insertion and the generation of a **nonsense codon** at or distal to the insertion, or perhaps reading through the normal termination codon. Following a deletion in a gene, an insertion (or vice versa) can reestablish the proper reading frame (example 4, Figure 38–5). The corresponding mRNA, when translated, would contain a garbled amino acid sequence between the insertion and deletion. Beyond the reestablishment of the reading frame, the amino acid sequence would be correct. One can imagine that different combinations of deletions, of insertions, or of both would result in formation of a protein wherein a portion is abnormal, but this portion is surrounded by the normal amino acid sequences. Such phenomena have been demonstrated convincingly in a number of diseases.

## Suppressor Mutations Can Counteract Some of the Effects of Missense, Nonsense, & Frameshift Mutations

The above discussion of the altered protein products of gene mutations is based on the presence of normally functioning tRNA molecules. However, in prokaryotic and lower eukaryotic organisms, abnormally functioning tRNA molecules have been discovered that are themselves the results of mutations. Some of these abnormal tRNA molecules are capable of binding to and decoding altered codons, thereby suppressing the effects of mutations in distant structural genes. These **suppressor tRNA molecules,** usually formed as the result of alterations in their anticodon regions, are capable of suppressing missense mutations, nonsense mutations, and frameshift mutations. However, since the suppressor tRNA molecules are not capable of distinguishing between a normal codon and one resulting from a gene mutation, their presence in a cell usually results in decreased viability. For instance, the nonsense suppressor tRNA molecules can suppress the normal termination signals to allow a read-through when it is not desirable. Frameshift suppressor tRNA molecules may read a normal codon plus a component of a juxtaposed codon to provide a frameshift, also when it is not desirable. Suppressor tRNA molecules may exist in mammalian cells, since read-through transcription occurs.

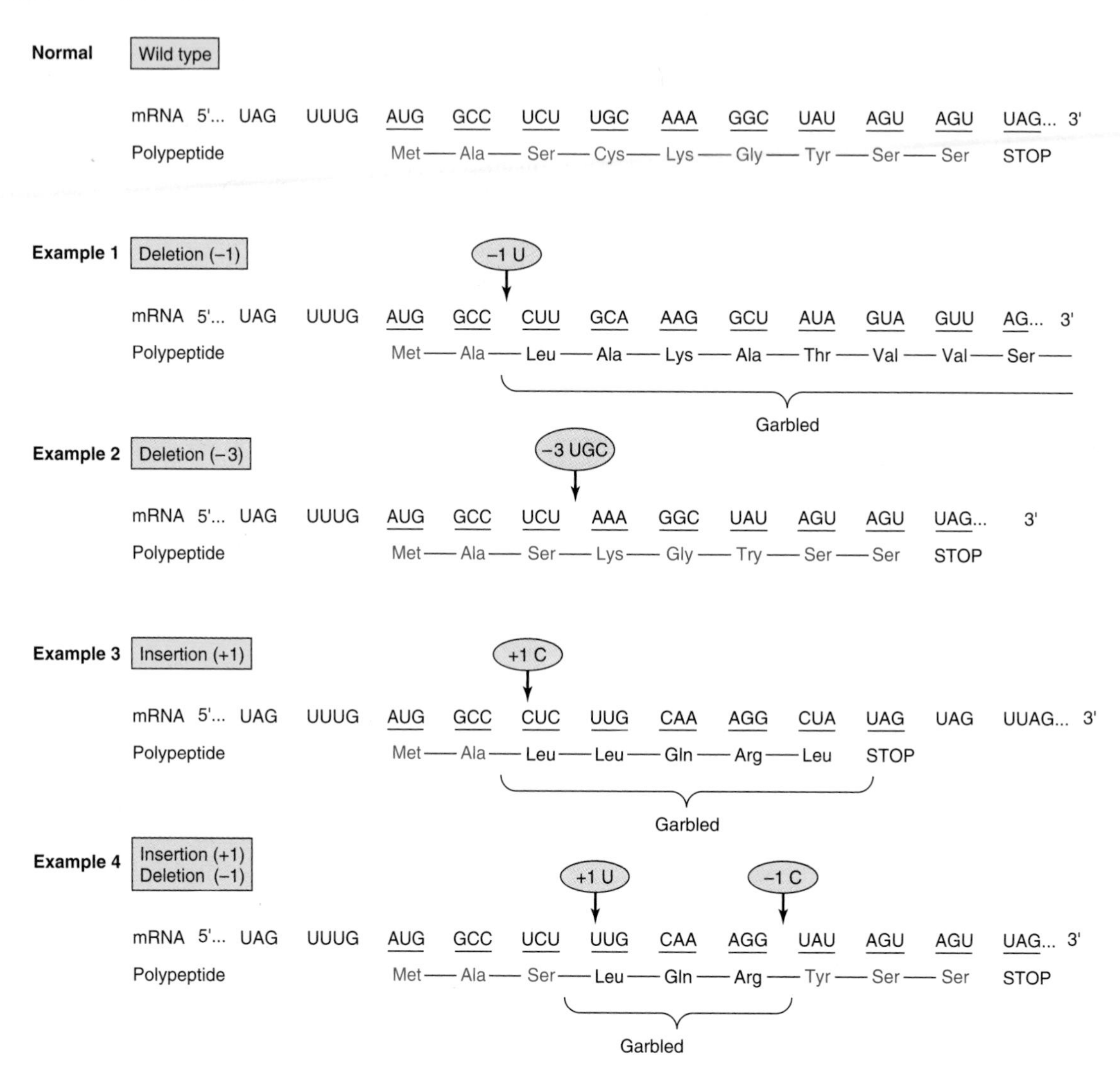


***Figure 38–5.*** Examples of the effects of deletions and insertions in a gene on the sequence of the mRNA transcript and of the polypeptide chain translated therefrom. The arrows indicate the sites of deletions or insertions, and the numbers in the ovals indicate the number of nucleotide residues deleted or inserted. Blue type indicates amino acids in correct order.

## LIKE TRANSCRIPTION, PROTEIN SYNTHESIS CAN BE DESCRIBED IN THREE PHASES: INITIATION, ELONGATION, & TERMINATION

The general structural characteristics of ribosomes and their self-assembly process are discussed in Chapter 37. These particulate entities serve as the machinery on which the mRNA nucleotide sequence is translated into the sequence of amino acids of the specified protein. The translation of the mRNA commences near its 5′ terminal with the formation of the corresponding amino terminal of the protein molecule. The message is read from 5′ to 3′, concluding with the formation of the carboxyl terminal of the protein. Again, the concept of **polarity** is apparent. As described in Chapter 37, the transcription of a gene into the corresponding mRNA or its precursor first forms the 5′ terminal of the RNA molecule. In prokaryotes, this allows for the beginning of mRNA translation before the transcription of the gene is completed. In eukaryotic organisms, the process of transcription is a nuclear one; mRNA translation occurs in the cytoplasm. This precludes simultaneous transcription and translation in eukaryotic organisms and makes possible the processing necessary to generate mature mRNA from the primary transcript—hnRNA.

## Initiation Involves Several Protein-RNA Complexes (Figure 38–6)

Initiation of protein synthesis requires that an mRNA molecule be selected for translation by a ribosome. Once the mRNA binds to the ribosome, the latter finds the correct reading frame on the mRNA, and translation begins. This process involves tRNA, rRNA, mRNA, and at least ten eukaryotic initiation factors (eIFs), some of which have multiple (three to eight) subunits. Also involved are GTP, ATP, and amino acids. Initiation can be divided into four steps: (1) dissociation of the ribosome into its 40S and 60S subunits; (2) binding of a ternary complex consisting of met-tRNA$^i$, GTP, and eIF-2 to the 40S ribosome to form a preinitiation complex; (3) binding of mRNA to the 40S preinitiation complex to form a 43S initiation complex; and (4) combination of the 43S initiation complex with the 60S ribosomal subunit to form the 80S initiation complex.

### A. Ribosomal Dissociation

Two initiation factors, eIF-3 and eIF-1A, bind to the newly dissociated 40S ribosomal subunit. This delays its reassociation with the 60S subunit and allows other translation initiation factors to associate with the 40S subunit.

### B. Formation of the 43S Preinitiation Complex

The first step in this process involves the binding of GTP by eIF-2. This binary complex then binds to met-tRNA$^i$, a tRNA specifically involved in binding to the initiation codon AUG. (There are two tRNAs for methionine. One specifies methionine for the initiator codon, the other for internal methionines. Each has a unique nucleotide sequence.) This ternary complex binds to the 40S ribosomal subunit to form the 43S preinitiation complex, which is stabilized by association with eIF-3 and eIF-1A.

eIF-2 is one of two control points for protein synthesis initiation in eukaryotic cells. eIF-2 consists of α, β, and γ subunits. eIF-2α is phosphorylated (on serine 51) by at least four different protein kinases (HCR, PKR, PERK, and GCN2) that are activated when a cell is under stress and when the energy expenditure required for protein synthesis would be deleterious. Such conditions include amino acid and glucose starvation, virus infection, misfolded proteins, serum deprivation, hyperosmolality, and heat shock. PKR is particularly interesting in this regard. This kinase is activated by viruses and provides a host defense mechanism that decreases protein synthesis, thereby inhibiting viral replication. Phosphorylated eIF-2α binds tightly to and inactivates the GTP-GDP recycling protein eIF-2B. This prevents formation of the 43S preinitiation complex and blocks protein synthesis.

### C. Formation of the 43S Initiation Complex

The 5′ terminals of most mRNA molecules in eukaryotic cells are "capped," as described in Chapter 37. This methyl-guanosyl triphosphate cap facilitates the binding of mRNA to the 43S preinitiation complex. A cap binding protein complex, eIF-4F (4F), which consists of eIF-4E and the eIF-4G (4G)-eIF4A (4A) complex, binds to the cap through the 4E protein. Then eIF-4A (4A) and eIF-4B (4B) bind and reduce the complex secondary structure of the 5′ end of the mRNA through ATPase and ATP-dependent helicase activities. The association of mRNA with the 43S preinitiation complex to form the 48S initiation complex requires ATP hydrolysis. eIF-3 is a key protein because it binds with high affinity to the 4G component of 4F, and it links this complex to the 40S ribosomal subunit. Following association of the 43S preinitiation complex with the mRNA cap and reduction ("melting") of the secondary structure near the 5′ end of the mRNA, the complex scans the mRNA for a suitable initiation codon. Generally this is the 5′-most AUG, but the precise initiation codon is determined by so-called **Kozak consensus sequences** that surround the AUG:

–3 –1 +4

GCCA / GCCAUGG

Most preferred is the presence of a purine at positions –3 and +4 relative to the AUG.

### D. Role of the Poly(A) Tail in Initiation

Biochemical and genetic experiments in yeast have revealed that the 3′ poly(A) tail and its binding protein, Pab1p, are required for efficient initiation of protein synthesis. Further studies showed that the poly(A) tail stimulates recruitment of the 40S ribosomal subunit to the mRNA through a complex set of interactions. Pab1p, bound to the poly(A) tail, interacts with eIF-4G, which in turn binds to eIF-4E that is bound to the cap structure. It is possible that a circular structure is formed and that this helps direct the 40S ribosomal subunit to the 5′ end of the mRNA. This helps explain how the cap and poly(A) tail structures have a synergistic effect on protein synthesis. It appears that a similar mechanism is at work in mammalian cells.

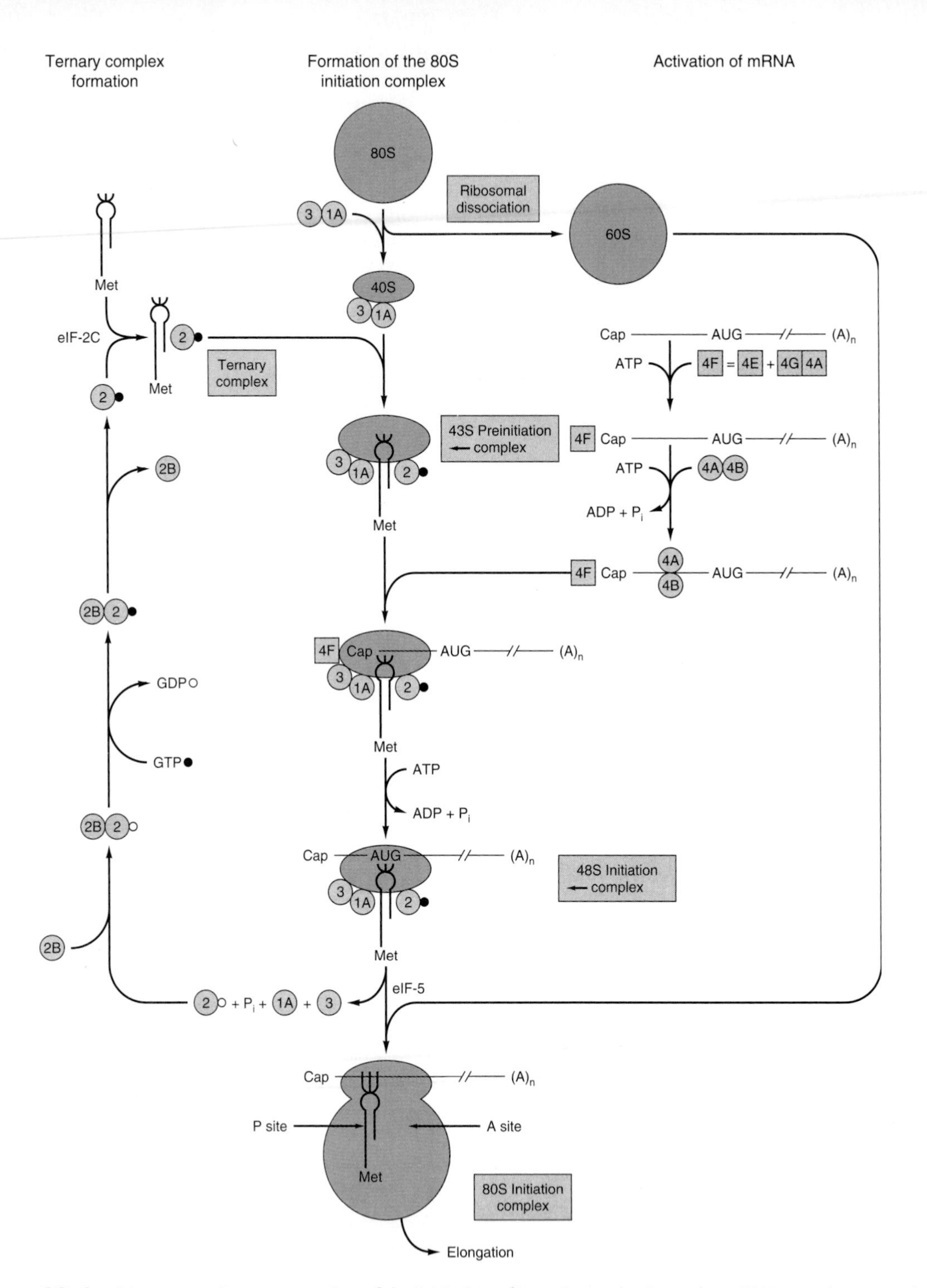


***Figure 38–6.*** Diagrammatic representation of the initiation of protein synthesis on the mRNA template containing a 5′ cap ($G^mTP$-5′) and 3′ poly(A) terminal [3′$(A)_n$]. This process proceeds in three steps: (1) activation of mRNA; (2) formation of the ternary complex consisting of $tRNAmet^i$, initiation factor eIF-2, and GTP; and (3) formation of the active 80S initiation complex. (See text for details.) GTP, ●; GDP, ○. The various initiation factors appear in abbreviated form as circles or squares, eg, eIF-3 (③), eIF-4F (4F). 4•F is a complex consisting of 4E and 4A bound to 4G (see Figure 38–7). The constellation of protein factors and the 40S ribosomal subunit comprise the 43S preinitiation complex. When bound to mRNA, this forms the 48S preinitiation complex.

### E. Formation of the 80S Initiation Complex

The binding of the 60S ribosomal subunit to the 48S initiation complex involves hydrolysis of the GTP bound to eIF-2 by eIF-5. This reaction results in release of the initiation factors bound to the 48S initiation complex (these factors then are recycled) and the rapid association of the 40S and 60S subunits to form the 80S ribosome. At this point, the met-tRNA$^{i}$ is on the P site of the ribosome, ready for the elongation cycle to commence.

## The Regulation of eIF-4E Controls the Rate of Initiation

The 4F complex is particularly important in controlling the rate of protein translation. As described above, 4F is a complex consisting of 4E, which binds to the m$^7$G cap structure at the 5′ end of the mRNA, and 4G, which serves as a scaffolding protein. In addition to binding 4E, 4G binds to eIF-3, which links the complex to the 40S ribosomal subunit. It also binds 4A and 4B, the ATPase-helicase complex that helps unwind the RNA (Figure 38–7).

4E is responsible for recognition of the mRNA cap structure, which is a rate-limiting step in translation. This process is regulated at two levels. Insulin and mitogenic growth factors result in the phosphorylation of 4E on ser 209 (or thr 210). Phosphorylated 4E binds to the cap much more avidly than does the nonphosphorylated form, thus enhancing the rate of initiation. A component of the MAP kinase pathway (see Figure 43–8) appears to be involved in this phosphorylation reaction.

The activity of 4E is regulated in a second way, and this also involves phosphorylation. A recently discovered set of proteins bind to and inactivate 4E. These proteins include 4E-BP1 (BP1, also known as PHAS-1) and the closely related proteins 4E-BP2 and 4E-BP3. BP1 binds with high affinity to 4E. The [4E]•[BP1] association prevents 4E from binding to 4G (to form 4F). Since this interaction is essential for the binding of 4F to the ribosomal 40S subunit and for correctly positioning this on the capped mRNA, BP-1 effectively inhibits translation initiation.

Insulin and other growth factors result in the phosphorylation of BP-1 at five unique sites. Phosphorylation of BP-1 results in its dissociation from 4E, and it cannot rebind until critical sites are dephosphorylated. The protein kinase responsible has not been identified, but it appears to be different from the one that phosphorylates 4E. A kinase in the mammalian target of rapamycin (mTOR) pathway, perhaps mTOR itself, is involved. These effects on the activation of 4E explain in part how insulin causes a marked posttranscriptional increase of protein synthesis in liver, adipose tissue, and muscle.

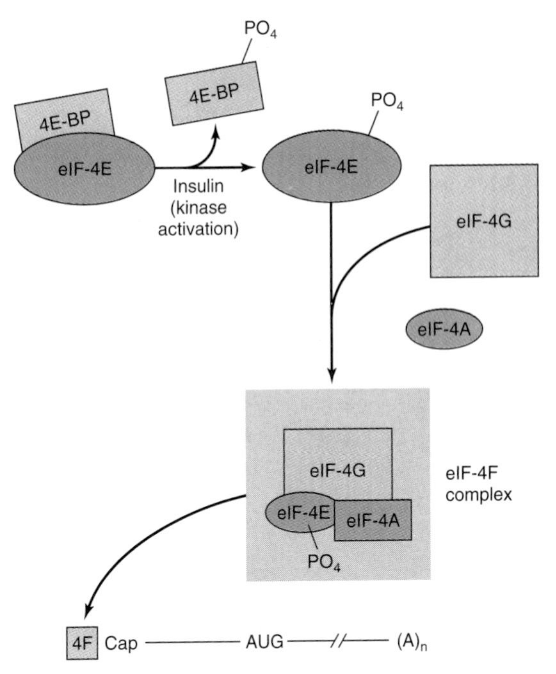


***Figure 38–7.*** Activation of eIF-4E by insulin and formation of the cap binding eIF-4F complex. The 4F-cap mRNA complex is depicted as in Figure 38–6. The 4F complex consists of eIF-4E (4E), eIF-4A, and eIF-4G. 4E is inactive when bound by one of a family of binding proteins (4E-BPs). Insulin and mitogenic factors (eg, IGF-1, PDGF, interleukin-2, and angiotensin II) activate a serine protein kinase in the mTOR pathway, and this results in the phosphorylation of 4E-BP. Phosphorylated 4E-BP dissociates from 4E, and the latter is then able to form the 4F complex and bind to the mRNA cap. These growth peptides also phosphorylate 4E itself by activating a component of the MAP kinase pathway. Phosphorylated 4E binds much more avidly to the cap than does nonphosphorylated 4E.

## Elongation Also Is a Multistep Process (Figure 38–8)

Elongation is a cyclic process on the ribosome in which one amino acid at a time is added to the nascent peptide chain. The peptide sequence is determined by the order of the codons in the mRNA. Elongation involves several steps catalyzed by proteins called elongation factors (EFs). These steps are (1) binding of aminoacyl-tRNA to the A site, (2) peptide bond formation, and (3) translocation.

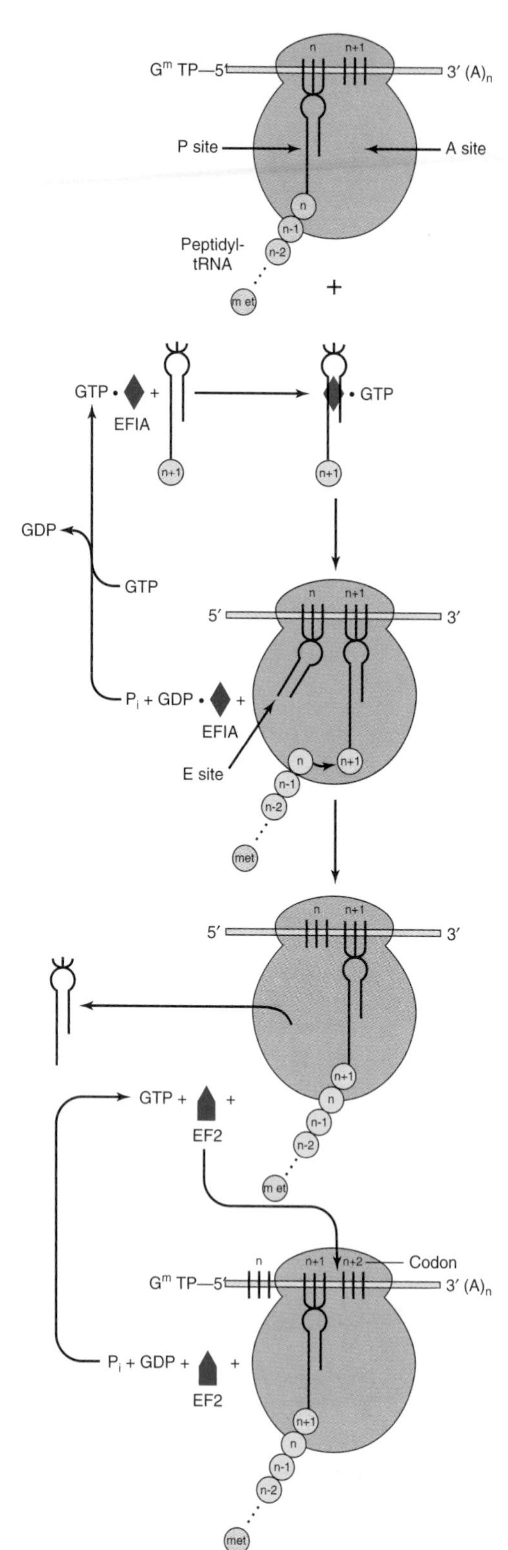


## A. Binding of Aminoacyl-tRNA to the A Site

In the complete 80S ribosome formed during the process of initiation, the A site (aminoacyl or acceptor site) is free. The binding of the proper aminoacyl-tRNA in the A site requires proper codon recognition. **Elongation factor EF1A** forms a ternary complex with GTP and the entering aminoacyl-tRNA (Figure 38–8). This complex then allows the aminoacyl-tRNA to enter the A site with the release of EF1A•GDP and phosphate. GTP hydrolysis is catalyzed by an active site on the ribosome. As shown in Figure 38–8, EF1A-GDP then recycles to EF1A-GTP with the aid of other soluble protein factors and GTP.

## B. Peptide Bond Formation

The α-amino group of the new aminoacyl-tRNA in the A site carries out a nucleophilic attack on the esterified carboxyl group of the peptidyl-tRNA occupying the P site (peptidyl or polypeptide site). At initiation, this site is occupied by aminoacyl-tRNA $met^i$. This reaction is catalyzed by a **peptidyltransferase,** a component of the 28S RNA of the 60S ribosomal subunit. This is another example of ribozyme activity and indicates an important—and previously unsuspected—direct role for RNA in protein synthesis (Table 38–3). Because the amino acid on the aminoacyl-tRNA is already "activated," no further energy source is required for this reaction. The reaction results in attachment of the growing peptide chain to the tRNA in the A site.

## C. Translocation

The now deacylated tRNA is attached by its anticodon to the P site at one end and by the open CCA tail to an **exit (E) site** on the large ribosomal subunit (Figure 38–8). At this point, **elongation factor 2 (EF2)** binds to and displaces the peptidyl tRNA from the A site to the P site. In turn, the deacylated tRNA is on the E site, from which it leaves the ribosome. The EF2-GTP complex is hydrolyzed to EF2-GDP, effectively moving the mRNA forward by one codon and leaving the A site open for occupancy by another ternary complex of amino acid tRNA-EF1A-GTP and another cycle of elongation.

***Figure 38–8.*** Diagrammatic representation of the peptide elongation process of protein synthesis. The small circles labeled n – 1, n, n + 1, etc, represent the amino acid residues of the newly formed protein molecule. EFIA and EF2 represent elongation factors 1 and 2, respectively. The peptidyl-tRNA and aminoacyl-tRNA sites on the ribosome are represented by P site and A site, respectively.

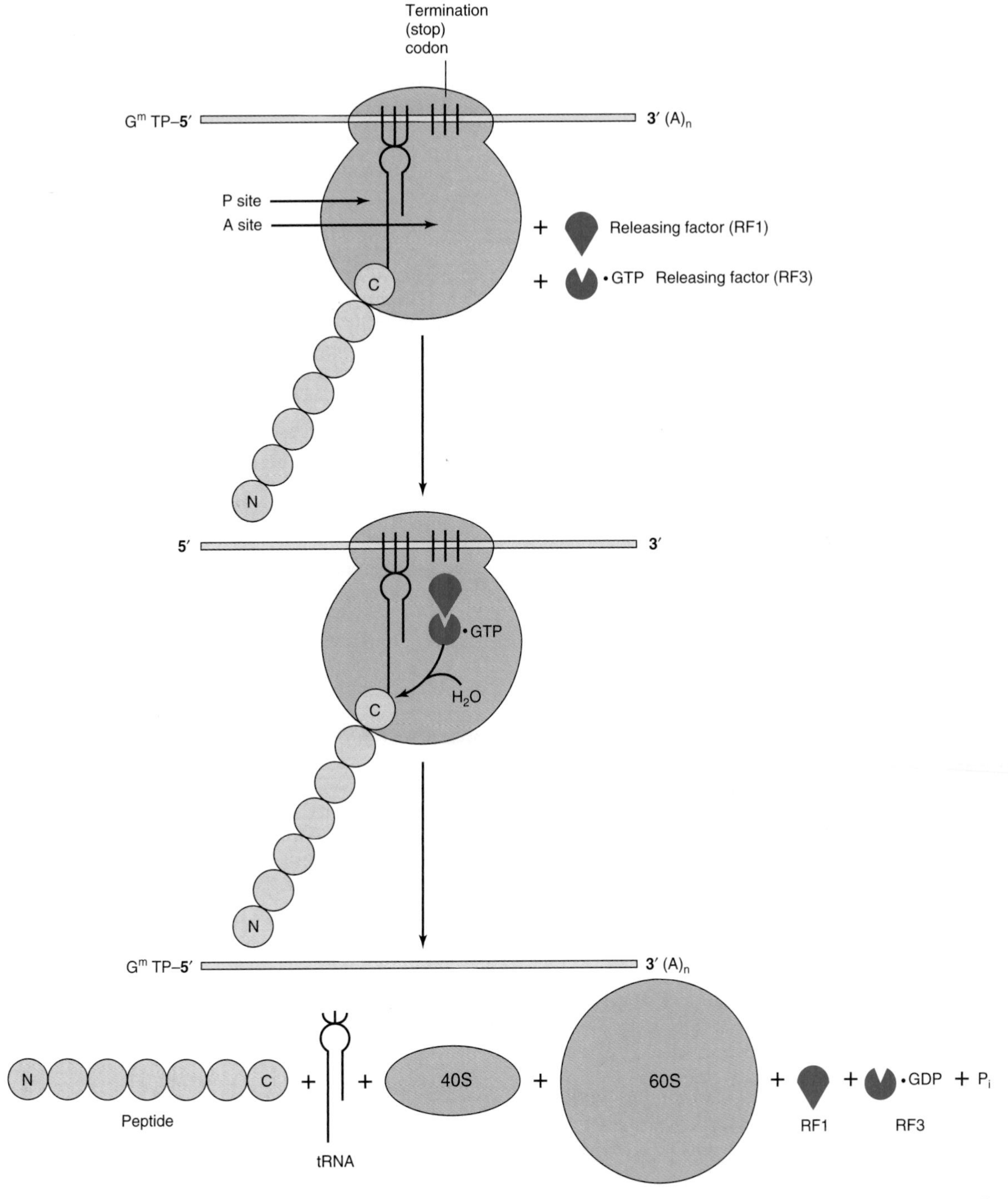


***Figure 38–9.*** Diagrammatic representation of the termination process of protein synthesis. The peptidyl-tRNA and aminoacyl-tRNA sites are indicated as P site and A site, respectively. The termination (stop) codon is indicated by the three vertical bars. Releasing factor RF1 binds to the stop codon. Releasing factor RF3, with bound GTP, binds to RF1. Hydrolysis of the peptidyl-tRNA complex is shown by the entry of $H_2O$. N and C indicate the amino and carboxyl terminal amino acids, respectively, and illustrate the polarity of protein synthesis.

***Table 38–3.*** Evidence that rRNA is peptidyltransferase.

- Ribosomes can make peptide bonds even when proteins are removed or inactivated.
- Certain parts of the rRNA sequence are highly conserved in all species.
- These conserved regions are on the surface of the RNA molecule.
- RNA can be catalytic.
- Mutations that result in antibiotic resistance at the level of protein synthesis are more often found in rRNA than in the protein components of the ribosome.

The charging of the tRNA molecule with the aminoacyl moiety requires the hydrolysis of an ATP to an AMP, equivalent to the hydrolysis of two ATPs to two ADPs and phosphates. The entry of the aminoacyl-tRNA into the A site results in the hydrolysis of one GTP to GDP. Translocation of the newly formed peptidyl-tRNA in the A site into the P site by EF2 similarly results in hydrolysis of GTP to GDP and phosphate. Thus, the energy requirements for the formation of one peptide bond include the equivalent of the hydrolysis of two ATP molecules to ADP and of two GTP molecules to GDP, or the hydrolysis of four high-energy phosphate bonds. A eukaryotic ribosome can incorporate as many as six amino acids per second; prokaryotic ribosomes incorporate as many as 18 per second. Thus, the process of peptide synthesis occurs with great speed and accuracy until a termination codon is reached.

## Termination Occurs When a Stop Codon Is Recognized (Figure 38–9)

In comparison to initiation and elongation, termination is a relatively simple process. After multiple cycles of elongation culminating in polymerization of the specific amino acids into a protein molecule, the stop or terminating codon of mRNA (UAA, UAG, UGA) appears in the A site. Normally, there is no tRNA with an anticodon capable of recognizing such a termination signal. **Releasing factor RF1** recognizes that a stop codon resides in the A site (Figure 38–9). RF1 is bound by a complex consisting of **releasing factor RF3** with bound GTP. This complex, with the peptidyl transferase, promotes hydrolysis of the bond between the peptide and the tRNA occupying the P site. Thus, a water molecule rather than an amino acid is added. This hydrolysis releases the protein and the tRNA from the P site. Upon hydrolysis and release, the **80S ribosome dissociates** into its 40S and 60S subunits, which are then recycled. Therefore, the releasing factors are proteins that hydrolyze the peptidyl-tRNA bond when a stop codon occupies the A site. The mRNA is then released from the ribosome, which dissociates into its component 40S and 60S subunits, and another cycle can be repeated.

## Polysomes Are Assemblies of Ribosomes

Many ribosomes can translate the same mRNA molecule simultaneously. Because of their relatively large size, the ribosome particles cannot attach to an mRNA any closer than 35 nucleotides apart. Multiple ribosomes on the same mRNA molecule form a **polyribosome,** or "polysome." In an unrestricted system, the number of ribosomes attached to an mRNA (and thus the size of polyribosomes) correlates positively with the length of the mRNA molecule. The mass of the mRNA molecule is, of course, quite small compared with the mass of even a single ribosome.

A single mammalian ribosome is capable of synthesizing about 400 peptide bonds each minute. Polyribosomes actively synthesizing proteins can exist as free particles in the cellular cytoplasm or may be attached to sheets of membranous cytoplasmic material referred to as **endoplasmic reticulum.** Attachment of the particulate polyribosomes to the endoplasmic reticulum is responsible for its "rough" appearance as seen by electron microscopy. The proteins synthesized by the attached polyribosomes are extruded into the cisternal space between the sheets of rough endoplasmic reticulum and are exported from there. Some of the protein products of the rough endoplasmic reticulum are packaged by the Golgi apparatus into zymogen particles for eventual export (see Chapter 46). The polyribosomal particles free in the cytosol are responsible for the synthesis of proteins required for intracellular functions.

## The Machinery of Protein Synthesis Can Respond to Environmental Threats

**Ferritin,** an iron-binding protein, prevents ionized iron ($Fe^{2+}$) from reaching toxic levels within cells. Elemental iron stimulates ferritin synthesis by causing the release of a cytoplasmic protein that binds to a specific region in the 5′ nontranslated region of ferritin mRNA. Disruption of this protein-mRNA interaction activates ferritin mRNA and results in its translation. This mechanism provides for rapid control of the synthesis of a protein that sequesters $Fe^{2+}$, a potentially toxic molecule.

## Many Viruses Co-opt the Host Cell Protein Synthesis Machinery

The protein synthesis machinery can also be modified in deleterious ways. **Viruses replicate by using host**

**cell processes,** including those involved in protein synthesis. Some viral mRNAs are translated much more efficiently than those of the host cell (eg, encephalomyocarditis virus). Others, such as reovirus and vesicular stomatitis virus, replicate abundantly, and their mRNAs have a competitive advantage over host cell mRNAs for limited translation factors. Other viruses inhibit host cell protein synthesis by preventing the association of mRNA with the 40S ribosome.

Poliovirus and other picornaviruses gain a selective advantage by disrupting the function of the 4F complex to their advantage. The mRNAs of these viruses do not have a cap structure to direct the binding of the 40S ribosomal subunit (see above). Instead, the 40S ribosomal subunit contacts an **internal ribosomal entry site (IRES)** in a reaction that requires 4G but not 4E. The virus gains a selective advantage by having a protease that attacks 4G and removes the amino terminal 4E binding site. Now the 4E-4G complex (4F) cannot form, so the 40S ribosomal subunit cannot be directed to capped mRNAs. Host cell translation is thus abolished. The 4G fragment can direct binding of the 40S ribosomal subunit to IRES-containing mRNAs, so viral mRNA translation is very efficient (Figure 38–10). These viruses also promote the dephosphorylation of BP1 (PHAS-1), thereby decreasing cap (4E)-dependent translation.

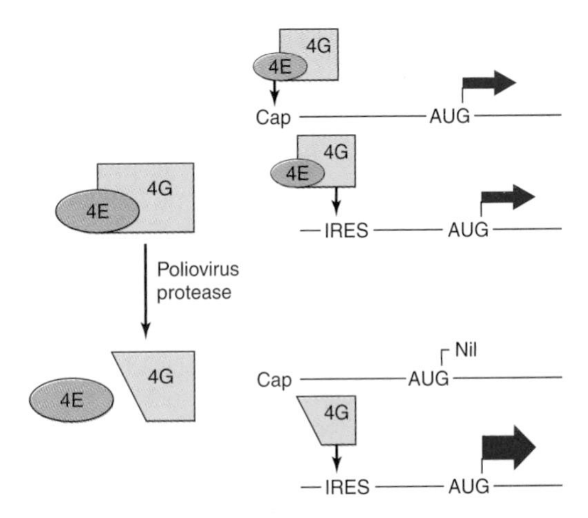


***Figure 38–10.*** Picornaviruses disrupt the 4F complex. The 4E-4G complex (4F) directs the 40S ribosomal subunit to the typical capped mRNA (see text). 4G alone is sufficient for targeting the 40S subunit to the internal ribosomal entry site (IRES) of viral mRNAs. To gain selective advantage, certain viruses (eg, poliovirus) have a protease that cleaves the 4E binding site from the amino terminal end of 4G. This truncated 4G can direct the 40S ribosomal subunit to mRNAs that have an IRES but not to those that have a cap. The widths of the arrows indicate the rate of translation initiation from the AUG codon in each example.

## POSTTRANSLATIONAL PROCESSING AFFECTS THE ACTIVITY OF MANY PROTEINS

Some animal viruses, notably poliovirus and hepatitis A virus, synthesize long polycistronic proteins from one long mRNA molecule. These protein molecules are subsequently cleaved at specific sites to provide the several specific proteins required for viral function. In animal cells, many proteins are synthesized from the mRNA template as a precursor molecule, which then must be modified to achieve the active protein. The prototype is insulin, which is a low-molecular-weight protein having two polypeptide chains with interchain and intrachain disulfide bridges. The molecule is synthesized as a single chain precursor, or **prohormone,** which folds to allow the disulfide bridges to form. A specific protease then clips out the segment that connects the two chains which form the functional insulin molecule (see Figure 42–12).

Many other peptides are synthesized as proproteins that require modifications before attaining biologic activity. Many of the posttranslational modifications involve the removal of amino terminal amino acid residues by specific aminopeptidases. Collagen, an abundant protein in the extracellular spaces of higher eukaryotes, is synthesized as procollagen. Three procollagen polypeptide molecules, frequently not identical in sequence, align themselves in a particular way that is dependent upon the existence of specific amino terminal peptides. Specific enzymes then carry out hydroxylations and oxidations of specific amino acid residues within the procollagen molecules to provide cross-links for greater stability. Amino terminal peptides are cleaved off the molecule to form the final product—a strong, insoluble collagen molecule. Many other posttranslational modifications of proteins occur. Covalent modification by acetylation, phosphorylation, methylation, ubiquitinylation, and glycosylation is common, for example.

## MANY ANTIBIOTICS WORK BECAUSE THEY SELECTIVELY INHIBIT PROTEIN SYNTHESIS IN BACTERIA

Ribosomes in bacteria and in the mitochondria of higher eukaryotic cells differ from the mammalian ribosome described in Chapter 35. The bacterial ribosome is smaller (70S rather than 80S) and has a different, somewhat simpler complement of RNA and protein molecules. This difference is exploited for clinical purposes because many effective antibiotics interact specifically with the proteins and RNAs of prokaryotic ribosomes and thus inhibit protein synthesis. This results in growth arrest or death of the bacterium. The most useful members of this class of antibiotics (eg, tetracyclines, lincomycin, erythromycin, and chloramphenicol) do not interact with components of eukaryotic ribosomal particles and thus are not toxic to eukaryotes. Tetracycline prevents the binding of aminoacyl-tRNAs to the A site. Chloramphenicol and the macrolide class of antibiotics work by binding to 23S rRNA, which is interesting in view of the newly appreciated role of rRNA in peptide bond formation through its peptidyltransferase activity. It should be mentioned that the close similarity between prokaryotic and mitochondrial ribosomes can lead to complications in the use of some antibiotics.

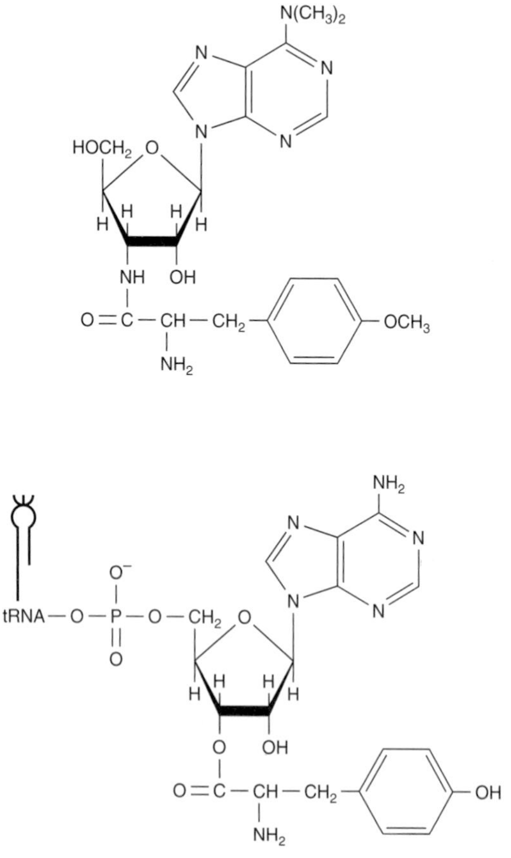

***Figure 38–11.*** The comparative structures of the antibiotic puromycin (*top*) and the 3′ terminal portion of tyrosinyl-tRNA (*bottom*).

Other antibiotics inhibit protein synthesis on all ribosomes **(puromycin)** or only on those of eukaryotic cells **(cycloheximide).** Puromycin (Figure 38–11) is a structural analog of tyrosinyl-tRNA. Puromycin is incorporated via the A site on the ribosome into the carboxyl terminal position of a peptide but causes the premature release of the polypeptide. Puromycin, as a tyrosinyl-tRNA analog, effectively inhibits protein synthesis in both prokaryotes and eukaryotes. Cycloheximide inhibits peptidyltransferase in the 60S ribosomal subunit in eukaryotes, presumably by binding to an rRNA component.

**Diphtheria toxin,** an exotoxin of *Corynebacterium diphtheriae* infected with a specific lysogenic phage, catalyzes the ADP-ribosylation of EF-2 on the unique amino acid diphthamide in mammalian cells. This modification inactivates EF-2 and thereby specifically inhibits mammalian protein synthesis. Many animals (eg, mice) are resistant to diphtheria toxin. This resistance is due to inability of diphtheria toxin to cross the cell membrane rather than to insensitivity of mouse EF-2 to diphtheria toxin-catalyzed ADP-ribosylation by NAD.

Ricin, an extremely toxic molecule isolated from the castor bean, inactivates eukaryotic 28S ribosomal RNA by providing the N-glycolytic cleavage or removal of a single adenine.

Many of these compounds—puromycin and cycloheximide in particular—are not clinically useful but have been important in elucidating the role of protein synthesis in the regulation of metabolic processes, particularly enzyme induction by hormones.

## SUMMARY

- The flow of genetic information follows the sequence DNA → RNA → protein.
- The genetic information in the structural region of a gene is transcribed into an RNA molecule such that the sequence of the latter is complementary to that in the DNA.
- Several different types of RNA, including ribosomal RNA (rRNA), transfer RNA (tRNA), and messenger RNA (mRNA), are involved in protein synthesis.
- The information in mRNA is in a tandem array of codons, each of which is three nucleotides long.
- The mRNA is read continuously from a start codon (AUG) to a termination codon (UAA, UAG, UGA).
- The open reading frame of the mRNA is the series of codons, each specifying a certain amino acid, that determines the precise amino acid sequence of the protein.
- Protein synthesis, like DNA and RNA synthesis, follows a 5′ to 3′ polarity and can be divided into three

processes: initiation, elongation, and termination. Mutant proteins arise when single-base substitutions result in codons that specify a different amino acid at a given position, when a stop codon results in a truncated protein, or when base additions or deletions alter the reading frame, so different codons are read.
- A variety of compounds, including several antibiotics, inhibit protein synthesis by affecting one or more of the steps involved in protein synthesis.

## REFERENCES

Crick F et al: The genetic code. Nature 1961;192:1227.

Green R, Noller HF: Ribosomes and translation. Annu Rev Biochem 1997;66:679.

Kozak M: Structural features in eukaryotic mRNAs that modulate the initiation of translation. J Biol Chem 1991;266:1986.

Lawrence JC, Abraham RT: PHAS/4E-BPs as regulators of mRNA translation and cell proliferation. Trends Biochem Sci 1997;22:345.

Sachs AB, Buratowski S: Common themes in translational and transcriptional regulation. Trends Biochem Sci 1997;22:189.

Sachs AB, Sarnow P, Hentze MW: Starting at the beginning, middle and end: translation initiation in eukaryotes. Cell 1997; 98:831.

Weatherall DJ et al: The hemoglobinopathies. In: *The Metabolic and Molecular Bases of Inherited Disease,* 8th ed. Scriver CR et al (editors). McGraw-Hill, 2001.

# Regulation of Gene Expression

# 39

Daryl K. Granner, MD, & P. Anthony Weil, PhD

## BIOMEDICAL IMPORTANCE

Organisms adapt to environmental changes by altering gene expression. The process of alteration of gene expression has been studied in detail and often involves modulation of gene transcription. Control of transcription ultimately results from changes in the interaction of specific binding regulatory proteins with various regions of DNA in the controlled gene. This can have a positive or negative effect on transcription. Transcription control can result in tissue-specific gene expression, and gene regulation is influenced by hormones, heavy metals, and chemicals. In addition to transcription level controls, gene expression can also be modulated by gene amplification, gene rearrangement, posttranscriptional modifications, and RNA stabilization. Many of the mechanisms that control gene expression are used to respond to hormones and therapeutic agents. Thus, a molecular understanding of these processes will lead to development of agents that alter pathophysiologic mechanisms or inhibit the function or arrest the growth of pathogenic organisms.

## REGULATED EXPRESSION OF GENES IS REQUIRED FOR DEVELOPMENT, DIFFERENTIATION, & ADAPTATION

The genetic information present in each somatic cell of a metazoan organism is practically identical. The exceptions are found in those few cells that have amplified or rearranged genes in order to perform specialized cellular functions. Expression of the genetic information must be regulated during ontogeny and differentiation of the organism and its cellular components. Furthermore, in order for the organism to adapt to its environment and to conserve energy and nutrients, the expression of genetic information must be cued to extrinsic signals and respond only when necessary. As organisms have evolved, more sophisticated regulatory mechanisms have appeared which provide the organism and its cells with the responsiveness necessary for survival in a complex environment. Mammalian cells possess about 1000 times more genetic information than does the bacterium *Escherichia coli.* Much of this additional genetic information is probably involved in regulation of gene expression during the differentiation of tissues and biologic processes in the multicellular organism and in ensuring that the organism can respond to complex environmental challenges.

In simple terms, there are only two types of gene regulation: **positive regulation** and **negative regulation** (Table 39–1). When the expression of genetic information is quantitatively increased by the presence of a specific regulatory element, regulation is said to be positive; when the expression of genetic information is diminished by the presence of a specific regulatory element, regulation is said to be negative. The element or molecule mediating negative regulation is said to be a negative regulator or **repressor;** that mediating positive regulation is a positive regulator or **activator.** However, a **double negative** has the effect of acting as a positive. Thus, an effector that inhibits the function of a negative regulator will appear to bring about a positive regulation. Many regulated systems that appear to be induced are in fact **derepressed** at the molecular level. (See Chapter 9 for explanation of these terms.)

## BIOLOGIC SYSTEMS EXHIBIT THREE TYPES OF TEMPORAL RESPONSES TO A REGULATORY SIGNAL

Figure 39–1 depicts the extent or amount of gene expression in three types of temporal response to an inducing signal. A **type A response** is characterized by an increased extent of gene expression that is dependent upon the continued presence of the inducing signal. When the inducing signal is removed, the amount of gene expression diminishes to its basal level, but the amount repeatedly increases in response to the reappearance of the specific signal. This type of response is commonly observed in prokaryotes in response to sudden changes of the intracellular concentration of a nutrient. It is also observed in many higher organisms after exposure to inducers such as hormones, nutrients, or growth factors (Chapter 43).

A **type B response** exhibits an increased amount of gene expression that is transient even in the continued presence of the regulatory signal. After the regulatory signal has terminated and the cell has been allowed to recover, a second transient response to a subsequent regulatory signal may be observed. This phenomenon of response-desensitization-recovery characterizes the action of many pharmacologic agents, but it is also a

***Table 39–1.*** Effects of positive and negative regulation on gene expression.

| | Rate of Gene Expression | |
|---|---|---|
| | **Negative Regulation** | **Positive Regulation** |
| Regulator present | Decreased | Increased |
| Regulator absent | Increased | Decreased |

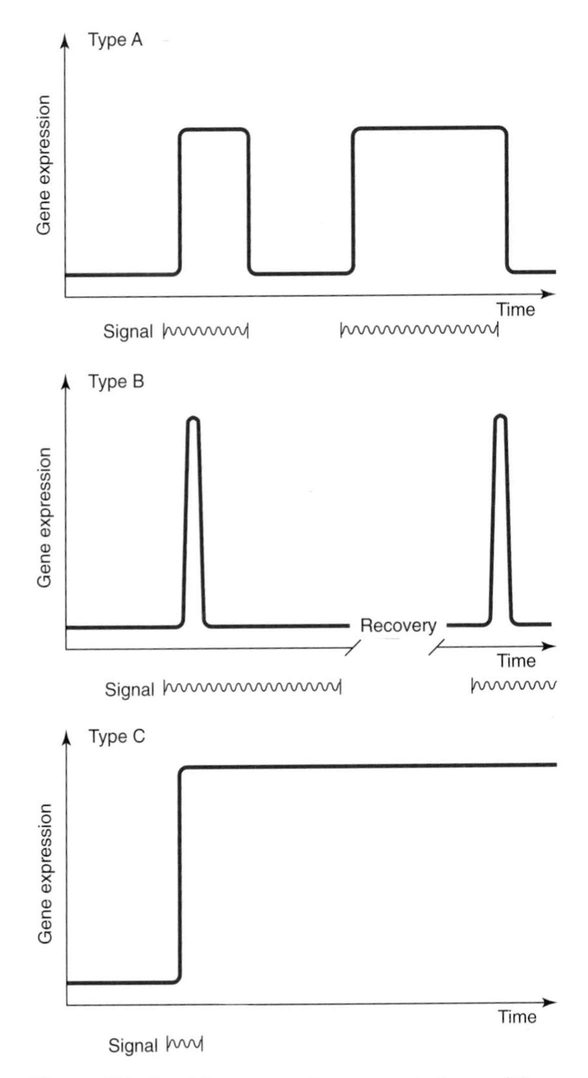


***Figure 39–1.*** Diagrammatic representations of the responses of the extent of expression of a gene to specific regulatory signals such as a hormone.

feature of many naturally occurring processes. This type of response commonly occurs during development of an organism, when only the transient appearance of a specific gene product is required although the signal persists.

The **type C response** pattern exhibits, in response to the regulatory signal, an increased extent of gene expression that persists indefinitely even after termination of the signal. The signal acts as a trigger in this pattern. Once expression of the gene is initiated in the cell, it cannot be terminated even in the daughter cells; it is therefore an irreversible and inherited alteration. This type of response typically occurs during the development of differentiated function in a tissue or organ.

## Prokaryotes Provide Models for the Study of Gene Expression in Mammalian Cells

Analysis of the regulation of gene expression in prokaryotic cells helped establish the principle that information flows from the gene to a messenger RNA to a specific protein molecule. These studies were aided by the advanced genetic analyses that could be performed in prokaryotic and lower eukaryotic organisms. In recent years, the principles established in these early studies, coupled with a variety of molecular biology techniques, have led to remarkable progress in the analysis of gene regulation in higher eukaryotic organisms, including mammals. In this chapter, the initial discussion will center on prokaryotic systems. The impressive genetic studies will not be described, but the physiology of gene expression will be discussed. However, nearly all of the conclusions about this physiology have been derived from genetic studies and confirmed by molecular genetic and biochemical studies.

## Some Features of Prokaryotic Gene Expression Are Unique

Before the physiology of gene expression can be explained, a few specialized genetic and regulatory terms must be defined for prokaryotic systems. In prokaryotes, the genes involved in a metabolic pathway are often present in a linear array called an **operon,** eg, the *lac* operon. An operon can be regulated by a single promoter or regulatory region. The **cistron** is the smallest unit of genetic expression. As described in Chapter 9, some enzymes and other protein molecules are composed of two or more nonidentical subunits. Thus, the "one gene, one enzyme" concept is not necessarily valid. The cistron is the genetic unit coding for the structure of the subunit of a protein molecule, acting as it does as the smallest unit of genetic expression. Thus, the one gene, one enzyme idea might more accurately be regarded as a **one cistron, one subunit** concept. A single mRNA that encodes more than one separately translated protein is referred to as a **polycistronic mRNA.** For example, the polycistronic *lac* operon mRNA is translated into three separate proteins (see below). Operons and polycistronic mRNAs are common in bacteria but not in eukaryotes.

An **inducible gene** is one whose expression increases in response to an **inducer** or **activator,** a specific positive regulatory signal. In general, inducible genes have relatively low basal rates of transcription. By contrast, genes with high basal rates of transcription are often subject to down-regulation by repressors.

The expression of some genes is **constitutive,** meaning that they are expressed at a reasonably constant rate and not known to be subject to regulation. These are often referred to as **housekeeping genes.** As a result of mutation, some inducible gene products become constitutively expressed. A mutation resulting in constitutive expression of what was formerly a regulated gene is called a **constitutive mutation.**

## Analysis of Lactose Metabolism in *E coli* Led to the Operon Hypothesis

Jacob and Monod in 1961 described their **operon model** in a classic paper. Their hypothesis was to a large extent based on observations on the regulation of lactose metabolism by the intestinal bacterium *E coli.* The molecular mechanisms responsible for the regulation of the genes involved in the metabolism of lactose are now among the best-understood in any organism. β-Galactosidase hydrolyzes the β-galactoside lactose to galactose and glucose. The structural gene for β-galactosidase *(lacZ)* is clustered with the genes responsible for the permeation of galactose into the cell *(lacY)* and for thiogalactoside transacetylase *(lacA).* The structural genes for these three enzymes, along with the *lac* promoter and *lac* operator (a regulatory region), are physically associated to constitute the ***lac* operon** as depicted in Figure 39–2. This genetic arrangement of the structural genes and their regulatory genes allows for **coordinate expression** of the three enzymes concerned with lactose metabolism. Each of these linked genes is transcribed into one large mRNA molecule that contains multiple independent translation start (AUG) and stop (UAA) codons for each cistron. Thus, each protein is translated separately, and they are not processed from a single large precursor protein. This type of mRNA molecule is called a **polycistronic mRNA.** Polycistronic mRNAs are predominantly found in prokaryotic organisms.

It is now conventional to consider that a gene includes regulatory sequences as well as the region that

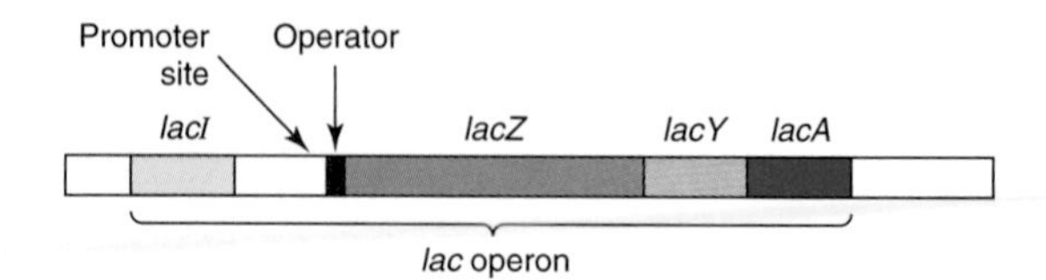


***Figure 39–2.*** The positional relationships of the structural and regulatory genes of the *lac* operon. *lacZ* encodes β-galactosidase, *lacY* encodes a permease, and *lacA* encodes a thiogalactoside transacetylase. *lacI* encodes the *lac* operon repressor protein.

encodes the primary transcript. Although there are many historical exceptions, a gene is generally italicized in lower case and the encoded protein, when abbreviated, is expressed in roman type with the first letter capitalized. For example, the gene *lacI* encodes the repressor protein LacI. When *E coli* is presented with lactose or some specific lactose analogs under appropriate nonrepressing conditions (eg, high concentrations of lactose, no or very low glucose in media; see below), the expression of the activities of β-galactosidase, galactoside permease, and thiogalactoside transacetylase is increased 100-fold to 1000-fold. This is a type A response, as depicted in Figure 39–1. The kinetics of induction can be quite rapid; *lac*-specific mRNAs are fully induced within 5–6 minutes after addition of lactose to a culture; β-galactosidase protein is maximal within 10 minutes. Under fully induced conditions, there can be up to 5000 β-galactosidase molecules per cell, an amount about 1000 times greater than the basal, uninduced level. Upon removal of the signal, ie, the inducer, the synthesis of these three enzymes declines.

When *E coli* is exposed to both lactose and glucose as sources of carbon, the organisms first metabolize the glucose and then temporarily stop growing until the genes of the *lac* operon become induced to provide the ability to metabolize lactose as a usable energy source. Although lactose is present from the beginning of the bacterial growth phase, the cell does not induce those enzymes necessary for catabolism of lactose until the glucose has been exhausted. This phenomenon was first thought to be attributable to repression of the *lac* operon by some catabolite of glucose; hence, it was termed catabolite repression. It is now known that catabolite repression is in fact mediated by a **catabolite gene activator protein (CAP)** in conjunction with **cAMP** (Figure 18–5). This protein is also referred to as the cAMP regulatory protein (CRP). The expression of many inducible enzyme systems or operons in *E coli* and other prokaryotes is sensitive to catabolite repression, as discussed below.

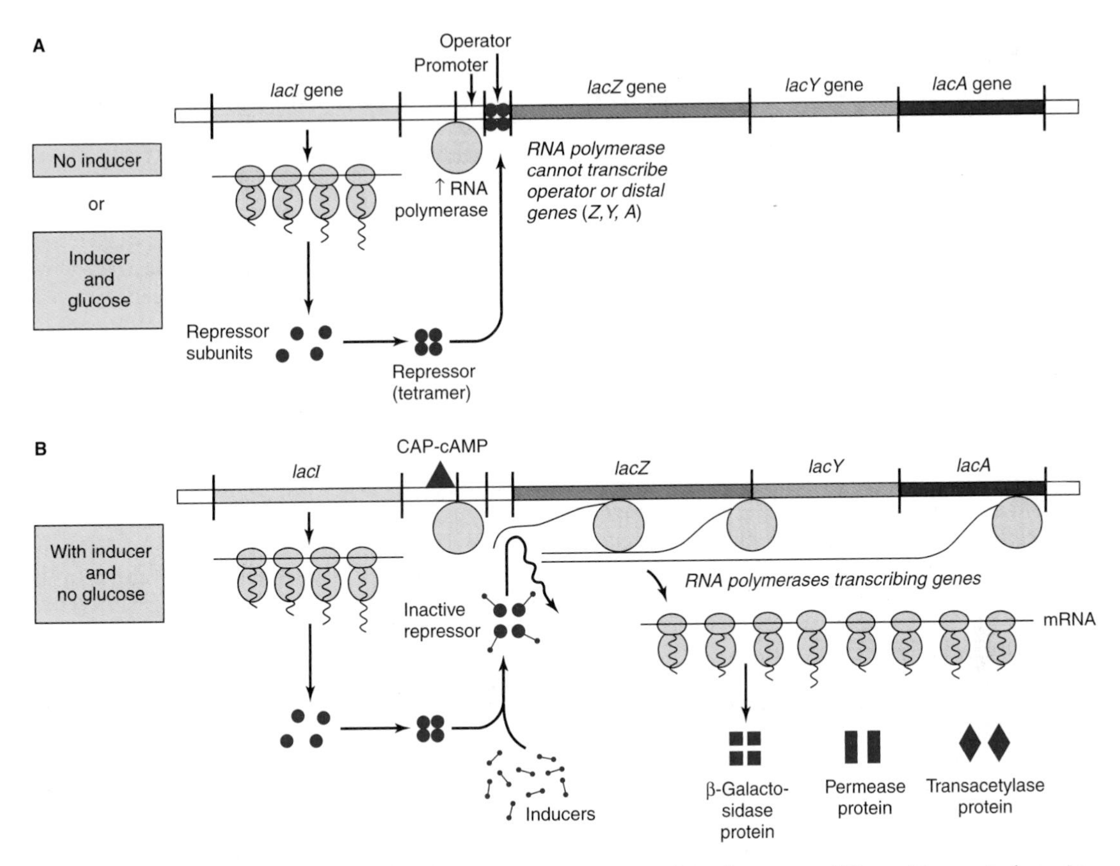


***Figure 39–3.*** The mechanism of repression and derepression of the *lac* operon. When either no inducer is present or inducer is present with glucose **(A),** the *lacI* gene products that are synthesized constitutively form a repressor tetramer molecule which binds at the operator locus to prevent the efficient initiation of transcription by RNA polymerase at the promoter locus and thus to prevent the subsequent transcription of the *lacZ, lacY,* and *lacA* structural genes. When inducer is present **(B),** the constitutively expressed *lacI* gene forms repressor molecules that are conformationally altered by the inducer and cannot efficiently bind to the operator locus (affinity of binding reduced > 1000-fold). In the presence of cAMP and its binding protein (CAP), the RNA polymerase can transcribe the structural genes *lacZ, lacY,* and *lacA,* and the polycistronic mRNA molecule formed can be translated into the corresponding protein molecules β-galactosidase, permease, and transacetylase, allowing for the catabolism of lactose.

The physiology of induction of the *lac* operon is well understood at the molecular level (Figure 39–3). Expression of the normal *lacI* gene of the *lac* operon is constitutive; it is expressed at a constant rate, resulting in formation of the subunits of the ***lac* repressor.** Four identical subunits with molecular weights of 38,000 assemble into a *lac* repressor molecule. The LacI repressor protein molecule, the product of *lacI*, has a high affinity ($K_d$ about $10^{-13}$ mol/L) for the operator locus. The **operator locus** is a region of double-stranded DNA 27 base pairs long with a twofold rotational symmetry and an inverted palindrome (indicated by solid lines about the dotted axis) in a region that is 21 base pairs long, as shown below:

:
5′ – AAT TGTGAGC G GATAACAATT
3′ – TTA ACACTCG C CTATTGTTAA
:

The minimum effective size of an operator for LacI repressor binding is 17 base pairs (boldface letters in the above sequence). At any one time, only two subunits of the repressors appear to bind to the operator, and within the 17-base-pair region at least one base of each base pair is involved in LacI recognition and binding. The binding occurs mostly in the **major groove** without interrupting the base-paired, double-helical nature of the operator DNA. The **operator locus** is between the **promoter site,** at which the DNA-dependent RNA polymerase attaches to commence transcription, and the transcription initiation site of the ***lacZ* gene,** the structural gene for β-galactosidase (Figure 39–2). When attached to the operator locus, the LacI repressor molecule prevents transcription of the operator locus as well as of the distal structural genes, *lacZ, lacY,* and *lacA.* Thus, the LacI repressor molecule is a **negative regulator;** in its presence (and in the absence of inducer; see below), expression from the *lacZ, lacY,* and *lacA* genes is prevented. There are normally 20–40 repressor tetramer molecules in the cell, a concentration of tetramer sufficient to effect, at any given time, > 95% occupancy of the one *lac* operator element in a bacterium, thus ensuring low (but not zero) basal *lac* operon gene transcription in the absence of inducing signals.

A lactose analog that is capable of inducing the *lac* operon while not itself serving as a substrate for β-galactosidase is an example of a **gratuitous inducer.** An example is isopropylthiogalactoside (IPTG). The addition of lactose or of a gratuitous inducer such as IPTG to bacteria growing on a poorly utilized carbon source (such as succinate) results in prompt induction of the *lac* operon enzymes. Small amounts of the gratuitous inducer or of lactose are able to enter the cell even in the absence of permease. The LacI repressor molecules—both those attached to the operator loci and those free in the cytosol—have a high affinity for the inducer. Binding of the inducer to a repressor molecule attached to the operator locus induces a conformational change in the structure of the repressor and causes it to dissociate from the DNA because its affinity for the operator is now $10^3$ times lower ($K_d$ about $10^{-9}$ mol/L) than that of LacI in the absence of IPTG. If DNA-dependent RNA polymerase has already attached to the coding strand at the promoter site, transcription will begin. The polymerase generates a polycistronic mRNA whose 5′ terminal is complementary to the template strand of the operator. In such a manner, **an inducer derepresses the *lac* operon** and allows transcription of the structural genes for β-galactosidase, galactoside permease, and thiogalactoside transacetylase. Translation of the polycistronic mRNA can occur even before transcription is completed. Derepression of the *lac* operon allows the cell to synthesize the enzymes necessary to catabolize lactose as an energy source. Based on the physiology just described, IPTG-induced expression of transfected plasmids bearing the *lac* operator-promoter ligated to appropriate bioengineered constructs is commonly used to express mammalian recombinant proteins in *E coli.*

In order for the RNA polymerase to efficiently form a PIC at the promoter site, there must also be present the **catabolite gene activator protein (CAP)** to which cAMP is bound. By an independent mechanism, the bacterium accumulates cAMP only when it is starved for a source of carbon. In the presence of glucose—or of glycerol in concentrations sufficient for growth—the bacteria will lack sufficient cAMP to bind to CAP because the glucose inhibits adenylyl cyclase, the enzyme that converts ATP to cAMP (see Chapter 42). Thus, in the presence of glucose or glycerol, cAMP-saturated CAP is lacking, so that the DNA-dependent RNA polymerase cannot initiate transcription of the *lac* operon. In the presence of the CAP-cAMP complex, which binds to DNA just upstream of the promoter site, transcription then occurs (Figure 39–3). Studies indicate that a region of CAP contacts the RNA polymerase α subunit and facilitates binding of this enzyme to the promoter. Thus, the CAP-cAMP regulator is acting as a **positive regulator** because its presence is required for gene expression. The *lac* operon is therefore controlled by two distinct, ligand-modulated DNA binding trans factors; one that acts positively (cAMP-CRP complex) and one that acts negatively (LacI repressor). Maximal activity of the *lac* operon occurs when glucose levels are low (high cAMP with CAP activation) and lactose is present (LacI is prevented from binding to the operator).

When the *lacI* gene has been mutated so that its product, LacI, is not capable of binding to operator DNA, the organism will exhibit **constitutive expression** of the *lac* operon. In a contrary manner, an organism with a *lacI* gene mutation that produces a LacI protein which prevents the binding of an inducer to the repressor will remain repressed even in the presence of the inducer molecule, because the inducer cannot bind to the repressor on the operator locus in order to derepress the operon. Similarly, bacteria harboring mutations in their *lac* operator locus such that the operator sequence will not bind a normal repressor molecule constitutively express the *lac* operon genes. Mechanisms of positive and negative regulation comparable to those described here for the *lac* system have been observed in eukaryotic cells (see below).

## The Genetic Switch of Bacteriophage Lambda (λ) Provides a Paradigm for Protein-DNA Interactions in Eukaryotic Cells

Like some eukaryotic viruses (eg, herpes simplex, HIV), some bacterial viruses can either reside in a dormant state within the host chromosomes or can replicate within the bacterium and eventually lead to lysis and killing of the bacterial host. Some *E coli* harbor such a "temperate" virus, bacteriophage lambda (λ). When lambda infects an organism of that species it injects its 45,000-bp, double-stranded, linear DNA genome into the cell (Figure 39–4). Depending upon the nutritional state of the cell, the lambda DNA will either **integrate** into the host genome **(lysogenic pathway)** and remain dormant until activated (see below), or it will commence **replicating** until it has made about 100 copies of complete, protein-packaged virus, at which point it causes lysis of its host **(lytic pathway).** The newly generated virus particles can then infect other susceptible hosts.

When integrated into the host genome in its dormant state, lambda will remain in that state until activated by exposure of its lysogenic bacterial host to DNA-damaging agents. In response to such a noxious stimulus, the dormant bacteriophage becomes "induced" and begins to transcribe and subsequently translate those genes of its own genome which are necessary for its excision from the host chromosome, its DNA replication, and the synthesis of its protein coat and lysis enzymes. This event acts like a trigger or type C (Figure 39–1) response; ie, once lambda has committed itself to induction, there is no turning back until the cell is lysed and the replicated bacteriophage released. This switch from a dormant or **prophage state** to a **lytic infection** is well understood at the genetic and molecular levels and will be described in detail here.

The switching event in lambda is centered around an 80-bp region in its double-stranded DNA genome referred to as the "right operator" ($O_R$) (Figure 39–5A). The **right operator** is flanked on its left side by the structural gene for the lambda repressor protein, the **cI protein,** and on its right side by the structural gene encoding another regulatory protein called **Cro.** When lambda is in its prophage state—ie, integrated into the host genome—the cI repressor gene is the *only* lambda gene cI protein that is expressed. When the bacteriophage is undergoing lytic growth, the cI repressor gene is not expressed, but the *cro* gene—as well as many other genes in lambda—is expressed. That is, **when the repressor gene is on, the *cro* gene is off, and when the *cro* gene is on, the repressor gene is off.** As we shall see, these two genes regulate each other's expression and thus, ultimately, the decision between lytic and lysogenic growth of lambda. **This decision between repressor gene transcription and *cro* gene transcription is a paradigmatic example of a molecular switch.**

The 80-bp λ right operator, $O_R$, can be subdivided into three discrete, evenly spaced, 17-bp cis-active DNA elements that represent the binding sites for either of two bacteriophage λ regulatory proteins. Impor-

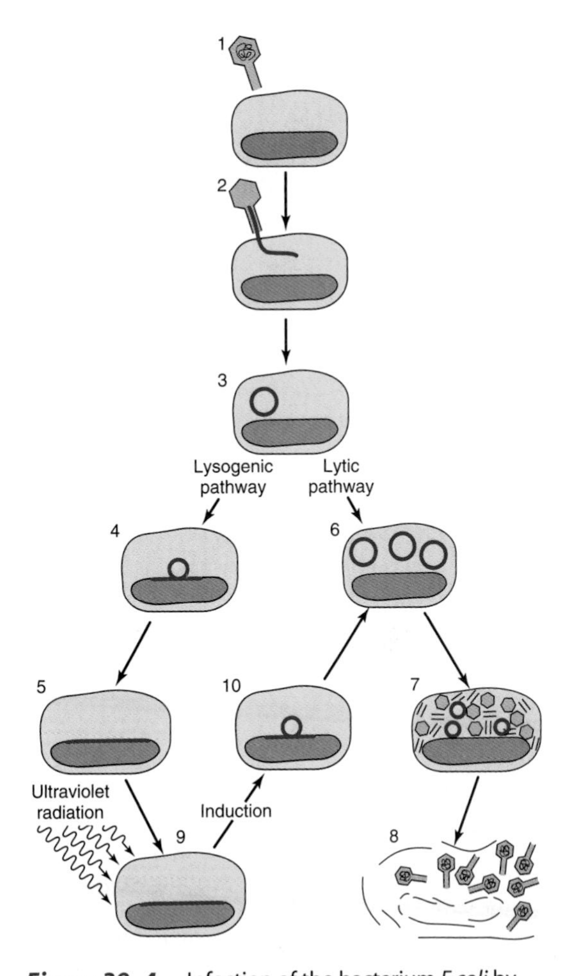


***Figure 39–4.*** Infection of the bacterium *E coli* by phage lambda begins when a virus particle attaches itself to the bacterial cell (1) and injects its DNA (shaded line) into the cell (2, 3). Infection can take either of two courses depending on which of two sets of viral genes is turned on. In the lysogenic pathway, the viral DNA becomes integrated into the bacterial chromosome (4, 5), where it replicates passively as the bacterial cell divides. The dormant virus is called a prophage, and the cell that harbors it is called a lysogen. In the alternative lytic mode of infection, the viral DNA replicates itself (6) and directs the synthesis of viral proteins (7). About 100 new virus particles are formed. The proliferating viruses lyse, or burst, the cell (8). A prophage can be "induced" by a DNA damaging agent such as ultraviolet radiation (9). The inducing agent throws a switch, so that a different set of genes is turned on. Viral DNA loops out of the chromosome (10) and replicates; the virus proceeds along the lytic pathway. (Reproduced, with permission, from Ptashne M, Johnson AD, Pabo CO: A genetic switch in a bacterial virus. Sci Am [Nov] 1982;247:128.)

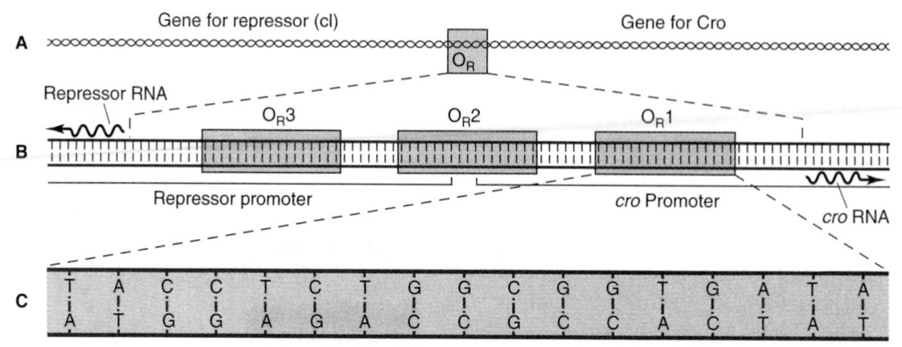


***Figure 39–5.*** Right operator ($O_R$) is shown in increasing detail in this series of drawings. The operator is a region of the viral DNA some 80 base pairs long **(A).** To its left lies the gene encoding lambda repressor (cI), to its right the gene *(cro)* encoding the regulator protein Cro. When the operator region is enlarged **(B),** it is seen to include three subregions, $O_R1$, $O_R2$, and $O_R3$, each 17 base pairs long. They are recognition sites to which both repressor and Cro can bind. The recognition sites overlap two promoters—sequences of bases to which RNA polymerase binds in order to transcribe these genes into mRNA (wavy lines), that are translated into protein. Site $O_R1$ is enlarged **(C)** to show its base sequence. Note that in this region of the λ chromosome, both strands of DNA act as a template for transcription (Chapter 39). (Reproduced, with permission, from Ptashne M, Johnson AD, Pabo CO: A genetic switch in a bacterial virus. Sci Am [Nov] 1982;247:128.)

tantly, the nucleotide sequences of these three tandemly arranged sites are similar but not identical (Figure 39–5B). The three related cis elements, termed operators $O_R1$, $O_R2$, and $O_R3$, can be bound by either cI or Cro proteins. However, the relative affinities of cI and Cro for each of the sites varies, and this differential binding affinity is central to the appropriate operation of the λ phage lytic or lysogenic "molecular switch." The DNA region between the *cro* and repressor genes also contains two promoter sequences that direct the binding of RNA polymerase in a specified orientation, where it commences transcribing adjacent genes. One promoter directs RNA polymerase to transcribe in the **rightward direction** and, thus, to transcribe *cro* and other distal genes, while the other promoter directs the transcription of the **repressor** gene in the **leftward direction** (Figure 39–5B).

The product of the repressor gene, the 236-amino-acid, 27 kDa **repressor protein,** exists as a **two-domain** molecule in which the **amino terminal domain binds to operator DNA** and the **carboxyl terminal domain promotes the association** of one repressor protein with another to form a dimer. A **dimer** of repressor molecules binds to **operator DNA** much more tightly than does the monomeric form (Figure 39–6A to 39–6C).

The product of the *cro* gene, the 66-amino-acid, 9 kDa **Cro protein,** has a single domain but also binds the operator DNA more tightly as a **dimer** (Figure 39–6D). The Cro protein's single domain mediates both operator binding and dimerization.

In a lysogenic bacterium—ie, a bacterium containing a lambda prophage—the lambda repressor dimer binds **preferentially to $O_R$**1 but in so doing, by a cooperative interaction, enhances the binding (by a factor of 10) of another repressor dimer to $O_R2$ (Figure 39–7). The affinity of repressor for $O_R3$ is the least of the three operator subregions. The binding of repressor to $O_R1$ has two major effects. The occupation of $O_R1$ by repressor **blocks the binding of RNA polymerase to the rightward promoter** and in that way prevents expression of *cro*. Second, as mentioned above, repressor dimer bound to $O_R1$ enhances the binding of repressor dimer to $O_R2$. The binding of repressor to $O_R2$ has the important added effect of **enhancing the binding of RNA polymerase to the leftward promoter** that overlaps $O_R2$ and thereby enhances transcription and subsequent expression of the repressor gene. This enhancement of transcription is apparently mediated through direct protein-protein interactions between promoter-bound RNA polymerase and $O_R2$-bound repressor. Thus, the lambda repressor is both a **negative regulator,** by preventing transcription of *cro,* and a **positive regulator,** by enhancing transcription of its own gene, the repressor gene. This dual effect of repressor is responsible for the stable state of the dormant lambda bacteriophage; not only does the repressor prevent expression of the genes necessary for lysis, but it also promotes expression of itself to

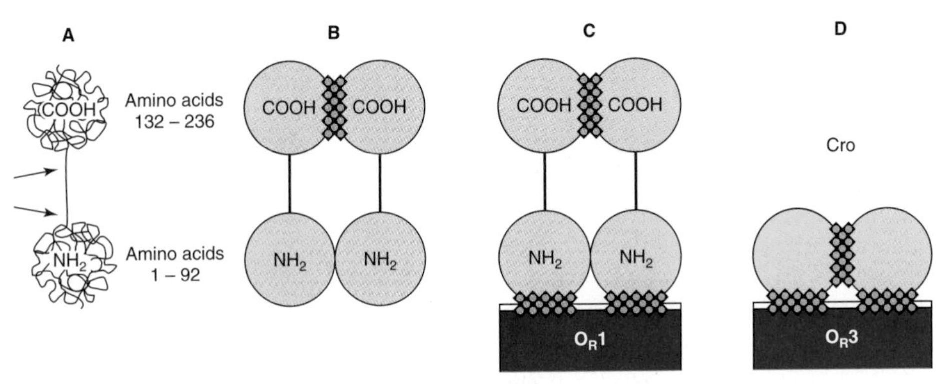


***Figure 39–6.*** Schematic molecular structures of cI (lambda repressor, shown in **A, B,** and **C**) and Cro **(D)**. Lambda repressor protein is a polypeptide chain 236 amino acids long. The chain folds itself into a dumbbell shape with two substructures: an amino terminal ($NH_2$) domain and a carboxyl terminal (COOH) domain. The two domains are linked by a region of the chain that is susceptible to cleavage by proteases (indicated by the two arrows in **A**). Single repressor molecules (monomers) tend to associate to form dimers **(B);** a dimer can dissociate to form monomers again. A dimer is held together mainly by contact between the carboxyl terminal domains (hatching). Repressor dimers bind to (and can dissociate from) the recognition sites in the operator region; they display the greatest affinity for site $O_R1$ **(C).** It is the amino terminal domain of the repressor molecule that makes contact with the DNA (hatching). Cro **(D)** has a single domain with sites that promote dimerization and other sites that promote binding of dimers to operator, preferentially to $O_R3$. (Reproduced, with permission, from Ptashne M, Johnson AD, Pabo CO: A genetic switch in a bacterial virus. Sci Am [Nov] 1982;247:128.)

stabilize this state of differentiation. In the event that intracellular repressor protein concentration becomes very high, this excess repressor will bind to $O_R3$ and by so doing diminish transcription of the repressor gene from the leftward promoter until the repressor concentration drops and repressor dissociates itself from $O_R3$.

With such a stable, repressive, cI-mediated, lysogenic state, one might wonder how the lytic cycle could ever be entered. However, this process does occur quite efficiently. When a DNA-damaging signal, such as ultraviolet light, strikes the lysogenic host bacterium, fragments of single-stranded DNA are generated that activate a specific **protease** coded by a bacterial gene and referred to as **recA** (Figure 39–7). The activated recA protease hydrolyzes the portion of the repressor protein that connects the amino terminal and carboxyl terminal domains of that molecule (see Figure 39–6A). Such cleavage of the repressor domains causes the **repressor dimers to dissociate,** which in turn causes **dissociation of the repressor molecules from $O_R2$** and eventually from $O_R1$. The effects of removal of repressor from $O_R1$ and $O_R2$ are predictable. RNA polymerase immediately has access to the rightward promoter and commences transcribing the ***cro* gene,** and the enhancement effect of the repressor at $O_R2$ on leftward transcription is lost (Figure 39–7).

The resulting newly synthesized Cro protein also binds to the operator region as a dimer, but its order of preference is opposite to that of repressor (Figure 39–7). That is, **Cro binds most tightly to $O_R3$**, but there is no cooperative effect of Cro at $O_R3$ on the binding of Cro to $O_R2$. At increasingly higher concentrations of Cro, the protein will bind to $O_R2$ and eventually to $O_R1$.

Occupancy of $O_R3$ by Cro immediately turns off transcription from the leftward promoter and in that way **prevents any further expression of the repressor gene.** The molecular switch is thus completely "thrown" in the lytic direction. The *cro* gene is now expressed, and the repressor gene is fully turned off. This event is irreversible, and the expression of other lambda genes begins as part of the lytic cycle. When Cro repressor concentration becomes quite high, it will eventually occupy $O_R1$ and in so doing reduce the expression of its own gene, a process that is necessary in order to effect the final stages of the lytic cycle.

The three-dimensional structures of Cro and of the lambda repressor protein have been determined by x-ray crystallography, and models for their binding and effecting the above-described molecular and genetic events have been proposed and tested. Both bind to DNA using helix-turn-helix DNA binding domain motifs (see below).

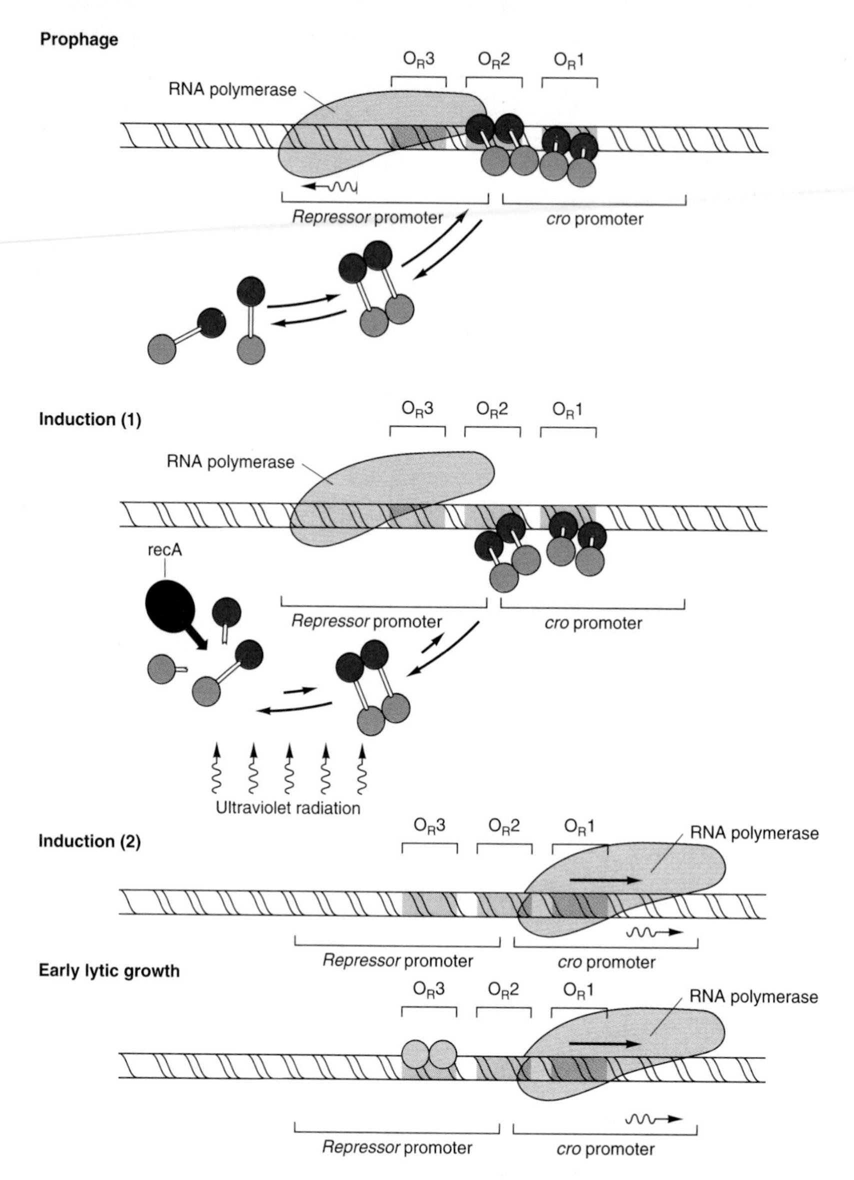


***Figure 39–7.*** Configuration of the switch is shown at four stages of lambda's life cycle. The lysogenic pathway (in which the virus remains dormant as a prophage) is selected when a repressor dimer binds to $O_R1$, thereby making it likely that $O_R2$ will be filled immediately by another dimer. In the prophage (top), the repressor dimers bound at $O_R1$ and $O_R2$ prevent RNA polymerase from binding to the rightward promoter and so block the synthesis of Cro (negative control). The repressors also enhance the binding of polymerase to the leftward promoter (positive control), with the result that the repressor gene is transcribed into RNA (wavy line) and more repressor is synthesized, maintaining the lysogenic state. The prophage is induced when ultraviolet radiation activates the protease recA, which cleaves repressor monomers. The equilibrium of free monomers, free dimers, and bound dimers is thereby shifted, and dimers leave the operator sites. RNA polymerase is no longer encouraged to bind to the leftward promoter, so that repressor is no longer synthesized. As induction proceeds, all the operator sites become vacant, and so polymerase can bind to the rightward promoter and Cro is synthesized. During early lytic growth, a single Cro dimer binds to $O_R3$ shaded circles, the site for which it has the highest affinity. Consequently, RNA polymerase cannot bind to the leftward promoter, but the rightward promoter remains accessible. Polymerase continues to bind there, transcribing *cro* and other early lytic genes. Lytic growth ensues. (Reproduced, with permission, from Ptashne M, Johnson AD, Pabo CO: A genetic switch in a bacterial virus. Sci Am [Nov] 1982;247:128.)

To date, this system provides the best understanding of the molecular events involved in gene regulation.

Detailed analysis of the lambda repressor led to the important concept that transcription regulatory proteins have several functional domains. For example, lambda repressor binds to DNA with high affinity. Repressor monomers form dimers, dimers interact with each other, and repressor interacts with RNA polymerase. The protein-DNA interface and the three protein-protein interfaces all involve separate and distinct domains of the repressor molecule. As will be noted below (see Figure 39–17), this is a characteristic shared by most (perhaps all) molecules that regulate transcription.

## SPECIAL FEATURES ARE INVOLVED IN REGULATION OF EUKARYOTIC GENE TRANSCRIPTION

Most of the DNA in prokaryotic cells is organized into genes, and the templates can always be transcribed. A very different situation exists in mammalian cells, in which relatively little of the total DNA is organized into genes and their associated regulatory regions. The function of the extra DNA is unknown. In addition, as described in Chapter 36, the DNA in eukaryotic cells is extensively folded and packed into the protein-DNA complex called chromatin. Histones are an important part of this complex since they both form the structures known as nucleosomes (see Chapter 36) and also factor significantly into gene regulatory mechanisms as outlined below.

### Chromatin Remodeling Is an Important Aspect of Eukaryotic Gene Expression

**Chromatin structure** provides an additional level of control of gene transcription. As discussed in Chapter 36, large regions of chromatin are transcriptionally inactive while others are either active or potentially active. With few exceptions, each cell contains the same complement of genes (antibody-producing cells are a notable exception). The development of specialized organs, tissues, and cells and their function in the intact organism depend upon the differential expression of genes.

Some of this differential expression is achieved by having different regions of chromatin available for transcription in cells from various tissues. For example, the DNA containing the β-globin gene cluster is in **"active" chromatin** in the reticulocyte but in **"inactive" chromatin** in muscle cells. All the factors involved in the determination of active chromatin have not been elucidated. The presence of nucleosomes and of complexes of histones and DNA (see Chapter 36) certainly provides a barrier against the ready association of transcription factors with specific DNA regions. The dynamics of the formation and disruption of nucleosome structure are therefore an important part of eukaryotic gene regulation.

**Histone acetylation and deacetylation** is an important determinant of gene activity. The surprising discovery that histone acetylase activity is associated with TAFs and the coactivators involved in hormonal regulation of gene transcription (see Chapter 43) has provided a new concept of gene regulation. Acetylation is known to occur on lysine residues in the amino terminal tails of histone molecules. This modification reduces the positive charge of these tails and decreases the binding affinity of histone for the negatively charged DNA. Accordingly, the acetylation of histone could result in disruption of nucleosomal structure and allow readier access of transcription factors to cognate regulatory DNA elements. As discussed previously, this would enhance binding of the basal transcription machinery to the promoter. Histone deacetylation would have the opposite effect. Different proteins with specific acetylase and deacetylase activities are associated with various components of the transcription apparatus. The specificity of these processes is under investigation, as are a variety of mechanisms of action. Some specific examples are illustrated in Chapter 43.

There is evidence that the **methylation of deoxycytidine residues** (in the sequence 5′-$^{m}$CpG-3′) in DNA may effect gross changes in chromatin so as to preclude its active transcription, as described in Chapter 36. For example, in mouse liver, only the unmethylated ribosomal genes can be expressed, and there is evidence that many animal viruses are not transcribed when their DNA is methylated. Acute demethylation of deoxycytidine residues in a specific region of the tyrosine aminotransferase gene—in response to glucocorticoid hormones—has been associated with an increased rate of transcription of the gene. However, it is not possible to generalize that methylated DNA is transcriptionally inactive, that all inactive chromatin is methylated, or that active DNA is not methylated.

Finally, the binding of specific transcription factors to cognate DNA elements may result in disruption of nucleosomal structure. Many eukaryotic genes have multiple protein-binding DNA elements. The serial binding of transcription factors to these elements—in a combinatorial fashion—may either directly disrupt the structure of the nucleosome or prevent its re-formation or recruit, via protein-protein interactions, multiprotein coactivator complexes that have the ability to covalently modify or remodel nucleosomes. These reactions result in chromatin-level structural changes that in the end increase DNA accessibility to other factors and the transcription machinery.

Eukaryotic DNA that is in an "active" region of chromatin can be transcribed. As in prokaryotic cells, a

**promoter** dictates where the RNA polymerase will initiate transcription, but this promoter cannot be neatly defined as containing a −35 and −10 box, particularly in mammalian cells (Chapter 37). In addition, the trans-acting factors generally come from other chromosomes (and so act in trans), whereas this consideration is moot in the case of the single chromosome-containing prokaryotic cells. Additional complexity is added by elements or factors that enhance or repress transcription, define tissue-specific expression, and modulate the actions of many effector molecules.

## Certain DNA Elements Enhance or Repress Transcription of Eukaryotic Genes

In addition to gross changes in chromatin affecting transcriptional activity, certain DNA elements facilitate or enhance initiation at the promoter. For example, in simian virus 40 (SV40) there exists about 200 bp upstream from the promoter of the early genes a region of two identical, tandem 72-bp lengths that can greatly increase the expression of genes in vivo. Each of these 72-bp elements can be subdivided into a series of smaller elements; therefore, some enhancers have a very complex structure. **Enhancer elements** differ from the promoter in two remarkable ways. They can exert their positive influence on transcription even when separated by thousands of base pairs from a promoter; they work when oriented in either direction; and they can work upstream (5′) or downstream (3′) from the promoter. Enhancers are promiscuous; they can stimulate any promoter in the vicinity and may act on more than one promoter. The SV40 enhancer element can exert an influence on, for example, the transcription of β-globin by increasing its transcription 200-fold in cells containing both the enhancer and the β-globin gene on the same plasmid (see below and Figure 39–8). The enhancer element does not produce a product that in turn acts on the promoter, since it is active only when it exists within the same DNA molecule as (ie, cis to) the promoter. Enhancer binding proteins are responsible for this effect. The exact mechanisms by which these transcription activators work are subject to much debate. Certainly, enhancer binding trans factors have been shown to interact with a plethora of other transcription proteins. These interactions include chromatin-modifying coactivators as well as the individual components of the basal RNA polymerase II transcription machinery. Ultimately, trans-factor-enhancer DNA binding events result in an increase in the binding of the basal transcription machinery to the promoter. Enhancer elements and associated binding proteins often convey nuclease hypersensitivity to those regions where they reside (Chapter 36). A summary of the properties of enhancers is presented in Table 39–2. One of the

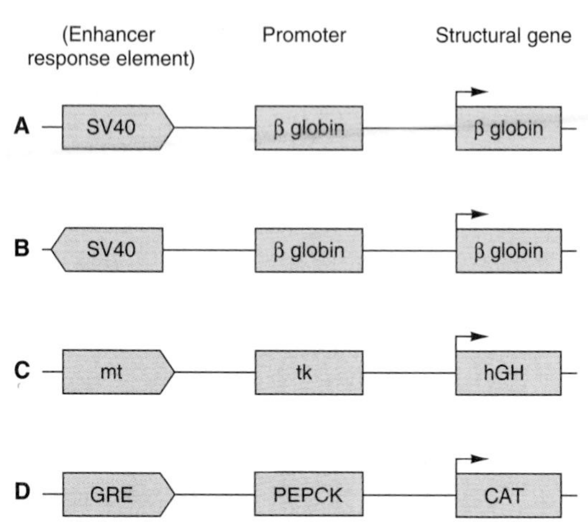


***Figure 39–8.*** A schematic explanation of the action of enhancers and other cis-acting regulatory elements. These model chimeric genes consist of a reporter (structural) gene that encodes a protein which can be readily assayed, a promoter that ensures accurate initiation of transcription, and the putative regulatory elements. In all cases, high-level transcription from the indicated chimeras depends upon the presence of enhancers, which stimulate transcription ≥ 100-fold over basal transcriptional levels (ie, transcription of the same chimeric genes containing just promoters fused to the structural genes). Examples **A** and **B** illustrate the fact that enhancers (eg, SV40) work in either orientation and upon a heterologous promoter. Example **C** illustrates that the metallothionein (mt) regulatory element (which under the influence of cadmium or zinc induces transcription of the endogenous mt gene and hence the metal-binding mt protein) will work through the thymidine kinase (tk) promoter to enhance transcription of the human growth hormone (hGH) gene. The engineered genetic constructions were introduced into the male pronuclei of single-cell mouse embryos and the embryos placed into the uterus of a surrogate mother to develop as transgenic animals. Offspring have been generated under these conditions, and in some the addition of zinc ions to their drinking water effects an increase in liver growth hormone. In this case, these transgenic animals have responded to the high levels of growth hormone by becoming twice as large as their normal litter mates. Example **D** illustrates that a glucocorticoid response element (GRE) will work through homologous (PEPCK gene) or heterologous promoters (not shown; ie, tk promoter, SV40 promoter, β-globin promoter, etc).

***Table 39–2.*** Summary of the properties of enhancers.

- Work when located long distances from the promoter
- Work when upstream or downstream from the promoter
- Work when oriented in either direction
- Work through heterologous promoters
- Work by binding one or more proteins
- Work by facilitating binding of the basal transcription complex to the promoter

best-understood mammalian enhancer systems is that of the β-interferon gene. This gene is induced upon viral infection of mammalian cells. One goal of the cell, once virally infected, is to attempt to mount an antiviral response—if not to save the infected cell, then to help to save the entire organism from viral infection. Interferon production is one mechanism by which this is accomplished. This family of proteins is secreted by virally infected cells. They interact with neighboring cells to cause an inhibition of viral replication by a variety of mechanisms, thereby limiting the extent of viral infection. The enhancer element controlling induction of this gene, located between nucleotides −110 and −45 of the β-interferon gene, is well characterized. This enhancer is composed of four distinct clustered cis elements, each of which is bound by distinct trans factors. One cis element is bound by the trans-acting factor NF-κB, one by a member of the IRF (interferon regulatory factor) family of trans factors, and a third by the heterodimeric leucine zipper factor ATF-2/c-Jun. The fourth factor is the ubiquitous, architectural transcription factor known as HMG I(Y). Upon binding to its degenerate, A+T-rich binding sites, HMG I(Y) induces a significant bend in the DNA. There are four such HMG I(Y) binding sites interspersed throughout the enhancer. These sites play a critical role in forming the enhanceosome, along with the aforementioned three trans factors, by inducing a series of critically spaced DNA bends. Consequently, HMG I(Y) induces the cooperative formation of a unique, stereospecific, three dimensional structure within which all four factors are active when viral infection signals are sensed by the cell. The structure formed by the cooperative assembly of these four factors is termed the β-interferon enhanceosome (see Figure 39–9), so named because of its obvious structural similarity to the nucleosome, also a unique three-dimensional protein DNA structure that wraps DNA about an assembly of proteins (see Figures 36–1 and 36–2). The enhanceosome, once formed, induces a large increase in β-interferon gene transcription upon virus infection. It is not simply the protein occupancy of the linearly apposed cis element sites that induces β-interferon gene transcription—rather, it is the formation of the enhanceosome proper that provides appropriate surfaces for the recruitment of coactivators that results in the enhanced formation of the PIC on the cis-linked promoter and thus transcription activation.

The cis-acting elements that decrease or **repress** the expression of specific genes have also been identified. Because fewer of these elements have been studied, it is not possible to formulate generalizations about their mechanism of action—though again, as for gene activation, chromatin level covalent modifications of histones and other proteins by (repressor)-recruited multisubunit corepressors have been implicated.

## Tissue-Specific Expression May Result From the Action of Enhancers or Repressors

Many genes are now recognized to harbor enhancer or activator elements in various locations relative to their coding regions. In addition to being able to enhance gene transcription, some of these enhancer elements clearly possess the ability to do so in a tissue-specific manner. Thus, the enhancer element associated with the immunoglobulin genes between the J and C regions enhances the expression of those genes preferentially in lymphoid cells. Similarly to the SV40 enhancer, which is capable of promiscuously activating a variety of cis-linked genes, enhancer elements associated with the genes for pancreatic enzymes are capable of enhancing even unrelated but physically linked genes preferentially in the pancreatic cells of mice into which the specifically engineered gene constructions were introduced microsurgically at the single-cell embryo stage. This **transgenic animal** approach has proved useful in studying tissue-specific gene expression. For example, DNA containing a pancreatic B cell tissue-specific enhancer (from the insulin gene), an oncogene, when ligated in a vector to polyoma large-T antigen, an oncogene, produced B cell tumors in transgenic mice. Tumors did not develop in any other tissue. Tissue-specific gene expression may therefore be mediated by enhancers or enhancer-like elements.

## Reporter Genes Are Used to Define Enhancers & Other Regulatory Elements

By ligating regions of DNA suspected of harboring regulatory sequences to various reporter genes (the **reporter** or **chimeric gene approach**) (Figures 39–10 and 39–11), one can determine which regions in the vicinity of structural genes have an influence on their expression. Pieces of DNA thought to harbor regulatory elements are ligated to a suitable reporter gene and

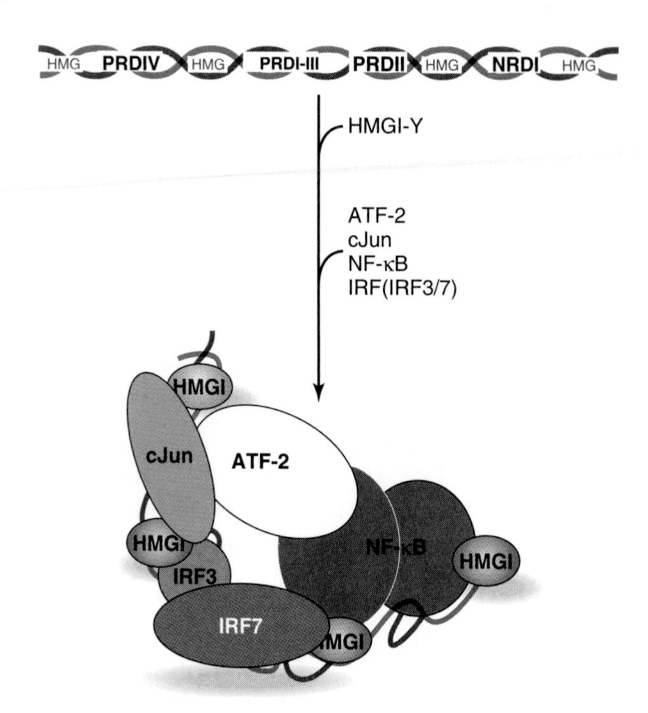


***Figure 39–9.*** Formation and putative structure of the enhanceosome formed on the human β-interferon gene enhancer. Diagramatically represented at the top is the distribution of the multiple cis-elements (HMG, PRDIV, PRDI-III, PRDII, NRDI) composing the β-interferon gene enhancer. The intact enhancer mediates transcriptional induction of the β-interferon gene (over 100-fold) upon virus infection of human cells. The cis-elements of this modular enhancer represent the binding sites for the trans-factors HMG I(Y), cJun-ATF-2, IRF3, IRF7, and NF-κB, respectively. The factors interact with these DNA elements in an obligatory, ordered, and highly cooperative fashion as indicated by the arrow. Initial binding of four HMG I(Y) proteins induces sharp DNA bends in the enhancer, causing the entire 70–80 bp region to assume a high level of curvature. This curvature is integral to the subsequent highly cooperative binding of the other trans-factors since this enables the DNA-bound factors to make important, direct protein-protein interactions that both contribute to the formation and stability of the enhanceosome and generate a unique three-dimensional surface that serves to recruit chromatin-modifying activities (eg, Swi/Snf and P/CAF) as well as the general transcription machinery (RNA polymerase II and GTFs). Although four of the five cis-elements (PRDIV, PRDI-III, PRDII, NRDI) independently can modestly stimulate (~tenfold) transcription of a reporter gene in transfected cells (see Figures 39–10 and 39–12), all five cis-elements, in appropriate order, are required to form an enhancer that can appropriately stimulate mRNA gene transcription (ie, ≥ 100-fold) in response to viral infection of a human cell. This distinction indicates the strict requirement for appropriate enhanceosome architecture for efficient trans-activation. Similar enhanceosomes, involving distinct cis- and trans-factors, are proposed to form on many other mammalian genes.

introduced into a host cell (Figure 39–10). Basal expression of the reporter gene will be increased if the DNA contains an enhancer. Addition of a hormone or heavy metal to the culture medium will increase expression of the reporter gene if the DNA contains a hormone or metal response element (Figure 39–11). The location of the element can be pinpointed by using progressively shorter pieces of DNA, deletions, or point mutations (Figure 39–11).

This strategy, **using transfected cells in culture and transgenic animals,** has led to the identification of dozens of enhancers, repressors, tissue-specific elements, and hormone, heavy metal, and drug-response elements. The activity of a gene at any moment reflects the interaction of these numerous cis-acting DNA elements with their respective trans-acting factors. The challenge now is to figure out how this occurs.

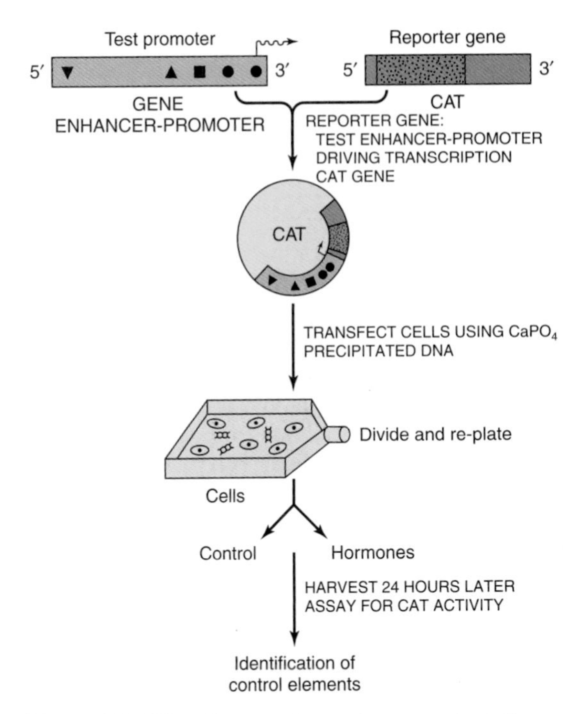


***Figure 39–10.*** The use of reporter genes to define DNA regulatory elements. A DNA fragment from the gene in question—in this example, approximately 2 kb of 5′-flanking DNA and cognate promoter—is ligated into a plasmid vector that contains a suitable reporter gene—in this case, the bacterial enzyme chloramphenicol transferase (CAT). The enzyme luciferase (abbreviated LUC) is another popular reporter gene. Neither LUC nor CAT is present in mammalian cells; hence, detection of these activities in a cell extract means that the cell was successfully transfected by the plasmid. An increase of CAT activity over the basal level, eg, after addition of one or more hormones, means that the region of DNA inserted into the reporter gene plasmid contains functional hormone response elements (HRE). Progressively shorter pieces of DNA, regions with internal deletions, or regions with point mutations can be constructed and inserted to pinpoint the response element (see Figure 39–11 for deletion mapping of the relevant HREs).

## Combinations of DNA Elements & Associated Proteins Provide Diversity in Responses

Prokaryotic genes are often regulated in an on-off manner in response to simple environmental cues. Some eukaryotic genes are regulated in the simple on-off manner, but the process in most, especially in mammals, is much more complicated. Signals representing a number of complex environmental stimuli may converge on a single gene. The response of the gene to these signals can have several physiologic characteristics. First, the response may extend over a considerable range. This is accomplished by having additive and synergistic positive responses counterbalanced by negative or repressing effects. In some cases, either the positive or the negative response can be dominant. Also required is a mechanism whereby an effector such as a hormone can activate some genes in a cell while repressing others and leaving still others unaffected. When all of these processes are coupled with tissue-specific element factors, considerable flexibility is afforded. These physiologic variables obviously require an arrangement much more complicated than an on-off switch. The array of DNA elements in a promoter specifies—with associated factors—how a given gene will respond. Some simple examples are illustrated in Figure 39–12.

## Transcription Domains Can Be Defined by Locus Control Regions & Insulators

The large number of genes in eukaryotic cells and the complex arrays of transcription regulatory factors presents an organizational problem. Why are some genes available for transcription in a given cell whereas others are not? If enhancers can regulate several genes and are not position- and orientation-dependent, how are they prevented from triggering transcription randomly? Part of the solution to these problems is arrived at by having the chromatin arranged in functional units that restrict patterns of gene expression. This may be achieved by having the chromatin form a structure with the nuclear matrix or other physical entity, or compartments within the nucleus. Alternatively, some regions are controlled by complex DNA elements called **locus control regions (LCRs).** An LCR—with associated bound proteins—controls the expression of a cluster of genes. The best-defined LCR regulates expression of the globin gene family over a large region of DNA. Another mechanism is provided by **insulators.** These DNA elements, also in association with one or more proteins, prevent an enhancer from acting on a promoter on the other side of an insulator in another transcription domain.

# SEVERAL MOTIFS MEDIATE THE BINDING OF REGULATORY PROTEINS TO DNA

The specificity involved in the control of transcription requires that regulatory proteins bind with high affinity to the correct region of DNA. Three unique motifs—the **helix-turn-helix,** the **zinc finger,** and the **leucine**

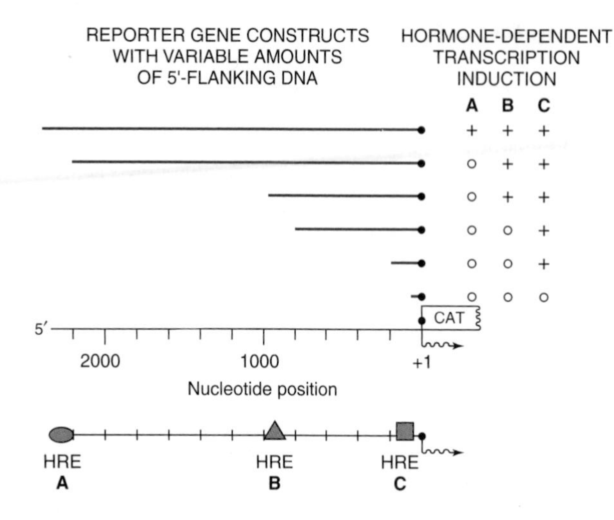


***Figure 39–11.*** Location of hormone response elements (HREs) **A, B,** and **C** using the reporter gene–transfection approach. A family of reporter genes, constructed as described in Figure 39–10, can be transfected individually into a recipient cell. By analyzing when certain hormone responses are lost in comparison to the 5′ deletion, specific hormone-responsive elements can be located.

**zipper**—account for many of these specific protein-DNA interactions. Examples of proteins containing these motifs are given in Table 39–3.

Comparison of the binding activities of the proteins that contain these motifs leads to several important generalizations.

(1) Binding must be of high affinity to the specific site and of low affinity to other DNA.

(2) Small regions of the protein make direct contact with DNA; the rest of the protein, in addition to pro-

***Table 39–3.*** Examples of transcription regulatory proteins that contain the various binding motifs.

| Binding Motif | Organism | Regulatory Protein |
|---|---|---|
| Helix-turn-helix | *E coli* | lac repressor |
| | | CAP |
| | Phage | λcI, cro, and tryptophan and 434 repressors |
| | Mammals | homeo box proteins |
| | | Pit-1, Oct1, Oct2 |
| Zinc finger | *E coli* | Gene 32 protein |
| | Yeast | Gal4 |
| | Drosophila | Serendipity, Hunchback |
| | Xenopus | TFIIIA |
| | Mammals | steroid receptor family, Sn1 |
| Leucine zipper | Yeast | GCN4 |
| | Mammals | C/EBP, fos, Jun, Fra-1, CRE binding protein, *c-myc*, *n-myc*, *l-myc* |

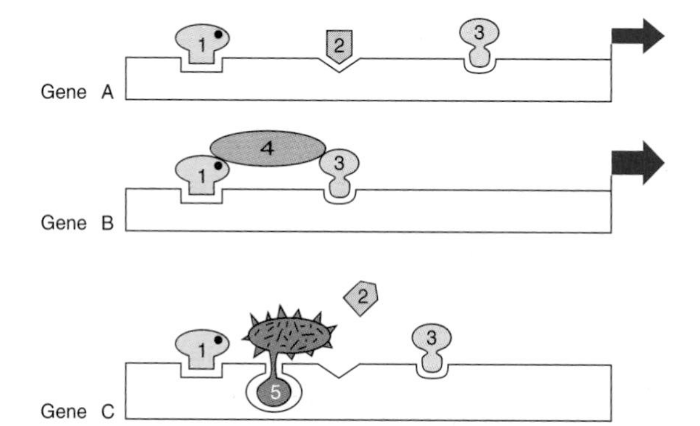


***Figure 39–12.*** Combinations of DNA elements and proteins provide diversity in the response of a gene. Gene A is activated (the width of the arrow indicates the extent) by the combination of activators 1, 2, and 3 (probably with coactivators, as shown in Figure 37–10). Gene B is activated, in this case more effectively, by the combination of 1, 3, and 4; note that 4 does not contact DNA directly in this example. The activators could form a linear bridge that links the basal machinery to the promoter, or this could be accomplished by looping out of the DNA. In either case, the purpose is to direct the basal transcription machinery to the promoter. Gene C is inactivated by the combination of 1, 5, and 3; in this case, factor 5 is shown to preclude the essential binding of factor 2 to DNA, as occurs in example A. If activator 1 helps repressor 5 bind and if activator 1 binding requires a ligand (solid dot), it can be seen how the ligand could activate one gene in a cell (gene A) and repress another (gene C).

viding the trans-activation domains, may be involved in the dimerization of monomers of the binding protein, may provide a contact surface for the formation of heterodimers, may provide one or more ligand-binding sites, or may provide surfaces for interaction with coactivators or corepressors.

(3) The protein-DNA interactions are maintained by hydrogen bonds and van der Waals forces.

(4) The motifs found in these proteins are unique; their presence in a protein of unknown function suggests that the protein may bind to DNA.

(5) Proteins with the helix-turn-helix or leucine zipper motifs form symmetric dimers, and their respective DNA binding sites are symmetric palindromes. In proteins with the zinc finger motif, the binding site is repeated two to nine times. These features allow for cooperative interactions between binding sites and enhance the degree and affinity of binding.

## The Helix-Turn-Helix Motif

The first motif described—and the one studied most extensively—is the helix-turn-helix. Analysis of the three-dimensional structure of the λ Cro transcription regulator has revealed that each monomer consists of three antiparallel β sheets and three α helices (Figure 39–13). The dimer forms by association of the antiparallel $\beta_3$ sheets. The $\alpha_3$ helices form the DNA recognition surface, and the rest of the molecule appears to be involved in stabilizing these structures. The average diameter of an α helix is 1.2 nm, which is the approximate width of the major groove in the B form of DNA. The DNA recognition domain of each Cro monomer interacts with 5 bp and the dimer binding sites span 3.4 nm, allowing fit into successive half turns of the major groove on the same surface (Figure 39–13). X-ray analyses of the λ cI repressor, CAP (the cAMP receptor

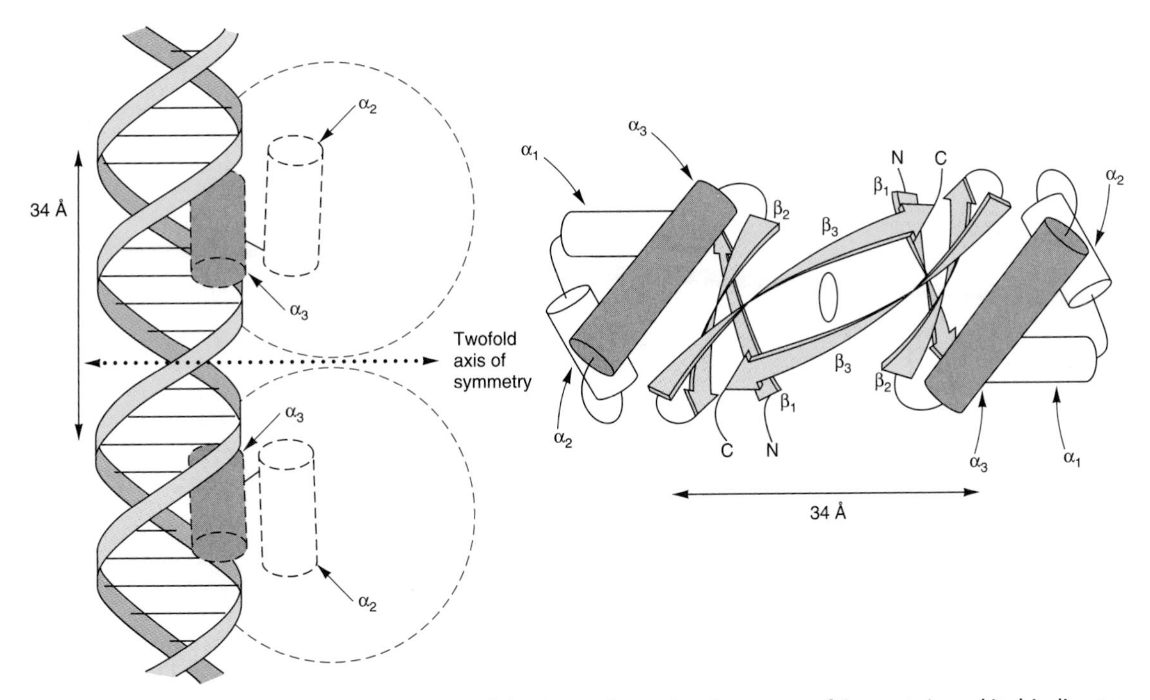


***Figure 39–13.*** A schematic representation of the three-dimensional structure of Cro protein and its binding to DNA by its helix-turn-helix motif. The Cro monomer consists of three antiparallel β sheets ($\beta_1$–$\beta_3$) and three α-helices ($\alpha_1$–$\alpha_3$). The helix-turn-helix motif is formed because the $\alpha_3$ and $\alpha_2$ helices are held at about 90 degrees to each other by a turn of four amino acids. The $\alpha_3$ helix of Cro is the DNA recognition surface (shaded). Two monomers associate through the antiparallel β3 sheets to form a dimer that has a twofold axis of symmetry (right). A Cro dimer binds to DNA through its $\alpha_3$ helices, each of which contacts about 5 bp on the same surface of the major groove. The distance between comparable points on the two DNA α-helices is 34 Å, which is the distance required for one complete turn of the double helix. (Courtesy of B Mathews.)

protein of *E coli*), tryptophan repressor, and phage 434 repressor all also display this dimeric helix-turn-helix structure that is present in eukaryotic DNA proteins as well (see Table 39–3).

### The Zinc Finger Motif

The zinc finger was the second DNA binding motif whose atomic structure was elucidated. It was known that the protein TFIIIA, a positive regulator of 5S RNA transcription, required zinc for activity. Structural and biophysical analyses revealed that each TFIIIA molecule contains nine zinc ions in a repeating coordination complex formed by closely spaced cysteine-cysteine residues followed 12–13 amino acids later by a histidine-histidine pair (Figure 39–14). In some instances—notably the steroid-thyroid receptor family—the His-His doublet is replaced by a second Cys-Cys pair. The protein containing zinc fingers appears to lie on one face of the DNA helix, with successive fingers alternatively positioned in one turn in the major groove. As is the case with the recognition domain in the helix-turn-helix protein, each TFIIIA zinc finger contacts about 5 bp of DNA. The importance of this motif in the action of steroid hormones is underscored by an "experiment of nature." A single amino acid mutation in either of the two zinc fingers of the $1,25(OH)_2$-$D_3$ receptor protein results in resistance to the action of this hormone and the clinical syndrome of rickets.

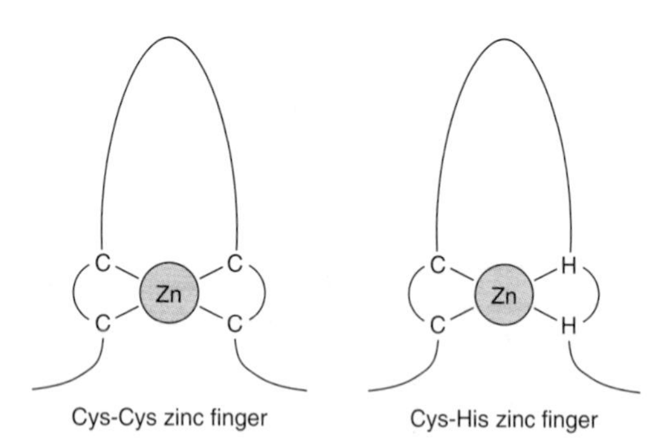


***Figure 39–14.*** Zinc fingers are a series of repeated domains (two to nine) in which each is centered on a tetrahedral coordination with zinc. In the case of TFIIIA, the coordination is provided by a pair of cysteine residues (C) separated by 12–13 amino acids from a pair of histidine (H) residues. In other zinc finger proteins, the second pair also consists of C residues. Zinc fingers bind in the major groove, with adjacent fingers making contact with 5 bp along the same face of the helix.

### The Leucine Zipper Motif

Careful analysis of a 30-amino-acid sequence in the carboxyl terminal region of the enhancer binding protein C/EBP revealed a novel structure. As illustrated in Figure 39–15, this region of the protein forms an α helix in which there is a periodic repeat of leucine residues at every seventh position. This occurs for eight helical turns and four leucine repeats. Similar structures have been found in a number of other proteins associated with the regulation of transcription in mammalian and yeast cells. It is thought that this structure allows two identical monomers or heterodimers (eg, Fos-Jun or Jun-Jun) to "zip together" in a coiled coil and form a tight dimeric complex (Figure 39–15). This protein-protein interaction may serve to enhance the association of the separate DNA binding domains with their target (Figure 39–15).

## THE DNA BINDING & TRANS-ACTIVATION DOMAINS OF MOST REGULATORY PROTEINS ARE SEPARATE & NONINTERACTIVE

DNA binding could result in a general conformational change that allows the bound protein to activate transcription, or these two functions could be served by separate and independent domains. Domain swap experiments suggest that the latter is the case.

The *GAL1* gene product is involved in galactose metabolism in yeast. Transcription of this gene is positively regulated by the GAL4 protein, which binds to an upstream activator sequence (UAS), or enhancer, through an amino terminal domain. The amino terminal 73-amino-acid DNA-binding domain (DBD) of GAL4 was removed and replaced with the DBD of LexA, an *E coli* DNA-binding protein. This domain swap resulted in a molecule that did not bind to the *GAL1* UAS and, of course, did not activate the *GAL1* gene (Figure 39–16). If, however, the *lexA* operator—the DNA sequence normally bound by the *lexA* DBD—was inserted into the promoter region of the *GAL* gene, the hybrid protein bound to this promoter (at the *lexA* operator) and it activated transcription of *GAL1*. This experiment, which has been repeated a number of times, affords solid evidence that the carboxyl terminal region of GAL4 causes transcriptional activation. These data also demonstrate that the DNA-binding DBD and trans-activation domains (ADs) are independent and noninteractive. The hierarchy involved in assembling gene transcription activating complexes includes proteins that bind DNA and trans-activate; others that form protein-protein complexes which bridge DNA-binding proteins to trans-activating proteins; and others that form protein-protein complexes with components of the basal transcription

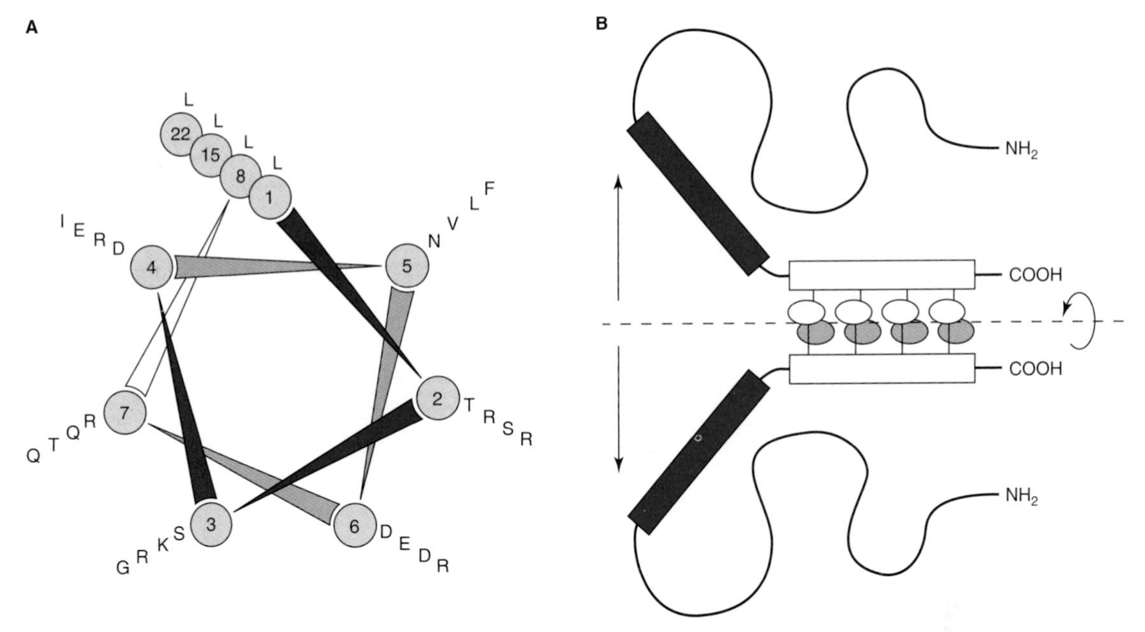


***Figure 39–15.*** The leucine zipper motif. **A** shows a helical wheel analysis of a carboxyl terminal portion of the DNA binding protein C/EBP. The amino acid sequence is displayed end-to-end down the axis of a schematic α-helix. The helical wheel consists of seven spokes that correspond to the seven amino acids that comprise every two turns of the α-helix. Note that leucine residues (L) occur at every seventh position. Other proteins with "leucine zippers" have a similar helical wheel pattern. **B** is a schematic model of the DNA binding domain of C/EBP. Two identical C/EBP polypeptide chains are held in dimer formation by the leucine zipper domain of each polypeptide (denoted by the rectangles and attached ovals). This association is apparently required to hold the DNA binding domains of each polypeptide (the shaded rectangles) in the proper conformation for DNA binding. (Courtesy of S McKnight.)

apparatus. A given protein may thus have several surfaces or domains that serve different functions (see Figure 39–17). As described in Chapter 37, the primary purpose of these complex assemblies is to facilitate the assembly of the basal transcription apparatus on the cis-linked promoter.

## GENE REGULATION IN PROKARYOTES & EUKARYOTES DIFFERS IN IMPORTANT RESPECTS

In addition to transcription, eukaryotic cells employ a variety of mechanisms to regulate gene expression (Table 39–4). The nuclear membrane of eukaryotic cells physically segregates gene transcription from translation, since ribosomes exist only in the cytoplasm. Many more steps, especially in RNA processing, are involved in the expression of eukaryotic genes than of prokaryotic genes, and these steps provide additional sites for regulatory influences that cannot exist in prokaryotes. These RNA processing steps in eukaryotes, described in detail in Chapter 37, include capping of the 5′ ends of the primary transcripts, addition of a polyadenylate tail to the 3′ ends of transcripts, and excision of intron regions to generate spliced exons in the mature mRNA molecule. To date, analyses of eukaryotic gene expression provide evidence that regulation occurs at the level of **transcription, nuclear RNA processing,** and **mRNA stability.** In addition, gene amplification and rearrangement influence gene expression.

Owing to the advent of recombinant DNA technology, much progress has been made in recent years in the understanding of eukaryotic gene expression. However, because most eukaryotic organisms contain so much more genetic information than do prokaryotes and because manipulation of their genes is so much more limited, molecular aspects of eukaryotic gene regulation are less well understood than the examples discussed earlier in this chapter. This section briefly describes a few different types of eukaryotic gene regulation.

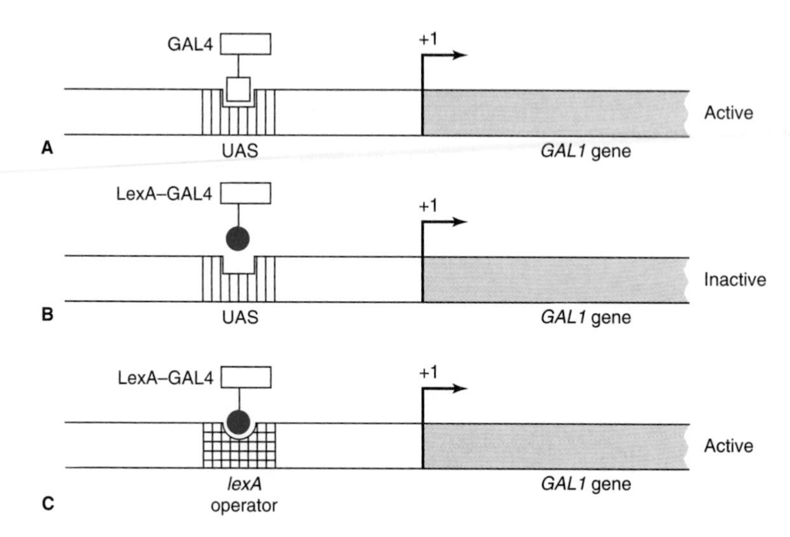


***Figure 39–16.*** Domain-swap experiments demonstrate the independent nature of DNA binding and transcription activation domains. The *GAL1* gene promoter contains an upstream activating sequence (UAS) or enhancer that binds the regulatory protein GAL4 **(A).** This interaction results in a stimulation of *GAL1* gene transcription. A chimeric protein, in which the amino terminal DNA binding domain of GAL4 is removed and replaced with the DNA binding region of the *E coli* protein LexA, fails to stimulate *GAL1* transcription because the LexA domain cannot bind to the UAS **(B).** The LexA–GAL4 fusion protein does increase *GAL1* transcription when the *lexA* operator (its natural target) is inserted into the *GAL1* promoter region **(C).**

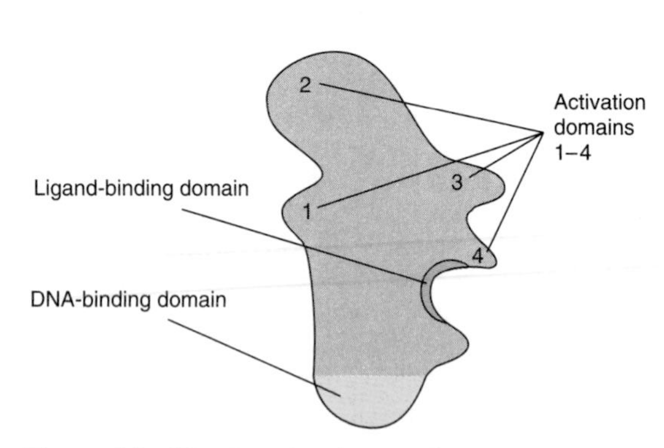


***Figure 39–17.*** Proteins that regulate transcription have several domains. This hypothetical transcription factor has a DNA-binding domain (DBD) that is distinct from a ligand-binding domain (LBD) and several activation domains (ADs) (1–4). Other proteins may lack the DBD or LBD and all may have variable numbers of domains that contact other proteins, including co-regulators and those of the basal transcription complex (see also Chapters 42 and 43).

## Eukaryotic Genes Can Be Amplified or Rearranged During Development or in Response to Drugs

During early development of metazoans, there is an abrupt increase in the need for specific molecules such as ribosomal RNA and messenger RNA molecules for proteins that make up such organs as the eggshell. One way to increase the rate at which such molecules can be formed is to increase the number of genes available for transcription of these specific molecules. Among the repetitive DNA sequences are hundreds of copies of ribosomal RNA genes and tRNA genes. These genes preexist repetitively in the genomic material of the gametes

***Table 39–4.*** Gene expression is regulated by transcription and in numerous other ways in eukaryotic cells.

- Gene amplification
- Gene rearrangement
- RNA processing
- Alternate mRNA splicing
- Transport of mRNA from nucleus to cytoplasm
- Regulation of mRNA stability

and thus are transmitted in high copy numbers from generation to generation. In some specific organisms such as the fruit fly *(drosophila),* there occurs during oogenesis an amplification of a few preexisting genes such as those for the chorion (eggshell) proteins. Subsequently, these amplified genes, presumably generated by a process of repeated initiations during DNA synthesis, provide multiple sites for gene transcription (Figures 36–16 and 39–18).

As noted in Chapter 37, the coding sequences responsible for the generation of specific protein molecules are frequently not contiguous in the mammalian genome. In the case of antibody encoding genes, this is particularly true. As described in detail in Chapter 50, immunoglobulins are composed of two polypeptides, the so-called heavy (about 50 kDa) and light (about 25 kDa) chains. The mRNAs encoding these two protein subunits are encoded by gene sequences that are subjected to extensive DNA sequence-coding changes. These DNA coding changes are integral to generating the requisite recognition diversity central to appropriate immune function.

IgG heavy and light chain mRNAs are encoded by several different segments that are tandemly repeated in the germline. Thus, for example, the IgG light chain is composed of variable ($V_L$), joining ($J_L$), and constant ($C_L$) domains or segments. For particular subsets of IgG light chains, there are roughly 300 tandemly repeated $V_L$ gene coding segments, five tandemly arranged $J_L$ coding sequences, and roughly ten $C_L$ gene coding segments. All of these multiple, distinct coding regions are located in the same region of the same chromosome, and each type of coding segment ($V_L$, $J_L$, and $C_L$) is tandemly repeated in head-to-tail fashion within the segment repeat region. By having multiple $V_L$, $J_L$, and $C_L$ segments to choose from, an immune cell has a greater repertoire of sequences to work with to develop both immunologic flexibility and specificity. However, a given functional IgG light chain transcription unit—like all other "normal" mammalian transcription units—contains only the coding sequences for a single protein. Thus, before a particular IgG light chain can be expressed, *single* $V_L$, $J_L$, and $C_L$ coding sequences must be recombined to generate a *single,* contiguous transcription unit excluding the multiple nonutilized segments (ie, the other approximately 300 unused $V_L$ segments, the other four unused $J_L$ segments, and the other nine unused $C_L$ segments). This deletion of unused genetic information is accomplished by selective DNA recombination that removes the unwanted coding DNA while retaining the required coding sequences: one $V_L$, one $J_L$, and one $C_L$ sequence. ($V_L$ sequences are subjected to additional point mutagenesis to generate even more variability—hence the name.) The newly recombined sequences thus form a single transcription unit that is competent for RNA polymerase II-mediated transcription. Although the IgG genes represent one of the best-studied instances of directed DNA rearrangement modulating gene expression, other cases of gene regulatory DNA rearrangement have been described in the literature. Indeed, as detailed below, drug-induced gene amplification is an important complication of cancer chemotherapy.

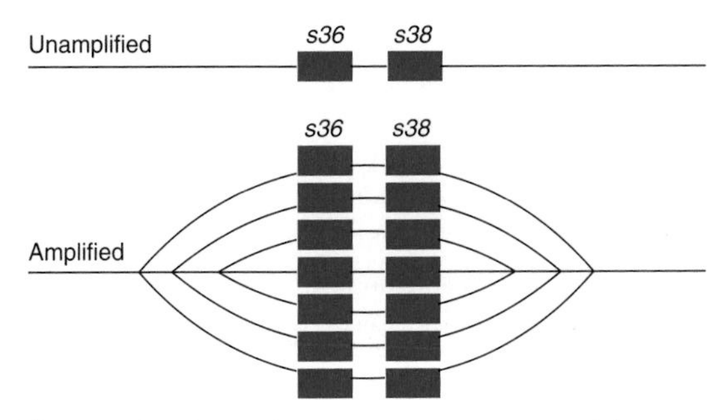


***Figure 39–18.*** Schematic representation of the amplification of chorion protein genes *s36* and *s38*. (Reproduced, with permission, from Chisholm R: Gene amplification during development. Trends Biochem Sci 1982;7:161.)

In recent years, it has been possible to promote the amplification of specific genetic regions in cultured mammalian cells. In some cases, a several thousand-fold increase in the copy number of specific genes can be achieved over a period of time involving increasing doses of selective drugs. In fact, it has been demonstrated in patients receiving methotrexate for cancer that malignant cells can develop **drug resistance** by increasing the number of genes for dihydrofolate reductase, the target of methotrexate. Gene amplification events such as these occur spontaneously in vivo—ie, in the absence of exogenously supplied selective agents—and these unscheduled extra rounds of replication can become "frozen" in the genome under appropriate selective pressures.

## Alternative RNA Processing Is Another Control Mechanism

In addition to affecting the efficiency of promoter utilization, eukaryotic cells employ alternative RNA processing to control gene expression. This can result when alternative promoters, intron-exon splice sites, or polyadenylation sites are used. Occasionally, heterogeneity within a cell results, but more commonly the same primary transcript is processed differently in different tissues. A few examples of each of these types of regulation are presented below.

The use of alternative **transcription start sites** results in a different 5′ exon on mRNAs corresponding to mouse amylase and myosin light chain, rat glucokinase, and drosophila alcohol dehydrogenase and actin. **Alternative polyadenylation sites** in the μ immunoglobulin heavy chain primary transcript result in mRNAs that are either 2700 bases long ($\mu_m$) or 2400 bases long ($\mu_s$). This results in a different carboxyl terminal region of the encoded proteins such that the $\mu_m$ protein remains attached to the membrane of the B lymphocyte and the $\mu_s$ immunoglobulin is secreted. **Alternative splicing and processing** results in the formation of seven unique α-tropomyosin mRNAs in seven different tissues. It is not clear how these processing-splicing decisions are made or whether these steps can be regulated.

## Regulation of Messenger RNA Stability Provides Another Control Mechanism

Although most mRNAs in mammalian cells are very stable (half-lives measured in hours), some turn over very rapidly (half-lives of 10–30 minutes). In certain instances, mRNA stability is subject to regulation. This has important implications since there is usually a direct relationship between mRNA amount and the translation of that mRNA into its cognate protein. Changes in the stability of a specific mRNA can therefore have major effects on biologic processes.

Messenger RNAs exist in the cytoplasm as ribonucleoprotein particles (RNPs). Some of these proteins protect the mRNA from digestion by nucleases, while others may under certain conditions promote nuclease attack. It is thought that mRNAs are stabilized or destabilized by the interaction of proteins with these various structures or sequences. Certain effectors, such as hormones, may regulate mRNA stability by increasing or decreasing the amount of these proteins.

It appears that **the ends of mRNA molecules are involved in mRNA stability** (Figure 39–19). The 5′ cap structure in eukaryotic mRNA prevents attack by 5′ exonucleases, and the poly(A) tail prohibits the action of 3′ exonucleases. In mRNA molecules with those structures, it is presumed that a single endonucleolytic cut allows exonucleases to attack and digest the entire molecule. Other structures (sequences) in the 5′ noncoding sequence (5′ NCS), the coding region, and the 3′ NCS are thought to promote or prevent this initial endonucleolytic action (Figure 39–19). A few illustrative examples will be cited.

Deletion of the 5′ NCS results in a threefold to fivefold prolongation of the half-life of c-*myc* mRNA. Shortening the coding region of histone mRNA results in a prolonged half-life. A form of autoregulation of mRNA stability indirectly involves the coding region. Free tubulin binds to the first four amino acids of a nascent chain of tubulin as it emerges from the ribosome. This appears to activate an RNase associated with the ribosome (RNP) which then digests the tubulin mRNA.

Structures at the 3′ end, including the poly(A) tail, enhance or diminish the stability of specific mRNAs. The absence of a poly(A) tail is associated with rapid degradation of mRNA, and the removal of poly(A) from some RNAs results in their destabilization. Histone mRNAs lack a poly(A) tail but have a sequence near the 3′ terminal that can form a stem-loop structure, and this appears to provide resistance to exonucleolytic attack. Histone H4 mRNA, for example, is degraded in the 3′ to 5′ direction but only after a single endonucleolytic cut occurs about nine nucleotides from the 3′ end in the region of the putative stem-loop structure. Stem-loop structures in the 3′ noncoding sequence are also critical for the regulation, by iron, of the mRNA encoding the transferrin receptor. Stem-loop structures are also associated with mRNA stability in bacteria, suggesting that this mechanism may be commonly employed.

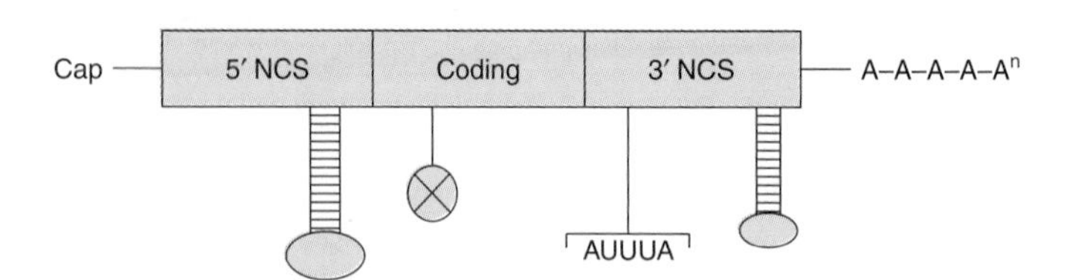


***Figure 39–19.*** Structure of a typical eukaryotic mRNA showing elements that are involved in regulating mRNA stability. The typical eukaryotic mRNA has a 5′ noncoding sequence (5′ NCS), a coding region, and a 3′ NCS. All are capped at the 5′ end, and most have a polyadenylate sequence at the 3′ end. The 5′ cap and 3′ poly(A) tail protect the mRNA against exonuclease attack. Stem-loop structures in the 5′ and 3′ NCS, features in the coding sequence, and the AU-rich region in the 3′ NCS are thought to play roles in mRNA stability.

Other sequences in the 3′ ends of certain eukaryotic mRNAs appear to be involved in the destabilization of these molecules. Of particular interest are AU-rich regions, many of which contain the sequence AUUUA. This sequence appears in mRNAs that have a very short half-life, including some encoding oncogene proteins and cytokines. The importance of this region is underscored by an experiment in which a sequence corresponding to the 3′ noncoding region of the short-half-life colony-stimulating factor (CSF) mRNA, which contains the AUUUA motif, was added to the 3′ end of the β-globin mRNA. Instead of becoming very stable, this hybrid β-globin mRNA now had the short-half-life characteristic of CSF mRNA.

From the few examples cited, it is clear that a number of mechanisms are used to regulate mRNA stability—just as several mechanisms are used to regulate the synthesis of mRNA. Coordinate regulation of these two processes confers on the cell remarkable adaptability.

## SUMMARY

- The genetic constitutions of nearly all metazoan somatic cells are identical.
- Phenotype (tissue or cell specificity) is dictated by differences in gene expression of this complement of genes.
- Alterations in gene expression allow a cell to adapt to environmental changes.
- Gene expression can be controlled at multiple levels by changes in transcription, RNA processing, localization, and stability or utilization. Gene amplification and rearrangements also influence gene expression.
- Transcription controls operate at the level of protein-DNA and protein-protein interactions. These interactions display protein domain modularity and high specificity.
- Several different classes of DNA-binding domains have been identified in transcription factors.
- Chromatin modifications are important in eukaryotic transcription control.

## REFERENCES


Albright SR, Tjian R: TAFs revisited: more data reveal new twists and confirm old ideas. Gene 2000;242:1.

Bird AP, Wolffe AP: Methylation-induced repression—belts, braces and chromatin. Cell 1999;99:451.

Berger SL, Felsenfeld G: Chromatin goes global. Mol Cell 2001; 8:263.

Busby S, Ebright RH: Promoter structure, promoter recognition, and transcription activation in prokaryotes. Cell 1994;79: 743.

Busby S, Ebright RH: Transcription activation by catabolite activator protein (CAP). J Mol Biol 1999;293:199.

Cowell IG: Repression versus activation in the control of gene transcription. Trends Biochem Sci 1994;1:38.

Ebright RH: RNA polymerase: structural similarities between bacterial RNA polymerase and eukaryotic RNA polymerase II. J Mol Biol 2000;304:687.

Fugman SD: RAG1 and RAG2 in V(D)J recombination and transposition. Immunol Res 2001;23:23.

Jacob F, Monod J: Genetic regulatory mechanisms in protein synthesis. J Mol Biol 1961;3:318.

Lemon B, Tjian R: Orchestrated response: a symphony of transcription factors for gene control. Genes Dev 2000;14:2551.

Letchman DS: Transcription factor mutations and disease. N Engl J Med 1996;334:28.

Merika M, Thanos D: Enhanceosomes. Curr Opin Genet Dev 2001;11:205.

Naar AM, Lemon BD, Tjian R: Transcriptional coactivator complexes. Annu Rev Biochem 2001;70:475.

Narlikar GJ, Fan HY, Kingston RE: Cooperation between complexes that regulate chromatin structure and transcription. Cell 2002;108:475.

Oltz EM: Regulation of antigen receptor gene assembly in lymphocytes. Immunol Res 2001;23:121.

Ptashne M: Control of gene transcription: an outline. Nat Med 1997;3:1069.

Ptashne M: *A Genetic Switch,* 2nd ed. Cell Press and Blackwell Scientific Publications, 1992.

Sterner DE, Berger SL: Acetylation of histones and transcription-related factors. Microbiol Mol Biol Rev 2000;64:435.

Wu R, Bahl CP, Narang SA: Lactose operator-repressor interaction. Curr Top Cell Regul 1978;13:137.

# Molecular Genetics, Recombinant DNA, & Genomic Technology

40

*Daryl K. Granner, MD, & P. Anthony Weil, PhD*

## BIOMEDICAL IMPORTANCE*

The development of recombinant DNA, high-density, high-throughput screening, and other molecular genetic methodologies has revolutionized biology and is having an increasing impact on clinical medicine. Much has been learned about human genetic disease from pedigree analysis and study of affected proteins, but in many cases where the specific genetic defect is unknown, these approaches cannot be used. The new technologies circumvent these limitations by going directly to the DNA molecule for information. Manipulation of a DNA sequence and the construction of chimeric molecules—so-called genetic engineering—provides a means of studying how a specific segment of DNA works. Novel molecular genetic tools allow investigators to query and manipulate genomic sequences as well as to examine both cellular mRNA and protein profiles at the molecular level.

Understanding this technology is important for several reasons: (1) It offers a rational approach to understanding the molecular basis of a number of diseases (eg, familial hypercholesterolemia, sickle cell disease, the thalassemias, cystic fibrosis, muscular dystrophy). (2) Human proteins can be produced in abundance for therapy (eg, insulin, growth hormone, tissue plasminogen activator). (3) Proteins for vaccines (eg, hepatitis B) and for diagnostic testing (eg, AIDS tests) can be obtained. (4) This technology is used to diagnose existing diseases and predict the risk of developing a given disease. (5) Special techniques have led to remarkable advances in forensic medicine. (6) Gene therapy for sickle cell disease, the thalassemias, adenosine deaminase deficiency, and other diseases may be devised.

* See glossary of terms at the end of this chapter.

## ELUCIDATION OF THE BASIC FEATURES OF DNA LED TO RECOMBINANT DNA TECHNOLOGY

### DNA Is a Complex Biopolymer Organized as a Double Helix

The fundamental organizational element is the sequence of purine (adenine [A] or guanine [G]) and pyrimidine (cytosine [C] or thymine [T]) bases. These bases are attached to the C-1′ position of the sugar deoxyribose, and the bases are linked together through joining of the sugar moieties at their 3′ and 5′ positions via a phosphodiester bond (Figure 35–1). The alternating deoxyribose and phosphate groups form the backbone of the double helix (Figure 35–2). These 3′–5′ linkages also define the orientation of a given strand of the DNA molecule, and, since the two strands run in opposite directions, they are said to be antiparallel.

### Base Pairing Is a Fundamental Concept of DNA Structure & Function

Adenine and thymine always pair, by hydrogen bonding, as do guanine and cytosine (Figure 35–3). These base pairs are said to be **complementary,** and the guanine content of a fragment of double-stranded DNA will always equal its cytosine content; likewise, the thymine and adenine contents are equal. Base pairing and hydrophobic base-stacking interactions hold the two DNA strands together. These interactions can be reduced by heating the DNA to denature it. The laws of base pairing predict that two complementary DNA strands will reanneal exactly in register upon renaturation, as happens when the temperature of the solution is slowly reduced to normal. Indeed, the degree of base-pair matching (or mismatching) can be estimated from the temperature required for denaturation-renaturation. Segments of DNA with high degrees of base-pair matching require more energy input (heat) to accomplish denaturation—or, to put it another way, a closely matched segment will withstand more heat before the strands separate. This reaction is used to determine whether there are significant differences between two DNA sequences, and it underlies the concept of **hybridization,** which is fundamental to the processes described below.

**There are about $3 \times 10^9$ base pairs (bp) in each human haploid genome.** If an average gene length is $3 \times 10^3$ bp (3 kilobases [kb]), the genome could consist of $10^6$ genes, assuming that there is no overlap and that transcription proceeds in only one direction. It is thought that there are $< 10^5$ genes in the human and that only 1–2% of the DNA codes for proteins. The exact function of the remaining ~98% of the human genome has not yet been defined.

**The double-helical DNA is packaged into a more compact structure by a number of proteins, most notably the basic proteins called histones.**[4] This condensation may serve a regulatory role and certainly has a practical purpose. The DNA present within the nucleus of a cell, if simply extended, would be about 1 meter long. The chromosomal proteins compact this long strand of DNA so that it can be packaged into a nucleus with a volume of a few cubic micrometers.

### DNA Is Organized Into Genes

In general, prokaryotic genes consist of a small regulatory region (100–500 bp) and a large protein-coding segment (500–10,000 bp). Several genes are often controlled by a single regulatory unit. Most mammalian genes are more complicated in that the coding regions are interrupted by noncoding regions that are eliminated when the primary RNA transcript is processed into mature **messenger RNA (mRNA).** The **coding regions** (those regions that appear in the mature RNA species) are called exons, and the **noncoding regions,** which interpose or intervene between the exons, are called introns (Figure 40–1). Introns are always removed from precursor RNA before transport into the cytoplasm occurs. The process by which introns are removed from precursor RNA and by which exons are ligated together is called **RNA splicing.** Incorrect processing of the primary transcript into the mature mRNA can result in disease in humans (see below); this underscores the importance of these posttranscriptional processing steps. The variation in size and complexity of some human genes is illustrated in Table 40–1. Although there is a 300-fold difference in the sizes of the genes illustrated, the mRNA sizes vary only about 20-fold. This is because most of the DNA in genes is present as introns, and introns tend to be much larger than exons. Regulatory regions for specific eukaryotic genes are usually located in the DNA that flanks the transcription initiation site at its 5′ end **(5′ flanking-sequence DNA).** Occasionally, such sequences are found within the gene itself or in the region that flanks the 3′ end of the gene. In mammalian cells, each gene has its own regulatory region. Many eukaryotic genes (and some viruses that replicate in mammalian cells) have special regions, called **enhancers,** that increase the rate of transcription. Some genes also have DNA sequences, known as **silencers,** that repress transcription. Mammalian genes are obviously complicated, multicomponent structures.

### Genes Are Transcribed Into RNA

Information generally flows from DNA to mRNA to protein, as illustrated in Figure 40–1 and discussed in more detail in Chapter 39. This is a rigidly controlled process involving a number of complex steps, each of which no doubt is regulated by one or more enzymes or factors; faulty function at any of these steps can cause disease.

## RECOMBINANT DNA TECHNOLOGY INVOLVES ISOLATION & MANIPULATION OF DNA TO MAKE CHIMERIC MOLECULES

Isolation and manipulation of DNA, including end-to-end joining of sequences from very different sources to make chimeric molecules (eg, molecules containing both human and bacterial DNA sequences in a sequence-independent fashion), is the essence of recombinant DNA research. This involves several unique techniques and reagents.

### Restriction Enzymes Cut DNA Chains at Specific Locations

Certain endonucleases—enzymes that cut DNA at specific DNA sequences within the molecule (as opposed to exonucleases, which digest from the ends of DNA molecules)—are a key tool in recombinant DNA research. These enzymes were called **restriction enzymes** because their presence in a given bacterium restricted the growth of certain bacterial viruses called bacteriophages. Restriction enzymes cut DNA of any source into short pieces in a sequence-specific manner—in contrast to most other enzymatic, chemical, or physical methods, which break DNA randomly. These defensive enzymes (hundreds have been discovered) protect the host bacterial DNA from DNA from foreign organisms (primarily infective phages). However, they are present only in cells that also have a companion enzyme which methylates the host DNA, rendering it an unsuitable substrate for digestion by the restriction enzyme. Thus,

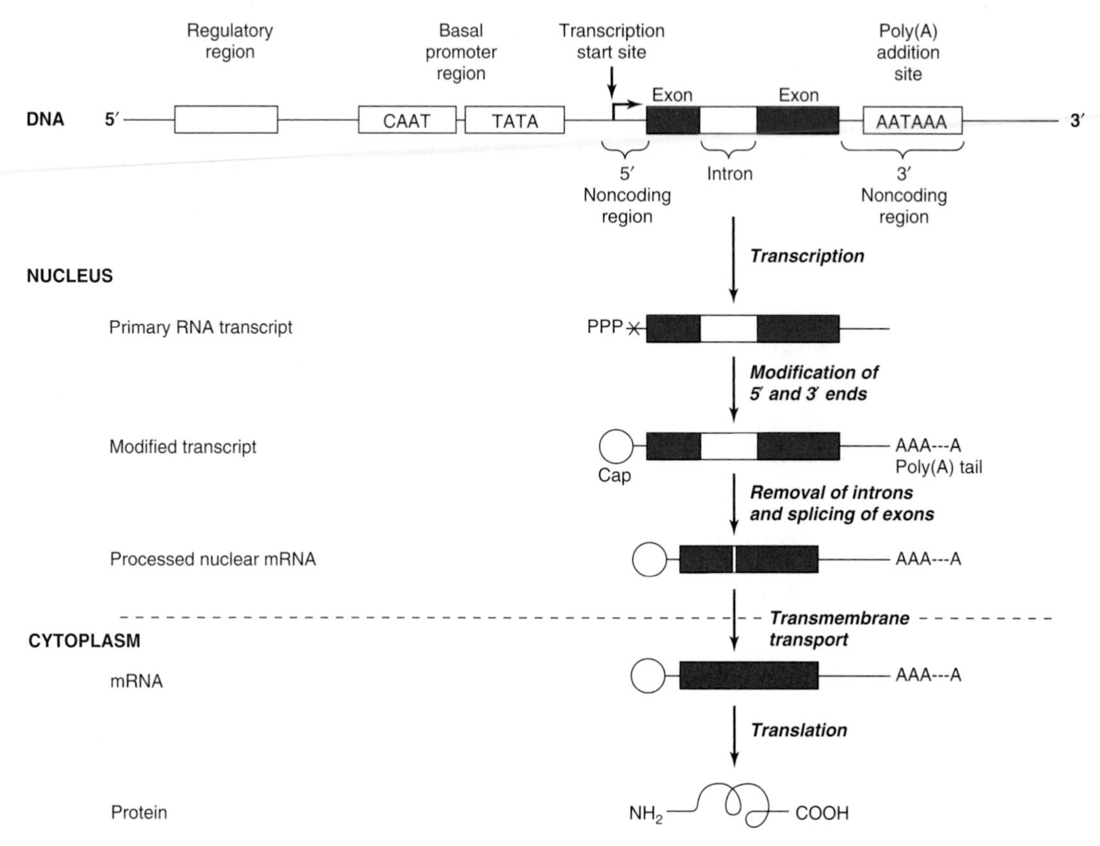


***Figure 40–1.*** Organization of a eukaryotic transcription unit and the pathway of eukaryotic gene expression. Eukaryotic genes have structural and regulatory regions. The structural region consists of the coding DNA and 5′ and 3′ noncoding DNA sequences. The coding regions are divided into two parts: (1) exons, which eventually are ligated together to become mature RNA, and (2) introns, which are processed out of the primary transcript. The structural region is bounded at its 5′ end by the transcription initiation site and at its 3′ end by the polyadenylate addition or termination site. The promoter region, which contains specific DNA sequences that interact with various protein factors to regulate transcription, is discussed in detail in Chapters 37 and 39. The primary transcript has a special structure, a cap, at the 5′ end and a stretch of As at the 3′ end. This transcript is processed to remove the introns; and the mature mRNA is then transported to the cytoplasm, where it is translated into protein.

**site-specific DNA methylases** and restriction enzymes always exist in pairs in a bacterium.

**Restriction enzymes are named after the bacterium from which they are isolated.** For example, *EcoRI* is from *Escherichia coli,* and *BamHI* is from *Bacillus amyloliquefaciens* (Table 40–2). The first three letters in the restriction enzyme name consist of the first letter of the genus (E) and the first two letters of the species (co). These may be followed by a strain designation (R) and a roman numeral (I) to indicate the order of discovery (eg, *EcoRI, EcoRII*). Each enzyme recognizes and cleaves a specific double-stranded DNA sequence that is 4–7 bp long. These DNA cuts result in **blunt ends** (eg, *HpaI)* or overlapping **(sticky) ends** (eg, *BamHI)* (Figure 40–2), depending on the mechanism used by the enzyme. Sticky ends are particularly useful in constructing hybrid or chimeric DNA molecules (see below). If the four nucleotides are distributed randomly in a given DNA molecule, one can calculate how frequently a given enzyme will cut a length of DNA. For each position in the DNA molecule, there are four possibilities (A, C, G, and T); therefore, a restriction enzyme that recognizes a 4-bp sequence cuts, on average, once every 256 bp ($4^4$), whereas another enzyme that recognizes a 6-bp sequence cuts once every 4096 bp ($4^6$). A given piece of DNA has a characteristic linear array of sites for

***Table 40–1.*** Variations in the size and complexity of some human genes and mRNAs.[1]

| Gene | Gene Size (kb) | Number of Introns | mRNA Size (kb) |
|---|---|---|---|
| β-Globin | 1.5 | 2 | 0.6 |
| Insulin | 1.7 | 2 | 0.4 |
| β-Adrenergic receptor | 3 | 0 | 2.2 |
| Albumin | 25 | 14 | 2.1 |
| LDL receptor | 45 | 17 | 5.5 |
| Factor VIII | 186 | 25 | 9.0 |
| Thyroglobulin | 300 | 36 | 8.7 |

[1]The sizes are given in kilobases (kb). The sizes of the genes include some proximal promoter and regulatory region sequences; these are generally about the same size for all genes. Genes vary in size from about 1500 base pairs (bp) to over $2 \times 10^6$ bp. There is also great variation in the number of introns and exons. The β-adrenergic receptor gene is intronless, and the thyroglobulin gene has 36 introns. As noted by the smaller difference in mRNA sizes, introns comprise most of the gene sequence.

the various enzymes dictated by the linear sequence of its bases; hence, a **restriction map** can be constructed. When DNA is digested with a given enzyme, the ends of all the fragments have the same DNA sequence. The fragments produced can be isolated by electrophoresis on agarose or polyacrylamide gels (see the discussion of blot transfer, below); this is an essential step in cloning and a major use of these enzymes.

A number of other enzymes that act on DNA and RNA are an important part of recombinant DNA technology. Many of these are referred to in this and subsequent chapters (Table 40–3).

## Restriction Enzymes & DNA Ligase Are Used to Prepare Chimeric DNA Molecules

Sticky-end ligation is technically easy, but some special techniques are often required to overcome problems inherent in this approach. Sticky ends of a vector may reconnect with themselves, with no net gain of DNA. Sticky ends of fragments can also anneal, so that tandem heterogeneous inserts form. Also, sticky-end sites may not be available or in a convenient position. To circumvent these problems, an enzyme that generates blunt ends is used, and new ends are added using the enzyme terminal transferase. If poly d(G) is added to the 3′ ends of the vector and poly d(C) is added to the 3′ ends of the foreign DNA, the two molecules can only anneal to each other, thus circumventing the problems listed above. This procedure is called homopolymer tailing. Sometimes, synthetic blunt-ended duplex oligonucleotide linkers with a convenient restriction enzyme se-

***Table 40–2.*** Selected restriction endonucleases and their sequence specificities.[1]

| Endonuclease | Sequence Recognized Cleavage Sites Shown | Bacterial Source |
|---|---|---|
| *BamHI* | G↓GATCC<br>CCTAG↑G | *Bacillus amyloliquefaciens* H |
| *BglII* | A↓GATCT<br>TCTAG↑A | *Bacillus glolbigii* |
| *EcoRI* | G↓AATTC<br>CTTAA↑G | *Escherichia coli* RY13 |
| *EcoRII* | ↓CCTGG<br>GGACC↑ | *Escherichia coli* R245 |
| *HindIII* | A↓AGCTT<br>TTCGA↑A | *Haemophilus influenzae* $R_d$ |
| *HhaI* | GCG↓C<br>C↑GCG | *Haemophilus haemolyticus* |
| *HpaI* | GTT↓AAC<br>CAA↑TTG | *Haemophilus parainfluenzae* |
| *MstII* | CC↓TNAGG<br>GGANT↑CC | Microcoleus strain |
| *PstI* | CTGCA↓G<br>G↑ACGTC | *Providencia stuartii* 164 |
| *TaqI* | T↓CGA<br>AGC↑T | *Thermus aquaticus* YTI |

[1]A, adenine; C, cytosine; G, guanine, T, thymine. Arrows show the site of cleavage; depending on the site, sticky ends *(BamHI)* or blunt ends *(HpaI)* may result. The length of the recognition sequence can be 4 bp *(TaqI)*, 5 bp *(EcoRII)*, 6 bp *(EcoRI)*, or 7 bp *(MstII)* or longer. By convention, these are written in the 5′ or 3′ direction for the upper strand of each recognition sequence, and the lower strand is shown with the opposite (ie, 3′ or 5′) polarity. Note that most recognition sequences are palindromes (ie, the sequence reads the same in opposite directions on the two strands). A residue designated N means that any nucleotide is permitted.

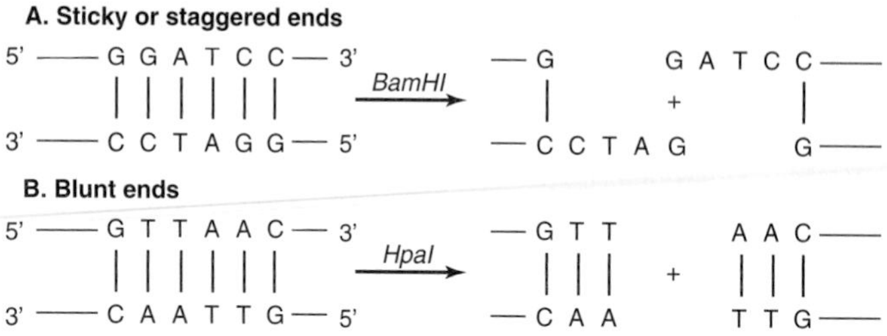

**A. Sticky or staggered ends**

5′ —— G G A T C C — 3′ → *BamHI* → — G + G A T C C ——

3′ —— C C T A G G — 5′ → — C C T A G + G ——

**B. Blunt ends**

5′ —— G T T A A C — 3′ → *HpaI* → — G T T + A A C ——

3′ —— C A A T T G — 5′ → — C A A + T T G ——

***Figure 40–2.*** Results of restriction endonuclease digestion. Digestion with a restriction endonuclease can result in the formation of DNA fragments with sticky, or cohesive, ends **(A)** or blunt ends **(B).** This is an important consideration in devising cloning strategies.

quence are ligated to the blunt-ended DNA. Direct blunt-end ligation is accomplished using the enzyme bacteriophage T4 DNA ligase. This technique, though less efficient than sticky-end ligation, has the advantage of joining together any pairs of ends. The disadvantages are that there is no control over the orientation of insertion or the number of molecules annealed together, and there is no easy way to retrieve the insert.

## Cloning Amplifies DNA

A **clone** is a large population of identical molecules, bacteria, or cells that arise from a common ancestor. Molecular cloning allows for the production of a large number of identical DNA molecules, which can then be characterized or used for other purposes. This technique is based on the fact that chimeric or hybrid DNA molecules can be constructed in **cloning vectors**—typically bacterial plasmids, phages, or cosmids—which then continue to replicate in a host cell under their own control systems. In this way, the chimeric DNA is amplified. The general procedure is illustrated in Figure 40–3.

Bacterial **plasmids** are small, circular, duplex DNA molecules whose natural function is to confer antibiotic resistance to the host cell. Plasmids have several properties that make them extremely useful as cloning vectors. They exist as single or multiple copies within the bacterium and replicate independently from the bacterial DNA. The complete DNA sequence of many plasmids is known; hence, the precise location of restriction enzyme

***Table 40–3.*** Some of the enzymes used in recombinant DNA research.[1]

| Enzyme | Reaction | Primary Use |
|---|---|---|
| Alkaline phosphatase | Dephosphorylates 5′ ends of RNA and DNA. | Removal of 5′-$PO_4$ groups prior to kinase labeling to prevent self-ligation. |
| BAL 31 nuclease | Degrades both the 3′ and 5′ ends of DNA. | Progressive shortening of DNA molecules. |
| DNA ligase | Catalyzes bonds between DNA molecules. | Joining of DNA molecules. |
| DNA polymerase I | Synthesizes double-stranded DNA from single-stranded DNA. | Synthesis of double-stranded cDNA; nick translation; generation of blunt ends from sticky ends. |
| DNase I | Under appropriate conditions, produces single-stranded nicks in DNA. | Nick translation; mapping of hypersensitive sites; mapping protein-DNA interactions. |
| Exonuclease III | Removes nucleotides from 3′ ends of DNA. | DNA sequencing; mapping of DNA-protein interactions. |
| λ exonuclease | Removes nucleotides from 5′ ends of DNA. | DNA sequencing. |
| Polynucleotide kinase | Transfers terminal phosphate (γ position) from ATP to 5′-OH groups of DNA or RNA. | $^{32}P$ labeling of DNA or RNA. |
| Reverse transcriptase | Synthesizes DNA from RNA template. | Synthesis of cDNA from mRNA; RNA (5′ end) mapping studies. |
| S1 nuclease | Degrades single-stranded DNA. | Removal of "hairpin" in synthesis of cDNA; RNA mapping studies (both 5′ and 3′ ends). |
| Terminal transferase | Adds nucleotides to the 3′ ends of DNA. | Homopolymer tailing. |

[1]Adapted and reproduced, with permission, from Emery AEH: Page 41 in: *An Introduction to Recombinant DNA.* Wiley, 1984.

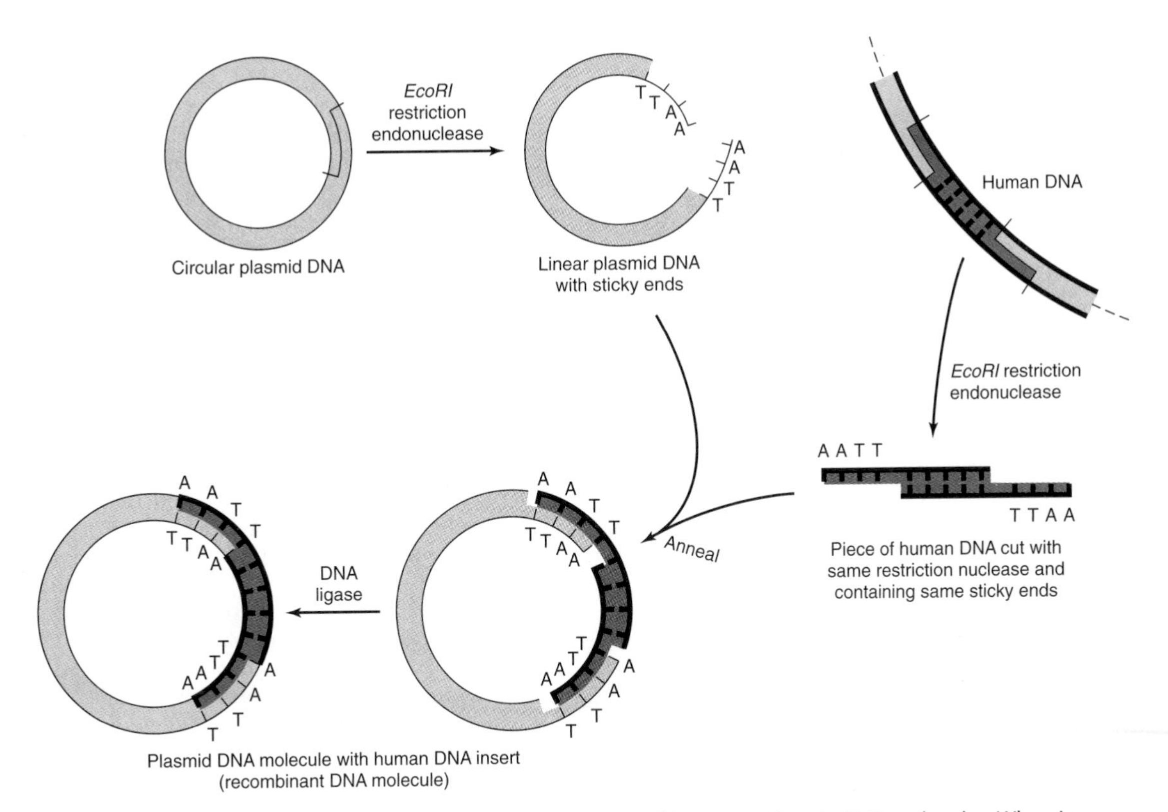


***Figure 40–3.*** Use of restriction nucleases to make new recombinant or chimeric DNA molecules. When inserted back into a bacterial cell (by the process called transformation), typically only a single plasmid is taken up by a single cell, and the plasmid DNA replicates not only itself but also the physically linked new DNA insert. Since recombining the sticky ends, as indicated, regenerates the same DNA sequence recognized by the original restriction enzyme, the cloned DNA insert can be cleanly cut back out of the recombinant plasmid circle with this endonuclease. If a mixture of all of the DNA pieces created by treatment of total human DNA with a single restriction nuclease is used as the source of human DNA, a million or so different types of recombinant DNA molecules can be obtained, each pure in its own bacterial clone. (Modified and reproduced, with permission, from Cohen SN: The manipulation of genes. Sci Am [July] 1975;233:34.)

cleavage sites for inserting the foreign DNA is available. Plasmids are smaller than the host chromosome and are therefore easily separated from the latter, and the desired plasmid-inserted DNA is readily removed by cutting the plasmid with the enzyme specific for the restriction site into which the original piece of DNA was inserted.

**Phages** usually have linear DNA molecules into which foreign DNA can be inserted at several restriction enzyme sites. The chimeric DNA is collected after the phage proceeds through its lytic cycle and produces mature, infective phage particles. A major advantage of phage vectors is that while plasmids accept DNA pieces about 6–10 kb long, phages can accept DNA fragments 10–20 kb long, a limitation imposed by the amount of DNA that can be packed into the phage head.

Larger fragments of DNA can be cloned in **cosmids,** which combine the best features of plasmids and phages. Cosmids are plasmids that contain the DNA sequences, so-called **cos sites,** required for packaging lambda DNA into the phage particle. These vectors grow in the plasmid form in bacteria, but since much of the unnecessary lambda DNA has been removed, more chimeric DNA can be packaged into the particle head. It is not unusual for cosmids to carry inserts of chimeric DNA that are 35–50 kb long. Even larger pieces of DNA can be incorporated into bacterial artificial chromosome (BAC), yeast artificial chromosome (YAC), or *E. coli* bacteriophage P1-based (PAC) vectors. These vectors will accept and propagate DNA inserts of several hundred kilobases or more and have largely re-

***Table 40–4.*** Cloning capacities of common cloning vectors.

| Vector | DNA Insert Size |
|---|---|
| Plasmid pBR322 | 0.01–10 kb |
| Lambda charon 4A | 10–20 kb |
| Cosmids | 35–50 kb |
| BAC, P1 | 50–250 kb |
| YAC | 500–3000 kb |

placed the plasmid, phage, and cosmid vectors for some cloning and gene mapping applications. A comparison of these vectors is shown in Table 40–4.

Because insertion of DNA into a functional region of the vector will interfere with the action of this region, care must be taken not to interrupt an essential function of the vector. This concept can be exploited, however, to provide a selection technique. For example, the common plasmid vector **pBR322** has both **tetracycline (tet)** and **ampicillin (amp)** resistance genes. A single *PstI* restriction enzyme site within the amp resistance gene is commonly used as the insertion site for a piece of foreign DNA. In addition to having sticky ends (Table 40–2 and Figure 40–2), the DNA inserted at this site disrupts the amp resistance gene and makes the bacterium carrying this plasmid amp-sensitive (Figure 40–4). Thus, the parental plasmid, which provides resistance to both antibiotics, can be readily separated from the chimeric plasmid, which is resistant only to tetracycline. YACs contain replication and segregation functions that work in both bacteria and yeast cells and therefore can be propagated in either organism.

In addition to the vectors described in Table 40–4 that are designed primarily for propagation in bacterial cells, vectors for mammalian cell propagation and insert gene (cDNA)/protein expression have also been developed. These vectors are all based upon various eukaryotic viruses that are composed of RNA or DNA genomes. Notable examples of such viral vectors are those utilizing adenoviral (DNA-based) and retroviral (RNA-based) genomes. Though somewhat limited in the size of DNA sequences that can be inserted, such mammalian viral cloning vectors make up for this shortcoming because they will efficiently infect a wide range of different cell types. For this reason, various mammalian viral vectors are being investigated for use in gene therapy experiments.

## A Library Is a Collection of Recombinant Clones

The combination of restriction enzymes and various cloning vectors allows the entire genome of an organism to be packed into a vector. A collection of these different recombinant clones is called a library. A **genomic library** is prepared from the total DNA of a cell line or tissue. A **cDNA library** comprises complementary DNA copies of the population of mRNAs in a tissue. Genomic DNA libraries are often prepared by performing partial digestion of total DNA with a restriction enzyme that cuts DNA frequently (eg, a four base cutter such as *TaqI*). The idea is to generate rather large fragments so that most genes will be left intact. The BAC, YAC, and P1 vectors are preferred since they can accept very large fragments of DNA and thus offer a better chance of isolating an intact gene on a single DNA fragment.

A vector in which the protein coded by the gene introduced by recombinant DNA technology is actually synthesized is known as an **expression vector.** Such vectors are now commonly used to detect specific cDNA molecules in libraries and to produce proteins by genetic engineering techniques. These vectors are specially constructed to contain very active inducible promoters, proper in-phase translation initiation codons, both transcription and translation termination signals, and appropriate protein processing signals, if needed. Some expression vectors even contain genes that code for protease inhibitors, so that the final yield of product is enhanced.

## Probes Search Libraries for Specific Genes or cDNA Molecules

A variety of molecules can be used to "probe" libraries in search of a specific gene or cDNA molecule or to define and quantitate DNA or RNA separated by electrophoresis through various gels. Probes are generally pieces of DNA or RNA labeled with a $^{32}$P-containing nucleotide—or fluorescently labeled nucleotides (more commonly now). Importantly, neither modification ($^{32}$P or fluorescent-label) affects the hybridization properties of the resulting labeled nucleic acid probes. The probe must recognize a complementary sequence to be effective. A cDNA synthesized from a specific mRNA can be used to screen either a cDNA library for a longer cDNA or a genomic library for a complementary sequence in the coding region of a gene. A popular technique for finding specific genes entails taking a short amino acid sequence and, employing the codon usage for that species (see Chapter 38), making an oligonucleotide probe that will detect the corresponding DNA fragment in a genomic library. If the sequences match exactly, probes 15–20 nucleotides long will hybridize. cDNA probes are used to detect DNA fragments on Southern blot transfers and to detect and quantitate RNA on Northern blot transfers. Specific antibodies can also be used as probes provided that the vector used synthesizes protein molecules that are recognized by them.

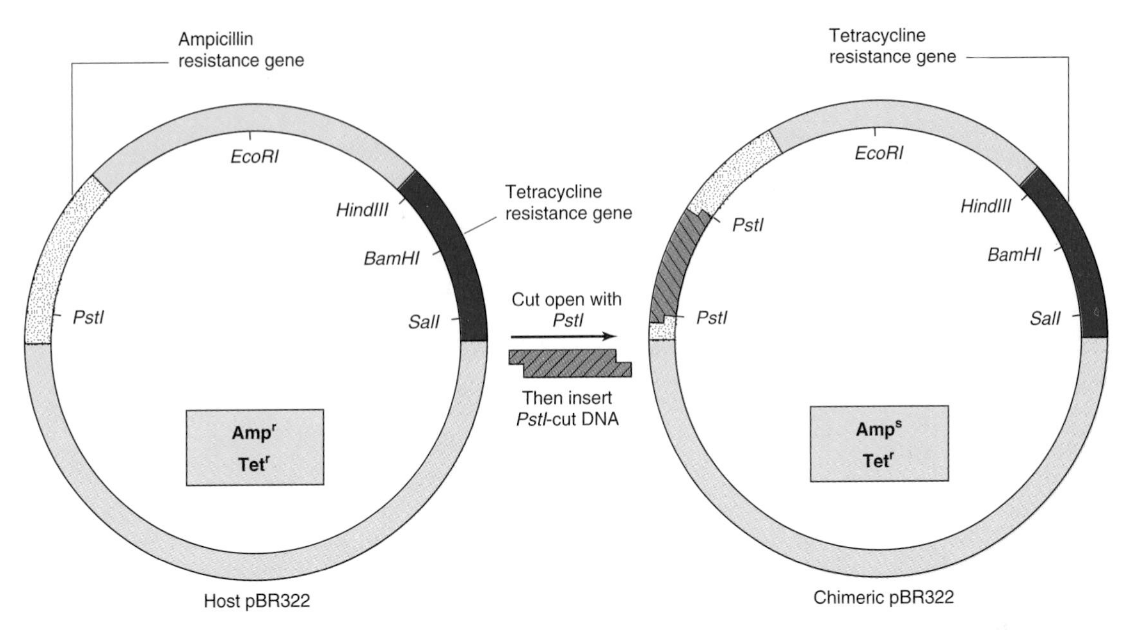


***Figure 40–4.*** A method of screening recombinants for inserted DNA fragments. Using the plasmid pBR322, a piece of DNA is inserted into the unique *PstI* site. This insertion disrupts the gene coding for a protein that provides ampicillin resistance to the host bacterium. Hence, the chimeric plasmid will no longer survive when plated on a substrate medium that contains this antibiotic. The differential sensitivity to tetracycline and ampicillin can therefore be used to distinguish clones of plasmid that contain an insert. A similar scheme relying upon production of an in-frame fusion of a newly inserted DNA producing a peptide fragment capable of complementing an inactive, deleted form of the enzyme β-galactosidase allows for blue-white colony formation on agar plates containing a dye hydrolyzable by β-galactoside. β-Galactosidase-positive colonies are blue.

## Blotting & Hybridization Techniques Allow Visualization of Specific Fragments

Visualization of a specific DNA or RNA fragment among the many thousands of "contaminating" molecules requires the convergence of a number of techniques, collectively termed **blot transfer.** Figure 40–5 illustrates the **Southern** (DNA), **Northern** (RNA), and **Western** (protein) blot transfer procedures. (The first is named for the person who devised the technique, and the other names began as laboratory jargon but are now accepted terms.) These procedures are useful in determining how many copies of a gene are in a given tissue or whether there are any gross alterations in a gene (deletions, insertions, or rearrangements). Occasionally, if a specific base is changed and a restriction site is altered, these procedures can detect a point mutation. The Northern and Western blot transfer techniques are used to size and quantitate specific RNA and protein molecules, respectively. A fourth hybridization technique, the Southwestern blot, examines protein•DNA interactions. Proteins are separated by electrophoresis, renatured, and analyzed for an interaction by hybridization with a specific labeled DNA probe.

**Colony** or **plaque hybridization** is the method by which specific clones are identified and purified. Bacteria are grown on colonies on an agar plate and overlaid with nitrocellulose filter paper. Cells from each colony stick to the filter and are permanently fixed thereto by heat, which with NaOH treatment also lyses the cells and denatures the DNA so that it will hybridize with the probe. A radioactive probe is added to the filter, and (after washing) the hybrid complex is localized by exposing the filter to x-ray film. By matching the spot on the autoradiograph to a colony, the latter can be picked from the plate. A similar strategy is used to identify fragments in phage libraries. Successive rounds of this procedure result in a clonal isolate (bacterial colony) or individual phage plaque.

All of the hybridization procedures discussed in this section depend on the specific base-pairing properties of complementary nucleic acid strands described above. Perfect matches hybridize readily and withstand high temperatures in the hybridization and washing reac-

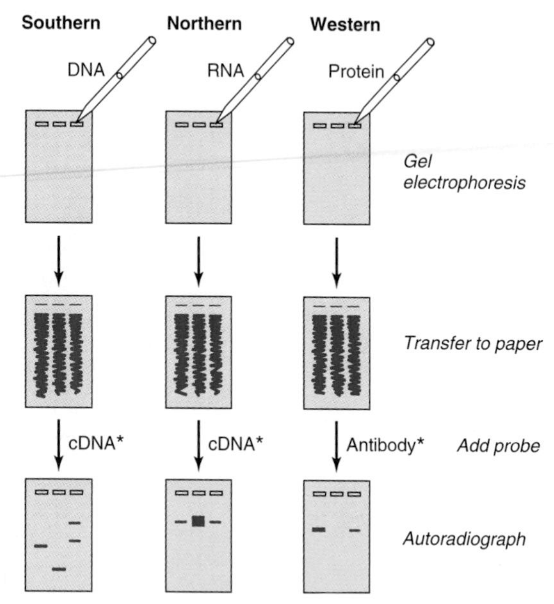


***Figure 40–5.*** The blot transfer procedure. In a Southern, or DNA, blot transfer, DNA isolated from a cell line or tissue is digested with one or more restriction enzymes. This mixture is pipetted into a well in an agarose or polyacrylamide gel and exposed to a direct electrical current. DNA, being negatively charged, migrates toward the anode; the smaller fragments move the most rapidly. After a suitable time, the DNA is denatured by exposure to mild alkali and transferred to nitrocellulose or nylon paper, in an exact replica of the pattern on the gel, by the blotting technique devised by Southern. The DNA is annealed to the paper by exposure to heat, and the paper is then exposed to the labeled cDNA probe, which hybridizes to complementary fragments on the filter. After thorough washing, the paper is exposed to x-ray film, which is developed to reveal several specific bands corresponding to the DNA fragment that recognized the sequences in the cDNA probe. The RNA, or Northern, blot is conceptually similar. RNA is subjected to electrophoresis before blot transfer. This requires some different steps from those of DNA transfer, primarily to ensure that the RNA remains intact, and is generally somewhat more difficult. In the protein, or Western, blot, proteins are electrophoresed and transferred to nitrocellulose and then probed with a specific antibody or other probe molecule. (Asterisks signify labeling, either radioactive or fluorescent.)

tions. Specific complexes also form in the presence of low salt concentrations. Less than perfect matches do not tolerate these **stringent conditions** (ie, elevated temperatures and low salt concentrations); thus, hybridization either never occurs or is disrupted during the washing step. Gene families, in which there is some degree of homology, can be detected by varying the stringency of the hybridization and washing steps. Cross-species comparisons of a given gene can also be made using this approach. Hybridization conditions capable of detecting just a single base pair mismatch between probe and target have been devised.

## Manual & Automatic Techniques Are Available to Determine the Sequence of DNA

The segments of specific DNA molecules obtained by recombinant DNA technology can be analyzed to determine their nucleotide sequence. This method depends upon having a large number of identical DNA molecules. This requirement can be satisfied by cloning the fragment of interest, using the techniques described above. The **manual enzymatic method (Sanger)** employs specific dideoxynucleotides that terminate DNA strand synthesis at specific nucleotides as the strand is synthesized on purified template nucleic acid. The reactions are adjusted so that a population of DNA fragments representing termination at every nucleotide is obtained. By having a radioactive label incorporated at the end opposite the termination site, one can separate the fragments according to size using polyacrylamide gel electrophoresis. An autoradiograph is made, and each of the fragments produces an image (band) on an x-ray film. These are read in order to give the DNA sequence (Figure 40–6). Another manual method, that of **Maxam and Gilbert,** employs **chemical methods** to cleave the DNA molecules where they contain the specific nucleotides. Techniques that do not require the use of radioisotopes are commonly employed in automated DNA sequencing. Most commonly employed is an automated procedure in which four different fluorescent labels—one representing each nucleotide—are used. Each emits a specific signal upon excitation by a laser beam, and this can be recorded by a computer.

## Oligonucleotide Synthesis Is Now Routine

The automated chemical synthesis of moderately long oligonucleotides (about 100 nucleotides) of precise sequence is now a routine laboratory procedure. Each synthetic cycle takes but a few minutes, so an entire molecule can be made by synthesizing relatively short segments that can then be ligated to one another. Oligonucleotides are now indispensable for DNA se-

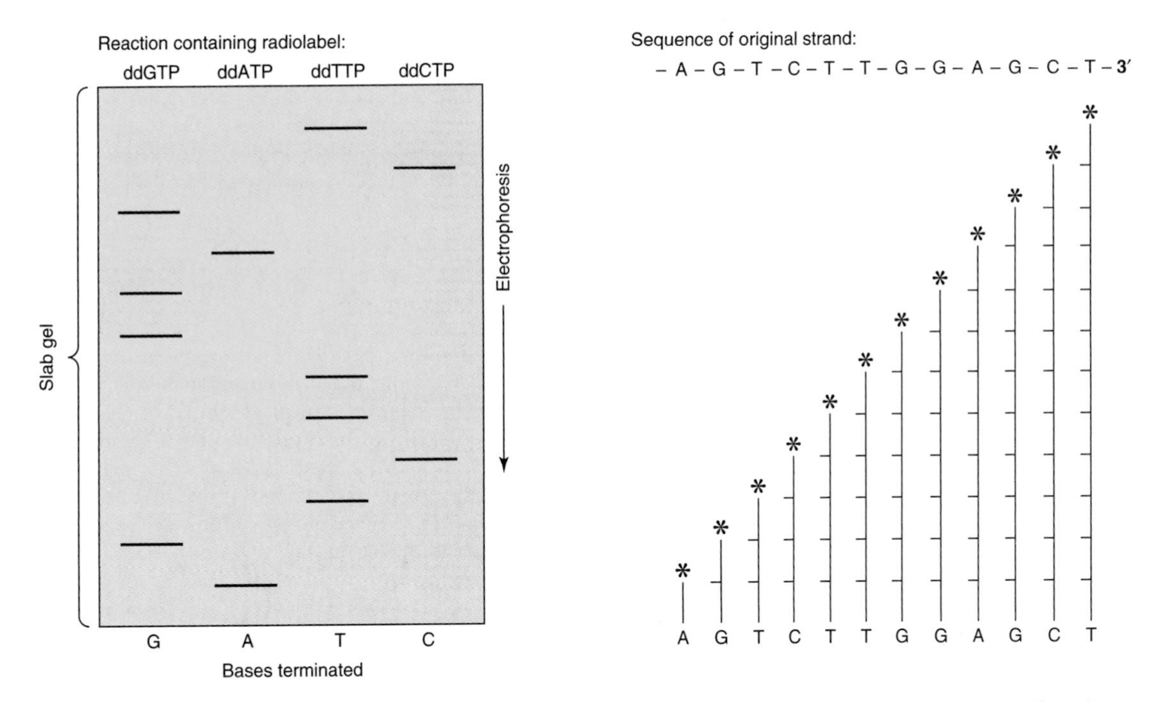


***Figure 40–6.*** Sequencing of DNA by the method devised by Sanger. The ladder-like arrays represent from bottom to top all of the successively longer fragments of the original DNA strand. Knowing which specific dideoxynucleotide reaction was conducted to produce each mixture of fragments, one can determine the sequence of nucleotides from the labeled end (asterisk) toward the unlabeled end by reading up the gel. Automated sequencing involves the reading of chemically modified deoxynucleotides. The base-pairing rules of Watson and Crick (A–T, G–C) dictate the sequence of the other (complementary) strand. (Asterisks signify radiolabeling.)

quencing, library screening, protein-DNA binding, DNA mobility shift assays, the polymerase chain reaction (see below), site-directed mutagenesis, and numerous other applications.

## The Polymerase Chain Reaction (PCR) Amplifies DNA Sequences

The polymerase chain reaction (PCR) is a method of amplifying a target sequence of DNA. PCR provides a sensitive, selective, and extremely rapid means of amplifying a desired sequence of DNA. Specificity is based on the use of two oligonucleotide primers that hybridize to complementary sequences on opposite strands of DNA and flank the target sequence (Figure 40–7). The DNA sample is first heated to separate the two strands; the primers are allowed to bind to the DNA; and each strand is copied by a DNA polymerase, starting at the primer site. The two DNA strands each serve as a template for the synthesis of new DNA from the two primers. Repeated cycles of heat denaturation, annealing of the primers to their complementary sequences, and extension of the annealed primers with DNA polymerase result in the exponential amplification of DNA segments of defined length. Early PCR reactions used an *E coli* DNA polymerase that was destroyed by each heat denaturation cycle. Substitution of a heat-stable DNA polymerase from *Thermus aquaticus* (or the corresponding DNA polymerase from other thermophilic bacteria), an organism that lives and replicates at 70–80 °C, obviates this problem and has made possible automation of the reaction, since the polymerase reactions can be run at 70 °C. This has also improved the specificity and the yield of DNA.

DNA sequences as short as 50–100 bp and as long as 10 kb can be amplified. Twenty cycles provide an amplification of $10^6$ and 30 cycles of $10^9$. The PCR allows the DNA in a single cell, hair follicle, or spermatozoon to be amplified and analyzed. Thus, the applications of PCR to forensic medicine are obvious. The PCR is also used (1) to detect infectious agents, especially latent viruses; (2) to make prenatal genetic diagnoses; (3) to detect allelic polymorphisms; (4) to establish precise tissue types for transplants; and (5) to study

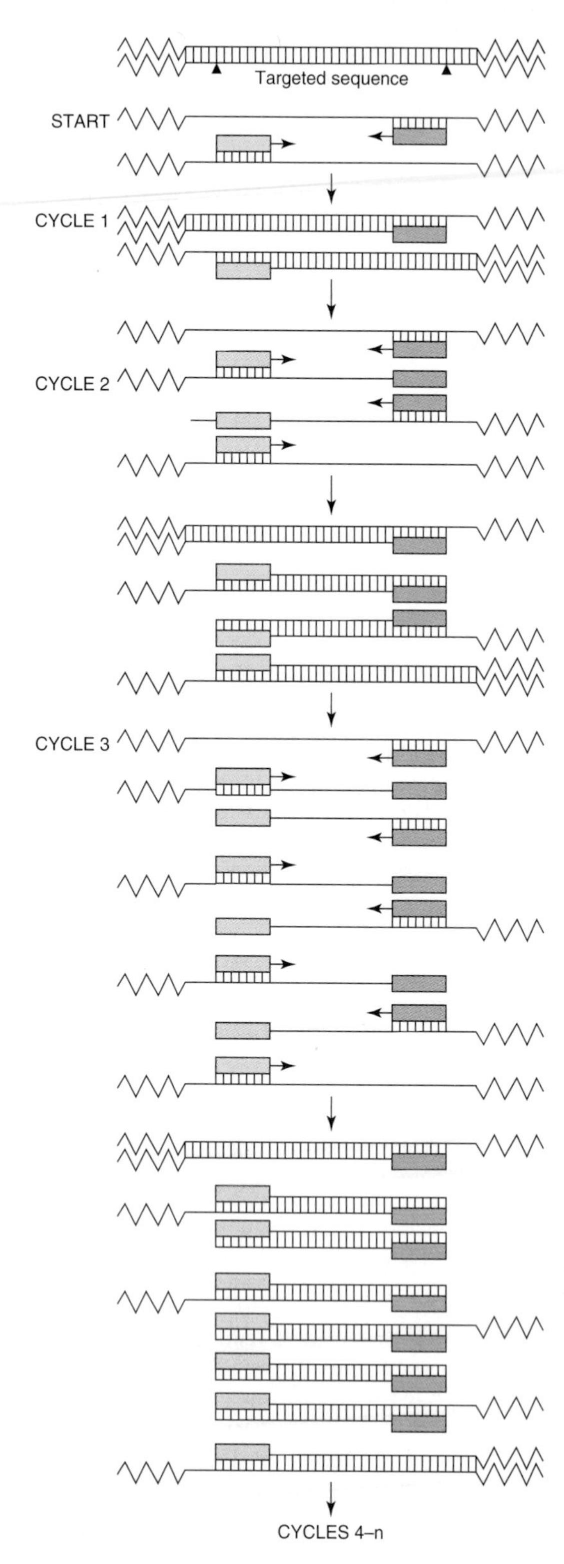


evolution, using DNA from archeological samples after RNA copying and mRNA quantitation by the so-called RT-PCR method (cDNA copies of mRNA generated by a retroviral reverse transcriptase). There are an equal number of applications of PCR to problems in basic science, and new uses are developed every year.

## PRACTICAL APPLICATIONS OF RECOMBINANT DNA TECHNOLOGY ARE NUMEROUS

The isolation of a specific gene from an entire genome requires a technique that will discriminate one part in a million. The identification of a regulatory region that may be only 10 bp in length requires a sensitivity of one part in $3 \times 10^8$; a disease such as sickle cell anemia is caused by a single base change, or one part in $3 \times 10^9$. Recombinant DNA technology is powerful enough to accomplish all these things.

### Gene Mapping Localizes Specific Genes to Distinct Chromosomes

Gene localizing thus can define a map of the human genome. This is already yielding useful information in the definition of human disease. Somatic cell hybridization and in situ hybridization are two techniques used to accomplish this. In **in situ hybridization,** the simpler and more direct procedure, a radioactive probe is added to a metaphase spread of chromosomes on a glass slide. The exact area of hybridization is localized by layering photographic emulsion over the slide and, after exposure, lining up the grains with some histologic identification of the chromosome. **Fluorescence in situ hybridization (FISH)** is a very sensitive technique that is also used for this purpose. This often places the gene at a location on a given band or region on the chromosome. Some of the human genes localized using these techniques are listed in Table 40–5. This table represents only a sampling, since thousands of genes have been mapped as a result of the recent sequencing of the

***Figure 40–7.*** The polymerase chain reaction is used to amplify specific gene sequences. Double-stranded DNA is heated to separate it into individual strands. These bind two distinct primers that are directed at specific sequences on opposite strands and that define the segment to be amplified. DNA polymerase extends the primers in each direction and synthesizes two strands complementary to the original two. This cycle is repeated several times, giving an amplified product of defined length and sequence. Note that the two primers are present in excess.

***Table 40–5.*** Localization of human genes.[1]

| Gene | Chromosome | Disease |
|---|---|---|
| Insulin | 11p15 | |
| Prolactin | 6p23-q12 | |
| Growth hormone | 17q21-qter | Growth hormone deficiency |
| α-Globin | 16p12-pter | α-Thalassemia |
| β-Globin | 11p12 | β-Thalassemia, sickle cell |
| Adenosine deaminase | 20q13-qter | Adenosine deaminase deficiency |
| Phenylalanine hydroxylase | 12q24 | Phenylketonuria |
| Hypoxanthine-guanine phosphoribosyltransferase | Xq26-q27 | Lesch-Nyhan syndrome |
| DNA segment G8 | 4p | Huntington's chorea |

[1]This table indicates the chromosomal location of several genes and the diseases associated with deficient or abnormal production of the gene products. The chromosome involved is indicated by the first number or letter. The other numbers and letters refer to precise localizations, as defined in McKusick VA: *Mendelian Inheritance in Man,* 6th ed. John Hopkins Univ Press, 1983.

human genome. Once the defect is localized to a region of DNA that has the characteristic structure of a gene (Figure 40–1), a synthetic gene can be constructed and expressed in an appropriate vector and its function can be assessed—or the putative peptide, deduced from the open reading frame in the coding region, can be synthesized. Antibodies directed against this peptide can be used to assess whether this peptide is expressed in normal persons and whether it is absent in those with the genetic syndrome.

## Proteins Can Be Produced for Research & Diagnosis

A practical goal of recombinant DNA research is the production of materials for biomedical application. This technology has two distinct merits: (1) It can supply large amounts of material that could not be obtained by conventional purification methods (eg, interferon, tissue plasminogen activating factor). (2) It can provide human material (eg, insulin, growth hormone). The advantages in both cases are obvious. Although the primary aim is to supply products—generally proteins—for treatment (insulin) and diagnosis (AIDS testing) of human and other animal diseases and for disease prevention (hepatitis B vaccine), there are other potential commercial applications, especially in agriculture. An example of the latter is the attempt to engineer plants that are more resistant to drought or temperature extremes, more efficient at fixing nitrogen, or that produce seeds containing the complete complement of essential amino acids (rice, wheat, corn, etc).

## Recombinant DNA Technology Is Used in the Molecular Analysis of Disease

### A. Normal Gene Variations

There is a normal variation of DNA sequence just as is true of more obvious aspects of human structure. Variations of DNA sequence, **polymorphisms,** occur approximately once in every 500 nucleotides, or about $10^7$ times per genome. There are without doubt deletions and insertions of DNA as well as single-base substitutions. In healthy people, these alterations obviously occur in noncoding regions of DNA or at sites that cause no change in function of the encoded protein. This heritable polymorphism of DNA structure can be associated with certain diseases within a large kindred and can be used to search for the specific gene involved, as is illustrated below. It can also be used in a variety of applications in forensic medicine.

### B. Gene Variations Causing Disease

Classic genetics taught that most genetic diseases were due to point mutations which resulted in an impaired protein. This may still be true, but if on reading the initial sections of this chapter one predicted that genetic disease could result from derangement of any of the steps illustrated in Figure 40–1, one would have made a proper assessment. This point is nicely illustrated by examination of the β-globin gene. This gene is located in a cluster on chromosome 11 (Figure 40–8), and an expanded version of the gene is illustrated in Figure 40–9. Defective production of β-globin results in a variety of diseases and is due to many

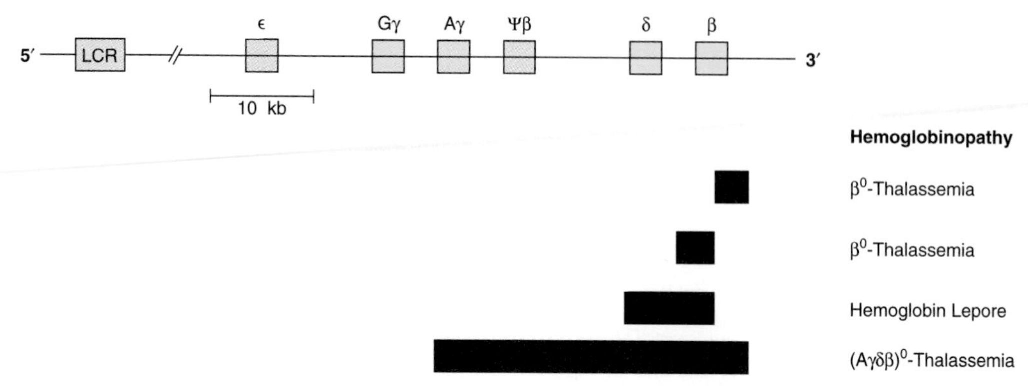


***Figure 40–8.*** Schematic representation of the β-globin gene cluster and of the lesions in some genetic disorders. The β-globin gene is located on chromosome 11 in close association with the two γ-globin genes and the δ-globin gene. The β-gene family is arranged in the order 5′-ε-Gγ-Aγ-ψβ-δ-β-3′. The ε locus is γ expressed in early embryonic life ($a_2\varepsilon_2$). The γ genes are expressed in fetal life, making fetal hemoglobin (HbF, $\alpha_2\gamma_2$). Adult hemoglobin consists of HbA ($\alpha_2\beta_2$) or $HbA_2$($\alpha_2\delta_2$). The ζβ is a pseudogene that has sequence homology with β but contains mutations that prevent its expression. A locus control region (LCR) located upstream (5′) from the ε gene controls the rate of transcription of the entire β-globin gene cluster. Deletions (solid bar) of the β locus cause β-thalassemia (deficiency or absence [$\beta^0$] of β-globin). A deletion of δ and β causes hemoglobin Lepore (only hemoglobin α is present). An inversion $(A\gamma\delta\beta)^0$ in this region (colored bar) disrupts gene function and also results in thalassemia (type III). Each type of thalassemia tends to be found in a certain group of people, eg, the $(A\gamma\delta\beta)^0$ deletion inversion occurs in persons from India. Many more deletions in this region have been mapped, and each causes some type of thalassemia.

different lesions in and around the β-globin gene (Table 40–6).

### C. Point Mutations

The classic example is **sickle cell disease,** which is caused by mutation of a single base out of the $3 \times 10^9$ in the genome, a T-to-A DNA substitution, which in turn results in an A-to-U change in the mRNA corresponding to the sixth codon of the β-globin gene. The altered codon specifies a different amino acid (valine rather than glutamic acid), and this causes a structural abnormality of the β-globin molecule. Other point mutations in and around the β-globin gene result in decreased production or, in some instances, no produc-

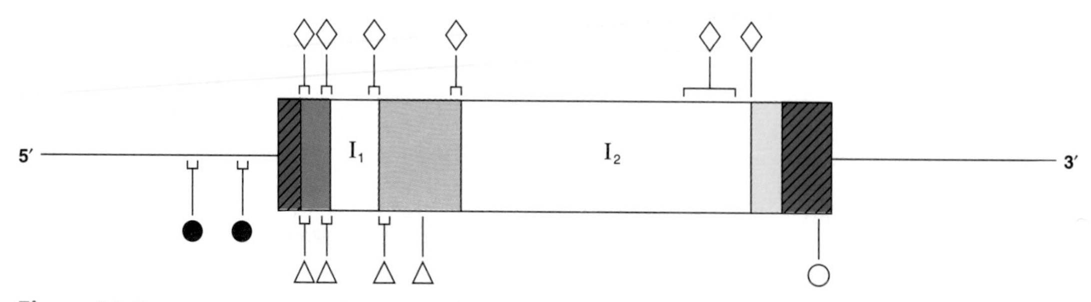


***Figure 40–9.*** Mutations in the β-globin gene causing β-thalassemia. The β-globin gene is shown in the 5′ to 3′ orientation. The cross-hatched areas indicate the 5′ and 3′ nontranslated regions. Reading from the 5′ to 3′ direction, the shaded areas are exons 1–3 and the clear spaces are introns 1 ($I_1$) and 2 ($I_2$). Mutations that affect transcription control (•) are located in the 5′ flanking-region DNA. Examples of nonsense mutations (△), mutations in RNA processing (◇), and RNA cleavage mutations (○) have been identified and are indicated. In some regions, many mutations have been found. These are indicated by the brackets.

***Table 40–6.*** Structural alterations of the β-globin gene.

| Alteration | Function Affected | Disease |
|---|---|---|
| Point mutations | Protein folding | Sickle cell disease |
| | Transcriptional control | β-Thalassemia |
| | Frameshift and nonsense mutations | β-Thalassemia |
| | RNA processing | β-Thalassemia |
| Deletion | mRNA production | $β^0$-Thalassemia<br>Hemoglobin Lepore |
| Rearrangement | mRNA production | β-Thalassemia type III |

tion of β-globin; β-thalassemia is the result of these mutations. (The thalassemias are characterized by defects in the synthesis of hemoglobin subunits, and so β-thalassemia results when there is insufficient production of β-globin.) Figure 40–9 illustrates that point mutations affecting each of the many processes involved in generating a normal mRNA (and therefore a normal protein) have been implicated as a cause of β-thalassemia.

### D. Deletions, Insertions, & Rearrangements of DNA

Studies of bacteria, viruses, yeasts, and fruit flies show that pieces of DNA can move from one place to another within a genome. The deletion of a critical piece of DNA, the rearrangement of DNA within a gene, or the insertion of a piece of DNA within a coding or regulatory region can all cause changes in gene expression resulting in disease. Again, a molecular analysis of β-thalassemia produces numerous examples of these processes—particularly deletions—as causes of disease (Figure 40–8). The globin gene clusters seem particularly prone to this lesion. Deletions in the α-globin cluster, located on chromosome 16, cause α-thalassemia. There is a strong ethnic association for many of these deletions, so that northern Europeans, Filipinos, blacks, and Mediterranean peoples have different lesions all resulting in the absence of hemoglobin A and α-thalassemia.

A similar analysis could be made for a number of other diseases. Point mutations are usually defined by sequencing the gene in question, though occasionally, if the mutation destroys or creates a restriction enzyme site, the technique of restriction fragment analysis can be used to pinpoint the lesion. Deletions or insertions of DNA larger than 50 bp can often be detected by the Southern blotting procedure.

### E. Pedigree Analysis

Sickle cell disease again provides an excellent example of how recombinant DNA technology can be applied to the study of human disease. The substitution of T for A in the template strand of DNA in the β-globin gene changes the sequence in the region that corresponds to the sixth codon from

```
↓
CCTGAGG      Coding strand
GGAC(T)CC    Template strand
       ↑
```

to

```
CCTGTGG      Coding strand
GGAC(A)CC    Template strand
```

and destroys a recognition site for the restriction enzyme *MstII* (CCTNAGG; denoted by the small vertical arrows; Table 40–2). Other *MstII* sites 5′ and 3′ from this site (Figure 40–10) are not affected and so will be cut. Therefore, incubation of DNA from normal (AA), heterozygous (AS), and homozygous (SS) individuals results in three different patterns on Southern blot transfer (Figure 40–10). This illustrates how a DNA pedigree can be established using the principles discussed in this chapter. Pedigree analysis has been applied to a number of genetic diseases and is most useful in those caused by deletions and insertions or the rarer instances in which a restriction endonuclease cleavage site is affected, as in the example cited in this paragraph. The analysis is facilitated by the PCR reaction, which can provide sufficient DNA for analysis from just a few nucleated red blood cells.

### F. Prenatal Diagnosis

If the genetic lesion is understood and a specific probe is available, prenatal diagnosis is possible. DNA from cells collected from as little as 10 mL of amniotic fluid (or by chorionic villus biopsy) can be analyzed by Southern blot transfer. A fetus with the restriction pattern AA in Figure 40–10 does not have sickle cell disease, nor is it a carrier. A fetus with the SS pattern will develop the disease. Probes are now available for this type of analysis of many genetic diseases.

### G. Restriction Fragment Length Polymorphism (RFLP)

The differences in DNA sequence cited above can result in variations of restriction sites and thus in the length of restriction fragments. An inherited difference in the pattern of restriction (eg, a DNA variation occurring in more than 1% of the general population) is known as a restriction fragment length polymorphism,

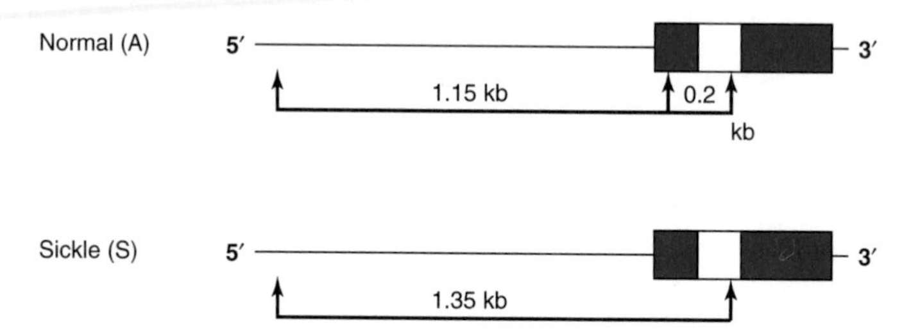


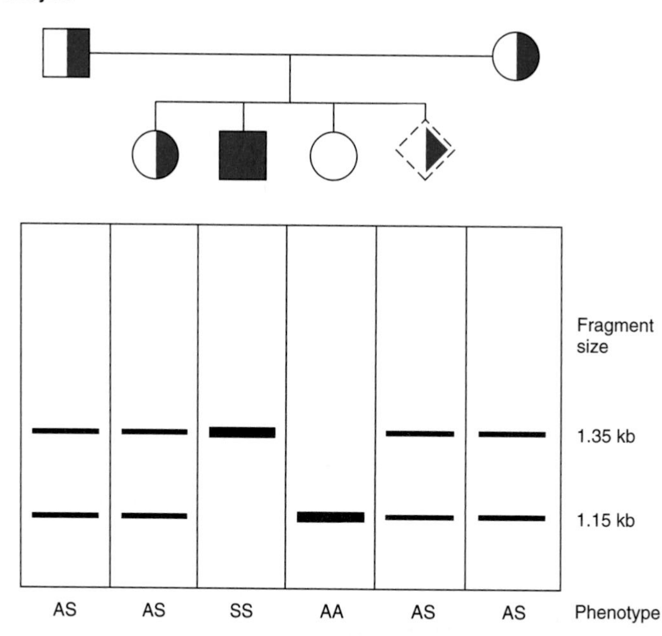


***Figure 40–10.*** Pedigree analysis of sickle cell disease. The top part of the figure **(A)** shows the first part of the β-globin gene and the *MstII* restriction enzyme sites in the normal (A) and sickle cell (S) β-globin genes. Digestion with the restriction enzyme *MstII* results in DNA fragments 1.15 kb and 0.2 kb long in normal individuals. The T-to-A change in individuals with sickle cell disease abolishes one of the three *MstII* sites around the β-globin gene; hence, a single restriction fragment 1.35 kb in length is generated in response to *MstII*. This size difference is easily detected on a Southern blot. (The 0.2-kb fragment would run off the gel in this illustration.) **(B)** Pedigree analysis shows three possibilities: AA = normal (open circle); AS = heterozygous (half-solid circles, half-solid square); SS = homozygous (solid square). This approach allows for prenatal diagnosis of sickle cell disease (dash-sided square with solid triangle on right).

or RFLP. An extensive RFLP map of the human genome has been constructed. This is proving useful in the human genome sequencing project and is an important component of the effort to understand various single-gene and multigenic diseases. RFLPs result from single-base changes (eg, sickle cell disease) or from deletions or insertions of DNA into a restriction fragment (eg, the thalassemias) and have proved to be useful diagnostic tools. They have been found at known gene loci and in sequences that have no known function; thus, RFLPs may disrupt the function of the gene or may have no biologic consequences.

RFLPs are inherited, and they segregate in a mendelian fashion. A major use of RFLPs (thousands are now known) is in the definition of inherited diseases in which the functional deficit is unknown. RFLPs can be used to establish linkage groups, which in turn, by the process of **chromosome walking,** will eventually define the disease locus. In chromosome walking (Figure 40–11), a fragment representing one end of a long piece of DNA is used to isolate another that overlaps but extends the first. The direction of extension is determined by restriction mapping, and the procedure is repeated sequentially until the desired sequence is obtained. The X chromosome-linked disorders are particularly amenable to this approach, since only a single allele is expressed. Hence, 20% of the defined RFLPs are on the X chromosome, and a reasonably complete linkage map of this chromosome exists. The gene for the X-linked disorder, Duchenne-type muscular dystrophy, was found using RFLPs. Likewise, the defect in Huntington's disease was localized to the terminal region of the short arm of chromosome 4, and the defect that causes polycystic kidney disease is linked to the α-globin locus on chromosome 16.

### H. Microsatellite DNA Polymorphisms

Short (2–6 bp), inherited, tandem repeat units of DNA occur about 50,000–100,000 times in the human genome (Chapter 36). Because they occur more frequently—and in view of the routine application of sensitive PCR methods—they are replacing RFLPs as the marker loci for various genome searches.

### I. RFLPs & VNTRs in Forensic Medicine

Variable numbers of tandemly repeated (VNTR) units are one common type of "insertion" that results in an RFLP. The VNTRs can be inherited, in which case they are useful in establishing genetic association with a disease in a family or kindred; or they can be unique to an individual and thus serve as a molecular fingerprint of that person.

### J. Gene Therapy

Diseases caused by deficiency of a gene product (Table 40–5) are amenable to replacement therapy. The strategy is to clone a gene (eg, the gene that codes for adenosine deaminase) into a vector that will readily be taken up and incorporated into the genome of a host cell. Bone marrow precursor cells are being investigated for this purpose because they presumably will resettle in the marrow and replicate there. The introduced gene would begin to direct the expression of its protein product, and this would correct the deficiency in the host cell.

### K. Transgenic Animals

The somatic cell gene replacement described above would obviously not be passed on to offspring. Other strategies to alter germ cell lines have been devised but have been tested only in experimental animals. A certain

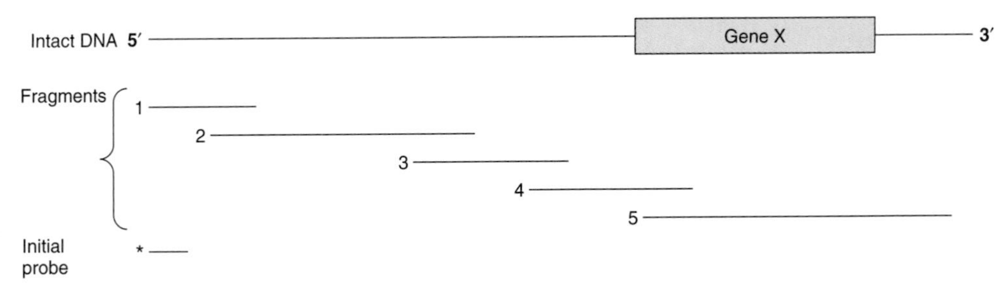


***Figure 40–11.*** The technique of chromosome walking. Gene X is to be isolated from a large piece of DNA. The exact location of this gene is not known, but a probe (*——) directed against a fragment of DNA (shown at the 5′ end in this representation) is available, as is a library containing a series of overlapping DNA fragments. For the sake of simplicity, only five of these are shown. The initial probe will hybridize only with clones containing fragment 1, which can then be isolated and used as a probe to detect fragment 2. This procedure is repeated until fragment 4 hybridizes with fragment 5, which contains the entire sequence of gene X.

percentage of genes injected into a fertilized mouse ovum will be incorporated into the genome and found in both somatic and germ cells. Hundreds of transgenic animals have been established, and these are useful for analysis of tissue-specific effects on gene expression and effects of overproduction of gene products (eg, those from the growth hormone gene or oncogenes) and in discovering genes involved in development—a process that heretofore has been difficult to study. The transgenic approach has recently been used to correct a genetic deficiency in mice. Fertilized ova obtained from mice with genetic hypogonadism were injected with DNA containing the coding sequence for the gonadotropin-releasing hormone (GnRH) precursor protein. This gene was expressed and regulated normally in the hypothalamus of a certain number of the resultant mice, and these animals were in all respects normal. Their offspring also showed no evidence of GnRH deficiency. This is, therefore, evidence of somatic cell expression of the transgene and of its maintenance in germ cells.

## Targeted Gene Disruption or Knockout

In transgenic animals, one is adding one or more copies of a gene to the genome, and there is no way to control where that gene eventually resides. A complementary—and much more difficult—approach involves the selective removal of a gene from the genome. Gene knockout animals (usually mice) are made by creating a mutation that totally disrupts the function of a gene. This is then used to replace one of the two genes in an embryonic stem cell that can be used to create a heterozygous transgenic animal. The mating of two such animals will, by mendelian genetics, result in a homozygous mutation in 25% of offspring. Several hundred strains of mice with knockouts of specific genes have been developed.

## RNA Transcript & Protein Profiling

The "-omic" revolution of the last several years has culminated in the determination of the nucleotide sequences of entire genomes, including those of budding and fission yeasts, various bacteria, the fruit fly, the worm *Caenorhabditis elegans,* the mouse and, most notably, humans. Additional genomes are being sequenced at an accelerating pace. The availability of all of this DNA sequence information, coupled with engineering advances, has lead to the development of several revolutionary methodologies, most of which are based upon **high-density microarray technology.** We now have the ability to deposit thousands of specific, known, definable DNA sequences (more typically now synthetic oligonucleotides) on a glass microscope-style slide in the space of a few square centimeters. By coupling such DNA microarrays with highly sensitive detection of hybridized fluorescently labeled nucleic acid probes derived from mRNA, investigators can rapidly and accurately generate profiles of gene expression (eg, specific cellular mRNA content) from cell and tissue samples as small as 1 gram or less. Thus entire **transcriptome information** (the entire collection of cellular mRNAs) for such cell or tissue sources can readily be obtained in only a few days. Transcriptome information allows one to predict the collection of proteins that might be expressed in a particular cell, tissue, or organ in normal and disease states based upon the mRNAs present in those cells. Complementing this high-throughput, transcript-profiling method is the recent development of high-sensitivity, high-throughput **mass spectrometry of complex protein samples.** Newer mass spectrometry methods allow one to identify hundreds to thousands of proteins in proteins extracted from very small numbers of cells (< 1 g). This critical information tells investigators which of the many mRNAs detected in transcript microarray mapping studies are actually translated into protein, generally the ultimate dictator of phenotype. Microarray techniques and mass spectrometric protein identification experiments both lead to the generation of huge amounts of data. Appropriate data management and interpretation of the deluge of information forthcoming from such studies has relied upon statistical methods; and this new technology, coupled with the flood of DNA sequence information, has led to the development of the field of **bioinformatics,** a new discipline whose goal is to help manage, analyze, and integrate this flood of biologically important information. Future work at the intersection of bioinformatics and transcript-protein profiling will revolutionize our understanding of biology and medicine.

## SUMMARY

- A variety of very sensitive techniques can now be applied to the isolation and characterization of genes and to the quantitation of gene products.
- In DNA cloning, a particular segment of DNA is removed from its normal environment using one of many restriction endonucleases. This is then ligated into one of several vectors in which the DNA segment can be amplified and produced in abundance.
- The cloned DNA can be used as a probe in one of several types of hybridization reactions to detect other related or adjacent pieces of DNA, or it can be used to quantitate gene products such as mRNA.
- Manipulation of the DNA to change its structure, so-called genetic engineering, is a key element in cloning (eg, the construction of chimeric molecules) and can

also be used to study the function of a certain fragment of DNA and to analyze how genes are regulated.

- Chimeric DNA molecules are introduced into cells to make transfected cells or into the fertilized oocyte to make transgenic animals.
- Techniques involving cloned DNA are used to locate genes to specific regions of chromosomes, to identify the genes responsible for diseases, to study how faulty gene regulation causes disease, to diagnose genetic diseases, and increasingly to treat genetic diseases.

## GLOSSARY

**ARS:** Autonomously replicating sequence; the origin of replication in yeast.

**Autoradiography:** The detection of radioactive molecules (eg, DNA, RNA, protein) by visualization of their effects on photographic film.

**Bacteriophage:** A virus that infects a bacterium.

**Blunt-ended DNA:** Two strands of a DNA duplex having ends that are flush with each other.

**cDNA:** A single-stranded DNA molecule that is complementary to an mRNA molecule and is synthesized from it by the action of reverse transcriptase.

**Chimeric molecule:** A molecule (eg, DNA, RNA, protein) containing sequences derived from two different species.

**Clone:** A large number of organisms, cells or molecules that are identical with a single parental organism cell or molecule.

**Cosmid:** A plasmid into which the DNA sequences from bacteriophage lambda that are necessary for the packaging of DNA (cos sites) have been inserted; this permits the plasmid DNA to be packaged in vitro.

**Endonuclease:** An enzyme that cleaves internal bonds in DNA or RNA.

**Excinuclease:** The excision nuclease involved in nucleotide exchange repair of DNA.

**Exon:** The sequence of a gene that is represented (expressed) as mRNA.

**Exonuclease:** An enzyme that cleaves nucleotides from either the 3′ or 5′ ends of DNA or RNA.

**Fingerprinting:** The use of RFLPs or repeat sequence DNA to establish a unique pattern of DNA fragments for an individual.

**Footprinting:** DNA with protein bound is resistant to digestion by DNase enzymes. When a sequencing reaction is performed using such DNA, a protected area, representing the "footprint" of the bound protein, will be detected.

**Hairpin:** A double-helical stretch formed by base pairing between neighboring complementary sequences of a single strand of DNA or RNA.

**Hybridization:** The specific reassociation of complementary strands of nucleic acids (DNA with DNA, DNA with RNA, or RNA with RNA).

**Insert:** An additional length of base pairs in DNA, generally introduced by the techniques of recombinant DNA technology.

**Intron:** The sequence of a gene that is transcribed but excised before translation.

**Library:** A collection of cloned fragments that represents the entire genome. Libraries may be either genomic DNA (in which both introns and exons are represented) or cDNA (in which only exons are represented).

**Ligation:** The enzyme-catalyzed joining in phosphodiester linkage of two stretches of DNA or RNA into one; the respective enzymes are DNA and RNA ligases.

**Lines:** Long interspersed repeat sequences.

**Microsatellite polymorphism:** Heterozygosity of a certain microsatellite repeat in an individual.

**Microsatellite repeat sequences:** Dispersed or group repeat sequences of 2–5 bp repeated up to 50 times. May occur at 50–100 thousand locations in the genome.

**Nick translation:** A technique for labeling DNA based on the ability of the DNA polymerase from *E coli* to degrade a strand of DNA that has been nicked and then to resynthesize the strand; if a radioactive nucleoside triphosphate is employed, the rebuilt strand becomes labeled and can be used as a radioactive probe.

**Northern blot:** A method for transferring RNA from an agarose gel to a nitrocellulose filter, on which the RNA can be detected by a suitable probe.

**Oligonucleotide:** A short, defined sequence of nucleotides joined together in the typical phosphodiester linkage.

**Ori:** The origin of DNA replication.

**PAC:** A high capacity (70–95 kb) cloning vector based upon the lytic *E. coli* bacteriophage P1 that replicates in bacteria as an extrachromosomal element.

**Palindrome:** A sequence of duplex DNA that is the same when the two strands are read in opposite directions.

**Plasmid:** A small, extrachromosomal, circular molecule of DNA that replicates independently of the host DNA.

**Polymerase chain reaction (PCR):** An enzymatic method for the repeated copying (and thus amplification) of the two strands of DNA that make up a particular gene sequence.

**Primosome:** The mobile complex of helicase and primase that is involved in DNA replication.

**Probe:** A molecule used to detect the presence of a specific fragment of DNA or RNA in, for instance, a bacterial colony that is formed from a genetic library or during analysis by blot transfer techniques; common probes are cDNA molecules, synthetic oligodeoxynucleotides of defined sequence, or antibodies to specific proteins.

**Proteome:** The entire collection of expressed proteins in an organism.

**Pseudogene:** An inactive segment of DNA arising by mutation of a parental active gene.

**Recombinant DNA:** The altered DNA that results from the insertion of a sequence of deoxynucleotides not previously present into an existing molecule of DNA by enzymatic or chemical means.

**Restriction enzyme:** An endodeoxynuclease that causes cleavage of both strands of DNA at highly specific sites dictated by the base sequence.

**Reverse transcription:** RNA-directed synthesis of DNA, catalyzed by reverse transcriptase.

**RT-PCR:** A method used to quantitate mRNA levels that relies upon a first step of cDNA copying of mRNAs prior to PCR amplification and quantitation.

**Signal:** The end product observed when a specific sequence of DNA or RNA is detected by autoradiography or some other method. Hybridization with a complementary radioactive polynucleotide (eg, by Southern or Northern blotting) is commonly used to generate the signal.

**Sines:** Short interspersed repeat sequences.

**SNP:** Single nucleotide polymorphism. Refers to the fact that single nucleotide genetic variation in genome sequence exists at discrete loci throughout the chromosomes. Measurement of allelic SNP differences is useful for gene mapping studies.

**snRNA:** Small nuclear RNA. This family of RNAs is best known for its role in mRNA processing.

**Southern blot:** A method for transferring DNA from an agarose gel to nitrocellulose filter, on which the DNA can be detected by a suitable probe (eg, complementary DNA or RNA).

**Southwestern blot:** A method for detecting protein-DNA interactions by applying a labeled DNA probe to a transfer membrane that contains a renatured protein.

**Spliceosome:** The macromolecular complex responsible for precursor mRNA splicing. The spliceosome consists of at least five small nuclear RNAs (snRNA; U1, U2, U4, U5, and U6) and many proteins.

**Splicing:** The removal of introns from RNA accompanied by the joining of its exons.

**Sticky-ended DNA:** Complementary single strands of DNA that protrude from opposite ends of a DNA duplex or from the ends of different duplex molecules (see also Blunt-ended DNA, above).

**Tandem:** Used to describe multiple copies of the same sequence (eg, DNA) that lie adjacent to one another.

**Terminal transferase:** An enzyme that adds nucleotides of one type (eg, deoxyadenonucleotidyl residues) to the 3′ end of DNA strands.

**Transcription:** Template DNA-directed synthesis of nucleic acids; typically DNA-directed synthesis of RNA.

**Transcriptome:** The entire collection of expressed mRNAs in an organism.

**Transgenic:** Describing the introduction of new DNA into germ cells by its injection into the nucleus of the ovum.

**Translation:** Synthesis of protein using mRNA as template.

**Vector:** A plasmid or bacteriophage into which foreign DNA can be introduced for the purposes of cloning.

**Western blot:** A method for transferring protein to a nitrocellulose filter, on which the protein can be detected by a suitable probe (eg, an antibody).

## REFERENCES


Lewin B: *Genes VII.* Oxford Univ Press, 1999.

Martin JB, Gusella JF: Huntington's disease: pathogenesis and management. N Engl J Med 1986:315:1267.

Sambrook J, Fritsch EF, Maniatis T: *Molecular Cloning: A Laboratory Manual.* Cold Spring Harbor Laboratory Press, 1989.

Spector DL, Goldman RD, Leinwand LA: *Cells: A Laboratory Manual.* Cold Spring Harbor Laboratory Press, 1998.

Watson JD et al: *Recombinant DNA,* 2nd ed. Scientific American Books. Freeman, 1992.

Weatherall DJ: *The New Genetics and Clinical Practice,* 3rd ed. Oxford Univ Press, 1991.

# SECTION V

# Biochemistry of Extracellular & Intracellular Communication

# Membranes: Structure & Function

# 41

*Robert K. Murray, MD, PhD, & Daryl K. Granner, MD*

## BIOMEDICAL IMPORTANCE

Membranes are highly viscous, plastic structures. Plasma membranes form closed compartments around cellular protoplasm to separate one cell from another and thus permit cellular individuality. The plasma membrane has **selective permeabilities** and acts as a barrier, thereby maintaining differences in composition between the inside and outside of the cell. The selective permeabilities are provided mainly by **channels** and **pumps** for ions and substrates. The plasma membrane also exchanges material with the extracellular environment by exocytosis and endocytosis, and there are special areas of membrane structure—the gap junctions—through which adjacent cells exchange material. In addition, the plasma membrane plays key roles in **cell-cell interactions** and in **transmembrane signaling.**

Membranes also form **specialized compartments** within the cell. Such intracellular membranes help shape many of the morphologically distinguishable structures (organelles), eg, mitochondria, ER, sarcoplasmic reticulum, Golgi complexes, secretory granules, lysosomes, and the nuclear membrane. Membranes localize **enzymes,** function as integral elements in **excitation-response coupling,** and provide sites of **energy transduction,** such as in photosynthesis and oxidative phosphorylation.

**Changes** in membrane structure (eg caused by ischemia) can affect water balance and ion flux and therefore every process within the cell. Specific deficiencies or alterations of certain membrane components lead to a variety of **diseases** (see Table 41–5). In short, normal cellular function depends on normal membranes.

## MAINTENANCE OF A NORMAL INTRA- & EXTRACELLULAR ENVIRONMENT IS FUNDAMENTAL TO LIFE

Life originated in an aqueous environment; enzyme reactions, cellular and subcellular processes, and so forth have therefore evolved to work in this milieu. Since mammals live in a gaseous environment, how is the aqueous state maintained? Membranes accomplish this by internalizing and compartmentalizing body water.

### The Body's Internal Water Is Compartmentalized

Water makes up about 60% of the lean body mass of the human body and is distributed in two large compartments.

#### A. Intracellular Fluid (ICF)

This compartment constitutes two-thirds of total body water and provides the environment for the cell (1) to make, store, and utilize energy; (2) to repair itself; (3) to replicate; and (4) to perform special functions.

#### B. Extracellular Fluid (ECF)

This compartment contains about one-third of total body water and is distributed between the plasma and interstitial compartments. The extracellular fluid is a delivery system. It brings to the cells nutrients (eg, glucose, fatty acids, amino acids), oxygen, various ions and trace minerals, and a variety of regulatory molecules (hormones) that coordinate the functions of widely separated cells. Extracellular fluid removes $CO_2$, waste products, and toxic or detoxified materials from the immediate cellular environment.

## The Ionic Compositions of Intracellular & Extracellular Fluids Differ Greatly

As illustrated in Table 41–1, the **internal environment** is rich in $K^+$ and $Mg^{2+}$, and phosphate is its major anion. **Extracellular fluid** is characterized by high $Na^+$ and $Ca^{2+}$ content, and $Cl^-$ is the major anion. Note also that the concentration of glucose is higher in extracellular fluid than in the cell, whereas the opposite is true for proteins. Why is there such a difference? It is thought that the primordial sea in which life originated was rich in $K^+$ and $Mg^{2+}$. It therefore follows that enzyme reactions and other biologic processes evolved to function best in that environment—hence the high concentration of these ions within cells. Cells were faced with strong selection pressure as the sea gradually changed to a composition rich in $Na^+$ and $Ca^{2+}$. Vast changes would have been required for evolution of a completely new set of biochemical and physiologic machinery; instead, as it happened, cells developed barriers—membranes with associated "pumps"—to maintain the internal microenvironment.

## MEMBRANES ARE COMPLEX STRUCTURES COMPOSED OF LIPIDS, PROTEINS, & CARBOHYDRATES

We shall mainly discuss the membranes present in eukaryotic cells, although many of the principles described also apply to the membranes of prokaryotes. The various cellular membranes have **different compositions**, as reflected in the ratio of protein to lipid (Figure 41–1). This is not surprising, given their divergent functions. Membranes are asymmetric sheet-like enclosed structures with distinct inner and outer surfaces.

***Table 41–1.*** Comparison of the mean concentrations of various molecules outside and inside a mammalian cell.

| Substance | Extracellular Fluid | Intracellular Fluid |
|---|---|---|
| $Na^+$ | 140 mmol/L | 10 mmol/L |
| $K^+$ | 4 mmol/L | 140 mmol/L |
| $Ca^{2+}$ (free) | 2.5 mmol/L | 0.1 μmol/L |
| $Mg^{2+}$ | 1.5 mmol/L | 30 mmol/L |
| $Cl^-$ | 100 mmol/L | 4 mmol/L |
| $HCO_3^-$ | 27 mmol/L | 10 mmol/L |
| $PO_4^{3-}$ | 2 mmol/L | 60 mmol/L |
| Glucose | 5.5 mmol/L | 0–1 mmol/L |
| Protein | 2 g/dL | 16 g/dL |

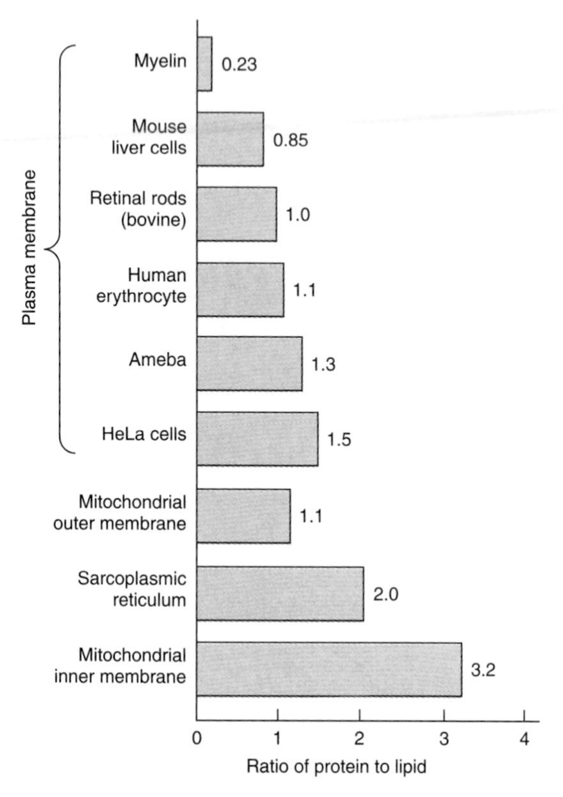


***Figure 41–1.*** Ratio of protein to lipid in different membranes. Proteins equal or exceed the quantity of lipid in nearly all membranes. The outstanding exception is myelin, an electrical insulator found on many nerve fibers.

These sheet-like structures are **noncovalent assemblies** that are thermodynamically stable and metabolically active. Numerous proteins are located in membranes, where they carry out specific functions of the organelle, the cell, or the organism.

## The Major Lipids in Mammalian Membranes Are Phospholipids, Glycosphingolipids, & Cholesterol

### A. Phospholipids

Of the two major phospholipid classes present in membranes, **phosphoglycerides** are the more common and consist of a glycerol backbone to which are attached two fatty acids in ester linkage and a phosphorylated alcohol (Figure 41–2). The fatty acid constituents are usually even-numbered carbon molecules, most commonly containing 16 or 18 carbons. They are unbranched and can be saturated or unsaturated. The simplest phosphoglyceride is phosphatidic acid, which is

Fatty acids

```
      O
      ||
R1—C—O—1CH2
            |
R2—C—O—2CH          O⁻
    ||       |         |
    O      3CH2—O—P—O—R3
                       ||
                       O
      Glycerol      Alcohol
```

***Figure 41–2.*** A phosphoglyceride showing the fatty acids ($R_1$ and $R_2$), glycerol, and phosphorylated alcohol components. In phosphatidic acid, $R_3$ is hydrogen.

1,2-diacylglycerol 3-phosphate, a key intermediate in the formation of all other phosphoglycerides (Chapter 24). In other phosphoglycerides, the 3-phosphate is esterified to an alcohol such as ethanolamine, choline, serine, glycerol, or inositol (Chapter 14).

The second major class of phospholipids is composed of **sphingomyelin,** which contains a sphingosine backbone rather than glycerol. A fatty acid is attached by an amide linkage to the amino group of sphingosine, forming ceramide. The primary hydroxyl group of sphingosine is esterified to phosphorylcholine. Sphingomyelin, as the name implies, is prominent in myelin sheaths.

The amounts and fatty acid compositions of the various phospholipids vary among the different cellular membranes.

### B. Glycosphingolipids

The glycosphingolipids (GSLs) are sugar-containing lipids built on a backbone of ceramide; they include galactosyl- and glucosylceramide (cerebrosides) and the gangliosides. Their structures are described in Chapter 14. They are mainly located in the plasma membranes of cells.

### C. Sterols

The most common sterol in membranes is **cholesterol** (Chapter 14), which resides mainly in the plasma membranes of mammalian cells but can also be found in lesser quantities in mitochondria, Golgi complexes, and nuclear membranes. Cholesterol intercalates among the phospholipids of the membrane, with its hydroxyl group at the aqueous interface and the remainder of the molecule within the leaflet. Its effect on the fluidity of membranes is discussed subsequently.

All of the above lipids can be separated from one another by techniques such as column, thin layer, and gas-liquid chromatography and their structures established by mass spectrometry.

Each eukaryotic cell membrane has a somewhat different lipid composition, though phospholipids are the major class in all.

## Membrane Lipids Are Amphipathic

All major lipids in membranes contain both hydrophobic and hydrophilic regions and are therefore termed **"amphipathic."** Membranes themselves are thus amphipathic. If the hydrophobic regions were separated from the rest of the molecule, it would be insoluble in water but soluble in oil. Conversely, if the hydrophilic region were separated from the rest of the molecule, it would be insoluble in oil but soluble in water. The amphipathic nature of a phospholipid is represented in Figure 41–3. Thus, the polar head groups of the phospholipids and the hydroxyl group of cholesterol interface with the aqueous environment; a similar situation applies to the sugar moieties of the GSLs (see below).

**Saturated fatty acids** have straight tails, whereas **unsaturated fatty acids,** which generally exist in the cis form in membranes, make kinked tails (Figure 41–3). As more kinks are inserted in the tails, the membrane becomes less tightly packed and therefore more fluid. **Detergents** are amphipathic molecules that are important in biochemistry as well as in the household. The molecular structure of a detergent is not unlike that of a phospholipid. Certain detergents are widely used to solubilize membrane proteins as a first step in their purification. The hydrophobic end of the detergent binds to

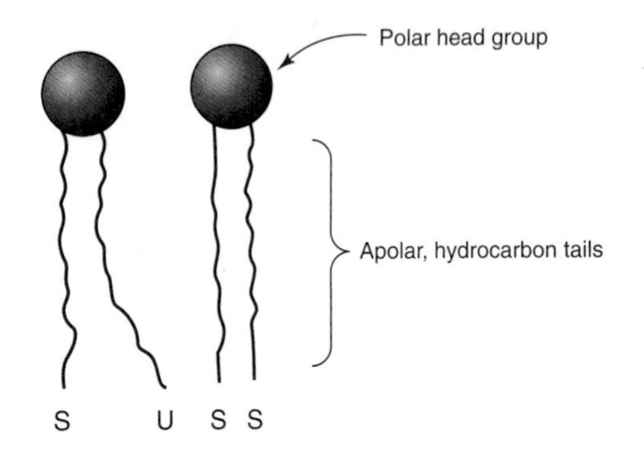


***Figure 41–3.*** Diagrammatic representation of a phospholipid or other membrane lipid. The polar head group is hydrophilic, and the hydrocarbon tails are hydrophobic or lipophilic. The fatty acids in the tails are saturated (S) or unsaturated (U); the former are usually attached to carbon 1 of glycerol and the latter to carbon 2. Note the kink in the tail of the unsaturated fatty acid (U), which is important in conferring increased membrane fluidity.

hydrophobic regions of the proteins, displacing most of their bound lipids. The polar end of the detergent is free, bringing the proteins into solution as detergent-protein complexes, usually also containing some residual lipids.

## Membrane Lipids Form Bilayers

The amphipathic character of phospholipids suggests that the two regions of the molecule have incompatible solubilities; however, in a solvent such as water, phospholipids organize themselves into a form that thermodynamically serves the solubility requirements of both regions. A **micelle** (Figure 41–4) is such a structure; the hydrophobic regions are shielded from water, while the hydrophilic polar groups are immersed in the aqueous environment. However, micelles are usually relatively small in size (eg, approximately 200 nm) and thus are limited in their potential to form membranes.

As was recognized in 1925 by Gorter and Grendel, a **bimolecular layer,** or **lipid bilayer,** can also satisfy the thermodynamic requirements of amphipathic molecules in an aqueous environment. **Bilayers,** not micelles, are indeed the key structures in biologic membranes. A bilayer exists as a sheet in which the hydrophobic regions of the phospholipids are protected from the aqueous environment, while the hydrophilic regions are immersed in water (Figure 41–5). Only the ends or edges of the bilayer sheet are exposed to an unfavorable environment, but even these exposed edges can be eliminated by folding the sheet back upon itself to form an enclosed vesicle with no edges. A bilayer can extend over relatively large distances (eg, 1 mm). The closed bilayer provides one of the most essential properties of membranes. It is impermeable to most water-soluble molecules, since they would be insoluble in the hydrophobic core of the bilayer.

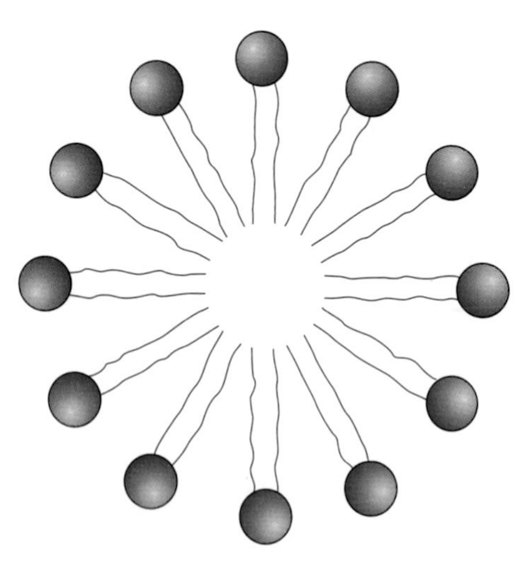

***Figure 41–4.*** Diagrammatic cross-section of a micelle. The polar head groups are bathed in water, whereas the hydrophobic hydrocarbon tails are surrounded by other hydrocarbons and thereby protected from water. Micelles are relatively small (compared with lipid bilayers) spherical structures.

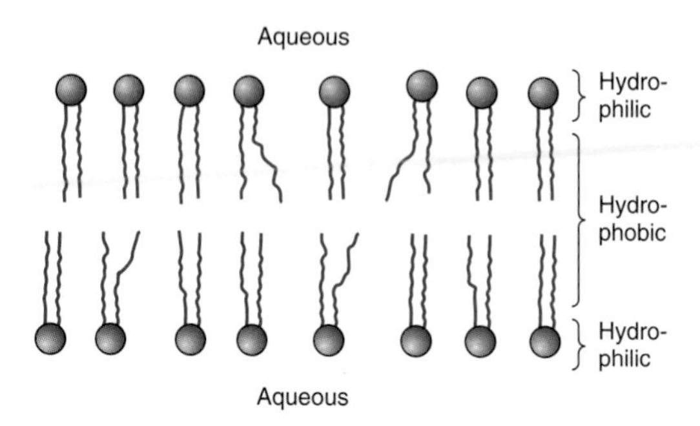


***Figure 41–5.*** Diagram of a section of a bilayer membrane formed from phospholipid molecules. The unsaturated fatty acid tails are kinked and lead to more spacing between the polar head groups, hence to more room for movement. This in turn results in increased membrane fluidity. (Slightly modified and reproduced, with permission, from Stryer L: Biochemistry, 2nd ed. Freeman, 1981.)

Lipid bilayers are formed by **self-assembly,** driven by the **hydrophobic effect.** When lipid molecules come together in a bilayer, the entropy of the surrounding solvent molecules increases.

Two questions arise from consideration of the above. First, how many biologic materials are **lipid-soluble** and can therefore readily enter the cell? Gases such as oxygen, $CO_2$, and nitrogen—small molecules with little interaction with solvents—readily diffuse through the hydrophobic regions of the membrane. The permeability coefficients of several ions and of a number of other molecules in a lipid bilayer are shown in Figure 41–6. The three electrolytes shown ($Na^+$, $K^+$, and $Cl^-$) cross the bilayer much more slowly than water. In general, the permeability coefficients of small molecules in a lipid bilayer correlate with their solubilities in nonpolar solvents. For instance, steroids more readily traverse the lipid bilayer compared with electrolytes. The high permeability coefficient of water itself is surprising but is partly explained by its small size and relative lack of charge.

The second question concerns **molecules that are not lipid-soluble:** How are the transmembrane concentration gradients for non-lipid-soluble molecules maintained? The answer is that membranes contain proteins,

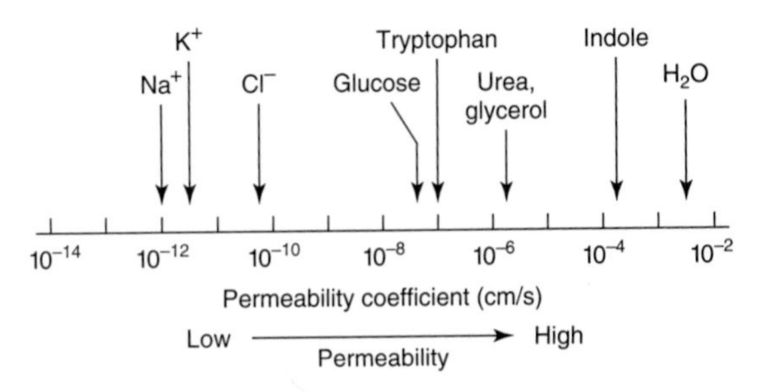


***Figure 41–6.*** Permeability coefficients of water, some ions, and other small molecules in lipid bilayer membranes. Molecules that move rapidly through a given membrane are said to have a high permeability coefficient. (Slightly modified and reproduced, with permission, from Stryer L: *Biochemistry,* 2nd ed. Freeman, 1981.)

and proteins are also amphipathic molecules that can be inserted into the correspondingly amphipathic lipid bilayer. Proteins form channels for the movement of ions and small molecules and serve as transporters for larger molecules that otherwise could not pass the bilayer. These processes are described below.

## Membrane Proteins Are Associated With the Lipid Bilayer

Membrane **phospholipids** act as a solvent for membrane proteins, creating an environment in which the latter can function. Of the 20 amino acids contributing to the primary structure of **proteins,** the functional groups attached to the α carbon are strongly hydrophobic in six, weakly hydrophobic in a few, and hydrophilic in the remainder. As described in Chapter 5, the α-helical structure of proteins minimizes the hydrophilic character of the peptide bonds themselves. Thus, proteins can be amphipathic and form an integral part of the membrane by having hydrophilic regions protruding at the inside and outside faces of the membrane but connected by a hydrophobic region traversing the hydrophobic core of the bilayer. In fact, those portions of membrane proteins that traverse membranes do contain substantial numbers of hydrophobic amino acids and almost invariably have either a high α-helical or β-pleated sheet content. For many membranes, a stretch of approximately 20 amino acids in an α helix will span the bilayer.

It is possible to calculate whether a particular sequence of amino acids present in a protein is consistent with a **transmembrane location.** This can be done by consulting a table that lists the hydrophobicities of each of the 20 common amino acids and the free energy values for their transfer from the interior of a membrane to water. Hydrophobic amino acids have positive values; polar amino acids have negative values. The total free energy values for transferring successive sequences of 20 amino acids in the protein are plotted, yielding a so-called **hydropathy plot.** Values of over 20 $kcal \cdot mol^{-1}$ are consistent with—but do not prove—a transmembrane location.

Another aspect of the interaction of lipids and proteins is that some proteins are anchored to one leaflet or another of the bilayer by **covalent linkages to certain lipids.** Palmitate and myristate are fatty acids involved in such linkages to specific proteins. A number of other proteins (see Chapter 47) are linked to glycophosphatidylinositol (GPI) structures.

## Different Membranes Have Different Protein Compositions

The **number of different proteins** in a membrane varies from less than a dozen in the sarcoplasmic reticulum to over 100 in the plasma membrane. Most membrane proteins can be separated from one another using sodium dodecyl sulfate polyacrylamide gel electrophoresis (SDS-PAGE), a technique that has revolutionized their study. In the absence of SDS, few membrane proteins would remain soluble during electrophoresis. Proteins are the **major functional molecules** of membranes and consist of enzymes, pumps and channels, structural components, antigens (eg, for histocompatibility), and receptors for various molecules. Because every membrane possesses a different complement of proteins, there is no such thing as a typical membrane structure. The enzymatic properties of several different membranes are shown in Table 41–2.

## Membranes Are Dynamic Structures

Membranes and their components are **dynamic structures.** The lipids and proteins in membranes undergo turnover there just as they do in other compartments of the cell. Different lipids have different turnover rates, and the turnover rates of individual species of membrane proteins may vary widely. The membrane itself can turn over even more rapidly than any of its constituents. This is discussed in more detail in the section on endocytosis.

## Membranes Are Asymmetric Structures

This asymmetry can be partially attributed to the irregular distribution of proteins within the membranes. An **inside-outside asymmetry** is also provided by the external location of the carbohydrates attached to membrane proteins. In addition, specific enzymes are lo-

***Table 41–2.*** Enzymatic markers of different membranes.[1]

| Membrane | Enzyme |
|---|---|
| Plasma | 5′-Nucleotidase<br>Adenylyl cyclase<br>$Na^+$-$K^+$ ATPase |
| Endoplasmic reticulum | Glucose-6-phosphatase |
| Golgi apparatus | |
| Cis | GlcNAc transferase I |
| Medial | Golgi mannosidase II |
| Trans | Galactosyl transferase |
| TGN | Sialyl transferase |
| Inner mitochondrial membrane | ATP synthase |

[1]Membranes contain many proteins, some of which have enzymatic activity. Some of these enzymes are located only in certain membranes and can therefore be used as markers to follow the purification of these membranes.
TGN, trans golgi network.

cated exclusively on the outside or inside of membranes, as in the mitochondrial and plasma membranes.

There are **regional asymmetries** in membranes. Some, such as occur at the villous borders of mucosal cells, are almost macroscopically visible. Others, such as those at gap junctions, tight junctions, and synapses, occupy much smaller regions of the membrane and generate correspondingly smaller local asymmetries.

There is also inside-outside (transverse) asymmetry of the **phospholipids.** The **choline-containing phospholipids** (phosphatidylcholine and sphingomyelin) are located mainly in the outer molecular layer; the **aminophospholipids** (phosphatidylserine and phosphatidylethanolamine) are preferentially located in the inner leaflet. Obviously, if this asymmetry is to exist at all, there must be limited transverse mobility (flip-flop) of the membrane phospholipids. In fact, phospholipids in synthetic bilayers exhibit an **extraordinarily slow rate of flip-flop;** the half-life of the asymmetry can be measured in several weeks. However, when certain membrane proteins such as the erythrocyte protein glycophorin are inserted artificially into synthetic bilayers, the frequency of phospholipid flip-flop may increase as much as 100-fold.

The mechanisms involved in the establishment of lipid asymmetry are not well understood. The enzymes involved in the synthesis of phospholipids are located on the cytoplasmic side of microsomal membrane vesicles. Translocases **(flippases)** exist that transfer certain phospholipids (eg, phosphatidylcholine) from the inner to the outer leaflet. Specific **proteins that preferentially bind** individual phospholipids also appear to be present in the two leaflets, contributing to the asymmetric distribution of these lipid molecules. In addition, phospholipid exchange proteins recognize specific phospholipids and transfer them from one membrane (eg, the endoplasmic reticulum [ER]) to others (eg, mitochondrial and peroxisomal). There is further asymmetry with regard to **GSLs** and also **glycoproteins;** the sugar moieties of these molecules all protrude outward from the plasma membrane and are absent from its inner face.

## Membranes Contain Integral & Peripheral Proteins (Figure 41–7)

It is useful to classify membrane proteins into two types: **integral** and **peripheral.** Most membrane proteins fall into the integral class, meaning that they interact extensively with the phospholipids and require the use of detergents for their solubilization. Also, they generally span the bilayer. Integral proteins are usually globular and are themselves amphipathic. They consist of two hydrophilic ends separated by an intervening hydrophobic region that traverses the hydrophobic core of the bilayer. As the structures of integral membrane proteins were being elucidated, it became apparent that certain ones (eg, transporter molecules, various receptors, and G proteins) span the bilayer many times (see Figure 46–5). Integral proteins are also asymmetrically distributed across the membrane bilayer. This asymmetric orientation is conferred at the time of their insertion in the lipid bilayer. The hydrophilic external region of an amphipathic protein, which is synthesized on polyribosomes, must traverse the hydrophobic core of its target membrane and eventually be found on the outside of that membrane. The molecular mechanisms involved in insertion of proteins into membranes and the topic of membrane assembly are discussed in Chapter 46.

**Peripheral proteins** do not interact directly with the phospholipids in the bilayer and thus do not require use of detergents for their release. They are weakly bound to the hydrophilic regions of specific integral proteins and can be released from them by treatment with salt solutions of high ionic strength. For example, ankyrin, a peripheral protein, is bound to the integral protein "band 3" of erythrocyte membrane. Spectrin, a cytoskeletal structure within the erythrocyte, is in turn bound to ankyrin and thereby plays an important role in maintenance of the biconcave shape of the erythrocyte. Many hormone receptor molecules are integral proteins, and the specific polypeptide hormones that bind to these receptor molecules may therefore be considered peripheral proteins. Peripheral proteins, such as polypeptide hormones, may help organize the distribu-

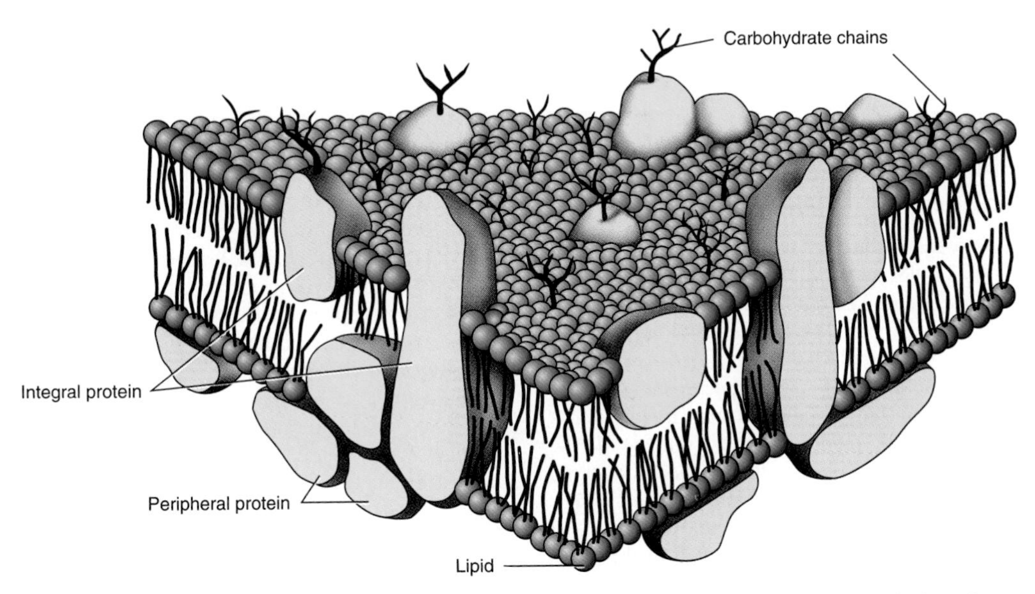


***Figure 41–7.*** The fluid mosaic model of membrane structure. The membrane consists of a bimolecular lipid layer with proteins inserted in it or bound to either surface. Integral membrane proteins are firmly embedded in the lipid layers. Some of these proteins completely span the bilayer and are called transmembrane proteins, while others are embedded in either the outer or inner leaflet of the lipid bilayer. Loosely bound to the outer or inner surface of the membrane are the peripheral proteins. Many of the proteins and lipids have externally exposed oligosaccharide chains. (Reproduced, with permission, from Junqueira LC, Carneiro J: *Basic Histology: Text & Atlas,* 10th ed. McGraw-Hill, 2003.)

tion of integral proteins, such as their receptors, within the plane of the bilayer (see below).

## ARTIFICIAL MEMBRANES MODEL MEMBRANE FUNCTION

Artificial membrane systems can be prepared by appropriate techniques. These systems generally consist of mixtures of one or more phospholipids of natural or synthetic origin that can be treated (eg, by using mild sonication) to form spherical vesicles in which the lipids form a bilayer. Such vesicles, surrounded by a lipid bilayer, are termed **liposomes.**

Some of the advantages and uses of artificial membrane systems in the study of membrane function can be briefly explained.

(1) The lipid content of the membranes can be varied, allowing systematic examination of the effects of varying lipid composition on certain functions. For instance, vesicles can be made that are composed solely of phosphatidylcholine or, alternatively, of known mixtures of different phospholipids, glycolipids, and cholesterol. The fatty acid moieties of the lipids used can also be varied by employing synthetic lipids of known composition to permit systematic examination of the effects of fatty acid composition on certain membrane functions (eg, transport).

(2) Purified membrane proteins or enzymes can be incorporated into these vesicles in order to assess what factors (eg, specific lipids or ancillary proteins) the proteins require to reconstitute their function. Investigations of purified proteins, eg, the $Ca^{2+}$ ATPase of the sarcoplasmic reticulum, have in certain cases suggested that only a single protein and a single lipid are required to reconstitute an ion pump.

(3) The environment of these systems can be rigidly controlled and systematically varied (eg, ion concentrations). The systems can also be exposed to known ligands if, for example, the liposomes contain specific receptor proteins.

(4) When liposomes are formed, they can be made to entrap certain compounds inside themselves, eg, drugs and isolated genes. There is interest in using liposomes to distribute drugs to certain tissues, and if components (eg, antibodies to certain cell surface molecules) could be incorporated into liposomes so that they would be targeted to specific tissues or tumors, the therapeutic impact would be considerable. DNA entrapped inside liposomes appears to be less sensitive to

attack by nucleases; this approach may prove useful in attempts at gene therapy.

## THE FLUID MOSAIC MODEL OF MEMBRANE STRUCTURE IS WIDELY ACCEPTED

The **fluid mosaic model** of membrane structure proposed in 1972 by Singer and Nicolson (Figure 41–7) is now widely accepted. The model is often likened to icebergs (membrane proteins) floating in a sea of predominantly phospholipid molecules. Early evidence for the model was the finding that certain species-specific integral proteins (detected by fluorescent labeling techniques) rapidly and randomly redistributed in the plasma membrane of an interspecies hybrid cell formed by the artificially induced fusion of two different parent cells. It has subsequently been demonstrated that phospholipids also undergo rapid redistribution in the plane of the membrane. This diffusion within the plane of the membrane, termed translational diffusion, can be quite rapid for a phospholipid; in fact, within the plane of the membrane, one molecule of phospholipid can move several micrometers per second.

The phase changes—and thus the **fluidity** of membranes—are largely dependent upon the lipid composition of the membrane. In a lipid bilayer, the hydrophobic chains of the fatty acids can be highly aligned or ordered to provide a rather stiff structure. As the temperature increases, the hydrophobic side chains undergo a transition from the ordered state (more gel-like or crystalline phase) to a disordered one, taking on a more liquid-like or fluid arrangement. The temperature at which the structure undergoes the transition from ordered to disordered (ie, melts) is called the **"transition temperature" ($T_m$)**. The longer and more saturated fatty acid chains interact more strongly with each other via their longer hydrocarbon chains and thus cause higher values of $T_m$—ie, higher temperatures are required to increase the fluidity of the bilayer. On the other hand, unsaturated bonds that exist in the cis configuration tend to increase the fluidity of a bilayer by decreasing the compactness of the side chain packing without diminishing hydrophobicity (Figure 41–3). The phospholipids of cellular membranes generally contain at least one unsaturated fatty acid with at least one cis double bond.

**Cholesterol** modifies the fluidity of membranes. At temperatures below the $T_m$, it interferes with the interaction of the hydrocarbon tails of fatty acids and thus increases fluidity. At temperatures above the $T_m$, it limits disorder because it is more rigid than the hydrocarbon tails of the fatty acids and cannot move in the membrane to the same extent, thus limiting fluidity. At high cholesterol:phospholipid ratios, transition temperatures are altogether indistinguishable.

The **fluidity** of a membrane significantly affects its **functions.** As membrane fluidity increases, so does its permeability to water and other small hydrophilic molecules. The lateral mobility of integral proteins increases as the fluidity of the membrane increases. If the active site of an integral protein involved in a given function is exclusively in its hydrophilic regions, changing lipid fluidity will probably have little effect on the activity of the protein; however, if the protein is involved in a transport function in which transport components span the membrane, lipid phase effects may significantly alter the transport rate. The insulin receptor is an excellent example of altered function with changes in fluidity. As the concentration of unsaturated fatty acids in the membrane is increased (by growing cultured cells in a medium rich in such molecules), fluidity increases. This alters the receptor so that it binds more insulin.

A state of fluidity and thus of translational mobility in a membrane may be confined to certain regions of membranes under certain conditions. For example, protein-protein interactions may take place within the plane of the membrane, such that the integral proteins form a rigid matrix—in contrast to the more usual situation, where the lipid acts as the matrix. Such regions of rigid protein matrix can exist side by side in the same membrane with the usual lipid matrix. Gap junctions and tight junctions are clear examples of such side-by-side coexistence of different matrices.

### Lipid Rafts & Caveolae Are Special Features of Some Membranes

While the fluid mosaic model of membrane structure has stood up well to detailed scrutiny, additional features of membrane structure and function are constantly emerging. Two structures of particular current interest, located in surface membranes, are lipid rafts and caveolae. The former are dynamic areas of the exoplasmic leaflet of the lipid bilayer enriched in cholesterol and sphingolipids; they are involved in signal transduction and possibly other processes. Caveolae may derive from lipid rafts. Many if not all of them contain the protein caveolin-1, which may be involved in their formation from rafts. Caveolae are observable by electron microscopy as flask-shaped indentations of the cell membrane. Proteins detected in caveolae include various components of the signal-transduction system (eg, the insulin receptor and some G proteins), the folate receptor, and endothelial nitric oxide synthase (eNOS). Caveolae and lipid rafts are active areas of research, and ideas concerning them and their possible roles in various diseases are rapidly evolving.

## MEMBRANE SELECTIVITY ALLOWS SPECIALIZED FUNCTIONS

If the plasma membrane is relatively impermeable, how do most molecules enter a cell? How is selectivity of this movement established? Answers to such questions are important in understanding how cells adjust to a constantly changing extracellular environment. Metazoan organisms also must have means of communicating between adjacent and distant cells, so that complex biologic processes can be coordinated. These signals must arrive at and be transmitted by the membrane, or they must be generated as a consequence of some interaction with the membrane. Some of the major mechanisms used to accomplish these different objectives are listed in Table 41–3.

### Passive Mechanisms Move Some Small Molecules Across Membranes

Molecules can passively traverse the bilayer down electrochemical gradients by **simple diffusion** or by **facilitated diffusion.** This spontaneous movement toward equilibrium contrasts with **active transport,** which requires energy because it constitutes movement against an electrochemical gradient. Figure 41–8 provides a schematic representation of these mechanisms.

As described above, some solutes such as gases can enter the cell by diffusing down an electrochemical gradient across the membrane and do not require metabolic energy. The simple **passive diffusion** of a solute across the membrane is limited by the thermal agitation of that specific molecule, by the concentration gradient across the membrane, and by the solubility of that solute (the permeability coefficient, Figure 41–6) in the hydrophobic core of the membrane bilayer. Solubility is inversely proportionate to the number of hydrogen bonds that must be broken in order for a solute in the external aqueous phase to become incorporated in the hydrophobic bilayer. Electrolytes, poorly soluble in lipid, do not form hydrogen bonds with water, but they do acquire a shell of water from hydration by electrostatic interaction. The size of the shell is directly proportionate to the charge density of the electrolyte. Electrolytes with a large charge density have a larger shell of hydration and thus a slower diffusion rate. $Na^+$, for example, has a higher charge density than $K^+$. Hydrated $Na^+$ is therefore larger than hydrated $K^+$; hence, the latter tends to move more easily through the membrane.

The following factors affect **net diffusion** of a substance: (1) Its concentration gradient across the membrane. Solutes move from high to low concentration. (2) The electrical potential across the membrane. Solutes move toward the solution that has the opposite charge. The inside of the cell usually has a negative charge. (3) The permeability coefficient of the substance for the membrane. (4) The hydrostatic pressure gradient across the membrane. Increased pressure will increase the rate and force of the collision between the molecules and the membrane. (5) Temperature. Increased temperature will increase particle motion and thus increase the frequency of collisions between external particles and the membrane. In addition, a multitude of channels exist in membranes that route the entry of ions into cells.

***Table 41–3.*** Transfer of material and information across membranes.

| |
|---|
| **Cross-membrane movement of small molecules** |
| Diffusion (passive and facilitated) |
| Active transport |
| **Cross-membrane movement of large molecules** |
| Endocytosis |
| Exocytosis |
| **Signal transmission across membranes** |
| Cell surface receptors |
| 1. Signal transduction (eg, glucagon → cAMP) |
| 2. Signal internalization (coupled with endocytosis, eg, the LDL receptor) |
| Movement to intracellular receptors (steroid hormones; a form of diffusion) |
| **Intercellular contact and communication** |

### Ion Channels Are Transmembrane Proteins That Allow the Selective Entry of Various Ions

In natural membranes, as opposed to synthetic membrane bilayers, there are transmembrane channels, pore-like structures composed of proteins that constitute selective **ion channels.** Cation-conductive channels have an average diameter of about 5–8 nm and are negatively charged within the channel. The permeability of a channel depends upon the size, the extent of hydration, and the extent of charge density on the ion. Specific channels for $Na^+$, $K^+$, $Ca^{2+}$, and $Cl^-$ have been identified; two such channels are illustrated in Figure 41–9. Both are seen to consist of four subunits. Each subunit consists of six α-helical transmembrane domains. The amino and carboxyl terminals of both ion channels are located in the cytoplasm, with both extracellular and intracellular loops being present. The actual pores in the channels through which the ions pass are not shown in the figure. They form the center (diameter about 5–8 nm) of a structure formed by apposition of the subunits. The channels are very selective, in most cases permitting the passage of only one type of ion ($Na^+$, $Ca^{2+}$, etc). Many variations on the above structural themes are found, but

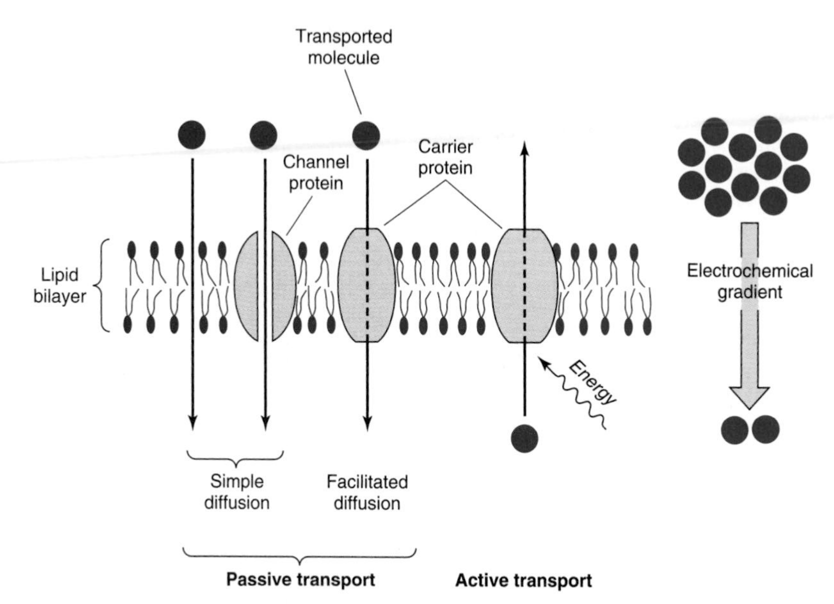


***Figure 41–8.*** Many small uncharged molecules pass freely through the lipid bilayer. Charged molecules, larger uncharged molecules, and some small uncharged molecules are transferred through channels or pores or by specific carrier proteins. Passive transport is always down an electrochemical gradient, toward equilibrium. Active transport is against an electrochemical gradient and requires an input of energy, whereas passive transport does not. (Redrawn and reproduced, with permission, from Alberts B et al: *Molecular Biology of the Cell.* Garland, 1983.)

all ion channels are basically made up of transmembrane subunits that come together to form a central pore through which ions pass selectively. The combination of x-ray crystallography (where possible) and site-directed mutagenesis affords a powerful approach to delineating the structure-function relationships of ion channels.

The membranes of **nerve cells** contain well-studied ion channels that are responsible for the action potentials generated across the membrane. The activity of some of these channels is controlled by neurotransmitters; hence, channel activity can be regulated. One ion can regulate the activity of the channel of another ion. For example, a decrease of $Ca^{2+}$ concentration in the extracellular fluid increases membrane permeability and increases the diffusion of $Na^+$. This depolarizes the membrane and triggers nerve discharge, which may explain the numbness, tingling, and muscle cramps symptomatic of a low level of plasma $Ca^{2+}$.

Channels are open transiently and thus are "gated." Gates can be controlled by opening or closing. In **ligand-gated channels,** a specific molecule binds to a receptor and opens the channel. **Voltage-gated channels** open (or close) in response to changes in membrane potential. Some properties of ion channels are listed in Table 41–4; other aspects of ion channels are discussed briefly in Chapter 49.

## Ionophores Are Molecules That Act as Membrane Shuttles for Various Ions

Certain microbes synthesize small organic molecules, **ionophores,** that function as shuttles for the movement of ions across membranes. These ionophores contain hydrophilic centers that bind specific ions and are surrounded by peripheral hydrophobic regions; this arrangement allows the molecules to dissolve effectively in the membrane and diffuse transversely therein. Others, like the well-studied polypeptide gramicidin, form channels.

Microbial toxins such as diphtheria toxin and activated serum complement components can produce large pores in cellular membranes and thereby provide macromolecules with direct access to the internal milieu.

## Aquaporins Are Proteins That Form Water Channels in Certain Membranes

In certain cells (eg, red cells, cells of the collecting ductules of the kidney), the movement of water by simple

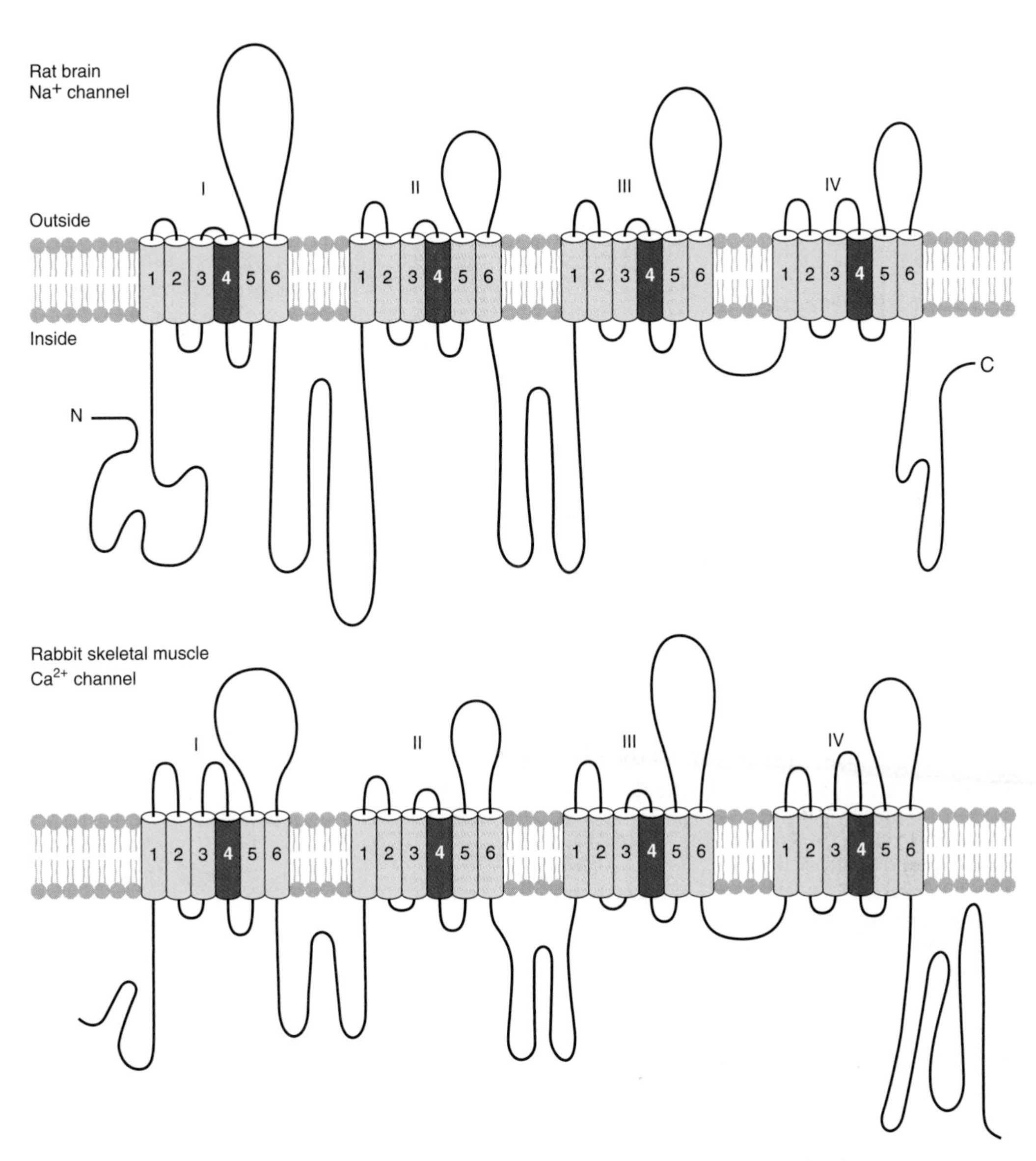


***Figure 41–9.*** Diagrammatic representation of the structures of two ion channels. The Roman numerals indicate the four subunits of each channel and the Arabic numerals the α-helical transmembrane domains of each subunit. The actual pores through which the ions pass are not shown but are formed by apposition of the various subunits. The specific areas of the subunits involved in the opening and closing of the channels are also not indicated. (After WK Catterall. Modified and reproduced from Hall ZW: *An Introduction to Molecular Neurobiology*. Sinauer, 1992.)

**Table 41–4.** Some properties of ion channels.

- They are composed of transmembrane protein subunits.
- Most are highly selective for one ion; a few are nonselective.
- They allow impermeable ions to cross membranes at rates approaching diffusion limits.
- They can permit ion fluxes of $10^6$–$10^7$/s.
- Their activities are regulated.
- The two main types are voltage-gated and ligand-gated.
- They are usually highly conserved across species.
- Most cells have a variety of $Na^+$, $K^+$, $Ca^{2+}$, and $Cl^-$ channels.
- Mutations in genes encoding them can cause specific diseases.[1]
- Their activities are affected by certain drugs.

[1]Some diseases caused by mutations of ion channels are briefly discussed in Chapter 49.

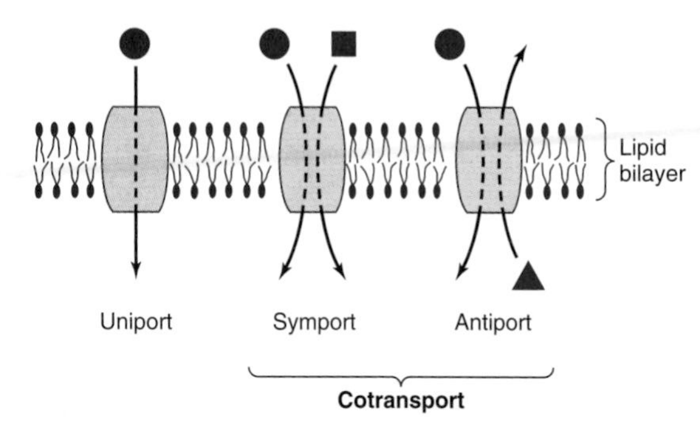


**Figure 41–10.** Schematic representation of types of transport systems. Transporters can be classified with regard to the direction of movement and whether one or more unique molecules are moved. (Redrawn and reproduced, with permission, from Alberts B et al: *Molecular Biology of the Cell*. Garland, 1983.)

diffusion is augmented by movement through water channels. These channels are composed of tetrameric transmembrane proteins named aquaporins. At least five distinct aquaporins (AP-1 to AP-5) have been identified. Mutations in the gene encoding AP-2 have been shown to be the cause of one type of nephrogenic diabetes insipidus.

## PLASMA MEMBRANES ARE INVOLVED IN FACILITATED DIFFUSION, ACTIVE TRANSPORT, & OTHER PROCESSES

Transport systems can be described in a functional sense according to the number of molecules moved and the direction of movement (Figure 41–10) or according to whether movement is toward or away from equilibrium. A **uniport** system moves one type of molecule bidirectionally. In **cotransport** systems, the transfer of one solute depends upon the stoichiometric simultaneous or sequential transfer of another solute. A **symport** moves these solutes in the same direction. Examples are the proton-sugar transporter in bacteria and the $Na^+$ - sugar transporters (for glucose and certain other sugars) and $Na^+$-amino acid transporters in mammalian cells. **Antiport** systems move two molecules in opposite directions (eg, $Na^+$ in and $Ca^{2+}$ out).

Molecules that cannot pass freely through the lipid bilayer membrane by themselves do so in association with carrier proteins. This involves two processes—**facilitated diffusion** and **active transport**—and highly specific transport systems.

Facilitated diffusion and active transport share many features. Both appear to involve **carrier proteins,** and both show **specificity** for ions, sugars, and amino acids. Mutations in bacteria and mammalian cells (including some that result in human disease) have supported these conclusions. Facilitated diffusion and active transport **resemble a substrate-enzyme reaction** except that no covalent interaction occurs. These points of resemblance are as follows: (1) There is a specific binding site for the solute. (2) The carrier is saturable, so it has a maximum rate of transport ($V_{max}$; Figure 41–11). (3) There is a binding constant ($K_m$) for the solute, and

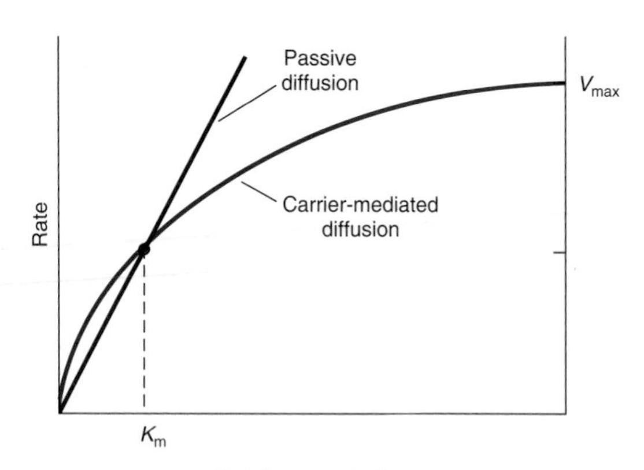


**Figure 41–11.** A comparison of the kinetics of carrier-mediated (facilitated) diffusion with passive diffusion. The rate of movement in the latter is directly proportionate to solute concentration, whereas the process is saturable when carriers are involved. The concentration at half-maximal velocity is equal to the binding constant ($K_m$) of the carrier for the solute. ($V_{max}$, maximal rate.)

so the whole system has a $K_m$ (Figure 41–11). (4) Structurally similar competitive inhibitors block transport.

Major **differences** are the following: (1) Facilitated diffusion can operate bidirectionally, whereas active transport is usually unidirectional. (2) Active transport always occurs against an electrical or chemical gradient, and so it requires energy.

## Facilitated Diffusion

Some specific solutes diffuse down electrochemical gradients across membranes more rapidly than might be expected from their size, charge, or partition coefficients. This **facilitated diffusion** exhibits properties distinct from those of simple diffusion. The rate of facilitated diffusion, a uniport system, can be saturated; ie, the number of sites involved in diffusion of the specific solutes appears finite. Many facilitated diffusion systems are stereospecific but, like simple diffusion, require no metabolic energy.

As described earlier, the inside-outside asymmetry of membrane proteins is stable, and mobility of proteins across (rather than in) the membrane is rare; therefore, transverse mobility of specific carrier proteins is not likely to account for facilitated diffusion processes except in a few unusual cases.

A **"Ping-Pong" mechanism** (Figure 41–12) explains facilitated diffusion. In this model, the carrier protein exists in two principal conformations. In the "pong" state, it is exposed to high concentrations of solute, and molecules of the solute bind to specific sites on the carrier protein. Transport occurs when a conformational change exposes the carrier to a lower concentration of solute ("ping" state). This process is completely reversible, and net flux across the membrane depends upon the concentration gradient. The rate at which solutes enter a cell by facilitated diffusion is determined by the following factors: (1) The concentration gradient across the membrane. (2) The amount of carrier available (this is a key control step). (3) The rapidity of the solute-carrier interaction. (4) The rapidity of the conformational change for both the loaded and the unloaded carrier.

**Hormones** regulate facilitated diffusion by changing the number of transporters available. **Insulin** increases glucose transport in fat and muscle by recruiting transporters from an intracellular reservoir. Insulin also enhances amino acid transport in liver and other tissues. One of the coordinated actions of **glucocorticoid hormones** is to enhance transport of amino acids into liver, where the amino acids then serve as a substrate for gluconeogenesis. **Growth hormone** increases amino acid transport in all cells, and estrogens do this in the uterus. There are at least five different carrier systems for amino acids in animal cells. Each is specific for a group of closely related amino acids, and most operate as $Na^+$-symport systems (Figure 41–10).

## Active Transport

The process of active transport differs from diffusion in that molecules are transported away from thermodynamic equilibrium; hence, **energy is required.** This energy can come from the hydrolysis of **ATP,** from **electron movement,** or from **light.** The maintenance of electrochemical gradients in biologic systems is so important that it consumes perhaps 30–40% of the total energy expenditure in a cell.

In general, cells maintain a low intracellular $Na^+$ concentration and a high intracellular $K^+$ concentration (Table 41–1), along with a net negative electrical potential inside. The pump that maintains these gradients is an **ATPase** that is activated by $Na^+$ and $K^+$ ($Na^+$-$K^+$ ATPase; see Figure 41–13). The ATPase is an integral

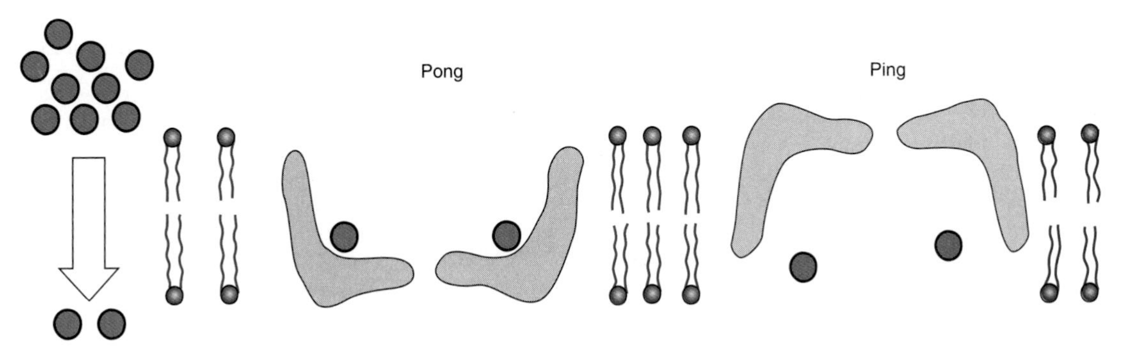


***Figure 41–12.*** The "Ping-Pong" model of facilitated diffusion. A protein carrier (gray structure) in the lipid bilayer associates with a solute in high concentration on one side of the membrane. A conformational change ensues ("pong" to "ping"), and the solute is discharged on the side favoring the new equilibrium. The empty carrier then reverts to the original conformation ("ping" to "pong") to complete the cycle.

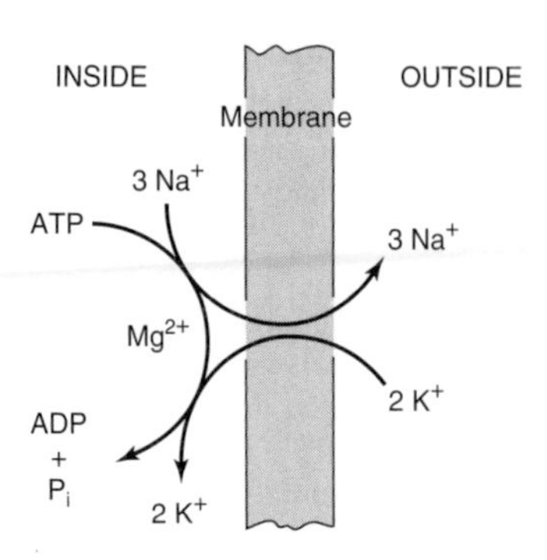


***Figure 41–13.*** Stoichiometry of the $Na^+$-$K^+$ ATPase pump. This pump moves three $Na^+$ ions from inside the cell to the outside and brings two $K^+$ ions from the outside to the inside for every molecule of ATP hydrolyzed to ADP by the membrane-associated ATPase. Ouabain and other cardiac glycosides inhibit this pump by acting on the extracellular surface of the membrane. (Courtesy of R Post.)

membrane protein and requires phospholipids for activity. The ATPase has catalytic centers for both ATP and $Na^+$ on the cytoplasmic side of the membrane, but the $K^+$ binding site is located on the extracellular side of the membrane. **Ouabain** or **digitalis** inhibits this ATPase by binding to the extracellular domain. Inhibition of the ATPase by ouabain can be antagonized by extracellular $K^+$.

## Nerve Impulses Are Transmitted Up & Down Membranes

The membrane forming the surface of **neuronal cells** maintains an asymmetry of inside-outside voltage (electrical potential) and is electrically excitable. When appropriately stimulated by a chemical signal mediated by a specific synaptic membrane receptor (see discussion of the transmission of biochemical signals, below), gates in the membrane are opened to allow the rapid influx of $Na^+$ or $Ca^{2+}$ (with or without the efflux of $K^+$), so that the voltage difference rapidly collapses and that segment of the membrane is depolarized. However, as a result of the action of the ion pumps in the membrane, the gradient is quickly restored.

When large areas of the membrane are **depolarized** in this manner, the electrochemical disturbance propagates in wave-like form down the membrane, generating a **nerve impulse. Myelin sheets,** formed by Schwann cells, wrap around nerve fibers and provide an electrical insulator that surrounds most of the nerve and greatly speeds up the propagation of the wave (signal) by allowing ions to flow in and out of the membrane only where the membrane is free of the insulation. The myelin membrane is composed of phospholipids, cholesterol, proteins, and GSLs. Relatively few proteins are found in the myelin membrane; those present appear to hold together multiple membrane bilayers to form the hydrophobic insulating structure that is impermeable to ions and water. Certain diseases, eg, multiple sclerosis and the Guillain-Barré syndrome, are characterized by demyelination and impaired nerve conduction.

## Glucose Transport Involves Several Mechanisms

A discussion of the **transport of glucose** summarizes many of the points made in this chapter. Glucose must enter cells as the first step in energy utilization. In adipocytes and muscle, glucose enters by a specific transport system that is enhanced by insulin. Changes in transport are primarily due to alterations of $V_{max}$ (presumably from more or fewer active transporters), but changes in $K_m$ may also be involved. Glucose transport involves different aspects of the principles of transport discussed above. Glucose and $Na^+$ bind to different sites on the glucose transporter. $Na^+$ moves into the cell down its electrochemical gradient and "drags" glucose with it (Figure 41–14). Therefore, the greater the $Na^+$ gradient, the more glucose enters; and if $Na^+$ in extracellular fluid is low, glucose transport stops. To maintain a steep $Na^+$ gradient, this $Na^+$-glucose symport is dependent on gradients generated by an $Na^+$-$K^+$ pump that maintains a low intracellular $Na^+$ concentration. Similar mechanisms are used to transport other sugars as well as amino acids.

The transcellular movement of sugars involves one additional component: a uniport that allows the glucose accumulated within the cell to move across a different surface toward a new equilibrium; this occurs in intestinal and renal cells, for example.

## Cells Transport Certain Macromolecules Across the Plasma Membrane

The process by which cells take up large molecules is called **"endocytosis."** Some of these molecules (eg, polysaccharides, proteins, and polynucleotides), when hydrolyzed inside the cell, yield nutrients. Endocytosis provides a mechanism for regulating the content of certain membrane components, hormone receptors being a case in point. Endocytosis can be used to learn more about how cells function. DNA from one cell type can be used to transfect a different cell and alter the latter's function or phenotype. A specific gene is often employed in these experiments, and this provides a unique way to study and analyze the regulation of that gene. DNA transfection depends upon endocytosis; endocy-

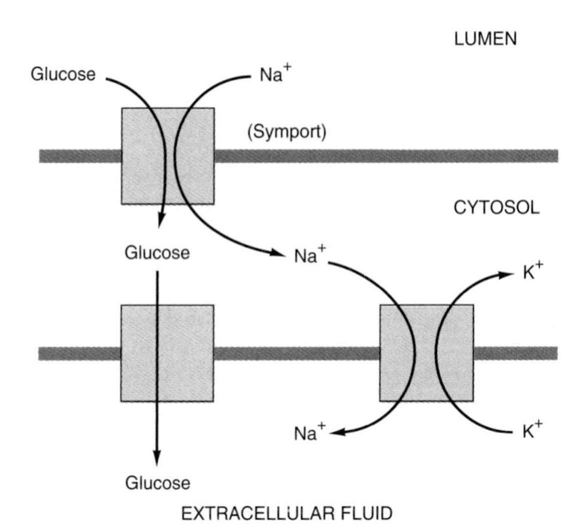


***Figure 41–14.*** The transcellular movement of glucose in an intestinal cell. Glucose follows $Na^+$ across the luminal epithelial membrane. The $Na^+$ gradient that drives this symport is established by $Na^+$-$K^+$ exchange, which occurs at the basal membrane facing the extracellular fluid compartment. Glucose at high concentration within the cell moves "downhill" into the extracellular fluid by facilitated diffusion (a uniport mechanism).

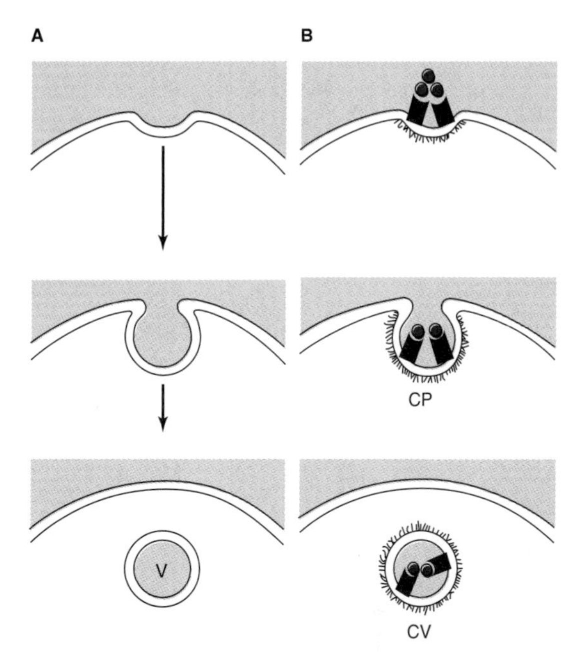


***Figure 41–15.*** Two types of endocytosis. An endocytotic vesicle (V) forms as a result of invagination of a portion of the plasma membrane. Fluid-phase endocytosis **(A)** is random and nondirected. Receptor-mediated endocytosis **(B)** is selective and occurs in coated pits (CP) lined with the protein clathrin (the fuzzy material). Targeting is provided by receptors (black symbols) specific for a variety of molecules. This results in the formation of a coated vesicle (CV).

tosis is responsible for the entry of DNA into the cell. Such experiments commonly use calcium phosphate, since $Ca^{2+}$ stimulates endocytosis and precipitates DNA, which makes the DNA a better object for endocytosis. Cells also release macromolecules by **exocytosis.** Endocytosis and exocytosis both involve vesicle formation with or from the plasma membrane.

### A. Endocytosis

All eukaryotic cells are continuously ingesting parts of their plasma membranes. Endocytotic vesicles are generated when segments of the plasma membrane invaginate, enclosing a minute volume of extracellular fluid and its contents. The vesicle then pinches off as the fusion of plasma membranes seals the neck of the vesicle at the original site of invagination (Figure 41–15). This vesicle fuses with other membrane structures and thus achieves the transport of its contents to other cellular compartments or even back to the cell exterior. Most endocytotic vesicles fuse with primary lysosomes to form secondary lysosomes, which contain hydrolytic enzymes and are therefore specialized organelles for intracellular disposal. The macromolecular contents are digested to yield amino acids, simple sugars, or nucleotides, and they diffuse out of the vesicles to be reused in the cytoplasm. Endocytosis requires (1) energy, usually from the hydrolysis of ATP; (2) $Ca^{2+}$ in extracellular fluid; and (3) contractile elements in the cell (probably the microfilament system) (Chapter 49).

There are two general types of endocytosis. **Phagocytosis** occurs only in specialized cells such as macrophages and granulocytes. Phagocytosis involves the ingestion of large particles such as viruses, bacteria, cells, or debris. Macrophages are extremely active in this regard and may ingest 25% of their volume per hour. In so doing, a macrophage may internalize 3% of its plasma membrane each minute or the entire membrane every 30 minutes.

**Pinocytosis** is a property of all cells and leads to the cellular uptake of fluid and fluid contents. There are two types. **Fluid-phase pinocytosis** is a nonselective process in which the uptake of a solute by formation of small vesicles is simply proportionate to its concentration in the surrounding extracellular fluid. The formation of these vesicles is an extremely active process. Fibroblasts, for example, internalize their plasma membrane at about one-third the rate of macrophages. This process occurs more rapidly than membranes are made. The surface area and volume of a cell do not change much, so membranes must be replaced by exocytosis or by being recycled as fast as they are removed by endocytosis.

The other type of pinocytosis, **absorptive pinocytosis,** is a receptor-mediated selective process primarily responsible for the uptake of macromolecules for which there are a finite number of binding sites on the plasma membrane. These high-affinity receptors permit the selective concentration of ligands from the medium, minimize the uptake of fluid or soluble unbound macromolecules, and markedly increase the rate at which specific molecules enter the cell. The vesicles formed during absorptive pinocytosis are derived from invaginations (pits) that are coated on the cytoplasmic side with a filamentous material. In many systems, the protein **clathrin** is the filamentous material. It has a three-limbed structure (called a triskelion), with each limb being made up of one light and one heavy chain of clathrin. The polymerization of clathrin into a vesicle is directed by **assembly particles**, composed of four **adapter proteins.** These interact with certain amino acid sequences in the receptors that become cargo, ensuring selectivity of uptake. The lipid **$PIP_2$** also plays an important role in vesicle assembly. In addition, the protein **dynamin**, which both binds and hydrolyzes GTP, is necessary for the pinching off of clathrin-coated vesicles from the cell surface. Coated pits may constitute as much as 2% of the surface of some cells.

As an example, the low-density lipoprotein (LDL) molecule and its receptor (Chapter 25) are internalized by means of coated pits containing the LDL receptor. These endocytotic vesicles containing LDL and its receptor fuse to lysosomes in the cell. The receptor is released and recycled back to the cell surface membrane, but the apoprotein of LDL is degraded and the cholesteryl esters metabolized. Synthesis of the LDL receptor is regulated by secondary or tertiary consequences of pinocytosis, eg, by metabolic products—such as cholesterol—released during the degradation of LDL. Disorders of the LDL receptor and its internalization are medically important and are discussed in Chapter 25.

Absorptive pinocytosis of **extracellular glycoproteins** requires that the glycoproteins carry specific carbohydrate recognition signals. These recognition signals are bound by membrane receptor molecules, which play a role analogous to that of the LDL receptor. A galactosyl receptor on the surface of hepatocytes is instrumental in the absorptive pinocytosis of asialoglycoproteins from the circulation (Chapter 47). Acid hydrolases taken up by absorptive pinocytosis in fibroblasts are recognized by their mannose 6-phosphate moieties. Interestingly, the mannose 6-phosphate moiety also seems to play an important role in the intracellular targeting of the acid hydrolases to the lysosomes of the cells in which they are synthesized (Chapter 47).

There is a **dark side** to receptor-mediated endocytosis in that viruses which cause such diseases as hepatitis (affecting liver cells), poliomyelitis (affecting motor neurons), and AIDS (affecting T cells) initiate their damage by this mechanism. Iron toxicity also begins with excessive uptake due to endocytosis.

### B. Exocytosis

Most cells release macromolecules to the exterior by exocytosis. This process is also involved in membrane remodeling, when the components synthesized in the Golgi apparatus are carried in vesicles to the plasma membrane. The signal for exocytosis is often a hormone which, when it binds to a cell-surface receptor, induces a local and transient change in $Ca^{2+}$ concentration. $Ca^{2+}$ triggers exocytosis. Figure 41–16 provides a comparison of the mechanisms of exocytosis and endocytosis.

Molecules released by exocytosis fall into three categories: (1) They can attach to the cell surface and become peripheral proteins, eg, antigens. (2) They can become part of the extracellular matrix, eg, collagen and glycosaminoglycans. (3) They can enter extracellular fluid and signal other cells. Insulin, parathyroid hormone, and the catecholamines are all packaged in gran-

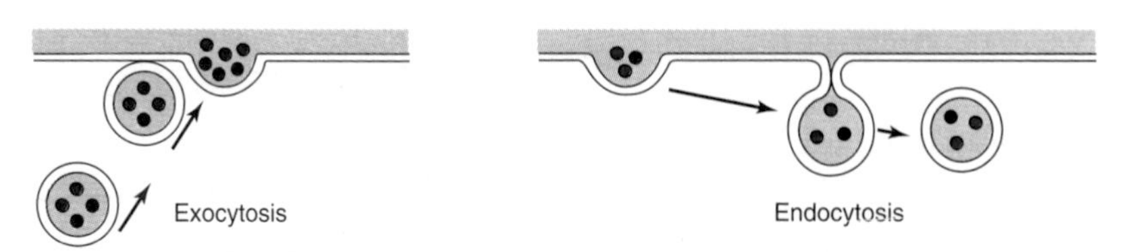


***Figure 41–16.*** A comparison of the mechanisms of endocytosis and exocytosis. Exocytosis involves the contact of two inside surface (cytoplasmic side) monolayers, whereas endocytosis results from the contact of two outer surface monolayers.

ules and processed within cells, to be released upon appropriate stimulation.

### Some Signals Are Transmitted Across Membranes

Specific biochemical signals such as neurotransmitters, hormones, and immunoglobulins bind to specific **receptors** (integral proteins) exposed to the outside of cellular membranes and transmit information through these membranes to the cytoplasm. This process, called **transmembrane signaling**, involves the generation of a number of signals, including cyclic nucleotides, calcium, phosphoinositides, and diacylglycerol. It is discussed in detail in Chapter 43.

### Information Can Be Communicated by Intercellular Contact

There are many areas of intercellular contact in a metazoan organism. This necessitates contact between the plasma membranes of the individual cells. Cells have developed specialized regions on their membranes for intercellular communication in close proximity. **Gap junctions** mediate and regulate the passage of ions and small molecules (up to 1000–2000 MW) through a narrow hydrophilic core connecting the cytosol of adjacent cells. These structures are primarily composed of the protein **connexin,** which contains four membrane-spanning α helices. About a dozen genes encoding different connexins have been cloned. An assembly of 12 connexin molecules forms a structure (a connexon) with a central channel that forms bridges between adjacent cells. Ions and small molecules pass from the cytosol of one cell to that of another through the channels, which open and close in a regulated fashion.

## MUTATIONS AFFECTING MEMBRANE PROTEINS CAUSE DISEASES

In view of the fact that membranes are located in so many organelles and are involved in so many processes, it is not surprising that **mutations** affecting their protein constituents should result in many diseases or disorders. Proteins in membranes can be classified as **receptors, transporters, ion channels, enzymes,** and **structural components.** Members of all of these classes are often **glycosylated,** so that mutations affecting this process may alter their function. Examples of diseases or disorders due to abnormalities in membrane proteins are listed in Table 41–5; these mainly reflect mutations in proteins of the **plasma membrane,** with one affecting lysosomal function (I-cell disease). Over 30 genetic diseases or disorders have been ascribed to mutations affecting various proteins involved in the transport of amino acids, sugars, lipids, urate, anions, cations, water, and vitamins across the plasma membrane. Mutations in genes encoding proteins in other membranes can also have harmful consequences. For example, mutations in genes encoding **mitochondrial membrane proteins** involved in oxidative phosphorylation can cause neurologic and other problems (eg, Leber's hereditary optic neuropathy; LHON). Membrane proteins can also be affected by conditions other than mutations. Formation of **autoantibodies** to the acetylcholine receptor in skeletal muscle causes myasthenia gravis. **Ischemia** can quickly affect the integrity of various ion channels in membranes. Abnormalities of membrane constituents other than proteins can also be harmful. With regard to **lipids,** excess of cholesterol (eg, in familial hypercholesterolemia), of lysophospholipid (eg, after bites by certain snakes, whose venom contains phospholipases), or of glycosphingolipids (eg, in a sphingolipidosis) can all affect membrane function.

### Cystic Fibrosis Is Due to Mutations in the Gene Encoding a Chloride Channel

Cystic fibrosis (CF) is a recessive genetic disorder prevalent among whites in North America and certain parts of northern Europe. It is characterized by chronic bacterial infections of the airways and sinuses, fat maldigestion due to pancreatic exocrine insufficiency, infertility in males due to abnormal development of the vas deferens, and elevated levels of chloride in sweat (> 60 mmol/L).

After a Herculean landmark endeavor, the gene for CF was identified in 1989 on chromosome 7. It was found to encode a protein of 1480 amino acids, named cystic fibrosis transmembrane regulator (CFTR), a cyclic AMP-regulated $Cl^-$ channel (see Figure 41–17). An abnormality of membrane $Cl^-$ permeability is believed to result in the increased viscosity of many bodily secretions, though the precise mechanisms are still under investigation. The commonest mutation (~70% in certain Caucasian populations) is deletion of three bases, resulting in loss of residue 508, a phenylalanine ($\Delta F_{508}$). However, more than 900 other mutations have been identified. These mutations affect CFTR in at least four ways: (1) its amount is reduced; (2) depending upon the particular mutation, it may be susceptible to misfolding and retention within the ER or Golgi apparatus; (3) mutations in the nucleotide-binding domains may affect the ability of the $Cl^-$ channel to open, an event affected by ATP; (4) the mutations may also reduce the rate of ion flow through a channel, generating less of a $Cl^-$ current.

The most serious and life-threatening complication is recurrent pulmonary infections due to overgrowth of various pathogens in the viscous secretions of the respi-

***Table 41–5.*** Some diseases or pathologic states resulting from or attributed to abnormalities of membranes.[1]

| Disease | Abnormality |
|---|---|
| Achondroplasia (MIM 100800) | Mutations in the gene encoding the fibroblast growth factor receptor 3 |
| Familial hypercholesterolemia (MIM 143890) | Mutations in the gene encoding the LDL receptor |
| Cystic fibrosis (MIM 219700) | Mutations in the gene encoding the CFTR protein, a $Cl^-$ transporter |
| Congenital long QT syndrome (MIM 192500) | Mutations in genes encoding ion channels in the heart |
| Wilson disease (MIM 277900) | Mutations in the gene encoding a copper-dependent ATPase |
| I-cell disease (MIM 252500) | Mutations in the gene encoding GlcNAc phosphotransferase, resulting in absence of the Man 6-P signal for lysosomal localization of certain hydrolases |
| Hereditary spherocytosis (MIM 182900) | Mutations in the genes encoding spectrin or other structural proteins in the red cell membrane |
| Metastasis | Abnormalities in the oligosaccharide chains of membrane glycoproteins and glycolipids are thought to be of importance |
| Paroxysmal nocturnal hemoglobinuria (MIM 311770) | Mutation resulting in deficient attachment of the GPI anchor to certain proteins of the red cell membrane |

[1]The disorders listed are discussed further in other chapters. The table lists examples of mutations affecting receptors, a transporter, an ion channel, an enzyme, and a structural protein. Examples of altered or defective glycosylation of glycoproteins are also presented. Most of the conditions listed affect the plasma membrane.

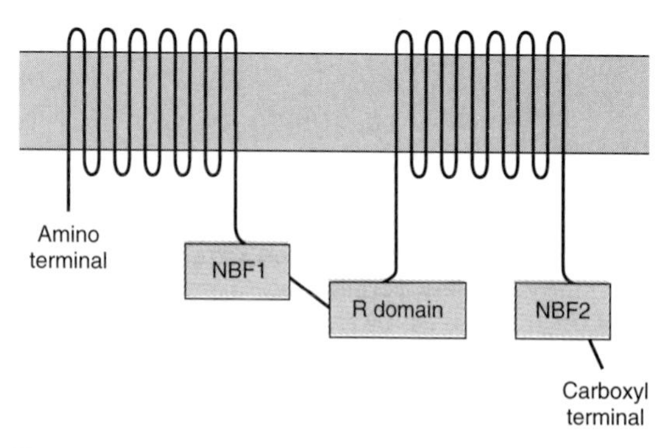


***Figure 41–17.*** Diagram of the structure of the CFTR protein (not to scale). The protein contains twelve transmembrane segments (probably helical), two nucleotide-binding folds or domains (NBF1 and NBF2), and one regulatory (R) domain. NBF1 and NBF2 probably bind ATP and couple its hydrolysis to transport of $Cl^-$. Phe 508, the major locus of mutations in cystic fibrosis, is located in NBF1.

ratory tract. Poor nutrition as a result of pancreatic insufficiency worsens the situation. The treatment of CF thus requires a comprehensive effort to maintain nutritional status, to prevent and combat pulmonary infections, and to maintain physical and psychologic health. Advances in molecular genetics mean that mutation analysis can be performed for prenatal diagnosis and for carrier testing in families in which one child already has the condition. Efforts are in progress to use gene therapy to restore the activity of CFTR. An aerosolized preparation of human DNase that digests the DNA of microorganisms in the respiratory tract has proved helpful in therapy.

## SUMMARY

- Membranes are complex structures composed of lipids, carbohydrates, and proteins.
- The basic structure of all membranes is the lipid bilayer. This bilayer is formed by two sheets of phospholipids in which the hydrophilic polar head groups

are directed away from each other and are exposed to the aqueous environment on the outer and inner surfaces of the membrane. The hydrophobic nonpolar tails of these molecules are oriented toward each other, in the direction of the center of the membrane.

- Membrane proteins are classified as integral if they are firmly embedded in the bilayer and as peripheral if they are loosely attached to the outer or inner surface.
- The 20 or so different membranes in a mammalian cell have intrinsic functions (eg, enzymatic activity), and they define compartments, or specialized environments, within the cell that have specific functions (eg, lysosomes).
- Certain molecules freely diffuse across membranes, but the movement of others is restricted because of size, charge, or solubility.
- Various passive and active mechanisms are employed to maintain gradients of such molecules across different membranes.
- Certain solutes, eg, glucose, enter cells by facilitated diffusion, along a downhill gradient from high to low concentration. Specific carrier molecules, or transporters, are involved in such processes.
- Ligand- or voltage-gated ion channels are often employed to move charged molecules ($Na^+$, $K^+$, $Ca^{2+}$, etc) across membranes.
- Large molecules can enter or leave cells through mechanisms such as endocytosis or exocytosis. These processes often require binding of the molecule to a receptor, which affords specificity to the process.
- Receptors may be integral components of membranes (particularly the plasma membrane). The interaction of a ligand with its receptor may not involve the movement of either into the cell, but the interaction results in the generation of a signal that influences intracellular processes (transmembrane signaling).
- Mutations that affect the structure of membrane proteins (receptors, transporters, ion channels, enzymes, and structural proteins) may cause diseases; examples include cystic fibrosis and familial hypercholesterolemia.

## REFERENCES

Doyle DA et al: The structure of the potassium channel: molecular basis of $K^+$ conductance and selectivity. Science 1998;280:69.

Felix R: Channelopathies: ion channel defects linked to heritable clinical disorders. J Med Genet 2000;37:729.

Garavito RM, Ferguson-Miller S: Detergents as tools in membrane biochemistry. J Biol Chem 2001;276:32403.

Gillooly DJ, Stenmark H: A lipid oils the endocytosis machine. Science 2001;291;993.

Knowles MR, Durie PR: What is cystic fibrosis? N Engl J Med 2002;347:439.

Longo N: Inherited defects of membrane transport. In: *Harrison's Principles of Internal Medicine,* 15th ed. Braunwald E et al (editors). McGraw-Hill, 2001.

Marx J: Caveolae: a once-elusive structure gets some respect. Science 2001;294;1862.

White SH et al: How membranes shape protein structure. J Biol Chem 2001:276:32395.

# The Diversity of the Endocrine System

42

Daryl K. Granner, MD

| | |
|---|---|
| **ACTH** | Adrenocorticotropic hormone |
| **ANF** | Atrial natriuretic factor |
| **cAMP** | Cyclic adenosine monophosphate |
| **CBG** | Corticosteroid-binding globulin |
| **CG** | Chorionic gonadotropin |
| **cGMP** | Cyclic guanosine monophosphate |
| **CLIP** | Corticotropin-like intermediate lobe peptide |
| **DBH** | Dopamine β-hydroxylase |
| **DHEA** | Dehydroepiandrosterone |
| **DHT** | Dihydrotestosterone |
| **DIT** | Diiodotyrosine |
| **DOC** | Deoxycorticosterone |
| **EGF** | Epidermal growth factor |
| **FSH** | Follicle-stimulating hormone |
| **GH** | Growth hormone |
| **IGF-I** | Insulin-like growth factor-I |
| **LH** | Luteotropic hormone |
| **LPH** | Lipotropin |
| **MIT** | Monoiodotyrosine |
| **MSH** | Melanocyte-stimulating hormone |
| **OHSD** | Hydroxysteroid dehydrogenase |
| **PNMT** | Phenylethanolamine-*N*-methyltransferase |
| **POMC** | Pro-opiomelanocortin |
| **SHBG** | Sex hormone-binding globulin |
| **StAR** | Steroidogenic acute regulatory (protein) |
| **TBG** | Thyroxine-binding globulin |
| **TEBG** | Testosterone-estrogen-binding globulin |
| **TRH** | Thyrotropin-releasing hormone |
| **TSH** | Thyrotropin-stimulating hormone |

## BIOMEDICAL IMPORTANCE

The survival of multicellular organisms depends on their ability to adapt to a constantly changing environment. Intercellular communication mechanisms are necessary requirements for this adaptation. The nervous system and the endocrine system provide this intercellular, organism-wide communication. The nervous system was originally viewed as providing a fixed communication system, whereas the endocrine system supplied hormones, which are mobile messages. In fact, there is a remarkable convergence of these regulatory systems. For example, neural regulation of the endocrine system is important in the production and secretion of some hormones; many neurotransmitters resemble hormones in their synthesis, transport, and mechanism of action; and many hormones are synthesized in the nervous system. The word "hormone" is derived from a Greek term that means to arouse to activity. As classically defined, a hormone is a substance that is synthesized in one organ and transported by the circulatory system to act on another tissue. However, this original description is too restrictive because hormones can act on adjacent cells (paracrine action) and on the cell in which they were synthesized (autocrine action) without entering the systemic circulation. A diverse array of hormones—each with distinctive mechanisms of action and properties of biosynthesis, storage, secretion, transport, and metabolism—has evolved to provide homeostatic responses. This biochemical diversity is the topic of this chapter.

## THE TARGET CELL CONCEPT

There are about 200 types of differentiated cells in humans. Only a few produce hormones, but virtually all of the 75 trillion cells in a human are targets of one or more of the over 50 known hormones. The concept of the target cell is a useful way of looking at hormone action. It was thought that hormones affected a single cell type—or only a few kinds of cells—and that a hormone elicited a unique biochemical or physiologic action. We now know that a given hormone can affect several different cell types; that more than one hormone can affect a given cell type; and that hormones can exert many different effects in one cell or in different cells. With the discovery of specific cell-surface and intracellular hormone receptors, the definition of a target has been expanded to include any cell in which the hormone (ligand) binds to its receptor, whether or not a biochemical or physiologic response has yet been determined.

Several factors determine the response of a target cell to a hormone. These can be thought of in two general ways: (1) as factors that affect the concentration of the hormone at the target cell (see Table 42–1) and (2) as factors that affect the actual response of the target cell to the hormone (see Table 42–2).

## HORMONE RECEPTORS ARE OF CENTRAL IMPORTANCE

### Receptors Discriminate Precisely

One of the major challenges faced in making the hormone-based communication system work is illustrated in Figure 42–1. Hormones are present at very low concentrations in the extracellular fluid, generally in the range of $10^{-15}$ to $10^{-9}$ mol/L. This concentration is much lower than that of the many structurally similar molecules (sterols, amino acids, peptides, proteins) and other molecules that circulate at concentrations in the $10^{-5}$ to $10^{-3}$ mol/L range. Target cells, therefore, must distinguish not only between different hormones present in small amounts but also between a given hormone and the $10^{6}$- to $10^{9}$-fold excess of other similar molecules. This high degree of discrimination is provided by cell-associated recognition molecules called receptors. Hormones initiate their biologic effects by binding to specific receptors, and since any effective control system also must provide a means of stopping a response, hormone-induced actions generally terminate when the effector dissociates from the receptor.

A target cell is defined by its ability to selectively bind a given hormone to its cognate receptor. Several biochemical features of this interaction are important in order for hormone-receptor interactions to be physiologically relevant: (1) binding should be specific, ie, displaceable by agonist or antagonist; (2) binding should be saturable; and (3) binding should occur within the concentration range of the expected biologic response.

***Table 42–1.*** Determinants of the concentration of a hormone at the target cell.

| |
|---|
| The rate of synthesis and secretion of the hormones. |
| The proximity of the target cell to the hormone source (dilution effect). |
| The dissociation constants of the hormone with specific plasma transport proteins (if any). |
| The conversion of inactive or suboptimally active forms of the hormone into the fully active form. |
| The rate of clearance from plasma by other tissues or by digestion, metabolism, or excretion. |

***Table 42–2.*** Determinants of the target cell response.

| |
|---|
| The number, relative activity, and state of occupancy of the specific receptors on the plasma membrane or in the cytoplasm or nucleus. |
| The metabolism (activation or inactivation) of the hormone in the target cell. |
| The presence of other factors within the cell that are necessary for the hormone response. |
| Up- or down-regulation of the receptor consequent to the interaction with the ligand. |
| Postreceptor desensitzation of the cell, including down-regulation of the receptor. |

### Both Recognition & Coupling Domains Occur on Receptors

All receptors have at least two functional domains. A recognition domain binds the hormone ligand and a second region generates a signal that couples hormone recognition to some intracellular function. Coupling (signal transduction) occurs in two general ways. Polypeptide and protein hormones and the catecholamines bind to receptors located in the plasma membrane and thereby generate a signal that regulates various intracellular functions, often by changing the activity of an enzyme. In contrast, steroid, retinoid, and thyroid hormones interact with intracellular receptors, and it is this ligand-receptor complex that directly provides the signal, generally to specific genes whose rate of transcription is thereby affected.

The domains responsible for hormone recognition and signal generation have been identified in the protein polypeptide and catecholamine hormone receptors. Steroid, thyroid, and retinoid hormone receptors have several functional domains: one site binds the hormone; another binds to specific DNA regions; a third is involved in the interaction with other coregulator proteins that result in the activation (or repression) of gene transcription; and a fourth may specify binding to one or more other proteins that influence the intracellular trafficking of the receptor.

The dual functions of binding and coupling ultimately define a receptor, and it is the coupling of hormone binding to signal transduction—so-called **receptor-effector coupling**—that provides the first step in amplification of the hormonal response. This dual purpose also distinguishes the target cell receptor from the

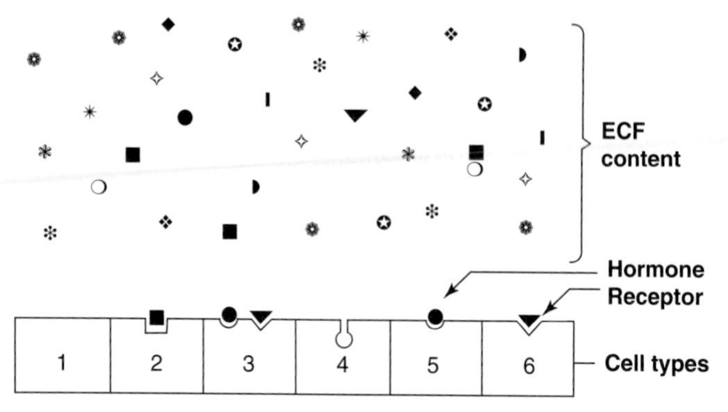


***Figure 42–1.*** Specificity and selectivity of hormone receptors. Many different molecules circulate in the extracellular fluid (ECF), but only a few are recognized by hormone receptors. Receptors must select these molecules from among high concentrations of the other molecules. This simplified drawing shows that a cell may have no hormone receptors (1), have one receptor (2+5+6), have receptors for several hormones (3), or have a receptor but no hormone in the vicinity (4).

plasma carrier proteins that bind hormone but do not generate a signal (see Table 42–6).

### Receptors Are Proteins

Several classes of peptide hormone receptors have been defined. For example, the insulin receptor is a heterotetramer ($\alpha_2\beta_2$) linked by multiple disulfide bonds in which the extracellular $\alpha$ subunit binds insulin and the membrane-spanning $\beta$ subunit transduces the signal through the tyrosine protein kinase domain located in the cytoplasmic portion of this polypeptide. The receptors for insulin-like growth factor I (IGF-I) and epidermal growth factor (EGF) are generally similar in structure to the insulin receptor. The growth hormone and prolactin receptors also span the plasma membrane of target cells but do not contain intrinsic protein kinase activity. Ligand binding to these receptors, however, results in the association and activation of a completely different protein kinase pathway, the Jak-Stat pathway. Polypeptide hormone and catecholamine receptors, which transduce signals by altering the rate of production of cAMP through G-proteins, are characterized by the presence of seven domains that span the plasma membrane. Protein kinase activation and the generation of cyclic AMP, (cAMP, 3′5′-adenylic acid; see Figure 20–5) is a downstream action of this class of receptor (see Chapter 43 for further details).

A comparison of several different steroid receptors with thyroid hormone receptors revealed a remarkable conservation of the amino acid sequence in certain regions, particularly in the DNA-binding domains. This led to the realization that receptors of the steroid or thyroid type are members of a large superfamily of nuclear receptors. Many related members of this family have no known ligand at present and thus are called orphan receptors. The nuclear receptor superfamily plays a critical role in the regulation of gene transcription by hormones, as described in Chapter 43.

## HORMONES CAN BE CLASSIFIED IN SEVERAL WAYS

Hormones can be classified according to chemical composition, solubility properties, location of receptors, and the nature of the signal used to mediate hormonal action within the cell. A classification based on the last two properties is illustrated in Table 42–3, and general features of each group are illustrated in Table 42–4.

The hormones in group I are lipophilic. After secretion, these hormones associate with plasma transport or carrier proteins, a process that circumvents the problem of solubility while prolonging the plasma half-life of the hormone. The relative percentages of bound and free hormone are determined by the binding affinity and binding capacity of the transport protein. The free hormone, which is the biologically active form, readily traverses the lipophilic plasma membrane of all cells and encounters receptors in either the cytosol or nucleus of target cells. The ligand-receptor complex is assumed to be the intracellular messenger in this group.

The second major group consists of water-soluble hormones that bind to the plasma membrane of the target cell. Hormones that bind to the surfaces of cells communicate with intracellular metabolic processes through intermediary molecules called **second messengers** (the hormone itself is the first messenger), which are generated as a consequence of the ligand-receptor interaction. The second messenger concept arose from an observation that epinephrine binds to the plasma membrane of certain cells and increases intracellular cAMP. This was followed by a series of experiments in which cAMP was found to mediate the effects of many hormones. Hormones that clearly employ this mechanism are shown in group II.A of Table 42–3. To date, only one hormone, atrial natriuretic factor (ANF), uses cGMP as its second messenger, but other hormones will probably be added to group II.B. Several hormones, many of which were previously thought to affect cAMP, appear to use ionic calcium ($Ca^{2+}$) or

***Table 42–3.*** Classification of hormones by mechanism of action.

***I. Hormones that bind to intracellular receptors***
- Androgens
- Calcitriol (1,25[OH]$_2$-D$_3$)
- Estrogens
- Glucocorticoids
- Mineralocorticoids
- Progestins
- Retinoic acid
- Thyroid hormones ($T_3$ and $T_4$)

***II. Hormones that bind to cell surface receptors***

**A. The second messenger is cAMP:**
- $\alpha_2$-Adrenergic catecholamines
- $\beta$-Adrenergic catecholamines
- Adrenocorticotropic hormone
- Antidiuretic hormone
- Calcitonin
- Chorionic gonadotropin, human
- Corticotropin-releasing hormone
- Follicle-stimulating hormone
- Glucagon
- Lipotropin
- Luteinizing hormone
- Melanocyte-stimulating hormone
- Parathyroid hormone
- Somatostatin
- Thyroid-stimulating hormone

**B. The second messenger is cGMP:**
- Atrial natriuretic factor
- Nitric oxide

**C. The second messenger is calcium or phosphatidylinositols (or both):**
- Acetylcholine (muscarinic)
- $\alpha_1$-Adrenergic catecholamines
- Angiotensin II
- Antidiuretic hormone (vasopressin)
- Cholecystokinin
- Gastrin
- Gonadotropin-releasing hormone
- Oxytocin
- Platelet-derived growth factor
- Substance P
- Thyrotropin-releasing hormone

**D. The second messenger is a kinase or phosphatase cascade:**
- Chorionic somatomammotropin
- Epidermal growth factor
- Erythropoietin
- Fibroblast growth factor
- Growth hormone
- Insulin
- Insulin-like growth factors I and II
- Nerve growth factor
- Platelet-derived growth factor
- Prolactin

***Table 42–4.*** General features of hormone classes.

| | Group I | Group II |
|---|---|---|
| Types | Steroids, iodothyronines, calcitriol, retinoids | Polypeptides, proteins, glycoproteins, catecholamines |
| Solubility | Lipophilic | Hydrophilic |
| Transport proteins | Yes | No |
| Plasma half-life | Long (hours to days) | Short (minutes) |
| Receptor | Intracellular | Plasma membrane |
| Mediator | Receptor-hormone complex | cAMP, cGMP, $Ca^{2+}$, metabolites of complex phosphoinositols, kinase cascades |

metabolites of complex phosphoinositides (or both) as the intracellular signal. These are shown in group II.C of the table. The intracellular messenger for group II.D is a protein kinase-phosphatase cascade. Several of these have been identified, and a given hormone may use more than one kinase cascade. A few hormones fit into more than one category, and assignments change as new information is brought forward.

## DIVERSITY OF THE ENDOCRINE SYSTEM

### Hormones Are Synthesized in a Variety of Cellular Arrangements

Hormones are synthesized in discrete organs designed solely for this specific purpose, such as the thyroid (triiodothyronine), adrenal (glucocorticoids and mineralocorticoids), and the pituitary (TSH, FSH, LH, growth hormone, prolactin, ACTH). Some organs are designed to perform two distinct but closely related functions. For example, the ovaries produce mature oocytes and the reproductive hormones estradiol and progesterone. The testes produce mature spermatozoa and testosterone. Hormones are also produced in specialized cells within other organs such as the small intestine (glucagon-like peptide), thyroid (calcitonin), and kidney (angiotensin II). Finally, the synthesis of some hormones requires the parenchymal cells of more than one organ—eg, the skin, liver, and kidney are required for the production of 1,25(OH)$_2$-D$_3$ (calcitriol). Examples of this diversity in the approach to hormone synthesis, each of which has evolved to fulfill a specific purpose, are discussed below.

### Hormones Are Chemically Diverse

Hormones are synthesized from a wide variety of chemical building blocks. A large series is derived from cholesterol. These include the glucocorticoids, mineralocorticoids, estrogens, progestins, and 1,25$(OH)_2$-$D_3$ (see Figure 42–2). In some cases, a steroid hormone is the precursor molecule for another hormone. For example, progesterone is a hormone in its own right but is also a precursor in the formation of glucocorticoids, mineralocorticoids, testosterone, and estrogens. Testosterone is an obligatory intermediate in the biosynthesis of estradiol and in the formation of dihydrotestosterone (DHT). In these examples, described in detail below, the final product is determined by the cell type and the associated set of enzymes in which the precursor exists.

The amino acid tyrosine is the starting point in the synthesis of the catecholamines and of the thyroid hormones tetraiodothyronine (thyroxine; $T_4$) and triiodothyronine ($T_3$) (Figure 42–2). $T_3$ and $T_4$ are unique in that they require the addition of iodine (as $I^-$) for bioactivity. Because dietary iodine is very scarce in many parts of the world, an intricate mechanism for accumulating and retaining $I^-$ has evolved.

Many hormones are polypeptides or glycoproteins. These range in size from thyrotropin-releasing hormone (TRH), a tripeptide, to single-chain polypeptides like adrenocorticotropic hormone (ACTH; 39 amino acids), parathyroid hormone (PTH; 84 amino acids), and growth hormone (GH; 191 amino acids) (Figure 42–2). Insulin is an AB chain heterodimer of 21 and 30 amino acids, respectively. Follicle-stimulating hormone (FSH), luteinizing hormone (LH), thyroid-stimulating hormone (TSH), and chorionic gonadotropin (CG) are glycoprotein hormones of $\alpha\beta$ heterodimeric structure. The $\alpha$ chain is identical in all of these hormones, and distinct $\beta$ chains impart hormone uniqueness. These hormones have a molecular mass in the range of 25–30 kDa depending on the degree of glycosylation and the length of the $\beta$ chain.

### Hormones Are Synthesized & Modified For Full Activity in a Variety of Ways

Some hormones are synthesized in final form and secreted immediately. Included in this class are the hormones derived from cholesterol. Others such as the catecholamines are synthesized in final form and stored in the producing cells. Others are synthesized from precursor molecules in the producing cell, then are processed and secreted upon a physiologic cue (insulin). Finally, still others are converted to active forms from precursor molecules in the periphery ($T_3$ and DHT). All of these examples are discussed in more detail below.

## MANY HORMONES ARE MADE FROM CHOLESTEROL

### Adrenal Steroidogenesis

The adrenal steroid hormones are synthesized from cholesterol. Cholesterol is mostly derived from the plasma, but a small portion is synthesized in situ from acetyl-CoA via mevalonate and squalene. Much of the cholesterol in the adrenal is esterified and stored in cytoplasmic lipid droplets. Upon stimulation of the adrenal by ACTH, an esterase is activated, and the free cholesterol formed is transported into the mitochondrion, where a **cytochrome P450 side chain cleavage enzyme (P450scc)** converts cholesterol to pregnenolone. Cleavage of the side chain involves sequential hydroxylations, first at $C_{22}$ and then at $C_{20}$, followed by side chain cleavage (removal of the six-carbon fragment isocaproaldehyde) to give the 21-carbon steroid (Figure 42–3, top). An ACTH-dependent **steroidogenic acute regulatory (StAR) protein** is essential for the transport of cholesterol to P450scc in the inner mitochondrial membrane.

All mammalian steroid hormones are formed from cholesterol via pregnenolone through a series of reactions that occur in either the mitochondria or endoplasmic reticulum of the adrenal cell. Hydroxylases that require molecular oxygen and NADPH are essential, and dehydrogenases, an isomerase, and a lyase reaction are also necessary for certain steps. There is cellular specificity in adrenal steroidogenesis. For instance, 18-hydroxylase and 19-hydroxysteroid dehydrogenase, which are required for aldosterone synthesis, are found only in the zona glomerulosa cells (the outer region of the adrenal cortex), so that the biosynthesis of this mineralocorticoid is confined to this region. A schematic representation of the pathways involved in the synthesis of the three major classes of adrenal steroids is presented in Figure 42–4. The enzymes are shown in the rectangular boxes, and the modifications at each step are shaded.

#### A. MINERALOCORTICOID SYNTHESIS

Synthesis of aldosterone follows the mineralocorticoid pathway and occurs in the zona glomerulosa. Pregnenolone is converted to progesterone by the action of two smooth endoplasmic reticulum enzymes, **3β-hydroxysteroid dehydrogenase (3β-OHSD)** and **$\Delta^{5,4}$-isomerase.** Progesterone is hydroxylated at the $C_{21}$ position to form 11-deoxycorticosterone (DOC), which is an active ($Na^+$-retaining) mineralocorticoid. The next hydroxylation, at $C_{11}$, produces corticosterone, which has glucocorticoid activity and is a weak mineralocorticoid (it has less than 5% of the potency of aldosterone). In some species (eg, rodents), it is the most potent glucocorticoid.

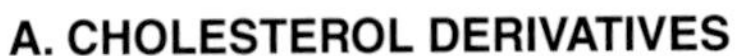

## A. CHOLESTEROL DERIVATIVES

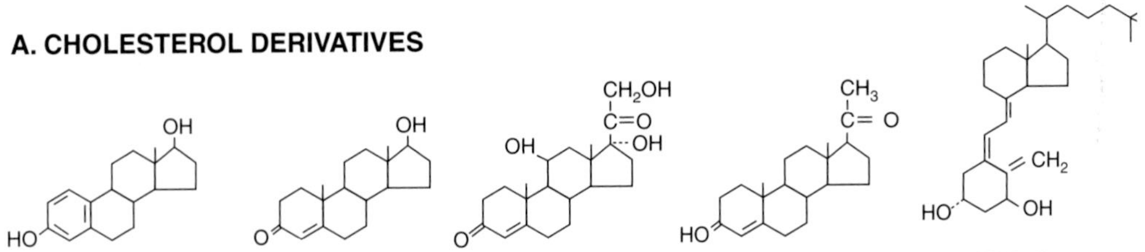

17β-Estradiol    Testosterone    Cortisol    Progesterone    $1{,}25(OH)_2$-$D_3$

## B. TYROSINE DERIVATIVES

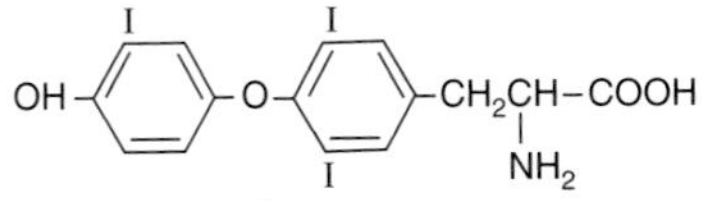

$T^3$

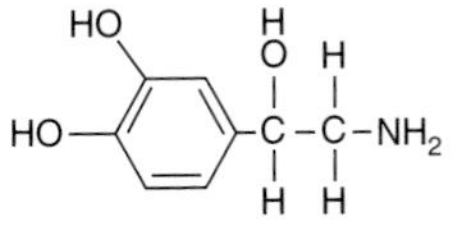

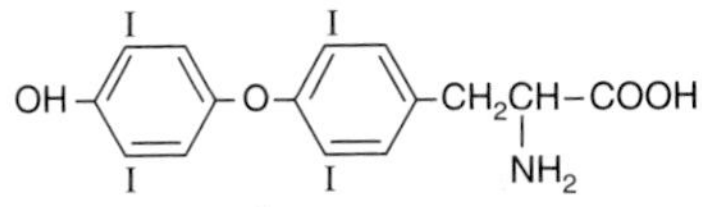

Norepinephrine

$T^4$

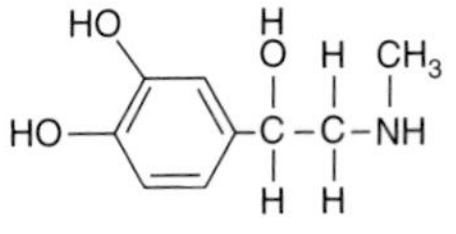

Epinephrine

## C. PEPTIDES OF VARIOUS SIZES

TRH

Structure of human ACTH.

ACTH

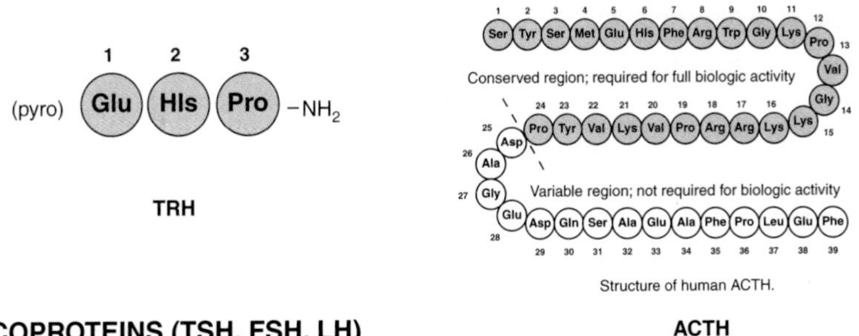

## D. GLYCOPROTEINS (TSH, FSH, LH)

common α subunits
unique β subunits

***Figure 42–2.*** Chemical diversity of hormones. **A.** Cholesterol derivatives. **B.** Tyrosine derivatives. **C.** Peptides of various sizes **D.** Glycoproteins (TSH, FSH, LH) with common α subunits and unique β subunits.

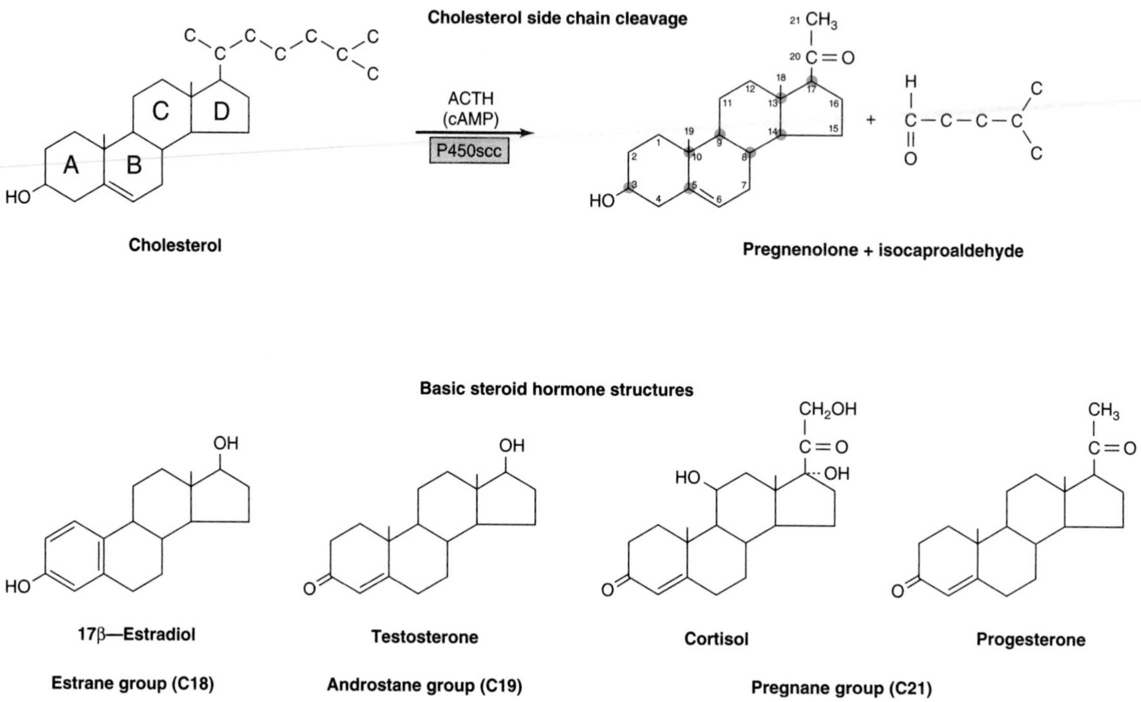

***Figure 42–3.*** Cholesterol side-chain cleavage and basic steroid hormone structures. The basic sterol rings are identified by the letters A–D. The carbon atoms are numbered 1–21 starting with the A ring. Note that the estrane group has 18 carbons (C18), etc.

$C_{21}$ hydroxylation is necessary for both mineralocorticoid and glucocorticoid activity, but most steroids with a $C_{17}$ hydroxyl group have more glucocorticoid and less mineralocorticoid action. In the zona glomerulosa, which does not have the smooth endoplasmic reticulum enzyme 17α-hydroxylase, a mitochondrial 18-hydroxylase is present. The **18-hydroxylase (aldosterone synthase)** acts on corticosterone to form 18-hydroxycorticosterone, which is changed to aldosterone by conversion of the 18-alcohol to an aldehyde. This unique distribution of enzymes and the special regulation of the zona glomerulosa by $K^+$ and angiotensin II have led some investigators to suggest that, in addition to the adrenal being two glands, the adrenal cortex is actually two separate organs.

## B. Glucocorticoid Synthesis

Cortisol synthesis requires three hydroxylases located in the fasciculata and reticularis zones of the adrenal cortex that act sequentially on the $C_{17}$, $C_{21}$, and $C_{11}$ positions. The first two reactions are rapid, while $C_{11}$ hydroxylation is relatively slow. If the $C_{11}$ position is hydroxylated first, the action of **17α-hydroxylase** is impeded and the mineralocorticoid pathway is followed (forming corticosterone or aldosterone, depending on the cell type). 17α-Hydroxylase is a smooth endoplasmic reticulum enzyme that acts upon either progesterone or, more commonly, pregnenolone. 17α-Hydroxyprogesterone is hydroxylated at $C_{21}$ to form 11-deoxycortisol, which is then hydroxylated at $C_{11}$ to form cortisol, the most potent natural glucocorticoid hormone in humans. 21-Hydroxylase is a smooth endoplasmic reticulum enzyme, whereas 11β-hydroxylase is a mitochondrial enzyme. Steroidogenesis thus involves the repeated shuttling of substrates into and out of the mitochondria.

## C. Androgen Synthesis

The major androgen or androgen precursor produced by the adrenal cortex is dehydroepiandrosterone (DHEA). Most 17-hydroxypregnenolone follows the glucocorticoid pathway, but a small fraction is subjected to oxidative fission and removal of the two-carbon side chain through the action of 17,20-lyase. The lyase activity is actually part of the same enzyme (P450c17) that catalyzes 17α-hydroxylation. This is therefore a **dual function protein.** The lyase activity is important in both the adrenals and

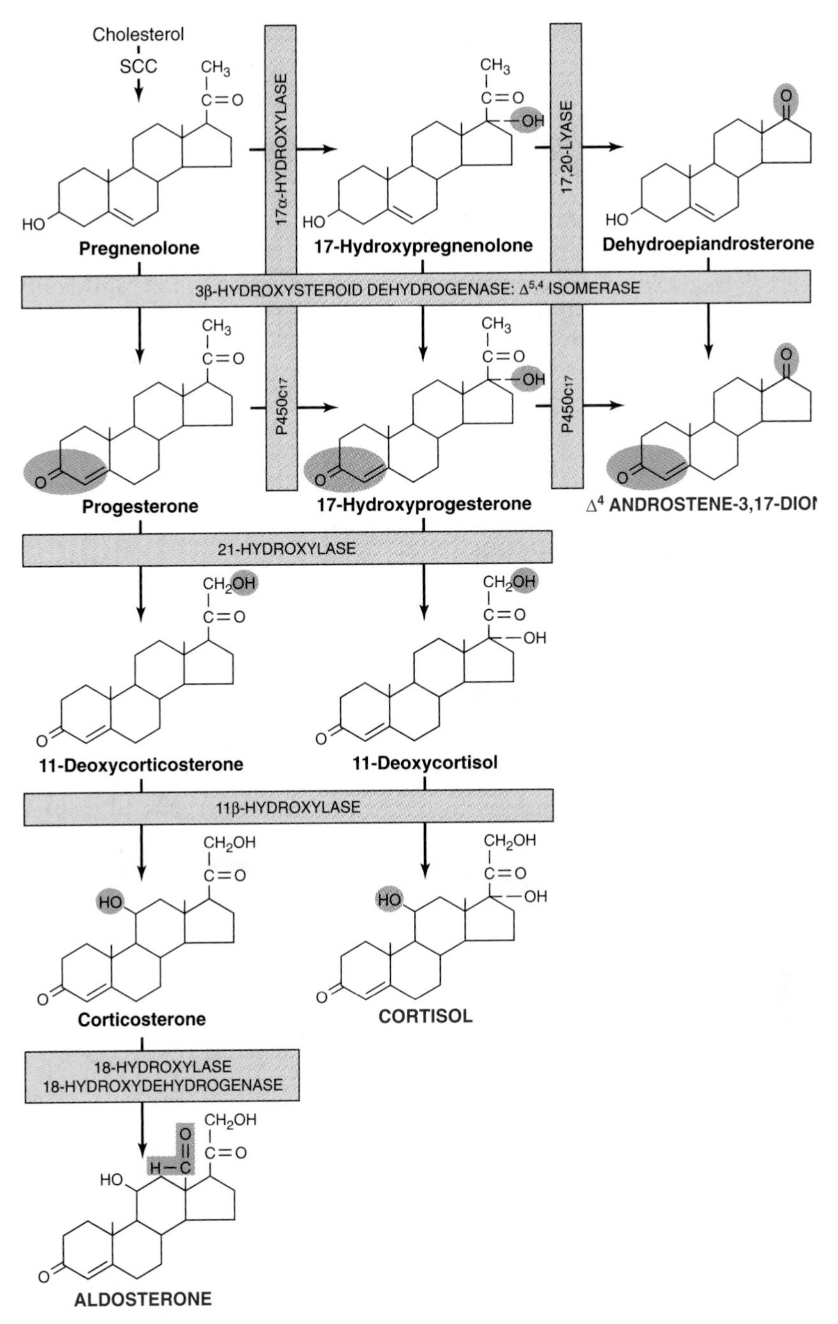


***Figure 42–4.*** Pathways involved in the synthesis of the three major classes of adrenal steroids (mineralocorticoids, glucocorticoids, and androgens). Enzymes are shown in the rectangular boxes, and the modifications at each step are shaded. Note that the 17α-hydroxylase and 17,20-lyase activities are both part of one enzyme, designated P450c17. (Slightly modified and reproduced, with permission, from Harding BW: In: *Endocrinology*, vol 2. DeGroot LJ [editor]. Grune & Stratton, 1979.)

the gonads and acts exclusively on 17α-hydroxy-containing molecules. Adrenal androgen production increases markedly if glucocorticoid biosynthesis is impeded by the lack of one of the hydroxylases **(adrenogenital syndrome).** DHEA is really a prohormone, since the actions of 3β-OHSD and $\Delta^{5,4}$-isomerase convert the weak androgen DHEA into the more potent **androstenedione.** Small amounts of androstenedione are also formed in the adrenal by the action of the lyase on 17α-hydroxyprogesterone. Reduction of androstenedione at the $C_{17}$ position results in the formation of **testosterone,** the most potent adrenal androgen. Small amounts of testosterone are produced in the adrenal by this mechanism, but most of this conversion occurs in the testes.

## Testicular Steroidogenesis

Testicular androgens are synthesized in the interstitial tissue by the Leydig cells. The immediate precursor of the gonadal steroids, as for the adrenal steroids, is cholesterol. The rate-limiting step, as in the adrenal, is delivery of cholesterol to the inner membrane of the mitochondria by the transport protein StAR. Once in the proper location, cholesterol is acted upon by the side chain cleavage enzyme P450scc. The conversion of cholesterol to pregnenolone is identical in adrenal, ovary, and testis. In the latter two tissues, however, the reaction is promoted by LH rather than ACTH.

The conversion of pregnenolone to testosterone requires the action of five enzyme activities contained in three proteins: (1) 3β-hydroxysteroid dehydrogenase (3β-OHSD) and $\Delta^{5,4}$-isomerase; (2) 17α-hydroxylase and 17,20-lyase; and (3) 17β-hydroxysteroid dehydrogenase (17β-OHSD). This sequence, referred to as the **progesterone (or $\Delta^4$) pathway,** is shown on the right side of Figure 42–5. Pregnenolone can also be converted to testosterone by the **dehydroepiandrosterone (or $\Delta^5$) pathway,** which is illustrated on the left side of Figure 42–5. The $\Delta^5$ route appears to be most used in human testes.

The five enzyme activities are localized in the microsomal fraction in rat testes, and there is a close functional association between the activities of 3β-OHSD and $\Delta^{5,4}$-isomerase and between those of a 17α-hydroxylase and 17,20-lyase. These enzyme pairs, both contained in a single protein, are shown in the general reaction sequence in Figure 42–5.

## Dihydrotestosterone Is Formed From Testosterone in Peripheral Tissues

Testosterone is metabolized by two pathways. One involves oxidation at the 17 position, and the other involves reduction of the A ring double bond and the 3-ketone. Metabolism by the first pathway occurs in many tissues, including liver, and produces 17-ketosteroids that are generally inactive or less active than the parent compound. Metabolism by the second pathway, which is less efficient, occurs primarily in target tissues and produces the potent metabolite dihydrotestosterone (DHT).

The most significant metabolic product of testosterone is DHT, since in many tissues, including prostate, external genitalia, and some areas of the skin, this is the active form of the hormone. The plasma content of DHT in the adult male is about one-tenth that of testosterone, and approximately 400 μg of DHT is produced daily as compared with about 5 mg of testosterone. About 50–100 μg of DHT are secreted by the testes. The rest is produced peripherally from testosterone in a reaction catalyzed by the NADPH-dependent **5α-reductase** (Figure 42–6). Testosterone can thus be considered a prohormone, since it is converted into a much more potent compound (dihydrotestosterone) and since most of this conversion occurs outside the testes. Some estradiol is formed from the peripheral aromatization of testosterone, particularly in males.

## Ovarian Steroidogenesis

The estrogens are a family of hormones synthesized in a variety of tissues. 17β-Estradiol is the primary estrogen of ovarian origin. In some species, estrone, synthesized in numerous tissues, is more abundant. In pregnancy, relatively more estriol is produced, and this comes from the placenta. The general pathway and the subcellular localization of the enzymes involved in the early steps of estradiol synthesis are the same as those involved in androgen biosynthesis. Features unique to the ovary are illustrated in Figure 42–7.

Estrogens are formed by the aromatization of androgens in a complex process that involves three hydroxylation steps, each of which requires $O_2$ and NADPH. The **aromatase enzyme complex** is thought to include a P450 monooxygenase. Estradiol is formed if the substrate of this enzyme complex is testosterone, whereas estrone results from the aromatization of androstenedione.

The cellular source of the various ovarian steroids has been difficult to unravel, but a transfer of substrates between two cell types is involved. Theca cells are the source of androstenedione and testosterone. These are converted by the aromatase enzyme in granulosa cells to estrone and estradiol, respectively. Progesterone, a precursor for all steroid hormones, is produced and secreted by the corpus luteum as an end-product hormone because these cells do not contain the enzymes necessary to convert progesterone to other steroid hormones (Figure 42–8).

Significant amounts of estrogens are produced by the peripheral aromatization of androgens. In human males, the peripheral aromatization of testosterone to estradiol ($E_2$) accounts for 80% of the production of the latter. In females, adrenal androgens are important

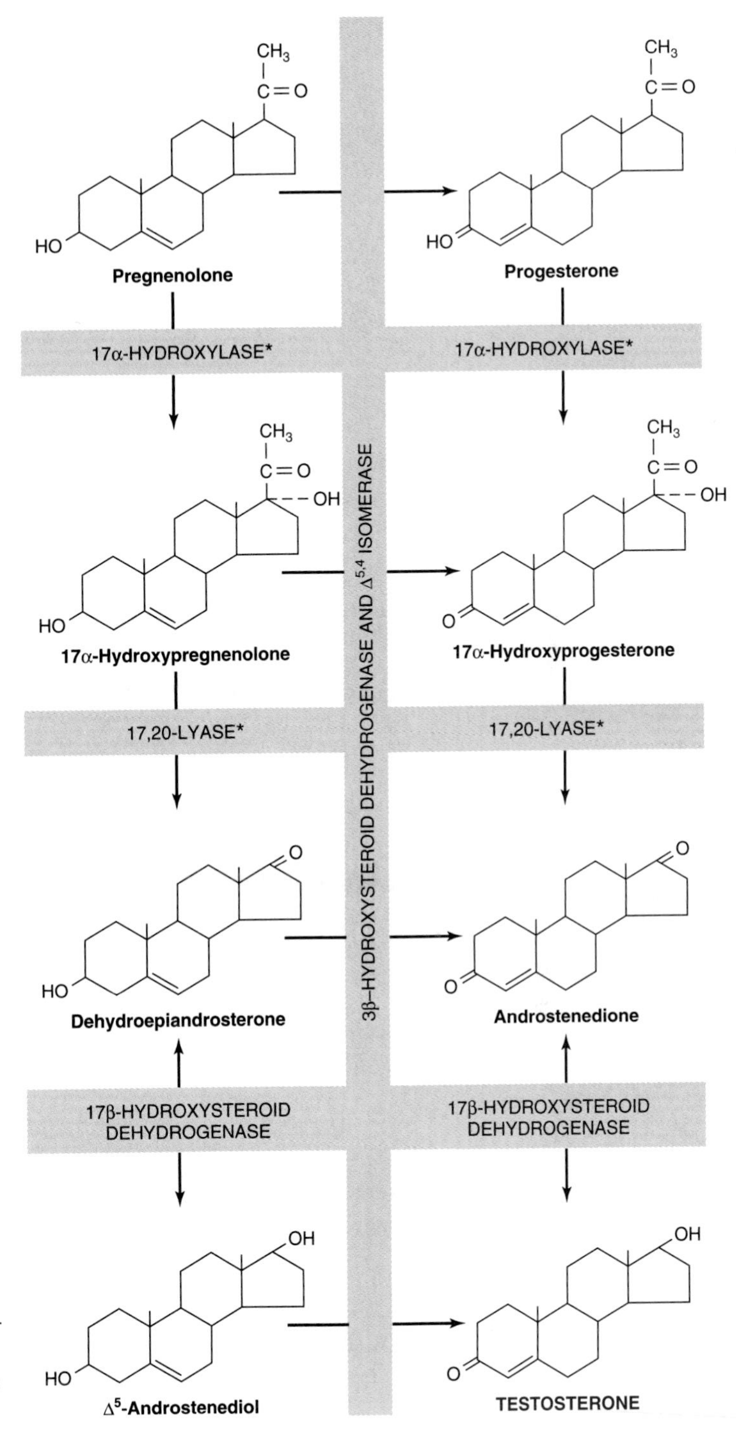


***Figure 42–5.*** Pathways of testosterone biosynthesis. The pathway on the left side of the figure is called the $\Delta^5$ or dehydroepiandrosterone pathway; the pathway on the right side is called the $\Delta^4$ or progesterone pathway. The asterisk indicates that the 17α-hydroxylase and 17,20-lyase activities reside in a single protein, P450c17.

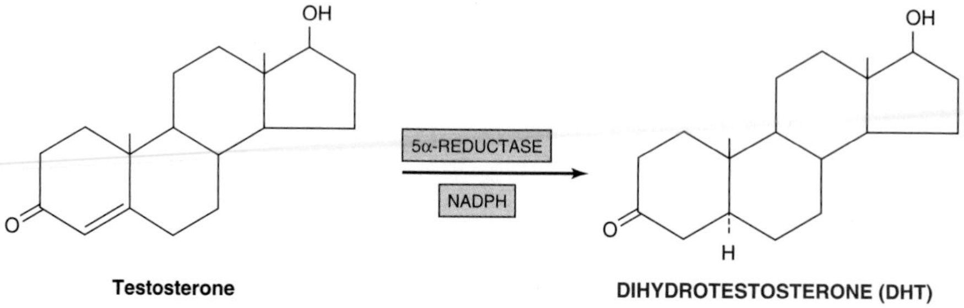

OH

5α-REDUCTASE

NADPH

O

H

Testosterone

DIHYDROTESTOSTERONE (DHT)

***Figure 42–6.*** Dihydrotestosterone is formed from testosterone through action of the enzyme 5α-reductase.

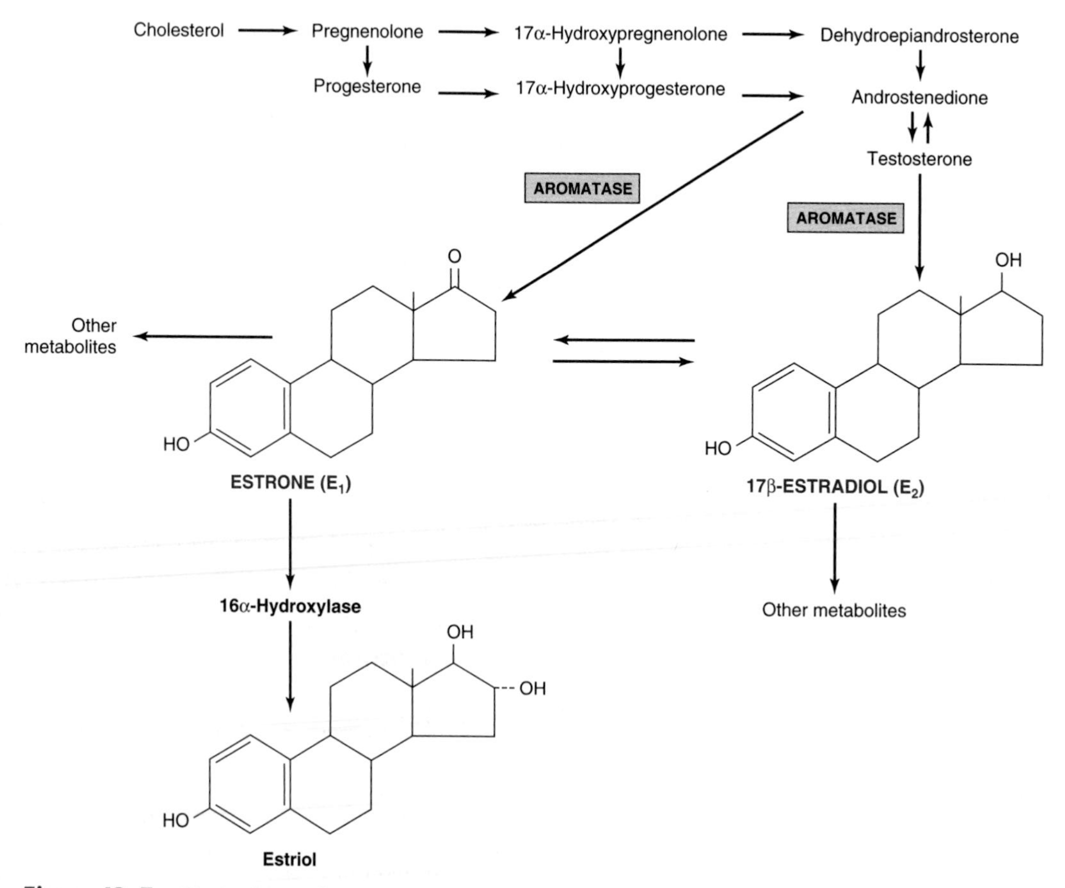


***Figure 42–7.*** Biosynthesis of estrogens. (Slightly modified and reproduced, with permission, from Ganong WF: *Review of Medical Physiology,* 20th ed. McGraw-Hill, 2001.)

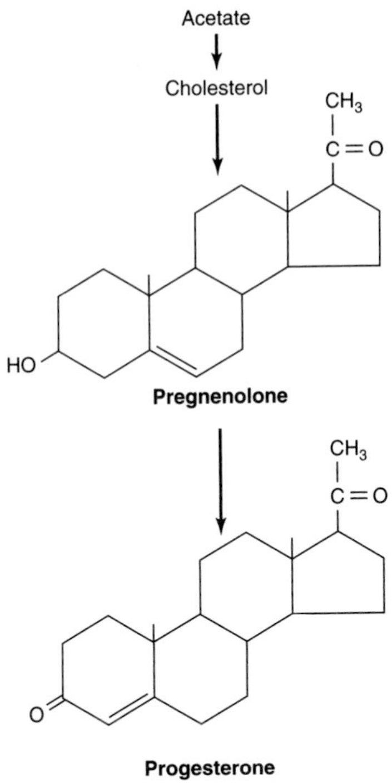

Acetate
Cholesterol
Pregnenolone
Progesterone

*Figure 42–8.* Biosynthesis of progesterone in the corpus luteum.

substrates, since as much as 50% of the $E_2$ produced during pregnancy comes from the aromatization of androgens. Finally, conversion of androstenedione to estrone is the major source of estrogens in postmenopausal women. Aromatase activity is present in adipose cells and also in liver, skin, and other tissues. Increased activity of this enzyme may contribute to the "estrogenization" that characterizes such diseases as cirrhosis of the liver, hyperthyroidism, aging, and obesity.

## $1,25(OH)_2$-$D_3$ (Calcitriol) Is Synthesized From a Cholesterol Derivative

$1,25(OH)_2$-$D_3$ is produced by a complex series of enzymatic reactions that involve the plasma transport of precursor molecules to a number of different tissues (Figure 42–9). One of these precursors is vitamin D—really not a vitamin, but this common name persists. The active molecule, $1,25(OH)_2$-$D_3$, is transported to other organs where it activates biologic processes in a manner similar to that employed by the steroid hormones.

### A. Skin

Small amounts of the precursor for $1,25(OH)_2$-$D_3$ synthesis are present in food (fish liver oil, egg yolk), but most of the precursor for $1,25(OH)_2$-$D_3$ synthesis is produced in the malpighian layer of the epidermis from 7-dehydrocholesterol in an ultraviolet light-mediated, nonenzymatic **photolysis** reaction. The extent of this conversion is related directly to the intensity of the exposure and inversely to the extent of pigmentation in the skin. There is an age-related loss of 7-dehydrocholesterol in the epidermis that may be related to the negative calcium balance associated with old age.

### B. Liver

A specific transport protein called the **vitamin D-binding protein** binds vitamin $D_3$ and its metabolites and moves vitamin $D_3$ from the skin or intestine to the liver, where it undergoes 25-hydroxylation, the first obligatory reaction in the production of $1,25(OH)_2$-$D_3$. 25-Hydroxylation occurs in the endoplasmic reticulum in a reaction that requires magnesium, NADPH, molecular oxygen, and an uncharacterized cytoplasmic factor. Two enzymes are involved: an NADPH-dependent cytochrome P450 reductase and a cytochrome P450. This reaction is not regulated, and it also occurs with low efficiency in kidney and intestine. The $25(OH)_2$-$D_3$ enters the circulation, where it is the major form of vitamin D found in plasma, and is transported to the kidney by the vitamin D-binding protein.

### C. Kidney

$25(OH)_2$-$D_3$ is a weak agonist and must be modified by hydroxylation at position $C_1$ for full biologic activity. This is accomplished in mitochondria of the renal proximal convoluted tubule by a three-component monooxygenase reaction that requires NADPH, $Mg^{2+}$, molecular oxygen, and at least three enzymes: (1) a flavoprotein, renal ferredoxin reductase; (2) an iron sulfur protein, renal ferredoxin; and (3) cytochrome P450. This system produces $1,25(OH)_2$-$D_3$, which is the most potent naturally occurring metabolite of vitamin D.

# CATECHOLAMINES & THYROID HORMONES ARE MADE FROM TYROSINE

## Catecholamines Are Synthesized in Final Form & Stored in Secretion Granules

Three amines—dopamine, norepinephrine, and epinephrine—are synthesized from tyrosine in the chromaffin cells of the adrenal medulla. The major product of the adrenal medulla is epinephrine. This compound constitutes about 80% of the catecholamines in the medulla, and it is not made in extramedullary tissue. In contrast, most of the norepinephrine present in organs innervated by sympathetic nerves is made in situ (about 80% of the total), and most of the rest is made in other nerve endings and reaches the target sites via the circu-

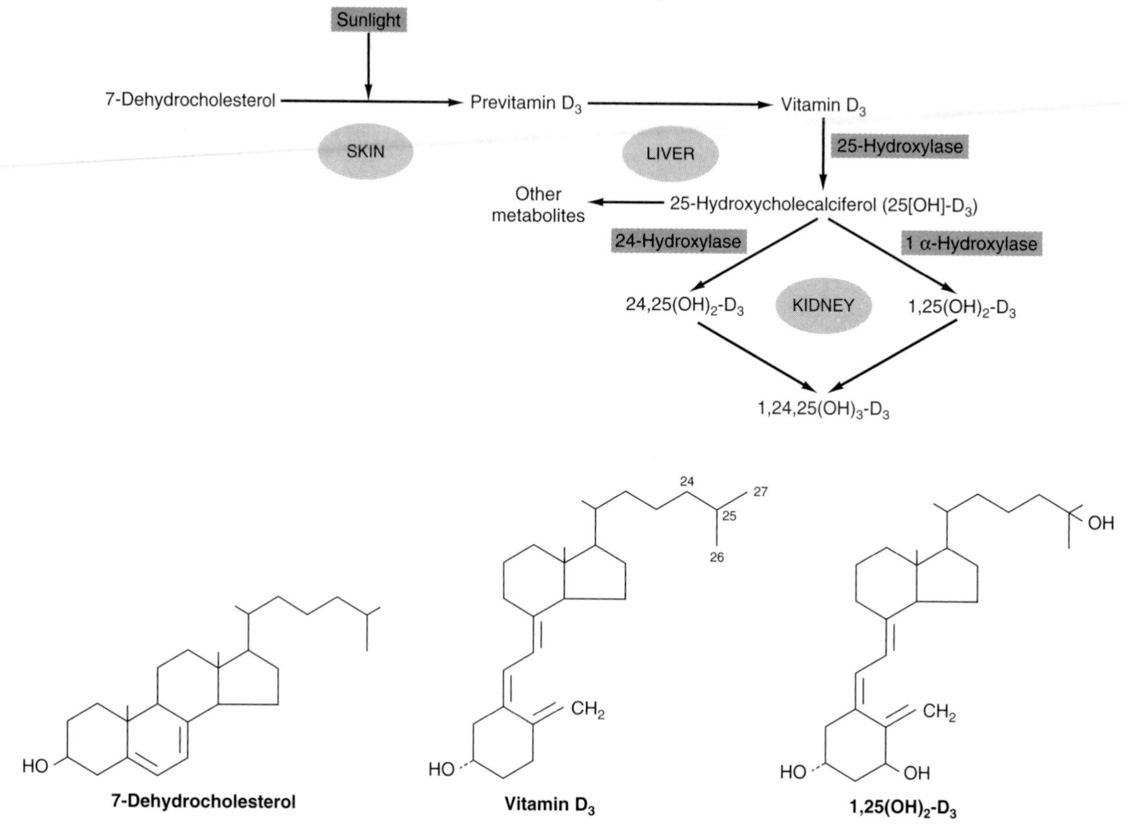


***Figure 42–9.*** Formation and hydroxylation of vitamin $D_3$. 25-Hydroxylation takes place in the liver, and the other hydroxylations occur in the kidneys. 25,26$(OH)_2$-$D_3$ and 1,25,26$(OH)_3$-$D_3$ are probably formed as well. The formulas of 7-dehydrocholesterol, vitamin $D_3$, and 1,25$(OH)_2$-$D_3$ are also shown. (Modified and reproduced, with permission, from Ganong WF: *Review of Medical Physiology,* 20th ed. McGraw-Hill, 2001.)

lation. Epinephrine and norepinephrine may be produced and stored in different cells in the adrenal medulla and other chromaffin tissues.

The conversion of tyrosine to epinephrine requires four sequential steps: (1) ring hydroxylation; (2) decarboxylation; (3) side chain hydroxylation to form norepinephrine; and (4) N-methylation to form epinephrine. The biosynthetic pathway and the enzymes involved are illustrated in Figure 42–10.

### A. Tyrosine Hydroxylase Is Rate-Limiting for Catecholamine Biosynthesis

**Tyrosine** is the immediate precursor of catecholamines, and **tyrosine hydroxylase** is the rate-limiting enzyme in catecholamine biosynthesis. Tyrosine hydroxylase is found in both soluble and particle-bound forms only in tissues that synthesize catecholamines; it functions as an oxidoreductase, with tetrahydropteridine as a cofactor, to convert L-tyrosine to L-dihydroxyphenylalanine **(L-dopa).** As the rate-limiting enzyme, tyrosine hydroxylase is regulated in a variety of ways. The most important mechanism involves feedback inhibition by the catecholamines, which compete with the enzyme for the pteridine cofactor. Catecholamines cannot cross the blood-brain barrier; hence, in the brain they must be synthesized locally. In certain central nervous system diseases (eg, Parkinson's disease), there is a local deficiency of dopamine synthesis. L-Dopa, the precursor of dopamine, readily crosses the blood-brain barrier and so is an important agent in the treatment of Parkinson's disease.

### B. Dopa Decarboxylase Is Present in All Tissues

This soluble enzyme requires pyridoxal phosphate for the conversion of L-dopa to 3,4-dihydroxyphenylethylamine **(dopamine).** Compounds that resemble L-dopa, such as α-methyldopa, are competitive inhibitors of this reaction. α-Methyldopa is effective in treating some kinds of hypertension.

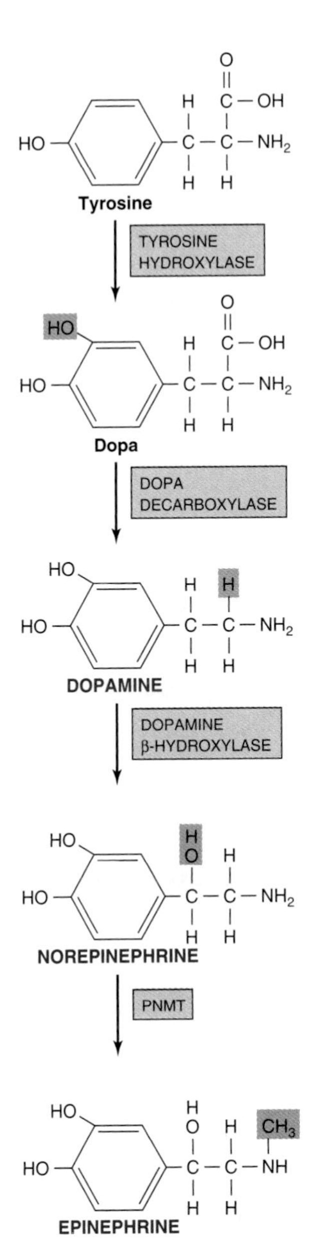


***Figure 42–10.*** Biosynthesis of catecholamines. (PNMT, phenylethanolamine-*N*-methyltransferase.)

### C. Dopamine β-Hydroxylase (DBH) Catalyzes the Conversion of Dopamine to Norepinephrine

DBH is a monooxygenase and uses ascorbate as an electron donor, copper at the active site, and fumarate as modulator. DBH is in the particulate fraction of the medullary cells, probably in the secretion granule; thus, the conversion of dopamine to **norepinephrine** occurs in this organelle.

### D. Phenylethanolamine-*N*-Methyltransferase (PNMT) Catalyzes the Production of Epinephrine

PNMT catalyzes the N-methylation of norepinephrine to form **epinephrine** in the epinephrine-forming cells of the adrenal medulla. Since PNMT is soluble, it is assumed that norepinephrine-to-epinephrine conversion occurs in the cytoplasm. The synthesis of PNMT is induced by glucocorticoid hormones that reach the medulla via the intra-adrenal portal system. This special system provides for a 100-fold steroid concentration gradient over systemic arterial blood, and this high intra-adrenal concentration appears to be necessary for the induction of PNMT.

## $T_3$ & $T_4$ Illustrate the Diversity in Hormone Synthesis

The formation of **triiodothyronine ($T_3$)** and **tetraiodothyronine (thyroxine; $T_4$)** (see Figure 42–2) illustrates many of the principles of diversity discussed in this chapter. These hormones require a rare element (iodine) for bioactivity; they are synthesized as part of a very large precursor molecule (thyroglobulin); they are stored in an intracellular reservoir (colloid); and there is peripheral conversion of $T_4$ to $T_3$, which is a much more active hormone.

The thyroid hormones $T_3$ and $T_4$ are unique in that iodine (as iodide) is an essential component of both. In most parts of the world, iodine is a scarce component of soil, and for that reason there is little in food. A complex mechanism has evolved to acquire and retain this crucial element and to convert it into a form suitable for incorporation into organic compounds. At the same time, the thyroid must synthesize thyronine from tyrosine, and this synthesis takes place in thyroglobulin (Figure 42–11).

**Thyroglobulin** is the precursor of $T_4$ and $T_3$. It is a large iodinated, glycosylated protein with a molecular mass of 660 kDa. Carbohydrate accounts for 8–10% of the weight of thyroglobulin and iodide for about 0.2–1%, depending upon the iodine content in the diet. Thyroglobulin is composed of two large subunits. It contains 115 tyrosine residues, each of which is a potential site of iodination. About 70% of the iodide in thyroglobulin exists in the inactive precursors, **monoiodotyrosine (MIT)** and **diiodotyrosine (DIT),** while 30% is in the **iodothyronyl residues,** $T_4$ and $T_3$. When iodine supplies are sufficient, the $T_4$:$T_3$ ratio is about 7:1. In **iodine deficiency,** this ratio decreases, as does the DIT:MIT ratio. Thyroglobulin, a large molecule of about 5000 amino acids, provides the confor-

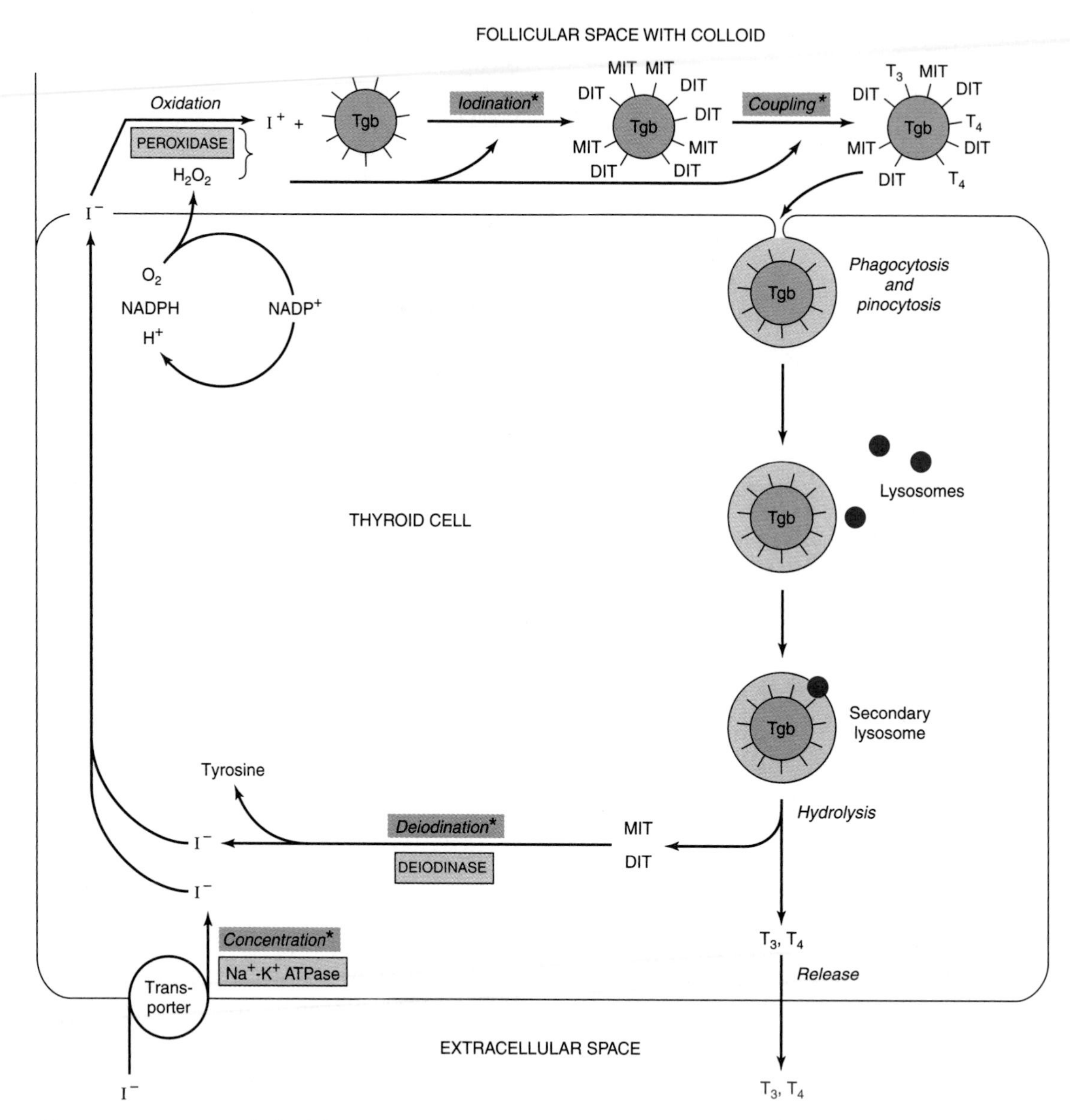


***Figure 42–11.*** Model of iodide metabolism in the thyroid follicle. A follicular cell is shown facing the follicular lumen (top) and the extracellular space (at bottom). Iodide enters the thyroid primarily through a transporter (bottom left). Thyroid hormone synthesis occurs in the follicular space through a series of reactions, many of which are peroxidase-mediated. Thyroid hormones, stored in the colloid in the follicular space, are released from thyroglobulin by hydrolysis inside the thyroid cell. (Tgb, thyroglobulin; MIT, monoiodotyrosine; DIT, diiodotyrosine; $T_3$, triiodothyronine; $T_4$, tetraiodothyronine.) Asterisks indicate steps or processes that are inherited enzyme deficiencies which cause congenital goiter and often result in hypothyroidism.

mation required for tyrosyl coupling and iodide organification necessary in the formation of the diaminoacid thyroid hormones. It is synthesized in the basal portion of the cell and moves to the lumen, where it is a storage form of $T_3$ and $T_4$ in the colloid; several weeks' supply of these hormones exist in the normal thyroid. Within minutes after stimulation of the thyroid by TSH, colloid reenters the cell and there is a marked increase of phagolysosome activity. Various acid proteases and peptidases hydrolyze the thyroglobulin into its constituent amino acids, including $T_4$ and $T_3$, which are discharged from the basal portion of the cell (see Figure 42–11). Thyroglobulin is thus a very large prohormone.

### Iodide Metabolism Involves Several Discrete Steps

The thyroid is able to concentrate $I^-$ against a strong electrochemical gradient. This is an energy-dependent process and is linked to the $Na^+$-$K^+$ ATPase-dependent thyroidal $I^-$ transporter. The ratio of iodide in thyroid to iodide in serum (T:S ratio) is a reflection of the activity of this transporter. This activity is primarily controlled by TSH and ranges from 500:1 in animals chronically stimulated with TSH to 5:1 or less in hypophysectomized animals (no TSH). The T:S ratio in humans on a normal iodine diet is about 25:1.

The thyroid is the only tissue that can oxidize $I^-$ to a higher valence state, an obligatory step in $I^-$ organification and thyroid hormone biosynthesis. This step involves a heme-containing peroxidase and occurs at the luminal surface of the follicular cell. Thyroperoxidase, a tetrameric protein with a molecular mass of 60 kDa, requires hydrogen peroxide as an oxidizing agent. The $H_2O_2$ is produced by an NADPH-dependent enzyme resembling cytochrome *c* reductase. A number of compounds inhibit $I^-$ oxidation and therefore its subsequent incorporation into MIT and DIT. The most important of these are the thiourea drugs. They are used as antithyroid drugs because of their ability to inhibit thyroid hormone biosynthesis at this step. Once iodination occurs, the iodine does not readily leave the thyroid. Free tyrosine can be iodinated, but it is not incorporated into proteins since no tRNA recognizes iodinated tyrosine.

The coupling of two DIT molecules to form $T_4$—or of an MIT and DIT to form $T_3$—occurs within the thyroglobulin molecule. A separate coupling enzyme has not been found, and since this is an oxidative process it is assumed that the same thyroperoxidase catalyzes this reaction by stimulating free radical formation of iodotyrosine. This hypothesis is supported by the observation that the same drugs which inhibit $I^-$ oxidation also inhibit coupling. The formed thyroid hormones remain as integral parts of thyroglobulin until the latter is degraded, as described above.

A deiodinase removes $I^-$ from the inactive mono- and diiodothyronine molecules in the thyroid. This mechanism provides a substantial amount of the $I^-$ used in $T_3$ and $T_4$ biosynthesis. A peripheral deiodinase in target tissues such as pituitary, kidney, and liver selectively removes $I^-$ from the 5′ position of $T_4$ to make $T_3$ (see Figure 42–2), which is a much more active molecule. In this sense, $T_4$ can be thought of as a prohormone, though it does have some intrinsic activity.

## SEVERAL HORMONES ARE MADE FROM LARGER PEPTIDE PRECURSORS

Formation of the critical disulfide bridges in insulin requires that this hormone be first synthesized as part of a larger precursor molecule, proinsulin. This is conceptually similar to the example of the thyroid hormones, which can only be formed in the context of a much larger molecule. Several other hormones are synthesized as parts of large precursor molecules, not because of some special structural requirement but rather as a mechanism for controlling the available amount of the active hormone. PTH and angiotensin II are examples of this type of regulation. Another interesting example is the POMC protein, which can be processed into many different hormones in a tissue-specific manner. These examples are discussed in detail below.

### Insulin Is Synthesized as a Preprohormone & Modified Within the β Cell

Insulin has an AB heterodimeric structure with one intrachain (A6–A11) and two interchain disulfide bridges (A7–B7 and A20–B19) (Figure 42–12). The A and B chains could be synthesized in the laboratory, but attempts at a biochemical synthesis of the mature insulin molecule yielded very poor results. The reason for this became apparent when it was discovered that insulin is synthesized as a **preprohormone** (molecular weight approximately 11,500), which is the prototype for peptides that are processed from larger precursor molecules. The hydrophobic 23-amino-acid pre-, or leader, sequence directs the molecule into the cisternae of the endoplasmic reticulum and then is removed. This results in the 9000-MW proinsulin molecule, which provides the conformation necessary for the proper and efficient formation of the disulfide bridges. As shown in Figure 42–12, the sequence of proinsulin, starting from the amino terminal, is B chain—connecting (C) peptide—A chain. The proinsulin molecule undergoes a series of site-specific peptide cleavages that result in the formation of equimolar amounts of mature insulin and C peptide. These enzymatic cleavages are summarized in Figure 42–12.

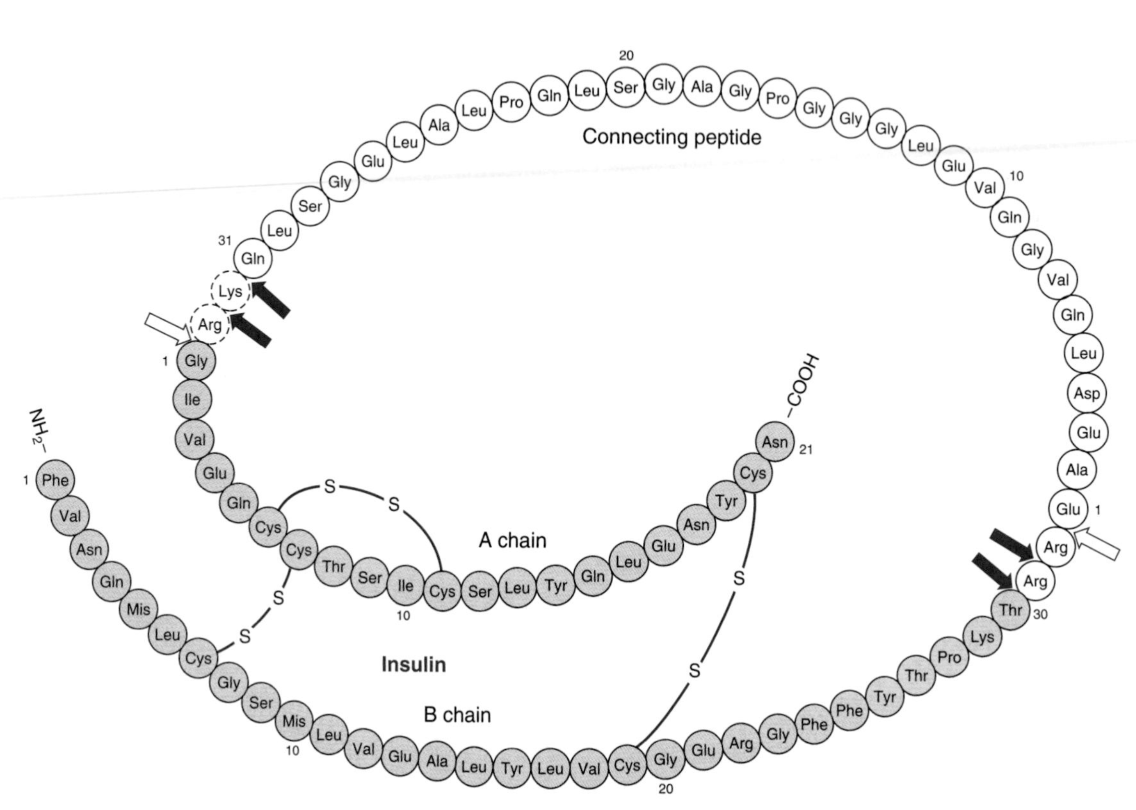


***Figure 42–12.*** Structure of human proinsulin. Insulin and C-peptide molecules are connected at two sites by dipeptide links. An initial cleavage by a trypsin-like enzyme (open arrows) followed by several cleavages by a carboxypeptidase-like enzyme (solid arrows) results in the production of the heterodimeric (AB) insulin molecule (light blue) and the C-peptide.

## Parathyroid Hormone (PTH) Is Secreted as an 84-Amino-Acid Peptide

The immediate precursor of PTH is **proPTH,** which differs from the native 84-amino-acid hormone by having a highly basic hexapeptide amino terminal extension. The primary gene product and the immediate precursor for proPTH is the 115-amino-acid **preproPTH.** This differs from proPTH by having an additional 25-amino-acid amino terminal extension that, in common with the other leader or signal sequences characteristic of secreted proteins, is hydrophobic. The complete structure of preproPTH and the sequences of proPTH and PTH are illustrated in Figure 42–13. $PTH_{1-34}$ has full biologic activity, and the region 25–34 is primarily responsible for receptor binding.

The biosynthesis of PTH and its subsequent secretion are regulated by the plasma ionized calcium ($Ca^{2+}$) concentration through a complex process. An acute decrease of $Ca^{2+}$ results in a marked increase of PTH mRNA, and this is followed by an increased rate of PTH synthesis and secretion. However, about 80–90% of the proPTH synthesized cannot be accounted for as intact PTH in cells or in the incubation medium of experimental systems. This finding led to the conclusion that most of the proPTH synthesized is quickly degraded. It was later discovered that this rate of degradation decreases when $Ca^{2+}$ concentrations are low, and it increases when $Ca^{2+}$ concentrations are high. Very specific fragments of PTH are generated during its proteolytic digestion (Figure 42–13). A number of proteolytic enzymes, including cathepsins B and D, have been identified in parathyroid tissue. Cathepsin B cleaves PTH into two fragments: $PTH_{1-36}$ and $PTH_{37-84}$. $PTH_{37-84}$ is not further degraded; however, $PTH_{1-36}$ is rapidly and progressively cleaved into di- and tripeptides. Most of the proteolysis of PTH occurs within the gland, but a number of studies confirm that PTH, once secreted, is proteolytically degraded in other tissues, especially the liver, by similar mechanisms.

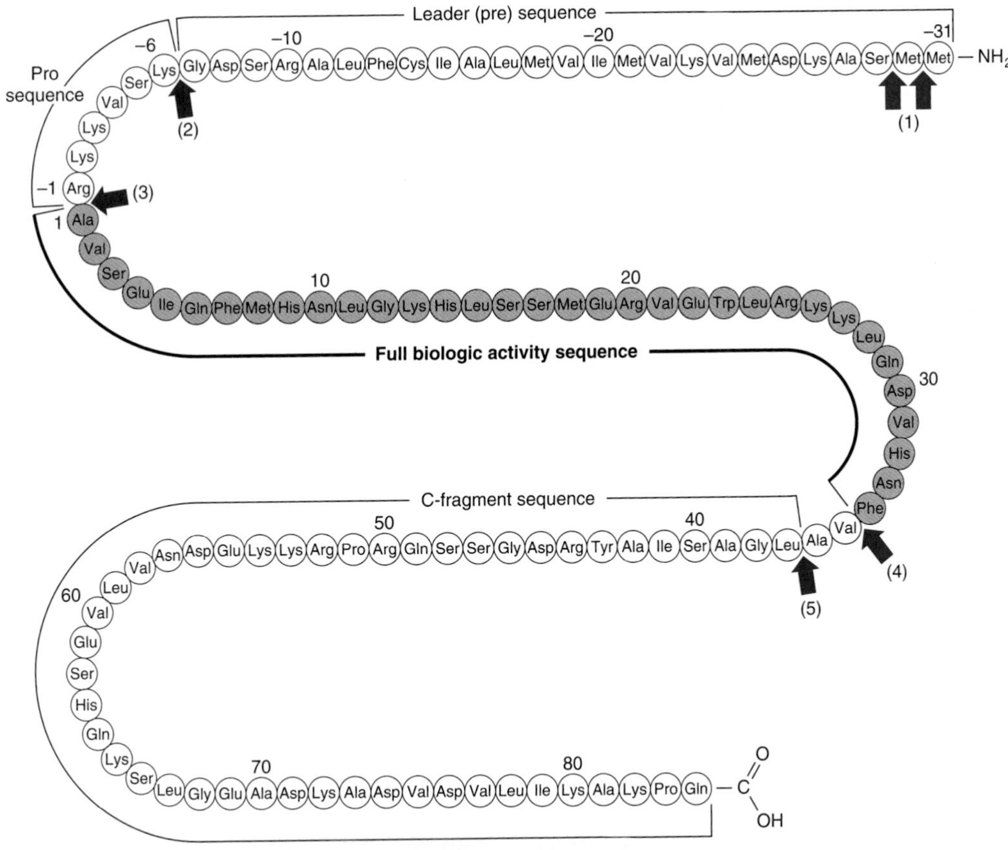


***Figure 42–13.*** Structure of bovine preproparathyroid hormone. Arrows indicate sites cleaved by processing enzymes in the parathyroid gland (1–5) and in the liver after secretion of the hormone (4–5). The biologically active region of the molecule is flanked by sequence not required for activity on target receptors. (Slightly modified and reproduced, with permission, from Habener JF: Recent advances in parathyroid hormone research. Clin Biochem 1981;14:223.)

## Angiotensin II Is Also Synthesized From a Large Precursor

The renin-angiotensin system is involved in the regulation of blood pressure and electrolyte metabolism (through production of aldosterone). The primary hormone involved in these processes is angiotensin II, an octapeptide made from angiotensinogen (Figure 42–14). Angiotensinogen, a large $\alpha_2$-globulin made in liver, is the substrate for renin, an enzyme produced in the juxtaglomerular cells of the renal afferent arteriole. The position of these cells makes them particularly sensitive to blood pressure changes, and many of the physiologic regulators of renin release act through renal baroreceptors. The juxtaglomerular cells are also sensitive to changes of $Na^+$ and $Cl^-$ concentration in the renal tubular fluid; therefore, any combination of factors that decreases fluid volume (dehydration, decreased blood pressure, fluid or blood loss) or decreases NaCl concentration stimulates renin release. Renal sympathetic nerves that terminate in the juxtaglomerular cells mediate the central nervous system and postural effects on renin release independently of the baroreceptor and salt effects, a mechanism that involves the β-adrenergic receptor. Renin acts upon the substrate angiotensinogen to produce the decapeptide angiotensin I.

Angiotensin-converting enzyme, a glycoprotein found in lung, endothelial cells, and plasma, removes two carboxyl terminal amino acids from the decapeptide angiotensin I to form angiotensin II in a step that is not thought to be rate-limiting. Various nonapeptide analogs of angiotensin I and other compounds act as competitive inhibitors of converting enzyme and are used to treat renin-dependent hypertension. These are

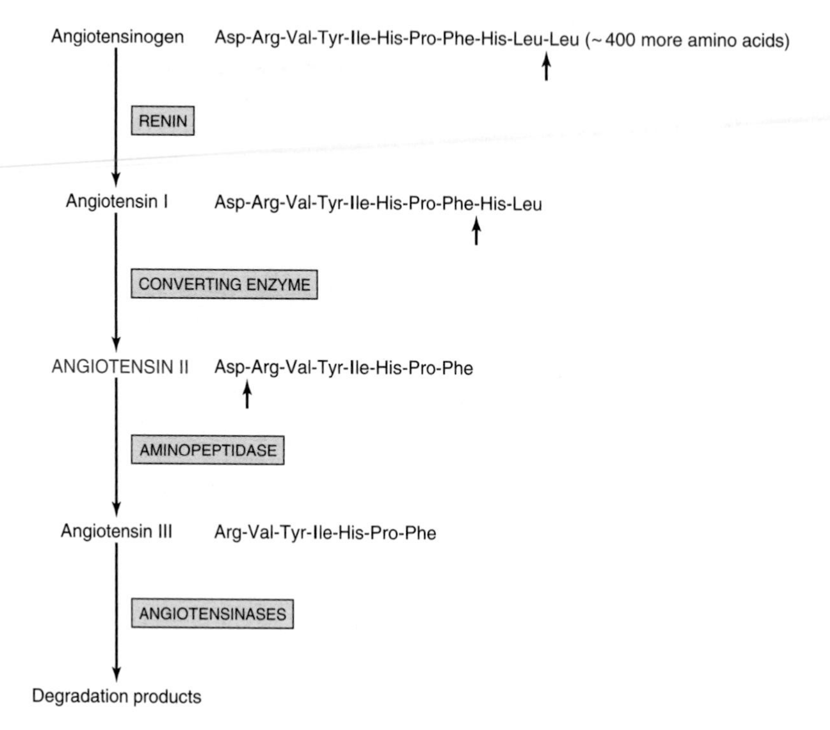


***Figure 42–14.*** Formation and metabolism of angiotensins. Small arrows indicate cleavage sites.

referred to as **angiotensin-converting enzyme (ACE) inhibitors.** Angiotensin II increases blood pressure by causing vasoconstriction of the arteriole and is a very potent vasoactive substance. It inhibits renin release from the juxtaglomerular cells and is a potent stimulator of aldosterone production. This results in $Na^+$ retention, volume expansion, and increased blood pressure.

In some species, angiotensin II is converted to the heptapeptide angiotensin III (Figure 42–14), an equally potent stimulator of aldosterone production. In humans, the plasma level of angiotensin II is four times greater than that of angiotensin III, so most effects are exerted by the octapeptide. Angiotensins II and III are rapidly inactivated by angiotensinases.

Angiotensin II binds to specific adrenal cortex glomerulosa cell receptors. The hormone-receptor interaction does not activate adenylyl cyclase, and cAMP does not appear to mediate the action of this hormone. The actions of angiotensin II, which are to stimulate the conversion of cholesterol to pregnenolone and of corticosterone to 18-hydroxycorticosterone and aldosterone, may involve changes in the concentration of intracellular calcium and of phospholipid metabolites by mechanisms similar to those described in Chapter 43.

## Complex Processing Generates the Pro-opiomelanocortin (POMC) Peptide Family

The POMC family consists of peptides that act as hormones (ACTH, LPH, MSH) and others that may serve as neurotransmitters or neuromodulators (endorphins) (see Figure 42–15). POMC is synthesized as a precursor molecule of 285 amino acids and is processed differently in various regions of the pituitary.

The POMC gene is expressed in the anterior and intermediate lobes of the pituitary. The most conserved sequences between species are within the amino terminal fragment, the ACTH region, and the β-endorphin region. POMC or related products are found in several other vertebrate tissues, including the brain, placenta, gastrointestinal tract, reproductive tract, lung, and lymphocytes.

The POMC protein is processed differently in the anterior lobe than in the intermediate lobe. The intermediate lobe of the pituitary is rudimentary in adult humans, but it is active in human fetuses and in pregnant women during late gestation and is also active in many animal species. Processing of the POMC protein in the peripheral tissues (gut, placenta, male reproductive tract) resem-

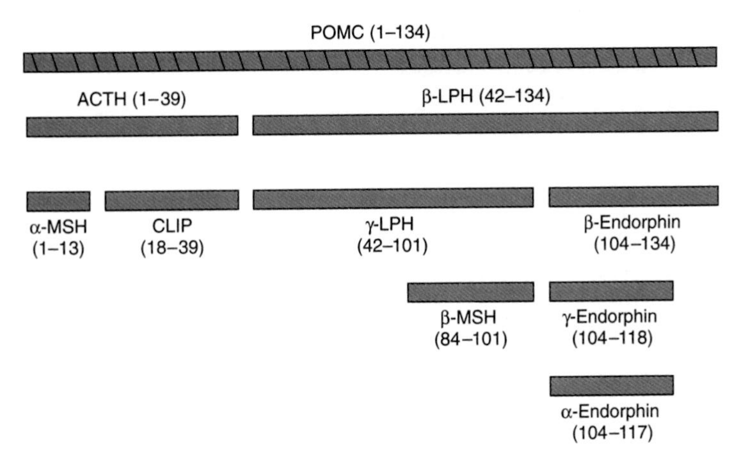


***Figure 42–15.*** Products of pro-opiomelanocortin (POMC) cleavage. (MSH, melanocyte-stimulating hormone; CLIP, corticotropin-like intermediate lobe peptide; LPH, lipotropin.)

bles that in the intermediate lobe. There are three basic peptide groups: (1) ACTH, which can give rise to α-MSH and corticotropin-like intermediate lobe peptide (CLIP); (2) β-lipotropin (β-LPH), which can yield γ-LPH, β-MSH, and β-endorphin (and thus α- and γ-endorphins); and (3) a large amino terminal peptide, which generates γ-MSH. The diversity of these products is due to the many dibasic amino acid clusters that are potential cleavage sites for trypsin-like enzymes. Each of the peptides mentioned is preceded by Lys-Arg, Arg-Lys, Arg-Arg, or Lys-Lys residues. After the prehormone segment is cleaved, the next cleavage, in both anterior and intermediate lobes, is between ACTH and β-LPH, resulting in an amino terminal peptide with ACTH and a β-LPH segment (Figure 42–15). $ACTH_{1-39}$ is subsequently cleaved from the amino terminal peptide, and in the anterior lobe essentially no further cleavages occur. In the intermediate lobe, $ACTH_{1-39}$ is cleaved into α-MSH (residues 1–13) and CLIP (18–39); β-LPH (42–134) is converted to γ-LPH (42–101) and β-endorphin (104–134). β-MSH (84–101) is derived from γ-LPH.

There are extensive additional tissue-specific modifications of these peptides that affect activity. These modifications include phosphorylation, acetylation, glycosylation, and amidation.

## THERE IS VARIATION IN THE STORAGE & SECRETION OF HORMONES

As mentioned above, the steroid hormones and 1,25$(OH)_2$-$D_3$ are synthesized in their final active form. They are also secreted as they are made, and thus there is no intracellular reservoir of these hormones. The catecholamines, also synthesized in active form, are stored in granules in the chromaffin cells in the adrenal medulla. In response to appropriate neural stimulation, these granules are released from the cell through exocytosis, and the catecholamines are released into the circulation. A several-hour reserve supply of catecholamines exists in the chromaffin cells.

Parathyroid hormone also exists in storage vesicles. As much as 80–90% of the proPTH synthesized is degraded before it enters this final storage compartment, especially when $Ca^{2+}$ levels are high in the parathyroid cell (see above). PTH is secreted when $Ca^{2+}$ is low in the parathyroid cells, which contain a several-hour supply of the hormone.

The human pancreas secretes about 40–50 units of insulin daily, which represents about 15–20% of the hormone stored in the B cells. Insulin and the C-peptide (see Figure 42–12) are normally secreted in equimolar amounts. Stimuli such as glucose, which provokes insulin secretion, therefore trigger the processing of proinsulin to insulin as an essential part of the secretory response.

A several-week supply of $T_3$ and $T_4$ exists in the thyroglobulin that is stored in colloid in the lumen of the thyroid follicles. These hormones can be released upon stimulation by TSH. This is the most exaggerated example of a prohormone, as a molecule containing approximately 5000 amino acids must be first synthesized, then degraded, to supply a few molecules of the active hormones $T_4$ and $T_3$.

The diversity in storage and secretion of hormones is illustrated in Table 42–5.

***Table 42–5.*** Diversity in the storage of hormones.

| Hormone | Supply Stored in Cell |
|---|---|
| Steroids and 1,25(OH)$_2$-D$_3$ | None |
| Catecholamines and PTH | Hours |
| Insulin | Days |
| T$_3$ and T$_4$ | Weeks |

## SOME HORMONES HAVE PLASMA TRANSPORT PROTEINS

The class I hormones are hydrophobic in chemical nature and thus are not very soluble in plasma. These hormones, principally the steroids and thyroid hormones, have specialized plasma transport proteins that serve several purposes. First, these proteins circumvent the solubility problem and thereby deliver the hormone to the target cell. They also provide a circulating reservoir of the hormone that can be substantial, as in the case of the thyroid hormones. Hormones, when bound to the transport proteins, cannot be metabolized, thereby prolonging their plasma half-life ($t_{1/2}$). The binding affinity of a given hormone to its transporter determines the bound versus free ratio of the hormone. This is important because only the free form of a hormone is biologically active. In general, the concentration of free hormone in plasma is very low, in the range of $10^{-15}$ to $10^{-9}$ mol/L. It is important to distinguish between plasma transport proteins and hormone receptors. Both bind hormones but with very different characteristics (Table 42–6).

The hydrophilic hormones—generally class II and of peptide structure—are freely soluble in plasma and do not require transport proteins. Hormones such as insulin, growth hormone, ACTH, and TSH circulate in the free, active form and have very short plasma half-lives. A notable exception is IGF-I, which is transported bound to members of a family of binding proteins.

***Table 42–6.*** Comparison of receptors with transport proteins.

| Feature | Receptors | Transport Proteins |
|---|---|---|
| Concentration | Very low (thousands/cell) | Very high (billions/μL) |
| Binding affinity | High (pmol/L to nmol/L range) | Low (μmol/L range) |
| Binding specificity | Very high | Low |
| Saturability | Yes | No |
| Reversibility | Yes | Yes |
| Signal transduction | Yes | No |

### Thyroid Hormones Are Transported by Thyroid-Binding Globulin

Many of the principles discussed above are illustrated in a discussion of thyroid-binding proteins. One-half to two-thirds of T$_4$ and T$_3$ in the body is in an extrathyroidal reservoir. Most of this circulates in bound form, ie, bound to a specific binding protein, **thyroxine-binding globulin (TBG).** TBG, a glycoprotein with a molecular mass of 50 kDa, binds T$_4$ and T$_3$ and has the capacity to bind 20 μg/dL of plasma. Under normal circumstances, TBG binds—noncovalently—nearly all of the T$_4$ and T$_3$ in plasma, and it binds T$_4$ with greater affinity than T$_3$ (Table 42–7). The plasma half-life of T$_4$ is correspondingly four to five times that of T$_3$. The small, unbound (free) fraction is responsible for the biologic activity. Thus, in spite of the great difference in total amount, the free fraction of T$_3$ approximates that of T$_4$, and given that T$_3$ is intrinsically more active than T$_4$, most biologic activity is attributed to T$_3$. TBG does not bind any other hormones.

### Glucocorticoids Are Transported by Corticosteroid-Binding Globulin

Hydrocortisone (cortisol) also circulates in plasma in protein-bound and free forms. The main plasma binding protein is an α-globulin called **transcortin,** or **corticosteroid-binding globulin (CBG).** CBG is produced in the liver, and its synthesis, like that of TBG, is increased by estrogens. CBG binds most of the hormone when plasma cortisol levels are within the normal range; much smaller amounts of cortisol are bound to albumin. The avidity of binding helps determine the biologic half-lives of various glucocorticoids. Cortisol binds tightly to CBG and has a $t_{1/2}$ of 1.5–2 hours, while corticosterone, which binds less tightly, has a $t_{1/2}$ of less than 1 hour (Table 42–8). The unbound (free) cortisol constitutes about 8% of the total and represents the biologically active fraction. Binding to CBG is not restricted to glucocorticoids. Deoxycorticosterone and

***Table 42–7.*** Comparison of T$_4$ and T$_3$ in plasma.

| | Total Hormone (μg/dL) | Free Hormone | | | $t^1/_2$ in Blood (days) |
|---|---|---|---|---|---|
| | | Percent of Total | ng/dL | Molarity | |
| T$_4$ | 8 | 0.03 | ~2.24 | $3.0 \times 10^{-11}$ | 6.5 |
| T$_3$ | 0.15 | 0.3 | ~0.4 | $\sim 0.6 \times 10^{-11}$ | 1.5 |

***Table 42–8.*** Approximate affinities of steroids for serum-binding proteins.

| | SHBG[1] | CBG[1] |
|---|---|---|
| Dihydrotestosterone | 1 | > 100 |
| Testosterone | 2 | > 100 |
| Estradiol | 5 | > 10 |
| Estrone | > 10 | > 100 |
| Progesterone | > 100 | ~ 2 |
| Cortisol | > 100 | ~ 3 |
| Corticosterone | > 100 | ~ 5 |

[1]Affinity expressed as $K_d$ (nmol/L).

progesterone interact with CBG with sufficient affinity to compete for cortisol binding. Aldosterone, the most potent natural mineralocorticoid, does not have a specific plasma transport protein. Gonadal steroids bind very weakly to CBG (Table 42–8).

## Gonadal Steroids Are Transported by Sex Hormone-Binding Globulin

Most mammals, humans included, have a plasma β-globulin that binds testosterone with specificity, relatively high affinity, and limited capacity (Table 42–8). This protein, usually called **sex hormone-binding globulin (SHBG)** or testosterone-estrogen-binding globulin (TEBG), is produced in the liver. Its production is increased by estrogens (women have twice the serum concentration of SHBG as men), certain types of liver disease, and hyperthyroidism; it is decreased by androgens, advancing age, and hypothyroidism. Many of these conditions also affect the production of CBG and TBG. Since SHBG and albumin bind 97–99% of circulating testosterone, only a small fraction of the hormone in circulation is in the free (biologically active) form. The primary function of SHBG may be to restrict the free concentration of testosterone in the serum. Testosterone binds to SHBG with higher affinity than does estradiol (Table 42–8). Therefore, a change in the level of SHBG causes a greater change in the free testosterone level than in the free estradiol level.

Estrogens are bound to SHBG and progestins to CBG. SHBG binds estradiol about five times less avidly than it binds testosterone or DHT, while progesterone and cortisol have little affinity for this protein (Table 42–8). In contrast, progesterone and cortisol bind with nearly equal affinity to CBG, which in turn has little avidity for estradiol and even less for testosterone, DHT, or estrone.

These binding proteins also provide a circulating reservoir of hormone, and because of the relatively large binding capacity they probably buffer against sudden changes in the plasma level. Because the metabolic clearance rates of these steroids are inversely related to the affinity of their binding to SHBG, estrone is cleared more rapidly than estradiol, which in turn is cleared more rapidly than testosterone or DHT.

## SUMMARY

- The presence of a specific receptor defines the target cells for a given hormone.
- Receptors are proteins that bind specific hormones and generate an intracellular signal (receptor-effector coupling).
- Some hormones have intracellular receptors; others bind to receptors on the plasma membrane.
- Hormones are synthesized from a number of precursor molecules, including cholesterol, tyrosine per se, and all the constituent amino acids of peptides and proteins.
- A number of modification processes alter the activity of hormones. For example, many hormones are synthesized from larger precursor molecules.
- The complement of enzymes in a particular cell type allows for the production of a specific class of steroid hormone.
- Most of the lipid-soluble hormones are bound to rather specific plasma transport proteins.

## REFERENCES


Bartalina L: Thyroid hormone-binding proteins: update 1994. Endocr Rev 1994;13:140.

Beato M et al: Steroid hormone receptors: many actors in search of a plot. Cell 1995;83:851.

Dai G, Carrasco L, Carrasco N: Cloning and characterization of the thyroid iodide transporter. Nature 1996;379:458.

DeLuca HR: The vitamin D story: a collaborative effort of basic science and clinical medicine. FASEB J 1988;2:224.

Douglass J, Civelli O, Herbert E: Polyprotein gene expression: Generation of diversity of neuroendocrine peptides. Annu Rev Biochem 1984;53:665.

Miller WL: Molecular biology of steroid hormone biosynthesis. Endocr Rev 1988;9:295.

Nagatsu T: Genes for human catecholamine-synthesizing enzymes. Neurosci Res 1991;12:315.

Russell DW, Wilson JD: Steroid 5 alpha-reductase: two genes/two enzymes. Annu Rev Biochem 1994;63:25.

Russell J et al: Interaction between calcium and 1,25-dihydroxyvitamin $D_3$ in the regulation of preproparathyroid hormone and vitamin D receptor mRNA in avian parathyroids. Endocrinology 1993;132:2639.

Steiner DF et al: The new enzymology of precursor processing endoproteases. J Biol Chem 1992;267:23435.

# Hormone Action & Signal Transduction

# 43

*Daryl K. Granner, MD*

## BIOMEDICAL IMPORTANCE

The homeostatic adaptations an organism makes to a constantly changing environment are in large part accomplished through alterations of the activity and amount of proteins. Hormones provide a major means of facilitating these changes. A hormone-receptor interaction results in generation of an intracellular signal that can either regulate the activity of a select set of genes, thereby altering the amount of certain proteins in the target cell, or affect the activity of specific proteins, including enzymes and transporter or channel proteins. The signal can influence the location of proteins in the cell and can affect general processes such as protein synthesis, cell growth, and replication, perhaps through effects on gene expression. Other signaling molecules—including cytokines, interleukins, growth factors, and metabolites—use some of the same general mechanisms and signal transduction pathways. Excessive, deficient, or inappropriate production and release of hormones and of these other regulatory molecules are major causes of disease. Many pharmacotherapeutic agents are aimed at correcting or otherwise influencing the pathways discussed in this chapter.

## HORMONES TRANSDUCE SIGNALS TO AFFECT HOMEOSTATIC MECHANISMS

The general steps involved in producing a coordinated response to a particular stimulus are illustrated in Figure 43–1. The stimulus can be a challenge or a threat to the organism, to an organ, or to the integrity of a single cell within that organism. Recognition of the stimulus is the first step in the adaptive response. At the organismic level, this generally involves the nervous system and the special senses (sight, hearing, pain, smell, touch). At the organismic or cellular level, recognition involves physicochemical factors such as pH, $O_2$ tension, temperature, nutrient supply, noxious metabolites, and osmolarity. Appropriate recognition results in the release of one or more hormones that will govern generation of the necessary adaptive response. For purposes of this discussion, the hormones are categorized as described in Chapter 42, ie, based on the location of their specific cellular receptors and the type of signals generated. Group I hormones interact with an intracellular receptor and group II hormones with receptor recognition sites located on the extracellular surface of the plasma membrane of target cells. The cytokines, interleukins, and growth factors should also be considered in this latter category. These molecules, of critical importance in homeostatic adaptation, are hormones in the sense that they are produced in specific cells, have the equivalent of autocrine, paracrine, and endocrine actions, bind to cell surface receptors, and activate many of the same signal transduction pathways employed by the more traditional group II hormones.

## SIGNAL GENERATION

### The Ligand-Receptor Complex Is the Signal for Group I Hormones

The lipophilic group I hormones diffuse through the plasma membrane of all cells but only encounter their specific, high-affinity intracellular receptors in target cells. These receptors can be located in the cytoplasm or in the nucleus of target cells. The hormone-receptor complex first undergoes an **activation reaction.** As shown in Figure 43–2, receptor activation occurs by at least two mechanisms. For example, glucocorticoids diffuse across the plasma membrane and encounter their cognate receptor in the cytoplasm of target cells. Ligand-receptor binding results in the dissociation of heat shock protein 90 (hsp90) from the receptor. This step appears to be necessary for subsequent nuclear localization of the glucocorticoid receptor. This receptor also contains nuclear localization sequences that assist in the translocation from cytoplasm to nucleus. The now activated receptor moves into the nucleus (Figure 43–2) and binds with high affinity to a specific DNA sequence called the **hormone response element (HRE).** In the case illustrated, this is a glucocorticoid response element, or GRE. Consensus sequences for HREs are shown in Table 43–1. The DNA-bound, liganded receptor serves as a high-affinity binding site for

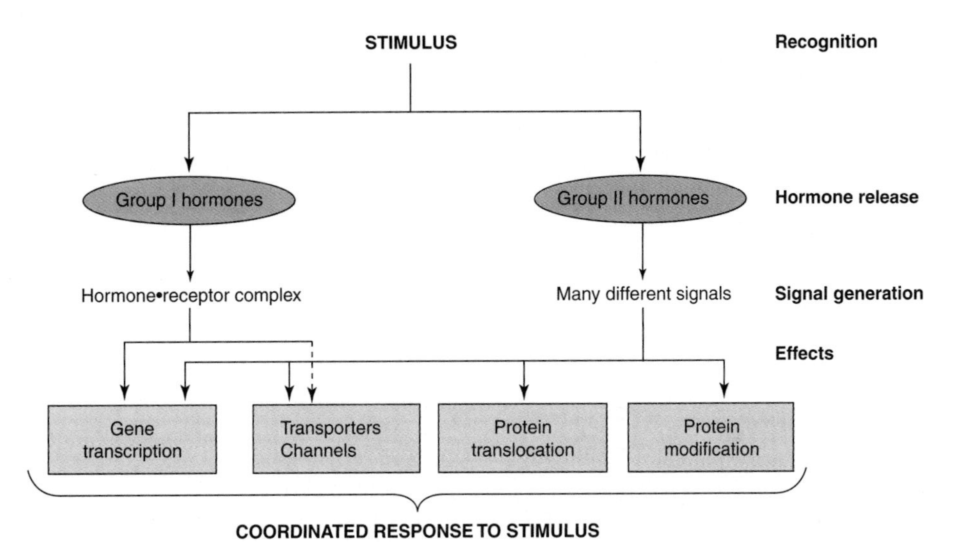


***Figure 43–1.*** Hormonal involvement in responses to a stimulus. A challenge to the integrity of the organism elicits a response that includes the release of one or more hormones. These hormones generate signals at or within target cells, and these signals regulate a variety of biologic processes which provide for a coordinated response to the stimulus or challenge. See Figure 43–8 for a specific example.

one or more coactivator proteins, and accelerated gene transcription typically ensues when this occurs. By contrast, certain hormones such as the thyroid hormones and retinoids diffuse from the extracellular fluid across the plasma membrane and go directly into the nucleus. In this case, the cognate receptor is already bound to the HRE (the thyroid hormone response element [TRE], in this example). However, this DNA-bound receptor fails to activate transcription because it is complexed with a corepressor. Indeed, this receptor-corepressor complex serves as an active repressor of gene transcription. The association of ligand with these receptors results in dissociation of the corepressor. The liganded receptor is now capable of binding one or more coactivators with high affinity, resulting in the activation of gene transcription. The relationship of hormone receptors to other nuclear receptors and to coregulators is discussed in more detail below.

By selectively affecting gene transcription and the consequent production of appropriate target mRNAs, the amounts of specific proteins are changed and metabolic processes are influenced. The influence of each of these hormones is quite specific; generally, the hormone affects less than 1% of the genes, mRNA, or proteins in a target cell; sometimes only a few are affected. The nuclear actions of steroid, thyroid, and retinoid hormones are quite well defined. Most evidence suggests that these hormones exert their dominant effect on modulating gene transcription, but they—and many of the hormones in the other classes discussed below—can act at any step of the "information pathway" illustrated in Figure 43–3. Direct actions of steroids in the cytoplasm and on various organelles and membranes have also been described.

## GROUP II (PEPTIDE & CATECHOLAMINE) HORMONES HAVE MEMBRANE RECEPTORS & USE INTRACELLULAR MESSENGERS

Many hormones are water-soluble, have no transport proteins (and therefore have a short plasma half-life), and initiate a response by binding to a receptor located in the plasma membrane (see Tables 42–3 and 42–4). The mechanism of action of this group of hormones can best be discussed in terms of the **intracellular signals** they generate. These signals include cAMP (cyclic AMP; 3′,5′-adenylic acid; see Figure 18–5), a nucleotide derived from ATP through the action of adenylyl cyclase; cGMP, a nucleotide formed by guanylyl cyclase; $Ca^{2+}$; and phosphatidylinositides. Many of these second messengers affect gene transcription, as described in the previous paragraph; but they also influ-

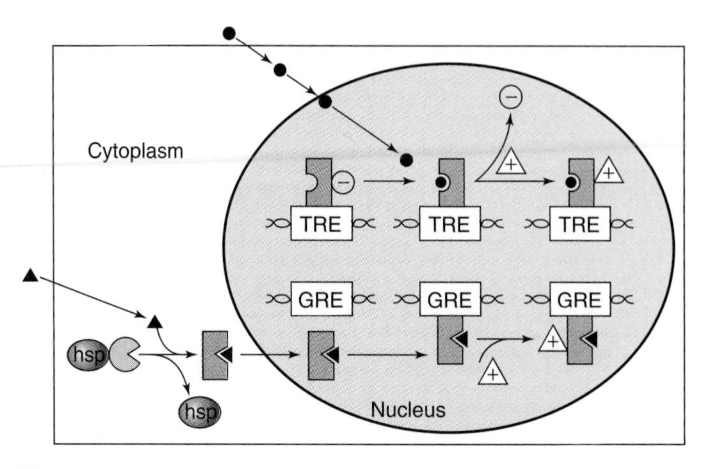


***Figure 43–2.*** Regulation of gene expression by class I hormones. Steroid hormones readily gain access to the cytoplasmic compartment of target cells. Glucocorticoid hormones (solid triangles) encounter their cognate receptor in the cytoplasm, where it exists in a complex with heat shock protein 90 (hsp). Ligand binding causes dissociation of hsp and a conformational change of the receptor. The receptor•ligand complex then traverses the nuclear membrane and binds to DNA with specificity and high affinity at a glucocorticoid response element (GRE). This event triggers the assembly of a number of transcription coregulators (⊕), and enhanced transcription ensues. By contrast, thyroid hormones and retinoic acid (●) directly enter the nucleus, where their cognate receptors are already bound to the appropriate response elements with an associated transcription repressor complex (⊖). This complex, which consists of molecules such as N-CoR or SMRT (see Table 43–6) in the absence of ligand, actively inhibits transcription. Ligand binding results in dissociation of the repressor complex from the receptor, allowing an activator complex to assemble. The gene is then actively transcribed.

ence a variety of other biologic processes, as shown in Figure 43–1.

## G Protein-Coupled Receptors (GPCR)

Many of the group II hormones bind to receptors that couple to effectors through a GTP-binding protein intermediary. These receptors typically have seven hydrophobic plasma membrane-spanning domains. This is illustrated by the seven interconnected cylinders extending through the lipid bilayer in Figure 43–4. Receptors of this class, which signal through guanine nucleotide-bound protein intermediates, are known as **G protein-coupled receptors,** or **GPCRs.** To date, over 130 G protein-linked receptor genes have been cloned from various mammalian species. A wide variety of responses are mediated by the GPCRs.

## cAMP Is the Intracellular Signal for Many Responses

Cyclic AMP was the first intracellular signal identified in mammalian cells. Several components comprise a system for the generation, degradation, and action of cAMP.

### A. Adenylyl Cyclase

Different peptide hormones can either stimulate (s) or inhibit (i) the production of cAMP from adenylyl cy-

***Table 43–1.*** The DNA sequences of several hormone response elements (HREs).[1]

| Hormone or Effector | HRE | DNA Sequence |
|---|---|---|
| Glucocorticoids<br>Progestins<br>Mineralocorticoids<br>Androgens | GRE<br>PRE<br>MRE<br>ARE | GGTACA NNN TGTTCT<br>⟵ ⟶ |
| Estrogens | ERE | AGGTCA --- TGA/TCCT<br>⟵ ⟶ |
| Thyroid hormone<br>Retinoic acid<br>Vitamin D | TRE<br>RARE<br>VDRE | AGGTCA N3,4,5, AGGTCA<br>⟶ ⟶ |
| cAMP | CRE | TGACGTCA |

[1]Letters indicate nucleotide; N means any one of the four can be used in that position. The arrows pointing in opposite directions illustrate the slightly imperfect inverted palindromes present in many HREs; in some cases these are called "half binding sites" because each binds one monomer of the receptor. The GRE, PRE, MRE, and ARE consist of the same DNA sequence. Specificity may be conferred by the intracellular concentration of the ligand or hormone receptor, by flanking DNA sequences not included in the consensus, or by other accessory elements. A second group of HREs includes those for thyroid hormones, estrogens, retinoic acid, and vitamin D. These HREs are similar except for the orientation and spacing between the half palindromes. Spacing determines the hormone specificity. VDRE (N=3), TRE (N=4), and RARE (N=5) bind to direct repeats rather than to inverted repeats. Another member of the steroid receptor superfamily, the retinoid X receptor (RXR), forms heterodimers with VDR, TR, and RARE, and these constitute the *trans*-acting factors. cAMP affects gene transcription through the CRE.

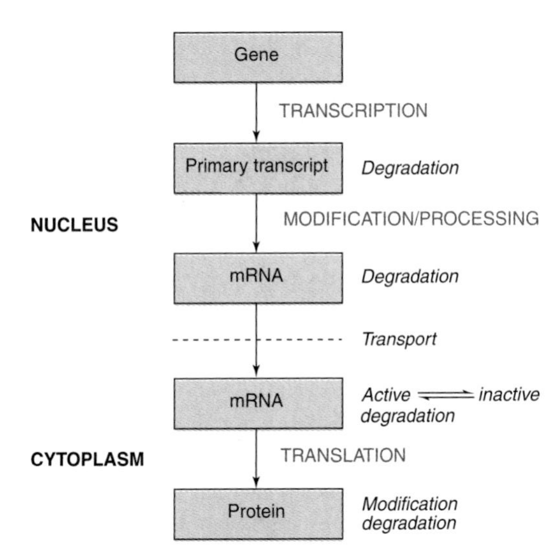


***Figure 43–3.*** The "information pathway." Information flows from the gene to the primary transcript to mRNA to protein. Hormones can affect any of the steps involved and can affect the rates of processing, degradation, or modification of the various products.

clase, which is encoded by at least nine different genes (Table 43–2). Two parallel systems, a stimulatory (s) one and an inhibitory (i) one, converge upon a single catalytic molecule (C). Each consists of a receptor, $R_s$ or $R_i$, and a regulatory complex, $G_s$ and $G_i$. $G_s$ and $G_i$ are each trimers composed of $\alpha$, $\beta$, and $\gamma$ subunits. Because the $\alpha$ subunit in $G_s$ differs from that in $G_i$, the proteins, which are distinct gene products, are designated $\alpha_s$ and $\alpha_i$. The $\alpha$ subunits bind guanine nucleotides. The $\beta$ and $\gamma$ subunits are always associated ($\beta\gamma$) and appear to function as a heterodimer. The binding of a hormone to $R_s$ or $R_i$ results in a receptor-mediated activation of G, which entails the exchange of GDP by GTP on $\alpha$ and the concomitant dissociation of $\beta\gamma$ from $\alpha$.

The $\alpha_s$ protein has intrinsic GTPase activity. The active form, $\alpha_s$•GTP, is inactivated upon hydrolysis of the GTP to GDP; the trimeric $G_s$ complex ($\alpha\beta\gamma$) is then re-formed and is ready for another cycle of activation. Cholera and pertussis toxins catalyze the ADP-ribosylation of $\alpha_s$ and $\alpha_{i\text{-}2}$ (see Table 43–3), respectively. In the case of $\alpha_s$, this modification disrupts the intrinsic GTP-ase activity; thus, $\alpha_s$ cannot reassociate with $\beta\gamma$ and is therefore irreversibly activated. ADP-ribosylation of $\alpha_{i\text{-}2}$ prevents the dissociation of $\alpha_{i\text{-}2}$ from $\beta\gamma$, and free $\alpha_{i\text{-}2}$ thus cannot be formed. $\alpha_s$ activity in such cells is therefore unopposed.

There is a large family of G proteins, and these are part of the superfamily of GTPases. The G protein family is classified according to sequence homology into four subfamilies, as illustrated in Table 43–3. There are 21 $\alpha$, 5 $\beta$, and 8 $\gamma$ subunit genes. Various combinations of these subunits provide a large number of possible $\alpha\beta\gamma$ and cyclase complexes.

The $\alpha$ subunits and the $\beta\gamma$ complex have actions independent of those on adenylyl cyclase (see Figure 43–4 and Table 43–3). Some forms of $\alpha_i$ stimulate $K^+$ channels and inhibit $Ca^{2+}$ channels, and some $\alpha_s$ molecules have the opposite effects. Members of the $G_q$ family activate the phospholipase C group of enzymes. The $\beta\gamma$ complexes have been associated with $K^+$ channel stimulation and phospholipase C activation. G proteins are involved in many important biologic processes in addition to hormone action. Notable examples include olfaction ($\alpha_{OLF}$) and vision ($\alpha_t$). Some examples are listed in Table 43–3. GPCRs are implicated in a number of diseases and are major targets for pharmaceutical agents.

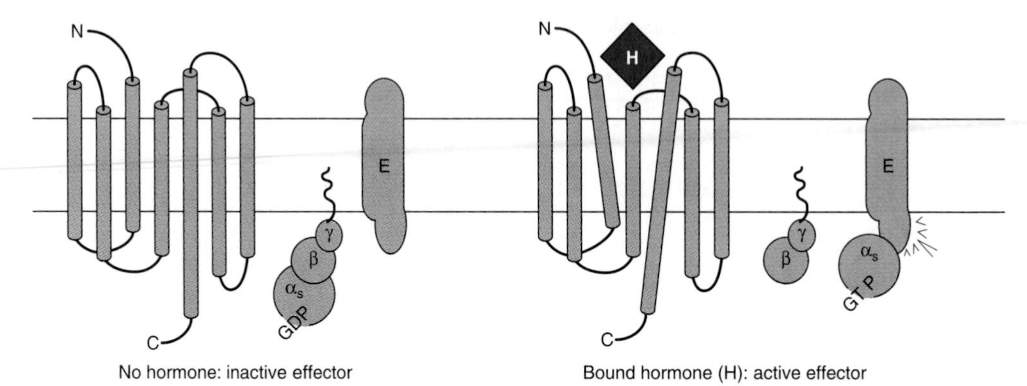


***Figure 43–4.*** Components of the hormone receptor–G protein effector system. Receptors that couple to effectors through G proteins (GPCR) typically have seven membrane-spanning domains. In the absence of hormone (left), the heterotrimeric G-protein complex (α, β, γ) is in an inactive guanosine diphosphate (GDP)-bound form and is probably not associated with the receptor. This complex is anchored to the plasma membrane through prenylated groups on the βγ subunits (wavy lines) and perhaps by myristoylated groups on α subunits (not shown). On binding of hormone (⟨H⟩) to the receptor, there is a presumed conformational change of the receptor—as indicated by the tilted membrane spanning domains—and activation of the G-protein complex. This results from the exchange of GDP with guanosine triphosphate (GTP) on the α subunit, after which α and βγ dissociate. The α subunit binds to and activates the effector (E). E can be adenylyl cyclase, $Ca^{2+}$, $Na^+$, or $Cl^-$ channels ($\alpha_s$), or it could be a $K^+$ channel ($\alpha_i$), phospholipase Cβ ($\alpha_q$), or cGMP phosphodiesterase ($\alpha_t$). The βγ subunit can also have direct actions on E. (Modified and reproduced, with permission, from Granner DK in: *Principles and Practice of Endocrinology and Metabolism,* 3rd ed. Becker KL [editor]. Lippincott, 2000.)

***Table 43–2.*** Subclassification of group II.A hormones.

| Hormones That Stimulate Adenylyl Cyclase ($H_s$) | Hormones That Inhibit Adenylyl Cyclase ($H_i$) |
|---|---|
| ACTH | Acetylcholine |
| ADH | $\alpha_2$-Adrenergics |
| β-Adrenergics | Angiotensin II |
| Calcitonin | Somatostatin |
| CRH | |
| FSH | |
| Glucagon | |
| hCG | |
| LH | |
| LPH | |
| MSH | |
| PTH | |
| TSH | |

### B. Protein Kinase

In prokaryotic cells, cAMP binds to a specific protein called catabolite regulatory protein (CRP) that binds directly to DNA and influences gene expression. In eukaryotic cells, cAMP binds to a protein kinase called **protein kinase A (PKA)** that is a heterotetrameric molecule consisting of two regulatory subunits (R) and two catalytic subunits (C). cAMP binding results in the following reaction:

$$4\ \text{cAMP} + R_2C_2 \rightleftharpoons R_2 \cdot (4\ \text{cAMP}) + 2C$$

The $R_2C_2$ complex has no enzymatic activity, but the binding of cAMP by R dissociates R from C, thereby activating the latter (Figure 43–5). The active C subunit catalyzes the transfer of the γ phosphate of ATP to a serine or threonine residue in a variety of proteins. The consensus phosphorylation sites are -Arg-Arg/Lys-X-Ser/Thr- and -Arg-Lys-X-X-Ser-, where X can be any amino acid.

Protein kinase activities were originally described as being "cAMP-dependent" or "cAMP-independent." This

***Table 43–3.*** Classes and functions of selected G proteins.[1,2]

| Class or Type | Stimulus | Effector | Effect |
|---|---|---|---|
| $G_s$ | | | |
| $\alpha_s$ | Glucagon, β-adrenergics | ↑ Adenylyl cyclase<br>↑ Cardiac $Ca^{2+}$, $Cl^-$, and $Na^+$ channels | Gluconeogenesis, lipolysis, glycogenolysis |
| $\alpha_{olf}$ | Odorant | ↑ Adenylyl cyclase | Olfaction |
| $G_i$ | | | |
| $\alpha_{i\text{-}1,2,3}$ | Acetylcholine, $\alpha_2$-adrenergics<br>$M_2$ cholinergics | ↓ Adenylyl cyclase<br>↑ Potassium channels<br>↓ Calcium channels | Slowed heart rate |
| $\alpha_0$ | Opioids, endorphins | ↑ Potassium channels | Neuronal electrical activity |
| $\alpha_t$ | Light | ↑ cGMP phosphodiesterase | Vision |
| $G_q$ | | | |
| $\alpha_q$ | $M_1$ cholinergics<br>$\alpha_1$-Adrenergics | ↑ Phospholipase C-β1 | ↑ Muscle contraction and |
| $\alpha_{11}$ | $\alpha_1$-Adrenergics | ↑ Phospholipase c-β2 | ↑ Blood pressure |
| $G_{12}$ | | | |
| $\alpha_{12}$ | ? | $Cl^-$ channel | ? |

[1]Modified and reproduced, with permission, from Granner DK in: *Principles and Practice of Endocrinology and Metabolism,* 3rd ed. Becker KL (editor). Lippincott, 2000.

[2]The four major classes or families of mammalian G proteins ($G_s$, $G_i$, $G_q$, and $G_{12}$) are based on protein sequence homology. Representative members of each are shown, along with known stimuli, effectors, and well-defined biologic effects. Nine isoforms of adenylyl cyclase have been identified (isoforms I–IX). All isoforms are stimulated by $\alpha_s$; $\alpha_i$ isoforms inhibit types V and VI, and $\alpha_0$ inhibits types I and V. At least 16 different α subunits have been identified.

classification has changed, as protein phosphorylation is now recognized as being a major regulatory mechanism. Several hundred protein kinases have now been described. The kinases are related in sequence and structure within the catalytic domain, but each is a unique molecule with considerable variability with respect to subunit composition, molecular weight, autophosphorylation, $K_m$ for ATP, and substrate specificity.

### C. Phosphoproteins

The effects of cAMP in eukaryotic cells are all thought to be mediated by protein phosphorylation-dephosphorylation, principally on serine and threonine residues. The control of any of the effects of cAMP, including such diverse processes as steroidogenesis, secretion, ion transport, carbohydrate and fat metabolism, enzyme induction, gene regulation, synaptic transmission, and cell growth and replication, could be conferred by a specific protein kinase, by a specific phosphatase, or by specific substrates for phosphorylation. These substrates help define a target tissue and are involved in defining the extent of a particular response within a given cell. For example, the effects of cAMP on gene transcription are mediated by the protein **cyclic AMP response element binding protein (CREB).** CREB binds to a cAMP responsive element (CRE) (see Table 43–1) in its nonphosphorylated state and is a weak activator of transcription. When phosphorylated by PKA, CREB binds the coactivator **CREB-binding protein CBP/p300** (see below) and as a result is a much more potent transcription activator.

### D. Phosphodiesterases

Actions caused by hormones that increase cAMP concentration can be terminated in a number of ways, including the hydrolysis of cAMP to 5′-AMP by phosphodiesterases (see Figure 43–5). The presence of these hydrolytic enzymes ensures a rapid turnover of the signal (cAMP) and hence a rapid termination of the biologic process once the hormonal stimulus is removed. There are at least 11 known members of the phosphodiesterase family of enzymes. These are subject to regulation by their substrates, cAMP and cGMP; by hormones; and by intracellular messengers such as calcium, probably acting through calmodulin. Inhibitors of phosphodiesterase, most notably methylated xanthine derivatives such as caffeine, increase intracellular cAMP and mimic or prolong the actions of hormones through this signal.

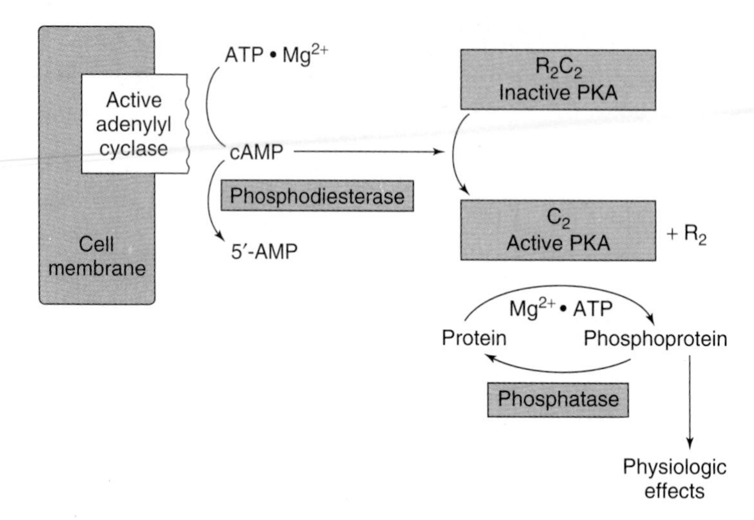


***Figure 43–5.*** Hormonal regulation of cellular processes through cAMP-dependent protein kinase (PKA). PKA exists in an inactive form as an $R_2C_2$ heterotetramer consisting of two regulatory and two catalytic subunits. The cAMP generated by the action of adenylyl cyclase (activated as shown in Figure 43–4) binds to the regulatory (R) subunit of PKA. This results in dissociation of the regulatory and catalytic subunits and activation of the latter. The active catalytic subunits phosphorylate a number of target proteins on serine and threonine residues. Phosphatases remove phosphate from these residues and thus terminate the physiologic response. A phosphodiesterase can also terminate the response by converting cAMP to 5′-AMP.

#### E. Phosphoprotein Phosphatases

Given the importance of protein phosphorylation, it is not surprising that regulation of the protein dephosphorylation reaction is another important control mechanism (see Figure 43–5). The phosphoprotein phosphatases are themselves subject to regulation by phosphorylation-dephosphorylation reactions and by a variety of other mechanisms, such as protein-protein interactions. In fact, the substrate specificity of the phosphoserine-phosphothreonine phosphatases may be dictated by distinct regulatory subunits whose binding is regulated hormonally. The best-studied role of regulation by the dephosphorylation of proteins is that of glycogen metabolism in muscle. Two major types of phosphoserine-phosphothreonine phosphatases have been described. Type I preferentially dephosphorylates the β subunit of phosphorylase kinase, whereas type II dephosphorylates the α subunit. Type I phosphatase is implicated in the regulation of glycogen synthase, phosphorylase, and phosphorylase kinase. This phosphatase is itself regulated by phosphorylation of certain of its subunits, and these reactions are reversed by the action of one of the type II phosphatases. In addition, two heat-stable protein inhibitors regulate type I phosphatase activity. Inhibitor-1 is phosphorylated and activated by cAMP-dependent protein kinases; and inhibitor-2, which may be a subunit of the inactive phosphatase, is also phosphorylated, possibly by glycogen synthase kinase-3.

## cGMP Is Also an Intracellular Signal

Cyclic GMP is made from GTP by the enzyme guanylyl cyclase, which exists in soluble and membrane-bound forms. Each of these isozymes has unique physiologic properties. The atriopeptins, a family of peptides produced in cardiac atrial tissues, cause natriuresis, diuresis, vasodilation, and inhibition of aldosterone secretion. These peptides (eg, atrial natriuretic factor) bind to and activate the membrane-bound form of guanylyl cyclase. This results in an increase of cGMP by as much as 50-fold in some cases, and this is thought to mediate the effects mentioned above. Other evidence links cGMP to vasodilation. A series of compounds, including nitroprusside, nitroglycerin, nitric oxide, sodium nitrite, and sodium azide, all cause smooth muscle relaxation and are potent vasodilators. These agents increase cGMP by activating the soluble form of guanylyl cyclase, and inhibitors of cGMP phosphodiesterase (the drug sildenafil [Viagra], for example) enhance and prolong these responses. The increased cGMP activates cGMP-dependent protein kinase (PKG), which in turn phosphorylates a number of smooth muscle proteins. Presumably, this is involved in relaxation of smooth muscle and vasodilation.

## Several Hormones Act Through Calcium or Phosphatidylinositols

Ionized calcium is an important regulator of a variety of cellular processes, including muscle contraction, stimulus-secretion coupling, the blood clotting cascade, enzyme activity, and membrane excitability. It is also an intracellular messenger of hormone action.

### A. Calcium Metabolism

The extracellular calcium ($Ca^{2+}$) concentration is about 5 mmol/L and is very rigidly controlled. Although substantial amounts of calcium are associated with intracellular organelles such as mitochondria and the endoplasmic reticulum, the intracellular concentration of free or ionized calcium ($Ca^{2+}$) is very low: 0.05–10 μmol/L. In spite of this large concentration gradient and a favorable transmembrane electrical gradient, $Ca^{2+}$ is restrained from entering the cell. A considerable amount of energy is expended to ensure that the intracellular $Ca^{2+}$ is controlled, as a prolonged elevation of $Ca^{2+}$ in the cell is very toxic. A $Na^+/Ca^{2+}$ exchange mechanism that has a high capacity but low affinity pumps $Ca^{2+}$ out of cells. There also is a $Ca^{2+}$/proton ATPase-dependent pump that extrudes $Ca^{2+}$ in exchange for $H^+$. This has a high affinity for $Ca^{2+}$ but a low capacity and is probably responsible for fine-tuning cytosolic $Ca^{2+}$. Furthermore, $Ca^{2+}$ ATPases pump $Ca^{2+}$ from the cytosol to the lumen of the endoplasmic reticulum. There are three ways of changing cytosolic $Ca^{2+}$: (1) Certain hormones (class II.C, Table 42–3) by binding to receptors that are themselves $Ca^{2+}$ channels, enhance membrane permeability to $Ca^{2+}$ and thereby increase $Ca^{2+}$ influx. (2) Hormones also indirectly promote $Ca^{2+}$ influx by modulating the membrane potential at the plasma membrane. Membrane depolarization opens voltage-gated $Ca^{2+}$ channels and allows for $Ca^{2+}$ influx. (3) $Ca^{2+}$ can be mobilized from the endoplasmic reticulum, and possibly from mitochondrial pools.

An important observation linking $Ca^{2+}$ to hormone action involved the definition of the intracellular targets of $Ca^{2+}$ action. The discovery of a $Ca^{2+}$-dependent regulator of phosphodiesterase activity provided the basis for a broad understanding of how $Ca^{2+}$ and cAMP interact within cells.

### B. Calmodulin

The calcium-dependent regulatory protein is calmodulin, a 17-kDa protein that is homologous to the muscle protein troponin C in structure and function. Calmodulin has four $Ca^{2+}$ binding sites, and full occupancy of these sites leads to a marked conformational change, which allows calmodulin to activate enzymes and ion channels. The interaction of $Ca^{2+}$ with calmodulin (with the resultant change of activity of the latter) is conceptually similar to the binding of cAMP to PKA and the subsequent activation of this molecule. Calmodulin can be one of numerous subunits of complex proteins and is particularly involved in regulating various kinases and enzymes of cyclic nucleotide generation and degradation. A partial list of the enzymes regulated directly or indirectly by $Ca^{2+}$, probably through calmodulin, is presented in Table 43–4.

In addition to its effects on enzymes and ion transport, $Ca^{2+}$/calmodulin regulates the activity of many structural elements in cells. These include the actin-myosin complex of smooth muscle, which is under β-adrenergic control, and various microfilament-mediated processes in noncontractile cells, including cell motility, cell conformation changes, mitosis, granule release, and endocytosis.

### C. Calcium Is a Mediator of Hormone Action

A role for $Ca^{2+}$ in hormone action is suggested by the observations that the effect of many hormones is (1) blunted by $Ca^{2+}$-free media or when intracellular calcium is depleted; (2) can be mimicked by agents that increase cytosolic $Ca^{2+}$, such as the $Ca^{2+}$ ionophore A23187; and (3) influences cellular calcium flux. The regulation of glycogen metabolism in liver by vasopressin and α-adrenergic catecholamines provides a good example. This is shown schematically in Figures 18–6 and 18–7.

***Table 43–4.*** Enzymes and proteins regulated by calcium or calmodulin.

| |
|---|
| Adenylyl cyclase |
| $Ca^{2+}$-dependent protein kinases |
| $Ca^{2+}$-$Mg^{2+}$ ATPase |
| $Ca^{2+}$-phospholipid-dependent protein kinase |
| Cyclic nucleotide phosphodiesterase |
| Some cytoskeletal proteins |
| Some ion channels (eg, L-type calcium channels) |
| Nitric oxide synthase |
| Phosphorylase kinase |
| Phosphoprotein phosphatase 2B |
| Some receptors (eg, NMDA-type glutamate receptor) |

A number of critical metabolic enzymes are regulated by $Ca^{2+}$, phosphorylation, or both, including glycogen synthase, pyruvate kinase, pyruvate carboxylase, glycerol-3-phosphate dehydrogenase, and pyruvate dehydrogenase.

### D. Phosphatidylinositide Metabolism Affects $Ca^{2+}$-Dependent Hormone Action

Some signal must provide communication between the hormone receptor on the plasma membrane and the intracellular $Ca^{2+}$ reservoirs. This is accomplished by products of phosphatidylinositol metabolism. Cell surface receptors such as those for acetylcholine, antidiuretic hormone, and $\alpha_1$-type catecholamines are, when occupied by their respective ligands, potent activators of phospholipase C. Receptor binding and activation of phospholipase C are coupled by the $G_q$ isoforms (Table 43–3 and Figure 43–6). Phospholipase C catalyzes the hydrolysis of phosphatidylinositol 4,5-bisphosphate to inositol trisphosphate ($IP_3$) and 1,2-diacylglycerol (Figure 43–7). Diacylglycerol is itself capable of activating **protein kinase C (PKC),** the activity of which also depends upon $Ca^{2+}$. $IP_3$, by interacting with a specific intracellular receptor, is an effective releaser of $Ca^{2+}$ from

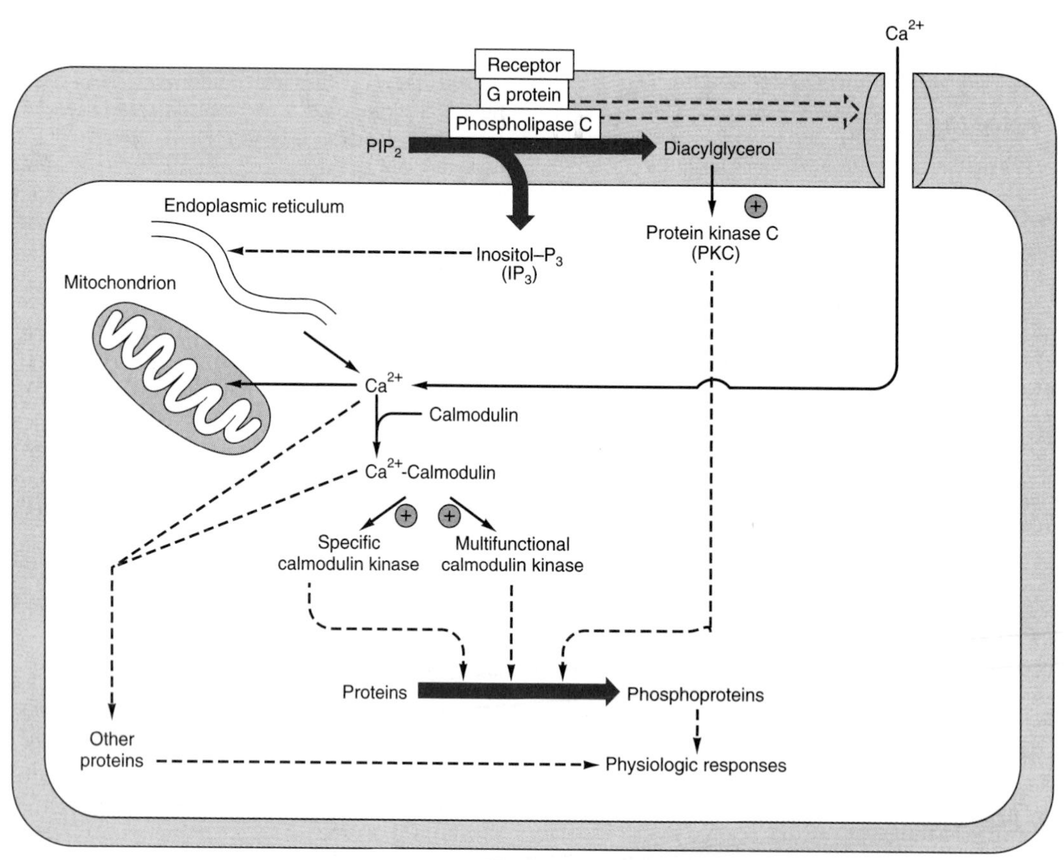


***Figure 43–6.*** Certain hormone-receptor interactions result in the activation of phospholipase C. This appears to involve a specific G protein, which also may activate a calcium channel. Phospholipase C results in generation of inositol trisphosphate ($IP_3$), which liberates stored intracellular $Ca^{2+}$, and diacylglycerol (DAG), a potent activator of protein kinase C (PKC). In this scheme, the activated PKC phosphorylates specific substrates, which then alter physiologic processes. Likewise, the $Ca^{2+}$-calmodulin complex can activate specific kinases, two of which are shown here. These actions result in phosphorylation of substrates, and this leads to altered physiologic responses. This figure also shows that $Ca^{2+}$ can enter cells through voltage- or ligand-gated $Ca^{2+}$ channels. The intracellular $Ca^{2+}$ is also regulated through storage and release by the mitochondria and endoplasmic reticulum. (Courtesy of JH Exton.)

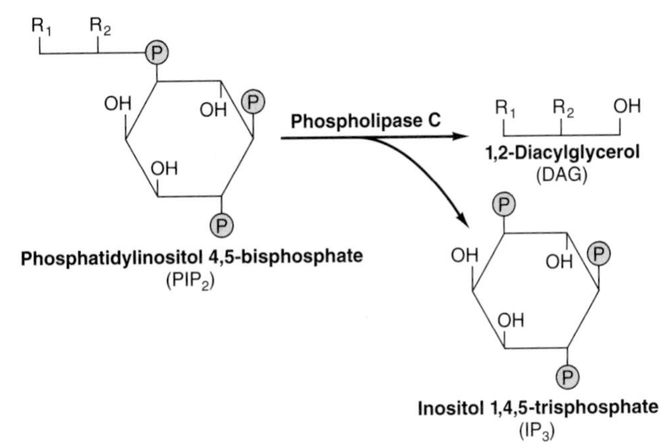


***Figure 43–7.*** Phospholipase C cleaves $PIP_2$ into diacylglycerol and inositol trisphosphate. $R_1$ generally is stearate, and $R_2$ is usually arachidonate. $IP_3$ can be dephosphorylated (to the inactive I-1,4-$P_2$) or phosphorylated (to the potentially active I-1,3,4,5-$P_4$).

intracellular storage sites in the endoplasmic reticulum. Thus, the hydrolysis of phosphatidylinositol 4,5-bisphosphate leads to activation of PKC and promotes an increase of cytoplasmic $Ca^{2+}$. As shown in Figure 43–4, the activation of G proteins can also have a direct action on $Ca^{2+}$ channels. The resulting elevations of cytosolic $Ca^{2+}$ activate $Ca^{2+}$–calmodulin-dependent kinases and many other $Ca^{2+}$–calmodulin-dependent enzymes.

Steroidogenic agents—including ACTH and cAMP in the adrenal cortex; angiotensin II, $K^+$, serotonin, ACTH, and cAMP in the zona glomerulosa of the adrenal; LH in the ovary; and LH and cAMP in the Leydig cells of the testes—have been associated with increased amounts of phosphatidic acid, phosphatidylinositol, and polyphosphoinositides (see Chapter 14) in the respective target tissues. Several other examples could be cited.

The roles that $Ca^{2+}$ and polyphosphoinositide breakdown products might play in hormone action are presented in Figure 43–6. In this scheme the activated protein kinase C can phosphorylate specific substrates, which then alter physiologic processes. Likewise, the $Ca^{2+}$-calmodulin complex can activate specific kinases. These then modify substrates and thereby alter physiologic responses.

## Some Hormones Act Through a Protein Kinase Cascade

Single protein kinases such as PKA, PKC, and $Ca^{2+}$-calmodulin (CaM)-kinases, which result in the phosphorylation of serine and threonine residues in target proteins, play a very important role in hormone action. The discovery that the EGF receptor contains an intrinsic tyrosine kinase activity that is activated by the binding of the ligand EGF was an important breakthrough. The insulin and IGF-I receptors also contain intrinsic ligand-activated tyrosine kinase activity. Several receptors—generally those involved in binding ligands involved in growth control, differentiation, and the inflammatory response—either have intrinsic tyrosine kinase activity or are associated with proteins that are tyrosine kinases. Another distinguishing feature of this class of hormone action is that these kinases preferentially phosphorylate tyrosine residues, and tyrosine phosphorylation is infrequent (< 0.03% of total amino acid phosphorylation) in mammalian cells. A third distinguishing feature is that the ligand-receptor interaction that results in a tyrosine phosphorylation event initiates a cascade that may involve several protein kinases, phosphatases, and other regulatory proteins.

### A. Insulin Transmits Signals by Several Kinase Cascades

The insulin, epidermal growth factor (EGF), and IGF-I receptors have intrinsic protein tyrosine kinase activities located in their cytoplasmic domains. These activities are stimulated when the receptor binds ligand. The receptors are then autophosphorylated on tyrosine residues, and this initiates a complex series of events (summarized in simplified fashion in Figure 43–8). The phosphorylated insulin receptor next phosphorylates insulin receptor substrates (there are at least four of these molecules, called IRS 1–4) on tyrosine residues. Phosphorylated IRS binds to the Src homology 2 (SH2) domains of a variety of proteins that are directly involved in mediating different effects of insulin. One of these proteins, PI-3 kinase, links insulin receptor activation to insulin action through activation of a number of molecules, including the kinase PDK1 (phosphoinositide-dependent kinase-1). This enzyme propagates the signal through several other kinases, including PKB (akt), SKG, and aPKC (see legend to Figure 43–8 for definitions and expanded abbreviations). An alternative

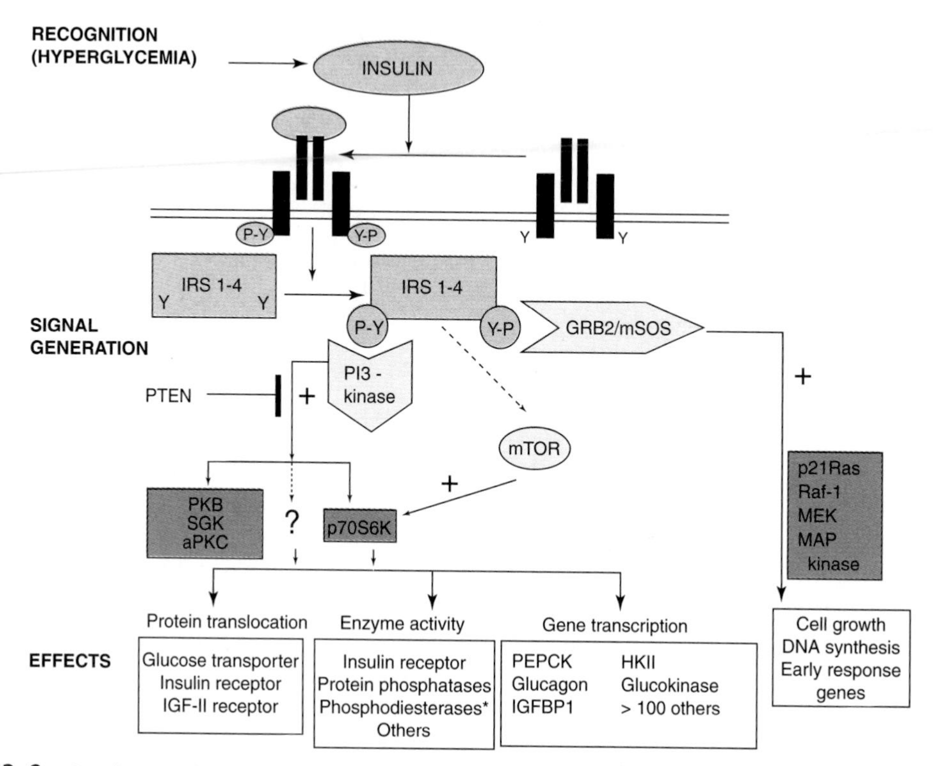


***Figure 43–8.*** Insulin signaling pathways. The insulin signaling pathways provide an excellent example of the "recognition → hormone release → signal generation → effects" paradigm outlined in Figure 43–1. Insulin is released in response to hyperglycemia. Binding of insulin to a target cell-specific plasma membrane receptor results in a cascade of intracellular events. Stimulation of the intrinsic tyrosine kinase activity of the insulin receptor marks the initial event, resulting in increased tyrosine (Y) phosphorylation (Y → Y-P) of the receptor and then one or more of the insulin receptor substrate molecules (IRS 1–4). This increase in phosphotyrosine stimulates the activity of many intracellular molecules such as GTPases, protein kinases, and lipid kinases, all of which play a role in certain metabolic actions of insulin. The two best-described pathways are shown. First, phosphorylation of an IRS molecule (probably IRS-2) results in docking and activation of the lipid kinase, PI-3 kinase, which generates novel inositol lipids that may act as "second messenger" molecules. These, in turn, activate PDK1 and then a variety of downstream signaling molecules, including protein kinase B (PKB or akt), SGK, and aPKC. An alternative pathway involves the activation of p70S6K and perhaps other as yet unidentified kinases. Second, phosphorylation of IRS (probably IRS-1) results in docking of GRB2/mSOS and activation of the small GTPase, p21RAS, which initiates a protein kinase cascade that activates Raf-1, MEK, and the p42/p44 MAP kinase isoforms. These protein kinases are important in the regulation of proliferation and differentiation of several cell types. The mTOR pathway provides an alternative way of activating p70S6K and appears to be involved in nutrient signaling as well as insulin action. Each of these cascades may influence different physiologic processes, as shown. Each of the phosphorylation events is reversible through the action of specific phosphatases. For example, the lipid phosphatase PTEN dephosphorylates the product of the PI-3 kinase reaction, thereby antagonizing the pathway and terminating the signal. Representative effects of major actions of insulin are shown in each of the boxes. The asterisk after phosphodiesterase indicates that insulin indirectly affects the activity of many enzymes by activating phosphodiesterases and reducing intracellular cAMP levels. (IGFBP, insulin-like growth factor binding protein; IRS 1–4, insulin receptor substrate isoforms 1–4); PI-3 kinase, phosphatidylinositol 3-kinase; PTEN, phosphatase and tensin homolog deleted on chromosome 10; PKD1, phosphoinositide-dependent kinase; PKB, protein kinase B; SGK, serum and glucocorticoid-regulated kinase; aPKC, atypical protein kinase C; p70S6K, p70 ribosomal protein S6 kinase; mTOR, mammalian target of rapamycin; GRB2, growth factor receptor binding protein 2; mSOS, mammalian son of sevenless; MEK, MAP kinase kinase and ERK kinase; MAP kinase, mitogen-activated protein kinase.)

pathway downstream from PKD1 involves p70S6K and perhaps other as yet unidentified kinases. A second major pathway involves mTOR. This enzyme is directly regulated by amino acids and insulin and is essential for p70S6K activity. This pathway provides a distinction between the PKB and p70S6K branches downstream from PKD1. These pathways are involved in protein translocation, enzyme activity, and the regulation, by insulin, of genes involved in metabolism (Figure 43–8). Another SH2 domain-containing protein is GRB2, which binds to IRS-1 and links tyrosine phosphorylation to several proteins, the result of which is activation of a cascade of threonine and serine kinases. A pathway showing how this insulin-receptor interaction activates the mitogen-activated protein (MAP) kinase pathway and the anabolic effects of insulin is illustrated in Figure 43–8. The exact roles of many of these docking proteins, kinases, and phosphatases remain to be established.

### B. The Jak/STAT Pathway Is Used by Hormones and Cytokines

Tyrosine kinase activation can also initiate a phosphorylation and dephosphorylation cascade that involves the action of several other protein kinases and the counterbalancing actions of phosphatases. Two mechanisms are employed to initiate this cascade. Some hormones, such as growth hormone, prolactin, erythropoietin, and the cytokines, initiate their action by activating a tyrosine kinase, but this activity is not an integral part of the hormone receptor. The hormone-receptor interaction promotes binding and activation of **cytoplasmic protein tyrosine kinases,** such as **Tyk-2, Jak1,** or **Jak2.** These kinases phosphorylate one or more cytoplasmic proteins, which then associate with other docking proteins through binding to SH2 domains. One such interaction results in the activation of a family of cytosolic proteins called **signal transducers and activators of transcription (STATs).** The phosphorylated STAT protein dimerizes and translocates into the nucleus, binds to a specific DNA element such as the interferon response element, and activates transcription. This is illustrated in Figure 43–9. Other SH2 docking events may result in the activation of PI 3-kinase, the MAP kinase pathway (through SHC or GRB2), or G protein-mediated activation of phospholipase C (PLCγ) with the attendant production of diacylglycerol and activation of protein kinase C. It is apparent that there is a potential for "cross-talk" when different hormones activate these various signal transduction pathways.

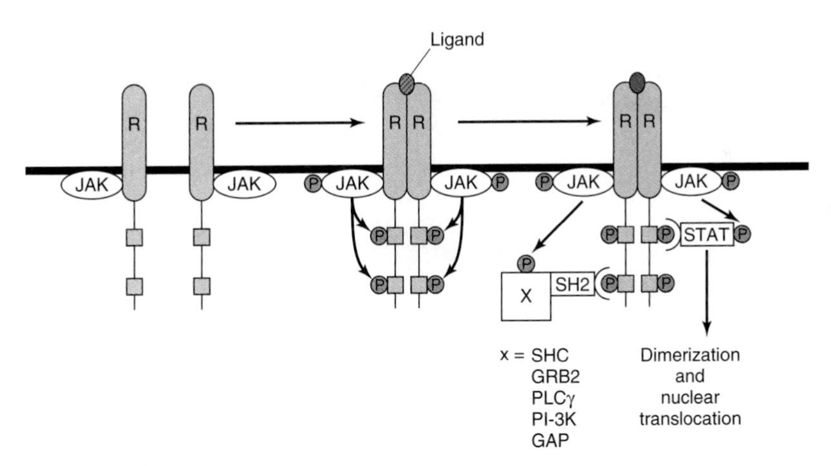


***Figure 43–9.*** Initiation of signal transduction by receptors linked to Jak kinases. The receptors (R) that bind prolactin, growth hormone, interferons, and cytokines lack endogenous tyrosine kinase. Upon ligand binding, these receptors dimerize and an associated protein (Jak1, Jak2, or TYK) is phosphorylated. Jak-P, an active kinase, phosphorylates the receptor on tyrosine residues. The STAT proteins associate with the phosphorylated receptor and then are themselves phosphorylated by Jak-P. STATⓅ dimerizes, translocates to the nucleus, binds to specific DNA elements, and regulates transcription. The phosphotyrosine residues of the receptor also bind to several SH2 domain-containing proteins. This results in activation of the MAP kinase pathway (through SHC or GRB2), PLCγ, or PI-3 kinase.

### C. The NF-κB Pathway Is Regulated by Glucocorticoids

The transcription factor NF-κB is a heterodimeric complex typically composed of two subunits termed p50 and p65 (Figure 43–10). Normally, NF-κB is kept sequestered in the cytoplasm in a transcriptionally inactive form by members of the inhibitor of NF-κB (IκB) family. Extracellular stimuli such as proinflammatory cytokines, reactive oxygen species, and mitogens lead to activation of the IκB kinase complex, IKK, which is a heterohexameric structure consisting of α, β, and γ subunits. IKK phosphorylates IκB on two serine residues, and this targets IκB for ubiquitination and subsequent degradation by the proteasome. Following IκB degradation, free NF-κB can now translocate to the nucleus, where it binds to a number of gene promoters and activates transcription, particularly of genes involved in the **inflammatory response.** Transcriptional regulation by NF-κB is mediated by a variety of coactivators such as CREB binding protein (CBP), as described below (Figure 43–13).

**Glucocorticoid hormones** are therapeutically useful agents for the treatment of a variety of inflammatory and immune diseases. Their anti-inflammatory and immunomodulatory actions are explained in part by the inhibition of NF-κB and its subsequent actions. Evidence for three mechanisms for the inhibition of NF-κB by glucocorticoids has been presented: (1) Glucocorticoids increase IκB mRNA, which leads to an increase of IκB protein and more efficient sequestration of NF-κB in the cytoplasm. (2) The glucocorticoid receptor competes with NF-κB for binding to coactivators. (3) The glucocorticoid receptor directly binds to the p65 subunit of NF-κB and inhibits its activation (Figure 43–10).

## HORMONES CAN INFLUENCE SPECIFIC BIOLOGIC EFFECTS BY MODULATING TRANSCRIPTION

The signals generated as described above have to be translated into an action that allows the cell to effectively adapt to a challenge (Figure 43–1). Much of this

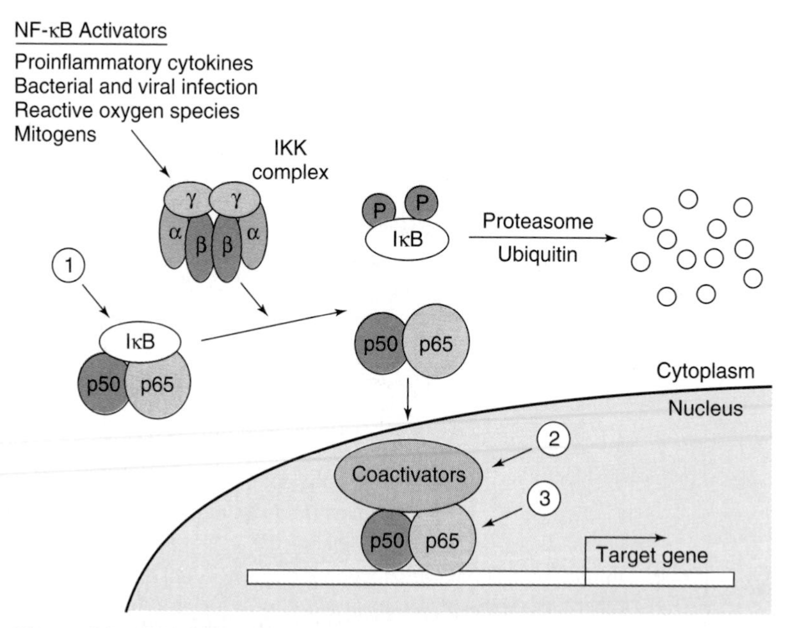


***Figure 43–10.*** Regulation of the NF-κB pathway. NF-κB consists of two subunits, p50 and p65, which regulate transcription of many genes when in the nucleus. NF-κB is restricted from entering the nucleus by IκB, an inhibitor of NF-κB. IκB binds to—and masks—the nuclear localization signal of NF-κB. This cytoplasmic protein is phosphorylated by an IKK complex which is activated by cytokines, reactive oxygen species, and mitogens. Phosphorylated IκB can be ubiquitinylated and degraded, thus releasing its hold on NF-κB. Glucocorticoids affect many steps in this process, as described in the text.

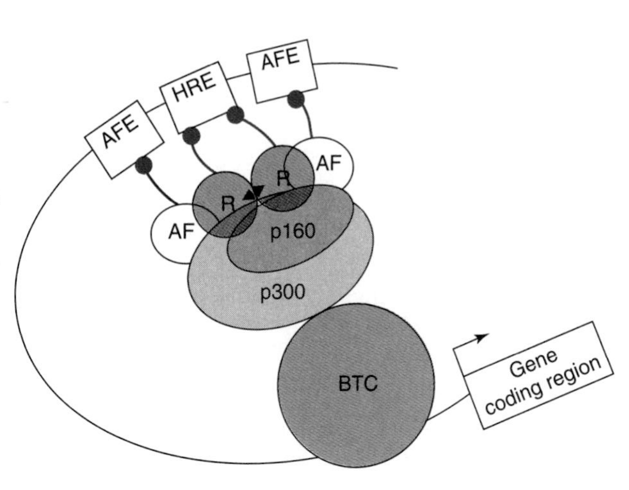


***Figure 43–11.*** The hormone response transcription unit. The hormone response transcription unit is an assembly of DNA elements and bound proteins that interact, through protein-protein interactions, with a number of coactivator or corepressor molecules. An essential component is the hormone response element which binds the ligand (▲)-bound receptor (R). Also important are the accessory factor elements (AFEs) with bound transcription factors. More than two dozen of these accessory factors (AFs), which are often members of the nuclear receptor superfamily, have been linked to hormone effects on transcription. The AFs can interact with each other, with the liganded nuclear receptors, or with coregulators. These components communicate with the basal transcription complex through a coregulator complex that can consist of one or more members of the p160, corepressor, mediator-related, or CBP/p300 families (see Table 43–6).

adaptation is accomplished through alterations in the rates of transcription of specific genes. Many different observations have led to the current view of how hormones affect transcription. Some of these are as follows: (1) Actively transcribed genes are in regions of "open" chromatin (defined by a susceptibility to the enzyme DNase I), which allows for the access of transcription factors to DNA. (2) Genes have regulatory regions, and transcription factors bind to these to modulate the frequency of transcription initiation. (3) The hormone-receptor complex can be one of these transcription factors. The DNA sequence to which this binds is called a hormone response element (HRE; see Table 43–1 for examples). (4) Alternatively, other hormone-generated signals can modify the location, amount, or activity of transcription factors and thereby influence binding to the regulatory or response element. (5) Members of a large superfamily of nuclear receptors act with—or in a manner analogous to—hormone receptors. (6) These nuclear receptors interact with another large group of coregulatory molecules to effect changes in the transcription of specific genes.

## Several Hormone Response Elements (HREs) Have Been Defined

Hormone response elements resemble enhancer elements in that they are not strictly dependent on position and location. They generally are found within a few hundred nucleotides upstream (5′) of the transcription initiation site, but they may be located within the coding region of the gene, in introns. HREs were defined by the strategy illustrated in Figure 39–11. The consensus sequences illustrated in Table 43–1 were arrived at through analysis of several genes regulated by a given hormone using simple, heterologous reporter systems (see Figure 39–10). Although these simple HREs bind the hormone-receptor complex more avidly than surrounding DNA—or DNA from an unrelated source—and confer hormone responsiveness to a reporter gene, it soon became apparent that the regulatory circuitry of natural genes must be much more complicated. Glucocorticoids, progestins, mineralocorticoids, and androgens have vastly different physiologic actions. How could the specificity required for these effects be achieved through regulation of gene expression by the same HRE (Table 43–1)? Questions like this have led to experiments which have allowed for elaboration of a very complex model of transcription regulation. For example, the HRE must associate with other DNA elements (and associated binding proteins) to function optimally. The extensive sequence similarity noted between steroid hormone receptors, particularly in their DNA-binding domains, led to discovery of the **nuclear receptor superfamily** of proteins. These—and a large number of **coregulator proteins**—allow for a wide variety of DNA-protein and protein-protein interactions and the specificity necessary for highly regulated physiologic control. A schematic of such an assembly is illustrated in Figure 43–11.

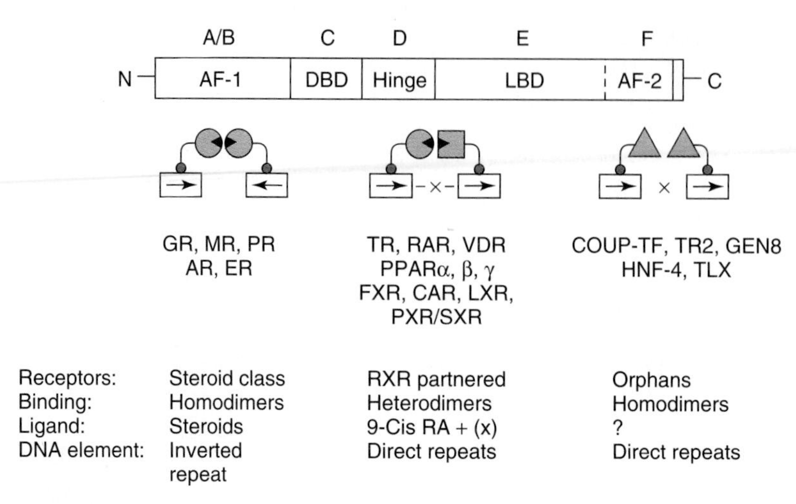


***Figure 43–12.*** The nuclear receptor superfamily. Members of this family are divided into six structural domains (A–F). Domain A/B is also called AF-1, or the modulator region, because it is involved in activating transcription. The C domain consists of the DNA-binding domain (DBD). The D region contains the hinge, which provides flexibility between the DBD and the ligand-binding domain (LBD, region E). The amino (N) terminal part of region E contains AF-2, another domain important for transactivation. The F region is poorly defined. The functions of these domains are discussed in more detail in the text. Receptors with known ligands, such as the steroid hormones, bind as homodimers on inverted repeat half-sites. Other receptors form heterodimers with the partner RXR on direct repeat elements. There can be nucleotide spacers of one to five bases between these direct repeats (DR1–5). Another class of receptors for which ligands have not been determined (orphan receptors) bind as homodimers to direct repeats and occasionally as monomers to a single half-site.

## There Is a Large Family of Nuclear Receptor Proteins

The nuclear receptor superfamily consists of a diverse set of transcription factors that were discovered because of a sequence similarity in their DNA-binding domains. This family, now with more than 50 members, includes the nuclear hormone receptors discussed above, a number of other receptors whose ligands were discovered after the receptors were identified, and many putative or orphan receptors for which a ligand has yet to be discovered.

These nuclear receptors have several common structural features (Figure 43–12). All have a centrally located **DNA-binding domain (DBD)** that allows the receptor to bind with high affinity to a response element. The DBD contains two zinc finger binding motifs (see Figure 39–14) that direct binding either as homodimers, as heterodimers (usually with a retinoid X receptor [RXR] partner), or as monomers. The target response element consists of one or two half-site consensus sequences arranged as an inverted or direct repeat. The spacing between the latter helps determine binding specificity. Thus, a direct repeat with three, four, or five nucleotide spacer regions specifies the binding of the vitamin D, thyroid, and retinoic acid receptors, respectively, to the same consensus response element (Table 43–1). A multifunctional **ligand-binding domain (LBD)** is located in the carboxyl terminal half of the receptor. The LBD binds hormones or metabolites with selectivity and thus specifies a particular biologic response. The LBD also contains domains that mediate the binding of heat shock proteins, dimerization, nuclear localization, and transactivation. The latter function is facilitated by the carboxyl terminal transcription activation function **(AF-2 domain),** which forms a surface required for the interaction with

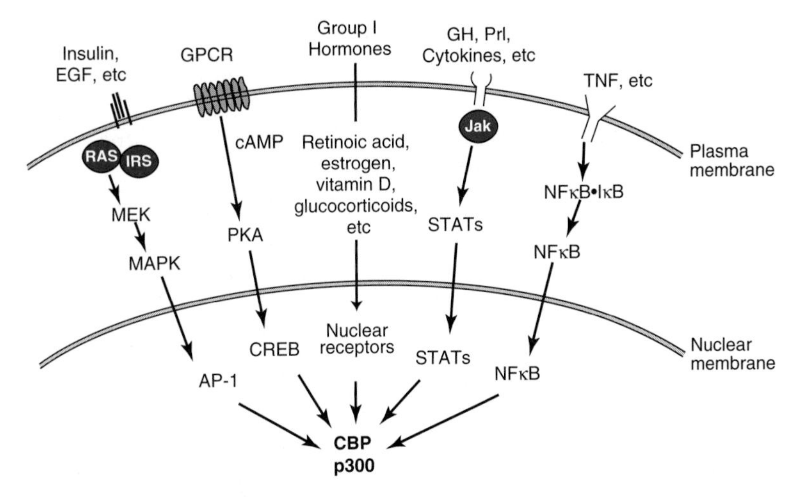


***Figure 43–13.*** Several signal transduction pathways converge on CBP/p300. Ligands that associate with membrane or nuclear receptors eventually converge on CBP/p300. Several different signal transduction pathways are employed. EGF, epidermal growth factor; GH, growth hormone; Prl, prolactin; TNF, tumor necrosis factor; other abbreviations are expanded in the text.

coactivators. A highly variable **hinge region** separates the DBD from the LBD. This region provides flexibility to the receptor, so it can assume different DNA-binding conformations. Finally, there is a highly variable amino terminal region that contains another transactivation domain referred to as **AF-1.** Less well defined, the AF-1 domain may provide for distinct physiologic functions through the binding of different coregulator proteins. This region of the receptor, through the use of different promoters, alternative splice sites, and multiple translation initiation sites, provides for receptor isoforms that share DBD and LBD identity but exert different physiologic responses because of the association of various coregulators with this variable amino terminal AF-1 domain.

It is possible to sort this large number of receptors into groups in a variety of ways. Here they are discussed according to the way they bind to their respective DNA elements (Figure 43–12). Classic hormone receptors for glucocorticoids (GR), mineralocorticoids (MR), estrogens (ER), androgens (AR), and progestins (PR) bind as homodimers to inverted repeat sequences. Other hormone receptors such as thyroid (TR), retinoic acid (RAR), and vitamin D (VDR) and receptors that bind various metabolite ligands such as PPAR $\alpha$ $\beta$, and $\gamma$, FXR, LXR, PXR/SXR, and CAR bind as heterodimers, with retinoid X receptor (RXR) as a partner, to direct repeat sequences (see Figure 43–12 and Table 43–5). Another group of orphan receptors that as yet have no known ligand bind as homodimers or monomers to direct repeat sequences.

As illustrated in Table 43–5, the discovery of the nuclear receptor superfamily has led to an important understanding of how a variety of metabolites and xenobiotics regulate gene expression and thus the metabolism, detoxification, and elimination of normal body products and exogenous agents such as pharmaceuticals. Not surprisingly, this area is a fertile field for investigation of new therapeutic interventions.

## A Large Number of Nuclear Receptor Coregulators Also Participate in Regulating Transcription

Chromatin remodeling, transcription factor modification by various enzyme activities, and the communication between the nuclear receptors and the basal transcription apparatus are accomplished by protein-protein interactions with one or more of a class of coregulator molecules. The number of these coregulator molecules now exceeds 100, not counting species variations and splice variants. The first of these to be described was the **CREB-binding protein, CBP.** CBP, through an amino terminal domain, binds to phosphorylated serine 137 of CREB and mediates transactivation in response to cAMP. It thus is described as a coactivator. CBP and

***Table 43–5.*** Nuclear receptors with special ligands.[1]

| Receptor | | Partner | Ligand | Process Affected |
|---|---|---|---|---|
| Peroxisome | PPARα | RXR (DR1) | Fatty acids | Peroxisome proliferation |
| Proliferator- | PPARβ | | Fatty acids | |
| activated | PPARγ | | Fatty acids | Lipid and carbohydrate metabolism |
| | | | Eicosanoids, thiazolidinediones | |
| Farnesoid X | FXR | RXR (DR4) | Farnesol, bile acids | Bile acid metabolism |
| Liver X | LXR | RXR (DR4) | Oxysterols | Cholesterol metabolism |
| Xenobiotic X | CAR | RXT (DR5) | Androstanes | |
| | | | Phenobarbital | Protection against certain drugs, toxic metabolites, and xenobiotics |
| | | | Xenobiotics | |
| | PXR | RXR (DR3) | Pregnanes | |
| | | | Xenobiotics | |

[1]Many members of the nuclear receptor superfamily were discovered by cloning, and the corresponding ligands were then identified. These ligands are not hormones in the classic sense, but they do have a similar function in that they activate specific members of the nuclear receptor superfamily. The receptors described here form heterodimers with RXR and have variable nucleotide sequences separating the direct repeat binding elements (DR1–5). These receptors regulate a variety of genes encoding cytochrome p450s (CYP), cytosolic binding proteins, and ATP-binding cassette (ABC) transporters to influence metabolism and protect cells against drugs and noxious agents.

its close relative, p300, interact directly or indirectly with a number of signaling molecules, including activator protein-1 (AP-1), signal transducers and activators of transcription (STATs), nuclear receptors, and CREB (Figure 39–11). **CBP/p300** also binds to the p160 family of coactivators described below and to a number of other proteins, including viral transcription factor Ela, the p90$^{rsk}$ protein kinase, and RNA helicase A. It is important to note that **CBP/p300 also has intrinsic histone acetyltransferase (HAT) activity.** The importance of this is described below. Some of the many actions of CBP/p300, which appear to depend on intrinsic enzyme activities and its ability to serve as a scaffold for the binding of other proteins, are illustrated in Figure 43–11. Other coregulators may serve similar functions.

Several other families of coactivator molecules have been described. Members of the **p160 family of coactivators,** all of about 160 kDa, include (1) SRC-1 and NCoA-1; (2) GRIP 1, TIF2, and NCoA-2; and (3) p/CIP, ACTR, AIB1, RAC3, and TRAM-1 (Table 43–6). The different names for members within a subfamily often represent species variations or minor splice variants. There is about 35% amino acid identity between members of the different subfamilies. The p160 coactivators share several properties. They (1) bind nuclear receptors in an agonist and AF-2 transactivation domain-dependent manner; (2) have a conserved amino terminal basic helix-loop-helix (bHLH) motif (see Chapter 39); (3) have a weak carboxyl terminal transactivation domain and a stronger amino terminal

***Table 43–6.*** Some mammalian coregulator proteins.

| | |
|---|---|
| **I. 300-kDa family of coactivators** | |
| CBP | CREB-binding protein |
| p300 | Protein of 300 kDa |
| **II. 160-kDa family of coactivators** | |
| A. SRC-1 | Steroid receptor coactivator 1 |
| NCoA-1 | Nuclear receptor coactivator 1 |
| B. TIF2 | Transcriptional intermediary factor 2 |
| GRIP1 | Glucocorticoid receptor-interacting protein |
| NCoA-2 | Nuclear receptor coactivator 2 |
| C. p/CIP | p300/CBP cointegrator-associated protein 1 |
| ACTR | Activator of the thyroid and retinoic acid receptors |
| AIB | Amplified in breast cancer |
| RAC3 | Receptor-associated coactivator 3 |
| TRAM-1 | TR activator molecule 1 |
| **III. Corepressors** | |
| NCoR | Nuclear receptor corepressor |
| SMRT | Silencing mediator for RXR and TR |
| **IV. Mediator-related proteins** | |
| TRAPs | Thyroid hormone receptor-associated proteins |
| DRIPs | Vitamin D receptor-interacting proteins |
| ARC | Activator-recruited cofactor |

transactivation domain in a region that is required for the CBP/p16O interaction; (4) contain at least three of the **LXXLL motifs** required for protein-protein interaction with other coactivators; and (5) often have HAT activity. The role of HAT is particularly interesting, as mutations of the HAT domain disable many of these transcription factors. Current thinking holds that these HAT activities acetylate histones and result in remodeling of chromatin into a transcription-efficient environment; however, other protein substrates for HAT-mediated acetylation have been reported. Histone acetylation/deacetylation is proposed to play a critical role in gene expression.

A small number of proteins, including NCoR and SMRT, comprise the **corepressor family.** They function, at least in part, as described in Figure 43–2. Another family includes the TRAPs, DRIPs, and ARC (Table 43–6). These so-called **mediator-related proteins** range in size from 80 kDa to 240 kDa and are thought to be involved in linking the nuclear receptor-coactivator complex to RNA polymerase II and the other components of the basal transcription apparatus.

The exact role of these coactivators is presently under intensive investigation. Many of these proteins have intrinsic enzymatic activities. This is particularly interesting in view of the fact that acetylation, phosphorylation, methylation, and ubiquitination—as well as proteolysis and cellular translocation—have been proposed to alter the activity of some of these coregulators and their targets.

It appears that certain combinations of coregulators—and thus different combinations of activators and inhibitors—are responsible for specific ligand-induced actions through various receptors.

## SUMMARY

- Hormones, cytokines, interleukins, and growth factors use a variety of signaling mechanisms to facilitate cellular adaptive responses.
- The ligand-receptor complex serves as the initial signal for members of the nuclear receptor family.
- Class II hormones, which bind to cell surface receptors, generate a variety of intracellular signals. These include cAMP, cGMP, $Ca^{2+}$, phosphatidylinositides, and protein kinase cascades.
- Many hormone responses are accomplished through alterations in the rate of transcription of specific genes.
- The nuclear receptor superfamily of proteins plays a central role in the regulation of gene transcription.
- These receptors, which may have hormones, metabolites, or drugs as ligands, bind to specific DNA elements as homodimers or as heterodimers with RXR. Some—orphan receptors—have no known ligand but bind DNA and influence transcription.
- Another large family of coregulator proteins remodel chromatin, modify other transcription factors, and bridge the nuclear receptors to the basal transcription apparatus.

## REFERENCES

Arvanitakis L et al: Constitutively signaling G-protein-coupled receptors and human disease. Trends Endocrinol Metab 1998;9:27.

Berridge M: Inositol triphosphate and calcium signalling. Nature 1993;361:315.

Chawla A et al: Nuclear receptors and lipid physiology: opening the X files. Science 2001;294:1866.

Darnell JE Jr, Kerr IM, Stark GR: Jak-STAT pathways and transcriptional activation in response to IFNs and other extracellular signaling proteins. Science 1994;264:1415.

Fantl WJ, Johnson DE, Williams LT: Signalling by receptor tyrosine kinases. Annu Rev Biochem 1993;62:453.

Giguère V: Orphan nuclear receptors: from gene to function. Endocr Rev 1999;20:689.

Grunstein M: Histone acetylation in chromatin structure and transcription. Nature 1997;389:349.

Hanoune J, Defer N: Regulation and role of adenylyl cyclase isoforms. Annu Rev Pharmacol Toxicol 2001;41:145.

Hermanson O, Glass CK, Rosenfeld MG: Nuclear receptor coregulators: multiple modes of receptor modification. Trends Endocrinol Metab 2002;13:55.

Jaken S: Protein kinase C isozymes and substrates. Curr Opin Cell Biol 1996;8:168.

Lucas P, Granner D: Hormone response domains in gene transcription. Annu Rev Biochem 1992;61:1131.

Montminy M: Transcriptional regulation by cyclic AMP. Annu Rev Biochem 1997;66:807

Morris AJ, Malbon CC: Physiological regulation of G protein-linked signaling. Physiol Rev 1999;79:1373.

Walton KM, Dixon JE: Protein tyrosine phosphatases. Annu Rev Biochem 1993;62:101.

# SECTION VI
## Special Topics

# Nutrition, Digestion, & Absorption 44

*David A. Bender, PhD, & Peter A. Mayes, PhD, DSc*

## BIOMEDICAL IMPORTANCE

Besides water, the diet must provide metabolic fuels (mainly carbohydrates and lipids), protein (for growth and turnover of tissue proteins), fiber (for roughage), minerals (elements with specific metabolic functions), and vitamins and essential fatty acids (organic compounds needed in small amounts for essential metabolic and physiologic functions). The polysaccharides, triacylglycerols, and proteins that make up the bulk of the diet must be hydrolyzed to their constituent monosaccharides, fatty acids, and amino acids, respectively, before absorption and utilization. Minerals and vitamins must be released from the complex matrix of food before they can be absorbed and utilized.

Globally, **undernutrition** is widespread, leading to impaired growth, defective immune systems, and reduced work capacity. By contrast, in developed countries, there is often excessive food consumption (especially of fat), leading to obesity and to the development of cardiovascular disease and some forms of cancer. Deficiencies of vitamin A, iron, and iodine pose major health concerns in many countries, and deficiencies of other vitamins and minerals are a major cause of ill health. In developed countries, nutrient deficiency is rare, though there are vulnerable sections of the population at risk. Intakes of minerals and vitamins that are adequate to prevent deficiency may be inadequate to promote optimum health and longevity.

Excessive secretion of gastric acid, associated with *Helicobacter pylori* infection, can result in the development of gastric and duodenal **ulcers;** small changes in the composition of bile can result in crystallization of cholesterol as **gallstones;** failure of exocrine pancreatic secretion (as in **cystic fibrosis**) leads to undernutrition and steatorrhea. **Lactose intolerance** is due to lactase deficiency leading to diarrhea and intestinal discomfort. Absorption of intact peptides that stimulate antibody responses causes **allergic reactions,** and **celiac disease** is an allergic reaction to wheat gluten.

## DIGESTION & ABSORPTION OF CARBOHYDRATES

The digestion of complex carbohydrates is by hydrolysis to liberate oligosaccharides, then free mono- and disaccharides. The increase in blood glucose after a test dose of a carbohydrate compared with that after an equivalent amount of glucose is known as the **glycemic index.** Glucose and galactose have an index of 1, as do lactose, maltose, isomaltose, and trehalose, which give rise to these monosaccharides on hydrolysis. Fructose and the sugar alcohols are absorbed less rapidly and have a lower glycemic index, as does sucrose. The glycemic index of starch varies between near 1 to near zero due to variable rates of hydrolysis, and that of nonstarch polysaccharides is zero. Foods that have a low glycemic index are considered to be more beneficial since they cause less fluctuation in insulin secretion.

### Amylases Catalyze the Hydrolysis of Starch

The hydrolysis of starch by salivary and pancreatic amylases catalyze random hydrolysis of $\alpha(1\rightarrow4)$ glycoside bonds, yielding dextrins, then a mixture of glucose, maltose, and isomaltose (from the branch points in amylopectin).

### Disaccharidases Are Brush Border Enzymes

The disaccharidases—maltase, sucrase-isomaltase (a bifunctional enzyme catalyzing hydrolysis of sucrose and isomaltose), lactase, and trehalase—are located on the brush border of the intestinal mucosal cells where the resultant monosaccharides and others arising from the diet are absorbed. In most people, apart from those of northern European genetic origin, lactase is gradually lost through adolescence, leading to **lactose intolerance.** Lactose remains in the intestinal lumen, where it is a substrate for bacterial fermentation to lactate, resulting in discomfort and diarrhea.

### There Are Two Separate Mechanisms for the Absorption of Monosaccharides in the Small Intestine

Glucose and galactose are absorbed by a sodium-dependent process. They are carried by the same transport protein (SGLT 1) and compete with each other for intestinal absorption (Figure 44–1). Other monosaccharides are absorbed by carrier-mediated diffusion. Because they are not actively transported, fructose and sugar alcohols are only absorbed down their concentration gradient, and after a moderately high intake some may remain in the intestinal lumen, acting as a substrate for bacterial fermentation.

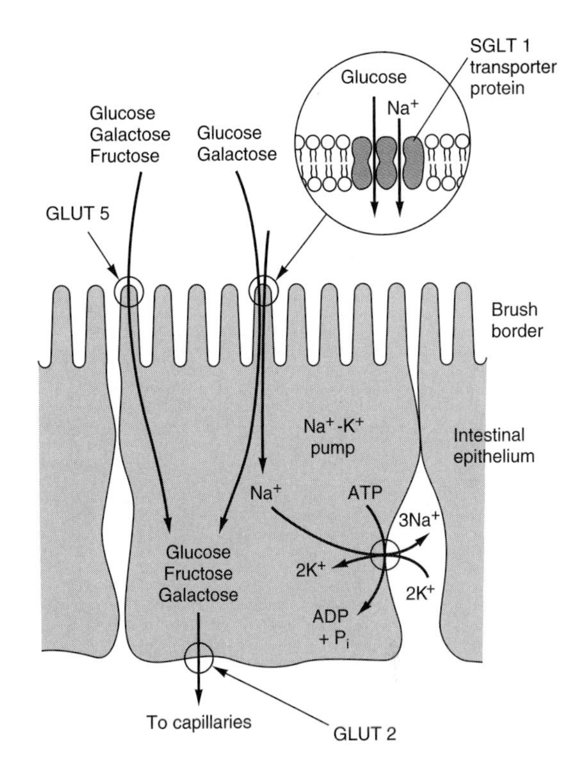


***Figure 44–1.*** Transport of glucose, fructose, and galactose across the intestinal epithelium. The SGLT 1 transporter is coupled to the $Na^+$-$K^+$ pump, allowing glucose and galactose to be transported against their concentration gradients. The GLUT 5 $Na^+$-independent facilitative transporter allows fructose as well as glucose and galactose to be transported with their concentration gradients. Exit from the cell for all the sugars is via the GLUT 2 facilitative transporter.

## DIGESTION & ABSORPTION OF LIPIDS

The major lipids in the diet are triacylglycerols and, to a lesser extent, phospholipids. These are hydrophobic molecules and must be hydrolyzed and emulsified to very small droplets (micelles) before they can be absorbed. The fat-soluble vitamins—A, D, E, and K—and a variety of other lipids (including cholesterol) are absorbed dissolved in the lipid micelles. Absorption of the fat-soluble vitamins is impaired on a very low fat diet.

Hydrolysis of triacylglycerols is initiated by lingual and gastric lipases that attack the *sn*-3 ester bond, forming 1,2-diacylglycerols and free fatty acids, aiding emulsification. Pancreatic lipase is secreted into the small intestine and requires a further pancreatic protein, colipase, for activity. It is specific for the primary ester links—ie, positions 1 and 3 in triacylglycerols—resulting in 2-monoacylglycerols and free fatty acids as the major end-products of luminal triacylglycerol digestion. Monoacylglycerols are hydrolyzed with difficulty to glycerol and free fatty acids, so that less than 25% of ingested triacylglycerol is completely hydrolyzed to glycerol and fatty acids (Figure 44–2). Bile salts, formed in the liver and secreted in the bile, enable emulsification of the products of lipid digestion into micelles and liposomes together with phospholipids and cholesterol from the bile. Because the micelles are soluble, they allow the products of digestion, including the fat-soluble vitamins, to be transported through the aqueous environment of the intestinal lumen and permit close contact with the brush border of the mucosal cells, allowing uptake into the epithelium, mainly of the jejunum. The bile salts pass on to the ileum, where most are absorbed into the **enterohepatic circulation** (Chapter 26). Within the intestinal epithelium, 1-monoacylglycerols are hydrolyzed to fatty acids and glycerol and 2-monoacylglycerols are re-acylated to triacylglycerols via the **monoacylglycerol pathway.** Glycerol released in the intestinal lumen is not reutilized but passes into the portal vein; glycerol released within the

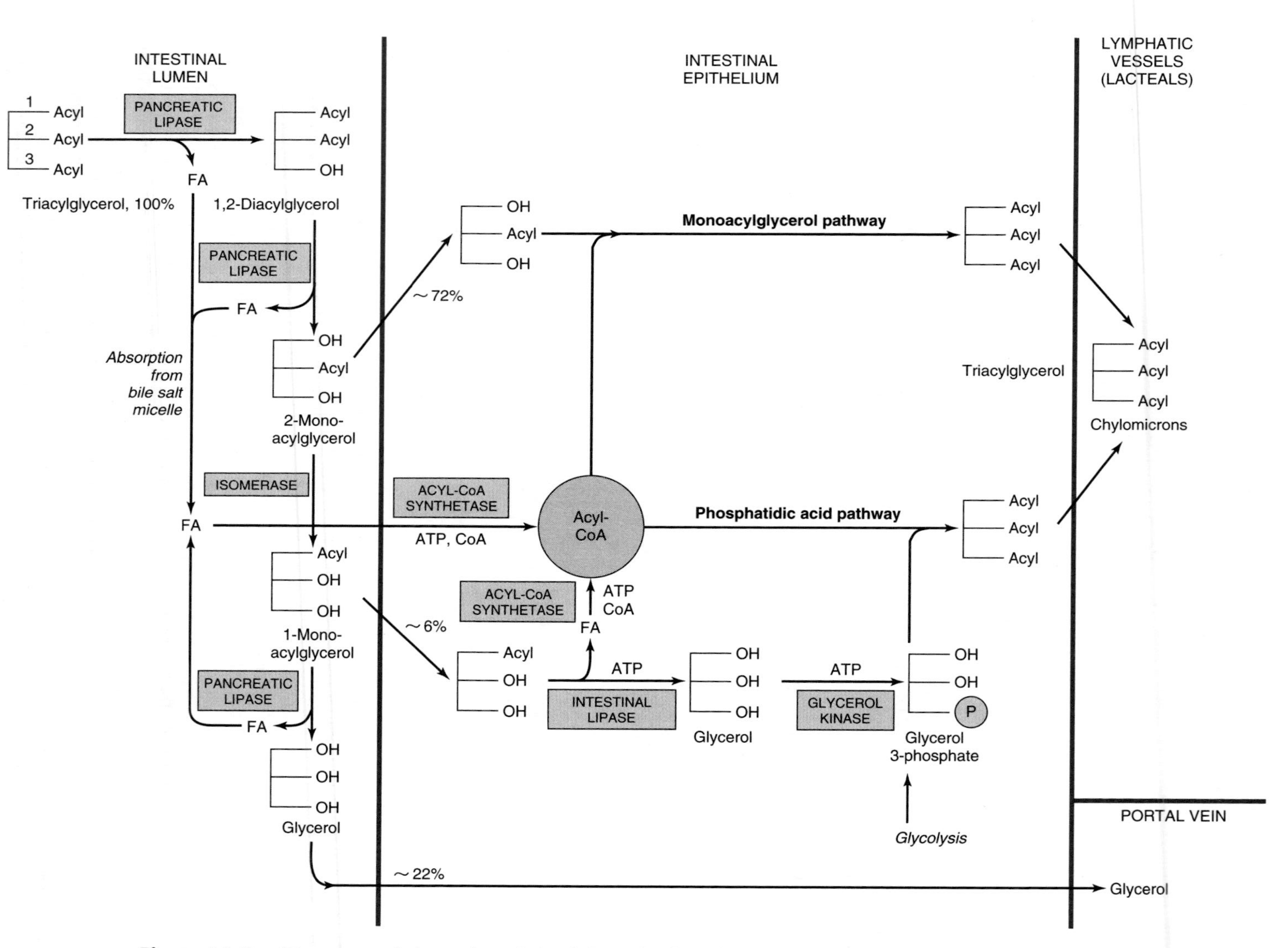


**Figure 44–2.** Digestion and absorption of triacylglycerols. The values given for percentage uptake may vary widely but indicate the relative importance of the three routes shown.

epithelium is reutilized for triacylglycerol synthesis via the normal phosphatidic acid pathway (Chapter 24). All long-chain fatty acids absorbed are converted to triacylglycerol in the mucosal cells and, together with the other products of lipid digestion, secreted as chylomicrons into the lymphatics, entering the blood stream via the thoracic duct (Chapter 25).

## DIGESTION & ABSORPTION OF PROTEINS

Few peptide bonds that are hydrolyzed by proteolytic enzymes are accessible without prior denaturation of dietary proteins (by heat in cooking and by the action of gastric acid).

### Several Groups of Enzymes Catalyze the Digestion of Proteins

There are two main classes of proteolytic digestive enzymes **(proteases),** with different specificities for the amino acids forming the peptide bond to be hydrolyzed. **Endopeptidases** hydrolyze peptide bonds between specific amino acids throughout the molecule. They are the first enzymes to act, yielding a larger number of smaller fragments, eg, **pepsin** in the gastric juice and **trypsin, chymotrypsin,** and **elastase** secreted into the small intestine by the pancreas. **Exopeptidases** catalyze the hydrolysis of peptide bonds, one at a time, from the ends of polypeptides. **Carboxypeptidases,** secreted in the pancreatic juice, release amino acids from the free carboxyl terminal, and **aminopeptidases,** secreted by the intestinal mucosal cells, release amino acids from the amino terminal. Dipeptides, which are not substrates for exopeptidases, are hydrolyzed in the brush border of intestinal mucosal cells by **dipeptidases.**

The proteases are secreted as inactive **zymogens;** the active site of the enzyme is masked by a small region of its peptide chain, which is removed by hydrolysis of a specific peptide bond. **Pepsinogen** is activated to pepsin by gastric acid and by activated pepsin (autocatalysis). In the small intestine, **trypsinogen,** the precursor of trypsin, is activated by **enteropeptidase,** which is secreted by the duodenal epithelial cells; trypsin can then activate **chymotrypsinogen** to chymotrypsin, **proelastase** to elastase, **procarboxypeptidase** to carboxypeptidase, and **proaminopeptidase** to aminopeptidase.

### Free Amino Acids & Small Peptides Are Absorbed by Different Mechanisms

The end product of the action of endopeptidases and exopeptidases is a mixture of free amino acids, di- and tripeptides, and oligopeptides, all of which are absorbed. Free amino acids are absorbed across the intestinal mucosa by sodium-dependent active transport. There are several different amino acid transporters, with specificity for the nature of the amino acid side chain (large or small; neutral, acidic, or basic). The various amino acids carried by any one transporter compete with each other for absorption and tissue uptake. Dipeptides and tripeptides enter the brush border of the intestinal mucosal cells, where they are hydrolyzed to free amino acids, which are then transported into the hepatic portal vein. Relatively large peptides may be absorbed intact, either by uptake into mucosal epithelial cells (transcellular) or by passing between epithelial cells (paracellular). Many such peptides are large enough to stimulate antibody formation—this is the basis of allergic reactions to foods.

## DIGESTION & ABSORPTION OF VITAMINS & MINERALS

Vitamins and minerals are released from food during digestion—though this is not complete—and the availability of vitamins and minerals depends on the type of food and, especially for minerals, the presence of chelating compounds. The fat-soluble vitamins are absorbed in the lipid micelles that result from fat digestion; water-soluble vitamins and most mineral salts are absorbed from the small intestine either by active transport or by carrier-mediated diffusion followed by binding to intracellular binding proteins to achieve concentration upon uptake. Vitamin $B_{12}$ absorption requires a specific transport protein, **intrinsic factor;** calcium absorption is dependent on vitamin D; zinc absorption probably requires a zinc-binding ligand secreted by the exocrine pancreas; and the absorption of iron is limited.

### Calcium Absorption Is Dependent on Vitamin D

In addition to its role in regulating calcium homeostasis, vitamin D is required for the intestinal absorption of calcium. Synthesis of the intracellular calcium-binding protein, **calbindin,** required for calcium absorption, is induced by vitamin D, which also affects the permeability of the mucosal cells to calcium, an effect that is rapid and independent of protein synthesis.

Phytic acid (inositol hexaphosphate) in cereals binds calcium in the intestinal lumen, preventing its absorption. Other minerals, including zinc, are also chelated by phytate. This is mainly a problem among people who consume large amounts of unleavened whole wheat products; yeast contains an enzyme, **phytase,** which dephosphorylates phytate, so rendering it inactive. High concentrations of fatty acids in the intestinal lumen—as a result of impaired fat absorption—can also reduce calcium absorption by forming insoluble calcium salts; a high intake of oxalate can sometimes cause deficiency, since calcium oxalate is insoluble.

### Iron Absorption Is Limited & Strictly Controlled but Is Enhanced by Vitamin C & Ethanol

Although iron deficiency is a common problem, about 10% of the population are genetically at risk of iron overload **(hemochromatosis),** and elemental iron can lead to nonenzymic generation of free radicals. Absorption of iron is strictly regulated. Inorganic iron is accumulated in intestinal mucosal cells bound to an intracellular protein, **ferritin.** Once the ferritin in the cell is saturated with iron, no more can enter. Iron can only leave the mucosal cell if there is **transferrin** in plasma to bind to. Once transferrin is saturated with iron, any that has accumulated in the mucosal cells will be lost when the cells are shed. As a result of this mucosal barrier, only about 10% of dietary iron is normally absorbed and only 1–5% from many plant foods.

Inorganic iron is absorbed only in the $Fe^{2+}$ (reduced) state, and for that reason the presence of reducing agents will enhance absorption. The most effective compound is **vitamin C,** and while intakes of 40–60 mg of vitamin C per day are more than adequate to meet requirements, an intake of 25–50 mg per meal will enhance iron absorption, especially when iron salts are used to treat iron deficiency anemia. Ethanol and fructose also enhance iron absorption. Heme iron from meat is absorbed separately and is considerably more available than inorganic iron. However, the absorption of both inorganic and heme iron is impaired by calcium—a glass of milk with a meal significantly reduces availability.

## ENERGY BALANCE: OVER- & UNDERNUTRITION

After the provision of water, the body's first requirement is for metabolic fuels—fats, carbohydrates, and amino acids from proteins (and ethanol) (Table 27–1). Food intake in excess of energy expenditure leads to **obesity,** while intake less than expenditure leads to emaciation and wasting, as in **marasmus** and **kwashiorkor.** Both obesity and severe undernutrition are associated with increased mortality. The **body mass index,** defined as weight in kilograms divided by height in meters squared, is commonly used as a way of expressing relative obesity to height. A desirable range is between 20 and 25.

### Energy Requirements Are Estimated by Measurement of Energy Expenditure

Energy expenditure can be determined directly by measuring heat output from the body but is normally estimated indirectly from the consumption of oxygen. There is an energy expenditure of 20 kJ/L of oxygen consumed regardless of whether the fuel being metabolized is carbohydrate, fat, or protein. Measurement of the ratio of the volume of carbon dioxide produced to volume of oxygen consumed (respiratory quotient; RQ) is an indication of the mixture of metabolic fuels being oxidized (Table 27–1). A more recent technique permits estimation of total energy expenditure over a period of 1–2 weeks using dual isotopically labeled water, $^{2}H_2{}^{18}O$. $^{2}H$ is lost from the body only in water, while $^{18}O$ is lost in both water and carbon dioxide; the difference in the rate of loss of the two labels permits estimation of total carbon dioxide production and thus oxygen consumption and energy expenditure.

**Basal metabolic rate (BMR)** is the energy expenditure by the body when at rest—but not asleep—under controlled conditions of thermal neutrality, measured at about 12 hours after the last meal, and depends on weight, age, and gender. Total energy expenditure depends on the basal metabolic rate, the energy required for physical activity, and the energy cost of synthesizing reserves in the fed state. It is therefore possible to calculate an individual's energy requirement from body weight, age, gender, and level of physical activity. Body weight affects BMR because there is a greater amount of active tissue in a larger body. The decrease in BMR with increasing age, even when body weight remains constant, is due to muscle tissue replacement by adipose tissue, which is metabolically much less active. Similarly, women have a significantly lower BMR than do men of the same body weight because women's bodies have proportionately more adipose tissue than men.

### Energy Requirements Increase With Activity

The most useful way of expressing the energy cost of physical activities is as a multiple of BMR. Sedentary activities use only about 1.1–1.2 × BMR. By contrast, vigorous exertion, such as climbing stairs, cross-country skiing, walking uphill, etc, may use 6–8 × BMR.

### Ten Percent of the Energy Yield of a Meal May Be Expended in Forming Reserves

There is a considerable increase in metabolic rate after a meal, a phenomenon known as **diet-induced thermogenesis.** A small part of this is the energy cost of secreting digestive enzymes and of active transport of the products of digestion; the major part is due to synthesizing reserves of glycogen, triacylglycerol, and protein.

### There Are Two Extreme Forms of Undernutrition

**Marasmus** can occur in both adults and children and occurs in vulnerable groups of all populations. **Kwash-**

**iorkor** only affects children and has only been reported in developing countries. The distinguishing feature of kwashiorkor is that there is fluid retention, leading to edema. Marasmus is a state of extreme emaciation; it is the outcome of prolonged negative energy balance. Not only have the body's fat reserves been exhausted, but there is wastage of muscle as well, and as the condition progresses there is loss of protein from the heart, liver, and kidneys. The amino acids released by the catabolism of tissue proteins are used as a source of metabolic fuel and as substrates for gluconeogenesis to maintain a supply of glucose for the brain and red blood cells. As a result of the reduced synthesis of proteins, there is impaired immune response and more risk from infections. Impairment of cell proliferation in the intestinal mucosa occurs, resulting in reduction in surface area of the intestinal mucosa and reduction in absorption of such nutrients as are available.

### Patients With Advanced Cancer & AIDS Are Malnourished

Patients with advanced cancer, HIV infection and AIDS, and a number of other chronic diseases are frequently undernourished—the condition is called **cachexia.** Physically, they show all the signs of marasmus, but there is considerably more loss of body protein than occurs in starvation. The secretion of cytokines in response to infection and cancer increases the catabolism of tissue protein. This differs from marasmus, in which protein synthesis is reduced but catabolism in unaffected. Patients are **hypermetabolic,** ie, there is a considerable increase in basal metabolic rate. Many tumors metabolize glucose anaerobically to release lactate. This is used for gluconeogenesis in the liver, which is energy-consuming with a net cost of six ATP for each mole of glucose cycled (Chapter 19). There is increased stimulation of **uncoupling proteins** by cytokines, leading to thermogenesis and increased oxidation of metabolic fuels. Futile cycling of lipids occurs because hormone-sensitive lipase is activated by a proteoglycan secreted by tumors, resulting in liberation of fatty acids from adipose tissue and ATP-expensive reesterification in the liver to triacylglycerols, which are exported in VLDL.

### Kwashiorkor Affects Undernourished Children

In addition to the wasting of muscle tissue, loss of intestinal mucosa, and impaired immune responses seen in marasmus, children with **kwashiorkor** show a number of characteristic features. The defining characteristic is **edema,** associated with a decreased concentration of plasma proteins. In addition, there is enlargement of the liver due to accumulation of fat. It was formerly believed that the cause of kwashiorkor was a lack of protein, with a more or less adequate energy intake; however, analysis of the diets of affected children shows that this is not so. Children with kwashiorkor are less stunted than those with marasmus, and the edema begins to improve early in treatment, when the child is still receiving a low-protein diet. Very commonly, an infection precipitates kwashiorkor. Superimposed on general food deficiency, there is probably a deficiency of the antioxidant nutrients such as zinc, copper, carotene, and vitamins C and E. The **respiratory burst** in response to infection leads to the production of oxygen and halogen **free radicals** as part of the cytotoxic action of stimulated macrophages. This added oxidant stress may well trigger the development of kwashiorkor.

## PROTEIN & AMINO ACID REQUIREMENTS

### Protein Requirements Can Be Determined by Measuring Nitrogen Balance

The state of protein nutrition can be determined by measuring the dietary intake and output of nitrogenous compounds from the body. Although nucleic acids also contain nitrogen, protein is the major dietary source of nitrogen and measurement of total nitrogen intake gives a good estimate of protein intake (mg N × 6.25 = mg protein, as nitrogen is 16% of most proteins). The output of nitrogen from the body is mainly in urea and smaller quantities of other compounds in urine and undigested protein in feces, and significant amounts may also be lost in sweat and shed skin.

The difference between intake and output of nitrogenous compounds is known as **nitrogen balance.** Three states can be defined: In a healthy adult, nitrogen balance is in **equilibrium** when intake equals output, and there is no change in the total body content of protein. In a growing child, a pregnant woman, or in recovery from protein loss, the excretion of nitrogenous compounds is less than the dietary intake and there is net retention of nitrogen in the body as protein, ie, **positive nitrogen balance.** In response to trauma or infection—or if the intake of protein is inadequate to meet requirements—there is net loss of protein nitrogen from the body, ie, **negative nitrogen balance.** The continual catabolism of tissue proteins creates the requirement for dietary protein even in an adult who is not growing, though some of the amino acids released can be reutilized. Nitrogen balance studies show that the average daily requirement is 0.6 g of protein per kilogram of body weight (the factor 0.75 should be used to allow for individual variation), or approximately 50 g/d. Average intakes of protein in developed countries are about 80–100 g/d, ie, 14–15% of energy intake. Because growing children are increasing the protein in the body, they have a proportionately greater requirement than adults and should be in positive nitrogen balance. Even so, the need is relatively small compared with the requirement for protein turnover. In some countries, protein intake may be inadequate to meet these requirements, resulting in stunting of growth.

### There Is a Loss of Body Protein in Response to Trauma & Infection

One of the metabolic reactions to major trauma, such as a burn, a broken limb, or surgery, is an increase in the net catabolism of tissue proteins. As much as 6–7% of the total body protein may be lost over 10 days. Prolonged bed rest results in considerable loss of protein because of atrophy of muscles. Protein is catabolized as normal, but without the stimulus of exercise it is not completely replaced. Lost protein is replaced during convalescence, when there is positive nitrogen balance. A normal diet is adequate to permit this replacement.

### The Requirement Is Not for Protein Itself but for Specific Amino Acids

Not all proteins are nutritionally equivalent. More of some than of others is needed to maintain nitrogen balance because different proteins contain different amounts of the various amino acids. The body's requirement is for specific amino acids in the correct proportions to replace the body proteins. The amino acids can be divided into two groups: **essential** and **nonessential.** There are nine essential or indispensable amino acids, which cannot be synthesized in the body: histidine, isoleucine, leucine, lysine, methionine, phenylalanine, threonine, tryptophan, and valine. If one of these is lacking or inadequate, then—regardless of the total intake of protein—it will not be possible to maintain nitrogen balance since there will not be enough of that amino acid for protein synthesis.

Two amino acids—cysteine and tyrosine—can be synthesized in the body, but only from essential amino acid precursors (cysteine from methionine and tyrosine from phenylalanine). The dietary intakes of cysteine and tyrosine thus affect the requirements for methionine and phenylalanine. The remaining 11 amino acids in proteins are considered to be nonessential or dispensable, since they can be synthesized as long as there is enough total protein in the diet—ie, if one of these amino acids is omitted from the diet, nitrogen balance can still be maintained. However, only three amino acids—alanine, aspartate, and glutamate—can be considered to be truly dispensable; they are synthesized from common metabolic intermediates (pyruvate, oxaloacetate, and α-ketoglutarate, respectively). The remaining amino acids are considered as nonessential, but under some circumstances the requirement for them may outstrip the organism's capacity for synthesis.

## SUMMARY

- Digestion involves hydrolyzing food molecules into smaller molecules for absorption through the gastrointestinal epithelium. Polysaccharides are absorbed as monosaccharides; triacylglycerols as 2-monoacylglycerols, fatty acids, and glycerol; and proteins as amino acids.
- Digestive disorders arise as a result of (1) enzyme deficiency, eg, lactase and sucrase; (2) malabsorption, eg, of glucose and galactose due to defects in the $Na^+$-glucose cotransporter (SGLT 1); (3) absorption of unhydrolyzed polypeptides, leading to immunologic responses, eg, as in celiac disease; and (4) precipitation of cholesterol from bile as gallstones.
- Besides water, the diet must provide metabolic fuels (carbohydrate and fat) for bodily growth and activity; protein for synthesis of tissue proteins; fiber for roughage; minerals for specific metabolic functions; certain polyunsaturated fatty acids of the n-3 and n-6 families for eicosanoid synthesis and other functions; and vitamins, organic compounds needed in small amounts for many varied essential functions.
- Twenty different amino acids are required for protein synthesis, of which nine are essential in the human diet. The quantity of protein required is affected by protein quality, energy intake, and physical activity.
- Undernutrition occurs in two extreme forms: marasmus in adults and children and kwashiorkor in children. Overnutrition from excess energy intake is associated with diseases such as obesity, type 2 diabetes mellitus, atherosclerosis, cancer, and hypertension.

## REFERENCES

Bender DA, Bender AE: *Nutrition: A Reference Handbook.* Oxford Univ Press, 1997.

Büller HA, Grand RJ: Lactose intolerance. Annu Rev Med 1990;41:141.

Fuller MF, Garlick PJ: Human amino acid requirements. Annu Rev Nutr 1994;14:217.

Garrow JS, James WPT, Ralph A: *Human Nutrition and Dietetics,* 10th ed. Churchill-Livingstone, 2000.

National Academy of Sciences report on diet and health. Nutr Rev 1989;47:142.

Nielsen FH: Nutritional significance of the ultratrace elements. Nutr Rev 1988;46:337.

# Vitamins & Minerals

# 45

*David A. Bender, PhD, & Peter A. Mayes, PhD, DSc*

## BIOMEDICAL IMPORTANCE

Vitamins are a group of organic nutrients required in small quantities for a variety of biochemical functions and which, generally, cannot be synthesized by the body and must therefore be supplied in the diet.

The lipid-soluble vitamins are apolar hydrophobic compounds that can only be absorbed efficiently when there is normal fat absorption. They are transported in the blood, like any other apolar lipid, in lipoproteins or attached to specific binding proteins. They have diverse functions, eg, vitamin A, vision; vitamin D, calcium and phosphate metabolism; vitamin E, antioxidant; vitamin K, blood clotting. As well as dietary inadequacy, conditions affecting the digestion and absorption of the lipid-soluble vitamins—such as steatorrhea and disorders of the biliary system—can all lead to deficiency syndromes, including: night blindness and xerophthalmia (vitamin A); rickets in young children and osteomalacia in adults (vitamin D); neurologic disorders and anemia of the newborn (vitamin E); and hemorrhage of the newborn (vitamin K). Toxicity can result from excessive intake of vitamins A and D. Vitamin A and β-carotene (provitamin A), as well as vitamin E, are antioxidants and have possible roles in atherosclerosis and cancer prevention.

The water-soluble vitamins comprise the B complex and vitamin C and function as enzyme cofactors. Folic acid acts as a carrier of one-carbon units. Deficiency of a single vitamin of the B complex is rare, since poor diets are most often associated with multiple deficiency states. Nevertheless, specific syndromes are characteristic of deficiencies of individual vitamins, eg, beriberi (thiamin); cheilosis, glossitis, seborrhea (riboflavin); pellagra (niacin); peripheral neuritis (pyridoxine); megaloblastic anemia, methylmalonic aciduria, and pernicious anemia (vitamin $B_{12}$); and megaloblastic anemia (folic acid). Vitamin C deficiency leads to scurvy.

Inorganic mineral elements that have a function in the body must be provided in the diet. When the intake is insufficient, deficiency symptoms may arise, eg, anemia (iron), cretinism and goiter (iodine). If present in excess as with selenium, toxicity symptoms may occur.

## THE DETERMINATION OF NUTRIENT REQUIREMENTS DEPENDS ON THE CRITERIA OF ADEQUACY CHOSEN

For any nutrient, particularly minerals and vitamins, there is a range of intakes between that which is clearly inadequate, leading to **clinical deficiency disease,** and that which is so much in excess of the body's metabolic capacity that there may be signs of **toxicity.** Between these two extremes is a level of intake that is adequate for normal health and the maintenance of metabolic integrity. Individuals do not all have the same requirement for nutrients even when calculated on the basis of body size or energy expenditure. There is a range of individual requirements of up to 25% around the mean. Therefore, in order to assess the adequacy of diets, it is necessary to set a reference level of intake high enough to ensure that no one will either suffer from deficiency or be at risk of toxicity. If it is assumed that individual requirements are distributed in a statistically normal fashion around the observed mean requirement, then a range of +/− 2 × the standard deviation (SD) around the mean will include the requirements of 95% of the population.

## THE VITAMINS ARE A DISPARATE GROUP OF COMPOUNDS WITH A VARIETY OF METABOLIC FUNCTIONS

A vitamin is defined as an organic compound that is required in the diet in small amounts for the maintenance of normal metabolic integrity. Deficiency causes a specific disease, which is cured or prevented only by restoring the vitamin to the diet (Table 45–1). However, **vitamin D,** which can be made in the skin after exposure to sunlight, and **niacin,** which can be formed from the essential amino acid tryptophan, do not strictly conform to this definition.

***Table 45–1.*** The vitamins.

| Vitamin | | Functions | Deficiency Disease |
|---|---|---|---|
| A | Retinol, β-carotene | Visual pigments in the retina; regulation of gene expression and cell differentiation; β-carotene is an antioxidant | Night blindness, xerophthalmia; keratinization of skin |
| D | Calciferol | Maintenance of calcium balance; enhances intestinal absorption of $Ca^{2+}$ and mobilizes bone mineral | Rickets = poor mineralization of bone; osteomalacia = bone demineralization |
| E | Tocopherols, tocotrienols | Antioxidant, especially in cell membranes | Extremely rare—serious neurologic dysfunction |
| K | Phylloquinone, menaquinones | Coenzyme in formation of γ-carboxyglutamate in enzymes of blood clotting and bone matrix | Impaired blood clotting, hemorrhagic disease |
| $B_1$ | Thiamin | Coenzyme in pyruvate and α-ketoglutarate, dehydrogenases, and transketolase; poorly defined function in nerve conduction | Peripheral nerve damage (beriberi) or central nervous system lesions (Wernicke-Korsakoff syndrome) |
| $B_2$ | Riboflavin | Coenzyme in oxidation and reduction reactions; prosthetic group of flavoproteins | Lesions of corner of mouth, lips, and tongue; seborrheic dermatitis |
| Niacin | Nicotinic acid, nicotinamide | Coenzyme in oxidation and reduction reactions, functional part of NAD and NADP | Pellagra—photosensitive dermatitis, depressive psychosis |
| $B_6$ | Pyridoxine, pyridoxal, pyridoxamine | Coenzyme in transamination and decarboxylation of amino acids and glycogen phosphorylase; role in steroid hormone action | Disorders of amino acid metabolism, convulsions |
| | Folic acid | Coenzyme in transfer of one-carbon fragments | Megaloblastic anemia |
| $B_{12}$ | Cobalamin | Coenzyme in transfer of one-carbon fragments and metabolism of folic acid | Pernicious anemia = megaloblastic anemia with degeneration of the spinal cord |
| | Pantothenic acid | Functional part of CoA and acyl carrier protein: fatty acid synthesis and metabolism | |
| H | Biotin | Coenzyme in carboxylation reactions in gluconeogenesis and fatty acid synthesis | Impaired fat and carbohydrate metabolism, dermatitis |
| C | Ascorbic acid | Coenzyme in hydroxylation of proline and lysine in collagen synthesis; antioxidant; enhances absorption of iron | Scurvy—impaired wound healing, loss of dental cement, subcutaneous hemorrhage |

# ■ LIPID-SOLUBLE VITAMINS

## RETINOIDS & CAROTENOIDS HAVE VITAMIN A ACTIVITY (Figure 45–1)

Retinoids comprise **retinol, retinaldehyde,** and **retinoic acid** (preformed vitamin A, found only in foods of animal origin); carotenoids, found in plants, comprise carotenes and related compounds, known as provitamin A, as they can be cleaved to yield retinaldehyde and thence retinol and retinoic acid. The α-, β-, and γ-carotenes and cryptoxanthin are quantitatively the most important provitamin A carotenoids. Although it would appear that one molecule of β-carotene should yield two of retinol, this is not so in practice; 6 μg of β-carotene is equivalent to 1 μg of preformed retinol. The total amount of vitamin A in foods is therefore expressed as micrograms of retinol equivalents. Beta-carotene and other provitamin A carotenoids are cleaved in the intestinal mucosa by **carotene dioxygenase,** yielding retinaldehyde, which is reduced to retinol, esterified, and secreted in chylomi-

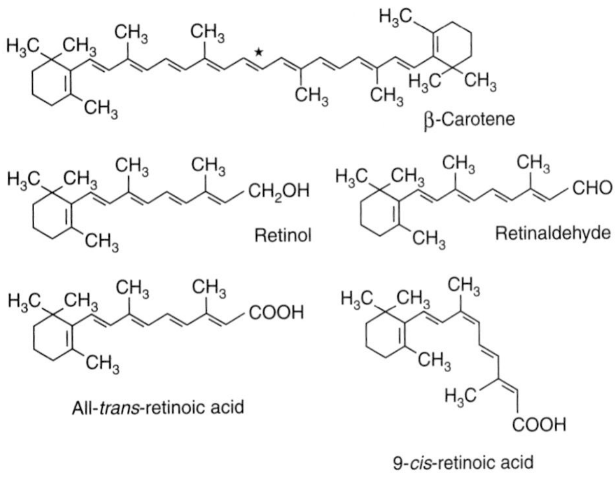

***Figure 45–1.*** β-Carotene and the major vitamin A vitamers. * Shows the site of cleavage of β-carotene into two molecules of retinaldehyde by carotene dioxygenase.

crons together with esters formed from dietary retinol. The intestinal activity of carotene dioxygenase is low, so that a relatively large proportion of ingested β-carotene may appear in the circulation unchanged. While the principal site of carotene dioxygenase attack is the central bond of β-carotene, asymmetric cleavage may also occur, leading to the formation of 8′-, 10′-, and 12′-apo-carotenals, which are oxidized to retinoic acid but cannot be used as sources of retinol or retinaldehyde.

## Vitamin A Has a Function in Vision

In the retina, retinaldehyde functions as the prosthetic group of the light-sensitive opsin proteins, forming **rhodopsin** (in rods) and **iodopsin** (in cones). Any one cone cell contains only one type of opsin and is sensitive to only one color. In the pigment epithelium of the retina, all-*trans*-retinol is isomerized to 11-*cis*-retinol and oxidized to 11-*cis*-retinaldehyde. This reacts with a lysine residue in opsin, forming the holoprotein rhodopsin. As shown in Figure 45–2, the absorption of light by rhodopsin causes isomerization of the retinaldehyde from 11-*cis* to all-*trans,* and a conformational change in opsin. This results in the release of retinaldehyde from the protein and the initiation of a nerve impulse. The formation of the initial excited form of rhodopsin, bathorhodopsin, occurs within picoseconds of illumination. There is then a series of conformational changes leading to the formation of metarhodopsin II, which initiates a guanine nucleotide amplification cascade and then a nerve impulse. The final step is hydrolysis to release all-*trans*-retinaldehyde and opsin. The key to initiation of the visual cycle is the availability of 11-*cis*-retinaldehyde, and hence vitamin A. In deficiency, both the time taken to adapt to darkness and the ability to see in poor light are impaired.

## Retinoic Acid Has a Role in the Regulation of Gene Expression & Tissue Differentiation

A most important function of vitamin A is in the control of cell differentiation and turnover. All-*trans*-retinoic acid and 9-*cis*-retinoic acid (Figure 45–1) regulate growth, development, and tissue differentiation; they have different actions in different tissues. Like the steroid hormones and vitamin D, retinoic acid binds to nuclear receptors that bind to response elements of DNA and regulate the transcription of specific genes. There are two families of nuclear retinoid receptors: the retinoic acid receptors (RARs) bind all-*trans*-retinoic acid or 9-*cis*-retinoic acid, and the retinoid X receptors (RXRs) bind 9-*cis*-retinoic acid.

## Vitamin A Deficiency Is a Major Public Health Problem Worldwide

Vitamin A deficiency is the most important preventable cause of blindness. The earliest sign of deficiency is a loss of sensitivity to green light, followed by impairment of adaptation to dim light, followed by night blindness. More prolonged deficiency leads to **xerophthalmia:** keratinization of the cornea and skin and blindness. Vitamin A also has an important role in differentiation of immune system cells, and mild deficiency leads to increased susceptibility to infectious diseases. Furthermore, the synthesis of retinol-binding protein in response to infection is reduced (it is a negative **acute phase protein**), decreasing the circulating vi-

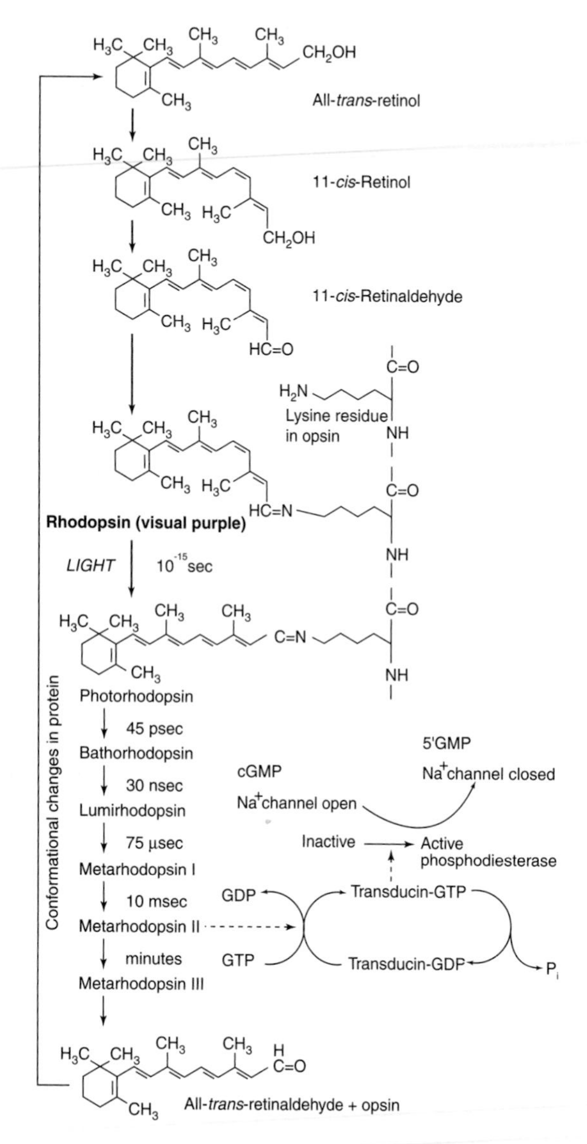


***Figure 45–2.*** The role of retinaldehyde in vision.

tamin, and therefore there is further impairment of immune responses.

### Vitamin A Is Toxic in Excess

There is only a limited capacity to metabolize vitamin A, and excessive intakes lead to accumulation beyond the capacity of binding proteins, so that unbound vitamin A causes tissue damage. Symptoms of toxicity affect the central nervous system (headache, nausea, ataxia, and anorexia, all associated with increased cerebrospinal fluid pressure), the liver (hepatomegaly with histologic changes and hyperlipidemia), calcium homeostasis (thickening of the long bones, hypercalcemia and calcification of soft tissues), and the skin (excessive dryness, desquamation, and alopecia).

## VITAMIN D IS REALLY A HORMONE

Vitamin D is not strictly a vitamin since it can be synthesized in the skin, and under most conditions that is its major source. Only when sunlight is inadequate is a dietary source required. The main function of vitamin D is in the regulation of calcium absorption and homeostasis; most of its actions are mediated by way of nuclear receptors that regulate gene expression. Deficiency—leading to rickets in children and osteomalacia in adults—continues to be a problem in northern latitudes, where sunlight exposure is poor.

### Vitamin D Is Synthesized in the Skin

7-Dehydrocholesterol (an intermediate in the synthesis of cholesterol that accumulates in the skin), undergoes a nonenzymic reaction on exposure to ultraviolet light, yielding previtamin D (Figure 45–3). This undergoes a further reaction over a period of hours to form the vitamin itself, cholecalciferol, which is absorbed into the bloodstream. In temperate climates, the plasma concentration of vitamin D is highest at the end of summer and lowest at the end of winter. Beyond about 40 degrees north or south in winter, there is very little ultraviolet radiation of appropriate wavelength.

### Vitamin D Is Metabolized to the Active Metabolite, Calcitriol, in Liver & Kidney

In the liver, cholecalciferol, which has been synthesized in the skin or derived from food, is hydroxylated to form the 25-hydroxy derivative calcidiol (Figure 45–4). This is released into the circulation bound to a vitamin D-binding globulin which is the main storage form of the vitamin. In the kidney, calcidiol undergoes either 1-hydroxylation to yield the active metabolite 1,25-dihydroxyvitamin D (calcitriol) or 24-hydroxylation to yield an inactive metabolite, 24,25-dihydroxyvitamin D (24-hydroxycalcidiol). Ergocalciferol from fortified foods undergoes similar hydroxylations to yield ercalcitriol.

### Vitamin D Metabolism Both Regulates & Is Regulated by Calcium Homeostasis

The main function of vitamin D is in the control of calcium homeostasis, and in turn vitamin D metabolism is

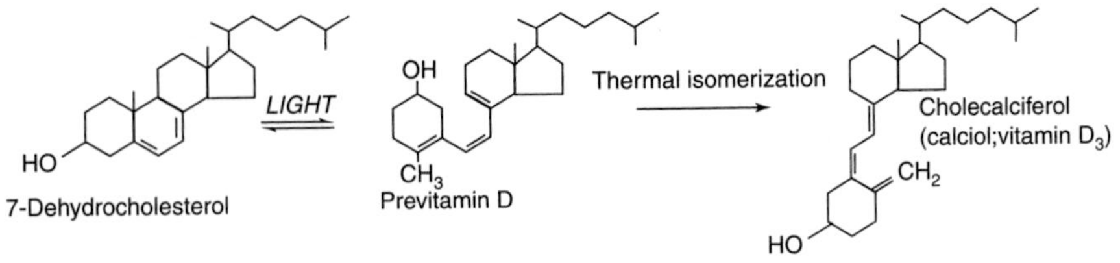

***Figure 45–3.*** Synthesis of vitamin D in the skin.

regulated by factors that respond to plasma concentrations of calcium and phosphate. Calcitriol acts to reduce its own synthesis by inducing the 24-hydroxylase and repressing the 1-hydroxylase in the kidney. Its principal function is to maintain the plasma calcium concentration. Calcitriol achieves this in three ways: it increases intestinal absorption of calcium, reduces excretion of calcium (by stimulating resorption in the distal renal tubules), and mobilizes bone mineral. In addition, calcitriol is involved in insulin secretion, synthesis and secretion of parathyroid and thyroid hormones, inhibition of production of interleukin by activated T lymphocytes and of immunoglobulin by activated B lymphocytes, differentiation of monocyte precursor cells, and modulation of cell proliferation. In its actions, it behaves like a **steroid hormone,** binding to a nuclear receptor protein.

## Vitamin D Deficiency Affects Children & Adults

In the vitamin D deficiency disease **rickets,** the bones of children are undermineralized as a result of poor absorption of calcium. Similar problems occur in adolescents who are deficient during their growth spurt. **Osteomalacia** in adults results from demineralization of bone in women who have little exposure to sunlight, often after several pregnancies. Although vitamin D is essential for prevention and treatment of osteomalacia in the elderly, there is little evidence that it is beneficial in treating **osteoporosis.**

## Vitamin D Is Toxic in Excess

Some infants are sensitive to intakes of vitamin D as low as 50 μg/d, resulting in an elevated plasma concen-

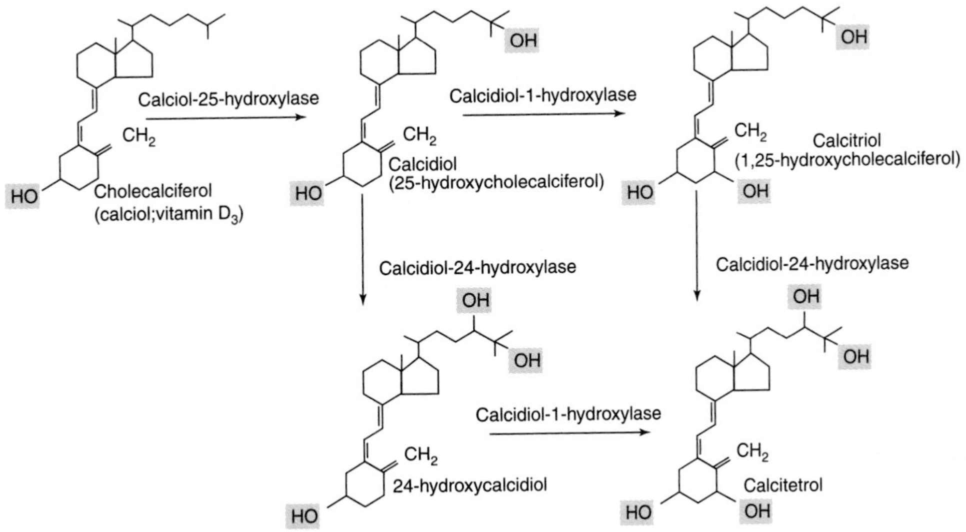

***Figure 45–4.*** Metabolism of vitamin D.

tration of calcium. This can lead to contraction of blood vessels, high blood pressure, and **calcinosis**—the calcification of soft tissues. Although excess dietary vitamin D is toxic, excessive exposure to sunlight does not lead to vitamin D poisoning because there is a limited capacity to form the precursor 7-dehydrocholesterol and to take up cholecalciferol from the skin.

## VITAMIN E DOES NOT HAVE A PRECISELY DEFINED METABOLIC FUNCTION

No unequivocal unique function for vitamin E has been defined. However, it does act as a lipid-soluble antioxidant in cell membranes, where many of its functions can be provided by synthetic antioxidants. Vitamin E is the generic descriptor for two families of compounds, the **tocopherols** and the **tocotrienols** (Figure 45–5). The different vitamers (compounds having similar vitamin activity) have different biologic potencies; the most active is D-α-tocopherol, and it is usual to express vitamin E intake in milligrams of D-α-tocopherol equivalents. Synthetic DL-α-tocopherol does not have the same biologic potency as the naturally occurring compound.

### Vitamin E Is the Major Lipid-Soluble Antioxidant in Cell Membranes & Plasma Lipoproteins

The main function of vitamin E is as a chain-breaking, free radical trapping antioxidant in cell membranes and plasma lipoproteins. It reacts with the lipid peroxide radicals formed by peroxidation of polyunsaturated fatty acids before they can establish a chain reaction. The tocopheroxyl free radical product is relatively unreactive and ultimately forms nonradical compounds. Commonly, the tocopheroxyl radical is reduced back to tocopherol by reaction with vitamin C from plasma (Figure 45–6). The resultant monodehydroascorbate free radical then undergoes enzymic or nonenzymic reaction to yield ascorbate and dehydroascorbate, neither of which is a free radical. The stability of the tocopheroxyl free radical means that it can penetrate farther into cells and, potentially, propagate a chain reaction. Therefore, vitamin E may, like other antioxidants, also have pro-oxidant actions, especially at high concentrations. This may explain why, although studies have shown an association between high blood concentrations of vitamin E and a lower incidence of atherosclerosis, the effect of high doses of vitamin E have been disappointing.

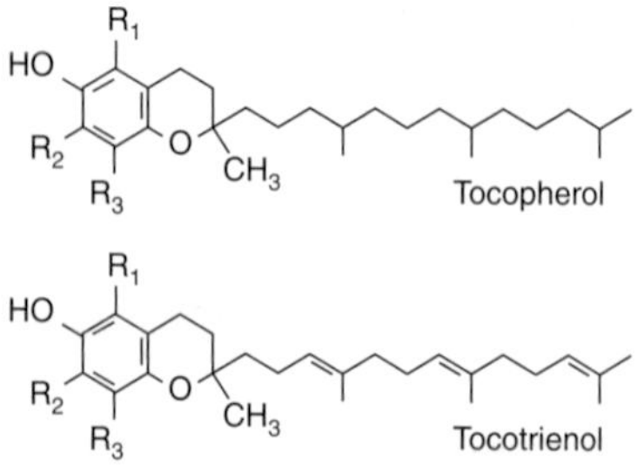

***Figure 45–5.*** The vitamin E vitamers. In α-tocopherol and tocotrienol $R_1$, $R_2$, and $R_3$ are all —$CH_3$ groups. In the β-vitamers $R_2$ is H; in the γ-vitamers $R_1$ is H, and in the δ-vitamers $R_1$ and $R_2$ are both H.

### Dietary Vitamin E Deficiency in Humans Is Unknown

In experimental animals, vitamin E deficiency results in resorption of fetuses and testicular atrophy. Dietary deficiency of vitamin E in humans is unknown, though patients with severe fat malabsorption, cystic fibrosis, and some forms of chronic liver disease suffer deficiency because they are unable to absorb the vitamin or transport it, exhibiting nerve and muscle membrane damage. Premature infants are born with inadequate reserves of the vitamin. Their erythrocyte membranes are abnormally fragile as a result of peroxidation, which leads to hemolytic anemia.

## VITAMIN K IS REQUIRED FOR SYNTHESIS OF BLOOD-CLOTTING PROTEINS

Vitamin K was discovered as a result of investigations into the cause of a bleeding disorder—hemorrhagic (sweet clover) disease—of cattle, and of chickens fed on a fat-free diet. The missing factor in the diet of the chickens was vitamin K, while the cattle feed contained **dicumarol,** an antagonist of the vitamin. Antagonists of vitamin K are used to reduce blood coagulation in patients at risk of thrombosis—the most widely used agent is **warfarin.**

Three compounds have the biologic activity of vitamin K (Figure 45–7): **phylloquinone,** the normal dietary source, found in green vegetables; **menaquinones,** synthesized by intestinal bacteria, with differing lengths of side-chain; **menadione, menadiol,** and **menadiol diacetate,** synthetic compounds that can be metabolized to phylloquinone. Menaquinones are absorbed to some extent but it is not clear to what extent they are biologically active as it is possible to induce signs of vitamin K deficiency simply by feeding a phylloquinone deficient diet, without inhibiting intestinal bacterial action.

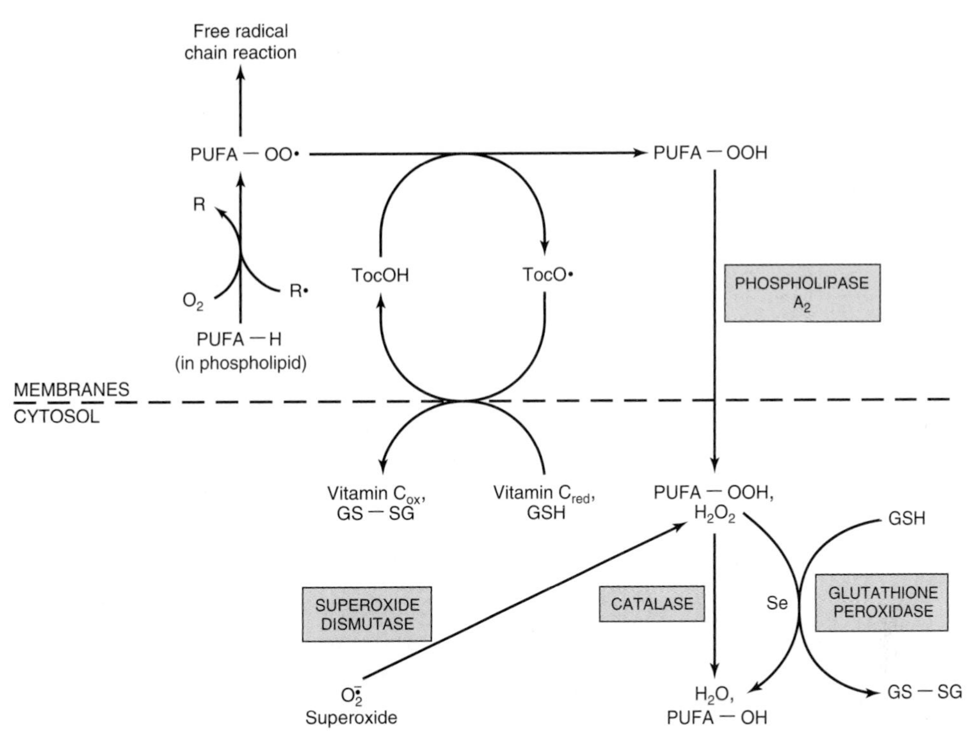


***Figure 45–6.*** Interaction and synergism between antioxidant systems operating in the lipid phase (membranes) of the cell and the aqueous phase (cytosol). (R•, free radical; PUFA-OO•, peroxyl free radical of polyunsaturated fatty acid in membrane phospholipid; PUFA-OOH, hydroperoxy polyunsaturated fatty acid in membrane phospholipid released as hydroperoxy free fatty acid into cytosol by the action of phospholipase $A_2$; PUFA-OH, hydroxy polyunsaturated fatty acid; TocOH, vitamin E ($\alpha$-tocopherol); TocO•, free radical of $\alpha$-tocopherol; Se, selenium; GSH, reduced glutathione; GS-SG, oxidized glutathione, which is returned to the reduced state after reaction with NADPH catalyzed by glutathione reductase; PUFA-H, polyunsaturated fatty acid.)

## Vitamin K Is the Coenzyme for Carboxylation of Glutamate in the Postsynthetic Modification of Calcium-Binding Proteins

Vitamin K is the cofactor for the carboxylation of glutamate residues in the post-synthetic modification of proteins to form the unusual amino acid $\gamma$-carboxyglutamate (Gla), which chelates the calcium ion. Initially, vitamin K hydroquinone is oxidized to the epoxide (Figure 45–8), which activates a glutamate residue in the protein substrate to a carbanion, that reacts nonenzymically with carbon dioxide to form $\gamma$-carboxyglutamate. Vitamin K epoxide is reduced to the quinone by a warfarin-sensitive reductase, and the quinone is reduced to the active hydroquinone by either the same warfarin-sensitive reductase or a warfarin-insensitive quinone reductase. In the presence of warfarin, vitamin K epoxide cannot be reduced but accumulates, and is excreted. If enough vitamin K (a quinone) is provided in the diet, it can be reduced to the active hydroquinone by the warfarin-insensitive enzyme, and carboxylation can continue, with stoichiometric utilization of vitamin K and excretion of the epoxide. A high dose of vitamin K is the antidote to an overdose of warfarin.

Prothrombin and several other proteins of the blood clotting system (Factors VII, IX and X, and proteins C and S) each contain between four and six $\gamma$-carboxyglutamate residues which chelate calcium ions and so permit the binding of the blood clotting proteins to membranes. In vitamin K deficiency or in the presence of warfarin, an abnormal precursor of prothrombin (preprothrombin) containing little or no $\gamma$-carboxyglutamate, and incapable of chelating calcium, is released into the circulation.

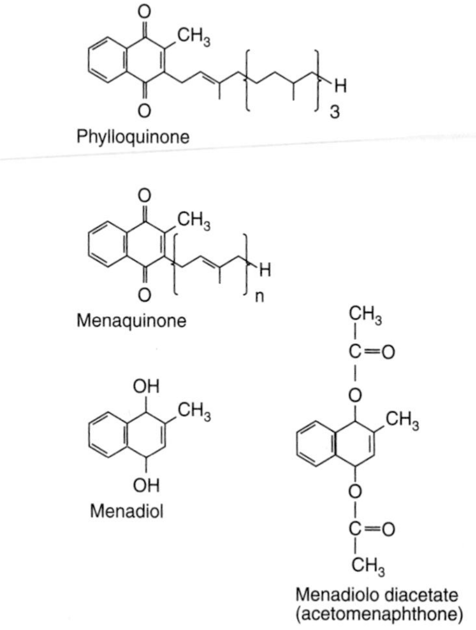

Phylloquinone

Menaquinone

Menadiol

Menadiolo diacetate (acetomenaphthone)

***Figure 45–7.*** The vitamin K vitamers. Menadiol (or menadione) and menadiol diacetate are synthetic compounds that are converted to menaquinone in the liver and have vitamin K activity.

### Vitamin K Is Also Important in the Synthesis of Bone Calcium-Binding Proteins

Treatment of pregnant women with warfarin can lead to fetal bone abnormalities (fetal warfarin syndrome). Two proteins are present in bone that contain γ-carboxyglutamate, osteocalcin and bone matrix Gla protein. Osteocalcin also contains hydroxyproline, so its synthesis is dependent on both vitamins K and C; in addition, its synthesis is induced by vitamin D. The release into the circulation of osteocalcin provides an index of vitamin D status.

# ■ WATER-SOLUBLE VITAMINS

## VITAMIN $B_1$ (THIAMIN) HAS A KEY ROLE IN CARBOHYDRATE METABOLISM

**Thiamin** has a central role in energy-yielding metabolism, and especially the metabolism of carbohydrate (Figure 45–9). **Thiamin diphosphate** is the coenzyme for three multi-enzyme complexes that catalyze oxidative decarboxylation reactions: pyruvate dehydrogenase in carbohydrate metabolism; α-ketoglutarate dehydro-

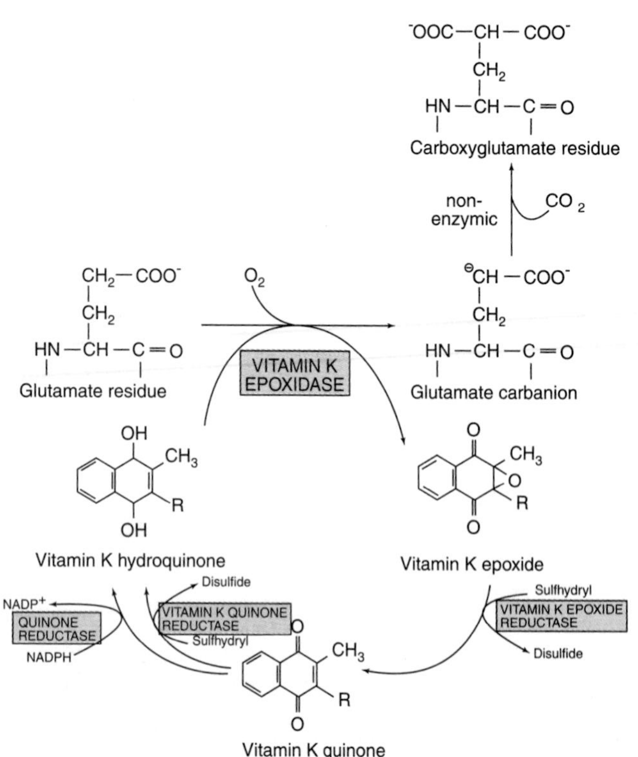


***Figure 45–8.*** The role of vitamin K in the biosynthesis of γ-carboxyglutamate.

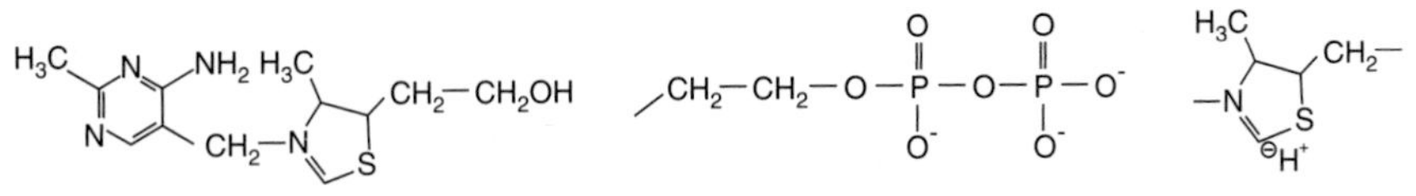

Thiamin Thiamin diphosphate Carbanion

***Figure 45–9.*** Thiamin, thiamin diphosphate, and the carbanion form.

genase in the citric acid cycle; and the branched-chain keto-acid dehydrogenase involved in the metabolism of leucine, isoleucine, and valine. It is also the coenzyme for transketolase, in the pentose phosphate pathway. In each case, the thiamin diphosphate provides a reactive carbon on the thiazole moiety that forms a carbanion, which then adds to the carbonyl group of, for instance, pyruvate. The addition compound then decarboxylates, eliminating $CO_2$. Electrical stimulation of nerve leads to a fall in membrane thiamin triphosphate and release of free thiamin. It is likely that thiamin triphosphate acts as a phosphate donor for phosphorylation of the nerve membrane sodium transport channel.

### Thiamin Deficiency Affects the Nervous System & Heart

Thiamin deficiency can result in three distinct syndromes: a chronic peripheral neuritis, **beriberi,** which may or may not be associated with **heart failure** and **edema;** acute pernicious (fulminating) beriberi (shoshin beriberi), in which heart failure and metabolic abnormalities predominate, without peripheral neuritis; and **Wernicke's encephalopathy** with **Korsakoff's psychosis,** which is associated especially with alcohol and drug abuse. The central role of thiamin diphosphate in pyruvate dehydrogenase means that in deficiency there is impaired conversion of pyruvate to acetyl CoA. In subjects on a relatively high carbohydrate diet, this results in increased plasma concentrations of lactate and pyruvate, which may cause life-threatening **lactic acidosis.**

### Thiamin Nutritional Status Can Be Assessed by Erythrocyte Transketolase Activation

The activation of apo-transketolase(the enzyme protein) in erythrocyte lysate by thiamin diphosphate added in vitro has become the accepted index of thiamin nutritional status.

## VITAMIN $B_2$ (RIBOFLAVIN) HAS A CENTRAL ROLE IN ENERGY-YIELDING METABOLISM

Riboflavin fulfills its role in metabolism as the coenzymes **flavin mononucleotide (FMN)** and **flavin adenine dinucleotide (FAD)** (Figure 45–10). FMN is formed by ATP-dependent phosphorylation of riboflavin, whereas FAD is synthesized by further reaction of FMN with ATP in which its AMP moiety is transferred to the

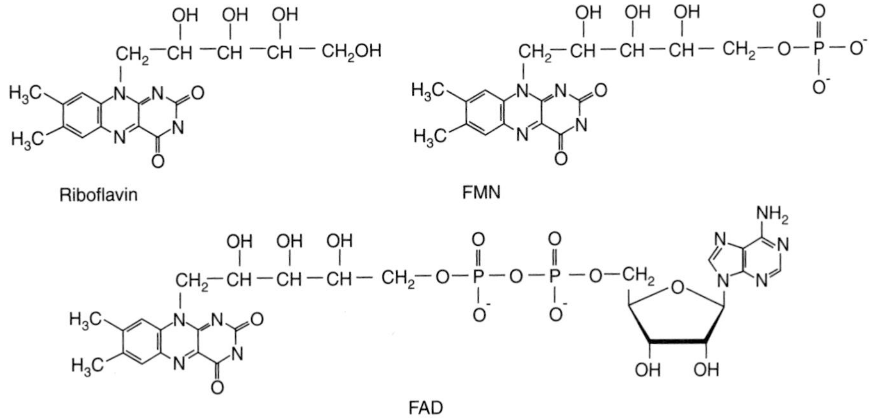

Riboflavin FMN

FAD

***Figure 45–10.*** Riboflavin and the coenzymes flavin mononucleotide (FMN) and flavin adenine dinucleotide (FAD).

FMN. The main dietary sources of riboflavin are milk and dairy products. In addition, because of its intense yellow color, riboflavin is widely used as a food additive.

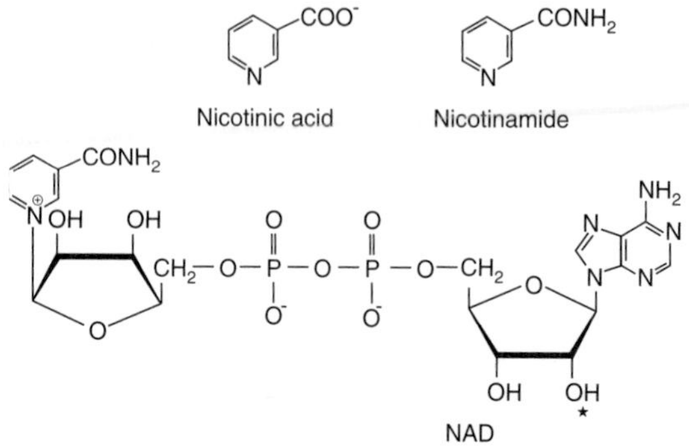

***Figure 45–11.*** Niacin (nicotinic acid and nicotinamide) and nicotinamide adenine dinucleotide (NAD). * Shows the site of phosphorylation in NADP.

## Flavin Coenzymes Are Electron Carriers in Oxidoreduction Reactions

These include the mitochondrial respiratory chain, key enzymes in fatty acid and amino acid oxidation, and the citric acid cycle. Reoxidation of the reduced flavin in oxygenases and mixed-function oxidases proceeds by way of formation of the flavin radical and flavin hydroperoxide, with the intermediate generation of superoxide and perhydroxyl radicals and hydrogen peroxide. Because of this, flavin oxidases make a significant contribution to the total oxidant stress of the body.

## Riboflavin Deficiency Is Widespread But Not Fatal

Although riboflavin is fundamentally involved in metabolism, and deficiencies are found in most countries, it is not fatal as there is very efficient conservation of tissue riboflavin. Riboflavin deficiency is characterized by cheilosis, lingual desquamation and a seborrheic dermatitis. Riboflavin nutritional status is assessed by measurement of the activation of erythrocyte glutathione reductase by FAD added in vitro.

# NIACIN IS NOT STRICTLY A VITAMIN

Niacin was discovered as a nutrient during studies of **pellagra.** It is not strictly a vitamin since it can be synthesized in the body from the essential amino acid tryptophan. Two compounds, **nicotinic acid** and **nicotinamide,** have the biologic activity of niacin; its metabolic function is as the nicotinamide ring of the coenzymes **NAD** and **NADP** in oxidation-reduction reactions (Figure 45–11). About 60 mg of tryptophan is equivalent to 1 mg of dietary niacin. The niacin content of foods is expressed as mg niacin equivalents = mg preformed niacin + 1/60 × mg tryptophan. Because most of the niacin in cereals is biologically unavailable, this is discounted.

## NAD Is the Source of ADP-Ribose

In addition to its coenzyme role, NAD is the source of ADP-ribose for the **ADP-ribosylation** of proteins and polyADP-ribosylation of nucleoproteins involved in the **DNA repair mechanism.**

## Pellagra Is Caused by Deficiency of Tryptophan & Niacin

Pellagra is characterized by a photosensitive dermatitis. As the condition progresses, there is dementia, possibly diarrhea, and, if untreated, death. Although the nutritional etiology of pellagra is well established and tryptophan or niacin will prevent or cure the disease, additional factors, including deficiency of riboflavin or vitamin $B_6$, both of which are required for synthesis of nicotinamide from tryptophan, may be important. In most outbreaks of pellagra twice as many women as men are affected, probably the result of inhibition of tryptophan metabolism by estrogen metabolites.

## Pellagra Can Occur as a Result of Disease Despite an Adequate Intake of Tryptophan & Niacin

A number of genetic diseases that result in defects of tryptophan metabolism are associated with the development of pellagra despite an apparently adequate intake of both tryptophan and niacin. **Hartnup disease** is a rare genetic condition in which there is a defect of the membrane transport mechanism for tryptophan, resulting in large losses due to intestinal malabsorption and failure of the renal resorption mechanism. In **carcinoid syndrome** there is metastasis of a primary liver tumor of enterochromaffin cells which synthesize 5-hydroxytryptamine. Overproduction of 5-hydroxytryptamine may account for as much as 60% of the body's tryptophan metabolism, causing pellagra because of the diversion away from NAD synthesis.

## Niacin Is Toxic in Excess

Nicotinic acid has been used to treat hyperlipidemia when of the order of 1–6 g/d are required, causing dilation of blood vessels and flushing, with skin irritation. Intakes of both nicotinic acid and nicotinamide in excess of 500 mg/d can cause liver damage.

## VITAMIN $B_6$ IS IMPORTANT IN AMINO ACID & GLYCOGEN METABOLISM & IN STEROID HORMONE ACTION

Six compounds have vitamin $B_6$ activity (Figure 45–12): **pyridoxine, pyridoxal, pyridoxamine,** and their 5′-phosphates. The active coenzyme is pyridoxal 5′-phosphate. Approximately 80% of the body's total vitamin $B_6$ is present as pyridoxal phosphate in muscle, mostly associated with glycogen phosphorylase. This is not available in $B_6$ deficiency but is released in starvation, when glycogen reserves become depleted, and is then available, especially in liver and kidney, to meet increased requirement for gluconeogenesis from amino acids.

### Vitamin $B_6$ Has Several Roles in Metabolism

Pyridoxal phosphate is a coenzyme for many enzymes involved in amino acid metabolism, especially in transamination and decarboxylation. It is also the cofactor of glycogen phosphorylase, where the phosphate group is catalytically important. In addition, vitamin $B_6$ is important in steroid hormone action where it removes the hormone-receptor complex from DNA binding, terminating the action of the hormones. In vitamin $B_6$ deficiency, this results in increased sensitivity to the actions of low concentrations of estrogens, androgens, cortisol, and vitamin D.

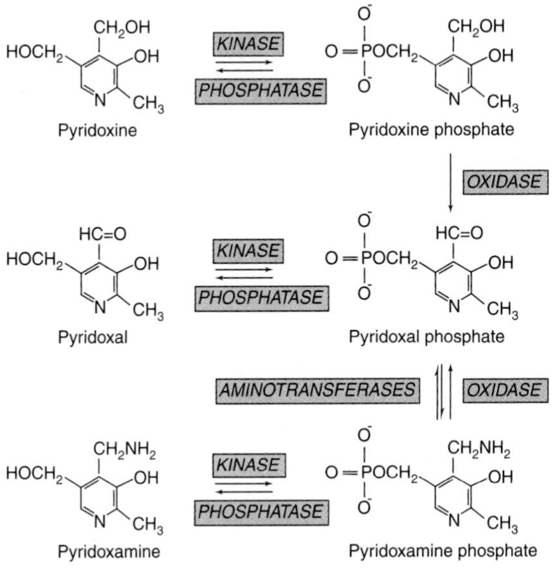

***Figure 45–12.*** Interconversion of the vitamin $B_6$ vitamers.

### Vitamin $B_6$ Deficiency Is Rare

Although clinical deficiency disease is rare, there is evidence that a significant proportion of the population have marginal vitamin $B_6$ status. Moderate deficiency results in abnormalities of tryptophan and methionine metabolism. Increased sensitivity to steroid hormone action may be important in the development of **hormone-dependent cancer** of the breast, uterus, and prostate, and vitamin $B_6$ status may affect the prognosis.

### Vitamin $B_6$ Status Is Assessed by Assaying Erythrocyte Aminotransferases

The most widely used method of assessing vitamin $B_6$ status is by the activation of erythrocyte aminotransferases by pyridoxal phosphate added in vitro, expressed as the activation coefficient.

### In Excess, Vitamin $B_6$ Causes Sensory Neuropathy

The development of sensory neuropathy has been reported in patients taking 2–7 g of pyridoxine per day for a variety of reasons (there is some slight evidence that it is effective in treating **premenstrual syndrome**). There was some residual damage after withdrawal of these high doses; other reports suggest that intakes in excess of 200 mg/d are associated with neurologic damage.

## VITAMIN $B_{12}$ IS FOUND ONLY IN FOODS OF ANIMAL ORIGIN

The term "vitamin $B_{12}$" is used as a generic descriptor for the **cobalamins**—those **corrinoids** (cobalt containing compounds possessing the corrin ring) having the biologic activity of the vitamin (Figure 45–13). Some corrinoids that are growth factors for microorganisms not only have no vitamin $B_{12}$ activity but may also be antimetabolites of the vitamin. Although it is synthesized exclusively by microorganisms, for practical purposes vitamin $B_{12}$ is found only in foods of animal origin, there being no plant sources of this vitamin. This means that strict vegetarians (vegans) are at risk of developing $B_{12}$ deficiency. The small amounts of the vitamin formed by bacteria on the surface of fruits may be adequate to meet requirements, but preparations of vitamin $B_{12}$ made by bacterial fermentation are available.

### Vitamin $B_{12}$ Absorption Requires Two Binding Proteins

Vitamin $B_{12}$ is absorbed bound to **intrinsic factor,** a small glycoprotein secreted by the parietal cells of the

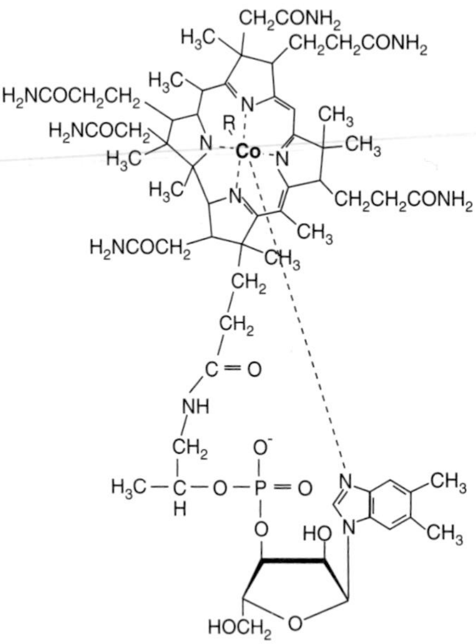

***Figure 45–13.*** Vitamin $B_{12}$ (cobalamin). R may be varied to give the various forms of the vitamin, eg, R = $CN^-$ in cyanocobalamin; R = $OH^-$ in hydroxocobalamin; R = 5′-deoxyadenosyl in 5′-deoxyadenosylcobalamin; R = $H_2O$ in aquocobalamin; and R = $CH_3$ in methylcobalamin.

gastric mucosa. Gastric acid and pepsin release the vitamin from protein binding in food and make it available to bind to **cobalophilin,** a binding protein secreted in the saliva. In the duodenum, cobalophilin is hydrolyzed, releasing the vitamin for binding to intrinsic factor. **Pancreatic insufficiency** can therefore be a factor in the development of vitamin $B_{12}$ deficiency, resulting in the excretion of cobalophilin-bound vitamin $B_{12}$. Intrinsic factor binds the various vitamin $B_{12}$ vitamers, but not other corrinoids. Vitamin $B_{12}$ is absorbed from the distal third of the ileum via receptors that bind the intrinsic factor-vitamin $B_{12}$ complex but not free intrinsic factor or free vitamin.

### There Are Three Vitamin $B_{12}$-Dependent Enzymes

**Methylmalonyl CoA mutase, leucine aminomutase,** and **methionine synthase** (Figure 45–14) are vitamin $B_{12}$-dependent enzymes. Methylmalonyl CoA is formed as an intermediate in the catabolism of valine and by the carboxylation of propionyl CoA arising in the catabolism of isoleucine, cholesterol, and, rarely, fatty acids with an odd number of carbon atoms—or directly from propionate, a major product of microbial fermentation in ruminants. It undergoes vitamin $B_{12}$-dependent rearrangement to succinyl-CoA, catalyzed by methylmalonyl-CoA isomerase (Figure 19–2). The activity of this enzyme is greatly reduced in vitamin $B_{12}$ deficiency, leading to an accumulation of methylmalonyl-CoA and urinary excretion of methylmalonic acid, which provides a means of assessing vitamin $B_{12}$ nutritional status.

### Vitamin $B_{12}$ Deficiency Causes Pernicious Anemia

Pernicious anemia arises when vitamin $B_{12}$ deficiency blocks the metabolism of folic acid, leading to functional folate deficiency. This impairs erythropoiesis, causing immature precursors of erythrocytes to be released into the circulation (megaloblastic anemia). The commonest cause of pernicious anemia is failure of the absorption of vitamin $B_{12}$ rather than dietary deficiency. This can be due to failure of intrinsic factor secretion caused by autoimmune disease of parietal cells or to generation of anti-intrinsic factor antibodies.

## THERE ARE MULTIPLE FORMS OF FOLATE IN THE DIET

The active form of folic acid (pteroyl glutamate) is tetrahydrofolate (Figure 45–15). The folates in foods may have up to seven additional glutamate residues linked by γ-peptide bonds. In addition, all of the one-carbon substituted folates in Figure 45–15 may also be present in foods.

The extent to which the different forms of folate can be absorbed varies, and this must be allowed for in calculating folate intakes.

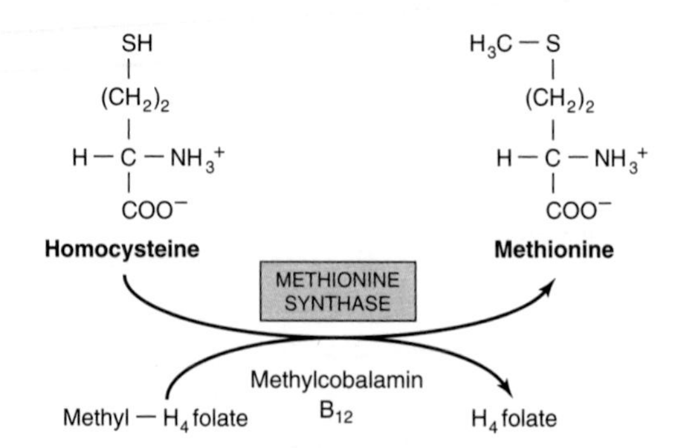


***Figure 45–14.*** Homocysteinuria and the folate trap. Vitamin $B_{12}$ deficiency leads to inhibition of methionine synthase activity causing homocysteinuria and the trapping of folate as methyltetrahydrofolate.

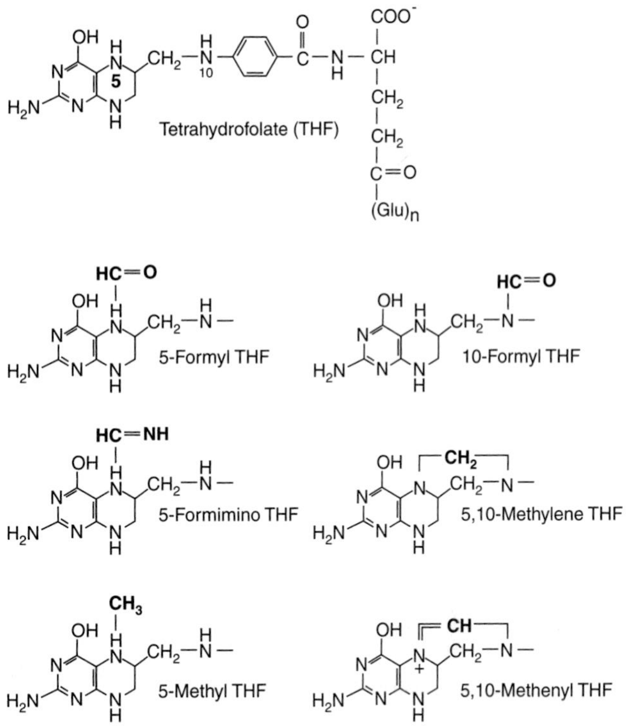

*Figure 45–15.* Tetrahydrofolic acid and the one-carbon substituted folates.

## Tetrahydrofolate Is a Carrier of One-Carbon Units

Tetrahydrofolate can carry one-carbon fragments attached to N-5 (formyl, formimino, or methyl groups), N-10 (formyl group), or bridging N-5 to N-10 (methylene or methenyl groups). 5-Formyl-tetrahydrofolate is more stable than folate and is therefore used pharmaceutically in the agent known as **folinic acid** and in the synthetic (racemic) compound **leucovorin.**

The major point of entry for one-carbon fragments into substituted folates is methylene tetrahydrofolate (Figure 45–16), which is formed by the reaction of glycine, serine, and choline with tetrahydrofolate. Serine is the most important source of substituted folates for biosynthetic reactions, and the activity of serine hy-

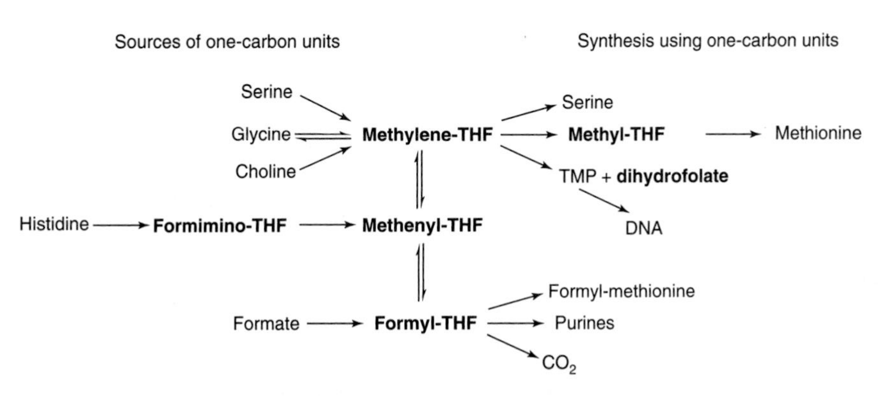


*Figure 45–16.* Sources and utilization of one-carbon substituted folates.

droxymethyltransferase is regulated by the state of folate substitution and the availability of folate. The reaction is reversible, and in liver it can form serine from glycine as a substrate for gluconeogenesis. Methylene, methenyl, and 10-formyl tetrahydrofolates are interconvertible. When one-carbon folates are not required, the oxidation of formyl tetrahydrofolate to yield carbon dioxide provides a means of maintaining a pool of free folate.

### Inhibitors of Folate Metabolism Provide Cancer Chemotherapy & Antibacterial & Antimalarial Drugs

The methylation of deoxyuridine monophosphate (dUMP) to thymidine monophosphate (TMP), catalyzed by thymidylate synthase, is essential for the synthesis of DNA. The one-carbon fragment of methylene-tetrahydrofolate is reduced to a methyl group with release of dihydrofolate, which is then reduced back to tetrahydrofolate by **dihydrofolate reductase.** Thymidylate synthase and dihydrofolate reductase are especially active in tissues with a high rate of cell division. **Methotrexate,** an analog of 10-methyl-tetrahydrofolate, inhibits dihydrofolate reductase and has been exploited as an anticancer drug. The dihydrofolate reductases of some bacteria and parasites differ from the human enzyme; inhibitors of these enzymes can be used as antibacterial drugs, eg, **trimethoprim,** and antimalarial drugs, eg, **pyrimethamine.**

### Vitamin $B_{12}$ Deficiency Causes Functional Folate Deficiency—the Folate Trap

When acting as a methyl donor, *S*-adenosylmethionine forms homocysteine, which may be remethylated by methyltetrahydrofolate catalyzed by methionine synthase, a vitamin $B_{12}$-dependent enzyme (Figure 45–14). The reduction of methylene-tetrahydrofolate to methyltetrahydrofolate is irreversible, and since the major source of tetrahydrofolate for tissues is methyl-tetrahydrofolate, the role of methionine synthase is vital and provides a link between the functions of folate and vitamin $B_{12}$. Impairment of methionine synthase in $B_{12}$ deficiency results in the accumulation of methyl-tetrahydrofolate—the "folate trap." There is therefore functional deficiency of folate secondary to the deficiency of vitamin $B_{12}$.

### Folate Deficiency Causes Megaloblastic Anemia

Deficiency of folic acid itself—or deficiency of vitamin $B_{12}$, which leads to functional folic acid deficiency—affects cells that are dividing rapidly because they have a large requirement for thymidine for DNA synthesis. Clinically, this affects the bone marrow, leading to megaloblastic anemia.

### Folic Acid Supplements Reduce the Risk of Neural Tube Defects & Hyperhomocysteinemia

Supplements of 400 μg/d of folate begun before conception result in a significant reduction in the incidence of neural tube defects as found in **spina bifida.** Elevated blood homocysteine is an associated risk factor for **atherosclerosis, thrombosis,** and **hypertension.** The condition is due to impaired ability to form methyl-tetrahydrofolate by **methylene-tetrahydrofolate reductase,** causing functional folate deficiency and resulting in failure to remethylate homocysteine to methionine. People with the causative abnormal variant of methylene-tetrahydrofolate reductase do not develop hyperhomocysteinemia if they have a relatively high intake of folate, but it is not yet known whether this affects the incidence of cardiovascular disease.

### Folate Enrichment of Foods May Put Some People at Risk

Folate supplements will rectify the megaloblastic anemia of vitamin $B_{12}$ deficiency but may hasten the development of the (irreversible) nerve damage found in $B_{12}$ deficiency. There is also antagonism between folic acid and the anticonvulsants used in the treatment of epilepsy.

## DIETARY BIOTIN DEFICIENCY IS UNKNOWN

The structures of **biotin, biocytin,** and **carboxy-biotin** (the active metabolic intermediate) are shown in Figure 45–17. Biotin is widely distributed in many foods as biocytin (ε-amino-biotinyl lysine), which is released on proteolysis. It is synthesized by intestinal flora in excess of requirements. Deficiency is unknown except among people maintained for many months on parenteral nutrition and a very small number who eat abnormally large amounts of uncooked egg white, which contains avidin, a protein that binds biotin and renders it unavailable for absorption.

### Biotin Is a Coenzyme of Carboxylase Enzymes

Biotin functions to transfer carbon dioxide in a small number of carboxylation reactions. A holocarboxylase synthetase acts on a lysine residue of the apoenzymes of acetyl-CoA carboxylase, pyruvate carboxylase, propionyl-CoA carboxylase, or methylcrotonyl-CoA carboxylase to react with free biotin to form the biocytin residue of the holoenzyme. The reactive intermediate is 1-*N*-carboxybiocytin, formed from bicarbonate in an ATP-dependent reaction. The carboxyl group is then transferred to the substrate for carboxylation.

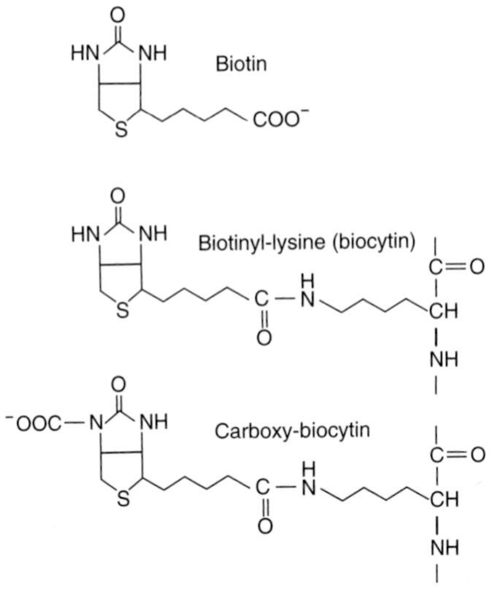

*Figure 45–17.* Biotin, biocytin, and carboxy-biocytin.

Biotin also has a role in regulation of the cell cycle, acting to biotinylate key nuclear proteins.

## AS PART OF CoA AND ACP, PANTOTHENIC ACID ACTS AS A CARRIER OF ACYL RADICALS

Pantothenic acid has a central role in acyl group metabolism when acting as the pantetheine functional moiety of coenzyme A or acyl carrier protein (ACP) (Figure 45–18). The pantetheine moiety is formed after combination of pantothenate with cysteine, which provides the —SH prosthetic group of CoA and ACP. CoA takes part in reactions of the citric acid cycle, fatty acid synthesis and oxidation, acetylations, and cholesterol synthesis. ACP participates in fatty acid synthesis. The vitamin is widely distributed in all foodstuffs, and deficiency has not been unequivocally reported in human beings except in specific depletion studies.

## ASCORBIC ACID IS A VITAMIN FOR ONLY SOME SPECIES

**Vitamin C** (Figure 45–19) is a vitamin for human beings and other primates, the guinea pig, bats, passerine birds, and most fishes and invertebrates; other animals synthesize it as an intermediate in the uronic acid pathway of glucose metabolism (Chapter 20). In those species for which it is a vitamin, there is a block in that pathway due to absence of gulonolactone oxidase. Both ascorbic acid and dehydroascorbic acid have vitamin activity.

### Vitamin C Is the Coenzyme for Two Groups of Hydroxylases

Ascorbic acid has specific roles in the copper-containing hydroxylases and the α-ketoglutarate-linked iron-containing hydroxylases. It also increases the activity of a number of other enzymes in vitro, though this is a nonspecific reducing action. In addition, it has a number of nonenzymic effects due to its action as a reducing agent and oxygen radical quencher.

**Dopamine β-hydroxylase** is a copper-containing enzyme involved in the synthesis of the catecholamines norepinephrine and epinephrine from tyrosine in the adrenal medulla and central nervous system. During hydroxylation, the $Cu^+$ is oxidized to $Cu^{2+}$; reduction back

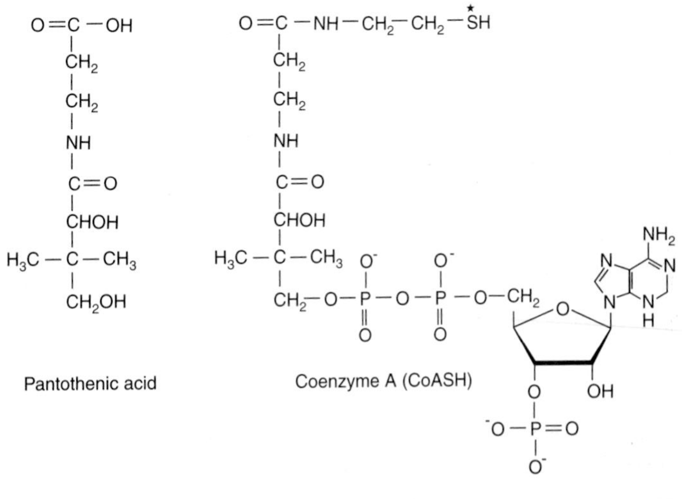

*Figure 45–18.* Pantothenic acid and coenzyme A. * Shows the site of acylation by fatty acids.

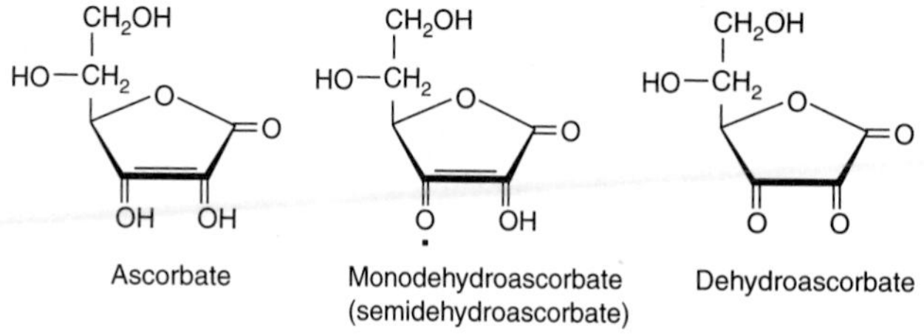

Ascorbate — Monodehydroascorbate (semidehydroascorbate) — Dehydroascorbate

***Figure 45–19.*** Vitamin C.

to $Cu^{+}$ specifically requires ascorbate, which is oxidized to monodehydroascorbate. Similar actions of ascorbate occur in tyrosine degradation at the *p*-hydroxyphenylpyruvate hydroxylase step and at the homogentisate dioxygenase step, which needs $Fe^{2+}$ (Figure 30–12).

A number of peptide hormones have a carboxyl terminal amide which is derived from a glycine terminal residue. This glycine is hydroxylated on the α-carbon by a copper-containing enzyme, **peptidylglycine hydroxylase,** which, again, requires ascorbate for reduction of $Cu^{2+}$.

A number of iron-containing, ascorbate-requiring hydroxylases share a common reaction mechanism in which hydroxylation of the substrate is linked to decarboxylation of α-ketoglutarate (Figure 28–11). Many of these enzymes are involved in the modification of precursor proteins. **Proline and lysine hydroxylases** are required for the postsynthetic modification of **procollagen** to **collagen,** and proline hydroxylase is also required in formation of **osteocalcin** and the C1q component of **complement.** Aspartate β-hydroxylase is required for the postsynthetic modification of the precursor of protein C, the vitamin K-dependent protease which hydrolyzes activated factor V in the blood clotting cascade. Trimethyllysine and γ-butyrobetaine hydroxylases are required for the synthesis of carnitine.

### Vitamin C Deficiency Causes Scurvy

Signs of vitamin C deficiency in scurvy include skin changes, fragility of blood capillaries, gum decay, tooth loss, and bone fracture, many of which can be attributed to deficient collagen synthesis.

### There May Be Benefits From Higher Intakes of Vitamin C

At intakes above approximately 100 mg/d, the body's capacity to metabolize vitamin C is saturated, and any further intake is excreted in the urine. However, in addition to its other roles, vitamin C enhances the absorption of iron, and this depends on the presence of the vitamin in the gut. Therefore, increased intakes may be beneficial. Evidence is unconvincing that high doses of vitamin C prevent the common cold or reduce the duration of its symptoms.

## MINERALS ARE REQUIRED FOR BOTH PHYSIOLOGIC & BIOCHEMICAL FUNCTIONS

Many of the essential minerals (Table 45–2) are widely distributed in foods, and most people eating a normal mixed diet are likely to receive adequate intakes. The

***Table 45–2.*** Classification of essential minerals according to their function.

| Function | Mineral |
|---|---|
| Structural function | Calcium, magnesium, phosphate |
| Involved in membrane function: principal cations of extracellular- and intracellular fluids, respectively | Sodium, potassium |
| Function as prosthetic groups in enzymes | Cobalt, copper, iron, molybdenum, selenium, zinc |
| Regulatory role or role in hormone action | Calcium, chromium, iodine, magnesium, manganese, sodium, potassium |
| Known to be essential, but function unknown | Silicon, vanadium, nickel, tin |
| Have effects in the body, but essentiality is not established | Fluoride, lithium |
| May occur in foods and known to be toxic in excess | Aluminum, arsenic, antimony, boron, bromine, cadmium, cesium, germanium, lead, mercury, silver, strontium |

amounts required vary from grams per day for sodium and calcium, through milligrams per day (eg, iron) to micrograms per day for the trace elements. In general, mineral deficiencies are encountered when foods come from one region, where the soil may be deficient in some minerals, eg, iodine deficiency. Where the diet comes from a variety of different regions, mineral deficiencies are unlikely. However, iron deficiency is a general problem because if iron losses from the body are relatively high (eg, from heavy menstrual blood loss), it is difficult to achieve an adequate intake to replace the losses. Foods from soils containing high levels of selenium cause toxicity, and increased intakes of common salt (sodium chloride) cause hypertension in susceptible individuals.

## SUMMARY


- Vitamins are organic nutrients with essential metabolic functions, generally required in small amounts in the diet because they cannot be synthesized by the body. The lipid-soluble vitamins (A, D, E, and K) are hydrophobic molecules requiring normal fat absorption for their efficient absorption and the avoidance of deficiency symptoms.
- Vitamin A (retinol), present in carnivorous diets, and the provitamin (β-carotene), found in plants, form retinaldehyde, utilized in vision, and retinoic acid, which acts in the control of gene expression. Vitamin D is a steroid prohormone yielding the active hormone derivative calcitriol, which regulates calcium and phosphate metabolism. Vitamin D deficiency leads to rickets and osteomalacia.
- Vitamin E (tocopherol) is the most important antioxidant in the body, acting in the lipid phase of membranes and protecting against the effects of free radicals. Vitamin K functions as cofactor to a carboxylase that acts on glutamate residues of clotting factor precursor proteins to enable them to chelate calcium.
- The water-soluble vitamins of the B complex act as enzyme cofactors. Thiamin is a cofactor in oxidative decarboxylation of α-keto acids and of transketolase in the pentose phosphate pathway. Riboflavin and niacin are important cofactors in oxidoreduction reactions, respectively present in flavoprotein enzymes and in NAD and NADP.
- Pantothenic acid is present in coenzyme A and acyl carrier protein, which act as carriers for acyl groups in metabolic reactions. Pyridoxine, as pyridoxal phosphate, is the coenzyme for several enzymes of amino acid metabolism, including the aminotransferases, and of glycogen phosphorylase. Biotin is the coenzyme for several carboxylase enzymes.
- Besides other functions, vitamin $B_{12}$ and folic acid take part in providing one-carbon residues for DNA synthesis, deficiency resulting in megaloblastic anemia. Vitamin C is a water-soluble antioxidant that maintains vitamin E and many metal cofactors in the reduced state.
- Inorganic mineral elements that have a function in the body must be provided in the diet. When insufficient, deficiency symptoms may arise, and if present in excess they may be toxic.


## REFERENCES


Bender DA, Bender AE: *Nutrition: A Reference Handbook.* Oxford Univ Press, 1997.

Bender DA: *Nutritional Biochemistry of the Vitamins.* Cambridge Univ Press, 1992.

Garrow JS, James WPT, Ralph A: *Human Nutrition and Dietetics,* 10th ed. Churchill-Livingstone, 2000.

Halliwell B, Chirico S: Lipid peroxidation: its mechanism, measurement, and significance. Am J Clin Nutr 1993;57(5 Suppl):715S.

Krinsky NI: Actions of carotenoids in biological systems. Annu Rev Nutr 1993;13:561.

Padh H: Vitamin C: newer insights into its biochemical functions. Nutr Rev 1991;49:65.

Shane B: Folylpolyglutamate synthesis and role in the regulation of one-carbon metabolism. Vitam Horm 1989;45:263.

Wiseman H, Halliwell B: Damage to DNA by reactive oxygen and nitrogen species: role in inflammatory disease and progression to cancer. Biochem J 1996;313:17.

# Intracellular Traffic & Sorting of Proteins

# 46

*Robert K. Murray, MD, PhD*

## BIOMEDICAL IMPORTANCE

Proteins must travel from polyribosomes to many different sites in the cell to perform their particular functions. Some are destined to be components of specific organelles, others for the cytosol or for export, and yet others will be located in the various cellular membranes. Thus, there is considerable intracellular traffic of proteins. Many studies have shown that the Golgi apparatus plays a major role in the sorting of proteins for their correct destinations. A major insight was the recognition that for proteins to attain their proper locations, they generally contain information (a signal or coding sequence) that targets them appropriately. Once a number of the signals were defined, it became apparent that certain diseases result from mutations that affect these signals. In this chapter we discuss the intracellular traffic of proteins and their sorting and briefly consider some of the disorders that result when abnormalities occur.

## MANY PROTEINS ARE TARGETED BY SIGNAL SEQUENCES TO THEIR CORRECT DESTINATIONS

The protein biosynthetic pathways in cells can be considered to be **one large sorting system.** Many proteins carry **signals** (usually but not always specific sequences of amino acids) that direct them to their destination, thus ensuring that they will end up in the appropriate membrane or cell compartment; these signals are a fundamental component of the sorting system. Usually the signal sequences are recognized and interact with complementary areas of proteins that serve as receptors for the proteins that contain them.

A major sorting decision is made early in protein biosynthesis, when specific proteins are synthesized either on free or on membrane-bound polyribosomes. This results in two sorting branches called the **cytosolic branch** and the **rough endoplasmic reticulum (RER) branch** (Figure 46–1). This sorting occurs because proteins synthesized on membrane-bound polyribosomes contain a **signal peptide** that mediates their attachment to the membrane of the ER. Further details on the signal peptide are given below. Proteins synthesized on **free polyribosomes** lack this particular signal peptide and are delivered into the cytosol. There they are directed to mitochondria, nuclei, and peroxisomes by specific signals—or remain in the cytosol if they lack a signal. Any protein that contains a targeting sequence that is subsequently removed is designated as a **preprotein.** In some cases a second peptide is also removed, and in that event the original protein is known as a **preproprotein** (eg, preproalbumin; Chapter 50).

Proteins synthesized and sorted in the **rough ER branch** (Figure 46–2) include many destined for various membranes (eg, of the ER, Golgi apparatus, lysosomes, and plasma membrane) and for secretion. Lysosomal enzymes are also included. Thus, such proteins may reside in the membranes or lumens of the ER or follow the major transport route of intracellular proteins to the Golgi apparatus. Further signal-mediated sorting of certain proteins occurs in the Golgi apparatus, resulting in delivery to lysosomes, membranes of the Golgi apparatus, and other sites. Proteins destined for the plasma membrane or for secretion pass through the Golgi apparatus but generally are not thought to carry specific sorting signals; they are believed to reach their destinations by default.

The entire pathway of ER → Golgi apparatus → plasma membrane is often called the **secretory** or **exocytotic pathway.** Events along this route will be given special attention. Most of the proteins reaching the Golgi apparatus or the plasma membrane are carried in **transport vesicles;** a brief description of the formation of these important particles will be given subsequently. Other proteins destined for secretion are carried in **secretory vesicles** (Figure 46–2). These are prominent in the pancreas and certain other glands. Their mobilization and discharge are regulated and often referred to as **"regulated secretion,"** whereas the secretory pathway involving transport vesicles is called **"constitutive."**

Experimental approaches that have afforded major insights to the processes described in this chapter include (1) use of yeast mutants; (2) application of recombinant DNA techniques (eg, mutating or eliminating particular sequences in proteins, or fusing new sequences onto them; and (3) development of in vitro

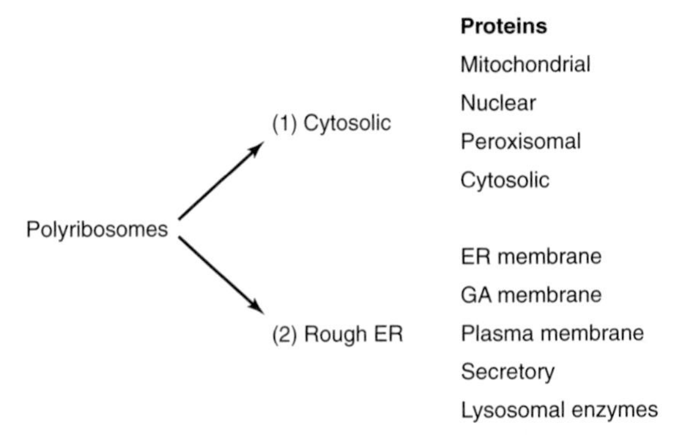


***Figure 46–1.*** Diagrammatic representation of the two branches of protein sorting occurring by synthesis on (1) cytosolic and (2) membrane-bound polyribosomes. The mitochondrial proteins listed are encoded by nuclear genes. Some of the signals used in further sorting of these proteins are listed in Table 46–4. (ER, endoplasmic reticulum; GA, Golgi apparatus.)

systems (eg, to study translocation in the ER and mechanisms of vesicle formation).

The sorting of proteins belonging to the cytosolic branch referred to above is described next, starting with mitochondrial proteins.

## THE MITOCHONDRION BOTH IMPORTS & SYNTHESIZES PROTEINS

Mitochondria contain many proteins. Thirteen proteins (mostly membrane components of the electron transport chain) are encoded by the **mitochondrial genome** and synthesized in that organelle using its own protein-synthesizing system. However, the majority (at least several hundred) are encoded by **nuclear genes,** are synthesized outside the mitochondria on cytosolic polyribosomes, and must be imported. **Yeast cells** have proved to be a particularly useful system for analyzing the mechanisms of import of mitochondrial proteins, partly because it has proved possible to generate a variety of mutants that have illuminated the fundamental processes involved. Most progress has been made in the study of proteins present in the mitochondrial matrix, such as the $F_1$ ATPase subunits. Only the pathway of import of matrix proteins will be discussed in any detail here.

**Matrix proteins** must pass from cytosolic polyribosomes through the outer and inner mitochondrial membranes to reach their destination. Passage through the two membranes is called **translocation.** They have an amino terminal leader sequence (presequence), about 20–80 amino acids in length, which is not highly conserved but contains many positively charged amino acids (eg, Lys or Arg). The presequence is equivalent to a signal peptide mediating attachment of polyribosomes to membranes of the ER (see below), but in this instance targeting proteins to the matrix; if the leader sequence is cleaved off, potential matrix proteins will not reach their destination.

Translocation is believed to occur **posttranslationally,** after the matrix proteins are released from the cytosolic polyribosomes. Interactions with a number of cytosolic proteins that act as **chaperones** (see below) and as targeting factors occur prior to translocation.

Two distinct **translocation complexes** are situated in the outer and inner mitochondrial membranes, referred to (respectively) as TOM (translocase-of-the-outer membrane) and TIM (translocase-of-the-inner membrane). Each complex has been analyzed and found to be composed of a number of proteins, some of which act as receptors for the incoming proteins and others as components of the transmembrane pores through which these proteins must pass. Proteins must be in the **unfolded state** to pass through the complexes, and this is made possible by **ATP-dependent binding to several chaperone proteins.** The roles of chaperone proteins in protein folding are discussed later in this chapter. In mitochondria, they are involved in translocation, sorting, folding, assembly, and degradation of imported proteins. A **proton-motive force** across the inner membrane is required for import; it is made up of the **electric potential** across the membrane (inside negative) and the **pH gradient** (see Chapter 12). The positively charged leader sequence may be helped through the membrane by the negative charge in the matrix. The presequence is split off in the matrix by a **matrix-processing peptidase (MPP).** Contact with **other chaperones** present in the matrix is essential to complete the overall process of import. Interaction with mt-Hsp70 (Hsp = heat shock protein) ensures proper import into the matrix and prevents misfolding or aggregation, while interaction with the mt-Hsp60-Hsp10 system ensures proper folding. The latter proteins resemble the bacterial GroEL chaperonins, a subclass of chaperones that form complex cage-like assemblies made up of heptameric ring structures. The interactions of imported proteins with the above chaperones require **hydrolysis of ATP** to drive them.

The details of how preproteins are translocated have not been fully elucidated. It is possible that the electric potential associated with the inner mitochondrial membrane causes a conformational change in the unfolded preprotein being translocated and that this helps to pull it across. Furthermore, the fact that the matrix is more negative than the intermembrane space may "attract" the positively charged amino terminal of the preprotein

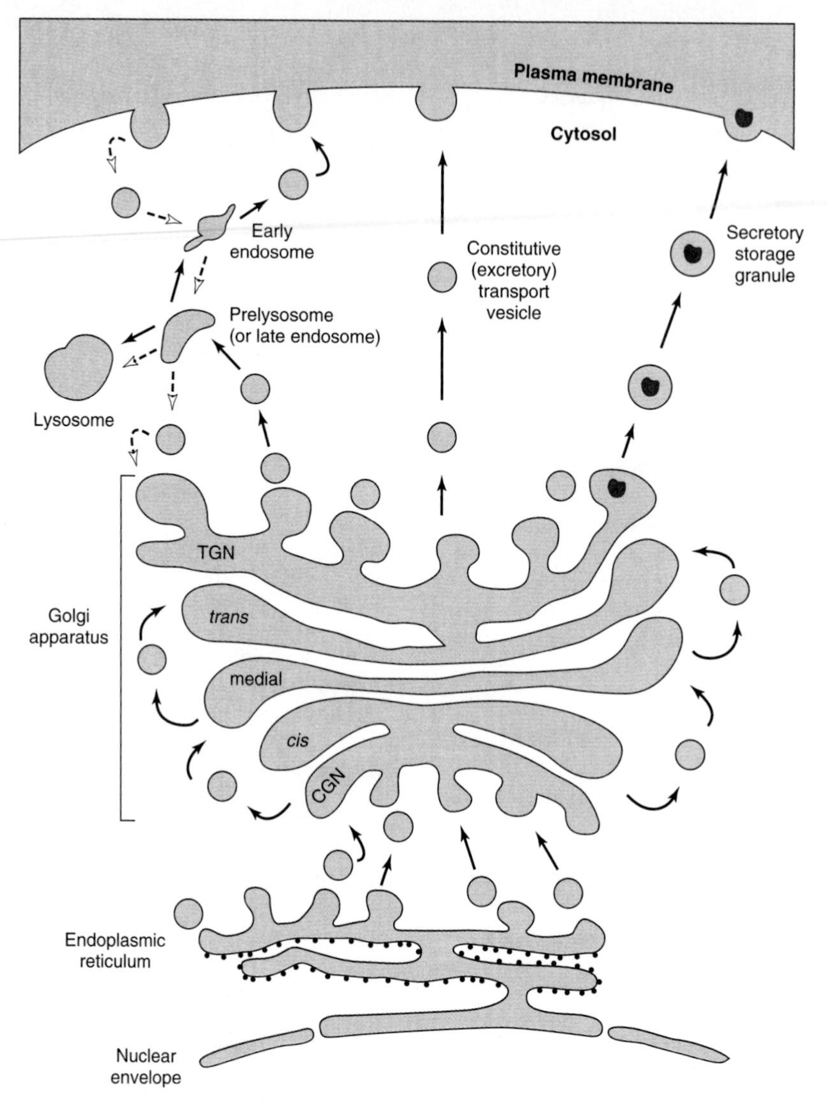


***Figure 46–2.*** Diagrammatic representation of the rough endoplasmic reticulum branch of protein sorting. Newly synthesized proteins are inserted into the ER membrane or lumen from membrane-bound polyribosomes (small black circles studding the cytosolic face of the ER). Those proteins that are transported out of the ER (indicated by solid black arrows) do so from ribosome-free transitional elements. Such proteins may then pass through the various subcompartments of the Golgi until they reach the TGN, the exit side of the Golgi. In the TGN, proteins are segregated and sorted. Secretory proteins accumulate in secretory storage granules from which they may be expelled as shown in the upper right-hand side of the figure. Proteins destined for the plasma membrane or those that are secreted in a constitutive manner are carried out to the cell surface in transport vesicles, as indicated in the upper middle area of the figure. Some proteins may reach the cell surface via late and early endosomes. Other proteins enter prelysosomes (late endosomes) and are selectively transferred to lysosomes. The endocytic pathway illustrated in the upper left-hand area of the figure is considered elsewhere in this chapter. Retrieval from the Golgi apparatus to the ER is not considered in this scheme. (CGN, *cis*-Golgi network; TGN, *trans*-Golgi network.) (Courtesy of E Degen.)

to enter the matrix. Close contact between the membrane sites in the outer and inner membranes involved in translocation is necessary.

The above describes the major pathway of proteins destined for the mitochondrial matrix. However, certain proteins insert into the **outer mitochondrial membrane** facilitated by the TOM complex. Others stop in the **intermembrane space,** and some insert into the **inner membrane.** Yet others proceed into the matrix and then return to the inner membrane or intermembrane space. A number of proteins contain two signaling sequences—one to enter the mitochondrial matrix and the other to mediate subsequent relocation (eg, into the inner membrane). Certain mitochondrial proteins do not contain presequences (eg, cytochrome *c,* which locates in the inter membrane space), and others contain **internal presequences.** Overall, proteins employ a variety of mechanisms and routes to attain their final destinations in mitochondria.

General features that apply to the import of proteins into organelles, including mitochondria and some of the other organelles to be discussed below, are summarized in Table 46–1.

## IMPORTINS & EXPORTINS ARE INVOLVED IN TRANSPORT OF MACROMOLECULES IN & OUT OF THE NUCLEUS

It has been estimated that more than a million macromolecules per minute are transported between the nucleus and the cytoplasm in an active eukaryotic cell.

***Table 46–1.*** Some general features of protein import to organelles.[1]

- Import of a protein into an organelle usually occurs in three stages: recognition, translocation, and maturation.
- Targeting sequences on the protein are recognized in the cytoplasm or on the surface of the organelle.
- The protein is unfolded for translocation, a state maintained in the cytoplasm by chaperones.
- Threading of the protein through a membrane requires energy and organellar chaperones on the trans side of the membrane.
- Cycles of binding and release of the protein to the chaperone result in pulling of its polypeptide chain through the membrane.
- Other proteins within the organelle catalyze folding of the protein, often attaching cofactors or oligosaccharides and assembling them into active monomers or oligomers.

[1]Data from McNew JA, Goodman JM: The targeting and assembly of peroxisomal proteins: some old rules do not apply. Trends Biochem Sci 1998;21:54.

These macromolecules include histones, ribosomal proteins and ribosomal subunits, transcription factors, and mRNA molecules. The transport is bidirectional and occurs through the nuclear pore complexes (NPCs). These are complex structures with a mass approximately 30 times that of a ribosome and are composed of about 100 different proteins. The diameter of an NPC is approximately 9 nm but can increase up to approximately 28 nm. Molecules smaller than about 40 kDa can pass through the channel of the NPC by diffusion, but special translocation mechanisms exist for larger molecules. These mechanisms are under intensive investigation, but some important features have already emerged.

Here we shall mainly describe **nuclear import** of certain macromolecules. The general picture that has emerged is that proteins to be imported (cargo molecules) carry a **nuclear localization signal (NLS).** One example of an NLS is the amino acid sequence $(Pro)_2$-$(Lys)_4$-Ala-Lys-Val, which is markedly rich in basic lysine residues. Depending on which NLS it contains, a cargo molecule interacts with one of a family of soluble proteins called **importins,** and the complex **docks** at the NPC. Another family of proteins called **Ran** plays a critical regulatory role in the interaction of the complex with the NPC and in its translocation through the NPC. Ran proteins are small monomeric nuclear GTPases and, like other GTPases, exist in either GTP-bound or GDP-bound states. They are themselves regulated by **guanine nucleotide exchange factors** (GEFs; eg, the protein RCC1 in eukaryotes), which are located in the nucleus, and Ran **guanine-activating proteins** (GAPs), which are predominantly cytoplasmic. The GTP-bound state of Ran is favored in the nucleus and the GDP-bound state in the cytoplasm. The conformations and activities of Ran molecules vary depending on whether GTP or GDP is bound to them (the GTP-bound state is active; see discussion of G proteins in Chapter 43). The **asymmetry** between nucleus and cytoplasm—with respect to which of these two nucleotides is bound to Ran molecules—is thought to be crucial in understanding the roles of Ran in transferring complexes unidirectionally across the NPC. When cargo molecules are released inside the nucleus, the importins recirculate to the cytoplasm to be used again. Figure 46–3 summarizes some of the principal features in the above process.

Other **small monomeric GTPases** (eg, ARF, Rab, Ras, and Rho) are important in various cellular processes such as vesicle formation and transport (ARF and Rab; see below), certain growth and differentiation processes (Ras), and formation of the actin cytoskeleton. A process involving GTP and GDP is also crucial in the transport of proteins across the membrane of the ER (see below).

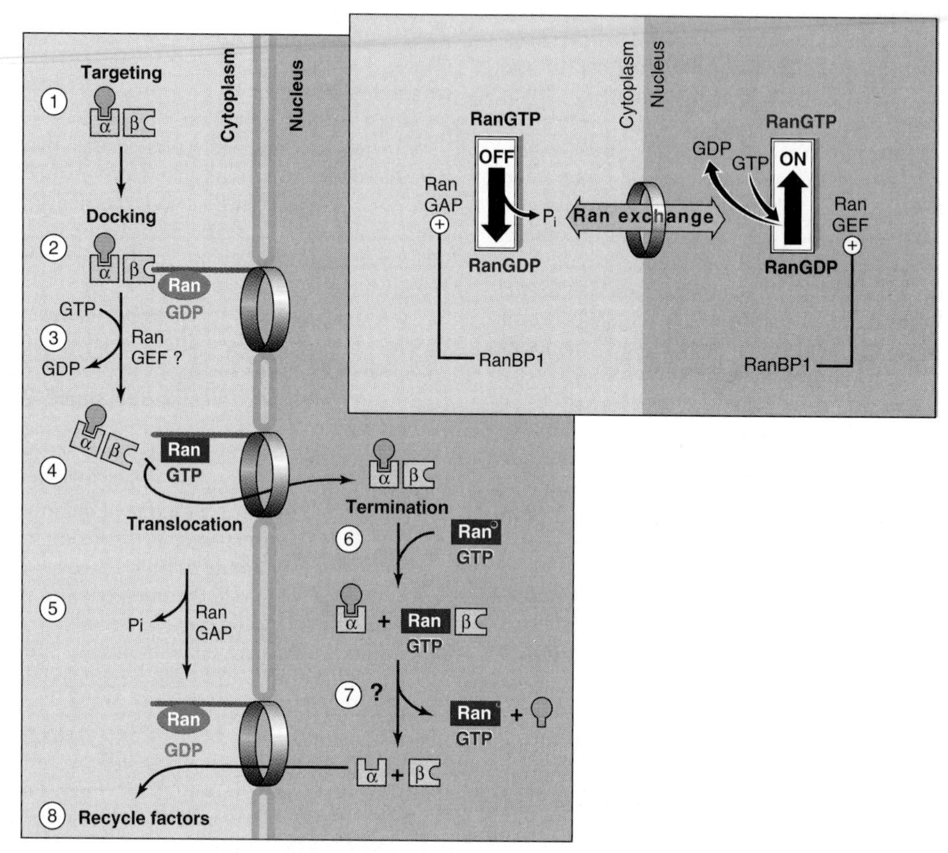


***Figure 46–3.*** Schematic representation of the proposed role of Ran in the import of cargo carrying an NLS signal. (1) The targeting complex forms when the NLS receptor (α, an importin) binds NLS cargo and the docking factor (β). (2) Docking occurs at filamentous sites that protrude from the NPC. Ran-GDP docks independently. (3) Transfer to the translocation channel is triggered when a RanGEF converts Ran-GDP to Ran-GTP. (4) The NPC catalyzes translocation of the targeting complex. (5) Ran-GTP is recycled to Ran-GDP by docked RanGAP. (6) Ran-GTP disrupts the targeting complex by binding to a site on β that overlaps with a binding site. (7) NLS cargo dissociates from α, and Ran-GTP may dissociate from β. (8) α and β factors are recycled to the cytoplasm. *Inset:* The Ran translocation switch is off in the cytoplasm and on in the nucleus. Ran-GTP promotes NLS- and NES-directed translocation. However, cytoplasmic Ran is enriched in Ran-GDP (OFF) by an active RanGAP, and nuclear pools are enriched in Ran-GTP (ON) by an active GEF. RanBP1 promotes the contrary activities of these two factors. Direct linkage of nuclear and cytoplasmic pools of Ran occurs through the NPC by an unknown shuttling mechanism. $P_i$, inorganic phosphate; NLS, nuclear localization signal; NPC, nuclear pore complex; GEF, guanine nucleotide exchange factor; GAP, guanine-activating protein; NES, nuclear export signal; BP, binding protein. (Reprinted, with permission, from Goldfarb DS: Whose finger is on the switch? Science 1997;276:1814.)

Proteins similar to importins, referred to as **exportins,** are involved in export of many macromolecules from the nucleus. Cargo molecules for export carry **nuclear export signals (NESs).** Ran proteins are involved in this process also, and it is now established that the processes of import and export share a number of common features.

## MOST CASES OF ZELLWEGER SYNDROME ARE DUE TO MUTATIONS IN GENES INVOLVED IN THE BIOGENESIS OF PEROXISOMES

The peroxisome is an important organelle involved in aspects of the metabolism of many molecules, including fatty acids and other lipids (eg, plasmalogens, cholesterol, bile acids), purines, amino acids, and hydrogen peroxide. The peroxisome is bounded by a single membrane and contains more than 50 enzymes; catalase and urate oxidase are marker enzymes for this organelle. Its proteins are synthesized on cytosolic polyribosomes and fold prior to import. The pathways of import of a number of its proteins and enzymes have been studied, some being **matrix components** and others **membrane components.** At least two **peroxisomal-matrix targeting sequences (PTSs)** have been discovered. One, PTS1, is a tripeptide (ie, Ser-Lys-Leu [SKL], but variations of this sequence have been detected) located at the carboxyl terminal of a number of matrix proteins, including catalase. Another, PTS2, consisting of about 26–36 amino acids, has been found in at least four matrix proteins (eg, thiolase) and, unlike PTS1, is cleaved after entry into the matrix. Proteins containing PTS1 sequences form complexes with a soluble receptor protein (PTS1R) and proteins containing PTS2 sequences complex with another, PTS2R. The resulting complexes then interact with a membrane receptor, Pex14p. Proteins involved in further transport of proteins into the matrix are also present. Most peroxisomal membrane proteins have been found to contain neither of the above two targeting sequences, but apparently contain others. The import system can handle **intact oligomers** (eg, tetrameric catalase). Import of matrix proteins requires ATP, whereas import of membrane proteins does not.

Interest in import of proteins into peroxisomes has been stimulated by studies on **Zellweger syndrome.** This condition is apparent at birth and is characterized by profound neurologic impairment, victims often dying within a year. The number of peroxisomes can vary from being almost normal to being virtually absent in some patients. Biochemical findings include an accumulation of very long chain fatty acids, abnormalities of the synthesis of bile acids, and a marked reduction of plasmalogens. The condition is believed to be due to mutations in genes encoding certain proteins—so called **peroxins**—involved in various steps of **peroxisome biogenesis** (such as the import of proteins described above), or in genes encoding certain peroxisomal enzymes themselves. Two closely related conditions are **neonatal adrenoleukodystrophy** and **infantile Refsum disease.** Zellweger syndrome and these two conditions represent a spectrum of overlapping features, with Zellweger syndrome being the most severe (many proteins affected) and infantile Refsum disease the least severe (only one or a few proteins affected). Table 46–2 lists some features of these and related conditions.

## THE SIGNAL HYPOTHESIS EXPLAINS HOW POLYRIBOSOMES BIND TO THE ENDOPLASMIC RETICULUM

As indicated above, the rough ER branch is the second of the two branches involved in the synthesis and sorting of proteins. In this branch, proteins are synthesized on membrane-bound polyribosomes and translocated into the lumen of the rough ER prior to further sorting (Figure 46–2).

The **signal hypothesis** was proposed by Blobel and Sabatini partly to explain the distinction between free and membrane-bound polyribosomes. They found that proteins synthesized on membrane-bound polyribosomes contained a peptide extension **(signal peptide)**

***Table 46–2.*** Disorders due to peroxisomal abnormalities.[1]

| | MIM Number[2] |
|---|---|
| Zellweger syndrome | 214100 |
| Neonatal adrenoleukodystrophy | 202370 |
| Infantile Refsum disease | 266510 |
| Hyperpipecolic acidemia | 239400 |
| Rhizomelic chondrodysplasia punctata | 215100 |
| Adrenoleukodystrophy | 300100 |
| Pseudo-neonatal adrenoleukodystrophy | 264470 |
| Pseudo-Zellweger syndrome | 261510 |
| Hyperoxaluria type 1 | 259900 |
| Acatalasemia | 115500 |
| Glutaryl-CoA oxidase deficiency | 231690 |

[1]Reproduced, with permission, from Seashore MR, Wappner RS: *Genetics in Primary Care & Clinical Medicine.* Appleton & Lange, 1996.

[2]MIM = *Mendelian Inheritance in Man.* Each number specifies a reference in which information regarding each of the above conditions can be found.

at their amino terminals which mediated their attachment to the membranes of the ER. As noted above, proteins whose entire synthesis occurs on free polyribosomes lack this signal peptide. An important aspect of the signal hypothesis was that it suggested—as turns out to be the case—that **all ribosomes have the same structure** and that the distinction between membrane-bound and free ribosomes depends solely on the former's carrying proteins that have signal peptides. Much evidence has confirmed the original hypothesis. Because many membrane proteins are synthesized on membrane-bound polyribosomes, the signal hypothesis plays an important role in concepts of membrane assembly. Some characteristics of signal peptides are summarized in Table 46–3.

Figure 46–4 illustrates the principal features in relation to the passage of a secreted protein through the membrane of the ER. It incorporates features from the original signal hypothesis and from subsequent work. The mRNA for such a protein encodes an amino terminal **signal peptide** (also variously called a leader sequence, a transient insertion signal, a signal sequence, or a presequence). The signal hypothesis proposed that the protein is inserted into the ER membrane at the same time as its mRNA is being translated on polyribosomes, so-called **cotranslational insertion.** As the signal peptide emerges from the large subunit of the ribosome, it is recognized by a **signal recognition particle (SRP)** that blocks further translation after about 70 amino acids have been polymerized (40 buried in the large ribosomal subunit and 30 exposed). The block is referred to as **elongation arrest.** The SRP contains six proteins and has a 7S RNA associated with it that is closely related to the Alu family of highly repeated DNA sequences (Chapter 36). The SRP-imposed block is not released until the SRP-signal peptide-polyribosome complex has bound to the so-called **docking protein** (SRP-R, a receptor for the SRP) on the ER membrane; the SRP thus guides the signal peptide to the SRP-R and prevents premature folding and expulsion of the protein being synthesized into the cytosol.

The SRP-R is an integral membrane protein composed of α and β subunits. The α subunit binds GDP and the β subunit spans the membrane. When the SRP-signal peptide complex interacts with the receptor, the exchange of GDP for GTP is stimulated. This form of the receptor (with GTP bound) has a high affinity for the SRP and thus releases the signal peptide, which binds to the translocation machinery (translocon) also present in the ER membrane. The α subunit then hydrolyzes its bound GTP, restoring GDP and completing a GTP-GDP cycle. The unidirectionality of this cycle helps drive the interaction of the polyribosome and its signal peptide with the ER membrane in the forward direction.

***Table 46–3.*** Some properties of signal peptides.

- Usually, but not always, located at the amino terminal
- Contain approximately 12–35 amino acids
- Methionine is usually the amino terminal amino acid
- Contain a central cluster of hydrophobic amino acids
- Contain at least one positively charged amino acid near their amino terminal
- Usually cleaved off at the carboxyl terminal end of an Ala residue by signal peptidase

The **translocon** consists of a number of membrane proteins that form a protein-conducting channel in the ER membrane through which the newly synthesized protein may pass. The channel appears to be open only when a signal peptide is present, preserving conductance across the ER membrane when it closes. The conductance of the channel has been measured experimentally. Specific functions of a number of components of the translocon have been identified or suggested. **TRAM** (translocating chain-associated membrane) protein may bind the signal sequence as it initially interacts with the translocon and the **Sec61p** complex (consisting of three proteins) binds the heavy subunit of the ribosome.

The insertion of the signal peptide into the conducting channel, while the other end of the parent protein is still attached to ribosomes, is termed **"cotranslational insertion."** The process of elongation of the remaining portion of the protein probably facilitates passage of the nascent protein across the lipid bilayer as the ribosomes remain attached to the membrane of the ER. Thus, the rough (or ribosome-studded) ER is formed. It is important that the protein be kept in an **unfolded state** prior to entering the conducting channel—otherwise, it may not be able to gain access to the channel.

Ribosomes remain attached to the ER during synthesis of signal peptide-containing proteins but are released and dissociated into their two types of subunits when the process is completed. The signal peptide is hydrolyzed by **signal peptidase,** located on the luminal side of the ER membrane (Figure 46–4), and then is apparently rapidly degraded by proteases.

Cytochrome P450 (Chapter 53), an integral ER membrane protein, does not completely cross the membrane. Instead, it resides in the membrane with its signal peptide intact. Its passage through the membrane is prevented by a sequence of amino acids called a halt- or stop-transfer signal.

Secretory proteins and proteins destined for membranes distal to the ER completely traverse the membrane bilayer and are discharged into the lumen of the ER. *N*-Glycan chains, if present, are added (Chapter 47) as these proteins traverse the inner part of the ER membrane—a process called "cotranslational glycosylation." Subsequently, the proteins are found in the

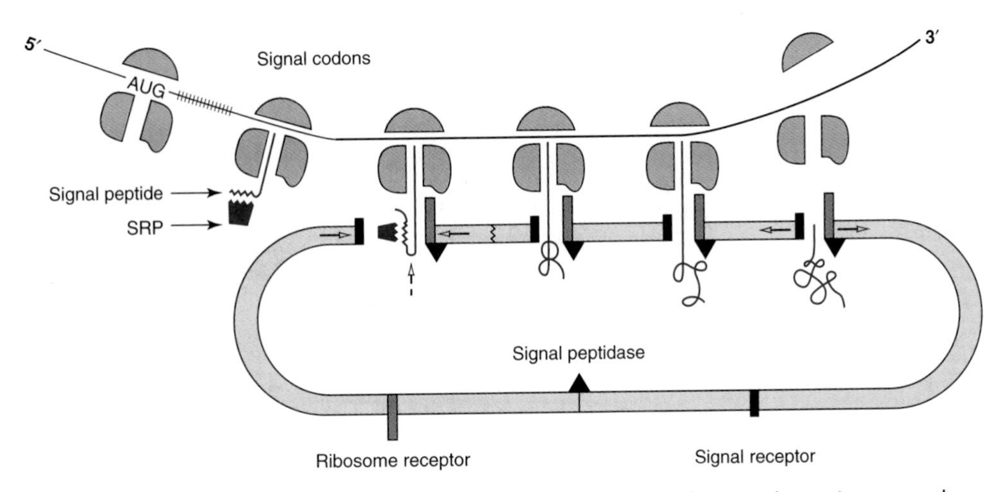


***Figure 46–4.*** Diagram of the signal hypothesis for the transport of secreted proteins across the ER membrane. The ribosomes synthesizing a protein move along the messenger RNA specifying the amino acid sequence of the protein. (The messenger is represented by the line between 5′ and 3′.) The codon AUG marks the start of the message for the protein; the hatched lines that follow AUG represent the codons for the signal sequence. As the protein grows out from the larger ribosomal subunit, the signal sequence is exposed and bound by the signal recognition particle (SRP). Translation is blocked until the complex binds to the "docking protein," also designated SRP-R (represented by the solid bar) on the ER membrane. There is also a receptor (open bar) for the ribosome itself. The interaction of the ribosome and growing peptide chain with the ER membrane results in the opening of a channel through which the protein is transported to the interior space of the ER. During translocation, the signal sequence of most proteins is removed by an enzyme called the "signal peptidase," located at the luminal surface of the ER membrane. The completed protein is eventually released by the ribosome, which then separates into its two components, the large and small ribosomal subunits. The protein ends up inside the ER. See text for further details. (Slightly modified and reproduced, with permission, from Marx JL: Newly made proteins zip through the cell. Science 1980;207:164. Copyright © 1980 by the American Association for the Advancement of Science.)

lumen of the Golgi apparatus, where further changes in glycan chains occur (Figure 47–9) prior to intracellular distribution or secretion. There is strong evidence that the signal peptide is involved in the process of protein insertion into ER membranes. Mutant proteins, containing altered signal peptides in which a hydrophobic amino acid is replaced by a hydrophilic one, are not inserted into ER membranes. Nonmembrane proteins (eg, α-globin) to which signal peptides have been attached by genetic engineering can be inserted into the lumen of the ER or even secreted.

There is considerable evidence that a second transposon in the ER membrane is involved in **retrograde transport** of various molecules from the ER lumen to the cytosol. These molecules include unfolded or misfolded glycoproteins, glycopeptides, and oligosaccharides. Some at least of these molecules are degraded in proteasomes. Thus, there is two-way traffic across the ER membrane.

## PROTEINS FOLLOW SEVERAL ROUTES TO BE INSERTED INTO OR ATTACHED TO THE MEMBRANES OF THE ENDOPLASMIC RETICULUM

The routes that proteins follow to be inserted into the membranes of the ER include the following.

### A. Cotranslational Insertion

Figure 46–5 shows a variety of ways in which proteins are distributed in the plasma membrane. In particular, the amino terminals of certain proteins (eg, the LDL receptor) can be seen to be on the extracytoplasmic face, whereas for other proteins (eg, the asialoglycoprotein receptor) the carboxyl terminals are on this face. To explain these dispositions, one must consider the initial biosynthetic events at the ER membrane. The **LDL receptor** enters the ER membrane in a manner analogous to a secretory protein (Figure 46–4); it partly traverses

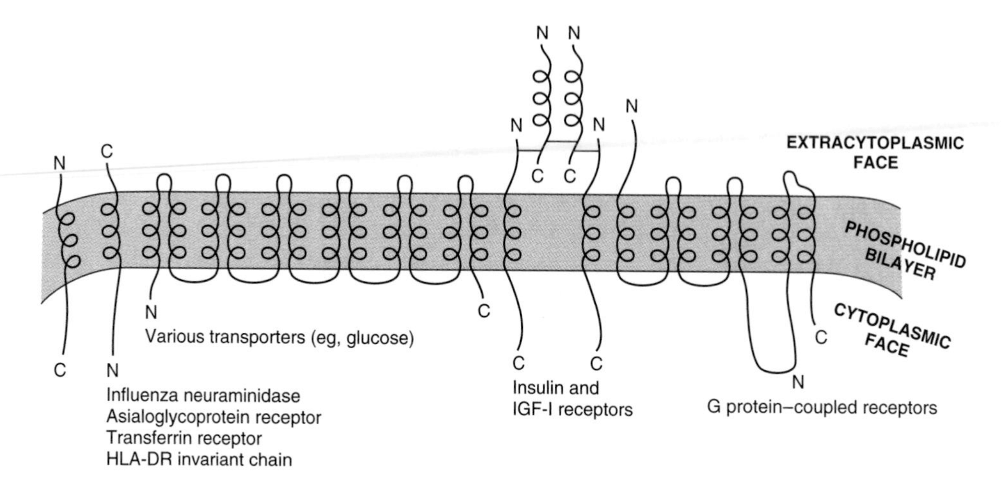


***Figure 46–5.*** Variations in the way in which proteins are inserted into membranes. This schematic representation, which illustrates a number of possible orientations, shows the segments of the proteins within the membrane as α-helices and the other segments as lines. The LDL receptor, which crosses the membrane once and has its amino terminal on the exterior, is called a type I transmembrane protein. The asialoglycoprotein receptor, which also crosses the membrane once but has its carboxyl terminal on the exterior, is called a type II transmembrane protein. The various transporters indicated (eg, glucose) cross the membrane a number of times and are called type III transmembrane proteins; they are also referred to as polytopic membrane proteins. (N, amino terminal; C, carboxyl terminal.) (Adapted, with permission, from Wickner WT, Lodish HF: Multiple mechanisms of protein insertion into and across membranes. Science 1985;230:400. Copyright © 1985 by the American Association for the Advancement of Science.)

the ER membrane, its signal peptide is cleaved, and its amino terminal protrudes into the lumen. However, it is retained in the membrane because it contains a highly hydrophobic segment, the **halt- or stop-transfer signal.** This sequence forms the single transmembrane segment of the protein and is its membrane-anchoring domain. The small patch of ER membrane in which the newly synthesized LDL receptor is located subsequently buds off as a component of a transport vesicle, probably from the transitional elements of the ER (Figure 46–2). As described below in the discussion of asymmetry of proteins and lipids in membrane assembly, the disposition of the receptor in the ER membrane is preserved in the vesicle, which eventually fuses with the plasma membrane. In contrast, the **asialoglycoprotein receptor** possesses an internal insertion sequence, which inserts into the membrane but is not cleaved. This acts as an anchor, and its carboxyl terminal is extruded through the membrane. The more complex disposition of the **transporters** (eg, for glucose) can be explained by the fact that alternating transmembrane α-helices act as uncleaved insertion sequences and as halt-transfer signals, respectively. Each pair of helical segments is inserted as a hairpin. Sequences that determine the structure of a protein in a membrane are called **topogenic sequences.** As explained in the legend to Figure 46–5, the above three proteins are examples of **type I, type II,** and **type III** transmembrane proteins.

### B. Synthesis on Free Polyribosomes & Subsequent Attachment to the Endoplasmic Reticulum Membrane

An example is cytochrome $b_5$, which enters the ER membrane spontaneously.

### C. Retention at the Luminal Aspect of the Endoplasmic Reticulum by Specific Amino Acid Sequences

A number of proteins possess the amino acid sequence **KDEL** (Lys-Asp-Glu-Leu) at their carboxyl terminal.

This sequence specifies that such proteins will be attached to the inner aspect of the ER in a relatively loose manner. The chaperone BiP (see below) is one such protein. Actually, KDEL-containing proteins first travel to the Golgi, interact there with a specific KDEL receptor protein, and then return in transport vesicles to the ER, where they dissociate from the receptor.

### D. Retrograde Transport From the Golgi Apparatus

Certain other non-KDEL-containing proteins destined for the membranes of the ER also pass to the Golgi and then return, by retrograde vesicular transport, to the ER to be inserted therein (see below).

The foregoing paragraphs demonstrate that a **variety of routes** are involved in assembly of the proteins of the ER membranes; a similar situation probably holds for other membranes (eg, the mitochondrial membranes and the plasma membrane). Precise targeting sequences have been identified in some instances (eg, KDEL sequences).

The topic of membrane biogenesis is discussed further later in this chapter.

## PROTEINS MOVE THROUGH CELLULAR COMPARTMENTS TO SPECIFIC DESTINATIONS

A scheme representing the possible flow of proteins along the ER → Golgi apparatus → plasma membrane route is shown in Figure 46–6. The horizontal arrows denote transport steps that may be independent of targeting signals, whereas the vertical open arrows represent steps that depend on specific signals. Thus, flow of certain proteins (including membrane proteins) from the ER to the plasma membrane (designated **"bulk flow,"** as it is nonselective) probably occurs without any targeting sequences being involved, ie, by default. On the other hand, insertion of resident proteins into the ER and Golgi membranes is dependent upon specific signals (eg, KDEL or halt-transfer sequences for the ER). Similarly, transport of many enzymes to **lysosomes** is dependent upon the Man 6-P signal (Chapter 47), and a signal may be involved for entry of proteins into **secretory granules.** Table 46–4 summarizes information on sequences that are known to be involved in targeting various proteins to their correct intracellular sites.

## CHAPERONES ARE PROTEINS THAT PREVENT FAULTY FOLDING & UNPRODUCTIVE INTERACTIONS OF OTHER PROTEINS

Exit from the ER may be the rate-limiting step in the secretory pathway. In this context, it has been found that certain proteins play a role in the assembly or proper folding of other proteins without themselves being components of the latter. Such proteins are called **molecular chaperones;** a number of important properties of these proteins are listed in Table 46–5, and the names of some of particular importance in the ER are listed in Table 46–6. Basically, they stabilize unfolded

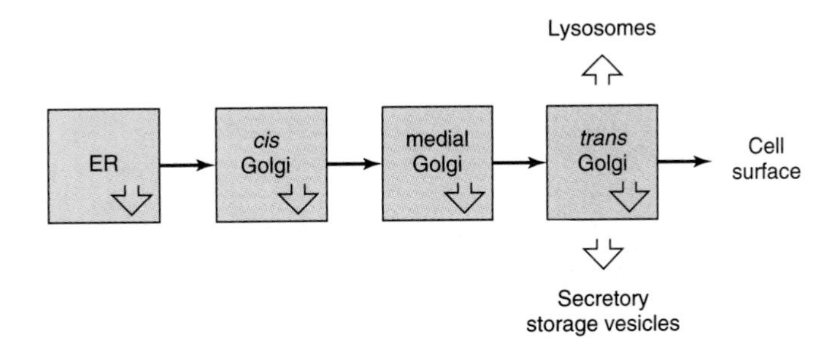


***Figure 46–6.*** Flow of membrane proteins from the endoplasmic reticulum (ER) to the cell surface. Horizontal arrows denote steps that have been proposed to be signal independent and thus represent bulk flow. The open vertical arrows in the boxes denote retention of proteins that are resident in the membranes of the organelle indicated. The open vertical arrows outside the boxes indicate signal-mediated transport to lysosomes and secretory storage granules. (Reproduced, with permission, from Pfeffer SR, Rothman JE: Biosynthetic protein transport and sorting by the endoplasmic reticulum and Golgi. Annu Rev Biochem 1987;56:829.)

***Table 46–4.*** Some sequences or compounds that direct proteins to specific organelles.

| Targeting Sequence or Compound | Organelle Targeted |
|---|---|
| Signal peptide sequence | Membrane of ER |
| Amino terminal KDEL sequence (Lys-Asp-Glu-Leu) | Luminal surface of ER |
| Amino terminal sequence (20–80 residues) | Mitochondrial matrix |
| NLS[1] (eg, $Pro_2$-$Lys_2$-Ala-Lys-Val) | Nucleus |
| PTS[1] (eg, Ser-Lys-Leu) | Peroxisome |
| Mannose 6-phosphate | Lysosome |

[1]NLS, nuclear localization signal; PTS, peroxisomal-matrix targeting sequence.

or partially folded intermediates, allowing them time to fold properly, and prevent inappropriate interactions, thus combating the formation of nonfunctional structures. Most chaperones exhibit **ATPase activity** and bind ADP and ATP. This activity is important for their effect on folding. The ADP-chaperone complex often has a high affinity for the unfolded protein, which, when bound, stimulates release of ADP with replacement by ATP. The ATP-chaperone complex, in turn, releases segments of the protein that have folded properly, and the cycle involving ADP and ATP binding is repeated until the folded protein is released.

***Table 46–5.*** Some properties of chaperone proteins.

- Present in a wide range of species from bacteria to humans
- Many are so-called heat shock proteins (Hsp)
- Some are inducible by conditions that cause unfolding of newly synthesized proteins (eg, elevated temperature and various chemicals)
- They bind to predominantly hydrophobic regions of unfolded and aggregated proteins
- They act in part as a quality control or editing mechanism for detecting misfolded or otherwise defective proteins
- Most chaperones show associated ATPase activity, with ATP or ADP being involved in the protein-chaperone interaction
- Found in various cellular compartments such as cytosol, mitochondria, and the lumen of the endoplasmic reticulum

***Table 46–6.*** Some chaperones and enzymes involved in folding that are located in the rough endoplasmic reticulum.

- BiP (immunoglobulin heavy chain binding protein)
- GRP94 (glucose-regulated protein)
- Calnexin
- Calreticulin
- PDI (protein disulfide isomerase)
- PPI (peptidyl prolyl cis-trans isomerase)

Several examples of chaperones were introduced above when the sorting of mitochondrial proteins was discussed. The **immunoglobulin heavy chain binding protein (BiP)** is located in the lumen of the ER. This protein will bind abnormally folded immunoglobulin heavy chains and certain other proteins and prevent them from leaving the ER, in which they are degraded. Another important chaperone is **calnexin,** a calcium-binding protein located in the ER membrane. This protein binds a wide variety of proteins, including mixed histocompatibility (MHC) antigens and a variety of serum proteins. As mentioned in Chapter 47, calnexin binds the monoglycosylated species of glycoproteins that occur during processing of glycoproteins, retaining them in the ER until the glycoprotein has folded properly. **Calreticulin,** which is also a calcium-binding protein, has properties similar to those of calnexin; it is not membrane-bound. Chaperones are not the only proteins in the ER lumen that are concerned with proper folding of proteins. Two enzymes are present that play an active role in folding. **Protein disulfide isomerase (PDI)** promotes rapid reshuffling of disulfide bonds until the correct set is achieved. **Peptidyl prolyl isomerase (PPI)** accelerates folding of proline-containing proteins by catalyzing the cis-trans isomerization of X-Pro bonds, where X is any amino acid residue.

## TRANSPORT VESICLES ARE KEY PLAYERS IN INTRACELLULAR PROTEIN TRAFFIC

Most proteins that are synthesized on membrane-bound polyribosomes and are destined for the Golgi apparatus or plasma membrane reach these sites inside transport vesicles. The precise mechanisms by which proteins synthesized in the rough ER are inserted into these vesicles are not known. Those involved in transport from the ER to the Golgi apparatus and vice versa—and from the Golgi to the plasma membrane—are mainly clathrin-free, unlike the coated vesicles involved in endocytosis (see discussions of the LDL receptor in Chapters 25 and 26). For the sake of clarity, the non-clathrin-coated vesicles will be referred to in this text as **transport vesicles.** There is evidence that proteins destined for the membranes of the Golgi apparatus contain specific signal sequences. On the other hand, most proteins destined for the plasma membrane or for secretion do not appear to contain specific signals, reaching these destinations by default.

## The Golgi Apparatus Is Involved in Glycosylation & Sorting of Proteins

The Golgi apparatus plays two important roles in membrane synthesis. First, it is involved in the **processing of the oligosaccharide chains** of membrane and other N-linked glycoproteins and also contains enzymes involved in O-glycosylation (see Chapter 47). Second, it is involved in the **sorting** of various proteins prior to their delivery to their appropriate intracellular destinations. All parts of the Golgi apparatus participate in the first role, whereas the trans-Golgi is particularly involved in the second and is very rich in vesicles. Because of their central role in protein transport, considerable research has been conducted in recent years concerning the formation and fate of transport vesicles.

## A Model of Non-Clathrin-Coated Vesicles Involves SNAREs & Other Factors

Vesicles lie at the heart of intracellular transport of many proteins. Recently, significant progress has been made in understanding the events involved in vesicle formation and transport. This has transpired because of the use of a number of approaches. These include establishment of **cell-free systems** with which to study vesicle formation. For instance, it is possible to observe, by electron microscopy, budding of vesicles from Golgi preparations incubated with cytosol and ATP. The development of genetic approaches for studying vesicles in yeast has also been crucial. The picture is complex, with its own nomenclature (Table 46–7), and involves a variety of cytosolic and membrane proteins, GTP, ATP, and accessory factors.

Based largely on a proposal by Rothman and colleagues, anterograde vesicular transport can be considered to occur in eight steps (Figure 46–7). The basic concept is that each transport vesicle bears a unique address marker consisting of one or more **v-SNARE proteins,** while each target membrane bears one or more **complementary t-SNARE proteins** with which the former interact specifically.

*Step 1:* **Coat assembly** is initiated when ARF is activated by binding GTP, which is exchanged for GDP. This leads to the association of GTP-bound ARF with its putative receptor (hatched in Figure 46–7) in the donor membrane.

***Table 46–7.*** Factors involved in the formation of non-clathrin-coated vesicles and their transport.

- ARF: ADP-ribosylation factor, a GTPase
- Coatomer: A family of at least seven coat proteins ($\alpha$, $\beta$, $\gamma$, $\delta$, $\varepsilon$, $\beta'$, and $\zeta$). Different transport vesicles have different complements of coat proteins.
- SNAP: Soluble NSF attachment factor
- SNARE: SNAP receptor
- v-SNARE: Vesicle SNARE
- t-SNARE: Target SNARE
- GTP-$\gamma$-S: A nonhydrolyzable analog of GTP, used to test the involvement of GTP
- NEM: *N*-Ethylmaleimide, a chemical that alkylates sulfhydryl groups
- NSF: NEM-sensitive factor, an ATPase
- Rab proteins: A family of ras-related proteins first observed in rat brain; they are GTPases and are active when GTP is found
- Sec1: A member of a family of proteins that attach to t-SNAREs and are displaced from them by Rab proteins, thereby allowing v-SNARE–t-SNARE interactions to occur.

*Step 2:* Membrane-associated ARF **recruits the coat proteins** that comprise the coatomer shell from the cytosol, forming a coated bud.

*Step 3:* The **bud pinches off** in a process involving acyl-CoA—and probably ATP—to complete the formation of the coated vesicle.

*Step 4:* **Coat disassembly** (involving dissociation of ARF and coatomer shell) follows hydrolysis of bound GTP; uncoating is necessary for fusion to occur.

*Step 5:* **Vesicle targeting** is achieved via members of a family of integral proteins, termed v-SNAREs, that tag the vesicle during its budding. v-SNAREs pair with cognate t-SNAREs in the target membrane to dock the vesicle.

It is presumed that steps 4 and 5 are closely coupled and that step 4 may follow step 5, with ARF and the coatomer shell rapidly dissociating after docking.

*Step 6:* The **general fusion machinery** then assembles on the paired SNARE complex; it includes an ATPase (NSF; NEM-sensitive factor) and the SNAP (soluble NSF attachment factor) proteins. SNAPs bind to the SNARE (SNAP receptor) complex, enabling NSF to bind.

*Step 7:* **Hydrolysis of ATP by NSF** is essential for fusion, a process that can be inhibited by NEM (*N*-ethylmaleimide). Certain other proteins and calcium are also required.

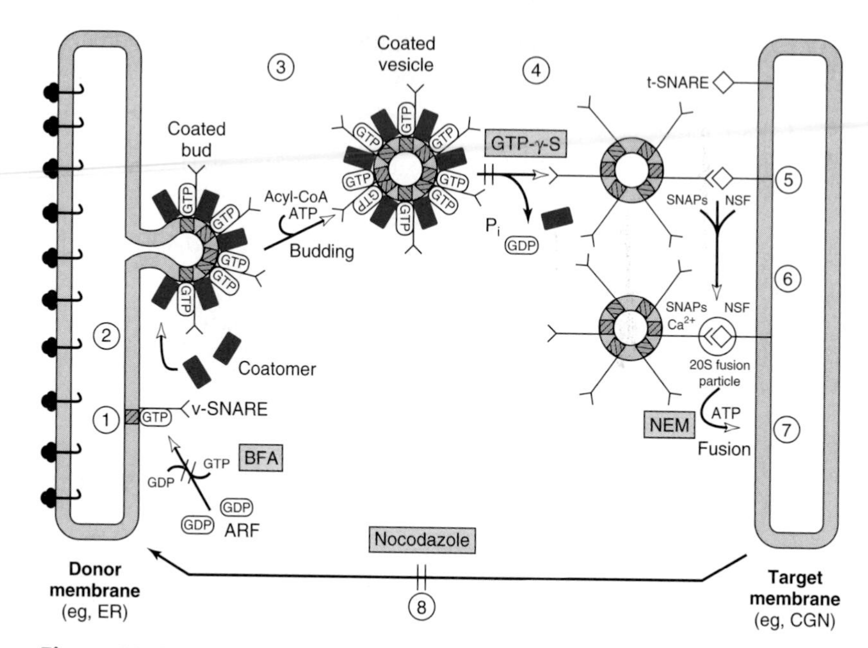


***Figure 46–7.*** Model of the steps in a round of anterograde vesicular transport. The cycle starts in the bottom left-hand side of the figure, where two molecules of ARF are represented as small ovals containing GDP. The steps in the cycle are described in the text. Most of the abbreviations used are explained in Table 46–7. The roles of Rab and Sec1 proteins (see text) in the overall process are not dealt with in this figure. (CGN, cis-Golgi network; BFA, Brefeldin A.) (Adapted from Rothman JE: Mechanisms of intracellular protein transport. Nature 1994;372:55.) (Courtesy of E Degen.)

***Step 8:*** **Retrograde transport** occurs to restart the cycle. This last step may retrieve certain proteins or recycle v-SNAREs. Nocodazole, a microtubule-disrupting agent, inhibits this step.

## Brefeldin A Inhibits the Coating Process

The following points expand and clarify the above.

**(a)** To participate in step 1, ARF must first be modified by addition of **myristic acid** (C14:0), employing myristoyl-CoA as the acyl donor. Myristoylation is one of a number of enzyme-catalyzed posttranslational modifications, involving addition of certain lipids to specific residues of proteins, that facilitate the binding of proteins to the cytosolic surfaces of membranes or vesicles. Others are addition of palmitate, farnesyl, and geranylgeranyl; the two latter molecules are polyisoprenoids containing 15 and 20 carbon atoms, respectively.

**(b)** At least three different types of coated vesicles have been distinguished: **COPI, COPII,** and **clathrin-coated vesicles;** the first two are referred to here as transport vesicles. Many other types of vesicles no doubt remain to be discovered. COPI vesicles are involved in bidirectional transport from the ER to the Golgi and in the reverse direction, whereas COPII vesicles are involved mainly in transport in the former direction. Clathrin-containing vesicles are involved in transport from the trans-Golgi network to prelysosomes and from the plasma membrane to endosomes, respectively. Regarding **selection** of cargo molecules by vesicles, this appears to be primarily a function of the coat proteins of vesicles. **Cargo molecules** may interact with coat proteins either directly or via intermediary proteins that attach to coat proteins, and they then become enclosed in their appropriate vesicles.

**(c)** The fungal metabolite **brefeldin A** prevents GTP from binding to ARF in step 1 and thus inhibits the entire coating process. In its presence, the Golgi apparatus appears to disintegrate, and fragments are lost. It may do this by inhibiting the guanine nucleotide exchanger involved in step 1.

**(d) GTP-γ-S** (a nonhydrolyzable analog of GTP often used in investigations of the role of GTP in biochemical processes) blocks disassembly of the coat from coated vesicles, leading to a build-up of coated vesicles.

(e) A family of Ras-like proteins, called the **Rab protein family,** are required in several steps of intracellular protein transport, regulated secretion, and endocytosis. They are small monomeric GTPases that attach to the cytosolic faces of membranes via geranylgeranyl chains. They attach in the GTP-bound state (not shown in Figure 46–7) to the budding vesicle. Another family of proteins **(Sec1)** binds to t-SNAREs and prevents interaction with them and their complementary v-SNAREs. When a vesicle interacts with its target membrane, Rab proteins displace Sec1 proteins and the v-SNARE-t-SNARE interaction is free to occur. It appears that the Rab and Sec1 families of proteins regulate the speed of vesicle formation, opposing each other. Rab proteins have been likened to throttles and Sec1 proteins to dampers on the overall process of vesicle formation.

(f) Studies using v- and t-SNARE proteins reconstituted into separate lipid bilayer vesicles have indicated that they form **SNAREpins,** ie, SNARE complexes that link two membranes (vesicles). SNAPs and NSF are required for formation of SNAREpins, but once they have formed they can apparently lead to spontaneous fusion of membranes at physiologic temperature, suggesting that they are the minimal machinery required for membrane fusion.

(g) The fusion of synaptic vesicles with the plasma membrane of **neurons** involves a series of events similar to that described above. For example, one v-SNARE is designated **synaptobrevin** and two t-SNAREs are designated **syntaxin** and **SNAP 25** (synaptosome-associated protein of 25 kDa). **Botulinum B toxin** is one of the most lethal toxins known and the most serious cause of food poisoning. One component of this toxin is a protease that appears to cleave only synaptobrevin, thus inhibiting release of acetylcholine at the neuromuscular junction and possibly proving fatal, depending on the dose taken.

(h) Although the above model describes non-clathrin-coated vesicles, it appears likely that many of the events described above apply, at least in principle, to clathrin-coated vesicles.

## THE ASSEMBLY OF MEMBRANES IS COMPLEX

There are many cellular membranes, each with its own specific features. No satisfactory scheme describing the assembly of any one of these membranes is available. How various proteins are initially inserted into the membrane of the ER has been discussed above. The transport of proteins, including membrane proteins, to various parts of the cell inside vesicles has also been described. Some general points about membrane assembly remain to be addressed.

### Asymmetry of Both Proteins & Lipids Is Maintained During Membrane Assembly

Vesicles formed from membranes of the ER and Golgi apparatus, either naturally or pinched off by homogenization, exhibit **transverse asymmetries** of both lipid and protein. These asymmetries are maintained during fusion of transport vesicles with the plasma membrane. The inside of the vesicles after fusion becomes the outside of the plasma membrane, and the cytoplasmic side of the vesicles remains the cytoplasmic side of the membrane (Figure 46–8). Since the transverse asymmetry of the membranes already exists in the vesicles of the ER well before they are fused to the plasma membrane, a major problem of membrane assembly becomes understanding how the integral proteins are inserted into the lipid bilayer of the ER. This problem was addressed earlier in this chapter.

**Phospholipids** are the major class of lipid in membranes. The enzymes responsible for the synthesis of phospholipids reside in the cytoplasmic surface of the cisternae of the ER. As phospholipids are synthesized at that site, they probably self-assemble into thermodynamically stable bimolecular layers, thereby expanding the membrane and perhaps promoting the detachment of so-called lipid vesicles from it. It has been proposed that these vesicles travel to other sites, donating their lipids to other membranes; however, little is known about this matter. As indicated above, cytosolic proteins that take up phospholipids from one membrane and release them to another (ie, phospholipid exchange proteins) have been demonstrated; they probably play a role in contributing to the specific lipid composition of various membranes.

### Lipids & Proteins Undergo Turnover at Different Rates in Different Membranes

It has been shown that the half-lives of the lipids of the ER membranes of rat liver are generally shorter than those of its proteins, so that the **turnover rates of lipids and proteins are independent.** Indeed, different lipids have been found to have different half-lives. Furthermore, the half-lives of the proteins of these membranes vary quite widely, some exhibiting short (hours) and others long (days) half-lives. Thus, individual lipids and proteins of the ER membranes appear to be inserted into it relatively independently; this is the case for many other membranes.

The biogenesis of membranes is thus a complex process about which much remains to be learned. One indication of the complexity involved is to consider the number of **posttranslational modifications** that membrane proteins may be subjected to prior to attaining their mature state. These include proteolysis, assembly

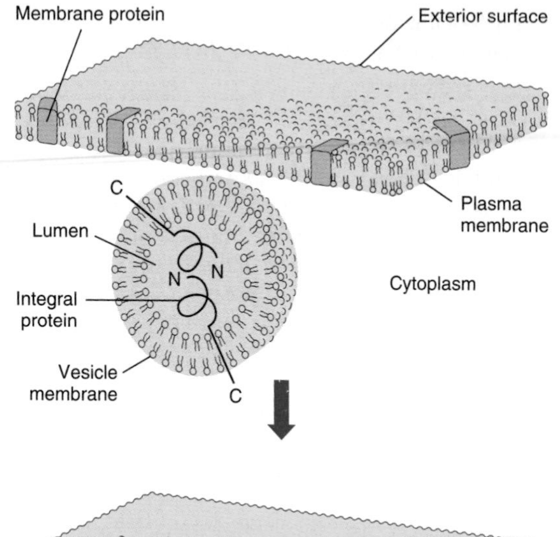


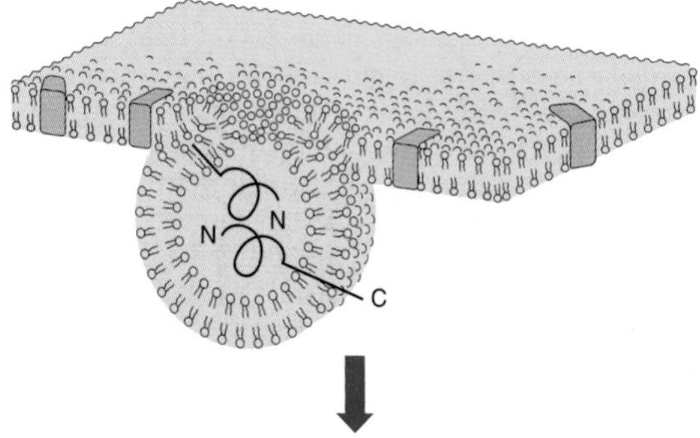


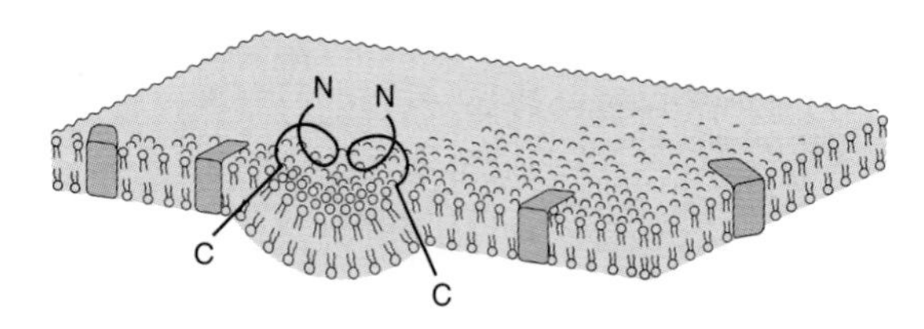


***Figure 46–8.*** Fusion of a vesicle with the plasma membrane preserves the orientation of any integral proteins embedded in the vesicle bilayer. Initially, the amino terminal of the protein faces the lumen, or inner cavity, of such a vesicle. After fusion, the amino terminal is on the exterior surface of the plasma membrane. That the orientation of the protein has not been reversed can be perceived by noting that the other end of the molecule, the carboxyl terminal, is always immersed in the cytoplasm. The lumen of a vesicle and the outside of the cell are topologically equivalent. (Redrawn and modified, with permission, from Lodish HF, Rothman JE: The assembly of cell membranes. Sci Am [Jan] 1979;240:43.)

***Table 46–8.*** Major features of membrane assembly.

- Lipids and proteins are inserted independently into membranes.
- Individual membrane lipids and proteins turn over independently and at different rates.
- Topogenic sequences (eg, signal [amino terminal or internal] and stop-transfer) are important in determining the insertion and disposition of proteins in membranes.
- Membrane proteins inside transport vesicles bud off the endoplasmic reticulum on their way to the Golgi; final sorting of many membrane proteins occurs in the trans-Golgi network.
- Specific sorting sequences guide proteins to particular organelles such as lysosomes, peroxisomes, and mitochondria.

into multimers, glycosylation, addition of a glycophosphatidylinositol (GPI) anchor, sulfation on tyrosine or carbohydrate moieties, phosphorylation, acylation, and prenylation—a list that is undoubtedly not complete. Nevertheless, significant progress has been made; Table 46–8 summarizes some of the major features of membrane assembly that have emerged to date.

***Table 46–9.*** Some disorders due to mutations in genes encoding proteins involved in intracellular membrane transport.[1]

| Disorder[2] | Protein Involved |
|---|---|
| Chédiak-Higashi syndrome, 214500 | Lysosomal trafficking regulator |
| Combined deficiency of factors V and VIII, 227300 | ERGIC-53, a mannose-binding lectin |
| Hermansky-Pudlak syndrome, 203300 | AP-3 adaptor complex β3A subunit |
| I-cell disease, 252500 | *N*-Acetylglucosamine 1-phosphotransferase |
| Oculocerebrorenal syndrome, 30900 | OCRL-1, an inositol polyphosphate 5-phosphatase |

[1]Modified from Olkonnen VM, Ikonen E: Genetic defects of intracellular-membrane transport. N Eng J Med 2000;343:1095. Certain related conditions not listed here are also described in this publication. I-cell disease is described in Chapter 47. The majority of the disorders listed above affect lysosomal function; readers should consult a textbook of medicine for information on the clinical manifestations of these conditions.
[2]The numbers after each disorder are the OMIM numbers.

### Various Disorders Result From Mutations in Genes Encoding Proteins Involved in Intracellular Transport

Some of these are listed in Table 46–9; the majority affect lysosomal function. A number of other mutations affecting intracellular protein transport have been reported but are not included here.

## SUMMARY

- Many proteins are targeted to their destinations by signal sequences. A major sorting decision is made when proteins are partitioned between cytosolic and membrane-bound polyribosomes by virtue of the absence or presence of a signal peptide.
- The pathways of protein import into mitochondria, nuclei, peroxisomes, and the endoplasmic reticulum are described.
- Many proteins synthesized on membrane-bound polyribosomes proceed to the Golgi apparatus and the plasma membrane in transport vesicles.
- A number of glycosylation reactions occur in compartments of the Golgi, and proteins are further sorted in the trans-Golgi network.
- Most proteins destined for the plasma membrane and for secretion appear to lack specific signals—a default mechanism.
- The role of chaperone proteins in the folding of proteins is presented, and a model describing budding and attachment of transport vesicles to a target membrane is summarized.
- Membrane assembly is discussed and shown to be complex. Asymmetry of both lipids and proteins is maintained during membrane assembly.
- A number of disorders have been shown to be due to mutations in genes encoding proteins involved in various aspects of protein traffic and sorting.

## REFERENCES

Fuller GM, Shields DL: *Molecular Basis of Medical Cell Biology.* McGraw-Hill, 1998.

Gould SJ et al: The peroxisome biogenesis disorders. In: *The Metabolic and Molecular Bases of Inherited Disease,* 8th ed. Scriver CR et al (editors). McGraw-Hill, 2001.

Graham JM, Higgins JA: *Membrane Analysis.* BIOS Scientific, 1997.

Griffith J, Sansom C: *The Transporter Facts Book.* Academic Press, 1998.

Lodish H et al: *Molecular Cell Biology,* 4th ed. Freeman, 2000. (Chapter 17 contains comprehensive coverage of protein sorting and organelle biogenesis.)

Olkkonen VM, Ikonen E: Genetic defects of intracellular-membrane transport. N Engl J Med 2000;343:1095.

Reithmeier RAF: Assembly of proteins into membranes. In: *Biochemistry of Lipids, Lipoproteins and Membranes.* Vance DE, Vance JE (editors). Elsevier, 1996.

Sabatini DD, Adesnik MB: The biogenesis of membranes and organelles. In: *The Metabolic and Molecular Bases of Inherited Disease,* 8th ed. Scriver CR et al (editors). McGraw-Hill, 2001.

# Glycoproteins

47

*Robert K. Murray, MD, PhD*

## BIOMEDICAL IMPORTANCE

**Glycoproteins** are proteins that contain oligosaccharide (glycan) chains covalently attached to their polypeptide backbones. They are one class of **glycoconjugate** or **complex carbohydrates**—equivalent terms used to denote molecules containing one or more carbohydrate chains covalently linked to protein (to form glycoproteins or proteoglycans) or lipid (to form glycolipids). (**Proteoglycans** are discussed in Chapter 48 and **glycolipids** in Chapter 14). Almost all the **plasma proteins** of humans—except albumin—are glycoproteins. Many **proteins of cellular membranes** (Chapter 41) contain substantial amounts of carbohydrate. A number of the **blood group substances** are glycoproteins, whereas others are glycosphingolipids. Certain **hormones** (eg, chorionic gonadotropin) are glycoproteins. A major problem in cancer is **metastasis,** the phenomenon whereby cancer cells leave their tissue of origin (eg, the breast), migrate through the bloodstream to some distant site in the body (eg, the brain), and grow there in an unregulated manner, with catastrophic results for the affected individual. Many cancer researchers think that alterations in the structures of glycoproteins and other glycoconjugates on the surfaces of cancer cells are important in the phenomenon of metastasis.

## GLYCOPROTEINS OCCUR WIDELY & PERFORM NUMEROUS FUNCTIONS

Glycoproteins occur in most organisms, from bacteria to humans. Many viruses also contain glycoproteins, some of which have been much investigated, in part because they are very suitable for biosynthetic studies. Numerous proteins with diverse functions are glycoproteins (Table 47–1); their carbohydrate content ranges from 1% to over 85% by weight.

Many studies have been conducted in an attempt to define the precise roles oligosaccharide chains play in the functions of glycoproteins. Table 47–2 summarizes results from such studies. Some of the functions listed are firmly established; others are still under investigation.

## OLIGOSACCHARIDE CHAINS ENCODE BIOLOGIC INFORMATION

An enormous number of glycosidic linkages can be generated between sugars. For example, three different hexoses may be linked to each other to form over 1000 different trisaccharides. The conformations of the sugars in oligosaccharide chains vary depending on their linkages and proximity to other molecules with which the oligosaccharides may interact. It is now established that certain oligosaccharide chains encode considerable **biologic information** and that this depends upon their constituent sugars, their sequences, and their linkages. For instance, mannose 6-phosphate residues target newly synthesized lysosomal enzymes to that organelle (see below).

## TECHNIQUES ARE AVAILABLE FOR DETECTION, PURIFICATION, & STRUCTURAL ANALYSIS OF GLYCOPROTEINS

A variety of methods used in the detection, purification, and structural analysis of glycoproteins are listed in Table 47–3. The conventional methods used to purify proteins and enzymes are also applicable to the purification of glycoproteins. Once a glycoprotein has been purified, the use of **mass spectrometry** and **high-resolution NMR spectroscopy** can often identify the structures of its glycan chains. Analysis of glycoproteins can be complicated by the fact that they often exist as **glycoforms;** these are proteins with identical amino acid sequences but somewhat different oligosaccharide compositions. Although linkage details are not stressed in this chapter, it is critical to appreciate that the precise natures of the linkages between the sugars of glycoproteins are of fundamental importance in determining the structures and functions of these molecules.

***Table 47–1.*** Some functions served by glycoproteins.

| Function | Glycoproteins |
|---|---|
| Structural molecule | Collagens |
| Lubricant and protective agent | Mucins |
| Transport molecule | Transferrin, ceruloplasmin |
| Immunologic molecule | Immunoglobulins, histocompatibility antigens |
| Hormone | Chorionic gonadotropin, thyroid-stimulating hormone (TSH) |
| Enzyme | Various, eg, alkaline phosphatase |
| Cell attachment-recognition site | Various proteins involved in cell-cell (eg, sperm-oocyte), virus-cell, bacterium-cell, and hormone-cell interactions |
| Antifreeze | Certain plasma proteins of cold water fish |
| Interact with specific carbohydrates | Lectins, selectins (cell adhesion lectins), antibodies |
| Receptor | Various proteins involved in hormone and drug action |
| Affect folding of certain proteins | Calnexin, calreticulin |

***Table 47–2.*** Some functions of the oligosaccharide chains of glycoproteins.[1]

- Modulate physicochemical properties, eg, solubility, viscosity, charge, conformation, denaturation, and binding sites for bacteria and viruses
- Protect against proteolysis, from inside and outside of cell
- Affect proteolytic processing of precursor proteins to smaller products
- Are involved in biologic activity, eg, of human chorionic gonadotropin (hCG)
- Affect insertion into membranes, intracellular migration, sorting and secretion
- Affect embryonic development and differentiation
- May affect sites of metastases selected by cancer cells

[1]Adapted from Schachter H: Biosynthetic controls that determine the branching and heterogeneity of protein-bound oligosaccharides. Biochem Cell Biol 1986;64:163.

***Table 47–3.*** Some important methods used to study glycoproteins.

| Method | Use |
|---|---|
| Periodic acid-Schiff reagent | Detects glycoproteins as pink bands after electrophoretic separation. |
| Incubation of cultured cells with glycoproteins as radioactive bands | Leads to detection of a radioactive sugar after electrophoretic separation. |
| Treatment with appropriate endo- or exoglycosidase or phospholipases | Resultant shifts in electrophoretic migration help distinguish among proteins with N-glycan, O-glycan, or GPI linkages and also between high mannose and complex N-glycans. |
| Sepharose-lectin column chromatography | To purify glycoproteins or glycopeptides that bind the particular lectin used. |
| Compositional analysis following acid hydrolysis | Identifies sugars that the glycoprotein contains and their stoichiometry. |
| Mass spectrometry | Provides information on molecular mass, composition, sequence, and sometimes branching of a glycan chain. |
| NMR spectroscopy | To identify specific sugars, their sequence, linkages, and the anomeric nature of glycosidic linkages. |
| Methylation (linkage) analysis | To determine linkages between sugars. |
| Amino acid or cDNA sequencing | Determination of amino acid sequence. |

## EIGHT SUGARS PREDOMINATE IN HUMAN GLYCOPROTEINS

About 200 monosaccharides are found in nature; however, only eight are commonly found in the oligosaccharide chains of glycoproteins (Table 47–4). Most of these sugars were described in Chapter 13. *N*-Acetylneuraminic acid (NeuAc) is usually found at the termini of oligosaccharide chains, attached to subterminal galactose (Gal) or *N*-acetylgalactosamine (GalNAc) residues. The other sugars listed are generally found in more internal positions. **Sulfate** is often found in glycoproteins, usually attached to Gal, GalNAc, or GlcNAc.

***Table 47–4.*** The principal sugars found in human glycoproteins. Their structures are illustrated in Chapter 13.

| Sugar | Type | Abbreviation | Nucleotide Sugar | Comments |
|---|---|---|---|---|
| Galactose | Hexose | Gal | UDP-Gal | Often found subterminal to NeuAc in N-linked glycoproteins. Also found in core trisaccharide of proteoglycans. |
| Glucose | Hexose | Glc | UDP-Glc | Present during the biosynthesis of N-linked glycoproteins but not usually present in mature glycoproteins. Present in some clotting factors. |
| Mannose | Hexose | Man | GDP-Man | Common sugar in N-linked glycoproteins. |
| *N*-Acetylneuraminic acid | Sialic acid (nine C atoms) | NeuAc | CMP-NeuAc | Often the terminal sugar in both N- and O-linked glycoproteins. Other types of sialic acid are also found, but NeuAc is the major species found in humans. Acetyl groups may also occur as O-acetyl species as well as N-acetyl. |
| Fucose | Deoxyhexose | Fuc | GDP-Fuc | May be external in both N- and O-linked glycoproteins or internal, linked to the GlcNAc residue attached to Asn in N-linked species. Can also occur internally attached to the OH of Ser (eg, in t-PA and certain clotting factors). |
| *N*-Acetylgalactosamine | Aminohexose | GalNAc | UDP-GalNAc | Present in both N- and O-linked glycoproteins. |
| *N*-Acetylglucosamine | Aminohexose | GlcNAc | UDP-GlcNAc | The sugar attached to the polypeptide chain via Asn in N-linked glycoproteins; also found at other sites in the oligosaccharides of these proteins. Many nuclear proteins have GlcNAc attached to the OH of Ser or Thr as a single sugar. |
| Xylose | Pentose | Xyl | UDP-Xyl | Xyl is attached to the OH of Ser in many proteoglycans. Xyl in turn is attached to two Gal residues, forming a link trisaccharide. Xyl is also found in t-PA and certain clotting factors. |

## NUCLEOTIDE SUGARS ACT AS SUGAR DONORS IN MANY BIOSYNTHETIC REACTIONS

The first nucleotide sugar to be reported was uridine diphosphate glucose (UDP-Glc); its structure is shown in Figure 18–2. The common nucleotide sugars involved in the biosynthesis of glycoproteins are listed in Table 47–4; the reasons some contain UDP and others guanosine diphosphate (GDP) or cytidine monophosphate (CMP) are obscure. Many of the glycosylation reactions involved in the biosynthesis of glycoproteins utilize these compounds (see below). The anhydro nature of the linkage between the phosphate group and the sugars is of the high-energy, high-group-transfer-potential type (Chapter 10). The sugars of these compounds are thus "activated" and can be transferred to suitable acceptors provided appropriate transferases are available.

Most nucleotide sugars are formed in the cytosol, generally from reactions involving the corresponding nucleoside triphosphate. CMP-sialic acids are formed in the nucleus. Formation of uridine diphosphate galactose (UDP-Gal) requires the following two reactions in mammalian tissues:

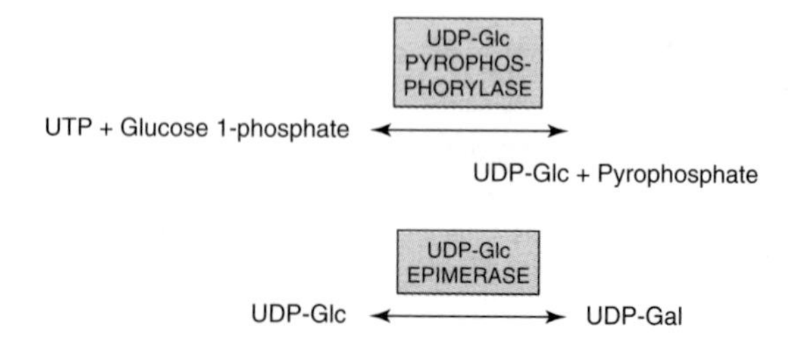

Because many glycosylation reactions occur within the lumen of the Golgi apparatus, **carrier systems** (permeases, transporters) are necessary to transport nucleotide sugars across the Golgi membrane. Systems transporting UDP-Gal, GDP-Man, and CMP-NeuAc into the cisternae of the Golgi apparatus have been described. They are **antiport** systems; ie, the influx of one molecule of nucleotide sugar is balanced by the efflux of one molecule of the corresponding nucleotide (eg, UMP, GMP, or CMP) formed from the nucleotide sugars. This mechanism ensures an adequate concentration of each nucleotide sugar inside the Golgi apparatus. UMP is formed from UDP-Gal in the above process as follows:

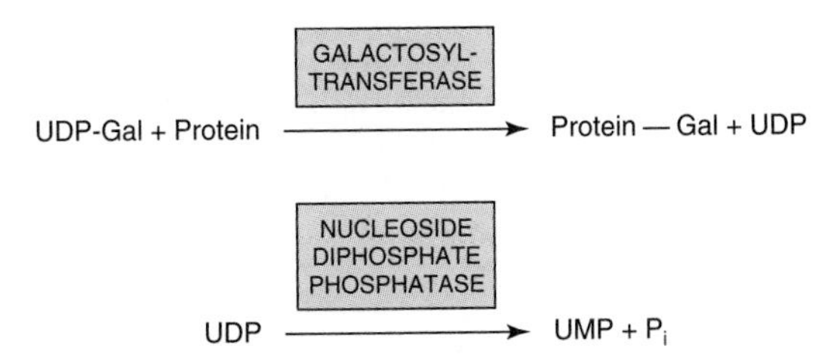


## EXO- & ENDOGLYCOSIDASES FACILITATE STUDY OF GLYCOPROTEINS

A number of **glycosidases** of defined specificity have proved useful in examining structural and functional aspects of glycoproteins (Table 47–5). These enzymes act at either external (exoglycosidases) or internal (endoglycosidases) positions of oligosaccharide chains. Examples of exoglycosidases are **neuraminidases** and **galactosidases;** their sequential use removes terminal NeuAc and subterminal Gal residues from most glycoproteins. **Endoglycosidases F** and **H** are examples of the latter class; these enzymes cleave the oligosaccharide chains at specific GlcNAc residues close to the polypeptide backbone (ie, at internal sites; Figure 47–5) and are thus useful in releasing large oligosaccharide chains for structural analyses. A glycoprotein can be treated with one or more of the above glycosidases to analyze the effects on its biologic behavior of removal of specific sugars.

***Table 47–5.*** Some glycosidases used to study the structure and function of glycoproteins.[1]

| Enzymes | Type |
|---|---|
| Neuraminidases | Exoglycosidase |
| Galactosidases | Exo- or endoglycosidase |
| Endoglycosidase F | Endoglycosidase |
| Endoglycosidase H | Endoglycosidase |

[1]The enzymes are available from a variety of sources and are often specific for certain types of glycosidic linkages and also for their anomeric natures. The sites of action of endoglycosidases F and H are shown in Figure 47–5. F acts on both high-mannose and complex oligosaccharides, whereas H acts on the former.

## THE MAMMALIAN ASIALOGLYCOPROTEIN RECEPTOR IS INVOLVED IN CLEARANCE OF CERTAIN GLYCOPROTEINS FROM PLASMA BY HEPATOCYTES

Experiments performed by Ashwell and his colleagues in the early 1970s played an important role in focusing attention on the functional significance of the oligosaccharide chains of glycoproteins. They treated rabbit ceruloplasmin (a plasma protein; see Chapter 50) with neuraminidase in vitro. This procedure exposed subterminal Gal residues that were normally masked by terminal NeuAc residues. Neuraminidase-treated radioactive ceruloplasmin was found to disappear rapidly from the circulation, in contrast to the slow clearance of the untreated protein. Very significantly, when the Gal residues exposed to treatment with neuraminidase were removed by treatment with a galactosidase, the clearance rate of the protein returned to normal. Further studies demonstrated that liver cells contain a **mammalian asialoglycoprotein receptor** that recognizes the Gal moiety of many desialylated plasma proteins and leads to their endocytosis. This work indicated that an individual sugar, such as Gal, could play an important role in governing at least one of the biologic properties (ie, time of residence in the circulation) of certain glycoproteins. This greatly strengthened the concept that oligosaccharide chains could contain biologic information.

## LECTINS CAN BE USED TO PURIFY GLYCOPROTEINS & TO PROBE THEIR FUNCTIONS

Lectins are carbohydrate-binding proteins that agglutinate cells or precipitate glycoconjugates; a number of lectins are themselves glycoproteins. Immunoglobulins that react with sugars are not considered lectins. Lectins contain at least two sugar-binding sites; proteins with a single sugar-binding site will not agglutinate cells or precipitate glycoconjugates. The specificity of a lectin is usually defined by the sugars that are best at inhibiting its ability to cause agglutination or precipitation. Enzymes, toxins, and transport proteins can be classified as lectins if they bind carbohydrate. Lectins were first discovered in plants and microbes, but many lectins of animal origin are now known. The mammalian asialoglycoprotein receptor described above is an important example of an animal lectin. Some important lectins are listed in Table 47–6. Much current research is centered on the roles of various animal lectins (eg, the selectins) in cell-cell interactions that occur in pathologic conditions such as inflammation and cancer metastasis (see below).

Numerous lectins have been purified and are commercially available; three plant lectins that have been widely used experimentally are listed in Table 47–7. Among many uses, lectins have been employed to purify specific glycoproteins, as tools for probing the glycoprotein profiles of cell surfaces, and as reagents for generating mutant cells deficient in certain enzymes involved in the biosynthesis of oligosaccharide chains.

## THERE ARE THREE MAJOR CLASSES OF GLYCOPROTEINS

Based on the nature of the linkage between their polypeptide chains and their oligosaccharide chains, glycoproteins can be divided into three major classes (Figure 47–1): (1) those containing an O-glycosidic linkage (ie, O-linked), involving the hydroxyl side chain of serine or threonine and a sugar such as *N*-acetylgalactosamine (GalNAc-Ser[Thr]); (2) those containing an N-glycosidic linkage (ie, N-linked), involving the amide nitrogen of asparagine and *N*-acetylglucosamine (GlcNAc-Asn); and (3) those linked to the carboxyl terminal amino acid of a protein via a phosphoryl-ethanolamine moiety joined to an oligosaccharide (glycan), which in turn is linked via glucosamine to phosphatidylinositol (PI). This latter class is referred to as **glycosylphosphatidylinositol-anchored** (**GPI-anchored,** or **GPI-linked**) glycoproteins. Other minor classes of glycoproteins also exist.

The number of oligosaccharide chains attached to one protein can vary from one to 30 or more, with the sugar chains ranging from one or two residues in length to much larger structures. Many proteins contain more than one type of linkage; for instance, **glycophorin,** an important red cell membrane glycoprotein (Chapter 52), contains both O- and N-linked oligosaccharides.

***Table 47–6.*** Some important lectins.

| Lectins | Examples or Comments |
|---|---|
| Legume lectins | Concanavalin A, pea lectin |
| Wheat germ agglutinin | Widely used in studies of surfaces of normal cells and cancer cells |
| Ricin | Cytotoxic glycoprotein derived from seeds of the castor plant |
| Bacterial toxins | Heat-labile enterotoxin of *E coli* and cholera toxin |
| Influenza virus hemagglutinin | Responsible for host-cell attachment and membrane fusion |
| C-type lectins | Characterized by a $Ca^{2+}$-dependent carbohydrate recognition domain (CRD); includes the mammalian asialoglycoprotein receptor, the selectins, and the mannose-binding protein |
| S-type lectins | β-Galactoside-binding animal lectins with roles in cell-cell and cell-matrix interactions |
| P-type lectins | Mannose 6-P receptor |
| I-type lectins | Members of the immunoglobulin superfamily, eg, sialoadhesin mediating adhesion of macrophages to various cells |

## GLYCOPROTEINS CONTAIN SEVERAL TYPES OF O-GLYCOSIDIC LINKAGES

At least four subclasses of O-glycosidic linkages are found in human glycoproteins: (1) The **GalNAc-Ser(Thr)** linkage shown in Figure 47–1 is the predominant linkage. Two typical oligosaccharide chains found in members of this subclass are shown in Figure 47–2. Usually a Gal or a NeuAc residue is attached to the GalNAc, but many variations in the sugar compositions and lengths of such oligosaccharide chains are found. This type of linkage is found in **mucins** (see below). (2) **Proteoglycans** contain a **Gal-Gal-Xyl-Ser** trisaccharide (the so-called link trisaccharide). (3) **Collagens** contain a **Gal-hydroxylysine (Hyl)** linkage. (Subclasses [2] and [3] are discussed further in Chapter 48.) (4) Many **nuclear proteins** (eg, certain transcription factors) and **cytosolic proteins** contain side chains consisting of a single GlcNAc attached to a serine or threonine residue **(GlcNAc-Ser[Thr]).**

***Table 47–7.*** Three plant lectins and the sugars with which they interact.[1]

| Lectin | Abbreviation | Sugars |
|---|---|---|
| Concanavalin A | ConA | Man and Glc |
| Soybean lectin | | Gal and GalNAc |
| Wheat germ agglutinin | WGA | Glc and NeuAc |

[1]In most cases, lectins show specificity for the anomeric nature of the glycosidic linkage (α or β); this is not indicated in the table.

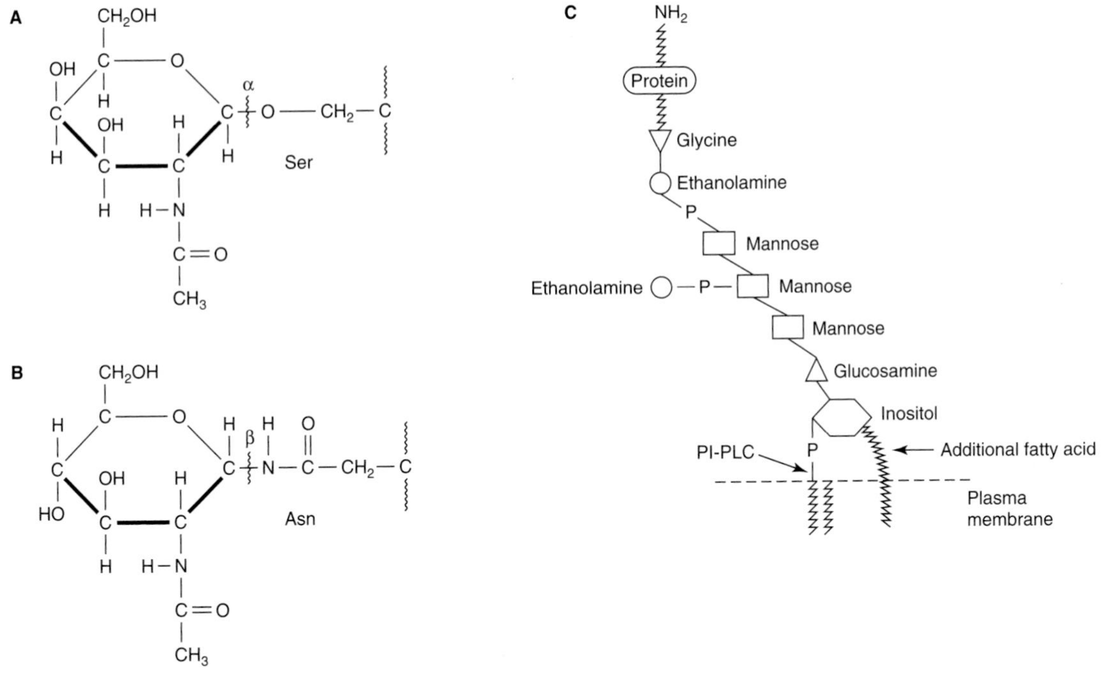

***Figure 47–1.*** Depictions of **(A)** an O-linkage (*N*-acetylgalactosamine to serine); **(B)** an N-linkage (*N*-acetylglucosamine to asparagine) and **(C)** a glycosylphosphatidylinositol (GPI) linkage. The GPI structure shown is that linking acetylcholinesterase to the plasma membrane of the human red blood cell. The carboxyl terminal amino acid is glycine joined in amide linkage via its COOH group to the $NH_2$ group of phosphorylethanolamine, which in turn is joined to a mannose residue. The core glycan contains three mannose and one glucosamine residues. The glucosamine is linked to inositol, which is attached to phosphatidic acid. The site of action of PI-phospholipase C (PI-PLC) is indicated. The structure of the core glycan is shown in the text. This particular GPI contains an extra fatty acid attached to inositol and also an extra phosphorylethanolamine moiety attached to the middle of the three mannose residues. Variations found among different GPI structures include the identity of the carboxyl terminal amino acid, the molecules attached to the mannose residues, and the precise nature of the lipid moiety.

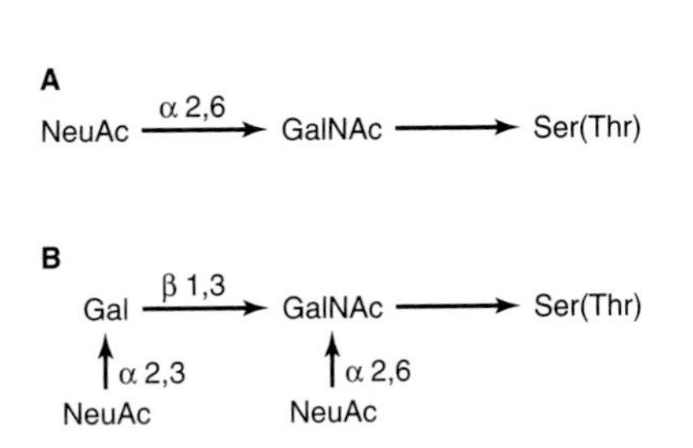


***Figure 47–2.*** Structures of two O-linked oligosaccharides found in **(A)** submaxillary mucins and **(B)** fetuin and in the sialoglycoprotein of the membrane of human red blood cells. (Modified and reproduced, with permission, from Lennarz WJ: ***The Biochemistry of Glycoproteins and Proteoglycans.*** Plenum Press, 1980.)

## Mucins Have a High Content of O-Linked Oligosaccharides & Exhibit Repeating Amino Acid Sequences

Mucins are glycoproteins with two major characteristics: (1) a high content of **O-linked oligosaccharides** (the carbohydrate content of mucins is generally more than 50%); and (2) the presence of **repeating amino acid sequences** (tandem repeats) in the center of their polypeptide backbones, to which the O-glycan chains are attached in clusters (Figure 47–3). These sequences are rich in serine, threonine, and proline. Although O-glycans predominate, mucins often contain a number of N-glycan chains. Both **secretory** and **membrane-bound** mucins occur. The former are found in the mucus present in the secretions of the gastrointestinal, respiratory, and reproductive tracts. **Mucus** consists of about 94% water and 5% mucins, with the remainder

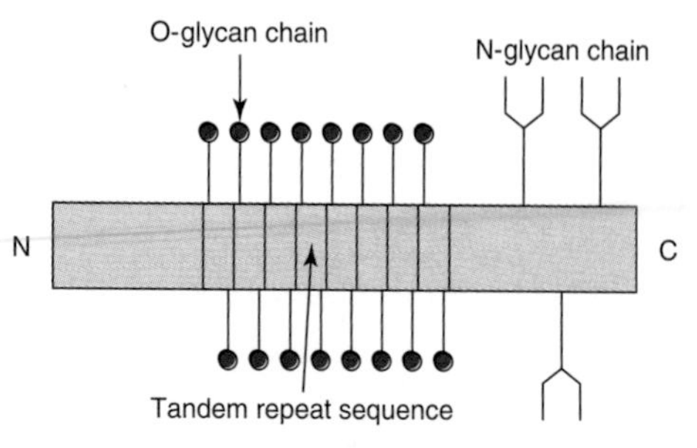


***Figure 47–3.*** Schematic diagram of a mucin. O-glycan chains are shown attached to the central region of the extended polypeptide chain and N-glycan chains to the carboxyl terminal region. The narrow rectangles represent a series of tandem repeat amino acid sequences. Many mucins contain cysteine residues whose SH groups form interchain linkages; these are not shown in the figure. (Adapted from Strous GJ, Dekker J: Mucin-type glycoproteins. Crit Rev Biochem Mol Biol 1992;27:57.)

being a mixture of various cell molecules, electrolytes, and remnants of cells. Secretory mucins generally have an oligomeric structure and thus often have a very high molecular mass. The oligomers are composed of monomers linked by disulfide bonds. Mucus exhibits a high **viscosity** and often forms a **gel.** These qualities are functions of its content of mucins. The high content of O-glycans confers an extended structure on mucins. This is in part explained by steric interactions between their GalNAc moieties and adjacent amino acids, resulting in a chain-stiffening effect so that the conformations of mucins often become those of rigid rods. Intermolecular noncovalent interactions between various sugars on neighboring glycan chains contribute to gel formation. The high content of **NeuAc** and **sulfate** residues found in many mucins confers a negative charge on them. With regard to function, mucins help **lubricate** and form a **protective physical barrier** on epithelial surfaces. Membrane-bound mucins participate in various **cell-cell interactions** (eg, involving selectins; see below). The density of oligosaccharide chains makes it difficult for **proteases** to approach their polypeptide backbones, so that mucins are often resistant to their action. Mucins also tend to "mask" certain surface antigens. Many cancer cells form excessive amounts of mucins; perhaps the mucins may mask certain surface antigens on such cells and thus protect the cells from immune surveillance. Mucins also carry cancer-specific peptide and carbohydrate epitopes (an epitope is a site on an antigen recognized by an antibody, also called an antigenic determinant). Some of these epitopes have been used to stimulate an immune response against cancer cells.

The **genes** encoding the polypeptide backbones of a number of mucins derived from various tissues (eg, pancreas, small intestine, trachea and bronchi, stomach, and salivary glands) have been cloned and sequenced. These studies have revealed new information about the polypeptide backbones of mucins (size of tandem repeats, potential sites of N-glycosylation, etc) and ultimately should reveal aspects of their genetic control. Some important properties of mucins are summarized in Table 47–8.

## The Biosynthesis of O-Linked Glycoproteins Uses Nucleotide Sugars

The polypeptide chains of O-linked and other glycoproteins are encoded by mRNA species; because most glycoproteins are membrane-bound or secreted, they are generally translated on membrane-bound polyribosomes (Chapter 38). Hundreds of different oligosaccharide chains of the O-glycosidic type exist. These glycoproteins are built up by the **stepwise donation of sugars from nucleotide sugars,** such as UDP-GalNAc, UDP-Gal, and CMP-NeuAc. The enzymes catalyzing this type of reaction are membrane-bound **glycoprotein glycosyltransferases.** Generally, synthesis of one specific type of linkage requires the activity of a correspondingly specific transferase. The factors that determine which specific serine and threonine residues are glycosylated have not been identified but are probably found in the peptide structure surrounding the glycosylation site. The enzymes assembling O-linked chains are located in the Golgi apparatus, sequentially arranged in an assembly line with terminal reactions occurring in the trans-Golgi compartments.

The major features of the biosynthesis of O-linked glycoproteins are summarized in Table 47–9.

***Table 47–8.*** Some properties of mucins.

- Found in secretions of the gastrointestinal, respiratory, and reproductive tracts and also in membranes of various cells.
- Exhibit high content of O-glycan chains, usually containing NeuAc.
- Contain repeating amino acid sequences rich in serine, threonine, and proline.
- Extended structure contributes to their high viscoelasticity.
- Form protective physical barrier on epithelial surfaces, are involved in cell-cell interactions, and may contain or mask certain surface antigens.

**Table 47–9.** Summary of main features of O-glycosylation.

- Involves a battery of membrane-bound glycoprotein glycosyltransferases acting in a stepwise manner; each transferase is generally specific for a particular type of linkage.
- The enzymes involved are located in various subcompartments of the Golgi apparatus.
- Each glycosylation reaction involves the appropriate nucleotide-sugar.
- Dolichol-P-P-oligosaccharide is not involved, nor are glycosidases; and the reactions are not inhibited by tunicamycin.
- O-Glycosylation occurs posttranslationally at certain Ser and Thr residues.

## N-LINKED GLYCOPROTEINS CONTAIN AN Asn-GlcNAc LINKAGE

N-Linked glycoproteins are distinguished by the presence of the Asn-GlcNAc linkage (Figure 47–1). It is the major class of glycoproteins and has been much studied, since the most readily accessible glycoproteins (eg, plasma proteins) mainly belong to this group. It includes both **membrane-bound** and **circulating** glycoproteins. The principal difference between this and the previous class, apart from the nature of the amino acid to which the oligosaccharide chain is attached (Asn versus Ser or Thr), concerns their biosynthesis.

### Complex, Hybrid, & High-Mannose Are the Three Major Classes of N-Linked Oligosaccharides

There are three major classes of N-linked oligosaccharides: **complex**, **hybrid**, and **high-mannose** (Figure 47–4). Each type shares a common pentasaccharide, $Man_3GlcNAc_2$—shown within the boxed area in Figure 47–4 and depicted also in Figure 47–5—but they differ in their outer branches. The presence of the **common pentasaccharide** is explained by the fact that all three classes share an initial common mechanism of biosynthesis. Glycoproteins of the complex type generally contain terminal NeuAc residues and underlying Gal and GlcNAc residues, the latter often constituting the disaccharide *N*-acetyllactosamine. Repeating ***N*-acetyllactosamine units**—$[Gal\beta 1–3/4GlcNAc\beta 1–3]_n$ (poly-*N*-acetyllactosaminoglycans)—are often found on N-linked glycan chains. I/i blood group substances belong to this class. The majority of complex-type oligosaccharides contain two, three, or four outer branches (Figure 47–4), but structures containing five branches have also been described. The oligosaccharide branches are often referred to as **antennae,** so that bi-, tri-, tetra-, and penta-antennary structures may all be found. A bewildering number of chains of the complex type exist, and that indicated in Figure 47–4 is only one of many. Other complex chains may terminate in Gal or Fuc. High-mannose oligosaccharides typically have two to six additional Man residues linked to the pentasaccharide core. Hybrid molecules contain features of both of the two other classes.

### The Biosynthesis of N-Linked Glycoproteins Involves Dolichol-P-P-Oligosaccharide

Leloir and his colleagues described the occurrence of a **dolichol-pyrophosphate-oligosaccharide (Dol-P-P-oligosaccharide),** which subsequent research showed to play a key role in the biosynthesis of N-linked glycoproteins. The oligosaccharide chain of this compound generally has the structure $R\text{-}GlcNAc_2Man_9Glc_3$ (R = Dol-P-P). The sugars of this compound are first assembled on the Dol-P-P backbone, and the oligosaccharide chain is then transferred en bloc to suitable Asn residues of acceptor apoglycoproteins during their synthesis on membrane-bound polyribosomes. All N-glycans have a common pentasaccharide core structure (Figure 47–5).

To form **high-mannose** chains, only the Glc residues plus certain of the peripheral Man residues are removed. To form an oligosaccharide chain of the **complex type,** the Glc residues and four of the Man residues are removed by glycosidases in the endoplasmic reticulum and Golgi. The sugars characteristic of complex chains (GlcNAc, Gal, NeuAc) are added by the action of individual glycosyltransferases located in the Golgi apparatus. The phenomenon whereby the glycan chains of N-linked glycoproteins are first partially degraded and then in some cases rebuilt is referred to as **oligosaccharide processing. Hybrid chains** are formed by partial processing, forming complex chains on one arm and Man structures on the other arm.

Thus, the initial steps involved in the biosynthesis of the N-linked glycoproteins differ markedly from those involved in the biosynthesis of the O-linked glycoproteins. The former involves Dol-P-P-oligosaccharide; the latter, as described earlier, does not.

The process of N-glycosylation can be broken down into two stages: (1) assembly of Dol-P-P-oligosaccharide and transfer of the oligosaccharide; and (2) processing of the oligosaccharide chain.

#### A. Assembly & Transfer of Dolichol-P-P-Oligosaccharide

**Polyisoprenol** compounds exist in both bacteria and eukaryotic cells. They participate in the synthesis of bacterial polysaccharides and in the biosynthesis of N-

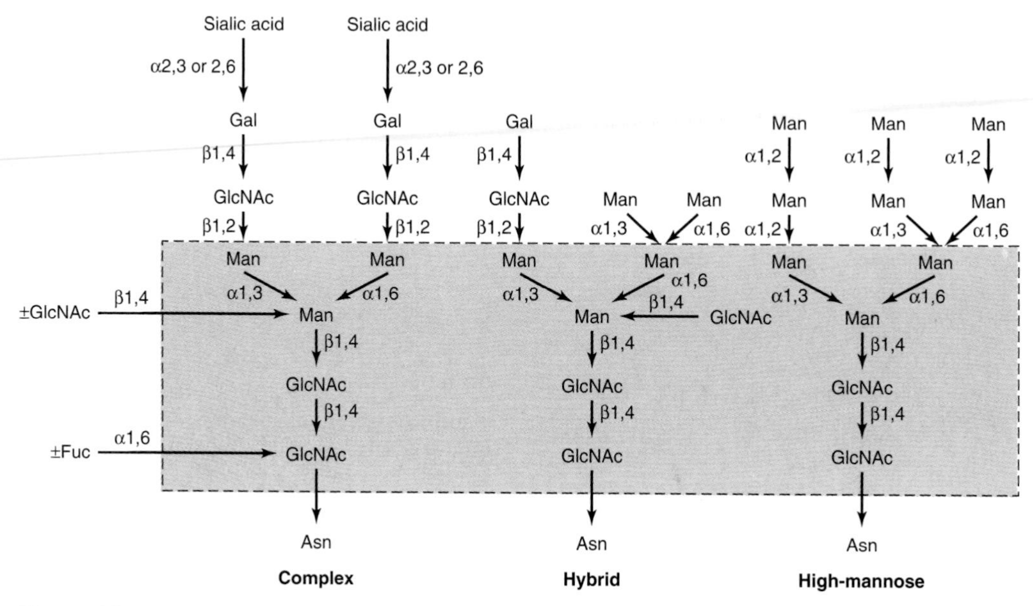


***Figure 47–4.*** Structures of the major types of asparagine-linked oligosaccharides. The boxed area encloses the pentasaccharide core common to all N-linked glycoproteins. (Reproduced, with permission, from Kornfeld R, Kornfeld S: Assembly of asparagine-linked oligosaccharides. Annu Rev Biochem 1985;54:631.)

linked glycoproteins and GPI anchors. The polyisoprenol used in eukaryotic tissues is **dolichol,** which is, next to rubber, the longest naturally occurring hydrocarbon made up of a single repeating unit. Dolichol is composed of 17–20 repeating isoprenoid units (Figure 47–6).

Before it participates in the biosynthesis of Dol-P-P-oligosaccharide, dolichol must first be phosphorylated to form dolichol phosphate (Dol-P) in a reaction catalyzed by **dolichol kinase** and using ATP as the phosphate donor.

**Dolichol-P-P-GlcNAc (Dol-P-P-GlcNAc)** is the key lipid that acts as an acceptor for other sugars in the assembly of Dol-P-P-oligosaccharide. It is synthesized

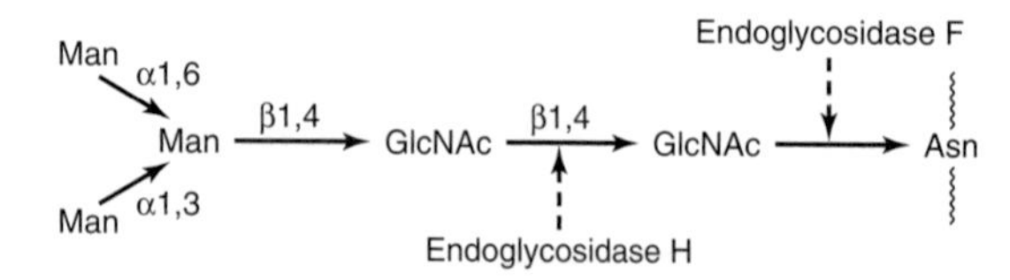


***Figure 47–5.*** Schematic diagram of the pentasaccharide core common to all N-linked glycoproteins and to which various outer chains of oligosaccharides may be attached. The sites of action of endoglycosidases F and H are also indicated.

in the membranes of the endoplasmic reticulum from Dol-P and UDP-GlcNAc in the following reaction, catalyzed by GlcNAc-P transferase:

$$\text{Dol-P} + \text{UDP-GlcNAc} \rightarrow \text{Dol-P-P-GlcNAc} + \text{UMP}$$

The above reaction—which is the first step in the assembly of Dol-P-P-oligosaccharide—and the other later reactions are summarized in Figure 47–7. The essential features of the subsequent steps in the assembly of Dol-P-P-oligosaccharide are as follows:

(1) A second GlcNAc residue is added to the first, again using UDP-GlcNAc as the donor.

(2) Five Man residues are added, using GDP-mannose as the donor.

(3) Four additional Man residues are next added, using Dol-P-Man as the donor. Dol-P-Man is formed by the following reaction:

$$\text{Dol-P} + \text{GDP-Man} \rightarrow \text{Dol-P-Man} + \text{GDP}$$

(4) Finally, the three peripheral glucose residues are donated by Dol-P-Glc, which is formed in a reaction analogous to that just presented except that Dol-P and UDP-Glc are the substrates.

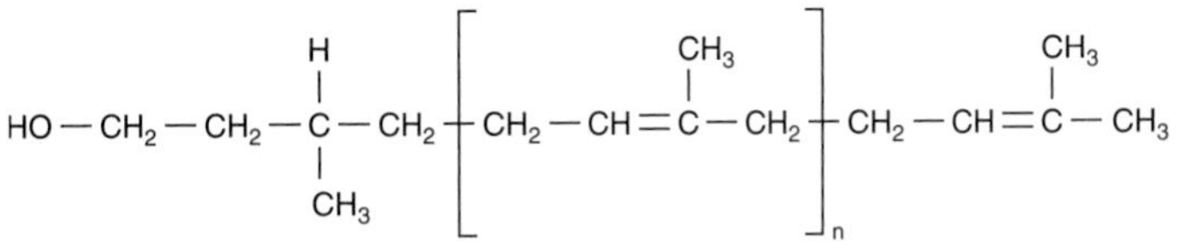

***Figure 47–6.*** The structure of dolichol. The phosphate in dolichol phosphate is attached to the primary alcohol group at the left-hand end of the molecule. The group within the brackets is an isoprene unit (n = 17–20 isoprenoid units).

$$HO-CH_2-CH_2-\underset{CH_3}{\overset{H}{C}}-CH_2-\left[CH_2-CH=\overset{CH_3}{C}-CH_2\right]_n-CH_2-CH=\overset{CH_3}{C}-CH_3$$

It should be noted that the first seven sugars (two GlcNAc and five Man residues) are donated by nucleotide sugars, whereas the last seven sugars (four Man and three Glc residues) added are donated by dolichol-P-sugars. The net result is assembly of the compound illustrated in Figure 47–8 and referred to in shorthand as Dol-P-P-GlcNAc$_2$Man$_9$Glc$_3$.

The oligosaccharide linked to dolichol-P-P is transferred en bloc to form an N-glycosidic bond with one or more specific Asn residues of an acceptor protein emerging from the luminal surface of the membrane of the endoplasmic reticulum. The reaction is catalyzed by **oligosaccharide:protein transferase,** a membrane-associated enzyme complex. The transferase will recognize and transfer any substrate with the general structure Dol-P-P-(GlcNAc)$_2$-R, but it has a strong preference for the Dol-P-P-GlcNAc$_2$Man$_9$Glc$_3$ structure. Glycosylation occurs at the Asn residue of an Asn-X-Ser/Thr tripeptide sequence, where X is any amino acid except proline, aspartic acid, or glutamic acid. A tripeptide site contained within a β turn is favored. Only about one-third of the Asn residues that are potential acceptor sites are actually glycosylated, suggesting that factors other than the tripeptide are also important. The acceptor proteins are of both the secretory and integral membrane class. Cytosolic proteins are rarely glycosylated. The transfer reaction and subsequent processes in the glycosylation of N-linked glycoproteins, along with their subcellular locations, are depicted in Figure 47–9. The other product of the oligosaccharide:protein transferase reaction is dolichol-P-P, which is subsequently converted to dolichol-P by a

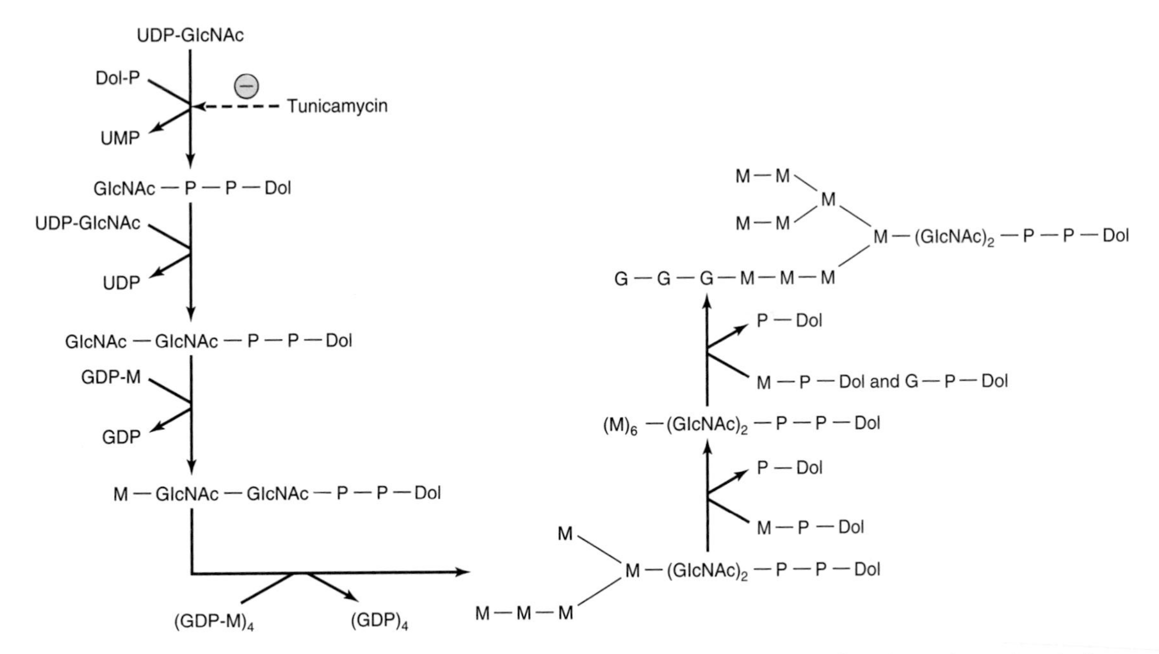


***Figure 47–7.*** Pathway of biosynthesis of dolichol-P-P-oligosaccharide. The specific linkages formed are indicated in Figure 47–8. Note that the first five internal mannose residues are donated by GDP-mannose, whereas the more external mannose residues and the glucose residues are donated by dolichol-P-mannose and dolichol-P-glucose. (UDP, uridine diphosphate; Dol, dolichol; P, phosphate; UMP, uridine monophosphate; GDP, guanosine diphosphate; M, mannose; G, glucose.)

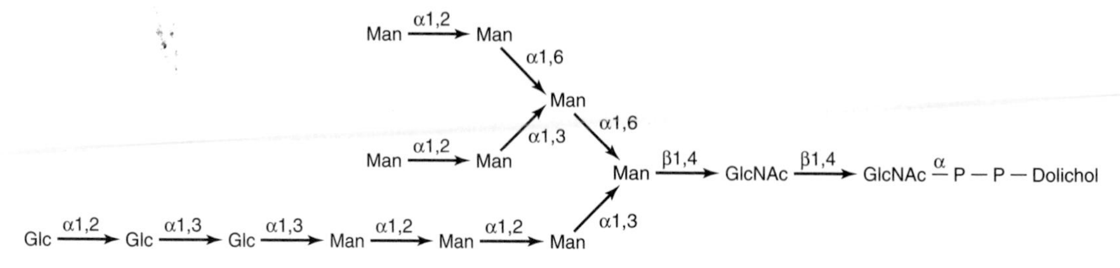


***Figure 47–8.*** **Structure of dolichol-P-P-oligosaccharide.** (Reproduced, with permission, from Lennarz WJ: *The Biochemistry of Glycoproteins and Proteoglycans.* Plenum Press, 1980.)

phosphatase. The dolichol-P can serve again as an acceptor for the synthesis of another molecule of Dol-P-P-oligosaccharide.

### B. Processing of the Oligosaccharide Chain

**1. Early phase**—The various reactions involved are indicated in Figure 47–9. The oligosaccharide:protein transferase catalyzes reaction 1 (see above). Reactions 2 and 3 involve the removal of the terminal Glc residue by glucosidase I and of the next two Glc residues by glucosidase II, respectively. In the case of **high-mannose** glycoproteins, the process may stop here, or up to four Man residues may also be removed. However, to form **complex** chains, additional steps are necessary, as follows. Four external Man residues are removed in reactions 4 and 5 by at least two different mannosidases. In reaction 6, a GlcNAc residue is added to the Man residue of the Manα1–3 arm by GlcNAc transferase I. The action of this latter enzyme permits the occurrence of reaction 7, a reaction catalyzed by yet another mannosidase (Golgi α-mannosidase II) and which results in a reduction of the Man residues to the core number of three (Figure 47–5).

An important additional pathway is indicated in reactions I and II of Figure 47–9. This involves enzymes destined for **lysosomes.** Such enzymes are targeted to the lysosomes by a specific chemical marker. In reaction I, a residue of GlcNAc-1-P is added to carbon 6 of one or more specific Man residues of these enzymes. The reaction is catalyzed by a GlcNAc phosphotransferase, which uses UDP-GlcNAc as the donor and generates UMP as the other product:

$$\text{UDP-GlcNAc + Man — Protein} \xrightarrow{\text{GlcNAc PHOSPHOTRANSFERASE}} \text{GlcNAc-1-P-6-Man — Protein + UMP}$$

In reaction II, the GlcNAc is removed by the action of a phosphodiesterase, leaving the Man residues phosphorylated in the 6 position:

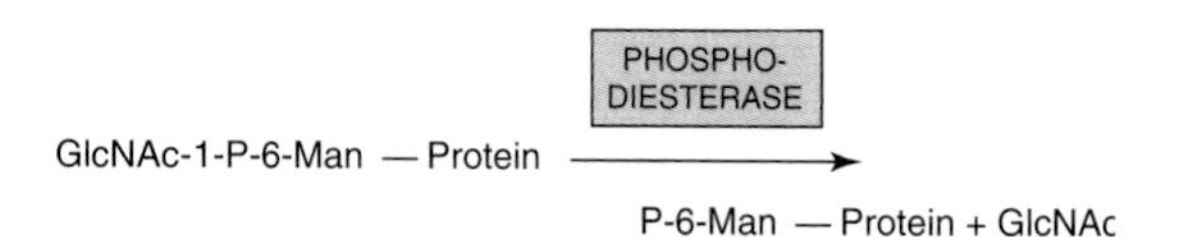


Man 6-P receptors, located in the Golgi apparatus, bind the Man 6-P residue of these enzymes and direct them to the lysosomes. Fibroblasts from patients with **I-cell disease** (see below) are severely deficient in the activity of the GlcNAc phosphotransferase.

**2. Late phase**—To assemble a typical complex oligosaccharide chain, additional sugars must be added to the structure formed in reaction 7. Hence, in reaction 8, a second GlcNAc is added to the peripheral Man residue of the other arm of the bi-antennary structure shown in Figure 47–9; the enzyme catalyzing this step is GlcNAc transferase II. Reactions 9, 10, and 11 involve the addition of Fuc, Gal, and NeuAc residues at the sites indicated, in reactions catalyzed by fucosyl, galactosyl, and sialyl transferases, respectively. The assembly of poly-*N*-acetyllactosamine chains requires additional GlcNAc transferases.

## The Endoplasmic Reticulum & Golgi Apparatus Are the Major Sites of Glycosylation

As indicated in Figure 47–9, the endoplasmic reticulum and the Golgi apparatus are the major sites involved in glycosylation processes. The assembly of Dol-P-P-oligosaccharide occurs on both the cytoplasmic and luminal surfaces of the ER membranes. Addition of the oligosaccharide to protein occurs in the rough endoplasmic reticulum during or after translation. Removal of the Glc and some of the peripheral Man residues also occurs in the endoplasmic reticulum. The Golgi apparatus is composed of cis, medial, and trans cisternae; these can be separated by appropriate centrifugation procedures. Vesicles containing glycoproteins appear to bud off in the endoplasmic reticulum and are transported to the cis Golgi. Various studies have shown

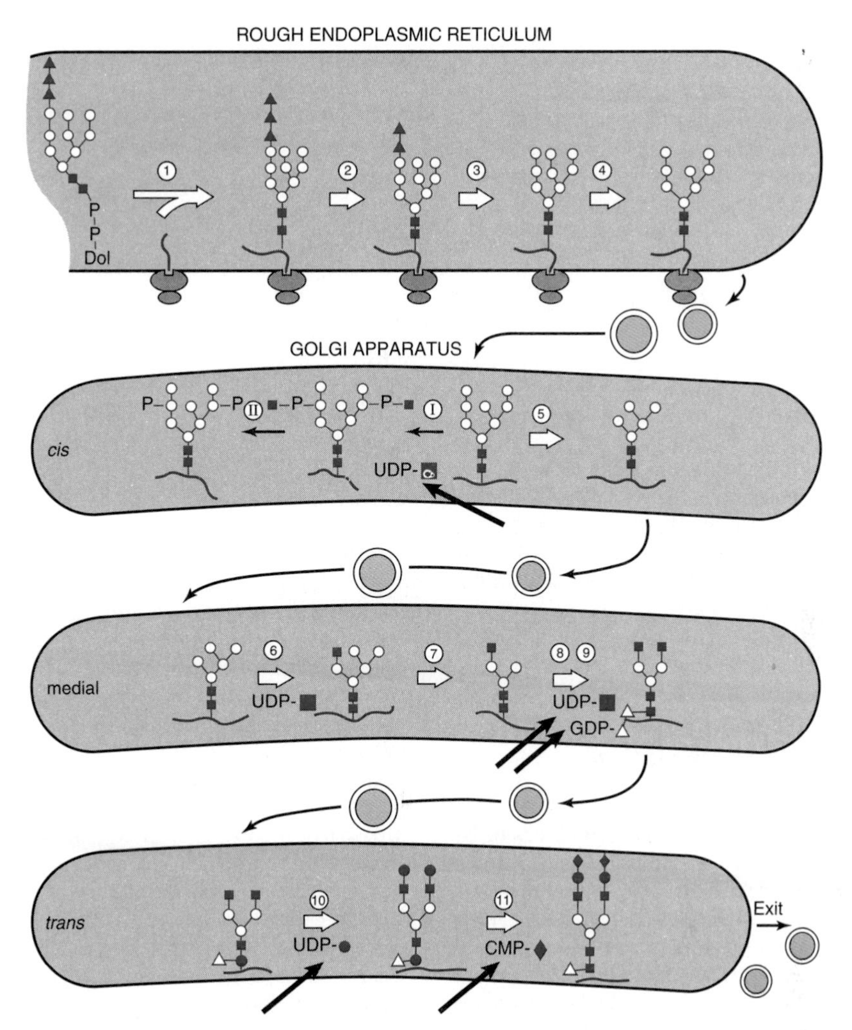


***Figure 47–9.*** Schematic pathway of oligosaccharide processing. The reactions are catalyzed by the following enzymes: ①, oligosaccharide:protein transferase; ②, α-glucosidase I; ③, α-glucosidase II; ④, endoplasmic reticulum α1,2-mannosidase; Ⓘ, *N*-acetylglucosaminylphosphotransferase; Ⓘⓘ, *N*-acetylglucosamine-1-phosphodiester α-*N*-acetylglucosaminidase; ⑤, Golgi apparatus α-mannosidase I; ⑥, *N*-acetylglucosaminyltransferase I; ⑦, Golgi apparatus α-mannosidase II; ⑧, *N*-acetylglucosaminyltransferase II; ⑨, fucosyltransferase; ⑩, galactosyltransferase; ⑪, sialyltransferase. The thick arrows indicate various nucleotide sugars involved in the oveall scheme. (Solid square, *N*-acetylglucosamine; open circle, mannose; solid triangle, glucose; open triangle, fucose; solid circle, galactose; solid diamond, sialic acid.) (Reproduced, with permission, from Kornfeld R, Kornfeld S: Assembly of asparagine-linked oligosaccharides. Annu Rev Biochem 1985;54:631.)

that the enzymes involved in glycoprotein processing show differential locations in the cisternae of the Golgi. As indicated in Figure 47–9, Golgi α-mannosidase I (catalyzing reaction 5) is located mainly in the cis Golgi, whereas GlcNAc transferase I (catalyzing reaction 6) appears to be located in the medial Golgi, and the fucosyl, galactosyl, and sialyl transferases (catalyzing reactions 9, 10, and 11) are located mainly in the trans Golgi. The major features of the biosynthesis of N-linked glycoproteins are summarized in Table 47–10 and should be contrasted with those previously listed (Table 47–9) for O-linked glycoproteins.

***Table 47–10.*** Summary of main features of N-glycosylation.

- The oligosaccharide $Glc_3Man_9(GlcNAc)_2$ is transferred from dolichol-P-P-oligosaccharide in a reaction catalyzed by oligosaccharide:protein transferase, which is inhibited by tunicamycin.
- Transfer occurs to specific Asn residues in the sequence Asn-X-Ser/Thr, where X is any residue except Pro, Asp, or Glu.
- Transfer can occur cotranslationally in the endoplasmic reticulum.
- The protein-bound oligosaccharide is then partially processed by glucosidases and mannosidases; if no additional sugars are added, this results in a high-mannose chain.
- If processing occurs down to the core heptasaccharide ($Man_5[GlcNAc]_2$), complex chains are synthesized by the addition of GlcNAc, removal of two Man, and the stepwise addition of individual sugars in reactions catalyzed by specific transferases (eg, GlcNAc, Gal, NeuAc transferases) that employ appropriate nucleotide sugars.

## Some Glycan Intermediates Formed During N-Glycosylation Have Specific Functions

The following are a number of specific functions of N-glycan chains that have been established or are under investigation. (1) The involvement of the **mannose 6-P signal** in targeting of certain lysosomal enzymes is clear (see above and discussion of I-cell disease, below). (2) It is likely that the large N-glycan chains present on newly synthesized glycoproteins may assist in keeping these proteins in a soluble state inside the lumen of the endoplasmic reticulum. (3) One species of N-glycan chains has been shown to play a role in the folding and retention of certain glycoproteins in the lumen of the endoplasmic reticulum. **Calnexin** is a protein present in the endoplasmic reticulum membrane that acts as a "chaperone" (Chapter 46). It has been found that calnexin will bind specifically to a number of glycoproteins (eg, the influenza virus hemagglutinin [HA]) that possess the **monoglycosylated core structure.** This species is the product of reaction 2 shown in Figure 47–9 but from which the terminal glucose residue has been removed, leaving only the innermost glucose attached. The release of fully folded HA from calnexin requires the enzymatic removal of this last glucosyl residue by α-glucosidase II. In this way, calnexin retains certain partly folded (or misfolded) glycoproteins and releases them when proper folding has occurred; it is thus an important component of the quality control systems operating in the lumen of the ER. The soluble protein **calreticulin** appears to play a similar function.

## Several Factors Regulate the Glycosylation of Glycoproteins

It is evident that glycosylation of glycoproteins is a complex process involving a large number of enzymes. One index of its complexity is that more than ten distinct GlcNAc transferases involved in glycoprotein biosynthesis have been reported, and many others are theoretically possible. Multiple species of the other glycosyltransferases (eg, sialyltransferases) also exist. Controlling factors of the first stage of N-linked glycoprotein biosynthesis (ie, **oligosaccharide assembly and transfer**) include (1) the presence of suitable acceptor sites in proteins, (2) the tissue level of Dol-P, and (3) the activity of the oligosaccharide:protein transferase.

Some factors known to be involved in the regulation of **oligosaccharide processing** are listed in Table 47–11. Two of the points listed merit further comment. First, **species variations** among processing enzymes have assumed importance in relation to production of glycoproteins of therapeutic use by means of recombinant DNA technology. For instance, recombinant erythropoietin (epoetin alfa; EPO) is sometimes administered to patients with certain types of chronic anemia in order to stimulate erythropoiesis. The half-life of EPO in plasma is influenced by the nature of its glycosylation pattern, with certain patterns being associated with a short half-life, appreciably limiting its period of therapeutic effectiveness. It is thus important to harvest EPO from host cells that confer a pattern of glycosylation consistent with a normal half-life in plasma. Second, there is great interest in analysis of the activities of glycoprotein-processing enzymes in various types of cancer cells. These cells have often been found to synthesize different oligosaccharide chains (eg, they often exhibit greater branching) from those made in control cells. This could be due to cancer cells containing different patterns of glycosyltransferases from those exhibited by corresponding normal cells, due to specific gene activation or repression. The differences in oligosaccharide chains could affect adhesive interactions between cancer cells and their normal parent tissue cells, contributing to metastasis. If a correlation could be found between the activity of particular processing enzymes and the **metastatic properties** of cancer cells, this could be important as it might permit synthesis of drugs to inhibit these enzymes and, secondarily, metastasis.

The genes encoding many glycosyltransferases have already been cloned, and others are under study. Cloning has revealed new information on both protein and gene structures. The latter should also cast light on

***Table 47–11.*** Some factors affecting the activities of glycoprotein processing enzymes.

| Factor | Comment |
|---|---|
| Cell type | Different cell types contain different profiles of processing enzymes. |
| Previous enzyme | Certain glycosyltransferases will only act on an oligosaccharide chain if it has already been acted upon by another processing enzyme.[1] |
| Development | The cellular profile of processing enzymes may change during development if their genes are turned on or off. |
| Intracellular location | For instance, if an enzyme is destined for insertion into the membrane of the ER (eg, HMG-CoA reductase), it may never encounter Golgi-located processing enzymes. |
| Protein conformation | Differences in conformation of different proteins may facilitate or hinder access of processing enzymes to identical oligosaccharide chains. |
| Species | Same cells (eg, fibroblasts) from different species may exhibit different patterns of processing enzymes. |
| Cancer | Cancer cells may exhibit processing enzymes different from those of corresponding normal cells. |

[1]For example, prior action of GlcNAc transferase I is necessary for the action of Golgi α-mannosidase II.

the mechanisms involved in their transcriptional control, and gene knockout studies are being used to evaluate the biologic importance of various glycosyltransferases.

### Tunicamycin Inhibits N- but Not O-Glycosylation

A number of compounds are known to inhibit various reactions involved in glycoprotein processing. **Tunicamycin, deoxynojirimycin,** and **swainsonine** are three such agents. The reactions they inhibit are indicated in Table 47–12. These agents can be used experimentally to inhibit various stages of glycoprotein biosynthesis and to study the effects of specific alterations upon the process. For instance, if cells are grown in the presence of tunicamycin, no glycosylation of their normally N-linked glycoproteins will occur. In certain cases, lack of glycosylation has been shown to increase the susceptibility of these proteins to proteolysis. Inhibition of glycosylation does not appear to have a consistent effect upon the secretion of glycoproteins that are normally secreted. The inhibitors of glycoprotein processing listed in Table 47–12 do not affect the biosynthesis of O-linked glycoproteins. The extension of O-linked chains can be prevented by GalNAc-benzyl. This compound competes with natural glycoprotein substrates and thus prevents chain growth beyond GalNAc.

***Table 47–12.*** Three inhibitors of enzymes involved in the glycosylation of glycoproteins and their sites of action.

| Inhibitor | Site of Action |
|---|---|
| Tunicamycin | Inhibits GlcNAc-P transferase, the enzyme catalyzing addition of GlcNAc to dolichol-P, the first step in the biosynthesis of oligosaccharide-P-P-dolichol |
| Deoxynojirimycin | Inhibitor of glucosidases I and II |
| Swainsonine | Inhibitor of mannosidase II |

## SOME PROTEINS ARE ANCHORED TO THE PLASMA MEMBRANE BY GLYCOSYLPHOSPHATIDYL-INOSITOL STRUCTURES

Glycosylphosphatidylinositol (GPI)-linked glycoproteins comprise the third major class of glycoprotein. The GPI structure (sometimes called a "sticky foot") involved in linkage of the enzyme acetylcholinesterase (ACh esterase) to the plasma membrane of the red blood cell is shown in Figure 47–1. GPI-linked proteins are anchored to the outer leaflet of the plasma membrane by the fatty acids of phosphatidylinositol (PI). The PI is linked via a $GlcNH_2$ moiety to a glycan chain that contains various sugars (eg, Man, $GlcNH_2$). In turn, the oligosaccharide chain is linked via phosphorylethanolamine in an amide linkage to the carboxyl terminal amino acid of the attached protein. The core of most GPI structures contains one molecule of phosphorylethanolamine, three Man residues, one molecule of $GlcNH_2$, and one molecule of phosphatidylinositol, as follows:

$$\text{Ethanolamine-phospho} \rightarrow 6\text{Man}\alpha1 \rightarrow$$
$$2\text{Man}\alpha1 \rightarrow 6\text{Man}\alpha1 \rightarrow \text{GlcN}\alpha1 \rightarrow$$
$$6 - \textit{myo}\text{-inositol-1-phospholipid}$$

Additional constituents are found in many GPI structures; for example, that shown in Figure 47–1 contains an extra phosphorylethanolamine attached to the middle of the three Man moieties of the glycan and an extra fatty acid attached to $GlcNH_2$. The functional significance of these variations among structures is not understood. This type of linkage was first detected by the use of bacterial PI-specific phospholipase C (PI-PLC), which was found to release certain proteins from the plasma membrane of cells by splitting the bond indicated in Figure 47–1. Examples of some proteins that are anchored by this type of linkage are given in Table 47–13. At least three possible functions of this type of linkage have been suggested: (1) The GPI anchor may allow greatly enhanced **mobility** of a protein in the plasma membrane compared with that observed for a protein that contains transmembrane sequences. This is perhaps not surprising, as the GPI anchor is attached only to the outer leaflet of the lipid bilayer, so that it is freer to diffuse than a protein anchored via both leaflets of the bilayer. Increased mobility may be important in facilitating rapid responses to appropriate stimuli. (2) Some GPI anchors may connect with **signal transduction** pathways. (3) It has been shown that GPI structures can target certain proteins to apical domains of the plasma membrane of certain epithelial cells. The biosynthesis of GPI anchors is complex and begins in the endoplasmic reticulum. The GPI anchor is assembled independently by a series of enzyme-catalyzed reactions and then transferred to the carboxyl terminal end of its acceptor protein, accompanied by cleavage of the preexisting carboxyl terminal hydrophobic peptide from that protein. This process is sometimes called **glypiation.** An acquired defect in an early stage of the biosynthesis of the GPI structure has been implicated in the causation of **paroxysmal nocturnal hemoglobinuria** (see below).

## GLYCOPROTEINS ARE INVOLVED IN MANY BIOLOGIC PROCESSES & IN MANY DISEASES

As listed in Table 47–1, glycoproteins have many different functions; some have already been addressed in this chapter and others are described elsewhere in this text (eg, transport molecules, immunologic molecules, and hormones). Here, their involvement in two specific processes—fertilization and inflammation—will be briefly described. In addition, the bases of a number of diseases that are due to abnormalities in the synthesis and degradation of glycoproteins will be summarized.

***Table 47–13.*** Some GPI-linked proteins.

- Acetylcholinesterase (red cell membrane)
- Alkaline phosphatase (intestinal, placental)
- Decay-accelerating factor (red cell membrane)
- 5′-Nucleotidase (T lymphocytes, other cells)
- Thy-1 antigen (brain, T lymphocytes)
- Variable surface glycoprotein (*Trypanosoma brucei*)

### Glycoproteins Are Important in Fertilization

To reach the plasma membrane of an oocyte, a sperm has to traverse the **zona pellucida (ZP),** a thick, transparent, noncellular envelope that surrounds the oocyte. The zona pellucida contains three glycoproteins of interest, ZP1–3. Of particular note is ZP3, an O-linked glycoprotein that functions as a receptor for the sperm. A protein on the sperm surface, possibly galactosyl transferase, interacts specifically with oligosaccharide chains of ZP3; in at least certain species (eg, the mouse), this interaction, by transmembrane signaling, induces the **acrosomal reaction,** in which enzymes such as proteases and hyaluronidase and other contents of the acrosome of the sperm are released. Liberation of these enzymes helps the sperm to pass through the zona pellucida and reach the plasma membrane (PM) of the oocyte. In hamsters, it has been shown that another glycoprotein, PH-30, is important in both the binding of the PM of the sperm to the PM of the oocyte and also in the subsequent fusion of the two membranes. These interactions enable the sperm to enter and thus fertilize the oocyte. It may be possible to inhibit fertilization by developing drugs or antibodies that interfere with the normal functions of ZP3 and PH-30 and which would thus act as contraceptive agents.

### Selectins Play Key Roles in Inflammation & in Lymphocyte Homing

Leukocytes play important roles in many inflammatory and immunologic phenomena. The first steps in many of these phenomena are interactions between circulating leukocytes and endothelial cells prior to passage of the former out of the circulation. Work done to identify specific molecules on the surfaces of the cells involved in such interactions has revealed that leukocytes and endothelial cells contain on their surfaces specific lectins, called **selectins,** that participate in their intercellular adhesion. Features of the three major classes of selectins are summarized in Table 47–14. Selectins are single-chain $Ca^{2+}$-binding transmembrane proteins that contain a number of domains (Figure 47–10). Their amino terminal ends contain the lectin domain, which is involved in binding to specific carbohydrate ligands.

The adhesion of neutrophils to endothelial cells of postcapillary venules can be considered to occur in four

***Table 47–14.*** Some molecules involved in leukocyte-endothelial cell interactions.[1]

| Molecule | Cell | Ligands |
|---|---|---|
| **Selectins** | | |
| L-selectin | PMN, lymphs | CD34, Gly-CAM-1[2]<br>Sialyl-Lewis$^x$ and others |
| P-selectin | EC, platelets | P-selectin glycoprotein ligand-1 (PSGL-1)<br>Sialyl-Lewis$^x$ and others |
| E-selectin | EC | Sialyl-Lewis$^x$ and others |
| **Integrins** | | |
| LFA-1 | PMN, lymphs | ICAM-1, ICAM-2 (CD11a/CD18) |
| Mac-1 | PMN | ICAM-1 and others (CD11b/CD18) |
| **Immunoglobulin superfamily** | | |
| ICAM-1 | Lymphs, EC | LFA-1, Mac-1 |
| ICAM-2 | Lymphs, EC | LFA-1 |
| PECAM-1 | EC, PMN, lymphs | Various platelets |

[1]Modified from Albelda SM, Smith CW, Ward PA: Adhesion molecules and inflammatory injury. FASEB J 1994;8:504.
[2]These are ligands for lymphocyte L-selectin; the ligands for neutrophil L-selectin have not been identified.
**Key:** PMN, polymorphonuclear leukocytes; EC, endothelial cell; lymphs, lymphocytes; CD, cluster of differentiation; ICAM, intercellular adhesion molecule; LFA-1, lymphocyte function-associated antigen-1; PECAM-1, platelet endothelial cell adhesion cell molecule-1.

stages, as shown in Figure 47–11. The initial baseline stage is succeeded by slowing or rolling of the neutrophils, mediated by selectins. Interactions between L-selectin on the neutrophil surface and CD34 and GlyCAM-1 or other glycoproteins on the endothelial surface are involved. These particular interactions are initially short-lived, and the overall binding is of relatively low affinity, permitting rolling. However, during

***Figure 47–10.*** Schematic diagram of the structure of human L-selectin. The extracellular portion contains an amino terminal domain homologous to C-type lectins and an adjacent epidermal growth factor-like domain. These are followed by a variable number of complement regulatory-like modules (numbered circles) and a transmembrane sequence (black diamond). A short cytoplasmic sequence (open rectangle) is at the carboxyl terminal. The structures of P- and E-selectin are similar to that shown except that they contain more complement-regulatory modules. The numbers of amino acids in L-, P-, and E- selectins, as deduced from the cDNA sequences, are 385, 789, and 589, respectively. (Reproduced, with permission, from Bevilacqua MP, Nelson RM: Selectins. J Clin Invest 1993;91:370.)

this stage, activation of the neutrophils by various chemical mediators (discussed below) occurs, resulting in a change of shape of the neutrophils and firm adhesion of these cells to the endothelium. An additional set of adhesion molecules is involved in firm adhesion, namely, LFA-1 and Mac-1 on the neutrophils and ICAM-1 and ICAM-2 on endothelial cells. LFA-1 and Mac-1 are CD11/CD18 integrins (see Chapter 52 for a discussion of integrins), whereas ICAM-1 and ICAM-2 are members of the immunoglobulin superfamily. The fourth stage is transmigration of the neutrophils across the endothelial wall. For this to occur, the neutrophils insert pseudopods into the junctions between endothelial cells, squeeze through these junctions, cross the basement membrane, and then are free to migrate in the extravascular space. Platelet-endothelial cell adhesion molecule-1 (PECAM-1) has been found to be localized at the junctions of endothelial cells and thus may have a role in transmigration. A variety of biomolecules have been found to be involved in activation of neutrophil and endothelial cells, including tumor necrosis factor α, various interleukins, platelet activating factor (PAF), leukotriene $B_4$, and certain complement fragments. These compounds stimulate various signaling pathways, resulting in changes in cell shape and function, and some are also chemotactic. One important functional change is recruitment of selectins to the cell surface, as in some cases selectins are stored in granules (eg, in endothelial cells and platelets).

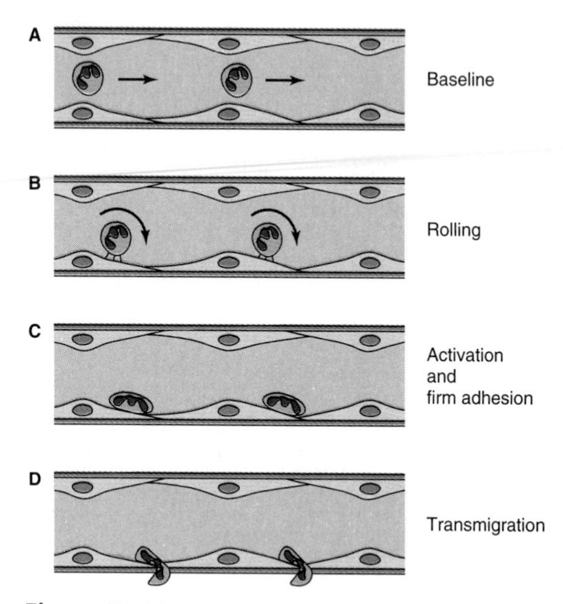


***Figure 47–11.*** Schematic diagram of neutrophil-endothelial cell interactions. **A:** Baseline conditions: Neutrophils do not adhere to the vessel wall. **B:** The first event is the slowing or rolling of the neutrophils within the vessel (venule) mediated by selectins. **C:** Activation occurs, resulting in neutrophils firmly adhering to the surfaces of endothelial cells and also assuming a flattened shape. This requires interaction of activated CD18 integrins on neutrophils with ICAM-1 on the endothelium. **D:** The neutrophils then migrate through the junctions of endothelial cells into the interstitial tissue; this requires involvement of PECAM-1. Chemotaxis is also involved in this latter stage. (Reproduced, with permission, from Albelda SM, Smith CW, Ward PA: Adhesion molecules and inflammatory injury. FASEB J 1994;8;504.)

The precise chemical nature of some of the ligands involved in selectin-ligand interactions has been determined. All three selectins bind **sialylated and fucosylated oligosaccharides,** and in particular all three bind **sialyl-Lewis$^X$** (Figure 47–12), a structure present on both glycoproteins and glycolipids. Whether this compound is the actual ligand involved in vivo is not established. Sulfated molecules, such as the sulfatides (Chapter 14), may be ligands in certain instances. This basic knowledge is being used in attempts to synthesize compounds that block selectin-ligand interactions and thus may inhibit the inflammatory response. Approaches include administration of specific monoclonal antibodies or of chemically synthesized analogs of sialyl-Lewis$^X$, both of which bind selectins. Cancer cells often exhibit sialyl-Lewis$^X$ and other selectin ligands on their surfaces. It is thought that these ligands play a role in the invasion and metastasis of cancer cells.

NeuAcα2 ⟶ 3Galβ1 ⟶ 4GlcNAc - - -

↑ α 1–3

Fuc

***Figure 47–12.*** Schematic representation of the structure of sialyl-Lewis$^X$.

## Abnormalities in the Synthesis of Glycoproteins Underlie Certain Diseases

Table 47–15 lists a number of conditions in which abnormalities in the synthesis of glycoproteins are of im-

***Table 47–15.*** Some diseases due to or involving abnormalities in the biosynthesis of glycoproteins.

| Disease | Abnormality |
|---|---|
| Cancer | Increased branching of cell surface glycans or presentation of selectin ligands may be important in metastasis. |
| Congenital disorders of glycosylation[1] | See Table 47–16. |
| HEMPAS[2] (MIM 224100) | Abnormalities in certain enzymes (eg, mannosidase II and others) involved in the biosynthesis of N-glycans, particularly affecting the red blood cell membrane. |
| Leukocyte adhesion deficiency, type II (MIM 266265) | Probably mutations affecting a Golgi-located GDP-fucose transporter, resulting in defective fucosylation. |
| Paroxysmal nocturnal hemoglobinuria (MIM 311770) | Acquired defect in biosynthesis of the GPI[3] structures of decay accelerating factor (DAF) and CD59. |
| I-cell disease (MIM 252500) | Deficiency of GlcNAc phosphotransferase, resulting in abnormal targeting of certain lysosomal enzymes. |

[1]The MIM number for congenital disorder of glycosylation type Ia is 212065.
[2]Hereditary erythroblastic multinuclearity with a positive acidified serum lysis test (congenital dyserythropoietic anemia type II). This is a relatively mild form of anemia. It reflects at least in part the presence in the red cell membranes of various glycoproteins with abnormal N-glycan chains, which contribute to the susceptibility to lysis.
[3]Glycosylphosphatidylinositol.

portance. As mentioned above, many **cancer cells** exhibit different profiles of oligosaccharide chains on their surfaces, some of which may contribute to metastasis. The **congenital disorders of glycosylation (CDG)** are a group of disorders of considerable current interest. The major features of these conditions are summarized in Table 47–16. **Leukocyte adhesion deficiency (LAD) II** is a rare condition probably due to mutations affecting the activity of a Golgi-located GDP-fucose transporter. It can be considered a congenital disorder of glycosylation. The absence of fucosylated ligands for selectins leads to a marked decrease in neutrophil rolling. Subjects suffer life-threatening, recurrent bacterial infections and also psychomotor and mental retardation. The condition appears to respond to oral fucose. **Hereditary erythroblastic multinuclearity with a positive acidified lysis test (HEMPAS)**—congenital dyserythropoietic anemia type II—is another disorder due to abnormalities in the processing of N-glycans. Some cases have been claimed to be due to defects in alpha–mannosidase II. I-cell disease is discussed further below. **Paroxysmal nocturnal hemoglobinuria** is an acquired mild anemia characterized by the presence of hemoglobin in urine due to hemolysis of red cells, particularly during sleep. This latter phenomenon may reflect a slight drop in plasma pH during sleep, which increases susceptibility to lysis by the complement system (Chapter 50). The basic defect in paroxysmal nocturnal hemoglobinuria is the acquisition of somatic mutations in the *PIG-A* (for phosphatidylinositol glycan class A) gene of certain hematopoietic cells. The product of this gene appears to be the enzyme that links glucosamine to phosphatidylinositol in the GPI structure (Figure 47–1). Thus, proteins that are anchored by a GPI linkage are deficient in the red cell membrane. Two proteins are of particular interest: **decay accelerating factor (DAF)** and another protein designated CD59. They normally interact with certain components of the complement system (Chapter 50) to prevent the hemolytic actions of the latter. However, when they are deficient, the complement system can act on the red cell membrane to cause hemolysis. Paroxysmal nocturnal hemoglobinuria can be diagnosed relatively simply, as the red cells are much more sensitive to hemolysis in normal serum acidified to pH 6.2 (Ham's test); the complement system is activated under these conditions, but normal cells are not affected. Figure 47–13 summarizes the etiology of paroxysmal nocturnal hemoglobinuria.

***Table 47–16.*** Major features of the congenital disorders of glycosylation.

- Autosomal recessive disorders
- Multisystem disorders that have probably not been recognized in the past
- Generally affect the central nervous system, resulting in psychomotor retardation and other features
- Type I disorders are due to mutations in genes encoding enzymes (eg, phosphomannomutase-2 [PMM-2], causing CDG Ia) involved in the synthesis of dolichol-P-P-oligosaccharide
- Type II disorders are due to mutations in genes encoding enzymes (eg, GlcNAc transferase-2, causing CDG IIa) involved in the processing of N-glycan chains
- About 11 distinct disorders have been recognized
- Isoelectric focusing of transferrin is a useful biochemical test for assisting in the diagnosis of these conditions; truncation of the oligosaccharide chains of this protein alters its isoelectric focusing pattern
- Oral mannose has proved of benefit in the treatment of CDG Ia

**Key:** CDG, congenital disorder of glycosylation.

## I-Cell Disease Results From Faulty Targeting of Lysosomal Enzymes

As indicated above, Man 6-P serves as a chemical marker to target certain lysosomal enzymes to that organelle. Analysis of cultured fibroblasts derived from patients with I-cell (inclusion cell) disease played a large part in revealing the above role of Man 6-P. I-cell disease is an uncommon condition characterized by severe progressive psychomotor retardation and a variety of physical signs, with death often occurring in the first decade. Cultured cells from patients with I-cell disease were found to lack almost all of the normal lysosomal enzymes; the lysosomes thus accumulate many different

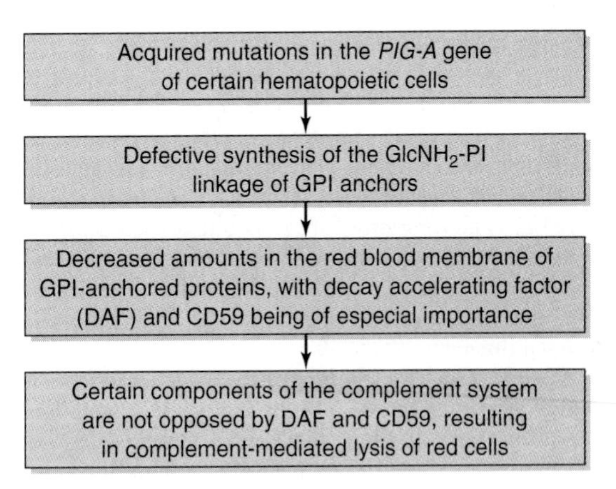


***Figure 47–13.*** Scheme of causation of paroxysmal nocturnal hemoglobinuria (MIM 311770).

types of undegraded molecules, forming inclusion bodies. Samples of plasma from patients with the disease were observed to contain very high activities of lysosomal enzymes; this suggested that the enzymes were being synthesized but were failing to reach their proper intracellular destination and were instead being secreted. Cultured cells from patients with the disease were noted to take up exogenously added lysosomal enzymes obtained from normal subjects, indicating that the cells contained a normal receptor on their surfaces for endocytic uptake of lysosomal enzymes. In addition, this finding suggested that lysosomal enzymes from patients with I-cell disease might lack a recognition marker. Further studies revealed that lysosomal enzymes from normal individuals carried the Man 6-P recognition marker described above, which interacted with a specific intracellular protein, the Man 6-P receptor. Cultured cells from patients with I-cell disease were then found to be deficient in the activity of the cis Golgi-located GlcNAc phosphotransferase, explaining how their lysosomal enzymes failed to acquire the Man 6-P marker. It is now known that there are two Man 6-P receptor proteins, one of high (275 kDa) and one of low (46 kDa) molecular mass. These proteins are lectins, recognizing Man 6-P. The former is cation-independent and also binds IGF-II (hence it is named the Man 6-P–IGF-II receptor), whereas the latter is cation-dependent in some species and does not bind IGF-II. It appears that both receptors function in the intracellular sorting of lysosomal enzymes into clathrin-coated vesicles, which occurs in the trans Golgi subsequent to synthesis of Man 6-P in the cis Golgi. These vesicles then leave the Golgi and fuse with a prelysosomal compartment. The low pH in this compartment causes the lysosomal enzymes to dissociate from their receptors and subsequently enter into lysosomes. The receptors are recycled and reused. Only the smaller receptor functions in the endocytosis of extracellular lysosomal enzymes, which is a minor pathway for lysosomal location. Not all cells employ the Man 6-P receptor to target their lysosomal enzymes (eg, hepatocytes use a different but undefined pathway); furthermore, not all lysosomal enzymes are targeted by this mechanism. Thus, biochemical investigations of I-cell disease not only led to elucidation of its basis but also contributed significantly to knowledge of how newly synthesized proteins are targeted to specific organelles, in this case the lysosome. Figure 47–14 summarizes the causation of I-cell disease.

**Pseudo-Hurler polydystrophy** is another genetic disease closely related to I-cell disease. It is a milder condition, and patients may survive to adulthood. Studies have revealed that the GlcNAc phosphotransferase involved in I-cell disease has several domains, including a catalytic domain and a domain that specifically recognizes and interacts with lysosomal enzymes. It has been proposed that the defect in pseudo-Hurler polydystrophy lies in the latter domain, and the retention of some catalytic activity results in a milder condition.

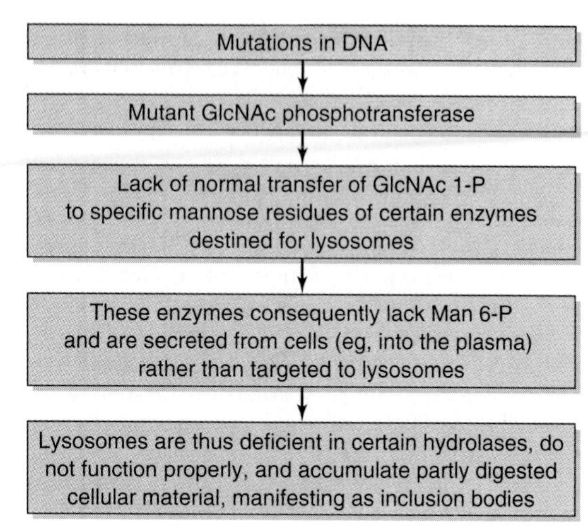


***Figure 47–14.*** Summary of the causation of I-cell disease (MIM 252500).

## Genetic Deficiencies of Glycoprotein Lysosomal Hydrolases Cause Diseases Such as α-Mannosidosis

Glycoproteins, like most other biomolecules, undergo both synthesis and degradation (ie, turnover). Degradation of the oligosaccharide chains of glycoproteins involves a battery of lysosomal hydrolases, including α-neuraminidase, β-galactosidase, β-hexosaminidase, α- and β-mannosidases, α-*N*-acetylgalactosaminidase, α-fucosidase, endo-β-*N*-acetylglucosaminidase, and aspartylglucosaminidase. The sites of action of the last two enzymes are indicated in the legend to Figure 47–5. Genetically determined defects of the activities of these enzymes can occur, resulting in abnormal degradation of glycoproteins. The accumulation in tissues of such abnormally degraded glycoproteins can lead to various diseases. Among the best-recognized of these diseases are mannosidosis, fucosidosis, sialidosis, aspartylglycosaminuria, and Schindler disease, due respectively to deficiencies of α-mannosidase, α-fucosidase, α-neuraminidase, aspartylglucosaminidase, and α-*N*-acetyl-galactosaminidase. These diseases, which are relatively uncommon, have a variety of manifestations; some of their major features are listed in Table 47–17. The fact that patients affected by these disorders all show signs referable to the central nervous sys-

***Table 47–17.*** Major features of some diseases (eg, α-mannosidosis, β-mannosidosis, fucosidosis, sialidosis, aspartylglycosaminuria, and Schindler disease) due to deficiencies of glycoprotein hydrolases.[1]

- Usually exhibit mental retardation or other neurologic abnormalities, and in some disorders coarse features or visceromegaly (or both)
- Variations in severity from mild to rapidly progressive
- Autosomal recessive inheritance
- May show ethnic distribution (eg, aspartylglycosaminuria is common in Finland)
- Vacuolization of cells observed by microscopy in some disorders
- Presence of abnormal degradation products (eg, oligosaccharides that accumulate because of the enzyme deficiency) in urine, detectable by TLC and characterizable by GLC-MS
- Definitive diagnosis made by assay of appropriate enzyme, often using leukocytes
- Possibility of prenatal diagnosis by appropriate enzyme assays
- No definitive treatment at present

[1]MIM numbers: α-mannosidosis, 248500; β-mannosidosis, 248510; fucosidosis, 230000; sialidosis, 256550; aspartylglycosaminuria, 208400; Schindler disease, 104170.

tem reflects the importance of glycoproteins in the development and normal function of that system.

From the above, it should be apparent that glycoproteins are involved in a wide variety of biologic processes and diseases. Glycoproteins play direct or indirect roles in a number of other diseases, as shown in the following examples.

(1) The **influenza virus** possesses a neuraminidase that plays a key role in elution of newly synthesized progeny from infected cells. If this process is inhibited, spread of the virus is markedly diminished. Inhibitors of this enzyme are now available for use in treating patients with influenza.

(2) **HIV-1,** thought by many to be the causative agent of AIDS, attaches to cells via one of its surface glycoproteins, gp120.

(3) **Rheumatoid arthritis** is associated with an alteration in the glycosylation of circulating immunoglobulin-γ (IgG) molecules (Chapter 50), such that they lack galactose in their Fc regions and terminate in GlcNAc. **Mannose-binding protein** (not to be confused with the mannose-6-P receptor), a C-lectin synthesized by liver cells and secreted into the circulation, binds mannose, GlcNAc, and certain other sugars. It can thus bind the agalactosyl IgG molecules, which subsequently activate the complement system, contributing to chronic inflammation in the synovial membranes of joints. This protein can also bind the above sugars when they are present on the surfaces of certain bacteria, fungi, and viruses, preparing these pathogens for opsonization or for destruction by the complement system. This is an example of innate immunity, not involving immunoglobulins. Deficiency of this protein in young infants, due to mutation, renders them very susceptible to **recurrent infections.**

Other disorders in which glycoproteins have been implicated include hepatitis B and C, Creutzfeldt-Jakob disease, and diarrheas due to a number of bacterial enterotoxins. It is hoped that basic studies of glycoproteins and other glycoconjugates (ie, the field of glycobiology) will lead to effective treatments for diseases in which these molecules are involved. Already, at least two disorders have been found to respond to oral supplements of sugars.

The fantastic progress made in relation to the human genome has stimulated intense interest in both **genomics** and **proteomics.** It is anticipated that the pace of research in **glycomics**—characterization of the entire complement of sugar chains found in cells (the glycome)—will also accelerate markedly. For a number of reasons, this field will prove more challenging than either genomics or proteomics. These reasons include the complexity of the structures of oligosaccharide chains due to linkage variations—in contrast to the generally uniform nature of the linkages between nucleotides and between amino acids. There are also significant variations in oligosaccharide structures among cells and at different stages of development. In addition, no simple technique exists for amplifying oligosaccharides, comparable to the PCR reaction. Despite these and other problems, it seems certain that research in this area will uncover many new important biologic interactions that are sugar-dependent and will provide targets for drug and other therapies.

## SUMMARY

- Glycoproteins are widely distributed proteins—with diverse functions—that contain one or more covalently linked carbohydrate chains.
- The carbohydrate components of a glycoprotein range from 1% to more than 85% of its weight and may be simple or very complex in structure.
- At least certain of the oligosaccharide chains of glycoproteins encode biologic information; they are also important to glycoproteins in modulating their solubility and viscosity, in protecting them against proteolysis, and in their biologic actions.

- The structures of many oligosaccharide chains can be elucidated by gas-liquid chromatography, mass spectrometry, and high-resolution NMR spectrometry.
- Glycosidases hydrolyze specific linkages in oligosaccharides and are used to explore both the structures and functions of glycoproteins.
- Lectins are carbohydrate-binding proteins involved in cell adhesion and other biologic processes.
- The major classes of glycoproteins are O-linked (involving an OH of serine or threonine), N-linked (involving the N of the amide group of asparagine), and glycosylphosphatidylinositol (GPI)-linked.
- Mucins are a class of O-linked glycoproteins that are distributed on the surfaces of epithelial cells of the respiratory, gastrointestinal, and reproductive tracts.
- The Golgi apparatus plays a major role in glycosylation reactions involved in the biosynthesis of glycoproteins.
- The oligosaccharide chains of O-linked glycoproteins are synthesized by the stepwise addition of sugars donated by nucleotide sugars in reactions catalyzed by individual specific glycoprotein glycosyltransferases.
- In contrast, the biosynthesis of N-linked glycoproteins involves a specific dolichol-P-P-oligosaccharide and various glycosidases. Depending on the glycosidases and precursor proteins synthesized by a tissue, it can synthesize complex, hybrid, or high-mannose types of N-linked oligosaccharides.
- Glycoproteins are implicated in many biologic processes. For instance, they have been found to play key roles in fertilization and inflammation.
- A number of diseases involving abnormalities in the synthesis and degradation of glycoproteins have been recognized. Glycoproteins are also involved in many other diseases, including influenza, AIDS, and rheumatoid arthritis.
- Developments in the new field of glycomics are likely to provide much new information on the roles of sugars in health and disease and also indicate targets for drug and other types of therapies.

## REFERENCES


Brockhausen I, Kuhns W: *Glycoproteins and Human Disease.* Chapman & Hall, 1997.

Kornfeld R, Kornfeld S: Assembly of asparagine-linked oligosaccharides. Annu Rev Biochem 1985;54:631.

Lehrman MA Oligosaccharide-based information in endoplasmic reticulum quality control and other biological systems. J Biol Chem. 2001;276:8623.

Perkel JM: Glycobiology goes to the ball. The Scientist 2002; 16:32.

Roseman S: Reflections on glycobiology. J Biol Chem 2001;276: 41527.

Schachter H: The clinical relevance of glycobiology. J Clin Invest 2001;108:1579.

Schwartz NB, Domowicz M: Chondrodysplasias due to proteoglycan defects. Glycobiology 2002;12:57R.

Science 2001;21(5512):2263. (This issue contains a special section entitled Carbohydrates and Glycobiology. It contains articles on the synthesis, structural determination, and functions of sugar-containing molecules and the roles of glycosylation in the immune system).

Scriver CR et al (editors): *The Metabolic and Molecular Bases of Inherited Disease,* 8th ed. McGraw-Hill, 2001.(Various chapters in this text give in-depth coverage of topics such as I-cell disease and disorders of glycoprotein degradation.)

Spiro RG: Protein glycosylation: nature, distribution, enzymatic formation, and disease implications of glycopeptide bonds. Glycobiology 2002;12:43R.

Varki A et al (editors): *Essentials of Glycobiology.* Cold Spring Harbor Laboratory Press, 1999.

Vestweber W, Blanks JE: Mechanisms that regulate the function of the selectins and their ligands. Physiol Rev 1999;79:181.

# The Extracellular Matrix

# 48

*Robert K. Murray, MD, PhD, & Frederick W. Keeley, PhD*

## BIOMEDICAL IMPORTANCE

Most mammalian cells are located in tissues where they are surrounded by a complex **extracellular matrix (ECM)** often referred to as **"connective tissue."** The ECM contains three major classes of biomolecules: (1) the **structural proteins,** collagen, elastin, and fibrillin; (2) certain **specialized proteins** such as fibrillin, fibronectin, and laminin; and (3) **proteoglycans,** whose chemical natures are described below. The ECM has been found to be involved in many normal and pathologic processes—eg, it plays important roles in development, in inflammatory states, and in the spread of cancer cells. Involvement of certain components of the ECM has been documented in both rheumatoid arthritis and **osteoarthritis.** Several diseases (eg, osteogenesis imperfecta and a number of types of the Ehlers-Danlos syndrome) are due to genetic disturbances of the synthesis of collagen. Specific components of proteoglycans (the glycosaminoglycans; GAGs) are affected in the group of genetic disorders known as the mucopolysaccharidoses. Changes occur in the ECM during the aging process. This chapter describes the basic biochemistry of the three major classes of biomolecules found in the ECM and illustrates their biomedical significance. Major biochemical features of two specialized forms of ECM—bone and cartilage—and of a number of diseases involving them are also briefly considered.

## COLLAGEN IS THE MOST ABUNDANT PROTEIN IN THE ANIMAL WORLD

Collagen, the major component of most connective tissues, constitutes approximately 25% of the protein of mammals. It provides an extracellular framework for all metazoan animals and exists in virtually every animal tissue. At least 19 distinct types of collagen made up of 30 distinct polypeptide chains (each encoded by a separate gene) have been identified in human tissues. Although several of these are present only in small proportions, they may play important roles in determining the physical properties of specific tissues. In addition, a number of proteins (eg, the C1q component of the complement system, pulmonary surfactant proteins SP-A and SP-D) that are not classified as collagens have collagen-like domains in their structures; these proteins are sometimes referred to as "noncollagen collagens."

Table 48–1 summarizes the types of collagens found in human tissues; the nomenclature used to designate types of collagen and their genes is described in the footnote.

The 19 types of collagen mentioned above can be subdivided into a number of classes based primarily on the structures they form (Table 48–2). In this chapter, we shall be primarily concerned with the fibril-forming collagens I and II, the major collagens of skin and bone and of cartilage, respectively. However, mention will be made of some of the other collagens.

## COLLAGEN TYPE I IS COMPOSED OF A TRIPLE HELIX STRUCTURE & FORMS FIBRILS

All collagen types have a **triple helical structure.** In some collagens, the entire molecule is triple helical, whereas in others the triple helix may involve only a fraction of the structure. Mature collagen type I, containing approximately 1000 amino acids, belongs to the former type; in it, each polypeptide subunit or alpha chain is twisted into a left-handed helix of three residues per turn (Figure 48–1). Three of these alpha chains are then wound into a right-handed superhelix, forming a rod-like molecule 1.4 nm in diameter and about 300 nm long. A striking characteristic of collagen is the occurrence of **glycine** residues at every third position of the triple helical portion of the alpha chain. This is necessary because glycine is the only amino acid small enough to be accommodated in the limited space available down the central core of the triple helix. This repeating structure, represented as $(Gly\text{-}X\text{-}Y)_n$, is an absolute requirement for the formation of the triple helix. While X and Y can be any other amino acids, about 100 of the X positions are proline and about 100 of the Y positions are hydroxyproline. Proline and hydroxyproline confer rigidity on the collagen molecule. **Hydroxyproline** is formed by the posttranslational hydroxylation of peptide-bound proline residues catalyzed by the enzyme **prolyl hydroxylase,** whose cofactors are **ascorbic acid** (vitamin C) and α-ketoglutarate. Lysines

***Table 48–1.*** Types of collagen and their genes.[1,2]

| Type | Genes | Tissue |
|---|---|---|
| I | *COL1A1, COL1A2* | Most connective tissues, including bone |
| II | *COL2A1* | Cartilage, vitreous humor |
| III | *COL3A1* | Extensible connective tissues such as skin, lung, and the vascular system |
| IV | *COL4A1–COL4A6* | Basement membranes |
| V | *COL5A1–COL5A3* | Minor component in tissues containing collagen I |
| VI | *COL6A1–COL6A3* | Most connective tissues |
| VII | *COL7A1* | Anchoring fibrils |
| VIII | *COL8A1–COL8A2* | Endothelium, other tissues |
| IX | *COL9A1–COL9A3* | Tissues containing collagen II |
| X | *COL10A1* | Hypertrophic cartilage |
| XI | *COL11A1, COL11A2, COL2A1* | Tissues containing collagen II |
| XII | *COL12A1* | Tissues containing collagen I |
| XIII | *COL13A1* | Many tissues |
| XIV | *COL14A1* | Tissues containing collagen I |
| XV | *COL15A1* | Many tissues |
| XVI | *COL16A1* | Many tissues |
| XVII | *COL17A1* | Skin hemidesmosomes |
| XVIII | *COL18A1* | Many tissues (eg, liver, kidney) |
| XIX | *COL19A1* | Rhabdomyosarcoma cells |

[1]Adapted slightly from Prockop DJ, Kivirrikko KI: Collagens: molecular biology, diseases, and potentials for therapy. Annu Rev Biochem 1995;64:403.

[2]The types of collagen are designated by Roman numerals. Constituent procollagen chains, called proα chains, are numbered using Arabic numerals, followed by the collagen type in parentheses. For instance, type I procollagen is assembled from two proα1(I) and one proα2(I) chain. It is thus a heterotrimer, whereas type 2 procollagen is assembled from three proα1(II) chains and is thus a homotrimer. The collagen genes are named according to the collagen type, written in Arabic numerals for the gene symbol, followed by an A and the number of the proα chain that they encode. Thus, the COL1A1 and COL1A2 genes encode the α1 and α2 chains of type I collagen, respectively.

***Table 48–2.*** Classification of collagens, based primarily on the structures that they form.[1]

| Class | Type |
|---|---|
| Fibril-forming | I, II, III, V, and XI |
| Network-like | IV, VIII, X |
| FACITs[2] | IX, XII, XIV, XVI, XIX |
| Beaded filaments | VI |
| Anchoring fibrils | VII |
| Transmembrane domain | XIII, XVII |
| Others | XV, XVIII |

[1]Based on Prockop DJ, Kivirrikko KI: Collagens: molecular biology, diseases, and potentials for therapy. Annu Rev Biochem 1995;64:403.

[2]FACITs = fibril-associated collagens with interrupted triple helices.

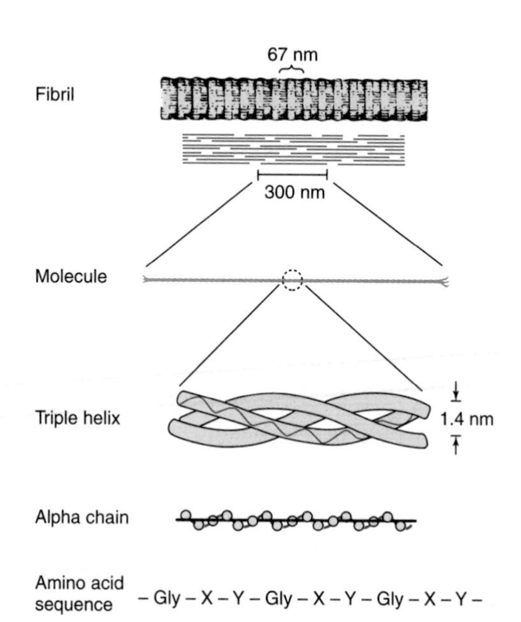


***Figure 48–1.*** **Molecular features of collagen structure from primary sequence up to the fibril.** (Slightly modified and reproduced, with permission, from Eyre DR: Collagen: Molecular diversity in the body's protein scaffold. Science 1980;207:1315. Copyright © 1980 by the American Association for the Advancement of Science.)

in the Y position may also be posttranslationally modified to hydroxylysine through the action of **lysyl hydroxylase,** an enzyme with similar cofactors. Some of these hydroxylysines may be further modified by the addition of galactose or galactosyl-glucose through an **O-glycosidic linkage,** a glycosylation site that is unique to collagen.

Collagen types that form long rod-like fibers in tissues are assembled by lateral association of these triple helical units into a "quarter staggered" alignment such that each is displaced longitudinally from its neighbor by slightly less than one-quarter of its length (Figure 48–1, upper part). This arrangement is responsible for the banded appearance of these fibers in connective tissues. Collagen fibers are further stabilized by the formation of **covalent cross-links,** both within and between the triple helical units. These cross-links form through the action of **lysyl oxidase,** a copper-dependent enzyme that oxidatively deaminates the ε-amino groups of certain lysine and hydroxylysine residues, yielding reactive aldehydes. Such aldehydes can form aldol condensation products with other lysine- or hydroxylysine-derived aldehydes or form Schiff bases with the ε-amino groups of unoxidized lysines or hydroxylysines. These reactions, after further chemical rearrangements, result in the stable covalent cross-links that are important for the tensile strength of the fibers. Histidine may also be involved in certain cross-links.

Several collagen types do not form fibrils in tissues (Table 48–2). They are characterized by interruptions of the triple helix with stretches of protein lacking Gly-X-Y repeat sequences. These non-Gly-X-Y sequences result in areas of globular structure interspersed in the triple helical structure.

**Type IV collagen,** the best-characterized example of a collagen with discontinuous triple helices, is an important component of **basement membranes,** where it forms a mesh-like network.

## Collagen Undergoes Extensive Posttranslational Modifications

Newly synthesized collagen undergoes extensive posttranslational modification before becoming part of a mature extracellular collagen fiber (Table 48–3). Like most secreted proteins, collagen is synthesized on ribosomes in a precursor form, **preprocollagen,** which contains a leader or signal sequence that directs the polypeptide chain into the lumen of the endoplasmic reticulum. As it enters the endoplasmic reticulum, this leader sequence is enzymatically removed. **Hydroxylation** of proline and lysine residues and **glycosylation** of hydroxylysines in the **procollagen** molecule also take place at this site. The procollagen molecule contains polypeptide extensions (extension peptides) of 20–35 kDa at both its amino and carboxyl terminal ends, neither of which is present in mature collagen. Both extension peptides contain cysteine residues. While the amino terminal propeptide forms only intrachain disulfide bonds, the carboxyl terminal propeptides form both intrachain and interchain disulfide bonds. Formation of these disulfide bonds assists in the registration of the three collagen molecules to form the triple helix, winding from the carboxyl terminal end. After formation of the triple helix, no further hydroxylation of proline or lysine or glycosylation of hydroxylysines can take place. **Self-assembly** is a cardinal principle in the biosynthesis of collagen.

Following **secretion** from the cell by way of the Golgi apparatus, extracellular enzymes called **procollagen aminoproteinase** and **procollagen carboxyproteinase** remove the extension peptides at the amino and carboxyl terminal ends, respectively. Cleavage of these propeptides may occur within crypts or folds in the cell membrane. Once the propeptides are removed, the triple helical collagen molecules, containing approximately 1000 amino acids per chain, spontaneously assemble into collagen fibers. These are further stabilized by the formation of inter- and intrachain cross-links through the action of lysyl oxidase, as described previously.

The same cells that secrete collagen also secrete **fibronectin,** a large glycoprotein present on cell surfaces, in the extracellular matrix, and in blood (see below). Fibronectin binds to aggregating precollagen fibers and alters the kinetics of fiber formation in the pericellular matrix. Associated with fibronectin and procollagen in

***Table 48–3.*** Order and location of processing of the fibrillar collagen precursor.

**Intracellular**

1. Cleavage of signal peptide
2. Hydroxylation of prolyl residues and some lysyl residues; glycosylation of some hydroxylysyl residues
3. Formation of intrachain and interchain S–S bonds in extension peptides
4. Formation of triple helix

**Extracellular**

1. Cleavage of amino and carboxyl terminal propeptides
2. Assembly of collagen fibers in quarter-staggered alignment
3. Oxidative deamination of ε-amino groups of lysyl and hydroxylysyl residues to aldehydes
4. Formation of intra- and interchain cross-links via Schiff bases and aldol condensation products

this matrix are the **proteoglycans** heparan sulfate and chondroitin sulfate (see below). In fact, type IX collagen, a minor collagen type from cartilage, contains attached proteoglycan chains. Such interactions may serve to regulate the formation of collagen fibers and to determine their orientation in tissues.

Once formed, collagen is relatively metabolically stable. However, its breakdown is increased during starvation and various inflammatory states. Excessive production of collagen occurs in a number of conditions, eg, hepatic cirrhosis.

## A Number of Genetic Diseases Result From Abnormalities in the Synthesis of Collagen

About 30 genes encode collagen, and its pathway of biosynthesis is complex, involving at least eight enzyme-catalyzed posttranslational steps. Thus, it is not surprising that a number of diseases (Table 48–4) are due to **mutations in collagen genes** or in **genes encoding some of the enzymes** involved in these posttranslational modifications. The diseases affecting bone (eg, osteogenesis imperfecta) and cartilage (eg, the chondrodysplasias) will be discussed later in this chapter.

**Ehlers-Danlos syndrome** comprises a group of inherited disorders whose principal clinical features are hyperextensibility of the skin, abnormal tissue fragility, and increased joint mobility. The clinical picture is variable, reflecting underlying extensive genetic heterogeneity. At least 10 types have been recognized, most but not all of which reflect a variety of lesions in the synthesis of collagen. **Type IV** is the most serious because of its tendency for spontaneous rupture of arteries or the bowel, reflecting abnormalities in type III collagen. Patients with **type VI,** due to a deficiency of lysyl hydroxylase, exhibit marked joint hypermobility and a tendency to ocular rupture. A deficiency of procollagen N-proteinase, causing formation of abnormal thin, irregular collagen fibrils, results in **type VIIC,** manifested by marked joint hypermobility and soft skin.

**Alport syndrome** is the designation applied to a number of genetic disorders (both X-linked and autosomal) affecting the structure of type IV collagen fibers, the major collagen found in the basement membranes of the renal glomeruli (see discussion of laminin, below). Mutations in several genes encoding type IV collagen fibers have been demonstrated. The presenting sign is hematuria, and patients may eventually develop end-stage renal disease. Electron microscopy reveals characteristic abnormalities of the structure of the basement membrane and lamina densa.

In **epidermolysis bullosa,** the skin breaks and blisters as a result of minor trauma. The dystrophic form is due to mutations in *COL7A1,* affecting the structure of type VII collagen. This collagen forms delicate fibrils that anchor the basal lamina to collagen fibrils in the dermis. These anchoring fibrils have been shown to be markedly reduced in this form of the disease, probably resulting in the blistering. Epidermolysis bullosa simplex, another variant, is due to mutations in keratin 5 (Chapter 49).

**Scurvy** affects the structure of collagen. However, it is due to a deficiency of ascorbic acid (Chapter 45) and is not a genetic disease. Its major signs are bleeding

***Table 48–4.*** Diseases caused by mutations in collagen genes or by deficiencies in the activities of posttranslational enzymes involved in the biosynthesis of collagen.[1]

| Gene or Enzyme | Disease[2] |
|---|---|
| *COL1A1, COL1A2* | Osteogenesis imperfecta, type 1[3] (MIM 1566200)<br>Osteoporosis[4] (MIM 166710)<br>Ehlers-Danlos syndrome type VII autosomal dominant (130060) |
| *COL2A1* | Severe chondrodysplasias<br>Osteoarthritis[4] (MIM 120140) |
| *COL3A1* | Ehlers-Danlos syndrome type IV (MIM 130050) |
| *COL4A3–COL4A6* | Alport syndrome (including both autosomal and X-linked forms) (MIM 104200) |
| *COL7A1* | Epidermolysis bullosa, dystrophic (MIM 131750) |
| *COL10A1* | Schmid metaphysial chondrodysplasia (MIM 156500) |
| Lysyl hydroxylase | Ehlers-Danlos syndrome type VI (MIM 225400) |
| Procollagen *N*-proteinase | Ehlers-Danlos syndrome type VII autosomal recessive (MIM 225410) |
| Lysyl hydroxylase | Menkes disease[5] (MIM 309400) |

[1]Adapted from Prockop DJ, Kivirrikko KI: Collagens: molecular biology, diseases, and potentials for therapy. Annu Rev Biochem 1995;64:403.
[2]Genetic linkage to collagen genes has been shown for a few other conditions not listed here.
[3]At least four types of osteogenesis imperfecta are recognized; the great majority of mutations in all types are in the *COL1A1* and *COL1A2* genes.
[4]At present applies to only a relatively small number of such patients.
[5]Secondary to a deficiency of copper (Chapter 50).

gums, subcutaneous hemorrhages, and poor wound healing. These signs reflect impaired synthesis of collagen due to deficiencies of prolyl and lysyl hydroxylases, both of which require ascorbic acid as a cofactor.

## ELASTIN CONFERS EXTENSIBILITY & RECOIL ON LUNG, BLOOD VESSELS, & LIGAMENTS

Elastin is a connective tissue protein that is responsible for properties of extensibility and elastic recoil in tissues. Although not as widespread as collagen, elastin is present in large amounts, particularly in tissues that require these physical properties, eg, lung, large arterial blood vessels, and some elastic ligaments. Smaller quantities of elastin are also found in skin, ear cartilage, and several other tissues. In contrast to collagen, there appears to be only one genetic type of elastin, although variants arise by alternative splicing (Chapter 37) of the hnRNA for elastin. Elastin is synthesized as a soluble monomer of 70 kDa called **tropoelastin.** Some of the prolines of tropoelastin are hydroxylated to **hydroxyproline** by prolyl hydroxylase, though hydroxylysine and glycosylated hydroxylysine are not present. Unlike collagen, tropoelastin is not synthesized in a pro- form with extension peptides. Furthermore, elastin does not contain repeat Gly-X-Y sequences, triple helical structure, or carbohydrate moieties.

After secretion from the cell, certain lysyl residues of tropoelastin are oxidatively deaminated to aldehydes by **lysyl oxidase,** the same enzyme involved in this process in collagen. However, the major cross-links formed in elastin are the **desmosines,** which result from the condensation of three of these lysine-derived aldehydes with an unmodified lysine to form a tetrafunctional cross-link unique to elastin. Once cross-linked in its mature, extracellular form, elastin is highly insoluble and extremely stable and has a very low turnover rate. Elastin exhibits a variety of random coil conformations that permit the protein to stretch and subsequently recoil during the performance of its physiologic functions.

Table 48–5 summarizes the main differences between collagen and elastin.

Deletions in the elastin gene (located at 7q11.23) have been found in approximately 90% of subjects with **Williams syndrome,** a developmental disorder affecting connective tissue and the central nervous system. The mutations, by affecting synthesis of elastin, probably play a causative role in the supravalvular aortic stenosis often found in this condition. A number of skin diseases (eg, scleroderma) are associated with accumulation of elastin. Fragmentation or, alternatively, a decrease of elastin is found in conditions such as pulmonary emphysema, cutis laxa, and aging of the skin.

***Table 48–5.*** Major differences between collagen and elastin.

| Collagen | Elastin |
|---|---|
| 1. Many different genetic types | One genetic type |
| 2. Triple helix | No triple helix; random coil conformations permitting stretching |
| 3. $(Gly\text{-}X\text{-}Y)_n$ repeating structure | No $(Gly\text{-}X\text{-}Y)_n$ repeating structure |
| 4. Presence of hydroxylysine | No hydroxylysine |
| 5. Carbohydrate-containing | No carbohydrate |
| 6. Intramolecular aldol cross-links | Intramolecular desmosine cross-links |
| 7. Presence of extension peptides during biosynthesis | No extension peptides present during biosynthesis |

## MARFAN SYNDROME IS DUE TO MUTATIONS IN THE GENE FOR FIBRILLIN, A PROTEIN PRESENT IN MICROFIBRILS

Marfan syndrome is a relatively prevalent inherited disease affecting connective tissue; it is inherited as an autosomal dominant trait. It affects the eyes (eg, causing dislocation of the lens, known as ectopia lentis), the skeletal system (most patients are tall and exhibit long digits [arachnodactyly] and hyperextensibility of the joints), and the cardiovascular system (eg, causing weakness of the aortic media, leading to dilation of the ascending aorta). Abraham Lincoln may have had this condition. Most cases are caused by mutations in the gene (on chromosome 15) for fibrillin; missense mutations have been detected in several patients with Marfan syndrome.

**Fibrillin** is a large glycoprotein (about 350 kDa) that is a structural component of microfibrils, 10- to 12-nm fibers found in many tissues. Fibrillin is secreted (subsequent to a proteolytic cleavage) into the extracellular matrix by fibroblasts and becomes incorporated into the insoluble microfibrils, which appear to provide a scaffold for deposition of elastin. Of special relevance to Marfan syndrome, fibrillin is found in the zonular fibers of the lens, in the periosteum, and associated with elastin fibers in the aorta (and elsewhere); these locations respectively explain the ectopia lentis, arachnodactyly, and cardiovascular problems found in the syndrome. Other proteins (eg, emelin and two microfibril-associated proteins) are also present in these microfibrils, and it appears likely that abnormalities of them may cause other connective tissue disorders. Another gene for fibrillin exists on chromosome 5; mutations in this gene are linked to causation of congenital contractural arachnodactyly but not to Marfan syndrome. The probable sequence of events leading to Marfan syndrome is summarized in Figure 48–2.

## FIBRONECTIN IS AN IMPORTANT GLYCOPROTEIN INVOLVED IN CELL ADHESION & MIGRATION

Fibronectin is a major glycoprotein of the extracellular matrix, also found in a soluble form in plasma. It consists of two identical subunits, each of about 230 kDa, joined by two disulfide bridges near their carboxyl terminals. The gene encoding fibronectin is very large, containing some 50 exons; the RNA produced by its transcription is subject to considerable alternative splicing, and as many as 20 different mRNAs have been detected in various tissues. Fibronectin contains three types of repeating motifs (I, II, and III), which are organized into functional **domains** (at least seven); functions of these domains include binding **heparin** (see below) and fibrin, collagen, DNA, and cell surfaces (Figure 48–3). The amino acid sequence of the fibronectin receptor of fibroblasts has been derived, and the protein is a member of the transmembrane integrin class of proteins (Chapter 51). The integrins are heterodimers, containing various types of α and β polypeptide chains. Fibronectin contains an Arg-Gly-Asp (RGD) sequence that binds to the receptor. The RGD sequence is shared by a number of other proteins present in the ECM that bind to integrins present in cell surfaces. Synthetic peptides containing the RGD sequence inhibit the binding of fibronectin to cell surfaces. Figure 48–4 illustrates the interaction of collagen, fibronectin, and laminin, all major proteins of the ECM, with a typical cell (eg, fibroblast) present in the matrix.

The fibronectin receptor interacts indirectly with **actin** microfilaments (Chapter 49) present in the cytosol (Figure 48–5). A number of proteins, collectively known as **attachment proteins,** are involved; these include talin, vinculin, an actin-filament capping protein, and α-actinin. Talin interacts with the receptor and vinculin, whereas the latter two interact with actin. The interaction of fibronectin with its receptor provides one route whereby the exterior of the cell can communicate with the interior and thus affect cell behavior. Via the interaction with its cell receptor, fibronectin plays an important role in the adhesion of cells to the ECM. It is also involved in cell migration by providing a binding site for cells and thus helping them to steer their way through the ECM. The amount of fibronectin around many transformed cells is sharply reduced, partly explaining their faulty interaction with the ECM.

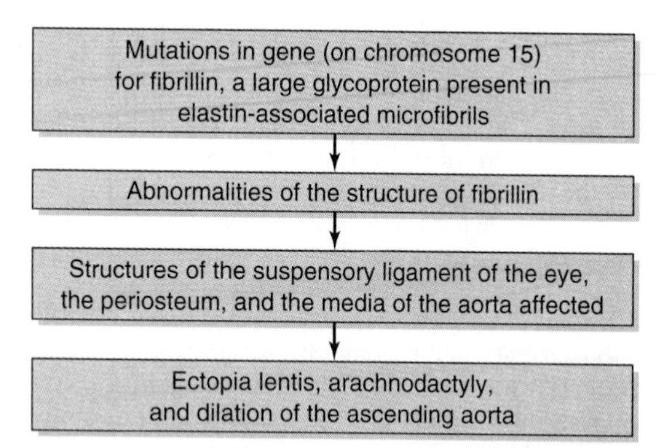


***Figure 48–2.*** Probable sequence of events in the causation of the major signs exhibited by patients with Marfan syndrome (MIM 154700).

## LAMININ IS A MAJOR PROTEIN COMPONENT OF RENAL GLOMERULAR & OTHER BASAL LAMINAS

Basal laminas are specialized areas of the ECM that surround epithelial and some other cells (eg, muscle cells); here we discuss only the laminas found in the **renal glomerulus.** In that structure, the basal lamina is contributed by two separate sheets of cells (one endothelial and one epithelial), each disposed on opposite sides of the lamina; these three layers make up the **glomerular membrane.** The primary components of the basal lamina are three proteins—laminin, entactin, and type IV collagen—and the GAG **heparin** or **heparan sulfate.** These components are synthesized by the underlying cells.

**Laminin** (about 850 kDa, 70 nm long) consists of three distinct elongated polypeptide chains (A, $B_1$, and $B_2$) linked together to form an elongated cruciform shape. It has binding sites for type IV collagen, heparin, and integrins on cell surfaces. The collagen interacts with laminin (rather than directly with the cell surface), which in turn interacts with integrins or other laminin receptor proteins, thus anchoring the lamina to the cells. **Entactin,** also known as "nidogen," is a glycoprotein containing an RGD sequence; it binds to laminin and is a major cell attachment factor. The relatively thick basal lamina of the renal glomerulus has an important role in **glomerular filtration,** regulating the passage of large molecules (most plasma proteins) across the glomerulus into the renal tubule. The glomerular membrane allows small molecules, such as **inulin** (5.2 kDa), to pass through as easily as water. On the other hand, only a small amount of the protein **albumin** (69

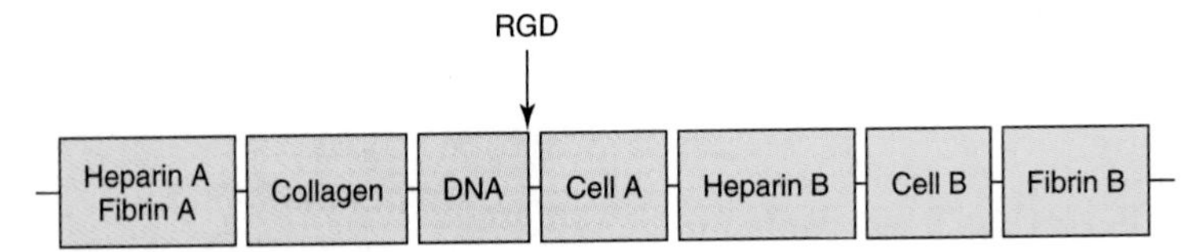


***Figure 48–3.*** Schematic representation of fibronectin. Seven functional domains of fibronectin are represented; two different types of domain for heparin, cell-binding, and fibrin are shown. The domains are composed of various combinations of three structural motifs (I, II, and III), not depicted in the figure. Also not shown is the fact that fibronectin is a dimer joined by disulfide bridges near the carboxyl terminals of the monomers. The approximate location of the RGD sequence of fibronectin, which interacts with a variety of fibronectin integrin receptors on cell surfaces, is indicated by the arrow. (Redrawn after Yamada KM: Adhesive recognition sequences. J Biol Chem 1991;266:12809.)

kDa), the major plasma protein, passes through the normal glomerulus. This is explained by two sets of facts: (1) The pores in the glomerular membrane are large enough to allow molecules up to about 8 nm to pass through. (2) Albumin is smaller than this pore size, but it is prevented from passing through easily by the negative charges of heparan sulfate and of certain sialic acid-containing glycoproteins present in the lamina. These negative charges repel albumin and most plasma proteins, which are negatively charged at the pH of blood. The normal structure of the glomerulus may be severely damaged in certain types of **glomerulonephritis** (eg, caused by antibodies directed against various components of the glomerular membrane). This alters the pores and the amounts and dispositions of the negatively charged macromolecules referred to above, and relatively massive amounts of albumin (and of certain

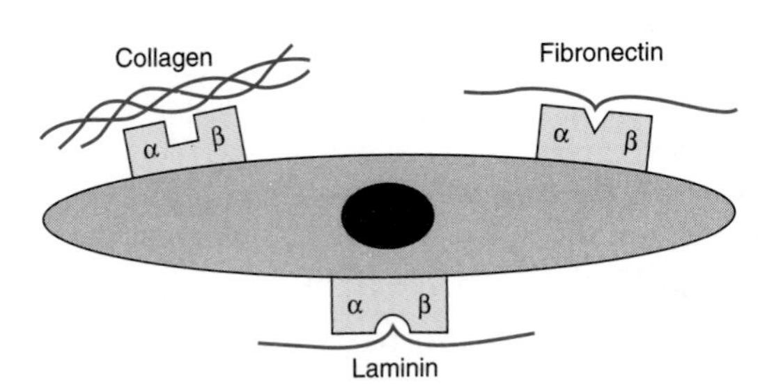


***Figure 48–4.*** Schematic representation of a cell interacting through various integrin receptors with collagen, fibronectin, and laminin present in the ECM. (Specific subunits are not indicated.) (Redrawn after Yamada KM: Adhesive recognition sequences. J Biol Chem 1991;266:12809.)

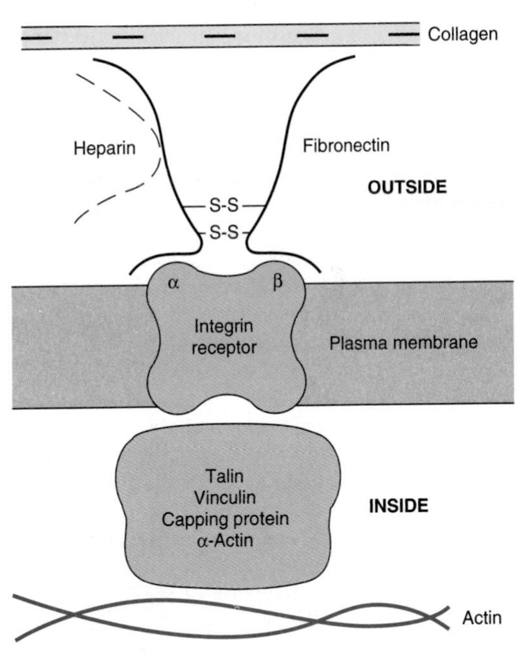


***Figure 48–5.*** Schematic representation of fibronectin interacting with an integrin fibronectin receptor situated in the exterior of the plasma membrane of a cell of the ECM and of various attachment proteins interacting indirectly or directly with an actin microfilament in the cytosol. For simplicity, the attachment proteins are represented as a complex.

other plasma proteins) can pass through into the urine, resulting in severe **albuminuria.**

## PROTEOGLYCANS & GLYCOSAMINOGLYCANS

### The Glycosaminoglycans Found in Proteoglycans Are Built Up of Repeating Disaccharides

**Proteoglycans** are proteins that contain covalently linked glycosaminoglycans. A number of them have been characterized and given names such as syndecan, betaglycan, serglycin, perlecan, aggrecan, versican, decorin, biglycan, and fibromodulin. They vary in tissue distribution, nature of the core protein, attached glycosaminoglycans, and function. The proteins bound covalently to glycosaminoglycans are called **"core proteins"**; they have proved difficult to isolate and characterize, but the use of recombinant DNA technology is beginning to yield important information about their structures. The amount of carbohydrate in a proteoglycan is usually much greater than is found in a glycoprotein and may comprise up to 95% of its weight. Figures 48–6 and 48–7 show the general structure of one particular proteoglycan, **aggrecan,** the major type found in cartilage. It is very large (about $2 \times 10^3$ kDa), with its overall structure resembling that of a bottle brush. It contains a long strand of hyaluronic acid (one type of GAG) to which link proteins are attached noncovalently. In turn, these latter interact noncovalently with core protein molecules from which chains of other GAGs (keratan sulfate and chondroitin sulfate in this case) project. More details on this macromolecule are given when cartilage is discussed below.

There are at least seven **glycosaminoglycans (GAGs):** hyaluronic acid, chondroitin sulfate, keratan sulfates I and II, heparin, heparan sulfate, and dermatan sulfate. A GAG is an unbranched polysaccharide made up of repeating disaccharides, one component of which is always an amino sugar (hence the name GAG), either D-glucosamine or D-galactosamine. The other component of the repeating disaccharide (except in the case of keratan sulfate) is a uronic acid, either L-glucuronic acid (GlcUA) or its 5′-epimer, L-iduronic acid (IdUA). With the exception of hyaluronic acid, all the GAGs contain sulfate groups, either as O-esters or as N-sulfate (in heparin and heparan sulfate). Hyaluronic acid affords another exception because there is no clear evidence that it is attached covalently to protein, as the definition of a proteoglycan given above specifies. Both GAGs and proteoglycans have proved difficult to work with, partly because of their complexity. However, they are major components of the ground substance; they have a number of important biologic roles; and they are involved in a number of disease processes—so that interest in them is increasing rapidly.

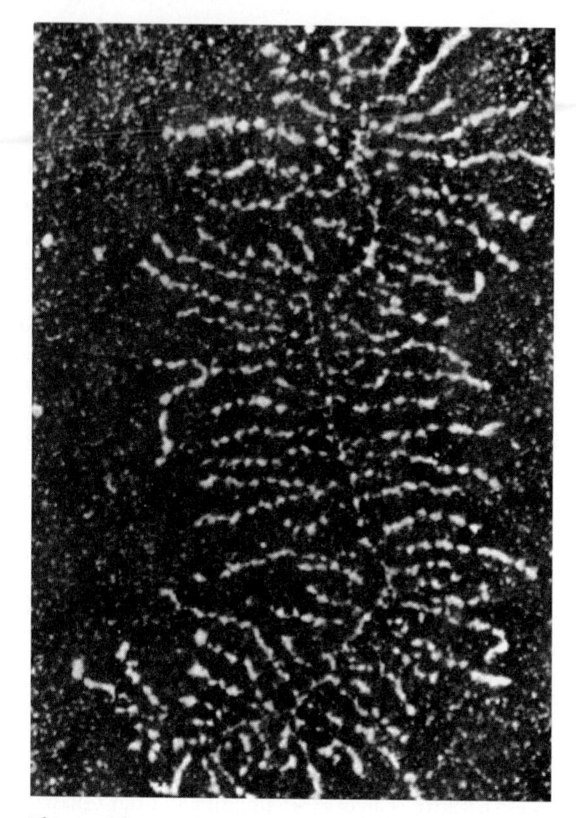

***Figure 48–6.*** Dark field electron micrograph of a proteoglycan aggregate in which the proteoglycan subunits and filamentous backbone are particularly well extended. (Reproduced, with permission, from Rosenberg L, Hellman W, Kleinschmidt AK: Electron microscopic studies of proteoglycan aggregates from bovine articular cartilage. J Biol Chem 1975;250:1877.)

### Biosynthesis of Glycosaminoglycans Involves Attachment to Core Proteins, Chain Elongation, & Chain Termination

#### A. Attachment to Core Proteins

The linkage between GAGs and their core proteins is generally one of three types.

**1.** An O-glycosidic bond between xylose (Xyl) and Ser, a bond that is unique to proteoglycans. This linkage is formed by transfer of a Xyl residue to Ser from UDP-xylose. Two residues of Gal are then added to the

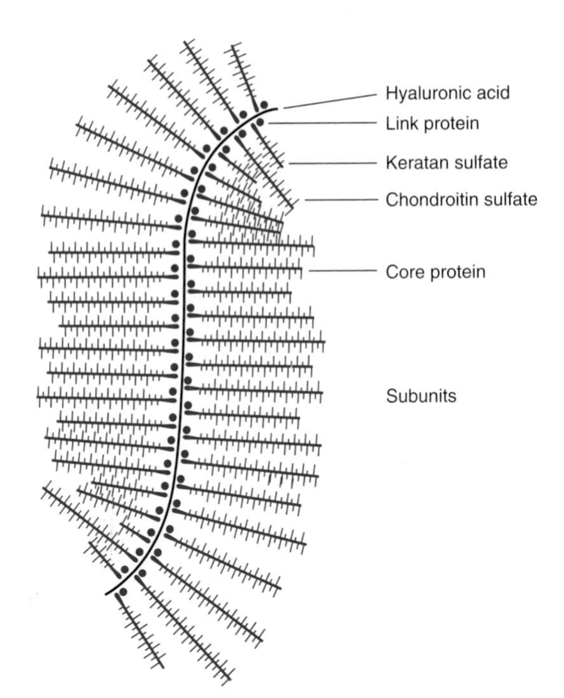


***Figure 48–7.*** Schematic representation of the proteoglycan aggrecan. (Reproduced, with permission, from Lennarz WJ: *The Biochemistry of Glycoproteins and Proteoglycans.* Plenum Press, 1980.)

Xyl residue, forming a **link trisaccharide,** Gal-Gal-Xyl-Ser. Further chain growth of the GAG occurs on the terminal Gal.

**2.** An O-glycosidic bond forms between GalNAc (*N*-acetylgalactosamine) and Ser (Thr) (Figure 47–1[a]), present in keratan sulfate II. This bond is formed by donation to Ser (or Thr) of a GalNAc residue, employing UDP-GalNAc as its donor.

**3.** An N-glycosylamine bond between GlcNAc (*N*-acetylglucosamine) and the amide nitrogen of **Asn,** as found in N-linked glycoproteins (Figure 47–1[b]). Its synthesis is believed to involve dolichol-P-P-oligosaccharide.

The synthesis of the core proteins occurs in the **endoplasmic reticulum,** and formation of at least some of the above linkages also occurs there. Most of the later steps in the biosynthesis of GAG chains and their subsequent modifications occur in the **Golgi apparatus.**

### B. Chain Elongation

Appropriate nucleotide sugars and highly specific Golgi-located glycosyltransferases are employed to synthesize the oligosaccharide chains of GAGs. The "one enzyme, one linkage" relationship appears to hold here, as in the case of certain types of linkages found in glycoproteins. The enzyme systems involved in chain elongation are capable of high-fidelity reproduction of complex GAGs.

### C. Chain Termination

This appears to result from (1) sulfation, particularly at certain positions of the sugars, and (2) the progression of the growing GAG chain away from the membrane site where catalysis occurs.

### D. Further Modifications

After formation of the GAG chain, numerous chemical modifications occur, such as the introduction of sulfate groups onto GalNAc and other moieties and the epimerization of GlcUA to IdUA residues. The enzymes catalyzing sulfation are designated **sulfotransferases** and use 3′-phosphoadenosine-5′-phosphosulfate (PAPS; active sulfate) as the sulfate donor. These Golgi-located enzymes are highly specific, and distinct enzymes catalyze sulfation at different positions (eg, carbons 2, 3, 4, and 6) on the acceptor sugars. An **epimerase** catalyzes conversions of glucuronyl to iduronyl residues.

## The Various Glycosaminoglycans Exhibit Differences in Structure & Have Characteristic Distributions

The seven GAGs named above differ from each other in a number of the following properties: amino sugar composition, uronic acid composition, linkages between these components, chain length of the disaccharides, the presence or absence of sulfate groups and their positions of attachment to the constituent sugars, the nature of the core proteins to which they are attached, the nature of the linkage to core protein, their tissue and subcellular distribution, and their biologic functions.

The structures (Figure 48–8) and the distributions of each of the GAGs will now be briefly discussed. The major features of the seven GAGs are summarized in Table 48–6.

### A. Hyaluronic Acid

Hyaluronic acid consists of an unbranched chain of repeating disaccharide units containing GlcUA and GlcNAc. Hyaluronic acid is present in bacteria and is widely distributed among various animals and tissues, including synovial fluid, the vitreous body of the eye, cartilage, and loose connective tissues.

### B. Chondroitin Sulfates (Chondroitin 4-Sulfate & Chondroitin 6-Sulfate)

Proteoglycans linked to chondroitin sulfate by the Xyl-Ser O-glycosidic bond are prominent components of **cartilage** (see below). The repeating disaccharide is similar to that found in hyaluronic acid, containing

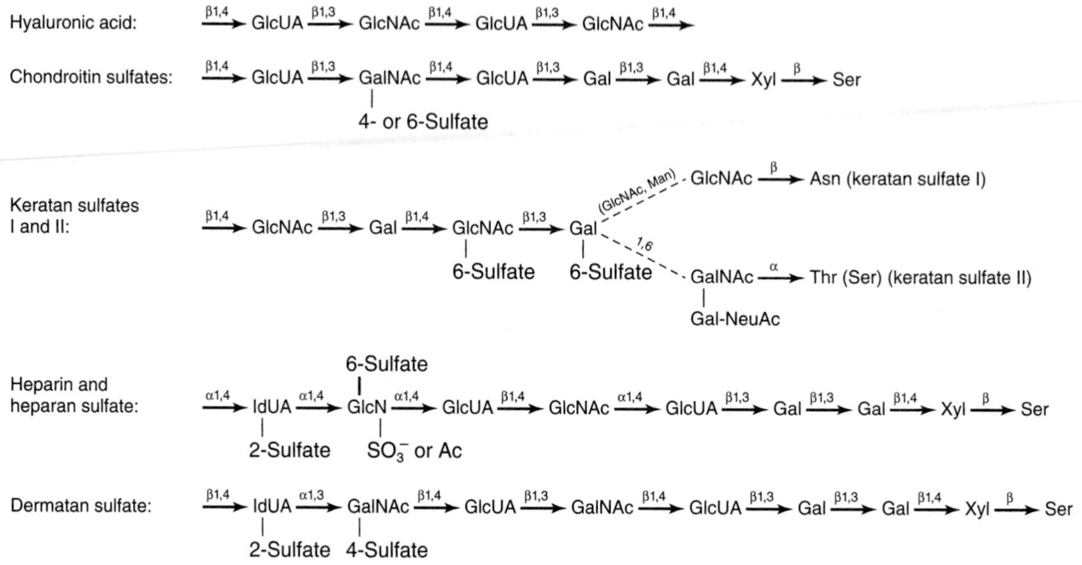

Hyaluronic acid: $\xrightarrow{\beta1,4}$ GlcUA $\xrightarrow{\beta1,3}$ GlcNAc $\xrightarrow{\beta1,4}$ GlcUA $\xrightarrow{\beta1,3}$ GlcNAc $\xrightarrow{\beta1,4}$

Chondroitin sulfates: $\xrightarrow{\beta1,4}$ GlcUA $\xrightarrow{\beta1,3}$ GalNAc $\xrightarrow{\beta1,4}$ GlcUA $\xrightarrow{\beta1,3}$ Gal $\xrightarrow{\beta1,3}$ Gal $\xrightarrow{\beta1,4}$ Xyl $\xrightarrow{\beta}$ Ser
(GalNAc: 4- or 6-Sulfate)

Keratan sulfates I and II: $\xrightarrow{\beta1,4}$ GlcNAc $\xrightarrow{\beta1,3}$ Gal $\xrightarrow{\beta1,4}$ GlcNAc $\xrightarrow{\beta1,3}$ Gal
(GlcNAc: 6-Sulfate; Gal: 6-Sulfate)
Gal (GlcNAc, Man) - GlcNAc $\xrightarrow{\beta}$ Asn (keratan sulfate I)
Gal 1,6 - GalNAc $\xrightarrow{\alpha}$ Thr (Ser) (keratan sulfate II)
(GalNAc: Gal-NeuAc)

Heparin and heparan sulfate: $\xrightarrow{\alpha1,4}$ IdUA $\xrightarrow{\alpha1,4}$ GlcN $\xrightarrow{\alpha1,4}$ GlcUA $\xrightarrow{\beta1,4}$ GlcNAc $\xrightarrow{\alpha1,4}$ GlcUA $\xrightarrow{\beta1,3}$ Gal $\xrightarrow{\beta1,3}$ Gal $\xrightarrow{\beta1,4}$ Xyl $\xrightarrow{\beta}$ Ser
(IdUA: 2-Sulfate; GlcN: 6-Sulfate, $SO_3^-$ or Ac)

Dermatan sulfate: $\xrightarrow{\beta1,4}$ IdUA $\xrightarrow{\alpha1,3}$ GalNAc $\xrightarrow{\beta1,4}$ GlcUA $\xrightarrow{\beta1,3}$ GalNAc $\xrightarrow{\beta1,4}$ GlcUA $\xrightarrow{\beta1,3}$ Gal $\xrightarrow{\beta1,3}$ Gal $\xrightarrow{\beta1,4}$ Xyl $\xrightarrow{\beta}$ Ser
(IdUA: 2-Sulfate; GalNAc: 4-Sulfate)

***Figure 48–8.*** Summary of structures of glycosaminoglycans and their attachments to core proteins. (GlcUA, D-glucuronic acid; IdUA, L-iduronic acid; GlcN, D-glucosamine; GalN, D-galactosamine; Ac, acetyl; Gal, D-galactose; Xyl, D-xylose; Ser, L-serine; Thr, L-threonine; Asn, L-asparagine; Man, D-mannose; NeuAc, *N*-acetylneuraminic acid.) The summary structures are qualitative representations only and do not reflect, for example, the uronic acid composition of hybrid glycosaminoglycans such as heparin and dermatan sulfate, which contain both L-iduronic and D-glucuronic acid. Neither should it be assumed that the indicated substituents are always present, eg, whereas most iduronic acid residues in heparin carry a 2′-sulfate group, a much smaller proportion of these residues are sulfated in dermatan sulfate. The presence of link trisaccharides (Gal-Gal-Xyl) in the chondroitin sulfates, heparin, and heparan and dermatan sulfates is shown. (Slightly modified and reproduced, with permission, from Lennarz WJ: *The Biochemistry of Glycoproteins and Proteoglycans.* Plenum Press, 1980.)

***Table 48–6.*** Major properties of the glycosaminoglycans.

| GAG | Sugars | Sulfate[1] | Linkage of Protein | Location |
|---|---|---|---|---|
| HA | GlcNAc, GlcUA | Nil | No firm evidence | Synovial fluid, vitreous humor, loose connective tissue |
| CS | GalNAc, GlcUA | GalNAc | Xyl-Ser; associated with HA via link proteins | Cartilage, bone, cornea |
| KS I | GlcNAc, Gal | GlcNAc | GlcNAc-Asn | Cornea |
| KS II | GlcNAc, Gal | Same as KS I | GalNAc-Thr | Loose connective tissue |
| Heparin | GlcN, IdUA | GlcN<br>GlcN<br>IdUA | Ser | Mast cells |
| Heparan sulfate | GlcN, GlcUA | GlcN | Xyl-Ser | Skin fibroblasts, aortic wall |
| Dermatan sulfate | GalNAc, IdUA, (GlcUA) | GalNAc<br>IdUa | Xyl-Ser | Wide distribution |

[1]The sulfate is attached to various positions of the sugars indicated (see Figure 48–7).

GlcUA but with GalNAc replacing GlcNAc. The GalNAc is substituted with sulfate at either its 4′ or its 6′ position, with approximately one sulfate being present per disaccharide unit.

#### C. Keratan Sulfates I & II

As shown in Figure 48–8, the keratan sulfates consist of repeating Gal-GlcNAc disaccharide units containing sulfate attached to the 6′ position of GlcNAc or occasionally of Gal. Type I is abundant in **cornea,** and type II is found along with chondroitin sulfate attached to hyaluronic acid in **loose connective tissue.** Types I and II have different attachments to protein (Figure 48–8).

#### D. Heparin

The repeating disaccharide contains **glucosamine** (GlcN) and either of the two uronic acids (Figure 48–9). Most of the amino groups of the GlcN residues are **N-sulfated,** but a few are acetylated. The GlcN also carries a $C_6$ sulfate ester.

Approximately 90% of the uronic acid residues are IdUA. Initially, all of the uronic acids are GlcUA, but a 5′-epimerase converts approximately 90% of the GlcUA residues to IdUA after the polysaccharide chain is formed. The protein molecule of the heparin proteoglycan is unique, consisting exclusively of serine and glycine residues. Approximately two-thirds of the serine residues contain GAG chains, usually of 5–15 kDa but occasionally much larger. Heparin is found in the granules of **mast cells** and also in liver, lung, and skin.

#### E. Heparan Sulfate

This molecule is present on many **cell surfaces** as a proteoglycan and is extracellular. It contains GlcN with fewer N-sulfates than heparin, and, unlike heparin, its predominant uronic acid is GlcUA.

#### F. Dermatan Sulfate

This substance is widely distributed in animal tissues. Its structure is similar to that of chondroitin sulfate, except that in place of a GlcUA in β-1,3 linkage to GalNAc it contains an IdUA in an α-1,3 linkage to GalNAc. Formation of the IdUA occurs, as in heparin and heparan sulfate, by 5′-epimerization of GlcUA. Because this is regulated by the degree of sulfation and because sulfation is incomplete, dermatan sulfate contains both IdUA-GalNAc and GlcUA-GalNAc disaccharides.

## Deficiencies of Enzymes That Degrade Glycosaminoglycans Result in Mucopolysaccharidoses

Both exo- and endoglycosidases degrade GAGs. Like most other biomolecules, GAGs are subject to turnover, being both synthesized and degraded. In adult tissues, GAGs generally exhibit relatively slow turnover, their half-lives being days to weeks.

Understanding of the degradative pathways for GAGs, as in the case of glycoproteins (Chapter 47) and glycosphingolipids (Chapter 24), has been greatly aided by elucidation of the specific enzyme deficiencies that occur in certain **inborn errors of metabolism.** When GAGs are involved, these inborn errors are called **mucopolysaccharidoses** (Table 48–7).

Degradation of GAGs is carried out by a battery of lysosomal hydrolases. These include certain endoglycosidases, various exoglycosidases, and sulfatases, generally acting in sequence to degrade the various GAGs. A number of them are indicated in Table 48–7.

The **mucopolysaccharidoses** share a common mechanism of causation, as illustrated in Figure 48–10. They are inherited in an autosomal recessive manner, with Hurler and Hunter syndromes being perhaps the most widely studied. None are common. In some cases, a family history of a mucopolysaccharidosis is obtained. Specific laboratory investigations of help in their diagnosis are urine testing for the presence of increased

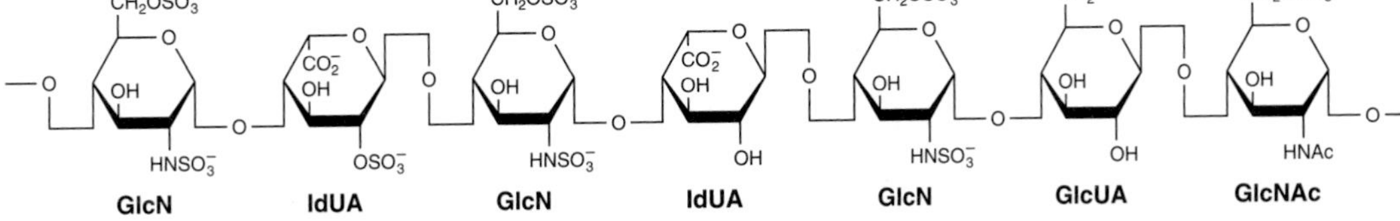

GlcN IdUA GlcN IdUA GlcN GlcUA GlcNAc

***Figure 48–9.*** Structure of heparin. The polymer section illustrates structural features typical of heparin; however, the sequence of variously substituted repeating disaccharide units has been arbitrarily selected. In addition, non-O-sulfated or 3-O-sulfated glucosamine residues may also occur. (Modified, redrawn, and reproduced, with permission, from Lindahl U et al: Structure and biosynthesis of heparin-like polysaccharides. Fed Proc 1977;36:19.)

***Table 48–7.*** Biochemical defects and diagnostic tests in mucopolysaccharidoses (MPS) and mucolipidoses (ML).[1]

| Name | Alternative Designation[2,3] | Enzymatic Defect | Urinary Metabolites |
|---|---|---|---|
| **Mucopolysaccharidoses** | | | |
| Hurler, Scheie, Hurler-Scheie (MIM 252800) | MPS I | α-L-Iduronidase | Dermatan sulfate, heparan sulfate |
| Hunter (MIM 309900) | MPS II | Iduronate sulfatase | Dermatan sulfate, heparan sulfate |
| Sanfilippo A (MIM 252900) | MPS IIIA | Heparan sulfate *N*-sulfatase (sulfamidase) | Heparan sulfate |
| Sanfilippo B (MIM 252920) | MPS IIIB | α-*N*-Acetylglucosaminidase | Heparan sulfate |
| Sanfilippo C (MIM 252930) | MPS IIIC | Acetyltransferase | Heparan sulfate |
| Sanfilippo D (MIM 252940) | MPS IIID | *N*-Acetylglucosamine 6-sulfatase | Heparan sulfate |
| Morquio A (MIM 253000) | MPS IVA | Galactosamine 6-sulfatase | Keratan sulfate, chondroitin 6-sulfate |
| Morquio B (MIM 253010) | MPS IVB | β-Galactosidase | Keratan sulfate |
| Maroteaux-Lamy (MIM 253200) | MPS VI | *N*-Acetylgalactosamine 4-sulfatase (arylsulfatase B) | Dermatan sulfate |
| Sly (MIM 253220) | MPS VII | β-Glucuronidase | Dermatan sulfate, heparan sulfate, chondroitin 4-sulfate, chondroitin 6-sulfate |
| **Mucolipidoses** | | | |
| Sialidosis (MIM 256550) | M LI | Sialidase (neuraminidase) | Glycoprotein fragments |
| I-cell disease (MIM 252500) | ML II | UDP-*N*-acetylglucosamine: glycoprotein *N*-acetylglu-cosamininylphosphotrans-ferase. (Acid hydrolases thus lack phosphoman-nosyl residues.) | Glycoprotein fragments |
| Pseudo-Hurler polydystrophy (MIM 252600) | ML III | As for ML II but deficiency is incomplete | Glycoprotein fragments |

[1]Modified and reproduced, with permission, from DiNatale P, Neufeld EF: The biochemical diagnosis of mucopolysaccharidoses, mucolipidoses and related disorders. In: *Perspectives in Inherited Metabolic Diseases,* vol 2. Barr B et al (editors). Editiones Ermes (Milan), 1979.
[2]Fibroblasts, leukocytes, tissues, amniotic fluid cells, or serum can be used for the assay of many of the above enzymes. Patients with these disorders exhibit a variety of clinical findings that may include cloudy corneas, mental retardation, stiff joints, cardiac abnormalities, hepatosplenomegaly, and short stature, depending on the specific disease and its severity.
[3]The term MPS V is no longer used. The existence of MPS VIII (suspected glucosamine 6-sulfatase deficiency: MIM 253230) has not been confirmed. At least one case of hyaluronidase deficiency (MPS IX; MIM 601492) has been reported.

amounts of GAGs and assays of suspected enzymes in white cells, fibroblasts, or sometimes in serum. In certain cases, a tissue biopsy is performed and the GAG that has accumulated can be determined by electrophoresis. DNA tests are increasingly available. Prenatal diagnosis can be made using amniotic cells or chorionic villus biopsy.

The term **"mucolipidosis"** was introduced to denote diseases that combined features common to both mucopolysaccharidoses and sphingolipidoses (Chapter 24). Three mucolipidoses are listed in Table 48–7. In **sialidosis** (mucolipidosis I, ML-I), various oligosaccharides derived from glycoproteins and certain gangliosides can accumulate in tissues. **I-cell disease** (ML-II)

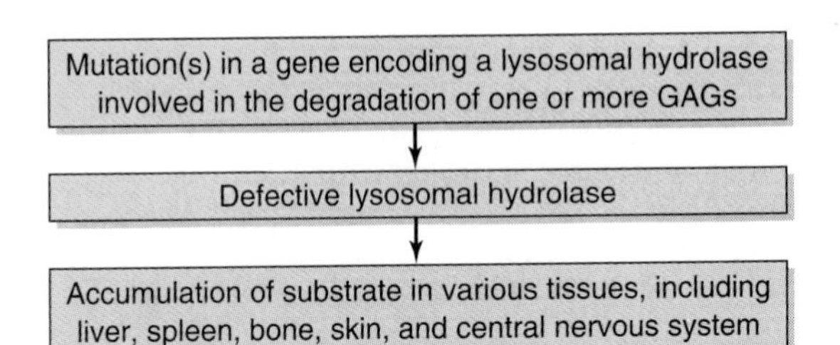


***Figure 48–10.*** Simplified scheme of causation of a mucopolysaccharidosis, such as Hurler syndrome (MIM 252800), in which the affected enzyme is α-L-iduronidase. Marked accumulation of the GAGs in the tissues mentioned in the figure could cause hepatomegaly, splenomegaly, disturbances of growth, coarse facies, and mental retardation, respectively.

and **pseudo-Hurler polydystrophy** (ML-III) are described in Chapter 47. The term "mucolipidosis" is retained because it is still in relatively widespread clinical usage, but it is not appropriate for these two latter diseases since the mechanism of their causation involves mislocation of certain lysosomal enzymes. Genetic defects of the catabolism of the oligosaccharide chains of glycoproteins (eg, mannosidosis, fucosidosis) are also described in Chapter 47. Most of these defects are characterized by increased excretion of various fragments of glycoproteins in the urine, which accumulate because of the metabolic block, as in the case of the mucolipidoses.

**Hyaluronidase** is one important enzyme involved in the catabolism of both hyaluronic acid and chondroitin sulfate. It is a widely distributed endoglycosidase that cleaves hexosaminidic linkages. From hyaluronic acid, the enzyme will generate a tetrasaccharide with the structure (GlcUA-β-1,3-GlcNAc-β-1,4)$_2$, which can be degraded further by a β-glucuronidase and β-*N*-acetylhexosaminidase. Surprisingly, only one case of an apparent genetic deficiency of this enzyme appears to have been reported.

## Proteoglycans Have Numerous Functions

As indicated above, proteoglycans are remarkably complex molecules and are found in every tissue of the body, mainly in the ECM or "ground substance." There they are associated with each other and also with the other major structural components of the matrix, collagen and elastin, in quite specific manners. Some proteoglycans bind to collagen and others to elastin. These interactions are important in determining the structural organization of the matrix. Some proteoglycans (eg, decorin) can also bind growth factors such as TGF-β, modulating their effects on cells. In addition, some of them interact with certain adhesive proteins such as fibronectin and laminin (see above), also found in the matrix. The GAGs present in the proteoglycans are polyanions and hence bind polycations and cations such as $Na^+$ and $K^+$. This latter ability attracts water by osmotic pressure into the extracellular matrix and contributes to its turgor. GAGs also gel at relatively low concentrations. Because of the long extended nature of the polysaccharide chains of GAGs and their ability to gel, the proteoglycans can act as sieves, restricting the passage of large macromolecules into the ECM but allowing relatively free diffusion of small molecules. Again, because of their extended structures and the huge macromolecular aggregates they often form, they occupy a large volume of the matrix relative to proteins.

### A. Some Functions of Specific GAGs & Proteoglycans

**Hyaluronic acid** is especially high in concentration in embryonic tissues and is thought to play an important role in permitting cell migration during morphogenesis and wound repair. Its ability to attract water into the extracellular matrix and thereby "loosen it up" may be important in this regard. The high concentrations of hyaluronic acid and chondroitin sulfates present in cartilage contribute to its compressibility (see below).

**Chondroitin sulfates** are located at sites of calcification in endochondral bone and are also found in cartilage. They are also located inside certain neurons and may provide an endoskeletal structure, helping to maintain their shape.

Both **keratan sulfate I** and **dermatan sulfate** are present in the cornea. They lie between collagen fibrils and play a critical role in corneal transparency. Changes in proteoglycan composition found in corneal scars disappear when the cornea heals. The presence of dermatan sulfate in the sclera may also play a role in maintaining the overall shape of the eye. Keratan sulfate I is also present in cartilage.

**Heparin** is an important anticoagulant. It binds with factors IX and XI, but its most important interaction is with **plasma antithrombin III** (discussed in Chapter 51). Heparin can also bind specifically to lipoprotein lipase present in capillary walls, causing a release of this enzyme into the circulation.

Certain proteoglycans (eg, **heparan sulfate**) are associated with the plasma membrane of cells, with their core proteins actually spanning that membrane. In it they may act as receptors and may also participate in the mediation of cell growth and cell-cell communication. The attachment of cells to their substratum in culture is mediated at least in part by heparan sulfate. This proteoglycan is also found in the basement membrane of the kidney along with type IV collagen and laminin (see above), where it plays a major role in determining the charge selectiveness of glomerular filtration.

Proteoglycans are also found in intracellular locations such as the nucleus; their function in this organelle has not been elucidated. They are present in some storage or secretory granules, such as the chromaffin granules of the adrenal medulla. It has been postulated that they play a role in release of the contents of such granules. The various functions of GAGs are summarized in Table 48–8.

### B. Associations With Major Diseases & With Aging

Hyaluronic acid may be important in permitting **tumor cells** to migrate through the ECM. Tumor cells can induce fibroblasts to synthesize greatly increased amounts of this GAG, thereby perhaps facilitating their own spread. Some tumor cells have less heparan sulfate at their surfaces, and this may play a role in the lack of adhesiveness that these cells display.

The intima of the **arterial wall** contains hyaluronic acid and chondroitin sulfate, dermatan sulfate, and heparan sulfate proteoglycans. Of these proteoglycans, dermatan sulfate binds plasma low-density lipoproteins. In addition, dermatan sulfate appears to be the major GAG synthesized by arterial smooth muscle cells. Because it is these cells that proliferate in atherosclerotic lesions in arteries, dermatan sulfate may play an important role in development of the atherosclerotic plaque.

**Table 48–8.** Some functions of glycosaminoglycans and proteoglycans.[1]

- Act as structural components of the ECM
- Have specific interactions with collagen, elastin, fibronectin, laminin, and other proteins such as growth factors
- As polyanions, bind polycations and cations
- Contribute to the characteristic turgor of various tissues
- Act as sieves in the ECM
- Facilitate cell migration (HA)
- Have role in compressibility of cartilage in weight-bearing (HA, CS)
- Play role in corneal transparency (KS I and DS)
- Have structural role in sclera (DS)
- Act as anticoagulant (heparin)
- Are components of plasma membranes, where they may act as receptors and participate in cell adhesion and cell-cell interactions (eg, HS)
- Determine charge-selectiveness of renal glomerulus (HS)
- Are components of synaptic and other vesicles (eg, HS)

[1]ECM, extracellular matrix; HA, hyaluronic acid; CS, chondroitin sulfate; KS I, keratan sulfate I; DS, dermatan sulfate; HS, heparan sulfate.

In various types of **arthritis,** proteoglycans may act as autoantigens, thus contributing to the pathologic features of these conditions. The amount of chondroitin sulfate in cartilage diminishes with age, whereas the amounts of keratan sulfate and hyaluronic acid increase. These changes may contribute to the development of **osteoarthritis.** Changes in the amounts of cer-

**Table 48–9.** The principal proteins found in bone.[1]

| Proteins | Comments |
|---|---|
| **Collagens** | |
| Collagen type I | Approximately 90% of total bone protein. Composed of two $\alpha$1(I) and one $\alpha$2(I) chains. |
| Collagen type V | Minor component. |
| **Noncollagen proteins** | |
| Plasma proteins | Mixture of various plasma proteins. |
| Proteoglycans[2] | |
| CS-PG I (biglycan) | Contains two GAG chains; found in other tissues. |
| CS-PG II (decorin) | Contains one GAG chain; found in other tissues. |
| CS-PG III | Bone-specific. |
| Bone SPARC[3] protein (osteonectin) | Not bone-specific. |
| Osteocalcin (bone Gla protein) | Contains $\gamma$-carboxyglutamate residues that bind to hydroxyapatite. Bone-specific. |
| Osteopontin | Not bone-specific. Glycosylated and phosphorylated. |
| Bone sialoprotein | Bone-specific. Heavily glycosylated, and sulfated on tyrosine. |
| Bone morphogenetic proteins (BMPs) | A family (eight or more) of secreted proteins with a variety of actions on bone; many induce ectopic bone growth. |

[1]Various functions have been ascribed to the noncollagen proteins, including roles in mineralization; however, most of them are still speculative. It is considered unlikely that the noncollagen proteins that are not bone-specific play a key role in mineralization. A number of other proteins are also present in bone, including a tyrosine-rich acidic matrix protein (TRAMP), some growth factors (eg, TGF$\beta$), and enzymes involved in collagen synthesis (eg, lysyl oxidase).
[2]CS-PG, chondroitin sulfate–proteoglycan; these are similar to the dermatan sulfate PGs (DS-PGs) of cartilage (Table 48–11).
[3]SPARC, secreted protein acidic and rich in cysteine.

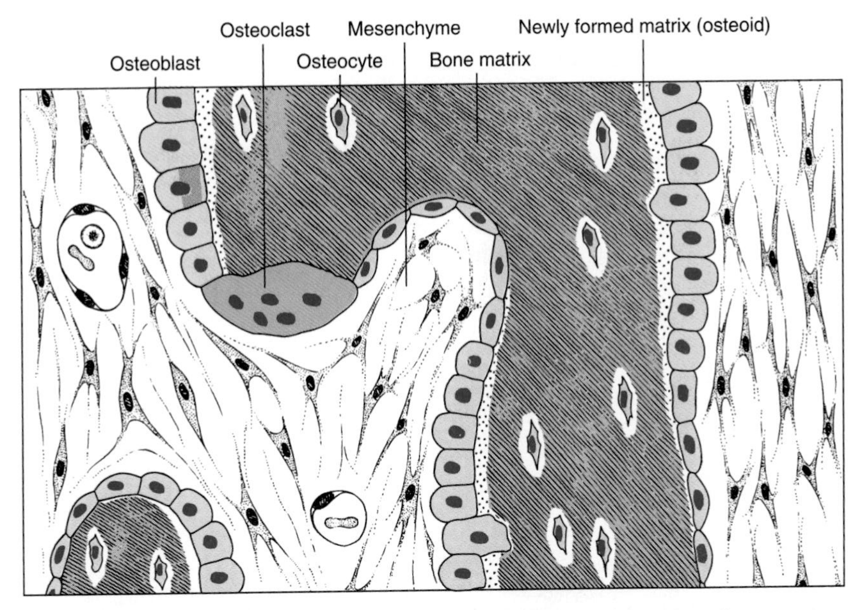


***Figure 48–11.*** Schematic illustration of the major cells present in membranous bone. Osteoblasts (lighter color) are synthesizing type I collagen, which forms a matrix that traps cells. As this occurs, osteoblasts gradually differentiate to become osteocytes. (Reproduced, with permission, from Junqueira LC, Carneiro J: *Basic Histology: Text & Atlas,* 10th ed. McGraw-Hill, 2003.)

tain GAGs in the skin are also observed with **aging** and help to account for the characteristic changes noted in this organ in the elderly.

An exciting new phase in proteoglycan research is opening up with the findings that mutations that affect individual proteoglycans or the enzymes needed for their synthesis alter the regulation of specific signaling pathways in drosophila and *Caenorhabditis elegans,* thus affecting development; it already seems likely that similar effects exist in mice and humans.

## BONE IS A MINERALIZED CONNECTIVE TISSUE

Bone contains both organic and inorganic material. The organic matter is mainly protein. The principal proteins of bone are listed in Table 48–9; **type I collagen** is the major protein, comprising 90–95% of the organic material. Type V collagen is also present in small amounts, as are a number of noncollagen proteins, some of which are relatively specific to bone. The inorganic or mineral component is mainly crystalline **hydroxyapatite**—$Ca_{10}(PO_4)_6(OH)_2$—along with sodium, magnesium, carbonate, and fluoride; approximately 99% of the body's calcium is contained in bone (Chapter 45). Hydroxyapatite confers on bone the strength and resilience required by its physiologic roles.

Bone is a dynamic structure that undergoes continuing cycles of remodeling, consisting of resorption followed by deposition of new bone tissue. This remodeling permits bone to adapt to both physical (eg, increases in weight-bearing) and hormonal signals.

The major cell types involved in bone resorption and deposition are **osteoclasts** and **osteoblasts** (Figure 48–11). The former are associated with resorption and the latter with deposition of bone. Osteocytes are descended from osteoblasts; they also appear to be involved in maintenance of bone matrix but will not be discussed further here.

Osteoclasts are multinucleated cells derived from pluripotent hematopoietic stem cells. Osteoclasts possess an apical membrane domain, exhibiting a ruffled border that plays a key role in bone resorption (Figure 48–12). A proton-translocating ATPase expels protons across the ruffled border into the resorption area, which is the microenvironment of low pH shown in the figure. This lowers the local pH to 4.0 or less, thus increasing the solubility of hydroxyapatite and allowing demineralization to occur. Lysosomal acid proteases are released that digest the now accessible matrix proteins.

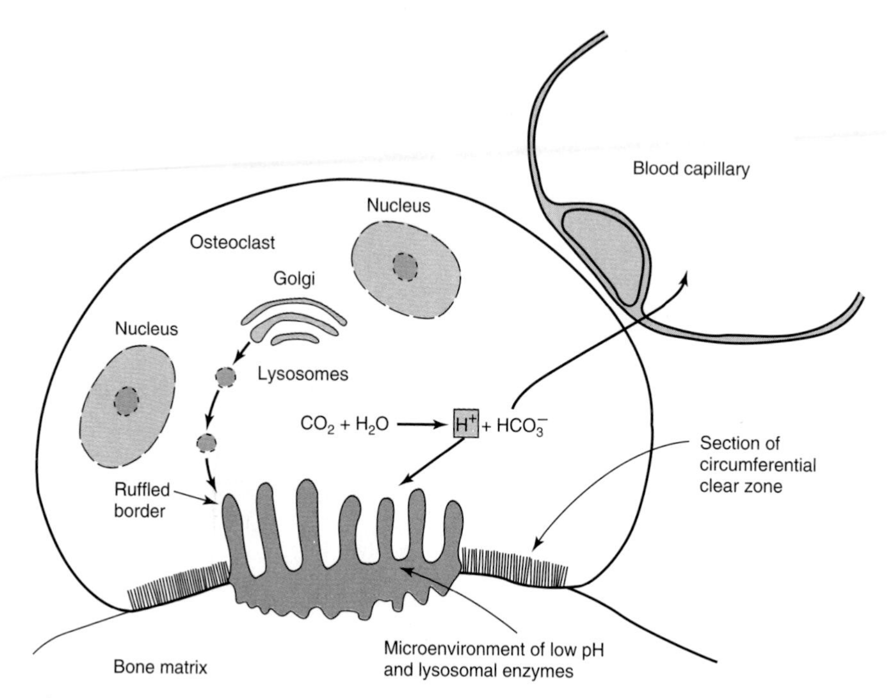


***Figure 48–12.*** Schematic illustration of some aspects of the role of the osteoclast in bone resorption. Lysosomal enzymes and hydrogen ions are released into the confined microenvironment created by the attachment between bone matrix and the peripheral clear zone of the osteoclast. The acidification of this confined space facilitates the dissolution of calcium phosphate from bone and is the optimal pH for the activity of lysosomal hydrolases. Bone matrix is thus removed, and the products of bone resorption are taken up into the cytoplasm of the osteoclast, probably digested further, and transferred into capillaries. The chemical equation shown in the figure refers to the action of carbonic anhydrase II, described in the text. (Reproduced, with permission, from Junqueira LC, Carneiro J: *Basic Histology: Text & Atlas,* 10th ed. McGraw-Hill, 2003.)

Osteoblasts—mononuclear cells derived from pluripotent mesenchymal precursors—synthesize most of the proteins found in bone (Table 48–9) as well as various growth factors and cytokines. They are responsible for the deposition of new bone matrix (osteoid) and its subsequent mineralization. Osteoblasts control mineralization by regulating the passage of calcium and phosphate ions across their surface membranes. The latter contain alkaline phosphatase, which is used to generate phosphate ions from organic phosphates. The mechanisms involved in mineralization are not fully understood, but several factors have been implicated. Alkaline phosphatase contributes to mineralization but in itself is not sufficient. Small vesicles (matrix vesicles) containing calcium and phosphate have been described at sites of mineralization, but their role is not clear. Type I collagen appears to be necessary, with mineralization being first evident in the gaps between successive molecules. Recent interest has focused on acidic phosphoproteins, such as bone sialoprotein, acting as sites of nucleation. These proteins contain motifs (eg, poly-Asp and poly-Glu stretches) that bind calcium and may provide an initial scaffold for mineralization. Some macromolecules, such as certain proteoglycans and glycoproteins, can also act as inhibitors of nucleation.

It is estimated that approximately 4% of compact bone is renewed annually in the typical healthy adult, whereas approximately 20% of trabecular bone is replaced.

Many factors are involved in the regulation of bone metabolism, only a few of which will be mentioned here. Some stimulate osteoblasts (eg, parathyroid hormone and 1,25-dihydroxycholecalciferol) and others inhibit them (eg, corticosteroids). Parathyroid hormone and 1,25-dihydroxycholecalciferol also stimulate osteoclasts, whereas calcitonin and estrogens inhibit them.

**Table 48–10.** Some metabolic and genetic diseases affecting bone and cartilage.

| Disease | Comments |
|---|---|
| Dwarfism | Often due to a deficiency of growth hormone, but has many other causes. |
| Rickets | Due to a deficiency of vitamin D during childhood. |
| Osteomalacia | Due to a deficiency of vitamin D during adulthood. |
| Hyperparathyroidism | Excess parathormone causes bone resorption. |
| Osteogenesis imperfecta (eg, MIM 166200) | Due to a variety of mutations in the *COL1A1* and *COL1A2* genes affecting the synthesis and structure of type I collagen. |
| Osteoporosis | Commonly postmenopausal or in other cases is more gradual and related to age; a small number of cases are due to mutations in the *COL1A1* and *COL1A2* genes and possibly in the vitamin D receptor gene (MIM 166710) |
| Osteoarthritis | A small number of cases are due to mutations in the *COL1A* genes. |
| Several chondrodysplasias | Due to mutations in *COL2A1* genes. |
| Pfeiffer syndrome[1] (MIM 100600) | Mutations in the gene encoding fibroblast growth receptor 1 (FGFR1). |
| Jackson-Weiss (MIM 123150) and Crouzon (MIM 123500) syndromes[1] | Mutations in the gene encoding FGFR2. |
| Achondroplasia (MIM 100800) and thanatophoric dysplasia[2] (MIM 187600) | Mutations in the gene encoding FGFR3. |

[1]The Pfeiffer, Jackson-Weiss, and Crouzon syndromes are craniosynostosis syndromes; craniosynostosis is a term signifying premature fusion of sutures in the skull.
[2]Thanatophoric (Gk *thanatos* "death" + *phoros* "bearing") dysplasia is the most common neonatal lethal skeletal dysplasia, displaying features similar to those of homozygous achondroplasia.

## MANY METABOLIC & GENETIC DISORDERS INVOLVE BONE

A number of the more important examples of metabolic and genetic disorders that affect bone are listed in Table 48–10.

**Osteogenesis imperfecta** (brittle bones) is characterized by abnormal fragility of bones. The scleras are often abnormally thin and translucent and may appear blue owing to a deficiency of connective tissue. Four types of this condition (mild, extensive, severe, and variable) have been recognized, of which the extensive type occurring in the newborn is the most ominous. Affected infants may be born with multiple fractures and not survive. Over 90% of patients with osteogenesis imperfecta have mutations in the *COL1A1* and *COL1A2* genes, encoding pro$\alpha$1(I) and pro$\alpha$2(I) chains, respectively. Over 100 mutations in these two genes have been documented and include partial gene deletions and duplications. Other mutations affect RNA splicing, and the most frequent type results in the replacement of glycine by another bulkier amino acid, affecting formation of the triple helix. In general, these mutations result in decreased expression of collagen or

**Table 48–11.** The principal proteins found in cartilage.

| Proteins | Comments |
|---|---|
| **Collagen proteins** | |
| Collagen type II | 90–98% of total articular cartilage collagen. Composed of three $\alpha$1(II) chains. |
| Collagens V, VI, IX, X, XI | Type IX cross-links to type II collagen. Type XI may help control diameter of type II fibrils. |
| **Noncollagen proteins** | |
| Proteoglycans | |
| Aggrecan | The major proteoglycan of cartilage. |
| Large non-aggregating proteoglycan | Found in some types of cartilage. |
| DS-PG I (biglycan)[1] | Similar to CS-PG I of bone. |
| DS-PG II (decorin) | Similar to CS-PG II of bone. |
| Chondronectin | May play role in binding type II collagen to surface of cartilage. |
| Anchorin C II | May bind type II collagen to surface of chondrocyte. |

[1]The core proteins of DS-PG I and DS-PG II are homologous to those of CS-PG I and CS-PG II found in bone (Table 48-9). A possible explanation is that osteoblasts lack the epimerase required to convert glucuronic acid to iduronic acid, the latter of which is found in dermatan sulfate.

in structurally abnormal pro$\alpha$ chains that assemble into abnormal fibrils, weakening the overall structure of bone. When one abnormal chain is present, it may interact with two normal chains, but folding may be prevented, resulting in enzymatic degradation of all of the chains. This is called "procollagen suicide" and is an example of a dominant negative mutation, a result often seen when a protein consists of multiple different subunits.

**Osteopetrosis** (marble bone disease), characterized by increased bone density, is due to inability to resorb bone. One form occurs along with renal tubular acidosis and cerebral calcification. It is due to mutations in the gene (located on chromosome 8q22) encoding carbonic anhydrase II (CA II), one of four isozymes of carbonic anhydrase present in human tissues. The reaction catalyzed by carbonic anhydrase is shown below:

67 nm

Fibril

Reaction II is spontaneous. In osteoclasts involved in bone resorption, CA II apparently provides protons to neutralize the $OH^-$ ions left inside the cell when $H^+$ ions are pumped across their ruffled borders (see above). Thus, if CA II is deficient in activity in osteoclasts, normal bone resorption does not occur, and osteopetrosis results. The mechanism of the cerebral calcification is not clear, whereas the renal tubular acidosis reflects deficient activity of CA II in the renal tubules.

**Osteoporosis** is a generalized progressive reduction in bone tissue mass per unit volume causing skeletal weakness. The ratio of mineral to organic elements is unchanged in the remaining normal bone. Fractures of various bones, such as the head of the femur, occur very easily and represent a huge burden to both the affected patients and to the health care budget of society. Among other factors, estrogens and interleukins-1 and -6 appear to be intimately involved in the causation of osteoporosis.

## THE MAJOR COMPONENTS OF CARTILAGE ARE TYPE II COLLAGEN & CERTAIN PROTEOGLYCANS

The principal proteins of hyaline cartilage (the major type of cartilage) are listed in Table 48–11. Type II collagen is the principal protein (Figure 48–13), and a number of other minor types of collagen are also present. In

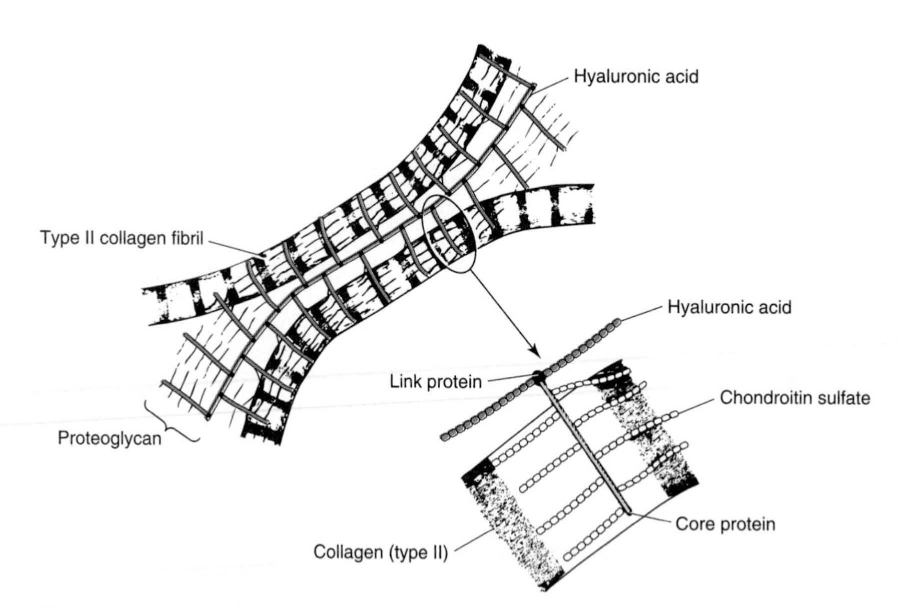


***Figure 48–13.*** Schematic representation of the molecular organization in cartilage matrix. Link proteins noncovalently bind the core protein (lighter color) of proteoglycans to the linear hyaluronic acid molecules (darker color). The chondroitin sulfate side chains of the proteoglycan electrostatically bind to the collagen fibrils, forming a cross-linked matrix. The oval outlines the area enlarged in the lower part of the figure. (Reproduced, with permission, from Junqueira LC, Carneiro J: ***Basic Histology***: *Text & Atlas,* 10th ed. McGraw-Hill, 2003.)

addition to these components, elastic cartilage contains elastin and fibroelastic cartilage contains type I collagen. Cartilage contains a number of proteoglycans, which play an important role in its compressibility. **Aggrecan** (about $2 \times 10^3$ kDa) is the major proteoglycan. As shown in Figure 48–14, it has a very complex structure, containing several GAGs (hyaluronic acid, chondroitin sulfate, and keratan sulfate) and both link and core proteins. The core protein contains three domains: A, B, and C. The hyaluronic acid binds noncovalently to domain A of the core protein as well as to the link protein, which stabilizes the hyaluronate–core protein interactions. The keratan sulfate chains are located in domain B, whereas the chondroitin sulfate chains are located in domain C; both of these types of GAGs are bound covalently to the core protein. The core protein also contains both O- and N-linked oligosaccharide chains.

The other proteoglycans found in cartilage have simpler structures than aggrecan.

**Chondronectin** is involved in the attachment of type II collagen to chondrocytes.

Cartilage is an avascular tissue and obtains most of its nutrients from synovial fluid. It exhibits slow but continuous turnover. Various **proteases** (eg, collagenases and stromalysin) synthesized by chondrocytes can degrade collagen and the other proteins found in cartilage. Interleukin-1 (IL-1) and tumor necrosis factor α (TNFα) appear to stimulate the production of such proteases, whereas transforming growth factor β (TGFβ) and insulin-like growth factor 1 (IGF-I) generally exert an anabolic influence on cartilage.

## THE MOLECULAR BASES OF THE CHONDRODYSPLASIAS INCLUDE MUTATIONS IN GENES ENCODING TYPE II COLLAGEN & FIBROBLAST GROWTH FACTOR RECEPTORS

Chondrodysplasias are a mixed group of hereditary disorders affecting cartilage. They are manifested by short-limbed dwarfism and numerous skeletal deformities. A number of them are due to a variety of mutations in the *COL2A1* gene, leading to abnormal forms of type II collagen. One example is **Stickler syndrome,** manifested by degeneration of joint cartilage and of the vitreous body of the eye.

The best-known of the chondrodysplasias is **achondroplasia,** the commonest cause of short-limbed dwarfism. Affected individuals have short limbs, nor-

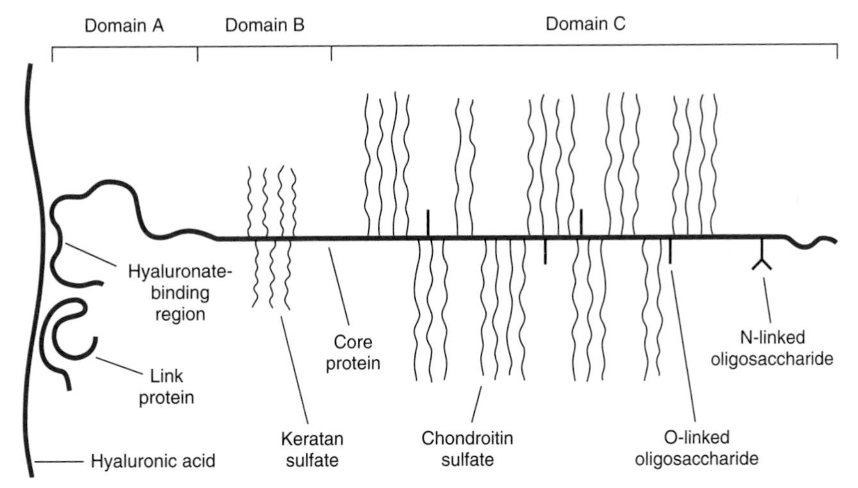


***Figure 48–14.*** Schematic diagram of the aggrecan from bovine nasal cartilage. A strand of hyaluronic acid is shown on the left. The core protein (about 210 kDa) has three major domains. Domain A, at its amino terminal end, interacts with approximately five repeating disaccharides in hyaluronate. The link protein interacts with both hyaluronate and domain A, stabilizing their interactions. Approximately 30 keratan sulfate chains are attached, via GalNAc-Ser linkages, to domain B. Domain C contains about 100 chondroitin sulfate chains attached via Gal-Gal-Xyl-Ser linkages and about 40 O-linked oligosaccharide chains. One or more N-linked glycan chains are also found near the carboxyl terminal of the core protein. (Reproduced, with permission, from Moran LA et al: *Biochemistry,* 2nd ed. Neil Patterson Publishers, 1994.)

mal trunk size, macrocephaly, and a variety of other skeletal abnormalities. The condition is often inherited as an autosomal dominant trait, but many cases are due to new mutations. The molecular basis of achondroplasia is outlined in Figure 48–15. Achondroplasia is not a collagen disorder but is due to mutations in the gene encoding **fibroblast growth factor receptor 3 (FGFR3). Fibroblast growth factors** are a family of at least nine proteins that affect the growth and differentiation of cells of mesenchymal and neuroectodermal origin. Their receptors are transmembrane proteins and form a subgroup of the family of receptor tyrosine kinases. FGFR3 is one member of this subgroup and mediates the actions of FGF3 on cartilage. In almost all cases of achondroplasia that have been investigated, the mutations were found to involve nucleotide 1138 and resulted in substitution of arginine for glycine (residue number 380) in the transmembrane domain of the protein, rendering it inactive. No such mutation was found in unaffected individuals. As indicated in Table 48–10, other skeletal dysplasias (including certain craniosynostosis syndromes) are also due to mutations in genes encoding FGF receptors. Another type of skeletal dysplasia (diastrophic dysplasia) has been found to be due to mutation in a sulfate transporter. Thus, thanks to recombinant DNA technology, a new era in understanding of skeletal dysplasias has begun.

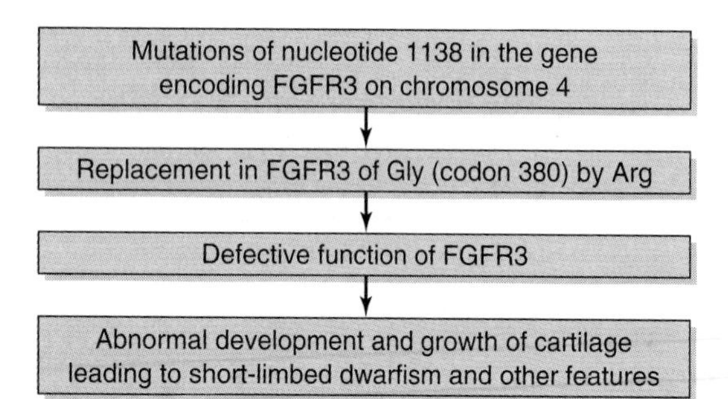


***Figure 48–15.*** Simplified scheme of the causation of achondroplasia (MIM 100800). In most cases studied so far, the mutation has been a G to A transition at nucleotide 1138. In a few cases, the mutation was a G to C transversion at the same nucleotide. This particular nucleotide is a real "hot spot" for mutation. Both mutations result in replacement of a Gly residue by an Arg residue in the transmembrane segment of the receptor. A few cases involving replacement of Gly by Cys at codon 375 have also been reported.

## SUMMARY


- The major components of the ECM are the structural proteins collagen, elastin, and fibrillin; a number of specialized proteins (eg, fibronectin and laminin); and various proteoglycans.
- Collagen is the most abundant protein in the animal kingdom; approximately 19 types have been isolated. All collagens contain greater or lesser stretches of triple helix and the repeating structure $(Gly\text{-}X\text{-}Y)_n$.
- The biosynthesis of collagen is complex, featuring many posttranslational events, including hydroxylation of proline and lysine.
- Diseases associated with impaired synthesis of collagen include scurvy, osteogenesis imperfecta, Ehlers-Danlos syndrome (many types), and Menkes disease.
- Elastin confers extensibility and elastic recoil on tissues. Elastin lacks hydroxylysine, Gly-X-Y sequences, triple helical structure, and sugars but contains desmosine and isodesmosine cross-links not found in collagen.
- Fibrillin is located in microfibrils. Mutations in the gene for fibrillin cause Marfan syndrome.
- The glycosaminoglycans (GAGs) are made up of repeating disaccharides containing a uronic acid (glucuronic or iduronic) or hexose (galactose) and a hexosamine (galactosamine or glucosamine). Sulfate is also frequently present.
- The major GAGs are hyaluronic acid, chondroitin 4- and 6-sulfates, keratan sulfates I and II, heparin, heparan sulfate, and dermatan sulfate.
- The GAGs are synthesized by the sequential actions of a battery of specific enzymes (glycosyltransferases, epimerases, sulfotransferases, etc) and are degraded by the sequential action of lysosomal hydrolases. Genetic deficiencies of the latter result in mucopolysaccharidoses (eg, Hurler syndrome).
- GAGs occur in tissues bound to various proteins (linker proteins and core proteins), constituting proteoglycans. These structures are often of very high molecular weight and serve many functions in tissues.
- Many components of the ECM bind to proteins of the cell surface named integrins; this constitutes one pathway by which the exteriors of cells can communicate with their interiors.
- Bone and cartilage are specialized forms of the ECM. Collagen I and hydroxyapatite are the major constituents of bone. Collagen II and certain proteoglycans are major constituents of cartilage.

- The molecular causes of a number of heritable diseases of bone (eg, osteogenesis imperfecta) and of cartilage (eg, the chondrodystrophies) are being revealed by the application of recombinant DNA technology.

## REFERENCES

Bandtlow CE, Zimmermann DR: Proteoglycans in the developing brain: new conceptual insights for old proteins. Physiol Rev 2000;80:1267.

Bikle DD: Biochemical markers in the assessment of bone diseases. Am J Med 1997;103:427.

Burke D et al: Fibroblast growth factor receptors: lessons from the genes. Trends Biochem Sci 1998;23:59.

Compston JE: Sex steroids and bone. Physiol Rev 2001;81:419.

Fuller GM, Shields D: *Molecular Basis of Medical Cell Biology.* Appleton & Lange, 1998.

Herman T, Horvitz HR: Three proteins involved in *Caenorhabditis elegans* vulval invagination are similar to components of a glycosylation pathway. Proc Natl Acad Sci U S A 1999;96:974.

Prockop DJ, Kivirikko KI: Collagens: molecular biology, diseases, and potential therapy. Annu Rev Biochem 1995;64:403.

Pyeritz RE: Ehlers-Danlos syndrome. N Engl J Med 2000;342:730.

Sage E: Regulation of interactions between cells and extracellular matrix: a command performance on several stages. J Clin Invest 2001;107:781. (This article introduces a series of six articles on cell-matrix interaction. The topics covered are cell adhesion and de-adhesion, thrombospondins, syndecans, SPARC, osteopontin, and Ehlers-Danlos syndrome. All of the articles can be accessed at www.jci.org.)

Scriver CR et al (editors): *The Metabolic and Molecular Bases of Inherited Disease,* 8th ed. McGraw-Hill, 2001 (This comprehensive four-volume text contains chapters on disorders of collagen biosynthesis and structure, Marfan syndrome, the mucopolysaccharidoses, achondroplasia, Alport syndrome, and craniosynostosis syndromes.)

Selleck SB: Genetic dissection of proteoglycan function in *Drosophila* and *C. elegans.* Semin Cell Dev Biol 2001;12:127.

# Muscle & the Cytoskeleton

49

*Robert K. Murray, MD, PhD*

## BIOMEDICAL IMPORTANCE

Proteins play an important role in movement at both the organ (eg, skeletal muscle, heart, and gut) and cellular levels. In this chapter, the roles of specific proteins and certain other key molecules (eg, $Ca^{2+}$) in **muscular contraction** are described. A brief coverage of **cytoskeletal proteins** is also presented.

Knowledge of the molecular bases of a number of conditions that affect muscle has advanced greatly in recent years. Understanding of the molecular basis of **Duchenne-type muscular dystrophy** was greatly enhanced when it was found that it was due to mutations in the gene encoding dystrophin. Significant progress has also been made in understanding the molecular basis of **malignant hyperthermia,** a serious complication for some patients undergoing certain types of anesthesia. **Heart failure** is a very common medical condition, with a variety of causes; its rational therapy requires understanding of the biochemistry of heart muscle. One group of conditions that cause heart failure are the **cardiomyopathies,** some of which are genetically determined. **Nitric oxide** (NO) has been found to be a major regulator of smooth muscle tone. Many widely used **vasodilators**—such as nitroglycerin, used in the treatment of angina pectoris—act by increasing the formation of NO. Muscle, partly because of its mass, plays major roles in the **overall metabolism** of the body.

## MUSCLE TRANSDUCES CHEMICAL ENERGY INTO MECHANICAL ENERGY

Muscle is the major biochemical transducer (machine) that converts potential (chemical) energy into kinetic (mechanical) energy. Muscle, the largest single tissue in the human body, makes up somewhat less than 25% of body mass at birth, more than 40% in the young adult, and somewhat less than 30% in the aged adult. We shall discuss aspects of the three types of muscle found in vertebrates: **skeletal, cardiac,** and **smooth.** Both skeletal and cardiac muscle appear **striated** upon microscopic observation; smooth muscle is **nonstriated.** Although skeletal muscle is under voluntary nervous control, the control of both cardiac and smooth muscle is involuntary.

### The Sarcoplasm of Muscle Cells Contains ATP, Phosphocreatine, & Glycolytic Enzymes

Striated muscle is composed of multinucleated muscle fiber cells surrounded by an electrically excitable plasma membrane, the **sarcolemma.** An individual muscle fiber cell, which may extend the entire length of the muscle, contains a bundle of many **myofibrils** arranged in parallel, embedded in intracellular fluid termed **sarcoplasm.** Within this fluid is contained glycogen, the high-energy compounds ATP and phosphocreatine, and the enzymes of glycolysis.

### The Sarcomere Is the Functional Unit of Muscle

An overall view of voluntary muscle at several levels of organization is presented in Figure 49–1.

When the myofibril is examined by electron microscopy, alternating dark and light bands (anisotropic bands, meaning birefringent in polarized light; and isotropic bands, meaning not altered by polarized light) can be observed. These bands are thus referred to as **A and I bands,** respectively. The central region of the A band (the H band) appears less dense than the rest of the band. The I band is bisected by a very dense and narrow **Z line** (Figure 49–2).

The **sarcomere** is defined as the region between two Z lines (Figures 49–1 and 49–2) and is repeated along the axis of a fibril at distances of 1500–2300 nm depending upon the state of contraction.

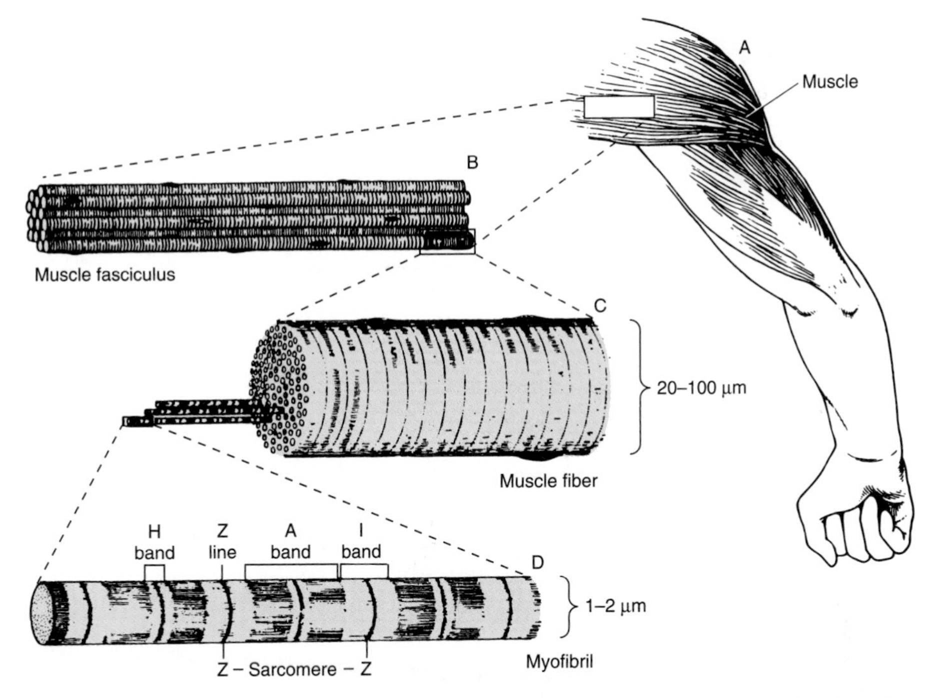


***Figure 49–1.*** **The structure of voluntary muscle. The sarcomere is the region between the Z lines.** (Drawing by Sylvia Colard Keene. Reproduced, with permission, from Bloom W, Fawcett DW: *A Textbook of Histology,* 10th ed. Saunders, 1975.)

The striated appearance of voluntary and cardiac muscle in light microscopic studies results from their high degree of organization, in which most muscle fiber cells are aligned so that their sarcomeres are in parallel register (Figure 49–1).

## Thick Filaments Contain Myosin; Thin Filaments Contain Actin, Tropomyosin, & Troponin

When myofibrils are examined by electron microscopy, it appears that each one is constructed of two types of longitudinal filaments. One type, the **thick filament,** confined to the A band, contains chiefly the protein myosin. These filaments are about 16 nm in diameter and arranged in cross-section as a hexagonal array (Figure 49–2, center; right-hand cross-section).

The **thin filament** (about 7 nm in diameter) lies in the I band and extends into the A band but not into its H zone (Figure 49–2). Thin filaments contain the proteins actin, tropomyosin, and troponin (Figure 49–3). In the A band, the thin filaments are arranged around the thick (myosin) filament as a secondary hexagonal array. Each thin filament lies symmetrically between three thick filaments (Figure 49–2, center; mid cross-section), and each thick filament is surrounded symmetrically by six thin filaments.

The thick and thin filaments interact via cross-bridges that emerge at intervals of 14 nm along the thick filaments. As depicted in Figure 49–2, the cross-bridges (drawn as arrowheads at each end of the myosin filaments, but not shown extending fully across to the thin filaments) have opposite polarities at the two ends of the thick filaments. The two poles of the thick filaments are separated by a 150-nm segment (the M band, not labeled in the figure) that is free of projections.

## The Sliding Filament Cross-Bridge Model Is the Foundation on Which Current Thinking About Muscle Contraction Is Built

This model was proposed independently in the 1950s by Henry Huxley and Andrew Huxley and their colleagues. It was largely based on careful morphologic observations on resting, extended, and contracting muscle. Basically, when muscle contracts, there is no change in the lengths of the thick and thin filaments, but the H zones and the I bands shorten (see legend to Fig-

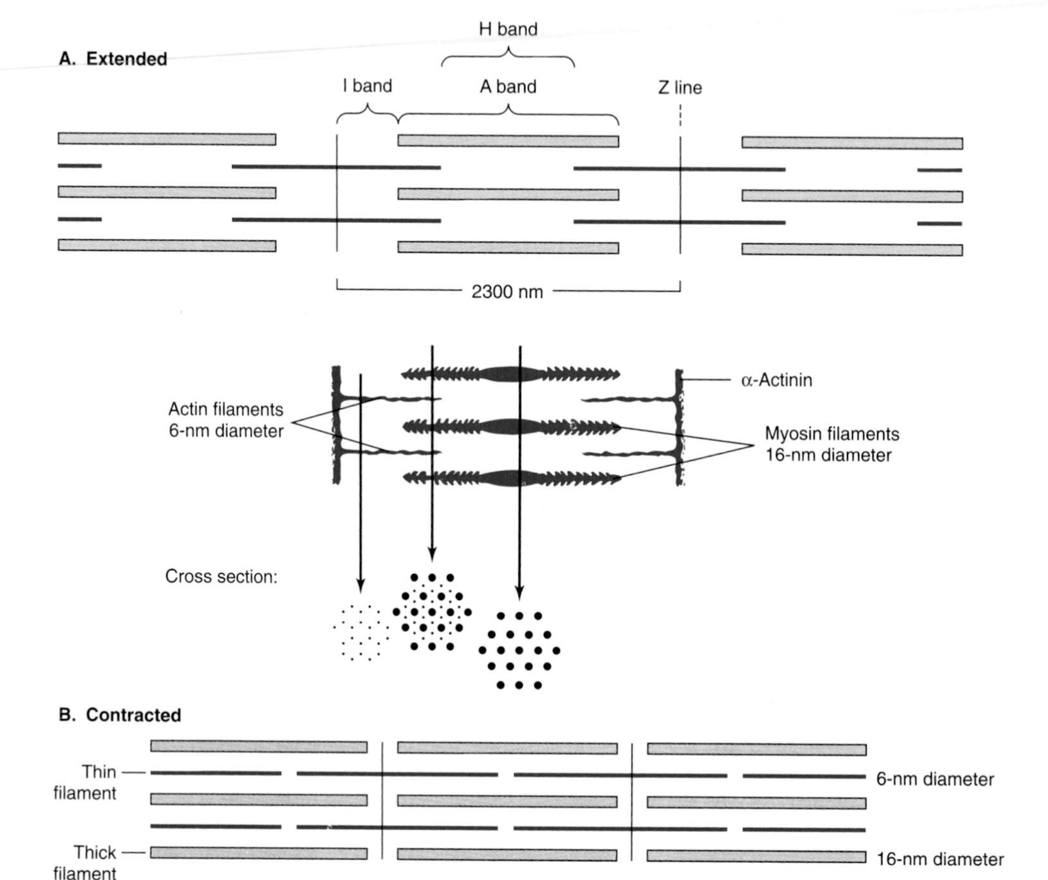


***Figure 49–2.*** Arrangement of filaments in striated muscle. **A:** Extended. The positions of the I, A, and H bands in the extended state are shown. The thin filaments partly overlap the ends of the thick filaments, and the thin filaments are shown anchored in the Z lines (often called Z disks). In the lower part of Figure 49–2A, "arrowheads," pointing in opposite directions, are shown emanating from the myosin (thick) filaments. Four actin (thin) filaments are shown attached to two Z lines via α-actinin. The central region of the three myosin filaments, free of arrowheads, is called the M band (not labeled). Cross-sections through the M bands, through an area where myosin and actin filaments overlap and through an area in which solely actin filaments are present, are shown. **B:** Contracted. The actin filaments are seen to have slipped along the sides of the myosin fibers toward each other. The lengths of the thick filaments (indicated by the A bands) and the thin filaments (distance between Z lines and the adjacent edges of the H bands) have not changed. However, the lengths of the sarcomeres have been reduced (from 2300 nm to 1500 nm), and the lengths of the H and I bands are also reduced because of the overlap between the thick and thin filaments. These morphologic observations provided part of the basis for the sliding filament model of muscle contraction.

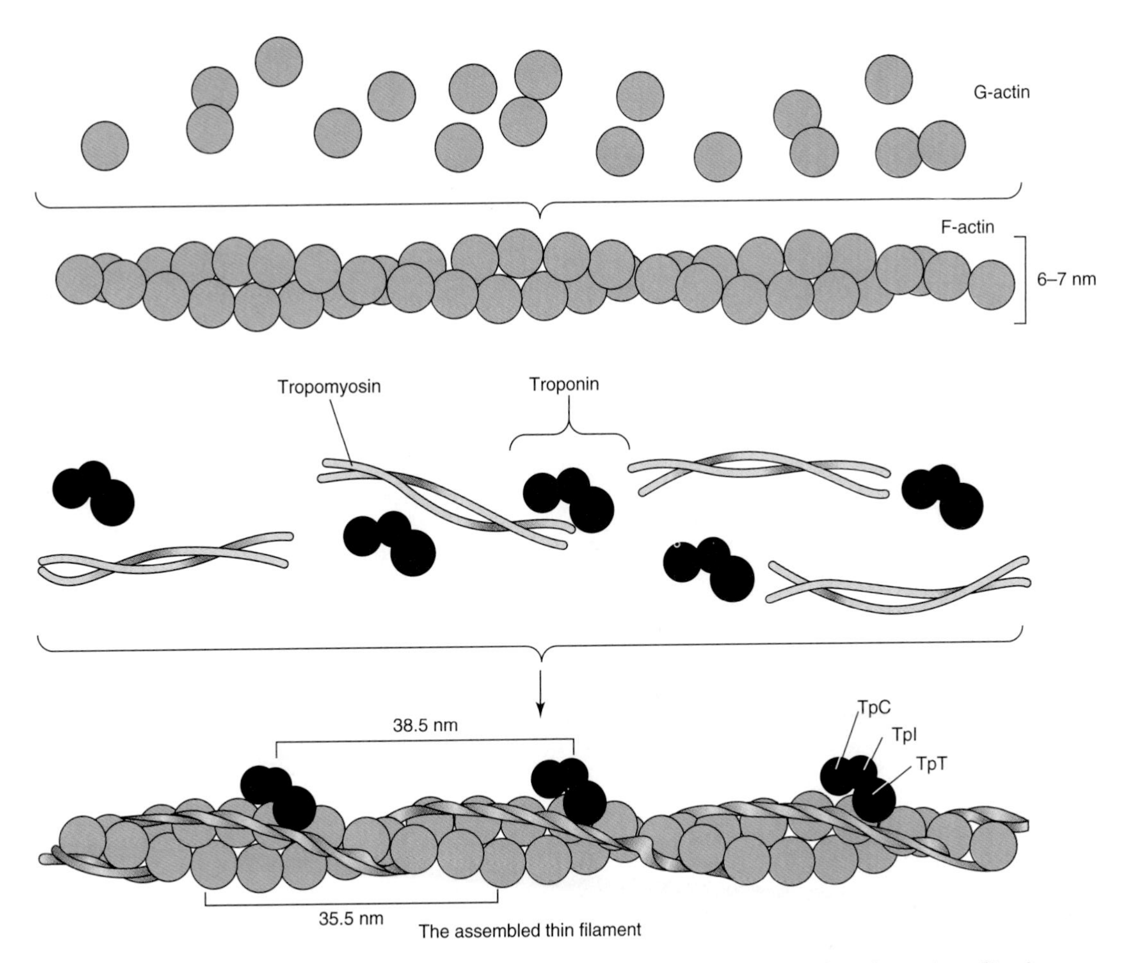


***Figure 49–3.*** Schematic representation of the thin filament, showing the spatial configuration of its three major protein components: actin, myosin, and tropomyosin. The upper panel shows individual molecules of G-actin. The middle panel shows actin monomers assembled into F-actin. Individual molecules of tropomyosin (two strands wound around one another) and of troponin (made up of its three subunits) are also shown. The lower panel shows the assembled thin filament, consisting of F-actin, tropomyosin, and the three subunits of troponin (TpC, TpI, and TpT).

ure 49–2). Thus, the arrays of interdigitating filaments must slide past one another during contraction. Cross-bridges that link thick and thin filaments at certain stages in the contraction cycle generate and sustain the tension. The tension developed during muscle contraction is proportionate to the filament overlap and to the number of cross-bridges. Each cross-bridge head is connected to the thick filament via a flexible fibrous segment that can bend outward from the thick filament. This flexible segment facilitates contact of the head with the thin filament when necessary but is also sufficiently pliant to be accommodated in the interfilament spacing.

## ACTIN & MYOSIN ARE THE MAJOR PROTEINS OF MUSCLE

The mass of a muscle is made up of 75% water and more than 20% protein. The two major proteins are actin and myosin.

Monomeric **G-actin** (43 kDa; G, globular) makes up 25% of muscle protein by weight. At physiologic ionic strength and in the presence of $Mg^{2+}$, G-actin polymerizes noncovalently to form an insoluble double helical filament called F-actin (Figure 49–3). The **F-actin** fiber is 6–7 nm thick and has a pitch or repeating structure every 35.5 nm.

**Myosins** constitute a family of proteins, with at least 15 members having been identified. The myosin discussed in this chapter is myosin-II, and when myosin is referred to in this text, it is this species that is meant unless otherwise indicated. Myosin-I is a monomeric species that binds to cell membranes. It may serve as a linkage between microfilaments and the cell membrane in certain locations.

Myosin contributes 55% of muscle protein by weight and forms the thick filaments. It is an asymmetric hexamer with a molecular mass of approximately 460 kDa. Myosin has a fibrous tail consisting of two intertwined helices. Each helix has a globular head portion attached at one end (Figure 49–4). The hexamer consists of one pair of **heavy (H) chains** each of approximately 200 kDA molecular mass, and two pairs of **light (L) chains** each with a molecular mass of approximately 20 kDa. The L chains differ, one being called the essential light chain and the other the regulatory light chain. Skeletal muscle myosin binds actin to form actomyosin (actin-myosin), and its intrinsic ATPase activity is markedly enhanced in this complex. Isoforms of myosin exist whose amounts can vary in different anatomic, physiologic, and pathologic situations.

The structures of actin and of the head of myosin have been determined by x-ray crystallography; these studies have confirmed a number of earlier findings concerning their structures and have also given rise to much new information.

## Limited Digestion of Myosin With Proteases Has Helped to Elucidate Its Structure & Function

When myosin is digested with trypsin, two myosin fragments (meromyosins) are generated. **Light meromyosin** (LMM) consists of aggregated, insoluble α-helical fibers from the tail of myosin (Figure 49–4). LMM

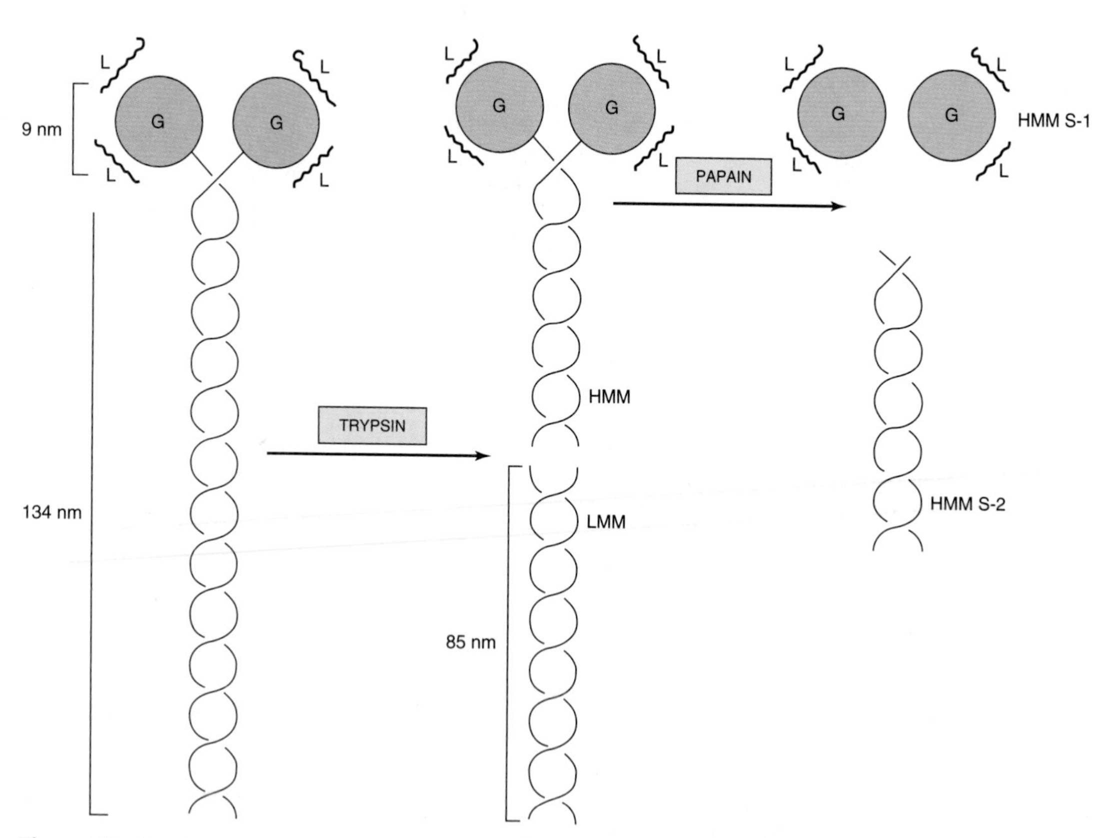


***Figure 49–4.*** Diagram of a myosin molecule showing the two intertwined α-helices (fibrous portion), the globular region or head (G), the light chains (L), and the effects of proteolytic cleavage by trypsin and papain. The globular region (myosin head) contains an actin-binding site and an L chain-binding site and also attaches to the remainder of the myosin molecule.

exhibits no ATPase activity and does not bind to F-actin.

**Heavy meromyosin** (HMM; molecular mass about 340 kDa) is a soluble protein that has both a fibrous portion and a globular portion (Figure 49–4). It exhibits ATPase activity and binds to F-actin. Digestion of HMM with papain generates two subfragments, S-1 and S-2. The S-2 fragment is fibrous in character, has no ATPase activity, and does not bind to F-actin.

S-1 (molecular mass approximately 115 kDa) does exhibit ATPase activity, binds L chains, and in the absence of ATP will bind to and decorate actin with "arrowheads" (Figure 49–5). Both S-1 and HMM exhibit ATPase activity, which is accelerated 100- to 200-fold by complexing with F-actin. As discussed below, F-actin greatly enhances the rate at which myosin ATPase releases its products, ADP and $P_i$. Thus, although F-actin does not affect the hydrolysis step per se, its ability to promote release of the products produced by the ATPase activity greatly accelerates the overall rate of catalysis.

## CHANGES IN THE CONFORMATION OF THE HEAD OF MYOSIN DRIVE MUSCLE CONTRACTION

How can hydrolysis of ATP produce macroscopic movement? Muscle contraction essentially consists of the cyclic attachment and detachment of the S-1 head of myosin to the F-actin filaments. This process can also be referred to as the making and breaking of cross-bridges. The attachment of actin to myosin is followed by conformational changes which are of particular importance in the S-1 head and are dependent upon which nucleotide is present (ADP or ATP). These changes result in the **power stroke,** which drives movement of actin filaments past myosin filaments. The energy for the power stroke is ultimately supplied by **ATP,** which is hydrolyzed to ADP and $P_i$. However, the power stroke itself occurs as a result of **conformational changes** in the myosin head when **ADP** leaves it.

The major biochemical events occurring during one cycle of muscle contraction and relaxation can be represented in the five steps shown in Figure 49–6:

**(1)** In the **relaxation phase** of muscle contraction, the S-1 head of myosin hydrolyzes ATP to ADP and $P_i$, but these products remain bound. The resultant ADP-$P_i$-myosin complex has been energized and is in a so-called high-energy conformation.

**(2)** When **contraction** of muscle is stimulated (via events involving $Ca^{2+}$, troponin, tropomyosin, and actin, which are described below), actin becomes accessible and the S-1 head of myosin finds it, binds it, and forms the actin-myosin-ADP-$P_i$ complex indicated.

**(3)** Formation of this complex **promotes the release of $P_i$,** which initiates the power stroke. This is followed by release of ADP and is accompanied by a large conformational change in the head of myosin in relation to its tail (Figure 49–7), pulling actin about 10 nm toward the center of the sarcomere. This is the power stroke. The myosin is now in a so-called low-energy state, indicated as actin-myosin.

**(4)** Another molecule of ATP binds to the S-1 head, forming an actin-myosin-ATP complex.

**(5)** Myosin-ATP has a low affinity for actin, and **actin is thus released.** This last step is a key component of relaxation and is dependent upon the binding of ATP to the actin-myosin complex.

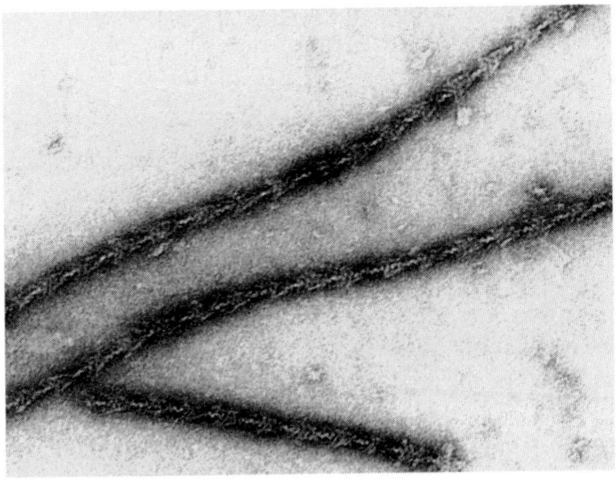

***Figure 49–5.*** The decoration of actin filaments with the S-1 fragments of myosin to form "arrowheads." (Courtesy of JA Spudich.)

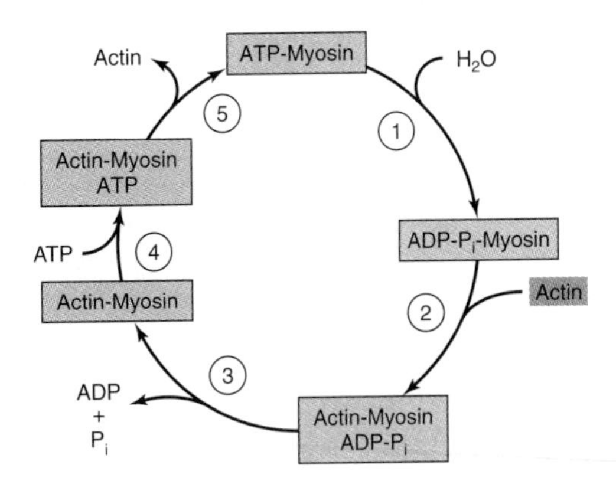


***Figure 49–6.*** The hydrolysis of ATP drives the cyclic association and dissociation of actin and myosin in five reactions described in the text. (Modified from Stryer L: *Biochemistry,* 2nd ed. Freeman, 1981.)

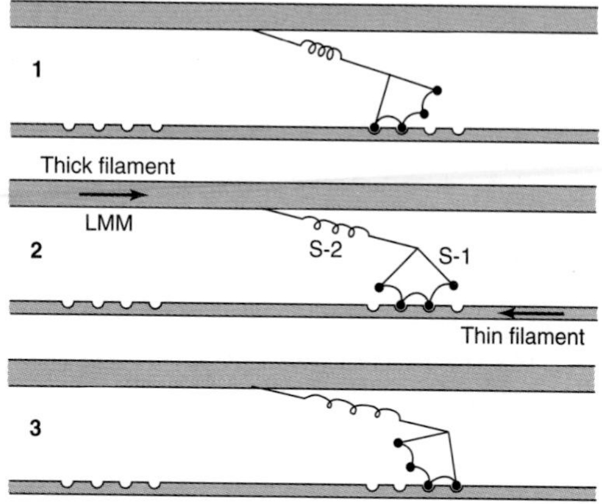


*Figure 49–7.* Representation of the active cross-bridges between thick and thin filaments. This diagram was adapted by AF Huxley from HE Huxley: The mechanism of muscular contraction. Science 1969;164:1356. The latter proposed that the force involved in muscular contraction originates in a tendency for the myosin head (S-1) to rotate relative to the thin filament and is transmitted to the thick filament by the S-2 portion of the myosin molecule acting as an inextensible link. Flexible points at each end of S-2 permit S-1 to rotate and allow for variations in the separation between filaments. The present figure is based on HE Huxley's proposal but also incorporates elastic (the coils in the S-2 portion) and stepwise-shortening elements (depicted here as four sites of interaction between the S-1 portion and the thin filament). (See Huxley AF, Simmons RM: Proposed mechanism of force generation in striated muscle. Nature [Lond] 1971;233:533.) The strengths of binding of the attached sites are higher in position 2 than in position 1 and higher in position 3 than position 2. The myosin head can be detached from position 3 with the utilization of a molecule of ATP; this is the predominant process during shortening. The myosin head is seen to vary in its position from about 90° to about 45°, as indicated in the text. (S-1, myosin head; S-2, portion of the myosin molecule; LMM, light meromyosin) (see legend to Figure 49–4). (Reproduced from Huxley AF: Muscular contraction. J Physiol 1974; 243:1. By kind permission of the author and the *Journal of Physiology*.)

Another cycle then commences with the hydrolysis of ATP (step 1 of Figure 49–6), re-forming the high-energy conformation.

Thus, hydrolysis of ATP is used to drive the cycle, with the actual power stroke being the conformational change in the S-1 head that occurs upon the release of ADP. The hinge regions of myosin (referred to as flexible points at each end of S-2 in the legend to Figure 49–7) permit the large range of movement of S-1 and also allow S-1 to find actin filaments.

If intracellular levels of ATP drop (eg, after death), ATP is not available to bind the S-1 head (step 4 above), actin does not dissociate, and relaxation (step 5) does not occur. This is the explanation for **rigor mortis,** the stiffening of the body that occurs after death.

Calculations have indicated that the efficiency of contraction is about 50%; that of the internal combustion engine is less than 20%.

## Tropomyosin & the Troponin Complex Present in Thin Filaments Perform Key Functions in Striated Muscle

In striated muscle, there are two other proteins that are minor in terms of their mass but important in terms of their function. **Tropomyosin** is a fibrous molecule that consists of two chains, alpha and beta, that attach to F-actin in the groove between its filaments (Figure 49–3). Tropomyosin is present in all muscular and muscle-like structures. The **troponin complex** is unique to striated muscle and consists of three polypeptides. **Troponin T** (TpT) binds to tropomyosin as well as to the other two troponin components. **Troponin I** (TpI) inhibits the F-actin-myosin interaction and also binds to the other components of troponin. **Troponin C** (TpC) is a calcium-binding polypeptide that is structurally and functionally analogous to **calmodulin,** an important calcium-binding protein widely distributed in nature. Four molecules of calcium ion are bound per molecule of troponin C or calmodulin, and both molecules have a molecular mass of 17 kDa.

## $Ca^{2+}$ Plays a Central Role in Regulation of Muscle Contraction

The contraction of muscles from all sources occurs by the general mechanism described above. Muscles from different organisms and from different cells and tissues within the same organism may have different molecular mechanisms responsible for the regulation of their contraction and relaxation. In all systems, $Ca^{2+}$ plays a key regulatory role. There are two general mechanisms of regulation of muscle contraction: **actin-based** and **myosin-based.** The former operates in skeletal and cardiac muscle, the latter in smooth muscle.

### Actin-Based Regulation Occurs in Striated Muscle

Actin-based regulation of muscle occurs in vertebrate skeletal and cardiac muscles, both striated. In the general mechanism described above (Figure 49–6), the only potentially limiting factor in the cycle of muscle contraction might be ATP. The skeletal muscle system is **inhibited** at rest; this inhibition is relieved to activate contraction. The inhibitor of striated muscle is the **troponin system,** which is bound to tropomyosin and F-actin in the thin filament (Figure 49–3). In striated muscle, there is no control of contraction unless the tropomyosin-troponin systems are present along with the actin and myosin filaments. As described above, **tropomyosin** lies along the groove of F-actin, and the three components of **troponin**—TpT, TpI, and TpC—are bound to the F-actin–tropomyosin complex. TpI prevents binding of the myosin head to its F-actin attachment site either by altering the conformation of F-actin via the tropomyosin molecules or by simply rolling tropomyosin into a position that directly blocks the sites on F-actin to which the myosin heads attach. Either way prevents activation of the myosin ATPase that is mediated by binding of the myosin head to F-actin. Hence, the TpI system blocks the contraction cycle at step 2 of Figure 49–6. This accounts for the inhibited state of relaxed striated muscle.

## The Sarcoplasmic Reticulum Regulates Intracellular Levels of $Ca^{2+}$ in Skeletal Muscle

In the sarcoplasm of resting muscle, the concentration of $Ca^{2+}$ is $10^{-8}$ to $10^{-7}$ mol/L. The resting state is achieved because $Ca^{2+}$ is pumped into the sarcoplasmic reticulum through the action of an active transport system, called the $Ca^{2+}$ ATPase (Figure 49–8), initiating relaxation. The sarcoplasmic reticulum is a network of fine membranous sacs. Inside the sarcoplasmic reticulum, $Ca^{2+}$ is bound to a specific $Ca^{2+}$-binding protein designated **calsequestrin.** The sarcomere is surrounded by an excitable membrane (the T tubule system) composed of transverse (T) channels closely associated with the sarcoplasmic reticulum.

When the sarcolemma is excited by a nerve impulse, the signal is transmitted into the T tubule system and a **$Ca^{2+}$ release channel** in the nearby sarcoplasmic reticulum opens, releasing $Ca^{2+}$ from the sarcoplasmic reticulum into the sarcoplasm. The concentration of $Ca^{2+}$ in the sarcoplasm rises rapidly to $10^{-5}$ mol/L. The $Ca^{2+}$-binding sites on TpC in the thin filament are quickly occupied by $Ca^{2+}$. The TpC-$4Ca^{2+}$ interacts with TpI and TpT to alter their interaction with tropomyosin. Accordingly, tropomyosin moves out of the way or alters the conformation of F-actin so that the myosin head-ADP-$P_i$ (Figure 49–6) can interact with F-actin to start the contraction cycle.

The $Ca^{2+}$ release channel is also known as the **ryanodine receptor** (RYR). There are two isoforms of this

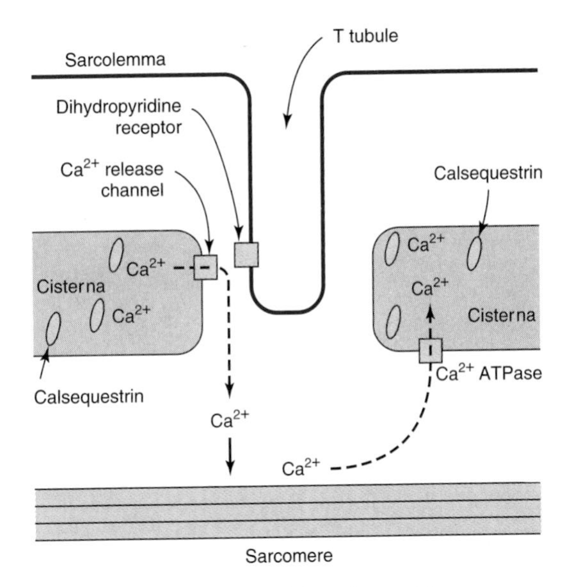


***Figure 49–8.*** Diagram of the relationships among the sarcolemma (plasma membrane), a T tubule, and two cisternae of the sarcoplasmic reticulum of skeletal muscle (not to scale). The T tubule extends inward from the sarcolemma. A wave of depolarization, initiated by a nerve impulse, is transmitted from the sarcolemma down the T tubule. It is then conveyed to the $Ca^{2+}$ release channel (ryanodine receptor), perhaps by interaction between it and the dihydropyridine receptor (slow $Ca^{2+}$ voltage channel), which are shown in close proximity. Release of $Ca^{2+}$ from the $Ca^{2+}$ release channel into the cytosol initiates contraction. Subsequently, $Ca^{2+}$ is pumped back into the cisternae of the sarcoplasmic reticulum by the $Ca^{2+}$ ATPase ($Ca^{2+}$ pump) and stored there, in part bound to calsequestrin.

receptor, RYR1 and RYR2, the former being present in skeletal muscle and the latter in heart muscle and brain. **Ryanodine** is a plant alkaloid that binds to RYR1 and RYR2 specifically and modulates their activities. The $Ca^{2+}$ release channel is a homotetramer made up of four subunits of kDa 565. It has transmembrane sequences at its carboxyl terminal, and these probably form the $Ca^{2+}$ channel. The remainder of the protein protrudes into the cytosol, bridging the gap between the sarcoplasmic reticulum and the transverse tubular membrane. The channel is ligand-gated, $Ca^{2+}$ and ATP working synergistically in vitro, although how it operates in vivo is not clear. A possible sequence of events leading to opening of the channel is shown in Figure 49–9. The channel lies very close to the **dihydropyridine receptor** (DHPR; a voltage-gated slow K type

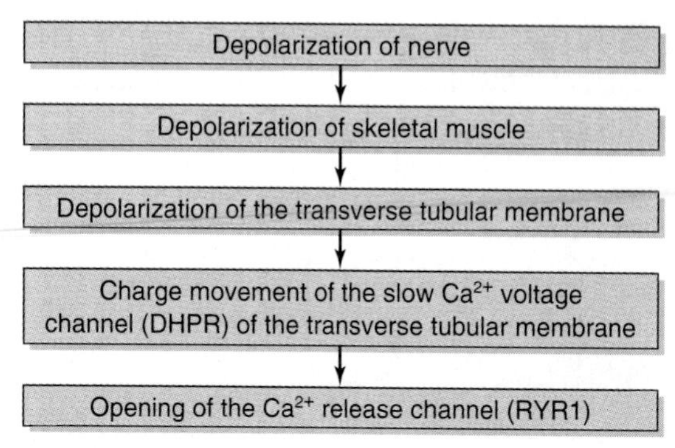


***Figure 49–9.*** Possible chain of events leading to opening of the $Ca^{2+}$ release channel. As indicated in the text, the $Ca^{2+}$ voltage channel and the $Ca^{2+}$ release channel have been shown to interact with each other in vitro via specific regions in their polypeptide chains. (DHPR, dihydropyridine receptor; RYR1, ryanodine receptor 1.)

$Ca^{2+}$ channel) of the transverse tubule system (Figure 49–8). Experiments in vitro employing an affinity column chromatography approach have indicated that a 37-amino-acid stretch in RYR1 interacts with one specific loop of DHPR.

**Relaxation** occurs when sarcoplasmic $Ca^{2+}$ falls below $10^{-7}$ mol/L owing to its resequestration into the sarcoplasmic reticulum by $Ca^{2+}$ ATPase. TpC.4$Ca^{2+}$ thus loses its $Ca^{2+}$. Consequently, troponin, via interaction with tropomyosin, inhibits further myosin head and F-actin interaction, and in the presence of ATP the myosin head detaches from the F-actin.

Thus, $Ca^{2+}$ controls skeletal muscle contraction and relaxation by an allosteric mechanism mediated by TpC, TpI, TpT, tropomyosin, and F-actin.

A decrease in the concentration of **ATP** in the sarcoplasm (eg, by excessive usage during the cycle of contraction-relaxation or by diminished formation, such as might occur in ischemia) has two major effects: (1) The $Ca^{2+}$ ATPase ($Ca^{2+}$ pump) in the sarcoplasmic reticulum ceases to maintain the low concentration of $Ca^{2+}$ in the sarcoplasm. Thus, the interaction of the myosin heads with F-actin is promoted. (2) The ATP-dependent detachment of myosin heads from F-actin cannot occur, and rigidity (contracture) sets in. The condition of **rigor mortis,** following death, is an extension of these events.

Muscle contraction is a delicate dynamic balance of the attachment and detachment of myosin heads to F-actin, subject to fine regulation via the nervous system.

Table 49–1 summarizes the overall events in contraction and relaxation of skeletal muscle.

## Mutations in the Gene Encoding the $Ca^{2+}$ Release Channel Are One Cause of Human Malignant Hyperthermia

Some genetically predisposed patients experience a severe reaction, designated malignant hyperthermia, on exposure to certain anesthetics (eg, halothane) and depolarizing skeletal muscle relaxants (eg, succinylcholine). The reaction consists primarily of rigidity of skeletal muscles, hypermetabolism, and high fever. A **high cytosolic concentration of $Ca^{2+}$** in skeletal muscle is a major factor in its causation. Unless malignant hyperthermia is recognized and treated immediately, patients may die acutely of ventricular fibrillation or survive to succumb subsequently from other serious complications. Appropriate treatment is to stop the anesthetic and administer the drug **dantrolene** intravenously. Dantrolene is a skeletal muscle relaxant that acts to inhibit release of $Ca^{2+}$ from the sarcoplasmic reticulum into the cytosol, thus preventing the increase of cytosolic $Ca^{2+}$ found in malignant hyperthermia.

***Table 49–1.*** Sequence of events in contraction and relaxation of skeletal muscle.[1]

| |
|---|
| **Steps in contraction** |
| (1) Discharge of motor neuron |
| (2) Release of transmitter (acetylcholine) at motor endplate |
| (3) Binding of acetylcholine to nicotinic acetylcholine receptors |
| (4) Increased $Na^+$ and $K^+$ conductance in endplate membrane |
| (5) Generation of endplate potential |
| (6) Generation of action potential in muscle fibers |
| (7) Inward spread of depolarization along T tubules |
| (8) Release of $Ca^{2+}$ from terminal cisterns of sarcoplasmic reticulum and diffusion to thick and thin filaments |
| (9) Binding of $Ca^{2+}$ to troponin C, uncovering myosin binding sites of actin |
| (10) Formation of cross-linkages between actin and myosin and sliding of thin on thick filaments, producing shortening |
| **Steps in relaxation** |
| (1) $Ca^{2+}$ pumped back into sarcoplasmic reticulum |
| (2) Release of $Ca^{2+}$ from troponin |
| (3) Cessation of interaction between actin and myosin |

[1]Reproduced, with permission, from Ganong WF: *Review of Medical Physiology,* 21st ed. McGraw-Hill, 2003.

Malignant hyperthermia also occurs in swine. Susceptible animals homozygous for malignant hyperthermia respond to stress with a fatal reaction **(porcine stress syndrome)** similar to that exhibited by humans. If the reaction occurs prior to slaughter, it affects the quality of the pork adversely, resulting in an inferior product. Both events can result in considerable economic losses for the swine industry.

The finding of a high level of cytosolic $Ca^{2+}$ in muscle in malignant hyperthermia suggested that the condition might be caused by abnormalities of the $Ca^{2+}$ ATPase or of the $Ca^{2+}$ release channel. No abnormalities were detected in the former, but sequencing of cDNAs for the latter protein proved insightful, particularly in swine. All cDNAs from swine with malignant hyperthermia so far examined have shown a substitution of T for C1843, resulting in the substitution of Cys for $Arg^{615}$ in the $Ca^{2+}$ release channel. The mutation affects the function of the channel in that it opens more easily and remains open longer; the net result is massive release of $Ca^{2+}$ into the cytosol, ultimately causing sustained muscle contraction.

The picture is more complex in humans, since malignant hyperthermia exhibits **genetic heterogeneity.** Members of a number of families who suffer from malignant hyperthermia have not shown genetic linkage to the *RYR1* gene. Some humans susceptible to malignant hyperthermia have been found to exhibit the same mutation found in swine, and others have a variety of point mutations at different loci in the *RYR1* gene. Certain families with malignant hypertension have been found to have mutations affecting the DHPR. Figure 49–10 summarizes the probable chain of events in malignant hyperthermia. The major promise of these findings is that, once additional mutations are detected, it will be possible to **screen,** using suitable DNA probes, for individuals at risk of developing malignant hyperthermia during anesthesia. Current screening tests (eg, the in vitro caffeine-halothane test) are relatively unreliable. Affected individuals could then be given alternative anesthetics, which would not endanger their lives. It should also be possible, if desired, to eliminate malignant hyperthermia from swine populations using suitable breeding practices.

Another condition due to mutations in the *RYR1* gene is **central core disease.** This is a rare myopathy presenting in infancy with hypotonia and proximal muscle weakness. Electron microscopy reveals an absence of mitochondria in the center of many type I (see below) muscle fibers. Damage to mitochondria induced by high intracellular levels of $Ca^{2+}$ secondary to abnormal functioning of *RYR1* appears to be responsible for the morphologic findings.

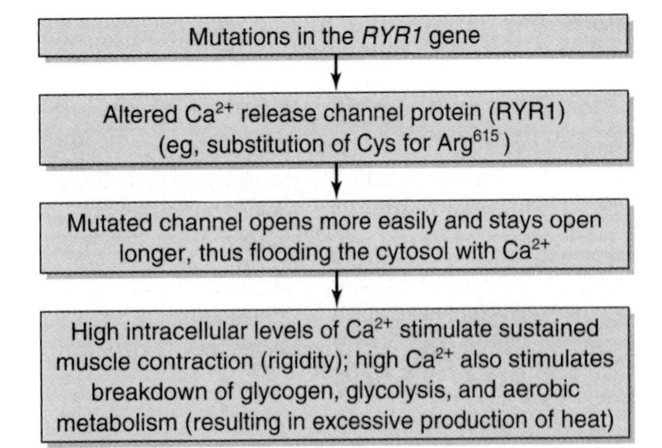


***Figure 49–10.*** Simplified scheme of the causation of malignant hyperthermia (MIM 145600). At least 17 different point mutations have been detected in the *RYR1* gene, some of which are associated with central core disease (MIM 117000). It is estimated that at least 50% of families with members who have malignant hyperthermia are linked to the *RYR1* gene. Some individuals with mutations in the gene encoding DHPR have also been detected; it is possible that mutations in other genes for proteins involved in certain aspects of muscle metabolism will also be found.

## MUTATIONS IN THE GENE ENCODING DYSTROPHIN CAUSE DUCHENNE MUSCULAR DYSTROPHY

A number of additional proteins play various roles in the structure and function of muscle. They include titin (the largest protein known), nebulin, α-actinin, desmin, dystrophin, and calcineurin. Some properties of these proteins are summarized in Table 49–2.

**Dystrophin** is of special interest. Mutations in the gene encoding this protein have been shown to be the cause of Duchenne muscular dystrophy and the milder Becker muscular dystrophy (see Figure 49–11). They are also implicated in some cases of dilated cardiomyopathy (see below). The gene encoding dystrophin is the largest gene known (≈ 2300 kb) and is situated on the X chromosome, accounting for the maternal inheritance pattern of Duchenne and Becker muscular dystrophies. As shown in Figure 49–12, dystrophin forms part of a large complex of proteins that attach to or interact with the plasmalemma. Dystrophin links the actin cytoskeleton to the ECM and appears to be needed for assembly of the synaptic junction. Impairment of these processes by formation of defective dystrophin is presumably critical in the causation of

***Table 49–2.*** Some other important proteins of muscle.

| Protein | Location | Comment or Function |
|---|---|---|
| Titin | Reaches from the Z line to the M line | Largest protein in body. Role in relaxation of muscle. |
| Nebulin | From Z line along length of actin filaments | May regulate assembly and length of actin filaments. |
| α-Actinin | Anchors actin to Z lines | Stabilizes actin filaments. |
| Desmin | Lies alongside actin filaments | Attaches to plasma membrane (plasmalemma). |
| Dystrophin | Attached to plasmalemma | Deficient in Duchenne muscular dystrophy. Mutations of its gene can also cause dilated cardiomyopathy. |
| Calcineurin | Cytosol | A calmodulin-regulated protein phosphatase. May play important roles in cardiac hypertrophy and in regulating amounts of slow and fast twitch muscles. |
| Myosin-binding protein C | Arranged transversely in sarcomere A-bands | Binds myosin and titin. Plays a role in maintaining the structural integrity of the sarcomere. |

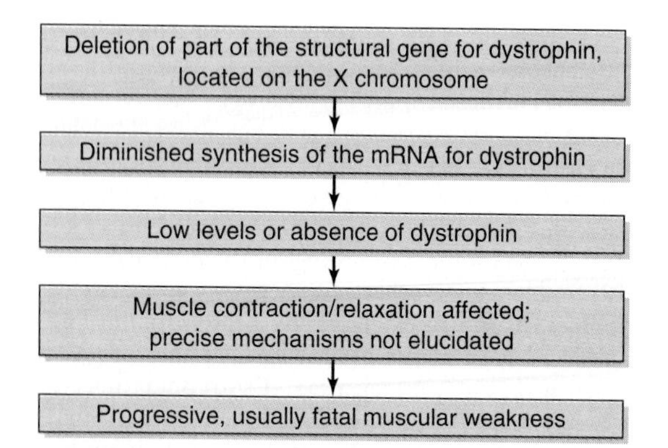


***Figure 49–11.*** Summary of the causation of Duchenne muscular dystrophy (MIM 310200).

Duchenne muscular dystrophy. Mutations in the genes encoding some of the components of the sarcoglycan complex shown in Figure 49–12 are responsible for limb-girdle and certain other congenital forms of muscular dystrophy.

## CARDIAC MUSCLE RESEMBLES SKELETAL MUSCLE IN MANY RESPECTS

The general picture of muscle contraction in the heart resembles that of skeletal muscle. Cardiac muscle, like skeletal muscle, is **striated** and uses the actin-myosin-tropomyosin-troponin system described above. Unlike skeletal muscle, cardiac muscle exhibits intrinsic rhythmicity, and individual myocytes communicate with each other because of its syncytial nature. The **T tubular system** is more developed in cardiac muscle, whereas the **sarcoplasmic reticulum** is less extensive and consequently the intracellular supply of $Ca^{2+}$ for contraction is less. Cardiac muscle thus relies on **extracellular $Ca^{2+}$** for contraction; if isolated cardiac muscle is deprived of $Ca^{2+}$, it ceases to beat within approximately 1 minute, whereas skeletal muscle can continue to contract without an extracellular source of $Ca^{2+}$. **Cyclic AMP** plays a more prominent role in cardiac than in skeletal muscle. It modulates intracellular levels of $Ca^{2+}$ through the activation of protein kinases; these enzymes phosphorylate various transport proteins in the sarcolemma and sarcoplasmic reticulum and also in the troponin-tropomyosin regulatory complex, affecting intracellular levels of $Ca^{2+}$ or responses to it. There is a rough correlation between the phosphorylation of TpI and the increased contraction of cardiac muscle induced by catecholamines. This may account for the **inotropic effects** (increased contractility) of β-adrenergic compounds on the heart. Some differences among skeletal, cardiac, and smooth muscle are summarized in Table 49–3.

### $Ca^{2+}$ Enters Myocytes via $Ca^{2+}$ Channels & Leaves via the $Na^+$-$Ca^{2+}$ Exchanger & the $Ca^{2+}$ ATPase

As stated above, extracellular $Ca^{2+}$ plays an important role in contraction of cardiac muscle but not in skeletal muscle. This means that $Ca^{2+}$ both enters and leaves myocytes in a regulated manner. We shall briefly consider three transmembrane proteins that play roles in this process.

#### A. $Ca^{2+}$ CHANNELS

$Ca^{2+}$ enters myocytes via these channels, which allow entry only of $Ca^{2+}$ ions. The major portal of entry is the

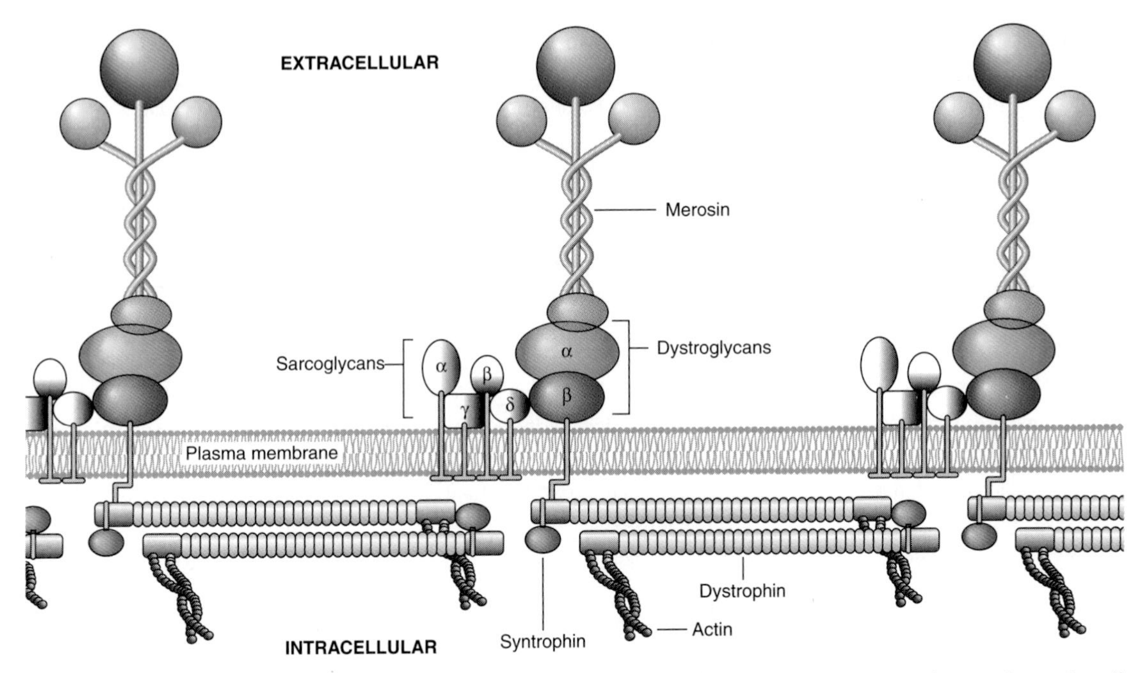


***Figure 49–12.*** Organization of dystrophin and other proteins in relation to the plasma membrane of muscle cells. Dystrophin is part of a large oligomeric complex associated with several other protein complexes. The dystroglycan complex consists of α-dystroglycan, which associates with the basal lamina protein merosin, and β-dystroglycan, which binds α-dystroglycan and dystrophin. Syntrophin binds to the carboxyl terminal of dystrophin. The sarcoglycan complex consists of four transmembrane proteins: α-, β-, γ-, and δ-sarcoglycan. The function of the sarcoglycan complex and the nature of the interactions within the complex and between it and the other complexes are not clear. The sarcoglycan complex is formed only in striated muscle, and its subunits preferentially associate with each other, suggesting that the complex may function as a single unit. Mutations in the gene encoding dystrophin cause Duchenne and Becker muscular dystrophy; mutations in the genes encoding the various sarcoglycans have been shown to be responsible for limb-girdle dystrophies (eg, MIM 601173). (Reproduced, with permission, from Duggan DJ et al: Mutations in the sarcoglycan genes in patients with myopathy. N Engl J Med 1997;336:618.)

L-type (long-duration current, large conductance) or slow $Ca^{2+}$ channel, which is voltage-gated, opening during depolarization induced by spread of the cardiac action potential and closing when the action potential declines. These channels are equivalent to the dihydropyridine receptors of skeletal muscle (Figure 49–8). Slow $Ca^{2+}$ channels are regulated by cAMP-dependent protein kinases (stimulatory) and cGMP-protein kinases (inhibitory) and are blocked by so-called calcium channel blockers (eg, verapamil). Fast (or T, transient) $Ca^{2+}$ channels are also present in the plasmalemma, though in much lower numbers; they probably contribute to the early phase of increase of myoplasmic $Ca^{2+}$.

The resultant increase of $Ca^{2+}$ in the myoplasm acts on the $Ca^{2+}$ release channel of the sarcoplasmic reticulum to open it. This is called $Ca^{2+}$-induced $Ca^{2+}$ release (CICR). It is estimated that approximately 10% of the $Ca^{2+}$ involved in contraction enters the cytosol from the extracellular fluid and 90% from the sarcoplasmic reticulum. However, the former 10% is important, as the rate of increase of $Ca^{2+}$ in the myoplasm is important, and entry via the $Ca^{2+}$ channels contributes appreciably to this.

### B. $Ca^{2+}$-$Na^+$ Exchanger

This is the principal route of exit of $Ca^{2+}$ from myocytes. In resting myocytes, it helps to maintain a low level of free intracellular $Ca^{2+}$ by exchanging one $Ca^{2+}$ for three $Na^+$. The energy for the uphill movement of $Ca^{2+}$ out of the cell comes from the downhill movement of $Na^+$ into the cell from the plasma. This exchange contributes to relaxation but may run in the re-

***Table 49–3.*** Some differences between skeletal, cardiac, and smooth muscle.

| Skeletal Muscle | Cardiac Muscle | Smooth Muscle |
|---|---|---|
| 1. Striated. | 1. Striated. | 1. Nonstriated. |
| 2. No syncytium. | 2. Syncytial. | 2. Syncytial. |
| 3. Small T tubules. | 3. Large T tubules. | 3. Generally rudimentary T tubules. |
| 4. Sarcoplasmic reticulum well-developed and $Ca^{2+}$ pump acts rapidly. | 4. Sarcoplasmic reticulum present and $Ca^{2+}$ pump acts relatively rapidly. | 4. Sarcoplasmic reticulum often rudimentary and $Ca^{2+}$ pump acts slowly. |
| 5. Plasmalemma lacks many hormone receptors. | 5. Plasmalemma contains a variety of receptors (eg, α- and β-adrenergic). | 5. Plasmalemma contains a variety of receptors (eg, α- and β-adrenergic). |
| 6. Nerve impulse initiates contraction. | 6. Has intrinsic rhythmicity. | 6. Contraction initiated by nerve impulses, hormones, etc. |
| 7. Extracellular fluid $Ca^{2+}$ not important for contraction. | 7. Extracellular fluid $Ca^{2+}$ important for contraction. | 7. Extracellular fluid $Ca^{2+}$ important for contraction. |
| 8. Troponin system present. | 8. Troponin system present. | 8. Lacks troponin system; uses regulatory head of myosin. |
| 9. Caldesmon not involved. | 9. Caldesmon not involved. | 9. Caldesmon is important regulatory protein. |
| 10. Very rapid cycling of the cross-bridges. | 10. Relatively rapid cycling of the cross-bridges. | 10. Slow cycling of the cross-bridges permits slow prolonged contraction and less utilization of ATP. |

verse direction during excitation. Because of the $Ca^{2+}$-$Na^+$ exchanger, anything that causes intracellular $Na^+$ ($Na^+_i$) to rise will secondarily cause $Ca^{2+}_i$ to rise, causing more forceful contraction. This is referred to as a positive inotropic effect. One example is when the drug **digitalis** is used to treat heart failure. Digitalis inhibits the sarcolemmal $Na^+$-$K^+$ ATPase, diminishing exit of $Na^+$ and thus increasing $Na^+_i$. This in turn causes $Ca^{2+}$ to increase, via the $Ca^{2+}$-$Na^+$ exchanger. The increased $Ca^{2+}_i$ results in increased force of cardiac contraction, of benefit in heart failure.

### C. $Ca^{2+}$ ATPase

This $Ca^{2+}$ pump, situated in the sarcolemma, also contributes to $Ca^{2+}$ exit but is believed to play a relatively minor role as compared with the $Ca^{2+}$-$Na^+$ exchanger.

It should be noted that there are a variety of **ion channels** (Chapter 41) in most cells, for $Na^+$, $K^+$, $Ca^{2+}$, etc. Many of them have been cloned in recent years and their dispositions in their respective membranes worked out (number of times each one crosses its membrane, location of the actual ion transport site in the protein, etc). They can be classified as indicated in Table 49–4. Cardiac muscle is rich in ion channels, and they are also important in skeletal muscle. Mutations in genes encoding ion channels have been shown to be responsible for a number of relatively rare conditions affecting muscle. These and other diseases due to mutations of ion channels have been termed **channelopathies;** some are listed in Table 49–5.

***Table 49–4.*** Major types of ion channels found in cells.

| Type | Comment |
|---|---|
| External ligand-gated | Open in response to a specific extracellular molecule, eg, acetylcholine. |
| Internal ligand-gated | Open or close in response to a specific intracellular molecule, eg, a cyclic nucleotide. |
| Voltage-gated | Open in response to a change in membrane potential, eg, $Na^+$, $K^+$, and $Ca^{2+}$ channels in heart. |
| Mechanically gated | Open in response to change in mechanical pressure. |

***Table 49–5.*** Some disorders (channelopathies) due to mutations in genes encoding polypeptide constituents of ion channels.[1]

| Disorder[2] | Ion Channel and Major Organs Involved |
|---|---|
| Central core disease (MIM 117000) | $Ca^{2+}$ release channel (RYR1)<br>Skeletal muscle |
| Cystic fibrosis (MIM 219700) | CFTR ($Cl^-$ channel)<br>Lungs, pancreas |
| Hyperkalemic periodic paralysis (MIM 170500) | Sodium channel<br>Skeletal muscle |
| Hypokalemic periodic paralysis (MIM 114208) | Slow $Ca^{2+}$ voltage channel (DHPR)<br>Skeletal muscle |
| Malignant hyperthermia (MIM 180901) | $Ca^{2+}$ release channel (RYR1)<br>Skeletal muscle |
| Myotonia congenita (MIM 160800) | Chloride channel<br>Skeletal muscle |

[1]Data in part from Ackerman NJ, Clapham DE: Ion channels—basic science and clinical disease. N Engl J Med 1997;336:1575.
[2]Other channelopathies include the long QT syndrome (MIM 192500); pseudoaldosteronism (Liddle syndrome, MIM 177200); persistent hyperinsulinemic hypoglycemia of infancy (MIM 601820); hereditary X-linked recessive type II nephrolithiasis of infancy (Dent syndrome, MIM 300009); and generalized myotonia, recessive (Becker disease, MIM 255700). The term "myotonia" signifies any condition in which muscles do not relax after contraction.

***Table 49–6.*** Biochemical causes of inherited cardiomyopathies.[1,2]

| Cause | Proteins or Process Affected |
|---|---|
| Inborn errors of fatty acid oxidation | Carnitine entry into cells and mitochondria<br>Certain enzymes of fatty acid oxidation |
| Disorders of mitochondrial oxidative phosphorylation | Proteins encoded by mitochondrial genes<br>Proteins encoded by nuclear genes |
| Abnormalities of myocardial contractile and structural proteins | β-Myosin heavy chains, troponin, tropomyosin, dystrophin |

[1]Based on Kelly DP, Strauss AW: Inherited cardiomyopathies. N Engl J Med 1994;330:913.
[2]Mutations (eg, point mutations, or in some cases deletions) in the genes (nuclear or mitochondrial) encoding various proteins, enzymes, or tRNA molecules are the fundamental causes of the inherited cardiomyopathies. Some conditions are mild, whereas others are severe and may be part of a syndrome affecting other tissues.

## Inherited Cardiomyopathies Are Due to Disorders of Cardiac Energy Metabolism or to Abnormal Myocardial Proteins

An inherited cardiomyopathy is any structural or functional abnormality of the ventricular myocardium due to an inherited cause. There are nonheritable types of cardiomyopathy, but these will not be described here. As shown in Table 49–6, the causes of inherited cardiomyopathies fall into two broad classes: (1) disorders of cardiac energy metabolism, mainly reflecting mutations in genes encoding enzymes or proteins involved in fatty acid oxidation (a major source of energy for the myocardium) and oxidative phosphorylation; and (2) mutations in genes encoding proteins involved in or affecting myocardial contraction, such as myosin, tropomyosin, the troponins, and cardiac myosin-binding protein C. Mutations in the genes encoding these latter proteins cause familial hypertrophic cardiomyopathy, which will now be discussed.

## Mutations in the Cardiac β-Myosin Heavy Chain Gene Are One Cause of Familial Hypertrophic Cardiomyopathy

Familial hypertrophic cardiomyopathy is one of the most frequent hereditary cardiac diseases. Patients exhibit hypertrophy—often massive—of one or both ventricles, starting early in life, and not related to any extrinsic cause such as hypertension. Most cases are transmitted in an autosomal dominant manner; the rest are sporadic. Until recently, its cause was obscure. However, this situation changed when studies of one affected family showed that a **missense mutation** (ie, substitution of one amino acid by another) in the β-myosin heavy chain gene was responsible for the condition. Subsequent studies have shown a number of missense mutations in this gene, all coding for highly conserved residues. Some individuals have shown other mutations, such as formation of an α/β-myosin heavy chain hybrid gene. Patients with familial hypertrophic cardiomyopathy can show great variation in clinical picture. This in part reflects genetic heterogeneity; ie, mutation in a number of **other genes** (eg, those encoding cardiac actin, tropomyosin, cardiac troponins I and T, essential and regulatory myosin light chains, and cardiac myosin-binding protein C) may also cause familial hypertrophic cardiomyopathy. In addition, mutations at different sites in the gene for β-myosin heavy chain may affect the function of the protein to a greater or lesser extent. The missense mutations are clustered in the head and head-rod regions of myosin heavy chain. One hypothesis is that the mutant polypeptides ("poison polypeptides") cause formation of abnormal myofibrils, eventually resulting in compensatory hypertrophy. Some mutations alter the charge of the amino acid (eg, substitution of arginine for glutamine), presumably affecting the conformation of the protein more markedly and thus affecting its function. Patients with these mutations have a significantly shorter life expectancy than patients in whom the mutation produced no alteration in charge. Thus, definition of the precise mutations involved in the genesis of FHC may prove to be of important prognostic value; it can be accomplished by appropriate use of the polymerase chain reaction on genomic DNA obtained from one sample of blood lymphocytes. Figure 49–13 is a simplified scheme of the events causing familial hypertrophic cardiomyopathy.

Another type of cardiomyopathy is termed **dilated cardiomyopathy.** Mutations in the genes encoding dystrophin, muscle LIM protein (so called because it was found to contain a cysteine-rich domain originally detected in three proteins: Lin-II, Isl-1, and Mec-3), and the cyclic response-element binding protein (CREB) have been implicated in the causation of this condition. The first two proteins help organize the contractile apparatus of cardiac muscle cells, and CREB is involved in the regulation of a number of genes in these cells. Current research is not only elucidating the molecular causes of the cardiomyopathies but is also disclosing mutations that cause **cardiac developmental disorders** (eg, septal defects) and arrhythmias (eg, due to mutations affecting ion channels).

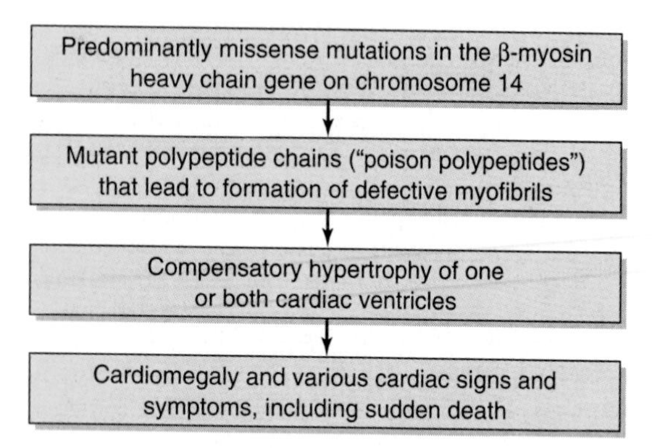


***Figure 49–13.*** Simplified scheme of the causation of familial hypertrophic cardiomyopathy (MIM 192600) due to mutations in the gene encoding β-myosin heavy chain. Mutations in genes encoding other proteins, such as the troponins, tropomyosin, and cardiac myosin-binding protein C can also cause this condition. Mutations in genes encoding yet other proteins (eg, dystrophin) are involved in the causation of dilated cardiomyopathy.

## $Ca^{2+}$ Also Regulates Contraction of Smooth Muscle

While all muscles contain actin, myosin, and tropomyosin, only vertebrate **striated** muscles contain the troponin system. Thus, the mechanisms that regulate contraction must differ in various contractile systems.

Smooth muscles have molecular structures similar to those in striated muscle, but the sarcomeres are not aligned so as to generate the striated appearance. Smooth muscles contain α-actinin and tropomyosin molecules, as do skeletal muscles. They do not have the troponin system, and the light chains of smooth muscle myosin molecules differ from those of striated muscle myosin. Regulation of smooth muscle contraction is **myosin-based,** unlike striated muscle, which is actin-based. However, like striated muscle, smooth muscle contraction is regulated by $Ca^{2+}$.

## Phosphorylation of Myosin Light Chains Initiates Contraction of Smooth Muscle

When smooth muscle myosin is bound to F-actin in the absence of other muscle proteins such as tropomyosin, there is no detectable ATPase activity. This absence of activity is quite unlike the situation described for striated muscle myosin and F-actin, which has abundant ATPase activity. Smooth muscle myosin contains light chains that prevent the binding of the myosin head to F-actin; they must be phosphorylated before they allow F-actin to activate myosin ATPase. The ATPase activity then attained hydrolyzes ATP about tenfold more slowly than the corresponding activity in skeletal muscle. The phosphate on the myosin light chains may form a chelate with the $Ca^{2+}$ bound to the tropomyosin-TpC-actin complex, leading to an increased rate of formation of cross-bridges between the myosin heads and actin. The phosphorylation of light chains initiates the attachment-detachment contraction cycle of smooth muscle.

## Myosin Light Chain Kinase Is Activated by Calmodulin-4$Ca^{2+}$ & Then Phosphorylates the Light Chains

Smooth muscle sarcoplasm contains a myosin light chain kinase that is calcium-dependent. The $Ca^{2+}$ activation of myosin light chain kinase requires binding of **calmodulin-4$Ca^{2+}$** to its kinase subunit (Figure 49–14).

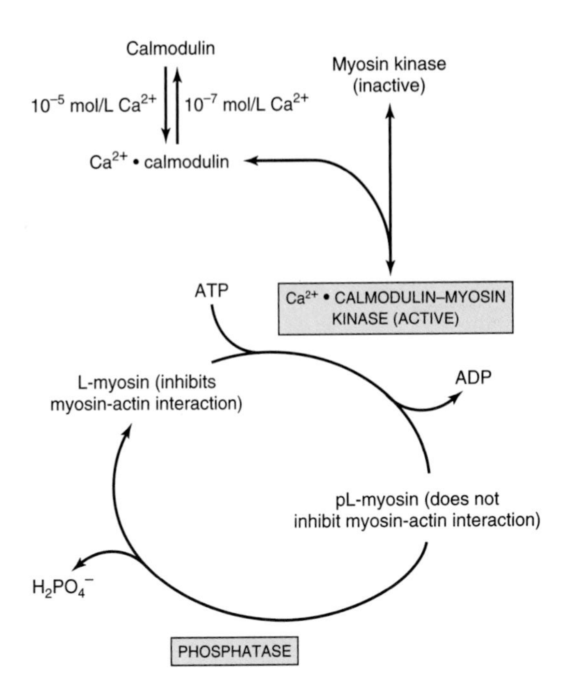


***Figure 49–14.*** **Regulation of smooth muscle contraction by $Ca^{2+}$. pL-myosin is the phosphorylated light chain of myosin; L-myosin is the dephosphorylated light chain.** (Adapted from Adelstein RS, Eisenberg R: Regulation and kinetics of actin-myosin ATP interaction. Annu Rev Biochem 1980;49:921.)

The calmodulin-$4Ca^{2+}$-activated light chain kinase phosphorylates the light chains, which then ceases to inhibit the myosin–F-actin interaction. The contraction cycle then begins.

## Smooth Muscle Relaxes When the Concentration of $Ca^{2+}$ Falls Below $10^{-7}$ Molar

Relaxation of smooth muscle occurs when sarcoplasmic $Ca^{2+}$ falls below $10^{-7}$ mol/L. The $Ca^{2+}$ dissociates from calmodulin, which in turn dissociates from the myosin light chain kinase, inactivating the kinase. No new phosphates are attached to the p-light chain, and light chain protein phosphatase, which is continually active and calcium-independent, removes the existing phosphates from the light chains. Dephosphorylated myosin p-light chain then inhibits the binding of myosin heads to F-actin and the ATPase activity. The myosin head detaches from the F-actin in the presence of ATP, but it cannot reattach because of the presence of dephosphorylated p-light chain; hence, relaxation occurs.

Table 49–7 summarizes and compares the regulation of actin-myosin interactions (activation of myosin ATPase) in striated and smooth muscles.

The myosin light chain kinase is not directly affected or activated by cAMP. However, cAMP-activated protein kinase can phosphorylate the myosin light chain kinase (not the light chains themselves). The phosphorylated myosin light chain kinase exhibits a significantly lower affinity for calmodulin-$Ca^{2+}$ and thus is less sensitive to activation. Accordingly, an increase in cAMP dampens the contraction response of smooth muscle to a given elevation of sarcoplasmic $Ca^{2+}$. This molecular mechanism can explain the relaxing effect of β-adrenergic stimulation on smooth muscle.

Another protein that appears to play a $Ca^{2+}$-dependent role in the regulation of smooth muscle contraction is **caldesmon** (87 kDa). This protein is ubiquitous in smooth muscle and is also found in nonmuscle tissue. At low concentrations of $Ca^{2+}$, it binds to tropomyosin and actin. This prevents interaction of actin with myosin, keeping muscle in a relaxed state. At higher concentrations of $Ca^{2+}$, $Ca^{2+}$-calmodulin binds caldesmon, releasing it from actin. The latter is then free to bind to myosin, and contraction can occur. Caldesmon is also subject to phosphorylation-dephosphorylation; when phosphorylated, it cannot bind actin, again freeing the latter to interact with myosin. Caldesmon may also participate in organizing the structure of the contractile apparatus in smooth muscle. Many of its effects have been demonstrated in vitro, and its physiologic significance is still under investigation.

As noted in Table 49–3, slow cycling of the cross-bridges permits slow prolonged contraction of smooth muscle (eg, in viscera and blood vessels) with less utilization of ATP compared with striated muscle. The ability of smooth muscle to maintain force at reduced velocities of contraction is referred to as the **latch state;** this is an important feature of smooth muscle, and its precise molecular bases are under study.

## Nitric Oxide Relaxes the Smooth Muscle of Blood Vessels & Also Has Many Other Important Biologic Functions

Acetylcholine is a vasodilator that acts by causing relaxation of the smooth muscle of blood vessels. However, it does not act directly on smooth muscle. A key observation was that if endothelial cells were stripped away from underlying smooth muscle cells, acetylcholine no longer exerted its vasodilator effect. This finding indicated that vasodilators such as acetylcholine initially interact with the endothelial cells of small blood vessels via receptors. The receptors are coupled to the phosphoinositide cycle, leading to the intracellular release of

***Table 49–7.*** Actin-myosin interactions in striated and smooth muscle.

| | Striated Muscle | Smooth Muscle (and Nonmuscle Cells) |
|---|---|---|
| Proteins of muscle filaments | Actin<br>Myosin<br>Tropomyosin<br>Troponin (TpI, TpT, TpC) | Actin<br>Myosin[1]<br>Tropomyosin |
| Spontaneous interaction of F-actin and myosin alone (spontaneous activation of myosin ATPase by F-actin | Yes | No |
| Inhibitor of F-actin–myosin interaction (inhibitor of F-actin–dependent activation of ATPase) | Troponin system (TpI) | Unphosphorylated myosin light chain |
| Contraction activated by | $Ca^{2+}$ | $Ca^{2+}$ |
| Direct effect of $Ca^{2+}$ | $4Ca^{2+}$ bind to TpC | $4Ca^{2+}$ bind to calmodulin |
| Effect of protein-bound $Ca^{2+}$ | TpC · $4Ca^{2+}$ antagonizes TpI inhibition of F-actin–myosin interaction (allows F-actin activation of ATPase) | Calmodulin · $4Ca^{2+}$ activates myosin light chain kinase that phosphorylates myosin p-light chain. The phosphorylated p-light chain no longer inhibits F-actin–myosin interaction (allows F-actin activation of ATPase). |

[1]Light chains of myosin are different in striated and smooth muscles.

$Ca^{2+}$ through the action of inositol trisphosphate. In turn, the elevation of $Ca^{2+}$ leads to the liberation of **endothelium-derived relaxing factor (EDRF),** which diffuses into the adjacent smooth muscle. There, it reacts with the heme moiety of a soluble guanylyl cyclase, resulting in activation of the latter, with a consequent elevation of intracellular levels of cGMP (Figure 49–15). This in turn stimulates the activities of certain cGMP-dependent protein kinases, which probably phosphorylate specific muscle proteins, causing relaxation; however, the details are still being clarified. The important coronary artery vasodilator **nitroglycerin,** widely used to relieve angina pectoris, acts to increase intracellular release of EDRF and thus of cGMP.

Quite unexpectedly, EDRF was found to be the gas **nitric oxide (NO).** NO is formed by the action of the enzyme NO synthase, which is cytosolic. The endothelial and neuronal forms of NO synthase are activated by $Ca^{2+}$ (Table 49–8). The substrate is arginine, and the products are citrulline and NO:

$$\text{Arginine} \xrightarrow{\text{NO SYNTHASE}} \text{Citrulline} + \text{NO}$$

NO synthase catalyzes a five-electron oxidation of an amidine nitrogen of arginine. L-Hydroxyarginine is an intermediate that remains tightly bound to the enzyme. NO synthase is a very complex enzyme, employing five redox cofactors: NADPH, FAD, FMN, heme, and tetrahydrobiopterin. NO can also be formed from **nitrite,** derived from vasodilators such as glyceryl trinitrate during their metabolism. NO has a very short half-life (approximately 3–4 seconds) in tissues because it reacts with oxygen and superoxide. The product of the reaction with superoxide is peroxynitrite ($ONOO^-$), which decomposes to form the highly reactive $OH^\bullet$ radical. NO is inhibited by hemoglobin and other heme proteins, which bind it tightly. Chemical inhibitors of NO synthase are now available that can markedly decrease formation of NO. Administration of such inhibitors to animals and humans leads to vasoconstriction and a marked elevation of blood pressure, indicating that NO is of major importance in the maintenance of blood pressure in vivo. Another important cardiovascular effect is that by increasing synthesis of cGMP, it acts as an inhibitor of platelet aggregation (Chapter 51).

Since the discovery of the role of NO as a vasodilator, there has been intense experimental interest in this substance. It has turned out to have a variety of physiologic roles, involving virtually every tissue of the body (Table 49–9). Three major isoforms of NO synthase have been identified, each of which has been cloned, and the chromosomal locations of their genes in hu-

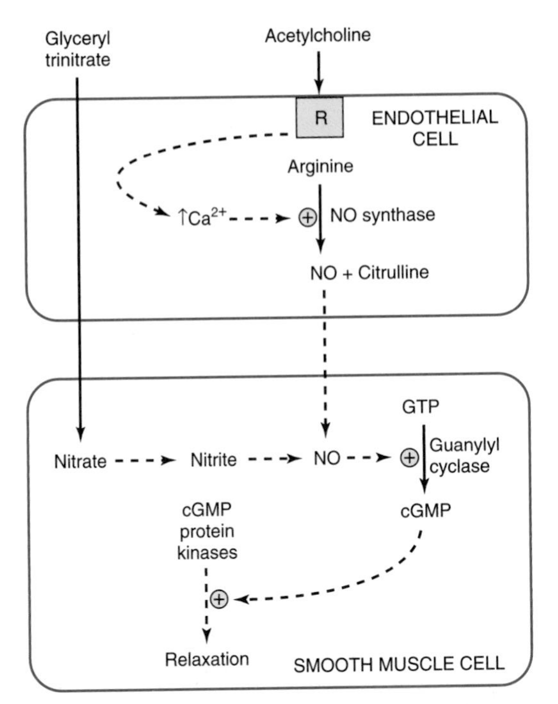


***Figure 49–15.*** Diagram showing formation in an endothelial cell of nitric oxide (NO) from arginine in a reaction catalyzed by NO synthase. Interaction of an agonist (eg, acetylcholine) with a receptor (R) probably leads to intracellular release of $Ca^{2+}$ via inositol trisphosphate generated by the phosphoinositide pathway, resulting in activation of NO synthase. The NO subsequently diffuses into adjacent smooth muscle, where it leads to activation of guanylyl cyclase, formation of cGMP, stimulation of cGMP-protein kinases, and subsequent relaxation. The vasodilator nitroglycerin is shown entering the smooth muscle cell, where its metabolism also leads to formation of NO.

mans have been determined. Gene knockout experiments have been performed on each of the three isoforms and have helped establish some of the postulated functions of NO.

To summarize, research in the past decade has shown that NO plays an important role in many physiologic and pathologic processes.

## SEVERAL MECHANISMS REPLENISH STORES OF ATP IN MUSCLE

The ATP required as the constant energy source for the contraction-relaxation cycle of muscle can be generated (1) by glycolysis, using blood glucose or muscle glycogen, (2) by oxidative phosphorylation, (3) from creatine phosphate, and (4) from two molecules of ADP in a reaction catalyzed by adenylyl kinase (Figure 49–16). The amount of ATP in skeletal muscle is only sufficient to provide energy for contraction for a few seconds, so that ATP must be constantly renewed from one or more of the above sources, depending upon metabolic conditions. As discussed below, there are at least two distinct types of fibers in skeletal muscle, one predominantly active in aerobic conditions and the other in anaerobic conditions; not unexpectedly, they use each of the above sources of energy to different extents.

### Skeletal Muscle Contains Large Supplies of Glycogen

The sarcoplasm of skeletal muscle contains large stores of glycogen, located in granules close to the I bands. The release of glucose from glycogen is dependent on a specific muscle glycogen phosphorylase (Chapter 18), which can be activated by $Ca^{2+}$, epinephrine, and AMP. To generate glucose 6-phosphate for glycolysis in skeletal muscle, glycogen phosphorylase b must be activated to phosphorylase a via phosphorylation by phosphorylase b kinase (Chapter 18). $Ca^{2+}$ promotes the activation of phosphorylase b kinase, also by phosphorylation. Thus, $Ca^{2+}$ both initiates muscle contraction and activates a pathway to provide necessary energy. The hormone epinephrine also activates glycogenolysis in muscle. AMP, produced by breakdown of ADP during muscular exercise, can also activate phosphorylase b without causing phosphorylation. Muscle glycogen phosphorylase b is inactive in **McArdle disease,** one of the glycogen storage diseases (Chapter 18).

### Under Aerobic Conditions, Muscle Generates ATP Mainly by Oxidative Phosphorylation

Synthesis of ATP via oxidative phosphorylation requires a supply of oxygen. Muscles that have a high demand for oxygen as a result of sustained contraction (eg, to maintain posture) store it attached to the heme moiety of **myoglobin.** Because of the heme moiety, muscles containing myoglobin are red, whereas muscles with little or no myoglobin are white. Glucose, derived from the blood glucose or from endogenous glycogen, and fatty acids derived from the triacylglycerols of adipose tissue are the principal substrates used for aerobic metabolism in muscle.

### Creatine Phosphate Constitutes a Major Energy Reserve in Muscle

Creatine phosphate prevents the rapid depletion of ATP by providing a readily available high-energy phosphate that can be used to regenerate ATP from ADP.

***Table 49–8.*** Summary of the nomenclature of the NO synthases and of the effects of knockout of their genes in mice.[1]

| Subtype | Name[2] | Comments | Result of Gene Knockout in Mice[3] |
|---|---|---|---|
| 1 | nNOS | Activity depends on elevated $Ca^{2+}$. First identified in neurons. Calmodulin-activated. | Pyloric stenosis, resistant to vascular stroke, aggressive sexual behavior (males). |
| 2 | iNOS[4] | Independent of elevated $Ca^{2+}$. Prominent in macrophages. | More susceptible to certain types of infection. |
| 3 | eNOS | Activity depends on elevated $Ca^{2+}$. First identified in endothelial cells. | Elevated mean blood pressure. |

[1]Adapted from Snyder SH: No endothelial NO. Nature 1995;377:196.
[2]n, neuronal; i, inducible; e, endothelial.
[3]Gene knockouts were performed by homologous recombination in mice. The enzymes are characterized as neuronal, inducible (macrophage), and endothelial because these were the sites in which they were first identified. However, all three enzymes have been found in other sites, and the neuronal enzyme is also inducible. Each gene has been cloned, and its chromosomal location in humans has been determined.
[4]iNOS is $Ca^{2+}$-independent but binds calmodulin very tightly.

Creatine phosphate is formed from ATP and creatine (Figure 49–16) at times when the muscle is relaxed and demands for ATP are not so great. The enzyme catalyzing the phosphorylation of creatine is creatine kinase (CK), a muscle-specific enzyme with clinical utility in the detection of acute or chronic diseases of muscle.

## SKELETAL MUSCLE CONTAINS SLOW (RED) & FAST (WHITE) TWITCH FIBERS

Different types of fibers have been detected in skeletal muscle. One classification subdivides them into type I (slow twitch), type IIA (fast twitch-oxidative), and type IIB (fast twitch-glycolytic). For the sake of simplicity, we shall consider only two types: type I (slow twitch, oxidative) and type II (fast twitch, glycolytic) (Table 49–10). The type I fibers are red because they contain myoglobin and mitochondria; their metabolism is aerobic, and they maintain relatively sustained contractions. The type II fibers, lacking myoglobin and containing few mitochondria, are white: they derive their energy from anaerobic glycolysis and exhibit relatively short durations of contraction. The proportion of these two types of fibers varies among the muscles of the body, depending on function (eg, whether or not a muscle is involved in sustained contraction, such as maintaining posture). The proportion also varies with training; for example, the number of type I fibers in certain leg muscles increases in athletes training for marathons, whereas the number of type II fibers increases in sprinters.

***Table 49–9.*** Some physiologic functions and pathologic involvements of nitric oxide (NO).

- Vasodilator, important in regulation of blood pressure
- Involved in penile erection; sildenafil citrate (Viagra) affects this process by inhibiting a cGMP phosphodiesterase
- Neurotransmitter in the brain and peripheral autonomic nervous system
- Role in long-term potentiation
- Role in neurotoxicity
- Low level of NO involved in causation of pylorospasm in infantile hypertrophic pyloric stenosis
- May have role in relaxation of skeletal muscle
- May constitute part of a primitive immune system
- Inhibits adhesion, activation, and aggregation of platelets

### A Sprinter Uses Creatine Phosphate & Anaerobic Glycolysis to Make ATP, Whereas a Marathon Runner Uses Oxidative Phosphorylation

In view of the two types of fibers in skeletal muscle and of the various energy sources described above, it is of interest to compare their involvement in a sprint (eg, 100 meters) and in the marathon (42.2 km; just over 26 miles) (Table 49–11).

The major sources of energy in the **100-m sprint** are **creatine phosphate** (first 4–5 seconds) and then **anaerobic glycolysis,** using muscle glycogen as the source of glucose. The two main sites of metabolic control are at glycogen phosphorylase and at PFK-1. The former is activated by $Ca^{2+}$ (released from the sarcoplasmic reticulum during contraction), epinephrine, and

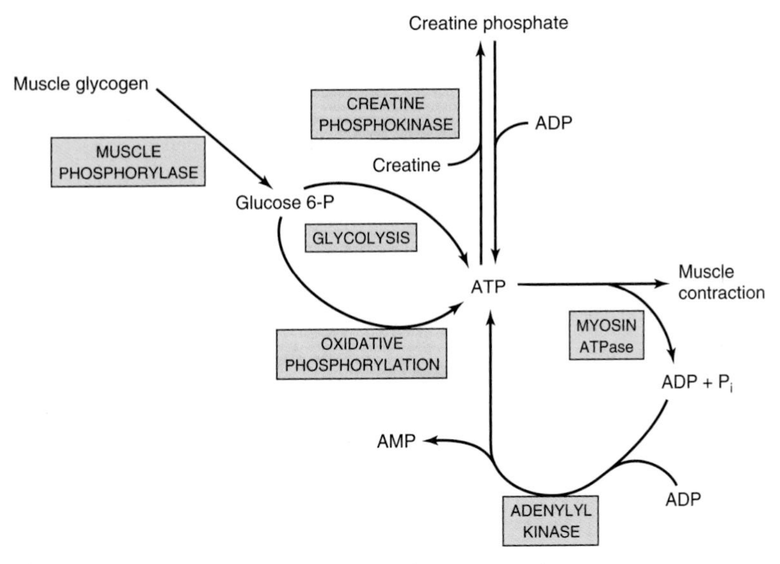


***Figure 49–16.*** The multiple sources of ATP in muscle.

AMP. PFK-1 is activated by AMP, $P_i$, and $NH_3$. Attesting to the efficiency of these processes, the flux through glycolysis can increase as much as 1000-fold during a sprint.

In contrast, in the **marathon,** aerobic metabolism is the principal source of ATP. The major fuel sources are blood glucose and free fatty acids, largely derived from the breakdown of triacylglycerols in adipose tissue, stimulated by epinephrine. Hepatic glycogen is degraded to maintain the level of blood glucose. Muscle glycogen is also a fuel source, but it is degraded much more gradually than in a sprint. It has been calculated that the amounts of glucose in the blood, of glycogen in the liver, of glycogen in muscle, and of triacylglycerol in adipose tissue are sufficient to supply muscle with energy during a marathon for 4 minutes, 18 minutes, 70 minutes, and approximately 4000 minutes, respectively. However, the rate of oxidation of fatty acids by muscle is slower than that of glucose, so that oxidation of glucose and of fatty acids are both major sources of energy in the marathon.

A number of procedures have been used by athletes to counteract muscle fatigue and inadequate strength. These include carbohydrate loading, soda (sodium bi-

***Table 49–10.*** Characteristics of type I and type II fibers of skeletal muscle.

| | Type I Slow Twitch | Type II Fast Twitch |
|---|---|---|
| Myosin ATPase | Low | High |
| Energy utilization | Low | High |
| Mitochondria | Many | Few |
| Color | Red | White |
| Myoglobin | Yes | No |
| Contraction rate | Slow | Fast |
| Duration | Prolonged | Short |

***Table 49–11.*** Types of muscle fibers and major fuel sources used by a sprinter and by a marathon runner.

| Sprinter (100 m) | Marathon Runner |
|---|---|
| Type II (glycolytic) fibers are used predominantly. | Type I (oxidative) fibers are used predominantly. |
| Creatine phosphate is the major energy source during the first 4–5 seconds. | ATP is the major energy source throughout. |
| Glucose derived from muscle glycogen and metabolized by anaerobic glycolysis is the major fuel source. | Blood glucose and free fatty acids are the major fuel sources. |
| Muscle glycogen is rapidly depleted. | Muscle glycogen is slowly depleted. |

carbonate) loading, blood doping (administration of red blood cells), and ingestion of creatine and androstenedione. Their rationales and efficacies will not be discussed here.

## SKELETAL MUSCLE CONSTITUTES THE MAJOR RESERVE OF PROTEIN IN THE BODY

In humans, skeletal muscle protein is the major nonfat source of stored energy. This explains the very large losses of muscle mass, particularly in adults, resulting from prolonged caloric undernutrition.

The study of tissue protein breakdown in vivo is difficult, because amino acids released during intracellular breakdown of proteins can be extensively reutilized for protein synthesis within the cell, or the amino acids may be transported to other organs where they enter anabolic pathways. However, actin and myosin are methylated by a posttranslational reaction, forming **3-methylhistidine.** During intracellular breakdown of actin and myosin, 3-methylhistidine is released and excreted into the urine. The urinary output of the methylated amino acid provides a reliable index of the rate of myofibrillar protein breakdown in the musculature of human subjects.

Various features of muscle metabolism, most of which are dealt with in other chapters of this text, are summarized in Table 49–12.

**Table 49–12.** Summary of major features of the biochemistry of skeletal muscle related to its metabolism.[1]

- Skeletal muscle functions under both aerobic (resting) and anaerobic (eg, sprinting) conditions, so both aerobic and anaerobic glycolysis operate, depending on conditions.
- Skeletal muscle contains myoglobin as a reservoir of oxygen.
- Skeletal muscle contains different types of fibers primarily suited to anaerobic (fast twitch fibers) or aerobic (slow twitch fibers) conditions.
- Actin, myosin, tropomyosin, troponin complex (TpT, TpI, and TpC), ATP, and $Ca^{2+}$ are key constituents in relation to contraction.
- The $Ca^{2+}$ ATPase, the $Ca^{2+}$ release channel, and calsequestrin are proteins involved in various aspects of $Ca^{2+}$ metabolism in muscle.
- Insulin acts on skeletal muscle to increase uptake of glucose.
- In the fed state, most glucose is used to synthesize glycogen, which acts as a store of glucose for use in exercise; "preloading" with glucose is used by some long-distance athletes to build up stores of glycogen.
- Epinephrine stimulates glycogenolysis in skeletal muscle, whereas glucagon does not because of absence of its receptors.
- Skeletal muscle cannot contribute directly to blood glucose because it does not contain glucose-6-phosphatase.
- Lactate produced by anaerobic metabolism in skeletal muscle passes to liver, which uses it to synthesize glucose, which can then return to muscle (the Cori cycle).
- Skeletal muscle contains phosphocreatine, which acts as an energy store for short-term (seconds) demands.
- Free fatty acids in plasma are a major source of energy, particularly under marathon conditions and in prolonged starvation.
- Skeletal muscle can utilize ketone bodies during starvation.
- Skeletal muscle is the principal site of metabolism of branched-chain amino acids, which are used as an energy source.
- Proteolysis of muscle during starvation supplies amino acids for gluconeogenesis.
- Major amino acids emanating from muscle are alanine (destined mainly for gluconeogenesis in liver and forming part of the glucose-alanine cycle) and glutamine (destined mainly for the gut and kidneys).

[1]This table brings together material from various chapters in this book.

## THE CYTOSKELETON PERFORMS MULTIPLE CELLULAR FUNCTIONS

Nonmuscle cells perform mechanical work, including self-propulsion, morphogenesis, cleavage, endocytosis, exocytosis, intracellular transport, and changing cell shape. These cellular functions are carried out by an extensive intracellular network of filamentous structures constituting the **cytoskeleton.** The cell cytoplasm is not a sac of fluid, as once thought. Essentially all eukaryotic cells contain three types of filamentous structures: actin filaments (7–9.5 nm in diameter; also known as microfilaments), microtubules (25 nm), and intermediate filaments (10–12 nm). Each type of filament can be distinguished biochemically and by the electron microscope.

### Nonmuscle Cells Contain Actin That Forms Microfilaments

G-actin is present in most if not all cells of the body. With appropriate concentrations of magnesium and potassium chloride, it spontaneously polymerizes to form double helical F-actin filaments like those seen in muscle. There are at least two types of actin in nonmuscle cells: β-actin and γ-actin. Both types can coexist in the same cell and probably even copolymerize in the same filament. In the cytoplasm, F-actin forms microfilaments of 7–9.5 nm that frequently exist as bundles of a tangled-appearing meshwork. These bundles are prominent just underlying the plasma membrane of many cells and are there referred to as stress fibers. The stress fibers disappear as cell motility increases or upon malignant transformation of cells by chemicals or oncogenic viruses.

Although not organized as in muscle, actin filaments in nonmuscle cells interact with myosin to cause cellular movements.

## Microtubules Contain α- & β-Tubulins

Microtubules, an integral component of the cellular cytoskeleton, consist of cytoplasmic tubes 25 nm in diameter and often of extreme length. Microtubules are necessary for the formation and function of the **mitotic spindle** and thus are present in all eukaryotic cells. They are also involved in the intracellular movement of endocytic and exocytic vesicles and form the major structural components of cilia and flagella. Microtubules are a major component of axons and dendrites, in which they maintain structure and participate in the axoplasmic flow of material along these neuronal processes.

Microtubules are cylinders of 13 longitudinally arranged protofilaments, each consisting of dimers of α-tubulin and β-tubulin, closely related proteins of approximately 50 kDa molecular mass. The tubulin dimers assemble into protofilaments and subsequently into sheets and then cylinders. A microtubule-organizing center, located around a pair of centrioles, nucleates the growth of new microtubules. A third species of tubulin, **γ-tubulin,** appears to play an important role in this assembly. GTP is required for assembly. A variety of proteins are associated with microtubules (microtubule-associated proteins [MAPs], one of which is tau) and play important roles in microtubule assembly and stabilization. Microtubules are in a state of dynamic instability, constantly assembling and disassembling. They exhibit polarity (plus and minus ends); this is important in their growth from centrioles and in their ability to direct intracellular movement. For instance, in axonal transport, the protein **kinesin,** with a myosin-like ATPase activity, uses hydrolysis of ATP to move vesicles down the axon toward the positive end of the microtubular formation. Flow of materials in the opposite direction, toward the negative end, is powered by **cytosolic dynein,** another protein with ATPase activity. Similarly, **axonemal dyneins** power ciliary and flagellar movement. Another protein, **dynamin,** uses GTP and is involved in endocytosis. Kinesins, dyneins, dynamin, and myosins are referred to as **molecular motors.**

An absence of dynein in cilia and flagella results in immotile cilia and flagella, leading to male sterility and chronic respiratory infection, a condition known as **Kartagener syndrome.**

Certain **drugs** bind to microtubules and thus interfere with their assembly or disassembly. These include colchicine (used for treatment of acute gouty arthritis), vinblastine (a vinca alkaloid used for treating certain types of cancer), paclitaxel (Taxol) (effective against ovarian cancer), and griseofulvin (an antifungal agent).

## Intermediate Filaments Differ From Microfilaments & Microtubules

An intracellular fibrous system exists of filaments with an axial periodicity of 21 nm and a diameter of 8–10 nm that is intermediate between that of microfilaments (6 nm) and microtubules (23 nm). Four classes of intermediate filaments are found, as indicated in Table 49–13. They are all elongated, fibrous molecules, with a central rod domain, an amino terminal head, and a carboxyl terminal tail. They form a structure like a rope, and the mature filaments are composed of tetramers packed together in a helical manner. They are important structural components of cells, and most are relatively stable components of the cytoskeleton, not undergoing rapid assembly and disassembly and not

***Table 49–13.*** Classes of intermediate filaments of eukaryotic cells and their distributions.

| Proteins | Molecular Mass | Distributions |
|---|---|---|
| Keratins | | |
| Type I (acidic) | 40–60 kDa | Epithelial cells, hair, |
| Type II (basic) | 50–70 kDa | nails |
| Vimentin-like | | |
| Vimentin | 54 kDa | Various mesenchymal cells |
| Desmin | 53 kDa | Muscle |
| Glial fibrillary acid protein | 50 kDa | Glial cells |
| Peripherin | 66 kDa | Neurons |
| Neurofilaments | | |
| Low (L), medium (M), and high (H)[1] | 60–130 kDa | Neurons |
| Lamins | | |
| A, B, and C | 65–75 kDa | Nuclear lamina |

[1]Refers to their molecular masses.

disappearing during mitosis, as do actin and many microtubular filaments. An important exception to this is provided by the lamins, which, subsequent to phosphorylation, disassemble at mitosis and reappear when it terminates.

**Keratins** form a large family, with about 30 members being distinguished. As indicated in Table 49–13, two major types of keratins are found; all individual keratins are heterodimers made up of one member of each class.

**Vimentins** are widely distributed in mesodermal cells, and desmin, glial fibrillary acidic protein, and peripherin are related to them. All members of the vimentin-like family can copolymerize with each other. Intermediate filaments are very prominent in nerve cells; neurofilaments are classified as low, medium, and high on the basis of their molecular masses. **Lamins** form a meshwork in apposition to the inner nuclear membrane. The distribution of intermediate filaments in normal and abnormal (eg, cancer) cells can be studied by the use of immunofluorescent techniques, using antibodies of appropriate specificities. These antibodies to specific intermediate filaments can also be of use to pathologists in helping to decide the origin of certain dedifferentiated malignant tumors. These tumors may still retain the type of intermediate filaments found in their cell of origin.

A number of **skin diseases,** mainly characterized by blistering, have been found to be due to mutations in genes encoding various keratins. Three of these disorders are epidermolysis bullosa simplex, epidermolytic hyperkeratosis, and epidermolytic palmoplantar keratoderma. The blistering probably reflects a diminished capacity of various layers of the skin to resist mechanical stresses due to abnormalities in microfilament structure.

## SUMMARY

- The myofibrils of skeletal muscle contain thick and thin filaments. The thick filaments contain myosin. The thin filaments contain actin, tropomyosin, and the troponin complex (troponins T, I, and C).
- The sliding filament cross-bridge model is the foundation of current thinking about muscle contraction. The basis of this model is that the interdigitating filaments slide past one another during contraction and cross-bridges between myosin and actin generate and sustain the tension.
- The hydrolysis of ATP is used to drive movement of the filaments. ATP binds to myosin heads and is hydrolyzed to ADP and $P_i$ by the ATPase activity of the actomyosin complex.
- $Ca^{2+}$ plays a key role in the initiation of muscle contraction by binding to troponin C. In skeletal muscle, the sarcoplasmic reticulum regulates distribution of $Ca^{2+}$ to the sarcomeres, whereas inflow of $Ca^{2+}$ via $Ca^{2+}$ channels in the sarcolemma is of major importance in cardiac and smooth muscle.
- Many cases of malignant hyperthermia in humans are due to mutations in the gene encoding the $Ca^{2+}$ release channel.
- A number of differences exist between skeletal and cardiac muscle; in particular, the latter contains a variety of receptors on its surface.
- Some cases of familial hypertrophic cardiomyopathy are due to missense mutations in the gene coding for β-myosin heavy chain.
- Smooth muscle, unlike skeletal and cardiac muscle, does not contain the troponin system; instead, phosphorylation of myosin light chains initiates contraction.
- Nitric oxide is a regulator of vascular smooth muscle; blockage of its formation from arginine causes an acute elevation of blood pressure, indicating that regulation of blood pressure is one of its many functions.
- Duchenne-type muscular dystrophy is due to mutations in the gene, located on the X chromosome, encoding the protein dystrophin.
- Two major types of muscle fibers are found in humans: white (anaerobic) and red (aerobic). The former are particularly used in sprints and the latter in prolonged aerobic exercise. During a sprint, muscle uses creatine phosphate and glycolysis as energy sources; in the marathon, oxidation of fatty acids is of major importance during the later phases.
- Nonmuscle cells perform various types of mechanical work carried out by the structures constituting the cytoskeleton. These structures include actin filaments (microfilaments), microtubules (composed primarily of α- tubulin and β-tubulin), and intermediate filaments. The latter include keratins, vimentin-like proteins, neurofilaments, and lamins.

## REFERENCES


Ackerman MJ, Clapham DE: Ion channels—basic science and clinical disease. N Engl J Med 1997;336:1575.

Andreoli TE: Ion transport disorders: introductory comments. Am J Med 1998;104:85. (First of a series of articles on ion transport disorders published between January and August, 1998. Topics covered were structure and function of ion channels, arrhythmias and antiarrhythmic drugs, Liddle syndrome, cholera, malignant hyperthermia, cystic fibrosis, the periodic paralyses and Bartter syndrome, and Gittelman syndrome.)

Fuller GM, Shields D: *Molecular Basis of Medical Cell Biology.* Appleton & Lange, 1998.

Geeves MA, Holmes KC: Structural mechanism of muscle contraction. Annu Rev Biochem 1999;68:728.

Hille B: *Ion Channels of Excitable Membranes.* Sinauer, 2001.

Howard J: *Mechanics of Motor Proteins and the Cytoskeleton.* Sinauer, 2001.

Lodish H et al (editors): *Molecular Cell Biology,* 4th ed. Freeman, 2000. (Chapters 18 and 19 of this text contain comprehensive descriptions of cell motility and cell shape.)

Loke J, MacLennan DH: Malignant hyperthermia and central core disease: disorders of $Ca^{2+}$ release channels. Am J Med 1998;104:470.

Mayer B, Hemmens B: Biosynthesis and action of nitric oxide in mammalian cells. Trends Biochem Sci 1998;22:477.

Scriver CR et al (editors): *The Metabolic and Molecular Bases of Inherited Disease,* 8th ed. McGraw-Hill, 2001. (This comprehensive four-volume text contains coverage of malignant hyperthermia [Chapter 9], channelopathies [Chapter 204], hypertrophic cardiomyopathy [Chapter 213], the muscular dystrophies [Chapter 216], and disorders of intermediate filaments and their associated proteins [Chapter 221].)

# Plasma Proteins & Immunoglobulins 50

*Robert K. Murray, MD, PhD*

## BIOMEDICAL IMPORTANCE

The fundamental role of blood in the maintenance of **homeostasis** and the ease with which blood can be obtained have meant that the study of its constituents has been of central importance in the development of biochemistry and clinical biochemistry. The basic properties of a number of plasma proteins, including the immunoglobulins (antibodies), are described in this chapter. Changes in the amounts of various plasma proteins and immunoglobulins occur in many diseases and can be monitored by electrophoresis or other suitable procedures. As indicated in an earlier chapter, alterations of the activities of certain **enzymes** found in plasma are of diagnostic use in a number of pathologic conditions.

## THE BLOOD HAS MANY FUNCTIONS

The functions of blood—except for specific cellular ones such as oxygen transport and cell-mediated immunologic defense—are carried out by plasma and its constituents (Table 50–1).

**Plasma** consists of water, electrolytes, metabolites, nutrients, proteins, and hormones. The water and electrolyte composition of plasma is practically the same as that of all extracellular fluids. Laboratory determinations of levels of $Na^+$, $K^+$, $Ca^{2+}$, $Cl^-$, $HCO_3^-$, $PaCO_2$, and of blood pH are important in the management of many patients.

## PLASMA CONTAINS A COMPLEX MIXTURE OF PROTEINS

The concentration of total protein in human plasma is approximately 7.0–7.5 g/dL and comprises the major part of the solids of the plasma. The proteins of the plasma are actually a complex mixture that includes not only simple proteins but also conjugated proteins such as **glycoproteins** and various types of **lipoproteins.** Thousands of **antibodies** are present in human plasma, though the amount of any one antibody is usually quite low under normal circumstances. The relative dimensions and molecular masses of some of the most important plasma proteins are shown in Figure 50–1.

The **separation** of individual proteins from a complex mixture is frequently accomplished by the use of solvents or electrolytes (or both) to remove different protein fractions in accordance with their solubility characteristics. This is the basis of the so-called salting-out methods, which find some usage in the determination of protein fractions in the clinical laboratory. Thus, one can separate the proteins of the plasma into three major groups—**fibrinogen, albumin,** and **globulins**—by the use of varying concentrations of sodium or ammonium sulfate.

The most common method of analyzing plasma proteins is by **electrophoresis.** There are many types of electrophoresis, each using a different supporting medium. In clinical laboratories, **cellulose acetate** is widely used as a supporting medium. Its use permits resolution, after staining, of plasma proteins into five bands, designated albumin, $\alpha_1$, $\alpha_2$, $\beta$, and $\gamma$ fractions, respectively (Figure 50–2). The stained strip of cellulose acetate (or other supporting medium) is called an electrophoretogram. The amounts of these five bands can be conveniently quantified by use of densitometric scanning machines. Characteristic changes in the amounts of one or more of these five bands are found in many diseases.

### The Concentration of Protein in Plasma Is Important in Determining the Distribution of Fluid Between Blood & Tissues

In arterioles, the **hydrostatic pressure** is about 37 mm Hg, with an interstitial (tissue) pressure of 1 mm Hg opposing it. The **osmotic pressure** (oncotic pressure) exerted by the plasma proteins is approximately 25 mm Hg. Thus, a net outward force of about 11 mm Hg drives fluid out into the interstitial spaces. In venules, the hydrostatic pressure is about 17 mm Hg, with the oncotic and interstitial pressures as described above; thus, a net force of about 9 mm Hg attracts water back into the circulation. The above pressures are often referred to as the **Starling forces.** If the concentration of plasma proteins is markedly diminished (eg, due to severe protein malnutrition), fluid is not attracted back into the intravascular compartment and accumulates in the extravascular tissue spaces, a condition known as **edema.** Edema has many causes; protein deficiency is one of them.

*Table 50–1.* Major functions of blood.

| | |
|---|---|
| (1) | **Respiration**—transport of oxygen from the lungs to the tissues and of $CO_2$ from the tissues to the lungs |
| (2) | **Nutrition**—transport of absorbed food materials |
| (3) | **Excretion**—transport of metabolic waste to the kidneys, lungs, skin, and intestines for removal |
| (4) | Maintenance of the normal **acid-base balance** in the body |
| (5) | Regulation of **water balance** through the effects of blood on the exchange of water between the circulating fluid and the tissue fluid |
| (6) | Regulation of **body temperature** by the distribution of body heat |
| (7) | **Defense** against infection by the white blood cells and circulating antibodies |
| (8) | Transport of **hormones** and regulation of metabolism |
| (9) | Transport of **metabolites** |
| (10) | **Coagulation** |

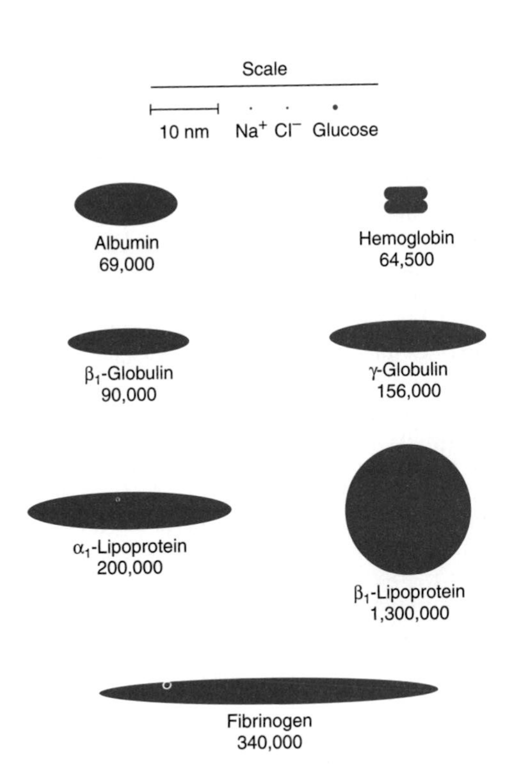


***Figure 50–1.*** Relative dimensions and approximate molecular masses of protein molecules in the blood (Oncley).

## Plasma Proteins Have Been Studied Extensively

Because of the relative ease with which they can be obtained, plasma proteins have been studied extensively in both humans and animals. Considerable information is available about the biosynthesis, turnover, structure, and functions of the major plasma proteins. Alterations of their amounts and of their metabolism in many disease states have also been investigated. In recent years, many of the genes for plasma proteins have been cloned and their structures determined.

The preparation of **antibodies** specific for the individual plasma proteins has greatly facilitated their study, allowing the precipitation and isolation of pure proteins from the complex mixture present in tissues or plasma. In addition, the use of **isotopes** has made possible the determination of their pathways of biosynthesis and of their turnover rates in plasma.

The following generalizations have emerged from studies of plasma proteins.

### A. Most Plasma Proteins Are Synthesized in the Liver

This has been established by experiments at the whole-animal level (eg, hepatectomy) and by use of the isolated perfused liver preparation, of liver slices, of liver homogenates, and of in vitro translation systems using preparations of mRNA extracted from liver. However, the γ-globulins are synthesized in plasma cells and certain plasma proteins are synthesized in other sites, such as endothelial cells.

### B. Plasma Proteins Are Generally Synthesized on Membrane-Bound Polyribosomes

They then traverse the major secretory route in the cell (rough endoplasmic membrane → smooth endoplasmic membrane → Golgi apparatus → secretory vesicles) prior to entering the plasma. Thus, most plasma proteins are synthesized as **preproteins** and initially contain amino terminal signal peptides (Chapter 46). They are usually subjected to various posttranslational modifications (proteolysis, glycosylation, phosphorylation, etc) as they travel through the cell. Transit times through the hepatocyte from the site of synthesis to the plasma vary from 30 minutes to several hours or more for individual proteins.

### C. Most Plasma Proteins Are Glycoproteins

Accordingly, they generally contain either N- or O-linked oligosaccharide chains, or both (Chapter 47). Albumin is the major exception; it does not contain sugar residues. The oligosaccharide chains have various functions (Table 47–2). Removal of terminal sialic acid

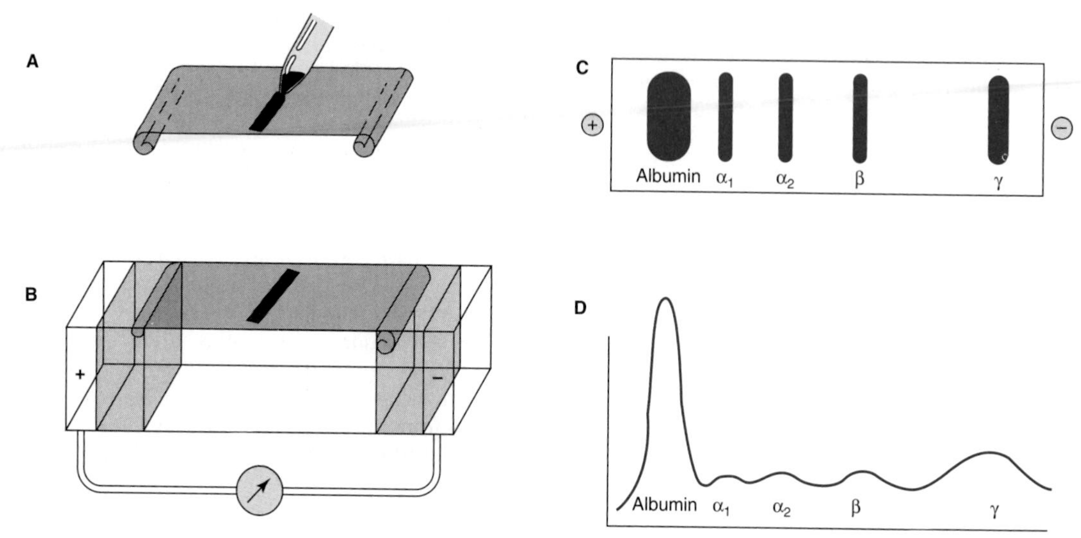


***Figure 50–2.*** Technique of cellulose acetate zone electrophoresis. **A:** A small amount of serum or other fluid is applied to a cellulose acetate strip. **B:** Electrophoresis of sample in electrolyte buffer is performed. **C:** Separated protein bands are visualized in characteristic positions after being stained. **D:** Densitometer scanning from cellulose acetate strip converts bands to characteristic peaks of albumin, $\alpha_1$-globulin, $\alpha_2$-globulin, β-globulin, and γ-globulin. (Reproduced, with permission, from Parslow TG et al [editors]: *Medical Immunology,* 10th ed. McGraw-Hill, 2001.)

residues from certain plasma proteins (eg, ceruloplasmin) by exposure to neuraminidase can markedly shorten their half-lives in plasma (Chapter 47).

### D. Many Plasma Proteins Exhibit Polymorphism

A polymorphism is a mendelian or monogenic trait that exists in the population in at least two phenotypes, neither of which is rare (ie, neither of which occurs with frequency of less than 1–2%). The ABO blood group substances (Chapter 52) are the best-known examples of human polymorphisms. Human plasma proteins that exhibit polymorphism include $\alpha_1$-antitrypsin, haptoglobin, transferrin, ceruloplasmin, and immunoglobulins. The polymorphic forms of these proteins can be distinguished by different procedures (eg, various types of electrophoresis or isoelectric focusing), in which each form may show a characteristic migration. Analyses of these human polymorphisms have proved to be of genetic, anthropologic, and clinical interest.

### E. Each Plasma Protein Has a Characteristic Half-Life in the Circulation

The half-life of a plasma protein can be determined by labeling the isolated pure protein with $^{131}I$ under mild, nondenaturing conditions. This isotope unites covalently with tyrosine residues in the protein. The labeled protein is freed of unbound $^{131}I$ and its specific activity (disintegrations per minute per milligram of protein) determined. A known amount of the radioactive protein is then injected into a normal adult subject, and samples of blood are taken at various time intervals for determinations of radioactivity. The values for radioactivity are plotted against time, and the half-life of the protein (the time for the radioactivity to decline from its peak value to one-half of its peak value) can be calculated from the resulting graph, discounting the times for the injected protein to equilibrate (mix) in the blood and in the extravascular spaces. The half-lives obtained for albumin and haptoglobin in normal healthy adults are approximately 20 and 5 days, respectively. In certain diseases, the half-life of a protein may be markedly altered. For instance, in some gastrointestinal diseases such as regional ileitis (Crohn disease), considerable amounts of plasma proteins, including albumin, may be lost into the bowel through the inflamed intestinal mucosa. Patients with this condition have a **protein-losing gastroenteropathy,** and the half-life of injected iodinated albumin in these subjects may be reduced to as little as 1 day.

### F. The Levels of Certain Proteins in Plasma Increase During Acute Inflammatory States or Secondary to Certain Types of Tissue Damage

These proteins are called **"acute phase proteins"** (or reactants) and include C-reactive protein (CRP, so-named because it reacts with the C polysaccharide of pneumococci), $\alpha_1$-antitrypsin, haptoglobin, $\alpha_1$-acid glycoprotein, and fibrinogen. The elevations of the levels of these proteins vary from as little as 50% to as much as 1000-fold in the case of CRP. Their levels are also usually elevated during chronic inflammatory states and in patients with cancer. These proteins are believed to play a role in the body's response to inflammation. For example, C-reactive protein can stimulate the classic complement pathway, and $\alpha_1$-antitrypsin can neutralize certain proteases released during the acute inflammatory state. CRP is used as a marker of tissue injury, infection, and inflammation, and there is considerable interest in its use as a predictor of certain types of cardiovascular conditions secondary to atherosclerosis. Interleukin-1 (IL-1), a polypeptide released from mononuclear phagocytic cells, is the principal—but not the sole—stimulator of the synthesis of the majority of acute phase reactants by hepatocytes. Additional molecules such as IL-6 are involved, and they as well as IL-1 appear to work at the level of gene transcription.

Table 50–2 summarizes the functions of many of the plasma proteins. The remainder of the material in this chapter presents basic information regarding selected plasma proteins: albumin, haptoglobin, transferrin, ceruloplasmin, $\alpha_1$-antitrypsin, $\alpha_2$-macroglobulin, the immunoglobulins, and the complement system. The lipoproteins are discussed in Chapter 25.

## Albumin Is the Major Protein in Human Plasma

Albumin (69 kDa) is the major protein of human plasma (3.4–4.7 g/dL) and makes up approximately 60% of the total plasma protein. About 40% of albumin is present in the plasma, and the other 60% is present in the extracellular space. The liver produces about 12 g of albumin per day, representing about 25% of total hepatic protein synthesis and half its secreted protein. Albumin is initially synthesized as a **preproprotein.** Its **signal peptide** is removed as it passes into the cisternae of the rough endoplasmic reticulum, and a **hexapeptide** at the resulting amino terminal is subsequently cleaved off farther along the secretory pathway. The synthesis of albumin is depressed in a variety of diseases, particularly those of the liver. The plasma of patients with liver disease often shows a decrease in the ratio of albumin to globulins (decreased albumin-globulin ratio). The synthesis of albumin decreases rela-

***Table 50–2.*** Some functions of plasma proteins.

| Function | Plasma Proteins |
|---|---|
| Antiproteases | Antichymotrypsin<br>$\alpha_1$-Antitrypsin ($\alpha_1$-antiproteinase)<br>$\alpha_2$-Macroglobulin<br>Antithrombin |
| Blood clotting | Various coagulation factors, fibrinogen |
| Enzymes | Function in blood, eg, coagulation factors, cholinesterase<br>Leakage from cells or tissues, eg, aminotransferases |
| Hormones | Erythropoietin[1] |
| Immune defense | Immunoglobulins, complement proteins, $\beta_2$-microglobulin |
| Involvement in inflammatory responses | Acute phase response proteins (eg, C-reactive protein, $\alpha_1$-acid glycoprotein [orosomucoid]) |
| Oncofetal | $\alpha_1$-Fetoprotein (AFP) |
| Transport or binding proteins | Albumin (various ligands, including bilirubin, free fatty acids, ions [$Ca^{2+}$], metals [eg, $Cu^{2+}$, $Zn^{2+}$], metheme, steroids, other hormones, and a variety of drugs<br>Ceruloplasmin (contains $Cu^{2+}$; albumin probably more important in physiologic transport of $Cu^{2+}$)<br>Corticosteroid-binding globulin (transcortin) (binds cortisol)<br>Haptoglobin (binds extracorpuscular hemoglobin)<br>Lipoproteins (chylomicrons, VLDL, LDL, HDL)<br>Hemopexin (binds heme)<br>Retinol-binding protein (binds retinol)<br>Sex hormone-binding globulin (binds testosterone, estradiol)<br>Thyroid-binding globulin (binds $T_4$, $T_3$)<br>Transferrin (transport iron)<br>Transthyretin (formerly prealbumin; binds $T_4$ and forms a complex with retinol-binding protein) |

[1]Various other protein hormones circulate in the blood but are not usually designated as plasma proteins. Similarly, ferritin is also found in plasma in small amounts, but it too is not usually characterized as a plasma protein.

tively early in conditions of protein malnutrition, such as kwashiorkor.

Mature human albumin consists of one polypeptide chain of 585 amino acids and contains 17 disulfide bonds. By the use of proteases, albumin can be subdivided into three **domains,** which have different functions. Albumin has an ellipsoidal shape, which means that it does not increase the viscosity of the plasma as much as an elongated molecule such as fibrinogen does. Because of its relatively low molecular mass (about 69 kDa) and high concentration, albumin is thought to be responsible for 75–80% of the **osmotic pressure** of human plasma. Electrophoretic studies have shown that the plasma of certain humans lacks albumin. These subjects are said to exhibit **analbuminemia.** One cause of this condition is a mutation that affects splicing. Subjects with analbuminemia show only moderate edema, despite the fact that albumin is the major determinant of plasma osmotic pressure. It is thought that the amounts of the other plasma proteins increase and compensate for the lack of albumin.

Another important function of albumin is its ability to **bind various ligands**. These include free fatty acids (FFA), calcium, certain steroid hormones, bilirubin, and some of the plasma tryptophan. In addition, albumin appears to play an important role in transport of copper in the human body (see below). A variety of drugs, including sulfonamides, penicillin G, dicumarol, and aspirin, are bound to albumin; this finding has important pharmacologic implications.

Preparations of human albumin have been widely used in the treatment of hemorrhagic shock and of burns. However, this treatment is under review because some recent studies have suggested that administration of albumin in these conditions may increase mortality rates.

## Haptoglobin Binds Extracorpuscular Hemoglobin, Preventing Free Hemoglobin From Entering the Kidney

Haptoglobin (Hp) is a plasma glycoprotein that binds extracorpuscular hemoglobin (Hb) in a tight noncovalent complex (Hb-Hp). The amount of haptoglobin in human plasma ranges from 40 mg to 180 mg of hemoglobin-binding capacity per deciliter. Approximately 10% of the hemoglobin that is degraded each day is released into the circulation and is thus extracorpuscular. The other 90% is present in old, damaged red blood cells, which are degraded by cells of the histiocytic system. The molecular mass of hemoglobin is approximately 65 kDa, whereas the molecular mass of the simplest polymorphic form of haptoglobin (Hp 1-1) found in humans is approximately 90 kDa. Thus, the Hb-Hp complex has a molecular mass of approximately 155 kDa. Free hemoglobin passes through the glomerulus of the kidney, enters the tubules, and tends to precipitate therein (as can happen after a massive incompatible blood transfusion, when the capacity of haptoglobin to bind hemoglobin is grossly exceeded) (Figure 50–3). However, the Hb-Hp complex is too large to pass through the glomerulus. The function of Hp thus appears to be to prevent loss of free hemoglobin into the kidney. This conserves the valuable iron present in hemoglobin, which would otherwise be lost to the body.

Human haptoglobin exists in three polymorphic forms, known as Hp 1-1, Hp 2-1, and Hp 2-2. Hp 1-1 migrates in starch gel electrophoresis as a single band, whereas Hp 2-1 and Hp 2-2 exhibit much more complex band patterns. Two genes, designated $Hp^1$ and $Hp^2$, direct these three phenotypes, with Hp 2-1 being the heterozygous phenotype. It has been suggested that the haptoglobin polymorphism may be associated with the prevalence of many inflammatory diseases.

The levels of haptoglobin in human plasma vary and are of some diagnostic use. Low levels of haptoglobin are found in patients with **hemolytic anemias.** This is explained by the fact that whereas the half-life of haptoglobin is approximately 5 days, the half-life of the Hb-Hp complex is about 90 minutes, the complex being rapidly removed from plasma by hepatocytes. Thus, when haptoglobin is bound to hemoglobin, it is cleared from the plasma about 80 times faster than normally. Accordingly, the level of haptoglobin falls rapidly in situations where hemoglobin is constantly being released from red blood cells, such as occurs in hemolytic anemias. Haptoglobin is an acute phase protein, and its plasma level is elevated in a variety of inflammatory states.

Certain other plasma proteins bind heme but not hemoglobin. Hemopexin is a $\beta_1$-globulin that binds free heme. Albumin will bind some metheme (ferric heme) to form methemalbumin, which then transfers the metheme to hemopexin.

## Absorption of Iron From the Small Intestine Is Tightly Regulated

Transferrin (Tf) is a plasma protein that plays a central role in transporting iron around the body to sites where

**A.** Hb → Kidney → Excreted in urine or precipitates in tubules; iron is lost to body
(MW 65,000)

**B.** Hb + Hp → Hb : Hp complex ↛ Kidney
(MW 65,000) (MW 90,000) (MW 155,000)
↓
Catabolized by liver cells; iron is conserved and reused

***Figure 50–3.*** Different fates of free hemoglobin and of the hemoglobin-haptoglobin complex.

it is needed. Before we discuss it further, certain aspects of iron metabolism will be reviewed.

Iron is important in the human body because of its occurrence in many hemoproteins such as hemoglobin, myoglobin, and the cytochromes. It is ingested in the diet either as heme or nonheme iron (Figure 50–4); as shown, these different forms involve separate pathways. Absorption of iron in the proximal duodenum is tightly regulated, as there is no physiologic pathway for its excretion from the body. Under normal circumstances, the body guards its content of iron zealously, so that a healthy adult male loses only about 1 mg/d, which is replaced by absorption. Adult females are more prone to states of iron deficiency because some may lose excessive blood during menstruation. The amounts of iron in various body compartments are shown in Table 50–3.

Enterocytes in the proximal duodenum are responsible for absorption of iron. Incoming iron in the $Fe^{3+}$ state is reduced to $Fe^{2+}$ by a **ferrireductase** present on the surface of enterocytes. Vitamin C in food also favors reduction of ferric iron to ferrous iron. The transfer of iron from the apical surfaces of enterocytes into their interiors is performed by a proton-coupled divalent metal transporter (DMT1). This protein is not specific for iron, as it can transport a wide variety of divalent cations.

Once inside an enterocyte, iron can either be stored as ferritin or transferred across the basolateral membrane into the plasma, where it is carried by transferrin (see below). Passage across the basolateral membrane appears to be carried out by another protein, possibly iron regulatory protein 1 (IREG1). This protein may interact with the copper-containing protein hephaestin, a protein similar to ceruloplasmin (see below). Hephaestin is thought to have a ferroxidase activity, which is important in the release of iron from cells. Thus, $Fe^{2+}$ is converted back to $Fe^{3+}$, the form in which it is transported in the plasma by transferrin.

**Overall regulation** of iron absorption is complex and not well understood mechanistically. It occurs at

***Table 50–3.*** Distribution of iron in a 70-kg adult male.[1]

| | |
|---|---|
| Transferrin | 3–4 mg |
| Hemoglobin in red blood cells | 2500 mg |
| In myoglobin and various enzymes | 300 mg |
| In stores (ferritin and hemosiderin) | 1000 mg |
| Absorption | 1 mg/d |
| Losses | 1 mg/d |

[1]In an adult female of similar weight, the amount in stores would generally be less (100–400 mg) and the losses would be greater (1.5–2 mg/d).

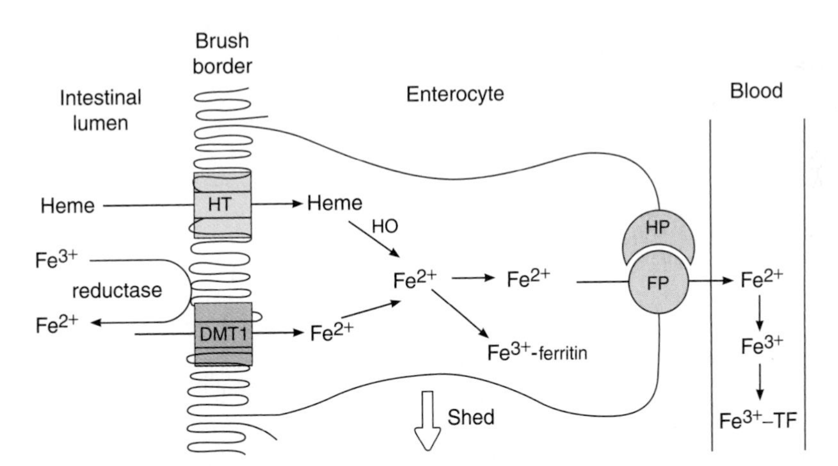


***Figure 50–4.*** Absorption of iron. $Fe^{3+}$ is converted to $Fe^{2+}$ by ferric reductase, and $Fe^{2+}$ is transported into the enterocyte by the apical membrane iron transporter DMT1. Heme is transported into the enterocyte by a separate heme transporter (HT), and heme oxidase (HO) releases $Fe^{2+}$ from the heme. Some of the intracellular $Fe^{2+}$ is converted to $Fe^{3+}$ and bound by ferritin. The remainder binds to the basolateral $Fe^{2+}$ transporter (FP) and is transported into the bloodstream, aided by hephaestin (HP). In plasma, $Fe^{3+}$ is bound to the iron transport protein transferrin (TF). (Reproduced, with permission, from Ganong WF: *Review of Medical Physiology,* 21st ed. McGraw-Hill, 2003.)

the level of the enterocyte, where further absorption of iron is blocked if a sufficient amount has been taken up (so-called dietary regulation exerted by "mucosal block"). It also appears to be responsive to the overall requirement of erythropoiesis for iron (erythropoietic regulation). Absorption is excessive in hereditary hemochromatosis (see below).

## Transferrin Shuttles Iron to Sites Where It Is Needed

Transferrin (Tf) is a $\beta_1$-globulin with a molecular mass of approximately 76 kDa. It is a glycoprotein and is synthesized in the liver. About 20 polymorphic forms of transferrin have been found. It plays a central role in the body's metabolism of iron because it transports iron (2 mol of $Fe^{3+}$ per mole of Tf) in the circulation to sites where iron is required, eg, from the gut to the bone marrow and other organs. Approximately 200 billion red blood cells (about 20 mL) are catabolized per day, releasing about 25 mg of iron into the body—most of which will be transported by transferrin.

There are receptors (TfRs) on the surfaces of many cells for transferrin. It binds to these receptors and is internalized by receptor-mediated endocytosis (compare the fate of LDL; Chapter 25). The acid pH inside the lysosome causes the iron to dissociate from the protein. The dissociated iron leaves the endosome via DMT1 to enter the cytoplasm. Unlike the protein component of LDL, apoTf is not degraded within the lysosome. Instead, it remains associated with its receptor, returns to the plasma membrane, dissociates from its receptor, reenters the plasma, picks up more iron, and again delivers the iron to needy cells.

Abnormalities of the glycosylation of transferrin occur in the congenital disorders of glycosylation (Chapter 47) and in chronic alcohol abuse. Their detection by, for example, isoelectric focusing is used to help diagnose these conditions.

## Iron Deficiency Anemia Is Extremely Prevalent

Attention to iron metabolism is **particularly important in women** for the reason mentioned above. Additionally, in **pregnancy,** allowances must be made for the growing fetus. Older people with poor dietary habits ("tea and toasters") may develop iron deficiency. Iron deficiency anemia due to inadequate intake, inadequate utilization, or excessive loss of iron is one of the most prevalent conditions seen in medical practice.

The concentration of transferrin in plasma is approximately 300 mg/dL. This amount of transferrin can bind 300 μg of iron per deciliter, so that this represents the **total iron-binding capacity** of plasma. However, the protein is normally only one-third saturated with iron. In **iron deficiency anemia,** the protein is even less saturated with iron, whereas in conditions of storage of excess iron in the body (eg, hemochromatosis) the saturation with iron is much greater than one-third.

## Ferritin Stores Iron in Cells

Ferritin is another protein that is important in the metabolism of iron. Under normal conditions, it stores iron that can be called upon for use as conditions require. In conditions of excess iron (eg, hemochromatosis), body stores of iron are greatly increased and much more ferritin is present in the tissues, such as the liver and spleen. Ferritin contains approximately 23% iron, and **apoferritin** (the protein moiety free of iron) has a molecular mass of approximately 440 kDa. Ferritin is composed of 24 subunits of 18.5 kDa, which surround in a micellar form some 3000–4500 ferric atoms. Normally, there is a little ferritin in human plasma. However, in patients with excess iron, the amount of ferritin in plasma is markedly elevated. The amount of ferritin in plasma can be conveniently measured by a sensitive and specific radioimmunoassay and serves as an index of body iron stores.

Synthesis of the **transferrin receptor (TfR)** and that of **ferritin** are reciprocally linked to cellular iron content. Specific untranslated sequences of the mRNAs for both proteins (named **iron response elements)** interact with a cytosolic protein sensitive to variations in levels of cellular iron (iron-responsive element-binding protein). When iron levels are high, cells use stored ferritin mRNA to synthesize ferritin, and the TfR mRNA is degraded. In contrast, when iron levels are low, the TfR mRNA is stabilized and increased synthesis of receptors occurs, while ferritin mRNA is apparently stored in an inactive form. This is an important example of control of expression of proteins at the **translational** level.

**Hemosiderin** is a somewhat ill-defined molecule; it appears to be a partly degraded form of ferritin but still containing iron. It can be detected by histologic stains (eg, Prussian blue) for iron, and its presence is determined histologically when excessive storage of iron occurs.

## Hereditary Hemochromatosis Is Due to Mutations in the *HFE* Gene

Hereditary (primary) hemochromatosis is a very prevalent autosomal recessive disorder in certain parts of the world (eg, Scotland, Ireland, and North America). It is characterized by excessive storage of iron in tissues, leading to tissue damage. Total body iron ranges between 2.5 g and 3.5 g in normal adults; in primary hemochromatosis it usually exceeds 15 g. The accumulated iron damages organs and tissues such as the liver, pancreatic islets, and heart, perhaps in part due to effects on free radical production (Chapter 52). Melanin and various amounts of iron accumulate in the skin, accounting for the slate-gray color often seen. The precise cause of melanin accumulation is not clear. The frequent coexistence of diabetes mellitus (due to islet damage) and the skin pigmentation led to use of the term **bronze diabetes** for this condition. In 1995, Feder and colleagues isolated a gene, now known as *HFE,* located on chromosome 6 close to the major histocompatibility complex genes. The encoded protein (HFE) was found to be related to MHC class 1 antigens. Initially, two different missense mutations were found in *HFE* in individuals with hereditary hemochromatosis. The more frequent mutation was one that changed cysteinyl residue 282 to a tyrosyl residue (CY282Y), disrupting the structure of the protein. The other mutation changed histidyl residue 63 to an aspartyl residue (H63D). Some patients with hereditary hemochromatosis have neither mutation, perhaps due to other mutations in *HFE* or because one or more other genes may be involved in its causation. Genetic screening for this condition has been evaluated but is not presently recommended. However, testing for *HFE* mutations in individuals with elevated serum iron concentrations may be useful.

HFE has been shown to be located in cells in the crypts of the small intestine, the site of iron absorption. There is evidence that it associates with $\beta_2$-microglobulin, an association that may be necessary for its stability, intracellular processing, and cell surface expression. The complex interacts with the transferrin receptor (TfR); how this leads to excessive storage of iron when *HFE* is altered by mutation is under close study. The mouse homolog of *HFE* has been knocked out, resulting in a potentially useful animal model of hemochromatosis.

A scheme of the likely main events in the causation of hereditary hemochromatosis is set forth in Figure 50–5.

**Secondary hemochromatosis** can occur after repeated transfusions (eg, for treatment of sickle cell anemia), excessive oral intake of iron (eg, by African Bantu peoples who consume alcoholic beverages fermented in containers made of iron), or a number of other conditions.

Table 50–4 summarizes laboratory tests useful in the assessment of patients with abnormalities of iron metabolism.

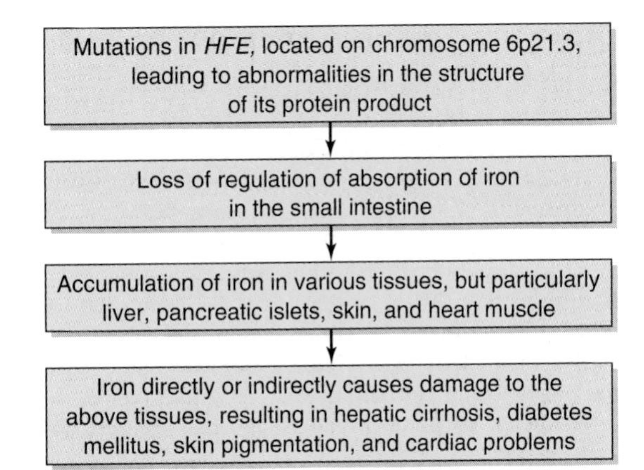


***Figure 50–5.*** Tentative scheme of the main events in causation of primary hemochromatosis (MIM 235200). The two principal mutations are CY282Y and H63D (see text). Mutations in genes other than *HFE* are also involved in some cases.

***Table 50–4.*** Laboratory tests for assessing patients with disorders of iron metabolism.

- Red blood cell count and estimation of hemoglobin
- Determinations of plasma iron, total iron-binding capacity (TIBC), and % transferrin saturation
- Determination of ferritin in plasma by radioimmunoassay
- Prussian blue stain of tissue sections
- Determination of amount of iron (μg/g) in a tissue biopsy

## Ceruloplasmin Binds Copper, & Low Levels of This Plasma Protein Are Associated With Wilson Disease

Ceruloplasmin (about 160 kDa) is an $\alpha_2$-globulin. It has a blue color because of its high copper content and carries 90% of the copper present in plasma. Each molecule of ceruloplasmin binds six atoms of copper very tightly, so that the copper is not readily exchangeable. Albumin carries the other 10% of the plasma copper but binds the metal less tightly than does ceruloplasmin. Albumin thus donates its copper to tissues more readily than ceruloplasmin and appears to be more important than ceruloplasmin in copper transport in the human body. Ceruloplasmin exhibits a copper-dependent **oxidase** activity, but its physiologic significance has not been clarified. The amount of ceruloplasmin in plasma is decreased in liver disease. In particular, low levels of ceruloplasmin are found in **Wilson disease** (hepatolenticular degeneration), a disease due to abnormal metabolism of copper. In order to clarify the description of Wilson disease, we shall first consider the metabolism of copper in the human body and then **Menkes disease**, another condition involving abnormal copper metabolism.

## Copper Is a Cofactor for Certain Enzymes

Copper is an essential trace element. It is required in the diet because it is the metal cofactor for a variety of enzymes (see Table 50–5). Copper accepts and donates electrons and is involved in reactions involving dismutation, hydroxylation, and oxygenation. However, excess copper can cause problems because it can oxidize proteins and lipids, bind to nucleic acids, and enhance the production of free radicals. It is thus important to have mechanisms that will maintain the amount of copper in the body within normal limits. The body of the normal adult contains about 100 mg of copper, located mostly in bone, liver, kidney, and muscle. The daily intake of copper is about 2–4 mg, with about 50% being absorbed in the stomach and upper small intestine and the remainder excreted in the feces. Copper is carried to the liver bound to albumin, taken up by liver cells, and part of it is excreted in the bile. Copper also leaves the liver attached to **ceruloplasmin,** which is synthesized in that organ.

## The Tissue Levels of Copper & of Certain Other Metals Are Regulated in Part by Metallothioneins

Metallothioneins are a group of small proteins (about 6.5 kDa), found in the cytosol of cells, particularly of liver, kidney, and intestine. They have a high content of cysteine and can bind copper, zinc, cadmium, and mercury. The SH groups of cysteine are involved in binding the metals. Acute intake (eg, by injection) of copper and of certain other metals increases the amount (induction) of these proteins in tissues, as does administration of certain hormones or cytokines. These proteins may function to store the above metals in a nontoxic form and are involved in their overall metabolism in the body. Sequestration of copper also diminishes the amount of this metal available to generate free radicals.

## Menkes Disease Is Due to Mutations in the Gene Encoding a Copper-Binding P-Type ATPase

Menkes disease ("kinky" or "steely" hair disease) is a disorder of copper metabolism. It is X-linked, affects only male infants, involves the nervous system, connective tissue, and vasculature, and is usually fatal in infancy. In 1993, it was reported that the basis of Menkes disease was mutations in the gene for a copper-binding P-type ATPase. Interestingly, the enzyme showed structural similarity to certain metal-binding proteins in microorganisms. This ATPase is thought to be responsible for directing the efflux of copper from cells. When altered by mutation, copper is not mobilized normally from the intestine, in which it accumulates, as it does in a variety of other cells and tissues, from which it cannot exit. Despite the accumulation of copper, the activities of many copper-dependent enzymes are decreased, perhaps because of a defect of its incorporation into the apoenzymes. Normal liver expresses very little of the ATPase, which explains the absence of hepatic involvement in Menkes disease. This work led to the suggestion that liver might contain a different copper-binding ATPase, which could be involved in the causation of Wilson disease. As described below, this turned out to be the case.

***Table 50–5.*** Some important enzymes that contain copper.

- Amine oxidase
- Copper-dependent superoxide dismutase
- Cytochrome oxidase
- Tyrosinase

## Wilson Disease Is Also Due to Mutations in a Gene Encoding a Copper-Binding P-Type ATPase

Wilson disease is a genetic disease in which copper fails to be excreted in the bile and accumulates in liver, brain, kidney, and red blood cells. It can be regarded as an inability to maintain a near-zero copper balance, resulting in **copper toxicosis.** The increase of copper in liver cells appears to inhibit the coupling of copper to apoceruloplasmin and leads to low levels of ceruloplasmin in plasma. As the amount of copper accumulates, patients may develop a hemolytic anemia, chronic liver disease (cirrhosis, hepatitis), and a neurologic syndrome owing to accumulation of copper in the basal ganglia and other centers. A frequent clinical finding is the **Kayser-Fleischer ring.** This is a green or golden pigment ring around the cornea due to deposition of copper in Descemet's membrane. The major laboratory tests of copper metabolism are listed in Table 50–6. If Wilson disease is suspected, a liver biopsy should be performed; a value for liver copper of over 250 μg per gram dry weight along with a plasma level of ceruloplasmin of under 20 mg/dL is diagnostic.

The cause of Wilson disease was also revealed in 1993, when it was reported that a variety of mutations in a gene encoding a copper-binding P-type ATPase were responsible. The gene is estimated to encode a protein of 1411 amino acids, which is highly homologous to the product of the gene affected in Menkes disease. In a manner not yet fully explained, a nonfunctional ATPase causes defective excretion of copper into the bile, a reduction of incorporation of copper into

***Table 50–6.*** Major laboratory tests used in the investigation of diseases of copper metabolism.[1]

| Test | Normal Adult Range |
|---|---|
| Serum copper | 10–22 μmol/L |
| Ceruloplasmin | 200–600 mg/L |
| Urinary copper | < 1 μmol/24 h |
| Liver copper | 20–50 μg/g dry weight |

[1]Based on Gaw A et al: *Clinical Biochemistry.* Churchill Livingstone, 1995.

apoceruloplasmin, and the accumulation of copper in liver and subsequently in other organs such as brain.

Treatment for Wilson disease consists of a diet low in copper along with lifelong administration of **penicillamine,** which chelates copper, is excreted in the urine, and thus depletes the body of the excess of this mineral.

Another condition involving ceruloplasmin is **aceruloplasminemia.** In this genetic disorder, levels of ceruloplasmin are very low and consequently its ferroxidase activity is markedly deficient. This leads to failure of release of iron from cells, and iron accumulates in certain brain cells, hepatocytes, and pancreatic islet cells. Affected individuals show severe neurologic signs and have diabetes mellitus. Use of a chelating agent or administration of plasma or ceruloplasmin concentrate may be beneficial.

## Deficiency of $\alpha_1$-Antiproteinase ($\alpha_1$-Antitrypsin) Is Associated With Emphysema & One Type of Liver Disease

$\alpha_1$-Antiproteinase (about 52 kDa) was formerly called **$\alpha_1$-antitrypsin,** and this name is retained here. It is a single-chain protein of 394 amino acids, contains three oligosaccharide chains, and is the major component (> 90%) of the $\alpha_1$ fraction of human plasma. It is synthesized by hepatocytes and macrophages and is the principal **serine protease inhibitor** (serpin, or Pi) of human plasma. It inhibits trypsin, elastase, and certain other proteases by forming complexes with them. At least 75 polymorphic forms occur, many of which can be separated by electrophoresis. The major genotype is MM, and its phenotypic product is PiM. There are two areas of clinical interest concerning $\alpha_1$-antitrypsin. A deficiency of this protein has a role in certain cases (approximately 5%) of **emphysema.** This occurs mainly in subjects with the **ZZ genotype,** who synthesize PiZ, and also in PiSZ heterozygotes, both of whom secrete considerably less protein than PiMM individuals. Considerably less of this protein is secreted as compared with PiM. When the amount of $\alpha_1$-antitrypsin is deficient and polymorphonuclear white blood cells increase in the lung (eg, during pneumonia), the affected individual lacks a countercheck to proteolytic damage of the lung by proteases such as elastase (Figure 50–6). It is of considerable interest that a particular methionine (residue 358) of $\alpha_1$-antitrypsin is involved in its binding to proteases. Smoking oxidizes this methionine to methionine sulfoxide and thus inactivates it. As a result, affected molecules of $\alpha_1$-antitrypsin no longer neutralize proteases. This is particularly devastating in patients (eg, PiZZ phenotype) who already have low levels of $\alpha_1$-antitrypsin. The further diminution in $\alpha_1$-antitrypsin brought about by smoking results in increased proteolytic destruction of lung tissue, accelerating the development of emphysema. Intravenous administration of $\alpha_1$-antitrypsin (augmentation therapy) has been used as an adjunct in the treatment of patients with emphysema due to $\alpha_1$-antitrypsin deficiency. Attempts are being made, using the techniques of protein engineering, to replace methionine 358 by another residue that would not be subject to oxidation. The resulting "mutant" $\alpha_1$-antitrypsin would thus afford protection against proteases for a much longer period of time than would native $\alpha_1$-antitrypsin. Attempts are also being made to develop **gene therapy** for this condition. One approach is to use a modified adenovirus (a pathogen of the respiratory tract) into which the gene for $\alpha_1$-antitrypsin has been inserted. The virus would then be introduced into the respiratory tract (eg, by an aerosol). The hope is that pulmonary epithelial cells would express the gene and secrete $\alpha_1$-antitrypsin locally. Experiments in animals have indicated the feasibility of this approach.

Deficiency of $\alpha_1$-antitrypsin is also implicated in one type of **liver disease** ($\alpha_1$-antitrypsin deficiency liver

**A.** Active elastase + $\alpha_1$–AT → Inactive elastase: $\alpha_1$–AT complex → No proteolysis of lung → No tissue damage

**B.** Active elastase + ↓ or no $\alpha_1$–AT → Active elastase → Proteolysis of lung → Tissue damage

***Figure 50–6.*** Scheme illustrating **(A)** normal inactivation of elastase by $\alpha_1$-antitrypsin and **(B)** situation in which the amount of $\alpha_1$-antitrypsin is substantially reduced, resulting in proteolysis by elastase and leading to tissue damage.

disease). In this condition, molecules of the ZZ phenotype accumulate and aggregate in the cisternae of the endoplasmic reticulum of hepatocytes. Aggregation is due to formation of polymers of mutant $\alpha_1$-antitrypsin, the polymers forming via a strong interaction between a specific loop in one molecule and a prominent $\beta$-pleated sheet in another (loop-sheet polymerization). By mechanisms that are not understood, **hepatitis** results with consequent **cirrhosis** (accumulation of massive amounts of collagen, resulting in fibrosis). It is possible that administration of a synthetic peptide resembling the loop sequence could inhibit loop-sheet polymerization. Diseases such as $\alpha_1$-antitrypsin deficiency, in which cellular pathology is primarily caused by the presence of aggregates of aberrant forms of individual proteins, have been named **conformational diseases.** Most appear to be due to the formation by conformationally unstable proteins of $\beta$ sheets, which in turn leads to formation of aggregates. Other members of this group of conditions include Alzheimer disease, Parkinson disease, and Huntington disease.

At present, severe $\alpha_1$-antitrypsin deficiency liver disease can be successfully treated by liver transplantation. In the future, introduction of the gene for normal $\alpha_1$-antitrypsin into hepatocytes may become possible, but this would not stop production of the PiZ protein. Figure 50–7 is a scheme of the causation of this disease.

## $\alpha_2$-Macroglobulin Neutralizes Many Proteases & Targets Certain Cytokines to Tissues

$\alpha_2$-Macroglobulin is a large plasma glycoprotein (720 kDa) made up of four identical subunits of 180 kDa. It comprises 8–10% of the total plasma protein in humans. Approximately 10% of the zinc in plasma is transported by $\alpha_2$-macroglobulin, the remainder being transported by albumin. The protein is synthesized by a variety of cell types, including monocytes, hepatocytes, and astrocytes. It is the major member of a group of plasma proteins that include complement proteins C3 and C4. These proteins contain a unique internal cyclic thiol ester bond (formed between a cysteine and a glutamine residue) and for this reason have been designated as the **thiol ester plasma protein family**.

$\alpha_2$-Macroglobulin binds many proteinases and is thus an important **panproteinase inhibitor.** The $\alpha_2$-macroglobulin-proteinase complexes are rapidly cleared from the plasma by a receptor located on many cell types. In addition, $\alpha_2$-macroglobulin binds many **cytokines** (platelet-derived growth factor, transforming growth factor-$\beta$, etc) and appears to be involved in targeting them toward particular tissues or cells. Once taken up by cells, the cytokines can dissociate from $\alpha_2$-macroglobulin and subsequently exert a variety of effects on cell growth and function. The binding of proteinases and cytokines by $\alpha_2$-macroglobulin involves different mechanisms that will not be considered here.

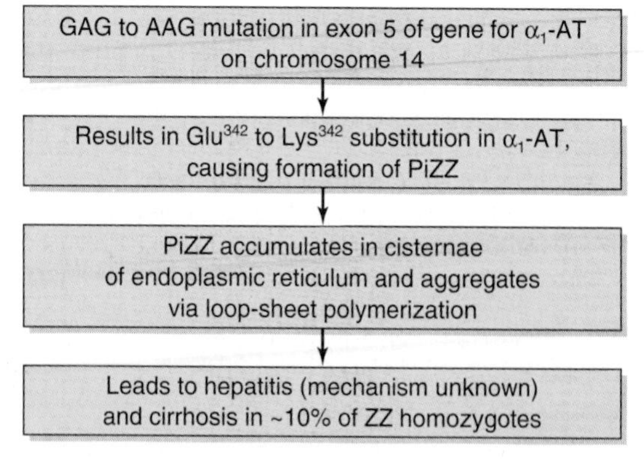


***Figure 50–7.*** Scheme of causation of $\alpha_1$-antitrypsin-deficiency liver disease. The mutation shown causes formation of PiZZ (MIM 107400). ($\alpha_1$-AT, $\alpha_1$-antitrypsin.)

## Amyloidosis Occurs by the Deposition of Fragments of Various Plasma Proteins in Tissues

Amyloidosis is the accumulation of various insoluble fibrillar proteins between the cells of tissues to an extent that affects function. The fibrils generally represent proteolytic fragments of various plasma proteins and possess a $\beta$-pleated sheet structure. The term "amyloidosis" is a misnomer, as it was originally thought that the fibrils were starch-like in nature. Among the most common **precursor proteins** are immunoglobulin light chains (see below), amyloid-associated protein derived from serum amyloid-associated protein (a plasma glycoprotein), and transthyretin (Table 50–2). The precursor proteins in plasma are generally either increased in amount (eg, immunoglobulin light chains in multiple myeloma or $\beta_2$-microglobulin in patients being maintained on chronic dialysis) or mutant forms (eg, of transthyretin in familial amyloidotic neuropathies). The precise factors that determine the deposition of proteolytic fragments in tissues await elucidation. Other proteins have been found in amyloid fibrils, such as calcitonin and amyloid $\beta$ protein (not derived from a plasma protein) in Alzheimer disease; a total of about 15 different proteins have been found. All fibrils have a **P component** associated with them, which is derived from serum amyloid P component, a plasma protein closely related to C-reactive protein. Tissue sections containing amyloid fibrils interact with Congo red stain and display striking green birefringence when viewed by polarizing microscopy. Deposition of amyloid occurs in patients with a variety of disorders; treatment of the underlying disorder should be provided if possible.

## PLASMA IMMUNOGLOBULINS PLAY A MAJOR ROLE IN THE BODY'S DEFENSE MECHANISMS

The immune system of the body consists of two major components: **B lymphocytes** and **T lymphocytes.** The B lymphocytes are mainly derived from bone marrow cells in higher animals and from the bursa of Fabricius in birds. The T lymphocytes are of thymic origin. The **B cells** are responsible for the synthesis of circulating, humoral antibodies, also known as **immunoglobulins.** The **T cells** are involved in a variety of important **cell-mediated immunologic processes** such as graft rejection, hypersensitivity reactions, and defense against malignant cells and many viruses. This section considers only the plasma immunoglobulins, which are synthesized mainly in **plasma cells.** These are specialized cells of B cell lineage that synthesize and secrete immunoglobulins into the plasma in response to exposure to a variety of **antigens.**

### All Immunoglobulins Contain a Minimum of Two Light & Two Heavy Chains

Immunoglobulins contain a minimum of two identical light (L) chains (23 kDa) and two identical heavy (H) chains (53–75 kDa), held together as a tetramer ($L_2H_2$) by disulfide bonds The structure of IgG is shown in Figure 50–8; it is Y-shaped, with binding of antigen occurring at both tips of the Y. Each chain can be divided conceptually into specific domains, or regions, that have structural and functional significance. The half of the light (L) chain toward the carboxyl terminal is referred to as the constant region ($C_L$), while the amino terminal half is the variable region of the light chain ($V_L$). Approximately one-quarter of the heavy (H) chain at the amino terminals is referred to as its variable region ($V_H$), and the other three-quarters of the heavy chain are referred to as the constant regions ($C_H1$, $C_H2$, $C_H3$) of that H chain. The portion of the immunoglobulin molecule that binds the specific antigen is formed by the amino terminal portions (variable regions) of both the H and L chains—ie, the $V_H$ and $V_L$ domains. The domains of the protein chains consist of two sheets of antiparallel distinct stretches of amino acids that bind antigen.

As depicted in Figure 50–8, digestion of an immunoglobulin by the enzyme **papain** produces two antigen-binding fragments (Fab) and one crystallizable fragment (Fc), which is responsible for functions of immunoglobulins other than direct binding of antigens. Because there are two Fab regions, IgG molecules bind two molecules of antigen and are termed divalent. The site on the antigen to which an antibody binds is termed an **antigenic determinant,** or **epitope.** The area in which papain cleaves the immunoglobulin molecule—ie, the region between the $C_H1$ and $C_H2$ domains—is referred to as the **"hinge region."** The hinge region confers flexibility and allows both Fab arms to move independently, thus helping them to bind to antigenic sites that may be variable distances apart (eg, on bacterial surfaces). Fc and hinge regions differ in the different classes of antibodies, but the overall model of antibody structure for each class is similar to that shown in Figure 50–8 for IgG.

### All Light Chains Are Either Kappa or Lambda in Type

There are two general types of light chains, kappa (κ) and lambda (λ), which can be distinguished on the basis of structural differences in their $C_L$ regions. A given immunoglobulin molecule always contains two κ or two λ light chains—never a mixture of κ and λ. In humans, the κ chains are more frequent than λ chains in immunoglobulin molecules.

### The Five Types of Heavy Chain Determine Immunoglobulin Class

Five classes of H chain have been found in humans (Table 50–7), distinguished by differences in their $C_H$ regions. They are designated γ, α, μ δ, and ε. The μ and ε chains each have four $C_H$ domains rather than the usual three. The type of H chain determines the class of immunoglobulin and thus its effector function. There are thus five immunoglobulin classes: **IgG, IgA, IgM, IgD,** and **IgE.** The biologic functions of these five classes are summarized in Table 50–8.

### No Two Variable Regions Are Identical

The variable regions of immunoglobulin molecules consist of the $V_L$ and $V_H$ domains and are quite heterogeneous. In fact, no two variable regions from different humans have been found to have identical amino acid sequences. However, amino acid analyses have shown that the variable regions are comprised of relatively invariable regions and other hypervariable regions (Figure 50–9). L chains have three hypervariable regions (in $V_L$) and H chains have four (in $V_H$). These hypervariable regions comprise the antigen-binding site (located at the tips of the Y shown in Figure 50–8) and dictate the amazing specificity of antibodies. For this reason, hypervariable regions are also termed **complementar-**

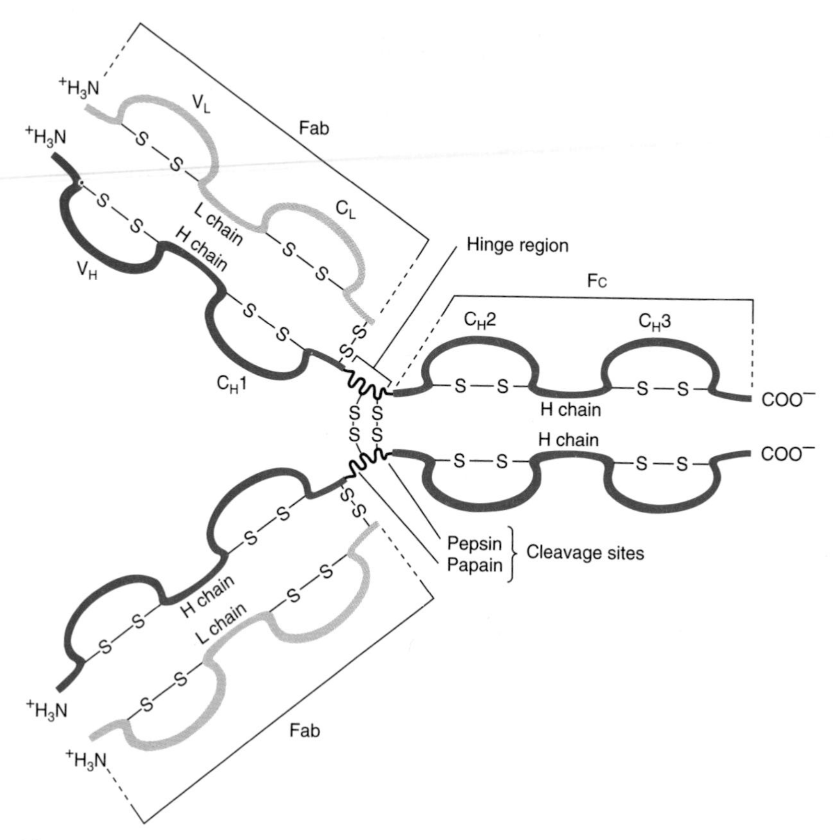


***Figure 50–8.*** Structure of IgG. The molecule consists of two light (L) chains and two heavy (H) chains. Each light chain consists of a variable ($V_L$) and a constant ($C_L$) region. Each heavy chain consists of a variable region ($V_H$) and a constant region that is divided into three domains ($C_H1$, $C_H2$, and $C_H3$). The $C_H2$ domain contains the complement-binding site and the $C_H3$ domain contains a site that attaches to receptors on neutrophils and macrophages. The antigen-binding site is formed by the hypervariable regions of both the light and heavy chains, which are located in the variable regions of these chains (see Figure 50–9). The light and heavy chains are linked by disulfide bonds, and the heavy chains are also linked to each other by disulfide bonds. (Reproduced, with permission, from Parslow TG et al [editors]: *Medical Immunology,* 10th ed. McGraw-Hill, 2001.)

**ity-determining regions (CDRs)**. About five to ten amino acids in each hypervariable region (CDR) contribute to the antigen-binding site. CDRs are located on small loops of the variable domains, the surrounding polypeptide regions between the hypervariable regions being termed **framework regions.** CDRs from both $V_H$ and $V_L$ domains, brought together by folding of the polypeptide chains in which they are contained, form a single hypervariable surface comprising the antigen-binding site. Various combinations of H and L chain CDRs can give rise to many antibodies of different specificities, a feature that contributes to the tremendous diversity of antibody molecules and is termed **combinatorial diversity.** Large antigens interact with all of the CDRs of an antibody, whereas small ligands may interact with only one or a few CDRs that form a pocket or groove in the antibody molecule. The essence of antigen-antibody interactions is mutual complementarity between the surfaces of CDRs and epitopes. The interactions between antibodies and antigens involve noncovalent forces and bonds (electrostatic and van der Waals forces and hydrogen and hydrophobic bonds).

***Table 50–7.*** Properties of human immunoglobulins.[1]

| Property | IgG | IgA | IgM | IgD | IgE |
|---|---|---|---|---|---|
| Percentage of total immunoglobulin in serum (approximate) | 75 | 15 | 9 | 0.2 | 0.004 |
| Serum concentration (mg/dL) (approximate) | 1000 | 200 | 120 | 3 | 0.05 |
| Sedimentation coefficient | 7S | 7S or 11S[2] | 19S | 7S | 8S |
| Molecular weight (× 1000) | 150 | 170 or 400[2] | 900 | 180 | 190 |
| Structure | Monomer | Monomer or dimer | Monomer or dimer | Monomer | Monomer |
| H-chain symbol | γ | α | μ | δ | ε |
| Complement fixation | + | – | + | – | – |
| Transplacental passage | + | – | – | ? | – |
| Mediation of allergic responses | – | – | – | – | + |
| Found in secretions | – | + | – | – | – |
| Opsonization | + | – | –[3] | – | – |
| Antigen receptor on B cell | – | – | + | ? | – |
| Polymeric form contains J chain | – | + | + | – | – |

[1]Reproduced, with permission, from Levinson W, Jawetz E: *Medical Microbiology and Immunology,* 7th ed. McGraw-Hill, 2002.
[2]The 11S form is found in secretions (eg, saliva, milk, tears) and fluids of the respiratory, intestinal, and genital tracts.
[3]IgM opsonizes indirectly by activating complement. This produces C3b, which is an opsonin.

## The Constant Regions Determine Class-Specific Effector Functions

The constant regions of the immunoglobulin molecules, particularly the $C_H2$ and $C_H3$ (and $C_H4$ of IgM and IgE), which constitute the Fc fragment, are responsible for the class-specific effector functions of the different immunoglobulin molecules (Table 50–7, bottom part), eg, complement fixation or transplacental passage.

Some immunoglobulins such as immune IgG exist only in the basic tetrameric structure, while others such as IgA and IgM can exist as higher order polymers of two, three (IgA), or five (IgM) tetrameric units (Figure 50–10).

The L chains and H chains are synthesized as separate molecules and are subsequently assembled within the B cell or plasma cell into mature immunoglobulin molecules, all of which are **glycoproteins.**

## Both Light & Heavy Chains Are Products of Multiple Genes

Each immunoglobulin light chain is the product of at least three separate structural genes: a variable region ($V_L$) gene, a joining region *(J)* gene (bearing no relationship to the J chain of IgA or IgM), and a constant region ($C_L$) gene. Each heavy chain is the product of at least four different genes: a variable region ($V_H$) gene, a diversity region *(D)* gene, a joining region *(J)* gene, and a constant region ($C_H$) gene. Thus, the "one gene, one protein" concept is not valid. The molecular mechanisms responsible for the generation of the single immunoglobulin chains from multiple structural genes are discussed in Chapters 36 and 39.

## Antibody Diversity Depends on Gene Rearrangements

Each person is capable of generating antibodies directed against perhaps 1 million different antigens. The generation of such immense antibody diversity depends upon a number of factors including the existence of multiple gene segments (V, C, J, and D segments), their recombinations (see Chapters 36 and 39), the combinations of different L and H chains, a high frequency of somatic mutations in immunoglobulin genes, and **junctional diversity.** The latter reflects the addi-

***Table 50–8.*** Major functions of immunoglobulins.[1]

| Immunoglobulin | Major Functions |
|---|---|
| IgG | Main antibody in the secondary response. Opsonizes bacteria, making them easier to phagocytose. Fixes complement, which enhances bacterial killing. Neutralizes bacterial toxins and viruses. Crosses the placenta. |
| IgA | Secretory IgA prevents attachment of bacteria and viruses to mucous membranes. Does not fix complement. |
| IgM | Produced in the primary response to an antigen. Fixes complement. Does not cross the placenta. Antigen receptor on the surface of B cells. |
| IgD | Uncertain. Found on the surface of many B cells as well as in serum. |
| IgE | Mediates immediate hypersensitivity by causing release of mediators from mast cells and basophils upon exposure to antigen (allergen). Defends against worm infections by causing release of enzymes from eosinophils. Does not fix complement. Main host defense against helminthic infections. |

[1]Reproduced, with permission, from Levinson W, Jawetz E: *Medical Microbiology and Immunology,* 7th ed. McGraw-Hill, 2002.

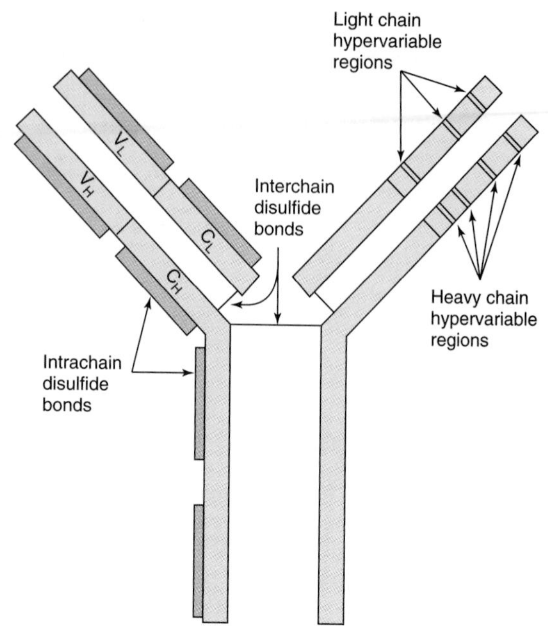


***Figure 50–9.*** Schematic model of an IgG molecule showing approximate positions of the hypervariable regions in heavy and light chains. The antigen-binding site is formed by these hypervariable regions. The hypervariable regions are also called complementarity-determining regions (CDRs). (Modified and reproduced, with permission, from Parslow TG et al [editors]: *Medical Immunology,* 10th ed. McGraw-Hill, 2001.)

tion or deletion of a random number of nucleotides when certain gene segments are joined together, and introduces an additional degree of diversity. Thus, the above factors ensure that a vast number of antibodies can be synthesized from several hundred gene segments.

## Class (Isotype) Switching Occurs During Immune Responses

In most humoral immune responses, antibodies with identical specificity but of different classes are generated in a specific chronologic order in response to the immunogen (immunizing antigen). For instance, antibodies of the IgM class normally precede molecules of the IgG class. The switch from one class to another is designated **"class or isotype switching,"** and its molecular basis has been investigated extensively. A single type of immunoglobulin light chain can combine with an antigen-specific μ chain to generate a specific IgM molecule. Subsequently, the same antigen-specific light chain combines with a γ chain with an identical $V_H$ region to generate an IgG molecule with antigen specificity identical to that of the original IgM molecule. The same light chain can also combine with an α heavy chain, again containing the identical $V_H$ region, to form an IgA molecule with identical antigen specificity. These three classes (IgM, IgG, and IgA) of immunoglobulin molecules against the same antigen have identical variable domains of both their light ($V_L$) chains and heavy ($V_H$) chains and are said to share an **idiotype.** (Idiotypes are the antigenic determinants formed by the specific amino acids in the hypervariable regions.) The different classes of these three immunoglobulins (called **isotypes**) are thus determined by their different $C_H$ regions, which are combined with the same antigen-specific $V_H$ regions.

## Both Over- & Underproduction of Immunoglobulins May Result in Disease States

Disorders of immunoglobulins include increased production of specific classes of immunoglobulins or even

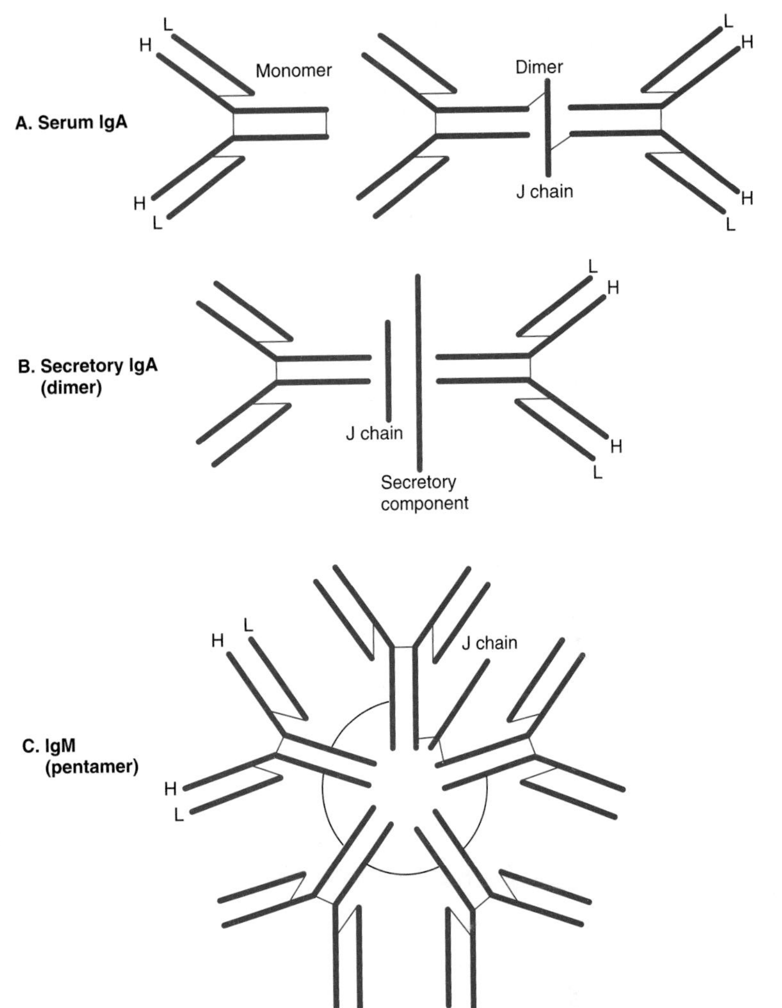


***Figure 50–10.*** **Schematic representation of serum IgA, secretory IgA, and IgM. Both IgA and IgM have a J chain, but only secretory IgA has a secretory component. Polypeptide chains are represented by thick lines; disulfide bonds linking different polypeptide chains are represented by thin lines.** (Reproduced, with permission, from Parslow TG et al [editors]: *Medical Immunology,* 10th ed. McGraw-Hill, 2001.)

specific immunoglobulin molecules, the latter by clonal tumors of plasma cells called myelomas. **Multiple myeloma** is a neoplastic condition; electrophoresis of serum or urine will usually reveal a large increase of one particular immunoglobulin or one particular light chain (the latter termed a Bence Jones protein). Decreased production may be restricted to a single class of immunoglobulin molecules (eg, IgA or IgG) or may involve underproduction of all classes of immunoglobulins (IgA, IgD, IgE, IgG, and IgM). A severe reduction in synthesis of an immunoglobulin class due to a genetic abnormality can result in a serious immunodeficiency disease—eg, **agammaglobulinemia,** in which production of IgG is markedly affected—because of impairment of the body's defense against microorganisms.

## Hybridomas Provide Long-Term Sources of Highly Useful Monoclonal Antibodies

When an antigen is injected into an animal, the resulting antibodies are polyclonal, being synthesized by a mixture of B cells. Polyclonal antibodies are directed against a number of different sites (epitopes or determinants) on the antigen and thus are not monospecific. However, by means of a method developed by Kohler and Milstein, large amounts of a single monoclonal antibody specific for one epitope can be obtained.

The method involves **cell fusion,** and the resulting permanent cell line is called a **hybridoma.** Typically, B cells are obtained from the spleen of a mouse (or other suitable animal) previously injected with an antigen or mixture of antigens (eg, foreign cells). The B cells are mixed with mouse **myeloma cells** and exposed to polyethylene glycol, which causes cell fusion. A summary of the principles involved in generating hybridoma cells is given in Figure 50–11. Under the conditions used, only the hybridoma cells multiply in cell culture. This involves plating the hybrid cells into hypoxanthine-aminopterin-thymidine (HAT)-containing medium at a concentration such that each dish contains approximately one cell. Thus, a **clone** of hybridoma cells multiplies in each dish. The culture medium is harvested and screened for antibodies that react with the original antigen or antigens. If the immunogen is a mixture of many antigens (eg, a cell membrane preparation), an individual culture dish will contain a clone of hybridoma cells synthesizing a monoclonal antibody to one specific antigenic determinant of the mixture. By harvesting the media from many culture dishes, a battery of monoclonal antibodies can be obtained, many of which are specific for individual components of the immunogenic mixture. The hybridoma cells can be frozen and stored and subsequently thawed when more of the antibody is required; this ensures its long-term supply. The hybridoma cells can also be grown in the abdomen of mice, providing relatively large supplies of antibodies.

Because of their specificity, monoclonal antibodies have become useful reagents in many areas of biology and medicine. For example, they can be used to measure the amounts of many individual proteins (eg, plasma proteins), to determine the nature of infectious agents (eg, types of bacteria), and to subclassify both normal (eg, lymphocytes) and tumor cells (eg, leukemic cells). In addition, they are being used to direct therapeutic agents to tumor cells and also to accelerate removal of drugs from the circulation when they reach toxic levels (eg, digoxin).

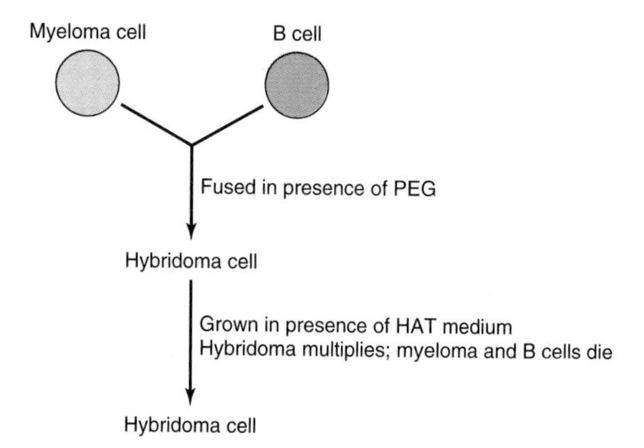


***Figure 50–11.*** Scheme of production of a hybridoma cell. The myeloma cells are immortalized, do not produce antibody, and are HGPRT$^-$ (rendering the salvage pathway of purine synthesis [Chapter 34] inactive). The B cells are not immortalized, each produces a specific antibody, and they are HGPRT$^+$. Polyethylene glycol (PEG) stimulates cell fusion. The resulting hybridoma cells are immortalized (via the parental myeloma cells), produce antibody, and are HGPRT$^+$ (both latter properties gained from the parental B cells). The B cells will die in the medium because they are not immortalized. In the presence of HAT, the myeloma cells will also die, since the aminopterin in HAT suppresses purine synthesis by the de novo pathway by inhibiting reutilization of tetrahydrofolate (Chapter 34). However, the hybridoma cells will survive, grow (because they are HGPRT$^+$), and—if cloned—produce monoclonal antibody. (HAT, hypoxanthine, aminopterin, and thymidine; HGPRT, hypoxanthine-guanine phosphoribosyl transferase.)

## The Complement System Comprises About 20 Plasma Proteins & Is Involved in Cell Lysis, Inflammation, & Other Processes

Plasma contains approximately 20 proteins that are members of the complement system. This system was discovered when it was observed that addition of fresh serum containing antibodies directed to a bacterium caused its lysis. Unlike antibodies, the factor was labile when heated at 56 °C. Subsequent work has resolved the proteins of the system and how they function; most have been cloned and sequenced. The major protein components are designated C1–9, with C9 associated with the C5–8 complex (together constituting the **membrane attack complex**) being involved in generating a lipid-soluble pore in the cell membrane that causes **osmotic lysis.**

The details of this system are relatively complex, and a textbook of immunology should be consulted. The basic concept is that the normally inactive proteins of the system, when triggered by a stimulus, become activated by proteolysis and interact in a specific sequence with one or more of the other proteins of the system. This results in cell lysis and generation of **peptide or polypeptide fragments** that are involved in various aspects of inflammation (chemotaxis, phagocytosis, etc). The system has other functions, such as clearance of antigen-antibody complexes from the circulation. Activation of the complement system is triggered by one of two routes, called the **classic** and the **alternative pathways.** The first involves interaction of C1 with antigen-antibody complexes, and the second (not involving antibody) involves direct interaction of bacterial cell surfaces or polysaccharides with a component designated C3b.

The complement system resembles blood coagulation (Chapter 51) in that it involves both conversion of inactive precursors to active products by proteases and a cascade with amplification.

## SUMMARY

- Plasma contains many proteins with a variety of functions. Most are synthesized in the liver and are glycosylated.
- Albumin, which is not glycosylated, is the major protein and is the principal determinant of intravascular osmotic pressure; it also binds many ligands, such as drugs and bilirubin.
- Haptoglobin binds extracorpuscular hemoglobin, prevents its loss into the kidney and urine, and hence preserves its iron for reutilization.
- Transferrin binds iron, transporting it to sites where it is required. Ferritin provides an intracellular store of iron. Iron deficiency anemia is a very prevalent disorder. Hereditary hemochromatosis has been shown to be due to mutations in *HFE,* a gene encoding the protein HFE, which appears to play an important role in absorption of iron.
- Ceruloplasmin contains substantial amounts of copper, but albumin appears to be more important with regard to its transport. Both Wilson disease and Menkes disease, which reflect abnormalities of copper metabolism, have been found to be due to mutations in genes encoding copper-binding P-type ATPases.
- $\alpha_1$-Antitrypsin is the major serine protease inhibitor of plasma, in particular inhibiting the elastase of neutrophils. Genetic deficiency of this protein is a cause of emphysema and can also lead to liver disease.
- $\alpha_2$-Macroglobulin is a major plasma protein that neutralizes many proteases and targets certain cytokines to specific organs.
- Immunoglobulins play a key role in the defense mechanisms of the body, as do proteins of the complement system. Some of the principal features of these proteins are described.

## REFERENCES

Andrews NC: Disorders of iron metabolism. N Engl J Med 1999;341:1986.

Carrell RW, Lomas DA: Alpha$_1$-antitrypsin deficiency—a model for conformational diseases. N Engl J Med 2002;346:45.

Gabay C, Kushner I: Acute-phase proteins and other systemic responses to inflammation. New Engl J Med 1999;340:448.

Harris ED: Cellular copper transport and metabolism. Annu Rev Nutr 2000;20:291.

Langlois MR, Delanghe JR: Biological and clinical significance of haptoglobin polymorphism in humans. Clin Chem 1996; 2:1589.

Levinson W, Jawetz E: *Medical Microbiology and Immunology,* 6th ed. Appleton & Lange, 2000.

Parslow TG et al (editors): *Medical Immunology,* 10th ed. Appleton & Lange, 2001.

Pepys MB, Berger A: The renaissance of C reactive protein. BMJ 2001;322:4.

Waheed A et al: Regulation of transferrin-mediated iron uptake by HFE, the protein defective in hereditary hemochromatosis. Proc Natl Acad U S A 2002;99:3117.

# Hemostasis & Thrombosis

51

*Margaret L. Rand, PhD, & Robert K. Murray, MD, PhD*

## BIOMEDICAL IMPORTANCE

Basic aspects of the proteins of the blood coagulation system and of fibrinolysis are described in this chapter. Some fundamental aspects of platelet biology are also presented. Hemorrhagic and thrombotic states can cause serious medical emergencies, and thromboses in the coronary and cerebral arteries are major causes of death in many parts of the world. Rational management of these conditions requires a clear understanding of the bases of blood clotting and fibrinolysis.

## HEMOSTASIS & THROMBOSIS HAVE THREE COMMON PHASES

Hemostasis is the cessation of bleeding from a cut or severed vessel, whereas thrombosis occurs when the endothelium lining blood vessels is damaged or removed (eg, upon rupture of an atherosclerotic plaque). These processes encompass blood clotting (coagulation) and involve blood vessels, platelet aggregation, and plasma proteins that cause formation or dissolution of platelet aggregates.

In hemostasis, there is initial vasoconstriction of the injured vessel, causing diminished blood flow distal to the injury. Then hemostasis and thrombosis share three phases:

(1) Formation of a loose and temporary platelet aggregate at the site of injury. Platelets bind to collagen at the site of vessel wall injury and are activated by thrombin (the mechanism of activation of platelets is described below), formed in the coagulation cascade at the same site, or by ADP released from other activated platelets. Upon activation, platelets change shape and, in the presence of fibrinogen, aggregate to form the hemostatic plug (in hemostasis) or thrombus (in thrombosis).

(2) Formation of a fibrin mesh that binds to the platelet aggregate, forming a more stable hemostatic plug or thrombus.

(3) Partial or complete dissolution of the hemostatic plug or thrombus by plasmin.

### There Are Three Types of Thrombi

Three types of thrombi or clots are distinguished. All three contain **fibrin** in various proportions.

(1) The **white** thrombus is composed of platelets and fibrin and is relatively poor in erythrocytes. It forms at the site of an injury or abnormal vessel wall, particularly in areas where blood flow is rapid (arteries).

(2) The **red** thrombus consists primarily of red cells and fibrin. It morphologically resembles the clot formed in a test tube and may form in vivo in areas of retarded blood flow or stasis (eg, veins) with or without vascular injury, or it may form at a site of injury or in an abnormal vessel in conjunction with an initiating platelet plug.

(3) A third type is a disseminated **fibrin deposit** in very small blood vessels or capillaries.

We shall first describe the coagulation pathway leading to the formation of fibrin. Then we shall briefly describe some aspects of the involvement of platelets and blood vessel walls in the overall process. This separation of clotting factors and platelets is artificial, since both play intimate and often mutually interdependent roles in hemostasis and thrombosis, but it facilitates description of the overall processes involved.

### Both Intrinsic & Extrinsic Pathways Result in the Formation of Fibrin

Two pathways lead to fibrin clot formation: the intrinsic and the extrinsic pathways. These pathways are not independent, as previously thought. However, this artificial distinction is retained in the following text to facilitate their description.

Initiation of the fibrin clot in response to tissue injury is carried out by the extrinsic pathway. How the intrinsic pathway is activated in vivo is unclear, but it involves a negatively charged surface. The intrinsic and extrinsic pathways converge in a **final common pathway** involving the activation of prothrombin to thrombin and the thrombin-catalyzed cleavage of fibrinogen to form the fibrin clot. The intrinsic, extrinsic, and final common pathways are complex and involve many different proteins (Figure 51–1 and Table 51–1). In

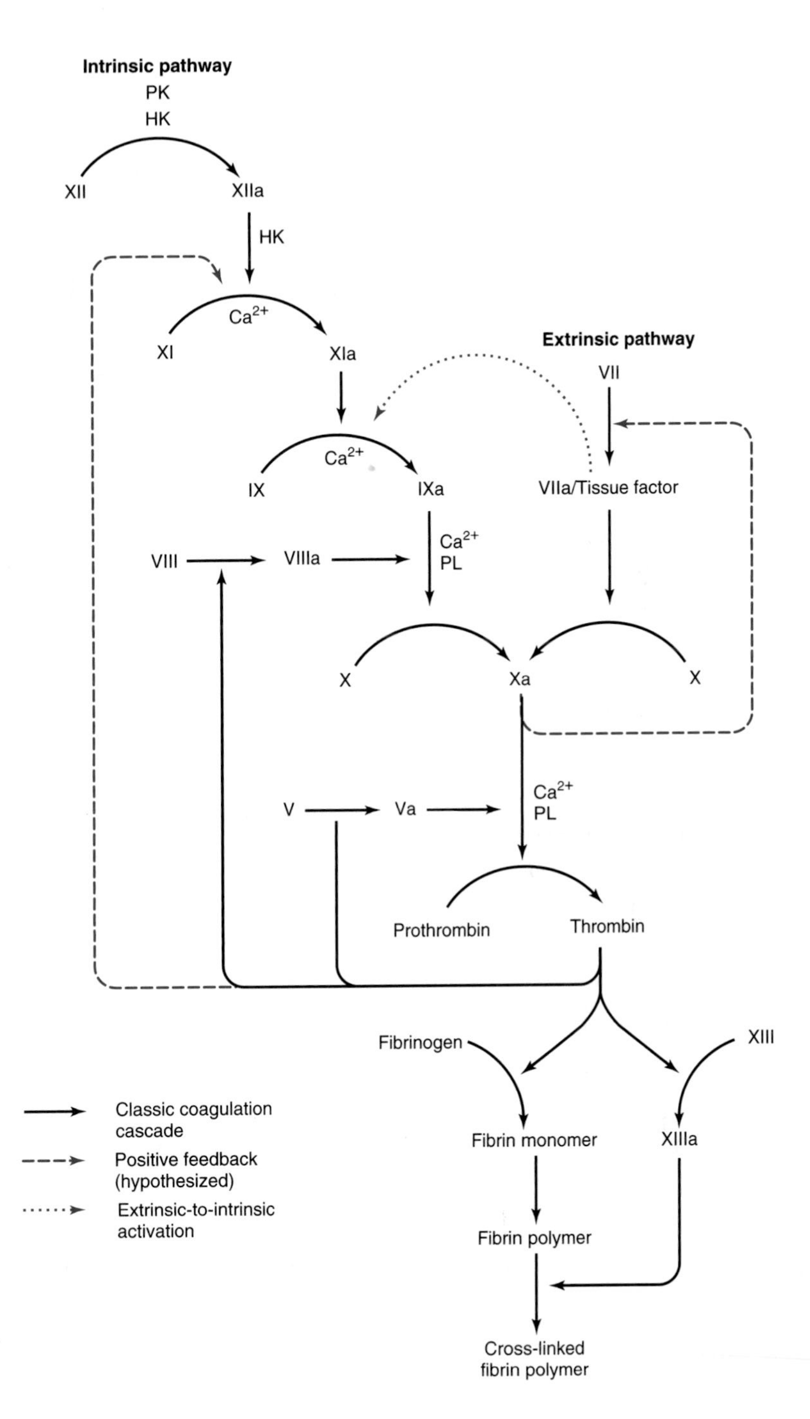


***Figure 51–1.*** The pathways of blood coagulation. The intrinsic and extrinsic pathways are indicated. The events depicted below factor Xa are designated the final common pathway, culminating in the formation of cross-linked fibrin. New observations (dotted arrow) include the finding that complexes of tissue factor and factor VIIa activate not only factor X (in the classic extrinsic pathway) but also factor IX in the intrinsic pathway. In addition, thrombin and factor Xa feedback-activate at the two sites indicated (dashed arrows). (PK, prekallikrein; HK, HMW kininogen; PL, phospholipids.) (Reproduced, with permission, from Roberts HR, Lozier JN: New perspectives on the coagulation cascade. Hosp Pract [Off Ed] 1992 Jan;27:97.)

***Table 51–1.*** Numerical system for nomenclature of blood clotting factors. The numbers indicate the order in which the factors have been discovered and bear no relationship to the order in which they act.

| Factor | Common Name |
|---|---|
| I | Fibrinogen — These factors are usually referred to by their common names. |
| II | Prothrombin — These factors are usually referred to by their common names. |
| III | Tissue factor — These factors are usually not referred to as coagulation factors. |
| IV | $Ca^{2+}$ — These factors are usually not referred to as coagulation factors. |
| V | Proaccelerin, labile factor, accelerator (Ac-) globulin |
| VII[1] | Proconvertin, serum prothrombin conversion accelerator (SPCA), cothromboplastin |
| VIII | Antihemophilic factor A, antihemophilic globulin (AHG) |
| IX | Antihemophilic factor B, Christmas factor, plasma thromboplastin component (PTC) |
| X | Stuart-Prower factor |
| XI | Plasma thromboplastin antecedent (PTA) |
| XII | Hageman factor |
| XIII | Fibrin stabilizing factor (FSF), fibrinoligase |

[1]There is no factor VI.

***Table 51–2.*** The functions of the proteins involved in blood coagulation.

| | |
|---|---|
| **Zymogens of serine proteases** | |
| Factor XII | Binds to negatively charged surface at site of vessel wall injury; activated by high-MW kininogen and kallikrein. |
| Factor XI | Activated by factor XIIa. |
| Factor IX | Activated by factor XIa in presence of $Ca^{2+}$. |
| Factor VII | Activated thrombin in presence of $Ca^{2+}$. |
| Factor X | Activated on surface of activated platelets by tenase complex ($Ca^{2+}$, factors VIIIa and IXa) and by factor VIIa in presence of tissue factor and $Ca^{2+}$. |
| Factor II | Activated on surface of activated platelets by prothrombinase complex ($Ca^{2+}$, factors Va and Xa). [Factors II, VII, IX, and X are Gla-containing zymogens.] (Gla = $\gamma$-carboxyglutamate.) |
| **Cofactors** | |
| Factor VIII | Activated by thrombin; factor VIIIa is a cofactor in the activation of factor X by factor IXa. |
| Factor V | Activated by thrombin; factor Va is a cofactor in the activation of prothrombin by factor Xa. |
| Tissue factor (factor III) | A glycoprotein expressed on the surface of injured or stimulated endothelial cells to act as a cofactor for factor VIIa. |
| **Fibrinogen** | |
| Factor I | Cleaved by thrombin to form fibrin clot. |
| **Thiol-dependent transglutaminase** | |
| Factor XIII | Activated by thrombin in presence of $Ca^{2+}$; stabilizes fibrin clot by covalent cross-linking. |
| **Regulatory and other proteins** | |
| Protein C | Activated to protein Ca by thrombin bound to thrombomodulin; then degrades factors VIIIa and Va. |
| Protein S | Acts as a cofactor of protein C; both proteins contain Gla ($\gamma$-carboxyglutamate) residues. |
| Thrombomodulin | Protein on the surface of endothelial cells; binds thrombin, which then activates protein C. |

general, as shown in Table 51–2, these proteins can be classified into five types: (1) zymogens of serine-dependent proteases, which become activated during the process of coagulation; (2) cofactors; (3) fibrinogen; (4) a transglutaminase, which stabilizes the fibrin clot; and (5) regulatory and other proteins.

## The Intrinsic Pathway Leads to Activation of Factor X

The intrinsic pathway (Figure 51–1) involves factors XII, XI, IX, VIII, and X as well as prekallikrein, high-molecular-weight (HMW) kininogen, $Ca^{2+}$, and platelet phospholipids. It results in the production of factor Xa (by convention, activated clotting factors are referred to by use of the suffix a).

This pathway commences with the "contact phase" in which prekallikrein, HMW kininogen, factor XII, and factor XI are exposed to a negatively charged activating surface. In vivo, the proteins probably assemble on endothelial cell membranes, whereas glass or kaolin can be used for in vitro tests of the intrinsic pathway. When the components of the contact phase assemble on the activating surface, factor XII is activated to factor XIIa upon proteolysis by kallikrein. This factor XIIa, generated by kallikrein, attacks prekallikrein to generate more kallikrein, setting up a reciprocal activation. Factor XIIa, once formed, activates factor XI to XIa and also releases bradykinin (a nonapeptide with potent vasodilator action) from HMW kininogen.

Factor XIa in the presence of $Ca^{2+}$ activates factor IX (55 kDa, a zymogen containing vitamin K-dependent $\gamma$-carboxyglutamate [Gla] residues; see Chapter 45), to the serine protease, factor IXa. This in turn cleaves an Arg-Ile bond in factor X (56 kDa) to produce the two-chain serine protease, factor Xa. This latter reaction requires the assembly of components, called **the tenase**

**complex,** on the surface of activated platelets: $Ca^{2+}$ and factor VIIIa, as well as factors IXa and X. It should be noted that in all reactions involving the Gla-containing zymogens (factors II, VII, IX, and X), the Gla residues in the amino terminal regions of the molecules serve as high-affinity binding sites for $Ca^{2+}$. For assembly of the tenase complex, the platelets must first be activated to expose the acidic (anionic) phospholipids, **phosphatidylserine** and **phosphatidylinositol,** that are normally on the internal side of the plasma membrane of resting, nonactivated platelets. Factor VIII (330 kDa), a glycoprotein, is not a protease precursor but a cofactor that serves as a receptor for factors IXa and X on the platelet surface. Factor VIII is activated by minute quantities of thrombin to form factor VIIIa, which is in turn inactivated upon further cleavage by thrombin.

## The Extrinsic Pathway Also Leads to Activation of Factor X But by a Different Mechanism

Factor Xa occurs at the site where the intrinsic and extrinsic pathways converge (Figure 51–1) and lead into the final common pathway of blood coagulation. The extrinsic pathway involves tissue factor, factors VII and X, and $Ca^{2+}$ and results in the production of factor Xa. It is initiated at the site of tissue injury with the exposure of **tissue factor** (Figure 51–1) on subendothelial cells. Tissue factor interacts with and activates factor VII (53 kDa), a circulating Gla-containing glycoprotein synthesized in the liver. Tissue factor acts as a cofactor for factor VIIa, enhancing its enzymatic activity to activate factor X. The association of tissue factor and factor VIIa is called **tissue factor complex.** Factor VIIa cleaves the same Arg-Ile bond in factor X that is cleaved by the tenase complex of the intrinsic pathway. Activation of factor X provides an important link between the intrinsic and extrinsic pathways.

Another important interaction between the extrinsic and intrinsic pathways is that complexes of tissue factor and factor VIIa also activate factor IX in the intrinsic pathway. Indeed, **the formation of complexes between tissue factor and factor VIIa is now considered to be the key process involved in initiation of blood coagulation in vivo.** The physiologic significance of the initial steps of the intrinsic pathway, in which factor XII, prekallikrein, and HMW kininogen are involved, has been called into question because patients with a hereditary deficiency of these components do not exhibit bleeding problems. Similarly, patients with a deficiency of factor XI may not have bleeding problems. The intrinsic pathway may actually be more important in fibrinolysis (see below) than in coagulation, since kallikrein, factor XIIa, and factor XIa can cleave plasminogen and kallikrein can activate single-chain urokinase.

**Tissue factor pathway inhibitor (TFPI)** is a major physiologic inhibitor of coagulation. It is a protein that circulates in the blood associated with lipoproteins. TFPI directly inhibits factor Xa by binding to the enzyme near its active site. This factor Xa-TFPI complex then inhibits the factor VIIa-tissue factor complex.

## The Final Common Pathway of Blood Clotting Involves Activation of Prothrombin to Thrombin

In the final common pathway, factor Xa, produced by either the intrinsic or the extrinsic pathway, activates **prothrombin** (factor II) to **thrombin** (factor IIa), which then converts fibrinogen to fibrin (Figure 51–1).

The activation of prothrombin, like that of factor X, occurs on the surface of activated platelets and requires the assembly of **a prothrombinase complex,** consisting of platelet anionic phospholipids, $Ca^{2+}$, factor Va, factor Xa, and prothrombin.

Factor V (330 kDa), a glycoprotein with homology to factor VIII and ceruloplasmin, is synthesized in the liver, spleen, and kidney and is found in platelets as well as in plasma. It functions as a cofactor in a manner similar to that of factor VIII in the tenase complex. When activated to factor Va by traces of thrombin, it binds to specific receptors on the platelet membrane (Figure 51–2) and forms a complex with factor Xa and prothrombin. It is subsequently inactivated by further action of thrombin, thereby providing a means of limiting the activation of prothrombin to thrombin. **Prothrombin** (72 kDa; Figure 51–3) is a single-chain glycoprotein synthesized in the liver. The amino terminal region of prothrombin (1 in Figure 51–3) contains ten Gla residues, and the serine-dependent active protease site (indicated by the arrowhead) is in the carboxyl terminal region of the molecule. Upon binding to the complex of factors Va and Xa on the platelet membrane, prothrombin is cleaved by factor Xa at two sites (Figure 51–2) to generate the active, two-chain thrombin molecule, which is then released from the platelet surface. The A and B chains of thrombin are held together by a disulfide bond.

## Conversion of Fibrinogen to Fibrin Is Catalyzed by Thrombin

Fibrinogen (factor I, 340 kDa; see Figures 51–1 and 51–4 and Tables 51–1 and 51–2) is a soluble plasma glycoprotein that consists of three nonidentical pairs of polypeptide chains $(A\alpha,B\beta\gamma)_2$ covalently linked by disulfide bonds. The $B\beta$ and $\gamma$ chains contain asparagine-linked complex oligosaccharides. All three

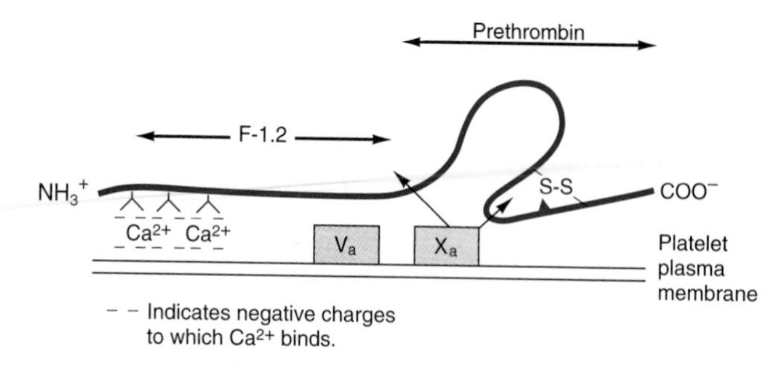


***Figure 51-2.*** Diagrammatic representation (not to scale) of the binding of factors Va, Xa, $Ca^{2+}$, and prothrombin to the plasma membrane of the activated platelet. The sites of cleavage of prothrombin by factor Xa are indicated by two arrows. The part of prothrombin destined to form thrombin is labeled prethrombin. The $Ca^{2+}$ is bound to anionic phospholipids of the plasma membrane of the activated platelet.

chains are synthesized in the liver; the three structural genes involved are on the same chromosome, and their expression is coordinately regulated in humans. The amino terminal regions of the six chains are held in close proximity by a number of disulfide bonds, while the carboxyl terminal regions are spread apart, giving rise to a highly asymmetric, elongated molecule (Figure 51–4). The A and B portions of the Aα and Bβ chains, designated **fibrinopeptides A (FPA) and B (FPB),** respectively, at the amino terminal ends of the chains, bear excess negative charges as a result of the presence of aspartate and glutamate residues, as well as an unusual tyrosine O-sulfate in FPB. These negative charges contribute to the solubility of fibrinogen in plasma and also serve to prevent aggregation by causing electrostatic repulsion between fibrinogen molecules.

**Thrombin** (34 kDa), a serine protease formed by the prothrombinase complex, hydrolyzes the four Arg-Gly bonds between the fibrinopeptides and the α and β portions of the Aα and Bβ chains of fibrinogen (Figure 51–5A). The release of the fibrinopeptides by thrombin generates fibrin monomer, which has the subunit structure $(\alpha, \beta, \gamma)_2$. Since FPA and FPB contain only 16 and 14 residues, respectively, the fibrin molecule retains 98% of the residues present in fibrinogen. The removal of the fibrinopeptides exposes binding sites that allow the molecules of fibrin monomers to aggregate spontaneously in a regularly staggered array, forming an insoluble fibrin clot. It is the formation of this insoluble fibrin polymer that traps platelets, red cells, and other components to form the white or red thrombi. This initial fibrin clot is rather weak, held together only by the noncovalent association of fibrin monomers.

In addition to converting fibrinogen to fibrin, thrombin also converts factor XIII to factor XIIIa. This factor is a highly specific **transglutaminase** that covalently cross-links fibrin molecules by forming peptide bonds between the amide groups of glutamine and the ε-amino groups of lysine residues (Figure 51–5B), yielding a more stable fibrin clot with increased resistance to proteolysis.

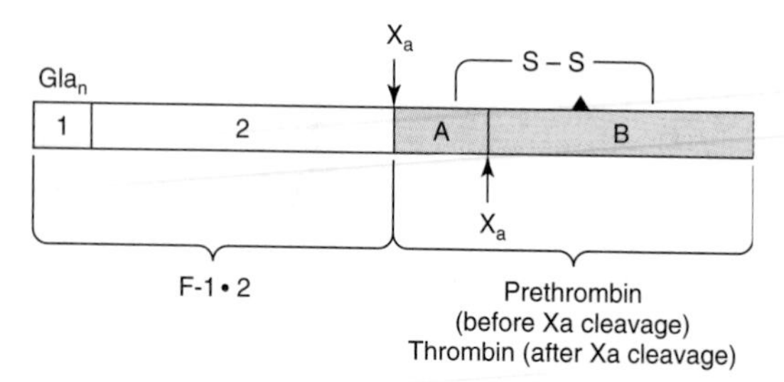


***Figure 51–3.*** Diagrammatic representation (not to scale) of prothrombin. The amino terminal is to the left; region 1 contains all ten Gla residues. The sites of cleavage by factor Xa are shown and the products named. The site of the catalytically active serine residue is indicated by the solid triangle. The A and B chains of active thrombin (shaded) are held together by the disulfide bridge.

## Levels of Circulating Thrombin Must Be Carefully Controlled or Clots May Form

Once active thrombin is formed in the course of hemostasis or thrombosis, its concentration must be carefully controlled to prevent further fibrin formation or platelet activation. This is achieved in two ways. Thrombin circulates as its inactive precursor, prothrombin, which is activated as the result of a cascade of enzymatic reactions, each converting an inactive zymogen to an active enzyme and leading finally to the conversion of prothrombin to thrombin (Figure 51–1). At each point in the cascade, **feedback mechanisms** produce a delicate balance of activation and inhibition. The concentration of factor XII in plasma is approximately 30 μg/mL, while that of fibrinogen is 3 mg/mL, with intermediate clotting factors increasing in concentration as one proceeds down the cascade, showing that the clotting cascade provides amplification. The second means of controlling thrombin activity is the inactivation of any thrombin formed by **circulating inhibi-**

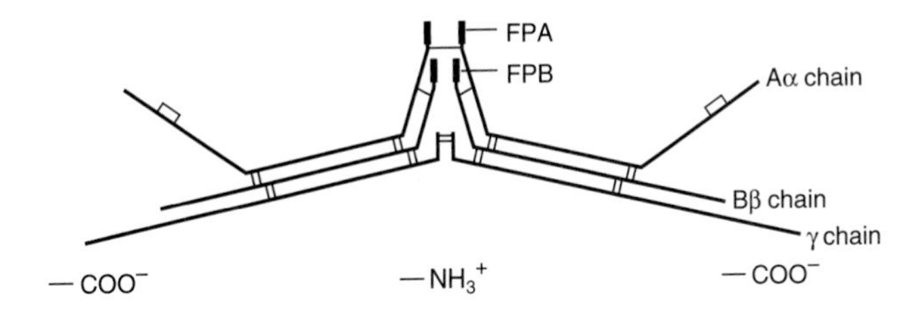


***Figure 51–4.*** Diagrammatic representation (not to scale) of fibrinogen showing pairs of Aα, Bβ, and γ chains linked by disulfide bonds. (FPA, fibrinopeptide A; FPB, fibrinopeptide B.)

**tors,** the most important of which is antithrombin III (see below).

## The Activity of Antithrombin III, an Inhibitor of Thrombin, Is Increased by Heparin

Four naturally occurring thrombin inhibitors exist in normal plasma. The most important is **antithrombin III** (often called simply antithrombin), which contributes approximately 75% of the antithrombin activity. Antithrombin III can also inhibit the activities of factors IXa, Xa, XIa, XIIa, and VIIa complexed with tissue factor. **$\alpha_2$-Macroglobulin** contributes most of the remainder of the antithrombin activity, with **heparin cofactor II** and **$\alpha_1$-antitrypsin** acting as minor inhibitors under physiologic conditions.

The endogenous activity of antithrombin III is greatly potentiated by the presence of acidic proteoglycans such as **heparin** (Chapter 48). These bind to a specific cationic site of antithrombin III, inducing a conformational change and promoting its binding to thrombin as well as to its other substrates. This is the basis for the use of heparin in clinical medicine to inhibit coagulation. The anticoagulant effects of heparin can be antagonized by strongly cationic polypeptides such as **protamine,** which bind strongly to heparin, thus inhibiting its binding to antithrombin III. Individuals with inherited deficiencies of antithrombin III are prone to develop venous thrombosis, providing evidence that antithrombin III has a physiologic function and that the coagulation system in humans is normally in a dynamic state.

Thrombin is involved in an additional regulatory mechanism that operates in coagulation. It combines with **thrombomodulin,** a glycoprotein present on the surfaces of endothelial cells. The complex activates **protein C.** In combination with **protein S,** activated protein C (APC) degrades factors Va and VIIIa, limiting their actions in coagulation. A genetic deficiency of either protein C or protein S can cause venous thrombosis. Furthermore, patients with **factor V Leiden** (which has a glutamine residue in place of an arginine at position 506) have an increased risk of venous thrombotic

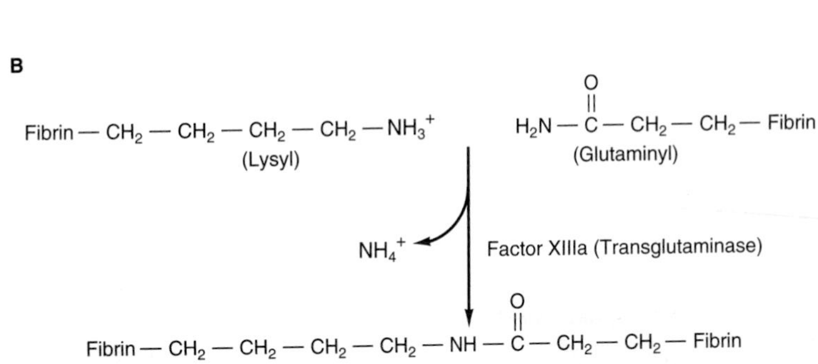

A

Thrombin

$NH_3^+$ — Arg — Gly — $COO^-$

Fibrinopeptide (A or B)

Fibrin chain (α or β)

B

Fibrin — $CH_2$ — $CH_2$ — $CH_2$ — $CH_2$ — $NH_3^+$ (Lysyl)

$H_2N$ — C(=O) — $CH_2$ — $CH_2$ — Fibrin (Glutaminyl)

$NH_4^+$

Factor XIIIa (Transglutaminase)

Fibrin — $CH_2$ — $CH_2$ — $CH_2$ — $CH_2$ — NH — C(=O) — $CH_2$ — $CH_2$ — Fibrin

***Figure 51-5.*** Formation of a fibrin clot. **A:** Thrombin-induced cleavage of Arg-Gly bonds of the Aα and Bβ chains of fibrinogen to produce fibrinopeptides (left-hand side) and the α and β chains of fibrin monomer (right-hand side). **B:** Cross-linking of fibrin molecules by activated factor XIII (factor XIIIa).

disease because factor V Leiden is resistant to inactivation by APC. This condition is termed APC resistance.

## Coumarin Anticoagulants Inhibit the Vitamin K-Dependent Carboxylation of Factors II, VII, IX, & X

The coumarin drugs (eg, warfarin), which are used as anticoagulants, inhibit the vitamin K-dependent carboxylation of Glu to Gla residues (see Chapter 45) in the amino terminal regions of factors II, VII, IX, and X and also proteins C and S. These proteins, all of which are synthesized in the liver, are dependent on the $Ca^{2+}$-binding properties of the Gla residues for their normal function in the coagulation pathways. The coumarins act by inhibiting the reduction of the quinone derivatives of vitamin K to the active hydroquinone forms (Chapter 45). Thus, the administration of vitamin K will bypass the coumarin-induced inhibition and allow maturation of the Gla-containing factors. Reversal of coumarin inhibition by vitamin K requires 12–24 hours, whereas reversal of the anticoagulant effects of heparin by protamine is almost instantaneous.

Heparin and warfarin are widely used in the treatment of thrombotic and thromboembolic conditions, such as deep vein thrombosis and pulmonary embolus. Heparin is administered first, because of its prompt onset of action, whereas warfarin takes several days to reach full effect. Their effects are closely monitored by use of appropriate tests of coagulation (see below) because of the risk of producing hemorrhage.

## Hemophilia A Is Due to a Genetically Determined Deficiency of Factor VIII

Inherited deficiencies of the clotting system that result in bleeding are found in humans. The most common is deficiency of factor VIII, causing **hemophilia A,** an X chromosome-linked disease that has played a major role in the history of the royal families of Europe. **Hemophilia B** is due to a deficiency of factor IX; its clinical features are almost identical to those of hemophilia A, but the conditions can be separated on the basis of specific assays that distinguish between the two factors.

The gene for human factor VIII has been cloned and is one of the largest so far studied, measuring 186 kb in length and containing 26 exons. A variety of mutations have been detected leading to diminished activity of factor VIII; these include partial gene deletions and point mutations resulting in premature chain termination. Prenatal diagnosis by DNA analysis after chorionic villus sampling is now possible.

In past years, treatment for patients with hemophilia A has consisted of administration of cryoprecipitates (enriched in factor VIII) prepared from individual donors or lyophilized factor VIII concentrates prepared from plasma pools of up to 5000 donors. It is now possible to prepare factor VIII by **recombinant DNA technology.** Such preparations are free of contaminating viruses (eg, hepatitis A, B, C, or HIV-1) found in human plasma but are at present expensive; their use may increase if cost of production decreases.

## Fibrin Clots Are Dissolved by Plasmin

As stated above, the coagulation system is normally in a state of dynamic equilibrium in which fibrin clots are constantly being laid down and dissolved. This latter process is termed **fibrinolysis. Plasmin,** the serine protease mainly responsible for degrading fibrin and fibrinogen, circulates in the form of its inactive zymogen, **plasminogen** (90 kDa), and any small amounts of plasmin that are formed in the fluid phase under physiologic conditions are rapidly inactivated by the fast-acting plasmin inhibitor, $\alpha_2$-antiplasmin. Plasminogen binds to fibrin and thus becomes incorporated in clots as they are produced; since plasmin that is formed when bound to fibrin is protected from $\alpha_2$-antiplasmin, it remains active. **Activators of plasminogen** of various types are found in most body tissues, and all cleave the same Arg-Val bond in plasminogen to produce the two-chain serine protease, plasmin (Figure 51–6).

**Tissue plasminogen activator (alteplase; t-PA)** is a serine protease that is released into the circulation from vascular endothelium under conditions of injury or

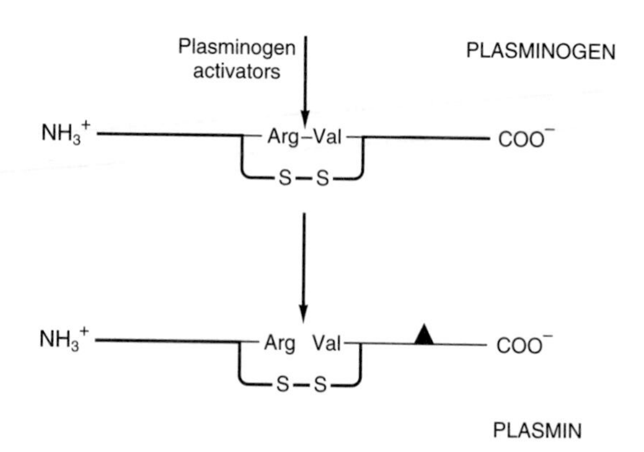


***Figure 51–6.*** Activation of plasminogen. The same Arg-Val bond is cleaved by all plasminogen activators to give the two-chain plasmin molecule. The solid triangle indicates the serine residue of the active site. The two chains of plasmin are held together by a disulfide bridge.

stress and is catalytically inactive unless bound to fibrin. Upon binding to fibrin, t-PA cleaves plasminogen within the clot to generate plasmin, which in turn digests the fibrin to form soluble degradation products and thus dissolves the clot. Neither plasmin nor the plasminogen activator can remain bound to these degradation products, and so they are released into the fluid phase, where they are inactivated by their natural inhibitors. Prourokinase is the precursor of a second activator of plasminogen, **urokinase.** Originally isolated from urine, it is now known to be synthesized by cell types such as monocytes and macrophages, fibroblasts, and epithelial cells. Its main action is probably in the degradation of extracellular matrix. Figure 51–7 indicates the sites of action of five proteins that influence the formation and action of plasmin.

## Recombinant t-PA & Streptokinase Are Used as Clot Busters

**Alteplase** (t-PA), produced by recombinant DNA technology, is used therapeutically as a fibrinolytic agent, as is **streptokinase.** However, the latter is less selective than t-PA, activating plasminogen in the fluid phase (where it can degrade circulating fibrinogen) as well as plasminogen that is bound to a fibrin clot. The amount of plasmin produced by therapeutic doses of streptokinase may exceed the capacity of the circulating $\alpha_2$-antiplasmin, causing fibrinogen as well as fibrin to be degraded and resulting in the bleeding often encountered during fibrinolytic therapy. Because of its **selectivity** for degrading fibrin, there is considerable therapeutic interest in the use of recombinant t-PA to restore the patency of coronary arteries following thrombosis. If administered early enough, before irreversible damage of heart muscle occurs (about 6 hours after onset of thrombosis), t-PA can significantly reduce the mortality rate from myocardial damage following coronary thrombosis. t-PA is more effective than streptokinase at restoring full patency and also appears to result in a slightly better survival rate. Table 51–3 compares some thrombolytic features of streptokinase and t-PA.

There are a number of disorders, including cancer and shock, in which **the concentrations of plasminogen activators increase.** In addition, the antiplasmin activities contributed by $\alpha_1$-antitrypsin and $\alpha_2$-antiplasmin may be impaired in diseases such as cirrhosis. Since certain bacterial products, such as streptokinase, are capable of activating plasminogen, they may be responsible for the diffuse hemorrhage sometimes observed in patients with disseminated bacterial infections.

## Activation of Platelets Involves Stimulation of the Polyphosphoinositide Pathway

Platelets normally circulate in an unstimulated disk-shaped form. During hemostasis or thrombosis, they become activated and help form hemostatic plugs or thrombi. Three major steps are involved: (1) adhesion to exposed collagen in blood vessels, (2) release of the contents of their granules, and (3) aggregation.

Platelets adhere to collagen via specific receptors on the platelet surface, including the glycoprotein complex GPIa–IIa ($\alpha_2\beta_1$ integrin; Chapter 52), in a reaction that involves **von Willebrand factor.** This is a glycoprotein, secreted by endothelial cells into the plasma, which stabilizes factor VIII and binds to collagen and the subendothelium. Platelets bind to von Willebrand factor via a glycoprotein complex (GPIb–V–IX) on the platelet surface; this interaction is especially important in platelet adherence to the subendothelium under conditions of high shear stress that occur in small vessels and stenosed arteries.

Platelets adherent to collagen change shape and spread out on the subendothelium. They release the contents of their storage granules (the dense granules and the alpha granules); secretion is also stimulated by thrombin.

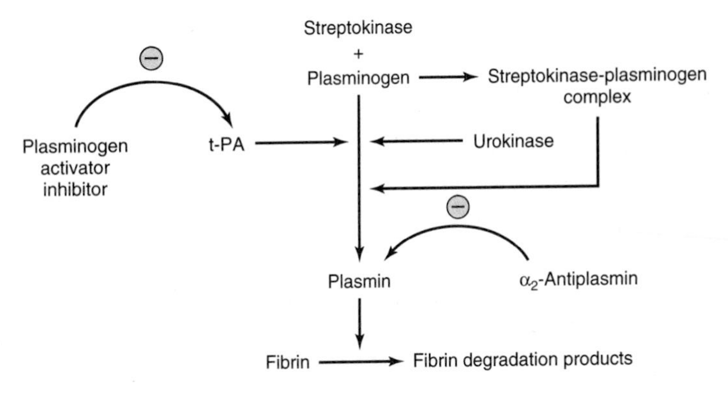


***Figure 51–7.*** Scheme of sites of action of streptokinase, tissue plasminogen activator (t-PA), urokinase, plasminogen activator inhibitor, and $\alpha_2$-antiplasmin (the last two proteins exert inhibitory actions). Streptokinase forms a complex with plasminogen, which exhibits proteolytic activity; this cleaves some plasminogen to plasmin, initiating fibrinolysis.

**Table 51–3.** Comparison of some properties of streptokinase (SK) and tissue plasminogen activator (t-PA) with regard to their use as thrombolytic agents.[1]

| | SK | t-PA |
|---|---|---|
| Selective for fibrin clot | – | + |
| Produces plasminemia | + | – |
| Reduces mortality | + | + |
| Causes allergic reaction | + | – |
| Causes hypotension | + | – |
| Cost per treatment (approximate) | Relatively low | Relatively high |

[1]Data from Webb J, Thompson C: Thrombolysis for acute myocardial infarction. Can Fam Physician 1992;38:1415.

**Thrombin,** formed from the coagulation cascade, is the most potent activator of platelets and initiates platelet activation by interacting with its receptor on the plasma membrane (Figure 51–8). The further events leading to platelet activation are examples of **transmembrane signaling,** in which a chemical messenger outside the cell generates effector molecules inside the cell. In this instance, thrombin acts as the external chemical messenger (stimulus or agonist). The interaction of thrombin with its receptor stimulates the activity of an intracellular **phospholipase Cβ.** This enzyme hydrolyzes the membrane phospholipid phosphatidylinositol 4,5-bisphosphate ($PIP_2$, a polyphosphoinositide) to form the two internal effector molecules, 1,2-diacylglycerol and 1,4,5-inositol trisphosphate.

Hydrolysis of $PIP_2$ is also involved in the action of many hormones and drugs. Diacylglycerol stimulates

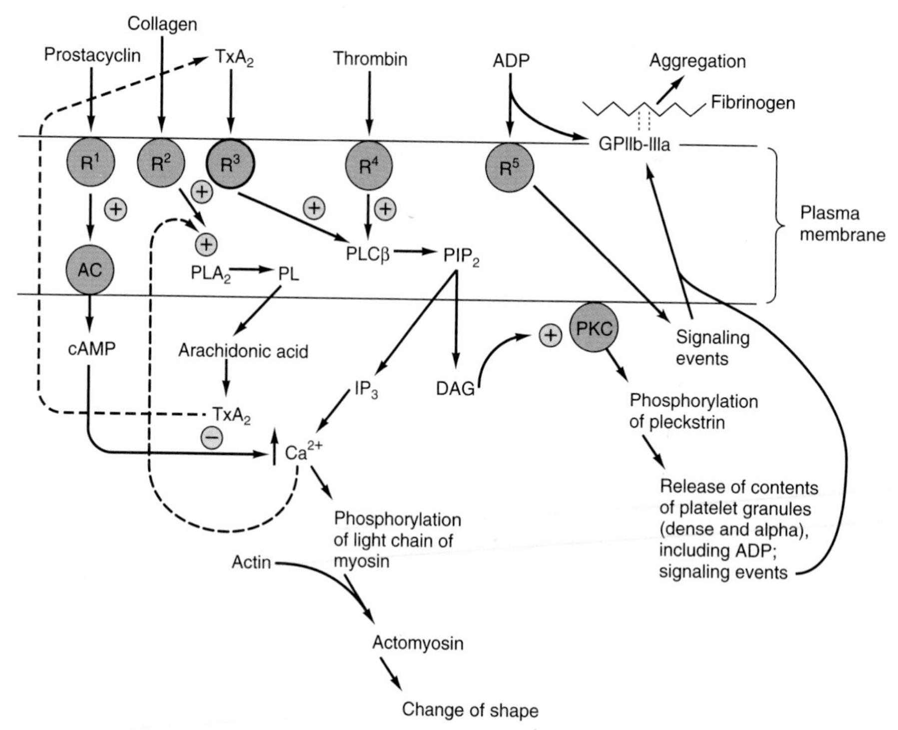


**Figure 51–8.** Diagrammatic representation of platelet activation. The external environment, the plasma membrane, and the inside of a platelet are depicted from top to bottom. Thrombin and collagen are the two most important platelet activators. ADP is considered a weak agonist; it causes aggregation but not granule release. (GP, glycoprotein; $R^1$–$R^5$, various receptors; AC, adenylyl cyclase; $PLA_2$, phospholipase $A_2$; PL, phospholipids; PLCβ, phospholipase Cβ; $PIP_2$, phosphatidylinositol 4,5-bisphosphate; cAMP, cyclic AMP; PKC, protein kinase C; $TxA_2$, thromboxane $A_2$; $IP_3$, inositol 1,4,5-trisphosphate; DAG, 1,2-diacylglycerol. The G proteins that are involved are not shown.)

**protein kinase C,** which phosphorylates the protein **pleckstrin** (47 kDa). This results in aggregation and release of the contents of the storage granules. ADP released from dense granules can also activate platelets, resulting in aggregation of additional platelets. $IP_3$ causes release of $Ca^{2+}$ into the cytosol mainly from the dense tubular system (or residual smooth endoplasmic reticulum from the megakaryocyte), which then interacts with calmodulin and myosin light chain kinase, leading to phosphorylation of the light chains of myosin. These chains then interact with actin, causing changes of platelet shape.

Collagen-induced activation of a platelet phospholipase $A_2$ by increased levels of cytosolic $Ca^{2+}$ results in liberation of arachidonic acid from platelet phospholipids, leading to the formation of **thromboxane $A_2$** (Chapter 23), which in turn, in a receptor-mediated fashion, can further activate phospholipase C, promoting platelet aggregation.

Activated platelets, besides forming a platelet aggregate, are required, via newly expressed anionic phospholipids on the membrane surface, for acceleration of the activation of factors X and II in the coagulation cascade (Figure 51–1).

All of the aggregating agents, including thrombin, collagen, ADP, and others such as platelet-activating factor, modify the platelet surface so that fibrinogen can bind to a glycoprotein complex, **GPIIb–IIIa** ($\alpha_{IIb}\beta_3$ integrin; Chapter 52), on the activated platelet surface. Molecules of divalent fibrinogen then link adjacent activated platelets to each other, forming a platelet aggregate. Some agents, including epinephrine, serotonin, and vasopressin, exert synergistic effects with other aggregating agents.

## Endothelial Cells Synthesize Prostacyclin & Other Compounds That Affect Clotting & Thrombosis

The endothelial cells in the walls of blood vessels make important contributions to the overall regulation of hemostasis and thrombosis. As described in Chapter 23, these cells synthesize **prostacyclin** ($PGI_2$), a potent inhibitor of platelet aggregation, opposing the action of thromboxane $A_2$. Prostacyclin acts by stimulating the activity of adenylyl cyclase in the surface membranes of platelets. The resulting increase of intraplatelet cAMP opposes the increase in the level of intracellular $Ca^{2+}$ produced by $IP_3$ and thus inhibits platelet activation (Figure 51–8). Endothelial cells play other roles in the regulation of thrombosis. For instance, these cells possess an ADPase, which hydrolyzes ADP, and thus opposes its aggregating effect on platelets. In addition, these cells appear to synthesize heparan sulfate, an anticoagulant, and they also synthesize plasminogen activators, which may help dissolve thrombi. Table 51–4 lists some molecules produced by endothelial cells that affect thrombosis and fibrinolysis. Endothelium-derived relaxing factor (nitric oxide) is discussed in Chapter 49.

Analysis of the mechanisms of uptake of atherogenic lipoproteins, such as LDL, by endothelial, smooth muscle, and monocytic cells of arteries, along with detailed studies of how these lipoproteins damage such cells is a key area of study in elucidating the mechanisms of **atherosclerosis** (Chapter 26).

## Aspirin Is an Effective Antiplatelet Drug

Certain drugs (antiplatelet drugs) modify the behavior of platelets. The most important is aspirin (acetylsalicylic acid), which irreversibly acetylates and thus inhibits the platelet cyclooxygenase system involved in formation of thromboxane $A_2$ (Chapter 14), a potent aggregator of platelets and also a vasoconstrictor. Platelets are very sensitive to aspirin; as little as 30 mg/d (one aspirin tablet usually contains 325 mg) effectively eliminates the synthesis of thromboxane $A_2$. Aspirin also inhibits production of prostacyclin ($PGI_2$, which opposes platelet aggregation and is a vasodilator) by en-

***Table 51–4.*** Molecules synthesized by endothelial cells that play a role in the regulation of thrombosis and fibrinolysis.[1]

| Molecule | Action |
|---|---|
| ADPase (an ectoenzyme) | Degrades ADP (an aggregating agent of platelets) to AMP + $P_i$ |
| Endothelium-derived relaxing factor (nitric oxide) | Inhibits platelet adhesion and aggregation by elevating levels of cGMP |
| Heparan sulfate (a glycosaminoglycan) | Anticoagulant; combines with antithrombin III to inhibit thrombin |
| Prostacyclin ($PGI_2$, a prostaglandin) | Inhibits platelet aggregation by increasing levels of cAMP |
| Thrombomodulin (a glycoprotein) | Binds protein C, which is then cleaved by thrombin to yield activated protein C; this in combination with protein S degrades factors Va and VIIIa, limiting their actions |
| Tissue plasminogen activator (t-PA, a protease) | Activates plasminogen to plasmin, which digests fibrin; the action of t-PA is opposed by plasminogen activator inhibitor-1 (PAI-1) |

[1]Adapted from Wu KK: Endothelial cells in hemostasis, thrombosis and inflammation. Hosp Pract (Off Ed) 1992 Apr; 27:145.

dothelial cells, but unlike platelets, these cells regenerate cyclooxygenase within a few hours. Thus, the overall balance between thromboxane $A_2$ and prostacyclin can be shifted in favor of the latter, opposing platelet aggregation. Indications for treatment with aspirin thus include management of angina and evolving myocardial infarction and also prevention of stroke and death in patients with transient cerebral ischemic attacks.

## Laboratory Tests Measure Coagulation & Thrombolysis

A number of laboratory tests are available to measure the phases of hemostasis described above. The tests include platelet count, bleeding time, activated partial thromboplastin time (aPTT or PTT), prothrombin time (PT), thrombin time (TT), concentration of fibrinogen, fibrin clot stability, and measurement of fibrin degradation products. The platelet count quantitates the number of platelets, and the bleeding time is an overall test of platelet function. aPTT is a measure of the intrinsic pathway and PT of the extrinsic pathway. PT is used to measure the effectiveness of oral anticoagulants such as warfarin, and aPTT is used to monitor heparin therapy. The reader is referred to a textbook of hematology for a discussion of these tests.

## SUMMARY

- Hemostasis and thrombosis are complex processes involving coagulation factors, platelets, and blood vessels.
- Many coagulation factors are zymogens of serine proteases, becoming activated during the overall process.
- Both intrinsic and extrinsic pathways of coagulation exist, the latter initiated by tissue factor. The pathways converge at factor Xa, embarking on the common final pathway resulting in thrombin-catalyzed conversion of fibrinogen to fibrin, which is strengthened by cross-linking, catalyzed by factor XIII.
- Genetic disorders of coagulation factors occur, and the two most common involve factors VIII (hemophilia A) and IX (hemophilia B).
- An important natural inhibitor of coagulation is antithrombin III; genetic deficiency of this protein can result in thrombosis.
- For activity, factors II, VII, IX, and X and proteins C and S require vitamin K-dependent $\gamma$-carboxylation of certain glutamate residues, a process that is inhibited by the anticoagulant warfarin.
- Fibrin is dissolved by plasmin. Plasmin exists as an inactive precursor, plasminogen, which can be activated by tissue plasminogen activator (t-PA). Both t-PA and streptokinase are widely used to treat early thrombosis in the coronary arteries.
- Thrombin and other agents cause platelet aggregation, which involves a variety of biochemical and morphologic events. Stimulation of phospholipase C and the polyphosphoinositide pathway is a key event in platelet activation, but other processes are also involved.
- Aspirin is an important antiplatelet drug that acts by inhibiting production of thromboxane $A_2$.

## REFERENCES

Bennett JS: Mechanisms of platelet adhesion and aggregation: an update. Hosp Pract (Off Ed) 1992;27:124.

Broze GJ: Tissue factor pathway inhibitor and the revised theory of coagulation. Annu Rev Med 1995;46:103.

Clemetson KJ: Platelet activation: signal transduction via membrane receptors. Thromb Haemost 1995;74:111.

Collen D, Lijnen HR: Basic and clinical aspects of fibrinolysis and thrombolysis. Blood 1991;78:3114.

Handin RI: Anticoagulant, fibrinolytic and antiplatelet therapy. In: *Harrison's Principles of Internal Medicine,* 15th ed. Braunwald E et al (editors). McGraw-Hill, 2001.

Handin RI: Disorders of coagulation and thrombosis. In: *Harrison's Principles of Internal Medicine,* 15th ed. Braunwald E et al (editors). McGraw-Hill, 2001.

Handin RI: Disorders of the platelet and vessel wall. In: *Harrison's Principles of Internal Medicine,* 15th ed. Braunwald E et al (editors). McGraw-Hill, 2001.

Kroll MH, Schafer AI: Biochemical mechanisms of platelet activation. Blood 1989;74:1181.

Roberts HR, Lozier JN: New perspectives on the coagulation cascade. Hosp Pract (Off Ed) 1992;27:97.

Roth GJ, Calverley DC: Aspirin, platelets, and thrombosis: theory and practice. Blood 1994;83:885.

Schmaier AH: Contact activation: a revision. Thromb Haemost 1997;78:101.

Wu KK: Endothelial cells in hemostasis, thrombosis and inflammation. Hosp Pract (Off Ed) 1992;27:145.

# Red & White Blood Cells

# 52

*Robert K. Murray, MD, PhD*

## BIOMEDICAL IMPORTANCE

Blood cells have been studied intensively because they are obtained easily, because of their functional importance, and because of their involvement in many disease processes. The structure and function of hemoglobin, the porphyrias, jaundice, and aspects of iron metabolism are discussed in previous chapters. Reduction of the number of red blood cells and of their content of hemoglobin is the cause of the anemias, a diverse and important group of conditions, some of which are seen very commonly in clinical practice. Certain of the blood group systems, present on the membranes of erythrocytes and other blood cells, are of extreme importance in relation to blood transfusion and tissue transplantation. Table 52–1 summarizes the causes of a number of important diseases affecting red blood cells; some are discussed in this chapter, and the remainder are discussed elsewhere in this text. Every organ in the body can be affected by inflammation; neutrophils play a central role in acute inflammation, and other white blood cells, such as lymphocytes, play important roles in chronic inflammation. Leukemias, defined as malignant neoplasms of blood-forming tissues, can affect precursor cells of any of the major classes of white blood cells; common types are acute and chronic myelocytic leukemia, affecting precursors of the neutrophils; and acute and chronic lymphocytic leukemias. Combination chemotherapy, using combinations of various chemotherapeutic agents, all of which act at one or more biochemical loci, has been remarkably effective in the treatment of certain of these types of leukemias. Understanding the role of red and white cells in health and disease requires a knowledge of certain fundamental aspects of their biochemistry.

## THE RED BLOOD CELL IS SIMPLE IN TERMS OF ITS STRUCTURE & FUNCTION

The major functions of the red blood cell are relatively simple, consisting of delivering oxygen to the tissues and of helping in the disposal of carbon dioxide and protons formed by tissue metabolism. Thus, it has a much simpler structure than most human cells, being essentially composed of a membrane surrounding a solution of hemoglobin (this protein forms about 95% of the intracellular protein of the red cell). There are no intracellular organelles, such as mitochondria, lysosomes, or Golgi apparatus. Human red blood cells, like most red cells of animals, are nonnucleated. However, the red cell is not metabolically inert. ATP is synthesized from glycolysis and is important in processes that help the red blood cell maintain its biconcave shape and also in the regulation of the transport of ions (eg, by the $Na^+$-$K^+$ ATPase and the anion exchange protein [see below]) and of water in and out of the cell. The biconcave shape increases the surface-to-volume ratio of the red blood cell, thus facilitating gas exchange. The red cell contains cytoskeletal components (see below) that play an important role in determining its shape.

### About Two Million Red Blood Cells Enter the Circulation per Second

The life span of the normal red blood cell is 120 days; this means that slightly less than 1% of the population of red cells (200 billion cells, or 2 million per second) is replaced daily. The new red cells that appear in the circulation still contain ribosomes and elements of the endoplasmic reticulum. The RNA of the ribosomes can be detected by suitable stains (such as cresyl blue), and cells containing it are termed reticulocytes; they normally number about 1% of the total red blood cell count. The life span of the red blood cell can be dramatically shortened in a variety of **hemolytic anemias.** The number of reticulocytes is markedly increased in these conditions, as the bone marrow attempts to compensate for rapid breakdown of red blood cells by increasing the amount of new, young red cells in the circulation.

### Erythropoietin Regulates Production of Red Blood Cells

Human erythropoietin is a glycoprotein of 166 amino acids (molecular mass about 34 kDa). Its amount in plasma can be measured by radioimmunoassay. It is the major regulator of human erythropoiesis. Erythropoietin is synthesized mainly by the kidney and is released in response to hypoxia into the bloodstream, in which it travels to the bone marrow. There it interacts with progenitors of red blood cells via a specific receptor. The receptor is a transmembrane protein consisting of two different subunits and a number of domains. It is not a tyrosine kinase, but it stimulates the activities of specific

***Table 52–1.*** Summary of the causes of some important disorders affecting red blood cells.

| Disorder | Sole or Major Cause |
|---|---|
| Iron deficiency anemia | Inadequate intake or excessive loss of iron |
| Methemoglobinemia | Intake of excess oxidants (various chemicals and drugs)<br>Genetic deficiency in the NADH-dependent methemoglobin reductase system (MIM 250800)<br>Inheritance of HbM (MIM 141800) |
| Sickle cell anemia (MIM 141900) | Sequence of codon 6 of the β chain changed from GAG in the normal gene to GTG in the sickle cell gene, resulting in substitution of valine for glutamic acid |
| α-Thalassemias (MIM 141800) | Mutations in the α-globin genes, mainly unequal crossing-over and large deletions and less commonly nonsense and frameshift mutations |
| β-Thalassemia (MIM 141900) | A very wide variety of mutations in the β-globin gene, including deletions, nonsense and frameshift mutations, and others affecting every aspect of its structure (eg, splice sites, promoter mutants) |
| Megaloblastic anemias<br>Deficiency of vitamin $B_{12}$ | Decreased absorption of $B_{12}$, often due to a deficiency of intrinsic factor, normally secreted by gastric parietal cells |
| Deficiency of folic acid | Decreased intake, defective absorption, or increased demand (eg, in pregnancy) for folate |
| Hereditary spherocytosis[1] | Deficiencies in the amount or in the structure of α or β spectrin, ankyrin, band 3 or band 4.1 |
| Glucose-6-phosphate dehydrogenase (G6PD) deficiency[1] (MIM 305900) | A variety of mutations in the gene (X-linked) for G6PD, mostly single point mutations |
| Pyruvate kinase (PK) deficiency[1] (MIM 255200) | Presumably a variety of mutations in the gene for the R (red cell) isozyme of PK |
| Paroxysmal nocturnal hemoglobinemia[1] (MIM 311770) | Mutations in the PIG-A gene, affecting synthesis of GPI-anchored proteins |

[1]The last four disorders cause hemolytic anemias, as do a number of the other disorders listed. Most of the above conditions are discussed in other chapters of this text. MIM numbers apply only to disorders with a genetic basis.

members of this class of enzymes involved in downstream signal transduction. Erythropoietin interacts with a red cell progenitor, known as the burst-forming unit-erythroid (BFU-E), causing it to proliferate and differentiate. In addition, it interacts with a later progenitor of the red blood cell, called the colony-forming unit-erythroid (CFU-E), also causing it to proliferate and further differentiate. For these effects, erythropoietin requires the cooperation of other factors (eg, interleukin-3 and insulin-like growth factor; Figure 52–1).

The availability of a cDNA for erythropoietin has made it possible to produce substantial amounts of this hormone for analysis and for therapeutic purposes; previously the isolation of erythropoietin from human urine yielded very small amounts of the protein. The major use of **recombinant erythropoietin** has been in the treatment of a small number of **anemic states,** such as that due to renal failure.

## MANY GROWTH FACTORS REGULATE PRODUCTION OF WHITE BLOOD CELLS

A large number of **hematopoietic growth factors** have been identified in recent years in addition to erythropoietin. This area of study adds to knowledge about the differentiation of blood cells, provides factors that may be useful in treatment, and also has implications for understanding of the abnormal growth of blood cells (eg, the leukemias). Like erythropoietin, most of the growth factors isolated have been glycoproteins, are very active in vivo and in vitro, interact with their target cells via specific cell surface receptors, and ultimately (via intracellular signals) affect gene expression, thereby promoting differentiation. Many have been cloned, permitting their production in relatively large amounts. Two of particular interest are **granulocyte-** and **granulocyte-macrophage colony-stimulating factors** (G-CSF and GM-CSF, respectively). G-CSF is relatively specific, inducing mainly granulocytes. GM-CSF affects a variety of progenitor cells and induces granulocytes, macrophages, and eosinophils. When the production of neutrophils is severely depressed, this condition is referred to as **neutropenia.** It is particularly likely to occur in patients treated with certain chemotherapeutic regimens and after bone marrow transplantation. These patients are liable to develop overwhelming infections. G-CSF has been administered to such patients to boost production of neutrophils.

## THE RED BLOOD CELL HAS A UNIQUE & RELATIVELY SIMPLE METABOLISM

Various aspects of the metabolism of the red cell, many of which are discussed in other chapters of this text, are summarized in Table 52–2.

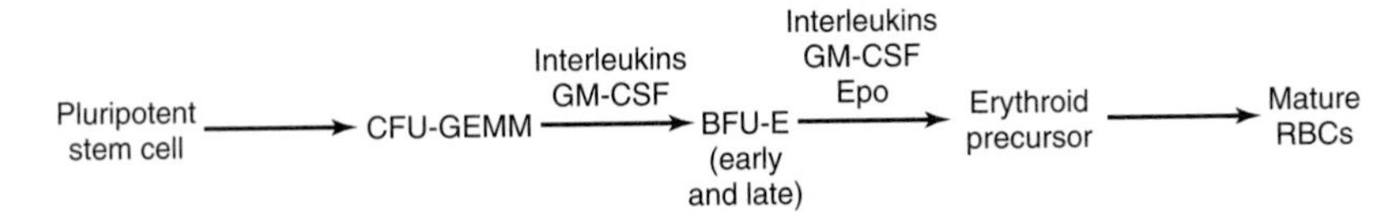


***Figure 52–1.*** Greatly simplified scheme of differentiation of stem cells to red blood cells. Various interleukins (ILs), such as IL-3, IL-4, IL-9, and IL-11, are involved at different steps of the overall process. Erythroid precursors include the pronormoblast, basophilic, polychromatophilic, and orthochromatophilic normoblasts, and the reticulocyte. Epo acts on basophilic normoblasts but not on later erythroid cells. (CFU-GEMM, colony-forming unit whose cells give rise to granulocytes, erythrocytes, macrophages, and megakaryocytes; BFU-E, burst-forming unit-erythroid; GM-CSF, granulocyte-macrophage colony-stimulating factor; Epo, erythropoietin; RBC, red blood cell.)

## The Red Blood Cell Has a Glucose Transporter in Its Membrane

The entry rate of glucose into red blood cells is far greater than would be calculated for simple diffusion. Rather, it is an example of **facilitated diffusion** (Chapter 41). The specific protein involved in this process is called the **glucose transporter** or glucose permease. Some of its properties are summarized in Table 52–3. The process of entry of glucose into red blood cells is of major importance because it is the major fuel supply for these cells. About seven different but related glucose transporters have been isolated from various tissues; unlike the red cell transporter, some of these are insulin-dependent (eg, in muscle and adipose tissue). There is considerable interest in the latter types of transporter because defects in their recruitment from intracellular sites to the surface of skeletal muscle cells may help explain the **insulin resistance** displayed by patients with type 2 diabetes mellitus.

## Reticulocytes Are Active in Protein Synthesis

The mature red blood cell cannot synthesize protein. Reticulocytes are active in protein synthesis. Once reticulocytes enter the circulation, they lose their intracellular organelles (ribosomes, mitochondria, etc) within about 24 hours, becoming young red blood cells and concomitantly losing their ability to synthesize protein. Extracts of rabbit reticulocytes (obtained by injecting rabbits with a chemical—phenylhydrazine—that causes a severe hemolytic anemia, so that the red cells are almost completely replaced by reticulocytes) are widely used as an in vitro system for synthesizing proteins. Endogenous mRNAs present in these reticulocytes are destroyed by use of a nuclease, whose activity can be inhibited by addition of $Ca^{2+}$. The system is then programmed by adding purified mRNAs or whole-cell extracts of mRNAs, and radioactive proteins are synthesized in the presence of $^{35}S$-labeled L-methionine or other radiolabeled amino acids. The radioactive proteins synthesized are separated by SDS-PAGE and detected by radioautography.

## Superoxide Dismutase, Catalase, & Glutathione Protect Blood Cells From Oxidative Stress & Damage

Several powerful oxidants are produced during the course of metabolism, in both blood cells and most other cells of the body. These include superoxide ($O_2^{\bar{\cdot}}$), hydrogen peroxide ($H_2O_2$), peroxyl radicals ($ROO^{\bullet}$), and hydroxyl radicals ($OH^{\bullet}$). The last is a particularly reactive molecule and can react with proteins, nucleic acids, lipids, and other molecules to alter their structure and produce tissue damage. The reactions listed in Table 52–4 play an important role in forming these oxidants and in disposing of them; each of these reactions will now be considered in turn.

Superoxide is formed (reaction 1) in the red blood cell by the auto-oxidation of hemoglobin to methemoglobin (approximately 3% of hemoglobin in human red blood cells has been calculated to auto-oxidize per day); in other tissues, it is formed by the action of enzymes such as cytochrome P450 reductase and xanthine oxidase. When stimulated by contact with bacteria, neutrophils exhibit a respiratory burst (see below) and produce superoxide in a reaction catalyzed by NADPH oxidase (reaction 2). Superoxide spontaneously dismutates to form $H_2O_2$ and $O_2$; however, the rate of this same reaction is speeded up tremendously by the action of the enzyme superoxide dismutase (reaction 3). Hydrogen peroxide is subject to a number of fates. The enzyme **catalase,** present in many types of cells, converts

***Table 52–2.*** Summary of important aspects of the metabolism of the red blood cell.

- The RBC is highly dependent upon glucose as its energy source; its membrane contains high affinity glucose transporters.
- Glycolysis, producing lactate, is the site of production of ATP.
- Because there are no mitochondria in RBCs, there is no production of ATP by oxidative phosphorylation.
- The RBC has a variety of transporters that maintain ionic and water balance.
- Production of 2,3-bisphosphoglycerate, by reactions closely associated with glycolysis, is important in regulating the ability of Hb to transport oxygen.
- The pentose phosphate pathway is operative in the RBC (it metabolizes about 5–10% of the total flux of glucose) and produces NADPH; hemolytic anemia due to a deficiency of the activity of glucose-6-phosphate dehydrogenase is common.
- Reduced glutathione (GSH) is important in the metabolism of the RBC, in part to counteract the action of potentially toxic peroxides; the RBC can synthesize GSH and requires NADPH to return oxidized glutathione (G-S-S-G) to the reduced state.
- The iron of Hb must be maintained in the ferrous state; ferric iron is reduced to the ferrous state by the action of an NADH-dependent methemoglobin reductase system involving cytochrome $b_5$ reductase and cytochrome $b_5$.
- Synthesis of glycogen, fatty acids, protein, and nucleic acids does not occur in the RBC; however, some lipids (eg, cholesterol) in the red cell membrane can exchange with corresponding plasma lipids.
- The RBC contains certain enzymes of nucleotide metabolism (eg, adenosine deaminase, pyrimidine nucleotidase, and adenylyl kinase); deficiencies of these enzymes are involved in some cases of hemolytic anemia.
- When RBCs reach the end of their life span, the globin is degraded to amino acids (which are reutilized in the body), the iron is released from heme and also reutilized, and the tetrapyrrole component of heme is converted to bilirubin, which is mainly excreted into the bowel via the bile.

***Table 52–3.*** Some properties of the glucose transporter of the membrane of the red blood cell.

- It accounts for about 2% of the protein of the membrane of the RBC.
- It exhibits specificity for glucose and related D-hexoses (L-hexoses are not transported).
- The transporter functions at approximately 75% of its $V_{max}$ at the physiologic concentration of blood glucose, is saturable and can be inhibited by certain analogs of glucose.
- At least seven similar but distinct glucose transporters have been detected to date in mammalian tissues, of which the red cell transporter is one.
- It is not dependent upon insulin, unlike the corresponding carrier in muscle and adipose tissue.
- Its complete amino acid sequence (492 amino acids) has been determined.
- It transports glucose when inserted into artificial liposomes.
- It is estimated to contain 12 transmembrane helical segments.
- It functions by generating a gated pore in the membrane to permit passage of glucose; the pore is conformationally dependent on the presence of glucose and can oscillate rapidly (about 900 times/s).

it to $H_2O$ and $O_2$ (reaction 4). Neutrophils possess a unique enzyme, myeloperoxidase, that uses $H_2O_2$ and halides to produce hypohalous acids (reaction 5); this subject is discussed further below. The selenium-containing enzyme glutathione peroxidase (Chapter 20) will also act on reduced glutathione (GSH) and $H_2O_2$ to produce oxidized glutathione (GSSG) and $H_2O$ (reaction 6); this enzyme can also use other peroxides as substrates. $OH^\bullet$ and $OH^-$ can be formed from $H_2O_2$ in a nonenzymatic reaction catalyzed by $Fe^{2+}$ (the **Fenton reaction,** reaction 7). $O_2^{\bar{\cdot}}$ and $H_2O_2$ are the substrates in the iron-catalyzed **Haber-Weiss reaction** (reaction 8), which also produces $OH^\bullet$ and $OH^-$. Superoxide can release iron ions from ferritin. Thus, production of $OH^\bullet$ may be one of the mechanisms involved in tissue injury due to iron overload (eg, hemochromatosis; Chapter 50).

Chemical compounds and reactions capable of generating potential toxic oxygen species can be referred to as **pro-oxidants.** On the other hand, compounds and reactions disposing of these species, scavenging them, suppressing their formation, or opposing their actions are **antioxidants** and include compounds such as NADPH, GSH, ascorbic acid, and vitamin E. In a normal cell, there is an appropriate pro-oxidant:antioxidant balance. However, this balance can be shifted toward the pro-oxidants when production of oxygen species is increased greatly (eg, following ingestion of certain chemicals or drugs) or when levels of antioxidants are diminished (eg, by inactivation of enzymes involved in disposal of oxygen species and by conditions that cause low levels of the antioxidants mentioned above). This state is called **"oxidative stress"** and can result in serious cell damage if the stress is massive or prolonged.

Oxygen species are now thought to play an important role in many types of **cellular injury** (eg, resulting from administration of various toxic chemicals or from ischemia), some of which can result in cell death. Indirect evidence supporting a role for these species in gen-

***Table 52–4.*** Reactions of importance in relation to oxidative stress in blood cells and various tissues.

| | |
|---|---|
| (1) Production of superoxide (by-product of various reactions) | $O_2 + e^- \rightarrow O_2^{\cdot-}$ |
| (2) NADPH-oxidase | $2\,O_2 + NADPH \rightarrow 2\,O_2^{\cdot-} + NADP + H^+$ |
| (3) Superoxide dismutase | $O_2^{\cdot-} + O_2^{\cdot-} + 2\,H^+ \rightarrow H_2O_2 + O_2$ |
| (4) Catalase | $H_2O_2 \rightarrow 2\,H_2O + O_2$ |
| (5) Myeloperoxidase | $H_2O_2 + X^- + H^+ \rightarrow HOX + H_2O$ ($X^- = Cl^-, Br^-, SCN^-$) |
| (6) Glutathione peroxidase (Se-dependent) | $2\,GSH + R\text{-}O\text{-}OH \rightarrow GSSG + H_2O + ROH$ |
| (7) Fenton reaction | $Fe^{2+} + H_2O_2 \rightarrow Fe^{3+} + OH^{\cdot} + OH^-$ |
| (8) Iron-catalyzed Haber-Weiss reaction | $O_2^{\cdot-} + H_2O_2 \rightarrow O_2 + OH^{\cdot} + OH^-$ |
| (9) Glucose-6-phosphate dehydrogenase (G6PD) | G6P + NADP → 6 Phosphogluconate + NADPH + $H^+$ |
| (10) Glutathione reductase | G-S-S-G + NADPH + $H^+$ → 2 GSH + NADP |

erating cell injury is provided if administration of an enzyme such as superoxide dismutase or catalase is found to protect against cell injury in the situation under study.

## Deficiency of Glucose-6-Phosphate Dehydrogenase Is Frequent in Certain Areas & Is an Important Cause of Hemolytic Anemia

NADPH, produced in the reaction catalyzed by the X-linked glucose-6-phosphate dehydrogenase (Table 52–4, reaction 9) in the pentose phosphate pathway (Chapter 20), plays a key role in supplying reducing equivalents in the red cell and in other cells such as the hepatocyte. Because the pentose phosphate pathway is virtually its sole means of producing NADPH, the red blood cell is very sensitive to oxidative damage if the function of this pathway is impaired (eg, by enzyme deficiency). One function of NADPH is to reduce GSSG to GSH, a reaction catalyzed by glutathione reductase (reaction 10).

Deficiency of the activity of glucose-6-phosphate dehydrogenase, owing to mutation, is extremely frequent in some regions of the world (eg, tropical Africa, the Mediterranean, certain parts of Asia, and in North America among blacks). It is the most common of all enzymopathies (diseases caused by abnormalities of enzymes), and over 300 genetic variants of the enzyme have been distinguished; at least 100 million people are deficient in this enzyme owing to these variants. The disorder resulting from deficiency of glucose-6-phosphate dehydrogenase is **hemolytic anemia.** Consumption of broad beans *(Vicia faba)* by individuals deficient in activity of the enzyme can precipitate an attack of hemolytic anemia (most likely because the beans contain potential oxidants). In addition, a number of drugs (eg, the antimalarial drug **primaquine** [the condition caused by intake of primaquine is called **primaquine-sensitive hemolytic anemia**] and **sulfonamides**) and chemicals (eg, naphthalene) precipitate an attack, because their intake leads to generation of $H_2O_2$ or $O_2^{\cdot-}$. Normally, $H_2O_2$ is disposed of by catalase and glutathione peroxidase (Table 52–4, reactions 4 and 6), the latter causing increased production of GSSG. GSH is regenerated from GSSG by the action of the enzyme glutathione reductase, which depends on the availability of NADPH (reaction 10). The red blood cells of individuals who are deficient in the activity of glucose-6-phosphate dehydrogenase cannot generate sufficient NADPH to regenerate GSH from GSSG, which in turn impairs their ability to dispose of $H_2O_2$ and of oxygen radicals. These compounds can cause oxidation of critical SH groups in proteins and possibly peroxidation of lipids in the membrane of the red cell, causing lysis of the red cell membrane. Some of the SH groups of hemoglobin become oxidized, and the protein precipitates inside the red blood cell, forming **Heinz bodies,** which stain purple with cresyl violet. The presence of Heinz bodies indicates that red blood cells have been subjected to oxidative stress. Figure 52–2 summarizes the possible chain of events in hemolytic anemia due to deficiency of glucose-6-phosphate dehydrogenase.

## Methemoglobin Is Useless in Transporting Oxygen

The ferrous iron of hemoglobin is susceptible to oxidation by superoxide and other oxidizing agents, forming methemoglobin, which cannot transport oxygen. Only a very small amount of methemoglobin is present in normal blood, as the red blood cell possesses an effective system (the NADH-cytochrome $b_5$ methemoglobin reductase system) for reducing heme $Fe^{3+}$ back to the $Fe^{2+}$ state. This system consists of NADH (generated by glycolysis), a flavoprotein named cytochrome $b_5$ reductase (also known as methemoglobin reductase), and cytochrome $b_5$. The $Fe^{3+}$ of methemoglobin is reduced

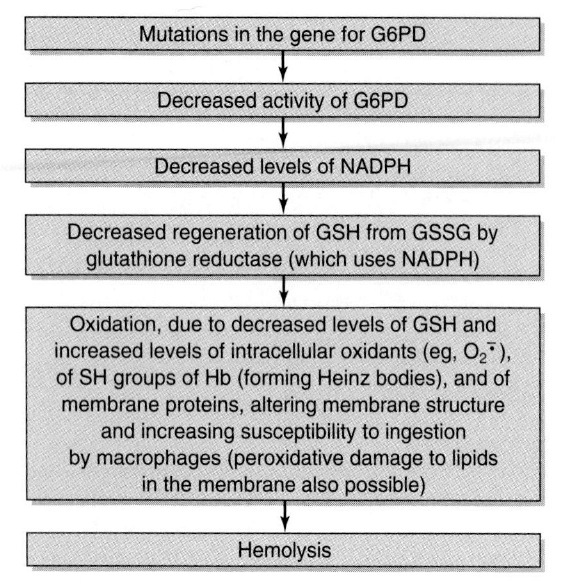


***Figure 52–2.*** Summary of probable events causing hemolytic anemia due to deficiency of the activity of glucose-6-phosphate dehydrogenase (G6PD) (MIM 305900).

back to the $Fe^{2+}$ state by the action of reduced cytochrome $b_5$:

$$\text{Hb-Fe}^{3+} + \text{Cyt } b_{5\,\text{red}} \rightarrow \text{Hb-Fe}^{2+} + \text{Cyt } b_{5\,\text{ox}}$$

Reduced cytochrome $b_5$ is then regenerated by the action of cytochrome $b_5$ reductase:

$$\text{Cyt } b_{5\,\text{ox}} + \text{NADH} \rightarrow \text{Cyt } b_{5\,\text{red}} + \text{NAD}$$

## Methemoglobinemia Is Inherited or Acquired

Methemoglobinemia can be classified as either inherited or acquired by ingestion of certain drugs and chemicals. Neither type occurs frequently, but physicians must be aware of them. The inherited form is usually due to deficient activity of methemoglobin reductase, transmitted in an autosomal recessive manner. Certain abnormal hemoglobins (Hb M) are also rare causes of methemoglobinemia. In Hb M, mutation changes the amino acid residue to which heme is attached, thus altering its affinity for oxygen and favoring its oxidation. Ingestion of certain drugs (eg, sulfonamides) or chemicals (eg, aniline) can cause acquired methemoglobinemia. Cyanosis (bluish discoloration of the skin and mucous membranes due to increased amounts of deoxygenated hemoglobin in arterial blood, or in this case due to increased amounts of methemoglobin) is usually the presenting sign in both types and is evident when over 10% of hemoglobin is in the "met" form. Diagnosis is made by spectroscopic analysis of blood, which reveals the characteristic absorption spectrum of methemoglobin. Additionally, a sample of blood containing methemoglobin cannot be fully reoxygenated by flushing oxygen through it, whereas normal deoxygenated blood can. Electrophoresis can be used to confirm the presence of an abnormal hemoglobin. Ingestion of methylene blue or ascorbic acid (reducing agents) is used to treat mild methemoglobinemia due to enzyme deficiency. Acute massive methemoglobinemia (due to ingestion of chemicals) should be treated by intravenous injection of methylene blue.

# MORE IS KNOWN ABOUT THE MEMBRANE OF THE HUMAN RED BLOOD CELL THAN ABOUT THE SURFACE MEMBRANE OF ANY OTHER HUMAN CELL

A variety of biochemical approaches have been used to study the membrane of the red blood cell. These include analysis of membrane proteins by SDS-PAGE, the use of specific enzymes (proteinases, glycosidases, and others) to determine the location of proteins and glycoproteins in the membrane, and various techniques to study both the lipid composition and disposition of individual lipids. Morphologic (eg, electron microscopy, freeze-fracture electron microscopy) and other techniques (eg, use of antibodies to specific components) have also been widely used. When red blood cells are lysed under specific conditions, their membranes will reseal in their original orientation to form **ghosts** (right-side-out ghosts). By altering the conditions, ghosts can also be made to reseal with their cytosolic aspect exposed on the exterior (inside-out ghosts). Both types of ghosts have been useful in analyzing the disposition of specific proteins and lipids in the membrane. In recent years, cDNAs for many proteins of this membrane have become available, permitting the deduction of their amino sequences and domains. All in all, more is known about the membrane of the red blood cell than about any other membrane of human cells (Table 52–5).

## Analysis by SDS-PAGE Resolves the Proteins of the Membrane of the Red Blood Cell

When the membranes of red blood cells are analyzed by SDS-PAGE, about ten major proteins are resolved

***Table 52–5.*** Summary of biochemical information about the membrane of the human red blood cell.

- The membrane is a bilayer composed of about 50% lipid and 50% protein.
- The major lipid classes are phospholipids and cholesterol; the major phospholipids are phosphatidylcholine (PC), phosphatidylethanolamine (PE), and phosphatidylserine (PS) along with sphingomyelin (Sph).
- The choline-containing phospholipids, PC and Sph, predominate in the outer leaflet and the amino-containing phospholipids (PE and PS) in the inner leaflet.
- Glycosphingolipids (GSLs) (neutral GSLs, gangliosides, and complex species, including the ABO blood group substances) constitute about 5–10% of the total lipid.
- Analysis by SDS-PAGE shows that the membrane contains about 10 major proteins and more than 100 minor species.
- The major proteins (which include spectrin, ankyrin, the anion exchange protein, actin, and band 4.1) have been studied intensively, and the principal features of their disposition (eg, integral or peripheral), structure, and function have been established.
- Many of the proteins are glycoproteins (eg, the glycophorins) containing O- or N-linked (or both) oligosaccharide chains located on the external surface of the membrane.

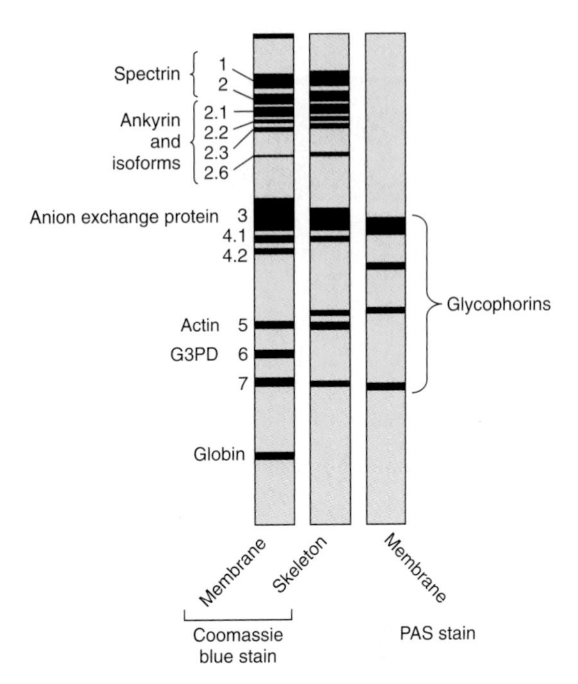


***Figure 52–3.*** Diagrammatic representation of the major proteins of the membrane of the human red blood cell separated by SDS-PAGE. The bands detected by staining with Coomassie blue are shown in the two left-hand channels, and the glycoproteins detected by staining with periodic acid-Schiff (PAS) reagent are shown in the right-hand channel. (Reproduced, with permission, from Beck WS, Tepper RI: Hemolytic anemias III: membrane disorders. In: *Hematology,* 5th ed. Beck WS [editor]. The MIT Press, 1991.)

(Figure 52–3), several of which have been shown to be **glycoproteins.** Their migration on SDS-PAGE was used to name these proteins, with the slowest migrating (and hence highest molecular mass) being designated band 1 or **spectrin.** All these major proteins have been isolated, most of them have been identified, and considerable insight has been obtained about their functions (Table 52–6). Many of their amino acid sequences also have been established. In addition, it has been determined which are integral or peripheral membrane proteins, which are situated on the external surface, which are on the cytosolic surface, and which span the membrane (Figure 52–4). Many minor components can also be detected in the red cell membrane by use of sensitive staining methods or two-dimensional gel electrophoresis. One of these is the glucose transporter described above.

## The Major Integral Proteins of the Red Blood Cell Membrane Are the Anion Exchange Protein & the Glycophorins

The anion exchange protein (band 3) is a transmembrane glycoprotein, with its carboxyl terminal end on the external surface of the membrane and its amino terminal end on the cytoplasmic surface. It is an example of a **multipass** membrane protein, extending across the bilayer at least ten times. It probably exists as a dimer in the membrane, in which it forms a tunnel, permitting the exchange of chloride for bicarbonate. Carbon dioxide, formed in the tissues, enters the red cell as bicarbonate, which is exchanged for chloride in the lungs, where carbon dioxide is exhaled. The amino terminal end binds many proteins, including hemoglobin, proteins 4.1 and 4.2, ankyrin, and several glycolytic enzymes. Purified band 3 has been added to lipid vesicles in vitro and has been shown to perform its transport functions in this reconstituted system.

**Glycophorins A, B, and C** are also transmembrane glycoproteins but of the **single-pass** type, extending across the membrane only once. A is the major glycophorin, is made up of 131 amino acids, and is heavily glycosylated (about 60% of its mass). Its amino terminal end, which contains 16 oligosaccharide chains (15 of which are O-glycans), extrudes out from the surface of

**Table 52–6.** Principal proteins of the red cell membrane.[1]

| Band Number[2] | Protein | Integral (I) or Peripheral (P) | Approximate Molecular Mass (kDa) |
|---|---|---|---|
| 1 | Spectrin (α) | P | 240 |
| 2 | Spectrin (β) | P | 220 |
| 2.1 | Ankyrin | P | 210 |
| 2.2 | " | P | 195 |
| 2.3 | " | P | 175 |
| 2.6 | " | P | 145 |
| 3 | Anion exchange protein | I | 100 |
| 4.1 | Unnamed | P | 80 |
| 5 | Actin | P | 43 |
| 6 | Glyceraldehyde-3-phosphate dehydrogenase | P | 35 |
| 7 | Tropomyosin | P | 29 |
| 8 | Unnamed | P | 23 |
| | Glycophorins A, B, and C | I | 31, 23, and 28 |

[1]Adapted from Lux DE, Becker PS: Disorders of the red cell membrane skeleton: hereditary spherocytosis and hereditary elliptocytosis. Chapter 95 in: *The Metabolic Basis of Inherited Disease,* 6th ed. Scriver CR et al (editors). McGraw-Hill, 1989.

[2]The band number refers to the position of migration on SDS-PAGE (see Figure 52–3). The glycophorins are detected by staining with the periodic acid-Schiff reagent. A number of other components (eg, 4.2 and 4.9) are not listed. Native spectrin is $\alpha_2\beta_2$.

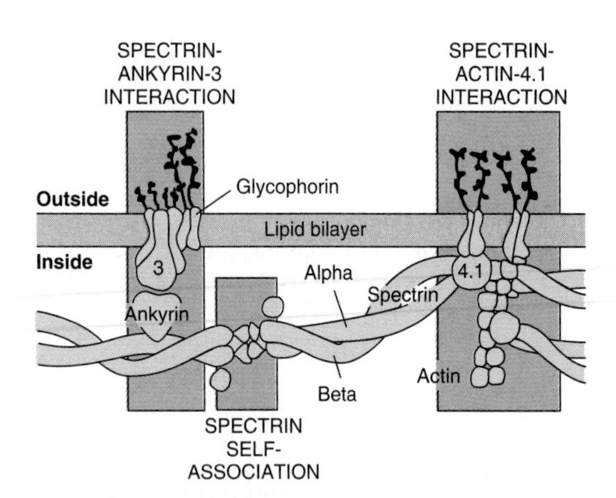


***Figure 52–4.*** Diagrammatic representation of the interaction of cytoskeletal proteins with each other and with certain integral proteins of the membrane of the red blood cell. (Reproduced, with permission, from Beck WS, Tepper RI: Hemolytic anemias III: membrane disorders. In: *Hematology,* 5th ed. Beck WS [editor]. The MIT Press, 1991.)

the red blood cell. Approximately 90% of the sialic acid of the red cell membrane is located in this protein. Its transmembrane segment (23 amino acids) is α-helical. The carboxyl terminal end extends into the cytosol and binds to protein 4.1, which in turn binds to spectrin. Polymorphism of this protein is the basis of the MN blood group system (see below). Glycophorin A contains binding sites for influenza virus and for *Plasmodium falciparum,* the cause of one form of malaria. Intriguingly, the function of red blood cells of individuals who lack glycophorin A does not appear to be affected.

## Spectrin, Ankyrin, & Other Peripheral Membrane Proteins Help Determine the Shape & Flexibility of the Red Blood Cell

The red blood cell must be able to squeeze through some tight spots in the microcirculation during its numerous passages around the body; the sinusoids of the spleen are of special importance in this regard. For the red cell to be easily and reversibly deformable, its membrane must be both fluid and flexible; it should also preserve its biconcave shape, since this facilitates gas exchange. Membrane lipids help determine membrane fluidity. Attached to the inner aspect of the membrane of the red blood cell are a number of peripheral cytoskeletal proteins (Table 52–6) that play important roles in respect to preserving shape and flexibility; these will now be described.

**Spectrin** is the major protein of the cytoskeleton. It is composed of two polypeptides: spectrin 1 (α chain) and spectrin 2 (β chain). These chains, measuring approximately 100 nm in length, are aligned in an antiparallel manner and are loosely intertwined, forming a dimer. Both chains are made up of segments of 106 amino acids that appear to fold into triple-stranded α-helical coils joined by nonhelical segments. One dimer interacts with another, forming a head-to-head tetramer. The overall shape confers flexibility on the protein and in turn on the membrane of the red blood cell. At least four binding sites can be defined in spectrin: (1) for self-association, (2) for ankyrin (bands 2.1, etc), (3) for actin (band 5), and (4) for protein 4.1.

**Ankyrin** is a pyramid-shaped protein that binds spectrin. In turn, ankyrin binds tightly to band 3, securing attachment of spectrin to the membrane. Ankyrin is sensitive to proteolysis, accounting for the appearance of bands 2.2, 2.3, and 2.6, all of which are derived from band 2.1.

**Actin** (band 5) exists in red blood cells as short, double-helical filaments of F-actin. The tail end of spectrin dimers binds to actin. Actin also binds to protein 4.1.

**Protein 4.1,** a globular protein, binds tightly to the tail end of spectrin, near the actin-binding site of the latter, and thus is part of a protein 4.1-spectrin-actin ternary complex. Protein 4.1 also binds to the integral proteins, glycophorins A and C, thereby attaching the ternary complex to the membrane. In addition, protein 4.1 may interact with certain membrane phospholipids, thus connecting the lipid bilayer to the cytoskeleton.

Certain other proteins (4.9, adducin, and tropomyosin) also participate in cytoskeletal assembly.

### Abnormalities in the Amount or Structure of Spectrin Cause Hereditary Spherocytosis & Elliptocytosis

Hereditary spherocytosis is a genetic disease, transmitted as an autosomal dominant, that affects about 1:5000 North Americans. It is characterized by the presence of spherocytes (spherical red blood cells, with a low surface-to-volume ratio) in the peripheral blood, by a hemolytic anemia, and by splenomegaly. The spherocytes are not as deformable as are normal red blood cells, and they are subject to destruction in the spleen, thus greatly shortening their life in the circulation. Hereditary spherocytosis is curable by splenectomy because the spherocytes can persist in the circulation if the spleen is absent.

The spherocytes are much more susceptible to osmotic lysis than are normal red blood cells. This is assessed in the **osmotic fragility test,** in which red blood cells are exposed in vitro to decreasing concentrations of NaCl. The physiologic concentration of NaCl is 0.85 g/dL. When exposed to a concentration of NaCl of 0.5 g/dL, very few normal red blood cells are hemolyzed, whereas approximately 50% of spherocytes would lyse under these conditions. The explanation is that the spherocyte, being almost circular, has little potential extra volume to accommodate additional water and thus lyses readily when exposed to a slightly lower osmotic pressure than is normal.

One cause of hereditary spherocytosis (Figure 52–5) is a deficiency in the amount of spectrin or abnormalities of its structure, so that it no longer tightly binds the other proteins with which it normally interacts. This weakens the membrane and leads to the spherocytic shape. Abnormalities of ankyrin and of bands 3 and 4.1 are involved in other cases.

**Hereditary elliptocytosis** is a genetic disorder that is similar to hereditary spherocytosis except that affected red blood cells assume an elliptic, disk-like shape, recognizable by microscopy. It is also due to abnormalities in spectrin; some cases reflect abnormalities of band 4.1 or of glycophorin C.

## THE BIOCHEMICAL BASES OF THE ABO BLOOD GROUP SYSTEM HAVE BEEN ESTABLISHED

At least 21 human blood group systems are recognized, the best known of which are the ABO, Rh (Rhesus), and MN systems. The term "blood group" applies to a defined system of red blood cell antigens (blood group substances) controlled by a genetic locus having a variable number of alleles (eg, A, B, and O in the ABO system). The term "blood type" refers to the antigenic phenotype, usually recognized by the use of appropriate

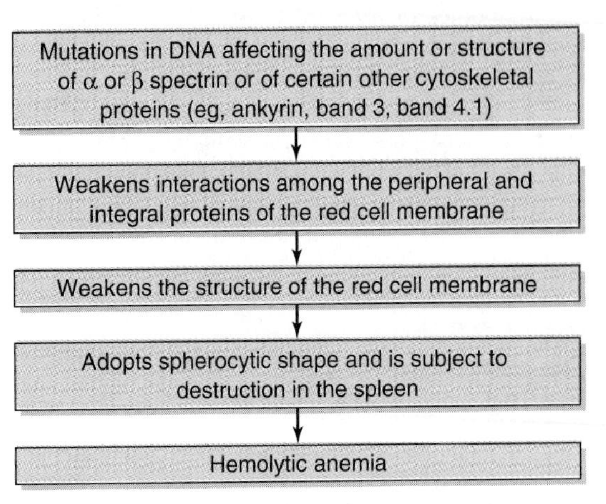


***Figure 52–5.*** Summary of the causation of hereditary spherocytosis (MIM 182900).

antibodies. For purposes of blood transfusion, it is particularly important to know the basics of the ABO and Rh systems. However, knowledge of blood group systems is also of biochemical, genetic, immunologic, anthropologic, obstetric, pathologic, and forensic interest. Here, we shall discuss only some key features of the ABO system. From a biochemical viewpoint, the major interests in the ABO substances have been in isolating and determining their structures, elucidating their pathways of biosynthesis, and determining the natures of the products of the A, B, and O genes.

## The ABO System Is of Crucial Importance in Blood Transfusion

This system was first discovered by Landsteiner in 1900 when investigating the basis of compatible and incompatible transfusions in humans. The membranes of the red blood cells of most individuals contain one blood group substance of type A, type B, type AB, or type O. Individuals of type A have anti-B antibodies in their plasma and will thus agglutinate type B or type AB blood. Individuals of type B have anti-A antibodies and will agglutinate type A or type AB blood. Type AB blood has neither anti-A nor anti-B antibodies and has been designated the **universal recipient.** Type O blood has neither A nor B substances and has been designated the **universal donor.** The explanation of these findings is related to the fact that the body does not usually produce antibodies to its own constituents. Thus, individuals of type A do not produce antibodies to their own blood group substance, A, but do possess antibodies to the foreign blood group substance, B, possibly because similar structures are present in microorganisms to which the body is exposed early in life. Since individuals of type O have neither A nor B substances, they possess antibodies to both these foreign substances. The above description has been simplified considerably; eg, there are two subgroups of type A: $A_1$ and $A_2$.

The genes responsible for production of the ABO substances are present on the long arm of chromosome 9. There are three alleles, two of which are codominant (A and B) and the third (O) recessive; these ultimately determine the four phenotypic products: the A, B, AB, and O substances.

## The ABO Substances Are Glycosphingolipids & Glycoproteins Sharing Common Oligosaccharide Chains

The ABO substances are complex oligosaccharides present in most cells of the body and in certain secretions. On membranes of red blood cells, the oligosaccharides that determine the specific natures of the ABO substances appear to be mostly present in **glycosphingolipids,** whereas in secretions the same oligosaccharides are present in **glycoproteins.** Their presence in secretions is determined by a gene designated ***Se*** (for **secretor**), which codes for a specific **fucosyl (Fuc) transferase** in secretory organs, such as the exocrine glands, but which is not active in red blood cells. Individuals of *SeSe* or *Sese* genotypes secrete A or B antigens (or both), whereas individuals of the *sese* genotype do not secrete A or B substances, but their red blood cells can express the A and B antigens.

## H Substance Is the Biosynthetic Precursor of Both the A & B Substances

The ABO substances have been isolated and their structures determined; simplified versions, showing only their nonreducing ends, are presented in Figure 52–6. It is important to first appreciate the structure of the H substance, since it is the precursor of both the A and B substances and is the blood group substance found in persons of type O. H substance itself is formed by the action of a **fucosyltransferase,** which catalyzes the addition of the terminal fucose in $\alpha 1 \rightarrow 2$ linkage onto the terminal Gal residue of its precursor:

$$\underset{\text{Precursor}}{\text{GDP-Fuc} + \text{Gal-}\beta\text{-R}} \rightarrow \underset{\text{H substance}}{\text{Fuc-}\alpha 1,2\text{-Gal-}\beta\text{-R}} + \text{GDP}$$

The H locus codes for this fucosyltransferase. The *h* allele of the H locus codes for an inactive fucosyltransferase; therefore, individuals of the *hh* genotype cannot generate H substance, the precursor of the A and B antigens. Thus, individuals of the *hh* genotype will have red blood cells of type O, even though they may possess the enzymes necessary to make the A or B substances (see below).

## The *A* Gene Encodes a GalNAc Transferase, the *B* Gene a Gal Transferase, & the *O* Gene an Inactive Product

In comparison with H substance (Figure 52–6), **A substance** contains an additional GalNAc and **B substance** an additional Gal, linked as indicated. Anti-A antibodies are directed to the additional GalNAc residue found in the A substance, and anti-B antibodies are directed toward the additional Gal residue found in the B substance. Thus, GalNAc is the **immunodominant sugar** (ie, the one determining the specificity of the antibody formed) of blood group A substance, whereas Gal is the immunodominant sugar of the B substance. In view of the structural findings, it is not surprising that A substance can be synthesized in vitro from O substance in a reaction catalyzed by a GalNAc transferase, employing UDP-GalNAc as the sugar donor. Similarly, blood

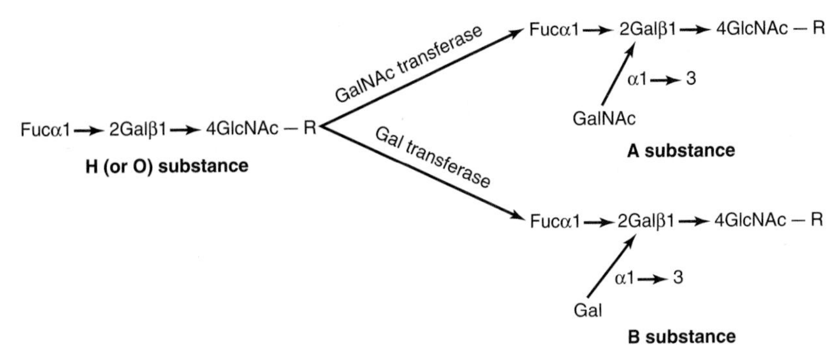


***Figure 52–6.*** Diagrammatic representation of the structures of the H, A, and B blood group substances. R represents a long complex oligosaccharide chain, joined either to ceramide where the substances are glycosphingolipids, or to the polypeptide backbone of a protein via a serine or threonine residue where the substances are glycoproteins. Note that the blood group substances are biantennary; ie, they have two arms, formed at a branch point (not indicated) between the GlcNAc—R, and only one arm of the branch is shown. Thus, the H, A, and B substances each contain two of their respective short oligosaccharide chains shown above. The AB substance contains one type A chain and one type B chain.

group B can be synthesized from O substance by the action of a Gal transferase, employing UDP-Gal. It is crucial to appreciate that the product of the *A* gene is the GalNAc transferase that adds the terminal GalNAc to the O substance. Similarly, the product of the *B* gene is the Gal transferase adding the Gal residue to the O substance. Individuals of type AB possess both enzymes and thus have two oligosaccharide chains (Figure 52–6), one terminated by a GalNAc and the other by a Gal. Individuals of type O apparently synthesize an inactive protein, detectable by immunologic means; thus, H substance is their ABO blood group substance.

In 1990, a study using cloning and sequencing technology described the nature of the differences between the glycosyltransferase products of the *A, B,* and *O* genes. A difference of four nucleotides is apparently responsible for the distinct specificities of the A and B glycosyltransferases. On the other hand, the *O* allele has a single base-pair mutation, causing a **frameshift mutation** resulting in a protein lacking transferase activity.

## HEMOLYTIC ANEMIAS ARE CAUSED BY ABNORMALITIES OUTSIDE, WITHIN, OR INSIDE THE RED BLOOD CELL MEMBRANE

Causes outside the membrane include hypersplenism, a condition in which the spleen is enlarged from a variety of causes and red blood cells become sequestered in it. Immunologic abnormalities (eg, transfusion reactions, the presence in plasma of warm and cold antibodies that lyse red blood cells, and unusual sensitivity to complement) also fall in this class, as do toxins released by various infectious agents, such as certain bacteria (eg, clostridium). Some snakes release venoms that act to lyse the red cell membrane (eg, via the action of phospholipases or proteinases).

Causes within the membrane include abnormalities of proteins. The most important conditions are hereditary spherocytosis and hereditary elliptocytosis, principally caused by abnormalities in the amount or structure of spectrin (see above).

Causes inside the red blood cell include **hemoglobinopathies** and **enzymopathies.** Sickle cell anemia is the most important hemoglobinopathy. Abnormalities of enzymes in the pentose phosphate pathway and in glycolysis are the most frequent enzymopathies involved, particularly the former. Deficiency of glucose-6-phosphate dehydrogenase is prevalent in certain parts of the world and is a frequent cause of hemolytic anemia (see above). Deficiency of pyruvate kinase is not frequent, but it is the second commonest enzyme deficiency resulting in hemolytic anemia; the mechanism appears to be due to impairment of glycolysis, resulting in decreased formation of ATP, affecting various aspects of membrane integrity.

Laboratory investigations that aid in the diagnosis of hemolytic anemia are listed in Table 52–7.

**Table 52–7.** Laboratory investigations that assist in the diagnosis of hemolytic anemia.

| |
|---|
| **General tests and findings** |
| Increased nonconjugated (indirect) bilirubin |
| Shortened red cell survival time as measured by injection of autologous $^{51}Cr$-labeled red cells |
| Reticulocytosis |
| Hemoglobinemia |
| Low level of plasma haptoglobin |
| **Specific tests and findings** |
| Hb electrophoresis (eg, HbS) |
| Red cell enzymes (eg, G6PD or PK deficiency) |
| Osmotic fragility (eg, hereditary spherocytosis) |
| Coombs test[1] |
| Cold agglutinins |

[1]The direct Coombs test detects the presence of antibodies on red cells, whereas the indirect test detects the presence of circulating antibodies to antigens present on red cells.

## NEUTROPHILS HAVE AN ACTIVE METABOLISM & CONTAIN SEVERAL UNIQUE ENZYMES & PROTEINS

The major biochemical features of neutrophils are summarized in Table 52–8. Prominent features are active aerobic glycolysis, active pentose phosphate pathway, moderately active oxidative phosphorylation (because mitochondria are relatively sparse), and a high content of lysosomal enzymes. Many of the enzymes listed in Table 52–4 are also of importance in the oxidative metabolism of neutrophils (see below). Table 52–9 summarizes the functions of some proteins that are relatively unique to neutrophils.

### Neutrophils Are Key Players in the Body's Defense Against Bacterial Infection

Neutrophils are motile phagocytic cells that play a key role in acute inflammation. When bacteria enter tissues, a number of phenomena result that are collectively known as the "acute inflammatory response." They include (1) increase of vascular permeability, (2) entry of activated neutrophils into the tissues, (3) activation of platelets, and (4) spontaneous subsidence (resolution) if the invading microorganisms have been dealt with successfully.

A variety of molecules are released from cells and plasma proteins during acute inflammation whose net overall effect is to increase vascular permeability, resulting in tissue edema (Table 52–10).

In acute inflammation, neutrophils are recruited from the bloodstream into the tissues to help eliminate the foreign invaders. The neutrophils are attracted into the tissues by **chemotactic factors,** including complement fragment C5a, small peptides derived from bacteria (eg, *N*-formyl-methionyl-leucyl-phenylalanine), and a number of leukotrienes. To reach the tissues, circulating neutrophils must pass through the capillaries. To achieve this, they marginate along the vessel walls and then adhere to endothelial (lining) cells of the capillaries.

**Table 52–8.** Summary of major biochemical features of neutrophils.

- Active glycolysis
- Active pentose phosphate pathway
- Moderate oxidative phosphorylation
- Rich in lysosomes and their degradative enzymes
- Contain certain unique enzymes (eg, myeloperoxidase and NADPH-oxidase) and proteins
- Contain CD 11/CD18 integrins in plasma membrane

### Integrins Mediate Adhesion of Neutrophils to Endothelial Cells

Adhesion of neutrophils to endothelial cells employs specific adhesive proteins (integrins) located on their surface and also specific receptor proteins in the endothelial cells. (See also the discussion of selectins in Chapter 47.)

The integrins are a superfamily of surface proteins present on a wide variety of cells. They are involved in the adhesion of cells to other cells or to specific components of the extracellular matrix. They are heterodimers, containing an $\alpha$ and a $\beta$ subunit linked noncovalently. The subunits contain extracellular, transmembrane, and intracellular segments. The extracellular segments bind to a variety of ligands such as specific proteins of the extracellular matrix and of the surfaces of other cells. These ligands often contain Arg-Gly-Asp (R-G-D) sequences. The intracellular domains bind to various proteins of the cytoskeleton, such as actin and vinculin. The integrins are proteins that link the outsides of cells to their insides, thereby helping to integrate responses of cells (eg, movement, phagocytosis) to changes in the environment.

Three subfamilies of integrins were recognized initially. Members of each subfamily were distinguished by containing a common $\beta$ subunit, but they differed in their $\alpha$ subunits. However, more than three $\beta$ subunits have now been identified, and the classification of integrins has become rather complex. Some integrins of specific interest with regard to neutrophils are listed in Table 52–11.

A deficiency of the $\beta_2$ subunit (also designated CD18) of LFA-1 and of two related integrins found in

***Table 52–9.*** Some important enzymes and proteins of neutrophils.[1]

| Enzyme or Protein | Reaction Catalyzed or Function | Comment |
|---|---|---|
| Myeloperoxidase (MPO) | $H_2O_2 + X^-$ (halide) $+ H^+ \rightarrow HOX + H_2O$ (where $X^- = Cl^-$, HOX = hypochlorous acid) | Responsible for the green color of pus<br>Genetic deficiency can cause recurrent infections |
| NADPH-oxidase | $2O_2 + NADPH \rightarrow 2O_2^{\bar{\cdot}} + NADP + H^+$ | Key component of the respiratory burst<br>Deficient in chronic granulomatous disease |
| Lysozyme | Hydrolyzes link between *N*-acetylmuramic acid and *N*-acetyl-D-glucosamine found in certain bacterial cell walls | Abundant in macrophages |
| Defensins | Basic antibiotic peptides of 20–33 amino acids | Apparently kill bacteria by causing membrane damage |
| Lactoferrin | Iron-binding protein | May inhibit growth of certain bacteria by binding iron and may be involved in regulation of proliferation of myeloid cells |
| CD11a/CD18, CD11b/CD18, CD11c/CD18[2] | Adhesion molecules (members of the integrin family) | Deficient in leukocyte adhesion deficiency type I (MIM 116920) |
| Receptors for Fc fragments of IgGs | Bind Fc fragments of IgG molecules | Target antigen-antibody complexes to myeloid and lymphoid cells, eliciting phagocytosis and other responses |

[1]The expression of many of these molecules has been studied during various stages of differentiation of normal neutrophils and also of corresponding leukemic cells employing molecular biology techniques (eg, measurements of their specific mRNAs). For the majority, cDNAs have been isolated and sequenced, amino acid sequences deduced, genes have been localized to specific chromosomal locations, and exons and intron sequences have been defined. Some important proteinases of neutrophils are listed in Table 52–12.
[2]CD = cluster of differentiation. This refers to a uniform system of nomenclature that has been adopted to name surface markers of leukocytes. A specific surface protein (marker) that identifies a particular lineage or differentiation stage of leukocytes and that is recognized by a group of monoclonal antibodies is called a member of a cluster of differentiation. The system is particularly helpful in categorizing subclasses of lymphocytes. Many CD antigens are involved in cell-cell interactions, adhesion, and transmembrane signaling.

neutrophils and macrophages, Mac-1 (CD11b/CD18) and p150,95 (CD11c/CD18), causes **type 1 leukocyte adhesion deficiency,** a disease characterized by recurrent bacterial and fungal infections. Among various results of this deficiency, the adhesion of affected white blood cells to endothelial cells is diminished, and lower numbers of neutrophils thus enter the tissues to combat infection.

Once having passed through the walls of small blood vessels, the neutrophils migrate toward the highest concentrations of the chemotactic factors, encounter the invading bacteria, and attempt to attack and destroy them. The neutrophils must be activated in order to turn on many of the metabolic processes involved in phagocytosis and killing of bacteria.

## Activation of Neutrophils Is Similar to Activation of Platelets & Involves Hydrolysis of Phosphatidylinositol Bisphosphate

The mechanisms involved in platelet activation are discussed in Chapter 51 (see Figure 51–8). The process involves interaction of the stimulus (eg, thrombin) with a receptor, activation of G proteins, stimulation of phospholipase C, and liberation from phosphatidylinositol

***Table 52–10.*** Sources of biomolecules with vasoactive properties involved in acute inflammation.

| Mast Cells and Basophils | Platelets | Neutrophils | Plasma Proteins |
|---|---|---|---|
| Histamine | Serotonin | Platelet-activating factor (PAF)<br>Eicosanoids (various prostaglandins and leukotrienes) | C3a, C4a, and C5a from the complement system<br>Bradykinin and fibrin split products from the coagulation system |

***Table 52–11.*** Examples of integrins that are important in the function of neutrophils, of other white blood cells, and of platelets.[1]

| Integrin | Cell | Subunit | Ligand | Function |
|---|---|---|---|---|
| VLA-1 (CD49a) | WBCs, others | $\alpha 1\beta 1$ | Collagen, laminin | Cell-ECM adhesion |
| VLA-5 (CD49e) | WBCs, others | $\alpha 5\beta 1$ | Fibronectin | Cell-ECM adhesion |
| VLA-6 (CD49f) | WBCs, others | $\alpha 6\beta 1$ | Laminin | Cell-ECM adhesion |
| LFA-1 (CD11a) | WBCs | $\alpha L\beta 2$ | ICAM-1 | Adhesion of WBCs |
| Glycoprotein IIb/IIIa | Platelets | $\alpha IIb\beta 3$ | ICAM-2<br>Fibrinogen, fibronectin, von Willebrand factor | Platelet adhesion and aggregation |

[1]LFA-1, lymphocyte function-associated antigen 1; VLA, very late antigen; CD, cluster of differentiation; ICAM, intercellular adhesion molecule; ECM, extracellular matrix. A deficiency of LFA-1 and related integrins is found in type I leukocyte adhesion deficiency (MIM 116290). A deficiency of platelet glycoprotein IIb/IIIa complex is found in Glanzmann thrombasthenia (MIM 273800), a condition characterized by a history of bleeding, a normal platelet count, and abnormal clot retraction. These findings illustrate how fundamental knowledge of cell surface adhesion proteins is shedding light on the causation of a number of diseases.

bisphosphate of inositol triphosphate and diacylglycerol. These two second messengers result in an elevation of intracellular $Ca^{2+}$ and activation of protein kinase C. In addition, activation of phospholipase $A_2$ produces arachidonic acid that can be converted to a variety of biologically active eicosanoids.

The process of activation of neutrophils is essentially similar. They are activated, via specific receptors, by interaction with bacteria, binding of chemotactic factors, or antibody-antigen complexes. The resultant rise in intracellular $Ca^{2+}$ affects many processes in neutrophils, such as assembly of microtubules and the actin-myosin system. These processes are respectively involved in secretion of contents of granules and in motility, which enables neutrophils to seek out the invaders. The activated neutrophils are now ready to destroy the invaders by mechanisms that include production of active derivatives of oxygen.

## The Respiratory Burst of Phagocytic Cells Involves NADPH Oxidase & Helps Kill Bacteria

When neutrophils and other phagocytic cells engulf bacteria, they exhibit a rapid increase in oxygen consumption known as the respiratory burst. This phenomenon reflects the rapid utilization of oxygen (following a lag of 15–60 seconds) and production from it of large amounts of reactive derivatives, such as $O_2^{\overline{\cdot}}$, $H_2O_2$, $OH^\bullet$, and $OCl^-$ (hypochlorite ion). Some of these products are potent microbicidal agents.

The **electron transport chain system** responsible for the respiratory burst (named NADPH oxidase) is composed of several components. One is **cytochrome *b*$_{558}$**, located in the plasma membrane; it is a heterodimer, containing two polypeptides of 91 kDa and 22 kDa. When the system is activated (see below), two cytoplasmic polypeptides of 47 kDa and 67 kDa are recruited to the plasma membrane and, together with cytochrome $b_{558}$, form the NADPH oxidase responsible for the respiratory burst. The reaction catalyzed by NADPH oxidase, involving formation of superoxide anion, is shown in Table 52–4 (reaction 2). This system catalyzes the one-electron reduction of oxygen to superoxide anion. The NADPH is generated mainly by the pentose phosphate cycle, whose activity increases markedly during phagocytosis.

The above reaction is followed by the spontaneous production (by spontaneous dismutation) of **hydrogen peroxide** from two molecules of superoxide:

$$O_2^{\overline{\cdot}} + O_2^{\overline{\cdot}} + 2\,H^+ \rightarrow H_2O_2 + O_2$$

The superoxide ion is discharged to the outside of the cell or into phagolysosomes, where it encounters ingested bacteria. Killing of bacteria within phagolysosomes appears to depend on the combined action of elevated pH, superoxide ion, or further oxygen derivatives ($H_2O_2$, $OH^\bullet$, and HOCl [hypochlorous acid; see below]) and on the action of certain bactericidal peptides (defensins) and other proteins (eg, cathepsin G and certain cationic proteins) present in phagocytic cells. Any superoxide that enters the cytosol of the phagocytic cell is converted to $H_2O_2$ by the action of **superoxide dismutase,** which catalyzes the same reaction as the spontaneous dismutation shown above. In turn, $H_2O_2$ is used by myeloperoxidase (see below) or disposed of by the action of glutathione peroxidase or catalase.

**NADPH oxidase** is inactive in resting phagocytic cells and is activated upon contact with various ligands (complement fragment C5a, chemotactic peptides, etc) with receptors in the plasma membrane. The events resulting in activation of the oxidase system have been much studied and are similar to those described above for the process of activation of neutrophils. They involve **G proteins,** activation of **phospholipase C,** and generation of **inositol 1,4,5-triphosphate** ($IP_3$). The last mediates a transient increase in the level of cytosolic $Ca^{2+}$, which is essential for induction of the respiratory burst. **Diacylglycerol** is also generated and induces the translocation of protein kinase C into the plasma membrane from the cytosol, where it catalyzes the **phosphorylation** of various proteins, some of which are components of the oxidase system. A second pathway of activation not involving $Ca^{2+}$ also operates.

## Mutations in the Genes for Components of the NADPH Oxidase System Cause Chronic Granulomatous Disease

The importance of the NADPH oxidase system was clearly shown when it was observed that the respiratory burst was defective in chronic granulomatous disease, a relatively uncommon condition characterized by recurrent infections and widespread granulomas (chronic inflammatory lesions) in the skin, lungs, and lymph nodes. The granulomas form as attempts to wall off bacteria that have not been killed, owing to genetic deficiencies in the NADPH oxidase system. The disorder is due to mutations in the genes encoding the four polypeptides that constitute the NADPH oxidase system. Some patients have responded to treatment with gamma interferon, which may increase transcription of the 91-kDa component if it is affected. The probable sequence of events involved in the causation of chronic granulomatous disease is shown in Figure 52–7.

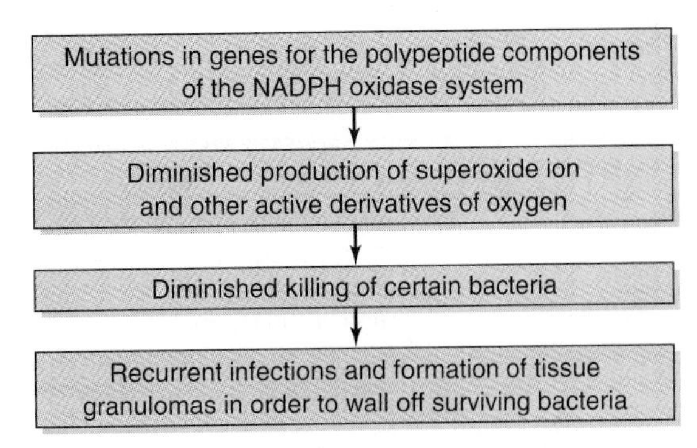


***Figure 52–7.*** Simplified scheme of the sequence of events involved in the causation of chronic granulomatous disease (MIM 306400). Mutations in any of the genes for the four polypeptides involved (two are components of cytochrome $b_{558}$ and two are derived from the cytoplasm) can cause the disease. The polypeptide of 91 kDa is encoded by a gene in the X chromosome; approximately 60% of cases of chronic granulomatous disease are X-linked, with the remainder being inherited in an autosomal recessive fashion.

## Neutrophils Contain Myeloperoxidase, Which Catalyzes the Production of Chlorinated Oxidants

The enzyme myeloperoxidase, present in large amounts in neutrophil granules and responsible for the green color of pus, can act on $H_2O_2$ to produce hypohalous acids:

$$H_2O_2 + X^- + H^+ \xrightarrow{\text{MYELOPEROXIDASE}} HOX + H_2O$$

($X^-$ = $Cl^-$, $Br^-$, $I^-$ or $SCN^-$; HOCl = hypochlorous acid)

The $H_2O_2$ used as substrate is generated by the NADPH oxidase system. $Cl^-$ is the halide usually employed, since it is present in relatively high concentration in plasma and body fluids. HOCl, the active ingredient of household liquid bleach, is a powerful oxidant and is highly microbicidal. When applied to normal tissues, its potential for causing damage is diminished because it reacts with primary or secondary amines present in neutrophils and tissues to produce various nitrogen-chlorine derivatives; these chloramines are also oxidants, though less powerful than HOCl, and act as microbicidal agents (eg, in sterilizing wounds) without causing tissue damage.

## The Proteinases of Neutrophils Can Cause Serious Tissue Damage If Their Actions Are Not Checked

Neutrophils contain a number of proteinases (Table 52–12) that can hydrolyze elastin, various types of collagens, and other proteins present in the extracellular matrix. Such enzymatic action, if allowed to proceed unopposed, can result in serious damage to tissues. Most of these proteinases are **lysosomal enzymes** and exist mainly as inactive precursors in normal neutrophils. Small amounts of these enzymes are released into normal tissues, with the amounts increasing markedly during inflammation. The activities of elastase and other proteinases are normally kept in check by a number of **antiproteinases** (also listed in Table 52–12) present in plasma and the extracellular fluid. Each of them can combine—usually forming a noncovalent complex—with one or more specific proteinases and thus cause inhibition. In Chapter 50 it was shown that a genetic deficiency of **$\alpha_1$-antiproteinase inhibitor** ($\alpha_1$-antitrypsin) permits elastase to act unopposed and digest pulmonary tissue, thereby participating in

***Table 52–12.*** Proteinases of neutrophils and antiproteinases of plasma and tissues.[1]

| Proteinases | Antiproteinases |
|---|---|
| Elastase | $\alpha_1$-Antiproteinase ($\alpha_1$-antitrypsin) |
| Collagenase | $\alpha_2$-Macroglobulin |
| Gelatinase | Secretory leukoproteinase inhibitor |
| Cathepsin G | $\alpha_1$-Antichymotrypsin |
| Plasminogen activator | Plasminogen activator inhibitor–1 |
| | Tissue inhibitor of metalloproteinase |

[1]The table lists some of the important proteinases of neutrophils and some of the proteins that can inhibit their actions. Most of the proteinases listed exist inside neutrophils as precursors. Plasminogen activator is not a proteinase, but it is included because it influences the activity of plasmin, which is a proteinase. The proteinases listed can digest many proteins of the extracellular matrix, causing tissue damage. The overall balance of proteinase:antiproteinase action can be altered by activating the precursors of the proteinases, or by inactivating the antiproteinases. The latter can be caused by proteolytic degradation or chemical modification, eg, Met-358 of $\alpha_1$-antiproteinase inhibitor is oxidized by cigarette smoke.

the causation of emphysema. **$\alpha_2$-Macroglobulin** is a plasma protein that plays an important role in the body's defense against excessive action of proteases; it combines with and thus neutralizes the activities of a number of important proteases (Chapter 50).

When increased amounts of chlorinated oxidants are formed during inflammation, they affect the proteinase:antiproteinase equilibrium, tilting it in favor of the former. For instance, certain of the proteinases listed in Table 52–12 are activated by HOCl, whereas certain of the antiproteinases are inactivated by this compound. In addition, the tissue inhibitor of metalloproteinases and $\alpha_1$-antichymotrypsin can be hydrolyzed by activated elastase, and $\alpha_1$-antiproteinase inhibitor can be hydrolyzed by activated collagenase and gelatinase. In most circumstances, an appropriate balance of proteinases and antiproteinases is achieved. However, in certain instances, such as in the lung when $\alpha_1$-antiproteinase inhibitor is deficient or when large amounts of neutrophils accumulate in tissues because of inadequate drainage, considerable tissue damage can result from the unopposed action of proteinases.

## RECOMBINANT DNA TECHNOLOGY HAS HAD A PROFOUND IMPACT ON HEMATOLOGY

Recombinant DNA technology has had a major impact on many aspects of hematology. The bases of the thalassemias and of many disorders of coagulation (Chapter 51) have been greatly clarified by investigations using cloning and sequencing. The study of oncogenes and chromosomal translocations has advanced understanding of the leukemias. As discussed above, cloning techniques have made available therapeutic amounts of erythropoietin and other growth factors. Deficiency of adenosine deaminase, which affects lymphocytes particularly, is the first disease to be treated by gene therapy. Like many other areas of biology and medicine, hematology has been and will continue to be revolutionized by this technology.

## SUMMARY

- The red blood cell is simple in terms of its structure and function, consisting principally of a concentrated solution of hemoglobin surrounded by a membrane.
- The production of red cells is regulated by erythropoietin, whereas other growth factors (eg, granulocyte- and granulocyte-macrophage colony-stimulating factors) regulate the production of white blood cells.
- The red cell contains a battery of cytosolic enzymes, such as superoxide dismutase, catalase, and glutathione peroxidase, to dispose of powerful oxidants generated during its metabolism.
- Genetically determined deficiency of the activity of glucose-6-phosphate dehydrogenase, which produces NADPH, is an important cause of hemolytic anemia.
- Methemoglobin is unable to transport oxygen; both genetic and acquired causes of methemoglobinemia are recognized. Considerable information has accumulated concerning the proteins and lipids of the red cell membrane. A number of cytoskeletal proteins, such as spectrin, ankyrin, and actin, interact with specific integral membrane proteins to help regulate the shape and flexibility of the membrane.
- Deficiency of spectrin results in hereditary spherocytosis, another important cause of hemolytic anemia.
- The ABO blood group substances in the red cell membrane are complex glycosphingolipids; the immunodominant sugar of A substance is *N*-acetylgalactosamine, whereas that of the B substance is galactose.
- Neutrophils play a major role in the body's defense mechanisms. Integrins on their surface membranes determine specific interactions with various cell and tissue components.
- Leukocytes are activated on exposure to bacteria and other stimuli; NADPH oxidase plays a key role in the process of activation (the respiratory burst). Mutations in this enzyme and associated proteins cause chronic granulomatous disease.

- The proteinases of neutrophils can digest many tissue proteins; normally, this is kept in check by a battery of antiproteinases. However, this defense mechanism can be overcome in certain circumstances, resulting in extensive tissue damage.
- The application of recombinant DNA technology is revolutionizing the field of hematology.

## REFERENCES

Borregaard N, Cowland JB: Granules of the human neutrophilic polymorphonuclear leukocyte. Blood 1997;89:3503.

Daniels G: A century of human blood groups. Wien Klin Wochenschr 2001;113:781.

Goodnough LT et al: Erythropoietin, iron, and erythropoiesis. Blood 2000;96:823.

Hirono A et al: Pyruvate kinase deficiency and other enzymopathies of the erythrocyte. In: *The Metabolic and Molecular Bases of Inherited Disease,* 8th ed. Scriver CR et al (editors). McGraw-Hill, 2001.

Israels LG, Israels ED: *Mechanisms in Hematology,* 2nd ed. Univ Manitoba Press, 1997. (Includes an excellent interactive CD.)

Jaffe ER, Hultquist DE: Cytochrome $b_5$ reductase deficiency and enzymopenic hereditary methemoglobinemia. In: *The Metabolic and Molecular Bases of Inherited Disease,* 8th ed. Scriver CR et al (editors). McGraw-Hill, 2001.

Lekstrom-Hunes JA, Gallin JI: Immunodeficiency diseases caused by defects in granulocytes. N Engl J Med 2000;343:1703.

Luzzato L et al: Glucose-6-phosphate dehydrogenase. In: *The Metabolic and Molecular Bases of Inherited Disease,* 8th ed. Scriver CR et al (editors). McGraw-Hill, 2001.

Rosse WF et al: New Views of Sickle Cell Disease Pathophysiology and Treatment. The American Society of Hematology. www.asheducationbook.org

Tse WT, Lux SE: Hereditary spherocytosis and hereditary elliptocytosis. In: *The Molecular Bases of Inherited Disease,* 8th ed. Scriver CR et al (editors). McGraw-Hill, 2001.

Weatherall DJ et al: The hemoglobinopathies. In: *The Metabolic and Molecular Bases of Inherited Disease,* 8th ed. Scriver CR et al (editors). McGraw-Hill, 2001.

# Metabolism of Xenobiotics 53

Robert K. Murray, MD, PhD

## BIOMEDICAL IMPORTANCE

Increasingly, humans are subjected to exposure to various foreign chemicals (xenobiotics)—drugs, food additives, pollutants, etc. The situation is well summarized in the following quotation from Rachel Carson: "As crude a weapon as the cave man's club, the chemical barrage has been hurled against the fabric of life." Understanding how xenobiotics are handled at the cellular level is important in learning how to cope with the chemical onslaught.

Knowledge of the metabolism of xenobiotics is basic to a rational understanding of pharmacology and therapeutics, pharmacy, toxicology, management of cancer, and drug addiction. All these areas involve administration of, or exposure to, xenobiotics.

## HUMANS ENCOUNTER THOUSANDS OF XENOBIOTICS THAT MUST BE METABOLIZED BEFORE BEING EXCRETED

A xenobiotic (Gk *xenos* "stranger") is a compound that is foreign to the body. The principal classes of xenobiotics of medical relevance are drugs, chemical carcinogens, and various compounds that have found their way into our environment by one route or another, such as polychlorinated biphenyls (PCBs) and certain insecticides. More than 200,000 manufactured environmental chemicals exist. Most of these compounds are subject to metabolism (chemical alteration) in the human body, with the liver being the main organ involved; occasionally, a xenobiotic may be excreted unchanged. At least 30 different enzymes catalyze reactions involved in xenobiotic metabolism; however, this chapter will only cover a selected group of them.

It is convenient to consider the metabolism of xenobiotics in two phases. In phase 1, the major reaction involved is **hydroxylation,** catalyzed by members of a class of enzymes referred to as **monooxygenases** or **cytochrome P450s.** Hydroxylation may terminate the action of a drug, though this is not always the case. In addition to hydroxylation, these enzymes catalyze a wide range of reactions, including those involving deamination, dehalogenation, desulfuration, epoxidation, peroxygenation, and reduction. Reactions involving hydrolysis (eg, catalyzed by esterases) and certain other non-P450-catalyzed reactions also occur in phase 1.

In phase 2, the hydroxylated or other compounds produced in phase 1 are converted by specific enzymes to various polar metabolites by **conjugation** with glucuronic acid, sulfate, acetate, glutathione, or certain amino acids, or by **methylation.**

The overall purpose of the two phases of metabolism of xenobiotics is to increase their **water solubility (polarity)** and thus **excretion** from the body. Very hydrophobic xenobiotics would persist in adipose tissue almost indefinitely if they were not converted to more polar forms. In certain cases, phase 1 metabolic reactions convert xenobiotics from **inactive** to **biologically active** compounds. In these instances, the original xenobiotics are referred to as **"prodrugs"** or **"procarcinogens."** In other cases, additional phase 1 reactions (eg, further hydroxylation reactions) convert the active compounds to less active or inactive forms prior to conjugation. In yet other cases, it is the conjugation reactions themselves that convert the active products of phase 1 reactions to less active or inactive species, which are subsequently excreted in the urine or bile. In a very few cases, conjugation may actually increase the biologic activity of a xenobiotic.

The term **"detoxification"** is sometimes used for many of the reactions involved in the metabolism of xenobiotics. However, the term is not always appropriate because, as mentioned above, in some cases the reactions to which xenobiotics are subject actually increase their biologic activity and toxicity.

## ISOFORMS OF CYTOCHROME P450 HYDROXYLATE A MYRIAD OF XENOBIOTICS IN PHASE 1 OF THEIR METABOLISM

**Hydroxylation** is the chief reaction involved in phase 1. The responsible enzymes are called **monooxygenases** or **cytochrome P450s;** the human genome encodes at least 14 families of these enzymes. Estimates of the number of distinct cytochrome P450s in human tissues range from approximately 35 to 60. The reaction catalyzed by a monooxygenase (cytochrome P450) is as follows:

$$RH + O_2 + NADPH + H^+ \rightarrow R—OH + H_2O + NADP$$

RH above can represent a very wide variety of xenobiotics, including drugs, carcinogens, pesticides, petroleum products, and pollutants (such as a mixture of PCBs). In addition, **endogenous compounds,** such as certain steroids, eicosanoids, fatty acids, and retinoids, are also substrates. The substrates are generally **lipophilic** and are rendered more **hydrophilic** by hydroxylation.

Cytochrome P450 is considered the **most versatile biocatalyst** known. The actual reaction mechanism is complex and has been briefly described previously (Figure 11–6). It has been shown by the use of $^{18}O_2$ that one atom of oxygen enters R—OH and one atom enters water. This dual fate of the oxygen accounts for the former naming of monooxygenases as **"mixed-function oxidases."** The reaction catalyzed by cytochrome P450 can also be represented as follows:

Reduced cytochrome P450 → Oxidized cytochrome P450

$$RH + O_2 \rightarrow R\text{—}OH + H_2O$$

The major monooxygenases in the endoplasmic reticulum are **cytochrome P450s**—so named because the enzyme was discovered when it was noted that preparations of microsomes that had been chemically reduced and then exposed to carbon monoxide exhibited a distinct peak at 450 nm. Among reasons that this enzyme is important is the fact that approximately 50% of the drugs humans ingest are metabolized by isoforms of cytochrome P450; these enzymes also act on various carcinogens and pollutants.

## Isoforms of Cytochrome P450 Make Up a Superfamily of Heme-Containing Enzymes

The following are important points concerning cytochrome P450s.

(1) Because of the large number of isoforms (about 150) that have been discovered, it became important to have a **systematic nomenclature** for isoforms of P450 and for their genes. This is now available and in wide use and is based on structural homology. The abbreviated root symbol CYP denotes a cytochrome P450. This is followed by an Arabic number designating the **family;** cytochrome P450s are included in the same family if they exhibit 40% or more sequence identity. The Arabic number is followed by a capital letter indicating the **subfamily,** if two or more members exist; P450s are in the same subfamily if they exhibit greater than 55% sequence identity. The **individual** P450s are then arbitrarily assigned Arabic numerals. Thus, CYP1A1 denotes a cytochrome P450 that is a member of family 1 and subfamily A and is the first individual member of that subfamily. The nomenclature for the **genes** encoding cytochrome P450s is identical to that described above except that italics are used; thus, the gene encoding CYP1A1 is *CYP1A1.*

(2) Like **hemoglobin,** they are hemoproteins.

(3) They are widely distributed across species. Bacteria possess cytochrome P450s, and $P450_{cam}$ (involved in the metabolism of camphor) of *Pseudomonas putida* is the only P450 isoform whose crystal structure has been established.

(4) They are present in highest amount in **liver** and **small intestine** but are probably present in all tissues. In liver and most other tissues, they are present mainly in the **membranes of the smooth endoplasmic reticulum,** which constitute part of the **microsomal fraction** when tissue is subjected to subcellular fractionation. In hepatic microsomes, cytochrome P450s can comprise as much as 20% of the total protein. P450s are found in most tissues, though often in low amounts compared with liver. In the **adrenal,** they are found in **mitochondria** as well as in the endoplasmic reticulum; the various hydroxylases present in that organ play an important role in cholesterol and steroid biosynthesis. The mitochondrial cytochrome P450 system differs from the microsomal system in that it uses an NADPH-linked flavoprotein, **adrenodoxin reductase,** and a nonheme iron-sulfur protein, **adrenodoxin.** In addition, the specific P450 isoforms involved in steroid biosynthesis are generally much more restricted in their substrate specificity.

(5) At least six isoforms of cytochrome P450 are present in the endoplasmic reticulum of human liver, each with wide and somewhat overlapping **substrate specificities** and acting on both xenobiotics and endogenous compounds. The genes for many isoforms of P450 (from both humans and animals such as the rat) have been isolated and studied in detail in recent years.

(6) **NADPH,** not NADH, is involved in the reaction mechanism of cytochrome P450. The enzyme that uses NADPH to yield the reduced cytochrome P450, shown at the left-hand side of the above equation, is called **NADPH-cytochrome P450 reductase.** Electrons are transferred from NADPH to NADPH-cytochrome P450 reductase and then to cytochrome P450. This leads to the **reductive activation of molecular oxygen,** and one atom of oxygen is subsequently inserted into the substrate. **Cytochrome $b_5$,** another hemoprotein found in the membranes of the smooth endoplasmic reticulum (Chapter 11), may be involved as an electron donor in some cases.

(7) **Lipids** are also components of the cytochrome P450 system. The preferred lipid is **phosphatidylcholine,** which is the major lipid found in membranes of the endoplasmic reticulum.

(8) Most isoforms of cytochrome P450 are **inducible.** For instance, the administration of phenobarbital or of many other drugs causes hypertrophy of the

smooth endoplasmic reticulum and a three- to fourfold increase in the amount of cytochrome P450 within 4–5 days. The mechanism of induction has been studied extensively and in most cases involves increased transcription of mRNA for cytochrome P450. However, certain cases of induction involve stabilization of mRNA, enzyme stabilization, or other mechanisms (eg, an effect on translation).

Induction of cytochrome P450 has important clinical implications, since it is a biochemical mechanism of **drug interaction.** A drug interaction has occurred when the effects of one drug are altered by prior, concurrent, or later administration of another. As an illustration, consider the situation when a patient is taking the anticoagulant **warfarin** to prevent blood clotting. This drug is metabolized by CYP2C9. Concomitantly, the patient is started on **phenobarbital** (an inducer of this P450) to treat a certain type of epilepsy, but the dose of warfarin is not changed. After 5 days or so, the level of CYP2C9 in the patient's liver will be elevated three- to fourfold. This in turn means that warfarin will be metabolized much more quickly than before, and its dosage will have become inadequate. Therefore, the dose must be increased if warfarin is to be therapeutically effective. To pursue this example further, a problem could arise later on if the phenobarbital is discontinued but the increased dosage of warfarin stays the same. The patient will be at risk of bleeding, since the high dose of warfarin will be even more active than before, because the level of CYP2C9 will decline once phenobarbital has been stopped.

Another example of enzyme induction involves **CYP2E1,** which is induced by consumption of **ethanol.** This is a matter for concern, because this P450 metabolizes certain widely used solvents and also components found in tobacco smoke, many of which are established **carcinogens.** Thus, if the activity of CYP2E1 is elevated by induction, this may increase the risk of carcinogenicity developing from exposure to such compounds.

**(9)** Certain isoforms of cytochrome P450 (eg, CYP1A1) are particularly involved in the metabolism of polycyclic aromatic hydrocarbons (PAHs) and related molecules; for this reason they were formerly called aromatic hydrocarbon hydroxylases (AHHs). This enzyme is important in the metabolism of PAHs and in carcinogenesis produced by these agents. For example, in the lung it may be involved in the conversion of inactive PAHs (procarcinogens), inhaled by smoking, to active carcinogens by hydroxylation reactions. Smokers have higher levels of this enzyme in some of their cells and tissues than do nonsmokers. Some reports have indicated that the activity of this enzyme may be elevated (induced) in the placenta of a woman who smokes, thus potentially altering the quantities of metabolites of PAHs (some of which could be harmful) to which the fetus is exposed.

**(10)** Certain cytochrome P450s exist in polymorphic forms (genetic isoforms), some of which exhibit low catalytic activity. These observations are one important explanation for the variations in drug responses noted among many patients. One P450 exhibiting polymorphism is CYP2D6, which is involved in the metabolism of debrisoquin (an antihypertensive drug; see Table 53–2) and sparteine (an antiarrhythmic and oxytocic drug). Certain polymorphisms of CYP2D6 cause poor metabolism of these and a variety of other drugs so that they can accumulate in the body, resulting in untoward consequences. Another interesting polymorphism is that of CYP2A6, which is involved in the metabolism of nicotine to conitine. Three *CYP2A6* alleles have been identified: a wild type and two null or inactive alleles. It has been reported that individuals with the null alleles, who have impaired metabolism of nicotine, are apparently protected against becoming tobacco-dependent smokers (Table 53–2). These individuals smoke less, presumably because their blood and brain concentrations of nicotine remain elevated longer than those of individuals with the wild-type allele. It has been speculated that inhibiting CYP2A6 may be a novel way to help prevent and to treat smoking.

Table 53–1 summarizes some principal features of cytochrome P450s.

## CONJUGATION REACTIONS PREPARE XENOBIOTICS FOR EXCRETION IN PHASE 2 OF THEIR METABOLISM

In phase 1 reactions, xenobiotics are generally converted to more polar, hydroxylated derivatives. In phase 2 reactions, these derivatives are conjugated with molecules such as glucuronic acid, sulfate, or glutathione. This renders them even more water-soluble, and they are eventually excreted in the urine or bile.

### Five Types of Phase 2 Reactions Are Described Here

#### A. Glucuronidation

The glucuronidation of bilirubin is discussed in Chapter 32; the reactions whereby xenobiotics are glucuronidated are essentially similar. UDP-glucuronic acid is the glucuronyl donor, and a variety of glucuronosyltransferases, present in both the endoplasmic reticulum and cytosol, are the catalysts. Molecules such as 2-acetylaminofluorene (a carcinogen), aniline, benzoic acid, meprobamate (a tranquilizer), phenol, and

***Table 53–1.*** Some properties of human cytochrome P450s.

- Involved in phase I of the metabolism of innumerable xenobiotics, including perhaps 50% of the drugs administered to humans
- Involved in the metabolism of many endogenous compounds (eg, steroids)
- All are hemoproteins
- Often exhibit broad substrate specificity, thus acting on many compounds; consequently, different P450s may catalyze formation of the same product
- Extremely versatile catalysts, perhaps catalyzing about 60 types of reactions
- However, basically they catalyze reactions involving introduction of one atom of oxygen into the substrate and one into water
- Their hydroxylated products are more water-soluble than their generally lipophilic substrates, facilitating excretion
- Liver contains highest amounts, but found in most if not all tissues, including small intestine, brain, and lung
- Located in the smooth endoplasmic reticulum or in mitochondria (steroidogenic hormones)
- In some cases, their products are mutagenic or carcinogenic
- Many have a molecular mass of about 55 kDa
- Many are inducible, resulting in one cause of drug interactions
- Many are inhibited by various drugs or their metabolic products, providing another cause of drug interactions
- Some exhibit genetic polymorphisms, which can result in atypical drug metabolism
- Their activities may be altered in diseased tissues (eg, cirrhosis), affecting drug metabolism
- Genotyping the P450 profile of patients (eg, to detect polymorphisms) may in the future permit individualization of drug therapy

many steroids are excreted as glucuronides. The glucuronide may be attached to oxygen, nitrogen, or sulfur groups of the substrates. Glucuronidation is probably the most frequent conjugation reaction.

### B. Sulfation

Some alcohols, arylamines, and phenols are sulfated. The **sulfate donor** in these and other biologic sulfation reactions (eg, sulfation of steroids, glycosaminoglycans, glycolipids, and glycoproteins) is **adenosine 3′-phosphate-5′-phosphosulfate (PAPS)** (Chapter 24); this compound is called "active sulfate."

### C. Conjugation With Glutathione

Glutathione (γ-glutamyl-cysteinylglycine) is a **tripeptide** consisting of glutamic acid, cysteine, and glycine (Figure 3–3). Glutathione is commonly abbreviated GSH (because of the sulfhydryl group of its cysteine, which is the business part of the molecule). A number of potentially toxic electrophilic xenobiotics (such as certain carcinogens) are conjugated to the nucleophilic GSH in reactions that can be represented as follows:

$$R + GSH \rightarrow R{-}S{-}G$$

where R = an electrophilic xenobiotic. The enzymes catalyzing these reactions are called **glutathione S-transferases** and are present in high amounts in liver cytosol and in lower amounts in other tissues. A variety of glutathione S-transferases are present in human tissue. They exhibit different substrate specificities and can be separated by electrophoretic and other techniques. If the potentially toxic xenobiotics were not conjugated to GSH, they would be free to combine covalently with DNA, RNA, or cell protein and could thus lead to serious cell damage. GSH is therefore an important **defense mechanism** against certain toxic compounds, such as some drugs and carcinogens. If the levels of GSH in a tissue such as liver are lowered (as can be achieved by the administration to rats of certain compounds that react with GSH), then that tissue can be shown to be more susceptible to injury by various chemicals that would normally be conjugated to GSH. Glutathione conjugates are subjected to further metabolism before excretion. The glutamyl and glycinyl groups belonging to glutathione are removed by specific enzymes, and an acetyl group (donated by acetyl-CoA) is added to the amino group of the remaining cysteinyl moiety. The resulting compound is a **mercapturic acid,** a conjugate of L-acetylcysteine, which is then excreted in the urine.

Glutathione has other important functions in human cells apart from its role in xenobiotic metabolism.

1. It participates in the decomposition of potentially toxic **hydrogen peroxide** in the reaction catalyzed by glutathione peroxidase (Chapter 20).
2. It is an important **intracellular reductant,** helping to maintain essential SH groups of enzymes in their reduced state. This role is discussed in Chapter 20, and its involvement in the hemolytic anemia caused by deficiency of glucose-6-phosphate dehydrogenase is discussed in Chapters 20 and 52.
3. A metabolic cycle involving GSH as a carrier has been implicated in the **transport of certain amino acids** across membranes in the kidney. The first reaction of the cycle is shown below.

$$\text{Amino acid} + GSH \rightarrow \gamma\text{-Glutamyl amino acid} + \text{Cysteinylglycine}$$

This reaction helps transfer certain amino acids across the plasma membrane, the amino acid being subsequently hydrolyzed from its complex with GSH and the GSH being resynthesized from cysteinylglycine. The enzyme catalyzing the above reaction is **γ-glutamyltransferase (GGT).** It is present in the plasma membrane of renal tubular cells and bile ductule cells, and in the endoplasmic reticulum of hepatocytes. The enzyme has diagnostic value because it is released into the blood from hepatic cells in various hepatobiliary diseases.

### D. Other Reactions

The two most important other reactions are acetylation and methylation.

**1. Acetylation**—Acetylation is represented by

$$X + \text{Acetyl-CoA} \rightarrow \text{Acetyl-X} + \text{CoA}$$

where X represents a xenobiotic. As for other acetylation reactions, **acetyl-CoA** (active acetate) is the acetyl donor. These reactions are catalyzed by **acetyltransferases** present in the cytosol of various tissues, particularly liver. The drug **isoniazid,** used in the treatment of tuberculosis, is subject to acetylation. **Polymorphic types** of acetyltransferases exist, resulting in individuals who are classified as **slow or fast acetylators,** and influence the rate of clearance of drugs such as isoniazid from blood. Slow acetylators are more subject to certain toxic effects of isoniazid because the drug persists longer in these individuals.

**2. Methylation**—A few xenobiotics are subject to methylation by methyltransferases, employing *S*-adenosylmethionine (Figure 30–17) as the methyl donor.

## THE ACTIVITIES OF XENOBIOTIC-METABOLIZING ENZYMES ARE AFFECTED BY AGE, SEX, & OTHER FACTORS

Various factors affect the activities of the enzymes metabolizing xenobiotics. The activities of these enzymes may differ substantially among species. Thus, for example, the possible toxicity or carcinogenicity of xenobiotics cannot be extrapolated freely from one species to another. There are significant differences in enzyme activities among individuals, many of which appear to be due to **genetic factors.** The activities of some of these enzymes vary according to **age** and **sex.**

Intake of various xenobiotics such as phenobarbital, PCBs, or certain hydrocarbons can cause **enzyme induction.** It is thus important to know whether or not an individual has been exposed to these inducing agents in evaluating biochemical responses to xenobiotics. Metabolites of certain xenobiotics can inhibit or stimulate the activities of xenobiotic-metabolizing enzymes. Again, this can affect the doses of certain drugs that are administered to patients. Various diseases (eg, cirrhosis of the liver) can affect the activities of drug-metabolizing enzymes, sometimes necessitating adjustment of dosages of various drugs for patients with these disorders.

## RESPONSES TO XENOBIOTICS INCLUDE PHARMACOLOGIC, TOXIC, IMMUNOLOGIC, & CARCINOGENIC EFFECTS

Xenobiotics are metabolized in the body by the reactions described above. When the xenobiotic is a drug, phase 1 reactions may produce its active form or may diminish or terminate its action if it is pharmacologically active in the body without prior metabolism. The diverse effects produced by drugs comprise the area of study of pharmacology; here it is important to appreciate that drugs act primarily through biochemical mechanisms. Table 53–2 summarizes four important reactions to drugs that reflect **genetically determined differences** in enzyme and protein structure among individuals—part of the field of study known as **pharmacogenetics** (see below).

***Table 53–2.*** Some important drug reactions due to mutant or polymorphic forms of enzymes or proteins.[1]

| Enzyme or Protein Affected | Reaction or Consequence |
|---|---|
| Glucose-6-phosphate dehydrogenase (G6PD) [mutations] (MIM 305900) | Hemolytic anemia following ingestion of drugs such as primaquine |
| $Ca^{2+}$ release channel (ryanodine receptor) in the sarcoplasmic reticulum [mutations] (MIM 180901) | Malignant hyperthermia (MIM 145600) following administration of certain anesthetics (eg, halothane) |
| CYP2D6 [polymorphisms] (MIM 124030) | Slow metabolism of certain drugs (eg, debrisoquin), resulting in their accumulation |
| CYP2A6 [polymorphisms] (MIM 122720) | Impaired metabolism of nicotine, resulting in protection against becoming a tobacco-dependent smoker |

[1]G6PD deficiency is discussed in Chapters 20 and 52 and malignant hyperthermia in Chapter 49. At least one gene other than that encoding the ryanodine receptor is involved in certain cases of malignant hypertension. Many other examples of drug reactions based on polymorphism or mutation are available.

Certain xenobiotics are very toxic even at low levels (eg, cyanide). On the other hand, there are few xenobiotics, including drugs, that do not exert some toxic effects if sufficient amounts are administered. The **toxic effects of xenobiotics** cover a wide spectrum, but the major effects can be considered under three general headings (Figure 53–1).

The first is **cell injury** (cytotoxicity), which can be severe enough to result in cell death. There are many mechanisms by which xenobiotics injure cells. The one considered here is **covalent binding to cell macromolecules** of reactive species of xenobiotics produced by metabolism. These macromolecular targets include **DNA, RNA,** and **protein.** If the macromolecule to which the reactive xenobiotic binds is essential for short-term cell survival, eg, a protein or enzyme involved in some critical cellular function such as oxidative phosphorylation or regulation of the permeability of the plasma membrane, then severe effects on cellular function could become evident quite rapidly.

Second, the reactive species of a xenobiotic may bind to a protein, altering its **antigenicity.** The xenobiotic is said to act as a **hapten,** ie, a small molecule that by itself does not stimulate antibody synthesis but will combine with antibody once formed. The resulting **antibodies** can then damage the cell by several immunologic mechanisms that grossly perturb normal cellular biochemical processes.

Third, reactions of activated species of chemical carcinogens with **DNA** are thought to be of great importance in **chemical carcinogenesis.** Some chemicals (eg, benzo[α]pyrene) require activation by monooxygenases in the endoplasmic reticulum to become carcinogenic (they are thus called **indirect carcinogens**). The activities of the monooxygenases and of other xenobiotic-metabolizing enzymes present in the endoplasmic reticulum thus help to determine whether such compounds become carcinogenic or are "detoxified." Other chemicals (eg, various alkylating agents) can react directly (direct carcinogens) with DNA without undergoing intracellular chemical activation.

The enzyme **epoxide hydrolase** is of interest because it can exert a protective effect against certain carcinogens. The products of the action of certain monooxygenases on some procarcinogen substrates are **epoxides.** Epoxides are highly reactive and mutagenic or carcinogenic or both. Epoxide hydrolase—like cytochrome P450, also present in the membranes of the endoplasmic reticulum—acts on these compounds, converting them into much less reactive dihydrodiols. The reaction catalyzed by epoxide hydrolase can be represented as follows:

$$-\underset{\diagdown}{C}-\underset{\diagup}{C}- \;(\text{bridged by } O) + H_2O \longrightarrow -\underset{HO}{\overset{|}{C}}-\underset{OH}{\overset{|}{C}}-$$

**Epoxide** → **Dihydrodiol**

## PHARMACOGENOMICS WILL DRIVE THE DEVELOPMENT OF NEW & SAFER DRUGS

As indicated above, **pharmacogenetics** is the study of the roles of genetic variations in the responses to drugs. As a result of the progress made in sequencing the

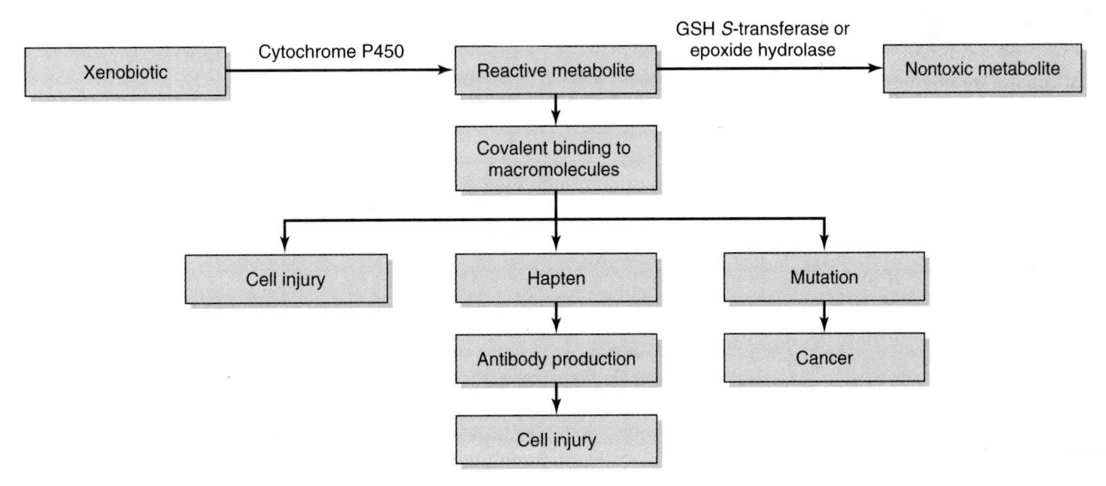


***Figure 53–1.*** Simplified scheme showing how metabolism of a xenobiotic can result in cell injury, immunologic damage, or cancer. In this instance, the conversion of the xenobiotic to a reactive metabolite is catalyzed by a cytochrome P450, and the conversion of the reactive metabolite (eg, an epoxide) to a nontoxic metabolite is catalyzed either by a GSH S-transferase or by epoxide hydrolase.

human genome, a new field of study—**pharmacogenomics**—has developed recently. It includes pharmacogenetics but covers a much wider sphere of activity. Information from genomics, proteomics, bioinformatics, and other disciplines such as biochemistry and toxicology will be integrated to make possible the synthesis of newer and safer drugs. As the sequences of all our genes and their encoded proteins are determined, this will reveal many new targets for drug actions. It will also reveal polymorphisms (this term is briefly discussed in Chapter 50) of enzymes and proteins related to drug metabolism, action, and toxicity. DNA probes capable of detecting them will be synthesized, permitting screening of individuals for potentially harmful polymorphisms prior to the start of drug therapy. As the structures of relevant proteins and their polymorphisms are revealed, model building and other techniques will permit the design of drugs that take into account both normal protein targets and their polymorphisms. At least to some extent, drugs will be tailor-made for individuals based on their genetic profiles. A new era of rational drug design built on information derived from genomics and proteomics has already commenced.

## SUMMARY

- Xenobiotics are chemical compounds foreign to the body, such as drugs, food additives, and environmental pollutants; more than 200,000 have been identified.
- Xenobiotics are metabolized in two phases. The major reaction of phase 1 is hydroxylation catalyzed by a variety of monooxygenases, also known as the cytochrome P450s. In phase 2, the hydroxylated species are conjugated with a variety of hydrophilic compounds such as glucuronic acid, sulfate, or glutathione. The combined operation of these two phases renders lipophilic compounds into water-soluble compounds that can be eliminated from the body.
- Cytochrome P450s catalyze reactions that introduce one atom of oxygen derived from molecular oxygen into the substrate, yielding a hydroxylated product. NADPH and NADPH-cytochrome P450 reductase are involved in the complex reaction mechanism.
- All cytochrome P450s are hemoproteins and generally have a wide substrate specificity, acting on many exogenous and endogenous substrates. They represent the most versatile biocatalyst known.
- Members of at least 11 families of cytochrome P450 are found in human tissue.
- Cytochrome P450s are generally located in the endoplasmic reticulum of cells and are particularly enriched in liver.
- Many cytochrome P450s are inducible. This has important implications in phenomena such as drug interaction.
- Mitochondrial cytochrome P450s also exist and are involved in cholesterol and steroid biosynthesis. They use a nonheme iron-containing sulfur protein, adrenodoxin, not required by microsomal isoforms.
- Cytochrome P450s, because of their catalytic activities, play major roles in the reactions of cells to chemical compounds and in chemical carcinogenesis.
- Phase 2 reactions are catalyzed by enzymes such as glucuronosyltransferases, sulfotransferases, and glutathione S-transferases, using UDP-glucuronic acid, PAPS (active sulfate), and glutathione, respectively, as donors.
- Glutathione not only plays an important role in phase 2 reactions but is also an intracellular reducing agent and is involved in the transport of certain amino acids into cells.
- Xenobiotics can produce a variety of biologic effects, including pharmacologic responses, toxicity, immunologic reactions, and cancer.
- Catalyzed by the progress made in sequencing the human genome, the new field of pharmacogenomics offers the promise of being able to make available a host of new rationally designed, safer drugs.

## REFERENCES


Evans WE, Johnson JA: Pharmacogenomics: the inherited basis for interindividual differences in drug response. Annu Rev Genomics Hum Genet 2001;2:9.

Guengerich FP: Common and uncommon cytochrome P450 reactions related to metabolism and chemical toxicity. Chem Res Toxicol 2001;14:611.

Honkakakoski P, Negishi M: Regulation of cytochrome P450 (CYP) genes by nuclear receptors. Biochem J 2000;347:321.

Kalow W, Grant DM: Pharmacogenetics. In: *The Metabolic and Molecular Bases of Inherited Disease,* 8th ed. Scriver CR et al (editors). McGraw-Hill, 2001.

Katzung BG (editor): *Basic & Clinical Pharmacology,* 8th ed. McGraw-Hill, 2001.

McLeod HL, Evans WE: Pharmacogenomics: unlocking the human genome for better drug therapy. Annu Rev Pharmacol Toxicol 2001;41:101.

Nelson DR et al: P450 superfamily: update on new sequences, gene mapping, accession numbers and nomenclature. Pharmacogenetics 1996;6:1.

# The Human Genome Project 54

*Robert K. Murray, MD, PhD*

## BIOMEDICAL SIGNIFICANCE

The information deriving from determination of the sequences of the human genome and those of other organisms will change biology and medicine for all time. For example, with reference to the human genome, new information on our origins, on disease genes, on diagnosis, and possible approaches to therapy are already flooding in. Progress in fields such as genomics, proteomics, bioinformatics, biotechnology, and pharmacogenomics is accelerating rapidly.

The aims of this chapter are to briefly summarize the major findings of the Human Genome Project (HGP) and indicate their implications for biology and medicine.

## THE HUMAN GENOME PROJECT HAS A VARIETY OF GOALS

The HGP, which started in 1990, is an international effort whose principal goals were to sequence the entire human genome and the genomes of several other model organisms that have been basic to the study of genetics (eg, *Escherichia coli, Saccharomyces cerevisiae* [a yeast], *Drosophila melanogaster* [the fruit fly], *Caenorhabditis elegans* [the roundworm], and *Mus musculus* [the common house mouse]). Most of these goals have been accomplished. In the United States, the National Center for Human Genome Research (NCHGR) was established in 1989, initially directed by James D. Watson and subsequently by Francis Collins. The NCHGR played a leading role in directing the United States effort in the HGP. In 1997, it became the National Human Genome Research Institute (NHGRI). The international collaboration—involving groups from the USA, UK, Japan, France, Germany, and China—came to be known as the International Human Genome Sequencing Consortium (IHGSC). Initially, a number of short-term goals were established for the United States effort—eg, producing a human genetic map with markers 2–5 centimorgans (cM) apart and constructing a physical map of all 24 human chromosomes (22 autosomal plus X and Y) with markers spaced at approximately 100,000 base pairs (bp). Figure 54–1 summarizes the differences between a genetic map, a cytogenetic map, and a physical map of a chromosome. These and other initial goals were achieved and surpassed by the mid-nineties. In 1998, new goals for the United States wing of the HGP were announced. These included the aim of completing the entire sequence by the end of 2003 or sooner. Other specific objectives concerned sequencing technology, comparative genomics, bioinformatics, ethical considerations, and other issues. By the fall of 1998, about 6% of the human genome sequence had been completed and the foundations for future work laid. Further progress was catalyzed by the announcement that a second group, the private company Celera Genomics, led by Craig Venter, had undertaken the objective of sequencing the human genome. Venter and colleagues had published in 1995 the entire genome sequences of *Haemophilus influenzae* and *Mycoplasma genitalium,* the first of many species to have their genomic sequences determined. An important factor in the success of these workers was the use of a shotgun approach, ie, sonicating the DNA, sequencing the fragments, and reassembling the sequence, based on overlaps. For comparison, a variety of approaches that have been used at different times to study normal and disease genes are listed in Table 54–1.

### A Draft Sequence of the Human Genome Was Announced in June 2000

In June 2000, leaders.of the IHGSC and the personnel at Celera Genomics announced completion of working drafts of the sequence of the human genome, covering more than 90% of it. The principal findings of the two groups were published separately in February 2001 in special issues of Nature (the IHGSC) and Science (Celera). The draft published by the Consortium was the product of at least 10 years of work involving 20 sequencing centers located in six countries. That published by Celera and associates was the product of some 3 years or less of work; it relied in part on data obtained by the IHGSC. The combined achievement has been hailed, among other descriptions, as providing a Library of Life, supplying a Periodic Table of Life, and finding the Holy Grail of Human Genetics.

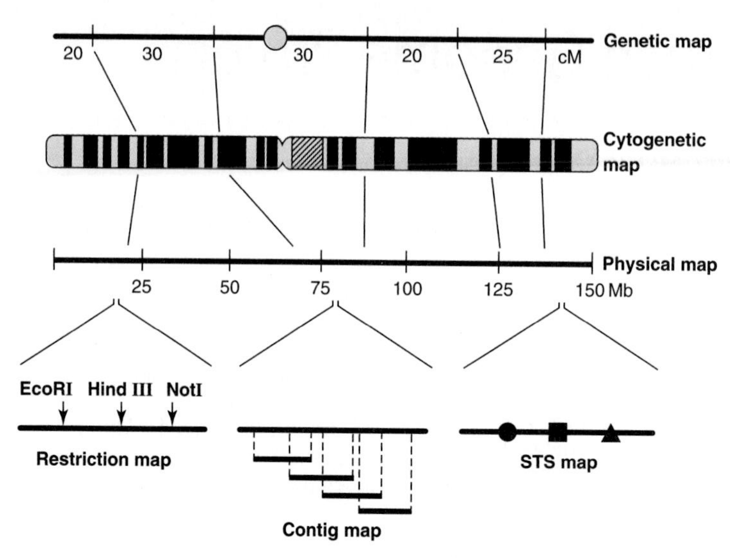


***Figure 54–1.*** Principal methods used to identify and isolate normal and disease genes. For the genetic map, the positions of several hypothetical genetic markers are shown, along with the genetic distances in centimorgans between them. The circle shows the position of the centromere. For the cytogenetic map, the classic banding pattern of a hypothetical chromosome is shown. For the physical map, the approximate physical positions of the above genetic markers are shown, along with the relative physical distances in megabase pairs. Examples of a restriction map, a contig mark, and an STS map are also shown. (Reproduced, with permission, from Green ED, Waterston RH: The Human Genome Project: Prospects and implications for clinical medicine. JAMA 1991;266: 1966. Copyright © 1991 by the American Medical Association.)

## Different Approaches Were Used by the Two Groups

We shall summarize the major findings reported in the two drafts and comment on their implications. While there are differences between the drafts, they will not be dwelt on here, as the areas of general agreement are much more extensive. It is worthwhile, however, to summarize the different approaches used by the two groups. Basically, the IHGSC employed a **map first, sequence later approach.** In part, this was because sequencing was a slow process when the public project started, and the strategy of the Consortium evolved over time as advances were made in sequencing and other techniques. The overall approach, referred to as **hierarchical shotgun sequencing,** consisted of fragmenting the entire genome into pieces of approximately 100–200 kb and inserting them into bacterial artificial chromosomes (BACs). The BACs were then positioned on individual chromosomes by looking for marker sequences known as sequence-tagged sites (STSs), whose locations had been already determined. STSs are short (usually < 500 bp), unique genomic loci for which a PCR assay is available. Clones of the BACs were then broken into small fragments (shotgunning). Each fragment was then sequenced, and computer algorithms were used that recognized matching sequence information from overlapping fragments to piece together the complete sequence.

Celera used the **whole genome shotgun approach,** in effect bypassing the mapping step. Shotgun fragments were assembled by algorithms onto large scaffolds, and the correct position of these scaffolds in the genome was determined using STSs. A scaffold is a series of "contigs" that are in the right order but not necessarily connected in one continuous sequence. Contigs are contiguous sequences of DNA made by assembling overlapping sequenced fragments of a natural chromosome or a BAC. The availability of **high-throughput sequenators, powerful computer programs,** the element of competition, and other factors accounted for the rapid progress made by both groups from 1998 onward.

***Table 54–1.*** Principal methods used to identify and isolate normal and disease genes.

| Procedure | Comments |
|---|---|
| Detection of specific cytogenetic abnormalities | For instance, a small deletion of band Xp21.2 was important in cloning the gene involved in Duchenne muscular dystrophy. |
| Extensive linkage studies | Large families with defined pedigrees are desirable. Dominant genes are easier to recognize than recessives. |
| Use of probes to define marker loci | Probes identify STSs, RFLPs, SNPs,[1] etc; thousands, covering all the chromosomes, are now available. It is desirable to flank the gene on both sides, clearly delineating it. |
| Radiation hybrid mapping[2] | Now the most rapid method of localizing a gene or DNA fragment to a subregion of a human chromosome and constructing a physical map. |
| Use of rodent or human somatic cell hybrids | Permits assignment of a gene to one specific chromosome but not to a subregion. |
| Fluorescence in situ hybridization | Permits localization of a gene to one chromosomal band. |
| Use of pulsed-field gel electrophoresis (PFGE) to separate large DNA fragments | Permits isolation of large DNA fragments obtained by use of restriction endonucleases (rare cutters) that result in very limited cutting of DNA. |
| Chromosome walking | Involves repeated cloning of overlapping DNA segments; the procedure is laborious and can usually cover only 100–200 kb. |
| Chromosome jumping | By cutting DNA into relatively large fragments and circularizing it, one can move more quickly and cover greater lengths of DNA than with chromosomal walking. |
| Cloning via YACs, BACs, cosmids, phages, and plasmids | Permits isolation of fragments of varying lengths. |
| Detection of expression of mRNAs in tissues by Northern blotting using one or more fragments of the gene as a probe | The mRNA should be expressed in affected tissues. |
| PCR | Can be used to amplify fragments of the gene; also many other applications. |
| DNA sequencing | Establishes the highest resolution physical map. Identifies open reading frame. Facilities with many high throughput instruments could sequence millions of base pairs per day. |
| Databases | Comparison of DNA and protein sequences obtained from unknown gene with known sequences in databases can facilitate gene identification. |

**Abbreviations:** STS, sequence tagged site; RFLP, restriction fragment linked polymorphism; SNP, single nucleotide polymorphism; YAC, yeast artificial chromosome; BAC, bacterial artificial chromosome; PCR, polymerase chain reaction.

[1]Many single nucleotide polymorphisms (SNPs) are being detected and catalogued. These are stable and frequent, and their detection can be automated. It is anticipated that they will be particularly useful for mapping complex traits such as diabetes mellitus.

[2]Radiation hybrid mapping (consult http://compgen.rutgers.edu/rhmap/ for a detailed bibliography of this technique) makes use of a panel of somatic cell hybrids, with each cell line containing a random set of irradiated human genomic DNA in a hamster background. Briefly, the radiation fragments the DNA into small pieces of variable length; if a gene is located close to another known gene, it is likely that the two will remain linked (compare genetic linkage) on the same fragment. An STS marker is typed against a radiation hybrid panel by using its two oligonucleotide primers to perform a PCR assay against the DNA from each hybrid cell line of the panel. If enough markers are typed on one panel, continuous linkage can be established along each arm of a chromosome, and the markers can be assembled into the map as a single linkage group.

## DETERMINATION OF THE SEQUENCE OF THE HUMAN GENOME HAS PRODUCED A WEALTH OF NEW FINDINGS

Only a small fraction of the findings can be covered here. The interested reader is referred to the original articles. Table 54–2 summarizes a number of the highlights, which can now be described.

### Most of the Human Genome Has Been Sequenced

Over 90% of the human genome had been sequenced by July 2000. This is by far the largest genome sequenced, with an estimated size of approximately 3.2 gigabases (Gb). Prior to the human genome, that of the fruit fly had been the largest (~180 Mb) sequenced. Gaps still exist, small and large, and the quality of some of the sequencing data will be refined since some of the findings are probably not exactly right.

### The Human Genome Is Estimated to Encode About 30,000–40,000 Proteins

The greatest surprise provided by the results to date has been the apparently low number of genes encoding proteins, estimated to lie between 30,000 and 40,000. The higher number could increase as new data are obtained. This number is approximately twice that found in the roundworm (19,099) and three times that of the fruit fly (13,061). The figures suggest that the complexity of humans compared with that of the two simpler organisms must have explanations other than strictly gene number.

**Table 54–2.** Major findings reported in the rough drafts of the human genome.

- More than 90% of the genome has been sequenced; gaps, large and small, remain to be filled in.
- Estimated number of protein-coding genes ranges from 30,000 to 40,000.
- Only 1.1–1.5% of the genome codes for proteins.
- There are wide variations in features of individual chromosomes (eg, in gene number per Mb, SNP density, GC content, numbers of transposable elements and CpG islands, recombination rate).
- Human genes do more work than those of the roundworm or fruit fly (eg, alternative splicing is used more frequently).
- The human proteome is more complex than that found in invertebrates.
- Repeat sequences probably constitute more than 50% of the genome.
- Approximately 100 coding regions have been copied and moved by RNA-based transposons.
- Approximately 200 genes may be derived from bacteria by lateral transfer.
- More than 3 million SNPs have been identified.

### Only 1.1–1.5% of the Human Genome Encodes Proteins

Analyses of the available data reveal that 1.1–1.5% of the genome consists of exons. About 24% consists of introns, and 75% of sequences lying between genes (intergenic). Comparisons with the data on the roundworm and fruit fly have shown that exon size across the three species is relatively constant (mean size of 145 bp in humans). However, intron size in humans is much more variable (mean size of over 3300 bp), resulting in great variation in gene size.

### The Landscape of Human Chromosomes Varies Widely

There are marked differences among individual chromosomes in many features, such as gene number per megabase, density of single nucleotide polymorphisms (SNPs), GC content, number of transposable elements and CpG islands, and recombination rate. To take one example, chromosome 19 has the richest gene content (23 genes per megabase), whereas chromosome 13 and the Y chromosome have the sparsest content (5 genes per megabase). Explanations for these variations are not apparent at this time.

### Human Genes Do More Work Than Those of Simpler Organisms

Alternative splicing appears to be more prevalent in humans, involving at least 35% of their genes. Data indicate that the average number of distinct transcripts per gene for chromosomes 22 and 19 were 2.6 and 3.2, respectively. These figures are higher than for the roundworm, where only 12.2% of genes appear to be alternatively spliced and only 1.34 splice variants per gene were noted.

### The Human Proteome Is More Complex Than That of Invertebrates

Relatively few new protein domains appear to have emerged among vertebrates. However, the number of distinct domain architectures (~1800) in human proteins is 1.8 times that of the roundworm or fruit fly. About 90 vertebrate-specific families of proteins have been identified, and these have been found to be enriched in proteins of the immune and nervous systems.

The results of the two drafts are rich in information about protein families and classes. One example is shown in Table 54–3, in which the major classes of proteins encoded by human genes are listed. As can be seen, the largest class is "unknown." Identification of these unknown proteins will be a major focus of activity for many laboratories.

## Repeat Sequences Probably Constitute More Than 50% of the Human Genome

Repeat sequences probably account for at least half of the genome. They fall into five classes: (1) transposon-derived repeats (interspersed repeats); (2) processed pseudogenes; (3) simple sequence repeats; (4) segmental duplications, made up of 10–300 kb that have been copied from one region of the genome into another; and (5) blocks of tandemly repeated sequences, found at centromeres, telomeres, and other areas. Considerable information on most of the above classes of repeat sequences—of great value in understanding the architecture and development of the human genome—is reported in the drafts. Only two points of interest will be mentioned here. It is speculated that Alu elements, the most prominent members (about 10% of the total genome) of the short interspersed elements (SINEs), may be present in GC-rich areas because of positive selection, implying that they are of benefit to the host.

***Table 54–3.*** Major classes of proteins encoded by human genes.[1]

| Class of Protein | Number (%)[2] |
|---|---|
| Unknown | 12,809 (41%) |
| Nucleic acid enzymes | 2,308 (7.5%) |
| Transcription factors | 1,850 (6%) |
| Receptors | 1,543 (5%) |
| Hydrolases | 1,227 (4.0%) |
| Select regulatory molecules (eg, G proteins, cell cycle regulators) | 988 (3.2%) |
| Protooncogenes | 902 (2.9%) |
| Cytoskeletal structural proteins | 876 (2.8%) |
| Kinases | 868 (2.8%) |

[1]Data from Venter JC et al: The sequence of the human genome. Science 2001;291:1304.

[2]The percentages are derived from a total of 26,383 genes reported in the rough draft by Celera Genomics. Classes containing more than 2.5% of the total proteins encoded by the genes identified when this rough draft was written are arbitrarily listed as major.

Segmental duplications have been found to be much more common than in the roundworm or fruit fly. It is possible that these structures may be involved in exon shuffling and the increased diversity of proteins found in humans.

## Other Findings of Interest

The last three major points of interest listed in Table 54–2 will be briefly described together.

Approximately 100 coding regions are estimated to have been copied and moved by RNA-based transposons (retrotransposons). It is possible that some of these genes may adopt new roles in the course of time. A surprising finding is that over 200 genes may be derived from bacteria by lateral transfer. None of these genes are present in nonvertebrate eukaryotes. More than 3 million SNPs have been identified. It is likely that they will prove invaluable for certain aspects of gene mapping.

It is stressed that the findings listed here are only a few of those reported in the drafts, and the reader is urged to consult the original reports (see References, below).

# FURTHER WORK IS PLANNED ON THE HUMAN & OTHER GENOMES

The IHGSC has indicated that it will determine the complete sequence, it is hoped, by 2003. The task involves filling in the gaps and identifying new genes, their locations, and functions. Regulatory regions will be identified, and the sequences of other large genomes (eg, of the house mouse; of *Rattus norvegicus,* the Norway rat; of *Danio rerio,* the zebra fish; of *Fugu rubripes,* the tiger puffer fish; and of one or more primates) will be obtained; indeed, a draft version of the genome of the tiger puffer fish was published in 2002. Additional SNPs will be identified; a complete catalog of these variants is expected to be of great value in mapping genes associated with complex traits and for other uses as well. Along with the above, existing databases will be added to as new information flows in, and new databases will probably be established to serve specific purposes. A variety of studies in functional genomics (ie, the study of genomes to determine the functions of all their genes and their products) will also be undertaken.

# IMPLICATIONS FOR PROTEOMICS, BIOTECHNOLOGY, & BIOINFORMATICS

Many fields will be influenced by knowledge of the human genome. Only a few are briefly discussed here.

**Proteomics** (see Chapter 4) in its broadest sense is the study of all the proteins encoded in an organism (ie, the proteome), including their structures, modifications, functions, and interactions. In a narrower sense, it involves the identification and study of multiple proteins linked through cellular actions—but not necessarily the entire proteome. With regard to humans, many individual proteins will be identified and characterized; their interactions and levels will be determined in physiologic and pathologic states, and the resulting information will be entered into appropriate databases. Techniques such as two-dimensional electrophoresis, a variety of modes of mass spectrometry, and antibody arrays will be central to expansion of this rapidly growing field. Overall, proteomics will greatly advance our knowledge of proteins at the basic level and will also nourish **biotechnology** as new proteins that are likely to have diagnostic, therapeutic, and other uses are discovered and methods for their economic production are developed. Specialists in **bioinformatics** will be in demand, as this field rapidly gears up to manage, analyze, and utilize the flood of data from genomic and proteomic studies.

## IMPLICATIONS FOR MEDICINE

Practically every area of medicine will be affected by the new information accruing from knowledge of the human genome. The **tracking of disease genes** will be enormously facilitated. As mentioned above, SNP maps should greatly assist determination of genes involved in complex diseases. Probes for any gene will be available if needed, leading to **improved diagnostic testing** for disease susceptibility genes and for genes directly involved in the causation of specific diseases. The field of **pharmacogenomics** (see Chapter 53) is already expanding greatly, and it is possible that in the future drugs will be tailored to accommodate the variations in enzymes and other proteins involved in drug action and metabolism found among individuals. Studies of genes involved in **behavior** may lead to new insights into the causation and possible treatment of psychiatric disorders. Many **ethical issues**—eg, privacy concerns and the use of genomic information for commercial purposes—will have to be addressed. It will also be important that medical and economic benefits accrue to individuals in **Third World countries** from the anticipated effects on health services and the diagnosis and treatment of disease.

## SUMMARY

- Determination of the complete sequence of the human genome, now almost completed, is one of the most significant scientific achievements of all time.
- Many important findings have already emerged. The one to date that has generated the most discussion is that the number of human genes may be only two to three times that estimated for the roundworm and the fruit fly.
- Information flowing from the Human Genome Project is having major influences in fields such as proteomics, bioinformatics, biotechnology, and pharmacogenomics as well as all areas of biology and medicine.
- It is hoped that the knowledge derived will be used wisely and fairly and that the benefits that will ensue regarding health, disease, and other matters will be made available to all people everywhere.

## REFERENCES

Collins FS, McKusick VA: Implications of the Human Genome Project for medical science. JAMA 2001;285:540. (The February 7, 2001, issue describes opportunities for medical research in the 21st century. Many articles of interest.)

Hedges SB, Kumar S: Vertebrate genomes compared. Science 2002;297:1283. (The same issue—No. 5585, August 23—contains a draft version of the genome of the tiger puffer fish.)

McKusick VA: The anatomy of the human genome: a neo-Vesalian basis for medicine in the 21st century. JAMA 2001;286: 2289. (The November 14, 2001, issue contains a number of other excellent articles—eg, on clinical proteomics, pharmacogenomics—relating to the Human Genome Project and its impact on medicine.)

Nature 2001;409(6822) (February 15), and Science 2001;291 (5507) (February 16). (These two issues present the rough drafts prepared by the IHGSC and Celera, respectively, along with many other articles analyzing the meaning and significance of the findings.)

Science 2001;294(5540) (October 5). (This issue contains a number of articles under the title Genome: Unlocking Biology's Storehouse. They describe new ideas, approaches, and research related to genome information.)

# APPENDIX

## SELECTED WORLD WIDE WEB SITES

The following is a list of Web sites that readers may find useful. The sites have been visited at various times by one of the authors (RKM). Most are located in the USA, but many provide extensive links to international sites and to databases (eg, for protein and nucleic acid sequences) and online journals. RKM would be grateful if readers who find other useful sites would notify him of their URLs by e-mail (rmurray6745@rogers. com) so that they may be considered for inclusion in future editions of this text.

Readers should note that URLs may change or cease to exist.

## ■ ACCESS TO THE BIOMEDICAL LITERATURE

HighWire Press: http://highwire.stanford.edu

(Extensive lists of various classes of journals—biology, medicine, etc—and offers also the most extensive list of journals with free online access.)

National Library of Medicine: http://www.nlm.nih.gov/

(Free access to Medline via PubMed.)

## GENERAL RESOURCE SITES

The Biology Project (from the University of Arizona): http://www.biology.arizona.edu/default.html

Harvard Department of Molecular & Cellular Biology Links: http://mcb.harvard.edu/BioLinks.html

## SITES ON SPECIFIC TOPICS

American Heart Association: http://www.americanheart.org

(Useful information on nutrition, on the role of various biomolecules—eg, cholesterol, lipoproteins—in heart disease, and on the major cardiovascular diseases.)

Cancer Genome Anatomy Project (CGAP): http://www.ncbi.nlm.nih.gov/ncicgap

(An interdisciplinary program to generate the information and technical tools needed to decipher the molecular anatomy of the cancer cell.)

European Bioinformatics Institute: http://www.ebi.ac.uk/ebi_home.html

(Maintains the EMBL Nucleotide and SWISS-PROT databases as well as other databases.)

GeneCards: http://bioinformatics.weizmann.ac.il/cards/

(A database of human genes, their products, and their involvements in diseases. From the Weizmann Institute of Science.)

GeneTestsGeneClinics: http://www.geneclinics.org/

(A medical genetics information resource with comprehensive articles on many genetic diseases.)

Genes and Disease: http://www.ncbi.nlm.nih.gov/disease/

(Coverage of the genetic basis of many different types of diseases.)

The Glycoscience Network (TGN): http://www.vei.co.uk/TGN/tgn_side.htm

(TGN is an informal worldwide grouping of scientists who share an interest in carbohydrates. The site contains considerable information on carbohydrates and an extensive list of links to other sites dealing with sugar-containing molecules).

Howard Hughes Medical Institute: http://www.hhmi.org/

(An excellent site for following current biomedical research. Contains a comprehensive Research News Archive.)

The Human Gene Mutation Database: http://archive.uwcm.ac.uk/uwcm/mg/hgmd0.html

(An extensive tabulation of mutations in human genes from the Institute of Medical Genetics in Cardiff, Wales.)

Human Genome Project Information: http://www.ornl.gov/hgmis/

(From the Human Genome Program of the United States Department of Energy.)

The Institute for Genetic Research: http://www.tigr.org/

(Sequences of various bacterial genomes and other information.)

Karolinska Institute Nutritional and Metabolic Diseases: http://www.mic.ki.se/Diseases/c18.html

(Access to information on many nutritional and metabolic disorders.)

MITOMAP: http://www.mitomap.org/

(A human mitochondrial genome database.)

National Center for Biotechnology Information: http://ncbi.nlm.nih.gov/

(Information on molecular biology and how molecular processes affect human health and disease.)

National Human Genome Research Institute: http://www.genome.gov/

(Extensive information about the Human Genome Project.)

National Institutes of Health (NIH): http://www.nih.gov/

(Includes links to the separate Institutes and Centers that constitute NIH, covering a wide range of biomedical research.)

Neuroscience (Biosciences): http://neuro.med.cornell.edu/VL/

(A comprehensive list of neuroscience resources; part of the World-Wide Web Virtual Library.)

Office of Rare Diseases: http://rarediseases.info.nih.gov/index_main.html

(Access to information on more than 7000 rare diseases, including current research.)

OMIM Home Page—Online Mendelian Inheritance in Man: http://www.ncbi.nlm.nih.gov/omim/

(An extensive catalog of human genetic disorders, updated daily. Lists access to various allied resources.)

Protein Data Bank (PDB): http://www.rcsb.org/pdb/

(A worldwide repository for the processing and distribution of three-dimensional biologic macromolecular structure data.)

The Protein Kinase Resource: http://pkr.sdsc.edu/html/index.shtml

(Information on the protein kinase family of enzymes.)

The Protein Society: http://www.faseb.org/protein/index.html

(An extensive list of Web resources for protein scientists.)

Signaling Update: http://www.signaling-update.org/

(A one-stop overview for the specialist or nonspecialist of what is happening in cell signaling.)

Society for Endocrinology: http://www.endocrinology.org

(Aims to promote advancement of public education in endocrinology. Contains a number of articles on endocrinology and a list of links to other relevant sites.)

tbase—the Transgenic/Targeted Mutation Database at the Jackson Laboratory, Bar Harbor, Maine: http://tbase.jax.org/

(An attempt to organize information on transgenic animals and targeted mutations generated and analyzed worldwide.)

The Wellcome Trust Sanger Institute: http://www.sanger.ac.uk

(A genome research center whose purpose it is to increase knowledge of genomes, particularly through large-scale sequencing and analysis,)

Whitehead Institute/MIT Center for Genome Research: http://www.genome.wi.mit.edu/

(Access to various databases and articles entitled "What's New in Genome Research.")

## Chapter 2

http://www.geocities.com/bioelectrochemistry/sorensen.htm

## Chapter 3

http: //www.bio.cmu.edu/Courses/03231

http: //www.bcbp.gu.se/~orjan/bmstruct/

## Chapter 4

http://www.lundberg.bcbp.gu.se/~orjan/bmstruct/

## Chapter 5

http://www.umass.edu/microbio/chime/explorer/

http://molvis.sdsc.edu/protexpl/index.htm

http://www.umass.edu/microbio/rasmol/

http://www.cryst.bbk.ac.uk/PPS2/course/section10/membrane.html

http://molbio.info.nih.gov/cgi-bin/pdb/

http://www.mc.vanderbilt.edu/peds/pidl/genetic/ehlers.htm

## Chapter 6

http://sickle.bwh.harvard.edu/

http://globin.cse.psu.edu/

## Chapter 7

http://s02.middlebury.edu/CH441A/EnzymeTutorials.html

http://www.i-a-s.de/IAS/botanik/e18/18.htm

## Chapter 8

http://www.indstate.edu/thcme/mwking/enzyme-kinetics.html

http://ntri.tamuk.edu/cell/kinetics.html

## Chapter 9

http://users.wmin.ac.uk/~mellerj/physiology/bernard.htm

http://www.cm.utexas.edu/academic/courses/Fall2001/CH369/LEC06/Lec6.htm

http://www.cellularsignaling.org

http://arethusa.unh.edu/bchm752/ppthtml/Jan27/sld015.htm

## Chapter 22

http://www.auhs.edu/netbiochem/NetWelco.htm

## Chapter 28

http://www.people.virginia.edu/~rjh9u/scurvy.html

http://www.mc.vanderbilt.edu/biolib/hc/journeys/scurvy.html

http://opbs.okstate.edu/~melcher/MG/MGW2/MG2411.html

## Chapter 29

http://www.nucdf.org/whatis.htm

## Chapter 30

http://www.pkunetwork.org/

http://www.rarediseases.org/search/rdblist.html

http://www.msud-support.org/overv.htm

## Chapter 34

http://www.rarediseases.org/search/rdblist.html

http://www.rheumatology.org/patients/factsheet/gout.html

http://www.merck.com/pubs/mmanual/section5/chapter55/55a.htm

http://www.nlm.nih.gov/medlineplus/goutandpseudogout.html

http://www.amg.gda.pl/~essppmm/ppd/ppd_py_umps.html

## BIOCHEMICAL JOURNALS AND REVIEWS

The following is a partial list of biochemistry journals and review series and of some biomedical journals that contain biochemical articles. Biochemistry and biology journals now usually have Web sites, often with useful links, and some journals are fully accessible without charge. The reader can obtain the URLs for the following by using a search engine.

Annual Reviews of Biochemistry, Cell and Developmental Biology, Genetics, Genomics and Human Genetics

Archives of Biochemistry and Biophysics (Arch Biochem Biophys)

Biochemical and Biophysical Research Communications (Biochem Biophys Res Commun)

Biochemical Journal (Biochem J)

Biochemistry (Biochemistry)

Biochemistry (Moscow) (Biochemistry [Mosc])

Biochimica et Biophysica Acta (Biochim Biophys Acta)

Biochimie (Biochimie)

European Journal of Biochemistry (Eur J Biochem)

Indian Journal of Biochemistry and Biophysics (Indian J Biochem Biophys)

Journal of Biochemistry (Tokyo) (J Biochem [Tokyo])

Journal of Biological Chemistry (J Biol Chem)

Journal of Clinical Investigation (J Clin Invest)

Journal of Lipid Research (J Lipid Res)

Nature (Nature)

Nature Genetics (Nat Genet)

Proceedings of the National Academy of Sciences USA (Proc Natl Acad Sci USA)

Science (Science)

Trends in Biochemical Sciences (Trends Biochem Sci)

# Index

*Note:* Page numbers in bold face type indicate a major discussion. A *t* following a page number indicates tabular material and an *f* following a page number indicates a figure.